BIOLOGICAL SCIENCE

WILLIAM T. KEETON
CORNELL UNIVERSITY

JAMES L. GOULD
PRINCETON UNIVERSITY

WITH CAROL GRANT GOULD

BIOLOGICAL SCIENCE

FOURTH EDITION

W · W · NORTON & COMPANY · NEW YORK · LONDON

Since this page cannot legibly accommodate all the copyright notices, pages
A20–A26 constitute an extension of the copyright page.

This book is composed in Aster. Composition by New England Typographic
Service, Inc. Manufacturing by R. R. Donnelley & Sons, Company. Book design
by Antonina Krass.

FOURTH EDITION

Library of Congress Cataloging in Publication Data

Keeton, William T.
    Biological science.
    Includes index.
    1. Biology. I. Gould, James L., 1945-
II. Gould, Carol Grant. III. Title.
QH308.K38 1986   574   86-5262
ISBN 0-393-95385-8

Cover photo by Nicholas Foster, The Image Bank.

W. W. Norton & Company, Inc., 500 Fifth Aveue, New York, N.Y. 10110
W. W. Norton & Company Ltd., 37 Great Russell Street, London WC1B 3NU
2 3 4 5 6 7 8 9 0

CONTENTS IN BRIEF

CONTENTS

PREFACE

Few challenges are more exciting and rewarding than teaching introductory biology. A teacher has the responsibility of presenting the essentials of a dynamic and critically relevant discipline to students of widely differing backgrounds and needs, of providing an informed lay understanding for many, and at the same time erecting a secure foundation for more advanced courses for others. Bill Keeton was a pioneer of the integrated approach to teaching introductory biology; his lucid and authoritative text has become the standard against which other books are judged. The success of this exceptional scientist and teacher was based both on his skill at exposition and on his conception of the field, in which he viewed the variety of living things in terms of the similarities and differences in the evolutionary adaptations they have undergone. This perspective led him to treat plants and animals together, comparing and contrasting the problems they face and the solutions that have evolved. For him, biology was never a series of special issues, a litany of social and environmental concerns, or an encyclopedia of premedical facts.

Despite the enduring excellence of *Biological Science,* it is no surprise that six years after the publication of the Third Edition an up-to-date presentation of modern biology is again needed. In preparing this Fourth Edition, we have been very fortunate to have a book of such obvious excellence from which to work.

We had three main objectives in revising: to improve the clarity of the presentation wherever possible, adding more intuitive explanations and more functional examples and thus making even the most complex subject matter accessible to a wider range of students; to keep the book manageably brief, which sometimes required the abbreviation or deletion of less essential topics; and to bring the book up to date in both depth and scope, so that it would reflect new discoveries and the shifting emphases in the advanced courses for which it may be the student's primary introduction. In particular, we wished to reinforce the evolutionary theme in all parts of the text, and to provide more intuitive molecular explanations of the mechanisms of biology in all chapters. From our own experience and the comments of other teachers of introductory biology, it was clear that the content of every chapter had to be reviewed for accuracy, emphasis, and effectiveness. In the end, every chapter benefited from this process: several were heavily revised and many of the longer chapters were divided into shorter, more manageable ones. Here, in brief, are some of the more important changes:

1. Chapter 1, the introduction, focuses new attention on the historical context in which modern biology developed, the role of intuition in formulating hypotheses, and the distinction between evolution as a process and natural selection as a mechanism.

2. Chapter 2, on simple chemistry, emphasizes the role of electrons in energy storage and release. The presentation of weak bonds—particularly the important concept of electronegativity—provides a groundwork for the later discussion of enzyme function.

3. Chapter 3, on the chemistry of life, now presents a brief, largely intuitive introduction to equilibrium constants that is particularly helpful in understanding how coupled reactions work. There is a new discussion of how the active sites of enzymes function to weaken the appropriate bonds in substrate molecules.

4. In Chapter 4, which focuses on the cell membrane, new descriptions of how the different sorts of microscopes work will enable students to interpret the many EM photographs in the book more easily. A more intuitive presentation of diffusion and osmosis complements the traditional explication. The biggest change has been in the added emphasis on the cell membrane itself, particularly the forces that stabilize it and the way its channels and gates work to regulate the movement of chemicals into and out of the cell. This discussion permits a more molecular presentation of nerve transmission, hormone action, organelle function (including the electron-transport chain), and fluid regulation in later chapters. The description of endocytosis has been enriched by a discussion of coated pits, with the LDL receptor as the primary example.

5. Chapter 5, on the interior of the cell, now presents an updated and expanded discussion of how newly synthesized proteins are labeled, packaged, and shipped by cellular organelles—particularly the ER and the Golgi. There is now explicit treatment of the microtrabecular lattice and a more extensive discussion of the endosymbiotic hypothesis.

6. In Chapter 6, on multicellular organization, the most important change has been the addition of a major discussion of cell-to-cell adhesion, including desmosomes and gap junctions.

7. In Chapters 7 and 8, on energy transformations, respiration is now discussed before photosynthesis; this order has an evolutionary logic, since fermentation probably appeared even before cyclic photosynthesis, but its primary advantage is pedagogical: respiration is easier to understand, and its basic features are virtually identical to those of the more complex chemistry of photosynthesis. It is now possible to teach glycolysis intuitively from a thermodynamic perspective, or as a series of chemical changes. New margin illustrations allow the interested student to follow the molecular alterations of glycolysis step by step, while the text outlines the thermodynamic considerations. The chemiosmotic hypothesis in respiration now receives central attention, and is presented both thermodynamically and anatomically. The treatment of photosynthesis has undergone similar changes, and focuses on its evolution, intuitive thermodynamics, and fascinating anatomy. Of the hundreds of illustrations added or modified to improve the visual presentation of ideas throughout the text, we are particularly pleased with the many new drawings in these two chapters.

8. The chapters on whole-organism physiology (Chapters 9–14) have undergone a number of small but important additions and improvements, most apparent in the new drawings of fluid flow and tissue growth in monocots and dicots, the sections on the evolution and comparative physiology of hearts and circulatory systems, and the treatments of liver and kidney function. As in all the cases in which we have added a molecular emphasis, we believe these details provide a more satisfying, unifying, and comprehensible picture of the life processes they elucidate.

9. The chapters dealing with hormonal control (Chapters 15–17) emphasize the molecular bases of hormone action, second-messenger strategies like that of the calcium-calmodulin system, local chemical mediators and other hormone-like molecules, and the evolution of hormones.

10. The chapters on neurobiology and effectors (Chapters 18–20) have been reorganized and rewritten to reflect a more modern emphasis on molecular mechanisms and the functional issues that relate to sensation and movement. The discussions of interneurons, habituation, and simple circuits use *Aplysia* as their primary example, to complement the extensive treatment of the human CNS. There is new emphasis on how sensory information is processed to extract patterns of shape and movement in the visual world. The discussion of brain evolution has been brought into line with current thinking.

11. Chapter 21, the first of two chapters on behavior, deals with the mechanisms of innate behavior and of learning, and ties these mechanisms to the neurobiology that underlies them; the chapter ends with a discussion of a prime example of the interplay of instinct and learning: bird navigation. Chapter 22 deals with the less mechanistic, more evolutionary aspects of behavior, often called behavioral ecology. Bringing together material formerly divided among several chapters, this discussion treats in one place the concepts of niche and habitat, kin selection and reciprocal altruism, communication, and the social behavior of insects, birds, and mammals (including humans).

12. The genetics and development chapters of Part III required extensive revision to keep pace with the most important new research findings in molecular biology; many more changes have been made in these chapters than can possibly

be mentioned here. Some alterations were strictly pedago-
gic, such as the reworking of the presentation and accompanying illustrations of mitosis and meiosis in Chapter 23, on
cellular reproduction. Also in that chapter, we have added a
discussion of the evolution of cell division and sexual recombination, as well as a molecular view of recombination. The
genetics chapters (Chapters 24 and 25) have undergone
many small changes to improve clarity. They now include
new discussions of complementation tests, trihybrid crosses,
statistical tests, and the molecular basis of mutation.

13. The chapters on the flow of information in cells
(Chapters 26–30) are the most heavily revised in Part III. Our
presentation of DNA replication and repair, transcription,
messenger RNA "processing" and translation is largely new,
and constitutes a modern molecular treatment of these subjects that is eminently suitable for an introductory course.
Here, as elsewhere, material most appropriate for majors or
advanced students, such as the discussion "Replication of
the *E. coli* chromosome," is presented in self-contained
boxes that supplement the more general treatment in the
text. Extrachromosomal inheritance is now presented in a
separate chapter (Chapter 28), which incorporates new or
heavily revised discussions of organelle heredity, the mechanisms of viral replication, and recombinant DNA technology.
The chapter on the control of gene expression (Chapter 29) is
almost entirely new, containing discussions of negative and
positive transcriptional control, a revised presentation of the
lac operon, two boxes on advanced topics ("How control
substances bind to DNA" and a particularly well understood
example of repression, "The lambda switch"), an exposition
of the curious organization of eucaryotic chromosomes, and
a modern discussion of cancer and oncogenes. The new
chapter on immunology (Chapter 30) features a well-inte-
grated discussion of how B cells and T cells work and interact to produce precisely modulated immune responses. This
chapter also includes important discussions of the genetic
basis of antibody diversity and of the various mechanisms of
genetic variability, including transposition, which may provide the variation necessary for the evolution of novel proteins. Intriguing hypotheses of gene evolution, like that of
Gilbert and Blake, are considered.

14. A variety of small improvements have been incorporated into Chapters 31 and 32, on development, but the main
revisions are those providing coverage of the mechanistic
bases of differentiation and pattern formation. In particular,
there is now major treatment of neural development.

15. The first two chapters of Part IV (Chapters 33 and 34)
now give more attention than the previous edition to the possible role of chance, and of genetic drift in particular, in evolution. They also provide necessary examples of sympatric
speciation and host specificity, and discuss several major
controversies among evolutionary biologists; these include
the debates over punctuated equilibria and cladistics.

16. The last two chapters of Part IV (Chapters 35 and 36)
deal with ecology and biogeography. Many teachers were
unhappy with the organization of this material in the Third
Edition, with the choice of examples in the sections on ecosystems and community ecology, and with the lack of emphasis on physiological ecology. These criticisms have been
taken to heart: the new organization puts material on population size and distribution first in Chapter 35; this is followed by sections on population growth and regulation, and
the chapter ends with a discussion of dominance, diversity,
stability, and succession. Chapter 36 now begins with a view
of the economy of ecosystems—the flow of energy and materials—followed by a new section on the role of the sun in
creating worldwide climatic zones. The discussion then focuses on more local factors like mountain ranges, and on the
different kinds of biomes. The chapter ends with a consideration of the evolution of biomes and the mechanisms and consequences of species dispersal.

17. The Fourth Edition continues to use Robert H. Whittaker's five-kingdom classification system, modified to accommodate the new findings on Archaebacteria. This system and
some alternatives are discussed in Chapter 37, which also includes a modernized account of current ideas about the origin and early evolution of life. In the Third Edition the
chapters on classification (now Chapters 38–43) were perhaps the most complete in an introductory book, and they
are therefore not much changed, except for Chapter 38, on
viruses and Monera, which now includes discussions of viroids, prions, and Archaebacteria. Numerous small changes
in the classification of higher animals have been necessary;
the sections dealing with the evolution of vertebrates, primates in particular, have also been carefully updated to reflect current thinking.

Throughout the Fourth Edition, our general pedagogical
strategy has been to begin a chapter—or, more often, a set of
chapters—with an overview of what is to come, thus providing an initial evolutionary or functional outline to help the
student place the sections that follow in a useful context. Several new summary diagrams and discussions reflect this approach.

We decided early on to keep the general chapter order of
the Third Edition. No single sequence is ideal, and the
present one has the advantage of familiarity. We have, however, made every effort to make the parts stand on their own,
so that they can be taught in different orders. At Princeton,

for example, we teach Part IV immediately after Part I, followed by Part III, Part V, and then Part II. It is a testament to the flexibility of the Third Edition that we have no difficulty using the book with our idiosyncratic order, and the Fourth Edition should prove even more adaptable. Teachers who wish to keep their presentations of plants and animals separate will find that the new chapter divisions facilitate this approach.

This brief listing of the major changes in the Fourth Edition should not be taken to imply that other parts of the text have escaped careful scrutiny. On the contrary, an intense re-evaluation has gone on throughout the book. Despite such exhaustive efforts, there will be parts of the book that do not reflect the most recent advances by the time students use the text. Not only will new developments have occurred in the interim between writing and publication but, lacking a crystal ball, we may have failed to recognize the significance of developments already taking place. For a text author, this is as unfortunate as it is unavoidable, but it is not all bad for students, since it reflects the vitality of the field.

A revision of this magnitude of a book with such high standards to maintain would have been impossible without the help of many reviewers. In particular we would like to thank Joseph M. Calvo, Cornell; Robert K. Colwell, Berkeley; Peter Grant, Princeton; Andre T. Jagendorf, Cornell; Carol H. McFadden, Cornell; C. O. Patterson, Texas A & M; Thomas Roos, Dartmouth; Daniel I. Rubenstein, Princeton; Peter M. Shugarman, Southern California; Malcolm Steinberg, Princeton; Volker M. Vogt, Cornell; and Timothy C. Williams, Swarthmore College, for help above and beyond the call of duty. Many other important criticisms were provided by Wayne Aspey, Ohio State; Robert A. Bender, Michigan; Anthony Blackler, Cornell; Robert W. Bouma, Cornell; George Bowes, Florida; George Cain, Iowa; W. Zacheus Cande, Berkeley; Steven Jay Gould (no relation), Harvard; Richard Hallberg, Iowa State; Steven Heidemann, Michigan State; Paul Hertz, Barnard; Robert P. Higgins, Smithsonian Institution; William Hodos, Maryland; Anthony R. Kaney, Bryn Mawr College; Joseph Levine, Boston College; William M. Lewis, Colorado; Karel F. Liem, Harvard; Ellis R. Loew, Cornell; Ross J. MacIntyre, Cornell; Peter Marler, Rockefeller; Mitch Masters, Oregon; Kenneth Miller, Brown; John Neess, Wisconsin; Michael Newlon, Iowa; Maggie T. Pennington, College of Charleston; David Pilbeam, Harvard; Patricia J. Pukilla, North Carolina; Robert Savage, Swarthmore College; John A. Schmitt, Ohio State; David Shappirio, Michigan; Steven D. Skopik, Delaware; Eric Skully, Towson State; James Smiley, College of Charleston; Douglas W. Smith, California, San Diego; Daryl Sweeney, Illinois; and Virginia Utermohlen, Cornell. The many new and uniformly excellent pictures and drawings in the Fourth Edition speak more eloquently than can we of the contribution of Ruth Mandel, our photo editor, and of Michael Reingold, the artist. The consistently high quality of the copy was maintained by the unparalleled vigilance and good judgment of Esther Jacobson, with help from Nancy Palmquist. Roy Tedoff and Fred Bidgood made the impossibly complicated task of book production at least seem feasible. Finally, the greatest contribution to both the rigor and the aesthetic appeal of this text was made by our indefatigable editor, James D. Jordan, whose remorseless insistence on clarity of exposition shows on every page. To all of these individuals, our heartfelt thanks.

J.L.G.
C.G.G.

Princeton, New Jersey
October 1985

BIOLOGICAL SCIENCE

Chapter 1

INTRODUCTION

LIFE

Every day, the sun radiates vast amounts of energy to the earth and our moon, and to all the other planets in our solar system (Fig. 1.1). And every day, the planets reradiate that energy into space. Only on earth is there a unique delay in the reemission of a tiny fraction of that energy: for a brief moment, a minute portion of it is trapped and stored. This minor delay in the flow of cosmic energy powers life. Plants capture sunlight and use its energy to build and maintain stems and leaves and seeds, while animals secure the energy of sunlight by eating plants, or by eating other animals that have eaten plants. At each stage in the many processes of living and dying, waste heat is produced, which joins the pool of energy being relayed back into space.

Biology is the study of this very special category of energy utilization—it is the science of living things. The world is teeming with life: millions of species of organisms of every description inhabit the earth, feeding directly or indirectly off lifeless sources of energy like the sun. What is there in this maze of diversity that unites all biology into one field? What, for instance, do amoebae, redwoods, and people have in common that lends unity to biological science? What differentiates the living from the nonliving? What, in short, *is* life?

Most dictionaries define life as the property that distinguishes the living from the dead, and define dead as deprived of life. These singularly circular and unsatisfactory definitions give us no clue to what we have in common with protozoans and plants. The difficulty for the scientist as well as for the writer of dictionaries is that life is not a separable, definable entity or property; it can't be isolated on a microscope slide or distilled into a test tube. To

1.1 The earth as seen from space

1

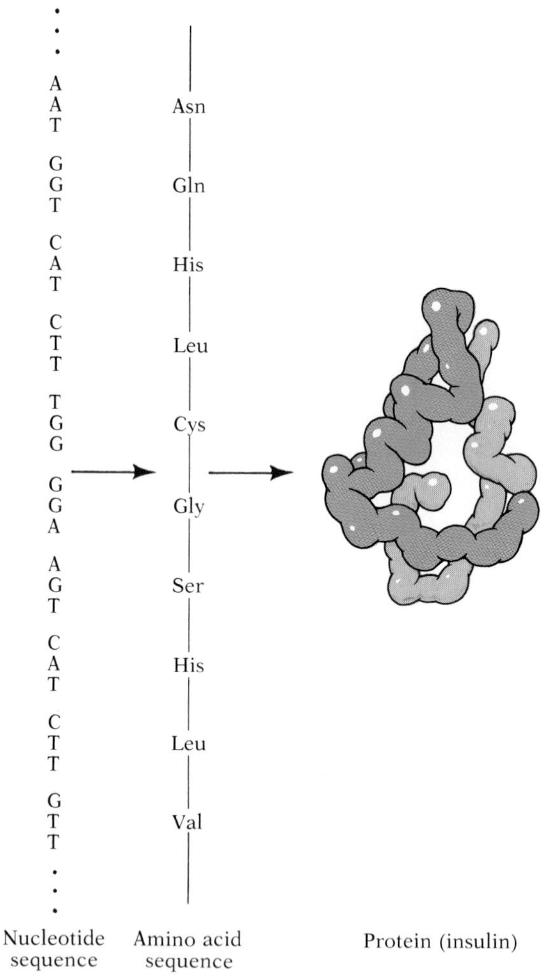

early "mechanistic" philosophers like Aristotle and Descartes, life was wholly explicable in terms of the natural laws of chemistry and physics. The "vitalists," philosophers of the opposing school, were convinced that there was a special property, a "vital force," absent in inanimate objects, that was unique to life. Though some scientists continue to think in terms of an unnamed and intangible special property, vitalism has been essentially dead in biology for at least half a century. The more we learn about living things, the clearer it becomes that life's processes are based on the same chemical and physical laws we see at work in a stone or a glass of water.

If life is not a special property, what is it? One answer may be found by comparing living and nonliving things. Organisms from bacteria to humans seem to have several attributes in common. For instance, all are chemically complex and highly organized. All use energy (metabolize), organize themselves (develop), and reproduce. All change (evolve) over generations. So far as we know, no nonliving thing possesses all these attributes. In addition, and perhaps most important, only the living organism has a set of instructions, or "program," resident in its genes, that directs its metabolism, organization, and reproduction, and is the raw material upon which evolution acts (Fig. 1.2).

But merely listing life's attributes is an inadequate and unsatisfying way of describing something so rich and complex, so varied and changing. This entire book is about life. When you have finished reading it, though a perfectly consistent and rigorous definition of life may still elude you (as it does most professional biologists), you will have a deeper understanding of life's mechanisms and the methods used to discern them, and perhaps a better appreciation for one of the most fascinating and revolutionary enterprises of our century. You will certainly have a greater appreciation of what it means to be alive.

THE SCIENTIFIC METHOD

We live in a science-conscious age. Almost every day the news media report some fresh development in agriculture, medicine, or space technology. Commercials use "scientists" in starched white lab coats to woo us with the latest products of "scientific" research. Yet even among well-educated people there is a surprising lack of understanding of what science really is. Some view it as akin to magic, as indeed it was in the days when many scientists were at least part-time alchemists or astrologers. The sinister scientist seeking forbidden knowledge whatever the cost to society pervades literature in such characters as Faustus, Dr. Jekyll, and, of course, Frankenstein. This image of science persists in some circles, manifesting itself as an uninformed phobia of, for instance, research on recombinant DNA or on nuclear fusion. Equally simplistic is the image of scientists as preoccupied, absentminded recluses shut contentedly in academic ivory towers, divorced from—and not greatly interested in—reality.

For the most part, though, scientists are neither sinister nor reclusive; rather, they are people whose native problem-solving mentality and curiosity about the natural world have simply deepened as they have grown older. It is this active curiosity about the living world, springing in most cases from a profound respect for nature, that binds together those who do re-

| Nucleotide sequence | Amino acid sequence | Protein (insulin) |

1.2 A simple model of the flow of chemical information in the cell

The internal chemistry of every cell is controlled ultimately by the cell's genes. These are long chains of DNA (deoxyribonucleic acid), composed of units called nucleotides. Information (instructions for the assembly of protein molecules) is carried in each gene in the sequence of four different nucleotides, represented in the model as A, C, G, and T. In the example shown here, each set of three nucleotides initiates (through intermediate steps that have been omitted) the production of a particular amino acid. The sequence of the sets of nucleotides determines the sequence of the amino acids, which constitute part of one chain of the protein insulin, a chemical that helps mediate the breakdown of complex food substances to release energy. We shall describe this flow of information in greater detail in later chapters.

search in the biological sciences. The unknown and perhaps unknowable instills in us humility and awe. This sense of mystery, accompanied by a belief in the existence of an underlying order, motivates virtually all scientists.

But motivation is only the first requirement for being a scientist. The second is the rigorous application of the ***scientific method***. So much has been said about the powers of the scientific method that many suspect it involves some formula too complicated for ordinary people to understand. It does not. Like so many truly great ideas, it is basically simple, and is used to some extent in fact by almost everyone every day. As the English biologist T. H. Huxley (1825–1895) put it, the scientific method is "nothing but trained and organized common sense." Its power in the hands of a scientist stems from the rigor of its application.

Formulating hypotheses   Science is concerned with the material universe, seeking to discover facts about it and to fit those facts into conceptual schemes, called theories or laws, that will clarify the relations between them. Science must therefore begin with observations of objects or events in the physical universe. The objects or events may occur naturally, or they may be the products of planned experiments; the important point is that they must be *observed*, either directly or indirectly. Science cannot deal with anything that cannot be observed.

Science rests on the philosophical assumption (well justified by its past successes) that virtually all events of the universe can be described by physical theories and laws, and that we get the data with which to formulate those theories and laws through our senses. Needless to say, natural laws are descriptive rather than prescriptive; they do not say how things *should* be, but instead how things are and probably will be. Scientists readily acknowledge the imperfection of human sensory perception: the major alterations our neural processing imposes upon our picture of the world around us are themselves a subject of scientific study. In addition, experience has shown that there is always a subtle interaction between phenomenon and observer: however careful we may be, our preconceived notions and even our physical presence may affect our observations and experiments. But to recognize the imperfection of sensory perception and observation is not to suggest that we may get scientific information from any other source (inspired or scriptural revelation, for example). No other means are open to us as scientists.

The first step in the scientific method is to formulate the question to be asked. This is not as simple as it sounds: scientists must decide which of the endless series of questions our escalating knowledge inspires are important and worth answering, and which are trivial. The next step is to make careful observations in an attempt to answer the question. Here too there are difficulties: the researcher must decide what to observe and, since measuring everything is impossible, what to ignore. The scientist must also decide how to make the measurements and how to record the data. This is no trivial matter: an oversight or a mistake can render years of work useless. Next, something must be done with the observations. Simply to amass data is not enough; the data must be analyzed and fitted into some sort of coherent pattern or generalization. A formal generalization, or ***hypothesis***, is a tentative causal explanation for a group of observations. The step from isolated bits of data to generalization can be taken with confidence only if enough observations have been made to give a firm basis for the generalization, and then only if the individual observations have been reliably made. But even when data have been carefully collected, a hypothesis does not automatically follow. Often data can be interpreted in several ways, or may appear to make no sense whatsoever.

Testing hypotheses   The making of a general statement, or hypothesis, is not the end of the process. Scientists must devise ways of testing their hypotheses by formulating predictions based on them and checking to see if these predictions are accurate. Again, this is often not easy: a hypothesis may supply the basis for many predictions, of various degrees of interest, but probably only one can be tested at a time. A scientist must decide which predictions can be most readily tested, and of those, which one provides the toughest challenge for the hypothesis.

Perhaps most difficult of all, when researchers find a discrepancy between a hypothesis and the results of their tests, they must be ready to change their generalization. If, however, all evidence continues to support the new idea, it may become widely accepted as probably true and be dignified by the appellation "theory." It is important to realize that scientists do not use the term "theory" as does the general public. To many people a theory is a highly tentative statement, a poor makeshift for fact. But when scientists dignify a statement by the name of theory, they imply that it has a very high degree of probability and that they have great confidence in it.

A *theory* is a hypothesis that has been repeatedly and extensively tested. It is supported by all the data that have been gathered, and helps order and explain those data. Many scientific theories, like the cell theory, are so well supported by essentially all the known facts that they themselves are "facts" in the nonscientific application of that term. But the testing of a theory never stops. No theory in science is ever absolutely and finally proven. Good scientists must be ready to alter or even abandon their most cherished generalizations when new evidence contradicts them. They must remember that all their theories, including the physical laws, are dependent on observable phenomena, and not vice versa. Even incorrect theories, however, can be enormously valuable in science. We usually think of mistaken hypotheses as just so much intellectual rubbish to be cleared away before science can progress, but tightly drawn, explicitly testable hypotheses, whether right or wrong, catalyze progress by focusing thought and experimentation.

The controlled experiment   In its simplest form the scientific method begins with careful observations, which must be shaped into a hypothesis. The hypothesis suggests predictions, which must be tested. The tests, furthermore, must be *controlled*, a matter of vital importance, since only a controlled test has any hope of illuminating anything. Controlling a test, or experiment, means making sure that the effects observed result from the phenomenon being tested, not from some tangential source. The most familiar way of controlling an experiment is to perform the same process time and again, varying only one minute part of it each time, so that if a difference appears in the result, we can easily trace the cause. When Louis Pasteur, for instance, took his memorable stand against the prevailing theory of spontaneous generation (the idea that life can arise spontaneously from nutrients), he designed his experiment so that there could be no doubt about the outcome, whatever it might be. He took identical flasks, filled them with identical nutrient solutions, subjected them to identical processes, but left some of them open to the air, while the others were effectively sealed. In time bacteria and mold developed in the open flasks, but the liquid in the sealed flasks remained clear. Since exposure to the outside air was the only variable in the procedure, it was then obvious that the bacteria and the mold-producing organisms, rather than being spontaneously generated from the broth itself, must have come from the air.

Intuition   We can see, then, that in actual practice science is neither easy nor mechanical. Most scientists will readily admit that every stage in the scientific process requires not just careful thought but a large measure of intuition and good fortune. At the end of the last century, for instance, physicists saw many small problems, or "anomalies," in Newtonian physics. Experience justifies living with small anomalies when a theory works well, and most minor problems prove in time to be irrelevant or mistaken—the result of observations that were faulty or interpretations that missed a point. Sometimes, though, anomalies in a theory may signal a conceptual error. Albert Einstein (Fig. 1.3) sensed that two of the many irritating anomalies in Newtonian physics were crucial difficulties, and rearranged the pieces of the puzzle to create a revolutionary new theory that accounted for those two anomalies. Like most new hypotheses, Einstein's theory of relativity did not win over the world of science at once. For one thing, it seemed far more complex than the Newtonian mechanics with which everyone was familiar, and it left at least as many (though different) loose ends dangling as had the theory it sought to replace. The potential appeal of Einstein's theory, however, was great: it made some very unlikely but testable predictions—that the light from stars, for instance, should bend as it passed near the sun. When these were subsequently proven to be correct, the theory of relativity was accepted rather quickly, loose ends and all, where a far more plausible but less dramatic hypothesis might well have been ignored.

Scientific investigation, then, depends on a combination of subjective judgments and objective tests, a delicate mixture of intuition and logic. Done well, scientific research is truly an art: the ability to make insightful guesses and imagine clever and critical ways in which to test them is usually the distinguishing characteristic of great scientists. But in the final analysis, the basic rules are the same for all: observations must be accurate and hypotheses testable. And as testing proceeds, hypotheses must be altered when necessary to conform to the evidence.

Limitations of the scientific method   The insistence on testability in science severely limits the range of its applications. For example, the idea —widely held by scientists and nonscientists alike—that there is a God working through the natural laws of the universe is simply not testable, and hence cannot be evaluated by science. Science seeks neither to confirm nor to refute it. Yet throughout history both scientists and theologians have

1.3   Albert Einstein

1.4   The Greek philosopher Aristotle
The central portion of *The School of Athens*, a mural by the Renaissance painter Raphael depicting philosophers and scholars of all ages. Aristotle is at right; the Greek philosopher Plato is at left.

made assertions about nature in the name of God, and have claimed that to deny these assertions is to deny their God. Those who make the existence of any deity stand or fall on some supposed fact about the universe risk having science destroy that deity. For this reason, such narrowly defined and dogmatic systems of belief have historically been enemies of free inquiry and freedom of thought.

Another limitation of science is that it cannot make value judgments: it cannot say, for example, that a painting or a sunset is beautiful. And science cannot make moral judgments: it cannot say that war is immoral. It cannot even say that a river should not be polluted. Science can, however, analyze responses to a painting; it can analyze the biological, social, and cultural implications of war; and it can demonstrate the consequences of pollution. It can, in short, try to predict what people will consider beautiful or moral, and it can provide them with information that may help them make value judgments or moral judgments about war or pollution. But the act of making the judgments is not itself science.

THE RISE OF MODERN BIOLOGICAL SCIENCE

EARLY SCIENCE

Science, as we have seen, is the endeavor to understand the natural world. Science must have originated with early man as he realized that his subjective observations could generate rules of practical utility—when to plant crops, for example, or how to recognize the approach of rain. Though a far cry from modern science and the scientific method, this intellectual observation of cause and effect marked the beginning of scientific thought.

The most important advance in early science came with the Greeks; instead of seeing the universe as ruled arbitrarily by a collection of gods who intervened frequently and capriciously in human events, they began to view the world as operating in a consistent, rule-governed fashion with a minimum of supernatural intervention. Philosophers could therefore set about attempting to discover the natural laws—philosophical principles, as they called them—that the universe obeyed.

The Greeks, particularly Aristotle (384–322 B.C.; Fig. 1.4), made systematic observations and from them formed generalizations, or hypotheses, divorced from strictly utilitarian goals. They developed and elaborated formal logic as a powerful intellectual tool, and employed it in their pioneering practice of making deductions from their hypotheses—what we would call making predictions, except that no effort was spent on experimental verification. Even Aristotle, the first great biologist, never tested a deduction. For instance, after observing and gathering information on an unprecedented variety of animals, he was faced with a number of loose ends. One of these was the issue of the sex of the various castes of honey bee. Though he very honestly pointed out the difficulties he had in reconciling his solution of the problem with what he had clearly observed, he nevertheless deduced that the queen and the workers had to be male because nature would never give weapons—stings, in this case—to females.

Clearly, the emphasis of Greek science was philosophical, its goal the creation of a unifying world picture rather than the working out of details. In addition to the grand philosophical system of Aristotle, there were others,

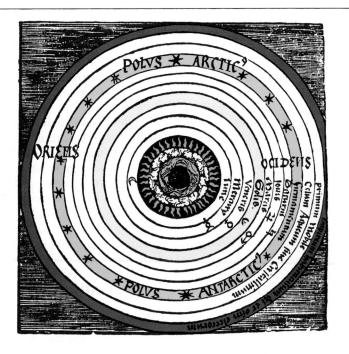

1.5 Aristotle's vision of the universe
The Aristotelian scheme of the universe had the earth
in the center, surrounded by a series of concentric
spheres containing water (the oceans), air, fire, the
moon, Mercury, Venus, the sun, Mars, Jupiter,
Saturn, the stars, and the region inhabited by the
divine beings who move the spheres. Ptolemaic
astronomy sought to explain the observed motion of
the planets and stars in terms of Aristotle's system of
concentric spheres.

notably that of Anaximander, who believed that everything in the universe
was composed of fire, earth, air, and water; that of Pythagoras, who was
convinced that all the secrets of the universe were contained in numerical
ratios; and that of Democritus, for whom all matter was composed of invisible atoms. Not all Greek science was quite so metaphysical, however, and
one particularly fortuitous combination of observation and deduction that
is still with us today is the method of medical diagnosis originated by Hip-
pocrates.

Roman culture, though marked by excellence in literature, history, and
the arts, added little to the scientific knowledge acquired by the Greeks,
and progress of every sort greatly declined after barbarians from northern
Europe sacked Rome in the fifth century and ushered in the Dark Ages.
Though the Arabs to the east continued to practice science according to the
Greek texts they had maintained, what little of Greek science survived in
Europe was preserved in remote or fortified monasteries until the slow
process of religious, cultural, and military conversion of the barbarians in
Europe by Charlemagne and his successors (beginning about A.D. 800) was
complete.

At first, scholars had all they could do simply to reabsorb Greek knowledge. The works of Aristotle in particular were enormously influential. His
logical, unifying, and aesthetically pleasing world view was especially attractive to a civilization emerging from centuries ruled by the irrational,
divisive, and uglier elements of man's nature. Aristotle's universe seemed
replete with perfection, beauty, and harmony. The earth lay at its center,
and the moon, sun, planets, and stars moved about the earth on a set of perfect, transparent, concentric spheres (Fig. 1.5). This idealized conception
became the basis of what we call Ptolemaic astronomy.

A

B

1.6 Albertus Magnus (A) and Thomas Aquinas (B)

Two very influential theologian-philosophers, Albertus Magnus (ca. 1193–1280; Fig. 1.6A) and Thomas Aquinas (1225–1274; Fig. 1.6B), both at the University of Paris, were largely responsible for incorporating the Aristotelian world view into Judaeo-Christian theology, and with it the critical distinction between the truth of nature and revealed, or scriptural, truth. Aristotle's dictum that man was at his best when he was using his rational powers fitted in particularly well with the biblical emphasis on free will and personal responsibility. This rationalistic approach prepared the way for the development of science. Even more important was the emergent Judaeo-Christian view of nature not as something arbitrary and to be feared, as it had been represented in many pagan mythologies, but as a creation by a just and merciful God for man's use and enjoyment—and, by extension, for man's study.

At the same time, the Franciscan monk Roger Bacon (ca. 1214–1294; Fig. 1.7) was enunciating the basic principles of modern science to students at a new and progressive university at Oxford. As he so succinctly put it, the purpose of science is "to work out the natures and properties of things." Bacon, himself an alchemist, appears to have been the first to champion the scientific method, including the then-revolutionary idea of the controlled experiment. To Bacon, accurate observation and experimental verification were the only bases for certainty.

EARLY DISCOVERIES IN ASTRONOMY AND PHYSICS

Though Albertus, Thomas Aquinas, Bacon, and others helped to prepare the intellectual climate for modern science, the dominant technique of scholarship for the next two centuries was scholasticism, a method of study that centered on the perusal and reperusal of the vast and fertile body of Scripture, and of other ancient writings that had achieved the status of doctrine. Bacon's plea to "cease to be ruled by dogmas and authorities; look at the world!" fell largely upon deaf ears. These very influential philosophers had legitimized science, but they had not really provided any compelling *reason* for scientific investigation. That critical motive was supplied in the sixteenth and seventeenth centuries by other intellectuals, among them Thomas More (1478–1535) and Francis Bacon (1561–1626). (It was Francis Bacon who formalized and popularized the scientific method first proposed by Roger Bacon three centuries earlier.) They argued persuasively that to study nature was to glorify the God who had created it. As Sir Thomas Browne (1605–1682) put it, "there is no danger to these profound mysteries" because God prefers a "devout and learned admiration" of his works to the "gross rusticity" that stares uncomprehendingly at nature, trembling at imagined portents. Generations of scientists—most of whom prior to the twentieth century were clerics—have found their calling in "reading the works of God in the book of nature."

Perhaps the first really important of these scientists was a Polish canon, Nicolaus Copernicus (1473–1543). He proposed that the earth orbited the sun, rather than the reverse. This hypothesis had the intuitively satisfying consequence of explaining why the more distant planets, such as Jupiter, reverse their direction of travel against the background of stars roughly once a year: the reversal is a simple consequence of the earth's "passing" them on its inside track around the sun. It also explained why Venus and Mercury never appear very far from the sun: they are circling the sun rather than the earth, and their orbits lie inside that of our planet.

Copernicus' idea was not immediately accepted, largely because the Aristotelian model had become so thoroughly integrated into Western thought and theology that the Copernican hypothesis appeared at first to threaten Western religion.

The really fatal blow to the Aristotelian model came from Galileo Galilei (1564–1642; Fig. 1.8), of the University of Padua. Galileo was probably the first person to apply the scientific method rigorously. He was able to show by careful measurement that much of Aristotelian physics was incorrect. For instance, Aristotle had reasoned that if an object weighed twice as much as another, it would fall twice as fast, but when Galileo actually measured the rates of movement (down an inclined plane, as he had no way to time free falls), he found them virtually identical. This finding meant that there was an enormous mistake in the accepted picture of the physical world. Galileo also discovered the principle of inertia: once moving, an object will continue at the same speed in the same direction unless acted upon by some other force (such as friction). The discovery of inertia ultimately had a profound effect upon the Western world view. Aristotle had thought that continued motion required continued force, and so someone—God, presumably—had to be continuously at work to keep the planets moving. Now, however, it appeared that God needed merely to have put the planets into place and set them in motion; hence in time the idea emerged of a God of Secondary Causes who had created the world, defined the natural laws, and then for the most part left things to take care of themselves. This concept became an underlying tenet for much of Western science in the eighteenth and nineteenth centuries.

Galileo seems also to have been the first scientist to point a telescope at the sky, and wherever he did so he saw that Aristotle's guesses were wrong: the moon was not a perfect sphere of some special celestial substance, but had mountains and craters; the sun was not perfect and immutable, but had spots that appeared, moved, and vanished; Venus had phases, which meant that it was a reflective rather than a luminous object, and that it did not follow the path required by the earth-centered system; Jupiter had four tiny moons circling it, in the same way that the planets, small compared to the sun, had been said by Copernicus to circle our star; and Saturn was not a sphere, but seen through Galileo's crude lenses appeared to have horns.

Galileo's observations provided strong evidence for the Copernican system, but the Catholic church concluded (unofficially) in 1616 that belief in the sun-centered universe was heretical. Galileo agreed to keep silent on the subject, but in 1632 he deliberately provoked a confrontation by publishing his *Dialogues Concerning the Two Principal Systems of the World*, in which two characters debate the very issue on which he had promised not to speak. The defender of the Aristotelian system (Simplicio, who often echoes the pope) is repeatedly made to look foolish, and Galileo was forced to recant by the Inquisition. The tragedy of Galileo is memorable because it has not been duplicated in Western history. Despite the periodic lapses of particular religious sects, the Western tradition of intellectual freedom has been especially strong in protecting science.

1.7 Roger Bacon

1.8 Galileo Galilei

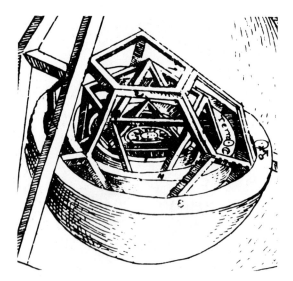

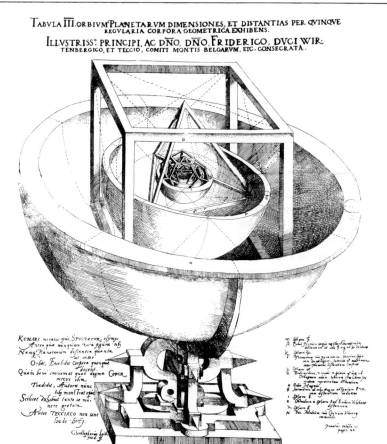

1.9 Kepler's view of the universe
Right: A model of the universe from Kepler's *Mystery of the Cosmos*. The outermost sphere is Saturn's. Above: The detail shows the spheres of Mars, the earth, Venus, and Mercury, with the sun in the center.

1.10   Isaac Newton

After Galileo's work, who could any longer doubt the value and wisdom of looking directly at nature? Perhaps the first and certainly the most interesting individual to be inspired by Copernicus and Galileo was the astrologer/astronomer Johannes Kepler (1571–1630), who, convinced both of the Copernican theory and of a God-given order and harmony in the universe, set out to find the divine plan. He began by trying to fit the orbits of the planets into a series of the five perfect geometric solids—the faces of a cube drawn inside the sphere of Saturn's orbit would just contain the sphere of Jupiter's, and a tetrahedron inside it would then just hold the Martian orbital sphere, and so on (Fig. 1.9). Later Kepler even tried to represent the planetary motions as musical chords. Along the way, he discovered the true order of the solar system: the planets move in elliptical orbits at speeds that vary according to their distance from the sun.

Kepler's work more than anyone else's demonstrated the value of a faith that order exists in nature. If there were no underlying order, science would be a waste of time. Faith that the physical world can be understood in terms of orderly relationships and universal laws, that physical events have comprehensible, impersonal causes, and that the whims of gods, magicians, or evil spirits need no longer be invoked to explain physical events, motivated that archetypal scientist Isaac Newton (1642–1727; Fig. 1.10). His discovery of the Law of Gravitation, the principles of optics, the so-called Newtonian mechanics, and the calculus fully justified that faith, and marked the birth of modern physics.

THE BEGINNINGS OF MODERN BIOLOGY

Newton's spectacular achievements began a 150-year alliance between science and religion that culminated in the discovery of evolution. Some clerics still argued endlessly over the seemingly unlimited interpretations of the Scriptures, but others, particularly the Protestants, saw in the testament of nature as explicated by science the eloquent and overwhelming evidence of a divine wisdom. This was the age of "natural religion"; nature was considered by many as more reliable evidence of God's existence than the Scriptures, with all their apparent contradictions and ambiguities.

The forerunners of modern biological investigation appeared at about the same time as Copernicus and Galileo. Three individuals set the basic course the life sciences were to follow. The earliest was Andreas Vesalius (1514–1564), who made the first serious studies of human anatomy by dissecting corpses. He discovered that the body is composed of numerous complex but beautiful subsystems, each with its own function, and he pioneered the comparative approach, using other animals to work out the purpose and organization of these anatomical units. A typical (if rather grisly) example is his demonstration that the nerve from the brain to the throat, common to so many animals, is responsible for controlling vocalization. When he took a squealing pig, dissected out the nerve, and cut it, the struggling animal instantly became mute even though its vocal apparatus remained intact (Fig. 1.11).

This powerful style of comparative and experimental study was carried forward by the English physician William Harvey (1578–1657), who showed conclusively that the heart pumps the blood, and the blood circulates. The heart, in short, is not in some metaphysical sense the seat of emotions, but a mechanical device with a clear function. As a result of these studies and the anatomical work that followed, an increasingly mechanistic point of view toward life began to develop.

The third of the pioneers was Antony van Leeuwenhoek (1632–1723; Fig. 1.12). Just as Galileo had the brilliant idea of pointing the newly invented telescope at the heavens, so Leeuwenhoek had the idea of using the microscope—with which he inspected cloth, as a draper's assistant—to look at living things. The most important of his many discoveries were microorganisms (including bacteria), sperm and the eggs they fertilized, and the cells of which all living things seemed to him to be composed.

For biology, unlike physics, centuries of painstaking observation were required to establish the science's fundamental generalizations. The cell theory, for instance, was not given its essentially modern form until 1858,

1.11 Cutting the vocal nerve
This is the initial letter in Vesalius' *Fabric of the Human Body* (1542).

1.12 Antony van Leeuwenhoek

1.13 Louis Pasteur

1.14 Charles Darwin

and only about 125 years ago, in 1862, did Louis Pasteur (Fig. 1.13) disprove the theory of spontaneous generation. With the realization that Leeuwenhoek's microorganisms might be responsible for disease, the English surgeon Joseph Lister proved the effectiveness of antiseptics (from *anti-*, "against," and *sepsis*, "decay") in 1865, and Pasteur greatly expanded the use of vaccination.

By far the most important figure in the history of biology, however, is Charles Darwin (1809–1882; Fig. 1.14). The publication in 1859 of his *The Origin of Species,* presenting the theory of evolution by natural selection, suddenly provided a coherent, organizing framework for the whole of biology. His work sparked the explosive growth of biological knowledge that continues today. As the most important unifying principle in biology, the theory of evolution underlies the logic of every chapter of this book.

DARWIN'S THEORY

The theory of evolution by natural selection, as modified since Darwin, will be treated in detail in Chapter 33. But since we shall be referring to it in earlier chapters, we must examine the essential concepts of Darwin's theory at the outset. It consists of two major parts: the concept of evolutionary change, for which Darwin presented a great deal of evidence, and the quite independent concept of **natural selection** as the agent of that change.

THE CONCEPT OF EVOLUTIONARY CHANGE

Until only two hundred years ago, it seemed self-evident that the world and the animals that fill it have not changed: robins look like robins and mice like mice year after year, generation after generation, at least within the short period of written history. This commonsense view is very like our untutored impression that the earth stands still and is circled by the sun, moon, planets, and stars: it accords well with day-to-day experience, and until evidence to the contrary appeared, it provided a satisfying picture of the living world. The idea of an unchanging world also corresponded to a literal reading of the powerfully poetic opening of the Book of Genesis, in which God is said to have created each species independently, simultaneously, and relatively recently—a little over six thousand years ago by reckonings based on Scripture.

But problems with the commonly held scriptural theory of creation arose from many sources; scientists attempted first, quite naturally, to discount the evidence and then, when that proved impossible, to construct a new explanation. Let's look at the evidence for evolution that confronted Darwin and his contemporaries.

The most dramatic findings came from geology. In the eighteenth century a picture of a changing earth had begun to emerge. Extinct volcanoes and their lava flows had been discovered; most geological strata were found to represent sedimentary deposits, laid down layer upon layer a millimeter at a time in columns three thousand meters or more deep; the gradual erosive action of wind and water were seen to have leveled entire mountains

A

B

and carved out valleys; unknown forces had caused mountains to rise where ocean floors had once been. This latter fact in particular was impressed upon Darwin when he discovered fossilized seashells high in the Andes. Each of these phenomena implied continuous change during vast periods of time.

Another problem for the static view of life was presented by the fossils themselves. Many represented plants and animals wholly unknown in Europe, and though theologians had argued that the organisms these fossils represented were alive in the New World, increasingly intensive exploration of the Americas indicated that the hundreds of species of dinosaurs, for instance, were really extinct. In addition, many previously unknown and often bizarre animals inhabited the Americas (Fig. 1.15). As the realization grew that the number of animal species for which evidence was accumulating ran at least into the hundreds of thousands, Noah's ark began to seem very small indeed. In fact, it appeared that the extinct species greatly outnumbered the living, and that new constellations of species had come and gone several times in the past. Moreover, the lowest, oldest rocks contained only the most primitive fossils—seashells, for instance—and these were followed in order by the more modern forms: fish appeared later, for example; reptiles still later; then birds and mammals. The hypothesis of a young earth populated almost overnight by a single bout of creation began to seem very unlikely.

Jean Baptiste de Lamarck (1744–1829; Fig. 1.16) was the first to offer the major alternative explanation of the fossil record: evolution. Lamarck had arranged fossils of various marine molluscs in order of increasing age; he saw clearly that certain species had slowly changed into others, and concluded that this process of slow change had continued right to the present day. As Lamarck put it in 1809, "it is no longer possible to doubt that nature has done everything little by little and successively," over a nearly infinite period of time. In Lamarck's view, the living world had begun with simple organisms in the sea, which eventually moved onto the land, and evolution

1.15 Challenges to traditional ideas on the origin of species

The discovery of fossils of now-extinct species brought into question the static view of life. Shown here are the remains of a baby mammoth that had been preserved in the permafrost in Siberia (A). The discovery in the New World of organisms unfamiliar to Europeans, such as the anteater (B), also required a reinterpretation of traditional ideas on the origin of species.

1.16 Jean Baptiste de Lamarck

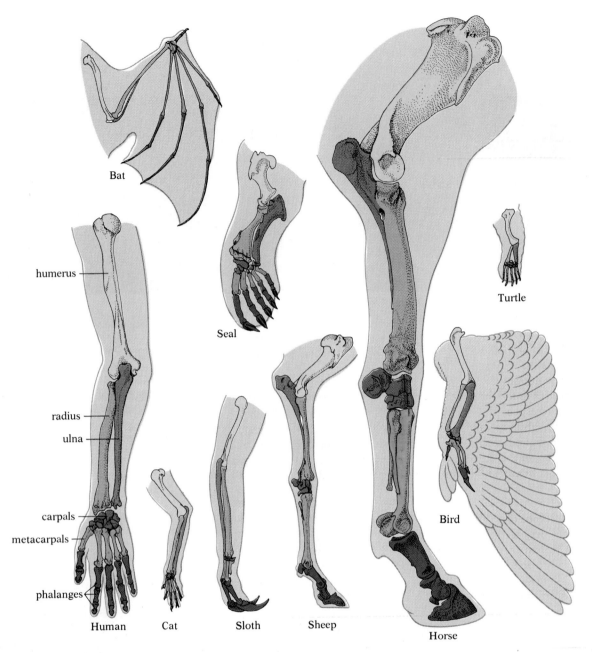

Bat

Seal

Turtle

humerus

radius

ulna

carpals

metacarpals

phalanges

Human   **Cat**   **Sloth**   **Sheep**

Bird

Horse

1.17   A comparison of the bones in some vertebrate forelimbs

The labeled and color-coded bones of the human arm at left permit identification of the same bones in the other forelimbs depicted. In the bat the metacarpals (hand bones) and phalanges (finger bones) are elongated as supports for the membranous wing. In the seal the bones are shortened and thickened in the flipper. The cat walks on its phalanges (toes), the metacarpals having come to form a part of the leg. The sloth normally hangs upside down from tree limbs—hence its recurved claws. The horse walks on the tip of one toe, which is covered by a hoof (specialized claw), and the sheep walks on the hoofed tips of two toes (it is therefore cloven-hoofed, though only one hoof can be seen in this side view). The carpals (wristbones) of both the horse and the sheep are elevated far off the ground, because the much-elongated metacarpals (hand bones) have become a section of the leg. Small splintlike bones that are vestiges of other ancestral metacarpals can be seen on the back of the upper portion of the functional metacarpals of both horse and sheep. All the animals mentioned so far—human, bat, seal, cat, sloth, horse, and sheep—are mammals, but the same bones can also be seen in the leg of a turtle and the wing of a bird. (All limbs are drawn to the same scale except that of the turtle, which is enlarged.)

14

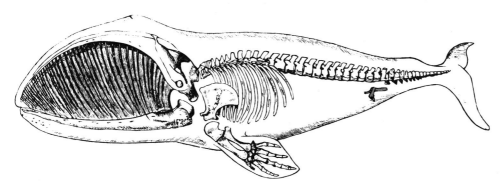

1.18   Rudimentary hind limb of a whale
Whales lost their hind limbs long ago, when they returned to the sea from the land, but they retain rudimentary bones that correspond to the pelvic girdle and the thighbone (color).

had culminated with the appearance of our species, the inevitable result of the gradual trend toward change and "increasing perfection."

Lamarck was basically on the right track, though as we shall see his mechanism for evolutionary change was incorrect. But he was ignored for the very understandable reason that he could not offer sufficient evidence for the *fact* of evolution. Darwin, only fifty years later, was in a far better position: there was much more evidence of the sort Lamarck had pointed to, and Darwin, a respected geologist, was well acquainted with it. Furthermore, he had the ability to spot important data in the midst of apparent chaos. He could find powerful support for the idea of evolution where Lamarck and others—if they looked at all—saw only irrelevancies.

One of the most important lines of evidence put forward by Darwin was the existence of morphological resemblances among living species (the findings of what we today call comparative anatomy). If, for example, we observe the forelimbs of a variety of different mammals, we see essentially the same bones arranged in the same order (Fig. 1.17). The basic bone structure of a human arm, a cat's front leg, and a seal's flipper is the same; the same bones are present even in a bird's wing. True, the size and shape of the individual bones vary from species to species, and some bones may be missing entirely in one species or another, but the basic construction is unmistakably the same. To Darwin the resemblance suggested that each of these species had descended from a common ancestor from which each had inherited the basic plan of its forelimb, modified to suit its present function. The observation that structures with important functions in some species appear in vestigial nonfunctional form in others further convinced Darwin of the reality of evolutionary change. Why otherwise would pigs, which walk on only two toes per foot, have two other toes that dangle uselessly well above the ground? Why would certain snakes, such as the boa constrictor, and many species of aquatic mammals, such as whales, have pelvic bones and small, internal hind-limb bones (Fig. 1.18)? Why would flightless birds such as penguins, ostriches, kiwis, and the cormorants of the Galápagos Islands still have rudimentary wings—or feathers, for that matter (Fig. 1.19)? Why would so many subterranean and cave-dwelling species have useless eyes buried under their skin?

1.19   Flightless cormorant on the Galápagos Islands

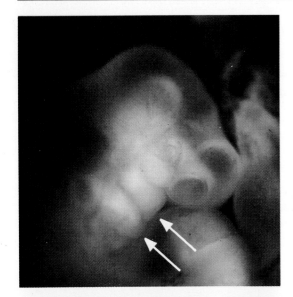

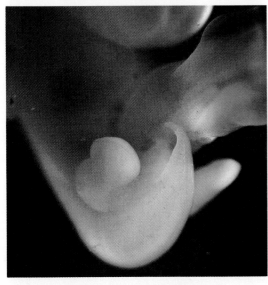

1.20 Embryological evidence of evolution
Top: Pharengeal ("gill") pouches (arrows) in a 4-week human embryo. Bottom: Tail in a 5-week human embryo.

Embryology—the study of how living things develop from eggs or seeds to their adult forms—also provided powerfully suggestive evidence. Darwin pointed out that in marine crustaceans as different as barnacles and lobsters the young larvae are virtually identical, implying a common descent. Telltale traces of their genealogy are obvious in vertebrates as well. Human embryos, for instance, have gill pouches and well-developed tails that disappear before the time of birth (Fig. 1.20). It seemed clear to Darwin that such inappropriate structures are inherited vestiges of structures that functioned in ancestral forms, and that may still function in other species descended from the same ancestors.

Another particularly convincing line of evidence offered by Darwin was the well-known ability of breeders to produce dramatic changes in both plants and animals. How could anyone contemplating the historical evidence of the alterations in domesticated species doubt that vast changes are possible, given sufficient time? Great Danes, sheep dogs, Irish setters, Yorkshire terriers, poodles, bulldogs, and dachschunds, for instance, are all members of the same species, bred from tamed wolves to look like almost anything breeders have fancied. Similarly, cabbage, brussels sprouts, cauliflower, broccoli, kohlrabi, rutabaga, curly greens, and savoy have all been bred from the same species, the wild form of which looks nothing like its domesticated progeny (Fig. 1.21). So, too, the many varieties of chickens, cattle, horses, flowers, grains, and so on have been bred over the years. Who could compare the colors and shapes of the wild rose or jonquil to the many colors and shapes of the far larger domesticated roses or daffodils, with new varieties bred each year, and doubt that a species has the capacity to change enormously even in a hundred years? In everything Darwin looked at—fossils, anatomy, embryology, and breeding—he saw the same message: species can and do change (Fig. 1.22).

THE CONCEPT OF NATURAL SELECTION

But what mechanism accounts for the changes? Lamarck's now-discredited hypothesis was one of the first attempts at a plausible explanation. Lamarck was impressed by how well suited each animal was to its particular position in the web of life, even though the environment had changed enormously again and again over countless millions of years. To account for this ability to adapt, he imagined that God had given each species a tendency toward perfection which allowed for small alterations in morphology, physiology, and behavior to accommodate changes in the environment, and that these alterations could be inherited by the offspring.

Belief in Lamarck's idea of a natural tendency toward perfection and the inheritance of acquired characteristics required no more faith in his day than did belief in other invisible everyday forces, such as gravity and magnetism. His hypothesis was a perfectly logical extension of the prevalent Western view that God had set things going by creating nature and nature's laws, and had then left things for the most part to run themselves. But where in the vestigial legs of whales or the dangling toes of pigs was there evidence of perfection? Instead the clear mark of compromise was everywhere. Plants and animals were well adapted to their places in the environment, but they were by no means perfect.

1.21 Selective breeding of the wild cabbage
The wild species *Brassica oleracea* (left) has been bred to create cauliflower (upper right), brussels sprouts (middle right), and cabbage (lower right). Each represents a selective exaggeration of one part of the wild plant—the flower heads for cauliflower, the side buds for brussels sprouts, and the leaves for cabbage. Despite the extreme morphological differences, however, the three domestic vegetables can be interbred.

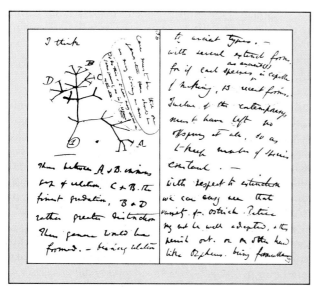

1.22 Pages from one of Darwin's notebooks
Scientists of the nineteenth century and earlier, among them Charles Darwin, often recorded in notebooks and journals their experiments, observations drawn from field trips, or simply their thoughts about hypotheses in the making. In addition to the many books he wrote, Darwin produced notebooks on a wide variety of subjects, including his trip to South America and the Galápagos (1831–1836), later described in the *Journal of Researches during the Voyage of the Beagle* (1840). The notebook from which this excerpt is taken was written in 1837, one year before Darwin conceived of natural selection as the likely mechanism for evolution. These pages show Darwin's first recorded drawing of an evolutionary tree, a metaphor for the diversity and interrelatedness of species that has continued in the scientific literature to the present time. The trunk represents a common ancestor, the limbs major groups, and the twigs particular species, either extinct (indicated by a crossbar) or living.

Darwin proposed a different mechanism—natural selection—requiring no internal tendency other than the one toward variation so obvious in nature. Darwin had conceived the idea of natural selection two years after his return from his voyage to the Americas on the *Beagle*, but was only goaded

1.23   Thomas Robert Malthus

into publishing, twenty years later, by the receipt in 1858 of A. R. Wallace's manuscript proposing essentially the same theory. (We normally associate Darwin's name with the theory because of the impressive evidence he presented—he had been collecting it for two decades—and because of his thorough exploration of the theory's many ramifications.)

In essence, Darwin put together two ideas. The first was that numerous variations exist within species, and that variations are largely heritable. Immersed as he was in the Victorian preoccupation with plant and animal breeding, Darwin knew that while cuttings produce plants identical to the parent, sexual reproduction produces individual offspring that differ both from their parents and from each other. Variation is a fact of life: breeders, as we know, are able to select for desirable traits and create new, morphologically distinct lines of plants and animals.

Darwin's second inspiration came on 28 September 1838, when he re-read the *Essay on the Principle of Population* by the economist Thomas Robert Malthus (Fig. 1.23). Malthus pointed out that humans produce far more offspring than can possibly survive; population growth always outruns any increase in the food supply and is held in check largely by war, disease, and famine, with vast numbers of people perpetually on the edge of starvation. Both Darwin and Wallace were struck at once by the consequence of applying the gloomy Malthusian logic to plants and animals: like humans, the creatures of each overpopulated generation must compete for the limited resources of their environment, and some—indeed most—must die. Each female frog, for example, produces thousands of eggs per year, and a fern produces tens of millions of spores, yet neither population is growing noticeably. Any organism with naturally occurring heritable variations that increase its chances in this life-or-death contest will be more likely than others to survive long enough to have offspring, some of which will inherit these variations. They in turn will have an above-average chance to survive the struggle, and so will form an increasingly large part of the population. As a result of this "selection," the population as a whole will become better adapted, and the never-ending struggle for existence will then turn on the possession of still better adaptations. To distinguish this process from the sort of directed selection practiced by agriculturalists, Darwin called it *natural* selection.

Nowhere is Darwin's attitude toward nature and his link with the tradition of natural religion and the idea of a God of Secondary Causes clearer than in the final sentence of later editions of *The Origin of Species:* "There is grandeur in this view of life, with its several powers, having been originally breathed by the Creator into a few forms or into one; and that, whilst this planet has gone cycling on according to the fixed law of gravity, from so simple a beginning endless forms most beautiful and most wonderful have been, and are being evolved."

The contrast between artificial and natural selection that served Darwin so well provides an instructive summary of the evolutionary process. In both, far more offspring are born than will reproduce; in both, differential reproduction, or selection, occurs, causing some inherited characteristics to become more frequent and prominent in the population and others to become less so as the generations pass. But in the breeding of domesticated plants and animals, selection results from the deliberate choice by the

1.24 Selective breeding of pigeons
By practicing rigorous selection, breeders can achieve major changes in relatively few generations. The ancestral rock dove is shown in the center. The domestic breeds (clockwise from upper left) are fantail, double-crested Saxon, Schmalkalden Moorhead, English carrier, pigmy pouter, Norwich cropper, and frillback.

breeder of which individuals to propagate. In nature, it takes place simply because individuals with different sets of inherited characteristics have unequal chances of surviving and reproducing. Notice, by the way, that selection does not change individuals. An individual cannot evolve. The changes are in the makeup of populations.

Artificial and natural selection also differ significantly in the *degree* of selection, and its effect on the rate of change. Breeders can practice rigorous selection, eliminating all unwanted individuals in every generation and allowing only a few of the most desirable to reproduce. They can thus bring about very rapid change (Fig. 1.24). Natural selection, which involves a large measure of chance, is usually much less rigorous: some poorly adapted individuals in each generation will be lucky enough to survive and reproduce, while some well-adapted members of the population will not. Hence evolutionary change is usually rather slow; major changes may take thousands or even millions of years, depending on the degree of selection pressure imposed by the environment and by other species.

Darwin's evidence for evolutionary change and the common descent of at least the major groups of organisms was widely accepted in his time, but the idea of natural selection by small steps remained controversial until the 1930s. Some biologists had difficulty seeing how an elaborate and specialized structure like an eye, for instance, could evolve, since the first rudimentary but necessary steps might lack obvious survival value. As we shall see in later chapters, an expanded understanding of the nature and organization of genes and their role in development has now made it clear that natural selection does explain most evolutionary change.

In summary, then, Darwin's explanation of evolution in terms of natural selection depends upon five basic assumptions:

1. Many more individuals are born in each generation than will survive and reproduce.
2. There is variation among individuals; they are not identical in all their characteristics.
3. Individuals with certain characteristics have a better chance of surviving and reproducing than individuals with other characteristics.
4. Some of the characteristics resulting in differential survival and reproduction are heritable.
5. Vast spans of time have been available for change.

All the known evidence supports the validity of these five assumptions.

MODERN BIOLOGY

The work in the latter half of the nineteenth century by scientists like Darwin, Pasteur, Gregor Mendel (the monk who discovered the basic principles of inheritance), and a number of developmental biologists combined to set the stage for the emergence of modern biology. Though we shall look briefly at the historical development of many of the central concepts of biology, this book will focus primarily on the mechanisms and logic behind life processes, the details of which are today being elucidated with breathtaking speed. Modern biology is approaching an understanding of the complex molecular, cellular, and evolutionary workings of life at a level undreamt of even a decade ago. Now more than ever, as we stand able to read out the genetic programs of organisms word for word, biology is the most exciting, intellectually stimulating, and promising discipline among the sciences.

PART I

THE CHEMICAL AND CELLULAR BASIS OF LIFE

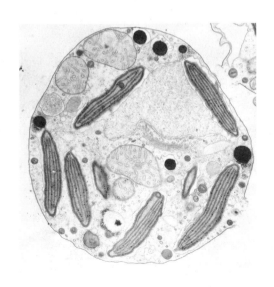

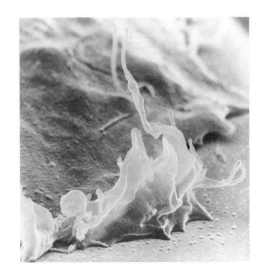

Chapter 2

SOME SIMPLE CHEMISTRY

The chemistry of life is both simple and complex. It is simple in the sense that the laws of chemistry can predict with great precision many chemical interactions. But, though often predictable, the chemistry of life is basically mysterious and intangible, depending on atoms we cannot see interacting by means of forces we cannot feel. Yet out of this invisible network of atomic interactions arises all that we see, hear, feel, taste, and smell of the world around us.

All life processes obey the laws of chemistry and physics. The same forces that assembled matter eons ago from the dust of the big bang regulate life today. In the behavior of molecules, atoms, and subatomic particles lies the key to such complex biological phenomena as the trapping and storing of solar energy by green plants, the extraction of usable energy from organic nutrients, the growth and development of organisms, the patterns of genetic inheritance, and the regulation of the activities of living cells. The study of biology, then, begins with—and continually returns to—the basic laws of chemistry and physics.

THE ELEMENTS

All the matter of the universe is composed of a limited number of basic substances called elements. There are 92 naturally occurring elements, and in addition many synthetic elements have been manufactured in the laboratory; the current total for both natural and artificial elements is well over a hundred.

Each element is designated by one or two letters that stand for its English or Latin name. Thus H is the symbol for hydrogen, O for oxygen, C for carbon, Cl for chlorine, Mg for magnesium, K for potassium (Latin, *kalium*),

TABLE 2.1 *Elements important to life*

Symbol	Element	Atomic number/ Mass number	Approximate percentage of earth's crust by weight	Approximate percentage of human body by weight
H	Hydrogen	1/1	0.14	9.5
B	Boron	5/11	Trace	Trace
C	Carbon	6/12	0.03	18.5
N	Nitrogen	7/14	Trace	3.3
O	Oxygen	8/16	46.6	65.0
F	Fluorine	9/19	0.07	Trace
Na	Sodium	11/23	2.8	0.2
Mg	Magnesium	12/24	2.1	0.1
Si	Silicon	14/28	27.7	Trace
P	Phosphorus	15/31	0.07	1.0
S	Sulfur	16/32	0.03	0.3
Cl	Chlorine	17/35	0.01	0.2
K	Potassium	19/39	2.6	0.4
Ca	Calcium	20/40	3.6	1.5
V	Vanadium	23/51	0.01	Trace
Cr	Chromium	24/52	0.01	Trace
Mn	Manganese	25/55	0.1	Trace
Fe	Iron	26/56	5.0	Trace
Co	Cobalt	27/59	Trace	Trace
Ni	Nickel	28/59	Trace	Trace
Cu	Copper	29/64	0.01	Trace
Zn	Zinc	30/65	Trace	Trace
Se	Selenium	34/79	Trace	Trace
Mo	Molybdenum	42/96	Trace	Trace
Sn	Tin	50/119	Trace	Trace
I	Iodine	53/127	Trace	Trace

TABLE 2.2 *Fundamental particles*

Particle	Mass (in units of 10^{-24} grams)	Charge (in electronic charge units)
Electron	0.001	−1
Proton	1.672	+1
Neutron	1.674	0

Na for sodium (Latin, *natrium*), and so on. Only a few of the 92 naturally occurring elements are important in life processes (Table 2.1).

Matter cannot be subdivided without limit. Progressive subdivision ultimately leads to units indivisible by ordinary chemical means. These units are called **atoms**. The atoms of a particular element are alike in many essential characteristics and differ in many measurable ways from the atoms of all other elements. A single atom is customarily represented by the chemical symbol for the element. N, for example, can represent either a single atom of nitrogen or the element itself.

ATOMIC STRUCTURE

Though atoms can be considered the basic chemical units of matter, they are themselves composed of still smaller particles. Many of these particles belong to the world of subatomic physics and are of little immediate concern to biologists. But three of them—the proton, the neutron, and the electron—play a central role in determining the biological activity of elements. In their interactions lie the power and the cohesion that make life possible.

The atomic nucleus All the positive charge and almost all the mass of an atom are concentrated in its center, or nucleus, which contains two kinds of so-called primary particles, the **proton** and the **neutron**. Each proton carries a charge of +1: one positive electronic charge unit.[1] The neutron, as its name implies, has no charge. The proton and the neutron have roughly the same mass, though strictly speaking the neutron is slightly heavier. In fact, the neutron is composed of a proton and an electron closely bound together and canceling each other's electrical charge (Table 2.2).

The number of protons in the nucleus is unique for each element. This number, called the **atomic number**, is sometimes written as a subscript immediately before the chemical symbol. Thus $_1$H indicates that the atomic number of hydrogen is 1; that is, its nucleus contains only one proton. Similarly, $_8$O indicates that each oxygen nucleus contains eight protons.

It is often desirable to indicate the total number of protons and neutrons in a nucleus; this number is called the **mass number**, because it approximates the total mass (commonly called the atomic weight) of the nucleus. The mass number is usually written as a superscript immediately preceding the chemical symbol. For example, most atoms of oxygen contain eight neutrons; the mass number is therefore 16, and the nucleus can be symbolized as ^{16}O or, if we wish to show both the atomic number and the mass number, as $^{16}_{8}$O.

Though the number of protons is the same for all atoms of the same element, the number of neutrons is not always the same, and neither, consequently, is the mass number. For example, most oxygen atoms, as we have seen, contain eight protons and eight neutrons and have a mass number of 16; some, however, contain nine neutrons and therefore have a mass number of 17 (symbolized as ^{17}O), and still others have ten neutrons and a mass number of 18 (symbolized as ^{18}O). Atoms of the same element that differ in mass, because they contain different numbers of neutrons, are called **isotopes**; ^{16}O, ^{17}O, and ^{18}O are three isotopes of oxygen. Some elements have as many as 20 naturally occurring isotopes; others have as few as two.

Figure 2.1 illustrates three different isotopes of hydrogen: $^1_1$H, which is the usual form of the element; $^2_1$H, a stable isotope generally called deuterium; and $^3_1$H, an unstable isotope called tritium. Both deuterium and tritium have been used extensively in tracing the movements of hydrogen in biochemical reactions. As we shall note again and again, isotopes of various elements are invaluable research tools for biologists.

The electrons The portion of the atom outside the nucleus contains the third kind of primary particle—the **electron**. Though electrons have very little mass (see Table 2.2), their behavior is the single most crucial factor in the chemistry of life. Each electron carries a charge of −1: one negative electronic charge unit—exactly the opposite of a proton's charge.

In a normal neutral atom, the number of electrons around the nucleus is exactly the same as the number of protons in the nucleus. The positive charges of the protons and the negative charges of the electrons cancel each

[1] For definitions of units of measurement, see Glossary.

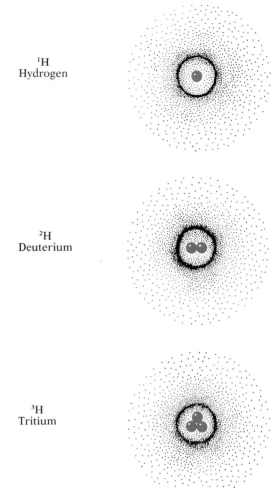

^{1}H
Hydrogen

^{2}H
Deuterium

^{3}H
Tritium

2.1 The three principal isotopes of hydrogen Each of the three isotopes has one proton in its nucleus and one electron. They differ in that ordinary hydrogen (^{1}H) has no neutrons in its nucleus, deuterium (^{2}H) has one, and tritium (^{3}H), which is unstable, has two. The single electron is not represented, but the volume within which it moves (a sphere) is indicated by stippling; the denser the stippling, the greater the likelihood that at any given moment the electron will be found in that portion of the sphere.

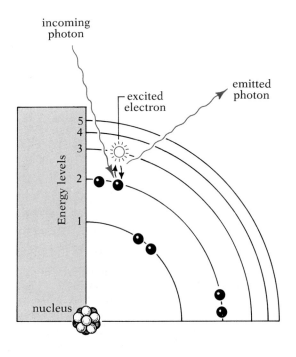

2.2   Energy levels of electrons
The electrons in an atom occupy discrete energy levels, here labeled 1 through 5. If an electron absorbs the right amount of energy (shown as a photon, a discrete particle of light) and there is a vacancy in a higher energy level, it can move to this level. Normally this "excited" electron quickly reemits the absorbed energy (here again shown as a photon) and returns to its original energy level.

other, making the total atom neutral. Consequently, in a neutral atom, the atomic number represents both the number of protons inside the nucleus and the number of electrons outside the nucleus. If, then, we see the symbol $^{35}_{17}Cl$, we can tell that a neutral atom of this isotope of chlorine has 17 protons, 18 neutrons, and 17 electrons. Similarly, the symbol $^{39}_{19}K$ means that this isotope of potassium contains 19 protons, 20 neutrons, and 19 electrons.

The electrons are not in fixed positions outside the nucleus. Each is in constant motion, and it is impossible to know exactly where a given electron is at any particular moment. For this reason some illustrations of atoms, such as Figure 2.1, do not show the electron itself, but indicate the region where the electron is likely to be.

The distance of an electron from the nucleus is a function of its energy; the higher its energy, the greater its probable distance from the nucleus. But in any particular atom, only certain discrete amounts, or "levels," of energy—like steps of a staircase—are possible. To occupy a certain step, or energy level, an electron must possess a specific amount of energy. To achieve a higher energy level an electron must absorb additional energy from some outside source. Conversely, when an electron falls into the next lower level, it emits the same amount of energy it previously took to move up from that level (Fig. 2.2). We refer to an electron occupying the lowest step available to it in the atom as being in the "ground state." Once it has absorbed enough energy to move up to the next energy level, it is said to be in an "excited state."

An electron in the excited state has a strong tendency to return to its ground state by emitting, in some form, the additional energy just acquired. Most often the energy is released as light. The decay of excited electrons in

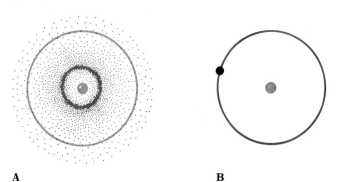

A   B

2.3   Two ways of representing the hydrogen atom
Since no one has ever seen the particles that make up an atom, all our knowledge of what atoms look like is indirect, and we can only picture them as models that fit the data. (A) The nucleus is shown here as a central gray area, with the "cloud" around it representing the region where the electron is likely to be. The circle encloses the orbital of the electron —the volume, a sphere, within which the electron will be found 90 percent of the time (see also Fig. 2.1). (B) Sometimes only the circle indicating the circumference of the orbital is shown; the electron may be represented by a small ball on the circle.

the lining of fluorescent tubes, for instance, helps us light our unnatural world. The fleeting moment of excitation, lasting 10^{-8} sec, is critical to living things: life on earth is based on an ability to capture and make use of the potential energy of excited electrons during that brief moment before they drop back down the energy staircase.

The volume within which an electron can be found 90 percent of the time is known as its **orbital**. In illustrations of atoms the orbital is often represented by a circle, the electrons sometimes being shown as round spots on the circle (Fig. 2.3). Such representations, of course, should not be taken to imply that the electrons are orbiting the nucleus at a precise distance from it, as envisioned in some older atomic models; the circle is merely a convention for indicating the border of the volume that probably contains the electrons.

The energy level of the hydrogen electron, which is the level nearest the nucleus, is often referred to as the K level (or K shell), and the orbital of the electron, which is spherical, is designated the $1s$ orbital. The K level can contain only two electrons. What happens, then, in an atom with more than two electrons, such as the oxygen atom, with its eight? Two of these electrons can be accommodated at the K level, but the other six must move at higher energy levels, farther from the nucleus. The next possible energy level is called the L level; it can contain a maximum of eight electrons. Since the most stable configuration for an atom is one in which its electrons have minimum energy, the six electrons outside the K level in an oxygen atom are all at the L energy level. Thus an oxygen atom has two K electrons and six L electrons.

The most likely distance from the nucleus of each of the six L electrons of an oxygen atom is roughly the same; as shown in Fig. 2.4, it is somewhat greater than the most likely distance of the K electrons. However, the orbitals of the L electrons are not all of the same shape. Two of these electrons have a spherical orbital (called the $2s$ orbital), which is like the $1s$ orbital of the K electrons except that it extends farther from the nucleus (Fig. 2.5). But the other L electrons have dumbbell-shaped orbitals (designated $2p$). The L energy level can contain three of these p orbitals, each oriented at right angles to the other two, so that each is aligned along a different one of

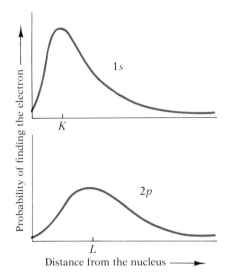

2.4 Graphs of the probability of finding electrons at various distances from the nucleus
Note that at no distance is the probability zero; the electrons could be at any distance, from right at the nucleus to infinity. But the most likely distance, K, for the $1s$ electron, which is at the first energy level, is less than the most likely distance, L, for the $2p$ electron, which is at the second energy level.

2.5 Representations of electron orbitals
The orbitals of s electrons are approximately spherical, those of p electrons roughly dumbbell-shaped. The numerals before s and p indicate the energy level. Thus the $1s$ electron is at the first energy level (K), that nearest the nucleus; the $2s$ and $2p$ electrons are at the L level, a higher energy level, and hence are at a greater average distance from the nucleus than the $1s$ electron. Note that, despite the very different shapes of their orbitals, the $2s$ and $2p$ electrons are at the same energy level—their most probable distances from the nucleus are the same.

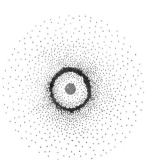

$1s$

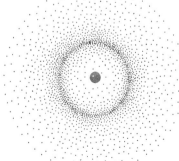

$2s$

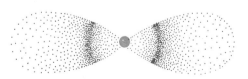

$2p$

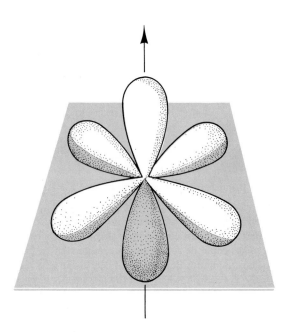

2.6   The three 2*p* electron orbitals
Each of the dumbbell-shaped orbitals is oriented in a different dimension of space, at right angles to the other two. (Imaginary plane and arrow inserted to aid visualization.)

the three dimensions of space (Fig. 2.6). The combination of the three 2*p* orbitals, which can contain a maximum of two electrons each, and the 2*s* orbital, which can also contain two electrons, accounts for the overall maximum of eight electrons for the *L* energy level. Obviously, not all these possible orbitals are filled in an oxygen atom.

When elements have more than 10 electrons (two at the *K* level and eight at the *L*), the additional electrons are accommodated at energy levels beyond *L*. Like the first two levels, each of these can contain only a limited number of electrons. Thus the third level *(M)* can contain a maximum of 18 electrons; the fourth *(N)* can contain 32; and so forth. In addition to *s* orbitals (spherical) and *p* orbitals (dumbbell-shaped), orbitals of other shapes may occur at these outer energy levels.

Though the third and successive levels can hold more than eight electrons, they are in a particularly stable configuration when they contain only eight. For our purposes, then, the first level can be considered complete when it holds two electrons, and every other level when it holds eight electrons.

Electron distribution and the chemical properties of elements   When the elements are arranged in sequence according to their atomic numbers —beginning with hydrogen, which has the atomic number 1, and proceeding to uranium, the last of the natural elements, with number 92—it can be seen that elements with very similar properties occur at regular intervals in the list (Fig. 2.7). For example, fluorine, number 9, is more like chlorine, number 17, bromine, number 35, and iodine, number 53, than like oxygen, number 8, or neon, number 10, the two elements immediately adjacent to it in the list. This tendency for chemical properties to recur periodically throughout the sequence of elements is called the Periodic Law.

The explanation for this periodicity is that the chemical properties of elements are largely determined by the number of electrons in their outermost shell (i.e., at their outermost energy level). If that shell is complete, as in helium (atomic number, 2), neon (10), or argon (18), the element has very little tendency to react chemically with other atoms (Fig. 2.7). If the outermost shell has one electron less than the full complement, the element has certain characteristic chemical properties; if it lacks two electrons, the element has somewhat different properties; if it lacks seven electrons, the element has very different properties.

We have said that fluorine is chemically similar to chlorine, bromine, and iodine. The critical characteristic shared by these four elements is that each has seven electrons in its outer shell. Oxygen, the element immediately preceding fluorine in the periodic table, has six electrons in its outer shell and hence is chemically quite different from fluorine. Neon, the element just after fluorine, has a full eight electrons in its outer shell and is therefore far less reactive than fluorine.

A convenient way to represent the electron configuration of the outer shell is to symbolize each electron by a dot placed near the chemical symbol for the element under consideration. Thus fluorine and chlorine, which, as we have said, have seven electrons at their outer energy level, would have the electron symbols

$$: \overset{\displaystyle ..}{\underset{\displaystyle ..}{F}} \cdot \qquad : \overset{\displaystyle ..}{\underset{\displaystyle ..}{Cl}} \cdot$$

Number of electrons in outer shell

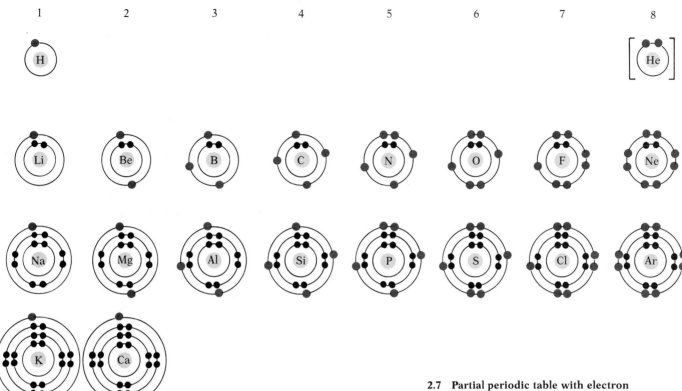

2.7 Partial periodic table with electron distributions
The first twenty elements are shown arranged according to their position in the periodic table. Elements in the same column share many chemical properties because they have the same number of electrons in their outer shell. (Helium is placed in column 8 even though it has only two outer electrons because, like neon, argon, and the other so-called noble gases, its outer shell is full, and its chemical properties are therefore those of a noble gas.)

Similarly, hydrogen with one electron in its shell, carbon with four in its outer shell, nitrogen with five, and oxygen with six would be shown as follows:

$$\mathrm{H}\cdot \qquad \cdot\overset{\displaystyle\cdot}{\underset{\displaystyle\cdot}{\mathrm{C}}}\cdot \qquad \cdot\overset{\displaystyle\cdot\cdot}{\underset{\displaystyle\cdot}{\mathrm{N}}}\cdot \qquad \cdot\overset{\displaystyle\cdot\cdot}{\underset{\displaystyle\cdot}{\mathrm{O}}}\!:$$

Of course the placement of the dots in no way indicates the actual positions of the electrons themselves.

Radioactive decay Atomic structure, as we shall see, is predictable. Protons, neutrons, and electrons all seek stability as atoms join together to complete their outer shells, or as unstable isotopes give off parts of themselves and reach a more stable state. Though the various isotopes of an element carry different numbers of neutrons, their identical electron distributions give them the same chemical properties. Their physical properties, however, differ in two ways that are important to biological research and to human health.

Unusual isotopes are taken up by tissues just as well as the more common forms of their respective elements, but since isotopes differ significantly in atomic weight, they can be distinguished from each other by such weight-sensitive techniques as mass spectroscopy and centrifugation. Matthew S. Meselson and Franklin W. Stahl, for instance, separated the rare isotope ^{15}N from the more common ^{14}N, and then provided it as the sole nitrogen source for DNA synthesis in bacteria. The resulting DNA, containing ^{15}N, was heavier than normal DNA, containing ^{14}N. Then they transferred the bacteria to a medium providing ^{14}N, so that any DNA synthesized thereafter would be of the lighter type. Because they could identify the DNA of earlier and later generations by weight as the cells grew and divided, Meselson and Stahl were able to show how DNA is replicated. Their work and its implications will be discussed in more detail in Chapter 26.

The other biologically significant physical property of some isotopes is their tendency to decay into a more stable form, giving off various particles on their way to physical stability. These so-called radioactive isotopes emit primarily ***alpha particles*** (each a unit of two protons and two neutrons), ***beta particles*** (electrons), and photons, often called ***gamma radiation***.

The tendency to decay is a consequence of balance between protons and neutrons in the nucleus. Protons, being each positively charged, strongly repel each other electrically. Protons and neutrons, however, attract each other by means of a nuclear force. The physical arrangement of the neutrons and protons, combined with the balance between these antagonistic forces, largely determines the stability of an isotope. That stability is measured by the ***half-life*** of the isotope: the time it takes half the atoms in a sample to decay. Tritium (^3H), for instance, has a half-life of about 12 years; ^{32}P, roughly 14 days; ^{40}K, 1.3×10^9 years; and so on.

Radioactive isotopes are extraordinarily useful in biology, since an isotope added to a sample emits radiation that scientists can track. With a "labeled" isotope of carbon dioxide (CO_2), for instance, we can trace how plants use carbon to build sugars. Because they are taken up by tissue as readily as their more stable counterparts, radioactive isotopes can also be used as "tracers" to help doctors locate circulatory blockages, pinpoint tumors, or predict potential problem areas. Naturally occurring isotopes make possible the dating of many rocks and fossils. The ratio of the radioactive isotope ^{14}C to the stable ^{12}C, for instance, is relatively constant in the CO_2 of the atmosphere, but once a plant has captured a CO_2 molecule and built it into a product like cellulose, the decay of ^{14}C atoms causes the ratio of ^{14}C to ^{12}C to decline steadily with time. Hence, the $^{14}C/^{12}C$ ratio in a sample provides a moderately accurate measure of age. With long-lived isotopes this "radioactive clock" technique can be extended far into the past: the ratio of uranium to the lead it decays into can reveal the age of rocks as much as 4 billion years old.

Isotopes, like most things, have their dark sides as well. A radioactive atom in a living cell poses two potential threats. First, it can, like uranium, decay into another element by losing protons, thereby altering the chemistry of its molecule completely. More often, radioactive isotopes produce highly reactive molecules that have too many or too few electrons to balance the electrical charge of their protons, and thus have a net charge. Beta decay, in which atoms throw off electrons, can produce this result directly. Gamma radiation, by bombarding nearby atoms with photons, can energize adjacent electrons so that they escape from their atoms. Since the behavior of electrons, as we shall see, determines the chemistry of life, such unpredictable and uncontrolled movement of electrons can disrupt the precisely ordered and carefully regulated workings of the cell. For instance, a change in a critical part of a cell's DNA, resulting from the redistribution of electrons in beta decay or gamma radiation, can trigger the complicated chain of events that leads to cancer. Occurring in the cells of the reproductive system, such a change in the DNA can cause defects in subsequent offspring. Changes of this sort arise in all of us every day from exposure to the sun's radiation and the natural decay of radioactive elements in the earth's atmosphere and crust, and each cell has a battery of defense mechanisms to counteract them. Our increasingly energy-dependent environment, however, may increase the odds of exposure to deleterious radioactivity.

CHEMICAL BONDS

The arrangement of electrons in the outer shell of the atoms of most elements gives those atoms an ability to bind to others to form new and more complex aggregates. When two or more atoms are bound together in this fashion, the force of attraction that holds them together is called a chemical bond. The atoms of each particular element can form only a limited number of such bonds; the arrangement of its electrons and the nature of the various charges they exhibit ensure that each element has its own characteristic bonding capacity.

IONIC BONDS

We have said that atoms are in a particularly stable configuration when the outer electron shell is complete—that is, in most cases, when it contains eight electrons. There is consequently a general tendency for atoms to form complete outer shells by reacting with other atoms. These reactions are the stuff of chemistry; the tendency of atoms to gain complete outer shells forms the basis upon which all chemistry is built.

Consider, for example, an atom of sodium (atomic number, 11). This atom has two electrons in its first shell, eight in the second, and only one in the third. One way sodium might gain a complete outer shell would be to acquire seven more electrons from some other atom or atoms. But the sodium atom would then have an enormous excess of negative charge, and since like charges repel each other, the electrons would tend to push each other away from the sodium. In point of fact, sodium cannot obtain a full outer shell by appropriating seven additional electrons. In nature it gives up the lone electron in its third shell to some electron acceptor, leaving the complete second shell as the new outer shell (Fig. 2.8).

Next, consider an atom of chlorine (atomic number, 17). This atom has two electrons in its first shell, eight in its second shell, and seven in its third shell. In other words, its outer shell is almost complete, lacking only a single electron. It cannot lose the seven electrons in its outer shell for reasons similar to those preventing sodium from gaining seven electrons. Only by

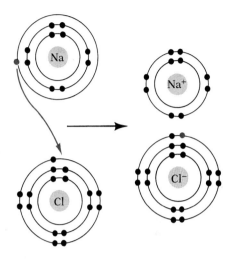

2.8 Ionic bonding of sodium and chlorine
Sodium has only one electron in its outer shell, while chlorine has seven. Sodium acts as an electron donor, giving up the one electron in its outer shell, whereupon the complete second shell functions as its new outer shell. Chlorine acts as an electron acceptor, picking up an additional electron to complete its outer shell. But after sodium has donated an electron to chlorine, the sodium, left with one more proton than it has electrons, has a positive charge. Conversely, the chlorine, with one more electron than it has protons, has a negative charge. The two charged atoms, called ions, are attracted to each other by their unlike charges. The result is sodium chloride (NaCl).

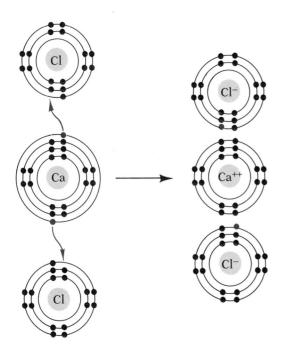

2.9   Ionic bonding of calcium and chlorine
Calcium has two electrons in its outer shell. It donates one to each of two chlorine atoms, and the two negatively charged chloride ions thus formed are attracted to the positively charged calcium ion to form calcium chloride (CaCl₂).

gaining an extra electron from some electron donor can chlorine acquire a complete outer shell.

If a strong electron donor like sodium (an atom with a strong tendency to get rid of an electron) and a strong electron acceptor like chlorine (an atom with a strong tendency to acquire an extra electron) come into contact, an electron may be completely transferred from the donor to the acceptor. The result, in the present example, is a sodium atom with one fewer electron than normal and a chlorine atom with one more electron than normal. Once it has lost an electron, the sodium is left with one more proton than it has electrons, and it therefore has a net charge of +1. Similarly, the chlorine atom that gained an electron has one more electron than it has protons and has a net charge of −1. Such charged atoms (or charged aggregates of atoms) are called *ions*, and are symbolized by the appropriate chemical symbol followed by a superscript indicating the charge. Sodium and chlorine ions are written Na⁺ and Cl⁻.

A sodium ion with its positive charge and a chlorine ion (usually called a chloride ion) with its negative charge tend to attract each other, since opposite charges attract. This important kind of electrical interaction is known as *electrostatic attraction*. In this instance electrostatic attraction holds the two ions together to form the compound we know as table salt, or sodium chloride, NaCl (Fig. 2.8). Such a bond, involving the complete transfer of an electron and the mutual electrostatic attraction of the two ions thus formed, is termed an *ionic bond*.

Ionic bonding may entail the transfer of more than one electron, as in calcium chloride, another common salt. Calcium (atomic number, 20) has two electrons in its outermost shell, and it loses both to form the calcium ion, Ca⁺⁺ (Fig. 2.9). Chlorine, however, requires only one electron to complete an octet in its outer shell, as we have already seen. Hence it takes two chlorine atoms to act as acceptors for the two electrons from a single calcium atom, and a total of three ions bond together to form calcium chloride, symbolized as CaCl₂ (the subscript 2 indicates that there are two chlorine atoms for each calcium atom in this compound). We can say, then, that calcium has a bonding capacity, or *valence*, of +2, while sodium has a valence of +1 and chlorine a valence of −1.

Ionic bonding occurs between strong electron donors and strong electron acceptors. It is not common between configurations that have intermediate numbers of electrons in the outer shells, or between two strong electron donors or electron acceptors.

Under some conditions the product of the binding of a calcium atom to two atoms of chlorine is one molecule of calcium chloride. A *molecule* is generally defined as an electrically neutral aggregate of atoms bonded together strongly enough to be regarded as a single entity. In many instances, however, *ionization* (the transfer of one or more electrons from one atom to another to form ions) occurs without true molecular formation. Substances like sodium chloride (NaCl) and calcium chloride (CaCl₂), for example, in which the bonds are almost exclusively ionic, have a pronounced tendency to dissociate into separate ions when in solution. (We shall look at why this happens shortly.) When they are ionized in solution, then, they do not exist as molecules; NaCl forms two separate entities, an Na⁺ ion and a

Cl⁻ ion (Fig. 2.10). Similarly, $CaCl_2$ in solution forms three separate entities, a Ca^{++} ion and two Cl⁻ ions. In nature, ionic compounds even in the solid state often do not form discrete molecules in the usual sense. In solid sodium chloride, many sodium and chlorine atoms are bound together into a large crystal (Fig. 2.11). There are no separate molecules composed of one sodium atom bonded to one chlorine atom, as the molecular symbol NaCl might seem to indicate. In a sense, the entire crystal can be conceived of as a single molecule.

Since ions are charged particles, they behave differently in living systems from neutral atoms or molecules, and substances wholly or partly ionized in water play many important roles in the functioning of biological systems. In later chapters we shall see the effects of charge on the movements of materials through the membranes of living cells, and the partitioning of positive and negative ions that gives rise to the differences in electrical potential essential for nerve and muscle activity.

Acids and bases   An *acid* can be characterized simply as a substance that increases the concentration of hydrogen ions (H⁺) in water, and a *base* as a substance that decreases the concentration of hydrogen ions, which in water is equivalent to increasing the concentration of hydroxyl ions (OH⁻). Note that H⁺ is simply a proton.

The degree of acidity or basicity (usually called alkalinity) of a solution is commonly measured in terms of a value known as *pH*, which is the negative logarithm of the concentration of hydrogen ions:

$$pH = \log\left(\frac{1}{[H^+]}\right) = -\log[H^+]$$

The pH scale generally ranges from 0 on the acidic end to 14 on the alkaline end. A solution is neutral, neither acidic nor alkaline (that is, it contains equal concentrations of H⁺ ions and OH⁻ ions), if its pH is exactly 7. Solutions with a pH of less than 7 are acidic (with a higher concentration of H⁺ ions than of OH⁻ ions); the lower the pH, the more acidic the solution. Conversely, solutions with a pH higher than 7 are alkaline (with a higher concentration of OH⁻ ions than of H⁺ ions); the higher the pH, the more alkaline the solution. You should realize that a change of one pH unit means a tenfold change in the concentration of hydrogen ions. Thus the concentration of H⁺ ions in the solution of a very strong acid may be as much as 100,000,000,000,000 (10^{14}) times greater than in the solution of a very

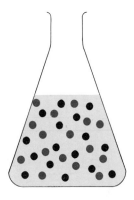

2.10   Ionization of sodium chloride
When in solution, the NaCl dissociates into separate Na⁺ (black) and Cl⁻ (color) ions.

2.11   The arrangement of ions in crystalline table salt (sodium chloride)
The imaginary lattice indicates the spatial arrangement of the Na⁺ and Cl⁻ ions.

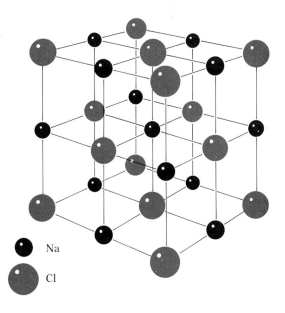

● Na

● Cl

	pH	H⁺	OH⁻	

Concentration of ions (moles/liter)

Substance	pH	H^+	OH^-	
Caustic soda (NaOH)	14	10^{-14}	10^0	A
	13			L
	12	10^{-12}	10^{-2}	K
Detergent	11			A
	10	10^{-10}	10^{-4}	L I
Baking soda	9			N
Seawater	8	10^{-8}	10^{-6}	E
Pure water	7	10^{-7}	10^{-7}	NEUTRAL
Saliva				
	6	10^{-6}	10^{-8}	
Rainwater Coffee	5			
Beer	4	10^{-4}	10^{-10}	A
Orange juice	3			C
Carbonated soft drink	2	10^{-2}	10^{-12}	I D
Stomach acid	1			I
Hydrochloric acid (HCl)	0	10^0	10^{-14}	C

strong base. Figure 2.12 illustrates the range of pHs we normally encounter.

Living matter is extraordinarily sensitive to pH. Except in parts of the animal digestive tract and a few other isolated areas, most cells function best when conditions are nearly neutral. Most of the interior material of living cells (excluding the nucleus and some vacuoles) has a pH of about 6.8. The blood plasma and other fluids that bathe the cells in our own bodies have a pH of 7.2–7.3. Numerous special mechanisms aid in stabilizing these fluids so that cells will not be subject to appreciable fluctuations in pH. Foremost among these mechanisms are certain chemical substances known as **buffers**, which have the capacity to bond to H⁺ ions, thereby removing them from solution whenever their concentration begins to rise, and conversely, to release H⁺ ions into solution whenever their concentration begins to fall. Buffers thus help minimize fluctuations in pH, which would otherwise be considerable since many of the biochemical reactions normally occurring in living organisms either release or use up H⁺ ions.

COVALENT BONDS

Ionic bonds, as we have seen, involve the complete transfer of electrons from one atom to another. But in most cases bonding occurs not by complete transfer, but by a sharing of electrons between the atoms involved. Bonds of this sort, based on shared electrons, are called **covalent bonds**. Take a hydrogen atom as an example. A complete first shell for hydrogen would contain two electrons, one more than each atom has normally. If the hydrogen gained an electron from some other atom it would have a full shell, but twice as much negative charge as positive charge (one proton and two electrons). Hydrogen does not, in fact, ionize in this manner. It tends to do the reverse; it loses its single electron, forming H⁺ ions, which are simply isolated protons since the hydrogen nucleus contains no neutrons. But suppose there is no strong electron acceptor available and the hydrogen cannot ionize. One possible reaction then is for two atoms of hydrogen to bond to each other and form what is called molecular hydrogen (H_2):

$$H \cdot + H \cdot \longrightarrow H : H$$

2.12 The pH scale

The concentration of hydrogen ions in a solution is measured by pH. At pH 7, the concentration of hydrogen ions (H⁺) exactly balances the concentration of hydroxyl ions (OH⁻), and so the solution is neutral. At lower pHs (corresponding to higher H⁺ concentrations) solutions are acidic; at higher pHs (corresponding to lower H⁺ concentrations) solutions are alkaline, or basic. Notice that the pH number matches the concentration of H⁺ in moles/liter—for example, pH 8 corresponds to an H⁺ concentration of 10^{-8}. A mole of any compound contains the same number of atoms—Avogadro's number, 6.02×10^{23}—as a mole of any other substance, and its weight in grams corresponds to the molecular weight (the summed atomic masses) of the substance.

In this molecule, each atom shares its electron with the other atom, so that each hydrogen has, in a sense, two electrons (Fig. 2.13).

Double and triple bonds   Covalent bonds are not limited to the sharing of one electron pair between two atoms. Sometimes two atoms share two or three electron pairs and form double or triple bonds. When two atoms of oxygen bond together, they form a double bond (remember that an oxygen atom needs two electrons to complete its outer shell), and when two atoms of nitrogen (atomic number, 7) bond together, they form a triple bond, because each nitrogen atom needs three additional electrons to fill its outer shell:

$$\overset{\cdot\cdot}{\underset{\cdot\cdot}{O}} :: \overset{\cdot\cdot}{\underset{\cdot\cdot}{O}} \qquad\qquad\qquad : N \vdots\vdots N :$$

A covalent bond may be represented simply by a line between two atoms, instead of a pair of dots; the other electrons in the outer shells are then ignored. Shown in this manner, H_2, O_2, and N_2 appear as follows:

$$H—H \qquad\qquad O{=}O \qquad\qquad N{\equiv}N$$

We have seen that hydrogen atoms tend to form only one bond; oxygen, two bonds; and nitrogen, three bonds. In other words, hydrogen has a covalent bonding capacity of 1; oxygen, a capacity of 2; and nitrogen, usually a bonding capacity of 3. Covalent bonding capacity is equivalent to valence: it corresponds to the number of vacancies in the outer shell of an atom with one to three vacancies, and to the number of sharable electrons in an atom with one to three outer electrons. The maximum covalent bonding capacity is 4: in an atom with four outer electrons, this is the number both of sharable electrons *and* of vacancies. Carbon is the most important example of an atom with a bonding capacity of 4; the ability of carbon to form so many bonds is in part what makes the diversity of life's chemistry possible.

Clearly, when two atoms of the same element bond together, as in these cases, one will not have more attraction for the shared electrons than the others; instead, the electrons will be shared equally by the two atoms. A bond in which the negatively charged electrons are likely to be no closer to one atom than to the other is said to be a ***nonpolar*** covalent bond.

Polar covalent bonds   Suppose, however, that instead of being bonded to each other, two hydrogen atoms are covalently bonded to an oxygen atom, forming water (H_2O):

$$H \cdot + H \cdot + \cdot \overset{\cdot\cdot}{\underset{\cdot}{O}} : \longrightarrow H : \overset{\cdot\cdot}{\underset{\ddot{H}}{O}} :$$

Oxygen, with six electrons in its outer shell (Fig. 2.7), needs two more. By sharing electrons with two hydrogen atoms, the oxygen atom can obtain a full outer octet, while at the same time each hydrogen obtains a complete first shell of two electrons. A covalent bond between a hydrogen atom and an oxygen atom is somewhat different from one between two hydrogen atoms or between two oxygen atoms, however. No two elements have exactly the same affinity for electrons. Consequently, when a covalent bond forms between two different elements, the shared electrons tend to be pulled closer to the more attractive element—the electron acceptor. Such a

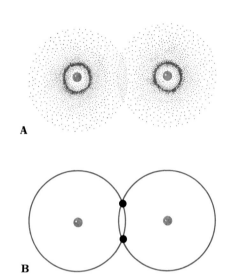

2.13   Covalent bonding of two hydrogen atoms
(A) The sharing of electrons is indicated by overlapping electron clouds. (B) Alternatively, the sharing may be indicated by interlocking orbital rings.

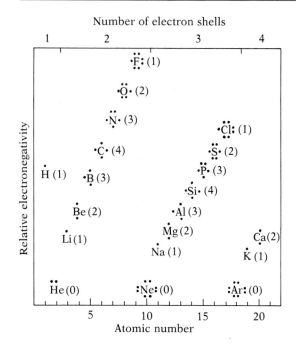

Number of electron shells

2.14 Electronegativity

The relative tendency of an atom to attract electrons depends on the number of spaces in the outer shell left to be filled. Hence, lithium (Li) with seven vacancies is less attractive to electrons (that is, less electronegative) than carbon (C), which has four. Oxygen (O), with only two missing electrons, is yet more electronegative. From this graph it is clear that in methane (CH_4) the shared electrons will be nearer the carbon atom, while in carbon dioxide (CO_2) they will be nearer the oxygens. The electronegativity of the noble gases (which have filled outer shells) cannot be measured, and so is simply estimated. This knowledge of relative electronegativity permits us to make important predictions about many biochemical reactions. (The covalent bonding capacity of each atom is shown in parentheses.)

bond is called a ***polar*** covalent bond because the charge is distributed asymmetrically.

The formal measure of an atom's attraction for free electrons is its ***electronegativity***; this depends upon the number of vacancies in the outer shell —an atom like oxygen with only one or two electron openings is generally more electronegative than one with three, and so on—and upon the distance of the outer shell from the nucleus (Fig. 2.14).

Since many covalent bonds are polar and since the degree of polarity of a bond varies over a wide range—from situations in which the shared electron is much closer to one of the atoms to situations in which it is only slightly closer—there is actually no sharp distinction between ionic bonds and covalent bonds. Ionic bonds represent one extreme, with the electrons pulled completely from one atom to the other, and nonpolar covalent bonds represent the other extreme, with the electrons pulled with equal force, and hence shared, by two atoms. Polar covalent bonds represent the usual case, a middle ground between these two extremes: the electrons are pulled closer to one atom than to the other, but not all the way.

The phenomenon of polarity helps explain many of the properties of various molecules in living systems. Whole molecules can be polar as a result of the polarity of bonds within them. One example is the water molecule. We can imagine how, even though the two hydrogen–oxygen bonds are polar, the atoms in the water molecule might be aligned in a straight line so that the charge would be distributed symmetrically within the molecule, which would then be nonpolar:

$$H : \overset{..}{\underset{..}{O}} : H$$

But this is not the actual arrangement. When the second energy level of oxygen has been filled by shared electrons, the resulting slight polarity of the covalent bonds induces the four pairs of electrons to adopt an arrangement in which the $2s$ orbital and the three $2p$ orbitals, their shapes highly modified, are oriented to the four corners of a tetrahedron (Fig. 2.15). Instead of the hypothetical linear arrangement, the three atoms form a bent-chain, or V-shaped, structure, with the oxygen at the apex of the V and the two hydrogen atoms as the arms:

$$+H : \overset{..}{\underset{..}{O}} : \overline{}$$
$$\overset{..}{H}$$
$$+$$

Since the electrons are drawn closer to the oxygen atom, there is a concentration of negative charge near the oxygen end of the molecule. Therefore, the molecule is polar (Fig. 2.16, right).

The carbon dioxide molecule, on the other hand, exhibits no polarity: its double bonds hold its atoms in rigid linear alignment (Fig. 2.16, left). Hence CO_2 is nonpolar.

As we shall see shortly when we look at the properties of water in detail, the polarity of molecules often has crucially important biological implications.

BIOLOGICALLY IMPORTANT WEAK BONDS

Strong versus weak bonds   To maintain internal stability, or homeostasis, living organisms must be able to change to meet the constant fluctuations of their environments. The changes all begin at the molecular level, and are powered by the liberated energy derived from the sorts of strong, energy-rich covalent bonds we have just discussed. Covalent bonds are called strong because breaking them is hard, usually requiring between 50 and 110 kilocalories of energy (most often supplied as heat) per mole.[2] Bond breakage usually results from collisions with rapidly moving molecules. Since the energy from even the most rapidly moving molecules at physiological temperatures is almost never above 10 kcal/mole, covalent bonds are stable and show little tendency to rupture spontaneously.

But life, as we have said, depends on a capacity for change, as well as on stability. The crucial sources of this ability to change are weak, noncovalent bonds, which can readily be broken and re-formed. Ionic bonds in aqueous solutions, for instance, are relatively weak, averaging abut 10 kcal/mole. The average duration of an ionic bond (the interval between formation of the bond and a collision with a molecule moving rapidly enough to break it) is quite short. The consequence, as we shall see, is that several weak bonds must act in concert to produce molecular stability sufficient for most of life's metabolic processes. The weak bonds (or interactions, as they are sometimes called) of biological significance include ionic bonds, hydrogen bonds, van der Waals interactions, and hydrophobic interactions.

[2] A calorie (spelled with a small *c*) is defined as the quantity of energy, in the form of heat, required to raise the temperature of one gram of pure water one degree, from 14.5°C to 15.5°C. One kilocalorie (kcal) is 1,000 calories. Nutritionists use a different scale to measure energy; their Calorie (spelled with a capital *C*) is equal to one kilocalorie on the standard scale. The number of calories required to break the bonds in a mole of particular compound tells us something about the strength of those bonds.

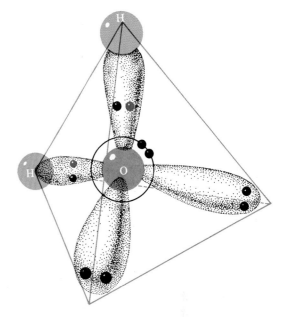

2.15   Structure of a water molecule
When oxygen bonds covalently with two hydrogen atoms, its second-level electrons are forced into hybrid orbitals oriented to the four corners of a tetrahedron. As a result, the angle between the two hydrogens is neither 90° nor 180°, as might be expected from the perpendicular arrangement of the 2p orbitals, but rather is 104.5°.

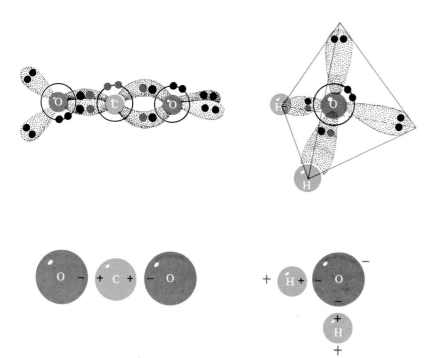

2.16   Polarities of two biologically important molecules
Left: In carbon dioxide, each of the four electrons of the carbon's outer orbital is shared between the carbon and one of the oxygens. The result is a linear molecule with no electrical polarity. Right: Two hydrogen atoms are bonded covalently to one oxygen atom, but the shared electrons are pulled closer to the oxygen than to the hydrogens. If the three atoms were arranged linearly, as in the carbon dioxide molecule, the charge distribution within the whole molecule would be symmetrical and the molecule would be nonpolar. But the atoms in water are arranged at an angle of 104.5°; the charge distribution is therefore asymmetrical, with negative charge concentrated at the oxygen end, and the molecule as a whole is polar.

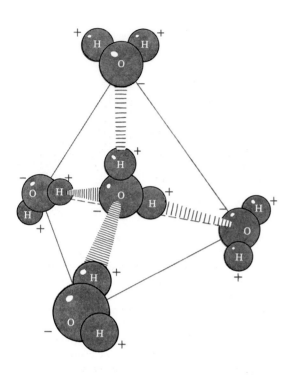

2.17 Hydrogen bonding between water molecules
Like the central H_2O molecule shown here, each
water molecule can form hydrogen bonds (color
bands) with four other water molecules. The array
then assumes the shape of a tetrahedron.

Hydrogen bonds   The electrostatic attraction between oppositely
charged portions of neighboring polar molecules results in *hydrogen
bonds*. Water molecules provide an excellent example. The hydrogen atoms
in each water molecule are covalently bonded to the oxygen atom, but because of the polarity of the bond—the electrons being closer to the oxygen
end than to the hydrogen ends—each hydrogen has a net positive charge. It
is therefore attracted by the oxygen atoms, with their net negative charge,
in other nearby water molecules. Since each of the hydrogens, while remaining covalently bonded to the oxygen atom of its own molecule, can
form a weak attachment with the oxygen of another water molecule, and
the oxygen can form a weak attachment with two external hydrogens, each
water molecule has the potential for being simultaneously linked by hydrogen bonds to four other water molecules (Fig. 2.17). In a sense, then, a volume of water is a continuous chemical entity, because of the hydrogen
bonding between the individual water molecules.

The distinction between hydrogen bonds and ionic bonds is clear: hydrogen bonds result from the electrostatic attraction between polar but electrically neutral molecules like water, while ionic bonds result from the
electrostatic attraction between oppositely charged atoms (ions). However,
an important bond also results from the electrostatic attraction between
ions and polar molecules, as we shall see when we discuss the hydration
sphere—the shell of polar water molecules drawn around an ion in solution. Pure hydrogen bonds usually have a bonding energy of about 4–5
kcal/mole; ionic bonds, about 10 kcal/mole; and polar/ionic bonds such as
those of the hydration sphere, about 7–8 kcal/mole.

Weaker interactions   Much weaker than ionic or hydrogen bonds are the
linkages known as *van der Waals interactions*, which have bonding energies of only 1–2 kcal/mole. These linkages occur between electrically neutral molecules (or parts of molecules) when they are so close to each other
that the electrons in their outer orbitals are set in synchronous, mutually
avoiding motion. The result of this momentary synchrony is that at one very
precise distance the normal repulsion between the two sets of electrons is
lessened and the atoms are able to bond weakly to each other. As the next
chapter will show, van der Waals interactions play a crucial role in the enzymatic reactions that control virtually all the processes of life.

Hydrophobic interactions are also very weak, their bonding energy
usually ranging between 1 and 3 kcal/mole. As implied by the name (from
the Greek roots *hydor*, "water," and *phobos*, "fear"), they occur between
groups of molecules that are insoluble in water. Such molecules, which are
nonpolar, tend to clump together in the presence of water, thus minimizing
their direct exposure to the water (Fig. 2.18).

The role of weak bonds   Weak bonds play a crucial role in stabilizing the
shape of many of the large molecules found in living matter—DNA and
proteins in particular—and they often hold together groups of such molecules in orderly arrays. Like the minute hooks and eyes of a Velcro fastening, the bonds are individually quite weak (their average duration is only
10^{-11} sec), but can be strong and stable when many act together (Fig. 2.19).

However, just as a Velcro patch is easy to unfasten when we begin at one end, attacking a few "bonds" at a time, so an array of molecules held together by weak bonds (as opposed to the more gripperlike covalent bonds) can be disassembled and rearranged with relative ease by forces that dissolve the fastening bonds one by one. These bonds are important because, as we shall see, an enormous number of life processes depend on just these sorts of changes.

SOME IMPORTANT INORGANIC MOLECULES

Chemists have traditionally referred to complex molecules containing the element carbon as *organic* compounds. All other compounds are called *inorganic*, a designation that should not mislead you into assuming that these compounds play no role in life processes. Many inorganic (non-carbon-based) substances are, in fact, basic to the chemistry of life. We shall examine a few of the most important here, for without some knowledge of them we can hardly understand the more complex organic compounds examined in the next chapter.

WATER

Life on earth is totally dependent on water. Between 70 and 90 percent of all living tissue is water, and the chemical reactions that characterize life all take place in a water-containing medium. Life based on some substance other than water could conceivably exist elsewhere in the universe, but such life would be vastly different from anything in our experience—so different that we might not recognize it as life even if we should stumble on it.

Water as a solvent   One of the main reasons water is so well adapted as the medium for life is that it is a superb solvent for many important classes of chemicals. It is a better solvent than most common liquids because of the marked polarity of the water molecule and the corresponding great propensity of water to form hydrogen bonds. Thanks to this polarity, both ionic substances and substances that are nonionic but polar are soluble in water. Let us consider the effect of water on each type of substance in turn.

We have mentioned that the ionic bonds linking the atoms of a salt such as NaCl are relatively weak when the salt is in an aqueous medium; but within a dry crystal of the same salt the bonds are comparatively strong. Why the difference? Within the dry crystal the electrostatic attractions between the positive sodium ions and the negative chloride ions is very strong; overcoming it requires a great deal of energy. But when the crystal is put into water, the attraction of the negatively charged oxygen ends of the water molecules for the positively charged sodium ions, and the similar attraction of the positively charged hydrogen ends of the water molecules for the negatively charged chloride ions, overcome by force of numbers the mutual attraction between the Na^+ and Cl^- ions. In water, then, the ionic bonds are broken with extreme ease because of the competitive attraction of the water; the Na^+ and Cl^- ions dissociate, and each of the ions becomes

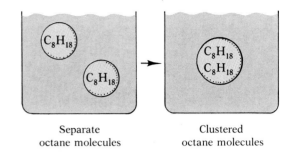

Separate   Clustered
octane molecules   octane molecules

2.18   Hydrophobic interactions
Left: Two molecules of octane, a strongly hydrophobic substance, are separately suspended in a volume of water. Each occupies its own cavity in the water and is directly exposed to it on all sides. Right: The octane molecules have clustered together and occupy a common cavity in the water. Because each octane has part of its surface in contact with the other, less total octane surface is exposed to water—an energetically more favorable arrangement.

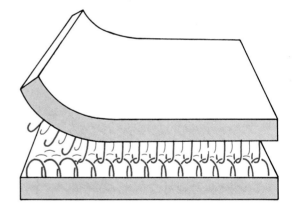

2.19   Stability from weak bonds
An array of many individually weak bonds—here represented as the many hooks and eyes of a Velcro fastener—can be surprisingly strong as a unit.

2.20 Hydration spheres of Na⁺ and Cl⁻

When dissolved in water, each of the Na⁺ and Cl⁻ ions is hydrated—that is, surrounded by water molecules electrostatically attracted to it. Note that the oxygen of the water molecules is attracted to the positively charged Na⁺, while the hydrogen of the water molecules is attracted to the negatively charged Cl⁻. Water molecules in a hydration sphere are called bound water. This bonding between ion and polar molecules (color-and-black bands) makes evident the common electrostatic basis of ionic bonds and polar (hydrogen) bonds.

surrounded by a sphere of regularly arranged water molecules that are electrostatically attracted to it—a process called **_hydration_** (Fig. 2.20).

From the point of view of some biological processes, an ion and its hydration spheres must be regarded as a single entity—as though the whole were a true molecule. For example, if the question is whether or not a given type of ion can move through tiny pores in cell membranes, then it is the size of the hydrated ion that must be compared with the dimensions of the pores.

Water is also an excellent solvent for nonionic molecules if they are polar. Such molecules are called **_hydrophilic_** ("water-loving"). They dissolve in water as a result of electrostatic attraction between the charged parts of the solute molecules and the oppositely charged parts of the water molecules. This occurs especially when, as in many biologically important compounds, the solute molecule has an oxygen with a hydrogen attached to it (—OH). As in water, the hydrogen in such a group has a net positive charge and is therefore attracted by the negatively charged oxygen end of a nearby water molecule, with the result that a hydrogen bond is formed. The solute molecules and the water molecules thus become weakly linked to each other (Fig. 2.21).

Substances that do not dissolve in water are electrically neutral and nonpolar. They therefore show no tendency to interact electrostatically with water and, indeed, are repulsed by it. A hydrophobic substance like oil or octane stirred into water will soon begin to separate out, because the water molecules tend slowly to reestablish the hydrogen bonds broken by the physical intrusion of the insoluble material and thus to push out that material. As a result, the nonpolar, insoluble molecules tend to coalesce to form droplets, which in general eventually fuse and form a separate layer outside the water. As we shall see, this basic chemical phenomenon, the tendency of hydrophobic molecules to be driven out of water, is the basis for the spontaneous formation of the cell membranes that protect organisms as they grow and develop.

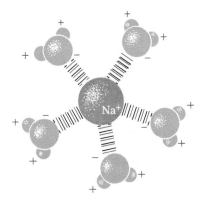

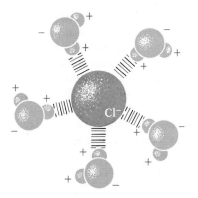

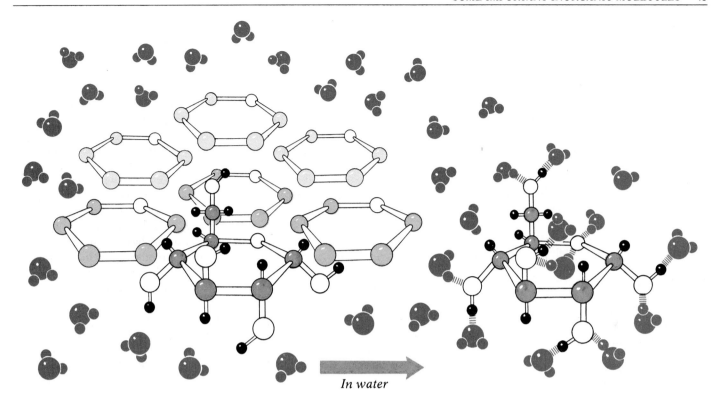

In water

Special physical properties of water   We have seen that substances dissolve in water if their molecules can form weak bonds or interactions with the water molecules. Such interactions have important implications for the water molecules themselves. As Figure 2.20 shows, water molecules in the hydration spheres of ions are arranged in orderly arrays; such water, referred to as ***bound water***, is essentially immobilized. The same is true of water molecules around polar groups of nonionic compounds. The orderly arrays of bound water are very different from those of pure water (Fig. 2.17), and the physical properties of bound water are consequently different from those of free water; the greater the proportion of bound water in a given volume, the lower the freezing point and the higher the boiling point of that volume of water. Since much of the water inside living cells is bound water, the physical properties of the cell contents are very different from those of pure water, even though water is the principal constituent of the cells.

The strong tendency of water molecules to form hydrogen bonds with each other, and the consequent ordering of the molecules, have important implications for life processes. For example, water has a high ***surface tension***: the surface of a volume of water is not easily broken. You witness the effects of surface tension when you watch a water strider or other insect

2.21   Polar basis of solubility

When a polar substance such as glucose, an energy-rich sugar (left), is placed in contact with water, the water molecules are attracted to the polar atoms of the sugar. The water forms hydrogen bonds with the substance, surrounding it with water molecules, and so dissolves it (right). In our bodies, only dissolved substances, like the glucose shown here, can be used to provide energy to run cellular processes.

2.22 A water strider on the surface of the water
Water striders can move rapidly across the surface of
still water, where they often congregate in large
numbers. Note the dimples in the water surface where
each foot rests.

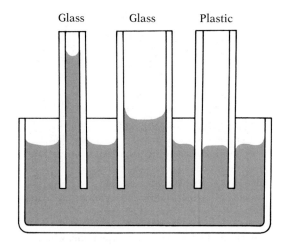

2.23 Capillarity
Water rises higher in a glass tube of small bore (left)
than in one of large bore (center) because in the
smaller tube a higher percentage of the water
molecules are in direct contact with the glass and
can form hydrogen bonds with charged groups on the
glass. By contrast, water cannot "stick" to the
surface of a plastic tube (right) because plastic is
uncharged.

walk on the surface of a pond without breaking the surface (Fig. 2.22) or
when you fill a glass slightly above the rim without spilling: the water has so
much cohesion that the extra water resists breaking away. Water has a high
surface tension because hydrogen bonds link the molecules at the surface
to each other and to the molecules below them. Before the legs of the water
strider (or any other object, for that matter) can penetrate the water's surface, they must break some of these hydrogen bonds and deform the orderly
array of water molcules. Similarly, in an overfilled glass the hydrogen
bonds that bind the extra water molecules to the water molecules below
them prevent water from spilling.

Just as water molecules are attracted electrostatically to areas of charge
on dissolved molecules, so also are they attracted to the charged groups
that characterize hydrophilic surfaces. Consequently such surfaces are
wettable—that is, water spreads over them and binds loosely to them. By
contrast, hydrophobic surfaces—those of most plastics and waxes, for example—lack surface charge and are not wettable; water on them will form
isolated droplets, but will not spread out.

The propensity of water to bind to hydrophilic surfaces explains the phenomenon of *capillarity*—the tendency of aqueous liquids to rise in narrow
tubes. If the end of a narrow glass tube is inserted below the surface of a
volume of water, water will rise in the tube to a level well above that of the
water outside (Fig. 2.23). Because glass is very hydrophilic, the water molecules, electrostatically attracted to the numerous charged groups on its
surface, tend to creep along the inside of the tube. As the ring of water molecules in contact with the inner surface creeps upward, it pulls along other
water molecules to which it is linked by hydrogen bonds. The larger the diameter of the tube, however, the smaller the percentage of water molecules
in direct contact with the glass and, correspondingly, the smaller the rise in
the water. Even though the relatively few molecules in contact with the
glass have a tendency to creep upward, they are held back by their cohesion
with the rest of the water in the tube.

Capillarity is by no means restricted to glass tubes. Water will climb any
charged surface. We are all familiar with the way it climbs up the fibers of
paper towels and spreads through the fibers of many sorts of cloth.

We have said that each water molecule has the potential for forming hydrogen bonds with four other water molecules (Fig. 2.17). In the liquid
state this potential is not fully realized because molecular motion prevents
stabilization, but as water is cooled the extent of hydrogen bonding increases. Hydrogen bonding reaches its full potential in water that has frozen into ice. When all four possible bonds have formed, each is oriented in
space at the greatest possible distance from the other three. Consequently
the bonds are directed toward the four corners of the tetrahedron. The resulting three-dimensional lattice of water molecules in ice is an open one
(Fig. 2.24A); the rigidly tetrahedral structure maintains space between the
molecules, so they can be only loosely packed.

When ice is warmed to the melting point, a few of the hydrogen bonds
rupture, and those that remain become less rigidly oriented. The resulting
deformation of the lattice and tighter packing of the molecules makes the
water denser. The density reaches its maximum when the water is at 4°C;
above this temperature, the further packing that might be expected as a

consequence of the increasing disruption of the lattice is more than offset by the expansion, shown by virtually all substances, that results from the increased molecular motion of heat energy. In summary, we see that unlike most other substances, which become increasingly dense as the temperature falls, water first becomes denser and then begins to expand again below 4°C. This means that ice, being less dense than cold water, floats and —further—that ponds and streams freeze from the top down rather than from the bottom up. The crust of ice that forms at the surface insulates the water below it from the cold air above and thereby often prevents the pond or stream from freezing solid, even in very cold weather. This special property of water makes life possible in the many ponds and streams that would otherwise freeze solid in the winter.

The role of water in regulating environmental temperature   The hydrogen bonds in water give it a high internal cohesion, which enables it to absorb much heat energy without undergoing a very large increase in temperature and to release much heat energy without undergoing a great drop in temperature. When most substances absorb heat, their molecules move more rapidly in relation to one another; temperature is an indication of the amount of such molecular motion. In water, by contrast, much of the absorbed heat energy is dissipated in increased vibration of the hydrogens, each of which is shared between the oxygen to which it is covalently bound and the oxygen of another water molecule, to which it is electrostatically bound. As a result, relatively little of the added heat energy is expressed as movement of whole water molecules, and the temperature increase is therefore modest. The high *heat capacity* of water (the amount of heat energy that must be added or subtracted to change the temperature by one degree), together with its high *heat of vaporization*—the amount of heat energy required for turning water from liquid to vapor (evaporation)— permits it to act as an effective buffer against extreme temperature fluctuations in the environment. In this way water helps stabilize the earth's temperatures within the range favorable to life.

Besides damping fluctuations in temperature, water plays an important role in determining the absolute temperature at the earth's surface, because the water vapor in the atmosphere exerts what has been called a greenhouse effect. The vapor absorbs much of the sunlight striking it from above and also much of the radiation reemitted by the earth. The absorbed radiation warms the atmosphere, which, in turn, warms the earth's surface.

CARBON DIOXIDE

As we have seen, carbon has only four electrons in its outer electron shell, and as a result it has a covalent bonding capacity of 4. Carbon dioxide (CO_2) is the compound formed when two atoms of oxygen bond to one atom of carbon. Though this substance contains carbon, it is generally thought of as inorganic, because it is simpler than all but a few of the compounds classified as organic.

Only a very small fraction of the atmosphere, roughly 0.033 percent, is CO_2; yet atmospheric carbon dioxide is the principal inorganic source of

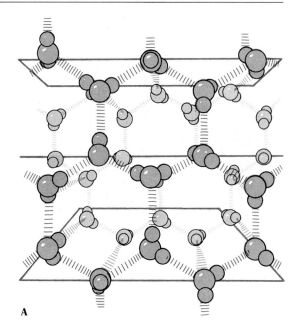

A

B

2.24   The molecular structure of ice
Because of the tetrahedral arrangement around each water molecule, the lattice is an open one, with considerable space between molecules (A). In liquid water the arrangement is not quite so rigid, and the packing of molecules is therefore slightly denser, but the general lattice arrangement is nonetheless largely preserved. (Planes have been added to help show the three-dimensional disposition of the molecules.) Note the hexagons created by the hydrogen bonding in this bit of ice. This conformation is the basis of the hexagonal shape of most snow crystals (B). Each "snowflake" contains about 10^{16} water molecules.

carbon, and carbon is the principal structural element of living tissue. Before CO_2 can take part in chemical reactions, it must usually first dissolve in water, which it does very readily, and then react with the water to form carbonic acid (H_2CO_3):

$$CO_2 + H_2O \longrightarrow H_2CO_3$$

This reaction involves so little energy change that it is easily reversible, and CO_2 can readily be released from water solution when conditions are appropriate:

$$H_2CO_3 \longrightarrow CO_2 + H_2O$$

Carbon dioxide and water are the raw materials from which green plants manufacture many complex organic compounds essential to life, as we shall see in detail in Chapter 8. When these complex compounds have run their course in the life system, they are broken down again to carbon dioxide and water, and the carbon dioxide is eventually released into the atmosphere. The simple compound carbon dioxide, then, is the beginning and the end of the immensely complex carbon cycle in nature.

OXYGEN

Molecular oxygen (O_2) constitutes approximately 21 percent of the atmosphere. It is necessary for the maintenance of life in most organisms, though a few can live without it. It can be utilized directly, without change, by both plants and animals in the process of extracting usable energy from nutrient molecules. Its role, as we shall see, is to serve as the ultimate acceptor of electrons. This is a crucial task: without oxygen to accept electrons, most cells can run at only 5 percent of their normal efficiency. Oxygen is not very soluble in water, but enough dissolves to supply the needs of aquatic organisms, provided (1) that the water is not too hot and (2) that the water's surface is exposed to the air or, alternatively, that green plants are growing in it, thus constantly releasing oxygen into it by the process of photosynthesis. Indeed, it is the production of oxygen by green plants that is the source of virtually all atmospheric oxygen.

Though water, oxygen, and carbon dioxide are truly basic to life as we know it, still other compounds are used to capture, store, transport, and utilize the energy that fuels life. In the next chapter we shall examine these complex, carbon-based compounds in an effort to understand the chemical reactions that make life possible.

Chapter 3

THE CHEMISTRY OF LIFE

Chemistry does what all good science does: it makes complex phenomena, if no less remarkable, at least easier to predict. Chemistry explains why certain reactions among molecules will take place, and why particular molecular combinations will be stable. Since the diversity of molecules in living material is great, and the possibilities for combining them are numerous, an understanding of chemistry is an indispensable predictive tool. Chemistry enables us to make molecular sense of the diversity around us; it lets us see how relatively few elements can constitute living matter in all its varied forms.

The source of the vast molecular diversity in living things is the bonding capacity of just one of the 92 elements—carbon. Carbon's power lies in its versatile structure: four unpaired electrons in its outer shell, which allow it to form covalent bonds with up to four other atoms, make possible enough different molecular connections to generate an almost endless variety of carbon-based—organic—molecules. In this chapter we shall look first at the important kinds of organic compounds—carbohydrates, lipids, proteins, and nucleic acids—and then at how some of the crucial well-ordered chemical changes inside cells are orchestrated and controlled.

SOME SIMPLE ORGANIC CHEMISTRY

Though carbon can and does bond to a variety of elements, its four unpaired electrons are most commonly bonded to hydrogen, oxygen, nitrogen, or more carbon. Compounds containing only carbon and hydrogen, the ***hydrocarbons***, are of central importance in organic chemistry; the number of different compounds of this kind is immense. The readiness with which carbon-to-carbon bonds can form and produce chains of varying

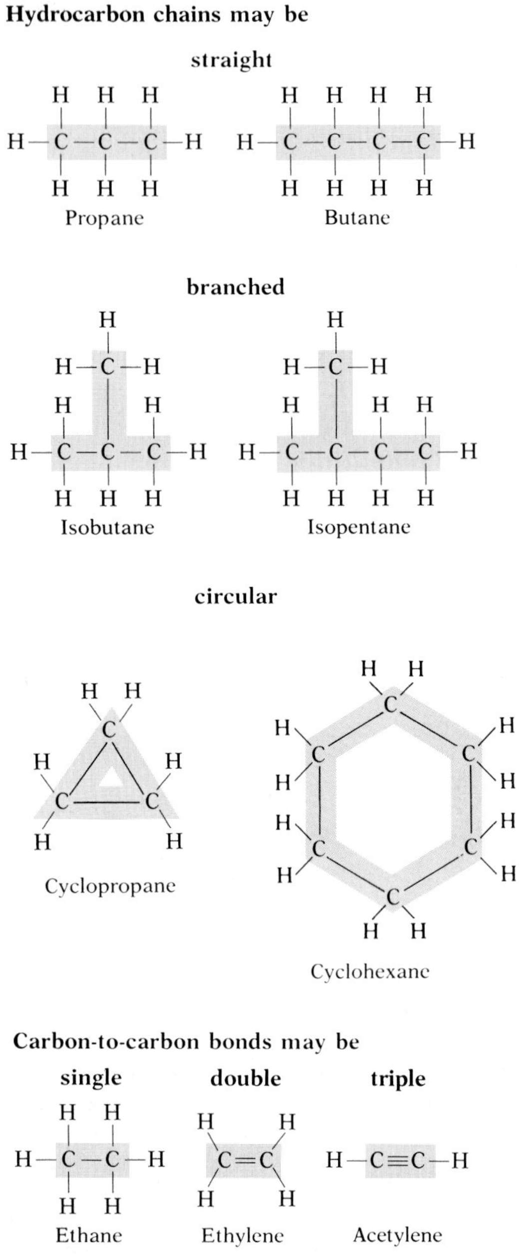

Hydrocarbon chains may be

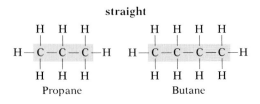

straight

Propane

Butane

branched

Isobutane

Isopentane

circular

Cyclopropane

Cyclohexane

Carbon-to-carbon bonds may be

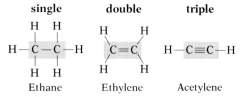

single   double   triple

Ethane   Ethylene   Acetylene

3.1 Examples of hydrocarbons
The molecules appear flat in these conventionally drawn structural diagrams, though they are, in fact, three-dimensional. The bonds around a carbon atom that forms only single bonds (all the carbons seen here except those in the last two molecules) are oriented toward the four corners of a tetrahedron.

lengths and shapes has generated a great variety of hydrocarbons (Fig. 3.1). Hydrocarbon chains may be simple, as in propane, butane, and substances with still longer molecules; they may be branched, as in isobutane and isopentane; or they may form circles of varying numbers of carbons, as in cyclopropane and cyclohexane. Obviously, the more atoms a molecule contains, the more different arrangements of those atoms will be possible. Compounds with the same atomic content and molecular formula but different atomic arrangements are called *isomers* (Fig. 3.2). Very large organic molecules may have hundreds of isomers, with many differing physical properties.

Another source of variety in hydrocarbon compounds is the capacity of adjacent carbon atoms to form single, double, or triple bonds (Fig. 3.1). And, of course, substitution of other elements or groups of elements for hydrogen atoms makes possible an almost infinite number of derivative compounds. The total number of hydrocarbons and derivatives that form in nature has been conservatively estimated at more than half a million. This great capacity for diverse atomic organization makes the hydrocarbon group ideal for building chemicals with unique properties, each precisely suited to the job at hand. For example, life-sustaining compounds like the sugars can be tailored by metabolic processes to suit the varying needs of the cell. From the sugars, cells can extract energy, derive building materials, or construct molecules that help direct cellular processes. On a very different scale, the organic carbon base allows one organism to use another, plant or animal, as food, and thus serves as a versatile common denominator connecting all organisms in an interdependent chain.

The four major classes of complex organic compounds are carbohydrates, lipids, proteins, and nucleic acids. Molecules in each of these classes are often identified on the basis of the subunits they contain. Each subunit, or *functional group*, has its own characteristic properties, which help determine solubility, reactivity, and other traits of the chemical "personality" of the whole molecule. We shall refer to many of the groups listed in Table 3.1 in this and later chapters.

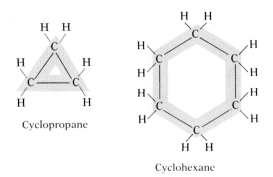

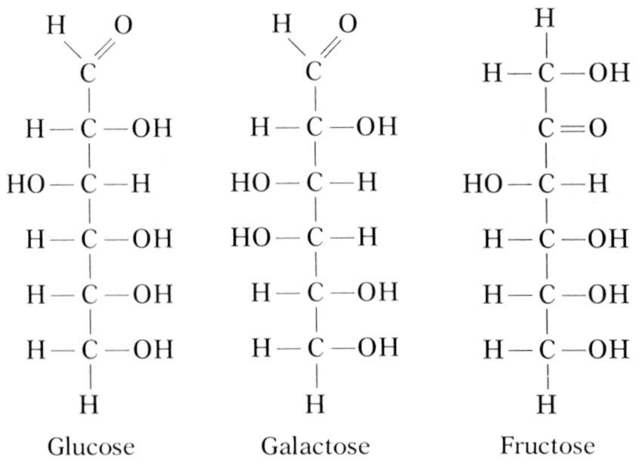

Glucose   Galactose   Fructose

3.2 Three isomeric hexoses
Each of these six-carbon sugars has the same molecular formula, $C_6H_{12}O_6$; hence each is an isomer of the others.

TABLE 3.1 *Important functional groups*

Group	Name	Properties
—OH	Hydroxyl, or alcohol	Polar (soluble, because it is able to form hydrogen bonds)
$-C\!\!\begin{smallmatrix}O\\OH\end{smallmatrix}$	Carboxyl	Polar (soluble); often loses its hydrogen, becoming negatively charged (an acid): $-C\!\!\begin{smallmatrix}O\\O^-\end{smallmatrix}$
$-N\!\!\begin{smallmatrix}H\\H\end{smallmatrix}$	Amino	Polar (soluble); often gains a hydrogen, becoming positively charged (a base): $-N\!\!\begin{smallmatrix}H\\H^+\\H\end{smallmatrix}$
$-C\!\!\begin{smallmatrix}O\\H\end{smallmatrix}$	Aldehyde	Polar (soluble)
C=O	Ketone	Polar (soluble)
$-C\!\!\begin{smallmatrix}H\\H\\H\end{smallmatrix}$	Methyl	Hydrophobic (insoluble); least reactive of the side groups
$-P\!\!\begin{smallmatrix}O\\OH\\OH\end{smallmatrix}$	Phosphate	Polar (soluble); usually loses its hydrogens, becoming negatively charged (an acid): $-P\!\!\begin{smallmatrix}O\\O^-\\O^-\end{smallmatrix}$

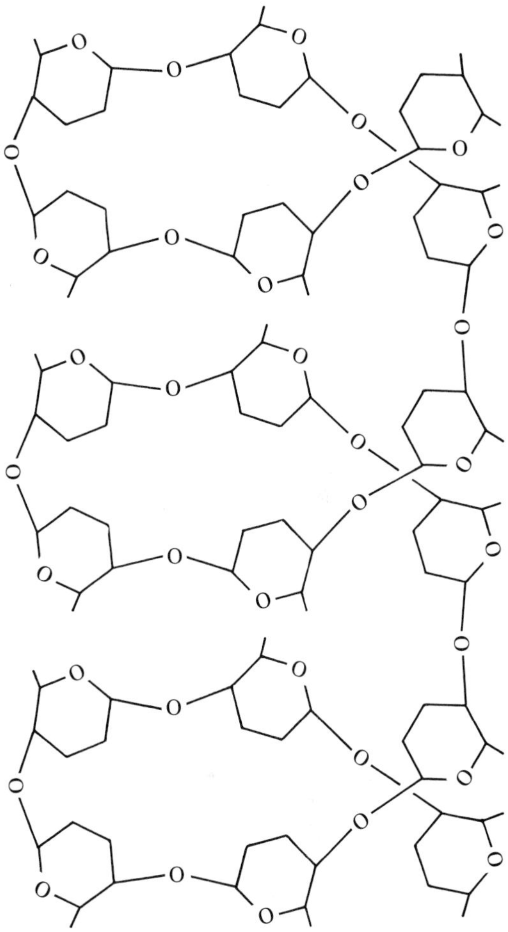

3.3 A moderately complex organic compound
One of the largest of the moderately complex organic compounds is starch, the main energy-storage molecule in plants. Despite its apparent complexity, the starch molecule is actually composed of a repetitive string of glucose units, each represented here as a hexagon. Only a small part of one starch chain is shown. A branched molecule composed of units of several different compounds would be far more complex.

CARBOHYDRATES

Carbohydrates are compounds composed of carbon, hydrogen, and oxygen. In simple carbohydrates the hydrogen and oxygen are characteristically present in the same proportions as in water: there are two hydrogen atoms and one oxygen atom for each carbon atom. Consequently the group —CH_2O recurs frequently in carbohydrate molecules; it is diagramed

$$H-C-OH$$

Some carbohydrates, such as starch and cellulose, are very large, complex molecules. But like most very large organic molecules, they are composed of many simpler "building-block" compounds bonded together (Fig. 3.3). Understanding the constituent, or building-block, compounds is the first step toward understanding the more complex substances.

STRUCTURAL ISOMERS

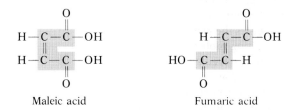

Ethyl alcohol   Dimethyl ether

3.4   Three types of isomerism

The two structural isomers differ in the basic grouping of their constituent atoms, one being an alcohol (characterized by an —OH group) and the other an ether (characterized by an oxygen bonded between two carbons). The two geometric stereoisomers are fixed in different spatial arrangements by their inability to rotate around the double bond between the middle two carbons. As Figure 3.5 shows, the two optical stereoisomers are asymmetric molecules that cannot be superimposed on each other.

GEOMETRIC STEREOISOMERS

Maleic acid   Fumaric acid

OPTICAL STEREOISOMERS

l-Lactic acid   d-Lactic acid

Simple sugars   The basic carbohydrate molecules are simple sugars, or *monosaccharides*. All sugars when in straight-chain form contain a —C$=$O group (Fig. 3.2). If the double-bonded O is attached to the terminal C of a chain, the combination is called an aldehyde group; if it is attached to a nonterminal C, the combination is called a ketone group (Table 3.1). The —OH (hydroxyl) groups, which are attached to all the carbons except those with a double-bonded oxygen, are polar. Hence sugars readily form hydrogen bonds with water and are soluble, unlike simple nonpolar hydrocarbon molecules, which tend to clump together in water.

The monosaccharide that forms the backbone of sugar can be of different lengths. Some monosaccharides contain as few as three carbons; others contain five carbons, six carbons, or more. Both three- and five-carbon sugars play important biological roles and will be mentioned in later chapters, but the six-carbon sugars (hexoses) are the most important building blocks for more complex carbohydrates.

There are many six-carbon sugars, glucose and fructose being two of the most important. Since they all have the proportions of oxygen and hydrogen typical of carbohydrates, all have the same molecular formula, $C_6H_{12}O_6$, and are therefore isomers of one another. Glucose and fructose are structural isomers (Fig. 3.2); the basic groupings of their constituent atoms are different, making one an aldehyde and the other a ketone sugar.

In addition to structural isomerism, which is readily understandable, there is another, more subtle, kind of isomerism called stereoisomerism. In a given pair of stereoisomers, identical groups are attached to the carbon atoms, but the spatial arrangements of the attached groups are different. The two middle compounds shown in Figure 3.4 are geometric stereoisomers; if the carbon-to-carbon bonds in these two molecules were single, the two would in fact be the same compound, because free rotation is possible around a single bond. A double bond, however, holds its atoms in a rigid configuration.

When considering optical stereoisomers (Fig. 3.4, bottom), keep in mind that molecules are not flat, even though they are often drawn that way. In a carbon atom the four unpaired electrons form the corners of a tetrahedron; if in two molecules the groups attached to corresponding electrons are different, the resulting molecules will be different (Fig. 3.5). Glucose and galactose are optical stereoisomers because the position of one —OH group is different in each (Fig. 3.2). Though the optical stereoisomers of a compound may appear very similar, they are usually quite different in their biological properties and behavior; as Figure 3.5 illustrates, for instance, subtle differences in shape may determine which isomer can bind to or react with a particular molecule and which cannot. Hence, different isomers can play very different roles in the chemistry of cells.

Glucose does not always exist as a straight-chain compound; indeed, this is probably its least common form. Usually it exists in ring form—most often, as a ring composed of five carbons and one oxygen (Fig. 3.6).

Glucose plays a unique role in the chemistry of life. As the primary prod-

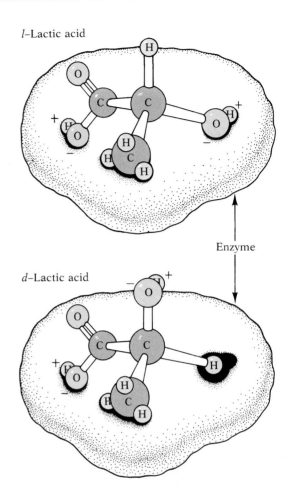

l–Lactic acid

Enzyme

d–Lactic acid

3.5 Optical stereoisomers and chemical specificity
Of these two stereoisomers of lactic acid, only one (top) can fit all the holes in the enzyme molecule, to which it binds. The other will not fit the holes no matter which way it is turned. Subtle distinctions of this kind are crucially important: they enable molecules of the substances known as enzymes to recognize and bind certain other particular molecules as part of the network of critical chemical reactions necessary for life. (The molecular structure of the enzyme has been omitted, and only its general shape indicated.)

3.6 Two forms of glucose
Glucose may exist in the straight-chain aldehyde form shown at left or as a ring structure, as shown in the center. The ring structure is the most common. (By convention, the unmarked corners of the hexagon signify carbon atoms.) Both representations fail to convey the true shape of the molecules, since the four bonds of each carbon atom are directed to the four corners of a tetrahedron. The illustration at right is a more realistic representation of the ring form, but such realism is impossible to preserve in representing any but the simplest organic molecules.

3.7 Two examples of derivative monosaccharides
Glucosamine is merely a glucose molecule with an amino group (—NH₂) substituted for an —OH group (see Fig. 3.6). Similarly glucose-6-phosphate is glucose with a substituted phosphate group.

uct of photosynthesis in plants, glucose becomes the ultimate source of all the carbon atoms in animal tissue. Moreover, the energy stored in its covalent bonds is usually, directly or indirectly, the source of the energy that powers cells. Other six-carbon monosaccharides, among them fructose and galactose, are constantly being converted into glucose or synthesized from glucose. And even such classes of compounds as fats and proteins can be converted into glucose or synthesized from glucose in the living body.

In addition to ordinary monosaccharides composed only of carbon, oxygen, and hydrogen, there are a variety of derivative monosaccharides containing other elements. For example, some have a phosphate group attached to one of the carbons, and others an *amino* group (a nitrogen with two hydrogens, —NH₂) (Fig. 3.7).

Disaccharides   The *disaccharides* are compound sugars composed of two simple sugars bonded together through a series of reactions that involves the removal of a molecule of water. This kind of reaction series is called a *condensation reaction* or a dehydration reaction.

Let us first examine the disaccharide *maltose*, or malt sugar. This compound is synthesized by a condensation reaction between two molecules of glucose. The reaction can be described by the following equation, which summarizes several intermediate steps:

$$2C_6H_{12}O_6 \longrightarrow C_{12}H_{22}O_{11} + H_2O$$

To understand specifically what is involved in the condensation synthesis of maltose, it is helpful to look at a diagram indicating the structures of the molecules. Figure 3.8 shows a slightly simplified version of the way in which the hydrogen atom from a hydroxyl group (—OH) of one molecule of glucose combines with a complete hydroxyl group from the other molecule of glucose to form water. The oxygen valence vacated by the removal of hydrogen and the carbon valence vacated by the removal of —OH are filled by the bonding together of the oxygen of one glucose molecule with the carbon of the other glucose molecule. As a result, the two glucose units are connected by an oxygen atom shared between them. The product is the disaccharide maltose.

Sucrose, our common table sugar, is also a disaccharide. It is synthesized by a condensation reaction between a molecule of glucose and a molecule of fructose. *Lactose*, or milk sugar, is a disaccharide composed of glucose and galactose joined by a condensation reaction (Fig. 3.8). In fact, it is a general (and important) rule that the synthesis of complex molecules from simpler units almost always produces water.

Synthesized by condensation reactions, disaccharides can be broken down to their constituent simple sugars by the reverse process. This reaction, called *hydrolysis*, involves addition of a water molecule through a series of steps which can be summarized as

$$C_{12}H_{22}O_{11} + H_2O \longrightarrow 2C_6H_{12}O_6$$

Hydrolysis reactions are of particular importance in digestion, as we shall see in a later chapter, because digestion breaks down complex molecules into simple building blocks, ready for subsequent use.

We can now define a monosaccharide more precisely than was hitherto

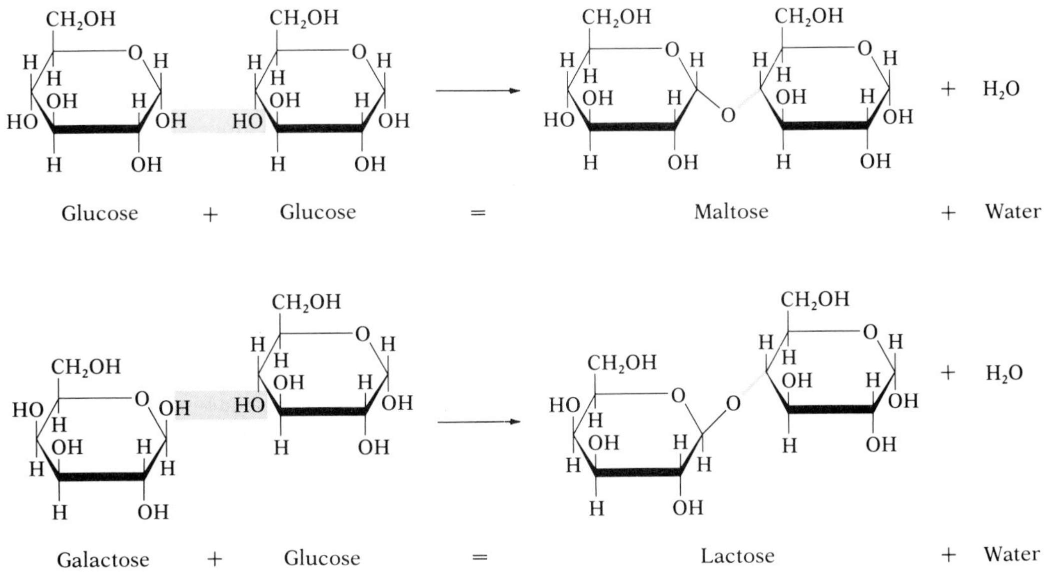

Glucose   +   Glucose   =   Maltose   +   Water

Galactose   +   Glucose   =   Lactose   +   Water

possible. By contrast with compound sugars, a monosaccharide is a sugar that cannot be hydrolyzed into smaller carbon-containing molecules.

Polysaccharides The prefix *poly-* means "many," and ***polysaccharides*** are complex carbohydrates composed of many simple-sugar building blocks bonded together in long chains (Fig. 3.9). They are synthesized by exactly the same kind of condensation reaction as the disaccharides and, like them, can be broken down into their constituent sugars by hydrolysis.

A number of complex polysaccharides are of great importance in biology. ***Starches***, for example, are the principal carbohydrate storage products of higher plants. They are composed of many hundreds of glucose units bonded together. In some forms of starch the chain of sugars is unbranched, and in others it is branched; both types are common in plant material. ***Glycogen*** is the principal carbohydrate storage product in animals and is sometimes called animal starch. Its molecules are much like those of starch; they have the same type of bond between adjacent glucose units, but

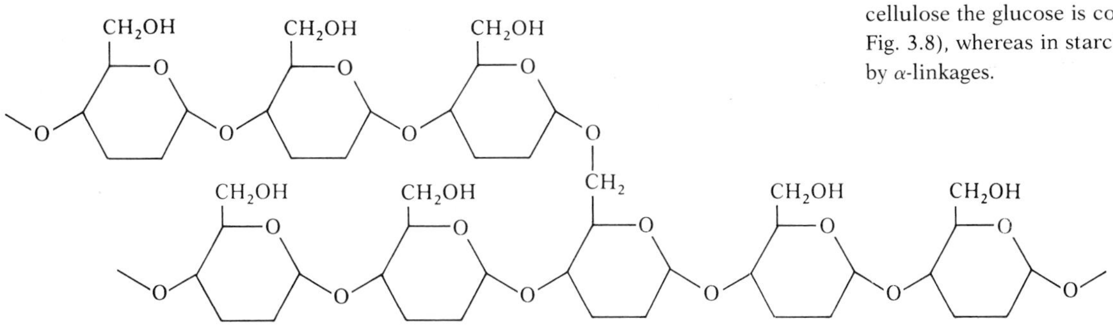

3.8 Synthesis of maltose and lactose
Removal of a molecule of water between two molecules of sugar results in formation of a bond (color) between the two. In the two examples shown here, intermediate steps are omitted. In the top example, a bottom-to-bottom (α) linkage forms between two glucose molecules, yielding maltose. In the bottom example, a top-to-bottom (β) linkage forms between galactose and glucose, yielding the milk sugar lactose. Many adults suffer from milk intolerance because they cannot digest this β-linkage.

3.9 Branched starch
Shown here is a small segment of a molecule of starch. This starch is branched, but some forms (as in Fig. 3.3) are unbranched. Like the cellulose of plants, starch is a polymer of glucose, but in cellulose the glucose is connected by β-linkages (see Fig. 3.8), whereas in starch the glucose is connected by α-linkages.

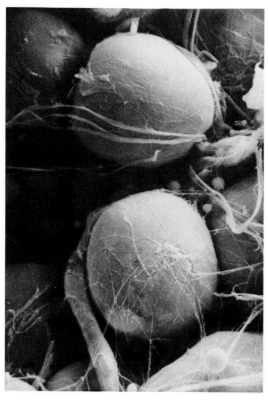

3.10   Fat-storage cells
Lipids are stored in spherical fat cells called
adipocytes, seen here. Small blood vessels
(capillaries) and support fibers (collagen) hold these
cells in place.

3.11   Beeswax production by bees
Honey bees extrude a waxy lipid from special glands
on the underside of the abdomen. These flakes are
then chewed until soft and worked to make honeycomb.

the chains are more extensively branched. ***Cellulose*** is a highly insoluble
unbranched polysaccharide common in plants, where it is a major supporting material. The bonds between its glucose units are β-linkages rather
than α-linkages (see Fig. 3.8); animals can digest the bonds of starch and
glycogen, but most animals are unable to hydrolyze those of cellulose.

Reactions like those that form polysaccharides—that is, reactions in
which small molecules bond together to form long chains—are called polymerization reactions. The products formed are called ***polymers***. If polymerization between molecules A, B, C, D, etc., is to occur, two requirements must be met: one end of molecule A must be capable of interacting
with one end of molecule B so as to split out some small group, such as
water; and each molecule must contain two functional groups, so that after
the first two combine, the free ends can continue to react, extending the
polymer. Polymers of several types play a critical role in biology, as we shall
see.

LIPIDS

Like carbohydrates, lipids, a second major group of biological compounds,
are composed principally of carbon, hydrogen, and oxygen, but they may
also contain other elements, particularly phosphorus and nitrogen. They
differ from carbohydrates in that they contain a much smaller proportion

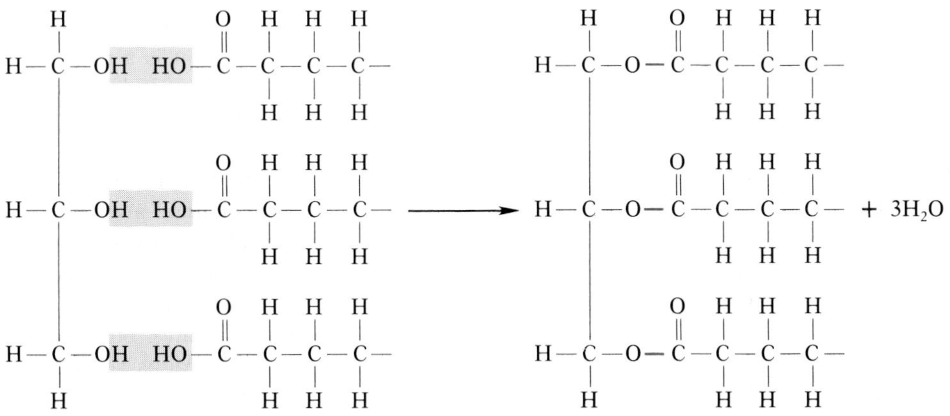

Glycerol + Fatty acids = Fat + Water

of oxygen. Unsubstituted lipids—that is, lipids without an ionic group attached to the carboxyl end—are primarily nonpolar, by virtue of their long hydrocarbon "tails." They are therefore relatively insoluble in water, but soluble in organic solvents such as ether.

Fats   Among the best-known lipids are the neutral fats. Important as energy-storage molecules in living organisms, the fats also provide insulation, cushioning, and protection for various parts of the body (Fig. 3.10). Honey bees even secrete beeswax, a substance similar to fat, from glands on the ventral (lower) side of the abdomen, and then use these flakes to build honeycomb (Fig. 3.11). Each molecule of fat is composed of two different types of building-block compounds: an alcohol called glycerol and fatty acids.

Glycerol (also sometimes called glycerin) has a backbone of three carbon atoms, each carrying a hydroxyl (—OH) group (Fig. 3.12). (Because the presence of an —OH group attached to a carbon atom characterizes a compound as an alcohol, the hydroxyl group is sometimes termed the alcohol functional group.)

Fatty acids, like all organic acids, contain a carboxyl (—COOH) group:

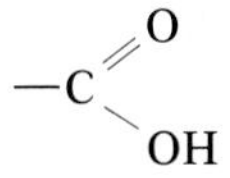

$$-C\diagup\!\!\!\!\!\!\raise2pt\hbox{}^{O}_{OH}$$

When both a double-bonded oxygen and an —OH group are attached to the same carbon atom, the double-bonded oxygen tends to cause the —OH part of this carboxyl to lose its hydrogen, making the group ionic and causing the compound to behave as an acid (see Table 3.1).

There are many different fatty acids, varying in carbon-chain length, in the number of single or double carbon-to-carbon bonds, and in other characteristics. The fatty acids in edible fats and oils contain an even number of carbon atoms, and most of them have relatively long carbon backbones, usually from 4 to 24 carbons, or more; three of the most common are

3.12   Synthesis of a fat
Removal of three molecules of water by condensation reactions results in the bonding of three molecules of fatty acid to a single molecule of glycerol. (Intermediate steps are omitted in this example.) Conversely, three molecules of water will be added by hydrolysis when this molecule of fat is digested. The carbon chains of the fatty acids are usually longer than shown here.

```
     HO    O                    HO    O
       \  //                      \  //
        C                          C
        |                          |
       CH₂                        CH₂
        |                          |
       CH₂                        CH₂
        |                          |
       CH₂                        CH₂
        |                          |
       CH₂                        CH₂
        |                          |
       CH₂                        CH₂
        |                          |
       CH₂                        CH₂
        |                          |
       CH₂                        CH₂
        |                          |
       CH₂                        CH
        |                          ‖
       CH₂                        CH
        |                          |
       CH₂                        CH₂
        |                          |
       CH₂                        CH
        |                          ‖
       CH₂                        CH
        |                          |
       CH₂                        CH₂
        |                          |
       CH₂                        CH₂
        |                          |
       CH₃                        CH₂
                                   |
                                  CH₂
                                   |
                                  CH₃

   Palmitic acid              Linoleic acid
```

3.13 Examples of saturated and unsaturated fatty acids

Palmitic acid is saturated with hydrogen—that is, it contains the maximum number of hydrogens possible. By contrast, linoleic acid, with its two inflexible carbon-to-carbon double bonds, accommodates four fewer than the maximum number of hydrogens.

stearic acid (18 carbons), palmitic acid (16 carbons) (Fig. 3.13), and oleic acid (18 carbons).

Organic acids and alcohols have a tendency to combine through condensation reactions. Since glycerol has three alcohol groups, it can combine with three molecules of fatty acid to form a molecule of fat (Fig. 3.12). Hence fats are sometimes also called triglycerides.

The various fats differ in the specific fatty acids, or types of fatty acids, composing them. You have doubtless read of the controversy in medical and nutritional circles concerning saturated and unsaturated fats. Saturated fats are simply those incorporating fatty acids with the maximum possible number of hydrogen atoms attached to each carbon, and hence no carbon-to-carbon double bonds (Fig. 3.13). The fatty acids in unsaturated fats (or perhaps we should say oils, since they are usually liquid at room temperature) have at least one carbon-to-carbon double bond—that is, they are not completely saturated with hydrogen. There is now good evidence that an elevated intake of saturated fats is one of many factors that predispose human beings to atherosclerosis—a disease of the arteries in which fatty deposits in the arterial walls cause partial obstruction and thus interfere with blood flow.

Since fats are synthesized by condensation reactions (removal of water), they, like complex carbohydrates, can be broken down into their building-block compounds by hydrolysis, as happens in digestion. And because fat contains 2.5 times as much usable energy per gram as monosaccharides, it is a good substance for long-term energy storage.

Phospholipids   Various lipids contain a phosphate group at the carboxyl end of the chain. Among the most common of these phospholipids are those composed of one unit of glycerol, two units of fatty acid, and a phosphate group often linked with a nitrogen-containing group (Fig. 3.14). The phosphate group is bonded to the glycerol at the point where the third fatty acid would be in a fat. Because the phosphate group has a marked tendency to lose a hydrogen ion, one of the oxygens becomes negatively charged; similarly, the nitrogen, being electronegative, tends to attract a hydrogen ion and thus to become positively charged. In short, the end of the phospholipid molecule with the phosphate and nitrogenous groups is strongly polar and hence soluble in water, whereas the other end, composed of two long hydrocarbon tails of the fatty acids, is nonpolar and insoluble. This curious property of solubility at one end but not at the other makes phospholipids especially well suited to function as major constituents of cellular membranes, as we shall see in Chapter 4.

Steroids   Though commonly classified as lipids because their solubility characteristics are similar to those of fats, oils, waxes, and phospholipids, the steroids differ markedly in structure from the other lipids we have discussed (Fig. 3.15). They are not based upon a bonding together of fatty acids and an alcohol. Instead, they are complex molecules composed of four interlocking rings of carbon atoms, with various side groups attached to the rings. Steroids are very important biologically. Some vitamins and hormones are steroids, and steroids often occur as structural elements in living cells, particularly in cellular membranes.

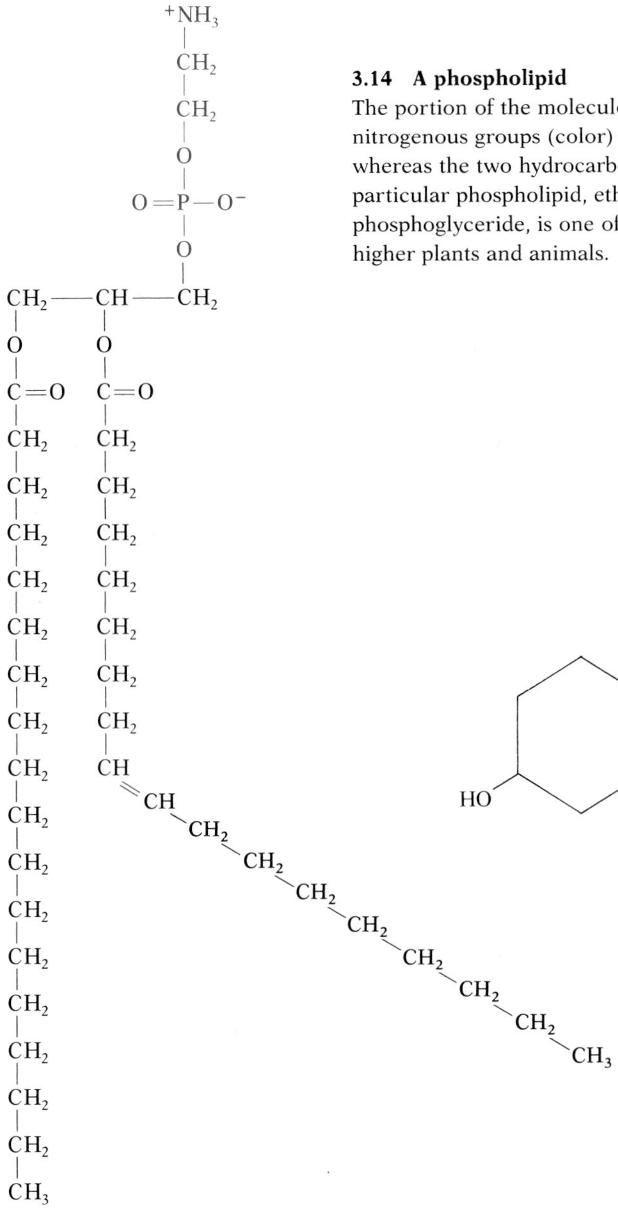

3.14 A phospholipid
The portion of the molecule with the phosphate and nitrogenous groups (color) is soluble in water, whereas the two hydrocarbon chains are not. This particular phospholipid, ethanolamine phosphoglyceride, is one of the two most abundant in higher plants and animals.

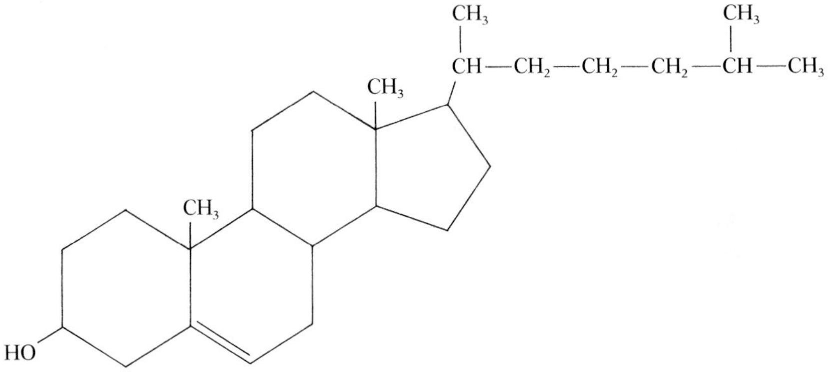

3.15 A steroid
All steroids have the same basic unit of four interlocking rings, but differ in their side groups. This particular steroid is cholesterol. (By convention, a hexagon signifies a six-carbon ring with its valences completed by hydrogens; see cyclohexane in Figure 3.1. A pentagon signifies a five-carbon ring, also with hydrogens attached to the carbons.)

PROTEINS

Far more complex than either carbohydrates or lipids, proteins are fundamental to both the structure and function of living material. Directly responsible for controlling the delicate chemistry of the cell, they exist in literally thousands of different forms. But like carbohydrates and lipids, proteins are composed of simple building-block compounds.

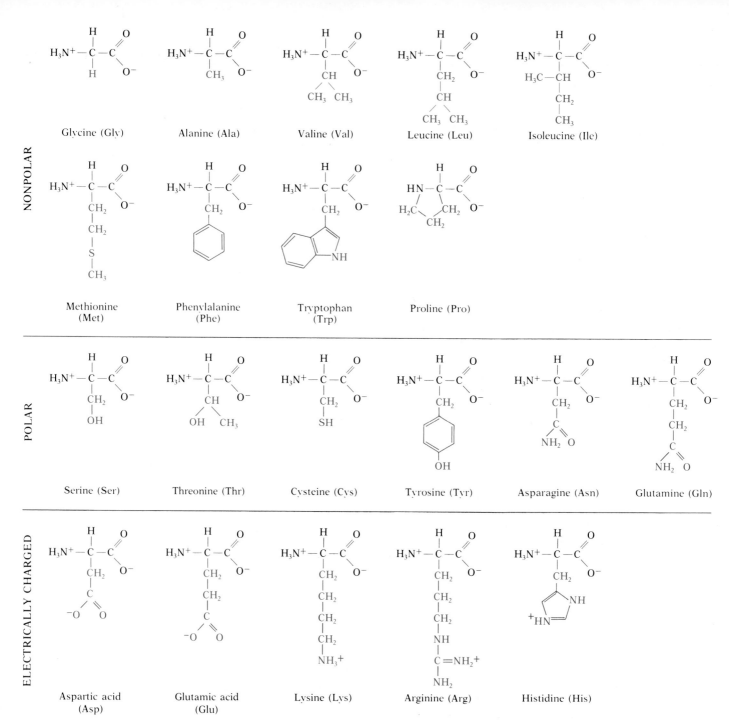

3.16 Structural formulas of the 20 amino acids common in proteins

The amino acids are shown in their ionized form. All have the same arrangement of a carboxyl group and an amino group attached to the same carbon; they differ in their R groups (color). Top two rows: These nine amino acids have nonpolar R groups and are relatively insoluble in water. (Glycine is an exception. Its R group, a

single hydrogen atom, is nonpolar, but is too small to outweigh the charge of the amino and carboxyl groups. The molecule therefore behaves more like a polar amino acid and is water-soluble. Proline is also unusual. It is technically not an amino acid, because the nitrogen is bonded to part of the R group. However, it is included because it is regularly incorporated into proteins along with the true amino acids.)

Third row: These six amino acids have polar R groups and are soluble. Bottom row: These five amino acids, with R groups ionized at intracellular pH levels, are electrically charged and thus water-soluble; the first two, being negatively charged, are acidic, whereas the last three, with a positive charge, are alkaline.

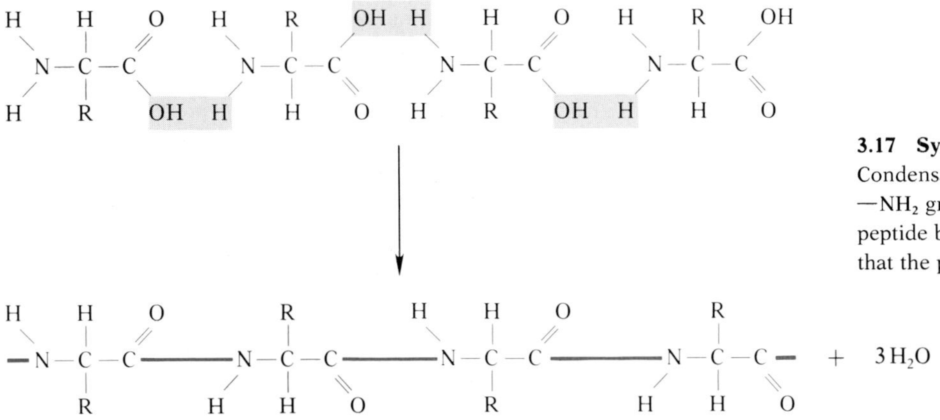

3.17 Synthesis of a polypeptide chain
Condensation reactions between the —COOH and
—NH$_2$ groups of adjacent amino acids result in
peptide bonds (color) between the acids. Notice again
that the process of combining units releases water.

The building blocks and primary structure of proteins All proteins
contain four essential elements: carbon, hydrogen, oxygen, and nitrogen;
most proteins also contain some sulfur. These elements are bonded together to form compounds called *amino acids*, which, being organic acids,
contain the carboxyl (—COOH) group. In addition, they each have an
amino (—NH$_2$) group. Both the —COOH and the —NH$_2$ group are attached
to the same carbon atom. Finally, each amino acid has a side chain, designated R:

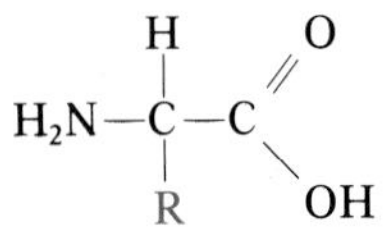

In the normally very slightly acidic pH within cells, a high percentage of the
amino acid molecules are ionized; the carboxyl group, having lost a hydrogen ion, is negatively charged, and the amino group, having attracted an
extra hydrogen ion, is positively charged:

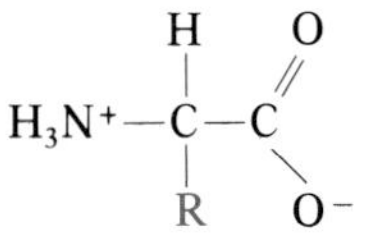

The various amino acids differ in their side chains, or R groups. R may be
very simple, as in glycine, where it is only a hydrogen atom, or it may be
very complex, as in tryptophan, where it includes two ring structures.
Twenty different amino acids are commonly found in proteins; their structural formulas are shown in Figure 3.16. The various R groups give each of
the amino acids different characteristics, which, in turn, greatly influence
the properties of the proteins incorporating them. For example, some
amino acids are relatively insoluble in water, owing to R groups that are
nonpolar at pH 6.5–7 (Fig. 3.16, top two rows), whereas other amino acids
are water-soluble, because their R groups are polar (third row) or electrically charged (bottom row).

Proteins are long and complex polymers of the twenty common amino
acids. The amino acid building blocks bond together by condensation reactions between the —COOH groups and the —NH$_2$ groups (Fig. 3.17). Such

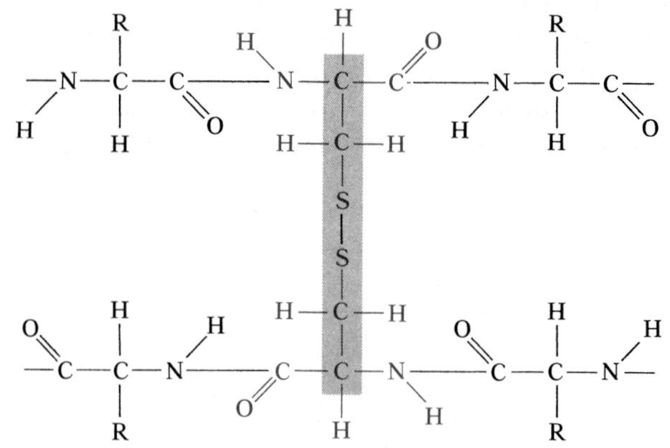

3.18   The structural formula of a cystine bridge
A cystine bridge (gray) is formed when two cysteine peptides (color) from different parts of a protein are linked by a disulfide bond.

bonds are called **peptide bonds**, and the chains they produce are called **polypeptide chains**. The amino acid units incorporated into a chain are called peptides. The number of peptides in a single polypeptide chain within a protein molecule is usually between 40 and 500, though shorter and longer chains sometimes occur. The variation, however, is between polypeptides of different kinds; for any given kind of polypeptide, the chain length is constant. As we shall see, these crucial molecules have a three-dimensional shape that is largely determined by the distribution of their polar, charged, and nonpolar R groups: the chain will tend to fold so that the nonpolar (hydrophobic) groups are inside the protein, where they bind to one another by means of hydrophobic interactions, as described in the preceding chapter, while the hydrophilic groups are exposed at the surface, where they interact with nearby polar molecules—particularly water.

Protein molecules often consist of more than one polypeptide chain. The chains may be held together by numerous weak bonds, especially hydrogen bonds; for example, a single molecule of hemoglobin, the red oxygen-carrying protein in blood, is composed of four polypeptide chains linked by hydrogen bonds. Insulin, an important hormone secreted by the pancreas in vertebrates, exemplifies a protein with polypeptide chains held together by both hydrogen and covalent bonds. The covalent bonds, called **disulfide bonds**, are between the sulfur atoms of two units of the amino acid cysteine (Fig. 3.16); two cysteines readily react with each other to form a symmetrical linkage called a cystine bridge (Fig. 3.18). As Figure 3.19 shows, disulfide bonds can also link two parts of a single polypeptide chain, maintaining it in a bent or folded shape. Clearly, disulfide bonds play a very important role in linking the constituent polypeptide chains of some complex proteins, whose folding patterns they thus help determine.

We have so far discussed the so-called **primary structure** of protein molecules—the number of polypeptide chains and the number and sequence of amino acids in each. Since these aspects of primary structure can vary among proteins of different kinds, the potential number of different proteins is enormous. For instance, for a relatively short polypeptide chain of 100 amino acids, 20^{100} sequences are possible. Each of the many millions of different species of organisms can therefore have its own peculiar proteins.

Determining the primary structure For a better understanding of the primary structure of proteins, let us trace some of the steps by which Frederick Sanger and his colleagues at Cambridge University determined the structural formula for insulin, the first protein for which a structural formula could be written (Fig. 3.19). Insulin was well suited to be the object of pioneering work on protein structure; it can easily be obtained pure from the bovine pancreas, and it is one of the smallest proteins known. Nevertheless, about ten years of exhaustive work were required before Sanger, in 1954, could feel confident that he finally knew the full sequence of amino acids in insulin. His achievement, a milestone in the history of biochemistry, won him the first of his two Nobel Prizes in 1958.

Sanger's method of determining the sequence of amino acids in the two polypeptide chains of insulin involved breaking the chains into fragments and then trying to establish how the pieces fitted together. If a polypeptide chain is heated in an acid solution for about 24 hours, all its peptide bonds are hydrolyzed (broken) by the addition of a water molecule at the site of each bond, the amino acids being uncoupled in the process. The results can then be analyzed.

The technique Sanger used for analyzing the amino acids was chromatography, a process that employs the different affinities of the unknown substances for two other materials. From this procedure he could find out which amino acids insulin contains and in what amounts. To learn the sequence of these amino acids, he had to use other approaches.

By less drastic treatment of insulin with hydrolyzing agents, he could preserve some of the peptide bonds and thus obtain many protein fragments consisting of two, three, four, five, or more amino acids. He analyzed these fragments for their amino acid content, utilizing particularly a technique that enabled him to determine which amino acid was on the end with the free amino group. After analyzing vast numbers of such pieces, he attempted to fit them together in proper sequence, by looking for fragments with regions of apparent overlap. For example, he found the following two fragments that seemed to have overlapping sequences at their ends:

Leu–Val–Cys–Gly–Glu–Arg–Gly–Phe–Phe
                             Gly–Phe–Phe–Tyr–Thr–Pro–Lys

He reasonably concluded that one part of the insulin molecule contained the sequence

     Leu–Val–Cys–Gly–Glu–Arg–Gly–Phe–Phe–Tyr–Thr–Pro–Lys

He then hunted for other fragments that overlapped this sequence, so that he could extend it. By laborious investigations of this sort, he finally determined the entire amino acid sequence of the two polypeptide chains. Later, by other techniques, he determined the positions of the three disulfide bonds.

Since Sanger's discovery of the structural formula of insulin—culminating more than a century of effort by scientists to learn the composition and

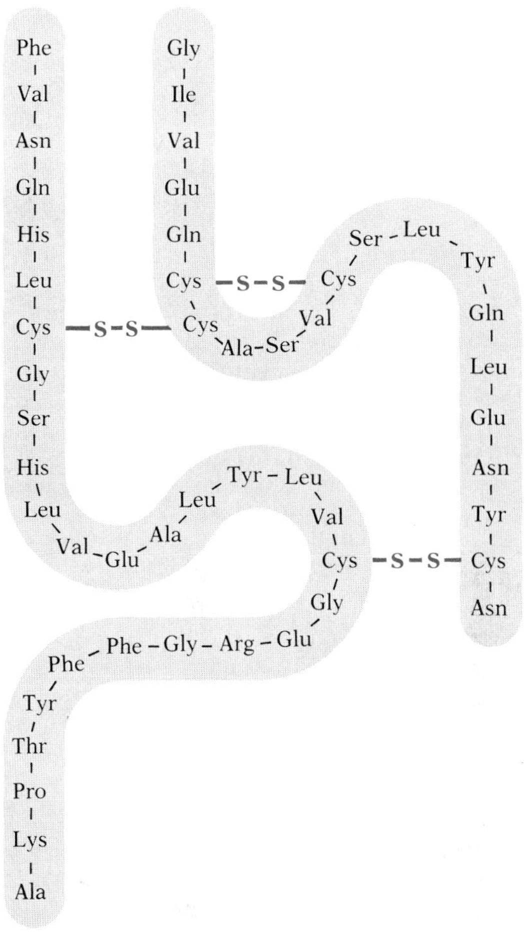

3.19 The structure of bovine insulin
The molecule consists of two polypeptide chains joined by two disulfide bonds. There is also one disulfide bond within the shorter chain (right). Hydrogen bonds (not shown) between the chains and between segments of the same chain are also present.

CHROMATOGRAPHY

Chromatography is one of the most valuable techniques available for separating the substances found in blood or cells, or separating a single substance into its constituent parts. For instance, after using chromatography to isolate a single protein, we can hydrolyze it and then use chromatography again to separate the amino acids from one another. The many kinds of chromatography all share the same fundamental mechanism—the simultaneous exposure of the mixture being studied to two different substances, such as two solvents that will not mix, or a solvent and an adsorbent solid. (An adsorbent is so named because molecules of a gas, dissolved solute, or liquid adhere to its surface.) Each solute in the mixture will distribute itself between the two substances in proportion to the relative affinities of the solute for those two substances. For example, if solute A, a material highly soluble in water but only minimally soluble in phenol, is shaken in a jar with water and phenol (which do not mix), the molecules of A will divide between the two solvents in such a fashion that most will be in the water and only a few in the phenol.

Paper chromatography is one of the simplest kinds of chromatography. Several drops of a mixture of unknown molecules are placed near one bottom corner of a piece of filter paper moistened with water (A, in the figure). The bottom edge of the paper is then dipped into a nonaqueous solvent such as phenol. As the solvent migrates up the paper by capillary action, those solutes in the mixture that have a much higher affinity for the solvent than for the water in the filter paper travel freely up the paper with the solvent. By contrast, those solutes that have a much higher affinity for the moisture in the filter paper do not travel far, but instead quickly transfer from the flowing solvent and bind to the stationary film of water on the filter paper. Materials with intermediate affinities for the solvent as compared to the water travel intermediate dis-

tances. At the end of a measured time interval, the various solutes in the original mixture have come to rest at different places along the filter paper (B). Frequently, further separation is achieved by using a second solvent and allowing the solutes to travel across the paper in a new direction (C and D). The technique is then known as two-axis chromatography. An example of the results obtainable by this method can be seen in the photograph.

Column chromatography works like simple paper chromatography, except that instead of filter paper, a glass column packed with some hydrated adsorbent material, such as starch or silica gel, is used. The mixture to be tested, in a nonaqueous solvent, is poured into the top of the column and allowed to filter downward (or is pulled down by a pump). Each component in the mixture tends to move at its own rate, which depends on its relative affinity for the flowing solvent and the stationary beads of adsorbent material. That is, substances with a much higher affinity for the solvent will flow straight through, while those with some affinity for the beads will be delayed. When enough solvent is poured into the column to keep all the materials moving, each of them emerges from the bottom of the column at a different time, and they are collected in separate containers, ready for further analysis.

Modern chromatographic equipment based on these and other separation techniques is highly sophisticated. Analyses that would have demanded years of work in Sanger's day can be performed in two or three hours. For example, in an amino acid analyzer each component of the substance being tested is caught by an automatic "fraction collector" as it emerges from the column, and a blue dye is added that reacts with amino acids. A photometer senses and records the intensity of the dye in each fraction, thus revealing what amino acids are present, and in what amounts.

structure of proteins—a number of other proteins have been similarly elucidated. The more recent work has been aided, as you would expect, by many technological advances, especially by rapid methods for determining the amino acid sequence in the relatively short chain fragments produced by partial hydrolysis. But the basic approach—complete hydrolysis and amino acid analysis, followed by partial hydrolysis and fragment matching —remains the same.

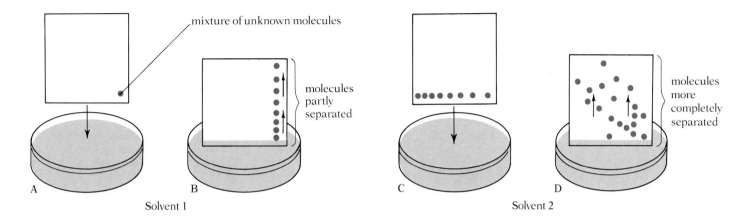

mixture of unknown molecules

molecules partly separated

A

B

Solvent 1

C

D

molecules more completely separated

Solvent 2

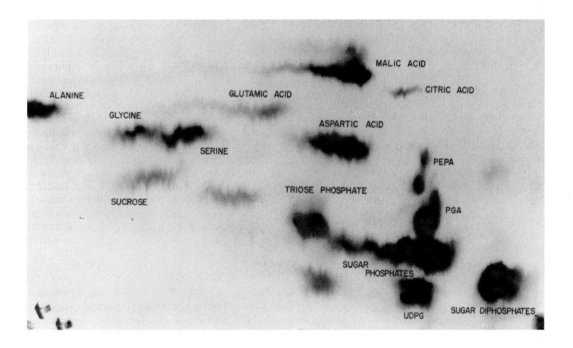

ALANINE

GLYCINE

SERINE

SUCROSE

GLUTAMIC ACID

MALIC ACID

CITRIC ACID

ASPARTIC ACID

PEPA

TRIOSE PHOSPHATE

PGA

SUGAR PHOSPHATES

UDPG

SUGAR DIPHOSPHATES

The spatial conformation of proteins Proteins are not laid out simply as straight chains of amino acids. Instead, they coil and fold into very complex spatial conformations, which play a crucial role in determining the distinctive biological properties of each protein. This three-dimensional character is a consequence of weak interactions between peptides in the protein. As Linus Pauling and Robert B. Corey of the California Institute of Technology showed in 1951, certain precise degrees of coiling allow internal hydrogen

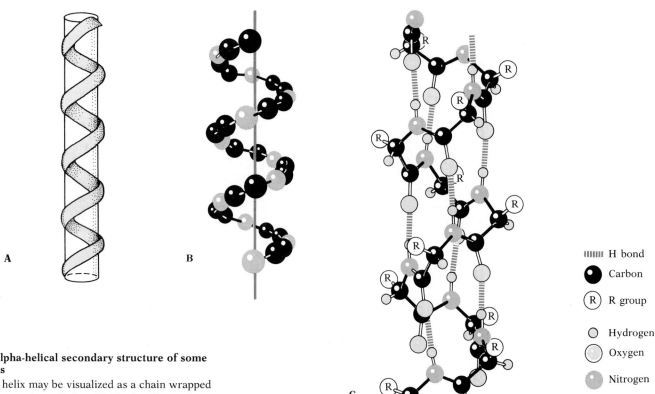

3.20   Alpha-helical secondary structure of some proteins
(A) The helix may be visualized as a chain wrapped around a regular cylinder. (B) The backbone of a polypeptide chain (the repeating sequence of N–C–C–N–C–C–N–C–C–) is shown coiled in a helix (all other atoms and R groups are omitted). It takes approximately 3.6 amino acid units (N–C–C) to form one complete turn of the helix. The vertical rod is meant to aid visualization. (C) A ball-and-stick model of an α-helical section of a protein, showing some of the intrachain hydrogen bonds that help stabilize the helix; the hydrogen bonds shown extend between the amino group of one amino acid and the oxygen of the third amino acid beyond it in the polypeptide chain.

bonds to form and stabilize what is called an ***alpha helix*** (Fig. 3.20). A helix can be visualized as a chain wound around a regular cylinder. In a protein, each complete turn of the helix takes up approximately 3.6 amino acid units of the polypeptide chain (Fig. 3.20B). The chain is held in this helical shape by hydrogen bonds formed between the amino group of one amino acid and the oxygen of the third amino acid beyond it in the polypeptide chain— which is the amino acid next to it in the axial direction of the helix (Fig. 3.20C). Regions displaying a regular α-helical arrangement are among those said to have a ***secondary structure***, above and beyond the linear sequence of amino acids that constitute the primary structure.

The helical pattern is seen at its simplest in some ***fibrous proteins***. One category of these insoluble proteins, extensively used in studies of protein structure, includes the ***keratins***, which provide the structural elements for many of the specialized derivatives of skin cells. Keratins with extensive α-helical secondary structure, such as nails, hooves, and horns, are hard and brittle. The hardness results from an extraordinarily large number of covalent cystine bridges: up to one amino acid in four is cysteine. Others, such as hair and wool, are soft and flexible and can easily be stretched (especially when moistened and warmed). The stretching is possible because there are many fewer cystine bridges. The intrachain hydrogen bonds are

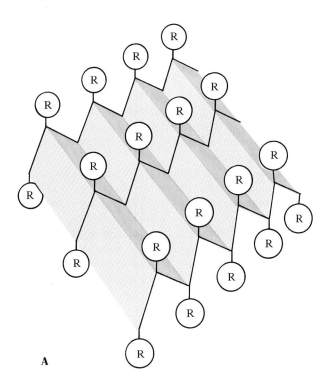

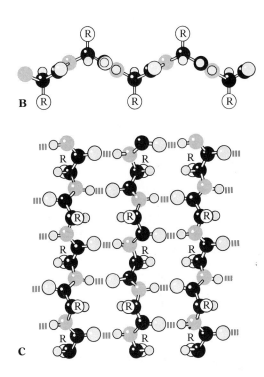

3.21 Pleated-sheet secondary structure of some proteins
(A) Diagrammatic representation of three parallel polypeptide chains in β conformation, with the imaginary pleated sheet between them shown in color. (B) Edge and (C) top views of a ball-and-stick model of polypeptide chains in β conformation.

easily broken, and the polypeptide chains can then be pulled out of their compact helical shapes into a more extended form. The chains tend to contract to their normal length, with re-formation of the hydrogen-bonded α helix, when the tension on them is released (or when they are dried and cooled).

Another stable arrangement of peptides within a polypeptide chain gives rise to the second major type of secondary structure. This type, designated beta (β) structure, is often called the ***pleated sheet***. In this conformation, also seen at its simplest among some of the keratins, many side-by-side polypeptide chains are cross-linked by interchain hydrogen bonds (Fig. 3.21). The resulting arrangement is flexible and strong, but resists stretching because the polypeptide chains are already almost fully extended. Probably the best-studied β-keratin is silk; other examples include spider webs, feathers (Fig. 3.22), and the scales, claws, and beaks of reptiles and birds.

In addition to the keratins, another kind of fibrous protein, with its own distinctive secondary structure, should be mentioned here because it is the

3.22 The feather of a bird, a β-keratin
This scanning electron micrograph shows the base of a parakeet's tail feather.

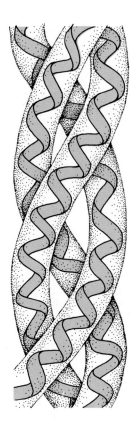

3.23 Model of a portion of a molecule of collagen
Three polypeptide chains, each helically coiled, are wound around one another to form a triple helix. The "sheaths" here and in Figures 3.24 and 3.25 are intended as a reminder that each molecule consists not merely of a backbone, but also of R groups, which give it volume.

most abundant protein in higher vertebrates. This is **collagen**, which may constitute one-third or more of all the body protein; it is especially abundant in skin, tendons, ligaments, and bones, and in the cornea of the eye. A molecule of collagen is composed of three polypeptide chains, each first helically coiled and then wound around the other two to form a triple helix (Fig. 3.23). What facilitates the intertwining of the three chains is that every third amino acid in the chains is glycine, whose R group, being only a single hydrogen atom (Fig. 3.16), takes up very little room. The chains are held together by hydrogen bonds. Collagen fibers are exceedingly strong and very resistant to stretching.

Far more complex in spatial conformation than the fibrous proteins are the globular proteins, whose polypeptide chains are folded into complicated spherical or globular shapes (Fig. 3.24). Because of charged and polar R groups on their exposed surfaces, globular proteins, which include enzymes, proteinaceous hormones, antibodies, and most blood proteins, are usually water-soluble. Typically, they are made up of sections of α helix interspersed with nonhelical regions; some globular proteins, however, have no obvious secondary structure at all. The protein myoglobin, which is the oxygen-storage protein in muscles, provides a more typical example. It consists of one polypeptide chain containing eight major sections of secondary structure—α helices—connected by short regions of irregular (nonhelical) coiling. At each nonhelical region, the three-dimensional orientation of the polypeptide chain changes, giving rise to the protein's characteristic folding pattern. This three-dimensional folding pattern, which is superimposed on the secondary structure, is called **tertiary structure**. In practice, tertiary structure is difficult to determine. The protein must first be crystallized. Then X rays are beamed through the crystals; deflected by the electrons of the thousands of atoms, they form a pattern which is then deciphered by a computer.

When a globular protein is composed of two or more independently folded polypeptide chains loosely held together, usually by weak bonds, the manner in which the already folded subunits fit together is called **quaternary structure** (Fig. 3.25).

Several aspects of a protein's primary structure (that is, its amino acid sequence) contribute to producing its tertiary and quaternary structure. If, for example, a polypeptide chain contains two cysteine units, the intrachain disulfide bond joining them may introduce a fold in the chain or stabilize one created in other ways (Fig. 3.19). The most common source of folding is proline. Wherever there is a proline, a kink or bend occurs, because the structure of proline is such that it cannot conform to the geometry of an α helix; as we have seen, proline, though one of the building-block units of protein, is not technically a true amino acid, since its R group circles around and links with its amino group (Fig. 3.16). Four of the eight bends in globular myoglobin, in fact, result from the presence of prolines in the chain.

The distinctive properties of the various R groups of the amino acids also impose constraints on the shape of the protein. For example, hydrophobic groups tend to be close to each other in the interior of the folded chains—as far away as possible from the water that suffuses living tissue—whereas hydrophilic groups tend to be on the outside, in contact with the water. Polar

R groups, such as that of tyrosine (Fig. 3.16), tend to assume positions where they can form hydrogen bonds with other polar R groups; similarly, electrically charged R groups can form ionic bonds with oppositely charged groups (aspartic acid and lysine, for example, form an ionic bond when they are forced into the interior of the protein). In myoglobin, again, all the hydrophobic peptides are in the interior, and all but two of the hydrophilic peptides are on the outside. (The two exceptions, both ionic amino acids, hold the heme group in place.) Thus the various kinds of weak bonds discussed earlier play crucially important roles in forming and stabilizing the tertiary structure of proteins.

As this discussion suggests, there are compelling reasons to believe that the primary structure of a protein determines its spatial conformation. More specifically, it appears that the primary structure determines the energetically most favorable, and therefore most stable, possible arrangement of the polypeptide chains. Hence the question, long perplexing to biochemists, of how conformation is specified when a protein is being synthesized becomes synonymous with the question of how amino acid sequence is specified—a question no longer so perplexing to scientists, as we shall see in a later chapter.

Further support for the idea that primary structure determines conformation comes from studies of *denatured* proteins—proteins that have lost most of their secondary, tertiary, and quaternary structure, and with it their normal biological activity, through exposure to high temperature or extreme pH. That a denatured protein should lack the characteristic biological activity of the natural protein is an indication that its conformation is functionally essential. Since conformation is dependent in large part on weak bonds (which are very sensitive to temperature and pH), it is easily disrupted by anything that breaks or alters those bonds; it is stable only within a limited range of temperature and pH. Even brief exposure to high temperatures (usually above 60°C) or to extremes of pH will cause denaturation of most globular proteins. But under favorable test-tube conditions some denatured proteins can spontaneously regain their native three-dimensional conformation; they can refold, and recover their normal biological activity. Since only the primary structure is available to dictate

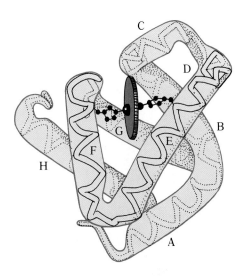

3.24 Spatial conformation of a molecule of myoglobin

Myoglobin, a globular protein related to hemoglobin and, like hemoglobin, characterized by a strong affinity for molecular oxygen, is a single complexly folded polypeptide chain of 151 amino acid units; attached to the chain is a nonproteinaceous "prosthetic" group called heme (represented by the disc). The polypeptide chain consists of eight sections of α helix (labeled A through H), with nonhelical regions between them. These nonhelical regions are a major factor in determining the tertiary structure of the molecule—that is, the way the helical sections are folded together. (Section D cannot be seen in this drawing, because it is oriented perpendicular to the plane of the page.)

3.25 Quaternary structure of hemoglobin

A single molecule of hemoglobin is composed of four independent polypeptide chains, each of which has a globular conformation and its own prosthetic group. The spatial relationship between these four—the way they fit together—is called the quaternary structure of the protein.

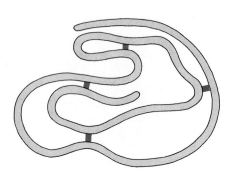 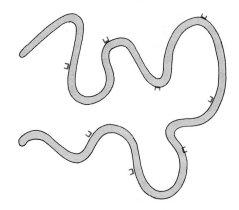 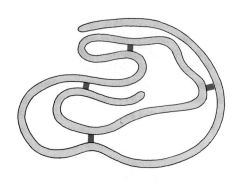

3.26   Denaturation and renaturation of ribonuclease

When ribonuclease, a normally globular protein (left), is denatured, with both its weak bonds and its four intrachain disulfide bonds (color) broken, it unfolds into an irregularly coiled state (middle). In this denatured condition the ribonuclease lacks its usual ability to digest RNA. When the denaturing agents are removed and favorable conditions restored, the protein spontaneously refolds into its native conformation (right) and regains its capacity for biological activity as an enzyme. Even the four disulfide bonds re-form correctly. Since there are 105 possible ways to join the cysteines but the enzyme folds only to bring the correct pairs together, we must conclude that the conformation does not result just from the positions of the cysteines; most disulfide bonds may serve to stabilize, rather than to determine structure.

the folding pattern in such cases, it alone must be sufficient to determine all other aspects of protein's structure (Fig. 3.26).

Conjugated proteins   Attached to some proteins are nonproteinaceous groups called *prosthetic groups*; an example is the heme group of myoglobin, a ringlike structure with an iron atom in its center (Fig. 3.24). Prosthetic groups may be as simple as a single metal ion bonded to the polypeptide chain; they may be sugars or other carbohydrate entities; or they may be of lipid form. Whatever the nature of the prosthetic group, its presence alters the properties of the protein in important ways. Without their heme groups, for example, myoglobin and hemoglobin lose their high affinity for molecular oxygen. All proteins that contain nonproteinaceous substances are called conjugated proteins.

NUCLEIC ACIDS

Nucleic acids constitute a fourth major class of organic compounds crucial to all life. They are the materials of which genes, the units of heredity, are composed. They are also the messenger substances that convey information from the genes in the nucleus to the rest of the cell, information that not only determines the structural attributes of the cell but also regulates its ongoing functional activities.

Like polysaccharides and proteins, nucleic acid molecules are long polymers of smaller building blocks. In this case the building blocks are called

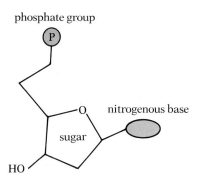

3.27   Diagram of a nucleotide

A phosphate group and a nitrogenous base are attached to a five-carbon sugar.

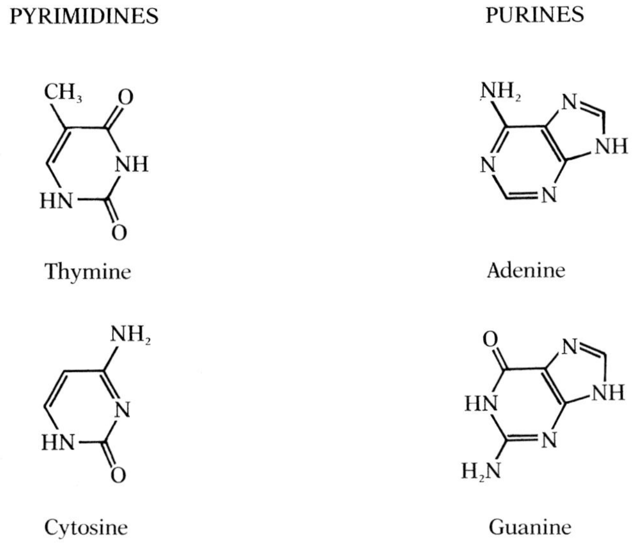

PYRIMIDINES   PURINES

Thymine   Adenine

Cytosine   Guanine

3.28 The four nitrogenous bases in DNA
The two single-ring bases, thymine and cytosine, are pyrimidines; the two double-ring bases, adenine and guanine, are purines.

nucleotides; they are themselves composed of still smaller constituent parts: a five-carbon sugar, a phosphate group, and an organic nitrogen-containing base. Both the phosphate group and the nitrogenous base are covalently bonded to the sugar (Fig. 3.27).

Deoxyribonucleic acid   This nucleic acid, commonly called *DNA*, is the one genes are made of. Four different kinds of nucleotide building blocks occur in DNA. All have, as the name suggests, deoxyribose as their sugar component, but they differ in their nitrogenous bases, which may be one of four different substances. Two of these, *adenine* and *guanine*, are double-ring structures of a class known as purines; the other two, *cytosine* and *thymine*, are single-ring structures known as pyrimidines (Fig. 3.28).

The nucleotides within a DNA molecule are bonded together in such a way that the sugar of one nucleotide is always attached to the phosphate group of the next nucleotide in the sequence (Fig. 3.29). Thus a long chain of alternating sugar and phosphate groups is established, with the nitrogenous bases oriented as side groups off this chain. The sequence in which the four different nucleotides occur is essentially constant in DNA molecules of the same species, but differs between species. It is this sequence that determines the specificity of each type of DNA. It is, in fact, the sequence of the nucleotides in DNA that encodes hereditary information, which is expressed through control of protein synthesis. More particularly, the sequence of nucleotides in DNA determines the sequencing of amino acids (the primary structure) in proteins; we shall examine this process in considerable detail in Chapter 27.

DNA molecules do not ordinarily exist in the single-chain form shown in Figure 3.29. Instead, two such chains, oriented in opposite directions, are arranged side by side like the uprights of a ladder with their nitrogenous

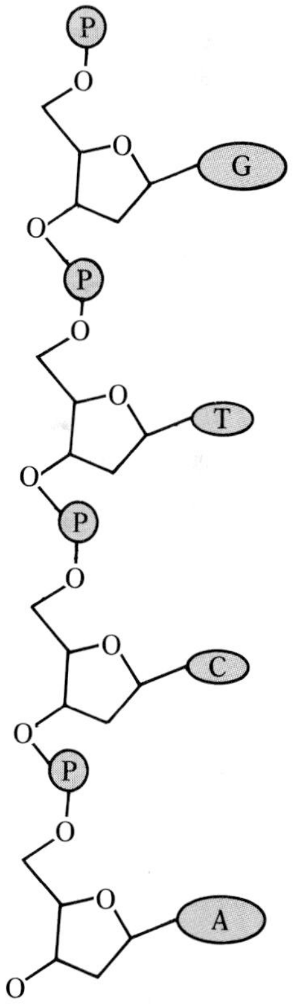

3.29 Portion of a single chain of DNA
Nucleotides are hooked together by bonds between their sugar and phosphate groups. The nitrogenous bases (G, guanine; T, thymine; C, cytosine; A, adenine) are side groups.

3.30 Portion of a DNA molecule uncoiled

The molecule has a ladderlike structure, with the two uprights composed of alternating sugar and phosphate groups and the cross rungs composed of paired nitrogenous bases. Each cross rung has one purine base (large oval) and one pyrimidine base (small oval). When the purine is guanine (G), the pyrimidine with which it is paired is always cytosine (C); when the purine is adenine (A), the pyrimidine is thymine (T). Adenine and thymine are linked by two hydrogen bonds (striped bands), guanine and cytosine by three. Note that the two chains run in opposite directions—that is, the free phosphate is at the upper end of the left chain and at the lower end of the right chain.

bases constituting the cross rungs of the ladder (Fig. 3.30). The two chains are held together by hydrogen bonds between adjacent bases. Finally, the entire double-chain molecule is coiled into a double helix (Fig. 3.31).

The regular helical coiling and the hydrogen bonding between bases impose two extremely important constraints on how the cross rungs of the ladderlike DNA molecule can be constructed. First, each rung must be composed of a purine (double ring) and a pyrimidine (single ring); only in this way will all cross rungs be of the same length and permit formation of a regular helix. Second, if the purine is adenine, the pyrimidine must be thymine, and, similarly, if the purine is guanine, the pyramidine must be cytosine; only these two pairs are capable of forming the required hydrogen bonds (Fig. 3.30). Since it does not matter, however, in which order the members of a pair appear (A–T or T–A; G–C or C–G), the double-chain molecule can have four different kinds of cross rungs, as shown in Figure 3.30. The biological significance of this arrangement is that the base sequence of one chain uniquely specifies the base sequence of the other, so that the two strands can be separated and exact copies made every time a cell divides.

Ribonucleic acid A second important category of nucleic acids comprises the ribonucleic acids, or **RNA**. There are several types of RNA, each with a different role in protein synthesis. Some act as messengers carrying instructions from the DNA of the genes to the sites of protein synthesis in the cell. Others are structural components of cytoplasmic organelles, called ribosomes, on which the process of protein synthesis takes place. Still others transport amino acids to the ribosomes, so that they may be incorporated into proteins. We shall discuss each of these types of RNA in much more detail in Chapter 27. Here let us note only that all types of RNA differ from DNA in three principal ways: (1) The sugar in RNA is ribose, whereas that in DNA is deoxyribose. (2) Instead of thymine, one of the four nitrogenous bases of DNA, RNA contains a very similar base called **uracil**. (3) RNA is ordinarily single-stranded, whereas DNA is usually double-stranded.

CHEMICAL REACTIONS

In previous sections we have mentioned several types of chemical reactions that take place within organisms: condensation reactions between simple sugars to form polysaccharides, and hydrolysis of polysaccharides back to simple sugars; condensation reactions of fatty acids and glycerol to form fats, and the reverse hydrolysis; condensation reactions of amino acids to form polypeptide chains and proteins, and the reverse hydrolysis. But we have said nothing about the conditions under which these reactions will take place. It is now time for a brief examination of those conditions.

All the processes of life depend on the ordered flow of energy. As we said in the preceding chapter, the behavior of electrons is the single most crucial factor in the chemistry of life. Virtually all the energy for living things comes as light from the sun and is captured by electrons, which are thereby excited into higher orbitals. The energy released by such electrons as they change orbitals, and as they move to more highly electronegative atoms in precisely ordered chemical reactions, is harvested to fuel all the processes of life. To understand biology, then, it is essential to understand how the transfer of energy in chemical reactions takes place as one set of covalent bonds is replaced by another, a subject known a thermodynamics.

FREE ENERGY

Instead of talking at this point about condensation or hydrolytic reactions or any other specific type of reaction, let's consider a generalized one. Suppose, for example, that two substances, A and B, can react with each other in solution to produce two new compounds, C and D:

$$A + B \longrightarrow C + D$$
$$\text{reactants} \qquad \text{products}$$

What determines whether this reaction will tend to take place spontaneously? The answer to this question turns on a concept of physics—*energy*, which is defined as *the capacity to do work*. *Free energy* (as the term is used in chemistry and biology) is the energy in a system available for doing work under conditions of constant temperature and pressure. Where there is energy to be tapped, whether it be in the weight of the water stored behind a dam, or in the covalent bonds of sugars like glucose, or in an electron that has been excited into a higher orbital by sunlight, or in the tightly bound nuclei of the atoms in a nuclear reactor, the potential for work is present.

The *First Law of Thermodynamics*—the Law of Conservation of Energy —tells us that *the total energy in the universe is constant:* energy needed to do work in a particular system—in a cell, for example—cannot be generated from nothing; it must be obtained from a source outside that system, which thereby loses a corresponding amount of energy to balance the books. The *Second Law of Thermodynamics* states that *in the universe as a whole the total amount of free energy*—that is, the energy actually available for doing work—*is declining*. This is because practically every energy transfer generates heat that is then no longer available for doing work. The magnitude of this waste is enormous, as the need for cooling towers in power plants and radiators in car engines to dissipate unused (and unusable) thermal energy makes evident. Even in our own bodies waste heat can be a serious problem, and highly specialized mechanisms have evolved for releasing this waste energy to the environment.

One simple and universal law of cellular chemistry holds that whether or not a reaction can proceed spontaneously depends on the net change in free energy that accompanies the reaction. In other words, the course of the reaction depends on whether the free energy of the set of covalent bonds in the reactants is greater or less than the free energy of the new set of covalent bonds in the products. In order to quantify free-energy changes, we shall use the symbol $\triangle G$ to denote changes in free energy under a defined set of conditions: temperature 25°C, pressure of 1 atmosphere, pH 7.0, with both the reactants and products at a concentration of 1 mole/liter.[1] With this in mind, let's rewrite the reaction from a thermodynamic perspective,

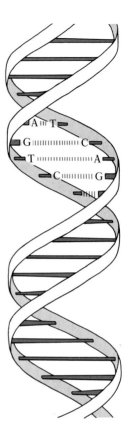

3.31 A model of the DNA molecule
The double-chained structure is coiled in a helix. As shown in detail in the second segment, it consists of two polynucleotide chains held together by hydrogen bonds (striped bands) between their adjacent bases.

---

[1] In thermodynamics texts the formal symbol for changes in free energy under these conditions is $\triangle G'°$. Since this is the only version of $\triangle G$ we shall discuss, we can dispense with the superscripts.

3.32 Endergonic versus exergonic reactions
Reactions either consume free energy from outside a system or liberate some of the energy from within it. Those that require energy (in the amount represented by ΔG) to go uphill from the initial state to the final state (left) are said to be endergonic, while those that go downhill and release energy (right) are said to be exergonic. (As we shall see, these diagrams describe ideal conditions and are therefore oversimplified.)

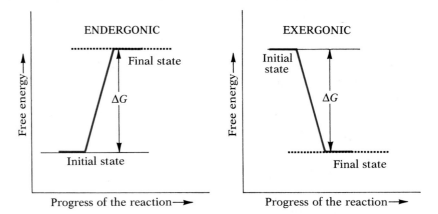

looking at the free energy of the initial reactants (G_i), the free energy of the final products (G_f), and the change in free energy engendered by the reaction (ΔG):

$$\begin{array}{ccc} \text{initial state} & & \text{final state} \\ \text{A + B} & \longrightarrow & \text{C + D} \\ G_i & \longrightarrow & G_f \qquad (\Delta G = G_f - G_i) \end{array}$$

The term ΔG, the change in free energy as a result of the reaction, is the crucial variable. *If the reaction results in products with less free energy in the covalent bonds than the reactants possessed, the reaction is "downhill" and can proceed spontaneously.* Since the free energy liberated from the covalent bonds of the reactants is usually released as heat, such reactions are said to be exothermic (heat-producing) or, more generally, **exergonic** (energy-releasing). If, on the other hand, the reaction requires an input of external energy, it cannot proceed spontaneously. An "uphill" reaction of this kind, in which the covalent bonds of the products have more free energy than those of the reactants, is said to be endothermic, or **endergonic**. These two alternatives are illustrated schematically in Figure 3.32.

In summary, then, we can say that all spontaneous reactions are downhill—exergonic. This fact is responsible for a general property of matter, both living and nonliving: all systems tend to lose free energy, and ultimately reach a state in which their free energy is as low as possible, just as rocks on a hill tend to move down to the bottom rather than up to the top.[2] Throughout this book we shall indicate the free-energy change, ΔG, resulting from biological reactions to the right of the reaction formula. You will need to remember that when ΔG is negative (meaning that the covalent bonds of the reactants had more energy than those of the products), the re-

---

[2] Physicists and chemists frequently relate free energy and "order," which is defined as an arrangement of components in a system that is unlikely to occur by chance. Consider the unlikely circumstance in which all the molecules of air in a room are on one side, with a vacuum on the other. As you can readily imagine, this orderly arrangement disintegrates as the molecules of air spread throughout the room, with a corresponding loss of free energy. The more orderly a system is, the more free energy it possesses; but since increasing disorder is inevitable, so also is the spontaneous loss of free energy. The amount of disorder in a system—the energy unavailable for doing work—is called entropy.

action is exergonic; when $\triangle G$ is positive, the covalent bonds of the products have gained energy, so the reaction is endergonic. This relationship holds even when factors like temperature do not correspond to the standard conditions listed above: though the exact magnitude of $\triangle G$ will shift as conditions change, the *relative* magnitude of the $\triangle G$s of different reactions in the same cell normally will not. In other words, if reaction I liberates twice the free energy of reaction II under standard conditions, the 2 to 1 ratio will persist under nonstandard conditions of temperature or pressure, provided they are the same for both reactions. Hence the $\triangle G$s of different reactions under the same conditions can be directly compared, and many aspects of cell chemistry can then be predicted.

THE EQUILIBRIUM CONSTANT

Our discussion so far has implied that two reactant molecules can combine spontaneously to create products only if free energy is liberated. By extension, this suggests that if a reaction is exergonic, all the reactants in a mixture will ultimately be turned into products. However, we have ignored the possibility of a **back reaction**, with the products C and D combining to regenerate the reactants A and B:

$$C + D \longrightarrow A + B$$

Though this reaction is endergonic (uphill), it can take place—very slowly —because there is a source of energy within every system. This source, which we have ignored so far, is the energy of motion, the **kinetic** or **thermal energy** of the molecules. Every molecule has both a certain characteristic amount of stored energy in its bonds *and* energy of motion, which depends on its speed. In any solution, some molecules move very fast, others more slowly. In our example, when two fast-moving product molecules, C and D, collide, their energy of motion will sometimes be converted into covalent bond energy to produce two (slow-moving) reactants, A and B. Though this back reaction is rare, it is very important when the forward reaction is near completion. At this point the reactants A and B are so scarce that the rare, kinetic-energy-dependent back reaction may be just as likely as the forward reaction.

When the forward reaction, slowed in consequence of the increasing scarcity of reactants, is just counterbalanced by the rare back reaction, the two processes are in equilibrium and no further change in the concentration of substances takes place. Chemists define this ultimate stable ratio of products to reactants as the equilibrium constant (K_{eq}):

$$K_{eq} = \frac{\text{products}}{\text{reactants}}$$

This relationship succinctly summarizes the results of any reaction, and so complements $\triangle G$ in describing chemical reactions. For instance, an equilibrium constant of 10 means that when equilibrium has been reached under conditions of stable temperature and pressure, the products of the reaction outnumber the reactants by a factor of 10:

$$A + B \xrightleftharpoons[]{K_{eq} = 10} C + D \qquad (\triangle G = -1.4 \text{ kcal/mole})$$

TABLE 3.2   *Relationship between* K_{eq} *and* $\triangle G$ *at 25°C*

$\triangle G$ (kcal/mole)	K_{eq}	
4.1	0.001	Endergonic reactions
2.7	0.01	
1.4	0.1	
0	1.0	
−1.4	10.0	Exergonic reactions
−2.7	100.0	
−4.1	1000.0	

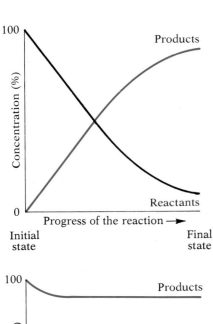

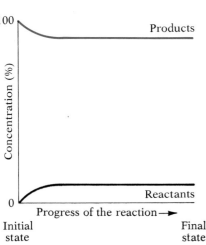

Such a reaction is downhill: the products are more abundant than the reactants, and the reaction liberates energy. The predominance of the forward reaction is indicated by having the arrow pointing from the reactants to the products longer than the one pointing back. In a reaction with an equilibrium constant of 0.1, on the other hand, the reactants outnumber the products by 10 to 1:

$$E + F \underset{}{\overset{K_{eq} = 0.1}{\rightleftharpoons}} G + H \qquad (\triangle G = +1.4 \text{ kcal/mole})$$

Such a reaction is uphill; it consumes outside energy. And finally, an equilibrium constant of 1.0 indicates that the products and the reactants are equally abundant:

$$I + J \underset{}{\overset{K_{eq} = 1.0}{\rightleftharpoons}} K + L \qquad (\triangle G = 0)$$

As we would expect, there is no free-energy change in this reaction. The numerical relationship between K_{eq} and $\triangle G$ at 25°C is summarized in Table 3.2.

Special note must be taken of one consequence of the relationship in cellular chemistry between the equilibrium constant and the concentration of products and reactants. Remember that the equilibrium constant of a reaction is a ratio—a simple empirical description of the outcome of the reaction—and depends, ultimately, on the $\triangle G$ of the reaction. Hence, the ultimate ratio of products to reactants in no way depends on the starting conditions. This point is illustrated in Figure 3.33, in which, whether we start with a great deal of reactant or none at all, we wind up with the ratio specified by the equilibrium constant.

ACTIVATION ENERGY

The question next arises, How do particular compounds follow particular pathways? For example, suppose the cell manufactures compounds D and Z, as follows:

$$A + B \xrightarrow{\quad} \overset{}{\rightleftharpoons} C + D \qquad W + X \xrightarrow{\quad} \overset{}{\rightleftharpoons} Y + Z$$

What is to prevent B from reacting with W, X, Y, or Z instead of with A? Some of these combinations may well have negative $\triangle G$s. What serves to prevent all but the "correct" reactants from combining?

For two molecules to combine, they must be brought unusually close to each other in a particular orientation and, frequently, one or more pre-existing bonds must be broken. This requires energy—specifically known as *activation energy* (E_a)—so even an exergonic reaction has an endergonic first step. The barrier that must be overcome by activation energy is illus-

3.33   The relationship between the equilibrium constant and concentration
The equilibrium constant, specifying the final ratio of products to reactants, is independent of the starting concentration. In the example shown here, the equilibrium constant is 10, which means that products (color curves) will be 10 times more common than reactants when the reaction ends. This is the outcome whether we begin with pure reactants (top) or pure products (bottom).

trated schematically in Figure 3.34. The only source of energy for this "priming" is the kinetic energy of colliding molecules.

Because of the strength of covalent bonds, a substantial amount of energy may be necessary to break the pre-existing bonds of reactants—often far more than the amounts listed for the $\triangle G$s in Table 3.2. Consider the following enormously exothermic reaction:

$$2H_2 + O_2 \longrightarrow 2H_2O$$

Even though this combination of oxygen and hydrogen can be explosive—for example, it provides much of the energy that pushes the space shuttle into orbit—the two reactants can coexist as a stable mixture almost indefinitely. A single spark, however, will initiate an explosive reaction. The same stability is a property of most reactants in living systems: the energy necessary to bring the reactants together and break their covalent bonds is far greater than the energy of all but the very few most rapidly moving molecules in a solution. The activation-energy barrier therefore prevents most reactions from taking place at a significant rate (Fig. 3.34). Without such a barrier the complex high-energy molecules (such as carbohydrates, lipids, proteins, and nucleic acids) on which life depends would be unstable, and would break down.

Once a reaction does get started, the combination of one pair of reactants may release enough energy (usually in the form of heat) to activate the next pair, and so on, in a chain reaction. This is precisely what happens in a rocket engine, and in the combustion of a dry piece of firewood. The wood, as we all know, can lie in a woodpile for years without bursting into flames spontaneously, but once set on fire, it literally consumes itself as the free energy liberated by the combining of carbon and oxygen into CO_2 supplies the activation energy to continue the burning. Put quite simply, heating a mixture will increase the *rate* of reaction (though it cannot affect the ultimate ratio of products to reactants, since that is a constant—K_{eq}). But cells literally cook at temperatures much above the 37°C of most mammals and birds, so they can use heat as a way of overcoming the activation-energy barrier only to a very limited extent. If the reactions necessary for life are to take place, cells must use some other method, and it must be one that lowers the barrier selectively, so that some exergonic reactions run, while others do not. Cellular chemistry, then, is essentially controlled by the selective lowering of particular activation-energy barriers. How is this crucial task managed?

The effect of catalysts   Chemists discovered years ago that certain chemicals speed up reactions between other chemicals. As we have seen, a simple mixture of hydrogen and oxygen does not react, but if we provide the initial activation energy (a spark) the mixture will explode. The same explosion will take place if we add instead a small quantity of platinum. After the reaction is over, the platinum will still be present, unchanged.

A substance that, like the platinum, speeds up a reaction but is itself unchanged when the reaction is over (even though it may have been temporarily changed during the reaction) is known as a ***catalyst***. A catalyst affects only the *rate* of reaction; it simply speeds up reactions that are thermodynamically possible to begin with. Like heat, a catalyst cannot alter the direction of a reaction, its final equilibrium, or the reaction energy involved.

In terms of our discussion, a catalyst decreases the activation energy

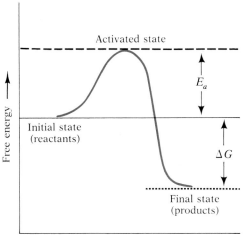

3.34 The energy changes in an exergonic reaction
Though the reactants are at a higher energy level than the products, the reaction cannot begin until the reactants have been raised from their initial energy state to an activated state by the addition of activation energy (E_a). It is the need for activation energy that ordinarily prevents high-energy substances from breaking down, and hence makes them stable; the higher the activation-energy barrier, the slower the reaction and hence the more stable the substance. When activation energy is available, the reactants form a temporary and unstable activated complex, which breaks down to yield the end products of the reaction; in the process both activation energy and free energy (in the amount represented by ΔG) are released.

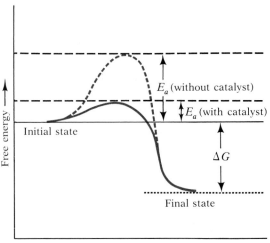

3.35   Reduction of necessary activation energy by catalysts

The activation energy (E_a) necessary to initiate the reaction is much less in the presence of a catalyst than in its absence. It is this lowering of the activation-energy barrier by enzyme catalysts that makes possible most of the chemical reactions of life. Note that the amount of free energy liberated by the reaction (ΔG) is unchanged by the catalyst—it is the same for both the catalyzed and the uncatalyzed reaction—and only the activation energy is changed.

needed for the reaction to take place (Fig. 3.35), thereby increasing the proportion of reactants energetic enough to react (Fig. 3.36). A catalyst does this by binding the reactants in an intermediate state in which the reactants are correctly oriented to each other and important internal bonds are weakened (Fig. 3.37). As a consequence of binding, then, conditions are highly favorable for the reaction.

An inorganic catalyst such as platinum is relatively unselective about the reactants it "helps." Natural selection has produced an enormous variety of highly specialized organic catalysts, called *enzymes*, which are included among the globular proteins described earlier in this chapter.

ENZYMES

Before we look at the way enzymes catalyze the selective chemistry of life, let's summarize what we know about the thermodynamics of biological reactions: (1) A chemical reaction can proceed spontaneously if it releases free energy. (2) Because the activation energy necessary for biochemical reactions is relatively high, they occur only very slowly without the intervention of catalysts. (3) Catalysts, including enzymes, alter neither the equilibrium constant of a reaction nor the net change in free energy—they cannot, by themselves, cause a reaction to run uphill, but they can make specific exergonic reactions run quickly.

Enzyme specificity and the active site   Unlike inorganic catalysts, enzymes are highly selective. A particular enzyme generally interacts with only one type of reactant or pair of reactants, customarily called the *substrates*. The enzyme thrombin, for instance, acts only on certain proteins, and only at very specific sites. It "recognizes" the bond between the amino acids arginine and glycine, which it then hydrolyzes (an important step in the formation of blood clots). Like all catalysts, enzymes lower the activation energy required (Fig. 3.35), so that the kinetic energy of many of the

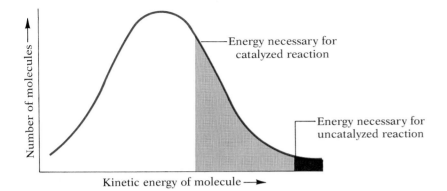

Energy necessary for catalyzed reaction

Energy necessary for uncatalyzed reaction

3.36 Effect of a catalyst on the ability of kinetic energy to activate a reaction
Reactant molecules exhibit a wide range of kinetic, or thermal, energy. Only a minute fraction (black area) have enough energy to overcome the activation-energy barrier for most reactions. In the presence of a catalyst, such as an enzyme, the barrier is lower, so a much larger proportion of the reactants (gray area) can combine to form products.

substrate molecules becomes sufficient to cause the reaction to take place (Fig. 3.36). As a result, enzymes vastly speed up the reactions they catalyze; a single molecule of enzyme may cause thousands or even hundreds of thousands of molecules of reactant to combine into product each second.

Because of their efficiency and specificity, enzymes can steer specific substrates into particular reaction pathways and block them from others, thus guiding the chemistry of life with great precision. In this and later chapters, we shall learn how enzymes can be so specific, how they lower the activation energy required for a reaction, and how they automatically control their own activity so as to best meet the cell's changing needs.

Biochemists have long held that the key to enzyme function—as well as to the operation of simpler inorganic catalysts—is surface activity. Enzymes, as we have said, are globular proteins, extremely complex molecules with intricate three-dimensional contours; each kind has its own distinctive surface geometry. It seems reasonable, then, that a given enzyme would interact only with substrates whose molecular configurations —conformation and location of charged groups—"fit" that enzyme's surface. Thus the specificity of enzymes can be viewed as depending on their three-dimensional molecular conformation.

3.37 Action of platinum as a catalyst
Because of its loosely bound outer electron (black dot), platinum is able to form weak temporary bonds with molecules of both hydrogen (left) and oxygen (right). This binding draws the hydrogen and oxygen electrons away from their covalent positions, thus weakening the bonds within their respective molecules. In addition, the spacing of the platinum atoms tends to align the hydrogen and oxygen atoms in such a way that new bonds, between hydrogen and oxygen, can be more easily formed. In the center portion of the figure, an oxygen and two hydrogens have bound to the platinum in a spatial relationship that is favorable for a water-producing reaction. (The other hydrogen and oxygen atoms in the original H_2 and O_2 molecules are bound to platinum atoms in an adjacent row, in front or behind the one shown here.) The complete reaction, of which essentially half is represented in the diagram, is $2H_2 + O_2 + 6Pt \longrightarrow 4HPt + 2OPt \longrightarrow 2H_2O + 6Pt$. Platinum is a catalyst in that it facilitates the reaction without itself being altered.

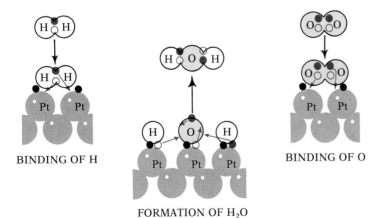

BINDING OF H

FORMATION OF H_2O

BINDING OF O

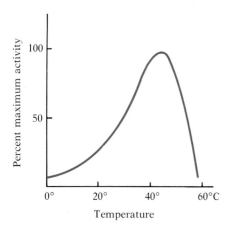

3.38 Enzyme activity as a function of temperature
Though temperature sensitivity varies somewhat
from one enzyme to another, the curve shown here
may be taken as applying to an "average" enzyme. Its
activity rises steadily with temperature
(approximately doubling for each 10°C increase)
until thermal denaturation causes a sudden sharp
decline, beginning between 40° and 45°. The enzyme
becomes completely inactivated at temperatures
above 60°, presumably because its three-dimensional
conformation has been severely disrupted.

The conclusion that the action of enzymes depends on their three-dimensional shape is consistent with the observation that when proteins are denatured—when their three-dimensional conformation is disrupted—they lose their characteristic biological activity; their enzymatic properties vanish (Fig. 3.38). Also consistent with this point of view is the observation that most enzymes are highly sensitive to changes in pH, and are active only within a limited pH range (Fig. 3.39). Apparently, change in pH results in the breakage of many of the weak bonds that help stabilize the conformation of proteins and, at the same time, leads to formation of new bonds, with consequent changes in the shape of the protein.

Enzyme and substrate have traditionally been visualized as fitting together like a lock and key, or like pieces of a puzzle. And it is true that the two must be roughly complementary if they are to combine. At least the reactive portion of the substrate molecule and the portion of the enzyme known as the ***active site*** must fit together in space intimately enough to become temporarily bonded, like the platinum and the hydrogen and oxygen in Figure 3.37. In this way they form a transient enzyme-substrate complex:

$$E + S \longrightarrow ES \longrightarrow E + P$$

(E stands for enzyme, S for substrate, and P for product.) But enzymes and their substrates probably do not always have to fit together exactly before ES (the enzyme-substrate complex) can form; according to the widely accepted ***induced-fit hypothesis***, an enzyme sometimes undergoes conformational changes in the course of bonding, which improve the fit and make ES more reactive (Fig. 3.40).

Spatial complementarity is only one of the prerequisites for enzyme-substrate interaction. Another is that E and S be chemically compatible and capable of forming numerous and precise weak bonds with each other. For though enzyme and substrate are sometimes held together by covalent bonds, the bonds are far more often weak ones of the same types—ionic, hydrogen, hydrophobic, and van der Waals—that stabilize protein conformation. They are bonds that can be made and broken rapidly in the collisions that result from random thermal motion at normal temperatures. The type of substrate to which a given enzyme molecule can become bonded

3.39 Enzyme activity as a function of pH
Most enzymes are very sensitive to pH, but they differ
markedly in their pH optima. Pepsin and trypsin are
both enzymes that digest protein, but the pH ranges
within which they are active overlap only slightly.
Pepsin is most active under strongly acidic conditions,
trypsin under neutral and slightly alkaline conditions.

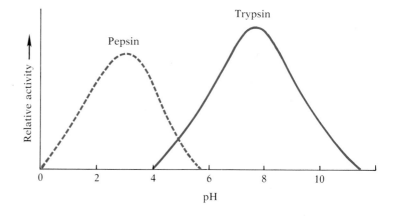

SUBSTRATE

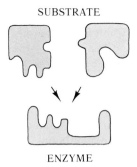

PRODUCT

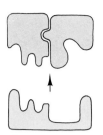

ENZYME   Enzyme-substrate complex   Enzyme resumes
original conformation

depends on the amino acids constituting its active site—more specifically, on the exposed R groups of these amino acids and the details of their arrangement relative to one another. Suppose the active site of a particular enzyme is a curving groove into which the reactive portion of the substrate must fit. Suppose further that most of the exposed R groups in this groove are electrically charged. It is obvious that the reactive portion of the substrate must be complementarily charged or polar; an electrically neutral nonpolar substrate molecule could not react with the active site of this enzyme, even if it could, by chance, fit into the groove. Conversely, only a hydrophobic substrate could interact with an active site made up largely of hydrophobic R groups; both electrically charged and polar molecules would be incompatible with such a site.

Figure 3.41 offers a model of what is currently known about the active site of one actual enzyme (carboxypeptidase, which catalyzes the removal

3.40 Induced-fit model of enzyme-substrate interaction
The enzyme molecule has an active site onto which the substrate molecules can fit (left), forming an enzyme-substrate complex (middle). The binding of the substrate induces conformational changes in the enzyme that maximize the fit and force the complex into a more reactive state. The enzyme molecule reverts to its original conformation when the product is released (right).

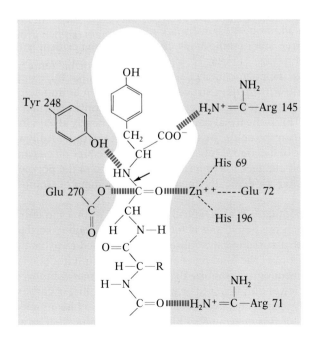

3.41 Model of an active site of an enzyme
Shown here in schematic form is the base of the cleft where the active site of carboxypeptidase is located. (The hydrophobic entrance to the cleft is out of the drawing, to the bottom. The entire enzyme, with the active site highlighted, is depicted in Figure 3.42). Part of a substrate molecule is shown in the cleft, linked to the enzyme by five weak bonds (striped bands). Seven of the amino acids of the active site are indicated by their abbreviated names; the numbers beside the names refer to the positions of the amino acids in the enzyme polypeptide chain. The function of this enzyme is to separate the terminal amino acid (top) from the amino acid chain (extending down out of the figure) at the covalent bond indicated by the arrow. The highly electronegative zinc and the charged oxygen draw away the electrons of this bond and so initiate its rupture.

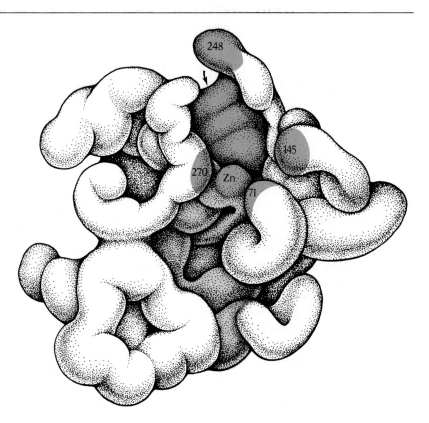

3.42 Location of the active site in an enzyme
The folding of the long chain of amino acids of
carboxypeptidase brings together the zinc atom and
three of the four amino acids that bind to the
substrate at the active site, even though—as their
numbers indicate—they are located at different
places in the chain. The zinc atom itself is held in
place by three different amino acids—nos. 69, 72, and
196 (not shown). The fifth part of the site folds into
place (arrow) when the substrate binds to the
enzyme. The region in dark color is the active site,
while the area in light color is the hydrophobic
entrance to the cleft.

of the terminal amino acid from one end of a polypeptide chain). As can be
seen, the site is visualized as a cleft into which the end of the substrate molecule can fit. The substrate is thought to form several weak bonds (five are
shown here) with the R groups of amino acids that constitute part of the active site. In addition, the substrate binds to a zinc held in place by the R
groups of three other amino acids. But note that the critical amino acids in
the active site (nos. 71, 196, 72, 69, 145, 248, and 270) are not adjacent to
each other in the polypeptide chain, which means that the complex folding
of the protein—its tertiary structure—has brought amino acids from several regions of the protein into close spatial proximity to form the active site
(Fig. 3.42). This is the usual pattern; active sites nearly always include some
nonadjacent amino acids. We can now understand more fully why elevated
temperature or a major pH change may greatly reduce an enzyme's activity:
anything that changes the precise folding pattern of the polypeptide chain
is likely to alter the critically important arrangement of amino acids in the
active site.

Carboxypeptidase is also typical of enzymes in that the entrance to the
cleft in which the active site is located is hydrophobic, so that the water
molecules that surround the substrate are stripped away as the substrate
enters the groove. The binding energy of the five simultaneous weak bonds
is just strong enough to stabilize the enzyme-substrate complex—a less perfect fit with fewer bonds would be unstable. The ability of carboxypeptidase
to form these particular five bonds accounts in part for its specificity. The
functioning of this enzyme also illustrates the induced-fit strategy: as the
binding begins at the other four parts of the active site, the part of the chain

containing tyrosine (no. 248) moves in from the periphery to trap the substrate (the polypeptide chain to be broken). The highly electronegative zinc atom and the charged oxygen of glutamate (no. 270) help activate the substrate by drawing away the electrons of the N—C bond that holds the terminal amino acid to the rest of the chain (Fig. 3.41). The resulting cascade of electron shifts between atoms of differing degrees of electronegativity makes the outcome—elimination of the terminal bond—inevitable. Finally, it should be clear why the two products—the terminal amino acid and the remainder of the chain—separate from the enzyme after the reaction: the new set of bonds in the products have redistributed their electrons, so the tenuous array of hydrogen bonds and van der Waals interactions that depended on precise electron interactions with the enzyme at the active site have been disturbed, and the products drift free.

Figure 3.41 illustrates another important point. Many enzymes contain a prosthetic group essential to their activity. A metal atom is often part of the prosthetic group; in carboxypeptidase, as we have seen, the metal is zinc.

Some enzymes that do not have a prosthetic group require a cofactor to which they bond only briefly and loosely during the reactions they catalyze. The cofactors may be metal ions, or they may be nonproteinaceous organic molecules; the latter are called *coenzymes*. Coenzyme molecules are much smaller and less complex than protein molecules. Like enzymes, they are not used up or permanently altered by the reactions in which they participate, and hence can be used over and over again. Only very tiny amounts are needed, therefore, but if the supply falls below normal, the health or even the life of the organism may be endangered. This is why vitamins, which act as parts of essential coenzymes, are so necessary in the diet.

Control of enzyme activity   Since enzymes control the myriad chemical reactions within living organisms, it is not surprising that a variety of mechanisms should have evolved for controlling the activity of enzymes themselves. These mechanisms depend not only on physical parameters such as temperature, pH, and substrate or enzyme concentration, but also on chemical agents, which mask, block, or alter the active sites of the enzymes they help regulate.

One common form of enzyme control, called *competitive inhibition*, involves an inhibitor substance sufficiently similar to the normal substrate of the enzyme to bind reversibly to its active site, but differing from the substrate in not being chemically changed in the process. By binding to the active site, the inhibitor (I) masks the site and prevents the normal substrate molecules from gaining access to it (Fig. 3.43). Thus the reaction

$$E + I \longrightarrow EI$$

competes with the reaction

$$E + S \longrightarrow ES \longrightarrow E + P$$

because both involve the same enzyme, which is present in only very small quantities. Which of the two reactions will predominate depends on their relative energetics, and, even more, on the relative concentrations of I and S. If there is much inhibitor and a low concentration of substrate molecules, a high percentage of the enzyme will be bound as EI and therefore unavailable; on the other hand, if there is much S and only a small concen-

Enzyme-substrate
complex

Competitive inhibitor
bound to enzyme

Noncompetitive inhibitor
bound to enzyme

3.43 Competitive and noncompetitive inhibition of an enzyme

Top: The substrate is bound to the catalytic site of the enzyme. Middle: The binding of a competitive-inhibitor molecule to the catalytic site prevents the substrate from binding. Bottom: A noncompetitive inhibitor bound to a different site on the enzyme induces an allosteric change that prevents the active site from catalyzing reactions.

tration of I, then most of the enzyme molecules will be free to catalyze the reaction of substrate molecules to form the product. Carbon monoxide poisoning is an example of competitive inhibition. The carbon monoxide competes with oxygen for the active sites in hemoglobin (Fig. 3.25), the enzymelike substance in the red blood corpuscles of vertebrates that carries oxygen to the body's cells. Carbon monoxide binds so strongly to the active sites that oxygen is effectively excluded. As a result of the oxygen deprivation that ensues, living tissue, particularly brain tissue, can be damaged or destroyed even if the concentration of carbon monoxide is relatively low.

A second category of reversible inhibition, called **noncompetitive inhibition**, depends on the operation of two kinds of binding sites in the same enzyme molecules—the usual active sites to which substrate can bind, and other sites to which inhibitors can bind. The most common kind of noncompetitive inhibition is **allosteric inhibition**. An allosteric enzyme is one that can exist in two distinct spatial conformations, which usually reflect alterations of tertiary structure. Often when the molecule is in one conformation the enzyme is active, and when it is in the other conformation it is inactive (or less active) because the substrate-binding site has been disrupted. Allosteric inhibition, the binding of inhibitor molecules—usually called negative **modulators** (or sometimes negative effectors)—stabilizes the enzyme in its inactive conformation (Fig. 3.43). Quite often the product itself (or the product of a later reaction in the same biochemical pathway) assumes the modulator role: when present in such high concentrations that no more is needed, it turns off the process responsible for its own synthesis. This self-limiting strategy is known as **feedback inhibition**.

Other types of allosteric enzymes have binding sites for positive modulators, which induce conformational changes enhancing enzyme reactivity.

Instead of having different kinds of binding sites, some allosteric enzymes have two or more sites of a single kind, and so can bind substrate at two or more locations simultaneously. The binding of substrate at one active site causes conformational changes that make the remaining sites more reactive. This phenomenon, called **cooperativity**, is exemplified in hemoglobin. A single molecule of hemoglobin is capable of carrying four oxygen molecules. The binding of the first oxygen molecule induces changes in the quaternary structure of the hemoglobin molecule that give the other three binding sites a higher affinity for oxygen. Thus in cooperativity, the first substrate molecule functions as the modulator. It stabilizes the allosteric protein in one of its possible conformations—in the case of hemoglobin, its most reactive conformation.

Because of the different effects that competitive inhibitors, noncompetitive inhibitors, and allosteric modulators have on the active sites of enzymes, biochemists find them exceedingly valuable in probing the nature of enzymes. For example, to gain a clearer idea of the active site—its physical shape and its reactive R groups—they may synthesize a series of slight variants of a competitive inhibitor and study the effectiveness of each in binding to the active site. Or, to investigate the roles of the various parts of an enzyme molecule, they may make use of irreversible inhibitors—chemicals that act as enzyme poisons by forming permanent covalent bonds with the functional groups necessary for catalysis.

Chapter 4

AT THE BOUNDARY OF THE CELL

We saw in the last chapter that organisms are composed of a great variety of chemicals, some simple and some complex. But these chemicals do not of themselves possess the properties we recognize as life. We cannot, for example, put a population of amoebae through a blender and expect that the resulting mixture of organic molecules will spontaneously reorganize itself into living entities.

Instead, life depends on a precise compartmentalization and organization of organic molecules. An intricate membrane must protect the interior of the cell—the nucleus, with its DNA, and the **cytoplasm**, which is composed of organic fluids, internal membranes, and a variety of specialized, self-contained entities known as organelles. The cell membrane separates the cell's delicate internal chemistry from the vagaries and dangers of the external environment, holding some chemicals in, passing some through, barring others from entering. And just as the cell membrane protects a favorable chemical atmosphere from the world outside, so the various structures inside separate themselves from the rest of the cell by means of their *own* membranes. In this way the cell's chemical processes are partitioned off—so that, for example, food is digested in membrane-lined compartments that prevent digestion of the contents of the rest of the cell.

In this chapter we shall look at the structure and function of the cell membrane and learn how it works, both actively and passively, to make life possible. In later chapters we shall see how the system of subunits inside the cell uses the same principles to direct cellular chemistry, to extract energy from food, to build new compounds and structures, and, in multicellular organisms, to orchestrate each cell's specialized role in a kind of society of cells.

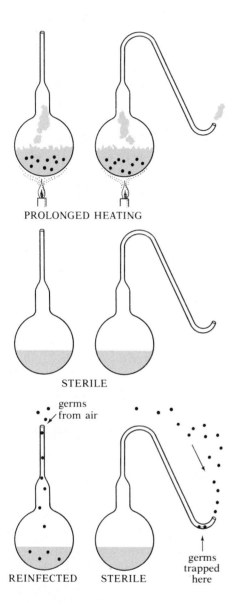

PROLONGED HEATING

STERILE

germs from air

REINFECTED   STERILE   germs trapped here

4.1 Pasteur's experiment
Nutrient broths in two kinds of flasks, one with a straight neck, the other with a bent neck, were boiled to kill any germs they might contain (top). The sterile broths were then allowed to sit in their open-mouthed containers for several weeks (middle). Microorganisms entering the straight-necked flask contaminated the broth, but those entering the bent neck of the other flask were trapped in films of moisture in the curves of the neck and did not contaminate the broth (bottom).

THE CELL THEORY

The discovery of cells and of their structure is linked to the development of magnifying lenses, particularly the microscope. Though some of the optical properties of curved surfaces had been known since 300 B.C., it was not until the seventeenth century that Antony van Leeuwenhoek and his contemporaries refined the production of lenses sufficiently to construct microscopes satisfactory for simple observations. Thus in 1665 Robert Hooke was able to report to the Royal Society of London on "the first microscopical pores I ever saw, and perhaps, that were ever seen," in a piece of cork; "these pores, or cells, were not very deep, but consisted of a great many little Boxes, separated out of one continued long pore, by certain Diaphragms." Hooke's microscopic examination of cork marks the beginning of the study of cells. Intensive work on cells was not pursued, however, until the early nineteenth century.

The idea that all living things are composed of cells—the *cell theory*—is commonly credited to two German investigators, the botanist Matthias Jakob Schleiden and the zoologist Theodor Schwann, who published their conclusions in 1838 and 1839 respectively. The idea that the cell is somehow integral to life had been considered before: the French naturalist Jean Baptiste de Lamarck, for instance, wrote in 1809 that "no body can have life if its constituent parts are not cellular tissue or are not formed by cellular tissue." But Schleiden and Schwann stated the principle with particular clarity, and they helped it gain general acceptance.

An important extension of the cell theory, proposed in 1858 by the German physician Rudolf Virchow, was that all living cells arise from pre-existing living cells (*"omnis cellula e cellula"*), and that there is therefore no spontaneous creation of cells from nonliving matter. The theory of *biogenesis*, life from life, contradicted the prevailing belief in spontaneous generation, then widely held not only by the general public but by scientists as well. It was Louis Pasteur in France who, a few years later (1862), supplied proof for Virchow's theory in a series of now classic experiments.

Pasteur's first step was to place various nutrient broths in long-necked flasks and then bend the necks of the flasks into curves (Fig. 4.1). Next, he boiled the broths in the flasks to kill any microorganisms (germs) that might be in them. While the flasks were left standing, germ-laden dust particles in the air moving into the flasks were trapped in the films of moisture on the humid curves of the necks; the curved necks acted as filters. Though the broths might be left standing in their swan-neck containers for months or even a year or more, no life appeared in them. Identical broths boiled in flasks with straight necks—the control solutions—did not remain free of microorganisms. They were soon teeming with life. Similarly, if the swan neck was broken off, the experimental broth rapidly developed colonies of molds and bacteria. The control solutions required by rigorous scientific procedure were crucial to the proof of Pasteur's theory: since handling of the control solutions differed in only one respect—exposure to air—the changed outcome had to be attributed to that difference. Thus Pasteur showed that the source of the microorganisms that fermented or putrefied such substances as milk, wine, and sugar-beet juice was the air. The organisms did not arise spontaneously from the nutrient media.

The two components of the cell theory—that all living things are composed of cells and that all cells arise from other cells—give us the basis for a working definition of living things: living things are chemical organizations composed of cells and capable of reproducing themselves.

VIEWING THE CELL

Much of our knowledge of subcellular organization has been made possible by the development of better and more powerful microscopes. In the detailed analysis of subcellular structure, three attributes of microscopes are of particular importance: magnification, resolution, and contrast. Magnification is a means of increasing the apparent size of the object being viewed until it provides an adequate stimulus to our eyes. Resolution is the capacity to show adjacent forms or objects as distinct. Contrast is important in distinguishing one part of a cell from another.

The ordinary light microscope has many features basic to the operation of all microscopes. Light passes through a specimen and is then captured, bent, and brought into focus by lenses (Fig. 4.2A). Depending on the magnification and the size of the specimen, a whole cell or only a tiny part may be in the field of view at any one time.

We vary the magnification by using lenses with different shapes that accept light from larger or smaller sections of the specimen: the higher the magnification, the smaller the amount of light that reaches the eyepiece. The limit on useful magnification in light microscopes is not a matter of exhausting the illumination, however. Instead it arises from the tendency of light to bend as it passes near an edge. This phenomenon, known as diffraction, spoils images by bending light out of the straight-line path as it moves from the its source, through the specimen, to the objective. The result is a degraded picture with decreased resolution. The amount of useful magnification possible is limited to about 1,000 times the actual size of the object in focus. Though magnification of 1,000 is an enormous improvement over the unaided eye, it is still not enough to let us see many of the smaller subcellular structures.

In the light microscope, contrast can be as important as magnification. Contrast is necessary if we are to distinguish structures from their backgrounds. Most cellular components are colorless and have essentially the same texture. But different parts of the cell often differ in their affinities for various dyes, so these areas can be stained by different color dyes or by dif-

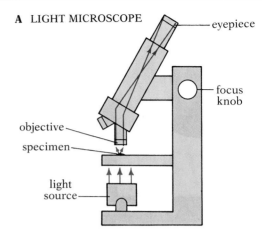

A LIGHT MICROSCOPE

eyepiece

focus knob

objective

specimen

light source

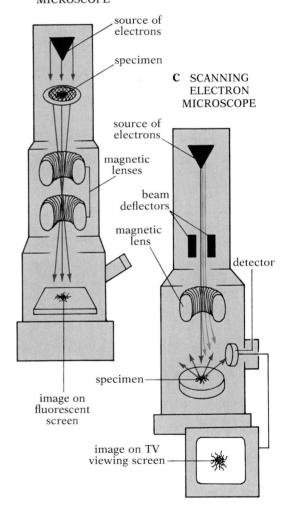

B TRANSMISSION ELECTRON MICROSCOPE

source of electrons

specimen

magnetic lenses

image on fluorescent screen

C SCANNING ELECTRON MICROSCOPE

source of electrons

beam deflectors

magnetic lens

detector

specimen

image on TV viewing screen

4.2 Microscopes
(A) In a simple light microscope, light passes through a specimen to the objective lens. Here the light is refracted and directed to the eyepiece, where it is focused for a camera or for our eyes. (B) In a transmission EM, electrons provide the illumination, which passes through the specimen and is then focused by magnets to form an image for a camera or on a fluorescent screen. (C) In a scanning EM, a focused beam of electrons moves back and forth across the specimen, while a detector monitors the consequent emission of secondary electrons and reconstructs an image.

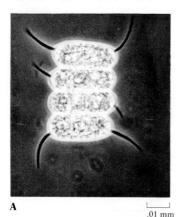
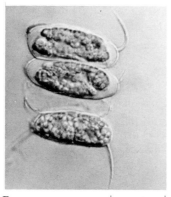
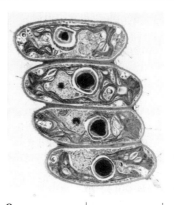
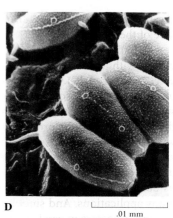

A   .01 mm   B   .01 mm   C   .01 mm   D   .01 mm

4.3   Views of the green alga *Scenedesmus* obtained with microscopes of various types.

(A) Photograph of an unstained specimen as seen with a phase-contrast light microscope. (B) Photograph taken by the Nomarski process. (C) Transmission electron micrograph. Many of the membranous and particulate intracellular organelles show up much more clearly here than under the light microscope. They are made visible by a stain containing heavy-metal atoms, which combines differentially with various structures. (D) Scanning electron micrograph, providing a three-dimensional view of surface features. The scale bars below the photographs in this and later figures indicate the dimensions of the organisms shown.

ferent intensities of dye to make them stand out from each other. Unfortunately, staining usually kills the cell, and may thereby change its internal structure. New techniques such as phase-contrast and Nomarski optics—both of which depend on elaborate optical manipulation—have greatly increased the value of the light microscope because they create contrast in cells optically without staining (Fig. 4.3A–B).

The electron microscope (EM) opened new vistas in the study of cells by using a beam of electrons instead of light as its source of illumination. Because resolution improves (that is, diffraction decreases) as the wavelength of the illumination becomes shorter, and because electron beams have much shorter wavelengths than visible light, electron microscopes can resolve objects about 10,000 times better than light microscopes (Fig. 4.3C–D). Many details of cellular structure would not be known but for the EM.

In a transmission EM (Fig. 4.2B) the electrons pass through the thinly sliced ("sectioned") specimen, are focused by magnets rather than by lenses, and fall on a photographic plate or a fluorescent screen, where they produce an image of the specimen. A specimen being prepared for the transmission EM must be differentially stained with an electron-dense chemical that binds to specific cell structures. Needless to say, the picture can be only as good as the staining technique used to create it, and for this reason there are many different techniques. One is to use stains containing heavy-metal atoms. These stains bind to different internal cellular structures, blocking the passage of electrons in these places (Fig. 4.3C). Another is to tilt the specimen to allow atoms of the electron-dense substance to fall onto it (Fig. 4.4). The shadows and highlights in the resulting EM picture create a three-dimensional effect—a kind of topographical map of the surface of the specimen.

A scanning EM can also produce a three-dimensional view. A specimen coated with atoms of metal is scanned from above by a moving beam of electrons (Fig. 4.2C). This focused probe does not penetrate the specimen, but instead causes so-called secondary electrons to be emitted from the surface. The intensity of the emission of secondary electrons depends on the angle at which the probe beam strikes the surface, and therefore varies with the contours of the specimen. Hence a point-by-point recording of the emission produces a three-dimensional picture (Fig. 4.3D). Though the resolution of the scanning EM does not approach that of the transmission EM, its ability to create a better three-dimensional effect is an advantage for many applications. And since it provides information by scanning surface features, a specimen can frequently be studied whole and intact. Moreover, since "shadowing" is not required, the same specimen can be turned repeatedly and observed from various perspectives.

FUNCTIONS OF THE CELL MEMBRANE

At one time, the cell membrane was considered little more than a bag to hold in all the organic chemicals that somehow combine to produce life. As we shall see, however, the cell membrane is far more than a passive envelope giving mechanical strength and shape to the cell. It bears the primary responsibility for regulating the chemical traffic between the precisely ordered interior of the cell and the essentially unfavorable and potentially disruptive outer environment. All substances moving into or out of a cell must pass through a membrane barrier, and the membrane of each cell can be quite specific about what is to pass through, and at what rate, and in which direction. The cell membrane exercises this control in two ways: by utilizing natural processes such as diffusion, and by transporting specific substances in and out.

DIFFUSION

Before we can understand how the membrane functions, we must return to the molecular level and consider the movement of materials in general. As we have seen, temperature affects the rates of chemical reactions; increased temperature is also conducive to more rapid particle movement.

Imagine a small stationary box containing 20 marbles in a tight cluster

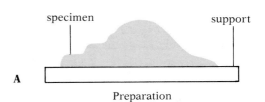

A   Preparation

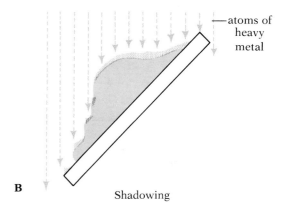

B   Shadowing

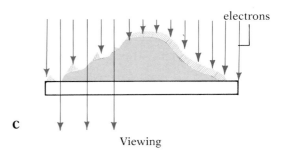

C   Viewing

4.4   Shadow staining for the transmission EM
The specimen (A) is tilted and dusted, or "shadowed," with atoms of a heavy metal such as platinum (B). (Another technique is to "shoot" the metal atoms from an angled source onto a horizontal specimen.) When an electron beam is aimed at the coated specimen (C), the unevenly distributed metal plating prevents most of the electrons from penetrating the shadowed areas, and produces a three-dimensional view of the surface of the specimen.

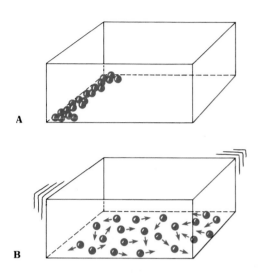

4.5 Mechanical model for diffusion
(A) All 20 marbles are placed in a cluster at one end of a rectangular box. (B) When the box is shaken to make the marbles move randomly, they become distributed throughout the box in nearly uniform density.

near one end (Fig. 4.5A). When we shake the box, the marbles disperse almost evenly over the bottom (Fig. 4.5B). Obvious as this result may seem, it is worth a closer look, because the marbles can be thought of as molecules, and the shaking as adding kinetic, or thermal, energy to the system.

First, it is immediately apparent that of all the possible directions in which a given marble might move, more lead *away* from the center of the cluster than toward it. Hence *random movement will tend to disrupt the cluster rather than to maintain it.* As we indicated in Chapter 3, in the absence of any counteracting external influence, a dynamic system will tend to move toward the more probable disorderly state rather than toward the less probable orderly state. This is precisely what happens, for instance, when a lump of sugar dissolves in a cup of warm coffee: the sugar molecules move from the region of high concentration (the sugar crystal) to regions of lower concentration; eventually the sugar molecules disperse throughout the liquid. The warmer the liquid, the more kinetic energy the molecules in solution will have on average, and the faster diffusion will take place.

Notice that the argument is statistical. It is possible that, as a result of random motion, all 20 scattered marbles will form a tight cluster at one end of the box. This result has a definite possibility of occurring, but one so slight that it can justifiably be disregarded. The kind of reasoning used here is typical of most scientific reasoning: the facts and laws of science are statistical rather than absolute. They describe natural phenomena in terms of *probable* outcomes; they do not assert that a certain outcome will occur 100 percent of the time.

We can now make a generalization based on our example of the marbles in the box and on others like it: all other factors being equal, *the net movement of the particles of a particular substance is from regions of higher to regions of lower concentration of that substance.* Note that we speak of *net* movement. There will always be some particles moving in the opposite direction, but, overall, the movement will be away from the centers of concentration. An obvious result is that the particles of a given substance tend to become relatively equidistant from one another within the available space. When this uniform density has been reached, the system is in equilibrium; the particles continue to move, but there is little net change in the system.

Movement of particles the size of molecules from one place to another in the manner we have been discussing is called *diffusion*. Diffusion is fastest by far in gases, where there is much space between the particles and hence relatively little chance of collisions which retard movement. Diffusion in liquids is much slower (Fig. 4.6); in the absence of convection currents a substance can take a very long time—years, in fact—to move in appreciable quantity only a few feet through cold water. Diffusion in solids is, of course, much slower still: there is very little space between the molecules of a solid, and collisions occur almost before the molecules get going. In all these instances, however, regardless of the rate of diffusion, the net effect is movement away from regions of higher concentration, as long as all regions are at the same temperature and pressure. In living organisms, where molecules are generally in a warm aqueous solution and the distances involved are measured in fractions of a millimeter, diffusion is a highly significant process.

So far, we have discussed diffusion in terms of movement from a higher to a lower concentration along a gradient. In the living world, however, diffusion is not strictly a function of concentration, since conditions are seldom constant where life processes are at work. It is therefore more useful for us to look at diffusion in terms of the free energy of the particles involved. A concentration of a substance is a relatively orderly and unlikely arrangement. We can see, for instance, that energy (among other things) would be needed to change a mixture of sugar molecules and coffee back into the original lump of sugar and unsweetened coffee. A random, disorderly arrangement of molecules has necessarily less potential for doing work than an orderly one, and has concomitantly less free energy. As you may recall from Chapter 3, the amount of disorder in a system is known as **entropy**. Since, as we know from the Second Law of Thermodynamics, the amount of free (useful) energy in the universe is always decreasing, entropy is always increasing.

Diffusion, then, is spontaneous because orderly molecules, concentrated together, have greater free energy than dispersed molecules: it is a downhill reaction from order to disorder. The mixture (or product) has less free energy than the separate original substances (the reactants). Like the chemical reactions discussed in Chapter 3, the rate of diffusion if we begin with two pure substances is fastest at the outset, and slows as the equilibrium of complete mixture is approached and the ratio of available reactants to product decreases. If we could observe diffusion at the molecular level we would see the sugar molecules speeding away from the lump during the early part of the reaction. Later, however, as the substances became more evenly mixed, the frequency of the "back reaction" returning sugar molecules to the location of the original lump would rise, until equilibrium was reached. In fact, diffusion is a chemical reaction with its own free energy, which depends on the characteristics of the substances involved:

$$\text{sugar} + \text{coffee} \rightleftharpoons \text{sweetened coffee} \qquad (\triangle G = -x)$$

Free energy is not only the more correct basis for understanding diffusion but also the more broadly applicable. Consider a situation where there is a slight concentration gradient from point Y to point Z and a pronounced temperature gradient in the reverse direction. If concentration alone were a factor, the net diffusion would be from Y, the region of higher concentration, to Z, the region of lower concentration. But temperature and pressure also play a role. In this case, Z has a higher temperature than Y. Now, the higher the temperature in a given system, the greater the thermal motion of the particles in that system; and the greater the thermal motion, the greater the free-energy content. Because the difference in free energy associated with the temperature gradient from Z to Y may outweigh the difference in free energy associated with the concentration gradient from Y to Z, net diffusion may be from Z, the region of lower concentration but higher temperature, to Y, the region of higher concentration but lower temperature. Thus the crucial factor in determining the movement of substances is not concentration (or temperature or pressure) but free energy.

The importance of diffusion and its basis in free energy with respect to cells is clear: the concentration of organic molecules and a select group of ions inside a cell is a very unlikely arrangement. Without the cell mem-

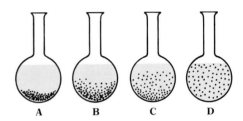

4.6 Diffusion in a liquid
Particles of solute are at the bottom of a flask of water (A). The particles slowly diffuse away from the cluster until (D) they are distributed with nearly uniform density through the water. If the water is cold and there are no convection currents to help move the particles, it may take a considerable period of time to reach the uniform distribution shown in D.

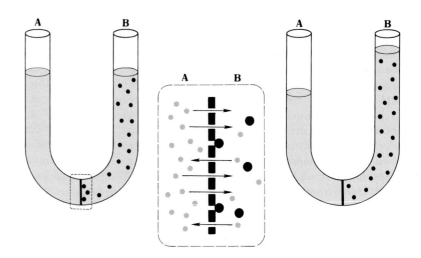

4.7   U-tube divided by a selectively permeable membrane
The membrane at the base of the U-tube is permeable to water, but not to sugar molecules (black balls). Left: Side A contains only water; side B contains a sugar solution. Initially the quantity of fluid in the two sides is the same. Center: A larger number of water molecules (colored balls) bump into the membrane per unit time on side A than on side B. Right: Because more water molecules move from A to B than from B to A, the level of fluid on side A falls while that on side B rises.

brane, the free energy of the cellular chemistry would be lost as the contents diffused into the environment. Two conclusions follow. First, there must be a barrier between the inside and the outside of the cell to maintain the integrity of the cellular chemistry. And second, the free-energy gradient across the cell membrane is available to do work.

OSMOSIS

To envision the way cell membranes can function, let's consider another model: a chamber divided into halves by a membrane partition. Let us assume, further, that particles of some substances can pass through the membrane while particles of other substances cannot. Such a membrane is said to be **differentially permeable** (or **selectively permeable**). How will the membrane affect the diffusion of materials between the two halves of the chamber? Suppose the chamber is a U-tube divided in half by a selectively permeable membrane (Fig. 4.7). Suppose side A contains pure water and side B an equal quantity of sugar solution (sugar dissolved in water), both sides being subject to the same initial temperature and pressure. If the membrane is permeable to water but not to sugar, water molecules will be able to pass in both directions, from A to B and from B to A.

This movement of a solvent (usually water) through a selectively permeable membrane is called **osmosis**. Biological membranes are selectively permeable, and the movement of water through them can be predicted on the basis of osmosis. As we shall see, some solutes, such as small lipid-soluble molecules, also pass through biological membranes freely.

Since water is already present on both sides of the membrane in the U-tube in Figure 4.7, it might at first seem that the movement of water molecules across the membrane would have no net effect. But consider the differences between the pure water and the sugar solution more carefully. On side A, all the molecules that bump into the membrane during a given interval are water molecules, and because the membrane is permeable to water, many of these will pass through the membrane from A to B. By contrast, on side B, some of the molecules bumping into the membrane during the same interval will be water molecules, which may pass through, and some will be sugar molecules, which cannot pass through because the membrane is impermeable to them. At any given instant, then, part of the membrane surface on side B is in contact with sugar molecules and part is in contact with water, whereas on side A all the membrane surface is in contact with water. Hence more water molecules will move across the membrane from side A to side B per unit time than in the opposite direction; the net osmosis will be from A to B.

We can think of the matter in another way. The arrangement of water molecules in pure water is orderly, in that every molecular location is occupied by a water molecule, whereas the arrangement in the sugar solution is disorderly, in the sense that any given molecular location may be occupied by either a water molecule or a sugar molecule. Now, we have said that, all other factors being equal, an orderly system possesses more free energy than a disorderly one. It follows that the orderly water molecules in the pure water (side A) have more free energy than the disorderly water molecules in the sugar solution (side B). There is a free-energy gradient for water from side A to side B, and, according to our generalization concerning diffusion, there will be a net movement of water down this gradient, from A to B.

We are now in a position to make some additional generalizations. *The free energy of water molecules is always decreased if osmotically active substances* (dissolved or colloidally suspended particles) *are present in the water.* (Colloidal particles are generally larger than the separated individual molecules of a dissolved substance, yet small enough so that—unlike the still larger particles of a true suspension—they do not settle out at an appreciable rate but remain dispersed within the fluid medium.) The **osmotic concentration** of a fluid—the number of osmotically active particles per unit volume—thus bears a direct relationship to that fluid's free energy. In the U-tube example, for instance, *the decrease in the free energy of water molecules is proportional to the osmotic concentration.* The reason for the diminution in free energy is that the osmotically active particles to some degree disrupt the orderly three-dimensional array of the water molecules (see Fig. 2.24, p. 45).

Each solution, then, has a certain free energy, depending on its osmotic concentration. Under conditions of constant temperature and pressure, this free energy can be calculated; it is called **osmotic potential**. (Pure water is arbitrarily assigned an osmotic-potential value of zero. Since osmotic potential decreases as osmotic concentration increases, all solutions have values of less than zero.) If two different solutions are separated by a membrane permeable only to water, and temperature and pressure are constant, *the net movement of water will be from the solution with the lower osmotic concentration to the solution with the higher osmotic concentration.* The steeper the osmotic-concentration gradient, the more rapid the movement. That is to say, the water flows from regions of higher osmotic potential to regions of lower potential at a rate proportional to the degree of difference in osmotic potential.

If the net movement of water in the U-tube is from side A to side B, the volume of liquid will increase on side B and and decrease on side A. Does the property of selective permeability cause this process to continue indefi-

OSMOTIC POTENTIAL, OSMOTIC PRESSURE, AND WATER POTENTIAL

As we have seen, osmotic potential is useful for thinking about how two solutions with differing osmotic concentrations interact. It is easily related to the underlying difference between them in free energy, which results in the movement of the solvent. Many researchers, however, prefer to think in terms of the pressure that must be exerted on a solution to keep it in equilibrium with pure water when the two are separated by a selectively permeable membrane. In our U-tube example, this pressure corresponds to the hydrostatic pressure exerted by the sugar solution at equilibrium; it is known as **osmotic pressure**. Clearly, *the osmotic pressure of a solution is a measure of the tendency of water to move by osmosis into it.* The more dissolved particles in a solution, the greater the tendency of water to move into it, and the higher the osmotic pressure of the solution. Thus, under constant temperature and pressure, water will move from the solution with the lower osmotic pressure to the solution with the higher osmotic pressure when the two solutions are separated by a selectively permeable membrane.

While the terms "osmotic pressure" and "osmotic potential" are regularly used by physiologists studying animals, plant physiologists more often refer to **water potential**, which is essentially the same as the free energy of water. At a pressure of one atmosphere, pure water is assigned a water potential of zero. Since the water potential decreases as the osmotic concentration increases, all solutions have values of less than zero. In this sense, water potential is like osmotic potential. But unlike osmotic potential, which is a function of solute concentration alone, water potential (like free energy) is also a function of temperature and pressure. When two solutions are separated by a selectively permeable membrane, water will move from the solution with the higher water potential to the solution with the lower water potential.

Familiarity with all of these terms is useful, because they are all common in the biological literature. In this book, however, we shall ordinarily use the term "osmotic potential."

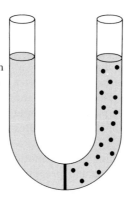

Lower osmotic concentration

Higher free energy of water

Higher osmotic potential

Higher water potential

Lower osmotic pressure

Higher osmotic concentration

Lower free energy of water

Lower osmotic potential

Lower water potential

Higher osmotic pressure

nitely, or will an equilibrium point be reached? Clearly, if the membrane is completely impermeable to sugar molecules, conditions on the two sides will never be equal, no matter how many water molecules move from A to B. The fluid in B will remain a sugar solution, though an increasingly weak one, and the fluid in A will remain pure water. Nevertheless, under normal conditions, the fluid level in B will rise to a certain point and then cease to rise further. Why? The column of fluid is, of course, being pulled downward by gravity. As the column rises, therefore, its weight exerts increasing downward hydrostatic pressure. As the pressure increases, the free energy of the water in the sugar solution rises, because pressure too is a form of free (useful) energy. Eventually the column of sugar solution becomes so high, and its pressure and free energy so great, that water molecules are pushed across the membrane from B to A as fast as they move into B from A.

When water is passing through the membrane in opposite directions at the same rate, the system is in dynamic equilibrium, with the free energy— the osmotic potential—of the pure water on one side of the membrane just matching the free energy—the osmotic potential and hydrostatic pressure —of the column of solution on the other side. Obviously, the greater the concentration difference across the membrane, the greater the difference in osmotic potential between the two sides and the higher the column of solution will rise before this difference is counterbalanced by the difference in hydrostatic pressure.

It is important to understand that osmotic concentration is not concentration by weight, but rather molecular or ionic concentration—the total number of solute particles per unit volume. If there are several kinds of solutes in the same solution, then the osmotic concentration of that solution is determined by the total (per unit volume) of *all* the particles of all kinds. If a dissolved substance ionizes, then each ion functions osmotically as a separate particle: one mole of sodium chloride (NaCl) dissolved in water produces two moles of particles—one of Na^+ ions and one of Cl^- ions. Colloidal particles may also contribute to the total osmotic concentration.

OSMOSIS AND THE CELL MEMBRANE

By now you probably realize that we have discussed diffusion and osmosis at such length because the cell membrane is selectively permeable, and the processes of diffusion and osmosis are fundamental to cell life. Though the membranes of different types of cells vary widely in their permeability characteristics—the membrane of a human red blood corpuscle,[1] for instance, is over a hundred times more permeable to water than the membrane of *Amoeba,* a single-celled organism—a few rough generalizations can be made: Cell membranes are relatively permeable to water and to certain simple sugars, amino acids, and lipid-soluble substances. They are relatively impermeable to polysaccharides, proteins, and other very large molecules. In short, cell membranes let pass only the building blocks of complex organic compounds, not the compounds themselves. The permeability of cell membranes to small inorganic ions varies greatly, depending on the particular ion, but in general negatively charged ions can cross more rapidly than positively charged ions, though neither can do so as readily as uncharged particles.[2]

What implications do these generalizations hold for life? On the one hand, selective permeability enables cells to retain the large organic mole-

---

[1] A red blood corpuscle begins as a cell but loses its nucleus as it matures and becomes specialized to transport oxygen. However, it remains sufficiently like true cells to be used as an example of many cellular properties.

[2] The process in which, in addition to solvent, some solutes selectively cross the membrane is often called dialysis.

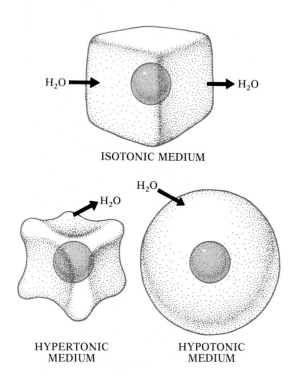

4.8 Osmotic relationships of a cell
In an isotonic medium water gain and water loss are equal; hence the cell neither shrinks nor swells. In a hypertonic medium there is a net loss of water from the cell, which therefore shrinks. In a hypotonic medium the cell has a net gain of water and swells.

cules they synthesize. On the other, the tendency of water to pass through selectively permeable membranes into regions of higher osmotic concentration can be harmful or even fatal. When a cell is in a medium that is *hypertonic* relative to it (a medium to which it loses water by osmosis, usually because the medium contains a higher concentration of osmotically active particles), the cell tends to shrink (Fig. 4.8); and if the process goes too far, it may die. Conversely, when a cell is in a medium *hypotonic* to it (a medium from which it gains water, usually because the medium contains a lower concentration of osmotically active particles), the cell tends to swell; and unless it has special mechanisms for expelling the excess water, or special structures that prevent excessive swelling (as most plant cells do), it may burst. A cell in an *isotonic* medium (one with which the cell is in osmotic balance, usually because it contains the same concentration of osmotically active particles) neither loses nor gains appreciable quantities of water by osmosis.

Obviously, the osmotic relationship between the cell and the medium surrounding it is a critical factor in the life of the cell. Some cells are normally bathed by an isotonic fluid and therefore have no serious osmotic problems. Human red blood corpuscles are an example; they are normally bathed by blood plasma, with which they are in relatively close osmotic balance. Most simpler oceanic plants and animals also exemplify cells in an isotonic medium; their cellular contents have an osmotic concentration close to that of seawater. All cells, however, have a higher osmotic concentration than fresh water. Freshwater organisms therefore live in a hypotonic medium and face the problem of accumulating excessive water within their cells by osmosis. Their very existence has depended on the evolution of ways of preventing their cells from becoming so turgid—so distended by their fluid content—that they would burst.

But controlling the flow of water is only one problem. Though the selective permeability of the membrane effectively traps large molecules inside, it does not provide any mechanism for concentrating the organic building blocks necessary for constructing substances like DNA, proteins, and polysaccharides in the first place. So while the cell membrane acts in many respects like an inert osmotic partition, it must also do more. For nutrients to be captured and retained, wastes expelled, and cell volume controlled, the cell membrane must have the capacity to pass many chemicals in only one direction. The secret of this critical ability lies beyond mere selective permeability, in the structure of the cell membrane itself.

STRUCTURE OF THE CELL MEMBRANE

Researchers have struggled for decades to explain the remarkable and apparently contradictory abilities of the cell membrane. The key to membrane function lies in understanding its structure, and a series of structural models have served to guide and focus the research that has led to our present understanding.

Despite the long-standing certainty that cells are bounded by a *plasma membrane*, only in the last three decades has direct proof of its existence been obtained. Most of the earlier conceptions of the membrane were deduced from the characteristics of cells, for the membrane is usually not visible even under the most powerful light microscopes. Though something believed to be the membrane could be isolated from red blood corpuscles, there was no conclusive proof that these corpuscular "ghosts" (Fig. 4.9) were really cell membranes and not artifacts of the procedures used to obtain them.

THE DAVSON-DANIELLI MODEL

Permeability studies had long shown that lipids and many substances soluble in lipids move with relative ease between the cell and the surrounding medium. From this fact it was deduced that the outer boundary of the cell, the cell membrane, must contain lipids, and that fat-soluble substances could move across the membrane by being dissolved in it. It was also observed that many small water-soluble molecules move quite freely between the inner portion of the cell and its external environment; it was therefore postulated that the cell membrane is a kind of sieve, containing pores or nonlipid patches. But still other observations had to be accounted for. Small water-soluble ions move through the cell membrane less freely than uncharged particles of roughly the same size; moreover, different ions do not all exhibit the same facility for crossing the cell boundary, some passing through rather freely, others doing so only in very limited numbers. It was therefore assumed that the cell membrane itself possesses charge, a property that tends to interfere with the movement of charged particles. Finally, the physical properties of the cell boundary, especially its wettability and elasticity, seemed to indicate the presence of protein in the membrane.

But this catalog of components does not by itself tell us much about how they are put together. The most useful early model was offered in the late 1930s by J. F. Danielli of Princeton University and H. Davson of University College, London. They formulated the idea that the membrane might be composed of two layers of phospholipids oriented with their polar (hydrophilic) ends exposed at the two surfaces of the membrane, and their nonpolar (hydrophobic) hydrocarbon chains buried in the interior, hidden from the surrounding water and tending to bind to each other through hydrophobic interactions (Fig. 4.10). A structure based on hydrophobic/hy-

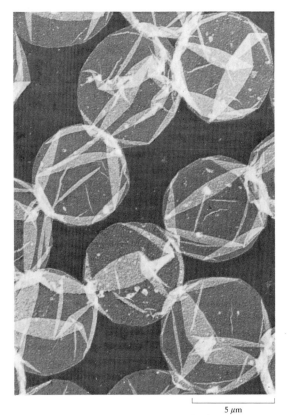

4.9 Ghosts of human red blood corpuscles
The "ghosts" are now known to be cell membranes. The whitish areas are places where the membrane is folded.

4.10 The Davson-Danielli model of the cell membrane
Two layers of lipids are sandwiched between two layers of protein. The phospholipids are oriented with their polar hydrophilic heads near the surfaces and their nonpolar hydrophobic tails projecting into the interior of the structure, at right angles to the surfaces. Pores penetrate the membrane at some points.

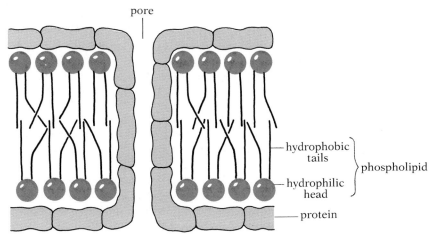

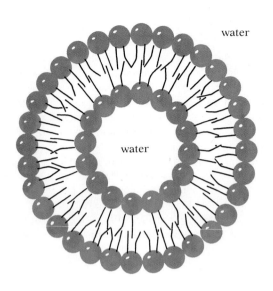

4.11 A liposome

Mixing phospholipids with water produces spherical phospholipid bilayers, each enclosing a droplet of water. The spontaneous formation of these spheres, called liposomes, is a result of the energetically favorable interaction of the hydrophilic ends of the phospholipids with the water molecules, and of the hydrophobic chains with each other. Cell membranes are structured in exactly this way. Hence, they are basically stable, forming almost automatically and requiring no energy to maintain.

drophilic interactions would be at once very stable and elastic. Indeed, as we now know, spheres—called *liposomes*—composed of phospholipid bilayers will spontaneously form when phospholipids are mixed with water (Fig. 4.11).

The Davson-Danielli model accounted for the stability, flexibility, and lipid-passing characteristics of the membrane with ease, and it also suggested how the proteins and pores might be arranged. Davson and Danielli imagined that both sides of the membrane were coated with protein, while charged protein-coated pores allowed small molecules and certain ions to pass through the membrane (Fig. 4.10). Their model thus predicted a symmetrical membrane, about 8 nm (nanometers) thick.

Confirmation of part of the Davson-Danielli model came from electron micrographs made in the 1950s by J. David Robertson, then of Harvard Medical School. These showed that the membrane is indeed about 8 nm thick and symmetrical, and that it consists of two electron-dense layers separated by a lighter area inside (Fig. 4.12). In the years since, the notion of a phospholipid bilayer has survived test after test. However, the idea of a protein coat has been repeatedly shown inadequate to account for the distribution and mobility of pores and specialized membrane proteins.

THE FLUID-MOSAIC MODEL

In 1972 S. J. Singer of the University of California, at San Diego, and G. L. Nicolson of the Salk Institute proposed the fluid-mosaic model, a hypothesis that is now almost universally accepted. The model incorporates the Davson-Danielli conception of a bilayer of phospholipids oriented with their hydrophobic tails toward the interior and their hydrophilic heads exposed to the aqueous environment on both surfaces. In the fluid-mosaic model, however, the arrangement of the proteins is dramatically different. Instead of coating the membrane, the various specialized proteins are now thought to be inserted in the membrane to mediate a wide range of critical functions, which we shall examine presently (Fig. 4.13).

4.12 Electron micrograph showing membrane of sectioned human red blood corpuscle

The cytoplasm of the corpuscle is in the bottom half of the picture. The membrane consists of two dark lines separated by a lighter area.

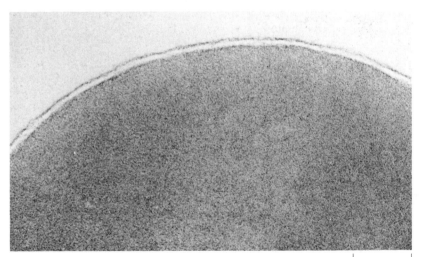

0.1 μm

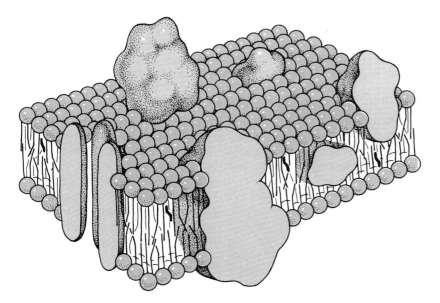

4.13 The fluid-mosaic model of the cell membrane
A double layer of lipids forms the main continuous part of the membrane; the lipids are mostly phospholipids, but in plasma membranes of higher organisms cholesterol (solid bars) is also present. Proteins occur in various arrangements. Some, called extrinsic proteins, are entirely on the surface of the membrane. Others, called intrinsic proteins, are wholly or partly embedded in the lipid layers; some of these may penetrate all the way through the membrane. The three units at left are joined by covalent bonds (not shown) to form part of a single protein molecule bounding a membrane-spanning pore.

Of the proteins confined to the surfaces (extrinsic proteins), those on the inner surface usually differ markedly from those on the outer surface; and some membranes have no extrinsic proteins at all. The proteins located wholly or partly within the lipid bilayer (intrinsic proteins) may exhibit one of several arrangements: some are entirely buried within the bilayer, whereas others have parts that project through the surface; some are confined to the outer half of the lipid core, and others to the inner half; some extend entirely through the bilayer, projecting into the watery medium on both sides. As would be expected, hydrophilic amino acids (those with polar or electrically charged R groups) predominate in the portions of the protein molecules that project out of the lipid bilayer into the water, whereas hydrophobic (nonpolar) amino acids are abundant in the portions buried in the lipid bilayer (Fig. 4.14). Indeed, the location of the hydrophilic and the hydrophobic amino acids in a membrane protein determines which part of the protein will be anchored in or to the membrane and whether the pro-

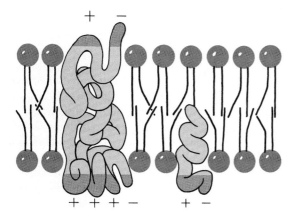

4.14 Orientation of proteins within membranes
The parts of the polypeptide chain containing most of the hydrophilic amino acids (polar or charged R groups) (dark color) tend to project into the watery medium outside the lipid layers, whereas the parts of the chain with hydrophobic amino acids (light color) tend to be folded into the inner, lipid portion of the membrane. The diameter of the protein strands has been reduced for clarity.

FREEZE-ETCHING

Freeze-etching, a technique for preparing specimens for electron microscopy, has become an indispensable tool for providing detailed confirmation of membrane structure. The specimen is first rapidly frozen and then fractured along the plane of its bilipid membrane (A–B). Some of the ice is then removed from the specimen by sublimation (conversion directly to vapor), which exposes the inside surface of the membrane and gives the specimen an etched appearance (C).

Carbon and a metal, usually platinum, are then applied at an angle to the specimen (D) so as to shadow any irregularities in the membrane. Next the original specimen is removed from the platinum cast or surface replica thus formed (E). The replica can now be examined by microscopy. EMs of freeze-etched cell membranes or other structures in cells usually have a striking three-dimensional appearance, as seen, for example, in Figure 5.5, p. 115.

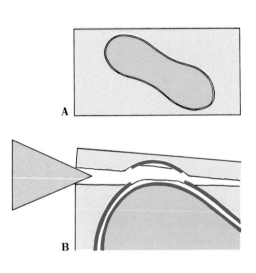

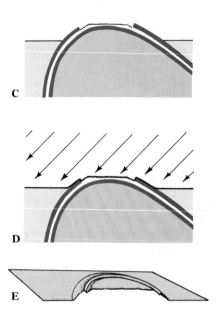

tein will be intrinsic or extrinsic. The existence of intrinsic proteins, which are predicted only by the fluid-mosaic model, has now been confirmed by freeze-etch microscopy (Fig. 4.15).

According to the fluid-mosaic model, the structure of the membrane is not static. The individual lipid molecules (which are linked to one another only by weak bonds, not by covalent bonds) can move laterally, in the plane of the membrane, so that a particular molecule found in one position at a given moment may be in an entirely different position only minutes later. Mobility of the lipids is greatest in membranes that are high in unsaturated phospholipids (Fig. 4.16) and that contain no cholesterol. Speeds of 2 μm (micrometers) per second are possible in such membranes—an astonishing mobility when we consider that many organisms (the bacterium *Escherichia coli*, for example) are only about 2 μm long. When cholesterol is present, it binds weakly to adjacent phospholipids, thereby joining them to-

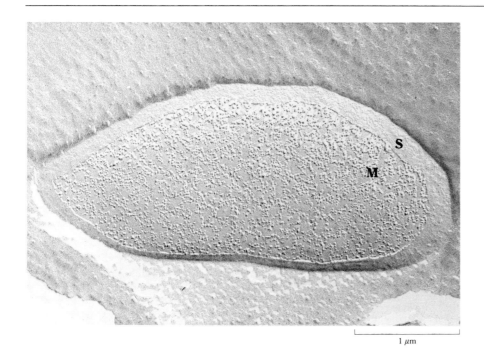

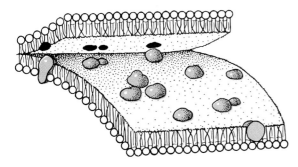

4.15   Electron micrograph of freeze-etched plasma membrane of red blood corpuscle

In this specimen the plasma membrane has been fractured along the plane between the two layers of lipids—that is, along the middle of the bimolecular lipid core (see sketch). The numerous spherical particles visible in the micrograph are interpreted as protein (see colored entities in sketch). They appear where the Davson-Danielli model predicts only lipid, but their presence is explained satisfactorily by the fluid-mosaic model. S: outer surface of membrane; M: interior of fractured membrane.

4.16   A phospholipid

The cell membrane is made up mostly of phospholipids. Phosphatidylcholine, a common membrane phospholipid, consists of a polar head (a positively charged choline, a negatively charged phosphate, and an uncharged glycerol) joined to two hydrophobic fatty acid chains. The "kink" in the right tail is created by a double carbon bond. Because this tail is unsaturated—that is, not every carbon has its full complement of hydrogen atoms—the phospholipid will be less tightly packed in the membrane, and hence will be more mobile.

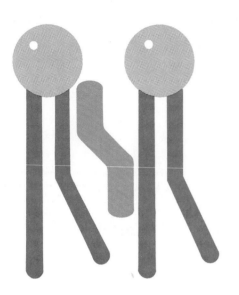

4.17 Cholesterol in the membrane
Cholesterol (color) binds weakly but effectively to two adjacent phospholipids, thereby partially immobilizing them. The result is a less-fluid and mechanically stronger membrane. The amount of cholesterol varies widely according to cell type, with the membranes of some cells possessing nearly as many cholesterol molecules as phospholipids while others lack cholesterol entirely. For the structural formula of cholesterol, see Figure 3.15, p. 57.

gether and reducing their mobility (Fig. 4.17). As a result, cholesterol stabilizes and strengthens the membrane.

The proteins, too, can move laterally to some extent, but much less than the lipids. Complete freedom of movement would be incompatible with the specific functional demands placed on the membrane proteins. Certain proteins in the membrane of nerve cells, for example, are essential to the transmission of nerve impulses from one cell to another; they are found only at points where one nerve cell is close to another—not in other positions, where they could not fulfill their function. Similarly, proteins responsible for pumping sodium ions out of the cells that line the intestine are located in the membrane on only one side of the cells (the side away from the intestinal cavity). In short, at least some of the membrane proteins are anchored in place, thereby limiting the fluidity of the membrane. In some cases the anchoring probably results when tight associations between two or more intrinsic proteins give rise to structural and functional complexes too large to move easily. In other cases intrinsic and extrinsic proteins may be bound weakly to each other. Even the lipid molecules may not all be free to move; there is evidence that the lipids immediately adjacent to intrinsic proteins may be loosely bound to the proteins and thus immobilized.

In the fluid-mosaic model, the pores in the membrane are depicted as channels through one or a group of protein molecules (Fig. 4.13). The ability of unanchored proteins to drift laterally in the lipid bilayer explains the observed mobility of many membrane pores. The distinctive properties of the various R groups of the amino acids in the proteins give the pores some selectivity; not all ions or molecules small enough to fit in the pores can actually move through them.

MEMBRANE CHANNELS AND PUMPS

The lipid bilayer that makes the cell membrane forms spontaneously, creating a flexible but effective barrier between the inside of the cell and the world outside. Moreover, the bilayer provides an anchoring plane for a variety of membrane proteins. The largely lipid composition of the membrane explains why small lipid-soluble molecules can diffuse into and out of cells, but the membrane's permeability to certain chemicals that do *not* dissolve in lipids must depend on the proteins in the bilayer. The ability of cells actively to transport specific substances against their osmotic-concentration gradients must also be accounted for by the properties of the membrane proteins.

It is easy to demonstrate that the membrane is highly selective. If a molecule that can readily enter a cell is slightly altered, but not in such a way as to change its solubility properties, it often loses its capacity to move through the membrane. This selectivity on the part of the membrane suggests that the transport agents, or "carriers," are enzymelike proteins—a hypothesis supported by a variety of experiments. For example, the movement of some substances through the membrane can be competitively inhibited by other structurally related substances. Inhibition would not occur if both substances were moving by simple diffusion; apparently the two substances compete for access to specific binding sites on enzymelike carrier molecules in the membrane.

We now know that the transport agents that control molecular traffic in and out of the cell are highly specialized channels and pumps. Each de-

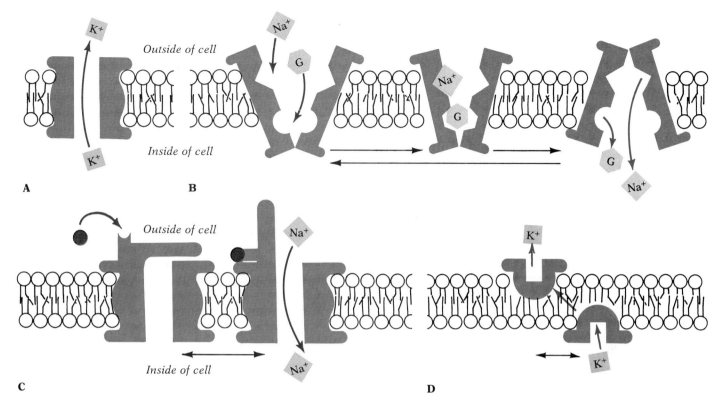

4.18 Models of membrane transport
Many strategies for moving substances across the cell membrane are known. (A) In facilitated diffusion of the simplest kind, a protein channel, or pore, embedded in the membrane provides a direct path for the chemical it passes down its osmotic-concentration gradient. The channel's diameter and the chemical environment it creates (hydrophilic or hydrophobic, for example) serve to prevent all but the correct substance from passing. (B) Other channels pass two substances cooperatively, or exchange two substances. Illustrated

here is a hypothetical mechanism for the channel that uses the highly favorable osmotic-concentration gradient for bringing sodium ions into the cell to overcome the osmotic-concentration gradient working against glucose (labeled G). When Na^+ binds to the channel, it may induce an allosteric change that enables glucose also to bind to the channel. This binding of glucose may then result in another change, which causes the channel to close to the outside and open to the inside. This change, which in turn may cause the channel to lose its affinity for glucose, releases the glucose as

well as the Na^+ into the interior. Having lost the Na^+ and glucose, the channel can reopen to the outside. (C) An allosteric interaction between a signal molecule (color circle) and the gated channel causes the gate to open, so that diffusion can take place down a favorable concentration gradient. Other molecular systems, not shown, then inactivate the signal molecule so that the channel can close again. (D) A mobile carrier would not provide transmembrane channels, but would itself migrate back and forth from one surface to the other. The existence of mobile carriers is controversial.

pends on membrane proteins for its operation; in fact, each is usually fabricated from several "cooperating" membrane proteins. Since they enable specific chemicals to permeate the membrane, channels and pumps are known collectively as *permeases*.

Membrane channels The simplest of these permeases, the ***membrane channels***, provide openings through which specific substances can diffuse across the membrane. These channels, though selective, are passive: they simply permit particular chemicals to move down their concentration gradients. This strategy, known as ***facilitated diffusion***, is the basis of the membrane's highly specific permeability. The protein channel for potassium ions (Fig. 4.18A) provides a good example of this strategy. Cellular processes result in the accumulation of K^+ inside most cells. As a charged par-

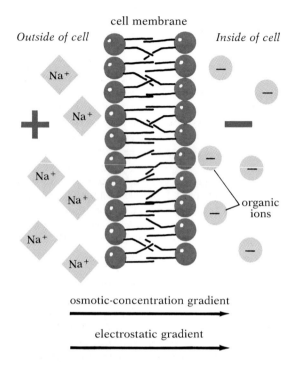

cell membrane

Outside of cell   *Inside of cell*

organic ions

osmotic-concentration gradient

electrostatic gradient

4.19   An electrochemical gradient
Cells have an electrical potential of about 70 mv negative with respect to the fluids that surround them. This gradient arises mainly because large numbers of negatively charged organic ions are trapped inside the cells, while a relatively high concentration of positively charged Na^+ exists outside. Sodium ions are therefore subject to both a strong osmotic-concentration gradient and a sizable electrostatic gradient. The effects of these two gradients combine to create an electrochemical gradient.

ticle, K^+ is insoluble in the lipid membrane, but the channel specific for K^+ allows it to leak out slowly at a controlled rate. Without such leakage the internal K^+ concentration would become too high for the cell to function properly. The specificity of the K^+ channel is a result of both its internal shape and its charge, but no one really understands in detail just what goes on inside these most simple of all membrane pores.

More complex channels, though passive, frequently move two specific substances in concert. Coordinated movement of this sort by a channel is important in the transport through the membrane of glucose, the most important energy source for most cells. Sodium ions are 11 times more concentrated outside the cell, and are therefore subject to a highly favorable osmotic gradient to the inside. Yet they must be accompanied by glucose to pass through the appropriate channel (Fig. 4.18B). The channel will not transmit either substance alone. It is as though both must bind to the outside of the channel before this special membrane pore will open. Thus the free energy of the osmotic-concentration gradient of Na^+ can be exploited to overcome the smaller unfavorable concentration gradient of glucose. In thermodynamic terms, the two diffusion "reactions" are linked: moving Na^+ "downhill" releases more free energy than is utilized in moving glucose "uphill," so the cooperative diffusion proceeds.

This description in terms of osmotic-concentration gradients accounts for the coordinated movement of Na^+ and glucose through a common channel, but the rate of movement is too great to be explained solely by the concentration difference across the membrane. In fact, there is a second important gradient that contributes to the diffusion of ions. As you know, oppositely charged ions attract one another electrostatically, while ions with the same charge repel one another. As a result, if a cell has, say, more negatively charged than positively charged ions, positive ions will be attracted to it from the surrounding fluid. (Most cells actually have a net negative charge of about 70 millivolts relative to the fluids surrounding them.) The difference in charge across the membrane generates an ***electrostatic gradient***, and when appropriate channels are open, positive ions tend to flow into the cell, while negative ions tend to flow out. For ions like Na^+, which are far more concentrated outside, the osmotic and electrostatic potentials combine to create a strong ***electrochemical gradient*** (Fig. 4.19). It is the free energy of the combined potentials that accounts for the particular effectiveness of Na^+ in moving glucose into cells. The mechanism by which the electrochemical gradient is maintained is the sodium-potassium pump, discussed below.

Another strategy for controlling movement across the membrane utilizes a gate across a membrane channel. This strategy is widely used to convert a molecular signal specialized for carrying information between cells into a second signal, more suitable for communicating inside the cell. When a molecular signal—a hormone or one of the transmitter substances that carry messages from one nerve to another—binds to an exposed part of a transmembrane protein, the ***receptor***, an allosteric change in conformation takes place. The change allows the gate to open, and the second signal, usually an ion like Na^+ or Ca^{++}, can then move across, carrying the message into the cell (Fig. 4.18C). This ***gated channel*** strategy underlies the transmission of many chemical messages in both plants and animals, and of the nerve impulses by which animals sense the outside world, move their muscles, and perhaps even think.

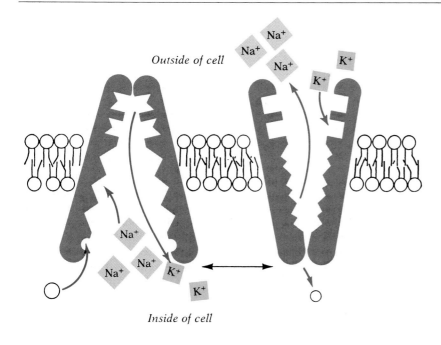

Outside of cell

Inside of cell

4.20 The sodium-potassium pump

In contrast to other methods of membrane transport, the pump strategy uses energy from the cell (rather than the free energy of a concentration gradient) and provides active transport of a substance against its gradient. In this case, three sodium ions are exchanged for two potassium ions; both kinds of ions are already more concentrated on the side to which they are being moved. In the model shown here, the release of K$^+$ ions brought in during the previous cycle is followed by the binding of three Na$^+$ ions and an energy source, ATP (circle), on the inside. The resulting conformational changes in the protein open it to the outside, reduce its affinity for Na$^+$, which is then released, and increase its affinity for K$^+$ ions. The binding of K$^+$ then causes the channel to open to the inside, increase its affinity for Na$^+$, and decrease its affinity for K$^+$, and the cycle begins again. The net ionic effect is to pump positive charges out of the cell, and the inside of the cell becomes negatively charged with respect to the outside. The electrical and osmotic potential created by the sodium-potassium pump ultimately makes possible the cooperative transport of glucose illustrated in Figure 4.18B.

Another way for molecules to get across the membrane would be for the permease to act as a *mobile carrier*, taking them through one by one (Fig. 4.18D). No example of such a permease is yet known, but valinomycin acts in just the way we would expect a mobile carrier to act. Valinomycin is a ring-shaped polymer with a hydrophobic exterior and a polar interior. The polar pocket, lined with six oxygen atoms, can hold a single potassium ion. Apparently the complex bobs back and forth randomly, carrying K$^+$ ions in both directions. The net transfer of K$^+$ is a statistical consequence of the electrochemical gradient: valinomycin more often picks up K$^+$ inside the cell and releases it outside simply because there is more K$^+$ inside than out. Valinomycin is not a normal membrane protein—in fact, technically it is not a protein at all. It is produced by one set of microorganisms to poison competing microorganisms by altering the selective permeability of their membranes. There is some evidence for mobile carriers in normal membranes, but their existence is still an open question.

Membrane pumps   Other permeases, known as *pumps*, do not depend on free-energy gradients. Instead, pumps use the cell's store of energy to move substances against their gradients. The process, known as *active transport*, is important in ridding the cell of accumulated substances that are insoluble in the membrane, and of molecules that are simply too big to escape. Pumps also transport many essential building blocks into the cell. The best-understood example of a membrane pump, however, is the complex responsible for maintaining the electrochemical gradient across the membrane: the *sodium-potassium pump* (Fig. 4.20). The free-energy source for this pump, and for many cellular processes, is the cellular energy carrier ATP, which we shall discuss in detail in later chapters. The pump

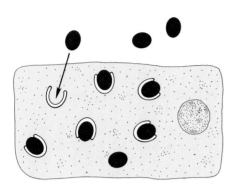

4.21 Formation of a complex inside the cell
As fast as G molecules (black ovals) enter the cell, they combine with molecules of X (white) already in the cell to form a new substance, XG. The concentration of free G inside the cell therefore remains low, and G continues to diffuse inward.

utilizes this energy to exchange potassium and sodium ions across the membrane, thereby maintaining the electrochemical gradient.

The effects of this pump are far-reaching: it is responsible for the electrical activity of nerves and muscles; it indirectly supplies the free energy for many osmotic transport systems, such as the one for glucose described earlier (which, you may recall, depend on the Na^+ electrochemical gradient); and it helps regulate the volume of many cells by controlling osmotic potential. Indeed, when the pump is destroyed by the Indian dart poison ouabain (a recent neurobiological research tool), cells swell uncontrollably with water until they burst. As we shall see, the sodium-potassium pump also contributes to many metabolic functions at the organismal level, among them maintenance of the electrical activity of nerves and muscles and the uptake of water by plant roots.

Mention should also be made of an important type of chemical manipulation, quite unrelated to channels and pumps, by which cells control some of the osmotic potentials across the membrane. In discussing the transport of glucose, we said that the Na^+ gradient can be used to "pull" glucose inside because the gradient against glucose is very small. Yet the world outside most cells has very little glucose to begin with, while most cells use it in large amounts. How can the concentration inside the cell be kept so low? The cell manages this trick by binding the glucose into another compound

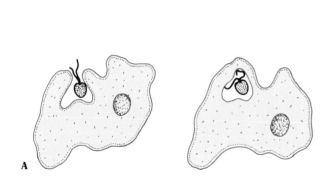

4.22 Phagocytosis
(A) In *Amoeba* pseudopodia flow around the prey until it is entirely enclosed within a vacuole. (B) White blood cells, or leukocytes, in the bloodstream use phagocytosis to capture foreign organisms; here the leukocyte is engulfing a dividing bacterium.

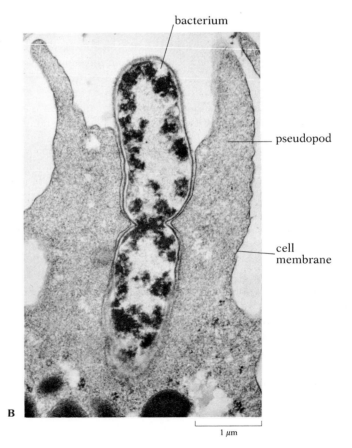

1 μm

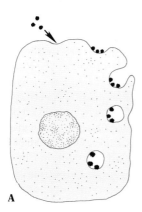

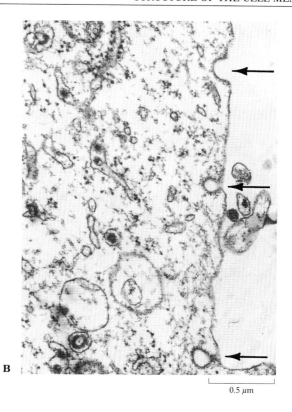

0.5 μm

4.23 Pinocytosis
(A) Particles adsorbed on the membrane are enclosed
in vesicles that detach directly from the cell surface
and move into the cell. (B) Three stages of pinocytosis
by a cultured nerve cell. The material being enclosed
is not visible in this photograph.

as soon as it is inside (Fig. 4.21). As a result, the free-glucose concentration
remains artificially low inside, and the osmotic potential against glucose
does not get out of hand.

ENDOCYTOSIS AND EXOCYTOSIS

As we have seen, permeases are the means by which substances enter and
leave cells through the cell membrane. But cells have ways of admitting
substances, usually in larger quantities, without having them actually pass
through the membrane. By an active process called ***endocytosis***, a cell encloses the substance in a membrane-bound vesicle that is pinched off from
the cell membrane. There are two types of endocytosis, both of which depend on specialized membrane proteins:

  1. When the material engulfed is in the form of large particles or chunks
of matter, the process is called ***phagocytosis*** (Fig 4.22). Usually, armlike
processes of the cell, called ***pseudopodia***, flow around the material, enclosing it within a vesicle, which then becomes detached from the plasma
membrane and migrates into the interior of the cell.

  2. When the engulfed material is liquid or consists of very small particles,
the process is termed ***pinocytosis***. The material first becomes adsorbed on
the cell membrane, usually at selective binding sites. The loaded vesicles
are formed, and detach from the membrane at the cell surface (Fig. 4.23).

  In some cases the selective binding site that collects a particular substance before trapping it in a vesicle appears as a "coated pit," with recep-

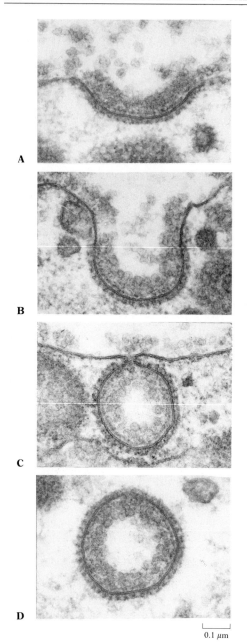

A

B

C

D

0.1 μm

4.24 Endocytosis by means of a coated pit
(A) Specialized receptors for lipoproteins have aggregated to form a coated pit in the membrane of an egg cell. (B–D) The pit is subsequently pinched off to form a vesicle. Most endocytotic vesicles are transported to lysosomes, organelles within the cell where the contents are enzymatically altered. The lipoprotein in the vesicle shown here will become part of the yolk.

tor molecules clustering in one spot in the membrane (Fig. 4.24). One example, which illustrates the importance and specificity of receptor-mediated endocytosis, involves cholesterol uptake by cells. Cholesterol is transported in the blood to cells by a carrier complex of low-density lipoprotein (LDL). When the cell needs cholesterol—usually for use in the manufacture of new membrane—LDL receptors are synthesized and incorporated into the cell membrane (Fig. 4.25). The LDL receptors aggregate spontaneously, particularly once they have bound LDL, and instigate endocytosis. The cholesterol is then transported in endocytotic vesicles for use in the cell's vast complex of membranes. One cause of atherosclerosis involves failure of the receptors to bind LDL; another involves failure of the receptors, having bound LDL, to aggregate and to initiate endocytosis.

Material enclosed within endocytotic vesicles has not yet entered the cell in the fullest sense. It is still separated from the cellular substance by a membrane, and it must eventually cross that membrane (or the membrane must disintegrate) if it is to become incorporated into the cell. Normally the vesicle membrane forms from cell membrane, and the vesicle is transported (by mechanisms not yet understood) to a cellular organelle known as a lysosome. The vesicle fuses with the lysosome, whose enzymes digest the substances that were transported. After digestion, many of the products can cross the lysosome membrane into the cytoplasm, while others remain trapped inside. As a result, the cell is able to ingest the substances it requires, while the unwanted parts of the miniature meal, and the lysosome's destructive enzymes, remain segregated from its delicate chemical interior.

In a process essentially the reverse of endocytosis, called *exocytosis*, materials contained in membranous vesicles are conveyed to the periphery of the cell, where the vesicular membrane fuses with the cell membrane and then bursts, releasing the materials to the surrounding medium (Fig. 4.26). Many glandular secretions are released from cells in this way; the hormone insulin, for example, is released by exocytosis from the pancreatic cells that synthesize it. Exocytosis also functions in the release of waste products from the cell. The undigested remains of materials brought in by endocytotic vesicles, for example, are normally disposed of through exocytosis. And in some cases a coupling of endocytosis and exocytosis moves a substance entirely across a cellular barrier, such as the wall of a blood vessel; the substance is picked up by endocytosis on one side of the cell, and the vesicle then moves through the cell to the other side, where the substance is released by exocytosis (see Fig. 13.14, p. 333).

CELL WALLS AND COATS

For as long as biologists have been examining cells under the microscope, they have been aware that plant cells are encased in conspicuous cell walls. These walls, which are located outside the plasma membrane, are composed primarily of carbohydrates. Biologists have also long known that the cells of fungi and most bacteria have strong, thick walls rich in carbohydrates. But only in recent years have they come to realize that most animal cells, too, have carbohydrates on the outer surface of their membranes. The carbohydrates do not form a wall, but are attached as independent side groups to some lipids and membrane proteins. Though not attached to one

4.25 Endocytosis of cholesterol

When a cell needs cholesterol, it synthesizes receptors for low-density lipoprotein (LDL) and incorporates them into the cell membrane, where they are free to migrate (A). The receptors soon bind LDL, a carrier complex that transports cholesterol in the blood (B). The LDL complex includes some 2,000 cholesterol molecules and an associated protein (called apoprotein) which binds to the LDL receptor. Having bound the LDL, the receptors stop drifting in the membrane and stick to one another (C). Even in the absence of bound LDL, many of the receptors eventually aggregate spontaneously. Aggregations of LDL receptors trigger endocytosis, the first step of which is the formation of a coated pit (D–E). The resulting vesicle (F) is transported to the site of membrane synthesis. This method of cholesterol uptake may demonstrate the usual strategy by which cells obtain nutrients that cannot pass through the membrane directly.

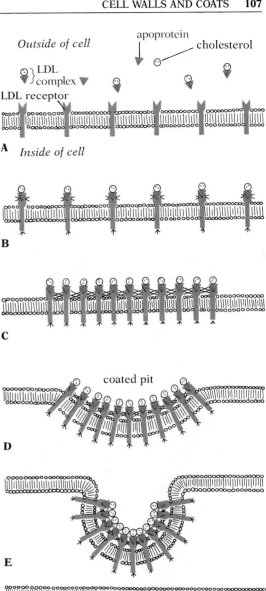

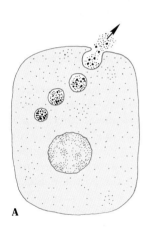

4.26 Exocytosis

(A) A membranous vesicle moves to the periphery of the cell, where it bursts, releasing its contents to the exterior. (B) The final steps of exocytosis are seen here, as a vesicle containing tear fluid fuses with the plasma membrane and bursts.

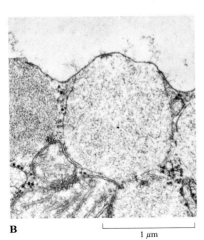

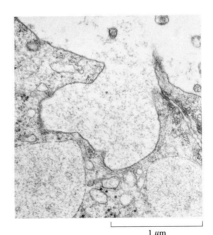

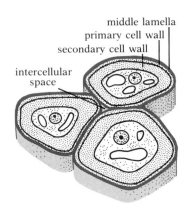

middle lamella
primary cell wall
secondary cell wall
intercellular
space

4.27   Cell walls and middle lamella of three adjacent plant cells

1 μm

4.28   Electron micrograph of cellulose microfibrils from the cell wall of a green alga
The microfibrils are laid out in parallel lines in two directions; each is about 20 nm wide. The specimen, from the green alga *Chaetomorpha melagonium*, was shadowed with palladium-gold.

another, these carbohydrate groups are usually described as a cell "coat," and the coat plays an important role in determining certain properties of the cells. The presence of carbohydrate materials on their outer surfaces, then, appears to be a general property of cells. Nonetheless, the conspicuous, thick, relatively rigid walls of plant, fungal, and bacterial cells, on the one hand, and the inconspicuous, thin, nonrigid coats of animal cells, on the other, remain among the most striking differences between these groups.

Cell walls of plants, fungi, and bacteria   Located outside the cell membrane, the plant cell wall is generally not considered part of the cytoplasm, though it is a product of the cell. The principal structural component of the cell wall is the complex polysaccharide ***cellulose***, which is generally present in the form of long threadlike structures called fibrils. The cellulose fibrils are cemented together by a matrix of other carbohydrate derivatives, including pectin and hemicellulose (a material not structurally related to cellulose). The spaces between the fibrils are not entirely filled with matrix, however, and they generally allow water, air, and dissolved materials to pass freely through the cell wall. The wall does not usually determine which materials can enter the cell and which cannot. This function is mostly reserved for the membrane located below the cell wall.

The first portion of the cell wall laid down by a young growing cell is the ***primary wall***. As long as the cell continues to grow, this somewhat elastic wall is the only one formed. Where the walls of two cells abut, a layer between them, known as the ***middle lamella***, binds them together. ***Pectin***, a complex polysaccharide generally present in the form of calcium pectate, is one of the principal constituents of the middle lamella. If the pectin is dissolved away, the cells become less tightly bound to one another. That is what happens, for example, when fruits ripen. The calcium pectate is partly converted into other, more soluble forms, the cells become looser, and the fruit becomes softer. Many of the bacteria and fungi that produce soft rots of the tissues of higher plants do so by first dissolving the pectin, reducing the tissue to a soft pulp which they can absorb.

Cells of the soft tissues of the plant have only primary walls and intercellular middle lamellae. After ceasing to grow, the cells that eventually form the harder, more woody portions of the plant add further layers to the cell wall, forming the ***secondary wall***. Since this wall, like the primary wall, is deposited by the cytoplasm of the cell, it is located inside the earlier-formed primary wall, lying between it and the membrane (Fig. 4.27). The secondary wall is often much thicker than the primary wall and is composed of a succession of compact layers, or lamellae. The cellulose fibrils in each lamella lie parallel to each other and are generally oriented at angles of 60–90 degrees to the fibrils of the adjacent lamellae (Fig. 4.28). This arrangement gives added strength to the cell wall. In addition to cellulose, secondary walls usually contain other materials, such as ***lignin***, which make them stiffer. Once deposition of the secondary wall is completed, many cells die, leaving the hard tube formed by their walls to function in mechanical support and internal transport for the body of the plant.

The cellulose of plant cell walls is commercially important as the main component of paper, cotton, flax, hemp, rayon, celluloid, and, obviously, wood itself. Lignin extracted from wood is sometimes used in the manufacture of synthetic rubber, adhesives, pigments, synthetic resins, and vanillin.

Plant cell walls generally do not form completely uninterrupted boundaries around the cells. There are often tiny holes in the walls through which delicate connections between adjacent cells may run. These connections, called *plasmodesmata*, are of two types. One type is a membrane-lined channel through which the cytoplasm of an individual cell in a multicellular plant body is in contact and communication with the cytoplasm of other cells (see Fig. 5.28, p. 133). The cytoplasm of cells interconnected by these plasmodesmata constitutes a continuous system called the *symplast*. A large portion of the intercellular exchange of such materials as sugars and amino acids probably takes place through the plasmodesmata of the symplast. The other type, usually called a pit, consists of a selectively permeable barrier formed by the primary walls (and middle lamella) of adjoining cells (see Fig. 12.12, p. 306).

The cell walls of both fungi and bacteria differ from those of plant cells. In fungi the main structural component of the wall is not cellulose but *chitin*, a polymer that is a derivative of the amino sugar glucosamine (see Fig. 3.7, p. 52). In bacteria the cell walls contain several kinds of organic substances, which vary from subgroup to subgroup. The distinctive responses of these organic substances to diagnostic stains are a regular means of identifying bacteria in the laboratory. Structurally, however, the cell walls of all bacterial groups are alike in one respect. Part of each bacterial wall has a rigid framework of polysaccharide chains covalently cross-linked by short chains of amino acids; the resulting structure can be regarded as a single enormous molecule, often called *murein*.

The presence of cell walls means that the cells of plants, fungi, and bacteria can withstand very dilute external media without bursting. In such media the cells are, of course, in a condition of turgor (distention). Water tends to move into them by osmosis, as a result of the high osmotic concentration of the cell contents. The cell swells, building up *turgor pressure* against the cell walls. The cell wall of a mature cell can usually be stretched only a minute amount. Equilibrium is reached when the resistance of the wall is so great that no further increase in the size of the cell is possible and, consequently, no more water can enter the cell. Thus the cells of plants, fungi, and bacteria are not so sensitive as animal cells to the difference in osmotic concentration between the cellular material and the surrounding medium. Because of their walls, these cells can withstand much wider fluctuations in the osmotic makeup of the surrounding medium than animal cells. Moreover, turgor pressure actually strengthens the mechanical structure of plants, just as inflating an initially limp balloon or tire produces a much stiffer and stronger structure.

The glycocalyx In plants, fungi, and bacteria, the cell wall is entirely separate from the membrane; if the cell shrinks in a hypertonic medium, the membrane separates from the much more rigid wall (see Fig. 14.2, p. 354). By contrast, the "coat" of an animal cell is not an independent entity. The carbohydrates (short chains of sugars called oligosaccharides) of which it is composed are covalently bonded to protein or lipid molecules in the plasma membrane (Fig. 4.29). The resulting complex molecules are termed

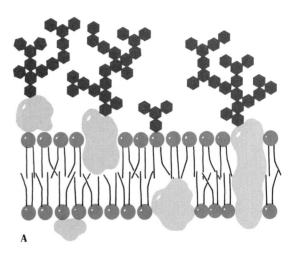

A

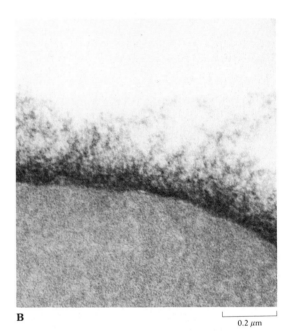

B $\llcorner\underline{\qquad}\lrcorner$ 0.2 μm

4.29 Plasma membrane with glycocalyx
(A) The glycocalyx of an animal cell is composed of oligosaccharides (branching structures) attached to some of the protein and lipid molecules of the outer surface of the membrane. (B) In this electron micrograph, the glycocalyx of a red blood corpuscle gives the outer surface of the membrane a fuzzy appearance.

glycoproteins and glycolipids, and the cell coat itself is often called the *glycocalyx.* It is important to realize that the membrane is strictly polarized: the glycolipids (which constitute about 50 percent of the lipids in the outer layer) and the carbohydrate-equipped ends of the glycoproteins are found *only* on the outside of the lipid bilayer.

According to recent research, the glycocalyx provides the recognition sites on the surface of the cell that enable it to interact with other cells. For example, if individual liver and kidney cells are mixed in a culture medium, the liver cells will recognize one another and reassociate; similarly, the kidney cells will seek out their own kind and reassociate. Apparently some property of the glycocalyx enables the cells to distinguish one from the other in such a situation. And indeed the nature of the carbohydrate markers varies consistently from tissue to tissue and from species to species. Cell recognition in the process of embryonic development must also depend, at least in part, on the glycocalyx, and the same is probably true for the control of cell growth. When normal cells grown in tissue culture touch each other, they cease moving and their growth slows down or stops altogether. This phenomenon of *contact inhibition* appears to be absent in cancer cells, which continue growing without restraint, probably because an abnormal glycocalyx prevents them from interacting normally. As a source of identity, the glycocalyx is also important in a variety of infectious diseases: malaria parasites, for instance, recognize their host (the red blood corpuscle) by a distinctive carbohydrate marker which the corpuscle probably manufactures for a completely different purpose. The recognition of host cells by invading viruses probably also depends on the carbohydrate markers of the glycocalyx. The markers in the glycocalyces of foreign cells probably provide the cues that the immune system's antibody molecules use to recognize invaders.

Chapter 5

INSIDE THE CELL

In the last chapter we saw how the cell membrane protects the cell from the world outside, and how it manages to trap some substances and to dispose of others. Some specialized membrane channels make use of the free energy of osmotic-concentration gradients, but most of the cell's selective permeability is the direct or indirect result of energy expenditure. The result is that a homeostatic chemical environment favorable to life is maintained inside the membrane, an environment with optimal ionic concentrations and pH, enough of the right sort of organic building blocks, and a collection of life-supporting enzymes. But the chemistry of many of the important processes of life is different from that of the cellular cytoplasm as a whole. Already we have seen, for example, that digestive enzymes are packaged in special structures called lysosomes, to protect the cell as a whole from these destructive but essential biological catalysts. In this chapter we shall examine a variety of subcellular *organelles*, each with its own specialized chemistry carefully isolated from the rest of the cell.

Organelles are organized like small cells, often complete with their own lipid-bilayer membrane. The membrane contains protein channels that maintain the organelle's own unique homeostasis, transporting in reactants, exporting products, and frequently even containing the enzymes necessary for many of the organelle's specialized functions. And as we shall see, current theory suggests a fascinating evolutionary history for some organelles.

SUBCELLULAR ORGANELLES

THE NUCLEUS

Within the cells of most organisms (though not of bacteria), the largest and one of the most conspicuous structural areas is the membrane-bounded *nucleus* (Fig. 5.1). The nucleus plays the central role in cellular reproduc-

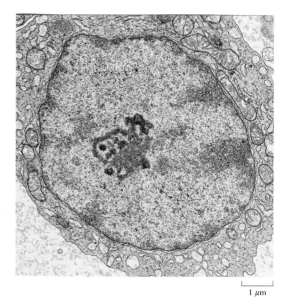

1 μm

5.1 Electron micrograph of an ovarian-cell nucleus from a rat
The dark area near the center of the nucleus is a nucleolus.

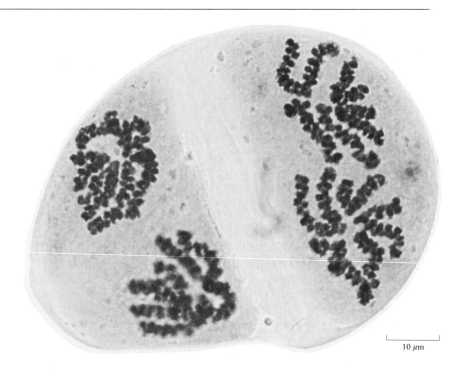

5.2   Chromosomes in a dividing cell of the reproductive organ of *Trillium*
The separate chromosomes in the two nuclei can easily be distinguished.

10 μm

tion, the process by which a single cell divides and forms two new cells. It also plays a crucial part, in conjunction with the environment, in determining what sort of differentiation a cell will undergo and what form it will exhibit at maturity. And the nucleus directs the metabolic activities of the living cell. In short, it is from the nucleus that the "instructions" emanate that guide the life processes of the cell as long as it lives.

We have said that the cells of bacteria differ from those of all other kinds of organisms in lacking a membrane-bounded nucleus (though they do possess genetic material that controls the cell's activities). Similarly, this group lacks many of the other cellular structures found in other organisms. These differences are so fundamental that bacteria are classified in a kingdom of their own (the Monera); and their cells are designated as ***procaryotic*** ("having a primitive nucleus"), whereas the cells of all other organisms are designated as ***eucaryotic*** ("having a true nucleus"). Therefore, the following discussion of the nucleus and other subcellular organelles concerns only eucaryotic cells; the characteristics of procaryotic cells will be discussed in a later section.

The eucaryotic nucleus contains two distinct types of structures, the chromosomes and the nucleoli. With an electron microscope we can see that both are embedded in a mass of amorphous, granular-appearing nucleoplasm. The entire nucleus is bounded by a nuclear membrane.

The ***chromosomes*** (Fig. 5.2) are elongate, threadlike bodies clearly visible only when the cell is undergoing division. They are composed of DNA and protein; the DNA is the substance of the basic units of heredity, called **genes**, while the protein provides spool-like supports, or cores, on which the DNA is wound to form structures called nucleosomes (Fig. 5.3). The genes transmit an elaborate message from generation to generation. They determine the characteristics of cells and act as the units of control in the day-to-day activities of living cells.

The hereditary information carried by the genes is written in the sequence of the nucleotide building blocks of the DNA molecules. Since the genes themselves remain in the nucleus, while most of the processes they control take place in the cytoplasm, some mechanism must exist for conveying the information outside the nucleus. The mechanism is transcription, the process by which the nucleotide sequence in the DNA gives rise to a corresponding nucleotide sequence in RNA. This RNA sequence is then somewhat modified, and the resulting messenger RNA (mRNA) can leave the nucleus and move to the sites of protein synthesis in the cytoplasm. There amino acids are linked by peptide bonds in a sequence corresponding to that of the mRNA nucleotides to form proteins, including enzymes. This process is known as translation. As we saw in Chapter 3, the sequence of amino acids—the primary structure of a protein—determines the three-dimensional conformation of the protein and the biological activity that this conformation bestows. Thus the genes are at the very hub of life; they encode all the information necessary for the synthesis of the enzymes regulating the myriad interdependent chemical reactions that determine the characteristics of cells and organisms. Later chapters will deal in more detail with the question of how genes encode information and direct protein synthesis, how genes are turned on or off by the cell as required, and how genes reproduce themselves during the process of cell division.

The other prominent structures in the nucleus, besides the chromosomes, are the **nucleoli**, the dark-staining, generally oval bodies usually visible within the nuclei of nondividing cells. There may be one or more nucleoli per nucleus, depending on the species of organism. Nucleoli form in association with particular regions of specific chromosomes; they are, in fact, simply specialized parts of the chromosome and, like the rest of the chromosome, are composed of DNA and protein. The DNA of the nucleoli includes multiple copies of the genes from which a type of RNA called ribosomal RNA (rRna) is transcribed. After this rRNA is synthesized, it combines with proteins, and the resulting complex detaches from the nucleolus, leaves the nucleus, and enters the cytoplasm, where it becomes a part of the protein-synthesizing organelles called ribosomes. Thus the nucleoli are responsible for manufacturing and exporting to the cytoplasm the precursors of the ribosomes on which proteins will be synthesized. Multiple copies of the genes for rRNA make possible rapid manufacture of the ribosomes necessary for active protein synthesis. Nucleoli tend to be small or absent in cells that carry out little protein synthesis.

The **nuclear membrane** helps maintain a chemical environment in the nucleus different from that provided by the surrounding cytoplasm. Unlike the cell membrane, the complete nuclear envelope consists of distinct inner and outer membranes, with space enclosed between them (Figs. 5.1 and 5.4).

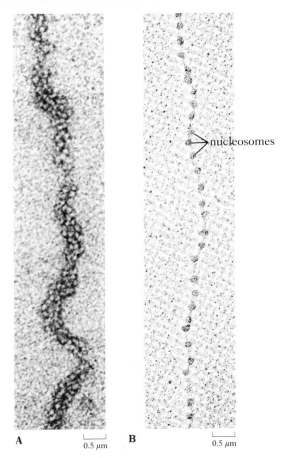

A 0.5 μm   B 0.5 μm

5.3 DNA on nucleosomes
The chromosomal DNA of eucaryotes is wound on protein spools, or cores, to form structures called nucleosomes. Normally the spools adhere to one another in a regular way, giving the DNA the appearance of a piece of yarn (A). When the DNA is treated to break the connections between nucleosomes, the individual protein spools and the thread of DNA can be seen (B).

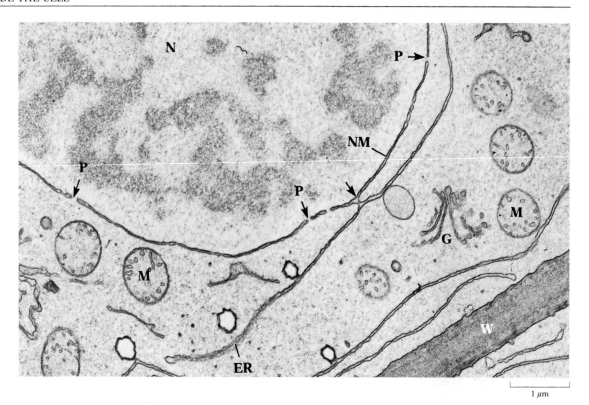

5.4   Electron micrograph showing nuclear membrane of a cell from corn root

The large structure filling the upper left quarter of the picture is the nucleus. The unlabeled arrow indicates a point where the endoplasmic reticulum and the double nuclear membrane interconnect. ER, endoplasmic reticulum; G, Golgi apparatus; M, mitochondrion; N, nucleus; NM, nuclear membrane; P, pore in nuclear membrane; W, cell wall.

Electron-microscope studies indicate that the double-membrane envelope is interrupted at intervals by fairly large and elaborate pores at points where the outer and inner membrane are continuous (Figs. 5.4, 5.5, and 5.6). Nevertheless, the membrane is highly selective. According to permeability experiments, some substances that can cross the cell membrane into the cytoplasm apparently cannot readily cross the nuclear membrane into the nucleus and are consequently restricted to the cytoplasm. Experiments with molecules much smaller than the pores confirm that simple unrestricted movement through the pores is definitely not possible. Yet some large molecules do pass readily through them. These seem primarily to be substances produced on the genes (such as mRNA) that are moving out of the nucleus, proteins moving into the nucleus to be incorporated into nuclear structures or to catalyze chemical reactions in the nucleus, and various substances from the cytoplasm that move into the nucleus and help regulate gene activity. There is thus a carefully controlled and highly selective two-way exchange through the pores between the nucleus and the cytoplasm. Apparently the pores recognize which substances are to be allowed to pass, and in which direction, on the basis of specific "passwords," chemical signal sequences attached to one end of the various molecules. When these passwords are found on RNA or protein, they must normally be removed before the substance can be used.

The electron microscope has revealed another particularly interesting

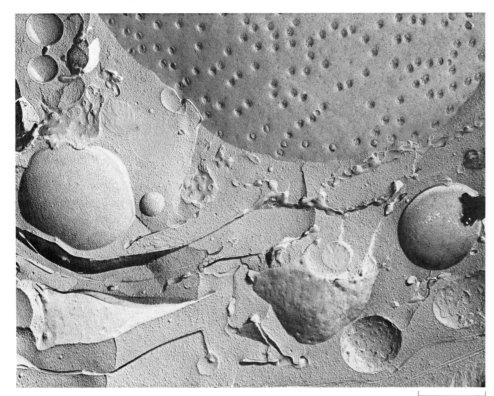

1 μm

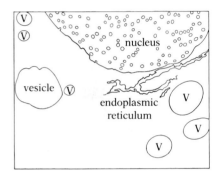

5.5   Electron micrograph of freeze-etched onion root-tip cell
At upper right is the surface of the nuclear membrane, with numerous pores. A variety of vesicles can be seen in the cytoplasm.

fact about the nuclear membrane—the continuity, at some points, of this double membrane with an extensive cytoplasmic membrane system called the endoplasmic reticulum (Fig. 5.4).

THE ENDOPLASMIC RETICULUM AND RIBOSOMES

Attempts to separate subcellular organelles from the cytoplasm for various research purposes often make use of centrifugation. In this technique, cells are first lysed (broken) by detergents that disrupt the lipids in the membrane, and then mixed with a viscous solution containing, for instance, glucose; finally, the samples are spun in a centrifuge to sort out the cell parts by weight.

In 1938 Albert Claude of the Rockefeller Institute isolated some cytoplasmic components which he called microsomes ("small bodies"). Claude's microsomes made up 15–20 percent of the total cell mass and could be isolated from almost any kind of cell—plant or animal. Chemical analysis showed that the microsomes had a very high nucleic acid content; in fact, they contained almost all the cytoplasmic nucleic acid. They also contained a high percentage of the cytoplasmic phospholipids. However, since the microsomes were not visible under the light microscope, there was much argument about whether they were actually discrete portions of

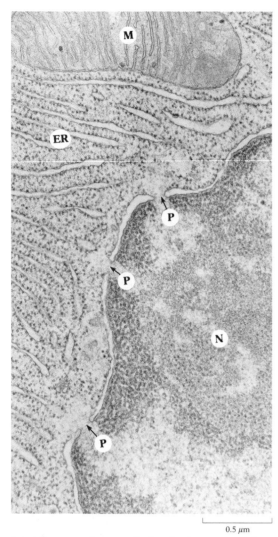

0.5 μm

5.6 Electron micrograph of endoplasmic reticulum
Thin section of a pancreatic cell from a bat, showing
many flattened cisternae of rough ER; the ribosomes
lining the ER membranes can be clearly
distinguished. In the lower right portion of the
micrograph is part of the nucleus (N); note the very
prominent pores (P) in the double nuclear membrane.
A mitochondrion (M) is at the top.

the living cells or simply artifacts produced by the breaking up and centrifugation of the cells.

Only the advent of the high-resolution phase-contrast microscope allowed Claude's microsomes to become established as part of the cell machinery. In 1945 Keith R. Porter, then of the Rockefeller Institute,
described a complex system of membranes that formed a network in the
cytoplasm. This sytem, which Porter named the **endoplasmic reticulum** (*reticulum* being Latin for "network"), has since been shown to be present to
some extent in all nucleated cells. Claude's microsomes were actually a
mixture of ribosomes and the fragmented endoplasmic reticulum.

Though the endoplasmic reticulum (ER) varies greatly in appearance in
different cells—its components may look like long tubules or round or ob-
long vesicles, or they may form stacks of flattened sacs—it is always a system of membrane-enclosed, fluid-filled spaces, or cisternae. And it seems
very likely that in most (and perhaps all) cells the ER forms a single continuous sheet. In most cells a portion of the ER membrane is lined on its outer
surface with **ribosomes**, the small protein-synthesizing organelles built in
part by the nucleolus. In this case, the ER is spoken of as "rough"; where no
ribosomes line its membrane, the ER is described as "smooth" (Fig. 5.7).
Just as the ER may exist without associated ribosomes, so too ribosomes
may occur independently of the ER.

Since the spaces between the inner and outer membranes of the nuclear
envelope are continuous with the membrane-enclosed channels and spaces
of the ER (Figs. 5.4 and 5.6), the nuclear membrane may be only a specialized part of the general ER system. If this proves true, one possible function
of the ER will be immediately apparent: its channels may serve as routes for
the transport of materials between the nucleus and various parts of the cytoplasm, forming a communication network between the nuclear control
center and the rest of the cell. This network would be independent of the
highly selective direct route, described earlier in this chapter, between the
nucleus and the cytoplasm by way of the pores in the nuclear membrane,
and would therefore probably serve a separate function.

Considerable evidence that substances do, in fact, move within the ER
has been provided by Philip Siekevitz and George Palade of the Rockefeller
Institute, who traced the movements of a particular zymogen (a substance
that is an inactive precursor of an enzyme, requiring a conformational
change to become active). They demonstrated that a digestive-enzyme zymogen synthesized in the pancreatic cells of guinea pigs by ribosomes on
the ER soon crosses the membranes of the cisternae, and then moves
within the ER to specialized areas for packaging in vesicles and transport to
other cellular targets.

In fact, the association of the ribosomes with the ER membrane appears
to be necessary for the protein they produce to penetrate the membrane to
the cisternae. Most or all of the proteins to be packaged and transported by
the ER are synthesized by the ribosomes of the rough ER. Proteins synthesized on free ribosomes in the cytoplasm are apparently not destined for
export from the cell or for incorporation into membranes, but rather are
released to function as enzymes in the **cytosol** (the more fluid part of the
cytoplasm). It seems clear that mRNA from the nucleus that is to produce
enzymes for the ER binds first to free ribosomes. This mRNA, however, sig-

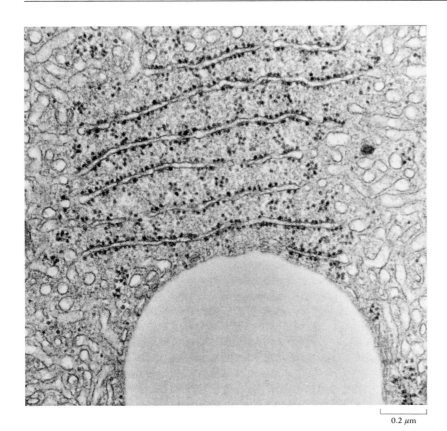

0.2 μm

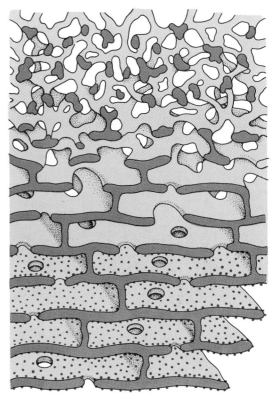

5.7 The endoplasmic reticulum

The endoplasmic reticulum from a steroid-producing cell of a guinea pig testis, shown in this electron micrograph, consists of a complex system of membranes including the rough ER, with associated ribosomes, and smooth ER. The relationship of rough and smooth ER varies from cell to cell; the sketch shows an association more typical of a cell in which macromolecules are synthesized in the rough ER and transported through the reticular channels to the smooth ER. Large openings through the rough ER allow mRNA and ribosomes access to the inner layers of the ER; small membrane channels under each bound ribosome allow the proteins synthesized by the ribosomes to be passed into the lumen of the ER.

nals the ribosome to bind to special channels in the rough ER. Once the ribosome binds, the translation begins and the enzyme being synthesized on the ribosome is threaded through the channels. Here again we see how membranes can serve to segregate the chemistry of one part of the cell from that of another.

Despite its importance in intracellular transport, it would be surprising if the ER functioned merely as a passageway. The ER membrane has a large protein content, and proteins, you will remember, may act both as structural elements in cells and as enzymes catalyzing chemical reactions. There is now abundant evidence that at least some of the many protein molecules of which the ER membranes are composed act as enzymes and that the ER functions as a cytoplasmic framework providing catalytic surfaces for some of the biochemical activity of the cell. Its complex folding provides an enormous surface for such activity. Some of the evidence has come from studies of the smooth ER found prominently in cells in which membrane phospholipids are synthesized in large quantities. Here the enzymes involved in lipid synthesis are firmly embedded in the ER membrane, arranged to process reactants in the appropriate sequence, and oriented toward the outside of the ER, where the reactants are to be found. The newly synthesized phospholipids are used in the manufacture of transport vesicles.

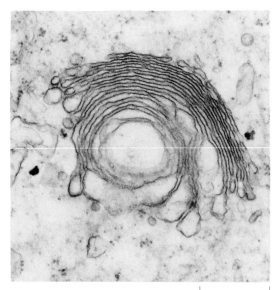

0.05 μm

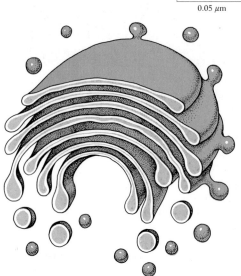

5.8   The Golgi apparatus
Electron micrograph of Golgi apparatus from an amoeba. Vesicles forming at the ends of some of the cisternae can be seen in the EM and in the interpretive drawing.

These two processes—enzymatic activity of the ER and compartmentalized vesicle transport—work together in the vertebrate liver to remove dangerous substances from the blood. The smooth ER of liver cells holds enzymes that detoxify many poisons, including barbiturates, amphetamines, and morphine. Poisons that appear in the bloodstream are transported to the liver. The rapid synthesis of more phospholipids causes the surface area of the liver cells' smooth ER to double. The positioning of detoxification enzymes in the ER takes maximum advantage of the cell's potential for compartmentalization to protect the cytosol from the poisons. The products of the detoxification process are probably packaged in vesicles that have been formed from the ER itself, and are then transported to other organelles, where they are further broken down.

THE GOLGI APPARATUS

In 1898 the Italian scientist Camillo Golgi first described a new "reticular apparatus" in certain cells of the vertebrate brain. This "apparatus," now recognized as yet another subcellular organelle, became visible under the light microscope only when treated with certain chemicals. Similar cytoplasmic regions have subsequently been found by numerous workers in a great variety of animal and plant cells. Though they vary in form and several different names were at first applied to them, all eventually came to be called *Golgi apparatus*. This organelle consists of a system of membrane-delimited compartments arranged approximately parallel to each other (Fig. 5.8).

The Golgi apparatus is particularly prominent in cells thought to be involved in the secretion of various chemical products; as the level of secretory activity of these cells changes, corresponding changes occur in the morphology of the organelle. In the preceding section we mentioned that in certain cells of the pancreas of guinea pigs, a zymogen synthesized on the ribosomes moves into the channels of the ER; it reaches the Golgi apparatus in vesicles budded off the ER. With the help of the electron microscope, we now know that the zymogen arrives at the compartment nearest the nucleus (the most highly curved cisterna in Figure 5.8), and then moves from one layer to the next until it reaches the farthest layer. Movement between layers is probably by means of vesicles that bud off from one cisterna and then fuse with another, farther from the nucleus. During this movement the zymogen is modified, and in the final cisterna it is concentrated and stored; it is eventually released from the cell via secretory vesicles that are produced by this outer compartment of the Golgi and move to the cell surface. Thus the role of the Golgi apparatus in secretion is clear: its functions include storage, modification (for example, removal of water or emulsification of lipids), and packaging of secretory products.

The Golgi apparatus is also the major director of macromolecular transport in cells. Though no protein synthesis takes place in the Golgi, polysaccharides are synthesized there from simple sugars and attached to proteins and lipids to create glycolipids and glycoproteins. Some of these are transported in as part of vesicle membranes to the glycocalyx. In addition, proteins already marked with carbohydrate groups (and most proteins transported to the Golgi apparatus from the ER are so marked) may have

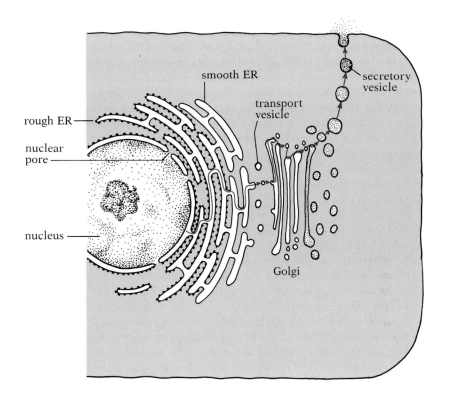

5.9 A probable path of phospholipid movement in the cell

Structural molecules of cellular membranes are constantly on the move. The path traced here shows the movement of a phospholipid from the point of synthesis in the nuclear envelope through the rough ER and into the smooth ER. There it receives a specific carbohydrate marker, which makes it a glycolipid, and becomes part of a vesicle transporting proteins to the Golgi apparatus; it is then incorporated into the membrane of that organelle. Subsequently the same molecule, perhaps with a new or modified carbohydrate marker, moves to the outer layer of the Golgi, where it becomes part of a secretory vesicle that is carried to the plasma membrane. There, in the process of exocytosis, the vesicle membrane (including the glycolipid molecule) fuses with the plasma membrane of the cell. Other membrane molecules follow different paths.

their sugar-based carbohydrate tags modified there. These carbohydrate tags may serve at least in part as intracellular "mailing labels," used to sort, package, and direct the various classes of chemicals from their sites of synthesis to their ultimate targets. Why proteins should need to be relabeled in the Golgi, though, is not fully understood.

The secretory vesicles produced by the Golgi apparatus probably play an important role in adding surface area to the cell membrane. When one of these vesicles moves to the cell surface, it becomes attached to the plasma membrane and then ruptures, releasing its contents to the exterior in the process of exocytosis. The membrane of the ruptured vesicle may remain as a permanent addition to the plasma membrane. Biochemical and morphological studies of the membranes show that the inner portion of the Golgi apparatus resembles the membranes of the nuclear envelope and endoplasmic reticulum, and that there is a progressive change within the Golgi apparatus until the outer portion, where secretory vesicles are produced, resembles the plasma membrane. It is likely, therefore, that a dynamic relationship exists between the different parts of the cellular membrane system (Fig. 5.9): a membrane molecule originating in the nuclear envelope could conceivably appear consecutively in the rough ER, the smooth ER, the Golgi apparatus, secretory or other vesicles, and finally the plasma membrane. From the plasma membrane it might eventually migrate back to the Golgi apparatus or some other organelle as part of an empty vesicle.

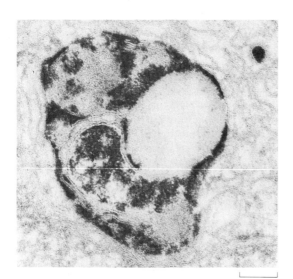

0.1 μm

5.10 Lysosomes in a connective-tissue cell from the vas deferens of a rat
The small dark body at upper right is a primary lysosome. The much larger body at left is a secondary lysosome (digestive vacuole) formed by fusion of a primary lysosome with a phagocytic or pinocytic vesicle. (The dark appearance of the lysosomes results from staining for acid phosphatase, a digestive enzyme whose presence is used as the definitive test for these organelles.)

LYSOSOMES

First described in the 1950s by the Belgian scientist Christian deDuve of the Catholic University of Louvain, *lysosomes* are membrane-enclosed bodies that function as storage vesicles for many powerful digestive (hydrolytic) enzymes (Fig. 5.10). The lysosome membrane, which is single, contains an ionic pump that maintains a highly acidic internal environment. The membrane permits desirable reaction products to pass through to the cytosol, but it is impermeable to the hydrolytic enzymes and capable of withstanding their digestive action. If the lysosome membrane is ruptured, the hydrolytic enzymes—no longer safely confined—are released into the surrounding cytoplasm and begin immediately to break down the interior of the cell.

As you have probably realized, lysosomes act as the digestive system of the cell, enabling it to process some of the bulk material taken in by endocytosis. Their hydrolytic enzymes are synthesized in the form of zymogens in the rough ER, packaged into transport vesicles in the smooth ER, and carried to the Golgi apparatus. There, receptor proteins on the inside of the Golgi membrane apparently form regions like coated pits to attract and trap these zymogens, probably recognized by their characteristic carbohydrate markers. When a region has received an appropriate supply of the zymogens, it buds off from the Golgi membrane as a lysosome and the enzymes are activated. Proteins mounted on the exterior of the lysosome's membrane must serve as recognition sites to assure that the enzymes it carries are delivered to the appropriate target. The targets of these so-called primary lysosomes include endocytotic vesicles newly arrived from the cell surface, and secondary lysosomes, also known as digestive vacuoles, which are the digestive vesicles already produced by the fusion of primary lysosomes and endocytotic vesicles. When digestion is complete, the useful products pass into the cytosol, while the residue is discharged by exocytosis (Fig. 5.11).

Several diseases are caused by lysosome disorders. In human inclusion cell disease, for instance, a sugar tag added to one of the digestive enzymes routes it not to a lysosome but to a fluid-filled vacuole, where it begins to digest the cell. The result is general debilitation, followed by death when one or more of the body's organs fail. In the devastating nervous disorder known as Tay-Sachs disease, lipid-digesting lysosomes lack a particular enzyme. When these deficient lysosomes fuse with lipid-containing vesicles, they do not fully digest their contents. The resulting defective secondary lysosomes can accumulate and block the long thin parts of nerve cells responsible for transmitting nerve impulses.

The elaborate organization of membrane movement in cells is vulnerable to many infectious diseases. Semliki Forest virus, for instance, infects a wide range of hosts from invertebrates to humans. The virus carries a marker in its protein coat that mimics in effect a substance for which one kind of coated pit has specific receptors. SFV is carried into the cell by endocytosis, and the vesicle thus formed is targeted for fusion with a particular type of lysosome. The SFV protein coat fuses with the secondary lysosome membrane, after which its RNA separates from it and escapes into the cytosol. This RNA uses the host cells' ribosomes, ER, Golgi apparatus, and the whole system of carbohydrate tags to reproduce itself and direct the

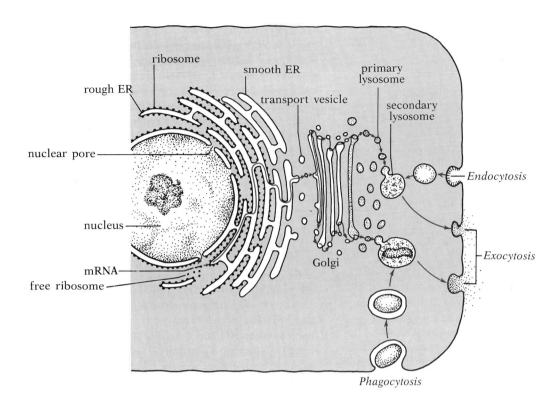

rough ER

ribosome

smooth ER

primary
lysosome

transport vesicle

secondary
lysosome

nuclear pore

Endocytosis

nucleus

Exocytosis

mRNA

Golgi

free ribosome

Phagocytosis

5.11   The hydrolytic enzyme cycle
The role of the ER and the Golgi apparatus
in cellular digestion is representative of
their functions in the cell. DNA in the cell
nucleus produces the mRNA that encodes
the hydrolytic enzymes. The mRNA is
transported to ribosomes, which then bind
to the rough ER. The enzymes are
synthesized there, passed into the ER, and
marked for later transport. The enzymes are
next collected by receptors in the smooth
ER and packaged into transport vesicles.
Markers on the outside of the transport
vesicles cause them to fuse with the
membrane of the Golgi apparatus, where
new vesicles (primary lysosomes) are
created. These packets of specific mixtures
of hydrolytic enzymes fuse with
appropriately marked endocytotic vesicles
or with secondary lysosomes (produced by
previous fusions of this kind) and help
digest the contents. Useful products of this
digestion pass through the membranes of
the secondary lysosomes into the cytosol,
while the residue is disposed of through
exocytosis. The same pattern of synthesizing
and marking enzymes in the ER,
transporting them to the Golgi apparatus,
sorting and repackaging them there, and
then dispatching them to various
intracellular targets seems to be the general
strategy for managing macromolecules in
cells.

subsequent exocytosis of its viral progeny. The virus thus ensures its own
propagation by exploiting much of the specialized membrane machinery of
the cell.

PEROXISOMES

Improved methods for separating cell contents have revealed that certain
membrane-bounded organelles previously confused with lysosomes actually have their own distinct identity; they are now known as microbodies.
The most thoroughly understood of these microbodies are the ***peroxisomes***

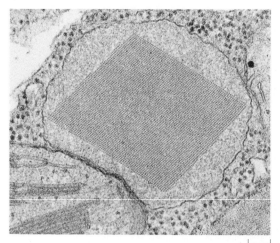

0.1 μm

5.12 Electron micrograph of a tobacco-leaf peroxisome
The tobacco-leaf cell has been treated with stain, which reveals the crystalline core of this peroxisome and the presence in it of the enzyme catalase. In mammalian peroxisomes, catalase functions to neutralize hydrogen peroxide, alcohol, and other potentially harmful substances.

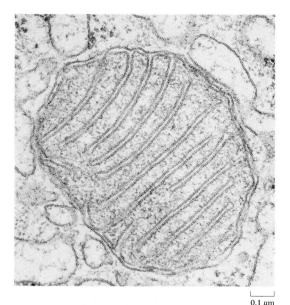

0.1 μm

5.13 Electron micrograph of a mitochondrion from an epithelial cell of a rat
Note the double outer membrane and the numerous cristae, which can be seen to arise as folds of the inner membrane.

(Fig. 5.12). Like lysosomes, peroxisomes contain an assortment of powerful enzymes. But where the enzymes of the lysosomes are hydrolytic (water-splitting), these enzymes are oxidative (water-producing) and catalyze condensation reactions, such as the oxidative removal of amino groups from amino acids, the detoxification of alcohol, the oxidation of the dangerous compound hydrogen peroxide into water and oxygen, and the reactions involved in the production of macromolecules used in respiration and other synthetic pathways. In the leaves of green plants, peroxisomes are involved in photorespiration (to be discussed in Chapter 8). Though similar in appearance to lysosomes, peroxisomes are not produced by budding from the Golgi apparatus. In fact, the source of their membrane is a mystery. The precursors of the enzymes they contain are produced in the cytoplasm rather than in the endoplasmic reticulum, and must somehow be captured and transported to these tiny organelles.

MITOCHONDRIA

We have now discussed the functioning of a variety of subcellular organelles, ranging from the nucleus to the lysosomes which fuse with the plasma membrane in exocytosis. But we have yet to discuss the organelles that provide the energy necessary to build and fuel all the others. These are the mitochondria and the chloroplasts.

Mitochondria, often thought of as the powerhouses of the cell, are the sites of the chemical reactions known as respiration that extract energy from food and make it available for innumerable energy-demanding activities. Each mitochondrion is bounded by a double membrane; the outer membrane is smooth, while the inner membrane has many inwardly directed folds (Fig. 5.13; see also Fig. 7.9, p. 185). These folds, called *cristae*, extend into an amorphous semifluid matrix. The outer mitochondrial membrane contains numerous gates and pumps, while the inner membrane is relatively impermeable. Reactants—fatty acids and pyruvic acid (an energy-rich product of glucose specially suitable for "burning" in mitochondria)—are concentrated in this organelle, and here, with the help of the appropriate enzymes, either of these fuels can combine with oxygen to generate water, carbon dioxide, and the energy that runs the cell. We shall discuss the operation of this important organelle more fully in Chapter 7.

PLASTIDS

Plastids are large cytoplasmic organelles found in the cells of most plants, but not in the cells of fungi or animals (except when symbiotic bacteria or algae have taken up residence there, as happens in some Protozoa, sea anemones, corals, and molluscs). Plastids are clearly visible through an ordinary light microscope. There are two principal categories of plastids: *chromoplasts* (colored plastids) and *leucoplasts* (white or colorless plastids).

Chloroplasts are chromoplasts that contain the green pigment *chlorophyll*, along with various yellow or orange pigments called *carotenoids*. We shall explore in detail in Chapter 8 how the radiant energy of sunlight is trapped in the chloroplasts by molecules of chlorophyll and is then used in the manufacture of complex organic molecules (particularly glucose) from

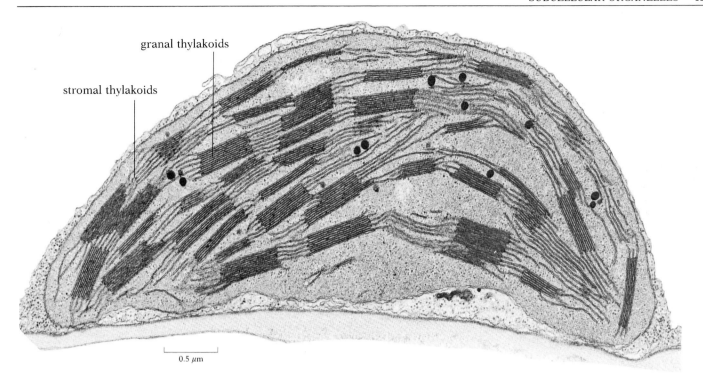

granal thylakoids

stromal thylakoids

0.5 μm

simple inorganic raw materials such as water and carbon dioxide. Oxygen is a by-product of this photosynthetic reaction upon which nearly all organisms depend.

The electron microscope reveals that the typical chloroplast is bounded by two concentric membranes and has, in addition, a complex internal membranous organization (Fig. 5.14). The fairly homogeneous internal proteinaceous matrix is called the *stroma*. Numerous flat compartments called *thylakoids*, or lamellae, are embedded in it. In most higher plants these thylakoids come in two varieties: separate thylakoids that run through the stroma, and stacks of platelike thylakoids forming regions known as *grana* (Fig. 5.14). Some other plants—the brown algae, for example—have no grana: all the thylakoids are stromal. The chlorophyll and carotenoids are bound to the proteins and lipids in the membranes of the thylakoids. The precise arrangement of the protein, lipid, and pigment components in the thylakoids is essential for complete photosynthesis. The membrane structure of chloroplasts will be considered in more detail in Chapter 8, when we examine the photosynthetic process.

Chloroplasts lacking chlorophyll are usually yellow or orange (occasionally red) because of the carotenoids they contain. It is these kinds of chloroplasts that give many flowers, ripe fruits, and autumn leaves their characteristic yellow or orange color. Some of these chloroplasts have never contained chlorophyll, while others are formed from chloroplasts whose chlorophyll has been lost. The latter are particularly common in ripe fruits and autumn leaves, structures that were once green.

The colorless plastids, or leucoplasts, are primarily organelles in which materials such as starch, oils, and protein granules are stored. Plastids

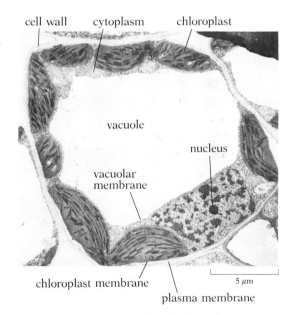

cell wall   cytoplasm   chloroplast

vacuole

nucleus

vacuolar membrane

chloroplast membrane

plasma membrane

5 μm

5.14 Electron micrographs of chloroplasts
Stacks of disclike thylakoids, forming grana, can be seen in this chloroplast of a corn leaf (top). Numerous chloroplasts lie close to the perimeter of a mature leaf cell of timothy grass (bottom).

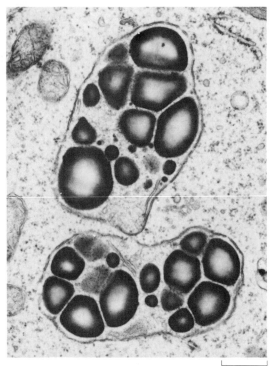

0.5 μm

5.15 Electron micrograph of leucoplasts from the root tip of *Arabidopsis*

Because they contain numerous prominent starch grains, these leucoplasts from the small desert plant *Arabidopsis* are called amyloplasts.

filled with starch (amyloplasts) are particularly common in seeds and in storage roots and stems, such as carrots and potatoes, but they also occur in the cells of many other parts of the plant. The starch, which you may remember from Chapter 3 is an energy-storage compound, is deposited as a grain or group of grains (Fig. 5.15); no starch is found in other parts of the cell.

All types of plastids form from small colorless bodies called proplastids. Once formed, many kinds of plastids can be converted into other types under appropriate conditions. It can be demonstrated, for example, that synthesis of chlorophyll is dependent on light and that under certain conditions leucoplasts exposed to light develop chlorophyll. When a leucoplast is converted into a chloroplast, the internal membranous thylakoids characteristic of a chloroplast develop from invaginations of the inner boundary membrane of the plastid.

VACUOLES

Membrane-enclosed, fluid-filled spaces called ***vacuoles*** are found in both animal and plant cells, though they have their greatest development in plant cells. There are various kinds of vacuoles, with a corresponding variety of functions. In some Protozoa, specialized vacuoles, called contractile vacuoles, play an important role in expelling excess water and some wastes from the cell; we shall discuss them in greater detail in a later chapter. Many Protozoa also possess food vacuoles, chambers that contain food particles. They are similar to the vesicles formed by many cells when they take in material by endocytosis. The distinction between vesicles and vacuoles, both of which are membrane-bound, is hard to draw with any precision, particularly since vesicles may fuse with or bud off from vacuoles. The most obvious differences are in permanence, activity, and size: vesicles are relatively short-lived transport vehicles, while vacuoles tend to be long-lived; vesicles usually move quickly, while vacuoles are relatively static; finally, vesicles are usually small, while vacuoles are most often quite large.

In most mature plant cells, a large vacuole occupies much of the volume of the cell. The immature cell usually contains many small vacuoles. As the cell matures, the vacuoles take in more water and become larger, eventually fusing to form the very large definitive vacuole of the mature cell (Fig. 5.16). This process pushes the cytoplasm to the periphery of the cell, where it forms a relatively thin layer.

The plant vacuole contains a liquid called cell sap—primarily water, with a variety of substances dissolved in it. Since the cell sap is generally hypertonic relative to the external medium, the vacuole tends to take in water by osmosis. As the vacuole swells, the vacuolar membrane (or tonoplast, as it is often called) pushes outward against the cytoplasm, which, being essentially fluid, resists compression and transmits the pressure to the cell wall. The wall is strong enough to limit the swelling and prevent the cell from bursting, but the outward push of the vacuolar membrane is sufficient to maintain cell turgidity and stiffness.

Many substances of importance in the life of the plant cell are stored in the vacuoles, among them high concentrations of soluble organic nitrogen compounds, including amino acids; vacuoles also store sugars, various organic acids, and some proteins.

 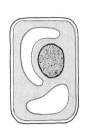 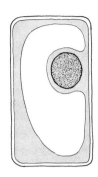

5.16 Development of the plant-cell vacuole
The immature cell (left) has many small vacuoles. As the cell grows, these vacuoles fuse and eventually form a single large vacuole, which occupies most of the volume of the mature cell, the cytoplasm having been pushed to the periphery (right).

The vacuoles also function as dumping sites for noxious wastes. Enzymes secreted into them degrade some of these wastes into simpler substances that can be reabsorbed into the cytosol and reused. The presence of poisonous wastes and powerful precipitating or denaturing agents in the vacuoles has greatly complicated studies of plant biochemistry, for when entire cells are disrupted, the mixing of these substances with the cytosol often results in the alteration or destruction of the compounds under investigation.

As might be expected, many of the substances accumulated in the vacuoles are selectively prevented from leaving by the vacuolar membrane, which must have its own distinctive permeability characteristics and must be capable of regulating the direction of movement of substances across it. If living beet cells are placed in distilled water, the pigment in the cell sap, one of a group of red pigments called **betacyanins**, does not diffuse out, even though it is in much higher concentration inside the vacuoles than outside. As soon as the beet cells die, the vacuolar membranes lose their selectivity and the betacyanin diffuses out.

Anthocyanins, another group of red pigments in the cell sap, are responsible for many of the purples, blues, and dark reds commonly seen in flowers, fruits, and autumn leaves (we have already seen that the carotenoids in the plastids are responsible for orange, yellow, and sometimes light red in these same structures). The relative amounts of anthocyanins and carotenoids in autumn leaves differ for different species of plants and also for the same species under different conditions. A high accumulation of sugars, low temperatures, and adequate light favor anthocyanin formation.

MICROFILAMENTS

As we have seen, the cell is full of specialized, membrane-lined compartments that mediate much of cellular chemistry. But the internal architecture of the cell also includes a variety of other important components, which help organize movement not only within the cell but by the cell itself, and which aid in defining and controlling cell shape. These protein-based components include microfilaments, microtubules, centrioles, cilia, and flagella.

Microfilaments are produced when molecules of a protein called actin polymerize spontaneously under conditions of elevated Ca^{++} and Mg^{++} concentration; the resulting long, extremely thin polymers, helically intertwined, form **actin** filaments (Fig. 5.17). Sometimes actin filaments are

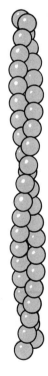

5.17 An actin filament
This portion of an actin filament shows the helically intertwined chains of protein subunits.

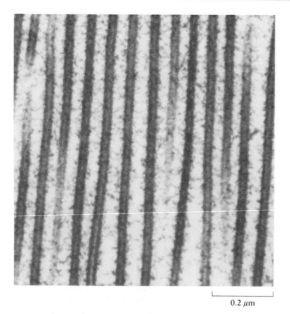

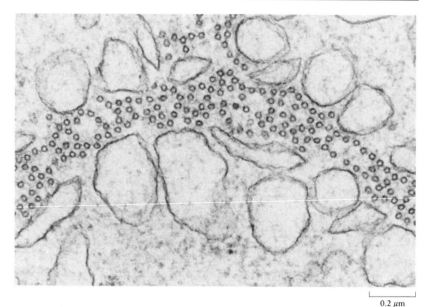

0.2 μm

0.2 μm

5.18   Electron micrographs of microtubules
Left: Longitudinal section, from bovine brain. Right:
Cross section, from hamster spermatid.

5.19   Structure of portion of a microtubule
The subunits of tubulin, each consisting of two
proteins (one kind shown colored, the other white),
are helically stacked to form the wall of the tubule,
which is usually 13 subunits in circumference.

found in association with **myosin** filaments, long threads very like actin fila-ments but with a hinged protein "foot" at one end. The unique feature of this actin-myosin combination is that the foot of a myosin filament can, when supplied with energy, literally climb along an adjacent actin filament as though it were a ladder. This ability of one filament to move with respect to another accounts for many types of movement, including the contrac-tion of muscles (which we shall examine in detail in Chapter 20), the amoe-boid movement of cells, the transport of some or all vesicles inside cells, and the constriction along the midline of cells as they divide. Each of these movements ceases if the cell is treated with cytochalasin B, a chemical that alters the properties of actin.

Microfilaments composed solely of actin play a purely structural role; because of their cross-links they provide necessary strength. Most com-monly they are found as a component of the **cytoskeleton**, a complex array of molecules that helps maintain cell shape. More specialized structural microfilaments provide reinforcement for various stiff cellular protuber-ances, including the dense forests of rodlike projections known as micro-villi that line the intestine.

MICROTUBULES

Microtubules are in many ways heavy-duty versions of microfilaments. They are long, hollow, cylindrical structures (Fig. 5.18) that, like microfila-ments, form spontaneously in response to the proper ionic signals. Mole-cules of a globular protein called tubulin, each molecule consisting of two proteins (α and β), polymerize to form a helical stack (Fig. 5.19). Microtu-bules radiate from organizing centers and play a critical role in general cell structure and in cell division. During cell division the microtubules radiate

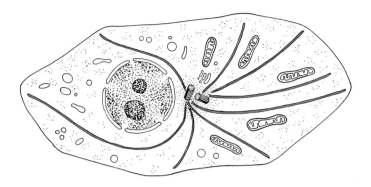

5.20 Microtubules in the cytoskeleton
It has been proposed that the microtubules (color), which often radiate outward from the region of the centrioles, help hold the cell in its characteristic shape, and provide a structural backbone for the other elements of the cytoskeleton.

5.21   The microtrabecular lattice
The still controversial microtrabecular lattice is thought by some researchers to be composed of a network of fibers that, joined to microtubules and microfilaments (not shown), help give shape to the cell and hold various cellular organelles in position. This interpretation suggests that microtrabecular fibers are anchored to the cell membrane, as shown in the interpretive drawing. But other researchers argue that the microtrabecular lattice is merely an artifact of the fixation technique used to prepare the specimen for the camera, and that cell movement, cell division, transport of vesicles along internal cell pathways, and similar phenomena are determined solely by the interactions of microtubules and microfilaments.

from organelles called centrioles at each end of the cell to form a basketlike arrangement (the spindle), which is instrumental in moving the chromosomes to the locations of the new nuclei (see Fig. 23.12D, p. 624). They also define the pathways to be followed by secretory vesicles and may, like microfilaments, have a part in their transport. Like microfilaments, they too help provide shape and support for the cell and its organelles as part of the cytoskeleton (Fig. 5.20). Indeed, when colchicine, a compound that prevents the stacking of tubulin, is added to a cell, it quickly loses its distinctive shape.

THE MICROTRABECULAR LATTICE

With the electron microscope, a faint, weblike network of fibers, the *microtrabecular lattice*, can be seen bridging the gaps between microtubules and microfilaments (Fig. 5.21). There is considerable controversy about

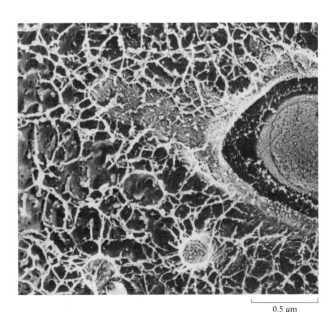

0.5 μm

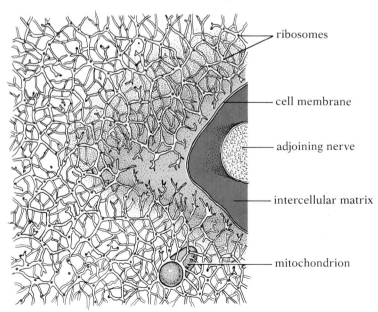

ribosomes

cell membrane

adjoining nerve

intercellular matrix

mitochondrion

whether the lattice is real, or a creation of the techniques by which cells are prepared for the EM. This is a familiar problem—researchers were divided for years about whether or not the Golgi apparatus was an artifact of Golgi's staining and fixation methods.

Unfortunately, the exact composition and the internal organization of the lattice fibers are not yet known, and so the information we have about the lattice can be interpreted in two quite different ways. For example, the proteins in the cytosol are concentrated in these fibers, while the surroundings are basically aqueous. Perhaps in living cells the proteins are mounted on the fibers, or perhaps they merely stick to the fibers when the specimen is undergoing fixation for viewing. Similarly, the free ribosomes of the cytoplasm are concentrated at lattice intersections, perhaps in living cells, perhaps as a result of fixation.

The fibers are thought by many researchers to be involved in the internal movements of the cell, though whether these movements arise from contractions of the lattice itself or from active movements of the microtubules and microfilaments to which the fibers are attached is not yet known. Other researchers, doubting that the fibers exist except as artifacts of fixation, see no need for any network beyond that of microtubules and microfilaments to account for the cell's internal movements. The debate about the lattice is fueling research that should establish its existence and role, if any, in the near future.

CENTRIOLES AND BASAL BODIES

Centrioles are found in pairs, oriented perpendicularly to each other, just outside the nuclei of many kinds of cells (Fig. 5.22). From the neighborhood

5.22 Centrioles
This electron micrograph shows newly replicated centrioles. Since the centrioles of each pair lie at right angles to each other, the sectioning of the specimen results in one from each pair being cut longitudinally and one being cut in cross section.

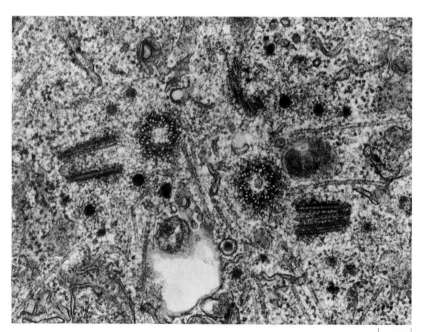

0.2 μm

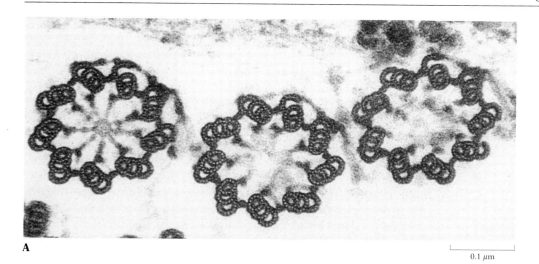

A

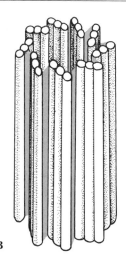

0.1 μm

B

5.23 Basal bodies and centrioles
Basal bodies (A), such as the three shown in cross section in this electron micrograph of a protozoan, are essentially identical to centrioles in structure. Centrioles (B) are composed of nine triplet microtubules.

of these organelles projects an array of microtubules (Fig. 5.20). When present, centrioles are also, as we have seen, the focus of the microtubule spindle during cell division. In cross section, centrioles display a remarkable uniformity of structure, with nine triplets arranged in a circle and each triplet composed of three fused microtubules (Fig. 5.23B).

Basal bodies anchor the many hairlike cilia and flagella (discussed next) to the cell membrane. As Figure 5.23A indicates, they have exactly the same structure as centrioles and are probably in truth the same organelle: basal bodies can become centrioles, and vice versa. During cell division in the mobile alga *Chlamydomonas*, for example, the basal bodies of the two flagella abandon their posts and migrate to the poles of the cell, where they take up their positions in the spindle. Similarly, the basal body of the flagellum in many kinds of sperm becomes one centriole of the egg after fertilization. And the centrioles of cells that differentiate to line the oviduct multiply to become the basal bodies of the cilia whose rhythmic beating moves eggs from the ovary to the site of fertilization.

In the past, centrioles have often been thought of as the organizing centers of the spindle, but now a great deal of evidence suggests a more passive role. As we have already seen, not all cells have centrioles (in fact, they are rare in plants), and yet virtually all cells have spindles during cell division. Moreover, in at least some cells destruction of the centrioles does not prevent subsequent spindle formation and cell division. It may be that centrioles are present merely to facilitate the production of basal bodies, and that they follow the spindle focus as a convenient way of segregating into the separating cells during cell division.

CILIA AND FLAGELLA

Some cells of both plants and animals have one or more movable hairlike structures projecting from their free surfaces. If there are only a few of these appendages and they are relatively long in proportion to the size of the cell, they are called *flagella*. If there are many and they are short, they are called *cilia* (Fig. 5.24). Actually, the basic structure of flagella and cilia is the same, and the terms are often used interchangeably. Both usually function either in moving the cell, or in moving liquids (or small particles) across the surface of the cell. They occur commonly on unicellular and small multicellular organisms and on the male reproductive cells of most animals and many plants, in both of which they may be the principal means of locomotion. They are also common on the cells lining many internal passageways and ducts in animals, where their beating aids in moving materials through the passageways. In the trachea they reach densities of a billion per square centimeter.

Electron-microscope studies of the flagella and cilia of eucaryotic cells have revealed a remarkable uniformity in their internal structure, regardless of the organism to which they belong, whether plant or animal, simple or complex. The slender cylindrical stalk, an extension of the cell membrane, contains a cytoplasmic matrix, with eleven groups of microtubules embedded in the matrix. Invariably, nine of these groups are fused pairs arranged around the periphery of the cylinder, while the other two are isolated microtubules lying in the center. Each cilium and flagellum is anchored to the cell by a basal body; the similarities between these organelles and the basal body in the arrangement of microtubules is obvious (Fig. 5.25). We shall discuss the mechanism of movement of cilia and flagella in Chapter 20.

EUCARYOTIC VS. PROCARYOTIC CELLS

In spite of the extreme variety among cells, they break down into two fundamental categories: eucaryotic cells, which compose the great majority of organisms, and procaryotic cells—the bacteria. The preceding discussion has been almost wholly concerned with eucaryotic cells. Before turning to the contrasting characteristics of procaryotic cells, let us summarize the

5.24 Scanning electron micrograph of ciliated surface of the trachea (windpipe) of a hamster
Coordinated movements of the many cilia function to sweep dust particles and other foreign material from the respiratory surfaces to the mouth.

10 μm

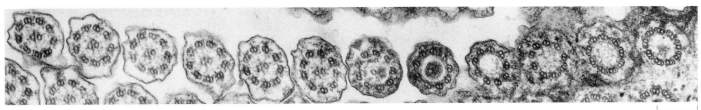

0.2 μm

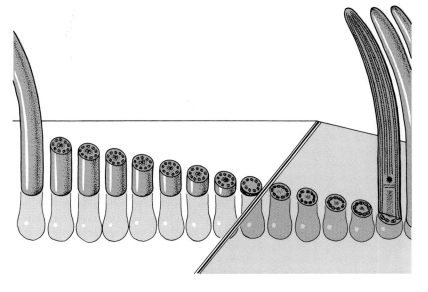

5.25 Cross section of cilia from the protozoan *Tetrahymena*
The electron micrograph of an oblique section of surface tissue reveals both the "nine plus two" arrangement of microtubules in cilia and the nine triplet microtubules characteristic of basal bodies. The interpretive drawing shows the plane of the section and the position of microtubules in a lateral section of an adjacent cilium.

main features of eucaryotic cells in an effort to achieve an integrated picture of cell structure and function.

"TYPICAL" EUCARYOTIC CELLS

How complex cells are was not truly appreciated until the advent of the electron microscope and modern biochemical techniques, which combined to change our whole picture of the cell. Four or five decades ago, biology books still regularly included a diagram of a so-called typical cell that showed only five or six simple internal components, a diagram easily memorized by students (Fig. 5.26). No simple diagram of this sort can be given today. In the first place, there is no such thing as a typical cell, not even a typical eucaryotic cell. Plant and animal cells differ from one another; cells of particular plants or animals differ from those of other plants or animals; and within the body of any one plant or animal the various cells often differ strikingly in shape, size, and function. This much, of course, has been known for a long time. But now that the number of known cellular components has grown so large and their great variability has been so well demonstrated, it becomes even more obvious that no single diagram, or even series of diagrams, can really portray a "typical" cell. Nevertheless, to help you visualize the arrangement of the organelles discussed in the preceding pages, two such diagrams (Figs. 5.27 and 5.28) are given here. As you examine them, keep in mind that not all the components shown always occur together in any one real cell.

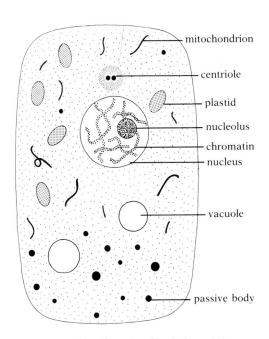

mitochondrion

centriole

plastid

nucleolus

chromatin

nucleus

vacuole

passive body

5.26 A "typical" cell as visualized about 1925

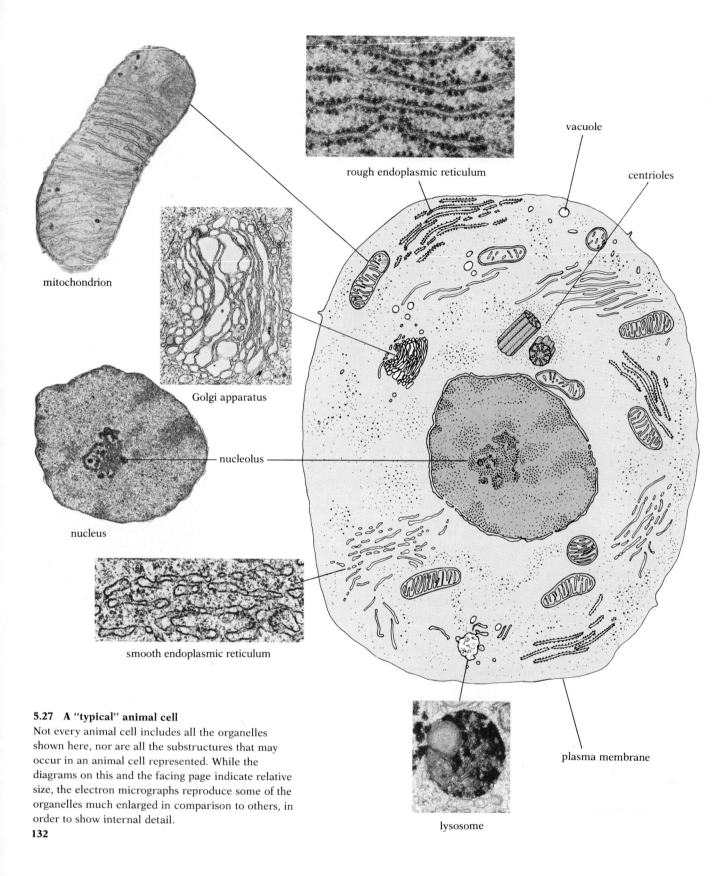

rough endoplasmic reticulum

vacuole

centrioles

mitochondrion

Golgi apparatus

nucleolus

nucleus

plasma membrane

smooth endoplasmic reticulum

lysosome

5.27 A "typical" animal cell
Not every animal cell includes all the organelles shown here, nor are all the substructures that may occur in an animal cell represented. While the diagrams on this and the facing page indicate relative size, the electron micrographs reproduce some of the organelles much enlarged in comparison to others, in order to show internal detail.

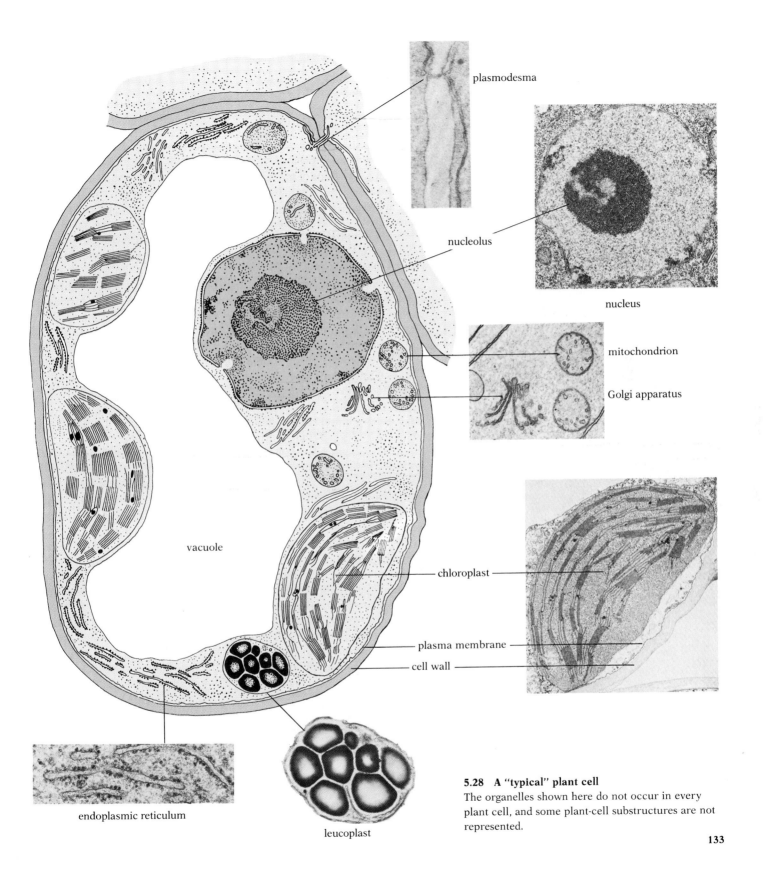

plasmodesma

nucleolus

nucleus

mitochondrion

Golgi apparatus

chloroplast

plasma membrane

cell wall

vacuole

endoplasmic reticulum

leucoplast

5.28   A "typical" plant cell
The organelles shown here do not occur in every plant cell, and some plant-cell substructures are not represented.

133

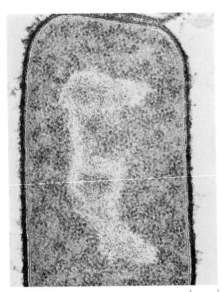

0.1 μm

5.29 Electron micrograph of part of a bacterial cell
The light area in the center of the cell, called the nucleoid, contains DNA, but is not bounded by a membrane. Note the prominent cell wall, with the plasma membrane visible just inside it.

PROCARYOTIC CELLS

Procaryotic cells lack most of the cytoplasmic organelles present in eucaryotic cells. We have already mentioned that they have no nuclear membrane; they also lack other membranous structures such as an endoplasmic reticulum, a Golgi apparatus, lysosomes, peroxisomes, and mitochondria (many of the functions of mitochondria are carried out by the inner surface of the plasma membrane). There is an exception, however, to the general rule that procaryotic cells lack intracellular membranous structures. Most photosynthetic bacteria contain chlorophyll, which is associated with membranous vesicles or lamellae; except in Cyanobacteria the lamellae are continuous with the cell membrane rather than being independent membrane-bounded plastids.

For a long time it was thought that procaryotic cells had no chromosomes. With the advent of the electron microscope, however, it became possible to detect in each procaryotic cell a nuclear region, or **nucleoid** (Fig. 5.29), containing a single large DNA molecule, which, though not tightly associated with proteins as DNA is in eucaryotic cells, may nonetheless be considered a chromosome. There may also be small independent pieces of DNA, called plasmids. Unlike eucaryotic chromosomes, which are usually linear, the procaryotic chromosome and plasmids are ordinarily circular (Fig. 5.30).

Like eucaryotic chromosomes, the procaryotic chromosome bears, in linear array, the genes that control both the hereditary traits of the cell and its ordinary activities. The DNA functions by directing protein synthesis on ribosomes via mRNA, in the way already described for eucaryotic cells. Note that ribosomes are the most prominent cytoplasmic organelles to occur in both eucaryotic and procaryotic cells. Those of procaryotic cells, however, are structurally different from those of eucaryotic cells, and they are somewhat smaller.

Some bacterial cells possess hairlike organelles used in swimming, and these have traditionally been called flagella. But in most procaryotic species these organelles do not have microtubules; their internal structure is completely different from that of eucaryotic flagella, and the mechanism of their movement must therefore also be different.

We have already (p. 109) contrasted the procaryotic cell walls with those of true plants and fungi. Suffice it to say here that the remarkable murein structure is unique to the major subgroup of procaryotes.

Table 5.1 gives a summary of some of the most important differences between procaryotic and eucaryotic cells, along with some material discussed in the next section.

THE ENDOSYMBIOTIC HYPOTHESIS

Though by no means unanimous, scientific opinion today is moving toward the view, presented most convincingly by Lynn Margulis of Boston University, that at least two organelles found only in eucaryotes—mitochondria and chloroplasts—are the descendants of procaryotic organisms that took up residence in the "hospitable" precursors of eucaryotes, often called urcaryotes. There are several lines of evidence for this so-called endosymbio-

TABLE 5.1   *A comparison of typical procaryotic cells, eucaryotic cells, and certain eucaryotic organelles*

Characteristic	Procaryotic cells	Eucaryotic cells	Mitochondria and chloroplasts
Size	1–10 μm	10–100 μm	1–10 μm
Nuclear membrane	Absent	Present	Absent
Chromosomes	Single, circular, contain no histone proteins	Multiple, linear, contain histone proteins	Single, circular, contain no histone proteins
Golgi apparatus	Absent	Present	Absent
Endoplasmic reticulum, lysosomes	Absent	Present	Absent
Mitochondria	Absent	Present	
Chlorophyll	Not in chloroplasts	In chloroplasts	
Ribosomes	Relatively small	Relatively large	Relatively small
Microtubules, microfilaments	Absent[a]	Present	Absent
Flagella	Lack 9–2 structure	Have 9–2 structure	
Cell wall	Contains murein	Lacks murein	Absent

[a] Microtubules have been reported in certain spirochete bacteria inhabiting the digestive tract of termites; these bacteria have true flagella.

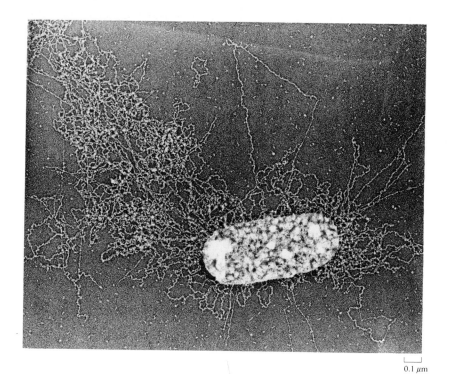

5.30   A disrupted cell of *Escherichia coli*
Exposure to detergent releases the DNA of the common intestinal bacterium *E. coli*, most of which can be seen outside the cell in this electron micrograph. Though not readily recognizable here, the main chromosome and the small accessory chromosomes of a bacterium each form a circle rather than individual strands in the cell. The mottled appearance of the surface of the bacterium results from the effects of alcohol, drying, and shadowing with platinum.

0.1 μm

tic hypothesis (the term is from *endo-*, "within," and *symbiosis*, "state of living together"); some of these are also summarized in Table 5.1.

1. Both mitochondria and chloroplasts contain their own ribosomes and their own chromosomes; the chromosomes code for their ribosomal RNA and ribosomal proteins, and for some (though not all) of their enzymes. Mitochondria and chloroplasts also build their own membranes.

2. The organelle chromosomes resemble those of procaryotes in that they are circular, not wound on special protein spools, and not enclosed by a nuclear membrane.

3. The internal organization of the organelle genes is similar to that of procaryotes, but very different from that of eucaryotes. We shall examine the details of this organization in later chapters.

4. The ribosomes of mitochondria and chloroplasts are more similar to those of procaryotes than to the ribosomes in the cytosol of the cells in which they live. Indeed, the large ribosomal subunit of the bacterium *Escherichia coli* and the large subunit of chloroplasts are so similar that they may be substituted for one another without affecting the hybrid ribosome's capacity to carry out protein synthesis.

5. Many present-day photosynthetic bacteria live inside eucaryotic hosts, providing them with food in return for shelter. Similarly, certain nonphotosynthetic bacteria live symbiotically within eucaryotes, extracting and sharing energy from foods that their hosts cannot themselves metabolize.

In summary, the endosymbiotic hypothesis assumes that an aerobic bacterium was captured through endocytosis about 1.5 billion years ago, resisted digestion, lived symbiotically inside its host, and divided independently of its host. Later, by a now-familiar genetic mechanism, some of the symbiont's genes were moved to the host nucleus, which took control from its guest. From this partnership all present-day plants, animals, fungi, and Protozoa must have evolved. Subsequently, according to this hypothesis, a photosynthetic bacterium (probably a cyanobacterium, also known as blue-green alga) met the same fate, the result being the unique evolutionary line we know as plants.

A much less well supported but thought-provoking speculation holds that basal bodies/centrioles and nematocysts (the devices by which jellyfish and similar creatures harpoon potential prey and careless swimmers) may also have had an endosymbiotic origin.

Chapter 6

MULTICELLULAR ORGANIZATION AND THE DIVERSITY OF ORGANISMS

Though single-celled organisms contribute roughly half of the earth's biomass (the total weight of all living things on earth), there are enormous benefits to being multicellular. Single cells cannot exploit the great advantages larger size gives an organism: the ability to capture or harvest smaller organisms efficiently, to move farther and faster, and so on. But large size cannot be achieved by simply increasing the size of a single-celled organism indefinitely. A cell must take in its nutrients and oxygen across the membrane. As a cell triples in volume, so does its need for nutrients and oxygen, and yet its membrane surface area does not even double. Since metabolic needs increase faster than the surface area through which they are satisfied, a point arrives at which the membrane can no longer support its contents. The need for efficient diffusion, therefore, puts a strict limit on the surface-to-volume ratio of a cell, and consequently limits cell size.

Many single-celled organisms are extraordinarily complex, for the one cell must do everything needed for survival. But in an assembly of cells, specialization is possible, and though simple aggregations of identical cells (like the 32-cell discs of some green algae) do exist in nature, the course of evolution has demonstrated that arrangements in which certain cells concentrate on particular functions (propulsion, feeding, reproduction, and so on) can be far more effective than those in which each cell pursues a "jack-of-all-trades" strategy.

The bodies of most multicellular organisms are organized on the basis of specialized tissues, organs, and systems. A *tissue* is composed of many cells, usually similar in both structure and function, bound together by intercellular material. An *organ*, in turn, is composed of various tissues (not necessarily similar) grouped together into a structural and functional unit. Similarly, a *system* is a group of interacting organs that cooperate as a functional complex in the life of the organism.

Before we survey the several kinds of tissues, organs, and systems (many of which will be examined in detail in later chapters), we must understand how the component cells manage to recognize each other, bind themselves together, and cooperate. Then we can turn to the tissues these recognition and adhesion processes make possible, and finally, in the last part of the chapter, we shall look briefly at the variety of organisms that have evolved, utilizing the cellular groups we know as tissues and organs.

CELLULAR ADHESION

There are two dramatically different strategies for giving form to muticellular aggregations, which would otherwise be amorphous lumps of cells. In certain animal tissues specialized cells called *fibroblasts* secrete fibrous proteins, among them collagen (see Fig. 3.23, p. 66) and elastin, which are components of an *intercellular matrix*. Cells are held in place by this structural network, and grow and function there. The resulting tissue, known as connective tissue, is described in detail later in this chapter.

The other strategy for providing shape and strength is for cells to adhere to one another. For this to take place the cells must specifically recognize which other cells are appropriate partners, and then secure their membranes to them. The mechanism of this recognition, a process especially critical during embryological development, when vast numbers of cells must be properly arranged, remains a mystery. There is some evidence that at least certain types of cells have characteristic carbohydrate markers on the lipids and proteins of their outer membranes that are recognized by specialized receptors on other cells. As we shall see when we discuss the immune system in a later chapter, several classes of cells have as their only function recognizing and attacking foreign cells that lack these molecular "passwords." In addition, a curious class of plant proteins, called lectins, recognize cells of specific tissue types on the basis of their carbohydrate identification tags. The role of lectins is unclear; apparently they help many plant cells to adhere to one another but there is also evidence that they act to glue together and thereby immobilize the bacterial and fungal cells that can otherwise cause plant diseases.

However cells locate each other, they frequently form strong junctions of several sorts. Most of the cellular attachment junctions utilized in multicellular organisms other than plants can be seen in the lining of the small intestine (Fig. 6.1). Intestinal cells are arranged with their actin-supported microvilli projecting into the intestine, where they absorb the nutrients released by digestion. They must adhere to one another not only to form the tubular channel of the gut but also to prevent the digestive enzymes (against which they are specifically armored) from leaking out and digesting the rest of the organism.

The general mechanical joining of two cells is accomplished by struc-

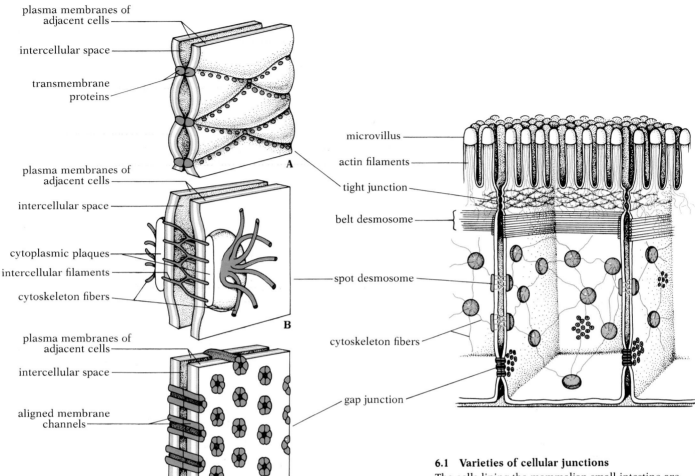

plasma membranes of adjacent cells

intercellular space

transmembrane proteins

A

plasma membranes of adjacent cells

intercellular space

cytoplasmic plaques
intercellular filaments
cytoskeleton fibers

B

plasma membranes of adjacent cells

intercellular space

aligned membrane channels

C

microvillus

actin filaments

tight junction

belt desmosome

spot desmosome

cytoskeleton fibers

gap junction

6.1 Varieties of cellular junctions
The cells lining the mammalian small intestine are attached by several types of specialized junctions. An example of each is shown in enlarged detail. (A) A tight junction is composed of rows of transmembrane proteins in adjacent cells that bind to each other. (B) A spot desmosome consists of a pair of cytoplasmic plaques, each just inside the cell membrane and connected to the other across the intermembrane space by specialized intercellular filaments. Each plaque is also attached to fibers of the cytoskeleton within its cell. (C) A gap junction is formed by a pair of membrane channels aligned and bound together to create a specialized pathway between cells.

tures known as *spot desmosomes*. A spot desmosome consists of two cytoplasmic plaques, one just inside each of two adjacent cells (Fig. 6.1B). The outside faces of the plaques are joined by intercellular filaments, acting like rivets, while the inside faces are firmly attached to the cellular cytoskeleton. There is a superficial resemblance between spot desmosomes and *belt desmosomes*, which are also composed of plaques and filaments (Fig. 6.1). Belt desmosomes, however, have no role in cell-to-cell adhesion. Instead, the circumferential plaque containing contractile fibers provides internal support for the cell.

Cells in the intestine are also joined by *tight junctions*, in which specific transmembrane proteins attach directly to their counterparts in the adjacent cell (Fig. 6.1A). The cells are thus drawn together in such intimate association that there is no intercellular space, and hence no possibility of leakage.

Finally, cells may be connected by *gap junctions*. A junction of this type is apparently formed by a pair of identical membrane channels in the two cells that line up with and bind to each other (Fig. 6.1C). The result is both mechanical strength and the ability to share certain particular substances between the cells. As we shall see in Chapter 18, gap junctions can connect two nerve cells electrically so that they behave as a single signaling element. Gap junctions are most common in developing tissue, and they may play a role in the arrangement and initial adhesion of cells.

The problems involved in achieving cellular adhesion and communication for most plant cells are very different from those for animal cells. In plants, rigid cell walls intervene between the plasma membranes of adjacent cells. Hence adhesion must be accomplished for the most part by relatively simple cross-linking between the polysaccharides of the cell walls. Effective adhesion and communication between plant cells is crucial, however, if water and inorganic nutrients are to be passed up from the roots, and the energy-rich products of photosynthesis are to be transmitted from the leaves to other parts of the plant. To accommodate this need, plant cell walls contain specialized openings—plasmodesmata—where the membranes of adjacent cells come into contact with each other. As we have seen (p. 109), some of these openings take the form of membrane-lined holes through which the cytoplasm of adjoining cells can mix directly. Others, more commonly known as pits, retain a double-membrane barrier; these play an important role in controlling the movement of both solutes and solvents between cells.

PLANT TISSUES

Plant tissues have been classified in a variety of ways by botanists. The system used here, while not necessarily better than other possible systems, is one of several acceptable ones. The lack of full agreement on any one classification springs from characteristics of the plant cells themselves. The different cell types are not perfectly distinct, but rather intergrade, and a given cell may even change from one type to another during the course of its life. Consequently the tissues formed from such cells may share structural and functional characteristics. Furthermore, plant tissues may contain cells of only one type, or they may be complex, containing a variety of cell types. In short, plant tissues cannot be fully characterized or distinguished on the basis of any single criterion such as structure, function, location, or mode of origin.

All plant tissues can be divided into two major categories: meristematic tissue and permanent tissue. *Meristematic tissues* are composed of immature cells and are regions of active cell division; *permanent tissues* are composed of more mature, differentiated cells. This distinction is not absolute, however, for some permanent tissues may revert to meristematic activity under certain conditions.

The permanent tissues fall into three subcategories: surface tissues, fundamental tissues, and vascular tissues. Each of these, in turn, contains several different tissue types. The classification used here can be summarized as follows:

I. Meristematic tissue
II. Permanent tissue
  A. Surface tissue
    1. Epidermis
    2. Periderm
  B. Fundamental tissue
    1. Parenchyma
    2. Collenchyma
    3. Sclerenchyma
    4. Endodermis
  C. Vascular tissue
    1. Xylem
    2. Phloem

It must be emphasized that this classification is based on the higher land plants—the vascular plants. It has little relevance for simpler plants like mosses and algae, in which multiple tissue types seldom occur.

MERISTEMATIC TISSUE

Meristematic tissues are composed of embryonic, undifferentiated cells capable of active cell division. Cell division occurs throughout the very early embryo, but as the young plant develops, many regions become specialized for other functions and cease playing a primary role in the production of new cells. Consequently cell division becomes restricted largely to certain undifferentiated tissues in localized regions; these tissues are the meristems.

It is hard to make any general statement on the characteristics of meristematic cells, because they show much variation. There is no such thing, in fact, as a typical meristematic cell. Nevertheless, we can say that these cells tend to be small, to have thin walls, and to be rich in cytoplasm (that is, to have only small vacuoles), and that meristematic tissues tend to lack intercellular spaces. New cells produced by a meristem are initially like those of the meristem itself, but as they grow and mature, their characteristics slowly change and they become differentiated as components of other tissues.

There are regions of meristematic tissue at the growing tips of roots and stems. These **apical meristems** are responsible for increase in length of the plant body. In many plants, there are also meristematic areas toward the periphery of the roots and stems, and these **lateral meristems** are responsible for increase in girth.

SURFACE TISSUE

As the term implies, surface tissues form the protective outer covering of the plant body. In young plants and adult plants that lack active lateral meristems the principal surface tissue of roots and stems is the *epidermis* (Fig. 6.2); epidermis is also the surface tissue of all leaves (Fig. 6.3). Often the epidermis is only one cell thick, though it may be thicker, as it is in some plants living in very dry habitats, where protection against water loss is critical.

Most epidermal cells have a very large vacuole and only a thin layer of cytoplasm. Often the outer and side walls of epidermal cells are thicker than the wall that faces the inside of the plant. Epidermal cells on the aerial parts of the plant often secrete a waxy, water-resistant *cuticle* on their outer surface; this, combined with the thick outer cell wall, aids in protecting against water loss, mechanical injury, and invasion by parasitic fungi. The irregularly shaped cells usually interlock tightly, like pieces of a puzzle, thus making the protective barrier even more complete.

Epidermal tissues of the aerial parts sometimes give rise to unicellular or multicellular hairs, spines, or glands. Some epidermal cells, particularly of the leaves, are specialized as guard cells and regulate the size of small holes in the epidermis through which gases can move into or out of the leaf (Fig. 6.3). Epidermal cells of the roots, which have no cuticle and function in water absorption, commonly bear long hairlike processes that greatly increase the total absorptive surface area (see Fig. 9.5, p. 229).

As the stems and roots of plants with active lateral meristems increase in diameter, the epidermis is slowly replaced by another surface tissue, the *periderm* (see Fig. 12.10, p. 304). This tissue constitutes the corky outer bark so characteristic of old trees. Functional cork cells are, in fact, dead; it is their waterproof cell walls that serve as the protective outer covering of the plant.

FUNDAMENTAL TISSUE

Most fundamental tissues are considered simple tissues, since each is usually composed of only one type of cell. Often these same types of cells also occur as components of the complex vascular tissues. The various fundamental tissues are not necessarily structurally similar, though they form from the same embryonic regions. They are often defined simply as those tissues that are neither surface tissues nor vascular tissues.

Parenchyma Parenchyma tissue occurs in roots, stems, and leaves. The parenchyma cells are relatively unspecialized, like those that make up almost the whole body of lower plants, and are relatively dormant. However, they retain the ability to break out of dormancy by beginning to divide actively, and to differentiate into specialized tissue—that is, they can take on meristematic activity or undergo further specialization, forming other cell types. Parenchyma cells usually have thin primary walls and no secondary walls. They generally have a large vacuole surrounded by a peripheral layer of cytoplasm. The cells are ordinarily loosely packed; consequently intercellular spaces are abundant in parenchyma tissue (Figs. 6.2 and 6.3). Most of the chloroplasts of leaves are in the cells of parenchyma tissue, and it is largely here that photosynthesis occurs. Parenchyma of stems and roots functions in storage of nutrients and water. When turgid, parenchyma is important in giving support and shape to the plant.

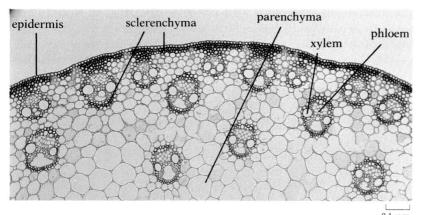

6.2 Photograph of portion of cross section of corn stem

0.1 mm

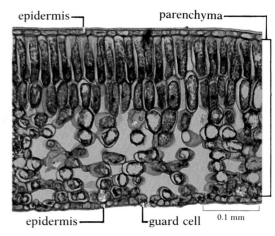

6.3 Photograph of cross section of ivy leaf

0.1 mm

Collenchyma Like parenchyma, collenchyma is a simple tissue whose cells remain alive during most of their functional existence. Though collenchyma cells are characteristically more elongate, they are structurally similar to parenchyma cells, except that their walls are irregularly thickened. The thickened areas are usually most prominent at the edges (the "corners" when viewed in cross section; Fig. 6.4). Collenchyma functions as an important supporting tissue in young plants, in the stems of non-woody older plants, and in leaves.

Sclerenchyma Sclerenchyma is a type of simple fundamental tissue that, like collenchyma, functions as support. However, sclerenchyma cells are far more specialized than collenchyma cells; at functional maturity most are dead, and their thick, hardened secondary walls give strength to the plant body. Often these walls are so thick that the lumen (internal space) of the cell has been nearly obliterated.

Sclerenchyma cells are customarily divided into two categories: fibers and sclereids (Fig. 6.5). *Fibers* are very elongate cells with tapered ends.

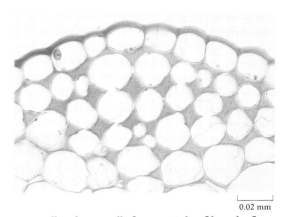

0.02 mm

6.4 Collenchyma cells from petiole of beet leaf
Notice the particularly thick walls at the corners of the cells.

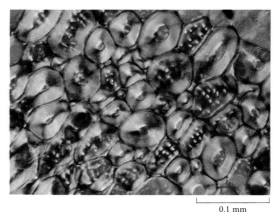

0.1 mm

6.5 Sclerenchyma
Left: Cross section of fibers from corn stem. Right: Stone cells of pear fruit.

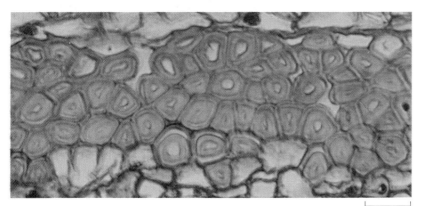

0.01 mm

They are tough and strong, but flexible; commercial flax and hemp are derived from strands of sclerenchyma fibers. *Sclereids* are of variable, often irregular, shape. The simpler, unbranched sclereids are frequently called stone cells; they are common in nutshells and the hard parts of seeds, and are scattered in the flesh of hard fruits. The gritty texture of pears, for instance, comes from small clusters of stone cells (Fig. 6.5).

Endodermis Endodermis is a type of tissue difficult to place in any classification. It occurs as a layer surrounding the vascular-tissue core of roots and, less frequently, of stems (see Fig. 9.4, p. 228). Young endodermal cells are much like elongate parenchyma cells, except that a band of chemically distinct thickening runs around each cell on its radial (side) and end walls. This reinforced, waterproof band is called the *Casparian strip*. In older endodermal cells the walls may become secondarily so thickened (sometimes almost obliterating the lumen) that the Casparian strip is obscured, but it can be detected by chemical tests. The cells of endodermal tissue occur in a single layer and are compactly arranged without intercellular spaces. Possible functions of the endodermis will be discussed in Chapter 9.

VASCULAR TISSUE

Vascular, or conductive, tissue is a distinctive feature of the higher plants, one that has made possible their extensive exploitation of the terrestrial environment. It incorporates cells that function as tubes or ducts through which water and numerous substances in solution move from one part of the plant body to another. There are two principal types of vascular tissue: xylem and phloem. Both of these are complex tissues—that is, they consist of more than one kind of cell.

Xylem Xylem is a vascular tissue that functions in the transport of water and dissolved substances upward in the plant body. It forms a continuous pathway running through the roots, the stem, and appendages of the stem such as leaves. In the flowering plants—the plants that contain an advanced form of xylem least like the ancestral form—the xylem cells are of two varieties: *tracheids* and *vessel elements*. The xylem of flowering plants also includes numerous parenchyma and sclerenchyma cells (mostly thick-walled fibers, but some sclereids as well). The parenchyma cells are the only living cells in mature functioning xylem: both the cytoplasm and the nuclei of tracheids, vessel elements, and sclerenchyma cells disintegrate at maturity, leaving the thick cell walls as the functional structures. In tracheids and vessels, the walls form passages or tubes in which vertical movement of materials can take place (see Figs. 12.11, 12.13, and 12.14, pp. 305–307).

Transport is not the only function of xylem. Another is support, particularly of the aerial parts of the plant. The numerous fibers in the xylem function almost exclusively in this way, and the thick-walled tracheids are also important as supportive elements. As a reminder of the enormous strength characteristic of xylem, keep in mind that its common name is wood.

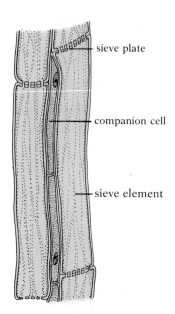

sieve plate

companion cell

sieve element

6.6 Longitudinal section of sieve elements and companion cells
The sieve elements lose their nuclei at maturity, but retain their cytoplasm.

Phloem   The second vascular tissue, phloem, is unlike xylem in that materials can move both up and down in it. Phloem functions particularly in the transport of organic materials such as carbohydrates and amino acids. For example, newly synthesized organic molecules are moved in the phloem from the leaves to the stem and roots for storage or to the growing points of the plant for immediate use. Like xylem, phloem is a complex tissue and contains both parenchyma and sclerenchyma cells in addition to those which are unique to it, in this instance sieve elements and companion cells. The *sieve elements* (Fig. 6.6) are the vertical transport units of phloem; at maturity their nuclei disintegrate, but their cytoplasm remains. ***Companion cells***, which retain both their nuclei and their cytoplasm at maturity, are closely associated with the sieve elements in the most advanced plants. Their possible function will be discussed in Chapter 12, where both xylem and phloem will be described in greater detail.

PLANT ORGANS

The body of the higher land plants is customarily divided into two major parts: the ***root*** and the ***shoot*** (Fig. 6.7). These can be distinguished on the basis of numerous morphological characteristics, especially the arrangement and mode of origin of the vascular tissue, the ways in which lateral roots and branch stems are formed, and the presence of leaves on the shoot but not on the root. Yet note that many of the tissues are essentially continuous throughout the entire axis, both root and shoot. For example, the vascular tissue of root and shoot, despite a somewhat different arrangement in each, forms an uninterrupted transport system.

The roots of a plant function mainly in procurement of inorganic nutrients such as minerals and water, in transport, in nutrient storage, and in anchoring the plant to the substrate.

The structurally somewhat more complex shoot consists of the stem and the appendages of the stem, particularly the foliage leaves and the reproductive organs. The stem, of course, functions in support and in internal transport, while the foliage leaves are organs in which the critical process of photosynthesis takes place.

You will notice that the number of distinct organs—root, stem, foliage leaf, and reproductive organs—is much smaller than it would be for higher animals. In general, the plant body is simply not so clearly subdivided into readily distinguishable functional components, or organs, as the animal body; the parts of the plant grade more imperceptibly into each other and, in some ways, form a more continuous whole. The arrangement of the tissues within these organs will be described in some detail in later chapters.

ANIMAL TISSUES

Animal tissues are traditionally divided into four categories: epithelium, connective tissue, muscle, and nerve. Each of these, and particularly connective tissue, is a diverse assemblage containing numerous subtypes. The subtype classification is based primarily on vertebrates (especially humans), and attempts to apply it to other animals, particularly the lower

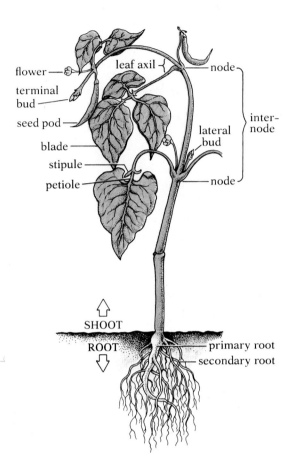

6.7   Diagram of flowering plant body
The vascular tissue (not shown) is continuous through all parts of the plant.

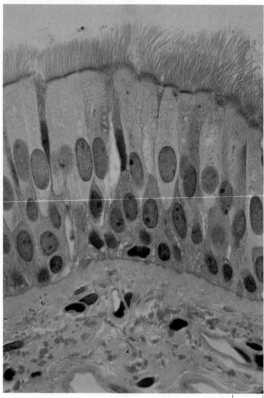

0.01 mm

6.8   Micrograph of pseudostratified columnar epithelium from human trachea

The dark basal bodies of the cilia and the long cones of fibrous rootlets are plainly visible. Note the basement membrane on which the epithelial cells rest.

invertebrates, are often not very useful. The classification of animal tissues employed here can be summarized as follows:

I. Epithelium
   A. Simple epithelium
      1. Squamous
      2. Cuboidal
      3. Columnar
   B. Stratified epithelium
      1. Stratified squamous
      2. Stratified cuboidal
      3. Stratified columnar
II. Connective tissue
   A. Vascular tissue
      1. Blood
      2. Lymph
   B. Connective tissue proper
      1. Loose connective tissue
      2. Dense connective tissue
   C. Cartilage
   D. Bone
III. Muscle
   A. Skeletal muscle
   B. Smooth muscle
   C. Cardiac muscle
IV. Nerve

EPITHELIUM

Epithelial tissue forms the covering or lining of all free body surfaces, both external and internal. The outer portion of the skin, for example, is epithelium, as are the linings of the digestive tract, the lungs, the blood vessels, the various ducts, the body cavity, and so on. Epithelial cells are packed tightly together, with only a small amount of cementing material between them and almost no intercellular spaces. Thus they provide a continuous barrier protecting the underlying cells from the external medium. Because anything entering or leaving the body must cross at least one layer of epithelium, the permeability characteristics of the cells of the various epithelia play an exceedingly important role in regulating the exchange of materials between different parts of the body and between the body and the external environment.

Since one surface of an epithelium is generally exposed to air or fluid and the opposite surface rests upon other cell layers, and since the epithelium plays a crucial part in the directional passage of materials, it is no surprise that epithelial cells should show significant differences between their free ends and their attached ends. Often highly specialized, the free ends commonly bear cilia, hairs, or short fingerlike processes; they may also have deep depressions and are sometimes covered with waxy or mucous secretions (Fig. 6.8). Inside the cells, pigments and such organelles as mitochondria are often more abundant at one end than at the other. Recent studies have shown that, as might be expected, the plasma membrane of an epithelial cell is not uniform in its permeability characteristics; the portion

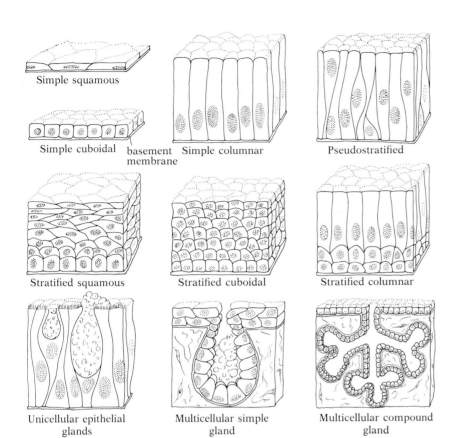

Simple squamous

Simple cuboidal   basement   Simple columnar
                   membrane

Pseudostratified

Stratified squamous

Stratified cuboidal

Stratified columnar

Unicellular epithelial
glands

Multicellular simple
gland

Multicellular compound
gland

6.9   Epithelial tissues

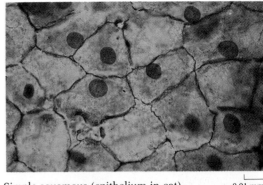

Simple squamous (epithelium in cat)   0.01 mm

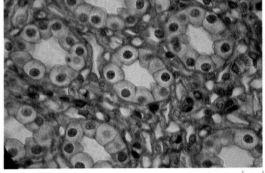

Simple cuboidal (in tubules   0.01 mm
of cat kidney)

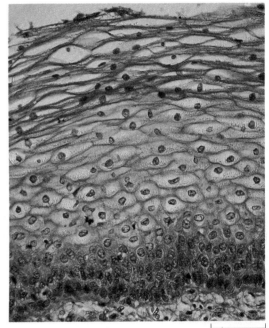

Stratified squamous   0.05 mm
(in human vagina)

6.10   Photographs of epithelial tissues

of the membrane on the outer surface of the cell, that exposed to the extra-cellular environment, is quite different from the portions of the membrane adjacent to other epithelial cells. When a fluorescent sodium dye is injected into a single epithelial cell, it can be detected within a few minutes in neighboring cells, but it does not appear in the external medium. Apparently the sodium dye can move across the plasma membrane at the junctions between epithelial cells, but cannot penetrate the plasma membrane where it is in contact with the external medium. Clearly, the chemical and/or physical properties of the membrane are not the same at those two locations.

Epithelial cells are usually divided into three categories: squamous, cuboidal, and columnar (Figs. 6.9 and 6.10). *Squamous cells* are much broader than they are thick and have the appearance of thin flat plates. *Cuboidal cells* are roughly as thick as they are wide and, as their name implies, have a rather square or cuboidal shape when viewed in a section perpendicular to the tissue surface; in surface view, however, they look like polygons, often with six sides. *Columnar cells* are much thicker than they are wide and, in vertical section, look like rectangles set on end.

Epithelial tissue may be only one cell thick, in which case it is called *simple epithelium*, or it may be two or more cells thick and is then known as *stratified epithelium*. There is, in addition, a third category, called *pseudostratified epithelium*, in which the tissue looks stratified, but actually is not; whereas in true stratified epithelium only the cells in the lowest layer are in contact with the underlying membrane, in pseudostratified epithelium all

6.11 Photographs of gland tissues

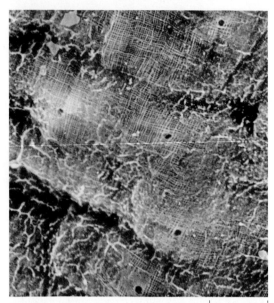

10 μm

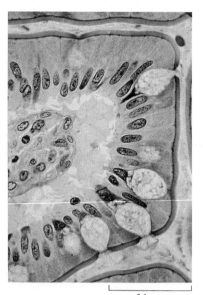

0.1 mm

Simple columnar, with unicellular
glands (in human small intestine)

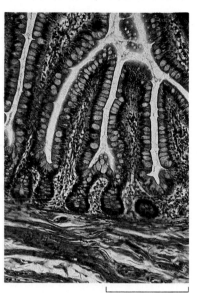

0.1 mm

Multicellular compound glands (in
villi of small intestine)

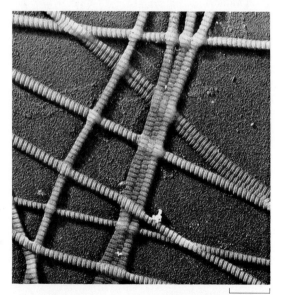

0.5 μm

6.12 Electron micrographs of collagen
Top: Scanning EM of the network of collagen fibers
in the skin of an earthworm. Bottom: Higher
magnification transmission EM of collagen fibrils
from calf skin.

the cells are in contact with it. The various types of epithelia are named on the basis of cell type and number of cell layers; in stratified epithelia the cells of the outermost layer determine the name. Epithelial tissue may be simple squamous, simple cuboidal, simple columnar, stratified squamous, stratified cuboidal, stratified columnar, and so on. Regardless of type, epithelium is usually separated from the underlying tissue by an extracellular **basement membrane** containing collagen fibers (Fig. 6.9).

Epithelial cells often become specialized as gland cells, secreting substances such as sweat, body oil, and mucus at the epithelial surface (Figs. 6.9 and 6.11). Sometimes a portion of the epithelial tissue becomes invaginated, and a multicellular gland is formed.

CONNECTIVE TISSUE

In connective tissue the cells are always embedded in an extensive intercellular matrix. Much of the total volume of connective tissue is matrix, the cells themselves often being widely separated. The matrix may be liquid, semisolid, or solid. Connective tissue is often divided into four main types: (1) blood and lymph, or vascular tissue; (2) connective tissue proper; (3) cartilage; (4) bone. The last three are sometimes collectively described as supporting tissues.

Blood and lymph Blood and lymph are rather atypical connective tissues with liquid matrixes. They will be discussed in some detail in Chapter 13.

Connective tissue proper Connective tissue proper is very variable, but its intercellular matrix always contains numerous fibers. These fibers are of three types:

Collagen fibers (or white fibers), which are very common, are composed of numerous fine fibrils of collagen, a protein that constitutes a very high percentage of the total protein in the animal body (Fig. 6.12). Such fibers are flexible, but resist stretching and confer considerable strength on the tissues containing them.

Elastic fibers (or yellow fibers) can, as their name implies, easily be stretched. When the stretching force ceases, the fibers return to their former length. Elastic fibers are often much thinner than collagen fibers. They are composed of the protein elastin.

Reticular fibers, as the term "reticular" indicates, branch and interlace to form complex networks. They are important at points where connective tissues and other tissues join, particularly in the basement membrane between epithelium and connective tissue.

Several kinds of cells are generally found in connective tissue proper (Fig. 6.13). They perform a variety of functions:

1. Fibroblasts secrete the proteins from which fibers form.

2. Macrophages, irregularly shaped cells particularly common near blood vessels, become mobilized when there is an inflammation. They can move by amoeboid motion and actively engulf particles such as dead red blood corpuscles and such foreign material as bacteria.

3. Mast cells produce a substance (heparin) that tends to prevent blood clotting and another substance (histamine) that increases the permeability of blood capillaries.

4. Fat cells are cells highly specialized for fat storage. When they are very numerous in a region of connective tissue, the tissue is often called adipose tissue.

5. The various kinds of white blood cells help fight infection. Some can move fairly easily between the blood or lymph and the connective tissue proper—a clear demonstration of the close interrelationship between these tissues.

Both cells and fibers are embedded in a rather amorphous ground substance, which is a mixture of water, proteins, carbohydrates, and lipids. Associated with the ground substance is the *tissue fluid*, a liquid derived from the blood.

Connective tissue proper is customarily subdivided into two basic types —loose connective tissue and dense connective tissue—though there is no rigid separation between them, and intermediate types sometimes occur.

Loose connective tissue is characterized by the loose, irregular arrangement of its fibers, the large amount of ground substance, and the presence of numerous cells of a variety of types (Fig. 6.13). It is very widely distributed in the animal body, no microscopic section of which is free of it. It has been said that if all other tissues were destroyed, the loose connective tissue alone would still show the exact contours of the body and the detailed shape of most internal organs. Much of the framework of the lymph glands, bone marrow, and liver is loose connective tissue; and loose connective tissue supports, surrounds, and connects the elements of all other tissues. For example, it binds muscle fibers together; attaches the skin to underlying tissues; forms the membranes that line the heart and abdominal cavities; forms the membranes, called mesenteries, that suspend the internal organs in their proper position and the membranes that bind together the parts of internal organs or that bind various organs together; functions as packing

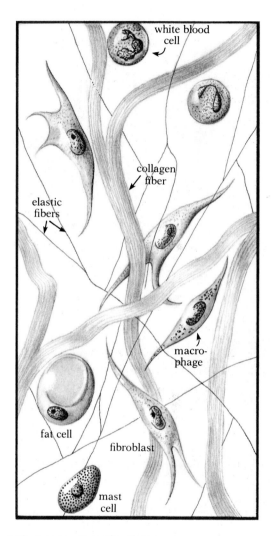

6.13 Loose connective tissue
The several varieties of cells are embedded in an extensive intercellular matrix of fibers and ground substance.

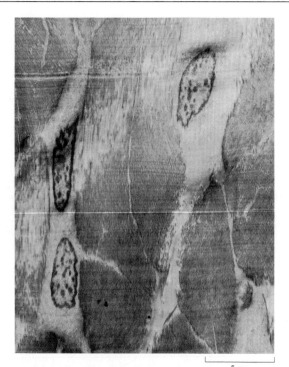

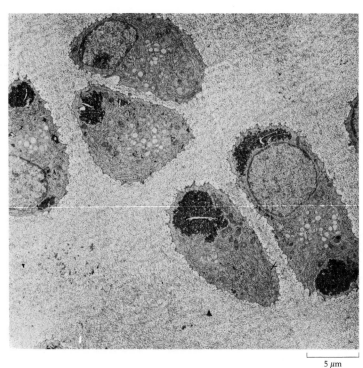

5 µm

5 µm

6.14 Photographs of tendon and cartilage
Left: Longitudinal section of tendon from the tail of a young rat, showing three cells surrounded by bundles of collagen fibrils. Right: Section of cartilage from a newborn mouse, showing five cells embedded in an intercellular matrix; the matrix looks almost homogeneous, but actually contains a dense network of very thin fibrils.

material in the spaces between organs; and forms a thin sheath around blood vessels, consequently penetrating with them into the interior of most organs. Because of its flexibility, loose connective tissue allows movement between the units it binds or connects.

Dense connective tissue is characterized by the compact arrangement of its many fibers, the limited amount of ground substance, and the relatively small number of cells. The fibers may be irregularly arranged into an interlacing network, as in the dermis of the skin or the sheaths (periostea) of bone; or they may be arranged in a definite pattern—usually parallel bundles oriented to withstand tension from one direction, as in tendons connecting muscle to bone (Fig. 6.14, left) or ligaments connecting bone to bone.

Cartilage   Cartilage (gristle) is a specialized form of dense fibrous connective tissue in which the intercellular matrix has a rubbery consistency (Fig. 6.14, right). The relatively few cells are located in cavities in the matrix. Cartilage can support great weight, yet it is often flexible. It can vary in texture, color, and elasticity.

Cartilage is found in the human body in such places as the nose and ears (where it forms pliable supports), the larynx and trachea (you can feel the rings of cartilage in the front of your throat), intervertebral discs, surfaces of skeletal joints, and ends of ribs. Most of the skeleton of the early vertebrate embryo is composed of cartilage; the developing bones follow this model and slowly replace it. Some vertebrate groups, the sharks, for example, retain a cartilaginous skeleton even in the adult.

Bone Bone has a hard, relatively rigid matrix. This matrix, which contains numerous collagen fibers and a surprising amount of water, is impregnated with inorganic salts such as calcium carbonate and calcium phosphate. This inorganic material may constitute as much as 65 percent of the dry weight of an adult bone. The few bone cells are widely separated and are located in spaces in the matrix (see Fig. 20.7, p. 532). We shall discuss the histology of bone in more detail in Chapter 20.

MUSCLE

The cells of muscle have greater capacity for contraction than most other cells, though all protoplasm probably possesses this capacity to some extent. Muscles are responsible for most movement in higher animals. The individual muscle cells are usually elongate and are bound together into sheets or bundles by connective tissue. Three principal types of muscle tissue (see Fig. 20.9, p. 534) are recognized in vertebrates: (1) skeletal or striated muscle, which is responsible for most voluntary movement; (2) smooth muscle, which is involved in most involuntary movements of internal organs; and (3) cardiac muscle, the tissue of which the heart is composed. In invertebrates one or more of the three muscle types may be missing entirely, and the distribution of the types within the body is nearly always different from that outlined here for vertebrates.

NERVE

To some extent, all protoplasm possesses the property of irritability—the ability to respond to stimuli—but nerve tissue is highly specialized for such responses. Nerve cells are easily stimulated and can transmit impulses very rapidly. Each cell is composed of a cell body, containing the nucleus, and one or more long thin extensions called fibers (Fig. 6.15). Nerve cells are thus admirably suited to serve as conductors of messages over long distances. An individual nerve cell may be a meter long, or even longer; no other kind of cell even approaches such length. Many nerve fibers bound together by connective tissue constitute a nerve.

The functional combination of nerve and muscle tissue is important for all multicellular animals except sponges. It is these tissues that confer on animals their characteristic ability to move rapidly in response to stimuli. We shall examine the functioning of nerves and muscles in detail in Chapters 18–20.

ANIMAL ORGANS

The bodies of the simpler multicellular animals generally show few clearly distinct organs, but most larger, more advanced animals characteristically have numerous organs, which in turn are organized into the functional complexes we call organ systems. Structural organization at the organ and organ-system levels is far more advanced in the higher animals than in the higher plants, a difference that probably reflects the different modes of life pursued by these two major groups of organisms.

An example of the complex integration of different types of cells and tis-

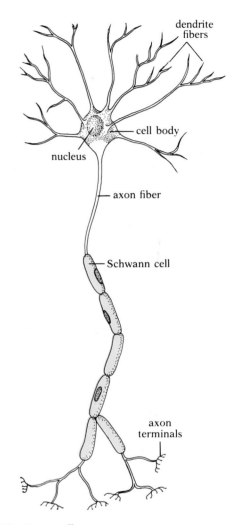

6.15 Nerve cell
The dendrites carry impulses toward the cell body; the axon carries impulses away from the cell body. The axons of vertebrates are often sheathed along much of their length by Schwann cells.

sues to form an animal organ is human skin (Fig. 6.16). It is clear from the illustration that skin is far more complex than a first impression might suggest. It contains elements of all four primary animal tissue types: epithelium, connective tissue, muscle, and nerve. Portions of these tissues, in turn, are organized into relatively complex structures like glands, ducts, hairs, blood vessels, and sensory devices. All these structural elements are integrated to form the functional organ. In this case the organ functions in many ways: as a protective covering for the body that resists penetration by many harmful substances and disease-producing organisms, and helps prevent excessive water loss; as a mechanism for excreting several different waste materials; as an aid in regulating the temperature of the body; as an instrument whose sensory nerve endings receive impulses from the outside environment; and as a depot in which reserve nutrients are stored.

The numerous organs, in addition to skin, that together constitute the body of a higher animal will be discussed in later chapters. They are commonly grouped into organ systems, of which the following will be of particular concern to us:

1. The digestive system, which functions in procuring and processing nutrients;
2. The respiratory system, functioning in the gas-exchange process by which oxygen is taken into the body and waste carbon dioxide released;
3. The circulatory system, the internal transport system of animals;
4. The excretory system, which not only functions in the release of certain metabolic wastes from the body, but acts as a critical regulator of the chemical makeup of the body fluids;
5. The endocrine system, whose glands, and the hormones they produce, play an important role in internal control;
6. The nervous system, a control system essential in coordinating the myriad functions of a complex multicellular animal;
7. The skeletal system, which provides support and determines shape in some animals;
8. The muscular system, of fundamental importance in the movement of animals;
9. The reproductive system, which functions in the production of new individuals.

THE DIVERSITY OF LIFE

In this chapter we have looked at how the cells of multicellular organisms differentiate into specialized tissues, each filling an essential role in the organism as a whole. Specialization can also be seen in the diversity of species in the world. As a result of natural selection, many species have differentiated and now exploit with great efficiency an enormous variety of habitats and food sources.

According to the fossil record, the most primitive organisms known—the bacteria—date back over 3 billion years. The first land plants and insects emerged over 400 million years ago, and the first birds and mammals over 180 million years ago. Since the simplest forms of life arose, innumerable

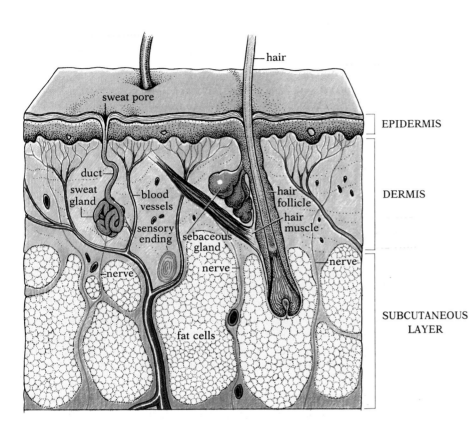

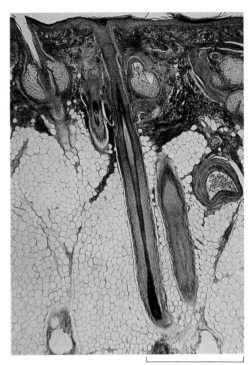

0.1 mm

6.16 Human skin in cross section

The skin shown in the photograph is from the human scalp. The outer portion of the skin, the epidermis, is composed of stratified squamous epithelial tissue. (Individual epithelial cells are not visible.) The outermost layer (*stratum corneum*) consists of hardened dead cells that are constantly being sloughed off. Active cell division in the deeper layers of the epidermis produces new cells that are pushed outward and take the place of those that are lost. Beneath the epidermis is a layer, called the dermis, composed chiefly of connective tissue (stained blue in the photograph). Blood vessels penetrate into the dermis but not into the epidermis. Sweat glands are embedded in the deeper layers of the dermis, and their ducts push outward through both dermis and epidermis to open onto the surface of the skin through sweat pores. Both the glands and their ducts are derived from the epidermis; they form initially as invaginations of the epidermis that push downward into the connective tissue of the dermis.

Hairs, and the inner layers of the hair follicles in which they are encased, are also derived from the epidermis and also develop as invaginations downward. When the hair follicle is fully developed, the bulbous base of the follicle and the hair root lie in the subcutaneous layer; the shaft of the hair extends at a slant from the root to the surface of the skin and beyond. A small muscle runs diagonally from the upper portion of the dermis to the hair follicle near its lower end; when this muscle contracts, it pulls the hair erect. One or more sebaceous (oil) glands empty into the hair follicle.

Numerous nerves penetrate into the dermis, and a few even penetrate into the epidermis. Among them are nerves to the hair muscles, sweat glands, and blood vessels, and also nerves terminating in the sensory structures for detecting touch, temperature, and pain.

Beneath the dermis, and not sharply delimited from it, is the subcutaneous layer, which is not considered a part of the skin itself. This is a layer of very loose connective tissue, usually with abundant fat cells. It is this layer that binds the skin to the body. The extent and form of its development determine the amount of possible skin movement.

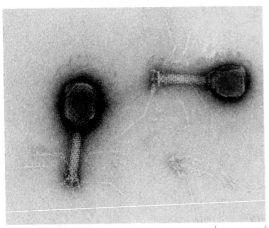

0.1 μm

6.17 T4 virus

T4 is one of many viruses that infect the common intestinal bacterium *E. coli*. Its six "legs" are equipped with proteins that recognize and bind to the host cell, bringing the base of the viral "tail" into contact with the bacterium. The tail contracts and the viral chromosome that fills the "head" is injected through the tail into the host. Once inside, the viral DNA exploits the bacterium's own enzymes and synthetic machinery to take over the cell, orchestrating the production of dozens or even hundreds of new virus particles.

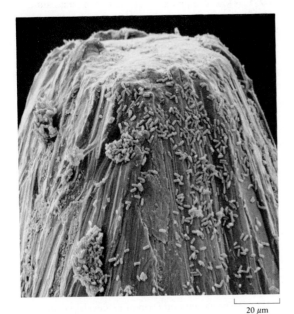

20 μm

6.18 Electron micrograph of *E. coli* bacteria on the point of a pin.

different kinds of organisms, increasingly complex and adapted to widely varying environments, have evolved. More than 2 million living species are recognized, and the best estimates put the number of extinct species at about 200 million. As we continue to explore the mechanisms and the evolution of life, we shall, of necessity, focus on only a tiny subset of "representative" species. The survey that follows will look at how these representatives are related to one another and to the hundreds of thousands of species we cannot explicitly discuss.

Living things are classified according to their probable evolutionary relationships. Though it is impossible to reconstruct the 3 billion years of evolution with great precision, a variety of techniques can be combined to provide a coherent and satisfactory overall picture. For instance, the course of evolution can be worked out by examining anatomical similarities and differences among organisms, or by charting the similarities and differences in the amino acid sequences of particular enzymes in various species. It is reassuring that the conclusions obtained through methods like these tend to support and reinforce each other. The grouping of organisms that appears most accurately to reflect the evolutionary relationships among living things is the five-kingdom system; it recognizes five broad categories: Monera, Protista, Plantae, Fungi, and Animalia. We shall refer to these major groups again and again in the course of this text, as we discuss the basic problems faced by all living things and the different adaptations these have evolved in response to them. The groups mentioned here, and many others, will be described in far greater detail in Part V.

One general category of enormous importance is omitted from this survey, that of the viruses and similar entities (Fig. 6.17). These encapsulated bits of nucleic acid are omitted from the classification system because they are not thought to represent an evolutionarily coherent line. Rather, the viruses are believed to have originated from the various organisms they now parasitize, and hence not to be related to one another. And as we shall see later, it is not even easy to decide whether, technically, viruses are living things, since they lack the machinery necessary for metabolism.

MONERA

The Monera—known more commonly as bacteria—are procaryotes. They differ from the members of the other four kingdoms—collectively known as the eucaryotes—in that their cells lack a membrane-enclosed nucleus, as well as other subcellular membranous structures like the mitochondria that are present in the cells of all other types of organisms. Most monerans, like plants, have strong cell walls.

Bacteria occupy a remarkable range of habitats: they live in the Antarctic ice, in the sands of the Sahara, and even in the boiling water of geysers and suboceanic fissures. Their numbers are enormous: large quantities of *Escherichia coli*, a bacterium that inhabits the digestive tracts of animals, are found in every gram of sand or soil on the earth (Fig. 6.18).

Eubacteria The Eubacteria are single-celled organisms visible only under high magnification (Fig. 6.19); not until the electron microscope was developed could their internal structure be elucidated. At least six broad

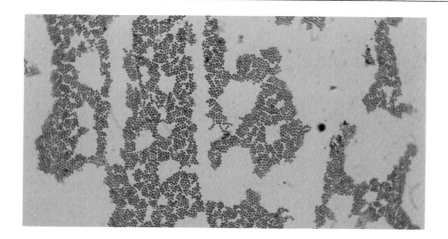

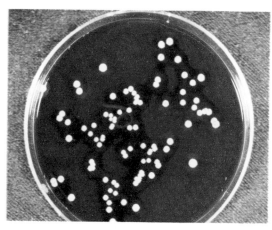

subgroups are generally recognized: the Gram-positive bacteria (so designated because a characteristic component in their cell walls reacts with a particular stain—Gram's solution), which include such disease-producing organisms as *Staphylococcus* and *Streptococcus;* the purple bacteria, some of which have an unusual photosynthetic pigment, while others, like *E. coli* (Fig. 6.20A), now lack all photosynthetic ability, and which are generally thought to have been the precursors of eucaryotic mitochondria; the Cyanobacteria, often called blue-green algae[1] (Fig. 6.20B), which have the same chlorophyll as plants and are believed to have been the precursors of eucaryotic chloroplasts; the spirochetes, which alone of the Monera have basal bodies and true flagella, and may have been the source of these organelles in eucaryotes; the Prochlorophyta; and the green photosynthetic bacteria.

[1] Since most algae have been found to be eucaryotes, the traditional term "blue-green algae" is now generally considered misleading; we shall refer to these organisms as Cyanobacteria.

6.19   The bacterium *Staphylococcus aureus*
Left: High magnification (here × 930) is needed—and the staining is helpful—in making this bacterium detectable by the eye. It is an agent for many infections, including boils and abscesses. Right: The visible colonies that *S. aureus* forms on a blood agar medium each contain many thousands of cells.

6.20   Representative Eubacteria
(A) The intestinal bacterium *E. coli.* Clearly visible in this electron micrograph are the stiff, helical flagella, which rotate like propellers to push the bacterium.
(B) Cyanobacteria (*Chroococcus*) from the Pine Barrens of New Jersey. Small groups of cells are enclosed within common gelatinous sheaths.

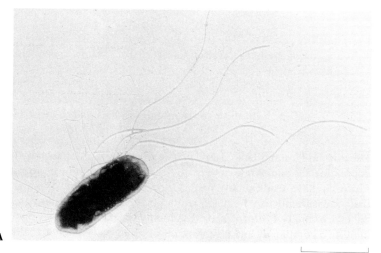

A

1 μm

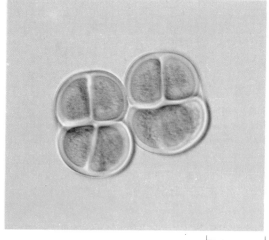

B

10 μm

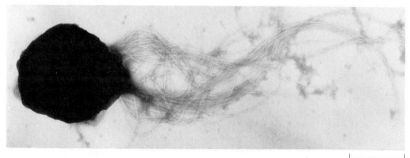

0.5 μm

6.21   Scanning electron micrograph of a thermoacidophilic archaebacterium
This chemosynthetic bacterium grows in the hot water of a submarine hydrothermal vent in the eastern Pacific.

Archaebacteria   The Archaebacteria form the other main division of the Monera. They include three main groups: the methanogens, which obtain energy by converting carbon dioxide into methane in bogs, stagnant ponds, and other anaerobic habitats; the halophiles, which live in extremely salty environments and photosynthesize by means of a pigment very like the visual pigment of vertebrates; and the thermoacidophiles, which inhabit extremely hot environments (Fig. 6.21). These three groups of hardy monerans were formerly classified as ordinary bacteria, but recent studies of their cell walls, cell membranes, gene structure, and biochemistry make it clear that they are very different from the Eubacteria. In fact, the differences already discovered are so great that Archaebacteria may deserve a phylogenetic kingdom of their own.

PROTISTA

The kingdom Protista includes a variety of groups whose members are predominantly unicellular, sometimes colonial, only rarely (and on a very rudimentary level) multicellular. Like true plants and animals, and unlike the unicellular Monera, protists have cells that contain a membrane-bounded nucleus and other subcellular membranous structures.

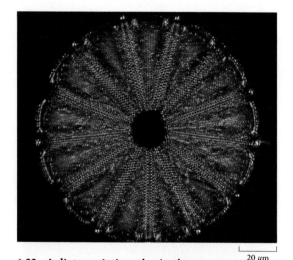

6.22   A diatom, *Actinosphenia elegans*   20 μm
The diatoms are single-celled organisms common both in freshwater habitats and in the oceans. They characteristically have elaborately ornamented glasslike cell walls.

The plantlike protists   The members of some protistan groups—the euglenoids, dinoflagellates, diatoms (Fig. 6.22), yellow-green algae, and golden-brown algae—have chlorophyll and are photosynthetic. In older classifications they are identified as primitive plants. Because they are indeed plantlike in many respects, it is still common practice to use the term "plant" to include both the photosynthetic protists and the true plants.

The animal-like protists   Many protistan groups lack chlorophyll and are not photosynthetic. Some of these are funguslike, but others, known as the **Protozoa**, have traditionally been viewed as unicellular animals, and we shall occasionally use the term "animal" to include both these groups and the true animals. Protozoa are often highly specialized, their single cell exhibiting a complexity and a separation of functions analogous to those observable in multicellular animals.

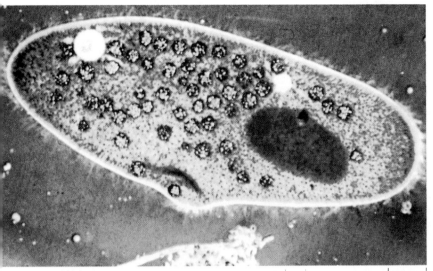

20 μm

6.23 A ciliate protozoan, *Paramecium caudatum*
The colors seen here, which were produced by optical staining with the interference microscope, help differentiate the component parts of this living specimen; they are not the organism's natural colors. Ciliate protozoans are unusual in having two kinds of nuclei—as seen here, a large brown structure with a small darker one overlapping it. The yellow cilia, which seem to edge the cell, actually cover all of it. The numerous small green circles are food vacuoles, chambers here containing yeast on which the protozoan has been feeding. The two white circles with radiating canals are contractile vacuoles, whose main function is eliminating water from the cell.

Protozoa are generally much larger than bacteria. They are very mobile, swimming rapidly in the water in which they live or crawling along the bottom or on submerged objects. Some propel themselves by the whiplike motion of their long hairlike flagella. Others, like *Paramecium* (Fig. 6.23), bear many cilia, the shorter hairlike structures that often function in both locomotion and feeding. Still others, like *Amoeba* and *Pelomyxa* (Fig. 6.24), have neither flagella nor cilia, but move by a complex flowing, as the cell, constantly changing shape, sends out extensions into which the rest of the cell contents flows. We shall refer to representative protozoans many times as we examine their fascinating evolutionary adaptations.

PLANTAE

Common to all plant groups are cells with rigid walls and subcellular membrane-bounded structures containing chlorophyll, the green pigment essential for photosynthesis. Thus plants, unlike fungi and animals, can themselves synthesize the high-energy compounds they need for maintenance and growth.

Brown algae and red algae   Both the brown algae and the red algae are primarily marine and are commonly known as seaweeds. They are especially prevalent in the intertidal zone along rocky coasts, where they can easily be observed at low tide. Their color results from brown and red pigments, which often mask the chlorophyll these algae also contain (Figs.

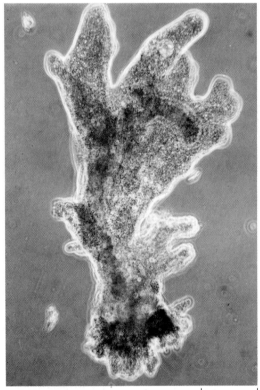

0.1 mm

6.24 An amoeboid protozoan, *Pelomyxa carolinensis*
"Amoeboid"—from the Greek word for "change"—describes a cell that can alter its shape as it thrusts out or withdraws many armlike extensions.

6.25   A brown alga, *Saccorhiza polyschides*
Brownish pigments mask the green pigment
chlorophyll also present in this seaweed. The
flattened leaflike blades are supported by the water
in which the plant is growing; when uncovered at
low tide, the blades lie flat on the substrate. Notice
the holdfasts, which anchor this alga to the substrate.

6.26   A red alga, *Rhodymenia palmata*
Like the brown algae, the red algae are mostly
seaweeds, but they generally grow at greater depths
than the brown algae.

6.25 and 6.26). They are always multicellular, but show relatively little of
the differentiation into distinct tissues found in the higher land plants.
Though some brown algae do show tissue differentiation and many other
characteristics recalling the higher land plants, it is believed that they and
the higher land plants evolved independently and are not closely related.
Indeed, even the brown and red algae are thought not to be closely related,
despite many superficial similarities.

Green algae   Green algae are relatively simple plants that live only in the
water or in very moist environments on land. Some are unicellular, others
multicellular. In multicellular green algae, most of the cells are similar to
one another and not highly specialized; they form what might be considered
a single continuous tissue, which is the plant body (Fig. 6.27). The plants are
green because their chlorophyll is not masked by accessory pigments. In
this and many other biochemical characteristics the green algae are similar
to the higher land plants, and most botanists agree that it was probably
from ancestral flagellated green algae that the land-plant groups arose.

Mosses, liverworts, and their relatives   Though these plants live on land,
they are not entirely independent of the ancestral aquatic environment;

thus they occur only in very moist habitats (Fig. 6.28). One reason is that parts of their reproductive cycle are dependent on abundant moisture. Another is that, unlike the vascular plants, they have not evolved an efficient internal transport system through which water from the soil can be carried to all parts of the plant.

Vascular plants   Of all members of the plant kingdom the vascular plants show the greatest internal specialization into tissues and organs—root stems, leaves, and reproductive organs (cones, flowers). Because they possess vascular tissue, they are less dependent than the other plant groups on abundant water directly available from the surrounding environment. They are the dominant plant group on land today.

The vascular-plant group is usually subdivided into several sections, three of which are probably familiar: the ferns, the conifers and their allies (gymnosperms), and the flowering plants (angiosperms). Of these three groups, the ferns are the most primitive; they appeared on the ancient earth before the other two groups, dominated the land for a long time, but eventually gave way to the other groups and are now largely overshadowed by them.

The gymnosperms and flowering plants are known collectively as the *seed plants*; they are more highly specialized for a terrestrial existence than the ferns. Some common gymnosperms are pine, cedar, spruce, fir, and hemlock; all of these bear cones and have needlelike leaves, though not all gymnosperms do.

The angiosperms, or flowering plants, are the most advanced of the three groups. The majority of the land plants familiar to you belong to this group, which includes plants of every shape and size from grasses to cacti, and

6.27   A marine green alga, *Cladophora rupestris*
The green algae include many unicellular and colonial forms. This most common species of multicellular green alga exhibits a branched and filamentous structure. Some less common multicellular algae are bladelike in form.

6.28   Liverworts, *Marchantia polymorpha*
Growing on a damp forest floor in England, these liverworts bear reproductive structures known as archegonia in receptacles at the ends of their stalks.

6.29 A monocot, *Lilium uchida* (left), and a dicot, *Rosa rubrifolia* (right)
Left: The garden lily has six petals. The leaves have parallel veins. Right: The wild rose has five petals and leaves with a network of veins.

from tiny herbs and wildflowers to large oaks and maple trees. This huge and very diverse group is customarily divided into two subgroups: the ***dicotyledons*** (dicots, for short), which include beans, buttercups, privets, dandelions, oak trees, maple trees, roses, potatoes, and a great variety of other plants; and the ***monocotyledons*** (monocots, for short), which include the grasses and grasslike plants such as corn, lilies, irises, and palm trees. The two subgroups differ in many characteristics, including two easily observable ones: (1) The leaves of dicots usually show a network of veins; those of monocots usually show parallel veins. (2) The petals of dicot flowers occur in fours or fives (or multiples of these); those of monocots occur in threes or multiples of three (Fig. 6.29).

FUNGI

It has been customary to include the fungi in the plant kingdom, because they are predominantly sedentary and because their cells, like plant cells, have walls. But unlike true plants, they lack chlorophyll and therefore cannot manufacture their own food. Like animals, they must obtain the complex high-energy nutrients they need in an already-synthesized form; unlike animals, however, they cannot ingest particulate food, and depend entirely on absorption of nutrient molecules; hence they must live in or on their nutrient sources, which are either other living organisms or the dead remains of other organisms. Since fungi differ from both plants and animals in so many ways, most recent classifications assign them to a kingdom of their own.

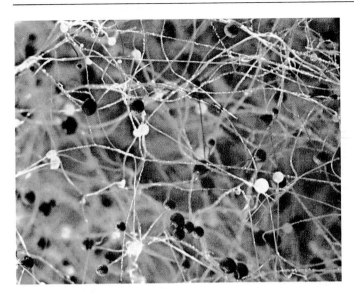

6.30 Two representative fungi
Left: The black bread mold *Rhizopus* is made up of numerous threadlike filaments hung with what look like black and white balls. These are spore-producing reproductive structures; the black ones are ripe and ready to release their spores. Right: The velvet-stemmed collybias growing among lichens on a tree trunk are composed of filaments so densely packed that the unaided eye cannot make them out individually.

Some fungi—yeast, for example—are unicellular, but most are multicellular. Among the latter are bread mold, fruit molds, mushrooms, toadstools, and bracket fungi (Fig. 6.30). However, these are not multicellular in the sense that plants and animals are, for the membranous partitions between adjacent fungal cells, unlike those between plant or animal cells, are usually absent or incomplete; thus the cell contents are essentially continuous within the long tubular body. Instead of calling these organisms multicellular, we might more appropriately describe them as multinucleate—having many nuclei.

ANIMALIA

Of the many characteristics that distinguish the animals from the two other main categories of multicellular organisms—the plants and the fungi—two are especially obvious: first, an animal cell differs from plant and fungal cells in that it lacks a rigid cell wall; second, the principal mode of nutrition in animals is ingestion of particulate food, while most plants depend on photosynthesis and most fungi absorb nutrients.

Coelenterates The coelenterates—also known as the cnidarians—constitute a large group of primitive aquatic animals, among them jellyfish, sea anemones (Fig. 6.31), and corals. The bodies of these sedentary animals are radially symmetrical: the body parts are arranged regularly around a central line, rather like the spokes on a wheel. They have for centuries frequently been mistaken for plants. The saclike bodies of coelenterates are

6.31 A sea anemone, *Epiactis*
A veritable thicket of tentacles surrounds the animal's mouth. Many newly budded young anemones are attached to the stalk portion of its body.

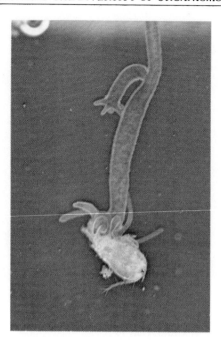

6.32 Hydra, feeding
The green hydra *Chlorhydra* owes its color to the cells of a green alga that it incorporates into its own cells, thus benefiting from the alga's photosynthetic activity. Note the buds with tentacles on the upper portion of the stalks; hydras reproduce asexually by budding. Left: The hydra is manipulating a small crustacean with its tentacles. Right: The food has been taken into the digestive cavity, where it makes a prominent bulge.

composed of two distinct tissue layers with a much less distinct third layer between them; and they have primitive nerves and muscles, though no circulatory system. They have a digestive cavity, but it only has one opening, which serves for both ingestion and excretion. The mouths of many coelenterates have tentacles, which they use to capture prey. The hydra (Fig. 6.32) is a freshwater coelenterate that has been the subject of much study, and will appear frequently in succeeding chapters.

Flatworms The flatworms are more complex than the coelenterates in some ways, but they, too, have a digestive tract with only one opening. The body is composed of three primary tissue layers, and the symmetry is bilateral: the body is elongated and has two similar sides, distinct dorsal (upper) and ventral (lower) surfaces, and distinct anterior (front) and posterior (rear) ends. Many flatworms, such as flukes and tapeworms, are parasites and show numerous interesting specializations for this mode of existence. Others, such as planarians (Fig. 6.33), small animals to which we shall make frequent reference, are free-living aquatic organisms.

Molluscs Fairly complex animals, most of which have shells, the molluscs include snails (Fig. 6.34), clams, oysters, and scallops, as well as slugs, octopuses, and squids, which do not have obvious shells. These animals are particularly abundant in the oceans, as anyone who has collected their shells along the seashore knows. They are also common in fresh water. Some snails and slugs have evolved lungs and become fully terrestrial.

Annelids The annelids are often called the segmented worms. As this term implies, the bodies of these highly evolved organisms are divided into a series of units, or segments, which are often clearly visible externally. Though most annelids are aquatic, some, like the earthworm, occur on land, always in moist places. We shall generally take the earthworm as representative of this group (Fig. 6.35); occasionally, however, we shall men-

6.33 Planarians, members of a group of free-living flatworms
These animals are free-living (nonparasitic) freshwater scavengers. Note in each the eyespots, which allow a primitive sort of vision, the much-branched digestive cavity (brown), and the centrally located tubular pharynx, which is extruded during feeding.

6.34 Florida tree snail
The majority of molluscs live in the water, but many snails, such as this one, are adapted for life on land. The snail moves by means of a muscular "foot" extruded from the shell. Note the eyes located on the ends of long retractable stalks.

6.35 An earthworm, *Lumbricus terrestris*
The animal's body is divided into numerous ringlike segments. Its front end, with the mouth, is at upper left. The prominent girdlelike structure is the clitellum—a group of glandular segments that secrete mucus during mating and produce a cocoon for the eggs.

6.36   A marine annelid worm, *Nereis*
This worm has a pair of large flaplike appendages on
each of its body segments, which function in
respiration and in locomotion.

6.37   Two representative arthropods
Left: The centiped has a segmented body
with a pair of jointed legs on each segment.
Two hornlike antennae curve outward at
lower right. Right: Like all insects, the
lubber grasshopper (*Romalea microptera*)
has only three pairs of legs. The abdominal
segments are legless.

6.38 Three representative echinoderms
Horse stars (*Hippasteria phrygiana*), right foreground and left background, a northern sea star (*Asterias vulgaris*), left foreground, and a sea urchin (*Strongylocentrotus droebachiensis*) in a New England tide pool. Sea stars anchor themselves against the ebb and flow of the tide by means of hundreds of tube feet; sea urchins often use their spines to anchor themselves against the tide in crevices .

tion *Nereis,* a marine worm with large lobes growing from both sides of the body segments (Fig. 6.36).

Arthropods The arthropods constitute an immense group of very advanced animals that includes more different species than all other animal groups combined. All arthropods have jointed legs and a hard outer skeleton. Spiders, scorpions, crabs, lobsters, crayfish, centipeds, millipeds, and insects all belong to this major group (Fig. 6.37). Of these, the insects are by far the largest subgroup; they are among the most successful of all land animals, rivaled only by the mammals, and particularly humans.

Echinoderms All echinoderms are strictly marine; they have apparently never been able to invade either the freshwater or the terrestrial habitats. Though their symmetry is radial, they are fairly advanced in many ways. The group includes sea stars (starfish), sand dollars, sea urchins, sea cucumbers, and a variety of other forms (Fig. 6.38). Though their appearance hardly suggests it, the echinoderms are regarded by most biologists as the group of animals most closely related to the next group, the chordates, which includes humans.

6.39 A frilled lizard, *Chlamydosaurus kingii*
By expanding its ribs, raising its frill, and opening its
mouth wide and hissing, this reptile performs a
threat display, a common example of a defensive
posture that actually serves to minimize physical
combat.

6.40 Japanese macaques (snow monkeys)
A member of a family of Old World monkeys that
includes mandrills and baboons, the Japanese
macaque is one of the few primates able to survive in
a cold climate.

Chordates Most members of this very important group are part of the
subgroup ***vertebrates***, which includes all animals possessing an internal
bony skeleton and, in particular, an articulated (or jointed) backbone. The
vertebrates are divided into eight classes: jawless fish, of which only a few,
such as the lamprey, survive; armored fish, all extinct; cartilaginous fish,
such as sharks and rays, with skeletons composed of cartilage rather than
bone; bony fish; amphibians, such as frogs and salamanders; reptiles, including snakes, lizards (Fig. 6.39), turtles, and alligators; birds; and mammals (Fig. 6.40), of which our species is a member. We shall pay special
attention to vertebrates throughout this book.

Chapter 7

ENERGY TRANSFORMATIONS: RESPIRATION AND OTHER CATABOLIC PATHWAYS

A basic principle of physics, as we saw in Chapter 3, is that all systems have a natural tendency toward disorder. The more orderly an arrangement of matter, the less probable it is, and the less likely to endure if energy is not expended to counteract the tendency toward disorder. Within a cell, therefore, energy is needed at every stage to drive the various chemical reactions that maintain life: to read, copy, and repair the genetic instructions in the chromosomes; to construct, repair, and move organelles; to bring in nutrients, expel wastes, preserve the proper pH and ionic balance; and so on. Without a constant supply of energy, these reactions cannot take place, the cell spontaneously decays, and life ceases. But where does the energy for maintaining life come from, and how is it used by the cell?

THE FLOW OF ENERGY

Virtually all the energy that fuels life today comes from the sun and is captured in the process known as photosynthesis by those organisms, notably plants, that are able to use it to build energy-rich compounds like glucose. Most creatures that do not capture the energy of sunlight directly obtain their energy by ingesting or absorbing photosynthetic organisms or by eating those that eat photosynthetic organisms.

The stored energy is usually released through a process known as respiration, in which, ultimately, oxygen and glucose are combined to yield CO_2 and water; the energy is then used to run all of the cell's enzyme-mediated

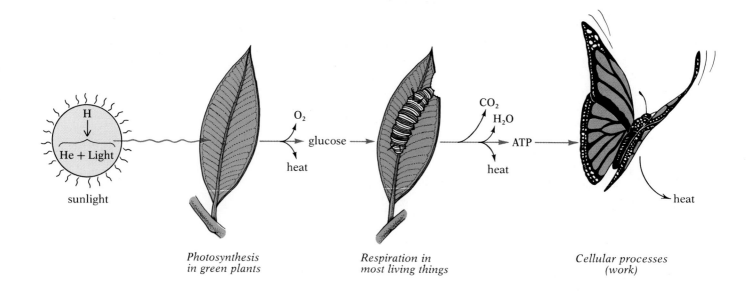

Photosynthesis in green plants   *Respiration in most living things*   *Cellular processes (work)*

7.1 Summary of biological energy flow

The energy for life originates in the sun, where hydrogen is converted by fusion into helium and light is produced. The process of photosynthesis, in green plants, converts the radiant energy of sunlight into chemical energy, most often stored initially in glucose. When cells need energy, the glucose is broken down and some of its chemical energy recovered, in most organisms by the process of respiration; the resulting product—ATP—supplies the energy in a more manageable form, making it available for muscular contraction, nerve conduction, active transport, and other work. Respiration utilizes oxygen, a by-product of photosynthesis, and photosynthesis utilizes carbon dioxide and water, by-products of respiration. With each of the transformations shown, much energy is lost as waste heat.

reactions (Fig. 7.1). These reactions, collectively known as **metabolism**, may be viewed as belonging to one of two phases—**anabolism**, encompassing the processes by which complex organic molecules are assembled, and **catabolism**, the processes by which living things extract energy from food.

EVOLUTION OF ENERGY TRANSFORMATIONS

A plausible picture of the early evolution of the chemical pathways of energy transformation and the organisms that embodied them is just now beginning to emerge. The weight of current evidence suggests that a limited supply of organic molecules developed abiotically on the early earth, and that the first living organisms arose at least 3.5 billion years ago and were capable of metabolizing this energy source. Because the supply of organic molecules available to the first living things was limited, those organisms probably gave rise to the first organisms capable of synthesizing their own food. To create organic compounds, this second group of early organisms used energy drawn not from the sun, but instead, for the most part, from the covalent bonds of molecular hydrogen. Organisms that obtain their energy in this way from inorganic energy sources are said to be **chemosynthetic**. Some of the early chemosynthetic organisms combined naturally occurring CO_2 with energy-rich H_2 to produce methane and water by the condensation reaction shown in the following equation, which summarizes several intermediate steps:

$$CO_2 + 4H_2 \longrightarrow CH_4 + 2H_2O$$

As a result of this reaction, some of the energy stored in the bonds of H_2 is released, and can be used to do work in the cell.

Another energy-liberating reaction uses sulfur to liberate the energy in H_2; indeed, many present-day bacteria thrive in the sulfur-rich waste found in sewers and bogs and (at over 100°C) near volcanic vents on the ocean floor. They produce the foul-smelling gas hydrogen sulfide (H_2S):

$$H_2 + S \longrightarrow H_2S + \text{energy}$$

Early organisms must have lived from hand to mouth, so to speak, eating whatever foods were available, but having little ability to store energy for later use. The evolution in chemosynthetic organisms of enzyme pathways that direct the synthesis of more complex organic molecules made possible long-term energy storage. The primary storage substance was probably glucose, or polysaccharides that could be converted into glucose. When needed, some of the energy of the glucose might then have been extracted, as energy still is today, by the series of enzyme-mediated reactions called glycolysis, or the glycolytic pathway.

About 3 billion years ago an enormously important biochemical event occurred: certain organisms acquired a primitive ability to capture the sun's energy directly and use it to synthesize glucose and other important organic compounds. Then, about 2.5 billion years ago, the forerunner of the chemical pathway we call respiration evolved. This pathway, which includes glycolysis as its first stage, extracts large amounts of energy from the end products of glycolysis, but requires molecular oxygen (O_2), which was in very short supply. However, oxygen became increasingly available when, after some further evolution, the descendants of the early photosynthetic organisms began producing it as a by-product of their now more advanced form of photosynthesis; thus, about 2.3 billion years ago, oxygen began to accumulate in the atmosphere. Since oxygen (in the form of atmospheric ozone, or O_3) filters out biologically destructive X rays and ultraviolet light from the sun, organisms could now emerge from beneath the radiation shielding provided by earth and water to colonize the surface of the land. In addition, the abundance of oxygen made respiration the dominant catabolic pathway: respiration is almost twenty times as efficient in extracting the stored energy of glucose as glycolysis alone.

OXIDATION AND REDUCTION

Chemical reactions transfer energy from one substance or set of substances (the reactants) to others (the products). Such a transfer, upon which much of life depends, occurs in the respiration of glucose, as shown in the general summary formula:

$$C_6H_{12}O_6 + 6O_2 \rightleftharpoons 6CO_2 + 6H_2O + \text{energy}$$

The initial storage of energy in substances like glucose, and the release of energy in reactions like the one shown here, involve two complementary processes, called oxidation and reduction. Originally these terms referred to the addition and removal of oxygen in a reaction. For example, when natural gas (methane, CH_4) is burned, the reactant methane is oxidized by the addition of oxygen, while the reactant oxygen is reduced by the removal of one atom of each oxygen molecule. The result is that the carbon of the methane combines with the added oxygen to form the product carbon dioxide, while the hydrogen of the methane combines with the remaining atoms of oxygen to form the other product, water:

$$CH_4 + 2O_2 \rightleftharpoons CO_2 + 2H_2O$$

TABLE 7.1 *Redox reactions*

Oxidation	Reduction
adds oxygen	removes oxygen
removes hydrogen	adds hydrogen
removes electron(s)	adds electron(s)
liberates energy	stores energy

Today these terms are applied also to reactions in which oxygen is not involved. All oxidation-reduction reactions have one thing in common: ***reduction*** always means the addition of one or more electrons (e^-)—either alone or in association with a proton (H^+) as a hydrogen atom—while ***oxidation*** means the removal of one or more electrons, again alone or with H^+. (The addition of oxygen has an equivalent effect because the oxygen, being strongly electronegative, attracts electrons away from their original atoms.) Since reduction and oxidation reactions inevitably occur together, with an electron/hydrogen added to one reactant and removed from the other, they are known as ***redox reactions*** (from *red*uction-*ox*idation).

These relationships are summarized in Table 7.1. The important point is that reduction *stores* energy in the reduced compound, while oxidation *liberates* energy from the oxidized substance. This redistribution is probably best illustrated by looking once again at the energy equation of life. The energy of high-energy photons from the sun is used by photosynthetic organisms to reduce carbon dioxide (that is, to remove oxygen from it and add hydrogen) and thereby create energy-rich carbohydrates such as glucose:

$$6CO_2 + 6H_2O + light \longrightarrow 6O_2 + C_6H_{12}O_6$$

The details of this photosynthetic reaction are examined in Chapter 8.

The energy stored in substances like glucose as a result of the reduction of their precursors can be partially recovered when the storage compound is in turn oxidized:

$$C_6H_{12}O_6 + 6O_2 \rightleftharpoons 6CO_2 + 6H_2O \qquad (\triangle G = -670 \text{ kcal/mole})$$

You may recall from page 72 that the $\triangle G$ of a reaction equals the difference in free energy between the products and the reactants, and a negative $\triangle G$ means that the reaction liberates free energy and so can proceed spontaneously. The stored energy is only partially recovered for use by the cell because, as in all reactions, some energy is lost as heat. The oxidation of glucose (the main subject of this chapter) provides energy for the creation of ATP—the energy currency of the cell—and other energy-storage compounds, thus providing energy to power other reactions; and it regenerates the raw materials (CO_2 and H_2O), which are then ready to be utilized again.

ADENOSINE TRIPHOSPHATE (ATP)

One of the essential substances of life is the compound ***adenosine triphosphate***, or ***ATP***, which plays a key role in nearly every transformation in every living thing.

The ATP molecule (Fig. 7.2) is composed of a nitrogen-containing compound (adenosine) plus three phosphate groups bonded in sequence:

$$adenosine—Ⓟ \sim Ⓟ \sim Ⓟ$$

According to convention, Ⓟ stands for the entire phosphate group, and the wavy lines between the first and second and the second and third phosphate groups represent so-called high-energy bonds:[1] if they are broken by hydrol-

---

[1] The term "high-energy bonds," though widely used by biologists, may be somewhat misleading. In fact, the energy is stored not in the bond, but in the phosphate group itself, and is liberated as the bond, represented in this book by a wavy line, is broken.

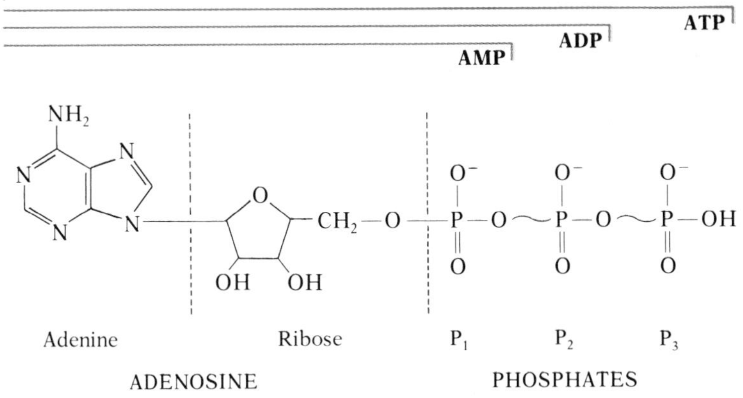

NH₂ ... Adenine — Ribose — P₁ P₂ P₃

Adenine   Ribose   P₁   P₂   P₃

ADENOSINE   PHOSPHATES

7.2 The ATP molecule
As its full name, adenosine triphosphate, implies, this molecule is composed of an adenosine unit (a complex of adenine and ribose sugar) and three phosphate groups arranged in sequence. The last two phosphates are attached by so-called high-energy bonds (wavy lines). The cell stores energy by adding a phosphate group to ADP (adenosine diphosphate) to make ATP, and later recovers some of this energy by hydrolyzing ATP into ADP and inorganic phosphate. Occasionally ADP is hydrolyzed to make AMP (adenosine monophosphate), one of the nucleotides of which RNA is composed.

lysis (the reaction in which a molecule is broken apart by the addition of water), far more free energy is released than if the other bonds in the ATP molecule are broken.

Actually, it is often only the terminal phosphate bond of ATP that is involved in energy conversions. The exergonic reaction by which this bond is hydrolyzed and the terminal phosphate group removed leaves a compound called adenosine diphosphate, or **ADP** (adenosine plus two phosphate groups), and inorganic phosphate (symbolized by P_i):

$$\text{ATP} + H_2O \xrightarrow{\text{enzyme}} \text{ADP} + P_i + \text{energy}$$

If both the second and third phosphate groups are removed from ATP, the resulting compound is adenosine monophosphate, or **AMP**.

New ATP can be synthesized from ADP and inorganic phosphate if adequate energy is available to force a third phosphate group onto the ADP. Addition of phosphate is termed **phosphorylation**:

$$\text{ADP} + P_i + \text{energy} \xrightarrow{\text{enzyme}} \text{ATP} + H_2O$$

ATP is often called the universal energy currency of living things, and this characterization is justified. Cells initially store energy in the form of carbohydrates, like glucose, and lipids, like the many kinds of fats. But the amount of energy in even a single glucose molecule is inconveniently large for driving most reactions: 670 kcal/mole. The hydrolysis of ATP to ADP releases a more useful amount of energy: 7.3 kcal/mole. Glucose molecules are the hundred-dollar bills of the cellular economy, while ATP molecules are the everyday denominations. Though some other compounds can supply energy, ATP is the one most often used by cells in the various kinds of work they perform: synthesis of more complex compounds, muscular contraction, nerve conduction, active transport across cell membranes, light production, and so on. The energy price of the work is paid through the energy-releasing hydrolysis of ATP to ADP.

CELLULAR METABOLISM

The energy stored in lipids and carbohydrates is not liberated through a single large reaction; rather, the universal catabolic process by which the molecules are broken down occurs as a series of smaller reactions, each catalyzed by its own specific enzyme. These result in the release of small amounts of energy, some of which is transferred to ATP by the phosphorylation which synthesizes ATP from ADP and inorganic phosphate. In this chapter we shall examine the most important steps in the catabolism of glucose, while in the next we shall look at the major pathway for the synthesis of glucose.

ANAEROBIC METABOLISM

Glycolysis (Stage I of respiration)   The complete catabolism of glucose involves five stages, divided between anaerobic and aerobic series of reactions. The anaerobic portion of the process, the breakdown of glucose to pyruvic acid, known as **glycolysis**, is the most ancient series of reactions in the pathway.

Glucose is a stable compound—one with little tendency to break down spontaneously into simpler products. If its energy is to be harvested, the glucose must first be made more reactive by the investment of a small amount of energy to "activate" the molecule. The first steps of glycolysis, therefore, are preparatory, enabling the later steps to extract the stored energy.

The energy for initiating glycolysis (Fig. 7.3) comes from ATP. The initial reaction, like the succeeding ones, is made possible by its own specific enzyme, which binds (via weak bonds) to the reactants, activates (by causing the electrons of the bound molecules to redistribute themselves), and joins

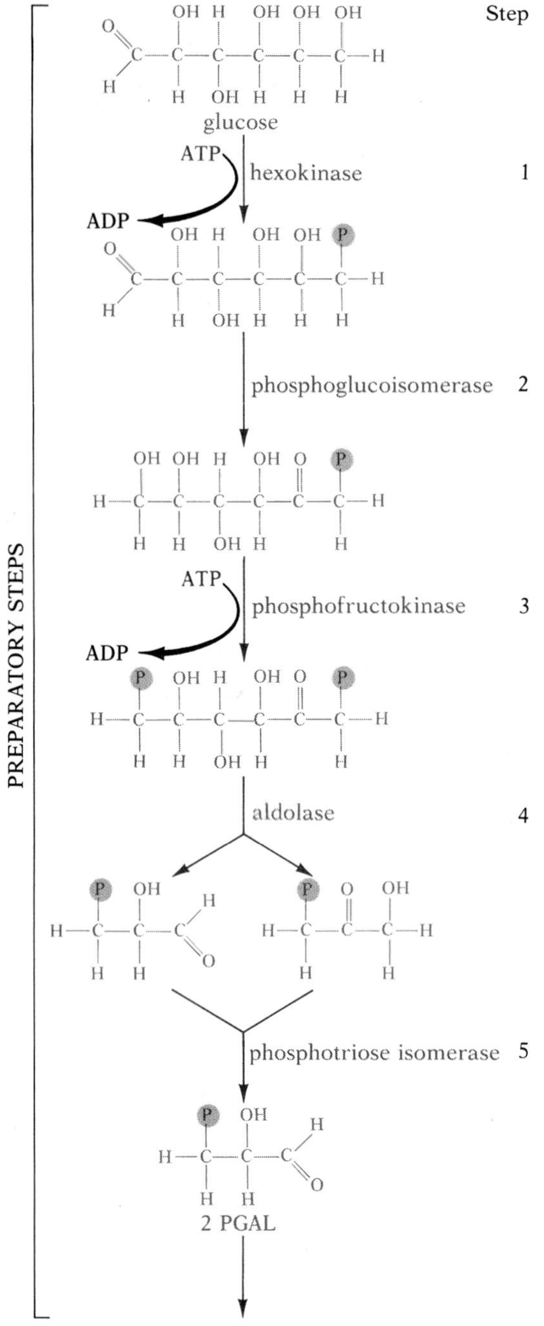

7.3 Glycolysis and fermentation
The entire reaction series for glycolysis is shown to illustrate how biochemical pathways work; there is no reason to try to memorize each step. Glucose in solution normally exists as a ring structure, but the straight-chain form is adopted here for simplicity. Energy to initiate the breakdown of glucose is supplied by two molecules of ATP (Steps 1–3). The resulting compound is then split into two molecules of PGAL (5). This completes the preparatory reactions. Next, the PGAL is oxidized by the removal of hydrogen, which is picked up by NAD_{ox} to form NAD_{re}; in the same reaction inorganic phosphate is added to each of the three-carbon molecules (6). A series of reactions then results in synthesis of four new molecules of ATP, for a net gain of two (7–10). The pyruvic acid produced by this anaerobic breakdown can be further oxidized if O_2 is present (by reactions not shown here). But in the absence of sufficient O_2, the pyruvic acid may accept hydrogen atoms from NAD_{re} to form lactic acid in some kinds of organisms or CO_2 and ethanol in others (11–12). At each step a particular enzyme catalyzes a specific redistribution of electrons, and thereby brings about changes in bonding. This step-by-step strategy is the only one by which enzymes can guide reactions. (In each diagram, bonds to be altered by enzymatic actions in the next step are shown in green. The fate of particular oxygens and hydrogens cannot always be traced from step to step because they are sometimes incorporated into units, such as phosphate groups, for which full diagrams are not given.)

or rearranges the reactants before releasing the products. In the first step, a molecule of ATP donates its terminal phosphate group to the glucose.

$$(1) \quad \underset{\text{glucose}}{C-C-C-C-C-C} + ATP \xrightarrow{\text{enzyme}} \underset{\text{glucose-6-phosphate}}{C-C-C-C-C-C-℗} + ADP$$

$$(\triangle G = -4.0 \text{ kcal/mole})$$

(The simplified equations given here show only the carbon skeleton; the more complete molecular structure is shown in Figure 7.3.)

Let's look carefully at what has happened in this reaction. An enzyme, hexokinase, has bound glucose and ATP, catalyzed the transfer of a phosphate group to the glucose, and released the products. The overall change in free energy of this reaction is −4.0 kcal/mole; the free energy is liberated primarily as heat. As always, the negative $\triangle G$ means that this is a strongly exergonic (downhill) reaction.[2] In fact, the reaction has an equilibrium constant (K_{eq}) of about 1000—that is, the products outnumber the reactants by 1000 to 1 (see Table 3.2, p. 74). The free energy for this reaction comes from ATP: the energy available from the terminal phosphate of the ATP is 7.3 kcal/mole. Only 4.0 kcal/mole is liberated to drive this first step of glycolysis, and the other 3.3 kcal/mole from the ATP is stored in the product —the activated glucose.

The next step in glycolysis converts glucose-6-phosphate into the nearly identical compound fructose-6-phosphate:

$$(2) \quad \underset{\text{glucose-6-phosphate}}{C-C-C-C-C-C-℗} \xrightarrow{\text{enzyme}} \underset{\text{fructose-6-phosphate}}{C-C-C-C-C-C-℗}$$

$$(\triangle G = +0.4 \text{ kcal/mole})$$

The positive $\triangle G$ indicates that this is an uphill, endergonic (energy-requiring) reaction, one that cannot proceed spontaneously. How is it, then, that glycolysis does not grind to a halt? The answer lies in the operation of **coupled reactions**: two reactions that share a common intermediate molecule —in this instance, glucose-6-phosphate, the product of Step 1 *and* the reactant of Step 2—can proceed thermodynamically as a single reaction. The −4.0 kcal/mole liberated by Step 1 is combined with the +0.4 kcal/mole consumed by Step 2 to yield a net $\triangle G$ of −3.6 kcal/mole. Taken together, the two steps are strongly exergonic, and so the reaction proceeds. The glycolytic pathway is a series of such coupled reactions, in which exergonic steps push or pull endergonic steps, with the favorable *net* free-energy change of the steps taken together enabling the sequence of reactions to proceed.

The conversion of glucose-6-phosphate into fructose-6-phosphate also provides a good illustration of how enzymatic pathways work. You may recall that when a substrate binds to an enzyme by means of weak bonds, a

[2] As noted in Chapter 3, we use $\triangle G$ (rather than the formal symbol $\triangle G'^{\circ}$) for the change in the free energy of a reaction under standard conditions. In addition, we divide $\triangle G$s simply between products and liberated free energy, even though, in fact, a more complicated but poorly understood division is probably occurring. The resulting approximation is reasonably accurate. The equilibrium constant (K_{eq}) is being treated for simplicity as though there is only a single product. This is possible because glycolysis has relatively little effect on the concentration of water, ATP, ADP, and so on; the concentration of these compounds is maintained through various homeostatic mechanisms in the cell.

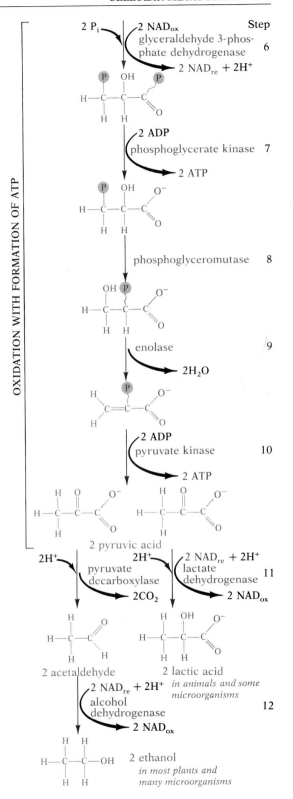

slight shift in the electron distribution of the substrate is induced, which lowers the activation energy required for a particular change in bonding and thereby catalyzes the reaction. The result of the redistribution of electrons can be seen in Step 2, in which two hydrogens, bonded to the fifth carbon and its oxygen, are transferred to the first carbon and its oxygen. This trivial change is essential to prepare for the next step in glycolysis. Once the change has taken place, the substrate no longer "fits" the enzyme, and so drifts away to be captured by the next enzyme in the series. Each step in the glycolytic pathway is therefore very small and is mediated by a highly specific enzyme.

After the formation of fructose-6-phosphate in Step 2, another molecule of ATP is consumed, to add a phosphate to the other end of the molecule. Of the ATP's energy, 3.9 kcal/mole is stored in the product, while the remaining 3.4 kcal/mole is liberated as heat; hence the reaction is exergonic:

$$(3) \quad C-C-C-C-C-C-\circledP + ATP \xrightarrow{\text{enzyme}}$$

$$\circledP-C-C-C-C-C-C-\circledP + ADP$$

fructose-6-phosphate

fructose-1,6-bisphosphate

$$(\triangle G = -3.4 \text{ kcal/mole})$$

Since in its turn this reaction is coupled to the previous one (with which it shares the intermediate compound fructose-6-phosphate), we can add the free energy of the separate steps along the way. The overall reaction chain so far has liberated 7.0 kcal/mole, and so has a highly favorable K_{eq} of more than 10^5.

Next, the fructose-1,6-bisphosphate is split between the third and fourth carbons, forming two essentially similar three-carbon molecules (Step 4). One is phosphoglyceraldehyde—**PGAL**—and the other, an intermediate compound, is usually converted immediately to PGAL, in Step 5. (The cell can also use this compound to synthesize fat if conditions warrant.) PGAL, a phosphorylated three-carbon sugar, is a key intermediate in both glycolysis and photosynthesis.

We can summarize these reactions simply:

$$(4-5) \quad \circledP-C-C-C-C-C-C-\circledP \xrightarrow{\text{enzyme}} 2\ C-C-C-\circledP$$

fructose-1,6-bisphosphate

PGAL

$$(\triangle G = +7.5 \text{ kcal/mole})$$

To this point, instead of releasing energy from glucose to form new ATP molecules, glycolysis has actually cost the cells two ATPs. Indeed, Steps 4 and 5 are so unfavorable energetically that the net change in free energy is now +0.5 kcal/mole. For subsequent reactions to proceed, significant amounts of free energy must be liberated to pull the reactants past the five preparatory steps.

The next step is a relatively complicated reaction (though it takes place on a single enzyme), and actually involves two separate molecular changes, which we summarize for simplicity in one step. The first change is oxidation of the PGAL by reduction of nicotinamide adenine dinucleotide, or **NAD**. (The characteristic function of NAD is temporary energy storage; it transports energy from one pathway to another, or from one step in a pathway to another step, elsewhere in the pathway.) Each NAD_{ox} accepts two hydrogens, keeping one and the electron of the other to produce NAD_{re} and an H^+ ion. The second change is phosphorylation of the PGAL:

(6) $2\ \circledP - C - C - C + 2\ NAD_{ox} + 2\ P_i \xrightarrow{\text{enzyme}}$
$$2\ \circledP - C - C - C \sim \circledP + 2\ NAD_{re} + 2H^+$$
$$(\triangle G = +3.0\ \text{kcal/mole})$$

The oxidation phase of this reaction, taken alone, is strongly exergonic, while the phosphorylation phase is strongly endergonic. Since the two processes occur together, the energy that would have been released by the oxidation (more than 100 kcal/mole) is conserved in the reduced NAD (NAD_{re}) and the phosphorylated PGAL. The consequence of Step 6 is to make the net change in free energy for the overall reaction chain even more unfavorable (+3.5 kcal/mole), but the next downhill reaction, to which Step 6 is coupled, begins once again to turn the balance, as the high-energy phosphate group is transferred to the ADP, to form ATP:

(7) $2\ \circledP - C - C - C \sim \circledP + 2\ ADP \xrightarrow{\text{enzyme}} 2\ \circledP - C - C - C + 2\ ATP$
$$(\triangle G = -9.0\ \text{kcal/mole})$$

At this point, then, the cell regains the two ATP molecules invested to activate the glucose in Steps 1 and 3, and the overall net change in free energy is again favorable: −5.5 kcal/mole. Moreover, a great deal of energy has been stored in NAD_{re}.

Next comes a reaction that energizes the remaining phosphate groups:

(8) $2\ \circledP - C - C - C \xrightarrow{\text{enzyme}} 2\ \circledP \sim C - C - C + H_2O$
$$(\triangle G = +1.5\ \text{kcal/mole})$$

After a reaction that rearranges the substrate and has a $\triangle G$ of −0.4 kcal/mole (Step 9), these energized phosphate groups are transferred to ADP; the products are **pyruvic acid** (also referred to as pyruvate) and ATP:

(10) $2\ \circledP \sim C - C - C + 2\ ADP \xrightarrow{\text{enzyme}} 2\ \underset{\text{pyruvic acid}}{C - C - C} + 2\ ATP$
$$(\triangle G = -15.0\ \text{kcal/mole})$$

Obviously this last reaction is overwhelmingly favorable; indeed, the K_{eq} is greater than 10^9. Moreover, a profit of two ATPs is generated by this step of glycolysis. Because the two ATP molecules used in Steps 1 and 3 have already been regained (in Step 7), the two additional molecules formed here represent a net gain in ATP for the cell. The highly exergonic last step results in an overall $\triangle G$ of −19.4 kcal/mole. It is the liberation of this energy, primarily as heat, that causes this series of coupled reactions to proceed.

Figure 7.4 shows graphically the changes in free-energy content at each successive step in glycolysis, from glucose to pyruvic acid (as well as in the process of fermentation, which will be discussed shortly).

We can now summarize the most important features of glycolysis:

1. Each molecule of glucose (a six-carbon compound) is broken down

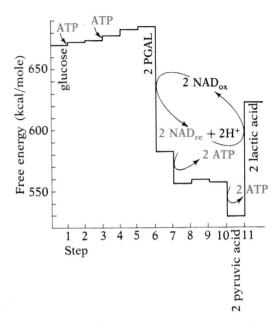

7.4   Changes in free-energy content at successive steps in glycolysis and fermentation
In the five preparatory steps, which convert glucose into PGAL, the free-energy content of the intermediate substances is slightly increased owing to the investment of two molecules of ATP. In Step 6 there is a sharp drop in free energy associated with the formation of two molecules of NAD_{re}. There are also major drops in Steps 7 and 10, each associated with the formation of two molecules of ATP. This summary includes both the free energy liberated (primarily as heat) or consumed, *and* the free energy stored in NAD_{re}, ATP, and the various intermediate products between glucose and pyruvic acid. The free-energy values listed to the right of the reactions in the text, on the other hand, reflect only the heat liberated by the particular reactions, or the energy required to drive them. The energy scale is relative to the total free energy of glucose, and shows how much energy remains untapped at the end of Steps 7, 10, and (in the absence of oxygen) 11.

COUPLED REACTIONS

As the process of glycolysis shows, cellular chemistry does not work by a uniform series of energetically downhill steps. Rather, many important reactions are endergonic; outside energy is required, for example, to build molecules like DNA, RNA, and protein, to form structures like cell membrane, and to maneuver energy sources like glucose and fat into positions where their stored energy can be utilized. This outside energy is supplied by a coupling of the unfavorable endergonic reaction to a strongly exergonic reaction. Let's look at the first two reactions of glycolysis as an example of how coupling works:

(1) glucose + ATP $\xrightleftharpoons{K_{eq} = 610}$

glucose-6-phosphate + ADP

($\triangle G = -4.0$ kcal/mole)

(2) glucose-6-phosphate $\xrightleftharpoons{K_{eq} = 0.54}$

fructose-6-phosphate

($\triangle G = +0.4$ kcal/mole)

(Changes in ADP, being affected also by reactions unrelated to glycolysis, need not be considered and are omitted in the equation for Step 2.)

Reaction 1 is downhill, with the products eventually outnumbering the reactants by 610 to 1, while reaction 2 is uphill, with the reactant eventually outnumbering the product by about 2 to 1. Notice that one of the products of reaction 1 is the reactant of reaction 2: the reactions are coupled—they share an **intermediate compound**, glucose-6-phosphate. The table, a molecular scorecard, shows these coupled reactions set in motion by the combination of 10^5 molecules of glucose and an equal number of molecules of ATP (a). The equilibrium constant of reaction 1 (610) means that it will proceed until a ratio of roughly 99,836 molecules of glucose-6-phosphate to 164 of glucose has been reached (b). Now, since glucose-6-phosphate is a reactant of reaction 2, some of it will be converted into fructose-6-phosphate. The equilibrium constant of this ratio is 0.54, so about 35,007 molecules of product will then be produced (c). The energy necessary for this uphill reaction comes from the free energy released by reaction 1; being highly exergonic, the downhill first reaction pushes the second reaction along. With the consequent reduction in the total amount of glucose-6-phosphate, reaction 1 is no longer at equilibrium, so additional glucose and ATP react to produce glucose-6-phosphate (d), only to have some of this intermediate promptly converted into fructose-6-phosphate (e). At this point equilibrium is reached, with the ratios of product to reactant for both reaction 1 and reaction 2 approximating the corresponding equilibrium constants. The addition of more glucose and ATP (which happens almost continuously in cells) or the removal of fructose-6-phosphate (which also happens continuously, as it is converted into another compound in the third step of glycolysis) will cause more molecules to move along this reaction

into two molecules of pyruvic acid (a three-carbon compound).
2. Two molecules of ATP are used to initiate the process. Later, four new molecules of ATP are synthesized, for a *net* gain of two molecules of ATP from each glucose molecule broken down. The energy stored in the new ATP molecules represents only about 2 percent of the energy initially present in the glucose molecule.
3. Two molecules of reduced NAD (NAD_{re}) are formed.
4. Because no molecular oxygen is used, glycolysis can occur whether or not O_2 is present. It is a process encountered in the cytoplasm of all living cells, whatever their mode of life.

Coupled Reactions

	Reactants	Reaction 1 $K_{eq} = 610$	Intermediate compound	Reaction 2 $K_{eq} = 0.54$	Product
	glucose + ATP ⇌		glucose-6-phosphate ⇌		fructose-6-phosphate
a	100,000		0		0
b	164		99,836		0
c	164		64,829		35,007
d	106		64,887		35,007
e	106		64,866		35,028

Addition of 10^4 molecules of glucose and 10^4 molecules of ATP

f	10,106		64,866		35,028
g	123		74,849		35,028
h	123		71,349		38,528
i	117		71,355		38,528
j	117		71,353		38,530

Removal of 10^4 molecules of fructose-6-phosphate

k	117		71,353		28,530
l	117		64,857		35,026
m	106		64,868		35,026
n	106		64,864		35,030

pathway (*f–j* and *k–n*).

Thus favorable and unfavorable reactions are coupled in the cell: reactants are pushed through the uphill steps as long as appropriate interme-diate compounds are present and the *net* ΔG is negative. In fact, cells function by coupling liter-ally thousands of reactions into long chains in this manner.

Fermentation   We have seen that in glycolysis two molecules of NAD_{ox} are reduced to NAD_{re}, and that NAD functions in the cell as an energy-transport compound, shuttling high-energy electrons between one substance and an-other. Thus NAD is only a temporary acceptor of electrons, promptly pass-ing its extra electrons to some other compound and then going back for another load. The cell has only a limited supply of NAD molecules, and these must be used over and over again. If the NAD_{re} molecules formed in glycolysis could not quickly unload electrons (that is, be oxidized back into NAD_{ox}), all the cell's NAD would soon be tied up. With Step 6 of glycolysis thus blocked, glycolysis would come to an end.

As we noted in the discussion of Step 6, more than 50 kcal/mole of free energy is stored in each of the two molecules of NAD_{re} produced during the glycolysis of a glucose molecule. For the cell to harvest this abundant energy, the NAD_{re} must transfer electrons to a lower energy level in some more electronegative acceptor molecule. In most cells, as we shall see, molecular oxygen becomes the ultimate acceptor of the transferred electrons. But under anaerobic conditions, with no oxygen present, the pyruvic acid formed by glycolysis accepts electrons from NAD_{re}. This reduction of pyruvic acid results in the formation of *lactic acid* in animal cells and some unicellular organisms, or of *ethanol* (ethyl alcohol) and carbon dioxide in most plants and many unicellular organisms:

(11) $\quad$ pyruvic acid $+ NAD_{re} + H^+ \xrightarrow{\text{enzyme}}$ lactic acid $+ NAD_{ox}$

or

(11) $\quad$ pyruvic acid $\xrightarrow{\text{enzyme}}$ acetaldehyde $+ CO_2$

(12) $\quad$ acetaldehyde $+ NAD_{re} + H^+ \xrightarrow{\text{enzyme}}$ ethanol $+ NAD_{ox}$

Thus, under anaerobic conditions, NAD shuttles back and forth, picking up electrons—becoming NAD_{re}—in Step 6 and giving up the electrons—becoming NAD_{ox}—in Step 11 or 12.

The process that begins with glycolysis and ends with the transformation of pyruvic acid into ethanol or lactic acid is called *fermentation*.[3] We can thus speak of alcoholic fermentation or lactic acid fermentation, depending on the end product of the process. (In a few organisms, fermentation leads to products other than ethanol or lactic acid, but these are of less general importance and will not be discussed here.)

Whatever the end product, fermentation enables a cell to continue synthesizing ATP by breakdown of nutrients under anaerobic conditions. But, because the electrons transferred from NAD_{re} remain at a relatively high energy level in the reduction of pyruvic acid, fermentation extracts only a very small portion (about 2 percent) of the energy present in the original glucose.

Fermentation by yeast cells and other microorganisms is, of course, the basis for the extensive and economically vital industry that produces both commercial alcohol and alcoholic beverages. Microbial fermentations are also essential to the production of most cheeses, yogurt, and a variety of other dairy products.

---

[3] The term "fermentation" has been used in countless ways in the scientific literature. It is often restricted to the breakdown of glucose to ethanol. It is also applied to the production of either ethanol or lactic acid by microorganisms—lactic acid production in animal cells being called glycolysis. Both these uses lead to confusion between the terms "fermentation" and "glycolysis," and both tend to obscure the general occurrence of the same basic fermentation process in all living cells. Accordingly, "fermentation" is here applied to the process of anaerobic production of ethanol or lactic acid (or other products of the reduction of pyruvic acid), whether by plant, animal, or microorganism, and the glycolytic pathway to pyruvic acid is taken both as a preparatory reaction sequence leading to the Krebs citric acid cycle when sufficient oxygen is present and as the initial portion of fermentation in the absence of sufficient oxygen. (A few microorganisms carry out fermentation in the presence of abundant oxygen.)

AEROBIC METABOLISM (RESPIRATION)

The more efficient process of energy extraction that occurs in the presence of abundant molecular oxygen was probably perfected in the tiny, single-celled organisms whose descendants, as noted in Chapter 5, may have given rise to the mitochondria of eucaryotic cells. In eucaryotic cells, aerobic respiration now takes place exclusively in the mitochondria. When O_2 is present, the metabolic breakdown of glucose initially proceeds, as it does under anaerobic conditions, along the glycolytic pathway to pyruvic acid. But the pyruvic acid need not then act as the electron acceptor and become converted into lactic acid or ethanol. Instead, oxygen acts as the electron acceptor; mitochondrial enzymes move the transient electrons to the oxygen, thereby releasing the free energy of NAD_{re}:

$$O_2 + 2\ NAD_{re} + 2H^+ \rightleftharpoons 2H_2O + 2\ NAD_{ox}$$
$$(\triangle G = -52.4 \text{ kcal/mole})$$

The free energy is then utilized in the formation of ATP. Moreover, the pyruvic acid (which still has 590 kcal/mole of free energy at Step 10) can be broken down to yield additional energy for the synthesis of still more ATP. Under aerobic conditions, then, ATP synthesis does not end with pyruvic acid. Indeed, if lactic acid has already been formed, it can be reconverted into pyruvic acid (with consequent regeneration of the lost NAD_{re}) when sufficient oxygen becomes available. This pyruvic acid too may then be oxidized.

The process of aerobic breakdown of nutrients with accompanying synthesis of ATP is called *cellular respiration*. Whereas anaerobic metabolism consists of glycolysis followed by fermentation, cellular respiration consists of glycolysis followed by the oxidation of pyruvic acid to acetyl-CoA, and then by the reactions of what is known as the Krebs citric acid cycle, the reactions involving the electron-transport chain, and finally the processes that culminate in the synthesis of ATP. The first three stages in the breakdown process yield the energy-rich compound NAD_{re}. It is the transfer of electrons from this compound to highly electronegative oxygen atoms via the respiratory electron-transport chain that ultimately produces most of the ATP from glucose.

Oxidation of pyruvic acid to acetyl-CoA (Stage II of respiration)   The aerobic oxidation of pyruvic acid begins with a complicated set of reactions whose net effect is to break down the three-carbon pyruvic acid to CO_2 and the two-carbon compound acetic acid, which is bonded to a coenzyme called coenzyme A, or CoA, for short; the complete compound is called *acetyl-CoA*. When a molecule of pyruvic acid is oxidized to acetyl-CoA and CO_2, hydrogen is removed and a molecule of NAD_{re} is formed. Since two molecules of pyruvic acid were formed from each glucose molecule, two molecules of NAD_{re} are formed here. This complicated series of reactions can be summarized by the following equation:

$$2 \text{ pyruvic acid} + 2\ CoA + 2\ NAD_{ox} \rightleftharpoons$$
$$2 \text{ acetyl-CoA} + 2CO_2 + 2\ NAD_{re} + 2H^+$$

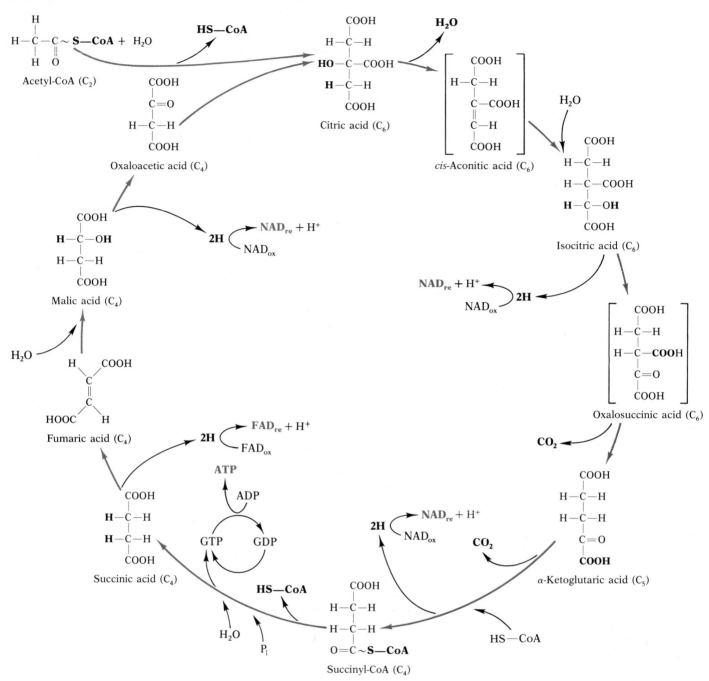

7.5 The Krebs citric acid cycle

The complete cycle is shown here only to help you appreciate the characteristic complexity of metabolic pathways; there is no point in your trying to learn all the reactions involved. As shown at upper left, the acetyl group (two carbons) from acetyl-CoA enters the cycle by combining with oxaloacetic acid (four carbons) to form citric acid (six carbons). During subsequent reactions two of the carbons are removed as CO_2 (between oxalosuccinic acid and α-ketoglutaric acid, and between α-ketoglutaric acid and succinyl-CoA), and a total of eight hydrogens are removed. These hydrogens are picked up by NAD (or by the related acceptor molecule FAD). One molecule of ATP is synthesized (bottom). Finally, oxaloacetic acid is regenerated and can combine with a new acetyl group to start the cycle over again. The cycle is completed twice for each molecule of glucose oxidized. (The atoms removed at each step are shown in boldface in the structural formulas. The two substances in brackets—cis-aconitic acid and oxalosuccinic acid—are enzyme-bound intermediates that seldom exist as free compounds.)

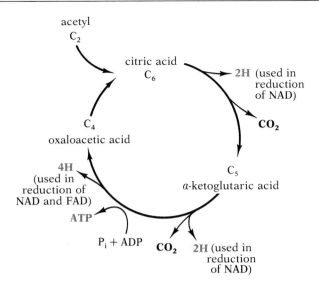

7.6 Simplified version of the Krebs citric acid cycle
The two carbons of the acetyl group combine with a four-carbon compound to form citric acid, a six-carbon compound. Removal of one carbon as CO_2 leaves a five-carbon compound. And removal of a second carbon as CO_2 leaves a four-carbon compound, which can combine with another acetyl group and start the cycle over again. In the course of the cycle, one molecule of ATP is synthesized and eight hydrogens are released; they are used in reduction of NAD and FAD. Since one molecule of glucose gives rise to two acetyl units, two turns of the cycle occur for each molecule of glucose oxidized, with production of four molecules of CO_2, two molecules of ATP, and 16 hydrogens.

Note that, at this stage, two of the six carbons present in the original glucose have been released as CO_2. Note also that the newly formed NAD_{re} must be oxidized if the breakdown process is to continue; we shall return to this problem shortly.

The Krebs citric acid cycle (Stage III of respiration)   The acetyl-CoA is next fed into a complex circular series of reactions called the Krebs citric acid cycle after the British scientist Sir Hans Krebs, who was awarded a Nobel Prize for his elucidation of this system. The cycle is shown in some detail in Figure 7.5; its essential features are outlined in Figure 7.6. Briefly, each of the two two-carbon acetyl-CoA molecules formed from one molecule of glucose is combined with a four-carbon compound (oxaloacetic acid) already present in the cell, to form a new six-carbon compound called **citric acid**. Each of the citric acid molecules is then oxidized to a five-carbon compound plus CO_2. The five-carbon unit, in turn, is oxidized to a four-carbon compound plus CO_2. This four-carbon compound is then converted into the four-carbon compound—oxaloacetic acid—to which acetyl-CoA was originally attached; it can now pick up more acetyl-CoA, forming new citric acid and beginning the cycle again.

We see, then, that two carbons are fed into the Krebs cycle as the acetyl group and two are released as CO_2. Since each glucose molecule being oxidized yields two molecules of acetyl-CoA, two turns of the cycle are required, and a total of four carbons are released as CO_2 during this stage of glucose breakdown. With the two carbons already released as CO_2 during the oxidation of pyruvic acid to acetyl-CoA, all six carbons of the original glucose are accounted for.

The oxidative breakdown of each molecule of acetyl-CoA via the Krebs citric acid cycle also involves the removal of eight hydrogens, which are picked up by NAD_{ox} (or by FAD_{ox}, the oxidized form of a related electron-

7.7   Summary of the most important products of Stages I, II, and III in the complete breakdown of one molecule of glucose

Stage I (glycolysis) begins with expenditure of two molecules of ATP to produce fructose-1,6-bisphosphate, which is broken down to two molecules of PGAL. After these preparatory steps, the two PGAL molecules are broken down to two molecules of pyruvic acid, in a process that first pays back the two ATP molecules originally invested and then yields two molecules each of ATP and NAD_{re} (color). Stage II (the breakdown of two molecules of pyruvic acid to two molecules of acetyl-CoA) yields two molecules each of CO_2 and NAD_{re}. Stage III, in which the two molecules of acetyl-CoA are fed into the Krebs cycle and further broken down, yields four CO_2 molecules, two ATP molecules, six NAD_{re} molecules, and two FAD_{re} molecules. (The H^+ ions liberated in the production of NAD_{re} and FAD_{re} are not shown.)

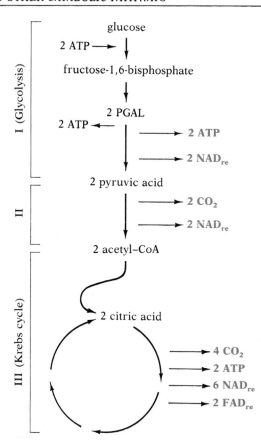

carrier compound, a flavoprotein called flavin adenine dinucleotide); four units of reduced carrier are thus formed (Fig. 7.5). Since the breakdown of one molecule of glucose leads to two turns of the Krebs cycle, a total of eight molecules of reduced carrier (six NAD_{re} and two FAD_{re}) are formed during this stage of the breakdown of glucose. Two molecules of ATP are also synthesized in the Krebs cycle.

Figure 7.7 summarizes the yield of ATP, NAD_{re}, FAD_{re}, and CO_2 from the three stages of the breakdown of glucose.

The respiratory electron-transport chain (Stage IV of respiration)   We pointed out earlier that glucose is energy-rich, and that its breakdown enables the cell to synthesize new ATP, the cell's energy currency. But in our examination of the three stages of catabolism—glycolysis, conversion of pyruvic acid into acetyl-CoA, and the Krebs cycle—we have so far seen a net gain of only four new ATP molecules (two in glycolysis and two in the Krebs cycle). These represent but a small fraction of the energy originally available in the glucose. Of the remainder, some is liberated during the first three stages (mostly as heat, which is essential for the reactions to proceed); the rest is stored in the high-energy intermediates NAD_{re} and FAD_{re}. Twelve of these molecules are synthesized in the breakdown of each molecule of

REGULATION OF GLUCOSE BREAKDOWN

There are several major points at which the interconnected enzyme pathways shown in the accompanying figure can be regulated. One of the most important is at the step in glycolysis where fructose-1,6-bisphosphate is produced. The allosteric enzyme that catalyzes this reaction is positively modulated (made more reactive) by ADP and AMP; it is negatively modulated (made less reactive) by ATP and by citric acid. Hence the enzyme is most active, and glucose breakdown is greatest, when there is a shortage of ATP and a buildup of ADP and AMP. The enzyme is least active, and glucose breakdown is slowed, when there is an accumulation of ATP and citric acid. Regulation of the enzyme for fructose-1,6-bisphosphate by citric acid is a good example of one type of *feedback inhibition*—the inhibition of an allosteric enzyme by one of the products of a later reaction in the biochemical pathway. The positive modulation by ADP and AMP, on the other hand, constitutes *activation*. Both types of control are illustrated in the figure.

Other steps of this pathway for glucose breakdown are also regulated. The product of the first step of glycolysis, glucose-6-phosphate, inhibits the enzyme that produces it, thereby assuring that additional glucose is not activated until the glucose-6-phosphate has been consumed in subsequent reactions. In the citric acid cycle, an excess of ATP or NAD_{re} inhibits the enzyme that converts isocitric acid into α-ketoglutaric acid, while an excess of ADP or NAD_{ox} activates it. Hence, the cycle continues only if there are elec-

tron acceptors ready to be charged. Each of these control mechanisms assures that the rate at which this glucose pathway operates is precisely matched to the current needs of the cell.

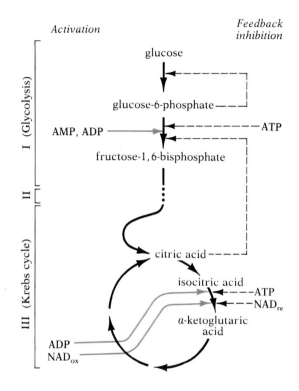

glucose: two NAD_{re} in glycolysis, two NAD_{re} in the breakdown of pyruvic acid to acetyl-CoA, and six NAD_{re} plus two FAD_{re} in the Krebs cycle (Fig. 7.7).

How is this energy used to synthesize ATP? We said earlier that under aerobic conditions the regeneration of NAD_{ox} from NAD_{re} is achieved by the passage of electrons from NAD_{re} to O_2, with oxygen thus acting as the ultimate acceptor of electrons:

$$O_2 + 2\ NAD_{re} + 2H \longrightarrow 2H_2O + 2\ NAD_{ox}$$

The NAD_{re} does not, however, pass its electron directly to the oxygen, as this summary equation might seem to indicate. The electrons and their associated proton reach their ultimate targets indirectly. In particular, the hydrogen electrons are passed down a "respiratory chain" of electron-transport compounds, many of which are iron-containing enzymes called

7.8   The respiratory electron-transport chain

The reactions summarized in this diagram take place in the inner membrane of the mitochondrion. NAD_{re} donates two electrons and a proton to the electron-transport chain; a second hydrogen ion is drawn from the medium. The electrons are passed from one acceptor substance to the next, step by step down an energy gradient from their initial high energy level in NAD_{re} to their final low energy level in H_2O. (When available, molecules of FAD_{re}, not shown, can also donate their electrons to the electron-transport chain; since these electrons have less energy than those of NAD_{re}, they enter lower down on the chain, at Q.) Each successive acceptor molecule is cyclically reduced when it receives the electrons and then oxidized when it passes them on to the next acceptor molecule. At three sites along the chain, some of the free energy released is used to pump H^+ ions into the compartment outside the inner membrane. Later the H^+ gradient generated by the electron-transport chain is used in the synthesis of ATP. The electron acceptors are a flavoprotein (FP); coenzyme Q; cytochromes a, a_3, b, b_2, c, and o; and two proteins containing iron and sulfur, FeS_a and FeS_b. For simplicity, steps have been combined wherever possible. A more complete sequence is shown as part of Figure 7.10.

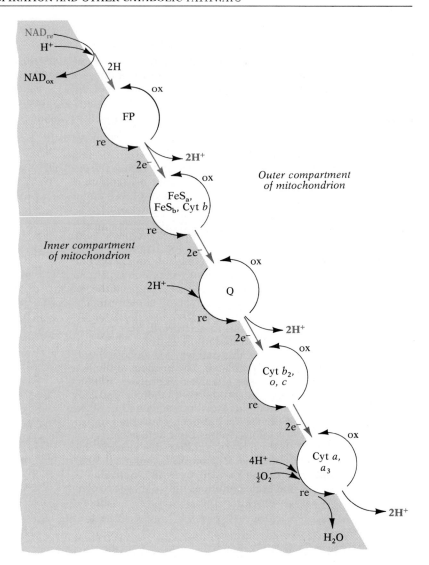

cytochromes (Fig. 7.8). The energy extracted by way of the electron-transport chain is then used for the production of ATP.

THE ANATOMY OF RESPIRATION

At this point the elaborate internal structure of the mitochondrion begins to make sense; indeed, it helps explain how respiration works. You may recall that an inner membrane divides each mitochondrion into two compartments, and that this membrane, being extensively folded, has a very large surface area (Fig 7.9). Since the usual role of biological membranes is to create and keep separate two different chemical environments and to serve as scaffolding for gates, pumps, and organized enzyme arrays, you can probably guess that the capacious inner membrane of the mitochondrion is the key to understanding how this organelle works.

To begin with, we now know that the inner membrane is clearly polarized; its inner face is studded with 9-nm spheres (thought to be the exposed parts of ATP-synthesizing F_1 enzyme complexes), and its outer face is smooth. And while the enzymes of the Krebs citric acid cycle are contained in the fluid of the inner compartment, the enzymes of the electron-transport chain are found in the inner membrane.

The anatomy of the electron-transport chain and of the other stages of aerobic respiration is shown in Figure 7.10. The sequence of biochemical events begins with glycolysis (upper left), and takes place in the cytosol of the cell, not the mitochondrion. The pyruvic acid generated in glycolysis is transported through both the outer and the inner membranes to the inner compartment of the mitochondrion. There it is converted into acetyl-CoA and enters the Krebs cycle (bottom center), which, as we have seen, yields CO_2, ATP, FAD_{re} (not shown), and NAD_{re}. The ATP is exported to the cytosol by means of a special exchange pump (far right), whose source of power we shall describe shortly.

The high-energy compound NAD_{re} carries its electrons to the inner membrane. Here NAD_{re} is oxidized by losing a hydrogen and an electron to FP, which is reduced; simultaneously, FP accepts an H^+ from the medium of the inner compartment. The hydrogen from NAD_{re} is split into a hydrogen ion (H^+) and an electron (e^-). The H^+ ion, along with the H^+ accepted from the medium, is deposited in the outer compartment; the two electrons are passed immediately to FeS_a. The result is that the H^+ concentration of the outer compartment is raised, high-energy-level electrons (two from each NAD_{re}) are inserted into the transport chain (center), where they will be used to do work, and NAD_{ox} is returned to the citric acid cycle to be "recharged."

The transfer of electrons from one acceptor to another proceeds because free energy is liberated (that is, lost by electrons) at each step. For each electron reaching enzyme Q (which shuttles back and forth across the membrane), another hydrogen ion enters the chain from the inner compartment, and is deposited in the outer compartment. (At this stage, the lower-energy electrons of FAD_{re} can also enter the electron-transport chain.) The electrons, whatever their origin, now move through the cytochrome series (left) to cytochrome a_3. This enzyme uses the energy of the electrons to split molecular oxygen (O_2) and catalyze the reaction

$$\frac{1}{2}O_2 + 4H^+ + 2e^- \xrightarrow{\text{cytochrome } a_3} H_2O + 2H^+$$

This reaction reduces by four the number of hydrogen ions in the inner compartment. Two of these hydrogen ions are exported to the outer compartment, while the other two are incorporated into the water molecule.

Chemiosmotic synthesis of ATP (Stage V of respiration)   All the energy stored in the NAD_{re} has now been used up without generating any new ATP. However, an electrostatic and osmotic-concentration gradient has been built up across the inner membrane, which is relatively impermeable to H^+ ions: the inner compartment, having lost H^+, has become negatively charged, while the outer compartment—the space between the inner and outer membrane—is filled with H^+ ions and therefore positively charged. The result is like a battery, and it is the flow of current (H^+ ions rather than the electrons of ordinary batteries) down the electrostatic and osmotic gradient of this continually charging mitochondrial battery that is subse-

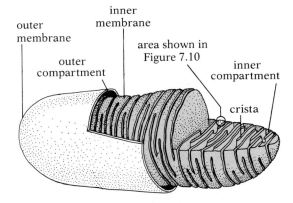

7.9   Structure of a mitochondrion
Much of the outer membrane has been cut away, and the interior has been sectioned to show how the inner membrane folds into cristae. The mitochondria of metabolically very active cells have more cristae than those of less active cells. The inner compartment is within the inner membrane, while the outer compartment is the space between the two membranes. As shown in Figure 7.10, much of the activity of cellular respiration, such as electron transport, occurs across the inner membrane between the inner and outer compartments.

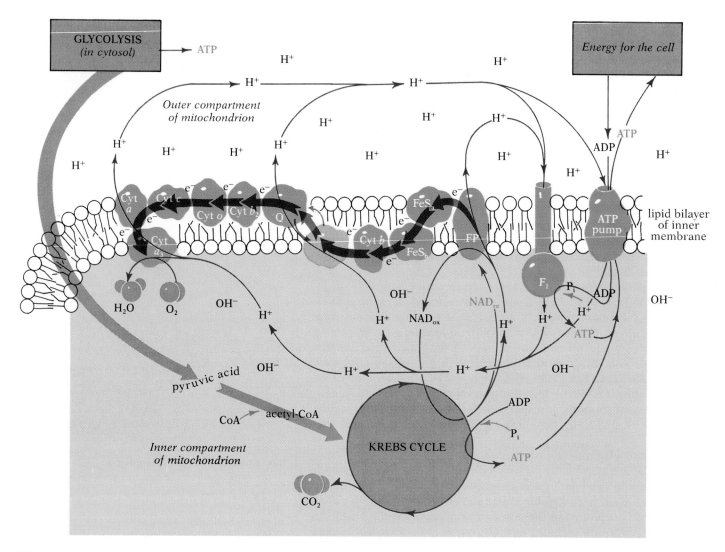

GLYCOLYSIS (in cytosol) → ATP

Energy for the cell

Outer compartment of mitochondrion

lipid bilayer of inner membrane

ATP pump

H_2O O_2

pyruvic acid

CoA   acetyl-CoA

Inner compartment of mitochondrion

KREBS CYCLE

CO_2

7.10 Anatomy of respiration

Glycolysis takes place in the cytosol of the cell, and supplies pyruvic acid, which is converted into acetyl-CoA in the inner compartment of the mitochondrion. The Krebs cycle operates in the inner compartment, producing CO_2, ATP, NAD_{re}, and FAD_{re}. Mounted in the inner membrane of the mitochondrion, which separates the inner and outer compartments, are the enzymes of the electron-transport chain. They extract the energy from NAD_{re} (and FAD_{re}, by a process not shown here) and use

it to pump H^+ ions from the inner to the outer compartment. Some of the ions are carried across by Q, which shuttles between the inner and outer surface of the membrane. The energy of the resulting electrochemical gradient is then used by the enzyme complex F_1 (right) to make ATP from ADP. Another membrane protein is responsible for exporting the ATP to the cell and importing ADP for phosphorylation. Some of the energy of the H^+ gradient is probably used to effect this exchange.

Approximately two H^+ ions must pass through the F_1 complex to create one ATP from ADP, and since the energy of the electrons donated by each NAD_{re} molecule is used to transport six H^+ into the outer compartment, the oxidation of an NAD_{re} produces about three ATPs. Two electrons from a molecule of FAD_{re}, which have less energy than those of NAD_{re}, can enter the electron-transport chain at Q; the oxidation of the lower-energy compound FAD_{re} produces about two ATPs.

quently used to make ATP from ADP. Most of the H^+ ions return from the outer to the inner compartment by way of the F_1 complex (Fig. 7.10), passing down what has come to be called the *chemiosmotic gradient*. As they do so, the energy of the combined potentials of the electrochemical gradient is

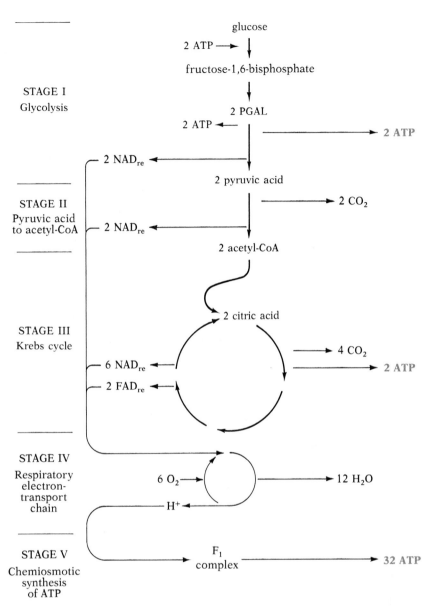

glucose

2 ATP ⟶

fructose-1,6-bisphosphate

STAGE I
Glycolysis

2 PGAL

2 ATP ⟵   ⟶ **2 ATP**

2 NAD$_{re}$ ⟵

2 pyruvic acid

STAGE II
Pyruvic acid
to acetyl-CoA

⟶ 2 CO$_2$

2 NAD$_{re}$ ⟵

2 acetyl-CoA

STAGE III
Krebs cycle

2 citric acid

⟶ 4 CO$_2$

6 NAD$_{re}$ ⟵

⟶ **2 ATP**

2 FAD$_{re}$ ⟵

STAGE IV
Respiratory
electron-
transport
chain

6 O$_2$ ⟶

⟶ 12 H$_2$O

H$^+$ ⟵

STAGE V
Chemiosmotic
synthesis
of ATP

F$_1$
complex

⟶ **32 ATP**

7.11 The ATP yield from complete breakdown of glucose to carbon dioxide and water
In the first three stages, four molecules of ATP are directly synthesized, along with ten of NAD$_{re}$ and two of FAD$_{re}$. In Stage IV, the NAD$_{re}$ and FAD$_{re}$ are fed to the electron-transport chain, and their stored energy is utilized to create a difference in charge across the inner membrane of the mitochondrion. Finally, in Stage V, the F$_1$ enzyme complex uses the energy of this chemiosmotic gradient across the membrane to make approximately 32 more molecules of ATP.

harvested: by processes yet to be understood, the abundant energy generated by this elegant system is converted and stored in the phosphate groups of ATP. The same H$^+$ gradient probably also powers the pump that exports ATP to the cytosol. Peter Mitchell of the Glynn Research Laboratories in England won the Nobel Prize in 1978 for working out this indirect route, by which mitochondria convert the energy of NAD$_{re}$ into ATP.

SUMMARY OF RESPIRATION ENERGETICS

Now that we have looked at all five stages in the aerobic breakdown of glucose, we can summarize their combined energy yield. The overall flow of energy in aerobic respiration is summarized in Figure 7.11. As we have seen, glycolysis generates two molecules of ATP per molecule of glucose,

TESTING THE MITCHELL HYPOTHESIS

Convincing evidence for Peter Mitchell's brilliant chemiosmotic hypothesis comes from several researchers, including (besides Mitchell) Efraim Racker, Andre T. Jagendorf, and Peter C. Hinkle, all of Cornell University. Much of their research involves artificially reconstructed vesicles. The technique, widely used in the study of membrane function in general, takes advantage of the ability of phospholipids in an aqueous solution to form spontaneously into spherical liposomes. When mitochondria are exposed to ultrasound (sound of very high frequency), both the outer and inner membranes break into sheet-like fragments; the inner membranes subsequently "heal" by forming spheres about 100 nm in diameter, as shown in the accompanying figure. Curiously enough, in these liposomes—called submitochondrial vesicles—the 9-nm spheres that are thought to be parts of the ATP-synthesizing F_1 complexes are found on the outer face, rather than on the inside, as in mitochondria. This characteristic inversion is the basis of a technique whereby researchers disrupt the mitochondria, let the submitochondrial vesicles form in one kind of solution—trapping that chemical milieu inside the vesicles—and then transfer the vesicles to a different chemical environment. In this way, the chemistry of the two sides of the membrane can be altered at will.

According to Mitchell's chemiosmotic battery model, in a normal mitochondrion the necessary gradient exists because the H^+ concentration is higher on the outside of the inner membrane than on the inside. If the model is correct, the inside-out patches of membrane should function when the H^+ concentration is higher on the inside. And in fact, when the vesicles are created with more H^+ inside than out, they begin producing ATP, but when the H^+ concentration is higher on the outside, they do nothing. The model also predicts that if these vesicles are supplied with NAD_{re} on the outside, they will increase the gradient by raising the H^+ concentration on the inside. This too does indeed happen. In addition, the model predicts that the insertion into these vesicles of channels permeable to H^+ should short-circuit the battery by allowing H^+ to cross the membrane freely, and should thereby prevent ATP synthesis. Again, this is exactly what happens.

Mitchell's chemiosmotic hypothesis has helped biologists understand a great deal. It clarifies the structure and functioning of the mitochondrial electron-transport chain. It suggests why the mitochondria are separate, highly organized organelles. The Mitchell hypothesis, as discussed in the next chapter, also accounts for the workings of chloroplasts, and adds credibility to the hypothesis that life evolved from anaerobic procaryotes to aerobic (mitochondrialike) procaryotes, and that photosynthetic chloroplastlike procaryotes evolved from aerobic procaryotes.

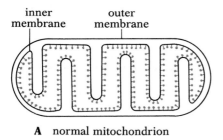

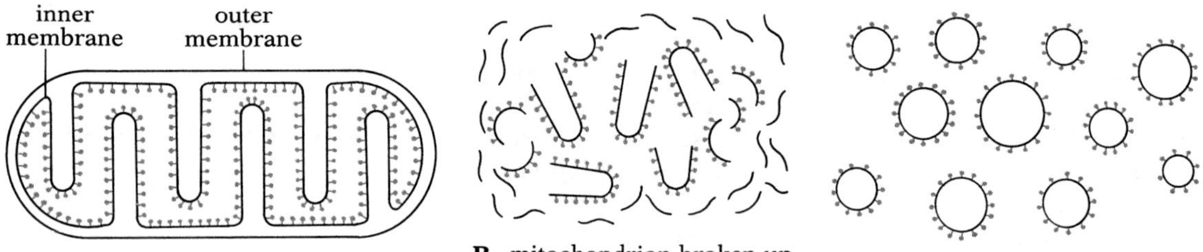

A normal mitochondrion

B mitochondrion broken up by ultrasound

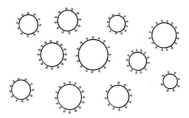

C submitochondrial vesicles formed when fragments "heal"

Formation of submitochondrial vesicles
Submitochondrial vesicles are made by breaking the mitochondrion into pieces with ultrasound (B), and then allowing the bits to "heal" (C). By controlling the medium in which vesicle formation takes place, and the solution into which the resulting inside-out membrane spheres are then put, the researcher can create virtually any desired combination of internal and external chemical environments.

along with two NAD_{re}; the conversion of pyruvic acid into acetyl-CoA yields another two NAD_{re}; the Krebs citric acid cycle produces an additional two ATP, six NAD_{re}, and two FAD_{re}; finally, and most critically, the charging of the mitochondrial battery—the expulsion of H^+ across the inner membrane by means of the electron-transport chain (driven by oxidation of the NAD_{re} and FAD_{re})—stores enough energy to synthesize roughly another 32 ATP. It should now be clear why atmospheric oxygen, and the unimpeded operation of the electron-transport chain, are so important to life as we know it: aerobic respiration extracts 18 times as much energy from glucose as does anaerobic metabolism (glycolysis followed by fermentation). We can see, then, why a metabolic poison like cyanide (which binds irreversibly with a cytochrome and thereby blocks the electron-transport chain) is fatal—the cell is suddenly denied 94 percent of its normal energy. This loss is normally disastrous. However, parts of our bodies *can* operate anaerobically for short periods. During intensive exercise, for instance, muscles often need so much energy that the oxygen supplied from breathing is insufficient. In such cases glycolysis and fermentation provide the needed energy for a time, but they are inefficient and fatigue soon results. Later, the oxygen debt is paid back by deep breathing or panting, and the lactic acid which accumulated in the muscles as a result of fermentation is removed to the liver and reconverted into glucose.

It is now simple to calculate the overall efficiency of aerobic respiration. We know that a molecule of glucose has a free-energy content of about 670 kcal/mole, while a molecule of ATP stores about 7.3 kcal/mole. Since the 36 ATPs that are generated therefore represent just under 270 kcal/mole, the cell has retained only 39 percent of the energy originally stored in the glucose; the other 61 percent is released, primarily as heat. This liberated energy is essential to the efficient shuttling of the reactants through the various chemical chains. And as we shall see in a later section, there is another use for some of the inevitable "waste" heat.

METABOLISM OF FATS AND PROTEINS

Cells can extract energy in the form of ATP not only from the carbohydrates we have focused on so far, but also from the two other major categories of nutrients: fats and proteins. As Figure 7.12 shows, early steps in the breakdown of fats and proteins create products that can be fed into the enzyme pathways we have already discussed.

Catabolism of fats begins with their hydrolysis to glycerol and fatty acids. The glycerol (a three-carbon compound) is then converted into PGAL and fed into the glycolytic pathway at the point where PGAL normally appears. The fatty acids, on the other hand, are broken down to a number of two-carbon fragments, which are converted into acetyl-CoA and fed into the respiratory pathway at the appropriate point. Since fats are more completely reduced compounds than carbohydrates (that is, they have a higher proportion of hydrogens), their full oxidation yields more energy per unit weight; one gram of fat yields slightly more than twice as much energy as one gram of carbohydrate.

The amino acids produced by hydrolysis of proteins are catabolized in a variety of ways. After removal of the amino group (deamination) in the

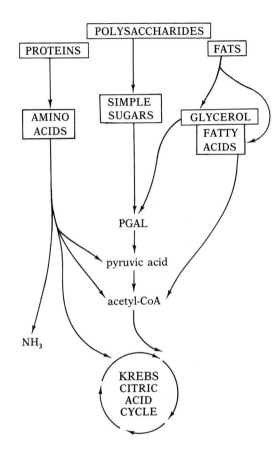

7.12 The relationships of the catabolism of proteins and fats to the catabolism of carbohydrates

form of ammonia (NH_3), some amino acids are converted into pyruvic acid, some into acetyl-CoA, and some into one or another compound of the citric acid cycle. Complete oxidation of a gram of protein yields roughly the same amount of energy as one gram of carbohydrate.

Such compounds as pyruvic acid, acetyl-CoA, and the compounds of the citric acid cycle, which are common to the catabolism of several different types of substances, not only play a crucial role in the oxidation of energy-rich compounds to carbon dioxide and water but also function in the anabolism of amino acids, sugars, and fats. They serve as biochemical crossroads, at which several enzyme pathways intersect. By investing energy, the cell can reverse the direction in which substances move along some of these pathways; for example, the PGAL and acetyl-CoA produced at different points in the breakdown of carbohydrate can be moved up the pathways to glycerol and fatty acids, for use in building fats. Similarly, many amino acids can be converted into carbohydrate via the common intermediates in their metabolic pathways. Not all pathways are two-way, however; most higher animals lack enzymes for converting fatty acids into carbohydrate.

The inability of animals to make carbohydrate from fatty acids, despite the convergence of the metabolic pathways for these two classes of compounds, shows that synthetic pathways are seldom simply the reverse of degradative pathways. For example, while it is technically true that the same enzymes can catalyze the back reactions as well as the forward reactions outlined in Figure 7.3, several of the reactions have equilibrium constants so high that for all practical purposes they are irreversible in the living cell. Hence the enzymes that catalyze the preparatory steps of glycolysis (those that convert a molecule of glucose into two molecules of PGAL) are not necessarily the same enzymes that catalyze the synthesis of glucose from PGAL in photosynthesis. Similarly, some of the reactions in the citric acid cycle are essentially irreversible, but the effect of the back reactions may be achieved by alternative pathways. It is the irreversibility of the reaction from pyruvic acid to acetyl-CoA, in combination with the absence of an alternative enzyme pathway, that makes higher animals unable to convert fatty acids into carbohydrate.

BODY TEMPERATURE AND METABOLIC RATE

We have seen that cellular metabolism captures some of the energy released by the oxidation of carbohydrates, fats, and proteins and converts it into the energy of high-energy phosphate groups in ATP. But this process fails to capture roughly 60 percent of the energy—energy that serves instead to make metabolism highly favorable thermodynamically. Most of this energy is released in the form of heat. The vast majority of animals, as well as all plants, promptly lose most of this thermal energy to their environment. Such animals are known as cold-blooded; more accurate terms are *poikilothermic* ("of variable temperature") and *ectothermic* ("externally heated"). Because the animals' heat comes largely from external sources, their body temperature fluctuates with the environmental temperature; when they are at rest, it is nearly the same as that of the surrounding medium, particularly if the medium is water.

The metabolism of an organism is closely tied to temperature. Within the narrow range of temperatures to which the active organism is tolerant, the metabolic rate—measured by the organism's rate of oxygen consumption and/or carbon dioxide production under standard conditions—increases with increasing temperature and decreases with decreasing temperature in a very regular fashion. The relationship between metabolic rate and temperature is often expressed in terms of a value, called the Q_{10}, which measures the rate increase for each 10°C rise in temperature. If the rate doubles for each 10° rise, the Q_{10} is said to be 2; if the rate triples for each 10° rise, the Q_{10} is said to be 3; and so forth. Metabolic rates frequently have a Q_{10} of about 2. Let us denote the metabolic rate of a given animal at 0°C by X. If its metabolic rate has a Q_{10} of 2, then at 10°C the rate will be $2X$, at 20° $4X$, at 30° $8X$, and at 40° $16X$. Notice that the rate increases more and more rapidly as the temperature increases; when represented graphically, as in Figure 7.13, this type of exponential increase produces a curve that becomes steeper and steeper as the temperature rises.

As would be expected, the activity of ectothermic animals is radically affected by temperature changes in their environment. As the temperature rises (with narrow limits), they become more active; as the temperature falls, they become sluggish and lethargic. Such animals, then, are restricted as to the habitats they can effectively occupy because they are at the mercy of the temperatures in those habitats.

A few animals, notably mammals and birds, can make use of the heat produced during the exergonic reactions of their metabolism, because they have evolved mechanisms—often including insulation by fat, hair, feathers, etc.—whereby heat loss to the environment is retarded. Such animals are commonly called warm-blooded; biologists use the term *homeothermic* ("of uniform temperature"). In recent years the term *endothermic* ("internally heated") has become increasingly accepted. The body temperature of animals that have internally produced heat is fairly high—usually higher than the environmental temperature—and relatively constant even when the environmental temperature fluctuates widely. The metabolic rate of endotherms can accordingly be maintained at a uniformly high level, and they remain very active. They are thus less dependent on environmental temperatures than ectotherms, and are freed for successful exploitation of more varied habitats.

It is not surprising that endotherms generally have body temperatures considerably higher than the average temperature of their environment. Not only does a high temperature produce a high metabolic rate and make possible a high activity level, but a constant temperature higher than that of the surroundings is much easier to maintain than one lower than that of the surroundings. The animal can have very effective insulation, such as the thick fur of polar bears or the massive fat layers of seals and whales; it can shunt blood away from blood vessels near the body surface; and it can speed up heat production by such responses as shivering, which is intensive muscle activity that rapidly uses up ATP energy and thus stimulates more cellular respiration and more heat production. In these ways an arctic mammal like the Eskimo dog, with a body temperature of 38.3°C, may be comfortable at −30°C or less, a temperature more than 60° below its body temperature.

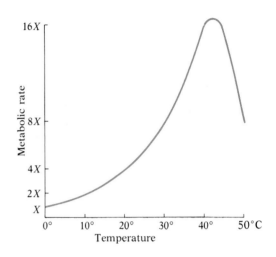

7.13   Graph of changes in metabolic rate with changes in temperature

The hypothetical organism has a Q_{10} of 2—that is, its metabolic rate doubles with every 10° rise in temperature. This rise is the result of the greater thermal energy of the reactants in the cell and the increasing effectiveness of the cellular enzymes. The abrupt decline above 40° represents the point at which the weak bonds that hold enzymes in their specific active conformations begin to break. As a result the enzymes become denatured and metabolic activity is severely disrupted.

But no mammal or bird could live more than a very short time in an environment 60° hotter than its body temperature. In fact, few endotherms can long withstand environmental temperatures more than a few degrees above their body temperature. Their cooling mechanisms are simply not effective enough. Their metabolism, with its unavoidable heat production, is by its very nature a furnace, not a refrigeration unit; it can easily be speeded up to counteract environmental cold, but it cannot be made into a cooling device. The external heat will tend, in fact, to do just the reverse of what is needed: speed up the metabolic rate according to the relationship we have already discussed. The animal can, of course, avoid the heat to some extent by seeking out a shady spot or retiring to a cool burrow. It may also have mechanisms for shunting much blood into surface capillaries, where heat loss is greatest, and for evaporative cooling such as sweating in humans, panting in dogs, and evaporation from licked parts in cats. But these methods are effective only in temperatures below, at, or just above body temperature; they cannot long counteract very high temperatures. In short, the most effective thermal regulatory devices available to animals are based largely on heat production and conservation, not on heat loss and cooling. Endothermy, then, involves mechanisms that can produce relative metabolic stability at a body temperature above that of the environment, but not at one much below that of the environment.

Under certain conditions many ectotherms can, like endotherms, take advantage of metabolic thermal energy. For example, on a cold day butterflies and moths are apt to vibrate their wings for several minutes before launching into the air. The heat produced by cellular respiration in the vibrating muscles may increase the muscle temperature by as much as 15°C in five or six minutes, the muscles thus getting into a condition in which they can contract fast enough to produce normal flight. Obvious analogies are the warming-up of an automobile or airplane engine and the warming-up exercises of athletes. Similarly, any rapidly moving ectotherm produces metabolic heat faster than that heat can radiate away into the surrounding medium, and the animal's body temperature consequently rises well above that of the environment. Honey bees provide a particularly interesting example of thermal regulation by ectotherms. When the temperature in their hive falls below a critical value, the bees become very active, releasing enough body heat to raise the hive temperature and maintain it at a level well above that of the outside environment. Conversely, when the hive temperature starts rising too high in the summer, the bees collect water, which they spread on the inner hive surfaces. They then create ventilating drafts by fanning their wings; despite the heat generated by this activity, the net result is cooling of the colony by evaporation. Since, as mentioned in Chapter 2, the evaporation of a small volume of water utilizes a relatively large amount of heat, this is an effective strategy; indeed, evaporation is used for cooling throughout the animal world.

In both endothermic and ectothermic animals and in plants, the normal metabolic rate is inversely related to body size; the smaller the organism, the higher the *relative* metabolic rate—that is, the higher the metabolic rate per gram of body tissue (Fig. 7.14). The reason for this is easily understood in the case of endotherms: smaller animals have a greater surface-to-volume ratio, and consequently a larger relative heat loss to the environment

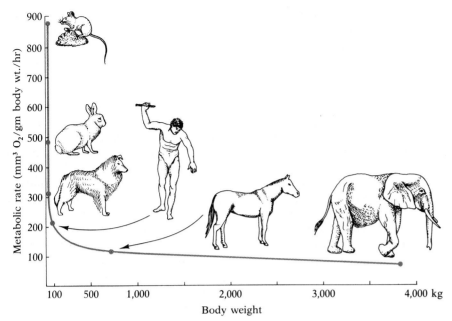

7.14 Graph showing inverse relationship of relative metabolic rate and body size in mammals

per unit time. To maintain a constant high body temperature despite rapid heat loss across its body surface, a small animal must oxidize food at a very high rate. Because the relative amount of food consumed and the pace of digestion, respiration, and so on must rise with decreasing size, there is a lower limit on the size of endotherms. The smallest living mammals are shrews, weighing only about 4 grams. They must eat nearly their own body weight of food every day, and can starve to death in a few hours if deprived of food.

It is more difficult to explain the inverse relationship between size and relative metabolic rate in ectothermic animals and in plants. Since cold-blooded organisms lose their metabolic heat to the environment and do not normally respond to heat loss by increased metabolism, larger size and its concomitant smaller surface-to-volume ratio should retard heat loss some-what, and the conserved heat ought then to speed up metabolism. In fact, why larger size in plants and ectotherms tends to be correlated with *lower* relative metabolic rates has never been fully explained. One likely factor is that increasing size generally involves a disproportionate increase of skeletal and other connective tissues in animals, and of supportive fibers and mature xylem in plants—a redwood requires a great deal more inactive support structure than a daisy, and an alligator more than a salamander. Since these tissues are relatively inactive metabolically, the average metabolic rate per unit weight for the organism as a whole may fall as the proportion of these less active but necessary structural tissues rises. This is seen in the course of development of an embryo: the early embryo, composed almost exclusively of metabolically active cells, has a high relative metabolic rate, while the later embryo has developed a larger proportion of the less active types of tissue, so its relative metabolic rate falls.

The relationship between metabolic rate and body size in endotherms has serious implications for small animals during the cold seasons of the year. Not only does the rate of heat loss rise at such times, but the food supply is generally low. Small mammals belonging to three groups—the insectivores, bats, and rodents—have evolved a mechanism that enables them partly to evade this problem. When winter comes, they *hibernate*: their body temperature falls far below its normal level, and their metabolism, heart rate, respiration, etc., are greatly depressed. The animals pass the winter in this dormant state, using up their energy reserves very slowly. Bats are of particular interest, because they not only hibernate in winter in cold climates, but go into a similar dormant state during the daylight hours every day, thus conserving the energy derived from the food they eat during the night. The reverse occurs in hummingbirds (which are very near the lower size limit for endotherms); they are active during the day and become torpid during the night. With a few exceptions, however, most birds do not hibernate during the winter; they either remain active, spending much of their time feeding, or avoid the cold by migrating to a warmer region.

Large mammals like bears are good heat conservers and, as we have seen, have a relatively low metabolic rate. Though they do not hibernate during the winter, they become relatively inactive and spend much of their time sleeping, while using up their extensive fat reserves; but their body temperature decreases by only a few degrees and they are not truly dormant.

Chapter 8

ENERGY TRANSFORMATIONS: PHOTOSYNTHESIS

In the last chapter we saw how cells metabolize glucose, first through glycolysis in the cytosol, and then through the remaining steps of aerobic respiration in the mitochondrion. We saw also that from the point of view of bioenergetics, the anaerobic pathway culminating in fermentation is wasteful, simply because oxygen is not available to act as the final acceptor of electrons in the reactions that yield up the energy stored in glucose. Though life existed on the earth for almost 2 billion years in the absence of significant quantities of molecular oxygen, and therefore in the absence of large-scale aerobic respiration, it can hardly be said to have flourished.

According to recent evolutionary evidence, an enormously important biochemical innovation occurred about 3 billion years ago: certain primitive organisms developed for the first time the ability to capture the sun's energy directly and to use it to synthesize foods such as glucose. In these organisms, photons interacting with electrons in special pigment molecules pushed the electrons up to higher levels, and the molecules thus activated passed the energy to still other molecules (Fig. 8.1). According to present evidence, special enzymes evolved to form an elaborate chemical pathway that made some of this energy from the photons available to drive cellular metabolism.

This process was probably the earliest form of photosynthesis, the transformation of light energy into chemical energy. Though it did not liberate oxygen into the early atmosphere, it did free organisms from dependence on inorganic food. Further evolution resulted in a form of photosynthesis that produced molecular oxygen as a by-product. As this highly electronegative substance began to accumulate in the atmosphere, high-efficiency aerobic respiration, with its own enzyme-mediated reac-

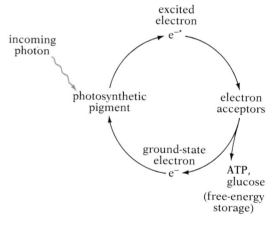

8.1 Energetic basis of life
The basic energy equation of life begins with a photon from the sun which excites an electron in a pigment molecule by moving the electron up to a higher level. The energy thus acquired ultimately excites an electron in an acceptor molecule. Subsequently, as the excited electron is returned to a low-energy level in a pigment molecule by way of a series of acceptors, its free energy is used for generating an energy-storage compound (usually glucose or ATP).

ENERGY PRODUCTION   METABOLISM

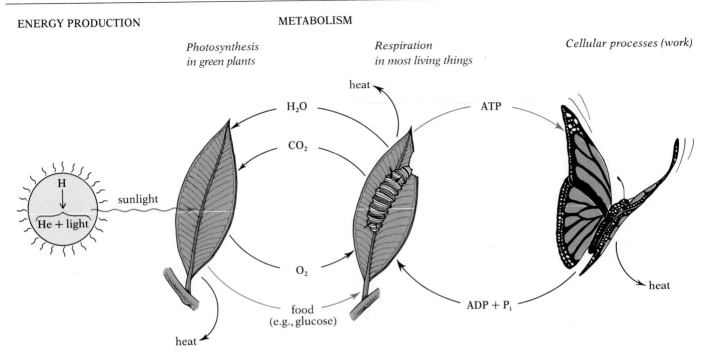

Photosynthesis in green plants   *Respiration in most living things*   *Cellular processes (work)*

8.2   Energy flow in aerobic pathways
With the evolution of so-called noncyclic photosynthesis approximately 2.3 billion years ago, oxygen, the electronegative element essential to the efficient operation of catabolic processes, began to accumulate in the atmosphere. This led in turn to the spread of organisms capable of high-efficiency aerobic respiration, and to the unification of the two great metabolic pathways—photosynthesis and respiration—that is characteristic of life processes today. The pathways are unified by shared products and by-products: organic food, H_2O, CO_2, and O_2. Note that both the caterpillar and the leaf respire.

tions, became the main mechanism for extracting energy from food, and the two major chemical pathways—photosynthesis and respiration—were joined through shared by-products (Fig. 8.2). Life became the dominant feature of the earth.

Today, virtually all organisms depend directly or indirectly on photosynthesis to fill their energy needs. *Autotrophs*, organisms like plants that are capable of making organic nutrients from inorganic materials, depend almost exclusively on photosynthesis, while *heterotrophs*, organisms like animals that must obtain organic nutrients from the environment, depend indirectly on photosynthesis, since they consume autotrophs, or heterotrophs that have eaten autotrophs, or both. Photosynthesis, then, is life's single most important biochemical process. Much about photosynthesis is still not known, but in the last three decades the chemical pathways have become much clearer. This chapter will acquaint you with the broad outlines and thermodynamic logic of photosynthesis.

EARLY RESEARCH IN PHOTOSYNTHESIS

Considering how much is now known of anabolic processes, it is easy to forget that for most of recorded history scientists had no idea that the sun supplies the surface of the earth with virtually all its energy, or that green plants trap that energy and produce the invisible gas we breathe. In fact, only in 1772 did the English clergyman Joseph Priestley demonstrate that green plants affect air in such a way as to reverse the effects of burning or of breathing. As he reported it:

> I flatter myself that I have accidentally hit upon a method of restoring air which has been injured by the burning of candles, and that I have discovered at

least one of the restoratives which nature employs for this purpose. It is vegetation. In what manner this process in nature operates, to produce so remarkable an effect, I do not pretend to have discovered; but a number of facts declare in favour of this hypothesis. I shall introduce my account of them, by reciting some of the observations which I made on the growing of plants in confined air, which led to this discovery.

One might have imagined that, since common air is necessary to vegetable, as well as to animal life, both plants and animals had affected it in the same manner, and I own I had that expectation, when I first put a sprig of mint into a glass-jar, standing inverted in a vessel of water; but when it had continued growing there for some months, I found that the air would neither extinguish a candle, nor was it at all inconvenient to a mouse, which I put into it.

Finding that candles burn very well in air in which plants had grown a long time, and having had some reason to think, that there was something attending vegetation, which restored air that had been injured by respiration, I thought it was possible that the same process might also restore the air that had been injured by the burning of candles.

Accordingly, on the 17th of August, 1771, I put a sprig of mint into a quantity of air, in which a wax candle had burned out, and found that, on the 27th of the same month, another candle burned perfectly well in it. This experiment I repeated, without the least variation in the event, not less than eight or ten times in the remainder of the summer. Several times I divided the quantity of air in which the candle had burned out, into two parts, and putting the plant into one of them, left the other in the same exposure, contained, also, in a glass vessel immersed in water, but without any plant; and never failed to find, that a candle would burn in the former, but not in the latter. I generally found that five or six days were sufficient to restore this air, when the plant was in its vigour; whereas I have kept this kind of air in glass vessels, immersed in water many months without being able to perceive that the least alteration had been made in it.

Priestley's important experiments were the first demonstration that plants produce oxygen, though he himself did not realize that this was what was happening; nor did he realize that light was essential for the process he observed. But his findings stimulated interest in photosynthesis, as we now call it, and led to further investigation. Only seven years later, the Dutch physician Jan Ingenhousz demonstrated the necessity of sunlight for oxygen production (though, like Priestley, he knew nothing about oxygen at that time and explained his results in other terms), and he also showed that only the green parts of the plant could photosynthesize. He reported his results in a book with the richly descriptive title *Experiments upon Vegetables, Discovering Their Great Power of Purifying the Common Air in the Sun-Shine, and of Injuring It in the Shade and at Night.* In 1782 a Swiss pastor and part-time scientist, Jean Senebier, showed that the process depended on a particular kind of gas, which he called "fixed air" (and we call carbon dioxide). Finally, in 1804, another Swiss researcher, Nicolas Théodore de Saussure, found that water is necessary for the photosynthetic production of organic materials.

Thus, early in the nineteenth century, all the important components of the photosynthetic process were at least vaguely known, and could be summarized by the following equation:

$$\text{carbon dioxide} + \text{water} + \text{light} \xrightarrow{\text{green plants}} \text{organic material} + \text{oxygen}$$

Later, scientists came to believe that light energy splits carbon dioxide, CO_2, and that the carbon is then combined with water, H_2O, to form the group —CH$_2$O, on which carbohydrates are based. According to this view, the oxygen released by the plant during photosynthesis comes from CO$_2$. This idea received a severe blow when, about 1930, C. B. van Niel of Stanford University showed that some photosynthetic bacteria, which use hydrogen sulfide, H$_2$S, instead of water as a raw material for photosynthesis, give off sulfur instead of oxygen as a by-product. Now, H$_2$S and H$_2$O have obvious chemical similarities, and if the sulfur produced by the bacteria during photosynthesis came from H$_2$S, it seemed reasonable to suppose that the oxygen produced by plants during photosynthesis might come from H$_2$O rather than CO$_2$. This was eventually shown by use of a heavy isotope of oxygen (^{18}O instead of the usual ^{16}O). If photosynthesizing plants are given normal carbon dioxide plus water containing heavy oxygen, the heavy isotope appears as molecular oxygen:

$$CO_2 + 2H_2{}^{18}O \longrightarrow {}^{18}O_2 + (CH_2O) + H_2O$$

(The parentheses indicate that CH$_2$O is not in itself a molecule.)

We now know that **_chlorophyll_**, the green pigment of plants, traps the energy required to split the water. The currently accepted equation for green-plant photosynthesis is given in more detail below; the dashed lines indicate the fates of all the atoms involved:

$$CO_2 + 2H_2O + light \xrightarrow{\text{chlorophyll}} O_2 + (CH_2O) + H_2O$$

Multiplying this summary equation by 6 has the advantage of showing that glucose, a six-carbon simple sugar, is often the end product:

$$6CO_2 + 12H_2O + light \xrightarrow{\text{chlorophyll}} 6O_2 + C_6H_{12}O_6 + 6H_2O$$
$$(\triangle G = -1300 \text{ kcal/mole})$$

It may seem curious that water should appear on both sides of the equation. The reason is that the water produced by the photosynthetic process is new; it is not the water used as a raw material. You may recall from the last chapter that the free energy stored in glucose is about 670 kcal/mole. The storage of this energy requires capturing roughly 1970 kcal/mole of light energy, the remaining 1300 kcal/mole being liberated during the process, which is thus strongly downhill, or exergonic.

Though the above equation is a convenient summary of photosynthetic carbohydrate synthesis, it tells us nothing about how the synthesis is actually achieved. The process is certainly not one gross chemical reaction, as the summary equation might imply. Many reactions are involved, some that require light, others that require it only indirectly—so-called "dark" reactions. The "light" reactions convert and store the energy from light in specialized energy-transfer molecules like ATP, while the dark reactions use that stored energy to convert (or "fix") carbon dioxide into carbohydrates like glucose.

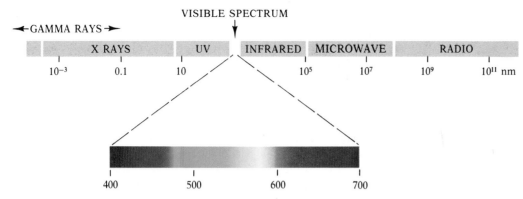

VISIBLE SPECTRUM

◄─GAMMA RAYS─►

X RAYS   UV   INFRARED   MICROWAVE   RADIO

10^{-3} 0.1   10 10^5 10^7 10^9 10^{11} nm

400   500   600   700

THE LIGHT REACTIONS: PHOTOPHOSPHORYLATION

The term *photophosphorylation* is often used to describe the light-dependent reactions of photosynthesis. Photophosphorylation means the use of light energy to phosphorylate (add inorganic phosphates to) a molecule, usually ADP:

$$ADP + P_i + energy \xrightarrow{\text{enzyme}} ATP + H_2O$$

Like so many terms, "photophosphorylation" became part of the scientific vocabulary before the process it describes was well understood. We now know that light-energy absorption and photophosphorylation are separate reactions in just the way the operation of the mitochondrial electron-transport chain and the subsequent synthesis of ATP by the F_1 complex are separate. In fact, as we shall see, the two parts of the light reactions, which occur in chloroplasts, directly parallel these two stages of respiration in the mitochondria.

LIGHT AND CHLOROPHYLL

Light waves constitute one small region of the spectrum of electromagnetic radiations (Fig. 8.3). Each radiation in this spectrum has a characteristic wavelength and energy content. These two characteristics are inversely related: the longer the wavelength, the smaller the energy content. Within the narrow band visible to human beings, the shortest light waves produce the sensation of violet and the longest produce the sensation of red. Radiations such as ultraviolet, X rays, and gamma rays, which are of shorter wavelengths than violet, are invisible to us, as are infrared, microwave, and radio-TV radiations, which are of longer wavelengths than red.

Is all light, regardless of wavelength, equally effective for photosynthesis? To answer this question, we must turn to the all-important green pigment chlorophyll, which, we have said, traps light energy and helps convert it into chemical energy. Of the several slightly different kinds of chloro-

8.3 Portion of the electromagnetic spectrum
Visible light constitutes only a very small portion of the total spectrum. Within the visible spectrum, light of different wavelengths stimulates different color sensations in us. Not only vision and photosynthesis, but also other radiation-dependent biological processes, rely on this same small portion of the electromagnetic spectrum (sometimes extended very slightly into the ultraviolet or infrared). Light of wavelengths shorter than about 300 nm (nanometers) is absorbed by the atmosphere, while wavelengths longer than 800 nm have too little energy to drive biological reactions.

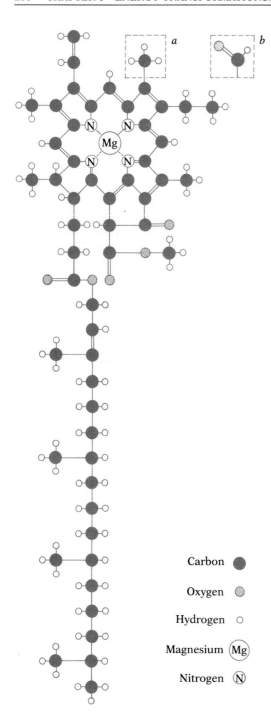

Carbon ●
Oxygen ◉
Hydrogen ○
Magnesium (Mg)
Nitrogen (N)

phyll, the most widespread is chlorophyll *a*, and our discussion will primarily refer to this compound (Fig. 8.4).

Light falling on an object may pass through the object (be transmitted), be absorbed by it, or be reflected from it (Fig. 8.5). We can see transmitted or reflected light, but not absorbed light. Now, if chlorophyll is the material that traps the incident light, and if it appears green to our eyes, several facts suggest themselves. First, chlorophyll cannot be absorbing much radiation of the wavelengths that produce in us the sensation of green, or we wouldn't see green color. Second, chlorophyll must be absorbing radiation of some wavelengths within the visible part of the spectrum, or the light transmitted or reflected to us would appear white (when all visible wavelengths are combined, they produce the sensation of white). At this point we have already partly answered the question of whether all light is equally effective for photosynthesis: green light is not as effective as light of some other colors, since it is not absorbed as readily by chlorophyll.

More precise information can be obtained if chlorophyll is extracted from the leaf and exposed to light of varying wavelengths to determine the amount of absorption at each wavelength. The absorption spectrum of chlorophyll *a* thus obtained (Fig. 8.6A) shows that primarily light in the violet, blue-violet, and red regions is absorbed, while green, yellow, and orange light is absorbed only very slightly. It must be noted that the action spectrum of photosynthesis—a measure of the effectiveness of light of various wavelengths in driving photosynthesis—is somewhat different from the absorption spectrum of chlorophyll *a*. That there is relatively high activity in parts of the spectrum where the chlorophyll absorbs very little light, as Figure 8.6B shows, is an indication that light of some wavelengths not readily absorbed by chlorophyll *a* is still effective in driving photosynthesis. Apparently other pigments that are present in green plants, principally the yellow and orange carotenoids and other forms of chlorophyll, absorb light in these regions of the spectrum and then pass the energy to the chlorophyll *a*. Accessory pigments like the carotenoids thus enable the plants to use light of more different wavelengths than could be trapped by chlorophyll *a* alone.

What happens when light of a proper wavelength strikes a chlorophyll molecule? The exact answer is not known, but many aspects of the process have been elucidated.

8.4 Molecular structure of chlorophyll

The structure of chlorophyll *a* is shown. The structure of the other major type of chlorophyll in green plants, chlorophyll *b*, differs only in the side group shown in the inset: a formyl group (—CHO) is substituted for the methyl group (—CH₃) of the *a* form. As you can probably guess, the long hydrophobic tail of the molecule serves to anchor the chlorophyll in the appropriate membrane. The electron that is excited by light energy is in the "head" region, near the magnesium atom.

In the chloroplasts of photosynthesizing cells in green plants, the chlorophyll and accessory pigments are organized into functional groups called ***photosynthetic units***. Each unit contains some 300 pigment molecules, including chlorophyll *a*, chlorophyll *b*, and carotenoid. One pigment molecule in each unit is distinct from all the rest; it is a specialized form of chlorophyll *a*, which acts as a reaction center. The other pigment molecules function something like antennas responsive to light energy.

As we have seen, light energy comes in discrete units called photons. When a photon strikes a chlorophyll (or a carotenoid) molecule and is absorbed, its energy is transferred to an electron of the pigment molecule; the

8.5 Light striking a leaf
Light striking an object, such as a leaf, may be reflected, absorbed, or transmitted.

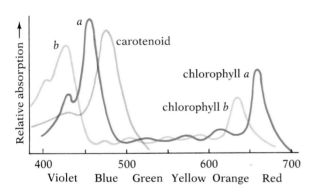

A　　　　　　Wavelength (nm)

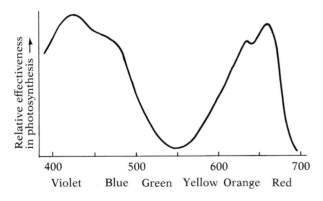

B　　　　　　Wavelength (nm)

8.6 Absorption and action spectra of photosynthetic pigments
(A) Absorption spectra of chlorophyll *a*, chlorophyll *b*, and a carotenoid. Taken together, the absorption spectra of the two chlorophylls and the carotenoid cover more of the range of wavelengths available to the plant than does the spectrum of chlorophyll *a* alone. (B) Action spectrum of photosynthesis. Light of intermediate wavelengths is more effective in driving photosynthesis than would be predicted on the basis of the absorption spectrum of chlorophyll *a* alone. Apparently other pigments, such as carotenoids, absorb light of these intermediate wavelengths to some extent and pass the energy to so-called reaction centers in chlorophyll *a* for photosynthesis.

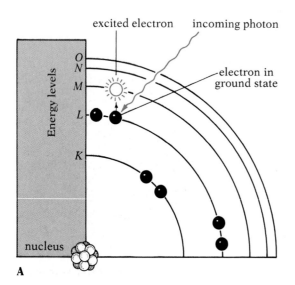

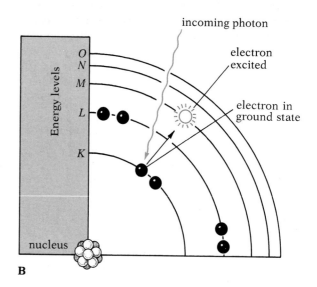

8.7   Effect of light on chlorophyll

(A) When a photon is absorbed by a chlorophyll molecule, the photon's energy raises an electron to a higher energy level. In this example we see a typical distribution of electrons at their lowest available energy levels, and the distribution as it is altered by absorption of a photon, which raises one electron from the second level (*L*) to the third (*M*). (B) As we saw in Figure 8.6A, the absorption spectrum of a particular pigment can have several peaks. These exist because not all electrons require the same amount of energy to be raised to an excited level. For example, an electron at a lower ground state than the one shown absorbing a photon in A can be excited to the same energy level by a photon of appropriate energy, and can then enter the photosynthetic pathway. The higher-energy, lower-wavelength photon is represented here in blue, while the less energetic photon in A is represented in red.

excited electron moves up to a higher, relatively unstable energy level (Fig. 8.7A). Not just any photon will do: because electrons occupy discrete energy levels, the photon must have a particular amount of energy to excite the pigment electron from its ground state to the higher level. Here is the explanation of the well-defined absorption peaks in Figure 8.6A: since, as noted earlier, the energy of a photon is related (inversely) to its wavelength, photons with the proper amount of energy come only from light within a certain range of wavelengths (Fig. 8.7B).

An excited and therefore unstable electron will spontaneously return to its inactivated state almost immediately, giving up its absorbed energy. Isolated chlorophyll in a test tube, for instance, promptly loses the energy it captures by reemitting it as visible light, in a process known as fluorescence: the chlorophyll pigment molecules alone, separated from their photosynthetic units, are incapable of converting light energy into chemical energy. But in the functioning chloroplast, once light energy has raised an electron in an antenna molecule to a high-energy state, the excited state is passed from pigment molecule to pigment molecule,[1] and may eventually reach the reaction-center molecule, which traps it (Fig. 8.8). The excited state is captured here because the free energy of the reaction center, even with this excitation energy, is *lower* than that of the antenna molecules. Hence, once the excited state has reached the reaction center, it cannot easily escape. In this molecule, the energized electron that

---

[1] The transfer of energy from one pigment molecule to an adjacent one does not appear to involve the physical transfer of an excited electron. Instead, when the excited electron in one molecule falls back to a lower energy level, an electron in an adjacent pigment molecule is boosted to a higher level, thus taking on the excited state. Some researchers refer to this process as a transfer of excitation energy.

characterizes the excited state does not decay back to a normal low-energy level. Instead, it is passed to an acceptor molecule and enters a series of enzyme-catalyzed reactions that convert the energy into a form more readily used by the cell.

CYCLIC PHOTOPHOSPHORYLATION

There are two general pathways by which the energy from excited electrons is harvested—the cyclic pathway and the noncyclic pathway. The cyclic pathway involves only one of the two types of photosynthetic units found in most plants; the noncyclic pathway involves both. We shall first examine the cyclic pathway, which is the simpler of the two.

Electron transport   We have seen that the specialized chlorophyll molecule that serves as the reaction center of a photosynthetic unit is capable of passing an energized electron to an acceptor molecule. Once cleared in this way, the reaction-center molecule is ready to be activated again, so photosynthesis can continue. In cyclic photophosphorylation the reaction-center molecule is designated P700 (because it cannot absorb light of wavelengths above 700 nm). The acceptor molecule to which it transfers the energized electron is an enzyme (FeS) containing iron and sulfur. The acceptor is reduced and P700 is oxidized by this electron transfer. The electron is then passed along by a series of membrane-mounted enzymes (Fig. 8.9) very like the electron-transport chain of the mitochondrion. Eventually, the electron reaches a molecule of the enzyme plastocyanin (PC); there it waits to fill an

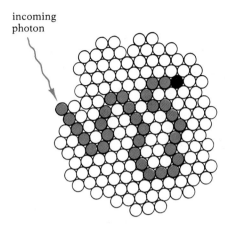

incoming photon

8.8   Flow of excited state within a photosynthetic unit

A photon strikes one of the antenna pigments (open circles) and raises an electron in the pigment to a higher energy level. This excited state is then passed from pigment molecule to pigment molecule in a random sequence (colored pathway) until it eventually reaches the reaction-center molecule (black circle), where it is trapped.

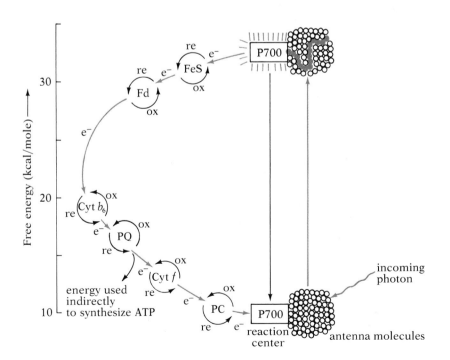

8.9   Cyclic photophosphorylation

A photon of light strikes a pigment molecule in the P700 antenna system. The excited state eventually reaches a molecule of P700, a specialized form of chlorophyll *a*, which is thereby energized. An energized electron from P700 then begins passage from one acceptor molecule to the next, releasing free energy at each step and ultimately returning to the ground state at which it began in P700. Only the energy released in the passage from plastoquinone (PQ) to cytochrome *f* is used by the cell. The other electron acceptors are FeS, an enzyme containing iron and sulfur; ferredoxin (Fd); cytochrome b_6; and plastocyanin (PC).

opening in the P700 molecule. When another electron is energized and transferred to FeS, the slot opens and our first electron fills it, thereby completing the cycle.

When a light-energized electron leaves a chlorophyll molecule, it is energy-rich; when it finally returns, it is energy-poor. The transition is gradual. As the electron is passed from transport molecule to transport molecule in the chain, it releases some of its extra energy with each transfer; in other words, it is eased down the energy gradient step by step from the excited state to the normal state. Hence when the electron finally falls back into the chlorophyll, it has discharged all its extra energy, but it has not discharged it all at once, as would happen with isolated chlorophyll in a test tube. Instead, the energy has been released in a series of small portions of manageable size. The whole process is called *cyclic photophosphorylation* because the electron is returned to the chlorophyll and some of its energy is used (indirectly, as we shall see) to phosphorylate ADP into ATP.

Cyclic photophosphorylation is thought to have been the first form of photosynthesis to evolve, and it is still the only form available to most photosynthetic bacteria. As Figure 8.9 indicates, however, the cyclic system is not very efficient: of the 25 kcal/mole of energy gained as a result of the excitation of P700, only the energy liberated in the passage from PQ to cytochrome *f*—3.4 kcal/mole—is actually made available to the cell. The energy released in the other steps is not utilized; it goes to waste. Though even 3.4 kcal/mole is far better than nothing (particularly since photons cost the cell nothing), most photosynthesis now follows a highly modified, noncyclic pathway, which is more efficient under many conditions. Indeed, just as cellular respiration represents an advance over the anaerobic processes of the glycolytic pathway to fermentation, so does noncyclic photophosphorylation represent an advance over the more primitive cyclic form. As we shall see, in most organisms today cyclic photophosphorylation is only a supplement to noncyclic photophosphorylation.

NONCYCLIC PHOTOPHOSPHORYLATION

Like the cyclic process already described, noncyclic photophosphorylation (Fig. 8.10) begins when a photon of light strikes an antenna molecule of chlorophyll and raises an electron to an excited state, which may be trapped by the reaction-center molecule P700. As before, an energized electron is then led away from the P700 molecule by electron acceptors, the first two being FeS and Fd. But here the similarity to cyclic photophosphorylation ends. Instead of continuing down the cyclic transport chain, the electron is passed from Fd to a different acceptor molecule: the flavoprotein FAD. FAD passes the electron to an extremely important substance called nicotinamide adenine dinucleotide phosphate, or *NADP*, which is closely related to the NAD in mitochondria. The antenna molecules and the P700 reaction center, together with the electron-transport chain from FeS to Fd to FAD, constitute *Photosystem I*.

Unlike the reduced electron-acceptor molecules in cyclic photophosphorylation, NADP does not promptly pass along the electrons it receives from the electron-transport chain to another acceptor molecule. Instead, it retains a pair of energized electrons and their associated protons. Eventu-

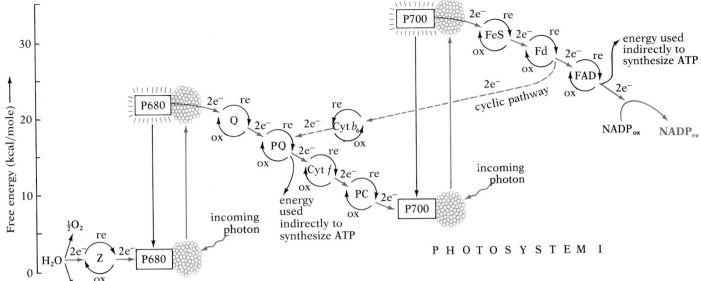

ally NADP$_{re}$ acts as an electron donor in the reduction of CO_2 to carbohydrate such as glucose, a process known as **carbon fixation**. Thus electrons move from the chlorophyll through an electron-transport chain to NADP to carbohydrate (via other intermediate compounds). In addition, some of the energy liberated in the passage from FAD to NADP is used, indirectly, in the synthesis of ATP.

Now, if the energized electrons from P700 are retained by NADP$_{re}$ and eventually incorporated into carbohydrate, it follows that Photosystem I is left short of electrons, that it is left with "electron holes." These electron holes are filled, indirectly, by electrons derived from water through a process we shall now examine.

At one time, it was thought that the electrons from the water passed, via a few transport molecules, directly to Photosystem I, and hence that there was only one light-driven event in photosynthesis—the one that initially excited the chlorophyll electrons of Photosystem I. Later, however, it became apparent that there are two light events, the one we have already discussed and a second one more intimately related to the splitting of water. This second light event involves a different type of photosynthetic unit, which contains about 200 molecules of chlorophyll a, about 200 molecules of chlorophyll b, c, or d, depending on the species of plant,[2] and one molecule of a specialized form of chlorophyll a, called P680, that acts as the reaction center for the unit. (As you may have guessed, the designation P680 indicates that this molecule cannot absorb light of wavelengths above 680 nm.) The antenna molecules and the P680 reaction center, plus its special set of electron-transport molecules, constitute **Photosystem II**.

[2] Chlorophyll b is found in green algae, bryophytes, and vascular plants. Chlorophyll c occurs in brown algae, and chlorophyll d in red algae.

8.10 Noncyclic photophosphorylation

As in cyclic photophosphorylation, electrons move along the pathway one at a time. However, for convenience in illustrating reactions at the two ends of the noncyclic pathway, the passage of two successive electrons is shown in the diagram. One of the two essential light events occurs when a photon of light strikes a pigment molecule in Photosystem II; the resulting excited state eventually reaches a molecule of P680, which then donates an energized electron to substance Q. The electron moves down the Photosystem II electron-transport chain from Q to PQ to cytochrome f to PC, and as in cyclic photophosphorylation, the energy released in the passage from PQ to cytochrome f is used by the cell. At PC the electron waits to fill an opening in the P700 molecule, which occurs when a photon strikes Photosystem I and P700 transfers an energized electron to the Photosystem I electron-transport chain. This electron passes from FeS to Fd to FAD, and finally to NADP. Energy released in the step down from FAD is used by the cell, and for each two electrons transported, one NADP$_{re}$ is generated to supply energy for carbon fixation. The vacancy in P680 that was created by the passage of an electron to Q is filled by an electron released in the splitting of water (lower left), with one molecule of water yielding two electrons, along with two H^+ ions and one atom of oxygen. As indicated by the changes in relative free energy, the electron that has moved through Photosystem II still has much of its original energy when it enters Photosystem I. The primitive cyclic pathway (which connects Fd to PQ via cytochrome b_6) is also shown.

When light of the proper wavelength strikes a pigment molecule of Photosystem II, the energy is passed around within the photosynthetic unit until it finally reaches the molecule of P680. This molecule, in turn, donates a high-energy electron to an acceptor, designated "Q" (Fig. 8.10). Substance Q then passes the electron to a chain of acceptor molecules, which transports the electron, by some of the same steps we saw in cyclic photophosphorylation, down an energy gradient to the electron hole in the P700 molecule of Photosystem I. As the electron moves down the transport chain, some of the energy released along the way is used by the cell indirectly to synthesize ATP.

Thus the electron holes created in Photosystem I by the first light event are refilled by electrons moved from Photosystem II by the second light event. But this process alone would leave electron holes in Photosystem II; the electron deficit would simply have been shifted from Photosystem I to Photosystem II. It is at this point that the electrons from water, mentioned earlier, play their role. As Figure 8.10 indicates, it is thought that P680 (with the aid of an enzyme complex referred to as "Z") pulls replacement electrons away from water, leaving behind free protons and molecular oxygen:

$$2H_2O \longrightarrow 4e^- + 4H^+ + O_2$$
$$\downarrow$$
$$\text{to P680}$$

The oxygen from H_2O is released as a gaseous by-product (note that two molecules of H_2O must be split to yield one molecule of O_2). The protons, as we shall see, also play an important role.

The electrons involved in the second light event move from water to the P680 molecule of Photosystem II to Q to the transport chain of Photosystem II and to Photosystem I (Fig. 8.10). If we combine these steps with the electron movement associated with the first light event, as traced above, we obtain the following abbreviated sequence showing the overall electron movement:

$H_2O \longrightarrow$ P680 $\longrightarrow$ Photosystem II transport chain $\longrightarrow$ P700 $\longrightarrow$
Photosystem I transport chain $\longrightarrow NADP_{re} \longrightarrow$ carbohydrate

This sequence shows that the electrons necessary to reduce carbon dioxide to carbohydrate come from water, and that the movement of electrons from water to carbohydrate is an indirect and complex process.

Since electrons are not passed in a circular chain in this process, some leaving the system via $NADP_{re}$ and others entering the system from water as replacements, this series of reactions is termed ***noncyclic photophosphorylation***. The whole process results in the formation of both ATP and $NADP_{re}$ and in the release of molecular oxygen. The reactions of photophosphorylation have traditionally been known as the "light" reactions of photosynthesis, but, as we have seen, only two steps are directly light-dependent.

We should point out that the details given above apply only to photosynthesis by green plants and a few bacteria. Some other bacteria possess a form of chlorophyll and can use light energy in synthesis of ATP and $NADP_{re}$, but do not use water as the source of electrons. Some of these use hydrogen sulfide (H_2S), which is very much like water, and they give off sulfur instead of oxygen. Others use inorganic compounds not much like water, but the basic processes are essentially the same, though oxygen is not

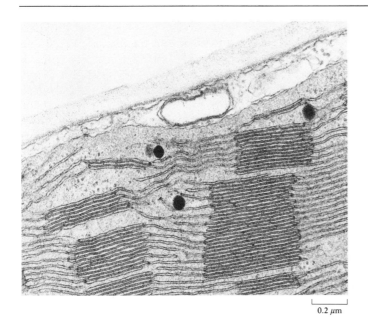

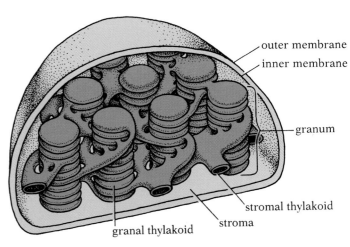

0.2 μm

8.11 Structure of a chloroplast
Left: Electron micrograph of a section of a
chloroplast of timothy grass showing several grana;
note the continuity between granal and stromal
thylakoids. Right: This cutaway view of a typical
chloroplast shows the inner and outer membranes
lying close together, enclosing the large compartment
known as the stroma. Inside the stroma can be seen
a third distinct membrane, which forms the
interconnected compartments called thylakoids. The
stacks of flat, disclike thylakoids are grana.
Chlorophyll molecules and most of the electron-
transport-chain molecules are located in the
thylakoid membrane.

a by-product. Green plants themselves can be experimentally induced to
use a source of electrons other than water. If oxidation of water is blocked
by chemical inhibitors, and then a strong electron donor is provided as a
substitute, noncyclic photophosphorylation can continue without production of oxygen; in other words, green-plant photosynthesis has been experimentally converted into an essentially bacterial type of photosynthesis.
Water is therefore only one of many possible electron sources for photosynthesis. It is not surprising, however, that the most successful photosynthetic
organisms, the green plants, evolved utilizing photosynthesis based on
water, that ubiquitous agent in the chemistry of life.

THE ANATOMY OF PHOTOPHOSPHORYLATION

By now you have probably noticed many similarities between the electron-
transport strategies of photosynthesis and those of respiration. We saw in
the last chapter that functioning of the respiratory electron-transport chain
depends on ordered arrays of enzyme molecules embedded in the highly
folded inner membrane of the mitochondrion. And we saw too that the energy made available by this stage is used in large part to transport H$^+$ out of
the inner compartment of the mitochondrion, with the resulting electrochemical gradient across the inner membrane acting like a battery to supply energy for the synthesis of ATP. The chloroplast is organized in a similar
way, and also establishes and maintains a chemiosmotic gradient.

You may recall from Chapter 5 that the chloroplast, like the mitochondrion, is a membrane-bound organelle bearing some of its own genes,
whose ancestors are thought to have evolved independently as procaryotes.
However, unlike the mitochondrion, which has a heavily folded inner
membrane containing electron-transport chains, the chloroplast has an
inner membrane that is relatively smooth and flat, and follows the contours
of the outer membrane (Fig. 8.11); it has many selective gates and pumps,

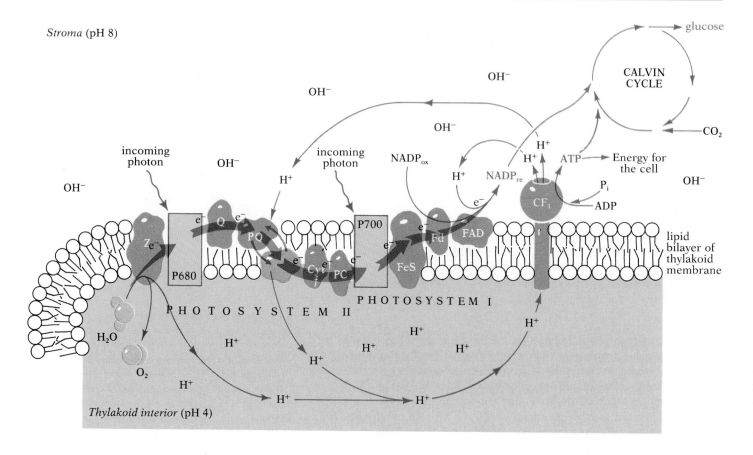

8.12 Anatomy of photophosphorylation

The enzyme molecules of photophosphorylation are shown here for simplicity as a linear chain; in the membrane the chain is thought to be folded back on itself, so that cytochrome b_6 (not shown) connects Fd and PQ, thereby making possible cyclic photophosphorylation. The antenna molecules of the two photosystems are omitted. This representation makes it clear how electrons from water are moved along the two electron-transport chains, and how the thylakoid battery is charged. Since the pH scale is logarithmic (see Fig. 2.12, p. 36), the values shown indicate that the concentration of H^+ ions is 10,000 times greater within the thylakoid than on the other side of the membrane. Note that PQ migrates from one side of the membrane to the other to transport H^+ ions. The CF_1 complex, which uses the energy of the electrochemical gradient to make ATP, is also shown.

but no electron-transport complexes. The antenna pigments, reaction centers, and electron-transport-chain molecules of the chloroplast are in fact embedded in a third membrane, which forms a series of flattened, interconnected compartments known as ***thylakoids*** that are often arranged in stacks called ***grana***.[3] The thylakoid membrane serves as a barrier between the interior of the thylakoid and the interior of the chloroplast as a whole, which is known as the ***stroma*** (Fig. 8.11). Like the inner mitochondrial membrane, the thylakoid membrane makes possible an electrochemical gradient, which, functioning like a battery, supplies energy for the synthesis of ATP. However, in the chloroplast the H^+ ions accumulate in the innermost compartment of the organelle—that is, the interior of the thylakoids—while the outer compartment, the stroma, becomes negatively charged.

How does the anatomy of the chloroplast clarify the details of photosynthesis? Let's review the flow of energy through the electron-transport chains, as shown in Figure 8.12. The process begins when a photon is absorbed by the antenna-molecule complex of Photosystem II, exciting an

[3] There is some evidence to suggest that the thylakoid membrane is continuous with the inner membrane.

electron; this excitation energy is passed to the reaction-center molecule, P680, which is thereby energized. P680 now transfers an energized electron to an acceptor, Q. The resulting electron vacancy in P680 is filled by the enzyme complex Z, which splits water, depositing one H^+ ion in the thylakoid interior for every electron delivered to P680.

Meanwhile, the original energized electron moves to PQ, a molecule that can shuttle between the stromal side of the membrane and the thylakoid side. Some of the energy of the electron is used here to move an H^+ ion from the stroma to the thylakoid interior, making a total of two H^+ ions added to the thylakoid. The electron now moves to cytochrome f and then to PC, where it waits for a vacancy in P700.

When a photon is absorbed by an antenna molecule in Photosystem I and excites an electron, the excitation energy is passed to the P700 molecule, which is thereby energized. The energized electron of P700 is immediately passed to an acceptor, FeS. The resulting vacancy in P700 can now be filled by the electron from Photosystem II waiting in PC.

The energized electron of Photosystem I, meanwhile, moves on to Fd, from which it can follow one of two pathways. Under normal conditions, the most likely fate for this electron is to reduce $NADP_{ox}$ via the FAD in the membrane. This process consumes an H^+ ion from the stroma, so that in all, two H^+ ions are now gone from the stroma while two H^+ ions have been added to the thylakoid. The alternative (not shown in Figure 8.12) is for cytochrome b_6 to accept the electron and pass it to PQ, in Photosystem II, where part of its energy is used to move an H^+ ion from the stroma to the thylakoid; this is the cyclic pathway described earlier. The cyclic alternative represents a less efficient use of energy, but is favored when $NADP_{ox}$ is in short supply, or when the cell is more in need of ATP than of $NADP_{re}$.

The result of this highly ordered flow of electrons is twofold: on the one hand, it generates a supply of the high-energy carrier molecule $NADP_{re}$, whose role in carbohydrate synthesis we shall examine shortly; on the other, it helps create, through the flow of H^+ ions, a powerful electrochemical gradient which is then used to generate ATP. For just as in the mitochondrion, the membrane in the thylakoid contains numerous enzyme complexes (here called CF_1) that can utilize the energy of the gradient to phosphorylate ADP.

A brief comparison of Figure 8.12 and Figure 7.10 (p. 186) shows that the organization and anatomy of the electron-transport chains of photophosphorylation and of respiration are very similar. In fact, there is little doubt that the anabolic and catabolic mechanisms have a common evolutionary origin. Current evidence indicates that a low-efficiency anaerobic pathway similar to glycolysis followed by fermentation evolved first, enabling chemosynthetic cells to store energy in carbohydrates for later use. Then followed a higher-efficiency membrane-embedded electron-transport chain, which used strong electron acceptors like sulfur and (when available) oxygen to yield greater supplies of free energy for the cell. Meanwhile, primitive kinds of photosynthesis had developed. Eventually, a major modification of the electron-transport system (previously involved only in catabolic processes) gave rise to cyclic photophosphorylation. When the more versatile noncyclic system evolved, providing energy for carbon fixation, atmospheric oxygen was produced and the further evolution and rapid

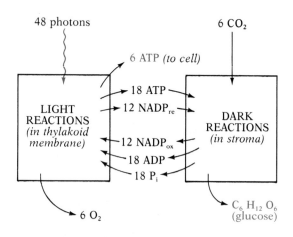

8.13 The light and dark reactions
Photosynthesis consists of two physically separate but interlocking sets of reactions. The light reactions—photophosphorylation—use light to generate the energy intermediates ATP and $NADP_{re}$. These reactions, as we have seen, take place in the thylakoid membrane. The dark reactions—carbon fixation—use these energy intermediates to turn carbon dioxide into carbohydrates like glucose. Noncyclic photophosphorylation, summarized in the left half of this figure, produces ATPs for the cell above and beyond those utilized in the synthesis of glucose. Cyclic photophosphorylation goes on simultaneously to provide additional ATP for the cell, but since it does not produce $NADP_{re}$ for carbon fixation, we have ignored it in this summary.

spread of the respiratory pathway using oxygen as the electron acceptor became possible. The plausibility of this scenario is bolstered by the recent discovery of procaryotes that may represent the "missing links." The gradual emergence of varied but related systems for obtaining the energy essential to life illustrates how evolution builds on modifications of what is already at hand, and how one change, such as the evolution of noncyclic photophosphorylation in the forerunners of chloroplasts, can suddenly make possible (or necessary) other rapid advances. Once the oxygen produced by the new photosynthetic strategy began to accumulate in the atmosphere, aerobic respiration permitted the evolution of vast numbers of highly efficient nonphotosynthetic organisms.

THE DARK REACTIONS: CARBON FIXATION

So far, we have seen how the energy of photons is captured and used to make ATP and $NADP_{re}$, but not how this energy in turn is used to transform the low-energy substance CO_2 into high-energy compounds like glucose. The entire process of photosynthesis, as we have noted, is frequently divided into the light reactions of photophosphorylation and the dark reactions of carbon fixation. This dichotomy is not quite accurate, since the energy stored by the thylakoid battery can be used to make ATP regardless of illumination, but it is a useful way of distinguishing between the energy-accumulating reactions of the thylakoids, which culminate in the production of ATP and $NADP_{re}$, and the energy-consuming carbon fixation that takes place in the stroma.

CARBOHYDRATE SYNTHESIS BY THE CALVIN CYCLE

Carbohydrates contain much chemical energy, while CO_2 contains very little. As might be predicted from what we have already learned about the chemistry of living cells, the reduction of CO_2 to form glucose proceeds by many steps, each catalyzed by an enzyme. In effect, CO_2 is pushed up an energy gradient through a series of intermediate compounds, some of them unstable, until the stable carbohydrate end product is formed. An analogy would be a man moving a large and very heavy chest up a flight of stairs from the first floor of his home to the second floor. The man might be able to lift the chest just high enough to get it up one step at a time, balancing it on each step long enough to marshal his strength before the next heave. If he let go (that is, stopped applying energy) at any point between the stable level of the first floor and the stable but higher energy level of the second floor, the chest would come crashing to the bottom. The steps, then, make it possible to move the chest up an energy gradient, but they themselves are unstable intermediate levels. In the case of the stepwise synthesis of carbohydrates from CO_2, the energy comes from light via ATP and $NADP_{re}$ (Fig. 8.13).

Since there are many sequential steps in the reduction of CO_2 to carbohydrates, and since many of the intermediate compounds occur also in other processes, leading to different end products, you may well wonder how the exact sequence of steps was ever discovered. The tool that made such discoveries possible was a radioactive isotope of carbon, designated ^{14}C. Sam-

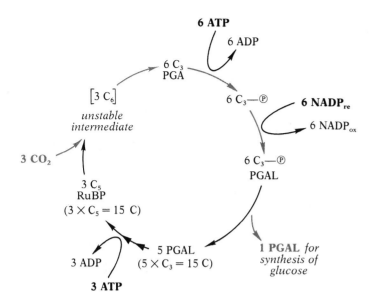

8.14   Synthesis of carbohydrate by the Calvin cycle
Each CO_2 molecule combines with a molecule of ribulose bisphosphate (RuBP), a five-carbon sugar, to form a highly unstable six-carbon intermediate, which promptly splits into two molecules of a three-carbon compound called PGA. Each PGA is phosphorylated by ATP and then reduced by $NADP_{re}$ to form PGAL, a three-carbon sugar. Thus each turn of the cycle produces two molecules of PGAL. Five of every six new PGAL molecules formed are used in synthesis of more RuBP by a complicated series of reactions (not shown separately here) driven by ATP. The sixth new PGAL molecule can be used in the synthesis of glucose. The path of carbon from CO_2 to glucose is here traced by color arrows. Since it takes three turns of the cycle to yield one PGAL for glucose synthesis, the diagram begins with three molecules of CO_2; it would require a total of six turns to produce one molecule of glucose, a six-carbon sugar. Note that the cycle is driven by energy from ATP and $NADP_{re}$, formed by the light reactions of photophosphorylation.

uel Ruben and Martin D. Kamen at the University of California, Berkeley, who discovered this isotope about 1940, immediately recognized its potential as a research tool in photosynthesis. They showed that plants exposed to carbon dioxide containing the radioactive isotope ($^{14}CO_2$ instead of the normal $^{12}CO_2$) incorporated the isotope into a variety of compounds. Later, in 1946, Melvin Calvin and his associates, also at Berkeley, began an intensive long-term investigation of carbon dioxide fixation in photosynthesis, using ^{14}C as their principal tool. They exposed algal cells to light in an atmosphere of $^{14}CO_2$ for a few seconds and then killed the cells by immersing them in alcohol. The alcohol not only killed the cells but also inactivated the enzymes that catalyze the reactions of photosynthesis. With the enzymes inactivated, whatever amount of each intermediate compound existed in the cell at the moment of inactivation was, in effect, locked in. Calvin and his co-workers could then determine which of these locked-in intermediate compounds contained ^{14}C. How long the algal cells were exposed to the $^{14}CO_2$ before being killed determined the number of compounds in which ^{14}C was detected: when the time was very short, the ^{14}C reached only the first few compounds in the synthetic sequence; when the time was longer, the isotope moved through more steps in the sequence and appeared in a great variety of compounds. After years of painstaking research, Calvin, who in 1961 was awarded the Nobel Prize for his critically important investigations, worked out the sequence of reactions now called the ***Calvin cycle***; we outline this sequence in very abbreviated form in Figure 8.14.

According to Calvin, the CO_2 first combines with a five-carbon sugar called ribulose bisphosphate, or ***RuBP***, to form a highly unstable six-carbon compound, which is promptly broken into two three-carbon molecules called phosphoglyceric acid, or PGA. Each molecule of PGA is then phosphorylated by ATP and reduced by hydrogen from $NADP_{re}$. The resulting energy-rich three-carbon compound is phosphoglyceraldehyde, or ***PGAL***.

VENATION   LEAF TYPE

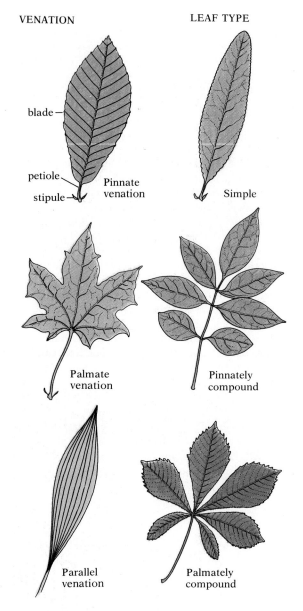

8.15 Leaf types
A leaf usually consists of a blade and of a petiole, which sometimes has stipules at its base. Veins run from the petiole into the blade. The main veins may branch in succession off the midvein (pinnate venation); or they may all branch from the base of the blade (palmate venation); or they may be parallel. The blade may be simple; or it may be compound—that is, divided into leaflets that may be pinnately or palmately arranged.

This compound, which we have already encountered in our study of glycolysis, is a true sugar and, in a sense, is the stable end product of photosynthesis. Because PGAL is a three-carbon compound, as are the intermediate compounds leading to its formation, the Calvin cycle is often called **C_3 photosynthesis**.

Five of every six molecules of PGAL are used in the formation of new RuBP (by a complicated series of reactions powered by ATP), with which more CO_2 can be processed. But one molecule out of six can be combined by a series of steps with another molecule of PGAL (produced in another turn of the cycle) to form the six-carbon sugar glucose.

Though glucose has traditionally been considered the end product of photosynthesis, free glucose is not present in significant amounts in most higher plants. Some of the PGAL produced by the Calvin cycle is at once utilized in the formation of lipids, amino acids, and nucleotides. Even when glucose is synthesized, it is normally used almost immediately as a building-block unit for double sugars, starch, cellulose, or other polysaccharides. As we mentioned in Chapter 3, carbohydrates are generally stored in higher plants in the form of starch. One of the most important advantages of storing carbohydrates in this form is that starch, which is insoluble in water, has much less osmotic activity than sugar. An excessive accumulation of sugar, which dissolves in the cellular cytoplasm, would raise the osmotic concentration of the cytoplasm relative to the environment and severely upset the osmotic balance between the cell and the surrounding fluid; the result would be the intake of too much water by the cell.

Photorespiration One property of the Calvin cycle is perplexing in that it has no obvious biological function: apparently RuBP carboxylase, the enzyme that catalyzes the carboxylation of ribulose bisphosphate (that is, the addition of CO_2 to RuBP) at the start of the Calvin cycle, can also catalyze the oxidation of RuBP by molecular oxygen (the addition of O_2 to RuBP). In other words, CO_2 and O_2 are alternative substrates that compete with each other for the same binding sites on this enzyme. When the concentration of CO_2 is high and that of O_2 is low, carboxylation is favored and carbohydrate synthesis by the Calvin cycle proceeds. But when the reverse conditions prevail—when the concentration of CO_2 is low and that of O_2 is high—oxidation is favored. Higher than normal temperatures also favor the alternative oxidation pathway.

Oxidation of RuBP results in formation of a two-carbon compound called phosphoglycolate, which can then be further broken down to CO_2. Under certain conditions, then, high-energy molecules such as RuBP, themselves produced by photosynthesis, are destroyed by a series of reactions initiated by the very same enzyme that under more favorable conditions would facilitate photosynthesis. This breakdown of photosynthetic intermediates to CO_2 is called **photorespiration**. Since it does not result in synthesis of ATP, as other types of respiration do, it would appear to be a wasteful process, short-circuiting the Calvin cycle to no purpose.

Things could be worse, however. By means of a complex series of reactions involving chloroplasts, mitochondria, and peroxisomes, plant cells salvage much of the energy they stand to lose from the breakdown of phosphoglycolate. Only one of every three carbons entering photorespiration is actually lost as CO_2.

Because photorespiration predominates over photosynthesis at low concentrations of CO_2, plants that depend exclusively on the Calvin cycle for CO_2 fixation cannot synthesize carbohydrates unless the CO_2 concentration in the air is above a critical level (commonly about 50 parts per million); even at normal levels much of the production of photosynthesis is undercut by concurrent photorespiration. At atmospheric CO_2 concentrations, net photosynthesis by such plants could be increased by as much as 50 percent if oxygen inhibition of photosynthesis and the associated photorespiration could be stopped. We shall return to this idea later in this chapter.

THE LEAF AS AN ORGAN OF PHOTOSYNTHESIS

As we have seen repeatedly, life depends both on the precisely catalyzed reactions of various biochemical pathways, with the accompanying interplay of their products, and on the particular anatomy of cells and their organelles. Nowhere is this intimate relationship between form and function more obvious than in the specialized tissues responsible for photosynthesis.

THE ANATOMY OF LEAVES

Photosynthesis can occur in all green parts of the plant, but in most vascular plants the leaves expose the greatest area of green tissue to the light and are therefore the principal organs of photosynthesis.

Figure 8.15 shows leaves of a variety of familiar land plants. Most dicot leaves consist of a stalk, or *petiole*, and a flattened *blade*. In addition, the petioles of some leaves bear small appendages, called stipules, at their bases. However, some leaves, particularly those of monocots (the grasslike plants), lack even petioles, the base of the blade being attached directly to the stem. The blades of most leaves are broad and flat and contain a complex system of veins. Because of the flatness of the blade, the leaf exposes to the light an area that is very large in relation to its volume.

When we examine a transverse section of a leaf under the microscope (see Fig. 8.16), it becomes clear that the outer surfaces are formed by layers of epidermis, usually only one cell thick, but sometimes two, three, or more cells thick. A waxy layer, the *cuticle*, usually covers the outer surfaces of both the upper and the lower epidermis, but is generally thicker on the former. The chief function of the epidermis is protection of the internal tissues of the leaf from excessive water loss, from invasion by fungi, and from mechanical injury. Most epidermal cells do not contain chloroplasts.

The entire region between the upper and lower epidermis contains parenchyma cells, which constitute the *mesophyll* portion of the leaf. The mesophyll is commonly (but by no means always) divided into two fairly distinct parts: an upper palisade mesophyll, consisting of cylindrical cells arranged vertically, and a lower spongy mesophyll, composed of irregularly shaped cells. The cells of both parts of the mesophyll are very loosely packed and have many intercellular air spaces between them. These spaces are interconnected and communicate with the atmosphere outside the leaf

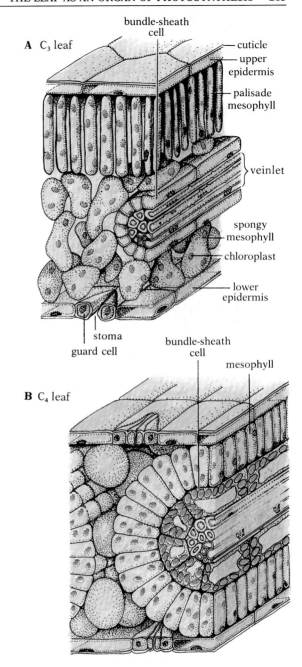

8.16 The anatomy of C₃ and C₄ (Kranz) leaves
In a C₃ leaf the palisade mesophyll cells typically form a layer in the upper part of the leaf; the corresponding mesophyll cells in a C₄ leaf are usually arranged in a ring around the bundle sheath. While the bundle-sheath cells of C₄ leaves have chloroplasts (dark green), those of C₃ leaves usually lack them.

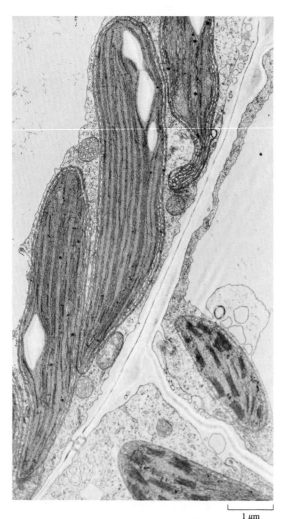

1 μm

8.17   Two kinds of chloroplasts in a C₄ leaf
At left in this electron micrograph of a segment of a
corn leaf is part of a bundle-sheath cell; its
chloroplasts have no grana and contain starch grains
(light areas). At right and bottom are parts of two
mesophyll cells; their chloroplasts are smaller and
contain numerous grana but no starch grains.

by way of holes in the epidermis called **stomata**. The size of the stomatal
openings is regulated by a pair of modified epidermal cells called **guard
cells**.

A conspicuous system of veins (also called vascular bundles) branches
into the leaf blade from the petiole (Fig. 8.15). The veins form a structural
framework for the blade and also act as transport pathways, being connected with the transport system of the rest of the plant. Each vein contains
cells of the two principal vascular tissues, xylem and phloem; and each is
usually surrounded by a **bundle sheath**, composed of parenchymatous cells
packed so tightly together that there are few spaces between them. In most
cases the branching of the veins is such that no mesophyll cell is far removed from a veinlet; in one study the veins were found to attain a combined length of 102 cm per square centimeter of leaf blade.

LEAVES WITH KRANZ ANATOMY

As early as 1904 German plant anatomists observed that the leaf anatomy of
some angiosperm plants of tropical origin—plants associated with bright,
hot, but especially xeric (dry) habitats—showed a combination of features
not generally found in plants native to the temperate zones. This unusual
complex of features came to be called Kranz anatomy. *Kranz*, which means
"wreath" in German, refers to the ringlike arrangement of photosynthetic
cells around the leaf veins of these plants.

The bundle-sheath cells of plants with Kranz anatomy (also called C₄
plants, for reasons we shall see shortly) contain numerous chloroplasts,
whereas those of other plants (C₃ plants) often do not. In plants with Kranz
anatomy the mesophyll cells that correspond to the palisade layer tend to be
clustered in a ringlike arrangement around the veins, just outside the bundle sheaths (Fig. 8.16B). These mesophyll cells contain numerous chloroplasts, but the spongy mesophyll cells outside the rings often have reduced
numbers of chloroplasts, or even none at all. In Kranz plants the chloroplasts of the bundle-sheath cells and mesophyll cells usually differ in a
number of ways. In the bundle-sheath cells the chloroplasts are bigger, they
accumulate large amounts of starch in the light, and the grana are few and
poorly developed; in the mesophyll cells the chloroplasts are smaller, they
usually do not accumulate much starch in the light, and they have numerous large grana (Fig. 8.17).[4]

Most suggestions concerning the functional significance of Kranz anatomy put forward between 1904 and 1965 turned either on enhanced conservation of water under dry conditions or on unusually rapid transport of
photosynthetic products away from the sites of synthesis. Though Kranz
anatomy may well serve both these functions, investigations of the late
1960s and early 1970s suggest that it is also associated with special biochemical adaptations that enhance the ability of the plants to carry out photosynthesis under conditions of high temperature, intense light, low
moisture, low CO₂ and high O₂ concentrations—all conditions far from op-

---

[4] The frequent use of "usually," "often," and other qualifiers, you probably have already gathered, indicates that there is considerable variation among species exhibiting Kranz anatomy, and
that the various Kranz characteristics do not always occur together.

timal for plants that depend entirely on the Calvin cycle for CO_2 fixation. We shall next examine the special photosynthetic pathways found in plants with Kranz anatomy.

C_4 PHOTOSYNTHESIS

As Figure 8.18A shows, corn, which originated in the tropics and has Kranz anatomy, can carry out photosynthesis at very low concentrations of CO_2, whereas bean plants, which are native to the temperate zone, cannot, because of photorespiration. When the CO_2 concentration falls below about 50 parts per million (in 21 percent O_2 at 20°C), bean plants become incapable of CO_2 fixation; and at CO_2 concentrations of 200 or 300 ppm, where corn approaches its maximum photosynthetic capacity, beans perform below their potential capacity. Again because of photorespiration, the concentration of O_2 that inhibits photosynthesis is far lower for beans than for corn (Fig. 8.18B). As was first realized in 1965, these contrasts between corn and beans are characteristic of the differences in photosynthetic capacity between Kranz plants and other plants.

Now, consider a plant exposed to great heat, dryness, and brilliant light, as in a desert or on a dry savanna. Under such conditions, when the moist walls of the mesophyll cells risk losing an excess of water by evaporation through the stomata, the guard cells close the stomata almost completely. Water loss is thus reduced, but now gases can no longer move freely between the atmosphere and the air spaces inside the leaf. As CO_2 is used up in photosynthesis, the nearly closed stomata prevent the supply inside the leaf from being fully replenished, and the CO_2 concentration in the air spaces around the mesophyll cells falls. Under such conditions, as we have seen, non-Kranz plants like the bean will carry out so much photorespiration that their ability to synthesize carbohydrate from C_2 will be greatly reduced. By contrast, corn and other Kranz plants *can* synthesize carbohydrate under

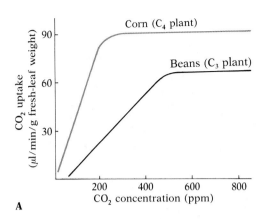

A

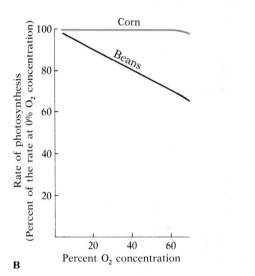

B

8.18 Comparison of photosynthetic efficiency in C_3 and C_4 plants
(A) Corn can fix CO_2 at CO_2 concentrations as low as one part per million, and it carries out photosynthesis at a very high rate at concentrations of 200–300 ppm (a normal concentration of CO_2 in the atmosphere is about 330 ppm). By contrast, beans perform no net CO_2 fixation at CO_2 concentrations below about 50 ppm, and their rate of photosynthesis at concentrations of 200–300 ppm is not very high.
(B) Photosynthesis in corn shows no inhibition at all at O_2 concentrations below 65 percent, whereas the photosynthetic rate of beans falls steadily as the O_2 concentration rises (a normal concentration of O_2 in the atmosphere is about 21 percent). Both A and B are for a temperature of 20°C and a light intensity of 2,000 foot-candles. Obviously, C_4 photosynthesis is superior under these conditions. However, when the temperature or illumination varies, while the O_2 and CO_2 concentrations remain at normal levels, a very different picture emerges. For example, as the temperature drops (C), C_3 plants clearly perform more efficiently, so they have the advantage in cooler (that is, more temperate) climates.

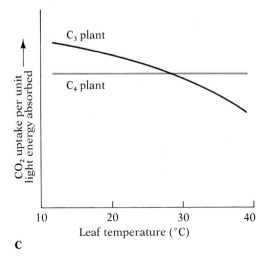

C

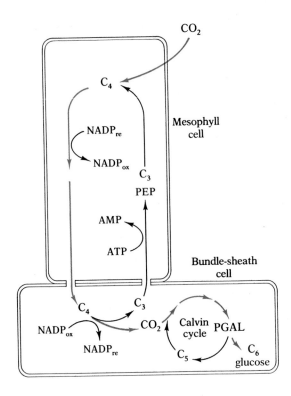

8.19   The Hatch-Slack pathway of C₄ photosynthesis
The path of carbon is here traced by color arrows. A
mesophyll cell absorbs CO_2 from the intercellular air
spaces (top). The CO_2 combines with PEP, a three-
carbon compound, to form a four-carbon compound.
After reduction by $NADP_{re}$, the C₄ substance moves
(probably via plasmodesmata) into an adjacent
bundle-sheath cell. There it is oxidized by $NADP_{ox}$ and
split into a C₃ compound and CO_2. The C₃ compound
moves back to the mesophyll cell and is converted
(by a reaction probably driven by energy from ATP)
into PEP. The CO_2 is fed into the Calvin cycle in the
bundle-sheath cell and is incorporated into
carbohydrate.

dry conditions; they are thus able to survive in climates that would be fatal
to other plants. Even under less extreme conditions they can often carry
out photosynthesis at a higher rate than other plants.

Surely it is not the distinctive anatomy of Kranz plants, by and of itself,
that enables them to carry out photosynthesis under conditions inhospitable to other plants. It must be, rather, some special biochemical capability
correlated with their anatomy—perhaps an ability to avoid the photorespiration that limits the photosynthetic ability of many plants under conditions of low CO_2 and high O_2. A finding in 1965 suggested that in the
presence of light, some plants with Kranz anatomy do not undergo photorespiration. But how do they avoid it? Here a third piece of the structure-
function puzzle must be added to the other two. This third piece is a special
way of fixing CO_2 initially.

In 1954 Hugo Kortschak of Hawaii, using the ¹⁴C tracer technique much
as Calvin did, found that one of the main early products of photosynthesis in
sugarcane is not one of the three-carbon intermediates (C₃) of the Calvin
cycle, but a four-carbon compound (C₄) instead. It was not until the late
1960s, however, that M. D. Hatch of Queensland, Australia, and C. R. Slack
of Liverpool, England, worked out the biochemical pathway responsible for
this C₄ type of photosynthesis. They found that in the mesophyll cells of sugarcane and other Kranz plants (where, you will recall, the mesophyll cells
richest in chloroplasts are arranged in rings around the veins) CO_2 is combined, not with ribulose bisphosphate as in the Calvin cycle, but rather with
a three-carbon compound called phosphoenolpyruvate or PEP, to form a
four-carbon compound. The enzyme that catalyzes this carboxylation of
PEP, unlike the one that catalyzes the carboxylation of RuBP in the Calvin
cycle, does not have O_2 as an alternative substrate and is not inhibited by
high O_2 concentrations. Thus it enables Kranz plants (which are therefore
also called C₄ plants) to fix CO_2 under conditions when photorespiration
would predominate over photosynthesis in C₃ plants, which use only the
Calvin cycle.

Curiously enough, the C₄ compound formed in the mesophyll cells is not
used for growth or nutrition by the plant. Instead, it is passed in reduced
form into the bundle-sheath cells (which in C₄ plants, you will remember,
are very well developed and contain chloroplasts), where it is decarboxy-
lated: the C₄ compound is broken down to CO_2 and a C₃ compound (Fig.
8.19). The C₃ residue moves back to the mesophyll cells, where it is reconverted into PEP and starts the C₄ cycle over again. But the CO_2 remains in
the bundle-sheath cells, where it is picked up by the RuBP carboxylase in
the chloroplasts and incorporated into carbohydrate via the Calvin cycle.

Note, then, that in both C₃ and C₄ plants the ultimate assimilation of CO_2
into carbohydrate is by the Calvin cycle. The difference is that in C₃ plants
the Calvin cycle is the only pathway of CO_2 fixation, whereas in C₄ plants
there is another, preliminary fixation pathway. It may at first seem strange
that a plant would have evolved a special mechanism for fixing CO_2 as C₄ in
the mesophyll, only to combine it with a mechanism for promptly breaking
off the CO_2 again and refixing it as carbohydrate in the bundle-sheath cells.
But remember that the C₄ fixation in the mesophyll cells, because it is insensitive to O_2 concentration, cannot be short-circuited by photorespira-

tion and can therefore operate under conditions where C_3 plants could not carry out net photosynthesis. The mesophyll cells can then pump enough CO_2 (via the C_4 intermediate) into the bundle-sheath cells to maintain an artificially high CO_2 concentration in which the Calvin cycle is able to function. Moreover, since the bundle sheath is surrounded by mesophyll cells, any CO_2 lost from the sheath cells as a result of photorespiration may be reclaimed by the mesophyll cells.

In summary, C_4 plants have an advantage over C_3 plants under conditions of high temperature and intense light, when stomatal closure results in low CO_2 and high O_2 in the air spaces inside the leaf. Under such conditions, C_3 plants are unable to use CO_2 effectively because O_2 competes for RuBP. But C_4 plants can fix CO_2, because the mesophyll cells, acting as CO_2 pumps, can elevate the CO_2 concentration in the bundle-sheath cells to a level where carboxylation of ribulose bisphosphate (leading into the Calvin cycle) exceeds its oxidation. The Kranz anatomy, with its concentric rings of mesophyll and bundle-sheath cells, facilitates the compartmentalization on which the process of CO_2 pumping depends.

The combination of Kranz anatomy and C_4 photosynthesis has evolved independently in a variety of unrelated plants, including both monocots like corn, sugarcane, sorghum, and crabgrass, and dicots like saltbush and portulaca. It is therefore an especially impressive illustration of the intimate relationship between structure and function in living systems.

CRASSULACEAN ACID METABOLISM (CAM)

Another interesting dry-climate variation of photosynthesis is found in succulents—plants that store water in fleshy leaves. Like C_4 plants, these succulents, called Crassulaceae, avoid water loss in their hot environment by closing their stomata during the day and opening them at night. The CO_2 necessary for photosynthesis is stored at night in the form of malic acid and isocitric acid, and then released in the cells during the day, to be fixed by C_3 photosynthesis. CAM evolved independently of C_4 photosynthesis; the existence of these variations illustrates again how natural selection can lead to several quite different solutions of the same problem.

PART II

THE BIOLOGY OF ORGANISMS

Chapter 9

NUTRIENT PROCUREMENT AND PROCESSING BY PLANTS AND OTHER AUTOTROPHS

The ability to synthesize new protoplasm, and the ability to break down energy-rich compounds by respiration, are vital to the well-being of almost all organisms. Both processes require the acquisition of two main types of molecules from the environment. These are (1) already synthesized high-energy compounds, or else the raw materials from which those compounds and new protoplasm can be synthesized, and (2) the oxygen used in cellular respiration. This chapter and the next will deal with the procurement and processing of materials of the first of these two categories, the **nutrients**.

We have already seen that organisms can be divided into two classes on the basis of their methods of nutrition. Fully **autotrophic** organisms can subsist in an exclusively inorganic environment because they can manufacture their own organic compounds from inorganic raw materials taken from the surrounding media. Since the molecules of these raw materials are small enough and soluble enough to pass through cell membranes, autotrophic organisms do not need to pretreat, or digest, their nutrients before taking them into their cells. Most autotrophs are photosynthetic, though a few are chemosynthetic. Photosynthetic plants are by far the most important of the earth's autotrophic organisms.

Heterotrophic organisms (most bacteria, fungi, Protozoa, and animals) are incapable of manufacturing energy-rich organic compounds from simple inorganic nutrients, and so must obtain prefabricated organic molecules from the environment. But many of the organic molecules found in nature are too large to be absorbed unaltered through cell membranes, and therefore must first be broken down into smaller, more easily absorbable molecular units—that is, they must be digested. Some of the products of digestion can then be used to supply the energy to alter and reassemble the others into organic macromolecules such as lipids, proteins, and nucleic acids.

It is clear, then, that autotrophic and heterotrophic organisms differ both in their nutrient requirements and in the problems associated with nutrient procurement. And it is not surprising that they should have evolved radically different adaptations in response to the different selection pressures acting upon them. But along with the wildly divergent adaptations, found even within each group, there are parallels in processing and procurement strategies that are intriguing and enlightening. Though the two great groups of organisms are discussed separately in this and the next chapter, we shall stress the similarities between them wherever possible.

NUTRIENT REQUIREMENTS OF GREEN PLANTS

RAW MATERIALS FOR PHOTOSYNTHESIS

The higher photosynthetic organisms need, as we have seen, carbon dioxide and water. These raw materials supply the carbon, oxygen, and hydrogen that are the predominant elements in the organic molecules they manufacture. Carbon dioxide, one of the constituent gases of the earth's atmosphere, is obtained directly from the air by the leaves of terrestrial plants; submerged aquatic plants absorb the dissolved gas from the surrounding water. Terrestrial plants obtain the other raw material, water, from the substrate in which they grow; most higher plants absorb water from the soil through their roots.

For many centuries it was assumed that the structural material of the plant body comes from the soil. Then, about 1450, the German cardinal Nicholas of Cusa suggested that the weight gained by a growing plant comes from water, not earth. He was sure that if one were to put a carefully weighed quantity of earth in a pot, plant some seeds in it, wait until the seeds had germinated and the plants had grown large and heavy, remove the plants, and again weigh the earth, very little loss of weight by the earth would be found. This, he thought, would prove that water had contributed the bulk of the weight of the plants. There is no evidence that the cardinal ever performed the experiment he suggested; in his time the gathering of empirical data by experiments was not as commonplace as it is today.

The experiment suggested by Nicholas of Cusa was finally performed by Jan Baptista van Helmont, a Flemish physician, and the results were published in 1648. Van Helmont described his experiment as follows:

> . . . I took an earthenware vessel, placed in it 200 pounds of soil dried in an oven, soaked this with rainwater, and planted in it a willow branch weighing 5 pounds. At the end of five years, the tree grown from it weighed 169 pounds and about 3 ounces. Now, the earthenware vessel was always moistened (when necessary) only with rainwater or distilled water, and it was large enough and embedded in the ground, and, lest dust flying about be mixed with the soil, an iron plate coated with tin and pierced by many holes covered the rim of the vessel. I

did not compute the weight of the fallen leaves of the four autumns. Finally, I dried the soil in the vessel again, and the same 200 pounds were found, less about 2 ounces. Therefore 164 pounds of wood, bark, and root had arisen from water only.[1]

Though van Helmont clearly demonstrated that most of the material of a plant's body does not come from the soil, he was not prepared to consider that it might have come from so weightless a thing as air. Finally, in 1727, Stephen Hales, an English clergyman, suggested that plants get at least part of their nourishment from the air, but the extent of such nourishment was not understood for a long time. And only with the advent of isotopic tracer techniques in the present century could it be conclusively demonstrated, as we saw in the last chapter, that CO_2 gas contributes the oxygen as well as the carbon to make glucose. The seeming implausibility of a tree's massive substance coming mainly from something as insubstantial as air was for centuries a serious impediment to research.

In fact, however, a very high percentage of the total dry body weight (the weight of the organic compounds) of a large tree is carbohydrate, and much of the rest is synthesized from carbohydrate. The formula for glucose, which may be taken as the central carbohydrate in protoplasm, is $C_6H_{12}O_6$, which means, you will recall, that one molecule of glucose contains six atoms of carbon, twelve of hydrogen, and six of oxygen. The atomic weight of carbon is 12, that of hydrogen is 1, and that of oxygen is 16; consequently the molecular weight of glucose is 180 (really 180.162 if isotopes are considered). Now, all the carbon and oxygen incorporated into glucose by photosynthesis comes from carbon dioxide, which in turn comes from the air. Since the combined weight of the six atoms of carbon and six atoms of oxygen in glucose is 168, which is about 93 percent of the total weight of glucose, it follows that about 93 percent of the dry weight of a large, immensely heavy tree comes initially from the air. The hydrogen in glucose comes from water, and hydrogen constitutes roughly 7 percent of the weight of glucose; hence about 7 percent of the dry weight of the tree comes initially from water. Thus the modern solution to the mystery of photosynthesis has proven that most of the dry weight of green plants comes from the least likely suspect, the air, rather than from water or from the solid earth in which they grow.

MINERAL NUTRITION

Despite their importance, carbon dioxide and water cannot be the only nutrient materials needed by a green plant. These two compounds provide only three elements: carbon, oxygen, and hydrogen; yet we know that other elements, too, enter into the composition of the plant. Nitrogen, for example, is always present in amino acids, the building-block units of the proteins, that are essential components of protoplasm; two very important amino acids also contain sulfur. Phosphorus is present in ATP, nucleic acids, and many other critically important compounds. Chlorophyll, the essential mediator of photosynthesis, contains magnesium (see Fig. 8.4, p. 200), and the cytochromes, so important in electron transport, contain

[1] Translated from the original Latin.

TABLE 9.1 *Essential minerals for higher plants.*[a]

Element	Approx. number of pounds needed to grow 100 bushels of corn	Function
MACRONUTRIENTS		
Nitrogen (N)	160	Structural component of amino acids, many hormones and coenzymes, etc.
Phosphorus (P)	40	Structural component of nucleic acids, phospholipids, ATP, coenzymes, etc.
Potassium (K)	125	Plays a role in the ionic balance of cells; cofactor for enzymes involved in protein synthesis and carbohydrate metabolism
Sulfur (S)	75	Structural component of two amino acids (cysteine and methionine) and of several vitamins
Magnesium (Mg)	50	Structural component of chlorophyll; cofactor for many enzymes involved in carbohydrate metabolism, nucleic acid synthesis, and the coupling of ATP with reactants
Calcium (Ca)	50	Influences permeability of membranes; component of pectic salts in middle lamellae and necessary for wall formation; activator for several enzymes
Iron (Fe)	2	Structural component of iron-porphyrins (hemes), which are incorporated into cytochromes, peroxidases, catalases, and some other enzymes; plays a role in the synthesis of chlorophyll
MICRONUTRIENTS		
Manganese (Mn)	0.3	Cofactor of many enzymes involved in cellular respiration, photosynthesis, and nitrogen metabolism
Boron (B)	0.06	Function unknown; may play a role in translocation of sugar; perhaps necessary for utilization of calcium in wall formation
Chlorine (Cl)	0.06	Plays an essential role in photosynthesis
Zinc (Zn)	Trace	Necessary for synthesis of tryptophan (a precursor of auxin); activator of many dehydrogenase enzymes; may play a role in protein synthesis
Copper (Cu)	Trace	Structural component of many enzymes that catalyze oxidation reactions, and of plastocyanin, which is important in electron transport in chloroplasts
Molybdenum (Mo)	Trace	Structural component of the enzyme that reduces nitrate to nitrite; essential for fixation of N_2 by nitrogen-fixing bacteria
Nickel (Ni)	Trace	Structural component of the enzyme that detoxifies waste nitrogen in plants

[a] An element must meet three criteria to be regarded as essential: (1) the element is needed for normal growth and reproduction in several different plants; (2) it is not replaceable by other elements; (3) its function is a direct one—in that it is not needed simply to correct a toxic condition induced by other substances.

iron. Where does the green plant obtain the nitrogen, sulfur, phosphorus, magnesium, iron, and other elements it needs? Obviously not from carbon dioxide or water. Here, finally, we see the role of the soil itself as a source of plant nutrients. From the soil the plant derives the minerals[2] essential to its life. Perhaps those two ounces of weight lost from the soil in van Helmont's experiment were more important than he knew.

During the nineteenth century, there was much interest in Europe in determining the mineral needs of crop plants and in devising ways of supplementing the amounts of essential mineral elements in the soil. By 1900, seven of these were known: nitrogen, sulfur, phosphorus, potassium, calcium, magnesium, and iron. Three of them—nitrogen, phosphorus, and potassium—were stressed particularly, as they are to this day in the manufacture of fertilizer. Modern commercial fertilizers are often designated by their N-P-K percentages; for example, the widely used garden fertilizer called 5-10-5 contains 5 percent nitrogen, 10 percent phosphoric acid, and 5 percent soluble potash (a potassium compound) by weight. These three are the elements most rapidly removed from the soil; consequently it is essential to replenish them if crops are to continue to flourish. Many modern fertilizers are also fortified by small amounts of some of the other essential minerals.

Much of the important research on the mineral requirements of plants was done by growing plants in distilled water to which measured amounts of minerals were added. This water-culture technique allowed a degree of control and a precision of measurement unattainable with plants growing in soil. Nevertheless, it was not until about 1920, after more than fifty years of water-culture research, that it became apparent that other elements, in addition to the seven already known, were essential to plants. These additional minerals (manganese, boron, chlorine, zinc, copper, molybdenum, and nickel) are required in such small amounts that the traces present as contaminants in the water or salts used in the early experiments were sufficient to meet the needs of the plants. Only with very elaborate purification procedures could their presence be controlled and their effects determined. Such elements, essential in minute amounts but sometimes toxic in excess, are now called trace elements or micronutrients.

Table 9.1 lists all the essential nutrients known at the present time and gives some indication of the relative amounts needed and of their known functions. Three other elements, vanadium, sodium, and cobalt, are under investigation and may someday be added to the list. Silicon and aluminum, two of the commonest elements in the earth's crust, often occur in quantity in plants, but they appear to be dispensable; however, the same was said of chlorine only a few years ago, and of other essential elements before that, so it seems advisable to reserve judgment.

The functions listed in Table 9.1 make it clear why the trace elements are needed in only minute amounts. Most of them are components of enzymes or coenzymes. You will remember that enzymes can be used over and over and that a very small quantity of each is sufficient. Only a small amount of a

---

[2] "Mineral" is a term applied to naturally occurring inorganic substances. As used here, it refers to an element in inorganic ionic form. Potassium and nitrogen, for example, are often available to a plant in soil in the form of potassium nitrate ($K^+NO_3^-$).

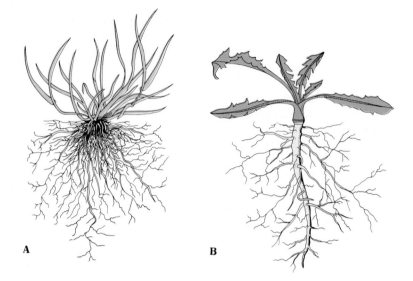

9.1 Two types of root systems
(A) Fibrous root system of grass. (B) Taproot system of dandelion.

A   B

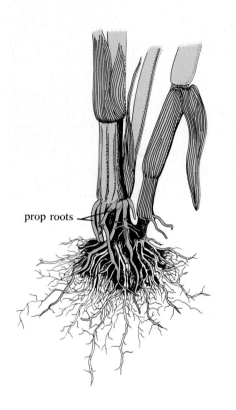

prop roots

9.2 The prop roots of corn
These roots arise from a portion of the stem and are therefore adventitious roots.

mineral is required, therefore, to synthesize the enzyme or coenzyme initially, and to replenish the supply as the enzyme molecules are slowly broken down.

NUTRIENT PROCUREMENT BY GREEN PLANTS

We have seen that three classes of nutrients are needed by green plants: carbon dioxide, water, and minerals. Carbon dioxide is absorbed by the leaves (and occasionally by stems that are green and carry out photosynthesis). This gas, together with the other components of air, moves into the internal spaces of the leaf through openings in the epidermis called stomata. Inside the leaf, the air circulates throughout the numerous intercellular spaces (see Fig. 11.3, p. 275). Carbon dioxide dissolves in the film of water on the surfaces of the leaf cells and diffuses into the cells, where it is used as a raw material for photosynthesis. Details of leaf structure and adaptations for gas exchange, both photosynthetic and respiratory, will be discussed in more detail in Chapter 11.

This chapter will be more concerned with the procurement of water and minerals. Aquatic plants such as algae absorb both of these from the surrounding medium. These plants usually lack specialized procurement organs and absorb their nutrients through the general body surface. Some nonaquatic plants, which grow on other plants, often on the branches of large trees, absorb all their nutrients through their leaves. Among these *epiphytes* are Spanish moss (really a flowering plant, not a moss) and many orchids. Their leaf-feeding strategy, known as foliar feeding, is also an effective method of providing nutrients for plants that normally obtain minerals through their roots, such as most house and garden plants; minerals in a water solution are sprayed on the leaves, where they are absorbed (minerals can also be lost from leaves by the leaching action of rainwater). But most higher land plants take in water and minerals primarily through their roots.

ROOTS AS ORGANS OF PROCUREMENT

Root structure   The first root formed by the young seedling is called the ***primary root***. Later, ***secondary roots*** (also called lateral roots) branch from the primary root, and a root system is formed. If the branching results in a system of numerous slender roots, with no single root predominating, as in clover and many grasses, the plant is said to have a ***fibrous root system*** (Fig. 9.1A). If, however, the primary root remains dominant, with smaller secondary roots branching from it, the arrangement is called a ***taproot system*** (Fig 9.1B). Dandelions, beets, and carrots, among others, are plants with taproots. As these examples suggest, taproots are frequently specialized as storage organs for the products of photosynthesis. Storage is a function of all roots, but particularly of taproots.

Obviously, procurement of water and minerals and storage of high-energy organic compounds are not the only functions of roots; they also serve to anchor the plant to the substrate. Usually the substrate is soil, but climbing vines commonly have, in addition to their normal root system in the ground, short specialized roots arising from the stem which fasten the plant to a vertical surface such as a tree trunk or the side of a building. These aerial roots of vines are examples of ***adventitious roots***; the term "adventitious" is applied to any root that arises after the embryo stage from a structure that is not a part of the root system. The prop roots of corn are also adventitious roots; they arise from the lower portion of the stem (Fig. 9.2), penetrate the soil, and become important components of the root system.

The root system of a plant is normally very extensive, far more extensive than is ordinarily realized. When we pull up a plant, we seldom get anything even approaching the entire root system, since most of the smaller roots are so firmly embedded in the soil that they break off and are lost. An extensive root system is important both in anchoring the plant and in providing sufficient absorptive surface. When we discussed possible limitations on potential cell size in an earlier chapter, we mentioned the problem of surface-to-volume ratio—that is, as a cell or an organism gets bigger, its volume increases much faster than its surface area. A large multicellular organism therefore faces a serious problem, particularly if most absorption is restricted to a limited region of the body, such as the roots; it needs an absorptive surface extensive enough to admit all the nutrients required to support its large volume. As an adaptation toward solving this problem, many organisms have evolved extensively subdivided absorptive surfaces, far greater in total area than those of an undivided system of the same volume. The manifold branching of a typical root system is an example of this kind of adaptation: a rye plant less than one meter tall has some 14 million branch roots with a combined length of over 600 kilometers.

Roots have evolved yet another adaptation that increases their absorptive capacity. Just behind the growing tip of each rootlet, there is usually an area bearing a dense cluster of tiny hairlike extensions of the epidermal cells (Fig. 9.3). The zone of these ***root hairs*** on each rootlet may be anywhere

9.3 Root of radish seedling with many prominent root hairs

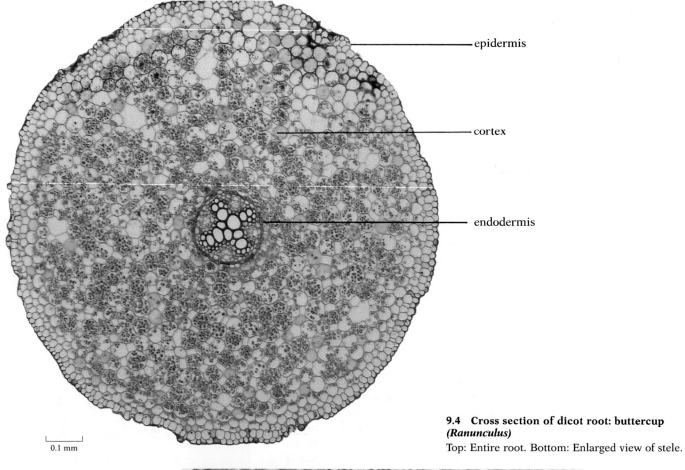

epidermis

cortex

endodermis

0.1 mm

9.4 Cross section of dicot root: buttercup (*Ranunculus*)
Top: Entire root. Bottom: Enlarged view of stele.

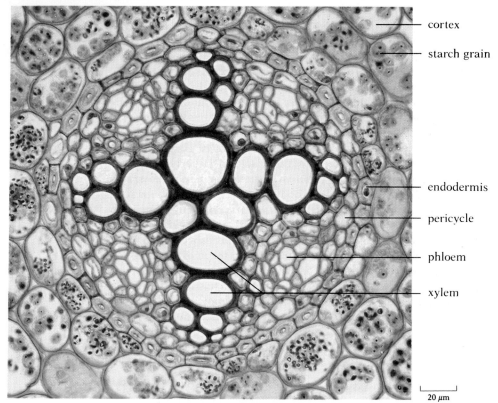

cortex

starch grain

endodermis

pericycle

phloem

xylem

20 μm

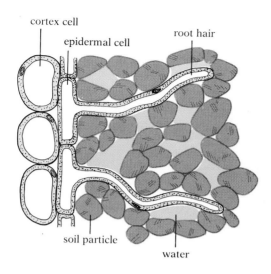

9.5 Root hairs penetrating soil
Each root hair, which is an extension of a single epidermal cell, is in contact with many soil particles (brown) and with soil spaces, some of which contain air, some water (blue).

from a centimeter long in some species to over a meter in length. It is in this region that most absorption of water and minerals takes place. Even if the root-hair zone on any one rootlet is not very long, the number of root hairs on all the many rootlets is so vast that the total absorptive surface they provide is enormous. For example, the rye plant just mentioned may have as many as 14 billion root hairs with a total surface area of more than 400 square meters.

If the root of a young dicot plant is viewed in cross section, the different tissue layers that form it become visible (Fig. 9.4). (As you may recall from Chapter 6, dicots and monocots are the two subgroups of flowering plants, and differ in many characteristics.) On the outer surface of the dicot root is a layer of *epidermis* one cell thick. Because young roots function as sponges to absorb water and pass it on to the rest of the plant, their epidermis usually lacks the waxy cuticle of the aerial parts of the plant, which need an efficient barrier against the desiccation that would result from the diffusion of water into the air. Each of the root hairs in the area just back of the growing tip of a rootlet arises from an epidermal cell (Fig. 9.5).

Beneath the epidermis is the *cortex*, a wide area composed primarily of parenchyma tissue, with numerous intercellular spaces. Sometimes sclerenchyma cells are also present; these are commonly in the outer portion of the cortex, adjacent to the epidermis. Large quantities of starch are often stored in cortex cells. The cortex, so prominent and important in young roots, is frequently much reduced or even lost in older roots, where both cortex and epidermis may be replaced by a corky periderm.

The innermost layer of the cortex, one cell thick, is the *endodermis* (Fig. 9.4). Endodermal cells are characterized by a waterproof band, the *Casparian strip*, which runs through their radial (side) and end walls (Fig. 9.6). The walls of mature endodermal cells are often very thick and hard. In some plants, however, a few of the endodermal cells, called passage cells, may

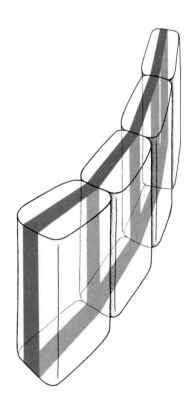

9.6 Endodermal cells with Casparian strip
The Casparian strip (color), located in the radial and end walls of each cell, forms a watertight barrier.

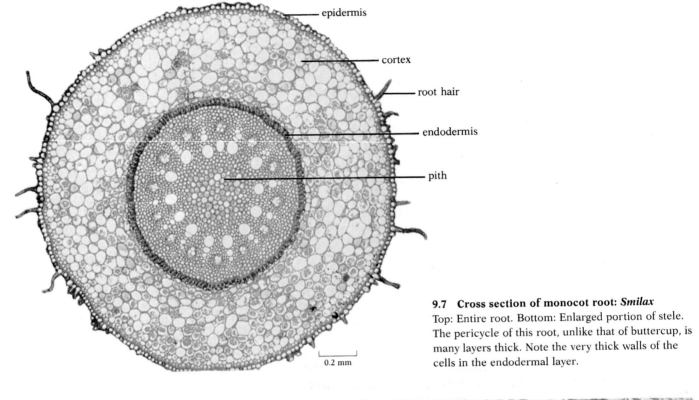

epidermis

cortex

root hair

endodermis

pith

0.2 mm

9.7 Cross section of monocot root: *Smilax*
Top: Entire root. Bottom: Enlarged portion of stele.
The pericycle of this root, unlike that of buttercup, is
many layers thick. Note the very thick walls of the
cells in the endodermal layer.

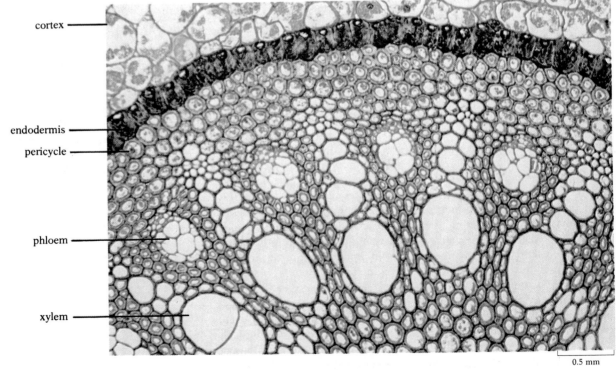

cortex

endodermis

pericycle

phloem

xylem

0.5 mm

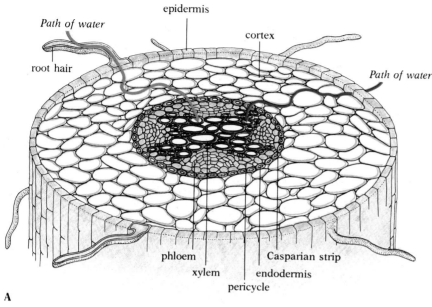

A

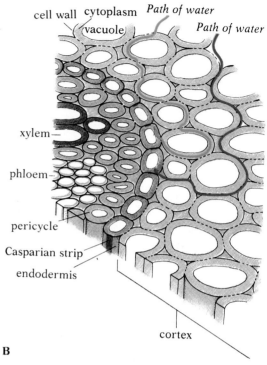

B

have relatively thin walls, though they still have Casparian strips. A well-dif-ferentiated endodermis is always present in roots, but occurs less regularly in stems.

The endodermis forms the outer boundary of a central core of the root that contains the vascular cylinder. This core is called the *stele*. Just inside the endodermis is a layer, often only one cell thick, of thin-walled paren-chymatous cells. The cells of this layer, called the *pericycle*, readily take on meristematic activity and may give rise to lateral, or secondary, roots (see Figs. 31.31 and 31.32, pp. 814–15).

The central portion of the dicot stele, surrounded by endodermis and pericycle, is filled with the two vascular tissues, *xylem* and *phloem*. The thick-walled xylem cells often form a cross- or star-shaped figure (Fig. 9.4). Bundles of phloem cells are located between the arms of the xylem. Thus, instead of forming a continuous cylinder like the epidermis, cortex, endo-dermis, and pericycle, the xylem and phloem alternate in this portion of the stele.

Large roots of monocots commonly have a region of parenchyma tissue, called *pith*, located at the very center of the stele (Fig. 9.7). The xylem there-fore does not form the star-shaped figure characteristic of dicots, but even in such roots the bundles of xylem and phloem alternate.

Absorption of nutrients   In the soil, rainwater generally becomes avail-able to plants as a loose film of water around soil particles, known as capil-lary water. The roots, and particularly the root hairs, are in contact with this water. Since root epidermis lacks a cuticle, the capillary water can eas-ily move into the roots, either by osmosis or by flowing along the continu-ous path provided by the cell walls and intercellular spaces (Fig. 9.8).

9.8   Movement of water from soil to xylem in root
(A) Cross section of a root. Some water (light-blue arrow) is absorbed by the epidermal cells, particularly the root hairs, and moves from cell to cell either by osmosis or by diffusion through plasmodesmata. Most of the water (dark-blue arrow) flows along cell walls and does not cross membranes of living cells until it reaches the endodermis. The Casparian strip (dark brown) of the endodermal cells prevents flow of water along their radial and end walls; hence all water entering the stele must move through the living cells of the endodermis. (B) A region near the endodermis of a root. The pathway along cell walls weaves through intercellular spaces.

The concentration of dissolved substances such as ions and organic compounds like sugar in an epidermal cell is normally higher than the concentration of dissolved substances in the soil water. With the osmotic concentration higher inside the cell than outside, a simple osmotic system is established and water can move across the membrane into the cell.

Once water has entered an epidermal cell, it must be moved to where it is needed—usually the leaves and growing part of the plant. This happens slowly but automatically as a consequence of osmosis: the water entering the epidermal cell dilutes the contents of that cell, with the result that the concentration of dissolved substances in it becomes lower than in the adjacent cell of the cortex, and water can move from it to the cortex cell by osmosis. But now a new concentration gradient is established, as water moving into the outermost cell of the cortex dilutes the contents of that cell and lowers its osmotic concentration to a point below that of the next cell of the cortex. As a result, water moves from the first cortex cell to the second cortex cell, following the concentration gradient. Again, dilution of the recipient cell occurs, a new gradient is established, and water moves on to the next cell. In this way, water can move slowly but quite easily from the capillary films of the soil into the epidermis and thence across the cortex to the stele. Once inside the xylem of the vascular cylinder, the water can rise to other parts of the plant body. Removal of water from the center of the root via the xylem maintains the concentration gradient from the epidermis to the xylem and allows the process of water absorption to continue.

Though water absorption through roots can be accomplished passively when water is abundant, active processes are probably also involved. In particular, the sodium-potassium pumps found in cell membranes act to maintain a strong osmotic gradient running from the soil to the epidermis of the root to the endodermal layer to the vascular system, with higher ionic concentrations in cells nearer the xylem. As a result, water absorption can be more rapid and can take place from drier soil than would be possible with passive osmosis alone.

Water can probably move from an epidermal cell to a cell in the cortex, and from one cortex cell to the next, by means other than osmosis through cell membranes. The cytoplasm of adjacent plant cells, you may recall, often forms a *symplast*, an association in which plasmodesmata interconnect the contents of the cells. Once water or some other substance has entered an epidermal cell, it can thus move to other cells through the plasmodesmata. Both the osmotic and symplast pathways through the cells are represented by the light-blue arrow in Figure 9.8.

Analogous to the symplast, in which the contents of adjacent cells are interconnected, is the *apoplast*, a network composed of the cell walls and intercellular spaces. The apoplast provides another means, besides osmosis, by which water can enter the root. Because cell walls are hydrophilic—their main component, cellulose, has a strong tendency to imbibe water—water from the soil can move in the apoplast across the epidermis and the entire cortex of a root without ever actually penetrating a membrane or entering a cell (this route is shown in dark blue in Figure 9.8). Indeed, recent experiments suggest that, in the normal uptake of water by roots, movement of water in the apoplast is much more important than movement in the symplast. But the water in the apoplast cannot flow across the

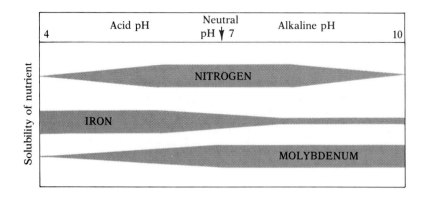

9.9 Solubility of three mineral nutrients as a function of pH
The changing width of each band indicates the relative solubility of the mineral between pH 4 and pH 10. Nitrogen is most soluble, and hence most available to plants, in the neutral pH range; iron is most soluble under acid conditions, and molybdenum under alkaline conditions. Other minerals have their own distinctive solubility curves. Since it is obviously impossible for soil to have a pH at which all minerals will be maximally available to plants, farmers or gardeners must adjust the soil pH according to the nutrient requirements of the particular plants they wish to grow.

endodermis because the Casparian strip, being hydrophobic, acts as a barrier. Consequently, all water entering the vascular cylinder must cross through the living cells of the endodermis. Thus the plant has an opportunity to control the movement into the stele of substances dissolved in water.

Plants usually absorb minerals in ionic form: nitrogen is absorbed as nitrate (NO_3^-) or ammonium (NH_4^+) ions; phosphorus as dihydrogen phosphate ($H_2PO_4^-$) or monohydrogen phosphate (HPO_4^{--}) ions; sulfur as sulfate ions (SO_4^{--}); and potassium, calcium, magnesium, and iron as their simple ions (K^+, Ca^{++}, Mg^{++}, and Fe^{++} or Fe^{+++}).

The ions available to plants for absorption are in solution in the soil water, their concentration varying according to the fertility and the acidity of the soil and other factors. When the soil minerals are not in solution, but are bound by ionic bonds to soil particles, they are not available to plants. Agricultural soil management often involves changing the soil acidity to free more such bound minerals for absorption by roots. For example, the addition of lime to very acid soil in order to raise the pH may increase the availability of phosphorus, potassium, and molybdenum, but an excess of lime may decrease the available iron, copper, manganese, and zinc (Fig. 9.9).

The rate of absorption of each mineral by roots is essentially independent of the rates of absorption of water and of other minerals. Each nutrient moves into the root at a rate determined by such factors as its concentration both inside and outside the root, the ease with which it can passively penetrate cell membranes, and the extent to which carrier molecules are involved. Though some of the inward movement of minerals, like that of water, is a result of passive diffusion along a concentration gradient—the mineral being in higher concentration in the soil solution than in the cells —numerous experiments have shown that simple diffusion alone cannot account for all the absorption of mineral nutrients by roots. Even when the concentration gradient favors inward movement, the rate of absorption is often greater than would be possible by passive diffusion alone—which means that facilitated diffusion is taking place. Moreover, plants can often take in a mineral that is in higher concentration inside the root cells than in the soil solution and that would, therefore, move the other way if simple

9.10 Photograph of roots of a legume (pea) showing nodules

diffusion alone were involved. Active transport is clearly a factor; the plant expends energy in the process of procuring the mineral nutrients essential to its continued existence. As will be evident throughout this book, active transport is the rule rather than the exception in most kinds of organisms, whether plant or animal, when substances are moving across the membranes of living cells.

Two factors help minerals enter roots by preventing their concentration in the root cells from becoming too high: as fast as a mineral enters a root, it may be removed and transported to some other part of the plant; also, it may be rapidly utilized in the synthesis of a different compound. For example, nitrogen, absorbed as nitrate, is quickly reduced and built into nitrogen-rich organic compounds such as amino acids and amides, which are then transported and stored. Much of the storage is in cell vacuoles, where the concentration of the nitrogen compounds is often much higher than in the cytoplasm itself—a clear indication that the vacuolar membrane, or tonoplast, acts selectively, admitting the compounds into the vacuole but preventing their escape from it.

Efficient as the root system of plants has become, there are still many soils in which plants cannot thrive unaided, either because proper nutrients are not present, or because they exist in forms that are inaccessible to the plants. Sometimes, however, fungi and plant roots form intimate associations known as **mycorrhizae**, which can greatly facilitate mineral uptake by the plants. The fungi, some of which are visible on the surface as mushrooms, actually penetrate the root cells of their hosts, and transport nutrients in directly. Apparently the fungi also obtain nutrients from the hosts, so that the relationship is mutually advantageous, or **mutualistic**.

NITROGEN FIXATION

Another mutualistic relationship between plant roots and other organisms is more completely understood. As Table 9.1 indicates, nitrogen is a mineral essential to plants. Nitrogen is abundant in the atmosphere that surrounds us: air consists of about 78 percent nitrogen, 21 percent oxygen, and 0.03 percent carbon dioxide. But although plants are incredibly successful at utilizing the carbon from the comparatively minute amount of carbon dioxide in the air, they have never evolved the ability to capture the much more abundant atmospheric nitrogen. Instead, most plants depend on nitrogen-containing compounds absorbed from the soil. When the soil is poor, farmers must add nitrogenous fertilizer for their crops to grow well.

The plants of one group, the legumes (which include the lupines, peas, peanuts, and alfalfa), have overcome dependence on soil nitrogen through mutualistic association with bacteria of the genus *Rhizobium*. These bacteria thrive in nodules on the plants' roots (Fig. 9.10). The bacteria have powerful enzymes capable of breaking the triple bonds of molecular nitrogen and trapping ("fixing") the nitrogen as ammonia (NH_3):

$$N_2 + 3H_2 \rightarrow 2N + 3H_2 \rightarrow 2NH_3 \qquad (\Delta G = -147 \text{ kcal/mole})$$

Since this reaction can take place only under anaerobic conditions, most oxygen-producing photosynthetic organisms like plants are unable to fix nitrogen directly. By supplying its host with nitrogen, *Rhizobium* completes

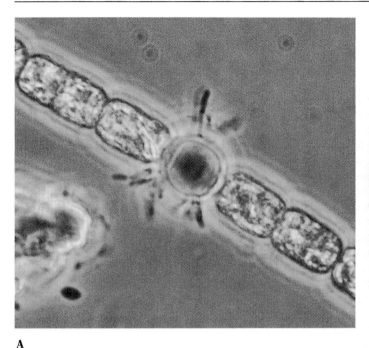

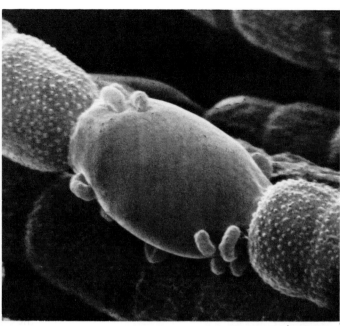

A   B

1 μm

the mutualistic cycle: the plants the anaerobic bacteria live on grow larger and are better able to produce sugars photosynthetically; some of the excess sugar feeds the *Rhizobium*. Modern agriculture makes use of this mutualism. Legumes such as clover and alfalfa are planted in rotation with other crops and plowed under to restore depleted nitrogen supplies, and legume crops like soybeans and peanuts are raised in less fertile soil.

Since nitrogen fixation produces a large free-energy profit, you may wonder why these bacteria do not use the extra energy to fix carbon directly into sugar, instead of depending on their hosts. The problem is that nitrogen fixation takes place only under anaerobic conditions, while direct carbon fixation from CO_2 and H_2O inevitably produces oxygen. Several groups of bacteria do use the free energy of nitrogen fixation and also fix carbon themselves, but the two processes are not connected. Carbon fixation in these systems is accomplished by cyclic photophosphorylation, while the energy from nitrogen fixation is used to fuel other processes. The most self-sufficient of these organisms are the Cyanobacteria (also called blue-green algae), which can survive in almost any habitat from volcanic springs to bare mountain rocks.

Some Cyanobacteria are able to have their cake and eat it too: they can fix nitrogen, but they can perform higher-efficiency aerobic respiration as well, without interfering with their nitrogen fixation. They manage this neat trick by sequestering the nitrogen-fixation system in separate (anaerobic) compartments known as heterocysts (Fig. 9.11). This strategy is very successful: up to 80 percent of the photosynthetic plankton of the oceans (which provide much of the oxygen in the atmosphere and virtually all of the food for marine organisms) are aerobic, nitrogen-fixing Cyanobacteria.

9.11 Nitrogen-fixing Cyanobacteria

Some species of Cyanobacteria form chains (A) consisting of photosynthetic cells and an occasional anaerobic nitrogen-fixing compartment known as a heterocyst. Such chains begin as a single cell. Planktonic Cyanobacteria benefit from the sharing of products between the photosynthetic cells and the heterocysts. The heterocysts anaerobically metabolize the bonds in molecular nitrogen to produce ammonia, which is then utilized to produce amino acids and proteins, while the photosynthetic cells produce sugars which the heterocysts utilize as a source of energy. In some species of Cyanobacteria, nitrogen fixation can be augmented indirectly by other bacteria, which congregate around the active heterocyst and ingest the excess amino acids that leak from it (B). This relationship appears to be of mutual benefit—in "return" for the amino acids, these bacteria reduce the oxygen concentration in the immediate environment of the heterocyst, thus increasing its efficiency in nitrogen fixation.

9.12 Pitcher plant (*Sarracenia purpurea*) from upstate New York
Several fruit flies are trapped within the leaf.

9.13 Leaf of Venus fly trap (*Dionaea muscipula*) from North Carolina
Top: One of the fly's legs is about to touch a trigger hair on the leaf surface. Bottom: The two halves of the leaf have quickly moved together, interlocking their marginal teeth and thus trapping the fly.

One of the most challenging goals for modern applied biology is to use the techniques of genetic engineering to develop crop plants that can fix atmospheric nitrogen themselves.

INSECTIVOROUS GREEN PLANTS

A few photosynthetic plants supplement their inorganic diet with organic compounds obtained by trapping and digesting insects and other small animals. Such plants are true autotrophs, since they can survive without capturing any prey, but when they do capture prey the nutrients thus obtained stimulate more rapid growth. Apparently it is the nitrogenous compounds of the animal's body that are of most benefit to insectivorous plants, which often grow in nitrogen-poor soils—particularly acid bogs and heavy volcanic clays—and have root systems that are not extensive. Their prey-capturing adaptations are interesting and worth examining here.

Pitcher plants (Fig. 9.12) have leaves modified into tubes or sacs, partly filled with water. The end of the leaf is further modified to form a hood, which partly covers the open mouth of the pitcher. Insects that fall into the sac are prevented from climbing out by numerous stiff downward-pointing hairs. The proteins of the trapped insects are digested by enzymes secreted into the water, and the products of this digestion are absorbed by the inner surface of the leaf.

The leaf of the Venus fly trap (Fig. 9.13) is composed of two lobes with a midrib between them. There is a row of long stiff teeth along the margin of each lobe. When an insect touches small sensitive hairs on the surface of the leaf, the lobes quickly change shape and come together with their teeth interlocked. (The plant guards against accidental triggering by responding only when two hairs are stimulated simultaneously.) Stimulation of the trigger hairs produces an electrical signal, much like a nerve impulse, which initiates a rapid pumping of H^+ ions through the cell membrane, consuming about 30 percent of the cells' available ATP in two seconds. The pH change in turn leads to a quick movement of water from the intercellular spaces into the cells at the base of the trap, and their consequent rapid enlargement causes the leaf to close on its victim. The trapped animal is then slowly digested by enzymes secreted from glands on the leaf surface, and the resulting amino acids are absorbed.

The leaves of sundews (Fig. 9.14) show still another type of modification for carnivorous activity. They bear numerous hairlike tentacles, each with a gland at its tip. The gland secretes a sticky fluid in which small insects, attracted by the plant's odor, become trapped. The stimulus from a trapped insect causes nearby tentacles to bend over the animal, further entangling it. As in the pitcher plants and Venus fly traps, the proteins of the insect are digested by enzymes and the amino acids are then absorbed.

Plants have evolved a variety of ways to procure from the environment the raw materials they need for growth. Adaptations and strategies as diverse as roots hundreds of kilometers long, mutualism, nitrogen fixation, and carnivorous feeding are found in the plant kingdom. At last we can begin to see how the integration of what biologists know about chemical reactions, cell structure, and such life-giving functions as respiration and photosynthesis provides a deeper understanding of how organisms actually work.

9.14 Leaf of sundew (*Drosera intermedia*) from the Pine Barrens of New Jersey
A damselfly is caught in the sticky fluid on the ends of the glandular hairs.

Chapter 10

NUTRIENT PROCUREMENT AND PROCESSING BY ANIMALS AND OTHER HETEROTROPHS

NUTRIENT REQUIREMENTS OF HETEROTROPHIC ORGANISMS

Heterotrophic organisms cannot manufacture their own high-energy compounds from low-energy inorganic raw materials. Yet they, like all living things, depend on high-energy compounds, and must extract from these complex molecules the energy necessary for both maintenance and growth. They must therefore obtain prefabricated high-energy organic nutrients.

There are four main groups of heterotrophic organisms: the nonphotosynthetic bacteria, the fungi, the nonphotosynthetic protozoans, and the animals. (In addition, some green plants—the insectivorous plants described in the last chapter, for example—are capable of heterotrophic as well as autotrophic nutrient procurement.) The first two groups, the bacteria and the fungi, lack internal digestive systems and hence depend mainly on absorption as their mode of feeding. They are usually either *saprophytic* (living and feeding on dead organic matter) or *parasitic* (living on or in other organisms and feeding on them). By contrast, the principal mode of feeding for animals and protozoans is ingestion—the taking in and digesting of particulate or bulk food. Animals and protozoans may be *herbivores*, in which case they obtain high-energy compounds by eating the plants[1] that have assembled those compounds from raw materials. Or they may be *car-*

[1] By "plants" in this chapter we mean the photosynthetic members of the kingdoms Monera, Protista, and Plantae.

nivores, in which case they eat other animals or protozoans that have eaten the plants. Some animals, the *omnivores*, need both plant and animal material to survive.

Much of the morphological diversity among living things simply reflects their different ways of employing the three major modes of nutrition—photosynthesis, absorption, and ingestion. Absorptive and ingestive organisms are as different from each other as each is from the green plants, but every heterotrophic organism, whether saprophytic or parasitic, or herbivorous, carnivorous, or omnivorous, depends on energy-yielding nutrients that came originally from photosynthetic organisms, which used radiant energy from the sun to make them.

NUTRIENTS REQUIRED IN BULK

Carbohydrates, fats, and proteins are the main energy sources for heterotrophic organisms. Of these, carbohydrates alone would suffice if organic nutrients functioned only as an energy source. But these nutrients perform another very important function—providing the carbon skeletons and functional groups necessary for the synthesis of new organic compounds. Assuming that enough inorganic minerals are included in the diet, can carbohydrates alone fulfill this second function?

For some heterotrophs, the answer is yes. Many bacteria and fungi, as well as a few Protozoa, can flourish on a diet consisting solely of carbohydrates and minerals. They need no protein in their diets because they, like green plants, can combine inorganic nitrogen with carbon skeletons from carbohydrates to make amino acids. They can also synthesize for themselves all the other classes of compounds necessary for life.

But for many other heterotrophs, such a limited diet cannot sustain life. Some bacteria and fungi have apparently lost the ability to synthesize certain organic compounds they need, and must obtain them as well as carbohydrates in their diet. Animals are especially deficient in synthetic ability. Among their extensive dietary requirements are proteins or the amino acids of which they are composed. Though a few heterotrophic flagellated Protozoa can survive with the inorganic compound ammonia (NH_3) as their only nitrogen source, organic nitrogen is indispensable to most animals, and for them a diet restricted to carbohydrates cannot support life.

The loss of a synthetic pathway may seem maladaptive and a contradiction to what we know about evolution by natural selection. But organisms may actually be at a competitive advantage if they do not need to use up their valuable raw materials and energy to fabricate organic compounds that are easily come by in the environment. Thus, the loss of unnecessary chemical pathways can actually be adaptive, and mutations in the genes coding for one or more of the enzymes necessary to synthesize readily available compounds would not be deleterious. They would, however, commit a species to an increasingly specific range of food. Hence, an organism's dependence on particular organic compounds implies that at some point in its evolution the compounds in question were readily available. Because nutritional needs tend to reflect in this way a species' past diet, they provide valuable clues about the course of evolution.

Most animals have lost the ability to synthesize certain amino acids and must get them from their diet. These are called *essential amino acids*—a somewhat misleading term, since it seems to imply that the other amino

10.1   A child suffering from kwashiorkor
Kwashiorkor is a severe protein-deficiency disease caused by eating protein from a single source, such as rice or corn. The most obvious external symptom is inflammation of the skin, evident here on the child's arms and legs.

acids commonly occurring in proteins are not essential. All are, of course, necessary, but only a few are essential *in the diet;* the others, which are also necessary for life, can be synthesized by the organism itself from other amino acids or organic nitrogen compounds. Though the essential amino acids vary for different species of animals, and even for different stages in the life history of the same species, the basic pattern is similar for all. Nine amino acids (histidine, isoleucine, leucine, lysine, methionine, phenylalanine, threonine, tryptophan, valine) are essential for almost all animals; the others may be essential for some, but not for all.

To accumulate all the essential amino acids, animals must normally ingest several different proteins, since a single protein may not contain them all. Zein, the main protein in corn, for example, is deficient in tryptophan and lysine. An animal that depended exclusively on a "poor quality" protein such as zein would suffer from a deficiency not only of these two amino acids but of other essential amino acids as well, since protein synthesis requires that all the essential amino acids be present simultaneously in the correct relative amounts. If there is not enough of one, utilization of the others is reduced proportionately, and since they cannot be stored they are lost through excretion.

To obtain all the essential amino acids, organisms must ingest a variety of different proteins, since it is unlikely that each will be deficient in the same amino acids. Specialists in human nutrition recommend, for instance, that an average adult male include in his daily diet at least 70 grams (about 2.5 ounces) of protein, of which at least half should be of animal origin. The proportions of the various amino acids in plant proteins are often quite different from those in animal proteins; hence plant proteins are less reliable than animal proteins as a source of essential amino acids for human beings. *Kwashiorkor* (Fig. 10.1), a protein-deficiency disease characterized by degeneration of the liver, severe anemia, and inflammation of the skin, is particularly common among children in countries where the diet consists primarily of a single plant material—as in Indonesia, where rice forms much of the diet, and in parts of Africa where corn is the principal staple.

A human following a vegetarian diet must take care to select a combination of plant proteins that will complement one another, making up for one another's deficiencies. For example, the proteins in beans are deficient in methionine, whereas those in wheat are deficient in lysine; if both beans and wheat are eaten at the same meal, they will complement each other and there will be no deficiency of either methionine or lysine. Since amino acids cannot be stored in the body, eating beans at one meal and wheat at the next would be futile: complementary proteins must be eaten *at the same meal.*

Some animals can survive and grow with little or no fat in their diets; because they can interconvert carbohydrate and fat, they may deposit much fat as a storage product in their bodies even when they do not eat any. But some animals (including rats and humans) cannot synthesize enough linoleic acid (a common fatty acid) for their needs, no matter how many other organic compounds are available to them. Severe disease symptoms or even death may result if such animals do not eat enough fat to provide presynthesized linoleic acid. For these animals, linoleic acid may be designated an ***essential fatty acid.***

The nutrient requirements for our species reinforce most current theories of human evolution, which indicate that early humans evolved as omnivorous hunter-gatherers. Regularly ingesting animal protein and fat along with plant material, these humans experienced no selection pressure to maintain the synthetic pathways for substances provided ready-made by the diet. Because these pathways have been lost, a strict vegetarian diet is now inherently artificial in our species, and care must be taken to balance the amino acids and to include animal fats (usually dairy products).

VITAMINS

Vitamins are organic compounds required in small quantities by organisms that cannot synthesize them and must therefore obtain them prefabricated in the diet. Note that a compound may be a vitamin for species A and not for species B, because B can synthesize it. Vitamins are necessary only in very small quantities, because they ordinarily function as coenzymes or as parts of coenzymes; enzymes and coenzymes, you may recall, are catalysts that can be reused many times and hence are not needed in large amounts.

That certain diseases are connected with dietary deficiencies, now identified as vitamin deprivation, was recognized long ago. By 1752 it was known that fresh fruit helps prevent *scurvy*, a painful disease common among sailors long at sea, the symptoms of which are bleeding gums and loosening of teeth, anemia (here a deficiency in hemoglobin level), delayed healing of wounds, and painful and swollen joints. Shortly before 1800, in an effort to control this debilitating disease, lime and lemon juice were made standard parts of rations for British sailors—hence their nickname "limeys." But at that time there was no way to know that a vitamin C deficiency was responsible for the disease. Again, our need for this vitamin reflects our evolution as omnivores: our species has always been able to ingest fresh vegetables, fruits, and berries. Carnivores do not suffer from scurvy despite the lack of fruit in their diets. Since they have never obtained much vitamin C from external sources, natural selection has worked to prevent their loss of the metabolic pathways necessary to synthesize it.

Another dietary-deficiency disease came to light almost a century after lemons and limes were discovered to be effective against scurvy. The Dutch government became concerned over the high incidence among troops in the East Indies of a crippling disease called *beriberi*, which is characterized by muscle atrophy, paralysis, mental confusion, and sometimes congestive heart failure. A team of investigators, aware of the discoveries of Pasteur, Robert Koch, and others implicating microorganisms as causative agents in disease, bent their efforts toward finding such an agent for beriberi. Two years of work met with no success. Then one member of the team, Christiaan Eijkman, discovered that chickens that were fed primarily on polished rice dropped in the kitchen and dining area of the military quarters developed symptoms similar to those of human beriberi. Because it was so cheap, polished rice was the main food supplied to the troops. Eijkman found that if unpolished rice was added to the diet neither chickens nor men developed the disease. He showed later that the antiberiberi factor was water-soluble and was removed by the process of polishing.

In 1906 F. G. Hopkins of Cambridge University demonstrated that, in addition to carbohydrates, fats, proteins, minerals, and water, normal foods contain minute traces of other substances essential to health. Hopkins called such substances accessory factors. Then in 1911 the American biochemist Casimir Funk isolated and crystallized the antiberiberi factor, which, because it was chemically an amine, he proposed naming "vitamine" (for "vital amine"). When it was found later that many accessory factors are not amines, some researchers wanted to abandon Funk's term; a compromise resulted in abandonment of only the final *e*, and the name became "vitamin." The antiberiberi factor is now called vitamin B_1 or thiamine.

It is often hard to demonstrate conclusively that a particular chemical compound is a vitamin, because a diet supposedly free of a given compound may contain trace amounts sufficient to prevent symptoms from developing in an experimental animal. Even the most elaborate purification techniques are not always successful. Various compounds that may be vitamins for human beings are under investigation, but reaching any certainty about them may take years. Determining reliable minimum daily requirements for known vitamins is even more difficult. The supposedly established requirements are still very much open to question. We know astoundingly little, for example, about how the requirements alter with an individual's age or with changing health. Though much research remains to be done, one thing can be asserted with reasonable confidence: healthy people who eat a varied diet including meats, fruits, and vegetables will probably get all the vitamins they need, numerous advertisements to the contrary notwithstanding.

The relation between the pathological symptoms of a vitamin deficiency and the actual biochemical function of the vitamin is often obscure. For example, the symptoms of beriberi, the vitamin B_1–deficiency disease, give no indication that vitamin B_1 functions in the conversion of pyruvic acid into acetic acid and carbon dioxide. In fact, the exact biochemical roles of many vitamins are still unknown, despite extensive clinical information on the symptoms that a lack of them will cause.

Water-soluble vitamins   Vitamins can be classified as water-soluble or fat-soluble. For humans the water-soluble vitamins include vitamin C and the vitamins of the so-called B complex, all of which tend to occur together. These substances function as coenzymes in metabolic reactions that take place in almost all animal cells. Some animals can synthesize one or more of these coenzymes, and for them, of course, such coenzymes are not vitamins and are not required in the diet.

Vitamin C, or ascorbic acid, is the factor in fruit that prevents scurvy. One of its most important functions is its role in the formation of collagen fibers, which, you may recall, are the chief components of connective tissue. When the diet is severely deficient in ascorbic acid, collagen formation ceases and the gravest symptoms of scurvy result. Severe scurvy is rare among adults in this country, but occurs occasionally in infants; very mild cases, which are difficult to recognize, are more frequent. Since fresh fruits or vegetables in the diet provide an ample supply of ascorbic acid, supplements are usually advisable only for infants, pregnant women, and the seriously ill. Citrus fruits are a good source of vitamin C but not necessarily the best. They got their good reputation because they travel well—

they last much longer than far richer sources of vitamin C like fresh cabbage, peas, and beans—and during a long sea voyage or a long winter, endurance counts.

The B complex includes a large number of compounds unrelated chemically, but somewhat similar in function and tending, as we have said, to occur together. Several of them are components of coenzymes functioning in cellular respiration. As already indicated, thiamine (vitamin B_1) is a part of the coenzyme that catalyzes the oxidation of pyruvic acid. Pantothenic acid is a component of coenzyme A, which, as we saw in Chapter 7, plays an essential role in carrying the acetyl group into the Krebs cycle. Riboflavin (vitamin B_2) is one of the carrier compounds in the respiratory electron-transport system. Pyridoxine (vitamin B_6) is a component of a coenzyme involved in transaminations—reactions transferring amino groups from one compound to another. Nicotinamide, another B vitamin, is a major component of both NAD and NADP (commercial vitamin preparations often contain niacin, which is converted into nicotinamide in the body); the nicotinamide-deficiency disease *pellagra* is a severe problem in many poor areas, but it can also be caused by chronic alcoholism, which interferes with nicotinamide uptake in the body. Clearly the B vitamins, as indispensable catalysts of the energy-releasing reactions of cellular respiration, are of prime importance in the diets of heterotrophs such as humans.

Some of the B vitamins—particularly B_{12} (cobalamin), a very important vitamin containing the element cobalt—seem to be involved in the formation of red blood corpuscles. Vitamin B_{12} deficiency results in *pernicious anemia*, a chronic disease most common to older people. This vitamin, like several others (vitamin E, vitamin K, niacin, pantothenic acid, and folic acid), is usually synthesized in mammals by microorganisms in the digestive tract, and may be absorbed from this source without having been present as such in the diet. When human beings develop pernicious anemia, the problem often is not insufficient vitamin B_{12} in the intestine, but rather an inability to absorb it or an inability to convert it into an active form once absorbed.

Folic acid, another of the B vitamins, is apparently also involved in red blood corpuscle formation. Its primary role, however, is in the synthesis of some of the nucleotides that are building blocks for nucleic acids, making it essential for cell division (Fig. 10.2).

Unlike the fat-soluble vitamins discussed next, water-soluble vitamins are not stored: we must replenish our supply of these substances frequently.

Fat-soluble vitamins   The compounds collectively known as fat-soluble vitamins are necessary only to vertebrate health. Of these, we shall concern ourselves just with the principal ones: A, D, E, and K. The same compounds, or very similar ones, occur in a great variety of other organisms, but they apparently function differently in these organisms and are not dietary essentials.[2] The development of changed functions for compounds present in ancestral organisms illustrates a common occurrence in the course of evolution: natural selection, instead of giving rise to completely new com-

10.2   Folic acid–deficient chick
Both birds are four weeks old. The one below was fed a diet deficient in folic acid, while the one above received a plentiful supply of the vitamin.

[2] A few invertebrates do require vitamins A and E.

10.3 Night blindness
Left: Road as seen by a normal individual. Right: Road approximately as seen under the same lighting conditions by an individual with a vitamin A deficiency. That person cannot see the road sign at all.

pounds, frequently acts upon those already existing, finding new uses for them. Let's look at the roles of the principal fat-soluble vitamins in vertebrates.

Symptoms of vitamin A deficiency include retarded growth, excessive keratinization of epithelia (hardening by deposition of keratins, the chief components of claws, nails, and horns), and degeneration of columnar and cuboidal epithelia into stratified squamous epithelia. But the most serious result of vitamin A deficiency is *xerophthalmia*—a keratinization of tissues of the eye that can lead to permanent blindness. Indeed, xerophthalmia is the most common cause of childhood blindness in many underdeveloped countries. A less extreme manifestation of vitamin A deficiency is night blindness: since vitamin A is a component of the light-sensitive pigment in the rod cells of the eye, lack of adequate vitamin A can cause a marked impairment of vision in dim light (Fig. 10.3). Vitamin A deficiency is not common in the United States because this vitamin can be synthesized in the animal body from the carotenoids in green and yellow vegetables and fruits (such as β-carotene, which gives carrots their characteristic color). Though carotenoids can be synthesized only by plants, these precursors of vitamin A are also abundant in such animal products as butter, cheese, milk, and egg yolk. Oddly enough, an excess of vitamin A, as of some other vitamins, can work to destroy the systems that in proper concentrations it helps.

Vitamin D is involved in calcium absorption and metabolism, and in children a deficiency results in the condition known as *rickets*. In an afflicted child, the growing skeleton becomes deformed because the bones, lacking sufficient calcium, are very soft. Exposure to sunlight is the best preventive for rickets, since the ultraviolet radiation in sunlight acts on sterols in the human skin to produce vitamin D. Rickets is therefore confined to the temperate zones, where people spend much time indoors and, when outdoors, wear clothing that shields most of the body from sunlight; it is almost unknown in the tropics, where our species evolved. Vitamin D is particularly abundant in egg yolk, milk, and fish oils.

Vitamin E is important in rats for maintaining good muscle and nerve condition, normal liver function, and male fertility, and for preventing rupture of red blood corpuscles. Since rats are omnivores, their dietary requirements are almost identical to ours, and it can be argued that vitamin E is likely to be equally important to human beings. However, a deficiency of this vitamin occurs only rarely in humans, and no clear deficiency symptoms are known. As a result, considerable controversy exists over whether it is actually essential for humans.

There is no question, however, that vitamin E has one beneficial effect. Many mutagenic ("gene-changing") and potentially cancer-causing substances have exposed oxygen atoms, each with a vacancy in its outer orbital. An example is the unstable compound hydrogen peroxide (H_2O_2). Because oxygen is highly electronegative, such compounds, known as oxygen radicals, are highly reactive and therefore potentially disruptive. Vitamin E, which is fat-soluble, reacts with and detoxifies radicals in lipid membranes; water-soluble vitamin C performs this function in the cytoplasm and extracellular fluids. The carotenoids found in vegetables, in addition to being a major source of vitamin A, are also able to detoxify oxygen radicals, particularly singlet oxygen (that is, unbonded atomic oxygen), which is highly reactive.

Vitamin K is essential for the formation of one of the chemicals necessary for blood clotting. A deficiency results in slow blood clotting, and sometimes in hemorrhages. In human beings enough vitamin K is normally synthesized by bacteria in the digestive tract, but a deficiency may develop if anything interferes with the absorption from the intestine of fats and fat-soluble material. Large doses of antibiotics can cause a temporary deficiency of this and other vitamins synthesized by intestinal bacteria: such nonspecific drugs kill not only disease-producing bacteria but the intestinal bacteria that keep us functioning as well.

Table 10.1 lists the main vitamins for human beings and indicates deficiency symptoms and important sources.

MINERAL NUTRITION

Like the autotrophs, heterotrophs require certain minerals, which are usually absorbed as ions. Some, like sodium, chlorine, potassium, phosphorus, magnesium, and calcium, are needed in relatively large amounts. In human beings the minimum daily requirement for these varies from about 0.35 gram for magnesium to nearly 3 grams for sodium chloride. Other minerals, like iron, manganese, and iodine, are needed in much smaller amounts. Copper, zinc, molybdenum, selenium, and cobalt, though essential to life, are needed only in trace amounts—no more than a few milligrams per day. Some elements, like vanadium, barium, tin, silicon, and nickel, are necessary in some species of animals, but have not been proven essential for human beings.

The function of some of the minerals is obvious. Calcium is a major constituent of bones and teeth in vertebrates and plays a variety of other roles in most organisms. Phosphorus is a component of nucleic acids and many

TABLE 10.1 *Some vitamins needed by human beings*

Vitamin	Some deficiency symptoms	Important sources
FAT-SOLUBLE		
Vitamin A (retinol)	Dry, brittle epithelia of skin, respiratory system, and urogenital tract; xerophthalmia and night blindness	Green and yellow vegetables and fruits, dairy products, egg yolk, fish-liver oil, liver, kidney, animal fat
Vitamin D (calciferol)	Rickets or osteomalacia (very low blood-calcium level, soft bones, distorted skeleton, poor muscular development)	Sunlight; egg yolk, milk, fish oils
Vitamin E (tocopherol) Need in humans not definitely established	In rats, malfunction of muscular and nervous systems; anemia (from rupture of red blood corpuscles); male sterility	Widely distributed in both plant and animal food, such as meat, egg yolk, green vegetables, seed oils, grains; intestinal bacteria
Vitamin K (phylloquinone, etc.)	Slow blood clotting and hemorrhage	Green vegetables; intestinal bacteria
WATER-SOLUBLE		
Thiamine (B_1)	Beriberi (muscle atrophy, paralysis, mental confusion, congestive heart failure)	Whole grains, yeast, nuts, liver, meat
Riboflavin (B_2)	Vascularization of the cornea, conjunctivitis, and disturbances of vision; sores on the lips and tongue; disorders of liver and nerves in experimental animals	Milk, cheese, eggs, yeast, liver, wheat germ, leafy vegetables, grains
Pyridoxine (B_6)	Convulsions, dermatitis, impairment of antibody synthesis	Whole grains, fresh meat, eggs, liver, fresh vegetables
Pantothenic acid	Impairment of adrenal cortex function, numbness and pain in toes and feet, impairment of antibody synthesis	Present in almost all foods, especially fresh vegetables and meat, whole grains, eggs; intestinal bacteria
Biotin	Clinical symptoms (dermatitis, conjunctivitis) extremely rare in humans, but can be produced by great excess of raw egg white in diet	Present in many foods, including liver, yeast, fresh vegetables
Nicotinamide	Pellagra (dermatitis, diarrhea, irritability, abdominal pain, numbness, mental disturbance)	Meat, yeast, grains; intestinal bacteria
Folic acid	Anemia, impairment of antibody synthesis, stunted growth in young animals	Leafy vegetables, liver, meat; intestinal bacteria
Cobalamin (B_{12})	Pernicious anemia	Liver, meat; intestinal bacteria
Ascorbic acid (C)	Scurvy (bleeding gums, loose teeth, anemia, painful and swollen joints, delayed healing of wounds, emaciation)	Fresh vegetables and fruits

high-energy organic compounds of critical importance. Iron is a constituent of the cytochromes and of hemoglobin. Sodium, chlorine, and potassium are important components of the body fluids, playing a role in osmotic phenomena and in such processes as nerve and muscle action. Iodine is a component of the hormones produced by the thyroid gland. But the function of some of the minerals, particularly those needed only in trace amounts, is less obvious. They may be components of enzymes or coenzymes; selenium, for example, is found at the active site of an enyzme that helps destroy mutagenic substances like hydrogen peroxide. Other minerals may be cofactors that help catalyze reactions without actually being incorporated into enzymes or coenzymes. In most cases we know only that without the necessary minerals, organisms fail to thrive. Like the vitamins, the minerals needed by each species come from the normal diet.

NUTRIENT PROCUREMENT BY FUNGI

The fungi constitute a large and diverse group of sedentary heterotrophic organisms that live on or in their food supply. Bread mold is a familiar example (see Fig. 6.30, p. 161). The bread on which it grows is composed mostly of starch, a rich source of energy. But starch is a polysaccharide, whose very large and insoluble molecules cannot move across the cell membranes of the mold. Before absorption can take place, the starch must be broken down to its constituent building-block compounds, the simple sugars; in short, the starch must be digested. **Digestion** is nothing more than enzymatic hydrolysis, which, you may recall, involves the addition of water (see p. 52). In bread mold the hydrolysis takes place outside the cells, in a process called **extracellular digestion.** Digestive enzymes synthesized inside the cells of the mold are released onto the bread and hydrolyze the starch. The simple sugars that are the products of this digestion are then absorbed, often by rootlike structures called **rhizoids** (Fig. 10.4A).

Mold living on bread exemplifies a saprophytic way of life, since it obtains its nutrients from dead organic matter. However, many fungi are parasitic. Indeed, bread mold itself is not restricted to saprophytic nutrition; it is one of the commonest destructive fungi on fresh fruit and vegetables. The various parasitic fungi differ in their relationships to their plant or animal hosts. Some small fungi grow between the cells of their host, but send out rootlike structures called **haustoria** which make deep invaginations in the host's cell membranes, through which they absorb nutrients from the host's cytoplasm (Fig. 10.4B). Still other filamentous types occupy many host cells simultaneously, penetrating through the cell membranes (and walls in the case of plant hosts) that divide one cell from the next. Whatever the details of their relationship to their hosts, the parasitic fungi employ basically the same mode of nutrition as the saprophytic fungi, such as mold on bread. Enzymes are secreted into the food supply on (or in) which the fungus lives; digestion takes place extracellularly; and the products of digestion are absorbed by the fungus. Note that fungi, unlike most animals, have no internal cavity where bulk food can be digested; they simply release digestive enzymes into their surroundings and absorb organic nutrients across the body surface, much as plant roots absorb inorganic nutrients.

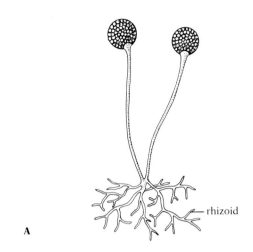

A

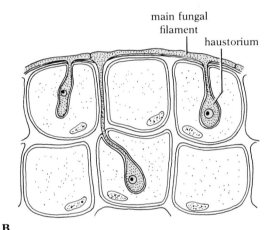

main fungal filament

haustorium

B

10.4 Nutrient-procurement structures of fungi
(A) The rootlike rhizoids of a saprophytic fungus. (B) The haustoria of a fungus parasitic on a multicellular plant. The body of the fungus (color) is filamentous and can grow between the cells of the plant host. The haustoria penetrate the cell walls and make deep invaginations in the membranes of host cells, through which they absorb nutrients. Note that the haustoria are not in direct contact with the cytoplasm of the host cells, because the haustorial invaginations are lined with host-cell membrane.

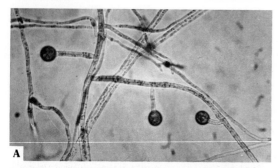

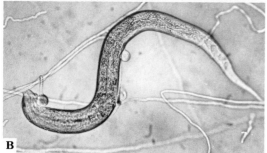

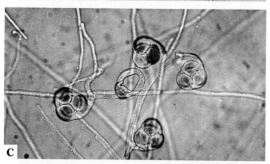

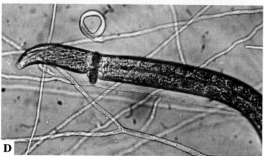

0.1 mm

10.5 Two fungi that trap nematode worms
Dactylella drechsleri has sticky knobs (A), which hold a worm that contacts them (B). *Arthrobotrys dactyloides* has rings formed of three cells (C); the three cells can be seen plainly in the closed rings. When a worm enters a ring, the cells swell and constrict the opening, trapping the worm (D).

A few fungi, departing somewhat from the usual pattern of feeding on the substrate, supplement their diets by trapping small animals such as nematode worms. When the prey has been trapped, branches of the fungus penetrate the victim's body and release digestive enzymes; extracellular digestion takes place, and the products are absorbed (Fig. 10.5).

One kind of predatory fungus, *Arthrobotrys*, provides an excellent illustration of the nutritional flexibility some fungi exhibit. *Arthrobotrys* grows readily on a variety of organic culture media, but in such cultures it seldom forms the traps it normally uses to capture prey. However, if a live nematode or an extract of nematode tissue (or even water in which nematodes have been living) is added to the culture, the fungus responds by developing traps. Apparently some chemical produced by the nematodes can induce the fungus to form traps; the prey stimulates the development of prey-capturing structures in the predator.

NUTRIENT PROCUREMENT BY ANIMALS AND PROTOZOANS

Nutrient procurement usually involves much more activity in animals and Protozoa than it does in plants. These organisms must often resort to elaborate methods of locating and trapping their food. Their incredibly varied feeding habits may be classified in any number of ways. We have already mentioned one possible classification, with identification as a carnivore, herbivore, or omnivore depending on whether the diet consists primarily of animals, of plants, or of both. Another possible criterion for classification is the size of the food. Thus we can recognize microphagous feeders, which strain microscopic organic materials from the surrounding water by an array of cilia, bristles, legs, nets, or the like. And we can recognize macrophagous feeders, which break up larger masses of food by means of teeth, jaws, pincers, or gizzards, or solely by the action of enzymes. Smaller groups would include the sucking animals, adapted to extract fluid from plants or from animal prey, and those parasitic animals and Protozoa that are bathed in the nutrients of the host and absorb them directly through the body surface.

Like the fungi already examined, most animals must digest their food before it can cross the membranes of their cells. Comparatively simple and diffusible compounds like glucose, glycerol, fatty acids, and amino acids are not usually available in nature. Rather, food material is likely to be in the form of large molecules such as polysaccharides, fats, and proteins, which must be hydrolyzed. Unlike the fungi, however, few animals secrete digestive enzymes directly onto their food. (Spiders and some insects are among these few. They inject their prey with digestive enzymes and drink the resulting "soup.") The vast majority ingest particles of food into some sort of digestive structure in which enzymatic action takes place. Often the structure is extracellular. In mammals, for instance, as we shall see later in the chapter, digestion proper takes place in the intestinal tube, and the nutrients released by the enzymatic breakdown are then absorbed into the surrounding cells. In some other organisms—the Protozoa in particular—food is ingested directly into a cell by phagocytosis or a similar process, and

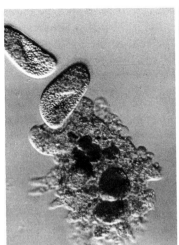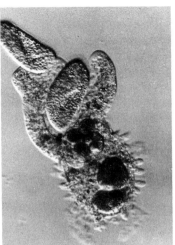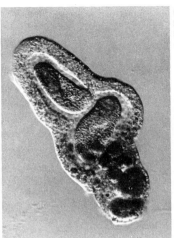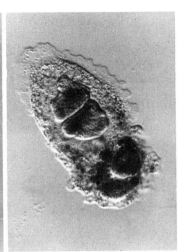

0.1 mm

then digested in a food vacuole. Though this process is classified as ***intracellular digestion***, the food material is actually separated from the rest of the cellular material by a membrane that it cannot cross until after digestion has occurred. Thus extracellular ingestion and intracellular ingestion are alike in that digestion always precedes the actual absorption of complex foods across a membrane.

Though both the nutritional requirements and the basic processes of digestion are essentially alike in protozoans and all types of animals from worms to human beings, the body plans of these organisms vary so greatly that the structures involved in food processing and the details of that processing are often very different. In the following sections we shall briefly examine the digestive mechanisms of protozoans and a variety of animals.

NUTRIENT PROCUREMENT BY PROTOZOANS

Since Protozoa, as single-celled organisms, have a body plan obviously very different from that of animals, which are almost always multicellular, we would expect their adaptations for food procurement to be likewise markedly different. And the differences are, in fact, considerable. But the similarities are often more striking and hold more biological interest than the differences.

Let us look first at an amoeboid protozoan, which constantly changes shape as its protoplasm flows along, pushing out new armlike pseudopodia and withdrawing others (see Fig. 6.24, p. 157). When an amoeba is stimulated by nearby food, some of the pseudopodia may flow around the food until they have completely surrounded it. This is the process known as phagocytosis. The food is completely engulfed by the cytoplasm and is enclosed in a ***food vacuole***, where is will be digested (Fig. 10.6). The amoeba is a good example of a protozoan without specialized permanent digestive structures, though its transitory food vacuoles are functional analogues of the extracellular digestive systems of high animals—the intestines of vertebrates, for example.

10.6 Phagocytosis of food by *Amoeba*
Pseudopodia flow around the prey (two cells of *Paramecium*) until it is entirely enclosed within a vacuole.

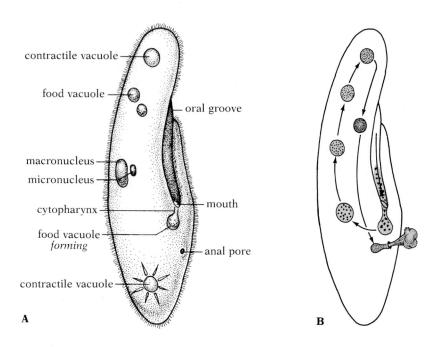

10.7   Paramecium

(A) Drawing showing major structures. (B) A food vacuole forms at lower end of the cytopharynx, then breaks off and moves toward the anterior end of the cell while enzymes are secreted into it; digestion takes place, and the products of digestion are absorbed into the general cytoplasm. The vacuole then moves toward the posterior end, attaches to the anal pore, and expels digestive wastes. The vacuole undergoes several changes in size and appearance as it moves.

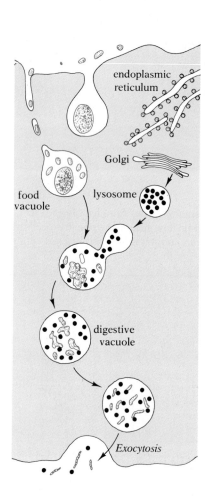

10.8   The role of lysosomes in intracellular digestion
Food material (color) that the cell takes in by phagocytosis is enclosed in a food vacuole, which fuses with a lysosome containing digestive enzymes. Digestion takes place within the composite structure thus formed (digestive vacuole), and the products of digestion are absorbed across the vacuolar membrane. The vacuole eventually fuses with the cell membrane and then ruptures, expelling digestive wastes to the outside.

The ciliates, another important group of Protozoa, of which *Paramecium* is an example, are characterized by hairlike cilia which cover the surface of the bodies (Fig. 10.7). Like all Protozoa, they are commonly regarded as unicellular. But though they lack actual subdivision into recognizable cellular units, the more complex ciliates show much of the internal specialization usually associated with multicelled organisms. For this reason many biologists prefer to regard them, not as single-celled, but as acellular—that is, as organisms whose bodies are not built of cells in the usual sense. Unlike the amoeba, *Paramecium* has a permanent structure, an organelle, that functions in feeding. Food particles are swept into an ***oral groove***, a ciliated channel located on one side of the cell (Fig. 10.7A), by water currents produced by the beating of the cilia, and are carried down the groove into a ***cytopharynx***. As food accumulates at the lower end of the cytopharynx, a food vacuole forms around it. Eventually the vacuole breaks off and begins to move toward the anterior end of the cell. Digestive enzymes are secreted into the vacuole and digestion begins. As digestion proceeds, the products (simple sugars, amino acids, and the like) diffuse across the membrane of the vacuole into the cytoplasm, and the vacuole begins to move back toward the posterior end of the cell. When the vacuole reaches a tiny specialized region of the cell surface called the anal pore, it becomes attached there and ruptures, expelling by exocytosis any remaining bits of indigestible material (Fig. 10.7B). Not only does the vacuole function as a digestive chamber, but by its movement it helps distribute the products of digestion to all parts of the cell.

We have said that digestive enzymes are secreted into the food vacuoles of both the amoeba and *Paramecium*. But if these powerful enzymes are capable of hydrolyzing such compounds as polysaccharides, fats, proteins, and nucleic acids, and if the cell itself is composed of these kinds of compounds, how can the cell contain the digestive enzymes without being destroyed by them? A partial answer was given in Chapter 5. Digestive enzymes are packaged in lysosomes, vesicles whose membranes are both impermeable to the enzymes and capable of resisting their hydrolytic action. The digestive enzymes, which are presumably synthesized on the ribosomes, move through the endoplasmic reticulum to the Golgi apparatus, and there become surrounded by a membrane to form the lysosome. When a food vacuole (sometimes also called a phagosome) is formed, a lysosome fuses with it (Fig. 10.8) and food materials and the digestive enzymes are mixed in the resulting digestive vacuole. As we have seen, this vacuole circulates in the cytoplasm, the products of digestion are absorbed, and indigestible materials are eventually expelled from the cell by exocytosis.

Though this description of lysosome activity pertains to digestion in protozoans, it applies equally well to intracellular digestion in any animal cell.

NUTRIENT PROCUREMENT BY COELENTERATES

With the evolution of multicellular organisms came a corresponding evolution of cellular specialization, resulting in a division of labor among cells. The coelenterates provide a comparatively simple example of this phenomenon. These radially symmetrical animals have a saclike body composed of two principal layers of cells (Fig. 10.9), with a jellylike layer, called mesoglea, between them. The cells of the outer layer function as a protective and sensory epithelium, while those of the inner layer, or gastrodermis, act as a nutritive epithelium. Some cells of both layers are specialized as muscle fibrils, and others as nerves. The central cavity of this saclike body functions as a digestive cavity. It has only one opening to the outside, which is surrounded by mobile tentacles. A digestive cavity of this sort, with a single opening that functions in both ingestion and excretion, is called a *gastrovascular cavity*.

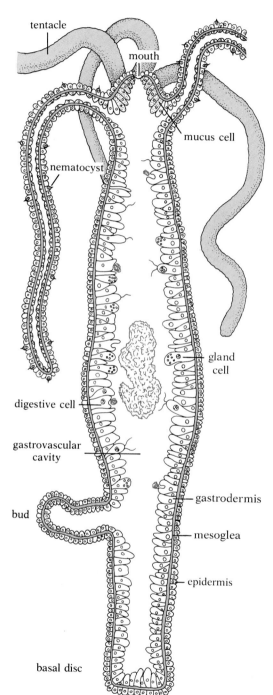

10.9 Hydra, showing gastrovascular cavity
The cavity contains food material (color). The mesoglea layer in the body wall is much more extensively developed in some other coelenterates.

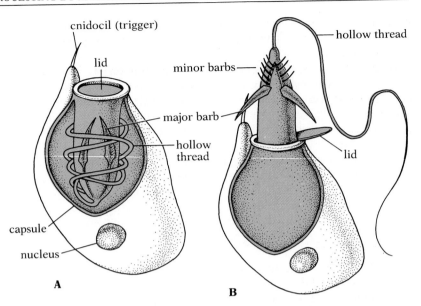

10.10 Nematocyst

Nematocysts are organelles contained in specialized cells known as cnidoblasts. This example of a nematocyst, shown before discharge (A) and after discharge (B), represents only one of more than a dozen general morphological types. The minor barbs are part of a hollow thread that is folded back inside itself before discharge. Most nematocysts fire only when the trigger has been stimulated simultaneously in two ways—for example, by touch and by exposure to chemicals characteristic of the coelenterate's prey. The exploding nematocyst propels the major barb at the target, and after it has struck, the hollow thread begins to be everted, and the minor barbs are exposed and dig themselves into the target. The rest of the hollow thread is similarly ejected, and the poison flows through it into the victim. This is the source of the jellyfish's painful sting.

Coelenterates are strictly carnivorous (see Fig. 6.32, p. 162). Embedded in their tentacles are numerous stinging structures called **nematocysts** (Fig. 10.10); some investigators think that nematocysts, like mitochondria and chloroplasts, originated as separate procaryotic organisms and now live in mutualistic symbiosis. Each nematocyst consists of a slender hollow thread coiled within a capsule, with a tiny hairlike trigger penetrating to the outside. When a coelenterate's prey comes into contact with the trigger, the nematocyst fires, the thread turns inside out, barbs on its surface unfold, and it either penetrates the body of the prey or entangles it in sticky loops. The nematocyst also ejects poison, which has a paralyzing action on the prey. The coelenterate then grasps its prey with a tentacle, and if it continues to struggle, neighboring tentacles may also become involved. The tentacles draw the inert prey toward the predator's mouth, which opens wide to receive it. Once the food is inside the gastrovascular cavity, digestive enzymes are secreted into the cavity by gastrodermal cells, and extracellular digestion begins. This extracellular digestion, largely limited to proteins in coelenterates, does not break down these substances completely to their constituent amino acids. As soon as the food has been reduced to small fragments, gastrodermal cells engulf them by phagocytosis, and digestion is completed intracellularly in digestive vacuoles. Indigestible remains of the food are expelled from the gastrovascular cavity via the mouth.

If phagocytosis and intracellular digestion are going to take place anyway, we can ask what adaptive advantage the evolution of the additional process of extracellular digestion might have. Why shouldn't coelenterates rely exclusively on intracellular digestion as the protozoans do? The problem with intracellular digestion is that it severely limits the size of the food the organism can handle. Extracellular digestion enables it to utilize much larger pieces of food; even whole multicellular animals become possible prey. Extracellular digestions is the rule rather than the exception in multicellular animals.

We have seen, then, that coelenterates exhibit a variety of interesting evolutionary adaptations for the capture and digestion of prey. As a result of cellular specialization and division of labor, certain cells, those of the gastrodermis, carry out digestion for the whole organism. The products of digestion can be distributed from the gastrodermal digestive cells to cells specialized for other functions, such as protection or movement or stimulus reception. Since the bodies of coelenterates are relatively small, and no cells are far removed from the gastrodermal layer, this distribution can be effected without any specialized transport system.

NUTRIENT PROCUREMENT BY FLATWORMS

Unlike the radially symmetrical coelenterates, the flatworms are bilaterally symmetrical; they have distinct anterior (front) and posterior (rear) ends, and also distinct dorsal (upper) and ventral (lower) surfaces. Their bodies are composed of three well-formed tissue layers. Many flatworms are parasitic on other animals, but some are free-living, and it is to these we shall turn first, using planaria as an example (Fig. 10.11; see also Fig. 6.33, p. 163).

The mouth of planaria, located on the ventral surface near the middle of the animal, opens into a muscular tubular *pharynx*, which planaria can protrude through its mouth directly onto its prey. The pharynx leads into a gastrovascular cavity, a cavity with only one opening to the outside. Though functionally similar to that of the coelenterates, this cavity branches elaborately throughout the animal's body. Literally gastrovascular (*gastro-* refers to the stomach and *vascular* to a circulatory vessel), it functions in both digestion and the transport of food to all parts of the body. The extensive branching has another important function: it greatly increases the total absorptive surface of the cavity. We saw in an earlier chapter that as organisms increase in size, and particularly as their volume increases, the problem of sufficient absorptive surface becomes more acute. Many organisms have evolved greatly subdivided absorptive surfaces, thereby compacting much total surface area into relatively little space. The root hairs of plants were one example, and the branched gastrovascular cavity of planaria is another; we shall encounter many more in this and later chapters.

Some extracellular digestion occurs in the gastrovascular cavity of planaria, but most of the food particles are engulfed by gastrodermal cells and digested intracellularly.

The members of one class of flatworms, the tapeworms, have become so highly specialized as parasites living in the digestive tracts of other animals that in the course of their evolution they have lost their own digestive systems. They are constantly bathed by the products of the host's digestion and can absorb them without having to carry out any digestion themselves. Evolutionary adaptation can involve the loss of structures, the acquisition of new structures, or the conversion of existing structures to new functions.

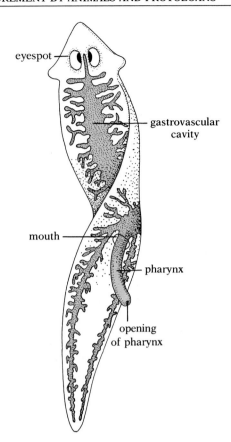

10.11   Planaria, showing much-branched gastrovascular cavity and extruded pharynx

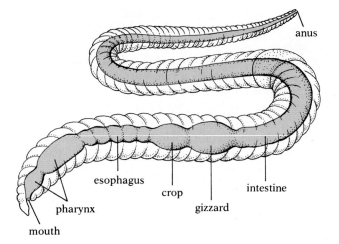

10.12   Digestive system of an earthworm

ANIMALS WITH COMPLETE DIGESTIVE TRACTS

Animals above the level of coelenterates and flatworms have a **complete digestive tract**—that is, one with two openings, a **mouth** and an **anus**. The advantages of such a system over a gastrovascular cavity are obvious. No longer must incoming food material and outgoing wastes pass through the same opening. Instead, food can be passed in one direction through a tubular system, which can be divided into a series of distinct sections or chambers, each specialized for a different function. As the food passes along this assembly line, it is acted upon in a different way in each section. The sections may be variously specialized for mechanical breakup of bulk food, temporary storage, enzymatic digestion, absorption of the products of digestion, reabsorption of water, storage of wastes, and so on. The overall result is a much more efficient digestive system, as well as a potential for special evolutionary modifications fitting different animals for different modes of existence.

The digestive system of an earthworm is a good example of division into specialized compartments (Fig. 10.12). Food, in the form of decaying organic matter mixed with soil, is drawn into the mouth by the sucking actions of a muscular chamber called the **pharynx**. It passes from the mouth through a short passageway into the pharynx and then through a connecting passage called the **esophagus**, after which it enters a relatively thin-walled **crop** that functions as a storage chamber. Next, it enters a compartment with thick muscular walls, the **gizzard**, where it is ground up by a churning action; the grinding is often facilitated by small stones in the gizzard. The pulverized food, suspended in water, passes into the long **intestine**, where enzymatic digestion and absorption take place. Finally, in the rear of the intestine, some of the water involved in the digestive process is reabsorbed, and the indigestible residue is eliminated from the body through the anus.

Notice that earthworms use extracellular digestion. Glandular cells in the epithelial lining of the intestine secrete hydrolytic enzymes into the intestinal cavity, and the end products of digestion—the simple building-

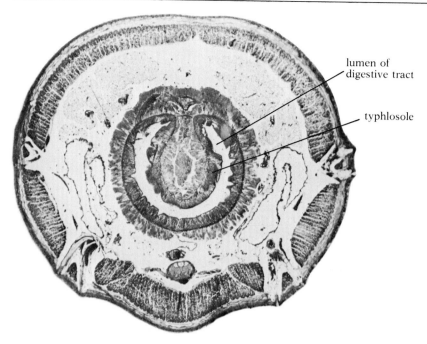

lumen of digestive tract

typhlosole

10.13 Photograph of cross section in the intestinal region of an earthworm
The typhlosole, which projects into the cavity of the intestine, greatly increases the surface area available for absorption of food.

block compounds—are absorbed. Here the problem of surface area arises again. The total interior surface presented by a plain tubular digestive tract would be inadequate in relation to the total volume of an animal the size of an earthworm. In point of fact, the inside of the earthworm's intestine is not a plain round tube (Fig. 10.13). A large dorsal fold, the typhlosole, projects downward into the digestive cavity, greatly increasing the total absorptive area exposed to the food without making the outer dimensions of the intestine prohibitively large. The typhlosole is therefore functionally analogous to the root hairs of higher plants and to the branching of the gastrovascular cavity in planaria.

We have already seen that extracellular digestion is an adaptation for eating sizable pieces of food; the gizzard of the earthworm is a crucial element in this digestive strategy. Mechanical breakup of bulk food is common among animals, and a variety of structures that serve this purpose have evolved. In our own case, food is torn and ground by the teeth. Many snails have a hard toothed pharyngeal plate, the radula, with which they rasp off small particles from larger pieces of food. Cockroaches and many other insects that feed on solid food have a chamber (the proventriculus) similar to the earthworm's gizzard except that its inner wall often bears several very hard ridges and teeth. The grinding or chewing device need not be in the first section of the digestive tract, as it is in humans; in both earthworms and cockroaches the grinding chamber comes after the crop, which is in some ways very like our stomach, and mechanical breakup thus follows temporary storage instead of preceding it. A similar arrangement exists even in some vertebrates; the muscular gizzard of birds, in which hard food is ground with rocks and pebbles (often called grit), is posterior to the less specialized stomach (Fig. 10.14).

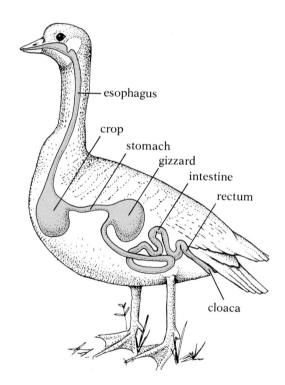

esophagus

crop

stomach

gizzard

intestine

rectum

cloaca

10.14 Digestive system of a bird
The chamber for mechanical breakup (gizzard) is located posterior to the stomach.

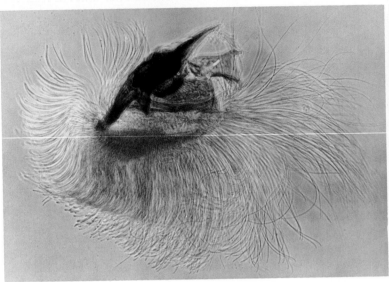

A

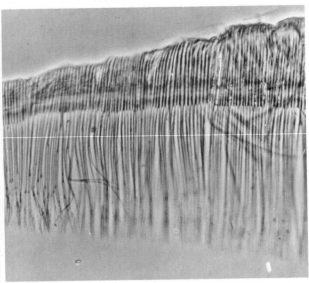

B

C

10.15 Structures used for filter feeding
(A) Isolated food brush of a mosquito larva. Its motion causes water to flow into the mouth. (B) Much-enlarged portion of comb from the pharyngeal filter of a mosquito larva. It strains food particles from water. (C) On a much larger scale, the strips of baleen hanging down from the top jaw of an Atlantic right whale, one of the filter-feeding whales. The baleen allows the cetacean to take a mouthful of water and strain out the organisms—primarily shrimp and small fish—that form its diet.

Not all complex multicellular animals eat large pieces of food and have special masticating devices. Some, such as the bloodsucking and sap-sucking insects, have liquid diets. Others are *filter feeders*, straining small particles of organic matter from water. Clams and many other molluscs filter water through tiny pores in their gills; microscopic food particles are trapped in streams of mucus that flow along the gills and enter the mouth, kept moving by beating cilia. In such molluscs digestion is largely intracellular, as might be expected in animals that eat microscopic food.

The larvae of mosquitoes are also filter feeders (Fig. 10.15A–B). They eat bacteria and other small particles of organic matter in the water where they live. Two small hair-covered brushes near the mouth of the larvae beat in a circular scooping motion, setting up water currents toward the mouth. The particles and water pass through the mouth and into the pharynx. If the larva swallowed all the water, the salt and water balance of its body fluids would be seriously disturbed. Instead, its pharynx eliminates the water while filtering out food particles. Muscles in the wall of the pharynx contract, expelling the water through two small canals. Tiny combs in the canals strain out the food while the water passes through, and the larva swallows the clump of food that remains.

Clams and mosquito larvae are only two of many possible examples of filter feeders. Some of the largest present-day vertebrates—certain species of whales—are filter feeders, straining small planktonic plants and animals from the vast quantities of water they take into their mouths (Fig 10.15C).

But let us return to the earthworm and examine the implications of another of the specialized compartments of its digestive tract, the crop, which, we have said, serves in food storage. The functional significance of a storage chamber should be clear after a moment's thought. It enables the animal to take in large amounts of food in a short time, when it is available and can be eaten in safety, and then to metabolize this food over a considerable period of time. Such discontinuous feeding makes it possible for the animal to devote much of its time to activities other than feeding, such as hiding, searching for a mate, mating, egg laying, and, in some cases, caring for young. Our own stomachs function as storage organs analogous to the earthworm's crop; they enable us to live well on only three or four meals a day and to devote the rest of our time to other pursuits. A man can survive if his stomach is removed surgically, but he can eat only a few bits at a sitting and must therefore eat frequently. It is not surprising that the vast majority of higher animals have evolved adaptations for discontinuous feeding, thereby gaining time for a behaviorally more varied existence.

Discontinuous feeding frequently results in adaptive advantages also in the feeding process itself. An animal that had to eat constantly to maintain its metabolic activity would be unable to spend time in searching for a new food supply, which might be at some distance, or in capturing more prey when its original supply had been depleted. In short, it would have to live in an essentially unlimited and continuous source of food; otherwise it would soon die or become inactive.

This is actually the case with tapeworms, nematode worms, and some other animals that lack storage ability; they must be almost constantly in contact with food. It would be a mistake, however, to assume that such animals are unsuccessful or poorly adapted. Their long evolutionary history and their large numbers today testify to the contrary. They are simply adapted—successfully adapted—for a different way of life. Biological success is not measured by structural complexity or by the possession of any particular organ. The earthworm with its crop and the nematode worm without one are both successful from the biological point of view, which equates success with survival.

Different kinds of food-storage organs occur in different species of animals. In many birds an expanded region of the esophagus anterior to the stomach forms a thin-walled chamber, the crop, which is functionally analogous to the earthworm's crop (Fig 10.14). Some birds also use the crop in carrying food to their young; they fill it with seeds, berries, fish, or whatever their food may be, and then fly to the nest, where they disgorge the food for their young. In many animals storage organs take the form of blind sacs, or diverticula, branching off the digestive tract. A good example is seen in adult female mosquitoes, which have a very large diverticulum (Fig. 10.16) that opens off the anterior portion of the digestive tract and runs

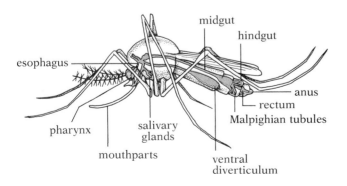

10.16 Digestive system of adult female mosquito, showing large diverticulum

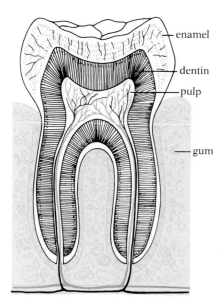

10.17 Internal structure of a human tooth
Blood vessels (color) and nerves penetrate into the pulp, but not into the outer harder layers.

10.18 Human teeth
(A) Lower jaw of adult. (B) Upper jaw of adult.
(C) Lower jaw of child, showing permanent teeth (color) in gums below milk teeth.

posteriorly, occupying much of the abdominal cavity. The female mosquito locates a suitable animal, pierces its skin with long needlelike mouthparts, and sucks blood until this diverticulum is filled. A single large blood meal may suffice to carry the female through the entire process of locating an egg-laying site and laying her eggs—a matter of four or five days.

THE DIGESTIVE SYSTEM OF HUMANS AND OTHER VERTEBRATES

Though an examination of the structure of the human digestive tract reveals little in the way of general principles that could not as easily be seen in an earthworm, natural interest in ourselves and our own species prompts a more detailed examination of human systems.

The oral cavity   The first chamber of the digestive tract is, of course, the oral cavity. Located here are the teeth, which function in the mechanical breakup of food by both biting and chewing. The internal structure of a human tooth is shown in Figure 10.17. Human teeth are of several different types, each adapted to a different function (Fig. 10.18). In front are the chisel-shaped *incisors,* four in the upper jaw and four in the lower, which are used for biting. Then come the more pointed *canine* teeth, one on each side in each jaw, which are specialized for tearing food. Behind each canine are two *premolars* and three *molars* in adults; these have flattened, ridged surfaces, and function in grinding, pounding, and crushing food. A child's first set of teeth does not include all those mentioned here; the first (or milk) teeth are lost as the child gets older and are replaced by the permanent teeth that have been growing in the gums (Fig. 10.18C).

The teeth of different species of vertebrates are specialized in a variety of ways and may be quite unlike human teeth in number, structure, arrangement, and function. For example, the teeth of snakes are very thin and sharp (Fig. 10.19A) and usually curve backward; they function in capturing prey, but not in mechanical breakup, for snakes do not chew their food, but swallow it whole. The teeth of carnivorous mammals, such as cats and dogs, are more pointed than human teeth (Fig 10.19C); the canines are long, and the premolars lack flat grinding surfaces, being more adapted to cutting and shearing (in many of these animals the more posterior molars are ab-

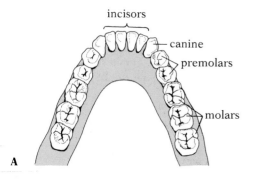

A

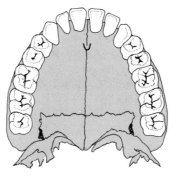

B

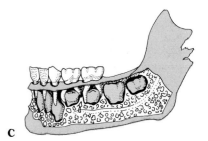

C

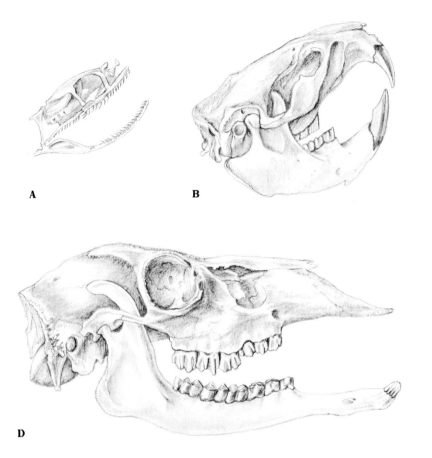

A

B

C

D

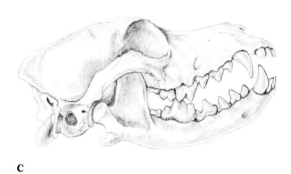

10.19 Structure and arrangement of teeth in different animals
(A) Snake: thin, sharp backward-curved teeth that have no chewing function (the snake skull is here shown disproportionately large in relation to the other three). (B) Beaver (gnawing herbivore): few but very large incisors, no canines, premolars and molars with flat grinding surfaces. (C) Dog (carnivore): large canines, premolars and molars adapted for cutting and shearing. (D) Deer (grazing and browsing herbivore): six lower incisors (three on each side), but no upper incisors (these are functionally replaced by a horny gum); premolars and molars with very large grinding surfaces. Notice the large gap between the incisors and premolars.

sent). On the other hand, such herbivores as cows and horses have very large flat premolars and molars with complex ridges and cusps; the canines are absent in many such animals.

Notice that sharp pointed teeth poorly adapted for chewing seem to characterize meat eaters like snakes, dogs, and cats, whereas broad flat teeth well adapted for chewing seem to characterize vegetarians. How can this difference be explained? Remember that plant cells are enclosed in a cellulose cell wall. Very few animals can digest cellulose; most of them must break up the cell walls of the plant they eat if the cell contents are to be exposed to the action of digestive enzymes. Animal cells, like those in meat, do not have any such nondigestible armor and can be acted upon directly by digestive enzymes. Therefore chewing is not as necessary for carnivores as for herbivores. You have probably seen how dogs gulp down their food, while cows and horses spend much time chewing. But carnivores have other problems. They must capture and kill their prey, and for this, sharp teeth capable of piercing, cutting, and tearing are well adapted. Humans, being omnivores, have teeth that belong, functionally and structurally, somewhere between the extremes of specialization attained by the teeth of

10.20   Human digestive system
The small intestine has been shortened for clarity.

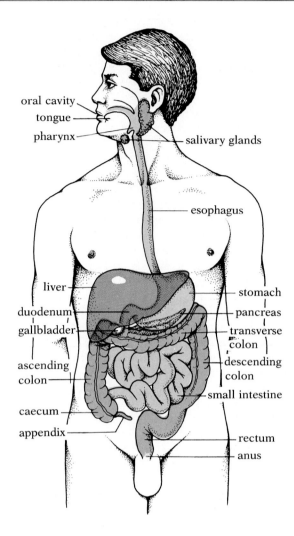

carnivores and herbivores. And as you might guess, the musculature of the jaws in each species has, like the teeth, evolved by and large to suit the needs of the species' diet.

The oral cavity has other functions besides those associated with the teeth. Here food is tasted and smelled, activities of great importance in food selection, and the food is mixed with saliva secreted by several sets of salivary glands. The saliva dissolves some of the food and acts as a lubricant, facilitating passage through the next portions of the digestive tract. Human saliva contains a starch-digesting enzyme, which initiates the process of enzymatic hydrolysis. It also contains an antimicrobial agent, the thiocyanate ion, together with a special enzyme that facilitates entry of the ion into microbial cells; these substances help prevent infection by the potentially harmful microbes that are regularly introduced into the mouth.

The muscular tongue manipulates the food during chewing and forms it into a mass, called a bolus, in preparation for swallowing, then pushes the bolus backward through a cavity called the ***pharynx*** and into the ***esophagus*** (Fig. 10.20; see also Fig. 11.18, p. 287). The pharynx functions also as a part of the respiratory passageway; the air and food passages cross here, in fact. Swallowing, therefore, involves a complex set of reflexes that close off the opening into the nasal passages and trachea (windpipe), thereby forcing the food to move into the esophagus. As you know, these reflexes occasionally fail to occur in proper sequence, and the food, entering the wrong passageway, makes you choke.

The esophagus and the stomach The esophagus is a long tube running downward through the throat and thorax and connecting to the stomach in the upper portion of the abdominal cavity (Fig. 10.20). Food moves quickly through the esophagus, pushed along by waves of muscular contraction in a process called ***peristalsis***. Circular muscles in the wall of the esophagus just behind the food bolus contract, squeezing the food forward (Fig. 10.21). As the food moves, the muscles it passes also contract, so that a region of contraction follows the bolus and constantly pushes it forward, much as though you were to keep a ball moving through a soft rubber tube by giving the tube a series of squeezes, with your hand always just behind the ball.

At the junction between the esophagus and the stomach is a special ring of muscle called a ***sphincter***, which, when it is contracted, closes the entrance to the stomach. It is normally closed, thus preventing the contents of the stomach from moving back into the esophagus when the stomach moves during digestion. It opens when a wave of peristaltic contraction coming down the esophagus reaches it.

The stomach lies slightly to the left side in the upper portion of the abdomen, just below the lower ribs. It is a large muscular sac, which, as we have already seen, functions as a storage organ, making possible discontinuous feeding. It has other functions, too. Its thick walls are composed of three layers: an inner mucous membrane composed of connective tissue and columnar epithelium with many glands, a thick middle layer of smooth muscle, and an outer layer of connective tissue. The muscle layer contains fibers running around the stomach, others running longitudinally, and still others oriented diagonally. Hence the stomach is capable of a great variety of movements. When it contains food, it is swept by powerful waves of contraction, which churn the food, mixing it and breaking the larger pieces. In this manner, the stomach supplements the action of the teeth in the mechanical breakup of food. The glands of the stomach lining are of several types. Some secrete mucus, which covers the stomach lining—hence the name ***mucosa*** or mucous membrane for the inner layer of the stomach wall; others secrete ***gastric juice***, a mixture of hydrochloric acid and digestive enzymes. Enzymatic digestion, then, is a third important function of the human stomach.

The small intestine Food leaves the stomach as a soupy mixture. It passes through the ***pyloric sphincter*** into the small intestine, which is the portion of the digestive tract where most of digestion and absorption takes place. The first section of the small intestine, attached to the stomach, is called the ***duodenum*** (Fig. 10.20). It leads into a very long coiled section lying lower

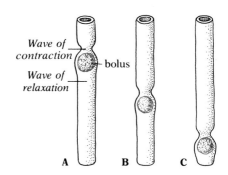

Wave of contraction

Wave of relaxation

bolus

A   B   C

10.21 Peristalsis
The wave of muscular contraction pushes the bolus of food ahead of it.

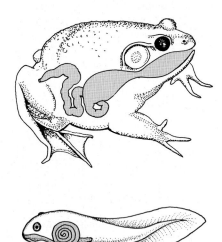

10.22 Intestines of adult frog and of tadpole
The much-coiled intestine of the tadpole is far longer relative to the size of the animal than the intestine of the adult frog.

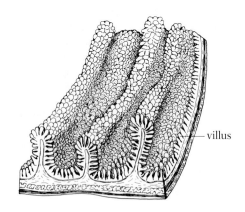

villus

in the abdominal cavity. The entire small intestine of an adult man is about 23 feet long and an inch in diameter.

The length of the small intestine shows interesting variations in different animals. The intestine is usually very long and much coiled in herbivores, much shorter in carnivores, and of medium length in omnivores like humans. These differences, like those of the teeth, are correlated with the difficulty of digesting plant material because of those cellulose cell walls. Even if the cellulose has been well broken up, it remains mixed with the digestible portions of the cells and tends to mask them from the digestive enzymes. This interference makes digestion and absorption of plant material much less efficient than the processing of animal material. A longer intestine is therefore an adaptive advantage, since it enables herbivores to extract a maximum amount of nutrition from their food. A striking example of adaptation of the small intestine is seen in frogs, where the immature stage, or tadpole, is herbivorous and has a long coiled small intestine, while the adult is carnivorous and has a relatively much shorter one (Fig. 10.22).

Since the small intestine is the place where absorption of the products of digestion occurs, we would expect it to have special structural adaptations to increase its absorptive surface area. Clearly, its great length plays a role here. But examination of its internal surface also reveals modifications that vastly increase its surface area over that of a smooth-walled tube of equal length and girth. First, the mucosa lining the intestine is arranged in numerous folds and ridges (Figs. 10.23 and 10.24). Second, small fingerlike outgrowths, called *villi*, cover the surface of the mucosa. And third, the electron microscope reveals that the individual epithelial cells covering the folds and villi have what is called, for obvious reasons, a brush border, consisting of countless closely packed cylindrical processes, the *microvilli* (Fig. 10.25). The total internal surface of the small intestine, including folds, villi, and microvilli, is incredibly large. Again, it is difficult not to think of the root hairs of plants and the rhizoids of fungi.

Some vertebrates show adaptations other than those seen in humans for increasing absorptive surface area. For instance, special blind sacs, called *caeca*, may branch from the anterior end of the small intestine; in many fish such caeca are present in the pyloric region. Another example is the spiral valve of many primitive fish and of sharks. The spiral valve is an epithelial fold extending the length of the intestine. Like a carpenter's bit tightly enclosed in a tube, it forms a spiral within the intestine, to whose wall its base is attached (Fig. 10.26); food cannot move in a straight path, but must follow the spiral of the valve and thus contact much more epithelial surface

10.23 Longitudinal section through the wall of the human intestine
Three folds of the lining of the intestine are shown, each bearing numerous villi, which greatly increase the absorptive surface area. The villi are not stationary; they have smooth muscle fibers which enable them to move back and forth. Such movement increases after a meal.

than it could by moving straight through a tubular intestine of the same length.

The large intestine   In humans the junction between the small intestine and the large intestine (colon) that follows it is usually in the lower right portion of the abdominal cavity. A blind sac, the *caecum*, projects from the large intestine near the point of juncture (Fig. 10.20). (Notice that most blind diverticula of the digestive tract are called caeca, even though their location and function may vary greatly. Thus the pyloric caeca of fish, mentioned above, are diverticula of the small intestine and often function in absorption, while the human caecum is a diverticulum of the large intestine and does not function in that way.) In humans there is a small fingerlike process, the *appendix*, at the top of the caecum. As you know, the appendix often becomes infected and must be surgically removed.

In humans the caecum is small and functionally unimportant (it may well be a relic of our long line of herbivorous primate ancestors), but in some mammals, particularly herbivorous ones, it is large and contains many microorganisms (bacteria and Protozoa) capable of digesting cellulose. Since the mammal cannot itself digest cellulose, it benefits from the

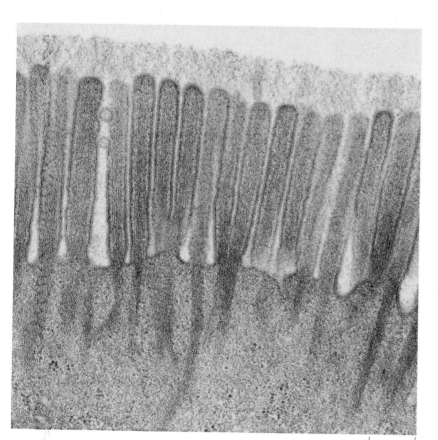

10.24   Cross section of intestine of calico bass, showing extensive folding

0.2 μm

10.25   Electron micrograph of microvilli on an epithelial cell of the intestinal lining of a cat
Notice the prominent glycocalyx covering the ends of the microvilli. A bundle of microfilaments (composed of actin) runs into each microvillus.

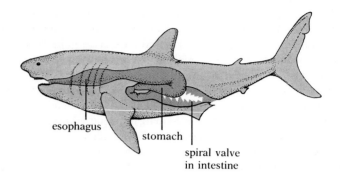

10.26   Digestive system of a shark
Because of the spiral valve of the intestine, food material must follow a winding course and is thus exposed to more surface area.

esophagus

stomach

spiral valve
in intestine

action of the microbes. The caecum, however, is not located where the mammal can derive maximal benefit from the microbial action. It is too far back in the digestive tract, behind the small intestine, where most of the digestion and absorption takes place. Hence even though horses have an enormous caecum, much coarse undigested plant material is expelled in their feces. A compensating adaptation has evolved in rabbits, which form two types of feces. One of these is material from the caecum; they re-ingest this, and it undergoes a second cycle of digestion and absorption.

Ruminants such as the cow also employ microbial digestion, but the microorganisms are not held in a posterior caecum. Instead, such animals have four different stomachlike chambers (Fig. 10.27), of which the first three are thought to be expanded sections of the esophagus. Vast numbers of bacteria and Protozoa live in the **rumen**, which is the largest of the four chambers, and in the reticulum. Swallowed food enters the rumen and reticulum, where the microbes begin digesting and fermenting it, breaking down not only protein, polysaccharides, and fats, but cellulose as well. The larger, coarser material is periodically regurgitated for further chewing, as the animal "chews its cud." This rechewed material is again swallowed and mixed with the fermenting material in the rumen. Slowly the products of the microbial action and some of the microbes themselves move on into the true stomach (the abomasum) and intestine, where the more usual type of digestion and absorption takes place. Thus, by using microbial digestion in the anterior portion of their digestive tracts rather than in a posterior caecum, ruminants derive maximal benefit from the microbial action.[3] Much less undigested plant material remains in their feces than in the feces of horses.

The rumen strategy is not perfect, however. Plant material must remain in the "fermentation vat" for a considerable period—hours, or even days

[3] It is worth noting—though the point has no direct bearing on digestion—that the microorganisms in the rumen can use ammonia and such comparatively simple organic compounds as urea as sources of nitrogen for synthesizing amino acids. By digesting the microorganisms, the ruminant obtains the amino acids, thus benefiting from the nitrogen metabolism of the microbes. Modern agriculture often takes advantage of this microbial metabolism by supplying nitrogen in the form of cheap ammonium salts in cattle feed, rather than in the more expensive form of protein.

—to give the microorganisms sufficient time to do their work. If the food is of low quality, a cow or an antelope can literally starve to death waiting. The less-efficient "afterburner" strategy of horses and zebras seems better adapted to handling large volumes of very poor quality food quickly. It enables zebras in Africa to thrive in almost desertlike habitats where the dozens of species of ruminant antelope cannot live.

Digestion of cellulose by symbiotic microorganisms is not limited to mammals. Various insects, notably the termites, feed on wood, which they could not use as food were it not for intestinal microbes that can ferment the cellulose. A few species of wood-eating beetles do, however, themselves secrete cellulase, an enzyme that digests cellulose, and such beetles do not have to rely on intestinal microbes.

But let us return to the large intestine in humans. From the caecum, the large intestine ascends on the right side to the mid-region of the abdominal cavity, then crosses to the left side, and descends again (Fig. 10.20). The three sections thus formed are frequently termed the ascending, transverse, and descending colons. One of the chief functions of the colon is reabsorption of much of the water used in the digestive process. If all the water in which enzymes are secreted into the digestive tract were lost with the feces, a severe problem of desiccation would result. Occasionally the intestine becomes irritated, and peristalsis moves material through it too fast for enough water to be reabsorbed; this condition is known as diarrhea. Conversely, if material moves too slowly, too much water is reabsorbed and constipation results. A proper amount of roughage (indigestible material, primarily cellulose) in the diet provides the bulk needed to stimulate enough peristalsis in the large intestine to prevent constipation. Our need for roughage is probably an artifact of our herbivorous ancestry.

A second function of the colon is the excretion of certain salts, such as those of calcium and iron, when their concentration in the blood is too high. The salts are excreted into the colon and are eliminated from the body in the feces. The large intestine also contains large numbers of bacteria, which live on the undigested food that reaches the colon. The significance of these bacteria in the life of a healthy person is not clearly understood. Approximately half the dry weight of the feces is made up of masses of these bacteria.

The last portion of the large intestine, the *rectum*, functions as a storage chamber for the feces until defecation. The feces are eliminated from the rectum through an opening called the anus.

ENZYMATIC DIGESTION IN HUMANS

Digestion by saliva   Having traced the human digestive tract from mouth to anus, let us next consider the chemical changes that occur in a food as it passes through this complex tubular system. We have said that enzymatic digestion starts in the mouth. The saliva contains an enzyme called *amylase*[4] (also known as ptyalin), which begins but does not complete the hy-

----

[4] The names of most enzymes end with the suffix *-ase*, which designates enzymes by international agreement. The first part of the name usually indicates the substrate upon which the enzyme acts; thus *amyl-* (from *amylum*, Latin for "starch") indicates that amylase acts upon starch.

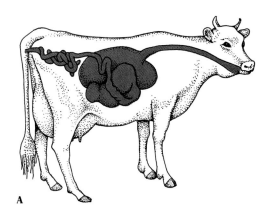

A

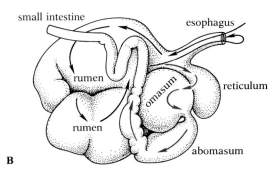

B

10.27   The digestive system of a ruminant
(A) A cow with the various chambers in approximately their correct locations. (B) Detailed path of food as it passes through the four "stomachs." Food moves initially (black arrows) into rumen and reticulum (the drawing shows it going only into the rumen), where it is digested and fermented by microorganisms. Then, as the cud, it is regurgitated for more chewing and later reswallowed and returned to the rumen. The fluids and finely divided particles produced in this second phase (color arrows) move through the reticulum into the omasum and then into the abomasum, which is the true stomach.

drolysis of starch to glucose. (As you may remember, starch is synthesized by water-producing condensation reactions, so its breakdown must involve hydrolysis.) Though amylase produces some glucose, it yields primarily the disaccharide maltose (Fig. 10.28) and, in lesser amounts, fragments three or four glucose units long, which must be further digested in the intestine.

Since the food remains in the mouth only a short time, the amylase has little opportunity to work there. Much of its actions occurs inside each bolus after it is swallowed and moved into the stomach. The acid of the stomach soon inactivates the enzyme, however, and salivary amylase actually digests only a small percentage of the starch in the food. In fact, the saliva of many mammals contains no amylase at all; dogs and their relatives, for example, have so little starch in their diets that there must have been little selection pressure for the evolution of a starch-digesting enzyme.

Digestion in the stomach   Once in the stomach, food is exposed to the action of gastric juice secreted by the numerous gastric glands of the mucosa of the stomach wall. This juice contains much hydrochloric acid and several enzymes. The acid makes the contents of the stomach very acidic (with a pH of about 1.5–2.5). Note that, advertisements for many patent medicines to the contrary, an acid stomach is both normal and necessary for proper function.

The principal enzyme of the gastric juice is *pepsin*, which digests protein. Unlike most proteolytic (protein-digesting) enzymes, it will function only in a strongly acid medium. Pepsin is a characteristic enzyme of vertebrates; and most invertebrates do not have any proteolytic enzymes active in strongly acid solutions. The evolution of pepsin may be correlated with feeding on animals with bones, since bones disintegrate more easily in acid.

Pepsin does not hydrolyze protein all the way to its amino acid components. It splits the peptide bonds adjacent to only a few amino acids, particularly tyrosine and phenylalanine (Fig. 10.29). The specificity of proteolytic enzymes is readily understandable. Since proteins are composed of a variety of building-block compounds, not just of one, the structural configura-

10.28   Digestion of starch
Amylase in the saliva and in the pancreatic juice hydrolyzes some of the bonds between glucose units, producing small amounts of free glucose, but much larger quantities of the disaccharide maltose. The maltose is then digested to glucose by maltase secreted by intestinal glands.

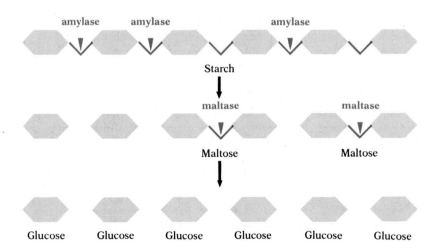

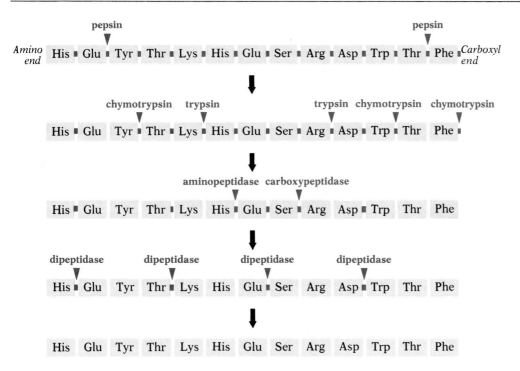

tion around the various peptide bonds varies, depending on which two amino acids the bond joins; some of the bonds may fit in the active side of a particular enzyme and others may not. Pepsin, for example, seems to have an active site complementary to peptide bonds at the amino end of amino acids whose R groups include a six-carbon ring (see Fig. 3.16, p. 58).

Any discussion of protein digestion immediately raises the question, Why isn't the wall of the digestive tract, which is composed primarily of protein, digested by the proteolytic enzymes? There are two reasons.

First, the wall of the digestive tract is covered with mucus, which apparently shields it from enzymes. When this defense breaks down, the digestive enzymes do begin to eat away a small portion of the lining; the resulting sore is known as an ulcer. Occasionally, an ulcer is so severe that a hole develops in the wall of the digestive tract, and the contents of the tract spill into the abdominal cavity. Why and how ulcers first develop is not clear.

Second, the gastric glands secrete, not the active enzyme pepsin, but an inactive precursor called pepsinogen. Inactive enzyme precursors of this sort are known as *zymogens*. Pepsinogen has no proteolytic activity and as long as it is stored in the glands of the stomach wall, poses no threat to the wall. It remains stored in the wall until needed, and can only be changed into active pepsin after exposure to the active pepsin already present in the lumen of the stomach. Activation results from the splitting off of 42 amino acids from one end of the pepsinogen molecule; the shorter polypeptide chain that remains is pepsin.[5]

[5] In addition to pepsin, the gastric juice of suckling ruminants like cattle and antelope contains another enzyme, rennin, which clumps milk proteins together.

10.29 Digestion of protein

Pepsin in the stomach hydrolyzes peptide bonds at the amino ends of tyrosine (Tyr) and phenylalanine (Phe). Then the food moves into the intestine, where trypsin from the pancreas hydrolyzes bonds at the carboxyl end of lysine and arginine, and chymotrypsin from the pancreas hydrolyzes bonds at the carboxyl end of tyrosine, tryptophan, and phenylalanine (and also bonds adjacent to methionine and leucine, when they are present). The remaining short chains of amino acids are digested in three ways: Aminopeptidase splits the bond connecting the peptide at the amino end of a chain to the rest of the fragment, while carboxypeptidase from the pancreas splits the bond connecting the peptide at the carboxyl end. In this way these two enzymes "nibble off" peptides from each end until only a dipeptide—a fragment consisting of two amino acids—remains. Bonds between these pairs of amino acids are split by dipeptidases. With digestion now complete, the amino acids may be absorbed through the cells of the intestinal wall.

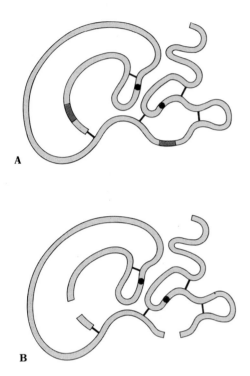

A

B

10.30 Conversion of chymotrypsinogen into chymotrypsin
The single polypeptide chain constituting a molecule of the enzyme precursor chymotrypsinogen is stabilized by disulfide bonds (black bars). The black circles indicate important parts of the catalytic site of the enzyme. The two portions of the polypeptide chain removed to produce the active enzyme chymotrypsin are shown in color before removal (A). Once these two portions are removed (B), the enzyme refolds into an active configuration (not shown).

Digestion in the small intestine   By far the most digestion takes place in the next section of the digestive tract, the small intestine. When partially digested food passes from the stomach into the duodenum, its acidity stimulates the release of a large number of different digestive enzymes into the lumen of the intestine. These enzymes are secreted from two principal sources, the *pancreas* and the *intestinal glands*. The pancreas, a large glandular organ lying just below the stomach (Fig. 10.20), forms during fetal development as an outgrowth of the digestive tract; it retains a connection to the duodenum called the *pancreatic duct*. When food enters the duodenum, the pancreas secretes a mixture of enzymes that flows through the pancreatic duct into the duodenum. Included in this mixture are enzymes that digest all three principal classes of foods—carbohydrates, fats, and proteins—as well as some that digest nucleic acids.

One of the pancreatic enzymes is *pancreatic amylase* (sometimes also called diastase or amylopsin), which, as its name implies, acts like salivary amylase, splitting starch into the disaccharide maltose. It is far more important than salivary amylase, for it carries out most of the starch digestion.

Lipase, also secreted by the pancreas, is the body's principal fat-digesting enzyme, but it completely hydrolyzes only a relatively small percentage of the fat to glycerol and fatty acids. Some of the fat is partly digested by removal of only one of the three fatty acids, and some is not digested at all. But since fats, and the products of the partial digestion of fats, are lipid-soluble, they can be absorbed across cell membranes.

Like pepsin, *trypsin* and *chymotrypsin*, two of the proteolytic enzymes of the pancreas, cleave only the peptide linkages adjacent to certain specific amino acids (Fig. 10.29). Trypsin splits the peptide bonds adjacent to lysine and arginine (both of which have positively charged R groups; see Fig. 3.16, p. 58); chymotrypsin splits those adjacent to tyrosine, phenylalanine, and tryptophan (which have six-carbon rings in their R groups) and, to a lesser extent, methionine and leucine. Chymotrypsin resembles pepsin in hydrolyzing bonds adjacent to tyrosine and phenylalanine, but this resemblance is superficial: the two enzymes do not split the same bonds. Pepsin cleaves the bonds on the amino side of tyrosine and phenylalanine, while chymotrypsin cleaves those on the carboxyl side. This specificity, far from being unusual, is typical of most enzymes. Indeed, the digestive enzymes are actually less specific than many others in that one enzyme will catalyze the reactions of several different substrates.

Again like pepsin, both trypsin and chymotrypsin are secreted in inactive (zymogen) forms, called trypsinogen and chymotrypsinogen respectively. In the intestine, trypsinogen is converted into active trypsin (through the removal of the terminal six amino acids from its polypeptide chain) by enterokinase, an enzyme secreted by the intestinal glands. Then the trypsin thus formed removes two small internal pieces of the chymotrypsinogen polypeptide chain; the remaining molecule, active chymotrypsin, consists of three separate polypeptide chains held together by disulfide bonds (Fig. 10.30). Note that all three zymogens we have discussed—pepsinogen, trypsinogen, and chymotrypsinogen—have polypeptide chains longer than those of the active enzymes; in each case activation simply involves cutting off one or more pieces of the chain so that the polypeptide can refold to create an active site.

Recent studies of the chemical structure of trypsin and chymotrypsin have helped clarify both their probable evolution and the basis for their catalytic specificity. These two enzymes are sufficiently alike—in amino acid sequence, in conformational folding pattern, and even in their catalytic sites—to make it reasonable to believe that they evolved from the same ancestral proteolytic enzyme. There is, however, a slight, but functionally crucial difference between them, which pertains to a pocketlike portion of the binding site that holds the R group of the substrate amino acid. In trypsin one of the amino acids that form this pocket is aspartic acid (which has a negatively charged R group); in chymotrypsin the same position is occupied by serine (which has an uncharged R group). This seemingly small variation accounts for the affinity of the two enzymes for different amino acids. That it should do so is a dramatic illustration of the critical importance of amino acid sequence (primary structure) in determining the functional properties of enzymes.

In summary, then, the action of the pepsin in the stomach and of trypsin and chymotrypsin from the pancreas results in a splitting of proteins into fragments of varying lengths, but does not produce many free amino acids. These three enzymes are known as *endopeptidases*—that is, enzymes that hydrolyze peptide bonds between amino acids located within the protein, not bonds linking terminal amino acids to the original chain. Another class of enzymes, called *exopeptidases*, hydrolyze off the terminal amino acids of the fragments produced by the endopeptidases, completing the digestive process. There is a great variety of exopeptidases, each highly specific in its action. One, for example (carboxypeptidase), hydrolyzes the linkage binding the terminal amino acid at the free carboxyl end of the fragment. Another (aminopeptidase) hydrolyzes the linkage of the terminal amino acid at the free amino end. Still others (dipeptidases) break apart pairs of amino acids; one breaks only the bond of a fragment consisting of glycine linked to leucine, another only the bond of a fragment consisting of two molecules of glycine linked together, and so on. These exopeptidases complete the job begun by the endopeptidases. Most of them are secreted by intestinal glands, but carboxypeptidase is produced in the pancreas.

The chemical action of the various proteolytic enzymes has been described at some length, not because it is important for you to remember in detail what bonds each enzyme hydrolyzes, but because the proteolytic mechanism provides a good example of enzyme specificity and of the way enzymes often work in teams. If we had simply said that proteins are digested by pepsin, trypsin, and a variety of other peptidases, you would have had no indication of the elaborate interplay that characterizes these seemingly commonplace processes.

Just as certain enzymes from the intestinal glands complete the digestion of protein, other intestinal enzymes complete the digestion of starch begun by salivary and pancreatic amylase. The amylases split starch molecules primarily into disaccharides. The intestinal enzymes split disaccharides into simple sugars. For example, *maltase* splits maltose (Fig. 10.28), *sucrase* splits sucrose, and *lactase* splits lactose. Digestion of nucleic acids is accomplished in an analogous manner. *Nucleases* secreted by the pancreas split nucleic acids into nucleotides. The latter are further digested by *phosphatases* from the intestinal glands, which break their phosphate bonds, and by *N-glycosylases,* which remove the ribose sugars.

Absorption of the products of digestion involves active transport, as you

THE DIGESTION OF LACTOSE BY HUMAN ADULTS

Digestive capabilities vary not only among different species (we have already mentioned the absence of salivary amylase in dogs), but also within particular species. The digestion of lactose, a sugar found only in milk, provides a striking example.

Secretion of milk by the mammary glands of female mammals evolved as a way of feeding the young. The only source of nourishment for very young mammals, milk is a nearly complete food, containing, in most species, carbohydrate (in the form of lactose),* fat, and protein, as well as important minerals. But except for humans, adult mammals do not use milk as a food. It is not surprising, then, that secretion of lactase, the lactose-digesting enzyme, usually greatly diminishes or even ceases altogether once an animal is past the age of weaning.

Only recently has it been realized, however, that this pattern applies also to most human beings; in most parts of the world, humans more than four years old secrete little or no lactase. Indeed, of the various peoples studied to date, only those of European ancestry and those belonging to a few pastoral tribes in Africa have been found to secrete enough of the enzyme to be able to digest the lactose in large quantities of milk (see graph). When people of other ancestries drink much milk they often become ill, getting a bloated feeling, cramps, and diarrhea, though they may have no such trouble with milk from which the lactose has been removed, or with milk products like yogurt and cheese in which the lactose has been broken down by microbial action. One reason for the illness is that the undigested sugar in the intestine upsets the normal osmotic balance and an excessive amount of water moves into the intestinal lumen from the cells; another is that fermentation of the lactose by bacteria in the large intestine produces large quantities of acids and carbon dioxide. In Europeans and pastoral Africans, lactose tolerance (resulting from the continued production of lactase in adults) must have evolved during the roughly 10,000 years since the milking of domestic animals began.

* There is no lactose in the milk of seals and their close relatives (the Pinnipedia).

How widely peoples living near one another may differ is shown by the major tribes of Nigeria. The Ibo and Yoruba live in the southern part of the country, where conditions are unfavorable for cattle; milk has not traditionally been a part of their diet after weaning, and they cannot tolerate lactose. By contrast, the nomadic Fulani in northern Nigeria have been raising milk cattle for thousands of years, and they are lactose-tolerant (see graph).

Most American blacks are descended from nonpastoral tribes of western Africa, and they are relatively intolerant of lactose, though not so much as native Africans. Their somewhat greater tolerance may be due, in part, to evolutionary change during the generations they have lived in dairying regions and, in part, to admixture of European genes.

might expect. Recent research on the "pumps" for amino acids and simple sugars has shown that both depend on a steep sodium ion (Na$^+$) gradient across the cell membrane, with a high concentration of Na$^+$ outside the cell and a low concentration inside. These active-transport pumps thus appear to be coupled to the sodium-potassium pump, which we discussed in Chapter 4.

Bile   One more secretion should be mentioned in this discussion of human digestion. The *liver*, a critically important organ about which much will be said in later chapters, produces a fluid called bile, which aids in fat digestion. The liver is very large, occupying much of the space in the upper part of the abdomen. On its surface is a small storage organ, the *gallbladder* (Fig. 10.20). Bile, produced throughout the liver, is collected by a series of branching ducts and emptied into the gallbladder. When food enters the duodenum, the muscular wall of the gallbladder is stimulated to contract, and the bile is forced down the bile duct into the duodenum.

Bile is not a digestive enzyme; it is not even a protein. It is a complex solution of salts, pigments, and cholesterol. The bile salts act as emulsifying agents, causing large fat droplets to be broken up into many tiny droplets suspended in water. This action is much like that of a good detergent. The many small fat droplets expose much more surface area to the digestive action of lipase than a few large droplets would. Bile salts apparently aid also in the absorption of fats. When insufficient bile salts are present in the intestine, both fat digestion and absorption are seriously impaired. The bile salts are reabsorbed by the large intestine, transported back to the liver, and used again.

The bile pigments and cholesterol play no perceptible role in digestion. The pigments are produced through the destruction of red blood cells in the liver and give the characteristic brown color to feces. The cholesterol, a relatively insoluble compound, sometimes causes trouble by becoming concentrated into hard gallstones, which may block the bile duct and interfere with the flow of bile.

Chapter 11

GAS EXCHANGE

We have seen that when nutrient compounds are broken down in the process of respiration, over 90 percent of the energy yield depends on the presence of oxygen, which makes possible the complete oxidation of the compounds to carbon dioxide and water. Thus a basic problem for the great majority of living organisms is the procurement of oxygen and the elimination of carbon dioxide.

It is true that a few unicellular organisms—notably the chemosynthetic microbes mentioned in Chapter 7—can subsist indefinitely in the total absence of oxygen, and exposure to it is even fatal to some. Other organisms can survive for limited periods under anaerobic conditions. But in these cases the respiratory degradation of nutrients stops far short of completion: the end products are usually lactic acid or ethyl alcohol, relatively large molecules that still contain much chemical energy. Anaerobic respiration thus appears very wasteful in comparison with aerobic respiration. Nevertheless, its importance should not be overlooked. By resorting to anaerobic metabolism, an organism (or part of an organism) may survive short periods of oxygen deprivation. For organisms capable of living anaerobically indefinitely, environments otherwise totally uninhabitable are open for colonization; many bacteria, for example, are anaerobic and live in habitats where no other type of organism could survive. Nonetheless, most environments suitable to life permit aerobic respiration, which has become the chief method of respiration in plants, animals, protists, and fungi.

It is a common misconception that oxygen procurement is a problem faced only by animals, and that gas exchange in green plants consists exclusively of intake of carbon dioxide and release of oxygen. Certainly this is the exchange that takes place in association with photosynthesis, but the carbohydrate products of photosynthesis are of little value to the plant if they cannot be respired to provide usable metabolic energy. Hence plants, like animals, are constantly taking in oxygen and releasing carbon dioxide as they carry out the process of cellular respiration. When a green plant is exposed to bright light, both photosynthetic and respiratory gas exchange are usually taking place; since the rate of photosynthesis then greatly exceeds the rate of respiration, the *net* effect is one of uptake of carbon dioxide and release of oxygen. The reverse is true, of course, when the green plant is in the dark or when it has no leaves in winter. Respiratory gas exchange, then, is as necessary for plants as it is for animals.

THE PROBLEM

Gas exchange between a living cell and its environment always takes place by diffusion across a moist cell membrane. The gases must be in solution if they are to move across the membrane. In unicellular organisms and many small multicellular ones, particularly those that are aquatic, this requirement poses no serious problem, because each cell is either in direct contact with the surrounding medium or only a few cells removed from that medium. Hence these organisms have usually not evolved special respiratory devices.

When large body size is predominantly in two dimensions, gas exchange can still take place chiefly by direct diffusion between the individual cells and the surrounding medium. Some brown algae, the kelps, may grow to a length of 50 m, but the blades of even the longest kelps remain very thin (Fig. 11.1). As a result, no cell is far from the surface of a blade, and the total gas-exchange area is fairly large in relation to the volume of the plant. The thicker stipe has numerous intercellular spaces filled with water that is continuous with the external medium. In these plants, then, which are large in only two dimensions, no special gas-exchange mechanism is needed.

When increase in body size involves three dimensions, as it generally does, the maintenance of a respiratory surface of adequate dimensions relative to the volume becomes a problem, because surface area (a square function) increases much more slowly than volume (a cube function). The problem is most acute for the more active animals, whose rapid utilization of energy demands a large amount of oxygen per unit of body volume per unit time.

An additional complicating factor is that many organisms have evolved relatively impermeable outer body coverings. The waxy epidermis of the leaves of terrestrial plants, for example, and animal skin with its derivative scales, feathers, or hair, function as protective barriers between the fragile internal tissues and organs and the often hostile outer environment. But their presence, though it confers many advantages, limits the gas-exchange surface to a restricted region of the body, making the problem of providing adequate exchange area even more critical.

Another complication brought on by large three-dimensional size in animals is that many cells are deep within the body of the organism, far from the gas-exchange surface. Diffusion alone is incapable of moving gases in adequate concentrations across the immense number of cells that may

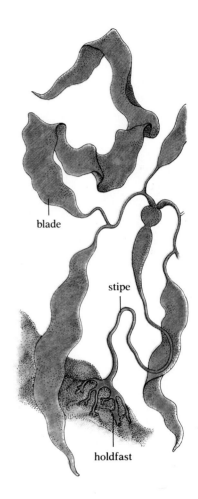

blade

stipe

holdfast

11.1 A kelp (*Pelagophycus*), one of the large brown algae
In a simple solution to the problem of gas exchange, the broad, flat kelp blades provide a relatively large surface for the diffusion of gases, with every cell in the organism close to the surface.

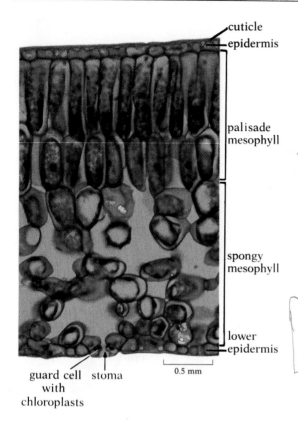

cuticle
epidermis

palisade
mesophyll

spongy
mesophyll

lower
epidermis

guard cell stoma
with
chloroplasts

0.5 mm

11.2 Cross section of part of a leaf of privet

intervene between these more distant cells and the exchange surface. In general, simple diffusion suffices for movement of substances through aqueous media only when the distances are less than one millimeter. Some other mechanism for conveying gases to the individual cells of the organism therefore becomes essential.

The need for direct contact between the moist membranes across which gas exchange occurs and the environmental medium also poses serious difficulties, especially for terrestrial organisms. The moist membranes must be exposed to the environment, but at the same time the chances of desiccation must be kept to a minimum. Since a large, thin, moist surface is often fragile and easily suffers mechanical damage, the tendency has been toward the evolution of protective devices, particularly when the respiratory surfaces protrude outward.

In general, specialized respiratory surfaces may be grouped in two categories: inward-oriented and outward-oriented extensions of the body surface (see Fig. 11.8). Each category embraces a diversity of form and detail, but the diversities become less bewildering if one bears in mind that each type of respiratory system represents merely one way of meeting the basic needs discussed above: (1) a respiratory surface of adequate dimensions; (2) for many organisms, methods of transporting gases between the area of exchange with the environment and the more internal cells; (3) means of protecting the fragile respiratory surface from mechanical injury; and (4) means of keeping the surface moist.

SOLUTIONS IN TERRESTRIAL PLANTS

As indicated earlier, most primitively aquatic plants, notably the algae, carry out gas exchange across almost the entire body surface; but in response to the problems of desiccation and large three-dimensional size, most terrestrial plants have evolved more elaborate mechanisms.

LEAVES

Gas exchange associated with both photosynthesis and cellular respiration takes place at a particularly high rate in green leaves, organs strikingly adapted for this process.

Recall that most of the visible outer surface of a leaf, covered as it is by a waxy cuticle, is more or less dry and impermeable, and hence ill suited for diffusion of gases. Exchange must therefore take place elsewhere. You may remember that the mesophyll parenchyma in a leaf is so arranged as to leave large intercellular spaces (Figs. 11.2 and 11.3; see also Fig. 6.3, p. 143). A high percentage of the total surface of each mesophyll cell is exposed to the air in these spaces, which are interconnected and continuous with the external atmosphere by way of openings in the epidermis: the *stomata*. Gases can thus move easily between the surrounding atmosphere and the internal spaces of the leaf. The actual gas exchange—the diffusion of gases into and out of living cells—takes place across the thin moist membranes of the cells inside the leaf.

Let us briefly consider how the structures of the leaf help it meet the four previously stated requirements for respiratory systems.

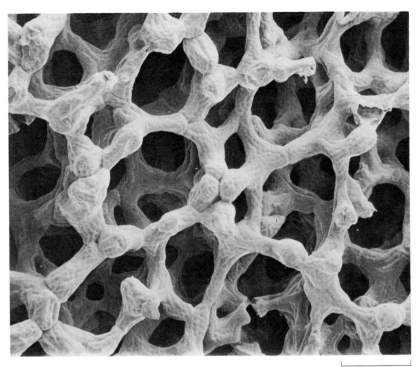

50 μm

11.3   Scanning electron micrograph of spongy mesophyll in a bean leaf
The micrograph gives an especially clear view of the extensive system of interconnecting air spaces.

1. The surface area available for gas exchange in the leaf is very large. By comparison with the outer area of the leaf, the total area of cell membrane exposed to the intercellular spaces is enormous. The principle involved is a very elementary one: a chamber irregularly shaped and greatly subdivided by partial partitions will have far more wall space than a round or square one of equal volume. The microvilli in the small intestine represent an analogous solution to the problem of increasing surface area within a small volume.

2. Internal transport of gases occurs in leaves without any special adaptations. Gases can reach each individual cell directly via the intercellular spaces.

3. The danger of mechanical injury is relatively minor for an internal exchange surface. The epidermis, with its hairs, spines, or other derivative structures, functions as a protective covering for the entire leaf.

4. The exchange surfaces remain moist because they are exposed to air only in intercellular spaces. With the humidity within those spaces nearly 100 percent, the membranes of the mesophyll cells always retain a thin film of water on their surfaces. Gases dissolve in this water before moving into the cells. The protective epidermal tissues and the layers of waxy cuticle on their outer surfaces act as barriers between the dry outside air and the moist inside air.

But these barriers are not complete; if they were, movement of gases between the outside and the inside could not take place. So the stomata, which provide the essential openings, in a sense constitute weak links in the

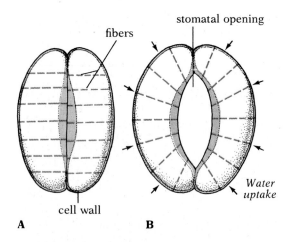

11.4 Guard cells
(A) Two unusual features of guard cells account for the way in which they regulate the stomatal opening: the cell wall is thicker on the side close to the stoma than on the side away from the opening; and bands of inelastic fibers (color) run around each cell.
(B) When uptake of water causes the cell to become turgid, expansion is limited to the side away from the opening. As a result the guard cells buckle, pulling apart, and the stoma opens. The degree of swelling is largely controlled by the availability of water and sunlight.

protective armor of the leaf. Here we see the sort of compromise that has been a constant feature of evolutionary adaptation. Few morphological structures or biochemical pathways, however beneficial in some way, are without possible deleterious effects in some other way. What determines the evolutionary fate of a particular trait is not whether it is exclusively beneficial or harmful, but whether the beneficial effects outweigh the harmful ones. In this case, stomata present advantages that outweigh the danger of desiccation; moreover, other adaptations help minimize this danger.

As you may recall from Chapter 8, the raw materials for noncyclic photophosphorylation and the Calvin cycle are light, water, and carbon dioxide. The function of the stomata is to balance the leaf's need to perform photosynthesis against the dangers of desiccation. When all three raw materials are available, the stomata must be open to allow gas exchange, but when one or more of them are missing, these openings must be closed to prevent water loss. Each opening, or stoma, in the epidermis is bounded by two highly specialized epidermal cells called **guard cells**, which are joined at their ends. Unlike most other epidermal cells, these bean-shaped cells contain chloroplasts (Fig. 11.2). The walls of each guard cell are of unequal thickness, being considerably thicker on the side next to the stoma than on the side away from it (Fig. 11.4). Moreover, the guard cells are wrapped by bundles of inelastic fibers that prevent them from expanding horizontally, in diameter. When the guard cells swell with water (which normally occurs only if sufficient water is available from the roots), the expansion of the cell surface to accommodate this increase in volume is restricted to the lengthening of the sides away from the stoma, where the wall is relatively thin. As a result the guard cells buckle (Fig. 11.4B) and the stoma opens (Fig. 11.5). Gas exchange can now take place, and the leaf can obtain carbon dioxide for photosynthesis. When water is not available, or is being lost from the plant too quickly, the leaf cells, including the guard cells, may become flaccid, causing the leaf to wilt and the guard cells to close.

At night, when the problems of water loss are rarely as severe, the stomata nevertheless usually narrow, since one of the raw materials for photosynthesis—sunlight—is not available. How the rapid changes in guard-cell turgidity are tied to the availability of sunlight, as well as water, is still a matter of dispute. One hypothesis takes its cue from the unusual starch metabolism of guard cells. By contrast with most cells, which convert starch into sugar mainly in the dark, guard cells do so in the light. The decreasing acidity (rise in pH) that results from the consumption of CO_2 in photosynthesis seems to make the enzyme responsible for converting starch into sugar in the guard cells progressively more active. The conversion of stored starch into sugar presumably brings about a rapid increase in the osmotic concentration of the guard cells' cytoplasm, so that water then moves into the guard cells by osmosis, causing them to swell and open the stomata.

Though the pH changes, the starch–sugar conversions, and the changes in sugar concentration unquestionably occur and no doubt play a role in guard-cell movement, this hypothesis does not fully explain the extremely rapid movements of guard cells often observed. A more recent hypothesis posits an additional factor that causes the guard cells to take up water: an ATP-driven pump that alters the osmotic balance between the cells and the surrounding fluid by moving potassium ions (K^+) into the guard cells. This mechanism is similar to the system by which the Venus fly trap closes its leaves. It has been demonstrated that there is indeed more K^+ in the guard cells of open than of closed stomata, and that rapid guard-cell movement can be inhibited by chemicals known to poison processes dependent on ATP. There is also evidence, at least under experimental conditions, that abscisic acid, a plant hormone, can induce stomatal closing. The various factors known to affect stomatal opening are summarized in Table 11.1.

Most plants lose large quantities of water by evaporation through the stomata every day in the process called ***transpiration***, but the humidity in the intercellular spaces of the leaf probably does not often drop appreciably, because the lost water is steadily replaced by water drawn up through the stem and distributed throughout the leaf by the many small veins. Indeed, the evolution of an effective transport system has involved the harnessing of the process of transpiration. Transpiration also serves as a cooling mechanism, preventing excessive buildup of heat inside the leaves even when they are exposed to the direct rays of the sun.

Closing of the stomata during the day when water is in short supply reduces photosynthesis, but does not altogether stop it. Their closing limits the supply of CO_2 to what is produced by cellular respiration; the rate of photosynthesis therefore declines until it is approximately equal to the rate of respiration. Respiration, in turn, is limited to the amount of oxygen released as a by-product of photosynthesis. The reduction in the supply of CO_2 may also bring about photorespiration (see Chapter 8, p. 212), with a resulting decrease in the synthesis of carbohydrate. As we saw in an earlier chapter, an alternative pathway for carbon fixation in the absence of abundant CO_2 has evolved in C_4 plants. This C_4 photosynthesis represents a metabolic adaptation complementary to the adaptation for preventing excessive water loss.

Before leaving the subject of stomata, let us note several other evolutionary adaptations associated with them. In most plants the stomata are located primarily, if not exclusively, in the lower epidermis—the side of the leaf usually turned away from the sun's rays, where the drying tendency is less severe. Further, the lower epidermis of many plants is covered with short hairs, which trap insulating air. This so-called boundary layer functions in reducing the flow of direct air currents across the stomatal openings and thereby slows down desiccation. Plant species adapted for life in particularly dry or particularly wet habitats often exhibit special adaptations of the stomata. In the oleander (*Nerium*), which lives in a very dry habitat, the stomata are located in deep hair-lined depressions in the lower

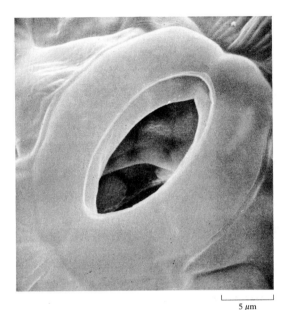

5 µm

11.5 Scanning electron micrograph of a stoma in a cucumber leaf
The stoma is open because the two large crescent-shaped guard cells have pulled away from each other. Mesophyll cells in the interior of the leaf can be glimpsed through the opening.

TABLE 11.1 *Environmental factors affecting stomatal opening and closing*

Conditions favoring opening	Conditions favoring closing
Abundant water	Lack of water
Abundant light	Darkness
Low internal CO_2	High internal CO_2
	Presence of abscisic acid

epidermis, an adaptation that eliminates convection currents across the stomata. In the pondweed *Potamogeton*, by contrast, the stomata of the floating leaves are located in the very thin upper epidermis and not in the lower epidermis (Fig. 11.6).

STEMS AND ROOTS

Do not infer from this discussion that leaves have the only gas-exchange areas in plants. We have simply chosen to examine these beautifully adapted organs in some detail in order to illustrate the types of problems and solutions characteristic of gas-exchange mechanisms.

The relatively impervious layer of bark on many old stems would effectively cut off most of their oxygen supply were it not for the development of numerous small areas of loosely arranged cells with many intercellular air spaces between them through which gases can move freely. Each such loose group of cells is called a **lenticel** (Fig. 11.7). The spongy streaks always seen in cork (an outer bark) are traces of the lenticels of the cork oak. Since most of the cells in the inner layers of large stems are dead, there is little need for oxygen in the intercellular air spaces to penetrate deep into the stem.

Roots also carry out gas exchange, though they usually possess no special structures for this function. Gases can diffuse readily across the moist membranes of root hairs and other epidermal cells. For roots to obtain enough oxygen, however, the soil in which they grow must be well aerated.

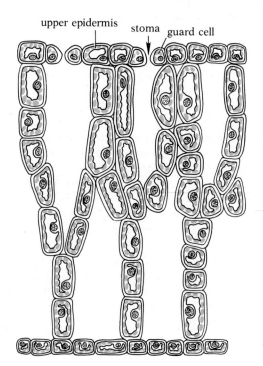

11.6 Cross section of portion of a floating leaf of pondweed (*Potamogeton*)

The stomata are in the upper surface, and the mesophyll has very large intercellular air spaces. There are chloroplasts in the epidermal cells.

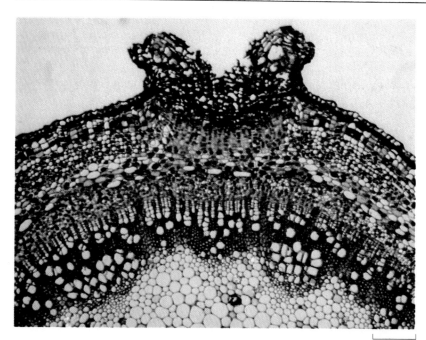

0.2 mm

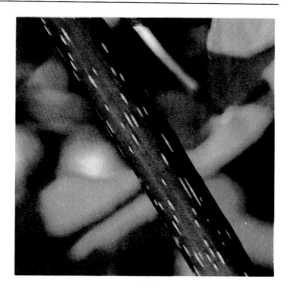

11.7 Lenticels

Left: The waterproof outer bark (layer of dark cells on the surface) on this section of elderberry (*Sambucus*) stem is interrupted at the center of the lenticel. Thus the more loosely arranged cell layers beneath, with their numerous intercellular air spaces, are exposed to the atmosphere. Right: The individual lenticels can be seen as white areas on the surface of a young sycamore maple (*Acer*) stem.

Different types of soils vary in their aeration characteristics, which depend on the amount of pore space, the affinity of the soil particles for water, and many other factors. Soils with very high percentages of clay particles, for example, have many pore spaces, but the tiny clay particles are so hydrophilic and absorb so much water during wet periods that the air spaces become filled with water, the result being a condition known as waterlogging. Frequent waterlogging may reduce the air content of the soil to the point at which many species of plants are stunted or cannot grow at all. In some poorly aerated soils, the circulation of gases between the pore spaces and the atmosphere above the soil is so slow that the air in the pore spaces becomes deficient in oxygen and often contains a deleteriously high concentration of carbon dioxide. One of the benefits of turning over the soil before planting is the increased air circulation this makes possible.

Plants, unlike animals, do not seem to need any special gas-transporting mechanisms. Most of the intercellular spaces in the tissues of land plants are filled with air, in contrast to those in animal tissues, which are filled with fluid. These air-filled spaces are interconnected to form an intercellular air-space system that opens to the outside through the stomata and lenticels and penetrates to the innermost parts of the plant body. Incoming gases can therefore move in gaseous form directly to the internal parts of the plant from the environmental atmosphere without having to cross membranous barriers, and they do not have to diffuse long distances through water or cell liquids because they do not go into solution until they reach the film of water on the surfaces of the individual cells. Since oxygen can diffuse some 10,000 times faster through air than through liquids, the intercellular air-space system ensures that all cells, even the more internal ones, are adequately supplied. If the oxygen had to diffuse through liquid

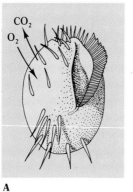

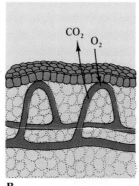

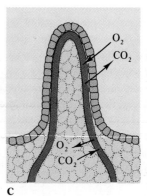

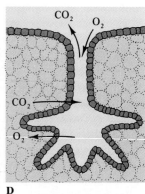

11.8   Types of gas-exchange systems in animals
(A) Unicellular organisms exchange gases with the surrounding water directly across the general cell membrane. (B) Some multicellular aquatic animals use the general body surface as an exchange surface; the blood (color) transports gases to and from the surface. (C) Many multicellular aquatic animals have specialized evaginated gas-exchange structures (gills). (D) A few aquatic animals, such as the sea cucumber, use invaginated exchange areas. (E) Most true air breathers have lungs, specialized invaginated areas that depend on a blood transport system. (F) Land arthropods have tracheal systems, invaginated tubes that carry air directly to the tissues without the intervention of a blood transport system.

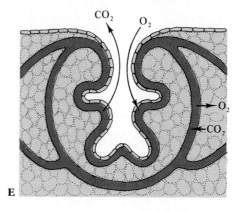

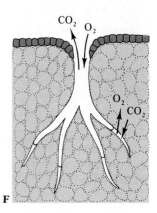

from the surface of a plant organ, it would penetrate less than one millimeter, and all the more internal cells would be deprived of oxygen and could not respire. Experiments show, in fact, that if the air-space system is blocked, the innermost cells soon start dying.

SOLUTIONS IN AQUATIC ANIMALS

As we have already indicated, unicellular animals have no special gas-exchange devices; simple diffusion across their cell membranes is sufficient (Fig. 11.8A). Some of the smaller and simpler multicellular animals like jellyfish, hydra, and planaria show little further development, although their multipurpose gastrovascular cavities do facilitate the exposure of the more internal cells to the environmental water (containing dissolved oxygen) they draw in through the mouth; no cell in these animals is far from the water medium. A few larger aquatic animals, particularly some of the marine segmented worms, lack special respiratory systems and use the skin of the general body surface, which is usually richly supplied with blood vessels (Fig. 11.8B). Most larger multicellular animals, however, have evolved true respiratory systems.

GILLS

With a few exceptions, the respiratory systems of multicellular aquatic animals involve evaginated (protuberant) exchange surfaces, usually known as gills. Gills vary in complexity all the way from the simple bumplike skin gills of some sea stars (Fig. 11.9), the flaplike parapodia of many segmented marine worms (Fig. 11.10), the mantle-protected gills of squids (Fig. 11.11), and the shell-protected gills of bivalves (Fig. 11.12), to the minutely subdivided gills of fish (Fig. 11.13). The gill system has evolved independently countless times in the history of animal life on earth, and such diverse animals as clams, lobsters, and salamanders use gills for respiration.

Most gills, particularly those of very active animals, have such finely subdivided surfaces that a few small gills may expose an immense total exchange surface to the water. Thus, though the gas-exchange surface takes up a very limited part of the animal body, most of which can be protected by relatively impermeable coverings, the surface-to-volume ratio of the exchange surface is high.

Another characteristic of most gills is that they contain a rich supply of blood vessels. Often the blood in these vessels is separated from the external water by only two cells: the single cell of the wall of the vessel and a cell of the gill surface. Sometimes even the vessel wall is eliminated, and only one cell remains between the blood and the water. Oxygen moves by diffusion from the water, across the intervening cell or cells, and into the blood, where it is ordinarily picked up by a carrier pigment. (Transport by the blood will be discussed in Chapter 13.) The blood then distributes the oxygen throughout the body to the individual cells. Carbon dioxide produced by cellular metabolism moves in the opposite direction: it is transported to the gills and discharged into the surrounding water.

One intriguing feature of the arrangement for exchange of oxygen and carbon dioxide between water and blood in fish gills deserves special men-

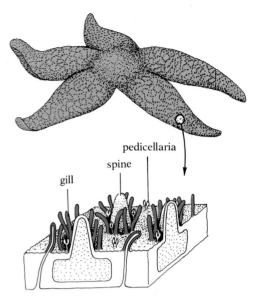

11.9 Sea star
The tiny skin gills are protected from damage by the spines and the pincerlike pedicellariae, which repel small animals that might otherwise settle on the surface of the sea star.

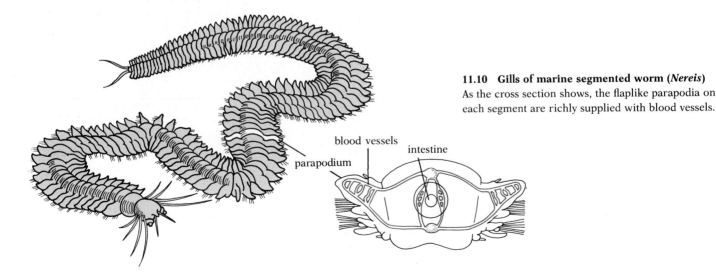

11.10 Gills of marine segmented worm (*Nereis*)
As the cross section shows, the flaplike parapodia on each segment are richly supplied with blood vessels.

tion. The water flows over the surface of a gill lamella in a direction opposite to the flow of blood in the vessels of the lamella (Fig. 11.14). Consequently blood just about to leave the gill, and already almost fully loaded with O_2, encounters water that has not yet given up any of its O_2, because it is just reaching the gill. The O_2 gradient therefore favors pickup of more O_2 by the blood. At the other end of the lamella, oxygen-poor blood entering the gill encounters water that has already lost much of its O_2, but because even this water contains more O_2 than blood that has not yet picked up any, the gradient here too favors diffusion of O_2 from the water to blood. In short, this **countercurrent exchange system**, thanks to a favorable gradient between blood and water at every point along the lamella, maximizes the amount of O_2 the blood can pick up from water. This would not be the case if the two fluids had the same direction of flow. This countercurrent strategy is by no means restricted to gills or to gas transfer. As we shall see in Chapter 14, the kidneys regulate body fluids by utilizing the same method to produce concentrated urine.

The fragile gills are easily damaged, and various structures have evolved for their protection. Frequently these are hard coverings like the carapace of lobsters and the operculum of fish (Fig. 11.13A), but sometimes they take other forms, such as the spines and pedicellariae that surround the skin gills of sea stars (Fig. 11.9).

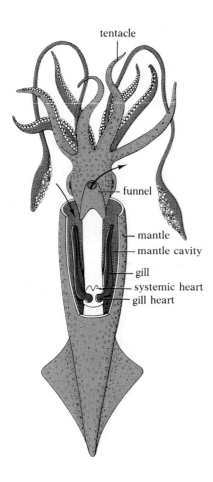

11.11 The gills of a squid
Part of the protective mantle has been cut away to expose the gills within the mantle cavity. Water (left-hand arrow) flows into the mantle cavity when the mantle is relaxed. When the mantle contracts, its collar seals the opening and the water is forced out of the funnel. The jetlike expulsion of water when the mantle is contracted propels the animal backward with great force.

tentacle

funnel

mantle
mantle cavity
gill
systemic heart
gill heart

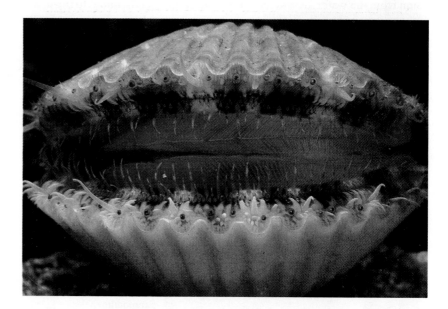

11.12 The calico scallop (*Argopecten gibbus*), a filter-feeding bivalve
The cilia on the gill move water through this bivalve. As the water crosses the gill, gas exchange—uptake of O_2 and release of CO_2—takes place. At the same time, a second set of cilia traps food particles and moves them to the labial palps near the mouth. Here yet another set of cilia separates out the coarse material, which is ejected; the remaining fine particles are then eaten.

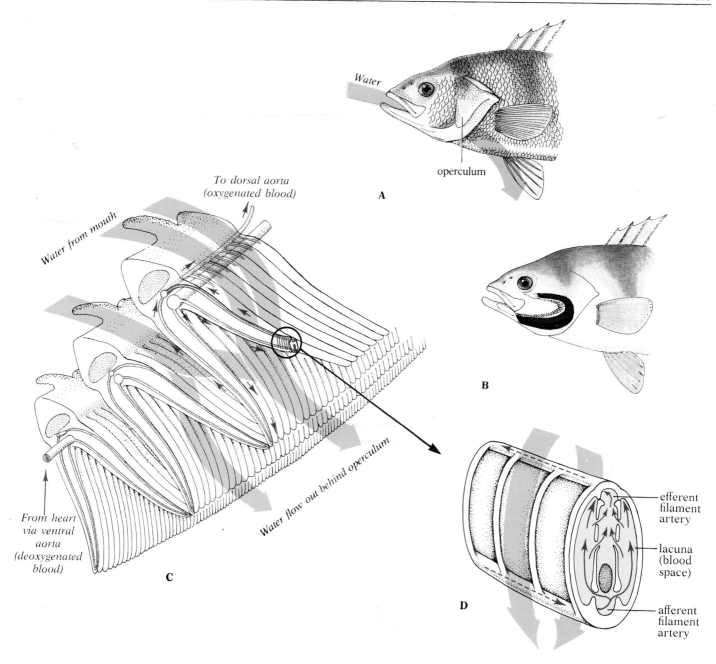

Water

operculum

A

To dorsal aorta
(oxygenated blood)

Water from mouth

B

From heart
via ventral
aorta
(deoxygenated
blood)

Water flow out behind operculum

C

efferent
filament
artery

lacuna
(blood
space)

afferent
filament
artery

D

11.13 Gills of fish

(A) Head with <u>operculum</u> covering gills. Water carrying O_2 is drawn in through the mouth, flows across the gills, and exits behind the operculum. (B) Head with the operculum cut away and the gills exposed. (C) Portions of three adjacent gill arches. Each arch bears two rows of primary filaments. The main paths of blood flow to and from the filaments are shown in red; the blue arrows trace the path of water across the gills. Each primary filament bears many disclike lamellae, which contain capillaries that run from the afferent artery to the efferent artery. The end of one filament has been cut off to show a lamella more clearly. (D) The lamella is magnified to show how the blood (red) in the lamellae and the water (blue) flowing between the lamellae are countercurrents, <u>moving in opposite</u> <u>directions. The lamellae are the actual sites</u> <u>of gas exchange.</u>

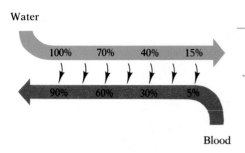

Water

Blood

11.14 Countercurrent exchange system in lamella of fish gill
Figures indicate the degree of oxygen saturation for both water and blood. Because of the counterflow, the O_2 gradient between water and blood always favors diffusion of O_2 (vertical arrows) from water to blood, and the blood can extract a high percentage of the O_2 from the water. If the flow were parallel, the blood could extract much less O_2 and would leave the gills far from fully loaded.

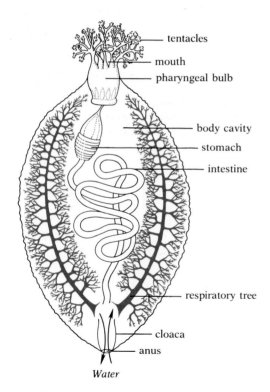

tentacles
mouth
pharyngeal bulb
body cavity
stomach
intestine
respiratory tree
cloaca
anus

Water

11.15 Respiratory tree of sea cucumber

Obtaining enough oxygen is a greater problem for aquatic animals than for air breathers, for two reasons: First, O_2 has a low solubility in water, constituting only about 0.004 percent of seawater (the percentage is usually slightly higher in fresh water, but far more variable) as compared with approximately 21 percent in air. Second, the diffusion of O_2 is many thousands of times slower in water than in air. Most aquatic animals must therefore move water across the exchange surfaces. If the water remained still, the O_2 in the vicinity of the exchange surfaces would soon be depleted, and it would not be renewed by diffusion fast enough to sustain the animal. Most fish actively pump water into their mouths, across the gill filaments, and out behind the operculum. Many fast-swimming fish keep their mouths open as they swim, so that their forward motion forces water across the gills; some species are so dependent on this method of ventilation that they will die for lack of O_2 if prevented from moving. Because water is far more viscous than air and therefore much harder to move, some aquatic organisms use up as much as 20 percent of their metabolic energy in moving water across their gills.

SOLUTIONS OTHER THAN GILLS

Biology is a science of exceptions. Not all aquatic animals with special respiratory systems use evaginated gills. For example, sea cucumbers—relatives of sea stars—have specialized invaginated (infolded) systems called respiratory trees (Fig. 11.15). These long branched tubes are diverticula of the cloaca; water is drawn into and expelled from the system by cloacal contractions. This system is an elaboration of a simpler one used by a few animals, in which gas exchange takes place as water is alternately drawn in and expelled through the enlarged thin-walled posterior portion of the digestive tract.

Many insects that live in water are not fully aquatic in that they must periodically come to the surface to breathe air. Of these, there are diving beetles (*Dytiscus*) that store air under their hard shell-like forewings when they surface, then dive with this air bubble and breathe from it. And there are spiders that construct an underwater web in which they store a large bubble of air (Fig. 11.16). You might think the O_2 in the bubble would very soon be depleted, but this is not the case. Remember that air contains high concentrations of other gases besides O_2. Since these gases are not used up, a gas bubble remains, and as the partial pressure of O_2 in the bubble falls, O_2 from the surrounding water tends to diffuse into the bubble, renewing the supply.[1] In short, the bubble acts as a gill. In a few insects this mechanism is so refined that their store of gases is permanent and they do not have to surface to renew it.

[1] The partial pressure of a gas is the total pressure of the mixture of gases in which it occurs multiplied by the percentage of the total volume that it occupies. Thus, if the total pressure of all the atmospheric gases is about 760 mm Hg (millimeters of mercury, a measure of barometric pressure), and if O_2 is about 20 percent of this mixture, then the partial pressure of O_2 in the atmosphere is equal to 760×0.20, or 152 mm Hg.

SOLUTIONS IN TERRESTRIAL ANIMALS

A few land animals have evolved highly modified gill-like respiratory structures that function in air. An example is the book lung of spiders, in which the gills are evaginated into an open body cavity and therefore resemble the pages of a book. But the hazards of desiccation are considerable for most exposed surfaces, and major structural problems are associated with an array of filaments or a branched structure that is at once sufficiently strong to maintain its shape against surface tension and gravity and sufficiently thin-walled to allow easy passage of gases. It is not surprising, therefore, that most terrestrial animals have evolved invaginated respiratory systems. These invaginated systems are of two principal types, **lungs** and **tracheae**. In both, the air inside the system is kept moist, and the cells of the exchange surface are covered by a film of water in which gases can dissolve. Thus the process of gas exchange has remained essentially aquatic in land animals, as it has in the leaf.

LUNGS

Lungs, which are invaginated gas-exchange organs limited to a particular region of the animal and dependent on a blood transport system, are most typical of two unrelated animal groups, the land snails and the higher vertebrates, including some fish, most amphibians, and all reptiles, birds, and mammals. In their simplest form, lungs are little more than chambers with slightly increased vascularization in their walls and with some sort of passageway leading to the outside. This simple type of lung is found, for example, in some snails that inhabit the lower levels of the ocean beach, where the necessity for air breathing is seldom pressing, because oxygen is available from the water. From such a rudimentary beginning, the evolution of the lung has tended toward a greatly increased surface area, by subdivision of its inner surface into many small pockets or folds, and toward increased vascularization of its exchange surfaces. The increase in vascularization is illustrated dramatically in four closely related species of snails, called periwinkles, which inhabit successively higher levels of the ocean beach. The extent of vascularization of the lung (or mantle cavity) in these snails is precisely correlated with the increased distance of the successive species from the ocean and the concomitant increased necessity for breathing air.

It is not surprising that terrestrial vertebrates have lungs, but it is odd that some relict species of fish (that is, species surviving essentially unchanged in structure from ancient times) have them also. Indeed, many biologists are now convinced that the ancestral fish from which both modern fish and the land vertebrates evolved had lungs that enabled them to live in stagnant, poorly aerated water for long periods of time when necessary. These primitive lungs were simple sacs that arose as ventral evaginations of the digestive tract in the pharyngeal region behind the gills. A few salamanders still have such simple lungs, but in most terrestrial vertebrates the inner surface has become increasingly folded and subdivided, providing a far higher surface-to-volume ratio for the exchange process. This evolutionary tendency reaches its culmination in mammals and birds, the two

11.16 The water spider *Argyoneta aquatica*
The male spider is beneath his underwater web, which holds a large bubble of air and to which he returns periodically to breathe. He transports the air there by trapping it among hairs on his abdomen (silvery area).

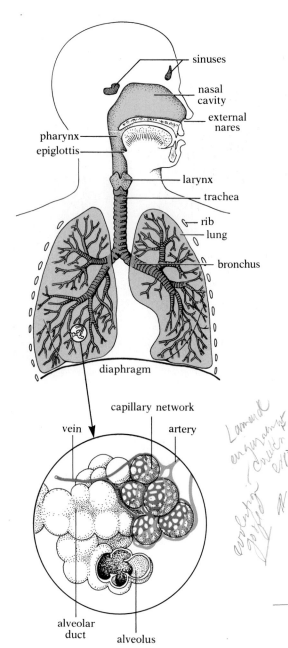

sinuses

nasal cavity

external nares

pharynx

epiglottis

larynx

trachea

rib

lung

bronchus

diaphragm

capillary network

vein

artery

alveolar duct

alveolus

11.17 The human respiratory system
The enlargement shows a few sectioned alveoli and other alveoli surrounded by blood vessels. Actually, all alveoli, including those lying along alveolar ducts, are surrounded by networks of capillaries.

warm-blooded classes, which must expend vast amounts of metabolic energy to maintain their stable high body temperatures, and hence have exceedingly high oxygen demands.

The human respiratory system (Fig. 11.17) provides a good example of the mammalian type. Air is drawn in through the **external nares**, or nostrils, and enters the **nasal cavities**, which function in warming and moistening the air, filtering out dust particles, and smelling. Bony ridges in the cavities cause eddies in the air stream, facilitating these processes. The mucous layer on the epithelium of the nasal passages, and the cilia of many of the epithelial cells, also increase the efficiency with which these processes occur; in addition, the mucus may have some bactericidal properties. Curiously enough, of all the processes, smelling is the most primitive one; the external nares and the nasal passages originally had nothing to do with respiration, but evolved as odor-sensing devices. Fish use their nostrils for smelling but not breathing; they take in water for respiratory purposes through the mouth. Since smelling and feeding are so intimately related, it is not surprising that in fish, amphibians, and many reptiles little or no separation exists between the nasal cavity and the mouth cavity. The separation has been developed furthest in mammals. In them a new "roof of the mouth" has evolved that consists of an anterior bony palate and a posterior soft palate (Fig. 11.18).

Even in mammals, however, the air and food passages ultimately join in the region known as the **pharynx**. During inhalation air leaves the pharynx via a ventral opening, the **glottis**, which leads into the larynx. (We are here using the terms "dorsal" and "ventral" as though the human were standing on four legs like other mammals.) Since air enters the pharynx dorsally and exits ventrally, and since food enters ventrally and exits dorsally into the esophagus, the air and food passages not only join but actually cross in the pharynx. (This rather inefficient arrangement is the price we pay because natural selection modified the already-existing smelling apparatus into respiratory passages instead of starting from scratch and building a totally new system. But this is typical of much evolution: the new is built from the old.) Elaborate mechanisms help ensure that when food is forced back into the pharynx it will not enter the nasal cavity or the larynx, but must be swallowed into the esophagus. Without attempting to describe the whole complex action in detail, we can point out that the internal nares, which connect the nasal cavities with the pharynx, are closed by the soft palate and that the glottis is closed by a flap of tissue called the **epiglottis** when the larynx is raised against it during swallowing.

After leaving the pharynx through the glottis, air enters the **larynx**, a chamber surrounded by a complex of cartilages (commonly called the Adam's apple). In many animals, including humans, the larynx functions as a voice box. It contains a pair of vocal cords—elastic ridges stretched across the laryngeal cavity that vibrate when air currents pass between them; changes in the tension of the cords result in changes in the pitch of the sounds emitted.

The **trachea** is an air duct leading from the larynx into the thoracic cavity. Its epithelial lining is ciliated; the cilia beat in waves that carry foreign particles and mucus up the trachea away from the lungs. A series of C-shaped rings of cartilage is embedded in the walls of the trachea and they

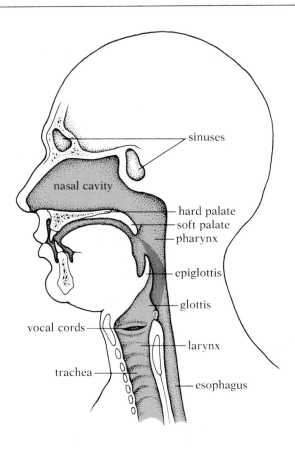

sinuses

nasal cavity

hard palate
soft palate
pharynx

epiglottis

glottis

vocal cords

larynx

trachea

esophagus

11.18 Detail of upper portion of human respiratory and digestive system
Path of air in color; path of food, gray. The two paths cross in the pharynx.

prevent it from collapsing upon inhalation. At its lower end, it divides into two **bronchi**, tubes that lead toward the two lungs. (It is at the lower end of the trachea, where the bronchi branch away, that the voice box of birds, the syrinx, is located. In birds each bronchus has its own set of vocal cords. Complex and beautiful songs are possible because birds can vary the output of the two sets of vocal cords independently, and so produce two separate melodies.) Each bronchus branches and rebranches, and the bronchioles thus formed branch repeatedly in their turn, forming smaller and smaller ducts that ultimately terminate in tiny air pockets, each of which has a series of small chamberlike bulges in its walls termed **alveoli**. The total alveolar surface is enormous: about 100 square meters—many times greater than the total area of the skin.

The walls of the alveoli are exceedingly thin, being usually only one cell thick, and each alveolus is surrounded by a dense bed of blood capillaries. The alveoli are the site of the actual gas exchange and may therefore be regarded as the primary functional units of the lungs. Oxygen entering an alveolus dissolves in the film of water on its wall and then moves by diffusion across the intervening cells to the blood. Experiments have demonstrated that both this movement and the reverse movement of CO_2 are cases of simple diffusion; no active transport across the cell barriers is involved. O_2 is in higher concentration in the air of the alveolus than in the blood, and CO_2 is

11.19   The mechanisms of human breathing
(A) Resting position. (B) Inhalation. The rib cage is raised up and out by the intercostal muscles, and the diaphragm is pulled downward. Both these motions increase the volume (color) of the thoracic cavity. The consequently reduced air pressure in the cavity causes more air to be drawn into the lungs.

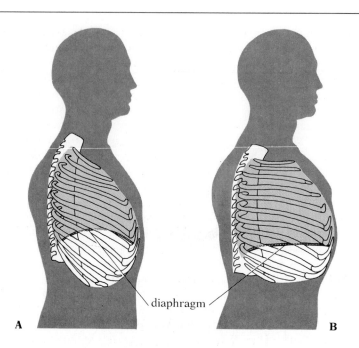

diaphragm

A   B

11.20   Respiratory system of a bird
Attached to the lungs are many air sacs (light brown), some of which even penetrate into the marrow cavities of the wing bones. As in a mammal, the respiratory system has bilateral symmetry (not obvious in this side view).

in higher concentration in the blood than in the alveolus. Each gas simply moves from the region of its higher concentration to the region of its lower concentration, in accordance with the principles of diffusion discussed in an earlier chapter. There are actually very few instances where diffusion takes place within the body totally unaided by active transport, but it has been demonstrated that the body is incapable of active transport of O_2 in the lungs. As a result, when the partial pressure of O_2 in the atmosphere falls below normal, as at high altitudes, symptoms of O_2 deprivation rapidly develop, because passive transport is insufficient to meet O_2 demands.

Air is drawn into and expelled from the lungs by the mechanical process called **breathing**. In mammals this process generally involves muscular contractions of two regions, the **rib cage** and the **diaphragm**. The latter is a muscular partition separating the thoracic and abdominal cavities (Fig. 11.19). Inhalation, or inspiration, occurs whenever the volume of the thoracic cavity, in which the lungs lie, is increased; such an increase reduces the air pressure within the chest below the atmospheric pressure and draws air into the lungs. The increase in thoracic volume is accomplished by contractions of the intercostal muscles that draw the rib cage up and out, and by contraction, or downward pull, of the normally upward-arched diaphragm; the first mechanism is popularly called chest breathing, while the second is called abdominal breathing. Normal exhalation, or expiration, is a passive process; the muscles relax, allowing the rib cage to fall back to its resting position and the diaphragm to arch upward. This reduction of thoracic volume, combined with the elastic recoil of the lungs themselves, causes a rise in the pressure inside the lungs to a level above that of the outside atmosphere and drives out the air.

The air moved by a single normal breath—the tidal air—represents only a small fraction of the total capacity of the lungs. Additional air (complementary air) can be forcibly inhaled, and, similarly, forcible exhalation can expel not only tidal and complementary air, but also additional air known as <u>reserve air</u>. The <u>total breathing capacity, or **vital capacity**, is the tidal air plus the complementary and reserve air;</u> though it varies greatly from person to person, 4 liters is probably a rough approximation of the average. Trained athletes usually develop larger lungs, and, as would be expected, their vital capacity is usually substantially greater than normal.

The pattern of air flow in the respiratory system of birds differs fundamentally from that of mammals. In addition to paired lungs, birds possess several (most commonly eight or nine) thin-walled air sacs that occupy much of the body cavity and even penetrate into the interior of some of the bones (Fig. 11.20). The air sacs are poorly supplied with blood vessels and do not themselves absorb O_2 or release CO_2. Their arrangement and bellowslike action, however, <u>make possible continuous unidirectional flow of air through the lungs.</u> Let us look at this process in more detail.

Like mammals, birds suck in air by increasing the volume of the body cavity. As Figure 11.21A shows, most of the air drawn in during inhalation does not go directly to the lungs, but flows through the bronchus to the posterior air sacs; simultaneously, air already in the lungs moves forward into the anterior air sacs via connecting passages called recurrent bronchi. During exhalation air from the posterior sacs moves into the lungs, primarily via recurrent bronchi while air from the anterior sacs empties to the outside (Fig. 11.21B). Thus air moves forward through the lungs during both inhalation and exhalation. Instead of alveoli, whose dead-end chambers would be incompatible with unidirectional air flow, bird lungs have tiny air ducts (parabronchi) running through the lung tissue, and it is across their walls that gas exchange takes place (Fig. 11.22).

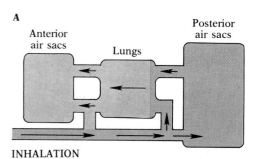

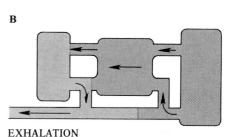

11.21   Respiratory cycle of a bird
(A) During inhalation, new air (color) is drawn into the posterior air sacs; a small amount also enters the posterior portion of the lungs. Air already in the system (gray) is simultaneously moved forward through the lungs and into the anterior air sacs.
(B) As air is exhaled from the anterior air sacs and air from the posterior air sacs moves forward into the lungs, the air sacs decrease in volume. The bolus of air (color) inhaled in (A) will be exhaled during the following respiratory cycle. Note that during both inhalation and exhalation oxygen-rich air is moving unidirectionally through the lungs.

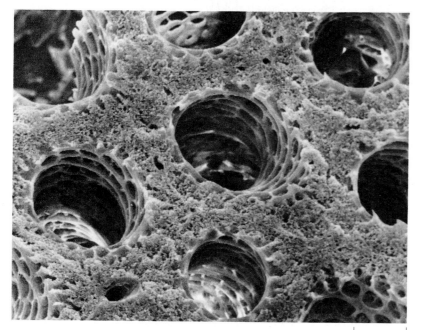

0.2 mm

11.22   Scanning electron micrograph of cross section of parabronchi in a two-week-old chick
It is through the parabronchial tubes that the unidirectional flow of air in the bird lung occurs. Exchange of gases takes place across the walls of these tubes, each less than 0.5 mm in diameter. Note the resemblance to the spongy mesophyll of a leaf (Fig. 11.3).

Birds are far more efficient than mammals in extracting O_2 from air, both because of the continuous unidirectional flow of air through their lungs, and because blood in the capillaries associated with the parabronchi flows at an angle to the flow of air and provides some of the same benefits as the countercurrent exchange system of fish gills. This superior efficiency enables birds to fly actively at high altitudes, where the partial pressure of O_2 is low. Vance Tucker of Duke University experimentally exposed sparrows and mice to an atmosphere simulating that at 6,000 m altitude (350 mm Hg), and found that the sparrows could fly vigorously while the mice were unable to stand up and could barely crawl.

The mammalian and avian method of breathing, in which air is drawn into the lungs, is known as ***negative-pressure breathing***. By contrast, in ***positive-pressure breathing*** air is forced into the lungs. Both methods are used by adult frogs. Mouth closed and nostrils open, the frog lowers the floor of its mouth, sucking air into the mouth cavity (negative-pressure method). Then it closes its nostrils and raises the mouth floor; this reduction in the volume of the mouth cavity exerts pressure on the imprisoned air and forces it into the lungs (positive-pressure method). The frog is also an excellent example of an animal that uses a variety of gas-exchange mechanisms. The lungs are only occasionally filled, because alternative mechanisms, notably the thin membrane of the mouth cavity and the creature's soft moist skin, provide ample gas-exchange surfaces.

Negative-pressure breathing works because the thorax, in which the lungs are suspended, is a sealed compartment filled with fluid that does not expand. As a result, when the thorax expands during inhalation, the walls of the lungs are pulled along, the volume of the lungs increases, and the resulting negative pressure draws in outside air. But if the seal separating the

11.23   Evolution of swim bladders
Swim bladder colored, esophagus uncolored. Each stage is shown in cross section (left) and longitudinal section (right). According to one hypothesis, the swim bladder of modern fish evolved from a primitive ventral lung (A). The lung may first have moved to a dorsal position while retaining a ventral attachment to the esophagus, as in some living lungfish (B). The attachment to the esophagus may then have moved laterally (C) and finally become dorsal, as in the typical modern swim bladder (D and E). In some modern fish the connection to the esophagus has been lost entirely (F).

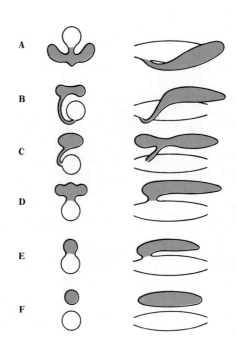

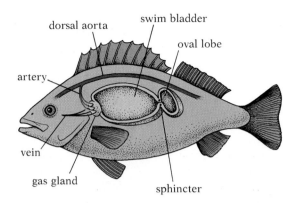

dorsal aorta   swim bladder

oval lobe

artery

vein

gas gland   sphincter

11.24 Functioning of a typical swim bladder
Gas is added to the bladder from the blood by means of the gas gland. Gas is removed from the bladder by absorption through the walls of the oval lobe. These two processes control the volume of the bladder, and hence the buoyancy of the fish. The sphincter is generally open only when gas is being removed.

thoracic cavity from the outside or from the rest of the body is broken, as in some chest injuries, the expansion of the thorax may draw in air or abdominal fluid and the lung may collapse. Positive pressure is then needed to reinflate the lung, and, of course, the thoracic cavity must be sealed to restore breathing. It should also be obvious that the discharge of fluid into either the thoracic cavity or the interior of the lung, a frequent consequence of certain diseases, will impair the efficiency of breathing.

The swim bladder of fish   Let's return to the hypothesis that the ancestral fish had lungs. Many researchers believe that the primitively ventral lung evolved into the swim bladder of modern fish by gradually shifting to a dorsal position (Fig. 11.23). Its attachment to the esophagus also gradually shifted to a dorsal position; it is still there in some species, but in other modern species the connecting duct has been lost and direct entrance or exit to the swim bladder no longer exists.

The swim bladder enables the fish to remain at a given level in the water without sinking by adjusting its density to that of the surrounding water. (As you might expect, bottom-dwelling fish seldom have swim bladders.) The volume of the gases in the bladder changes as the fish shifts to a water level with a different pressure. If the fish swims upward, the swim bladder swells in response to the reduced pressure, and gases must be removed from it to restore it to its normal size; if the fish swims downward, gases must be added to the swim bladder. The addition of gases is accomplished by a region of specialized glandular cells (the gas gland), associated with many blood vessels, in the walls of the swim bladder. The removal of gases generally occurs elsewhere in the bladder, or even in a separate specialized area, the oval lobe (Fig. 11.24).

Analysis of the gases in the swim bladder reveals a surprisingly high concentration of O_2—a concentration often far above that in the surrounding water. Still more surprising is a high concentration of molecular nitrogen (which is chemically inactive in the bodies of animals) and sometimes even of such inert gases as argon. Normally, the accumulation of substances against a concentration gradient can be attributed to active transport. But

11.25 Tracheae and spiracle of an insect
At top center is a spiracle (brown), from which numerous branching tubes—the tracheae—can be seen running to many parts of the insect's body. The spiracles are usually located on the sides of the body segments, the number varying in different kinds of insects. See Figure 11.26 for a drawing of terminal branches, or tracheoles.

active transport of O_2—a feat the lungs cannot perform—would be very unusual. Moreover, active transport requires that the transported substances enter into chemical reactions, which such chemically inactive gases as nitrogen and argon would be very unlikely to do. How, then, might gases be accumulated in the swim bladders of fish? A partial answer has come from the investigations of Johan B. Steen of the University of Oslo, Norway. Steen succeeded in taking minute blood samples from the tiny capillaries that enter and leave the gas gland in eels. His analyses indicate that gases are released from the blood, despite the apparent adverse concentration gradients, because lactic acid secreted into the blood by the gas gland greatly reduces the blood's capacity for carrying the gases. In short, no active transport is involved; the process of gas secretion is more physical than chemical.

TRACHEAL SYSTEMS

The second principal type of invaginated respiratory system evolved for air breathing is the tracheal system. It is typical of most terrestrial arthropods, in which it has evolved independently many times. Here we find no localized respiratory organ and little or no significant transport of gases by the blood. Instead, the system is composed of many small tubes, called *tracheae*, that branch throughout the body (Fig. 11.25). The tracheae and the smaller tracheoles into which they branch carry air directly to the individual cells, where diffusion across the cell membranes takes place (Fig. 11.26).

Air enters the tracheae by way of *spiracles*, apertures in the body wall that usually open and close by valves (Fig. 11.27). Some of the larger insects actively ventilate their tracheal systems by muscular contraction, but most small insects and some fairly large ones apparently do not. Calculations have shown that the rate of diffusion of oxygen in air is rapid enough to maintain at the tracheal endings an O_2 concentration only slightly below that of the external atmosphere. This type of respiratory system, however, has doubtless been a factor in limiting the size attainable by insects.

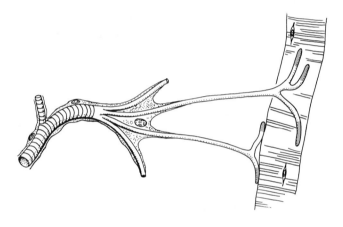

11.26 Terminal tracheoles
Unlike the larger tracheae (segments at left), the tracheoles do not have thickened supportive rings in their walls. The branching ends of the tracheoles lie on the surface of the target cell, to which they bring oxygen. The ends usually contain a small amount of liquid (color); when the oxygen requirements of the target cell rise, some of the liquid is withdrawn from the tracheoles, so that more air comes into direct contact with the cell.

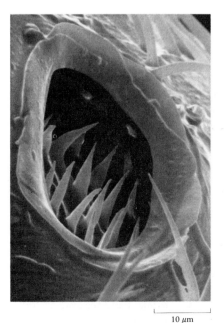

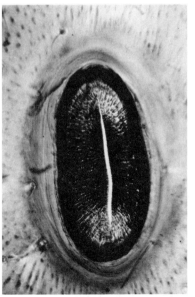

10 μm   10 μm

11.27 Spiracles of two insects
Left: Scanning electron micrograph of a fully open ant spiracle. The pointed projections are sensory hairs that monitor external conditions and can trigger spiracle closing when necessary. Right: A nearly closed grasshopper spiracle; the black areas are the valves. Note the resemblance to the stoma of a leaf (Fig. 11.5).

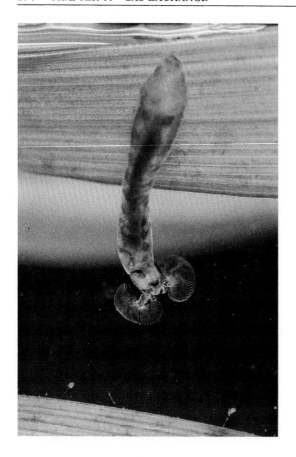

The aquatic larvae of some insects, such as damselflies and mayflies, have tracheal gills—platelike or feathery structures richly supplied with tracheae (Fig. 11.28). Whereas in the other gills that we have discussed absorbed O_2 is transported in the blood, the O_2 absorbed by tracheal gills comes out of solution and moves in gaseous form into the general tracheal system.

We have seen, then, the various ways in which plants and animals have adapted to their different environments, responding to environmental changes with new solutions to the problem of gas exchange. To meet the demand for greater respiratory-system surface area necessitated by increased body size, organisms have evolved complex branched and folded respiratory structures and more efficient systems of oxygen procurement. In most plants and animals, however, the oxygen thus obtained must still be transported to tissues not in direct contact with the air or water. The next two chapters will examine how the transport of oxygen and nutrients is accomplished in multicellular organisms.

11.28   Larva of a buffalo fly (*Siphona exigua*) with tracheal gills extended
Like this juvenile buffalo fly, many aquatic insect larvae breathe by means of tracheal gills. These structures serve a dual purpose: they provide a fine surface for the absorption of dissolved oxygen, and also function as filter-feeding combs, sifting food from the environment.

Chapter 12

INTERNAL TRANSPORT IN UNICELLULAR ORGANISMS AND PLANTS

THE PROBLEM

Every living cell, whether it exists alone as a single-cell organism or is a component of a multicellular one, must perform its own metabolic activities. It must synthesize its own ATP by cellular respiration (and/or photosynthesis) and carry out for itself those activities necessary for its growth and maintenance. To support its metabolism, the cell must obtain raw materials—nutrients—and, if it uses aerobic respiration, oxygen. At the same time, it must rid itself of metabolic wastes such as carbon dioxide and, in animals, nitrogenous compounds. In short, every cell must have access to a medium from which it can extract raw materials and into which it can dump wastes. In unicellular organisms and some of the structurally simple multicellular ones, each cell is either in direct contact with the environmental medium or only a short distance from it. But in the larger and structurally more complex multicellular plants and animals, the more internal cells are far from the body surface and from the general environmental medium. We have already seen that in such organisms nutrient procurement, gas exchange, and waste expulsion take place in certain restricted regions of the body specialized for those functions. Obviously, these organisms must also possess mechanisms for transporting substances between the specialized systems of procurement, synthesis, and elimination and the individual living cells throughout the body.

This chapter will discuss transport in unicellular organisms, simple multicellular organisms, and vascular plants. The next will examine transport in multicellular animals.

ORGANISMS WITHOUT SPECIAL TRANSPORT SYSTEMS

In bacteria, Protozoa, and unicellular algae (and within single living cells in general), diffusion plays an important role in the movement of materials. Since the individual particles of all substances within the cell exhibit random thermal motion, they tend to become evenly distributed throughout the cell when they are not prevented from doing so by specialized intracellular membranes. Diffusion is also important in movement of materials from cell to cell within the body of a multicellular organism; we have examined its part in the procurement of water by plant roots, for example. Such intercellular diffusion may be facilitated in plants by the plasmodesmata, which interconnect the cytoplasmic contents of adjacent cells. We have also seen the importance in plants of diffusion of water along cell walls and of diffusion of gases through intercellular spaces.

But diffusion is a very slow process. If only diffusion were involved, a substance would take a long time to move from one cell to another, or even from one end of a single large cell to the other end. It is not surprising, therefore, that even in unicellular and small multicellular organisms diffusion is supplemented by other transport mechanisms. We have seen, for example, that food vacuoles commonly move along a fairly precise path within the cell, thereby distributing the products of digestion to all parts of the cytoplasm. We have also seen that the endoplasmic reticulum may provide a pathway for intracellular movement of some substances. And the cytoplasm itself is seldom motionless; it frequently exhibits rapid massive flow within the cell. The flowing cytoplasm of an active amoeba is an example. The cytoplasm of many plant cells undergoes a characteristic movement called *cytoplasmic streaming*, or cyclosis, in which the cytoplasm flows in definite currents along the surface of the cell vacuole (Fig. 12.1). Sometimes the streaming is restricted to local regions of the cell; at other times most of the cytoplasm becomes involved, and a general circulation results. Such mass flow, which is accomplished by the active movement of cytoskeleton elements (probably the actin microfilaments), can transport substances from one part of a cell to another many times faster than simple diffusion.

Virtually all multicellular plants have a specialized internal-transport system to distribute water, essential gases, and nutrients to all their cells. However, many algae, particularly the red algae and most brown algae, lack conducting tissue, which is unnecessary because of their anatomy and their environment. As we have already seen, the cells of such plants are seldom far from the surrounding water or from water in intercellular spaces continuous with the external medium. Furthermore, nutrient and gas procurement is not limited to specialized restricted regions of the body, and photosynthesis is seldom localized in specific structures. Consequently each cell gets ample supplies locally, and long-distance transport is rarely

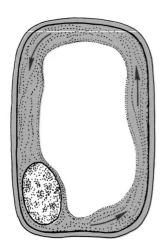

12.1 Cytoplasmic streaming in a plant cell
The cytoplasm flows around the large central vacuole.

necessary. Nevertheless, there is evidence that certain substances, particularly hormones, sometimes move over long distances surprisingly fast in some large nonvascular plants. How this happens is not clear.

SOLUTIONS IN VASCULAR PLANTS

Vascular plants are so named because their transport systems depend, like ours, on vessels. They incorporate two principal types of conducting tissue: the xylem (from the Greek *xylon*, "wood") and the phloem (from the Greek *phloios*, "bark," a reference to the location of this tissue in the outer growing perimeter of the plant); both were described briefly in Chapter 9. These specialized internal-transport tissues have enabled vascular plants to evolve bodies large in all dimensions and to develop far greater specialization of parts and more complete integration of function than other plant groups. Water and minerals can be taken up by specialized roots and distributed to the rest of the organism by the xylem, while photosynthesis can be restricted largely to the leaves, and its products transported throughout the plant by the phloem. In some very tall forest trees the distance between the roots and the leaves may be enormous; yet the xylem and phloem form continuous pathways between them, and they can exchange materials with relative ease (see Fig. 12.8). Plants could not have exploited the land environment successfully without the evolution of such a transport system.

STRUCTURE OF STEMS

Stems of plants serve many functions. Some contain chlorophyll and carry out photosynthesis. Most also store nutrients in their pith; indeed, some stems are highly specialized as storage organs: potato tubers, which are underground stems, are an example. Here, however, we shall concentrate on stems (usually called trunks in larger plants) as organs of transport and support, and examine in some detail the structural adaptations associated with these two functions. Keep in mind that, although our discussion of transport is based on stems, the vascular tissue of the stem is continuous with that in the roots and leaves.

From the standpoint of transport strategies, there are three basic kinds of higher plants: monocots, herbaceous dicots, and woody dicots.[1] Monocots, of which there are about 50,000 species, are mostly annuals (that is, they live for only a single season). All the grasses (which include wheat, oats, rice, and corn) are monocots, as are tulips, lilies, daffodils, and palms. As we saw in Chapter 6, monocots can be recognized by their morphology: the major veins in their leaves are roughly parallel, whereas those of most dicot leaves have a netlike arrangement; monocot flowers have petals in threes or multiples of three, rather than the fours or fives of dicot flowers; and (the source of their name) germinating monocot seeds have a single leaf (cotyledon) rather than the two of dicots.

Herbaceous dicots are plants whose stems remain soft and succulent.

---

[1] The vascular plants, or Tracheophyta, also include about 650 species of conifers, such as pine trees, which have vascular systems similar to those of dicots, and about 12,000 species of ferns and their relatives.

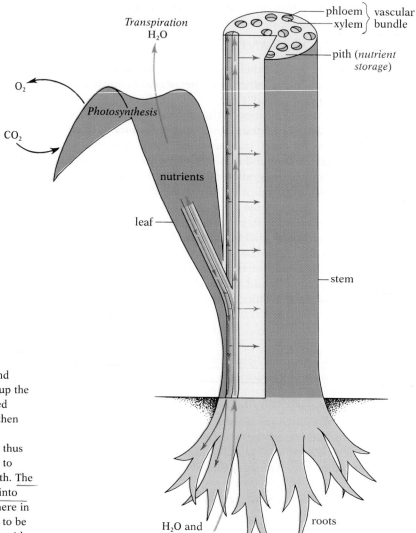

12.2 Summary of fluid flow in a monocot
In this stylized model of a grass plant, water and
minerals are gathered by the roots and pulled up the
xylem to the leaves by a process to be described
later. Some of the water and the minerals are then
used in photosynthesis and for making organic
compounds. The nutrients and other materials thus
generated are transported through the phloem to
build or nourish cells, or to be stored in the pith. The
phloem and xylem of monocots are organized into
discrete bundles (only one of which is shown here in
vertical section) surrounded by pith. Nutrients to be
stored simply diffuse out of the phloem into the pith,
as indicated by the red arrows. There is no distinct
outer cortex in monocots.

They too are mostly annuals: the entire plant dies after one season of
growth. In perennial herbaceous dicots, the part of the plant that remains
above ground dies each year, but the roots survive. The bean plant is an example of an economically important herbaceous dicot.

Woody dicots are all perennials, and both their top growth and their root
systems survive for many seasons. Most trees and many flowering plants,
such as the raspberry and the rose, are woody dicots. Altogether there are
about 200,000 species of herbaceous and woody dicots.

Monocot organization The simplest vascular organization is that of monocots. The nutrient- and water-conducting (vascular) tissue of monocots is organized in discrete vertical bundles scattered throughout the living structural and supportive cells of the stems (Figs. 12.2 and 12.3). Each bundle contains both xylem and phloem, and is surrounded by a supportive bundle sheath (Fig. 12.4). As they grow in diameter, monocot stems make new bundles, so no tissue that requires nutrients is too far away from a source. This means that most monocot stems exhibit no clear separation between outer cortex and central pith.

Herbaceous dicot organization Herbaceous dicots are superficially similar to young monocots; indeed, many researchers believe monocots to have evolved from herbaceous dicots. As in young monocots, the phloem

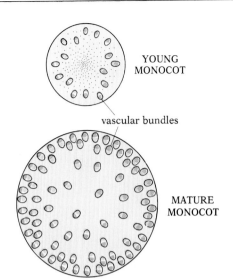

YOUNG MONOCOT

vascular bundles

MATURE MONOCOT

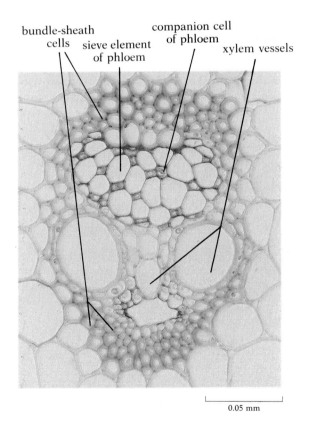

bundle-sheath cells

sieve element of phloem

companion cell of phloem

xylem vessels

0.05 mm

12.4 Cross section of a monocot vascular bundle (corn)
A single bundle is shown here. Each one contains a few xylem vessels in a mass of phloem, and is surrounded by a supportive bundle sheath.

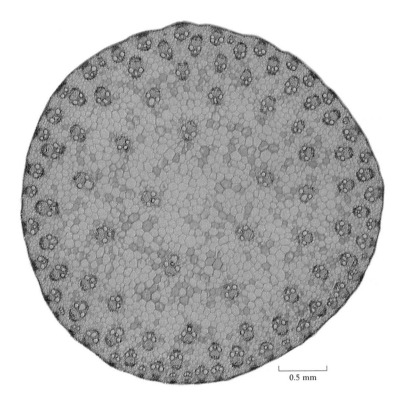

0.5 mm

12.3 Cross sections of monocot stems
A young monocot has a ring of discrete vascular bundles, each with phloem and xylem vessels (top). As it grows, new vascular bundles develop, until they are distributed throughout the stem (middle). The rest of the stem is pith, the tissue used for nutrient storage. At bottom is a photograph showing a cross section of a monocot stem (corn).

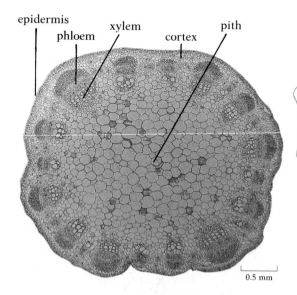

epidermis
phloem
xylem
cortex
pith

0.5 mm

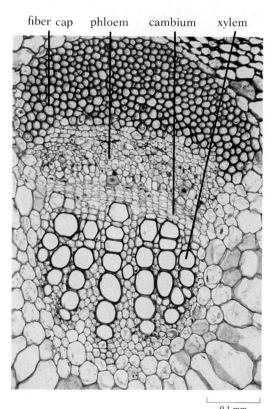

fiber cap   phloem   cambium   xylem

0.1 mm

12.5   Cross sections of herbaceous dicot stems
Top: Whole stem of sunflower (*Helianthus*). Bottom: Vascular bundle of sunflower. The vascular bundles in dicot stems are arranged in a circle.

and xylem of herbaceous dicots are organized in discrete bundles (Fig. 12.5). However, the differences between the two groups of plants are striking. The ring of vascular bundles in dicots separates the central pith from an outer cortex, and new vascular tissue, if any, is added by the growth of the existing bundles rather than by the addition of new bundles. This is because in dicots the actively growing (meristematic) tissue responsible for new vascular cells, a tissue called *vascular cambium*, lies between the xylem and phloem in each bundle. (In monocots the analogous tissue is at the periphery of the stem.) Nutrients from the phloem reach the pith through gaps between the bundles.

In many herbaceous dicots, such as the buttercup, the cambium never becomes active and never produces additional phloem or xylem cells. All the vascular tissue in such plants is said to be *primary tissue*—tissue derived from the apical meristem (the bud or root tip) as the stem or root grows in length. But in some species of herbaceous dicots, such as alfalfa, the cambium does become active. As the cambial cells divide, they give rise to new cells both to the inside and to the outside. The new cells formed on the outer side of the cambium differentiate as *secondary phloem*; those formed on the inner side of the cambium differentiate as *secondary xylem*.

Secondary vascular tissue, then, is tissue derived from the cambium, and its production results in growth in diameter rather than growth in length. The secondary phloem pushes the older, primary phloem farther and farther away from the cambium toward the outside of the stem. Similarly, as secondary xylem is produced, the cambium becomes increasingly distant from the primary xylem, which is left in the inner portion of the stem. In a

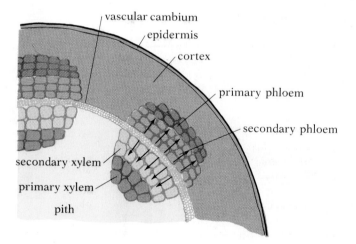

vascular cambium
epidermis
cortex
primary phloem
secondary phloem
secondary xylem
primary xylem
pith

12.6   Cross section of part of a herbaceous dicot stem after secondary growth
The vascular tissue of herbaceous dicots is arranged in discrete bundles, with the xylem separated from the phloem by vascular cambium. As the stem grows, the cambium produces new (secondary) xylem and phloem cells, which push away the original (primary) xylem and phloem. The arrows indicate the directions of growth.

stem that has undergone secondary growth, therefore, the sequence of tissues (moving from the outside toward the center through a vascular bundle) is: epidermis, cortex, primary phloem, secondary phloem, cambium, secondary xylem, primary xylem, pith (Fig. 12.6).

The organization of the tissues in the central vascular core, or stele, of a dicot stem differs in two obvious ways from that in a typical young dicot root (see Fig. 9.4, p. 228): (1) The bundles of the phloem and xylem in a dicot are arranged in concentric rings, whereas the two tissues alternate side by side in a circle around the central xylem in the young root. (2) Dicot stems characteristically have pith, whereas most dicot roots do not.

Woody dicot organization   Young woody dicot stems have the same organization as herbaceous dicot stems: the xylem and phloem, separated by the vascular cambium, are located in discrete bundles near the cortex and surrounding the pith (Fig. 12.7, top). Since herbaceous dicot stems are never more than one season old, this kind of organization remains functional throughout the life of the plant. But woody dicot stems survive the winter, and with continued growth their vascular anatomy changes. As additional tissue becomes necessary to transport the increasing quantities of water and nutrients required by the growing plant, some of the cells lying between the vascular bundles become vascular cambium. Like the cambium within the bundles, this new cambium produces xylem and phloem, and eventually the vascular bundles fuse, forming concentric rings of xylem, cambium, and phloem (Fig. 12.7, middle). As in herbaceous dicots, the youngest tissues lie closest to the cambium: the newest secondary

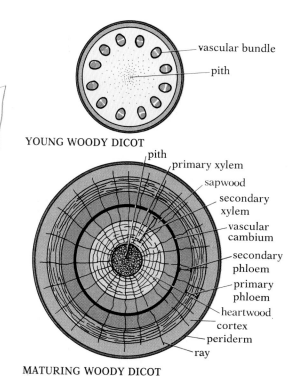

YOUNG WOODY DICOT

MATURING WOODY DICOT

12.7   Cross sections of woody dicot stems
In a young woody dicot the phloem and xylem are located in discrete vascular bundles (top). After several seasons the primary xylem and phloem of the first year have been supplanted by new cells—secondary xylem and phloem—produced by the cambium, and the vascular bundles have fused to form a cylinder. Rays carry nutrients from the phloem to other parts of the stem (middle). At bottom is a photograph showing a cross section of a woody dicot stem (basswood) at the end of three years of growth. Only the most recently produced xylem—the sapwood—is still active in transport, while older xylem—the heartwood—functions in support.

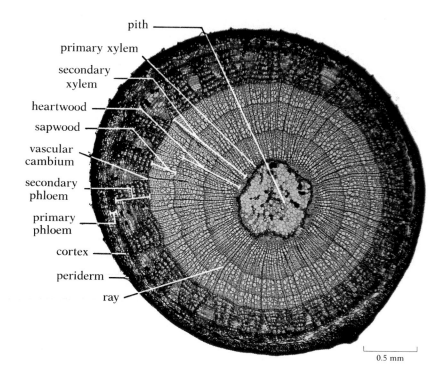

0.5 mm

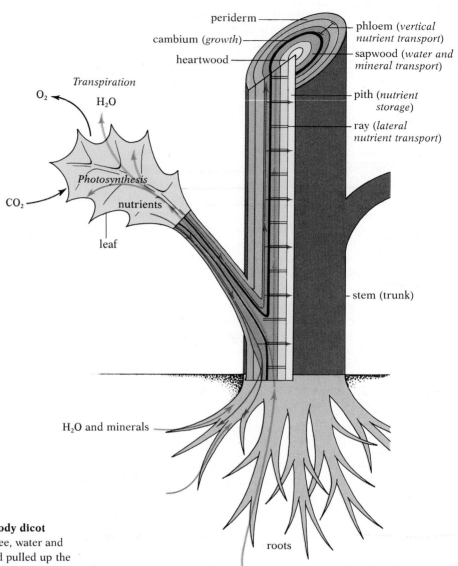

12.8   Summary of fluid flow in a woody dicot
In this stylized model of a tall forest tree, water and minerals are gathered by the roots and pulled up the xylem to the leaves by a process to be described later. Some of the water and the minerals are then used in photosynthesis and for making new cell materials. The nutrients and other materials are transported through the phloem to build or nourish cells. Rays carry the nutrients laterally to active cells and to storage areas in the plant. The width of some of the layers shown in this model has been exaggerated for clarity.

xylem lies just inside the cambium, while the newest secondary phloem is added just outside it. The secondary xylem becomes thicker and thicker, until almost the entire stem of an older plant is xylem tissue—commonly called wood. Though the formation of rings of vascular tissue increases the growth and transport capacity of the stem, it also creates a serious problem: how can the nutrients from the phloem be transported to the central pith for storage? As we shall see later in this chapter, in woody dicots a horizontal system of vessels called rays develops to carry out lateral transport (Fig. 12.7, middle). The rays carry nutrients from the phloem to the pith and also

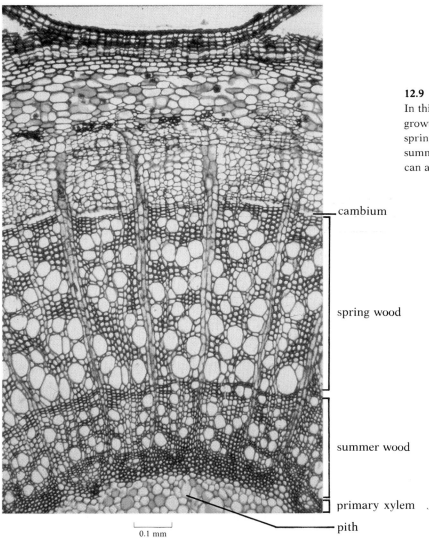

12.9   Cross section of ivy (*Hedera*)
In this photograph taken during the plant's spring growth, the larger, thinner-walled xylem cells are spring wood, and the smaller, thick-walled cells are summer wood from the previous year. Several rays can also be seen.

cambium

spring wood

summer wood

primary xylem

pith

0.1 mm

nourish active cells. They thus play an essential role in fluid flow in a woody dicot (Fig. 12.8).

Since new xylem cells produced early in the growing season, when conditions are best, grow larger than cells produced later in the season, a series of concentric annual rings, clearly visible in cross sections of the stem, is formed. Each ring is made up of an inner area of spring wood with large cells and an outer area of summer wood with smaller cells (Fig. 12.9). A fairly accurate estimate of the age of a tree can be made by counting the annual rings. The width of the rings may be affected by such factors as the vigor of the tree and the climatic conditions during the growing season. Since variations in ring size from year to year tend to reflect climatic changes during the life of a tree, study of the rings of a large sample of very old trees can give a clue to the climate of an area in past ages. Trees frequently live for many centuries, and some, like the bristlecone pine, may be

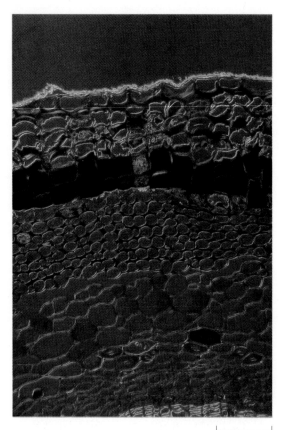

0.2 mm

12.10 Periderm of young stem of elderberry (*Sambucus*)
The periderm layer, here consisting of the cork cambium and several rows of flattened cork cells overlying collenchyma tissue, provides a protective layer on roots and stems that undergo secondary growth.

as much as 4,500 years old, so they can provide a valuable historical record. By matching the inner rings of living trees with the outer rings of older dead trees, and then the inner rings of these trees with the outer rings of still older dead trees, researchers may be able to date the older trees, and thus trace climatic conditions back as much as 10,000 years. Tree-ring dating has even been used in archeological studies; by matching the rings in wooden artifacts found among ancient ruins with cross sections of very old trees from the area, archeologists can make a reliable estimate of the age of the artifacts.

In older trees of most species, chemical and physical changes occur in the older rings of xylem toward the center of the stem. The conducting cells become plugged, the unspecialized parenchyma cells between the vessels die, and pigments, resins, tannins, and gums (all of which provide protection against insect pests) are deposited. As these changes take place, the older xylem ceases to function in transport, and the current year's xylem performs most or all of the transport from the roots. The older inactive xylem cells nevertheless remain important as a strong supportive component of the tree. The rings in which the changes have occurred are known as *heartwood*, or inactive xylem, while the newest outer ring or rings—which still function in transport—constitute the *sapwood*, or active xylem (Fig. 12.7, bottom). A tree can continue to live after its heartwood has burned or rotted away, but it is much weakened and cannot withstand strong winds.

We have discussed changes in the aging woody stem internal to the cambium; now let us examine the portions of the stem outside the cambium. As woody stems (or roots) grow in diameter, a layer of cells outside the phloem takes on meristematic activity of its own and becomes the *cork cambium*. In the second year this cambium forms *cork cells* just under the epidermis, and this growth causes the original epidermis and cortex to flake off. In subsequent years, the cork cambium can either continue to form outside the primary phloem, or can form outside the newest ring of secondary phloem cells. The cork cambium and the cell layers derived from it are collectively termed the *periderm* (Fig. 12.10). The walls of most cork cells usually develop layers of the same fatty substance, called suberin, that coats the root, and often layers of waxes; they thus provide a waterproof coating for the plant. As more and more layers of cork cells are produced, the cells in the older, outermost layers usually die and may begin to flake off. Because the breaking of the cork tends to be patchy, the outer bark of some species of trees is very rough and uneven.

The layers of cork cells constitute the outer bark of the older stem or root. The inner bark is the phloem tissue. Annual growth rings are very difficult, if not impossible, to detect in it, since the phloem layer never becomes thick like the xylem layer. There are a number of reasons for this: fewer phloem than xylem cells are produced by the vascular cambium; the phloem cells have thinner walls and are easily crushed; moreover, the new layers of cork that often form internal to the older layers of phloem push these older phloem cells to the outside, where they are periodically sloughed off. Unlike the xylem, therefore, the phloem of an older woody plant does not function as an important supportive tissue, but its role in transport is very important, as we shall see.

In summary, then, the old woody stem of a tree has no epidermis or cortex. Its surface is covered by an outer bark of cork tissue. Beneath the cork cambium is the thin layer of phloem, or inner bark, and beneath this is the vascular cambium, which is usually only one cell thick. The rest of the stem is mostly xylem, of which only the outer annual rings, or sapwood, still function in transport. It should be clear now why girdling a tree (cutting a ring around it) will kill it: the thin phloem system, which encircles the tree, is severed, so nutrients produced by the photosynthetic activity of the leaves cannot reach and sustain the roots.

Xylem   Xylem is a complex of conducting and supporting tissue formed by several types of cells. Two of these, tracheids and vessel elements, are important in the transport of water and minerals. When tracheids and vessel elements mature, their cellular contents, both cytoplasm and nucleus, disintegrate, leaving only their support structure. The main transport in the xylem occurs in these tubular remnants of cells, not in living cells themselves. The distinction between active and inactive xylem—between sapwood and heartwood—is therefore based only on whether or not these elements of the vascular system are functional. All tracheids and vessel elements in mature xylem, in both sapwood and heartwood, are dead.

Tracheids are elongate and tapering with thick, hard, secondary cell walls; the walls are particularly heavy in summer wood and are important as supportive elements. Tracheids of the first-matured primary xylem are stretched during their development and their secondary walls are usually in the form of rings or spirals (Fig. 12.11A–B). Those of secondary xylem arise after all lengthwise growth has ceased, and they are not stretched during their development; their secondary walls are more continuous, but are interrupted by numerous *pits* (Fig. 12.11C). As we noted in Chapter 4, pits are a type of plasmodesma. They may occur anywhere on the cell wall, but they are often particularly numerous on the tapered ends of the cell, where it abuts on the next cell beyond it. Water and dissolved substances move from tracheid to tracheid through the pits.

Most tracheids have *bordered pits*, which are of rather intricate structure (Fig. 12.12). At these pits, the secondary walls of two adjacent cells are interrupted and their edges overhang the pit chamber, forming the pit borders. The primary walls and middle lamella are continuous through the pit and form the pit membrane, which is generally very thin and selectively permeable to water and dissolved substances. The bordered pit is a singu-

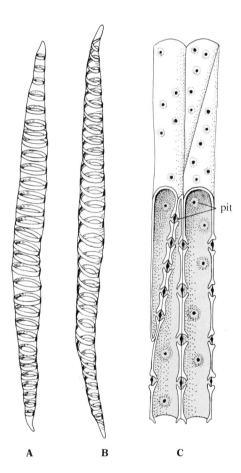

pit

12.11   Tracheids
(A) Primary tracheid with secondary walls in the form of rings. (B) Primary tracheid with spiral secondary wall. (C) Secondary tracheids. Parts of four cells are shown, with one wall cut away from portions of three of the cells to expose their lumina and give a clearer view of the junction between cells. Notice the pits, which are particularly abundant along the tapering ends of the cells.

A   B   C

12.12   Diagrams of pit structure
(A) Bordered pit pair without torus. The secondary
walls overhang the pit chamber. The primary walls of
any two adjacent cells, and the middle lamella
between them, are continuous through the pit and
constitute the pit membrane. (B–C) Bordered pit pair
with torus, in pine. In C the torus has been pushed
against the pit borders on one side; thus movement of
materials through the pit is impeded. (D) Three-
dimensional representation, showing one bordered
pit in section and another in surface view.

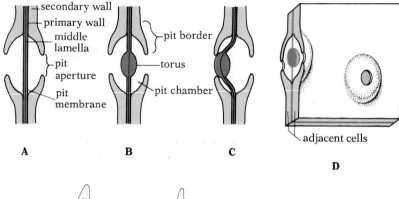

12.13   Vessel elements
Five different types of vessel elements are shown—
those thought to be the more primitive on the left,
those thought to be the more advanced on the right.
The last example (E) shows a single vessel element
on top, three elements linked in sequence to form a
vessel below. The evolutionary trend seems to have
been toward shorter and wider elements, larger
perforations in the end walls until no end walls
remained, and less oblique, more nearly horizontal
ends.

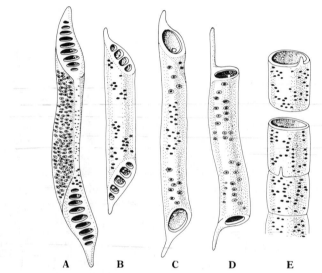

larly well-adapted structure: it provides extra tissue surface for permeability, without reducing structural rigidity as much as it would be reduced by a
nonbordered pit providing the same surface area. The bordered pits of conifers and a few other plants are particularly interesting in that the pit membrane, though very thin toward its edges, is thickened centrally to form a
buttonlike *torus*. If the pressure in one of the cells becomes much greater
than in the adjacent cell—perhaps because air entering the first cell has
formed a bubble—the torus, forced against the pit borders of the adjacent
cell, blocks the pit aperture and obstructs the flow of materials. In such
plants the pit membrane with its torus functions as a check valve between
the cells, stopping the flow of sap from a wound, for instance, or preventing
moisture from ebbing in a drought.

   Vessel elements are more highly specialized for transport than tracheids,
from which they probably evolved; fluid can flow directly through them, as
through a pipe. They are characteristic of the flowering plants and do not
occur in most conifers. In general, vessel elements are shorter and wider
than tracheids. They have bordered pits in their sides, through which some

lateral movement of substances may take place, but movement occurs chiefly through their ends, which are extensively perforated or may even be entirely open (Figs. 12.13 and 12.14). Since the perforations lack the membrane (primary walls plus middle lamella) found in pits, material moving vertically from one vessel element to the next forms a continuous column, and transport is more efficient. A vertical series of vessel elements is called a *vessel*.

In addition to tracheids and vessel elements, xylem contains fiber cells and parenchyma cells. The fibers are elongate, very thick-walled cells that function as supportive elements. They apparently evolved from tracheids since numerous intermediate cell types still exist in some species.

Some of the parenchyma cells of the xylem are scattered among the other cells, but many of them are grouped together to form the *rays* that radiate

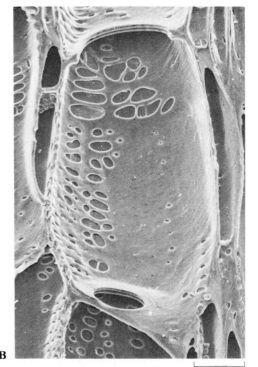

B

0.2 mm

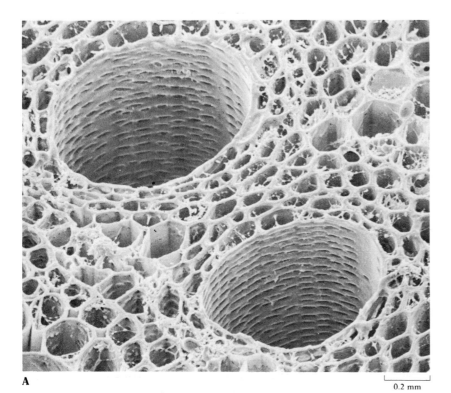

A

0.2 mm

12.14 Scanning electron micrographs of xylem vessels
(A) Portion of cross section of corn root showing two large xylem vessels. (B) Inner surface of a single vessel element. Numerous pits can be seen in the side walls. The large opening at the lower end is the perforation through which water moves vertically from one element to the next. (C) Lateral view of four vessel elements stacked one above the other.

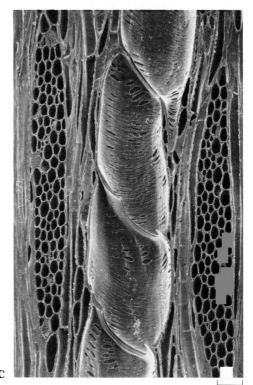

C

0.2 mm

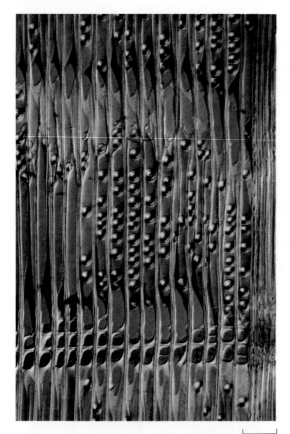

0.1 mm

12.15 Radial section of pine wood
Note the oblique junctions between successive tracheids. Bordered pits are clearly visible. Part of a ray about three cells high is shown in the lower quarter of the photograph.

through the xylem horizontally, perpendicular to the vessels (Fig. 12.7), and function as pathways for the lateral movement of materials and as storage areas. Rays may be small (Fig. 12.15) or large, depending on the species of plant. Without the rays to transport nutrients to them, cells far from the active phloem cells would die. Rays are particularly important in older trees, where a continuous layer of cambium separates the active xylem, or sapwood, from the phloem cells, and layer after layer of inactive xylem, or heartwood, intervenes between the phloem and the pith.

The number, form, and distribution of tracheids, vessels, fibers, and parenchyma cells vary from species to species, and this variation causes the woods of different species to differ in appearance and properties. The rays in oak, for instance, are typically macroscopic, so their presence greatly affects the appearance of the wood. The wood of pine, a conifer, has rays, but lacks vessels and is thus very different from that of oak, which has vessels (Fig. 12.16); oak wood, with its relatively few vessels, is, in turn, different from elm wood or tulip-tree wood, both of which are very porous and have numerous vessels. These differences have both aesthetic and mechanical consequences when the wood is used in construction.

Phloem   Like xylem, phloem is a complex tissue. It contains supportive fibers and also parenchyma, but its principal function is to transport the nutrient products of photosynthesis throughout the plant. The principal vertical conductive elements in phloem are the **sieve elements**, which in vertical series form **sieve tubes**. In their most advanced form, sieve elements are elongate cells with specialized areas on their end walls called **sieve plates** (Fig. 12.17). As the name implies, a sieve plate is a surface with numerous perforations or pores. These openings are etched into the cell's end as it matures and through them strands of cytoplasm connect the contents of one cell with those of the next. Unlike the tracheids and vessels of xylem, the sieve element retains its cytoplasm at maturity, though the nucleus disintegrates.

Closely associated with the sieve elements of most flowering plants are usually one or more mysterious, elongate parenchymatous cells. These **companion cells**, which are derived from the same cell that gave rise to the associated sieve element, retain both cytoplasm and nucleus. Some biologists have suggested that the nucleus of the companion cell may control both its own cytoplasm and that of the adjoining sieve element after the nucleus of the latter has disintegrated. Such an association, they think, would help explain why the mature sieve element can continue to carry out many of the normal activities of a living cell even though it has no nucleus of its own. An intimate association clearly exists between sieve elements and companion cells, but its exact nature and a full explanation for the activity of the sieve elements remain to be discovered.

TRANSPORT OF SAP

Sap—the water and dissolved elements that the roots absorb—moves upward through the plant body in the mature tracheids and vessels of the xylem. That the upward movement is primarily in the xylem can be easily

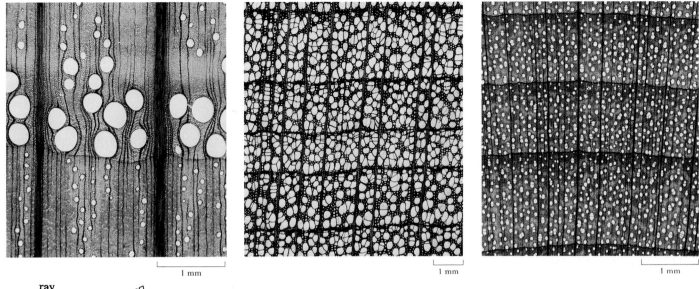

12.16   Cross sections of woods
Left: Red oak. Middle: Tulip tree. Right: Sugar maple.
Note the differences in the number and size of vessels.

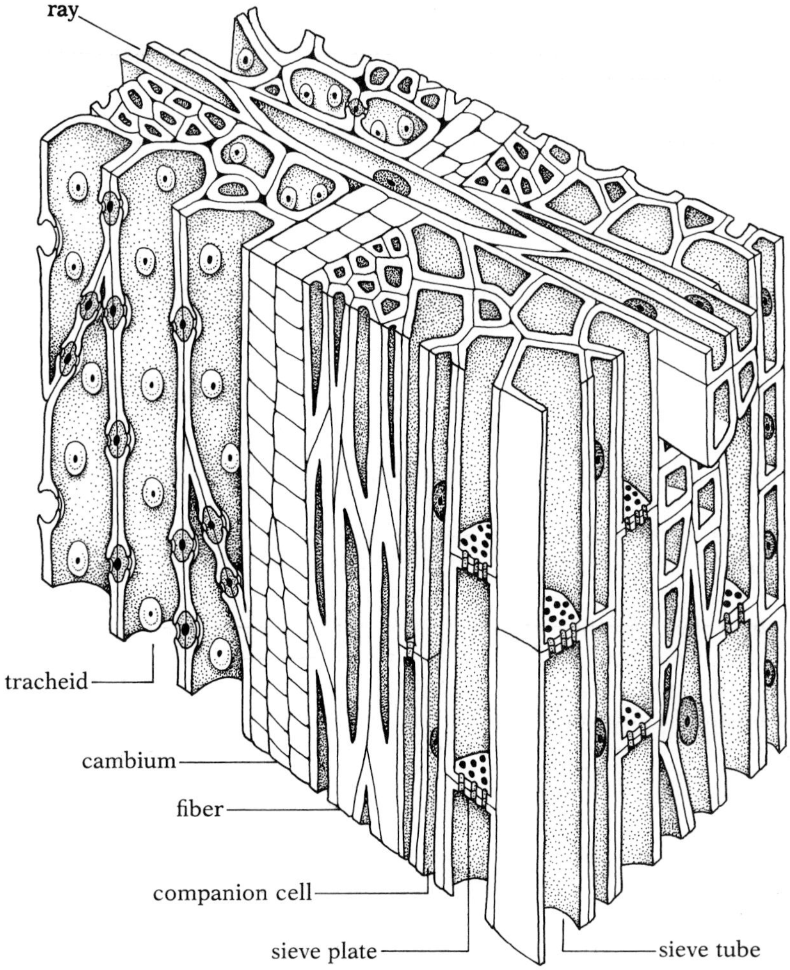

ray

tracheid

cambium

fiber

companion cell

sieve plate

sieve tube

12.17   Phloem
Some nearby xylem tissue is also shown.

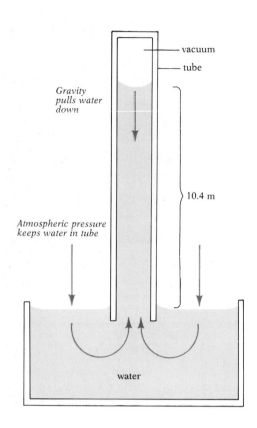

12.18   Role of atmospheric pressure in supporting a column of water
Atmospheric pressure—the weight of the air above us pressing down—is sufficient to support a column of water 10.4 m high. In the same way, the weight of the atmosphere supports the column of mercury in the familiar mercury barometer, but because mercury is so much denser than water, the column reaches only about 76 cm.

demonstrated by ringing experiments; if the cork, phloem, and cambium are removed from a ring around the trunk of a tree, the leaves will still remain turgid, even though they are connected to the roots only by xylem. Furthermore, the movement does not depend on the few cells that remain alive in mature xylem; if heat or poisons kill these cells, the rise of sap continues unabated.

The problem   Any general explanation for the ascent of sap in xylem must identify the forces capable of raising water to the tops of the tallest trees, which may be 90–120 m high. This is not a simple problem. Imagine filling a tube with water, sealing the top end, and leaving the bottom in an uncovered pool of water (Fig. 12.18). Gravity pulls down the water in the tube, and if the tube is tall enough, a vacuum forms at the top. The sap in the xylem of a tree is subject to this same gravitational force.

What keeps the water in the tube at all is the pressure of air on the surface of the surrounding pool of water. The existence of air pressure is difficult for most of us to imagine, living as we do at the bottom of an ocean of air. Nevertheless, the pressure is quite real. The weight of miles of atmosphere above us—designated 1 atmosphere, or atm—supports a column of water 10.4 m high at sea level. Therefore, a pressure of about 12 atm would be needed to support a column of 120 m, the height of the tallest trees. But the column must be more than supported: the fluid must be moved upward at a rate that may sometimes be as fast as a meter or more a minute, and this movement must take place in a system—the xylem—that offers far more frictional resistance than the smooth tube of Figure 12.18. Under these conditions, the movement of fluid to the highest branches of the tallest trees would require the force of at least 30 atm.

Any general theory of fluid movement in the xylem, then, must explain how a force of this magnitude arises, and this question has been a topic of hot debate among botanists for generations. Many ideas have been proposed and examined. Until recently, no one was even sure whether the sap was being pushed from below or pulled from above.

Capillarity   For a long time capillarity (which, you may recall from Chapter 2, is the tendency of water to rise in a thin wettable tube) was a popular candidate as the source of this fluid movement. Only an accurate assessment of the magnitude of the forces involved proved that capillarity, in and of itself, is far too weak to account for the massive phenomenon of the ascent of sap.

Root pressure   Another proposed explanation is root pressure. When the stems of certain species of plants are cut, sap flows from the surface of the stump for some time, and if a tube is attached to the stump, a column of fluid a meter high or more may rise in it. Similarly, when conditions are optimal for water absorption by the roots, but the humidity is so high that little water is lost by transpiration, water under pressure may be forced out at the ends of the leaf veins, forming droplets along the edges of the leaves (Fig. 12.19). This process of water secretion is called *guttation*. When the fluid in the xylem is under pressure, as in these instances of bleeding and guttation, the pushing force involved is apparently in the roots, and is called *root pressure*. Until this force was measured accurately, root pressure was a favored explanation for sap flow.

How is root pressure built up? We have already seen that water moves from the soil, through the epidermis, cortex, endodermis, and pericycle of the root to the xylem, in which it then flows upward to the rest of the plant. But a tall column of water under positive pressure in the xylem would exert a strong downward hydrostatic force by virtue of its own weight, and this force might be expected to drive water out of the xylem in the roots. Yet water is not only held in the stele of the root, but also continues to move into the stele in sufficient quantity to build up a force capable of pushing the column upward. However, if the roots are killed, all root pressure disappears; or if the roots are simply deprived of oxygen, the root pressure ceases—an indication that respiratory production of ATP is necessary to provide the energy for development of root pressure.

Apparently the energy from ATP drives active absorption of ions from the apoplast by cells of the cortex. The ions then move from cell to cell in the cortex through plasmodesmata until they have crossed the endodermal layer of the root and entered the stele. Now, the cells of the stele, being metabolically less active than those of the cortex, apparently allow the ions to leak out of their symplast into the surrounding apoplast. In this way an ion-concentration gradient may be built up across the root, from a low level in the apoplast of the cortex (where the ions are being removed from the apoplast by the cells) to a high level in the apoplast of the stele (where ions are being released into the apoplast). This gradient produces a corresponding osmotic gradient, so that the osmotic concentration is always higher toward the center of the root. Since, as we have seen, water moves from regions of lower osmotic concentration to regions of higher concentration, water in roots will move by osmosis into the stele. In short, energy is expended in the active pumping of ions into the stele, and water follows passively in such quantity as to build up root pressure in the xylem.

But root pressure cannot reach the magnitude required for the movement of sap. In fact, some plants, particularly the conifers and their relatives, are incapable of developing much root pressure at all. Attempts to measure the root pressure in species in which it does occur have rarely yielded values exceeding 1 or 2 atm, though isolated root tips of tomatoes growing in a tissue-culture medium can develop root pressure as high as 6–10 atm.

The low values for root pressure found by most investigators are not the only reason for doubting that it is the principal motive force for the ascent of sap, at least in tall trees. If we puncture a xylem vessel during the summer, water does not gush out, as it would if it were under pressure. On the contrary, there may be a short hissing sound as air is drawn into the vessel. Yet it is in summer that much of the upward movement of water occurs.

In summary, water does move into the roots primarily by means of osmosis, but the resulting root pressure cannot provide the explanation we are seeking for the movement of sap in the shoot. At best it may be involved in the ascent of sap in some plants, particularly very young plants in early spring.

12.19 Guttation by strawberry leaves

The cohesion theory We know now that as water in the walls of parenchyma cells in the leaves (or other parts of the shoot) is lost by transpiration, it is replaced by water from the cell contents. With the removal of water the osmotic concentration of leaf cells rises, and they take up water from adjoining cells—which, in turn, withdraw water from cells adjacent to them. In this way a gradient extends to the xylem in the veins of the leaf, and the parenchyma cells next to the xylem withdraw water from the column in the xylem, in the process helping to pull the column upward. Notice that this is not a matter of pulling the column up by air pressure or vacuum; the mechanism is not analogous to sucking up a liquid through a straw. Air pressure could raise water only 10.4 m, but we are dealing with a mechanism presumed to be the major factor involved in moving a water column that may be over 100 m high. What is required is a continuity between the water on the evaporating surfaces of the cell wall and the water in the xylem, and a continuity between the water at the top of the xylem and that in the roots. If this continuity of water all the way from leaf cell to root were broken by the entrance of air into the system, that particular xylem pathway would cease to function.

This idea of pull from above as a result primarily of transpiration was stated in tentative form early in the eighteenth century by the English clergyman and pioneering botanist Stephen Hales. It was given a more complete formulation in 1894 by the Irish botanist H. H. Dixon and his physicist collaborator J. Joly. Its validity hinges on the existence of great cohesion between individual water molecules. Water molecules moving out of the top of the xylem must pull other water molecules behind them; there can be no break, no separation between the molecules.

As we have already seen, water molecules do indeed exhibit great cohesion, as a result of hydrogen bonds. The theoretical cohesive strength of the entire column of water—a continuous system of hydrogen-bonded molecules from roots to transpiration surfaces in the leaves—is as high as 15,000 atm. Actual experimental values are lower, but may reach 300 atm or more. If the ascent of sap requires a pressure of 30 atm or more, as we found previously, these values are compatible with Dixon and Joly's thesis —commonly called the cohesion theory, or the transpiration theory. The source of energy for the transpiration system is the sun, which causes the evaporation of water from the leaves that pulls replacement water up the xylem.

Though the pull of transpiration, transmitted throughout the entire column, may be the principal force in the ascent of sap, capillarity—the electrostatic attraction of the water molecules to the charged hydrophilic walls of the thin tubes formed by tracheids and vessels—would provide additional support for the column.

The cohesion theory has not been tested under conditions duplicating those in very tall trees, but in the 1890s Josef Böhm in Austria and E. Askenasy in Germany performed some interesting experiments on a smaller scale. They showed that if water that has been boiled to remove all dissolved air is evaporated from the top of a thin glass tube whose lower end is immersed in mercury, a column of mercury will be pulled up the tube by the evaporating water. If the base of a cut branch is inserted tightly in the upper end of the tube and the leaves of the branch become the site of evaporation, the same thing happens. Clearly, the water molecules adhere to each other (and to the walls of the tube) tightly enough to pull up a heavy column of mercury (Fig. 12.20).

Support for the cohesion theory comes from other types of experiments too. In 1935 the German botanist Bruno Huber inserted small electric elements into the xylem and heated the sap. He then measured the time it took for the warmed sap to pass a thermocouple placed a short distance higher on the tree. He found that water begins to move in the upper parts of the tree earlier in the morning than it does in the lower parts of the trunk—an indication that the upward movement of the sap is initiated at the top of the tree, not at the bottom.

More evidence comes from precise measurements of the girth of tree trunks at different times of day. During the daylight hours, when the xylem should be under maximum tension because of high rates of transpiration, the trunk diminishes in girth—that is, it shrinks, just as any closed, flexible, liquid-filled container will do when its contents are being pulled upward. The suction, as we have seen, comes from transpiration, and measurements indicate that in most plants fully 90 percent of the water that ascends the xylem system is lost through transpiration rather than being consumed in photosynthesis or growth.

Though the cohesion theory has gained wide acceptance among plant physiologists, it leaves a number of problems unresolved. The theory requires the maintenance of continuity within the water column in the xylem; yet breaks in the column occur rather frequently. For example, during times of drought some of the gases dissolved in the sap may form gas bubbles; and when sap freezes in winter the dissolved gases are forced out, forming bubbles that break the water column. It is not entirely clear how these difficulties are met. Perhaps broken columns rejoin at night when the tension is relaxed; and perhaps the gases forced out of solution during freezing redissolve under more favorable environmental conditions. In addition, the pits of the tracheids or vessels probably allow fluid to bypass gas-filled or damaged parts of the xylem. At any rate, the breaks do not appear to affect water movement seriously; but they remain a challenge to further research.

Another puzzle is how the water column that enables sap to move upward in the xylem is originally established, how it gets up there in the first place. The answer seems to be that the water, in a manner of speaking, grows there. The cambium produces a new layer of potential xylem cells every year. This layer lies just outside the previous year's xylem. The as-yet-undifferentiated and still-living cells draw water laterally from the older xylem as they grow. These cells may, in fact, absorb so much water from the older xylem that their contents are under considerable positive pressure early in the growing season, and if they are punctured liquid may exude from them. As these cells mature, their end walls become more permeable and their contents can move more freely; a conductive system has been established. As the new spring leaves grow larger and the transpiration rate begins to rise, the liquid contents of the newly formed xylem elements are pulled upward. Tension develops in the column of liquid, and the positive pressure of the early part of the growing season is lost.

Notice that if the water in the xylem is under tension, the question of how

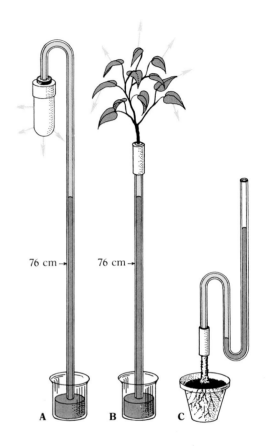

12.20   Demonstrations of rise of water by pull from above (A, B) and root-pressure push from below (C)

(A) Water is evaporated from a clay pot attached to the top of a tin tube whose lower end is in a beaker of mercury. The water in the tube rises and pulls a column of mercury to a point well above the 76 cm that atmospheric pressure can support. (B) The same results are obtained when transpiration from the leaves of a shoot is substituted for evaporation from a clay pot. (C) In some plants root pressure can raise a column of mercury.

water moves into the xylem in the roots presents less difficulty. We noticed earlier that a water column under positive pressure would exert a downward hydrostatic force opposing the entrance of more water into the stele. But if the water in the xylem is under tension, it will exert no such downward force, and the movement of water into the roots by osmosis will be correpondingly easier.

TRANSPORT OF SOLUTES

Transport of organic solutes   Two principal classes of solutes are transported—or translocated, as plant physiologists generally call it—within the plant body: organic solutes and inorganic solutes. Let us consider the organic solutes first. We can divide these conveniently into two principal types: carbohydrates (usually transported as sucrose) and organic nitrogen compounds. (Plant hormones are also transported.)

The classical picture of the translocation of solutes was that all upward movement was through the xylem and all downward movement through the phloem. About 1920, however, this view had to be revised. It became apparent that most of the movement of carbohydrates, whether up or down, was through the phloem.

Most of the early work on the path of movement involved ringing experiments. It could be demonstrated that if a ring of bark (which includes, of course, the phloem) was removed from the trunk of a tree, the supply of carbohydrates to all parts of the plant below the ring was cut off, and those parts died when they had depleted their stored reserves. Downward movement of carbohydrates was clearly through the phloem, not through the xylem, which was left intact in these experiments. But it could also be demonstrated that if a branch was ringed a short distance behind the growing bud, the supply of carbohydrates moving to the bud was cut off. Again, the movement must have been in the phloem, but in this case the movement was upward. Numerous experiments such as these led the majority of botanists to conclude that most carbohydrate movement is through the phloem.

But not all botanists accepted this view. Some objected that far too much carbohydrate moves within the plant body, and moves too rapidly, for the phloem to be the exclusive channel for this movement. After all, the total amount of functional phloem tissue in the trunk of a large tree is rather small. Surely, the argument ran, it is physically inconceivable that so much material should pass through so few sieve tubes, particularly since these tubes are not open pathways like xylem vessels. Numerous botanists, however, most notably T. G. Mason and E. J. Maskell of the Cotton Research Station, Trinidad, showed by careful ringing experiments that, hard to conceive as it is, the phloem is indeed the principal pathway of sugar movement, and that this movement is amazingly rapid. Another kind of demonstration was provided by Susann and Orlin Biddulph of the State University of Washington. They grew plants in an atmosphere containing carbon dioxide made from radioactive carbon. When a thin section was cut from the stem of one of these plants and placed in contact with photographic film, the resulting exposure showed that the sugar synthesized from the radioactive carbon had clearly traveled only in the phloem.

The situation is less clear with organic nitrogen compounds. It was for-

merly thought that nitrogen, absorbed by the roots primarily as nitrate, was carried upward in inorganic form through the xylem to the leaves, to be used in synthesis of organic compounds which were then transported through the phloem. This sequence probably holds true for some plants. But there is now good evidence that many species promptly incorporate incoming nitrogen into organic compounds such as amides and amino acids in the roots. In some species, especially herbaceous ones, these organic nitrogen compounds move upward in the phloem, but in other species, especially trees, they move in the xylem. We see, then, that not *all* upward transport of organic compounds is in the phloem, though most of it appears to be. Virtually all *downward* transport of organic compounds seems to be in the phloem.

Transport of inorganic solutes   Inorganic ions such as calcium, sulfur, and phosphorus ions are translocated upward from the roots to the leaves primarily through the xylem. However, experiments using radioactive forms of these minerals indicate that some are quite mobile in the plant, traveling rapidly back down in the phloem, or moving out of the older leaves through the phloem and being transferred to the newer, more actively growing leaves. Phosphorus, for example, easily moves upward in the xylem and downward in the phloem, often circulating rapidly throughout the plant in this manner. If a plant is grown for a short time in a solution containing radioactive phosphorus, and the plant is then placed against a photographic plate, the younger leaves will be found to contain the greatest concentrations of radioactive phosphorus (Fig. 12.21). If the plant is then moved into a normal solution (one without the radioactive tracer), allowed

12.21   Movement of radioactive phosphorus in a growing plant
The plant was grown for one hour in a nutrient solution containing ^{32}P. It was then removed to a nonradioactive solution. At the end of 6 hours (left), the ^{32}P was particularly concentrated in the youngest leaves. At the end of 96 hours (right), much of the ^{32}P had moved from the leaves in which it was formerly most concentrated to new leaves that had developed above them. (The darker the area, the more ^{32}P it contains.)

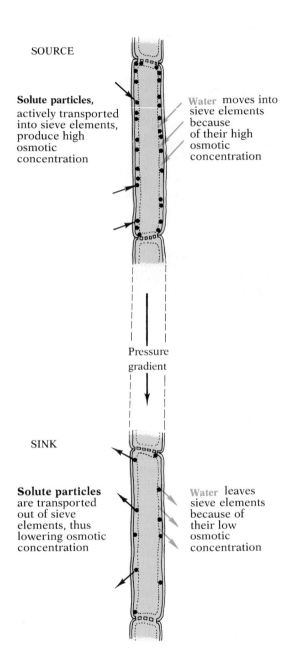

SOURCE

Solute particles,
actively transported
into sieve elements,
produce high
osmotic
concentration

Water moves into
sieve elements
because
of their high
osmotic
concentration

Pressure
gradient

SINK

Solute particles
are transported
out of sieve
elements, thus
lowering osmotic
concentration

Water leaves
sieve elements
because of
their low
osmotic
concentration

12.22 Pressure-flow model of phloem transport

to grow for a day or so, and again placed against a photographic plate, the resulting pictures will show that the radioactive phosphorus has moved from the leaves in which it was first concentrated to the new leaves just beginning to develop. Calcium, on the other hand, is not mobile in the phloem, and therefore cannot move from old leaves to newer ones. Consequently plants must obtain a steady supply of new calcium from the soil, whereas they can easily survive with only intermittent feedings of phosphorus, since this element can shift from place to place within the plant and be reused many times. Well-designed fertilization programs take into account such differences in the properties of the different mineral nutrients.

Hypotheses of phloem function   We have seen that most transport of organic solutes, both up and down, is through the phloem; most downward transport of minerals is also through the phloem. How phloem functions in transport is a problem that has been under investigation for a very long time. A number of hypotheses have been put forward, but none is fully convincing.

Any hypothesis about the transport mechanism in phloem must account for several facts: (1) The movement is often rapid, much more so than simple diffusion alone could make it. In fact, it has been estimated that sugar moves through the phloem of a cotton plant more than 40,000 times faster than it diffuses in a liquid. (2) The speed of movement through the sieve tubes differs for different substances. (3) The direction of movement may be reversed periodically within a given sieve tube. (4) The movement in neighboring sieve tubes may be in opposite directions. (5) The movement takes place through sieve elements that, unlike xylem, retain their cytoplasm (though the cytoplasm is not exactly like that of most other cells). (6) Unlike the ends of xylem vessels, the ends of the individual sieve elements are not broadly open, but are penetrated only by the tiny pores of the sieve plates. Clearly, we are dealing with transport through active cells, not merely with movement through dead tubes by purely mechanical processes, as in mature xylem.

One hypothesis is that materials are carried the length of each sieve element by cytoplasmic streaming. The suggestion is that materials diffusing into one end of a sieve element through the sieve plate are picked up by the streaming cytoplasm and carried to the other end; there they diffuse across the sieve plates at that end and, on entering the next element in the tube, are again picked up by streaming cytoplasm. In this way, by alternately streaming within elements and diffusing between them, the materials would move long distances through the sieve elements of the phloem. The diffusion across the sieve plates might, of course, involve active transport. One argument against this hypothesis is that there is little evidence that cytoplasmic streaming occurs in mature sieve elements. Furthermore, measurements in other cells where the process does occur yield velocities of streaming much lower than the known rates of solute movement through sieve tubes. At present, not many botanists accept the streaming hypothesis.

A second hypothesis, probably the one most widely supported by botanists today, invokes *pressure flow*, or mass flow—the mass flow of water and solutes through the sieve tubes along a turgor-pressure gradient. The process begins with the active transport of photosynthetically produced sugars into phloem cells in the leaves. Because these cells then contain high concentrations of sugar, which is osmotically active, water tends to diffuse into them. This passive osmotic movement of water following the sugar causes the turgor pressure of the cells to rise. The pressure then tends to force substances from these cells into the cells next to them. Thus, under hydrostatic pressure, substances are forced en masse from cell to cell along the sieve tubes. At the same time, in storage organs or actively growing tissues, where sugars are being used up and actively removed from the sieve tubes, the osmotic concentration in the tubes falls. The tubes therefore tend to lose water, and their turgor pressure drops.

The contents of the sieve tubes, then, are under considerable turgor pressure in one portion of the plant (the "source") and under lower turgor pressure in another portion of the plant (the "sink"). The result is a mass flow of the contents of the sieve tubes from the region under high pressure (usually the leaves, but sometimes storage organs when reserves are being mobilized for use, as in early spring) to the region under lower pressure (usually actively growing regions or storage depots). The whole process depends on massive uptake of water by cells at the source end, because of their high osmotic concentrations, and massive loss of water by cells at the sink end, because their osmotic concentrations are lowered by their loss of sugar (Fig. 12.22). Phloem transport is thus seen by the mass-flow hypothesis as push-powered, while by contrast xylem flow, according to the cohesion theory, is pull-powered by transpiration.

The mass-flow hypothesis is open to an obvious objection. It assumes that material can flow with relative freedom from one sieve element to the next. But the openings in the sieve plates between successive sieve elements are very tiny indeed. For many years, too, investigations showed the cytoplasm of sieve elements, particularly in the vicinity of the sieve plates, to be extremely viscous. Until the 1950s, however, all attempts to sample the cytoplasm involved invasive methods. Inserting tubes, however fine, inevitably damaged the sieve element, and when such damage occurs, the pores in the sieve plates are quickly plugged. This phenomenon, though it seems to suggest the presence of a very viscous fluid inside, probably demonstrates rather the efficiency of the plant's wound-healing system. Interestingly enough, aphids, small insects that suck the phloem juices out of stems, are able to insert their delicate stylets without triggering any wound-response reaction (Fig. 12.23). Recent research has focused on cutting a sucking aphid away from its stylet, which is then used to sample the phloem. The results obtained so far demonstrate that the fluid is not very viscous, indicating that the mass-flow hypothesis is the most likely explanation for vertical transport of solutes.

12.23 An aphid sucking phloem
The stylet of this aphid, on the underside of a branch of basswood, has penetrated the bark to reach the phloem in a sieve element.

Chapter 13

INTERNAL TRANSPORT IN ANIMALS

THE PROBLEM

Like the vascular plants and unicellular organisms discussed in the last chapter, animals face the problem of getting nutrients to every cell in their bodies and carrying off metabolic wastes. We saw that single-celled organisms can rely on simple diffusion, supplemented with active transport across membranes, because their surface area is large relative to their cellular volume. Even multicellular organisms may find diffusion and active transport sufficient if their bodies are so thin that every cell is exposed to the surrounding medium, as is the case for those algae with large thin leaves only two cells thick. But most larger organisms, like the vascular plants, must have specialized internal-transport systems.

Unlike plants, animals are generally adapted for active locomotion. The more rapid metabolism required by their active lives makes them correspondingly less able to rely on such a slow process as diffusion, even when it is supplemented by the intracellular processes we have already discussed. Furthermore, because of their way of life, animals are much less likely than plants to have bodies large in one or two dimensions but flat and thin in the third. There are exceptions to this rule, however. Some relatively large and sedentary animals can, like some large aquatic plants, keep one dimension thin. Sea fans, a curious group of large thin coelenterates, are an example of such plantlike animals. Tapeworms are another example; they may be 20 m long or more, but they are always flat and thin, and no cell is far from the food supply that surrounds them in the host's digestive tract.

In general, however, only very small animals lack circulatory systems, and even among small animals in which diffusion plays a major role, the body plan often exhibits some adaptations for transport. The hydra's body wall (see Fig. 10.9, p. 251) is basically only two cells thick, but even the cells of the inner layer are exposed directly.to water containing dissolved oxygen, because such water is drawn into the gastrovascular cavity. It might seem at first glance that the cells in the tentacles would be far removed from the food supply, but closer examination shows that branches of the gastrovascular cavity penetrate into each tentacle and that food particles can be absorbed directly from this cavity by the tentacle cells. Planaria (see Fig. 10.11, p. 253) is another example. We have already seen that its gastrovascular cavity branches into all parts of the body and functions as a primitive transport system. Small animals like hydra and planaria, then, usually have adaptations that supplement diffusion and intracellular transport. But these small animals, like the larger sea fans and tapeworms, are the exceptions rather than the rule among animals. Body size, body shape, and activity dictate that most animals, like the vascular plants, must have true circulatory systems to accomplish efficient internal transport of nutrients and wastes.

Animal circulatory systems usually include some sort of pumping device, called a **heart**. There may be only one heart, as in our own case, or a number of separate hearts, as in earthworms, where five blood vessels on each side of the animal pulsate, pumping blood from the main dorsal longitudinal vessel into the main ventral longitudinal vessel (Fig. 13.1). Many insects have both a large general heart and a series of smaller accessory hearts at the bases of their legs and wings.

The one-way pumping action of the heart, usually combined with a system of one-way valves, moves the blood in a regular fashion through the circuit. This circuit may be rigidly encompassed in well-defined channels or vessels, in which case it is called a **closed circulatory system**. Or the circuit may have some sections where definite vessels are absent and the blood flows through large open spaces known as sinuses; such a system is called an **open circulatory system**. Closed circulatory systems are characteristic of a great variety of animals, including the earthworms and all vertebrates. Open circulatory systems are characteristic of most molluscs (snails, oysters, clams, etc.) and all arthropods (insects, spiders, crabs, crayfish, etc.).

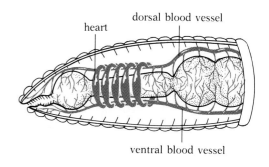

13.1 The circulatory system of an earthworm
Ten hearts, five on each side, pump blood through the longitudinal vessels, which themselves pulsate and help move the blood.

CIRCULATION IN INSECTS

Since movement of the blood through an open system is not as fast, orderly, or efficient as through a closed system, it may seem surprising that such active animals as insects, which must have relatively high metabolic rates and precise internal regulation, should have open circulatory systems. But you will recall that insects do not rely on the blood to carry oxygen to their tissues; that function is fulfilled by the much-branched tracheal system. Consequently it is not vital for insects that their blood flow very fast and in a precise pathway. This is a good example of the way the various systems of a living creature are interrelated in adaptation to its needs.

The circulatory systems of insects are even more reduced than those of

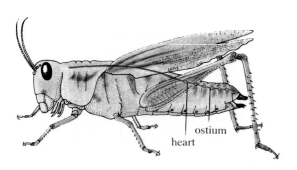

13.2 The dorsal heart of a grasshopper
Blood enters the heart through the ostia and is
pumped forward and out at its open anterior end.

most other arthropods. Ordinarily the only definite blood vessel in an insect is a longitudinal vessel, often designated the heart, which runs through
the dorsal portion of the animal's thorax and abdomen (Fig 13.2). The posterior portion of the heart is pierced by a series of openings, or ostia, each
regulated by a valve that will allow movement of blood only into the vessel.
When the heart contracts it forces blood out of its open anterior end into
the head region. When it relaxes again, blood is drawn in through the ostia.
Once outside the heart, the blood is no longer in vessels; there are no veins,
capillaries, or arteries, other than the heart itself with its valve segments
and the short so-called artery that forms its anterior end. The blood simply
fills the spaces between the internal organs of the insect, thus bathing them
directly.

The action of the heart causes the blood to move sluggishly through the
body spaces from the anterior end, where it was released, to the posterior
end, where it will again enter the heart. The movement of the blood is accelerated by the stirring and mixing action of the muscles of the body wall and
gut during activity. Thus, when the animal is most active, as in running or
flying, and its organs are in most need of rapid delivery of nutrients and removal of wastes, the blood moves relatively fast because of the activity itself.

CIRCULATION IN VERTEBRATES

THE CIRCULATORY PATH

All vertebrates have a closed circulatory system, which consists basically of
a heart and numerous arteries, capillaries, and veins. *Arteries* are blood
vessels that carry blood away from the heart, while *veins* carry blood back
toward the heart. Note that, contrary to a common impression, the definitions of these two types of vessels are not based on the condition of the
blood carried. Though most arteries carry oxygenated blood and most
veins carry deoxygenated blood, oxygen content is not always a reliable way
to distinguish them; indeed, as we shall see, the vein that leads from the
lungs to the heart carries the most highly oxygenated blood in the body.
Capillaries are tiny blood vessels that interconnect the arteries and the
veins; they usually run from very small arteries, called arterioles, to very
small veins, called venules. It is across the thin walls of the capillaries that
most of the exchange of materials between the blood and the other tissues
takes place.

The idea that blood circulates may seem perfectly obvious. After all, the
constant beating of the heart is one of the most conspicuous of body functions; and, in humans, blood vessels are clearly visible through the wrist or
the back of the hand, and the pulse in these vessels can hardly be missed. Yet
for centuries this idea was anything but obvious even to the well-informed.
The pumping action of the heart went unrecognized, and even after the notion of the heart as a pump prevailed, the idea of circulation remained
alien; it was thought that blood ebbed and flowed in the veins until it seeped
into the tissues.

A major turning point in understanding the functioning of the human
body came in 1628, when the great English biologist William Harvey pub-

lished a short work founded on his extensive examinations of many different species of animals, from worms and insects to human beings, in which he clearly enunciated the idea of circulation of blood. Though he had never actually seen a capillary, he succeeded in outlining the basic components of the circulatory system as we know them today. Harvey's work not only improved knowledge of the circulatory system, it marked the beginning of the modern science of physiology, the attempt to understand bodily processes in terms of physics and chemistry.

The circuit in humans   Let us trace the movement of blood through the human circulatory system, beginning with the blood returning to the heart from the legs or arms. Such blood enters the upper right chamber of the heart, called the ***right atrium***, or auricle (Figs. 13.3 and 13.4). This chamber then contracts, forcing the blood through a valve (the tricuspid valve) into the ***right ventricle***, the lower right chamber of the heart. Now this blood, having just returned to the heart from its circulation through tissues, con-

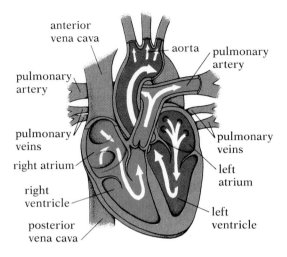

13.3   The human heart
The arrows show the direction of blood flow (oxygenated blood, red; deoxygenated blood, blue).

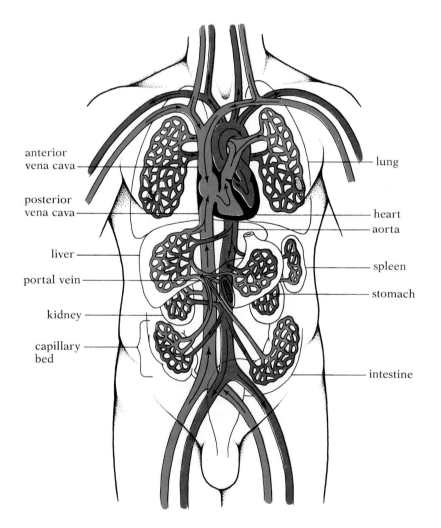

13.4   Diagram of the human circulatory system
Red vessels contain oxygenated blood; blue vessels, deoxygenated blood. The transfer of oxygen and nutrients from the blood to tissues, as well as the transfer of carbon dioxide and other wastes from cells to the blood, takes place in each of the capillary beds. The capillaries, which are of necessity represented here as being relatively thick, are actually microscopic. Only a very few of the vast number of arteries and the capillary beds they supply are shown.

tains little oxygen and much carbon dioxide. It would be of little value to the body simply to pump this deoxygenated blood back out to the general body tissues. Instead, contraction of the right ventricle sends the blood through a valve (the pulmonary semilunar valve) into the *pulmonary artery*, which soon divides into two branches, one going to each lung. In the lungs the pulmonary arteries branch repeatedly, and each terminal branch connects with dense beds of capillaries lying in the walls of the alveoli. Here gas exchange takes place, as carbon dioxide is discharged from the blood into the air in the alveoli and oxygen is picked up by the hemoglobin in the corpuscles of the blood. (Though these bodies are often referred to as "red blood cells," the term "corpuscles" is more accurate: they are not true cells, since they lack nuclei and organelles.) From the capillaries, the blood passes into venules, which soon join to form the large *pulmonary veins* that run back toward the heart from the lungs. The four pulmonary veins (two from each lung) empty into the upper left chamber of the heart, called the *left atrium*, or auricle. When the left atrium contracts, it forces the blood through a valve (the bicuspid, or mitral, valve) into the *left ventricle*, which is the lower left chamber of the heart. The left ventricle, then, is a pump for recently oxygenated blood. When it contracts, it pushes the blood through a valve (the aortic semilunar valve) into a very large artery called the *aorta*.

After the aorta emerges from the anterior portion of the heart (the upper portion, in humans standing erect), it forms a prominent arch and runs posteriorly along the middorsal wall of the thorax and abdomen (Fig. 13.4). Numerous branch arteries arise from the aorta along its length, and these arteries carry blood to all parts of the body. For example, the first branch of the aorta is the coronary artery, which carries blood to the muscular wall of the heart itself. Other early branches of the aorta, which arise in the region of the aortic arch, are the arteries that supply the head, neck, and arms. As the aorta runs posteriorly, arteries to the body wall, stomach, intestines, liver, pancreas, spleen, kidneys, legs, etc., arise from it. Each of these arteries, in turn, branches into smaller arteries, until eventually the smallest arterioles connect with the numerous tiny capillaries embedded in the tissues. Here oxygen, nutrients, hormones, and other substances move out of the blood into the tissues; such waste products as carbon dioxide and nitrogenous wastes are picked up by the blood, and substances to be transported, such as hormones secreted by the tissues, or nutrients from the intestine and liver, are also picked up. The blood then runs from the capillary bed into tiny veins, which fuse to form larger and larger veins, until eventually one or more large veins exit from the organ in question. These veins, in turn, empty into one of two very large veins that empty into the right atrium of the heart: the *anterior vena cava* (sometimes called the superior vena cava), which drains the head, neck, and arms, and the *posterior vena cava* (inferior vena cava), which drains the rest of the body.

Very little, if any, exchange of materials between the blood and the other tissues occurs across the walls of the arteries and veins. The walls of these vessels are apparently impermeable to the substances in the blood and tissue fluid. Walls of arteries and veins are composed of three layers: (1) an outer connective-tissue layer with numerous fibers, which give the vessels their characteristic elasticity; (2) a middle layer of smooth muscle, which can change the size of the vessels (some of the largest arteries also have

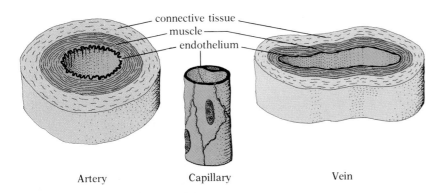

connective tissue
muscle
endothelium

Artery   Capillary   Vein

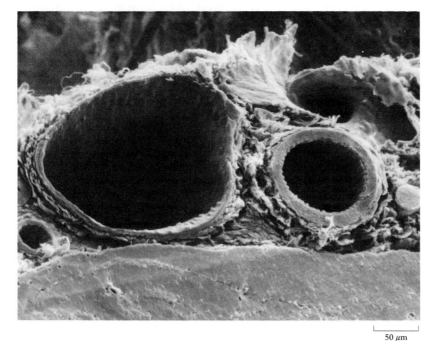

50 μm

13.5 Walls of artery, vein, and capillary compared
Arteries and veins have the same three layers in their
walls, but the walls of veins are much less rigid and
they readily change shape when muscles press
against them. Capillaries have walls composed only
of a thin endothelium. In the scanning electron
micrograph of tissue surrounding a human vas
deferens, a medium-size vein (left) and a medium-size
artery (right) can be seen embedded in connective
tissue. Note that the vein has a thinner wall and a
larger lumen than the artery.

many fibers in this middle layer); and (3) an inner layer of connective tissue
lined with simple squamous endothelium (Fig. 13.5). The two outer layers
and the connective-tissue portion of the inner layer terminate at the ends of
the arterioles and venules, leaving the capillaries with walls composed of
only the one-cell-thick endothelium. Across these very thin walls of the capillaries the exchange of materials takes place. As we would expect by analogy with roots, microvilli, and other structures specialized for the exchange
of materials, the capillaries provide an enormous amount of absorptive
surface. Indeed, some estimates place the combined length of the capillary
routes at over 100 km.

Let us now retrace the complete circuit traveled by the blood. It enters
the right side of the heart and is pumped to the lungs, where it picks up oxy-

Pulmonary circulation of adult

Pulmonary circulation of fetus

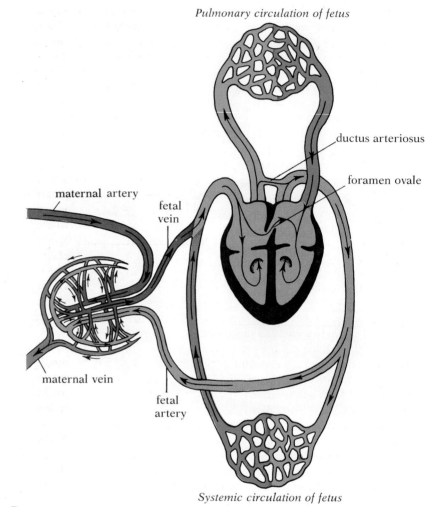

ductus arteriosus

foramen ovale

maternal artery

fetal vein

maternal vein

fetal artery

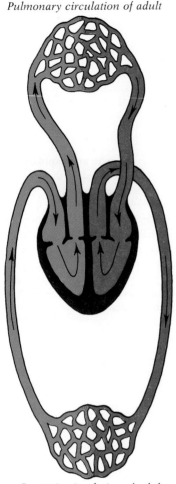

Systemic circulation of adult

A

Systemic circulation of fetus

B

13.6   Diagrams of mammalian circulation Red vessels contain oxygenated blood; blue vessels, deoxygenated blood. (A) In an adult, blood passes from the right ventricle to the lungs, where it picks up oxygen. It then returns to the left side of the heart, from which it is pumped into the systemic circulation. Finally, having largely given up its oxygen to cells lining the capillary beds, the blood returns to the right side of the heart. (B) In a fetus, by contrast, the blood is oxygenated by a countercurrent system in the placenta, in which fetal blood passes through capillaries alongside those containing maternal blood. The transfer takes place in thousands of cuplike structures known as microcotyledons; this figure shows a diagrammatic version of a single cup. As we shall see later in this chapter, a special sort of hemoglobin is needed to accomplish efficient gas exchange. The oxygenated fetal blood then enters the systemic veins leading to the heart, and forms a mixture (purple) with deoxygenated blood from the systemic circulation. Once the blood has reached the right atrium, the foramen ovale and the ductus arteriosus shunt most of it away from the nonfunctional lungs and into the systemic arteries.

gen and gives up carbon dioxide; then it returns to the left side of the heart. This portion of the circulatory system is called *pulmonary circulation* (Fig. 13.6A). Note that in the pulmonary circuit the arteries carry deoxygenated blood and the veins carry oxygenated blood. From the left side of the heart, the blood is pumped into the aorta and its numerous branches, from which it moves into capillaries, and then into veins, and finally back in the anterior or posterior vena cava to the right side of the heart. This portion of the circulatory system is called *systemic circulation*. The arteries of the systemic circuit carry oxygenated blood and the veins carry deoxygenated blood—a reversal of their roles in the pulmonary circuit.

You may have wondered how the developing tissues of a mammalian fetus are supplied with oxygenated blood, since its lungs do not work. Gas, nutrient, and waste transfers between the mother and the fetus take place in the placenta, a large adjoining structure that contains a specialized pair of capillary beds, one carrying fetal blood and the other carrying maternal blood (Fig. 13.6B). The umbilical cord brings part of the fetal blood supply from the fetal leg arteries to the fetal capillaries in the placenta (where CO_2 and other wastes diffuse into the maternal capillaries, while O_2 and nutrients move into the fetal system), and then back to the posterior vena cava. Very little of this freshly oxygenated blood needs to flow to the lungs, so there are two alternate pathways, unique to fetuses. The first is a valve—the foramen ovale—which joins the two atria; the second is the ductus arteriosus, which connects the pulmonary artery to the aorta. Both serve to transfer oxygenated blood to the systemic arteries.

At birth, the placental exchange with the maternal capillaries comes to an end, and these two connections must be closed. This is accomplished within seconds of birth as a result of the expansion of the lungs with the first breath. As the capillary beds in the lungs of the newborn open out, blood suddenly begins to flow in large quantities for the first time through the pulmonary arteries and veins. As a result, the pressure in the right atrium abruptly drops, while the pressure in the left atrium rises sharply, and the flap between the two is forced shut. At the same time, the increased flow of blood into the pulmonary artery causes it to expand and close off the now-useless ductus arteriosus. Subsequently both the duct and the flap are permanently sealed by new cell growth.

The circuit in other vertebrates   We have seen that the human heart, which lies in the thoracic cavity just beneath the sternum (breastbone), is, in effect, two hearts in one, since blood in the left side of a normal heart is completely separated from blood in the right side. This type of heart—four-chambered, with complete separation of sides—is characteristic of mammals and birds, the two groups of vertebrates commonly termed warm-blooded. These animals, which maintain a relatively constant high body temperature regardless of fluctuations in the environmental temperature, have a high metabolic rate and very precise internal control mechanisms. Constant perfusion of the tissues with blood rich in oxygen is clearly essential to them. It would be highly disadvantageous to such animals if the oxygen-rich blood from the pulmonary circulation were mixed with the oxygen-poor blood returning from the systemic circulation.

The hearts of primitive vertebrates apparently had only one atrium and

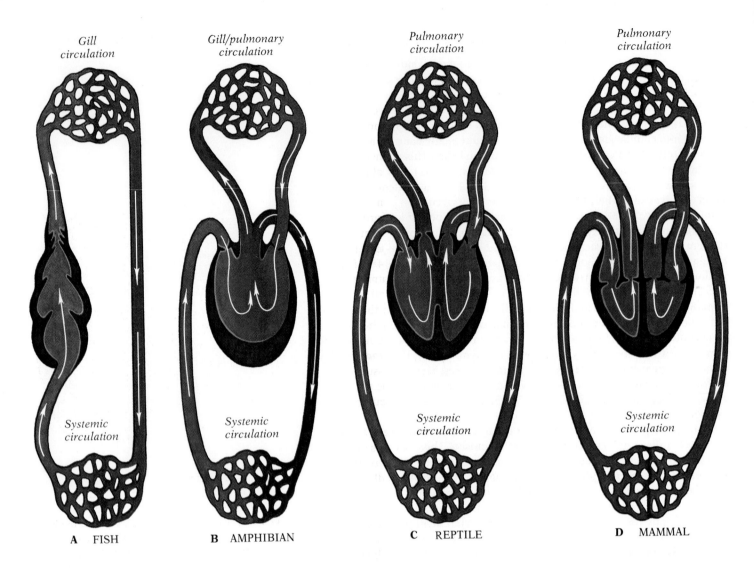

Gill
circulation

Gill/pulmonary
circulation

Pulmonary
circulation

Pulmonary
circulation

Systemic
circulation

Systemic
circulation

Systemic
circulation

Systemic
circulation

A   FISH

B   AMPHIBIAN

C   REPTILE

D   MAMMAL

13.7   A schematic comparison of vertebrate hearts
(A) In modern fish the blood is pumped through a linear, multi-chambered heart to the gills, where it picks up oxygen. The oxygenated blood (red) then passes without further pumping to the systemic circulation, where it gives up its oxygen before returning to the heart. (B) In the amphibians the blood that has picked up oxygen in the gills and/or lungs returns to the heart, from which it is pumped into the systemic circulation. Since the ventricle is not divided, some mixing of the pulmonary and systemic flows occurs in the heart. In most reptiles (C) the ventricles are partially divided, so less mixing takes place. (D) In mammals and birds the two halves of the heart are effectively separated.

one ventricle, arranged linearly. Modern fish display an elaboration of this linear design (Fig. 13.7A). In them, no mixing of oxygenated and deoxygenated blood occurs because blood aerated in the capillaries of the gills goes straight from the gills to the systemic circulation without first returning to the heart. This one-pump strategy has a drawback: because of the high resistance of the gill capillaries, blood leaving the gills to enter the systemic circulation is under relatively low pressure, and so moves through the systemic capillaries sluggishly.

This problem is overcome in other vertebrates by the addition of a second pump, between the gill/pulmonary and systemic circuits, to boost the pressure. In amphibians and reptiles, unlike birds and mammals, the two pumps are not fully separated. In amphibians, the heart has a distinct right atrium and left atrium, but the ventricle is essentially an undivided structure, whereas in reptiles the ventricles are more distinct (Fig. 13.7B–D). Though mixing of oxygenated and deoxygenated blood is inevitable in amphibians, the muscular ridges that divide the ventricles of reptiles allow relatively little mixing to take place.

PUMPING OF THE BLOOD

The heart   Even though the human heart is double, the two halves beat essentially in unison. The beating is inherent in the heart itself and not dependent on stimulation from the central nervous system. If all nerve connections to the heart are cut, the heart will continue to beat in a normal manner, though the rate of beat may change slightly. As you probably know, the heart of a frog or turtle can continue to beat even after removal from the animal's body, if it is placed in a solution with the proper osmotic concentration. But while the initiation of the beat and the beat itself are intrinsic properties of the heart, the rate of beat is partly regulated by stimulation from two sets of nerves of the involuntary nervous system; one set tends to accelerate the heartbeat and the other to decelerate it (see pp. 486–87).

The initiation of the heartbeat normally comes from the sino-atrial node, or **S-A node**, often called the pacemaker of the heart; it is a small mass of **nodal tissue** on the wall of the right atrium near the point where the anterior vena cava empties into it. Nodal tissue is unique to the heart; it has the contractile properties of muscle and can transmit impulses like nerve. A second mass of nodal tissue, the atrio-ventricular node, or **A-V node**, is located in the partition between the two atria. A bundle of nodal-tissue fibers (the bundle of His) extends from the A-V node into the walls of the two ventricles, with branches (**Purkinje fibers**) in all parts of the ventricular musculature (Fig. 13.8).

At regular intervals, a wave of contraction spreads from the S-A node across the walls of the atria. When this wave of contraction reaches the A-V node, the node is stimulated and excitatory impulses are rapidly transmitted from it to all parts of the ventricles via the fibers of the bundle of His. These impulses stimulate the ventricles to contract.

The heart rate—the alternation of systole (contraction) and diastole (relaxation)—is inversely related to body size and thus varies from species to species. In the Asiatic elephant, for example, a normal rate is 30 beats per minute; in the tiny masked shrew the reported average is 780 beats per minute; in a normal human being at rest it is about 70 beats per minute, but there is much individual variation.

In the course of the beat, the heart emits several characteristic sounds, which can be heard easily through a stethoscope placed against the chest. First, there is a long, low-pitched sound produced by the closing of the valves between the atria and the ventricles and by the contraction of the ventricles. Then there is a shorter, louder, higher-pitched sound produced by the closing of the valves between the ventricles and the arteries leading from them. Changes in these sounds often indicate to a physician that the heart is defective. A normal heart valve opens when the pressure in front of it is greater than the pressure behind it. For example, when the atria start contracting, they put pressure on the blood they contain, and as soon as this pressure is greater than the pressure in the ventricles, the tricuspid and bicuspid valves are forced open and the blood can flow into the ventricles. As soon as the atria begin to relax and the pressure in them falls below that in the ventricles, the valves snap shut. Similarly, when the ventricles contract and the pressure in them exceeds the pressure in the arteries leading from them, the semilunar valves open and the blood is forced into the arteries; as

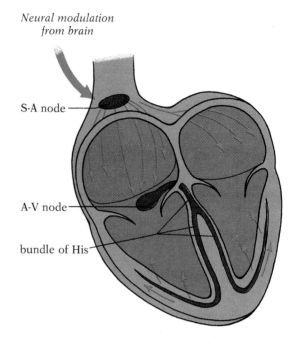

Neural modulation from brain

S-A node

A-V node

bundle of His

13.8   Electrical control of the heart
The pacemaker of the heart is the S-A node. Its spontaneous rate can be accelerated or decelerated by nerve impulses from the brain. Each time the S-A node fires, a signal spreads across the two atria, generating a beat and stimulating the A-V node. From there, after a brief delay, the signal is relayed rapidly down the fibers of the bundle of His to generate a beat in the ventricles.

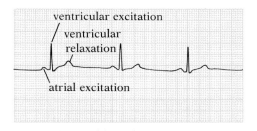

13.9 An electrocardiogram
By recording the electrical activity of the heart under various conditions of rest and exercise, electrocardiograms can sometimes provide important information about potential heart problems. Notice the regular order, strength, and spacing of events in this normal EKG. Since atrial relaxation normally takes place during ventricular excitation, it is not visible in this recording.

soon as the ventricles start to relax, the valves snap shut and prevent the blood in the arteries from flowing backward into the ventricles. The normal heart sounds are an indication that all these valves are functioning properly. If, however, a valve has been damaged and cannot shut completely, a hissing or murmuring sound can be heard as blood leaks backward through the damaged valve; this condition is called a diastolic heart murmur. Sometimes a damaged valve partly obstructs blood flow during systole, and the resulting sound (due to the turbulence in the blood flow) is called a systolic murmur. Heart murmurs are a common result of rheumatic fever and some other diseases. The more extensive the damage to the valve, the less efficient the heart action and the greater the strain placed on the heart.

In the course of contraction, the heart muscle undergoes a series of electrical changes. These changes can be detected by electrodes attached to the skin and can be graphed by an electrocardiograph (Fig. 13.9). Abnormalities in heart action alter the pattern of the resulting electrocardiogram.

The heart of a resting human adult pumps about 5 liters of blood every minute, which is approximately equal to the total amount of blood in the body. This does not mean, of course, that each individual drop of blood passes through the heart every minute; blood that happens to flow into one of the shorter circuits, such as those supplying the neck or chest, may return to the heart quickly and make several rounds in a minute, while blood going to more distant parts of the body, such as the legs, may take several minutes to return to the heart of a resting person. During exercise both the rate of contraction and the amount of blood pumped per beat (the stroke volume) increase greatly. The combination of elevated heart rate and increased stroke volume may raise cardiac output (total amount of blood pumped per minute) to a level four to seven times the resting level. Under such conditions a given drop of blood may pass through the heart many times every minute.

Blood pressure and rate of flow   When the left ventricle contracts, it forces blood under high pressure into the aorta, and blood surges forward in each of the arteries. The walls of the arteries are elastic, and the pulse wave stretches them. During diastole, the relaxation phase of the heart cycle, the heart is not exerting pressure on the blood in the arteries and the pressure in them falls, but elastic recoil of the previously stretched artery walls maintains some pressure on the blood. There is thus a regular cycle of pressure in the larger arteries, which reaches its high point during systole and its low point during diastole.

In humans arterial blood pressure in the systemic circuit is usually measured in the upper arm, where systolic values of about 120 mm Hg and diastolic values of about 80 mm are normal in young adult males at rest. Note that these pressures apply to the upper arm only; the values would not be the same for the lower arm, the leg, or any other part of the body. The blood pressure decreases continuously as the blood moves farther and farther away from the heart. Greatest in the part of the aorta close to the heart, it falls off steadily in the more distant parts of the aorta and its branches, and falls even more rapidly in the arterioles and capillaries. It continues to de-

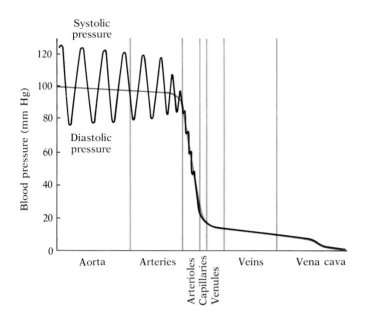

13.10 Graph of blood pressure in different parts of the human circulatory system
The color curve traces the mean-pressure values. As shown by the black curve, there is considerable fluctuation between the systolic and the diastolic pressure in the arteries. This fluctuation diminishes in the arterioles and no longer occurs in the capillaries and veins, because the elasticity of the vessel walls has effectively dampened out the oscillations. The most rapid fall in pressure is in the arterioles and capillaries.

cline, though more slowly, in the veins, reaching its lowest point in the veins nearest the heart (Fig. 13.10). The decline of the blood pressure in successive parts of the circuit is the result of friction between the flowing blood and the walls of the vessels. Such a gradient of pressure is essential, of course, if the blood is to continue to flow; the fluid can only move from a region of higher pressure toward a region of lower pressure. Now we can see more clearly the problem with the original vertebrate heart: the extreme drop in pressure across the capillaries shown in Figure 13.10 takes place in a fish's gills, with yet another set of high-resistance capillaries, in the body tissues, still to be traversed. For such a system to work, the heart must be quite powerful and the capillaries must be relatively wide (to provide less resistance) and hence less efficient for the exchange of gases, nutrients, and wastes.

Several other changes occur along the route of human blood flow. First, with increased distance from the heart, the difference between systolic and diastolic pressures diminishes, because the elasticity of the artery walls tends to damp out the fluctuations in blood pressure. This means that the cyclic, surging type of flow characteristic in the arteries is replaced by a constant rate of flow by the time the blood reaches the capillaries and veins. Second, the rate of flow tends to fall as the blood moves through the branching arteries and arterioles; the rate is lowest in the capillaries and increases again in the venules and veins. These changes in rate of flow result from changes in the total cross section of the vessel system. Linear rate of flow is inversely proportional to a cross-sectional area. In other words, if a fluid is flowing through a tube that has a smaller cross section in some regions than in others, the fluid will move faster in the regions with the smaller cross sections and more slowly in the regions with the larger cross

MEASURING BLOOD PRESSURE

Both the systolic pressure and the diastolic pressure are important diagnostic indicators to the physician, as you know. Ordinarily these pressures are measured in the artery of the arm with a sphygmomanometer (from the Greek *sphygmos*, "pulse"). This instrument has a rubber cuff, which is attached around the upper arm and exerts more and more pressure as air is pumped into it, until finally all blood flow through the artery is blocked. A graduated column of mercury attached to the cuff indicates the pressure the cuff is exerting on the arm. Next, the bell of a stethoscope is placed against the artery just beyond the cuff, and the air in the cuff is gradually released, with consequent reduction of pressure on the arm. When the pressure of the cuff has fallen to a value slightly lower than the maximum systolic pressure in the artery, a small stream of blood can squirt through the artery for an instant during each pulse, producing vibrations that can be heard through the stethoscope. The value shown on the column of mercury the instant before this sound is first detected is taken as the systolic pressure. As the pressure in the cuff is gradually lowered beneath this value, the sound produced as more and more blood surges through the artery at each systole becomes louder and louder. Eventually the pressure in the cuff drops to a value only slightly above that in the artery at diastole, and at this point the sound becomes muffled; the flow of blood through the artery is now continuous. The value shown on the column of mercury at this point is taken as the diastolic pressure. The systolic and diastolic pressures are frequently written together as a fraction—120/80, for example.

sections. The same rule applies if the tube is divided into many branches; the greater the effective cross section—the total cross-sectional area of all the branches in any given region—the slower the flow. As the arteries break up into arterioles and the arterioles break up into capillaries, the total cross-sectional area increases, and the rate of flow becomes slower. As the capillaries unite to form venules and these join to form veins, the total cross-sectional area diminishes again, and the rate of flow increases (Fig. 13.11).

The hydrostatic pressure is so low by the time the blood reaches the veins that some other mechanisms besides pressure from the beating heart must be at work in moving the blood. The walls of veins have the same three layers as arteries, but because the muscle layer is much less developed and there is more connective tissue, the walls are easily collapsible. When nearby muscles contract as the body moves, they put pressure on the veins, compressing their walls and forcing the fluid in them forward. The fluid can move only toward the heart, because the numerous one-way valves with which veins are equipped prevent it from flowing backward into the section from which it has just come (Fig. 13.12). If, when you stand still for a long period, your feet begin to swell and you have a feeling of fatigue, the reason is that the muscle action in your legs is not sufficient to push the body fluids upward against the pull of gravity. The unpleasant symptoms will not be so pronounced if you can manage, while standing, to keep moving your feet and legs, or to contract and relax the leg muscles regularly. Similarly, to prevent pooling of blood in the lower extremities, modern medical practice encourages hospitalized patients to begin walking as soon as they can, often only a day or two after surgery; for such pooling of blood may lead to formation of a clot in a leg vein (thrombophlebitis)—a potentially fatal development all too common in bedridden patients.

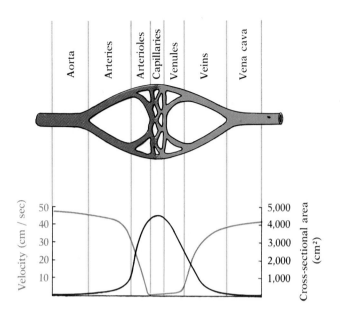

13.11   Changes in the velocity of blood flow in the various parts of a systemic circulatory pathway
As the color curve shows, the velocity falls precipitously as the blood flows through the arterioles, where the total cross-sectional area (black curve) rises. The velocity remains low in the capillaries and venules, but rises again in the veins as the total cross-sectional area falls.

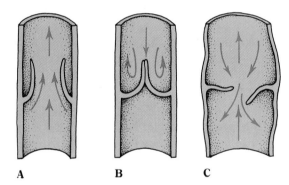

13.12   Valves in the veins
(A) When blood is flowing toward the heart, it forces the flaps open and moves through. (B) When the pressure declines and the blood is pulled down by gravity, the flaps automatically act as a check valve, closing to prevent a backward flow. (C) Damaged valves fail to prevent this backflow. The result is varicose veins, which are especially common on the legs.

The motions of the chest during breathing also aid in moving blood in the veins. When the chest expands during inhalation, the pressure in the thorax falls. There is thus a pressure gradient from other parts of the body to the thorax, and blood tends to be drawn into the large vessels of the thorax and into the heart.

CAPILLARY FUNCTION

The capillaries are so numerous that they penetrate into all parts of every tissue; no cell is far removed from at least one capillary. It is estimated that muscle tissue contains as many as 60,000 capillaries per square centimeter of cross section. The diameters of the capillaries are very small, being seldom much larger than those of the blood corpuscles that must pass through

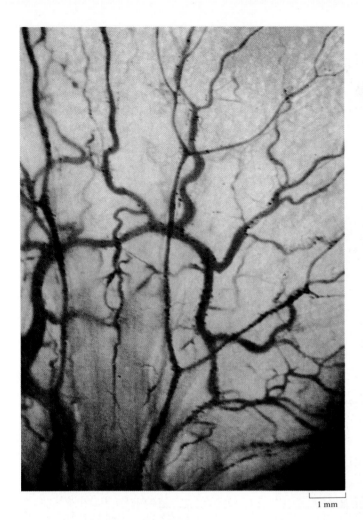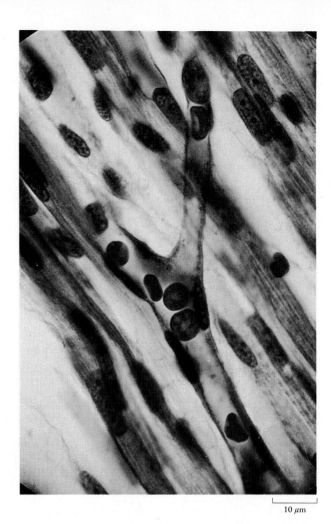

13.13   Blood vessels and capillaries
Left: The vessels in the web of a living frog's foot
branch repeatedly. The smallest capillaries (seen here
as very thin, faintly red lines) reach all cells. Right:
In this longitudinal section of a human capillary,
individual erythrocytes are seen moving single file.

them (Fig. 13.13). The extensive branching and the small diameters of individual capillaries are functionally important in several respects. They ensure not only that all portions of the tissues will be supplied with capillaries, but also that a very great capillary surface area will be available for the exchange process. It has been estimated that every cubic centimeter of blood contacts nearly one square meter of capillary surface each time it passes through a capillary bed. As we have seen, the branching also increases the total cross-sectional area of the system and thus makes blood flow more slowly in the capillaries than in the arteries or veins (Fig. 13.11). This slower flow allows more time for the exchange process. Furthermore, the very small bore of the capillaries makes for high frictional resistance to blood flow; the resulting considerable drop in blood pressure in the capillary bed plays an important role in the exchange process.

Exchange of materials between the blood in the capillaries and the tissue fluids outside the capillaries can occur in at least three ways: (1) The materials may move entirely by diffusion through the membrane of an endothelial cell in the wall of the capillary, across the cytoplasm of the cell to the other side, and out through the cell membrane on that side. (2) Electron-microscope studies of the endothelial cells of capillaries have revealed large numbers of vesicles that apparently pick up materials by endocytosis on one side of the cell, move across the cell, and then expel the materials by exocytosis on the other side (Fig. 13.14). (3) Electron-microscope studies

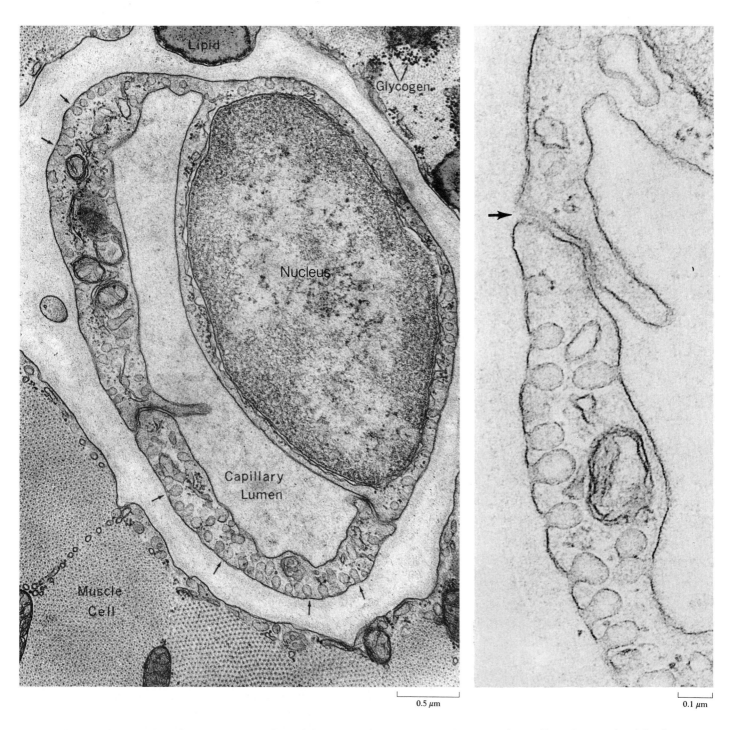

Lipid

Glycogen

Nucleus

Capillary
Lumen

Muscle
Cell

0.5 μm

0.1 μm

13.14 Electron micrographs of cross section of capillary

Left: Two endothelial cells (the section shows the large nucleus of one of them) make up the capillary wall. Note the clefts at each of the two junctions between the cells and the numerous pinocytic vesicles (arrows) in the cytoplasm. These vesicles may transport materials from outside the capillary, across the endothelial cell, into the lumen of the capillary, or they may take the reverse route. Right: Enlarged view of wall of a capillary, showing the cleft where two endothelial cells join (arrow) and numerous pinocytic vesicles opening on the outer face of the lower cell.

333

Thrombosis is the formation of a solid mass or plug of blood constituents in a blood vessel. The mass, or thrombus, may block (wholly or only in part) the vessel in which it forms (see photograph), or it may become dislodged and be carried to some other location in the circulatory system, in which case it is called an embolus. Thromboembolisms are among the leading causes of serious illness and death in Western civilizations.

Many factors can predispose a person to formation of a thrombus. These include irritation or infection of the lining of a blood vessel, or a reduced rate of blood flow through a vessel, which may result from disease or merely from long periods of inactivity. For example, formation of a thrombus in a vein, especially of the leg, a condition known as thrombophlebitis, is particularly common in postoperative patients who remain immobilized in bed; it is also common in the elderly, whose leg muscles have lost much of their pumping action. The thrombus often leads to an inflammatory reaction and pain.

A thrombus that becomes detached from its site of formation and moves in the bloodstream as an embolus is extremely dangerous, because it may become lodged in a vessel of some essential organ such as a lung, the heart, or the brain and cut off its blood supply. Such emboli often lodge in the lung (pulmonary embolism) and cause

death (necrosis) of a portion of the lung tissue. After pneumonia, pulmonary embolism is the most common acute disease of the lungs seen in hospitalized patients in the United States.

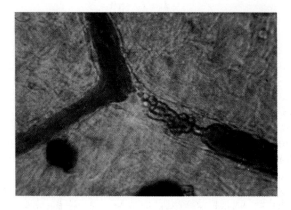

A thrombus in a small blood vessel
The thrombus (tangled red mass) has blocked blood flow near a point where the vessel branches. The blood has pulled away from the left end of the thrombus and is beginning to pull away from the right end also.

have also shown that in the capillaries of most parts of the body (the central nervous system is an exception) there are clefts between adjacent endothelial cells wide enough to permit filtration of water and most dissolved molecules, but not proteins (Fig. 13.14).

We shall now examine the third mechanism in more detail as it operates in human beings. At the arteriole end of a representative capillary bed, the hydrostatic blood pressure is, on the average, about 36 mm Hg higher than the hydrostatic pressure of the tissue fluid outside the capillaries (Fig. 13.15). The pressure differential has fallen to about 15 by the time the blood reaches the venule end of the capillary bed. The hydrostatic blood pressure tends to force materials out of the capillaries into the surrounding tissue fluid. If this were the only force involved, there would be a steady loss from the blood by filtration of both water and those dissolved substances that can readily be carried by the water through the clefts in the capillary walls. It can be demonstrated, however, that normally there is relatively little net loss of water from the blood in the capillaries. Clearly, some other force must act in opposition to the hydrostatic force.

When an embolus (or a locally formed thrombus) blocks a blood vessel in the brain and causes necrosis of the surrounding neural tissue from lack of oxygen, the condition is known as a stroke, or cerebral infarction. The symptoms of stroke vary, depending on the part of the brain that has been damaged. When the infarction is in the cerebral hemispheres, there is usually some loss of muscular control in part of the body, sometimes sensory impairment, and often some loss of language ability, which may be a difficulty of expression (expressive dysphasia) or a difficulty of understanding (receptive dysphasia), or both.

The structure of the circulatory system helps guard against frequent damage from small emboli. Most parts of the body are reached by capillaries from two or more arterioles—a strategy known as collateral circulation. Hence, to cause damage an embolus must be of sufficient size to block the circulation upstream, in the relatively large blood vessel that has not yet branched into alternative pathways, or separate small emboli must block each of the branches independently. However, some parts of the body lack effective collateral circulation; these include the retina and, unfortunately, the heart.

Blockage of a blood vessel in the heart by an embolus (or by a locally formed thrombus) causes necrosis of a portion of the heart muscle, a condition familiarly known as a heart attack (more technically, as a myocardial infarction). A high percentage of the deaths that occur in the first few hours after a heart attack result from disruptions of the control system of the heart, with accompanying arrhythmias, especially ventricular fibrillation.

In the vast majority of cases of cerebral or myocardial infarction, the patient already suffers from atherosclerosis, a condition in which fatty deposits in the arteries and thickening of their walls diminish the size of the lumen; both the reduced size and the consequent reduced blood flow make it easier for an embolus to become lodged in the vessel. Indeed, conditions caused wholly or in part by atherosclerosis are the leading cause of death in the United States; they are responsible for more deaths than the next two leading causes, cancer and accidents, combined.

Atherosclerosis is not the only major condition predisposing a person to a heart attack. Another extremely common precondition is damage to the lining of the arteries due to prolonged high blood pressure (hypertension). Hypertension can also lead eventually to weakening of the heart muscle (which has thickened from the continuing strain imposed on it) and to declining efficiency of its pumping action. Blood may then back up in the heart and lungs, an often fatal condition called congestive heart failure.

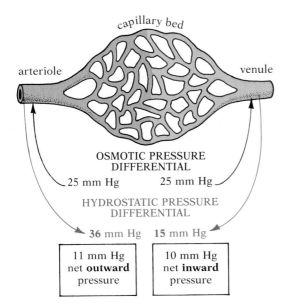

13.15 Diagram of forces involved in the filtration of materials across capillary walls

The blood in the capillaries has both a greater hydrostatic and a greater osmotic pressure than the surrounding tissue fluid. The hydrostatic pressure differential tends to force water and dissolved materials out of the capillaries into the tissue fluid; the osmotic pressure differential has the opposite effect, causing the capillaries to take up water and dissolved materials from the tissue fluid. At the arteriole end of a characteristic capillary bed, the hydrostatic pressure differential of 36 mm Hg is greater than the osmotic pressure differential of 25 mm; the difference of 11 mm favors the outflow of materials. At the venule end, the osmotic pressure differential, which remains 25 mm, is greater than the hydrostatic pressure differential, which has dropped to 15 mm; here the difference of 10 mm favors the pickup of materials from the tissue fluid.

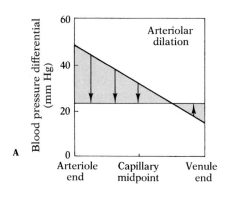

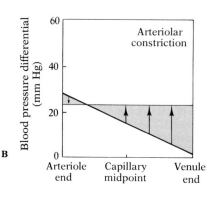

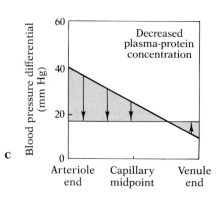

13.16 Conditions that alter the balance between blood pressure and osmotic pressure in the capillaries

(A) When the arterioles are dilated, and much blood is flowing into the capillary bed, the hydrostatic pressure differential (the difference—here represented by the diagonal line—between the hydrostatic pressure of the blood and that of the fluid in the surrounding tissue) is much greater at the arteriole end of the capillary than the osmotic pressure differential (horizontal line); hence flow of water out of the capillary (left color area) much exceeds reabsorption near the venule end (right color area). (B) When the arterioles are constricted, and little blood is entering the capillary bed, the hydrostatic blood pressure in the capillary is low; hence reabsorption of water into the blood much exceeds outward filtration. (C) When the plasma-protein concentration is low, the osmotic pressure of the blood falls (in this case lowering the differential —horizontal line—to below 20 mm Hg); hence the balance is shifted toward outward filtration of water.

This other force derives from the difference in osmotic concentration between the blood and the tissue fluid. The blood of mammals contains a relatively high concentration of proteins, and these large molecules cannot easily pass through the capillary walls. The same kinds of proteins occur in the tissue fluids, but in much lower concentration. Because of the difference in protein concentration on the two sides of the capillary wall, the blood and tissue fluids have different osmotic pressures. Normally, the osmotic pressure of the blood is about 25 mm Hg higher than that of the tissue fluid, with the result that water tends to move into the capillaries from the tissue fluid by osmosis.

We have, then, a system in which hydrostatic pressure developed by the heart tends to force water out of the capillaries and osmotic pressure reflecting differences in protein concentration tends to force water into the capillaries. Obviously, the net movement of water will be determined by the relative magnitudes of these two opposing forces. Notice that at the arteriole end of our representative capillary bed the hydrostatic pressure differential is 36 and the osmotic pressure differential is 25 (Fig. 13.15). Subtracting one from the other, we find that there is a net pressure of 11 tending to force water out of the capillaries. At the venule end of the capillary bed, the hydrostatic pressure differential has fallen to 15, while the osmotic pressure differential has not changed greatly.[1] Therefore, there is now a net pressure of at least 10 tending to force water into the capillaries. In summary, the balance between hydrostatic blood pressure and osmotic pressure is such that water is forced out of the capillaries at the arteriole end and into the capillaries at the venule end. The net effect, in a capillary bed such as the one we have just described, is that nearly all (about 99 percent) of the water filtered out of the capillaries at the arteriole end is reabsorbed at the venule end. Since the water carries with it molecules of many dissolved substances, we can say that the blood in the capillaries first unloads materials for the tissues at the arteriole end and then picks up materials for transport at the venule end. In the process, there is normally only a slight net loss of water from the blood.

The balance of hydrostatic and osmotic pressures in the capillaries is very delicate. Since it plays such an important role in the exchange of

---

[1] Loss of water from the blood has, of course, slightly increased the concentration of protein in the blood and raised the osmotic pressure accordingly, but this change is relatively slight and can be ignored for our purposes here.

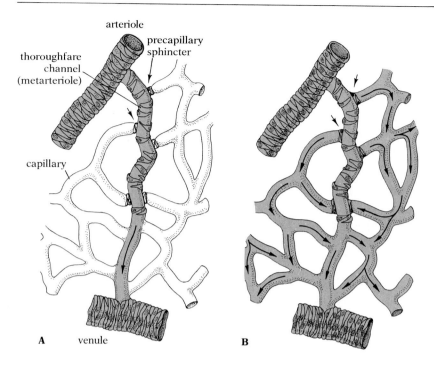

arteriole

precapillary
sphincter

thoroughfare
channel
(metarteriole)

capillary

A venule

B

13.17 The vessels of a capillary bed
When the precapillary sphincters are closed (A),
blood flows only through the capillaries known as
thoroughfare channels. When the sphincters are open
(B), blood flows through all the capillaries.

materials between the blood and the tissue fluid, any disturbance of it may
have profound effects on the condition of the organism. For example, an
increase in blood pressure in a capillary would tend to increase loss of fluid
from the blood, while a decrease in blood pressure would have the opposite
effect (Fig. 13.16A–B). Such changes in blood pressure could be produced
by one or more of a variety of factors, such as changes in rate or strength of
heart action, increase or decrease in total blood volume, changes in the
elasticity of the walls of the arteries, or increased dilation or constriction of
arterioles and capillary sphincters.

The last-mentioned factor—degree of dilation or constriction of capillary sphincters—is also important in determining the extent of the blood
supply to any given tissue at a given time. The capillaries of the body are
never fully open all at the same time; many capillaries are usually closed by
constriction of rings of sphincter muscle at their bases. In a resting muscle,
for example, only certain capillaries—called thoroughfare channels, or
metarterioles—are generally open; but once the muscle becomes active,
and its need for oxygen and nutrients increases, the numerous smaller
branch capillaries become dilated, and the local blood supply is greatly increased (Fig. 13.17). Similarly, the capillaries in the wall of the intestine are
extensively dilated following a large meal, when a major portion of the
blood supply is channeled into this region where much absorption of digestive products is occurring. Increased dilation of skin capillaries often gives
the skin a reddish hue, seen in blushing, while constriction of these same
capillaries gives the skin a bleached, whitish look. A major form of heat loss
from the body is by radiation from the blood in the superficial capillaries of
the skin (Fig. 13.18); changes in dilation of these capillaries, by altering the
amount of blood flow, are an important factor in helping to regulate heat
loss.

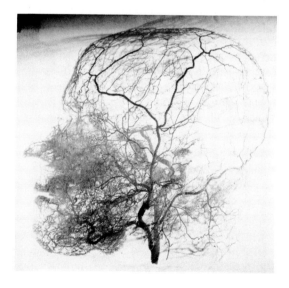

13.18 Blood vascular network in the human head
The dense array of capillaries in the human face,
notably the lips, suggests why this part of the body is
often affected severely by heat loss.

Clearly, simultaneous dilation of a high percentage of the body's capillaries, such as those of the muscles, intestine, and skin, tends to lower the blood pressure in any given capillary, because the same amount of blood is now distributed in a greater total space. If all the capillaries were fully open at the same time, which never actually happens, they would contain all the blood in the body, and arterial blood pressure would fall to zero or below. When blood pressure falls because of vasodilation (extensive dilation of the vessels, producing a condition known as vascular shock) or because of loss of blood by hemorrhage, the consequent increased absorption of tissue fluid by the capillaries increases the total blood volume (though not the total number of blood cells) and tends to compensate partially for the deficiency. At such times, the supply of circulating blood is also augmented by reserve blood previously stored in the *spleen*. This organ stores blood in large cavities connected to its capillaries. Contractions of the smooth muscles in the walls of the spleen can expel this blood into the general circulation.

Changes in the relative concentration of proteins in the blood and in the tissue fluid can also severely alter the balance of forces operating in the capillaries (Fig. 13.16C). Experiments in which the protein concentration in the blood supply to an animal's limb is artificially regulated have shown that increasing the protein concentration in the blood decreases loss of fluid from the blood and increases absorption of tissue fluid. Conversely, decreasing the protein concentration in the blood increases loss of fluid from the blood and decreases reabsorption of fluid from the tissues. The result is an abnormal accumulation of fluid in the tissues that causes swelling, a condition known as edema.

THE LYMPHATIC SYSTEM

We have seen that approximately 99 percent of the water that leaves the capillaries at the arteriole end is normally reabsorbed at the venule end. But what about the remaining 1 percent? Can this fluid return to the blood by any other means than direct reabsorption into the blood capillaries? The answer is yes. Vertebrates have a special system of vessels that function in returning materials from the tissues to the blood. These vessels are called lymph vessels, and together they constitute the lymphatic system, which includes lymph veins and lymph capillaries, but no arteries. The lymph capillaries, which like the blood capillaries are distributed throughout most of the body, are closed at one end. Their walls, like those of blood capillaries, are composed of a single layer of endothelium. Tissue fluid is absorbed into the lymph capillaries (whereupon it is called *lymph*) and slowly flows through the capillaries into small lymph veins, which unite to form larger and larger veins until finally two very large lymph ducts empty into veins of the blood circulatory system in the upper portion of the thorax near the heart.

Though the walls of the lymph capillaries are structurally similar to those of blood capillaries, their permeability characteristics are different. Lymph often contains a small concentration of proteins of the same types as those in the blood; these proteins can apparently move easily across the walls of the lymph capillaries. Since the lymphatic system steadily carries small

13.19 Elephantiasis
Elephantiasis is a condition of extreme edema that occurs when lymph vessels become blocked by filarial worms. Here the left leg is swollen with the fluids accumulated in the tissues as a result of the blockage.

quantities of proteins from the tissue fluid to the blood, but the protein concentration in the tissue fluid does not normally decrease, there must ordinarily be some slight leakage of proteins from the blood capillaries, even though the walls of these capillaries are highly impermeable to proteins. The lymph vessels, whose walls are very permeable to proteins, return such proteins to the blood. The importance of this process in maintaining the normal osmotic balance between the blood and the tissue fluid is very great. Under certain conditions major lymph vessels may become blocked; the protein concentration in the tissue fluid then steadily rises, and the difference in osmotic concentration between it and the blood steadily diminishes, which means that less and less fluid is reabsorbed by the blood capillaries. The result is severe edema (Fig. 13.19).

The lymphatic system performs many other functions besides returning excess tissue fluid and proteins to the blood. For example, much of the fat absorbed from the intestine is picked up by lymph vessels rather than by blood capillaries. Absorption of fats thus differs from the absorption of sugars and amino acids, which are picked up by blood capillaries.

Lymph nodes are present in the lymphatic systems of mammals and some birds, but are absent in most lower vertebrates. Located along major lymph vessels and composed of a meshwork of connective tissue harboring many phagocytic cells, they act as filters and are sites of formation of certain types of white blood cells. As the lymph trickles through the nodes, it is filtered and such particles as dead cells, cell fragments, and invading bacteria are destroyed by phagocytic cells. Nondigestible particles such as dust and soot, which the phagocytic cells cannot destroy, are stored in the nodes. Since the nodes are particularly active during an infection, they often become swollen and sore, as the lymph nodes at the base of the jaw are apt to be during a throat infection.

Since the lymphatic system is not connected to the arterial portion of the blood circulatory system, it is obvious that lymph is not moved by pressure developed by the heart. Its movement, like that of blood in the veins, is due to the contractions of skeletal muscles that press on the lymph vessels and push the lymph forward past one-way valves (Fig. 13.20). Though this mechanism of movement holds for all mammals, including humans, many other vertebrates have lymph hearts, which are pumping devices located along major lymph vessels. Most animals that have lymph hearts lack valves in their lymph vessels.

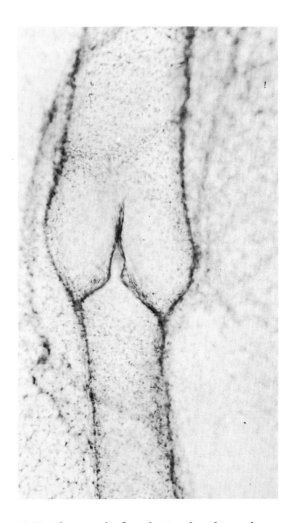

13.20 Photograph of a valve in a lymph vessel

HUMAN BLOOD

We have so far discussed the routes followed by the circulating blood, the mechanism of circulation, and the process of exchange of certain materials with the tissue fluid. We should now examine the blood itself in more detail. It is one of the most important and unusual tissues in the animal body.

In Chapter 6 we classified blood as a type of connective tissue with a liquid matrix. The intercellular liquid matrix of blood is called *plasma*. Suspended in the plasma are the formed elements, which are of three major types in vertebrates: (1) the red blood corpuscles, or *erythrocytes*; (2) the white blood cells, or *leukocytes*; (3) the *platelets*, which are small disc-shaped bodies that arise as cell fragments.[2] If whole blood, treated to prevent clotting, is left standing in a test tube, the formed elements will settle slowly to the bottom, leaving the fluid plasma above. The specific gravity of the formed elements does not greatly exceed that of the plasma, however, and the agitation associated with normal circulation is sufficient to prevent separation within the circulatory system. Normally, the formed elements constitute about 40–50 percent of the volume of whole blood, while the plasma constitutes the other 50–60 percent.

COMPOSITION OF THE PLASMA

The basic solvent of the plasma is, of course, water, which constitutes roughly 90 percent of the plasma. A great variety of substances are dissolved in the water; the relative concentrations of these vary with the condition of the organism and with the portion of the system under examination. For convenience, let us divide these solutes into six categories: (1) inorganic ions and salts; (2) plasma proteins; (3) organic nutrients; (4) nitrogenous waste products; (5) special products being transported; and (6) dissolved gases.

1. The principal inorganic cations (positively charged ions) in the plasma are sodium (Na^+), calcium (Ca^{++}), potassium (K^+), and magnesium (Mg^{++}). The chief inorganic anions (negatively charged ions) are chloride (Cl^-), bicarbonate (HCO_{3-}), phosphate (HPO_4^{--} and $H_2PO_4^-$), and sulfate (SO_4^{--}); of these, chloride and bicarbonate are by far the most abundant. Together, the inorganic ions and salts (ionic compounds composed of positively and negatively charged ions) make up about 0.9 percent of the plasma of mammals by weight; more than two-thirds of this amount is sodium chloride, ordinary table salt.

The concentrations of the individual ions remain relatively stable, regulated as they are by a variety of agencies, particularly the kidneys and other excretory organs, as well as a number of hormones. This stability, this maintenance of equilibrium—which, you may remember, is called ***homeostasis***—is essential to the normal function of the organism. Any appreciable shift in the concentrations of sodium chloride ($NaCl$) and sodium bicarbonate ($NaHCO_3$), for example, would cause severe disturbance, and even death, to the cells, for these compounds (together with plasma proteins) help determine the osmotic balance between the plasma and the fluids bathing the cells. Even if the total concentration of dissolved substances remains the same, shifts in the concentrations of particular ions in the plasma, leading to corresponding shifts in the tissue fluids, can create serious disturbances. Nerves and muscles, for example, are highly sensitive to changes in the concentrations of K^+ and Ca^{++}. Similarly, the integrity of cell membranes depends on proper balance of Ca^{++}, Mg^{++}, K^+, and Na^+ in the intercellular medium. The concentrations of certain ions are also very important in determining the pH of the body fluids, and even the slightest changes in pH (normally slightly alkaline in plasma) may kill the organism.

---

[2] True platelets are found only in mammals. The blood of most other vertebrates contains cells called thrombocytes, which function in blood clotting in a manner similar to platelets.

2. The plasma proteins, which constitute 7–9 percent by weight of the plasma, are of three types—fibrinogen, albumins, and globulins. Most of these proteins are synthesized in the liver, though some of the globulins are synthesized in lymphoid tissue.

We have already discussed the great importance of proteins in determining the osmotic pressure of the plasma and the influence they consequently have on the exchange of materials in the capillary beds and on the general water balance of the body. These proteins are also instrumental in stabilizing the pH of the plasma, as well as in the crucial task of controlling the viscosity of the plasma (viscosity is a measure of the internal friction of a fluid—that is, the friction between molecules as they slide past each other). The heart can maintain normal blood pressure only if the viscosity of the blood is nearly normal. Injection of an isotonic salt solution into the circulatory system as an emergency measure following extensive hemorrhage can restore the blood volume to normal and thus raise the blood pressure somewhat, but it cannot raise the pressure to the normal level, because the saline solution has too low a viscosity.

The plasma proteins have other functions as well. By binding to certain hormones, fatty acids and other lipids, some vitamins, and various minerals, they greatly facilitate the transport of such substances by the blood. In addition, fibrinogen and certain globulins function in blood clotting (see box, p. 342) and gamma globulin plays a role in antibody reactions to infections, as discussed in a later chapter.

3. Organic nutrients in the blood include glucose, fats, phospholipids, amino acids, and lactic acid. Some of these are picked up from the intestine; others enter the blood from storage areas such as the liver and the fat depots. Lactic acid is a product of glycolysis, especially in muscles; it is transported by the blood to the liver, where some of it may be used in resynthesis of carbohydrates and some may be further oxidized to carbon dioxide and water.

Another substance found in the plasma is cholesterol. It is metabolized to some extent as a source of energy and is a stiffening component of the cell membrane, but plays its major role as the precursor of most other important steroids, such as the bile acids and the steroid hormones.

4. The plasma also carries nitrogenous waste products from their sites of formation to such organs of excretion as the kidneys. In mammals this waste is primarily in the form of urea, with small amounts of ammonia and uric acid.

5. Among the special products carried by the plasma, the hormones are of particular significance. These substances, synthesized by the endocrine tissues, are important regulatory chemicals; we shall look at them more closely in another chapter.

6. Three principal gases are found dissolved in the plasma. One of these, nitrogen, which diffuses into the blood in the lungs, seems to be physiologically inert, and will be disregarded here. The other two, oxygen and carbon dioxide, are of critical importance, and we shall discuss the details of their transport in a later section. Actually, in vertebrates most of the oxygen and much of the carbon dioxide are transported in the red blood corpuscles, so only small quantities are present in the plasma.

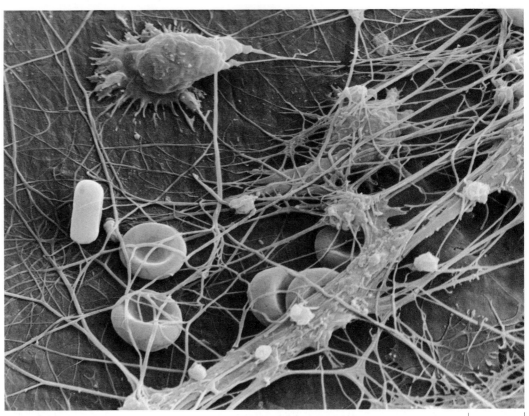

2 µm

13.21 Blood clotting
Blood clots as the various components of the blood
become trapped in a network of fibrin. Several
erythrocytes and a mobile macrophage (top left),
which plays a central role in eliminating dead
erythrocytes and bacteria, are visible in this electron
micrograph.

BLOOD CLOTTING

Normally, the plasma of the circulating blood remains a liquid. Under certain conditions, however, when a blood vessel has been ruptured or otherwise damaged, or when certain kinds of foreign substances have gained entrance into the circulating blood, or when the blood has been removed from the body, one of the plasma proteins, *fibrinogen*, comes out of solution and converts into *fibrin*, which forms a hard lump, or clot (Fig. 13.21). In this way a small hole in a vessel may be plugged, or a weakened place in a vessel wall may be strengthened. Blood clotting is a powerful evolutionary adaptation for emergency repair of the circulatory system and for preventing excessive blood loss (see box). Clotting occurs in all vertebrates and in some invertebrates. Some invertebrates have an alternative adaptation serving the same basic function: powerful muscles contract and close off any hole or damaged area.

It would perhaps be well to clarify here the meanings of the terms "whole blood," "blood plasma," and "blood serum," which occur frequently in references to medical procedures, particularly blood transfusions. Whole blood is blood just as it exists in the circulatory system, with none of its constituents removed. Blood plasma is whole blood minus the formed elements. Blood serum is plasma minus fibrinogen.

HOW DOES BLOOD CLOT?

Consider the following series of experiments.

1. Blood removed from a living blood vessel, but carefully prevented from touching the portion of the vessel that has been damaged, is put into an open dish lined with nonwettable plastic. The blood is thus simultaneously subjected to two conditions traditionally thought to result in clotting—exposure to air and cessation of flowing—yet it remains liquid for hours. Nor does it clot when it is held stationary in a portion of a blood vessel that has been tied off.

2. When the same procedure is repeated, but the blood is allowed to come into contact with the damaged vessel wall through which it is removed, clotting begins at once. Since clotting occurs when a vessel is damaged, and since it is initiated at the site of the damage, might the damaged tissue release some chemical that initiates the clotting process?

3. If we prepare an extract from tissue cells and add this to the liquid blood in the plastic-lined dish of the first experiment, a clot promptly starts to form. We seem to be on the right track. But when we remove blood from a blood vessel as we did in the first experiment, preventing it from touching damaged tissue, and put it into a glass dish instead of one lined with nonwettable plastic, the blood promptly clots even without the addition of tissue extract. Where does this leave us? Might there be some other source of the clot-initiating chemical we have postulated?

4. If we allow fresh blood to touch glass and watch it under a high-powered microscope, we can see that blood platelets tend to disintegrate on contact with the glass. Perhaps the disintegrating platelets release a clot-initiating chemical similar to that released by damaged tissues. Platelets seem to disintegrate much more readily on contact with wettable foreign surfaces like glass than on contact with nonwettable surfaces; this would explain why clotting did not occur in the dish lined with nonwettable plastic. It turns out that platelets also disintegrate when they touch damaged tissue.

The hypotheses suggested by the various experiments outlined here have been corroborated: both damaged tissues and disintegrating platelets release a complex substance, called *thromboplastin*, that initiates blood clotting. For reasons not well understood, however, thromboplastin is not effective unless calcium ions are present. Some of the anticoagulants used in storing blood for later use in transfusions are chemicals that remove calcium ions from the blood and thus prevent clotting.

Two substances essential for normal blood clotting, then, are thromboplastin and calcium ions; a third is the plasma protein fibrinogen. But if we mix these three substances in a dish, no clotting occurs. Clearly, something else must be involved. That something else seems to be one of the globulin proteins of the plasma, known as *prothrombin*. If prothrombin is added to the mixture of fibrinogen, thromboplastin, and calcium ions, a clot will form. It can be demonstrated, however, that prothrombin itself has no effect on clotting; it must first be converted into *thrombin*, the substance that converts fibrinogen into its crystalized form, fibrin, of which the clot is made.

Now we have identified the main ingredients in the clotting process. Thromboplastin, produced by disintegrating platelets or damaged tissue, converts the plasma protein prothrombin into thrombin; this is the reaction in which the calcium ions participate. The thrombin then converts another plasma protein, fibrinogen, into fibrin. The fibrin fibers form a meshwork, which begins to shrink; finally, the fluid blood serum is squeezed out, and a hardened clot is left in place. The reactions can be summarized as follows:

$$(1) \qquad \text{prothrombin} \xrightarrow[\text{Ca}^{++}]{\text{thromboplastin}} \text{thrombin}$$

$$(2) \qquad \text{fibrinogen} \xrightarrow{\text{thrombin}} \text{fibrin}$$

These simplified summary equations show the relations between the essential substances. Actually, numerous other substances—accelerators, inhibitors, and the like—also play roles in the clotting process. A reaction series of this sort in which the first step releases another, which then triggers yet another, and so on, is referred to as a cascade reaction.

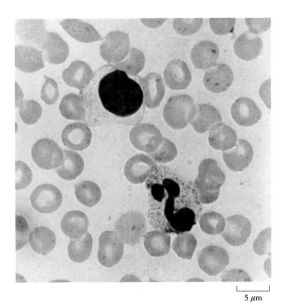

13.22 Photograph of human blood
The bodies without nuclei are erythrocytes. Two types of leukocytes can be seen: the cell with the very large round nucleus is a lymphocyte, and the cell with the segmented nucleus is a polymorphonuclear neutrophil.

LEUKOCYTES AND THEIR FUNCTIONS

Human leukocytes, or white blood cells, have large, often irregularly shaped nuclei (Fig. 13.22). At least five different types can be distinguished on the basis of the shape of the nucleus and the density of granules in the cytoplasm. The granular types are formed in the red bone marrow, while the agranular, or clear, types are formed in lymphoid tissues such as those of the lymph nodes, spleen, tonsils, adenoids, and thymus. Leukocytes are not restricted to the blood; the majority of our 10^{13} leukocytes are actually found in the lymphatic system. They also wander free in loose connective tissues and occasionally in other tissues as well. Leukocytes are capable of amoeboid movement, and can escape from the blood and lymph vessels by squeezing through the vessel walls at the points of contact between endothelial cells. In essence, then, all connective tissues, including blood and lymph, form one continuous system as far as the leukocytes are concerned.

The leukocytes play a variety of very important roles in the body's defenses against disease and infection. Apparently both damaged tissues and invading bacteria release chemicals that attract leukocytes. Some kinds of leukocytes act as phagocytes, engulfing and destroying bacteria and remnants of damaged tissue cells. Other kinds produce powerful enzymes that help detoxify foreign proteins and other potentially dangerous substances. In a severe infection the leukocyte count in the blood and lymph increases enormously, and vast numbers may invade the infected area. The resulting accumulation of dead tissue, bacterial cells, and living and dead leukocytes at the site of the infection is commonly known as pus. Abnormal dilation of the blood vessels in the infected area produces the local increase in temperature and reddening known as inflammation. Local swelling occurs, because the blood vessels are often more permeable than normal and lose fluid to the tissues. Both the vasodilation and the increased capillary permeability are stimulated by a substance called **histamine**, which is secreted by one kind of leukocyte and also by some other cells in the infected area. It is the excessive secretion of histamine and the consequent massive vasodilation that creates the potentially fatal symptoms of severe allergic reactions. The rare individuals allergic to honey-bee venom, for instance, respond to stings with increasing swelling of the tissues of the throat and bronchial passages. As a result, breathing becomes difficult. Compounding this problem is the vasodilation, which increases capillary capacity and thus lowers blood pressure and reduces flow through the lungs. Repeated exposure to the venom can cause allergic reactions severe enough to lead to death by asphyxiation.

Though the leukocytes of other vertebrates are often not morphologically identical with those of human beings, and the corresponding cells of invertebrates differ even more, most animals have blood cells that serve similar phagocytic, detoxifying, and histamine-secreting functions.

Leukocytes have still another way of fighting disease and infection. Certain nonphagocytic leukocytes, known as **lymphocytes**, give rise to specialized cells that play a central role in immunologic reactions. These cells respond to the presence of certain kinds of foreign substances, called **antigens**, by making **antibodies** that help destroy or inactivate the antigens.

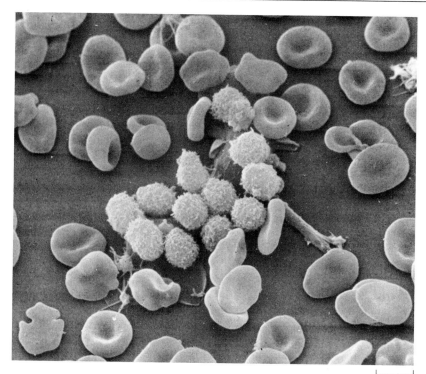

5 μm

13.23 Scanning electron micrograph of blood cells and corpuscles
The erythrocytes look like biconcave discs. In the center is a cluster of lymphocytes, a type of white blood cell important in immunologic defense against diseases. A large phagocytic macrophage can be seen under the lymphocytes.

Each type of antibody is usually very specific and will inactivate only the antigen that stimulated its synthesis. The antibodies are globulin proteins in the blood plasma. Whereas phagocytosis and the detoxification of foreign proteins are particularly important as first defenses against acute infection, the production of antibodies confers a degree of active immunity against the infection and is critical in fighting long-term chronic infections and in building resistance against further infection. We shall discuss the topic of immunologic competence at greater length in a later chapter.

ERYTHROCYTES AND THEIR FUNCTION

Human erythrocytes, or red blood corpuscles, are biconcave, disc-shaped bodies that resemble cells without nuclei (Fig. 13.23). There are normally about 5 million of them per cubic millimeter of blood. Though the number of corpuscles remains amazingly constant from day to day, continual destruction of some corpuscles and formation of new ones goes on; the normal survival time of an erythrocyte is 120 days. More than 2 million erythrocytes are destroyed every second, chiefly in the liver and the spleen, where they are engulfed by large phagocytic cells. Also, phagocytic cells in the lymph nodes destroy any erythrocytes that escape from the blood and get into the lymph. The common supposition that the phagocytic cells of the liver and spleen destroy only old, worn-out corpuscles has not been proven.

The erythrocytes of adults are formed in the red bone marrow, which fills the interior of the upper ends of the long bones and the shafts of flat bones

13.24 Structure of the heme group
A single molecule of hemoglobin has four of these
iron-containing prosthetic groups.

like those of the skull, ribs, and pelvis. Though the mature erythrocytes of
mammals are devoid of nuclei, mitochondria, Golgi apparatus, and other
subcellular organelles, and therefore lack many of the characteristics of living cells, they arise from normally nucleated, rapidly dividing connective-tissue cells of the bone marrow. Toward the end of their development, they lose their nuclei and acquire the red oxygen-carrying pigment
hemoglobin, a protein with iron containing prosthetic (nonproteinaceous)
groups. They then enter the circulating blood. In vertebrates other than
mammals the mature erythrocytes retain their nuclei. The evolutionary
loss of the nucleus from the mammalian erythrocyte might conceivably
have the adaptive advantage of leaving room for more hemoglobin. The
desirability of extra hemoglobin might, in turn, be correlated with the high
metabolic rates—and therefore high oxygen demands—of the tissues of endothermic animals. (However, since birds are also endotherms but retain
nucleated erythrocytes, this explanation is not fully convincing.) Analysis
of the hemoglobin content of the erythrocytes of a variety of vertebrates
reveals that the red corpuscles of mammals, though smaller than the red
cells of many of the lower vertebrates, do, in fact, contain more
hemoglobin.

Not all invertebrates have hemoglobin. In some of those that do, it is
located within cells, as in vertebrates, but in many it is simply dissolved in
the plasma. Though hemoglobin molecules might function just as well in
the plasma as in erythrocytes, their location in cells or corpuscles has a
decided adaptive advantage in animals with high metabolic rates; more
pigment molecules can then be carried per unit volume of blood, and
the oxygen-transporting capacity is correspondingly increased. A single human erythrocyte usually contains about 280 million molecules of hemoglobin. If all this hemoglobin were loose in the plasma, the concentration
of plasma protein would be about three times higher than at present,
with profound effects on the osmotic balance between the blood and the
tissue fluid. Erythrocytes, then, are a convenient method of packaging
large amounts of hemoglobin with relatively little disturbance of the
osmotic concentration of the blood.

Many of the invertebrates that lack hemoglobin have different oxygen-
transporting pigments, which, like hemoglobin, combine a metal with protein. For example, many molluscs and arthropods have a pigment called
hemocyanin, which contains copper instead of iron; when oxygenated, it is
blue instead of red. Hemocyanin never occurs in cells, but is dissolved in
the plasma.

HEMOGLOBIN

Hemoglobin and its role in the transport of oxygen   The hemoglobin
molecule is a globulin protein composed of four independent polypeptide
chains (see Fig. 3.24, p. 67). Each of the four chains enfolds a complex prosthetic group called heme, which has an iron atom at its center (Fig. 13.24).

Human hemoglobin was one of the proteins used in developing methods
for analysis of protein structure. The sequence of amino acids in the four

chains (two pairs, one of type α, the other of type β) was determined by Gerhardt Braunitzer of the Max Planck Institute for Biochemistry in Munich, Germany, and by William H. Konigsberg and Robert J. Hill of the Rockefeller Institute in New York. Similar determinations for the hemoglobin of other vertebrates have demonstrated that they differ from one another in the number and sequence of amino acids. Nevertheless, as shown by Nobel Prize winner M. F. Perutz of Cambridge University through X-ray crystallographic studies, all normal hemoglobins have essentially the same three-dimensional structure. In some hereditary blood diseases, however, which involve changes in only one or two amino acids, conformational alterations occur—alterations that severely impair the oxygen-transporting capability of the hemoglobin molecule. The dysfunction that causes sickle-cell anemia is a single amino acid substitution (the hydrophobic amino acid valine replaces glutamate, which is polar). This alteration reduces the solubility of the hemoglobin, so it tends to crystallize. Once crystallized, the hemoglobin can no longer load and unload oxygen efficiently; worse yet, the red corpuscles tend to collapse, taking on a sickle shape and forming clumps that can obstruct the capillaries (see Fig. 25.2, p. 673).

As would be expected, the more closely related two animals are, the more similar their hemoglobins tend to be; those of humans and apes, for instance, are much more alike than those of humans and fish. Sometimes hemoglobins differ within one and the same species. Thus the hemoglobin of human embryos is slightly different from adult hemoglobin, which replaces it shortly after birth.

Each of the four iron atoms in a hemoglobin molecule can, by virtue of its structural relationships within the molecule, combine loosely with one molecule of oxygen. The compound formed by the union of one molecule of hemoglobin (Hb for short) with four molecules of oxygen (Hb + $4O_2$) is called oxyhemoglobin.

Hemoglobin exemplifies cooperativity in an allosteric molecule (see p. 82). When it binds with the first O_2 molecule, conformational changes occur that enhance its affinity for oxygen and greatly facilitate the binding of the other three O_2 molecules. X-ray analysis shows that the important changes pertain, not to the tertiary structure of the four polypeptide chains, but to the way the chains fit together (the quarternary structure); the heme groups of the two β chains move farther apart, while those of the α chains move closer together. Just how these conformational changes influence the affinity of hemoglobin for oxygen is not yet fully clear. Presumably the altered positions of the heme groups relative to the R groups of neighboring amino acids permit easier access of O_2 to the iron atoms. An incidental effect of the formation of oxyhemoglobin, and the concomitant conformational changes, is to make the blood redder; it is on account of the oxyhemoglobin that arterial blood in the systemic circulation is more crimson than venous blood.

We have said that the combination of hemoglobin with O_2 is a loose one. Under certain conditions the combination will form, and under other conditions it will break down. Clearly, conditions in the lungs must favor formation of oxyhemoglobin, and conditions in the capillary beds of the systemic circulation must favor release of O_2 and re-formation of hemoglobin. The critical condition in determining whether hemoglobin will load or

13.25 Dissociation curves of adult human hemoglobin at normal and low blood pH
The nonlinear shape of the curves is a consequence of cooperativity, which results in the enhanced affinity of the hemoglobin molecule for oxygen once it has bound its first molecule of O_2. In the lungs the hemoglobin becomes fully loaded with oxygen, whereas oxygen is unloaded in the tissue capillaries. And as both curves make clear, when the oxygen concentration in the tissues is especially low, as during exercise, more of the oxygen will be unloaded to meet the needs of the tissues. Notice also that when CO_2 rises, and so acidifies the blood (black curve), the change in pH alters the binding properties of hemoglobin, causing it to deliver more oxygen than usual.

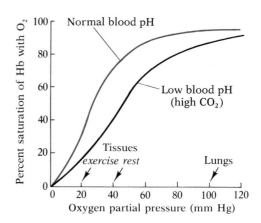

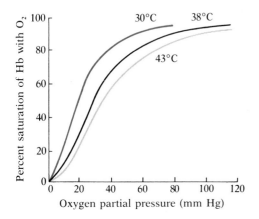

13.26 The effect of temperature on the affinity of human hemoglobin for oxygen
At higher temperatures the dissociation curve of hemoglobin is displaced to the right. In physiological terms, the hemoglobin has a lower affinity for O_2 and will therefore unload it at higher partial pressures of O_2, thus making it more readily available to the tissues.

unload O_2 is the partial pressure[3] of O_2 in the medium to which the hemoglobin is exposed. When the partial pressure of O_2 is high, the hemoglobin picks up O_2; when the partial pressure of O_2 is low, the hemoglobin releases O_2. This is simply another way of saying that hemoglobin loads O_2 when there is a relatively high percentage of O_2 in the surrounding medium, and unloads O_2 when the percentage is relatively low. There is, of course, a relatively high partial pressure of O_2 in the air in the alveoli of the lungs and a relatively low partial pressure of O_2 in the tissues serviced by the systemic circulation, where the O_2 is being consumed in cellular respiration. Consequently hemoglobin tends to pick up O_2 in the capillaries of the lungs and to release O_2 in the capillaries of the systemic circulation.

Figure 13.25 shows the percentage of O_2 saturation of human hemoglobin at different partial pressures of O_2 in the blood; the lower the partial pressure, the greater the tendency for oxyhemoglobin to dissociate into hemoglobin and O_2—hence the name "dissociation curve" for a graph of this type. As you can see, when the blood pH is normal (color curve), the hemoglobin is about 98 percent saturated with O_2 at the partial pressure of O_2 typical of the tissue fluid of the lungs (100 mm Hg), while it is only about 68 percent saturated at the partial pressure of O_2 typical of the fluid of tissues at rest (40 mm). The difference (30 percent) represents the approximate percentage of the O_2 carried by hemoglobin that is actually released to the tissues at rest. We see, then, that the oxyhemoglobin releases less than half its O_2 to the tissues and that venous blood still contains much O_2. During exercise the more rapid utilization of O_2 by the muscle tissues causes a drop in its partial pressure in the tissues to a level of 20 mm or even lower. As a result, the oxyhemoglobin releases more of its O_2 to the tissues, and the saturation of venous blood may fall as low as 25 percent.

The S shape of the dissociation curve of hemoglobin is a result of the cooperativity factor in the binding of O_2. At very low O_2 pressures the curve

---

[3] The partial pressure of a gas dissolved in a liquid refers to the amount of pressure that must be exerted on the gas for it to go into solution in the liquid in the concentration observed. For a general definition of the partial pressure of a gas, see p. 284.

rises slowly, because the binding of the first O_2 molecule to a molecule of hemoglobin is difficult. But after those first molecules are bound, the binding of additional O_2 is easy, and the curve rises steeply until it finally levels off when the hemoglobin is nearly saturated with O_2. That hemoglobin saturation should vary with O_2 partial pressure according to an S-shape curve has important implications for the release of O_2 to the tissues. Notice that the steepest portion of the curve is in the range of O_2 partial pressures prevalent in tissue fluid. Only a slight drop in the pressure here, as occurs in the transition from rest to exercise, results in a very sizable increase in the amount of O_2 released from the oxyhemoglobin. The drop in pressure from 40 to 20 mm discussed in the preceding paragraph, for instance, means an increase in the total percentage of O_2 released from about 30 to about 73 percent.

The affinity of hemoglobin for O_2 is markedly influenced by pH, because H^+ ions act as negative allosteric modulators for hemoglobin. CO_2, produced as metabolic waste by the cells, is abundant in the tissue fluid of most parts of the body; it combines with water to form carbonic acid (H_2CO_3), which in turn dissociates into H^+ ions and bicarbonate ions (HCO_3^-). In short, the result is increased acidity—that is, an increased concentration of H^+:

$$CO_2 + H_2O \rightleftharpoons H^+ + HCO_3^-$$

Hence hemoglobin in capillaries of the systemic circulation is exposed to an acid environment, where its affinity for O_2 is reduced, and it therefore readily unloads its O_2 (Fig. 13.25, black curve). Conversely, in the pulmonary arteries, where CO_2 is released to the lungs and acidity is consequently lower, the hemoglobin is in an environment in which its affinity for O_2 is high, and it therefore readily loads O_2. Thus the waste product CO_2 plays an important regulatory role in shifting the condition of the hemoglobin back and forth between a propensity for loading O_2 in the lungs and for unloading it in the other tissues.

Temperature, as well as pH, affects the O_2 affinity of blood pigments. As shown by Figure 13.26, hemoglobin releases O_2 more easily at higher temperatures, such as occur in muscles during strenuous exercise, when extra O_2 is needed.

Species-specific differences appear in the hemoglobin dissociation curves of mammals, the curves for smaller animals generally being to the right of those for larger animals (Fig. 13.27). It is, of course, advantageous to the smaller animals, which have a higher metabolic rate and whose tissues therefore require more O_2 per unit time, that their oxyhemoglobin should dissociate more readily.

The curves for birds, which have very high metabolic rates, as would be expected in such active animals, tend to be to the right of those for mammals. Those for cold-blooded animals such as fish and amphibians are generally to the left of those for warm-blooded animals; on account of their lower metabolic rates, the rate of demand for O_2 by their tissues is lower, and they have less need for oxyhemoglobin that dissociates very easily.

We have already mentioned that human fetal hemoglobin is slightly different chemically from adult hemoglobin. Figure 13.28 shows that fetal he-

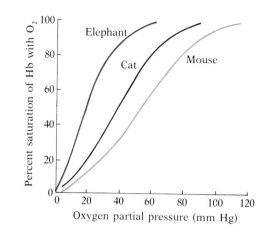

13.27 Dissociation curves for the hemoglobin of different animals
In general, the smaller the animal, the farther to the right its curve will be located. This means that small animals, with high metabolic rates and correspondingly high O_2 requirements, have hemoglobin that tends to unload more readily.

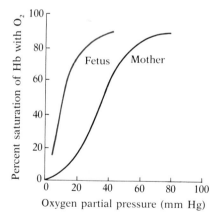

13.28 Dissociation curves for fetal and maternal hemoglobin in the cow
Because fetal homoglobin has a higher affinity for O_2 than maternal hemoglobin, it can take O_2 from the maternal hemoglobin.

TABLE 13.1 *Percentage of blood by volume occupied by corpuscles*[a]

Animal	Altitude	Corpuscles
Human	Sea level	46.0%
	5,360 m	59.9
Sheep	Sea level	35.3
	4,700 m	50.2
Dog	Sea level	34.6
	4,540 m	50.0
Rabbit	Sea level	35
	5,340 m	57
Vicuña	Sea level	29.8
	4,700 m	31.9

[a] Data from C. L. Prosser and F. A. Brown, *Comparative Animal Physiology*, Saunders, 1961.

moglobin has a higher affinity for O_2 than adult hemoglobin.[4] The adaptive significance of this difference is readily apparent. The fetus gets its O_2 from the mother's blood, not directly from the air. If the hemoglobin of the fetus is to take O_2 from the hemoglobin of the mother, it must have a greater affinity for O_2; it must be able to compete successfully with the mother's hemoglobin.

The human fetus exemplifies an organism that must get its O_2 from a medium in which the partial pressure of O_2 is lower than in the atmosphere we ordinarily breathe. Animals like the South American llama and vicuña live at very high altitudes in the mountains, where the partial pressure of O_2 is appreciably lower than it is nearer sea level (as you know if you have ever experienced shortness of breath at high altitudes). It is not surprising, therefore, to find that these animals, like the human fetus, have hemoglobin with a relatively high affinity to O_2; their dissociation curves are to the left of those of the average mammal, meaning that their hemoglobin loads more easily.

The llama and the vicuña have evolved a genetically determined type of hemoglobin that adapts them for life at high altitudes. But what of an animal such as a human, a sheep, a dog, or a rabbit that is moved to a high altitude? We know that eventually such an animal will become acclimatized to its new environment and will no longer experience the severe shortness of breath it experienced at first. Does its hemoglobin change? The answer is no. An individual animal's genes, and the hemoglobin they determine, are not changed simply by changing the environment; acclimatization involves other processes. For example, Table 13.1 shows that acclimatized humans, sheep, dogs, and rabbits at high altitudes have more erythrocytes per unit volume of blood than they normally have at sea level. This is apparently the result of at least two different reactions of the body to the decreased amount of O_2 reaching the tissues. First, erythrocytes stored in such areas as the spleen and skin capillaries are released into the general circulation. Second, in response to a hormonelike substance produced in the kidneys, the red bone marrow becomes more active and produces erythrocytes at a faster rate. Thus, through an increase in the number of corpuscles, these animals partly compensate for the reduced percentage of saturation of the hemoglobin resulting from the lower partial pressure of O_2 at high altitudes. Notice, however, that the vicuña, which has hemoglobin adapted to high altitudes, shows little difference in red-corpuscle count between sea level and high altitudes.

There are other gases besides O_2 that will bind loosely to hemoglobin. One that binds even more readily than O_2 is carbon monoxide (CO). This gas, common in coal gas used for heating and cooking, in the exhaust from automobiles, and in tobacco smoke, is a dangerous poison because, even when its partial pressure in the air is relatively low, such a high percentage of the hemoglobin may bind with it that the remainder cannot carry sufficient O_2 to the tissues. Severe symptoms of asphyxiation (impairment of vision, hearing, and thought) or even death may therefore result from exposure to carbon monoxide.

[4] In isolation, adult and fetal hemoglobins have the same affinity for O_2. But in the living animal adult hemoglobin has a much greater tendency to bind a substance called 2,3-diphosphoglycerate, and as a result, a lesser tendency to bind O_2.

The role of hemoglobin in transport of carbon dioxide   The blood not only transports O_2 from the lungs to the tissues but also has the very important function of transporting CO_2 in the reverse direction, from the tissues to the lungs. As we shall see, some of this CO_2 is carried as gas dissolved in the plasma and some in loose combination with hemoglobin in the red corpuscles, but most of it is carried as bicarbonate ions in the red corpuscles and plasma.

Relatively little of the CO_2 released from the tissue cells remains in the form of dissolved gas. Instead, as we saw earlier, it tends to combine with water to form carbonic acid. This takes place particularly fast within the erythrocytes because they contain an enzyme that accelerates the reaction. But the blood must transport much CO_2, and if it were all converted into carbonic acid and transported in this form, the pH of the blood would drop considerably. The difficulty created by CO_2 transport arises because, as we saw, carbonic acid in the blood greatly increases the concentration of H^+ ions. Cells are very sensitive to pH changes and can live only within a very narrow pH range, so any major drop in pH would clearly be very harmful to the organism.

However, if the H^+ ions could somehow be combined tightly with something else, the acidity would not increase so much. This is exactly what happens in the blood—and hemoglobin and other proteins play the critical role in the process. Much of the hemoglobin (Hb) is normally present as an almost completely ionized potassium salt (K^+Hb^-). Carbonic acid reacts with the potassium hemoglobin to form acid hemoglobin (HHb) and potassium bicarbonate ($KHCO_3$), which is usually completely ionized ($K^+HCO_3^-$):

$$H^+ + HCO_3^- + K^+ + Hb^- \longrightarrow K^+HCO_3^- + HHb$$

You might well question the advantage of exchanging carbonic acid for acid hemoglobin; after all, both are acids. But there is a big difference between them: acid hemoglobin is a much weaker acid than carbonic acid. Within the pH range of blood, very little of the acid hemoglobin is ionized. The reaction that would free H^+ ions does not occur at a significant rate; instead, the back reaction prevails:

$$HHb \rightleftharpoons H^+ + Hb^-$$

In other words, the equilibrium constant of this reaction favors the formation of acid hemoglobin, which results in the removal of free H^+ ions from solution in the blood. This is an excellent example of the potential buffering action of proteins. Hemoglobin is the principal buffer substance within the erythrocytes; it even indirectly buffers the plasma and increases its bicarbonate-transporting capacity. The plasma proteins also act as buffers in the plasma. Transport of lactic acid, for example, involves buffering action by plasma proteins.

In summary, CO_2 released by the tissues and picked up by the blood reacts with water to form carbonic acid, most of which ionizes into H^+ ions and bicarbonate ions. It is in the form of bicarbonate that most of the transport of CO_2 takes place; the excess H^+ ions are bound to hemoglobin. In the lungs the situation is reversed. Here the CO_2 pressure is less than in the blood, and the gradient therefore favors release of the CO_2 by a reversal of the chemical reactions outlined above.

Chapter 14

REGULATION OF BODY FLUIDS

Many sorts of evidence have led biologists to the conclusion that life had its origin in the ancient seas. Of the major environmental media of the earth —seawater, fresh water, air—seawater exhibits by far the greatest stability. In such crucial characteristics as temperature, acidity, and salt concentration, the seas must have fluctuated remarkably little over the immense spans of time, their vast bulk ensuring a stable environment for growth and development. Moreover, the ancient seas must have shielded the fragile pioneers of life from the damaging effects of ultraviolet radiation until a protective layer of ozone (O_3) came to surround the earth.

It is not surprising, therefore, that the protoplasm of the early cells had many characteristics in common with the seawater that bathed them, and that life processes evolved a close dependence on the stable conditions existing in seawater. Nor is it surprising that the evolution of complex multicellular marine animals involved the development of body fluids—tissue fluid, blood, and the like—that could provide even the innermost body cells with a relative nonfluctuating aquatic environment, and that the internal body fluids of those primitive marine animals resembled in many important ways the seawater that had been the cradle of life.

As the ages passed and evolution continued, the body fluids of different organisms evolved in different ways, just as their other characteristics did. Present-day marine animals, for example, differ noticeably in the chemical makeup of their body fluids (Table 14.1), though these fluids are more similar to one another and to seawater than they are to the body fluids of freshwater or terrestrial organisms, not to mention those of plants. Nonetheless, all these fluids have much in common, and, as Ernest Baldwin of Cambridge University has said, "The conditions under which cell life is possible are very restricted indeed and have not changed substantially since life first began."

All living things must maintain within themselves a fluid environment favorable to the continued life of their cells. The reason was stated succinctly by the great nineteenth-century physiologist Claude Bernard: *"La fixité du milieu intérieur est la condition de la vie libre"*; or, to paraphrase: A constant internal fluid environment is the prerequisite to survival under varying external conditions. Hence the evolutionary development of immense diversity among living organisms has necessarily involved the concomitant evolution of diverse mechanisms for maintaining homeostasis in their body fluids. These mechanisms include and supplement the activities of the cell membrane which, as we have seen, are directed at maintaining chemical homeostasis within cells.

THE EXTRACELLULAR FLUIDS OF PLANTS

Multicellular marine algae differ markedly from multicellular marine animals in the sort of fluid environment to which their cells are exposed. Roughly 50 percent of the water in the body of a complex animal is extracellular, taking the form of tissue fluid, lymph, or blood plasma. This extracellular fluid, which bathes most of the cells, is separated from the environmental water by cellular barriers and has a characteristic composition which differs both from that of the intracellular fluid and from that of the surrounding water. By contrast, most of the fluid content of a multicellular alga is intracellular. The fluid filling its intercellular spaces is essentially continuous with the environmental water and cannot be regarded as

TABLE 14.1   *Concentrations of ions in seawater and in body fluids (millimoles/liter)[a]*

	Na$^+$	K$^+$	Ca^{++}	Mg^{++}	Cl$^-$
Seawater	470.2	9.9	10.2	53.6	548.3
Marine invertebrates					
Jellyfish (*Aurelia*)	454.0	10.2	9.7	51.0	554.0
Sea urchin (*Echinus*)	444.0	9.6	9.9	50.2	522.0
Lobster (*Homarus*)	472.0	10.0	15.6	6.8	470.0
Crab (*Carcinus*)	468.0	12.1	17.5	23.6	524.0
Freshwater invertebrates					
Mussel (*Anodonta*)	13.9	0.3	11.0	0.3	12.0
Crayfish (*Cambarus*)	146.0	3.9	8.1	4.3	139.0
Terrestrial animals					
Cockroach (*Periplaneta*)	161.0	7.9	4.0	5.6	144.0
Honey bee (*Apis*)	11.0	31.0	18.0	21.0	?
Japanese beetle (*Popillia*)	20.0	10.0	16.0	39.0	19.0
Chicken	154.0	6.0	5.6	2.3	122.0
Human	145.0	5.1	2.5	1.2	103.0

[a] This table is based on a larger one in C. L. Prosser and F. A. Brown, *Comparative Animal Physiology*, Saunders, 1961.

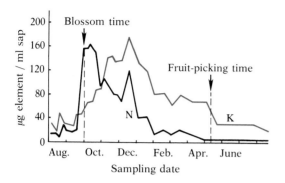

14.1   Yearly fluctuations in nitrogen and potassium concentrations in the xylem sap of apple trees in New Zealand
The fluctuations of both substances are very great—far greater than could be tolerated by animal cells.

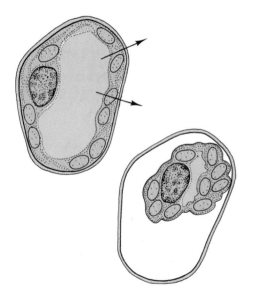

14.2   Plasmolysis
A plant cell in a hypertonic medium will lose so much water (left) that, as it shrinks, it will pull away from its more rigid wall (right).

separate or distinct. The alga thus has no fluid fully analogous to the tissue fluid and blood of an animal. Hence, unlike the animal, which must regulate the composition of both intracellular and extracellular fluids, the alga must regulate the composition of its intracellular fluids alone.

A similar contrast appears between an animal and a large vascular land plant. Such a plant obviously contains much extracellular fluid in the form of xylem sap and the water in the apoplast. But this fluid is not as fully distinct from the environmental water as the tissue fluid and blood of an animal. You will recall that water can penetrate far into the cortex of a root by flowing along cell walls without having to cross any membranous barrier. Thus, much of the fluid that directly bathes the plant cells, even those far inside the plant body, is essentially continuous with the environmental water and is therefore not fully analogous to animal tissue fluid, which is separated from the environmental medium by a membranous barrier. This means, of course, that the composition of much of the extracellular fluid of the plant cannot be so well regulated as the tissue fluid and blood of animals. Even the composition of the xylem sap, which is separated from external water by the membranous barrier of the endodermis, fluctuates widely, since it depends on such factors as the environmental conditions, the health of the plant, and the season of the year (Fig. 14.1).

It is easy to understand why marine algae need not regulate the composition of the fluid that bathes their cells; that fluid, after all, is essentially the same as seawater, the nonfluctuating medium in which life arose. (Though seawater today has a composition rather different from that of the ancient seas, the changes have come about so gradually that the organisms living in the seas have had ample time to adapt to their changing environment.) But what about the typical plant living in fresh water or on land, whose extracellular fluids fluctuate much more than the tissue fluids of animals and the seawater medium of marine algae?

Why the cells of terrestrial and freshwater plants are able to withstand great fluctuations in the composition of the fluids bathing them is not yet well understood, but a partial explanation can be given. Unlike the typical animal cell, which thrives in fluids with an osmotic concentration like its own, the cell of a land or freshwater plant almost always exists in a medium that is much more dilute than the cell's contents. In other words, the plant cell is decidedly hypertonic relative to the fluid that bathes it. In such a situation an animal cell would take in so much water by osmosis that it would burst, unless it had some special mechanism for expelling the excess water. But the plant cell is surrounded by its cell wall, and as the cell takes in more water and becomes turgid, the wall resists further expansion. Eventually the resistance of the wall is as great as the tendency of water to enter the cell by osmosis, and no further net gain of water by the cell is possible.

We see, then, that the plant cell can withstand rather pronounced changes in the osmotic concentration of the surrounding fluids as long as those fluids remain more dilute than the cell's contents—that is, as long as the fluids remain appreciably hypotonic relative to the cell. If the external fluids become decidedly hypertonic relative to the cell, the cell may lose so much water and shrink so grievously that it pulls away from its more rigid wall; such a cell is said to be plasmolyzed, and the phenomenon is called *plasmolysis* (Fig. 14.2). The presence of the cell wall in plants and its absence in animals thus make the problem of salt and water balance quite different in each.

To say that plants can tolerate much greater changes in the osmotic concentration of the fluids that bathe their cells is not to say that they are unaffected by changes in the concentration of individual ions in the surrounding medium. Such changes sometimes have pronounced effects on their health and growth. But the effects are usually attributable to an alteration in the chemical makeup of the plant, rather than to a disruption of its osmotic balance.

Land plants, like land animals, are often exposed to conditions that may cause excessive water loss by evaporation. We have already examined some of the adaptations whereby plants resist desiccation—cuticle on their exposed surfaces, regulation of stomatal openings by guard cells, stomata in deep hair-lined pits in some plants living in very dry regions, and so on. Some plants show little tolerance of drought; the cells lose their turgidity and the plant wilts and eventually dies when the soil moisture becomes deficient. Many plants that grow only in the shade are in this category. Some plants, including many mosses, lichens, and ferns, are drought-tolerant because their cells can be dehydrated without permanent injury. Other plants can survive through long periods of drought because they store very large quantities of water, and because they lose little by evaporation, thanks to very thick cuticles, few stomata, and low surface-to-volume ratio: cacti and other succulent desert plants are good examples. Other plants are only moderately well equipped to withstand drought; with a limited ability to endure dehydration, they frequently combine some adaptations for preventing water loss with large deep-penetrating root systems that increase absorptive capacity. Study of adaptations that enable plants to withstand drought is an interesting field in its own right.

THE VERTEBRATE LIVER

As a first example of the problems attendant on keeping the internal fluids of a complex vertebrate animal, such as a human, relatively constant in composition, consider the blood leaving the intestinal capillaries shortly after a meal. Digestion is taking place in the small intestine, and the products of digestion are moving in large quantities into the capillaries of the intestinal villi. This means that the blood leaving these capillaries contains high concentrations of such compounds as simple sugars and amino acids —concentrations considerably greater than those normally found in the blood in most parts of the circulatory system. The wholesale addition of these materials to the blood, if not controlled, would drastically alter the composition of the blood and other body fluids and make impossible the maintenance of a relatively nonfluctuating fluid environment for the cells.

Vertebrates overcome this difficulty with the help of a very important homeostatic organ, the liver. Blood from the intestine and stomach is collected in the **portal vein**, which does not empty into the vena cava, as might be expected, but goes to the liver, where it breaks up into a network of capillaries (or, more precisely, sinuses) in the liver tissue (Fig. 14.3). The liver is one of only three places in the mammalian body where blood passes

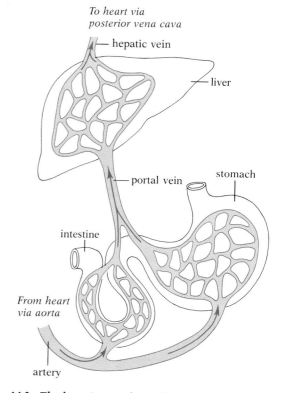

14.3 The hepatic portal circulation
Blood from the capillaries of the digestive tract and stomach is carried by the portal vein to a second bed of capillaries in the liver. It then flows via the hepatic vein into the posterior vena cava, which takes it to the heart.

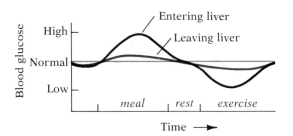

14.4 Regulation of blood sugar by the vertebrate liver
The liver acts to maintain constant levels of many substances in the blood. In this example, the liver removes and stores excess glucose carried by the blood from the intestine after a meal (left), but later liberates glucose into the blood during exercise, when muscles begin using up this source of energy at an unusually high rate (right).

through a second set of capillaries before returning to the heart;[1] other blood circuits involve only a single capillary bed.

The liver's role in regulation of the blood-sugar level   After a meal the blood coming to the liver via the portal vein has a higher than normal concentration of glucose (Fig. 14.4, left). Under these conditions the liver removes most of the excess, converting it into the insoluble polysaccharide glycogen, which is the principal storage form of carbohydrate in animal cells. Therefore, the blood leaving the liver via the ***hepatic vein*** (a vein that leads into the posterior vena cava) contains a concentration of glucose only slightly higher than that normally found in the arteries. After this blood is mixed in the vena cava with blood from other parts of the body—blood that contains a lower concentration of sugar because it has given up glucose to the tissues through which it has passed—the blood entering the heart has a glucose concentration within the normal tolerance range.

If the incoming supply of glucose from the intestine exceeds all the body's immediate needs, and the liver has stored its full capacity of glycogen, the liver begins converting glucose into fat, which can then be stored in the various regions of adipose tissue throughout the body. Thus, in spite of the great quantities of glucose absorbed by the intestine, the blood-sugar level in most of the circulatory system is not greatly raised and homeostasis is maintained.

The whole process is reversed when exercise depletes the supply of glucose in the blood. At such times, blood in the muscle capillaries gives up glucose to the muscle cells. This means that the blood reaching the liver via the portal vein is poor in glucose. Under these conditions the liver converts some of its stored glycogen into glucose and adds it to the blood. The result is that blood leaving the liver in the hepatic vein has its normal glucose concentration (Fig. 14.4, right), and the blood entering the heart, a mixture of blood from the liver and venous blood from other parts of the body, has a nearly normal blood-sugar level. We see, then, that the liver functions in helping maintain a homeostatic blood-sugar concentration: it converts a highly variable input into a virtually constant output by actively adding and subtracting substances from the blood.

The human liver is capable of storing enough glycogen to supply glucose to the blood for a period of about 4 hours. What happens if at the end of that period no new glucose has come to the liver from the intestine? A drop in the blood-sugar concentration to a level much below normal would soon be fatal; the brain cells are particularly sensitive to such a drop, because they cannot store adequate amounts of glucose themselves or use fats or amino acids as energy sources, and are thus wholly dependent on a regular supply of glucose from the blood. Under such conditions the liver begins converting other substances, such as amino acids, into glucose, and in this way maintains the normal blood-sugar level. Some other tissues of the body, particularly muscle, can store glucose as glycogen, but muscle glycogen apparently serves as a local fuel deposit only and is not generally available for maintaining the blood-glucose level.

The liver's activity in carbohydrate metabolism is regulated in a complex

[1]The others, to be discussed later, are the kidney and the pituitary gland.

fashion by several hormones, as will be described in Chapter 16. An abnormal balance of these hormones may result in an unusually high blood-sugar level or an unusually low one. Either condition can be dangerous.

The liver's role in the metabolism of lipids and amino acids   The liver is not only responsible for carbohydrate conversions essential to the maintenance of a relatively constant blood-sugar level, it is also important in the processing and modification of fatty acids and other lipid materials. For example, much of the fatty acid mobilized from adipose tissues during periods of starvation is incorporated into lipoprotein by the liver before it is used as a source of metabolic energy by other tissues of the body.

The liver's role in amino acid metabolism is more complicated. Like glucose, the amino acids absorbed by the villi of the intestine pass into the portal vein and thence to the liver. The liver removes many of these amino acids from the blood, temporarily storing small quantities and later gradually returning them to the blood, which carries them to other tissues for use in the synthesis of enzymes, hormones, or new protoplasm. But the usual diet contains far more amino acid than can be used in such syntheses. Unlike the plant, the animal body is capable of very little long-term storage of amino acids, proteins, or other nitrogenous compounds; those used as an energy source when supplies of carbohydrates and fats are exhausted are not stored products, but the actual structural material of living cells. Hence excess amino acids from the diet must be converted into other substances, such as glucose, glycogen, or fat. Such conversions take place in the liver.

You will recall that amino acids differ from most carbohydrates and fats in containing nitrogen, in the form of an amino group ($-NH_2$). It is not surprising, therefore, that the first step in converting amino acids into these other substances is *deamination*, or the removal of the amino group. In the deamination reaction the amino group is converted into *ammonia* (NH_3).[2] In some animals the liver simply releases this ammonia, which is a waste material, into the blood, and it is soon removed from the blood and from the body by excretory mechanisms. In many other animals, including humans, the liver first combines the ammonia with carbon dioxide to form a more complex but less toxic nitrogenous compound, called *urea* (Fig. 14.5), and then releases the urea into the blood. In still other animals, the liver converts the waste ammonia into a compound more complex than urea, called *uric acid*, which it releases into the blood. In short, whether the nitrogenous waste product is ammonia, urea, uric acid, or some other compound, the liver dumps it into the blood, and another regulatory system of the body acts to prevent the wastes from reaching too high a concentration in the body fluids.

Notice that animals differ markedly from green plants in being unable to reuse much of the nitrogen from the amino acids they metabolize. Green plants can shift nitrogen-containing groups from one organic compound to another more freely than can animals, and they can also use inorganic nitrogen to synthesize organic nitrogen-containing compounds. Hence excretion of nitrogenous wastes is essentially an animal activity.

[2] At physiological pH, most of the ammonia is rapidly converted into the ammonium ion (NH_4^+).

Ammonia   Urea   Uric acid

14.5   Three important nitrogenous waste compounds
Because the ammonia produced as a result of deamination of amino acids can be toxic in relatively low concentrations and requires considerable water for its elimination, some animals combine it with CO_2 to produce the less toxic urea. Other animals go a step further by converting ammonia into the insoluble and nontoxic uric acid. Though the conversion of ammonia to uric acid requires a considerably greater expenditure of ATP than conversion to urea, the greater energy investment may be worthwhile for certain land animals, since less water is required for uric acid elimination.

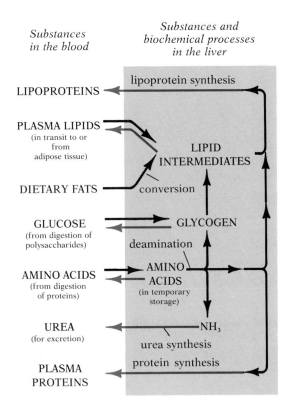

*Substances
in the blood*

*Substances and
biochemical processes
in the liver*

lipoprotein synthesis

LIPOPROTEINS

PLASMA LIPIDS
(in transit to or
from
adipose tissue)

LIPID
INTERMEDIATES

DIETARY FATS

conversion

GLUCOSE
(from digestion of
polysaccharides)

GLYCOGEN

deamination

AMINO ACIDS
(from digestion
of proteins)

AMINO
ACIDS
(in temporary
storage)

UREA
(for excretion)

NH₃

urea synthesis

protein synthesis

PLASMA
PROTEINS

14.6 Chemical pathways in the human liver
The liver removes excess sugar, fat, and amino acids
(among other things) from the blood when their
concentration is too high, and releases them into the
blood when their circulating levels fall too low. The
liver also synthesizes lipoproteins, plasma proteins,
and plasma lipids. Finally, the liver detoxifies many
substances, and produces urea. Some of the chemical
pathways involved are diagrammed here.

A summary of the liver's functions   Over and over, as we have examined
the physiology of vertebrate animals, we have mentioned some important
role played by the liver. Before we leave the present discussion of this vital
and versatile organ, let us compile a list of its functions. Though far from
complete, the list will convey how extensive and varied a part the liver plays
in the maintenance of life.

1. The liver removes excess glucose from the blood and stores it as
   glycogen, and reconverts glycogen into glucose to maintain the
   blood-sugar level when the incoming supply is insufficient (Fig.
   14.4).
2. It resynthesizes glycogen from some of the lactic acid produced by
   muscles during glycolysis.
3. It plays a major role in the interconversion of various nutrients,
   such as the conversion of carbohydrates into fats, of incoming fats
   into fats more typical of the organism's own body, of amino acids
   into carbohydrates or fats.
4. It deaminates amino acids, converts the ammonia thus obtained
   into urea, uric acid, or some other compound, and releases the nitrogenous wastes into the blood.
5. It detoxifies a great variety of injurious chemical compounds and is
   thus one of the body's most important defenses against poisons
   such as alcohol and barbiturates.
6. It manufactures many of the plasma proteins, including fibrinogen, prothrombin, albumin, and some globulin.
7. It manufactures some plasma lipids, including cholesterol, and
   plays an important role in processing and modifying lipids mobilized from adipose tissues.
8. It stores various important substances such as vitamins and iron.
9. It forms erythrocytes in the embryo.
10. It destroys red blood corpuscles.
11. It excretes bile pigments.
12. It synthesizes bile salts.

A few of these functions are summarized in Figure 14.6.

THE PROBLEM OF EXCRETION AND
SALT AND WATER BALANCE IN ANIMALS

We have seen that animals need mechanisms for ridding their bodies of
metabolic wastes—particularly nitrogenous ones, but many others as well.
The process of releasing such useless, even poisonous, substances is called
excretion. It should not be confused with elimination (defecation).
Whereas excretion is release of wastes that have been inside the cells, tissue
fluids, or blood of the organism, elimination is the release of unabsorbed
wastes by the digestive tract.

In general, excretory mechanisms also serve a second very important
function: they help regulate the water and salt balance of the organism. Our
examination of excretion will focus on both these aspects, which are in
most instances inextricably intertwined.

THE PROBLEM FOR AQUATIC ANIMALS

As we saw, the first nitrogenous waste formed by deamination of amino acids is ammonia. Since ammonia is an exceedingly poisonous compound, no organism can survive if its concentration in the body fluids gets very high. But the small, highly soluble molecules of ammonia readily diffuse across cell membranes, and there is no great difficulty in getting rid of them if an adequate supply of water is available. The water keeps the solution dilute while the ammonia is in the body, acts as a vehicle for its expulsion from the body, and flushes it rapidly away from the vicinity of the animal. In view of the plentiful supply of water available to aquatic animals, it is not surprising that for many of these the characteristic nitrogenous excretory product is ammonia.

Marine invertebrates   Many marine invertebrates lack special excretory systems, relying instead on release of wastes across the general surface membranes. Such organisms seldom have any problem with water balance, because they are essentially isotonic with the surrounding seawater, and hence neither take in much excess water nor lose too much. Some of these organisms supplement the excretory process by phagocytic excretion: certain cells pick up solid particles of waste material by phagocytosis and then move to the outer body surface or to the surface of the digestive cavity, where the particles are released.

Maintenance of the proper nonfluctuating internal fluid environment is relatively simple for marine invertebrates as long as they remain in the sea; it is quite a different matter when they move into a hypotonic environment such as the brackish water of estuaries or the fresh waters of rivers and lakes. Many marine animals are incapable of moving into such habitats, because their body fluids always lose salts until they have about the same salinity and osmotic concentration as their environment. Since their cells generally cannot tolerate much change in the makeup of the fluids bathing them, these animals soon die when they are put into brackish or fresh water. An example is the spider crab (*Maia*) (Fig. 14.7).

Some marine animals, however, have evolved adaptations that enable them to move into hypotonic media. The adaptations may be of an evasive character, as in oysters and clams, which simply close their shells and exclude the external water during those parts of the tidal cycle when the water in the estuaries is very dilute. But by far the most important adaptations for survival in dilute media—and the ones that have played the principal role in the evolutionary movement of animals into fresh water—are those that enable animals to regulate the osmotic concentrations of their body fluids and keep them constant despite fluctuations in the external medium. Such organisms are said to have the power of ***osmoregulation***.

The shore crab (*Carcinus*) is an example of a marine invertebrate that has evolved a degree of osmoregulation enabling it to live in both seawater and brackish water (Fig. 14.7). In seawater the crab's body fluids are in osmotic equilibrium, but in brackish water they are hypertonic relative to the surrounding medium. To maintain the internal fluids near their normal concentration in brackish water, cells on the crab's gills remove salt from

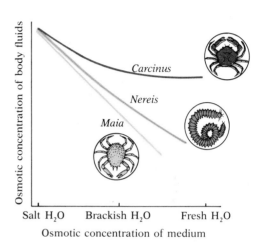

14.7   Variation of internal osmotic concentration with external osmotic concentration in three marine invertebrates

The spider crab (*Maia*) has no osmoregulatory capacity, and the concentration of its body fluids falls in direct proportion to the fall in the concentration of the external medium. The clam worm (*Nereis*) has very slight osmoregulatory capacity; the concentration of its body fluids does not bear a straight-line relationship to the concentration of the external medium. The shore crab (*Carcinus*) has considerable osmoregulatory capability and can maintain relatively concentrated body fluids even in a very dilute external medium.

TABLE 14.2   *Concentrations of ions in the blood of freshwater animals compared with seawater and fresh water (millimoles/liter)*[a]

	Na^+	K^+	Ca^{++}	Mg^{++}	Cl^-
Seawater	470.2	9.9	10.2	53.6	548.3
Brown trout (*Salmo*)	144.0	6.0	5.3	?	151.0
Crayfish (*Cambarus*)	146.0	3.9	8.1	4.3	139.0
Mussel (*Anodonta*)	13.9	0.3	11.0	0.3	12.0
Fresh water[b]	0.65	0.01	2.0	0.2	0.5

[a] This table is based on a larger one in C. L. Prosser and F. A. Brown, *Comparative Animal Physiology*, Saunders, 1961.

[b] The values for fresh water are representative only; actual values vary greatly, as the water ranges from "soft" (low concentrations of dissolved minerals, e.g., 0.22 Ca^{++}) to "hard" (high concentrations of dissolved minerals, e.g., 5.0 Ca^{++}).

the surrounding water and actively secrete it into the blood, while the excretory organs eliminate the excess water that constantly pours in.

Freshwater animals   Once the ancestors of the modern freshwater animals had made the transition to the freshwater environment, presumably by way of the estuaries, there was no longer any great advantage to their descendants in continuing to maintain body fluids as concentrated as seawater, as long as they remained in their new environment. Such excessively hypertonic internal conditions simply aggravated the problems of obtaining enough salt and bailing out excess water. Hence it is understandable that natural selection should have favored a reduction of the osmotic concentration of the body fluids within the bounds possible for the continuance of the life of the tissues, and that modern freshwater animals, both invertebrate and vertebrate, should have osmotic concentrations decidedly lower than seawater (Table 14.2). It seems incompatible with cellular existence, however, for the body fluids to be as dilute as fresh water, for no organisms are actually isotonic with their freshwater medium; the body fluids of freshwater animals are typically hypotonic relative to seawater, but hypertonic relative to fresh water (Table 14.2).

Now, if freshwater animals are hypertonic relative to the surrounding environmental medium, there will be a strong tendency for water to move into the organism and for salts to be lost from the organism to the surrounding water. At first glance, the obvious evolutionary solution to this problem might seem to be the development of completely impermeable membranes covering the entire body, but further thought shows that this solution would have been impracticable, since a truly aquatic organism must maintain some permeable membranes exposed to the water for gas exchange. Because mammals that live in the water breathe air and hence need never expose permeable respiratory membranes to the water, they can maintain an impermeable barrier between their body fluids and the water in which they live. But fully aquatic freshwater animals cannot use the method of "evasion" exclusively. They must also be able to carry out active osmoregulation, which usually involves excretory organs that can pump out the water as fast as it floods in—preferably through the production of urine more dilute than the body fluids—and/or special secretory cells somewhere on the body that can absorb salts from the environment and release them into the blood. Both corrective measures—production of dilute urine and absorption of salts—entail movement of materials against concentration gradients and therefore necessitate the expenditure of energy.

An examination of the water and salt regulation typical of modern freshwater bony fish provides a good example of the above-mentioned processes. The blood and tissue fluids of these fish are more concentrated than the environmental water (Table 14.2). The method of evasion is used to the extent that much of the body is covered by relatively impermeable skin and scales, and that the fish almost never drink. There is, however, a constant osmotic intake of water across the membranes of the gills and mouth, and a constant loss of salts across the same membranes. Correction of the resulting imbalance occurs in two ways: the excess water is eliminated in the form of very dilute and copious urine produced by the kidneys, and salts are actively absorbed by specialized cells in the gills (Fig. 14.8).

Marine vertebrates   Curiously enough, bony fish living in the sea have the reverse problem: they live in water, yet they steadily lose water to their environment and are in constant danger of dehydration. The explanation is that the ancestors of the bony fish apparently lived in fresh water, not in the sea, and when some of their descendants moved to the marine environment they retained their dilute body fluids. Marine bony fish are therefore hypotonic relative to the surrounding water, and have the problem of excessive water loss and excessive salt intake. Besides benefiting from the evasive adaptation of relatively impermeable skin and scales, they use two corrective measures: they drink almost continuously to replace the water they are constantly losing, and, by means of specialized cells in the gills, they actively excrete the salts they unavoidably take in with the water (Fig. 14.8). Most of the nitrogenous wastes are excreted as ammonia through the gills; hence only a small quantity of urine is produced by the kidneys, and little water need be lost in this manner. Apparently the kidneys of fish have not evolved the capacity to produce urine more concentrated than their blood, and are consequently of no help in salt elimination.

The marine elasmobranch fish (sharks and their relatives) probably also evolved from freshwater ancestors, but they solved the osmotic problem in a very different way. Their blood contains about the same concentrations of salt as the blood of marine bony fishes, but their blood also contains high concentrations of urea, to which they are much more tolerant than most vertebrates. By conserving urea instead of excreting it, the marine elasmobranchs maintain a total osmotic concentration in their blood slightly greater than that of seawater. They therefore have no problem of water loss. Excess salt is excreted by special glandular cells in the rectum.

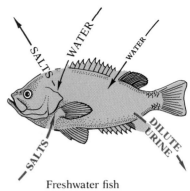

Freshwater fish
(hypertonic relative to medium)

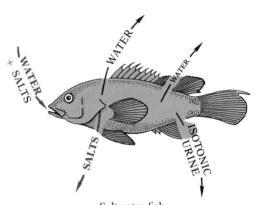

Saltwater fish
(hypotonic relative to medium)

14.8   Osmoregulation in bony fish
Freshwater fish tend to take in excessive amounts of water and to lose too much salt. They compensate by seldom drinking, by actively absorbing salts through specialized cells in their gills, and by excreting copious dilute urine. Saltwater fish tend to lose too much water and to take in too much salt. They compensate by drinking constantly and by actively excreting salts across their gills. They cannot produce hypertonic urine; hence the kidneys are of little aid to them in osmoregulation.

THE PROBLEM FOR TERRESTRIAL ANIMALS

We have already seen that one of the conditions of animal life in fresh water is a relatively impermeable covering for all but certain portions of the body surface, as an aid in preventing excessive absorption of water. For this reason freshwater animals had a important preadaptive advantage over primitive marine animals in colonizing the terrestrial environment. The evidence strongly supports the view that the movement to land was by way of fresh water, not directly from the sea.

On land the greatest threat to life is desiccation. Water is lost by evaporation from the respiratory surfaces (lungs, tracheae, etc.), by evaporation from the general body surface, by elimination in the feces, and by excretion in the urine. The lost water must obviously be replaced if life is to continue. It is replaced by drinking, by eating foods containing water, and by the oxidation of nutrients (remember that water is one of the products of cellular respiration).

We saw that ammonia is a satisfactory nitrogenous excretory product for aquatic animals. It is far from satisfactory for terrestrial ones, because of the difficulty of getting rid of this highly toxic substance on land, where an unlimited water supply is not available. Amphibians and mammals, therefore, rapidly convert ammonia to urea, a compound that, though very soluble, is relatively nontoxic. Urea can remain in the body for some time before being excreted, and we can regard its production as an adaptation to the conditions of water shortage characteristic of terrestrial existence.

Though urea is a far more satisfactory excretory product than ammonia for land animals, it has the disadvantage of draining away some of the critically needed water, for, being highly soluble, it must be released in an aqueous solution. If, however, uric acid, a very insoluble compound, is excreted instead of urea, almost no water need be lost. It is not surprising, therefore, that many terrestrial animals—most reptiles, birds, insects, and land snails—excrete uric acid or its salts. The excretion of this substance not only allows them to conserve water, but has another advantage, which may have been even more important in the evolution of uric acid metabolism. All these animals lay eggs enclosed within a relatively impermeable shell or membrane. If the embryos excreted ammonia, they would rapidly be poisoned, and if they produced urea, the concentration in the egg by the latter part of development would become decidedly harmful. Uric acid, on the other hand, is so insoluble that it can be precipitated in almost solid form and stored in the egg without exerting harmful toxic or osmotic effects. In the nitrogen metabolism of fully terrestrial animals, uric acid excretion is correlated with egg laying, while urea excretion is correlated with viviparity (giving birth to living young).

EXCRETORY MECHANISMS IN ANIMALS

CONTRACTILE VACUOLES

Many unicellular and simple multicellular animals have no special excretory structures. Nitrogenous wastes are simply excreted across the general cell membranes into the surrounding water. Some Protozoa do, however, have a special excretory organelle, the contractile vacuole. Each vacuole goes through a regular cycle consisting of a stage in which it fills with liquid and becomes larger and larger, followed by a contraction stage in which the contents of the vacuole are ejected from the cell. Though there is now evidence that contractile vacuoles excrete some nitrogenous wastes, it seems clear that their primary function is elimination of excess water. As might be predicted, they are much more common in freshwater Protozoa than in marine forms, and their rate of fluid elimination becomes slower as the osmotic concentration of the environmental medium increases (Fig. 14.9).

If the principal function of contractile vacuoles is expulsion of water from the cell, then the fluid in the vacuoles should have a lower osmotic concentration than the cytosol. That it actually does, during all stages of the cycle, has been confirmed by examination of samples of the fluid withdrawn from the vacuoles with micropipettes. But how can the vacuole take in fluid when, according to the rules of osmosis, water should move out of it, not in? Since water itself cannot be moved by active transport, there must be some other explanation.

In most Protozoa the vacuole is surrounded by a layer of tiny vesicles, and these, in turn, are surrounded by a layer of mitochondria (Fig. 14.10). The vesicles initially contain a fluid isotonic with the cytosol, but later actively pump out ions, using energy from ATP manufactured in the mitochondria. Eventually, when the osmotic concentration of the vesicular fluid has fallen to about one-third that of the cytosol, the vesicles move to the contractile vacuole and fuse with it. The contractile vacuole grows larger as more and more vesicles merge with it and empty their fluid into it. The membrane of the vacuole itself must be nearly impermeable to water, which is held in the vacuole until sudden contraction expels it from the cell.

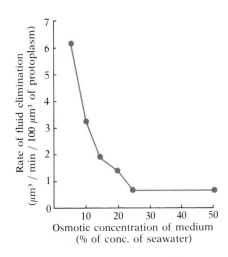

14.9 Rate of fluid elimination by contractile vacuole of *Amoeba lacerata* as a function of the osmotic concentration of the medium
Contractile-vacuole activity falls precipitously as the concentration of the medium goes up.

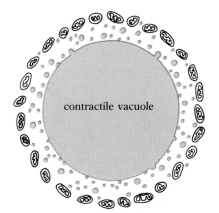

14.10   Contractile vacuole of *Amoeba proteus*
The numerous tiny vesicles around the vacuole fill with fluid; after most of the ions have been pumped out, the vesicles fuse with the vacuole and empty their contents into it. The layer of mitochondria just outside the vesicle layer presumably provides the ATP necessary to pump the ions out of the vesicles and then to expel the contents of the vacuole from the cell.

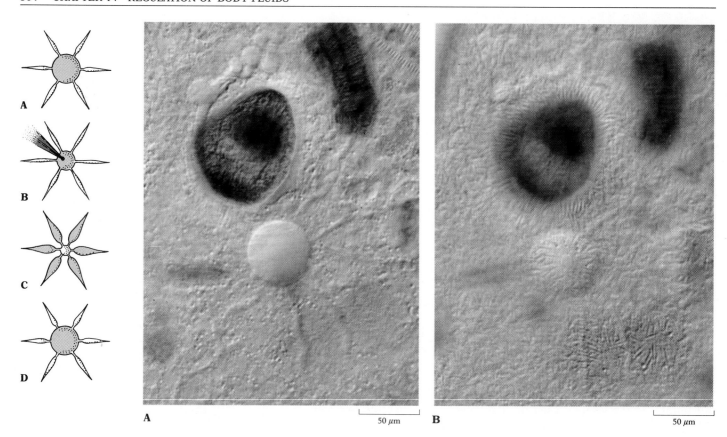

A   50 μm   B   50 μm

14.11 Contractile vacuole of *Paramecium caudatum*
The sequence shown diagrammatically above can
also be observed in the three photographs of a live
organism taken in the course of the expansion-
contraction cycle of its contractile vacuoles (only
one vacuole is seen here). The large green object and
the brown ones are remains of other microorganisms
ingested by the *Paramecium*. (A) The vacuole is full.
As shown in the diagram, a system of radiating canals
brings fluid from the cytoplasm to the vacuole.
(B) The vacuole is in the process of expelling its
contents; in the photograph the opening to the
outside can be seen as a small circle on the surface
of the vacuole. (C) The vacuole is nearly empty, but
the radiating canals are collecting more fluid from
the cytoplasm and fill the reservoir again (D).
Photographs taken by the Nomarski process.

In a few ciliate Protozoa fluid enters the contractile vacuole from radially oriented feeder canals rather than from vesicles (Fig. 14.11). The
feeder canals, in turn, appear to collect fluid from a network of tiny tubules
probably derived from the endoplasmic reticulum. Despite these anatomical differences, the mechanism of production of dilute fluid is probably
similar to the one already described, in which ions are pumped out before
the fluid reaches the contractile vacuole itself.

FLAME-CELL SYSTEMS

The beginnings of a tubular excretory system can be seen in the flatworms
(planaria, flukes, tapeworms, and the like). These animals are relatively
small and lack a functional body cavity—that is, there is no major break in
the tissue mass between the outer epithelium of the body and the gastrovascular cavity. They do not have a circulatory system.

Flatworm excretory systems usually consist of two or more longitudinal
branching tubules running the length of the body (Fig. 14.12). In planaria
and its relatives, the tubules open to the body surface through a number of
tiny pores. In some other flatworms, such as the flukes, the tubules unite to
form an enlarged bladder that opens to the outside. The critical portions of
the systems are many small bulblike structures located at the ends of side

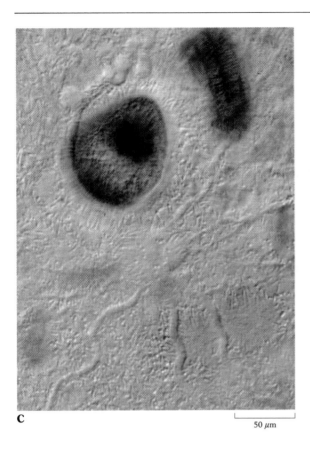

C

50 μm

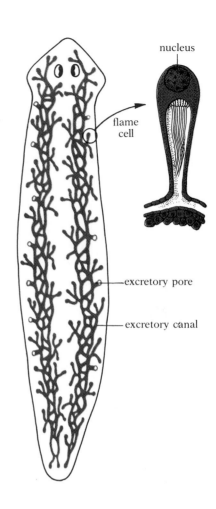

branches of the tubules. Each bulb has a hollow center into which a tuft of long cilia projects. The hollow centers of the bulbs are continuous with the cavities of the tubules. Water and some waste materials move from the tissue fluids into the bulbs. The constant undulating movement of the cilia creates a current that moves the collected liquid through the tubules to the excretory pores, where it leaves the body. The motion of the tuft of cilia resembles the flickering of a flame, and for this reason this type of excretory system is often called a flame-cell system.

Like the contractile vacuoles discussed earlier, flame-cell systems seem to function primarily in the regulation of water balance; most metabolic wastes of flatworms are excreted from the tissues into the gastrovascular cavity and eliminated from the body through the mouth.

14.12 Flame-cell system of planaria
Each of the two excretory canals consists of a longitudinal network of tubules, some ending in flame cells (one is shown enlarged at right) and others in excretory pores. The cilia in the flame cells create currents that move water and waste materials through the canals and out through the pores.

NEPHRIDIA OF EARTHWORMS

Flame-cell systems, because they function in animals without a circulatory system, pick up substances only from the tissue fluids. In animals that have evolved a closed circulatory system, the blood vessels have become intimately associated with the excretory organs, making possible direct exchange of materials between the blood and the excretory system.

The critical role of the circulatory system in excretion can be observed in

14.13 Nephridia of an earthworm

Each segment of the worm's body contains a pair of nephridia, one on each side. The open nephrostome of each nephridium is located in the segment ahead of the one containing the rest of the nephridium. The tubule from the nephrostome penetrates through the membranous partition between the segments and is then thrown into a series of coils, with which a network of blood capillaries is closely associated (the capillaries are shown here only on the nephridia of one segment). The coiled tubule empties into a storage bladder that opens to the outside through a nephridiopore. (The bladder and nephridiopore can be seen in the photograph of a cross section of an earthworm, Figure 10.13, p. 255).

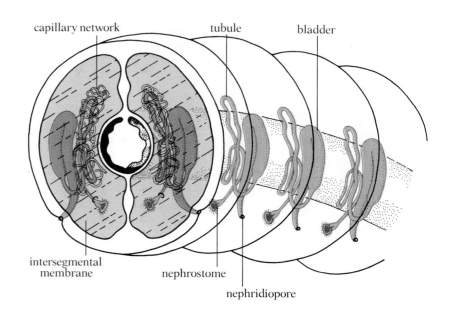

14.14 Malpighian tubules of an insect

These excretory organs arise as diverticula of the digestive system at the junction between the midgut and the hindgut.

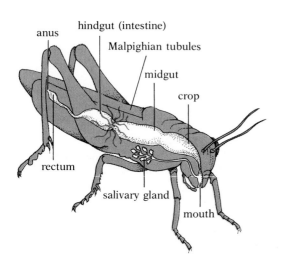

the earthworm. The earthworm's body is composed of a series of segments internally partitioned from each other by membranes. In general, each of the compartments thus formed has its own pair of excretory organs, called the nephridia, which open independently to the outside. A typical nephridium (Fig. 14.13) consists of an open ciliated funnel, or nephrostome (which corresponds functionally to the bulb of a flame-cell system), a coiled tubule running from the nephrostome, an enlarged bladder into which the tubule empties, and a nephridiopore through which materials are expelled from the bladder to the outside. Blood capillaries form a network around the coiled tubule. Materials move from the body fluids into the nephridium through the open nephrostome, but some materials are also picked up by the coiled tubule directly from the blood in the capillaries. There is probably also some reabsorption of materials from the tubule into the blood capillaries. The principal advance of this type of excretory system over the flame cell, then, is the association of blood vessels with the coiled tubule.

MALPIGHIAN TUBULES

Insects probably evolved from an ancestral form similar to the ancestor of the modern segmented worms, and this ancestor, like the earthworm, probably had nephridia. Yet insects do not have nephridia, nor have their excretory organs evolved from nephridia. The evolution of an open circulatory system in insects, with their consequent lack of blood capillaries, probably accounts for the evolutionary loss of nephridia, which are dependent on capillaries. Instead, insects and many of their relatives have evolved an entirely new excretory system, one that functions well in association with an open circulatory system.

The excretory organs of insects are called Malpighian tubules. They are diverticula of the digestive tract located at the junction between the midgut and the hindgut (Fig. 14.14). These blind sacs, variable in number, are bathed directly by the blood in the open sinuses of the animal's body. Fluid is absorbed from the blood into the blind distal end of the Malpighian tubules. As the fluid moves through the proximal portion of the tubules, the nitrogenous material is precipitated as uric acid and much of the water and various salts are reabsorbed. The concentrated, but still fluid, urine next passes into the hindgut and then into the rectum. The rectum has very powerful water-reabsorptive capacities, and the urine and feces leave the rectum as very dry material.

The highly effective role played in water conservation by the insect rectum is similar to that of the cloaca, a chamber in birds and some other vertebrates through which the urine is placed, in effect, in the posterior portion of the digestive tract. There it is subjected to the powerful reabsorption action of the rectum; the uric acid is therefore eliminated as a nearly dry powder or hard mass.

THE VERTEBRATE KIDNEY

Structure of the kidney   Like the nephridial system of earthworms, the excretory systems of vertebrates are closely associated with the closed circulatory system. When an efficient circulatory system can bring wastes to the excretory organs, the functional excretory units no longer have to be scattered throughout the body tissues, as in planaria. And the absence of internal segmentation of the body obviates the need for a series of individual excretory organs, as in earthworms. Higher vertebrates have typically evolved compact discrete organs, the kidneys, in which the functional units are massed. In humans the kidneys are located in the back of the abdominal cavity.

The functional units of the kidneys of higher vertebrates are called *nephrons*. Each nephron consists of a fully invaginated bulb called a *Bowman's capsule*, or renal capsule, and a fairly long coiled tubule consisting of three sections: the proximal convoluted tubule, the loop of Henle, and the distal convoluted tubule. The tubules of the various nephrons empty into collecting tubules, which in turn empty into the central cavity of the kidney, the pelvis. From the pelvis, a large duct leaves each kidney and runs posteriorly. As we have noted, in some animals—frogs and birds, for example— these ducts empty into the *cloaca*, a common chamber through which pass materials from the digestive, excretory, and reproductive systems. In mammals, which have no cloaca, the ducts, called *ureters*, empty into the *urinary bladder*. This storage organ drains to the outside via another duct, the *urethra* (Fig. 14.15).

Blood capillaries and the capsules and tubules of the nephrons are intimately associated in the modern vertebrate kidney. No longer are materials picked up from the general body fluids; exchange of substances takes place almost exclusively between blood capillaries and nephrons. Blood reaches

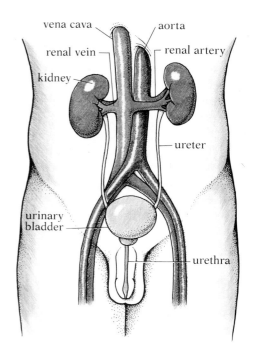

14.15   The human excretory system
The organs and vessels are shown larger relative to the body than they actually are.

THE KIDNEYS AS A FACTOR
IN HIGH BLOOD PRESSURE

It has long been known that high blood pressure, or hypertension, is commonly associated with kidney malfunction. But it was not until 1934 that H. Goldblatt and his associates at Western Reserve University showed that constriction of the renal arteries consistently causes pronounced and permanent hypertension. They fastened tiny adjustable clamps on the renal arteries of dogs which enabled them to reduce the blood flow to the kidneys by any desired amount. The dogs consistently developed hypertension to a degree correlated with the restriction in blood flow.

The result of restricting renal blood flow is oxygen deficiency in the kidney. It was later shown that the kidney responds to the deficiency by secreting an enzyme (renin), which reacts in the blood with a protein secreted by the liver to form the vasoconstrictor **angiotensin**. Angiotensin stimulates the smooth muscles in the walls of small blood vessels to contract. The resulting constriction of the vessels causes the blood pressure to rise, both because the constricted vessels offer more resistance to flow and because the heart compensates for the lessened flow by increased output. The higher blood pressure can force more blood through the partly blocked renal arteries into the glomeruli of the kidneys.

Thus the kidneys have a way of compensating for the reduced blood flow caused by constrictions or other obstructions in their arteries, but that very compensation may prove to be a critical causal element in the onset of hypertension, a very common and dangerous pathological condition in our society.

each kidney via a **renal artery,** a short vessel leading directly from the aorta to the kidney (Fig. 14.16). The renal artery enters the kidney at its median depression and then breaks up into many smaller branches that run through the inner portion of the kidney (the medulla) into the outer kidney layer (the cortex), where each of the many tiny branch arterioles penetrates into a cuplike depression in the wall of a Bowman's capsule. Within each capsule, the arteriole breaks up into a tuft of capillaries called the **glomerulus** (Fig. 14.17). Blood leaves the glomerulus via an arteriole formed by

14.16 Sections of the human kidney
(A) The blood circulation of the kidney. (B) The cortex and the medulla, and the large renal pelvis into which the collecting tubules of the nephrons empty. One nephron is shown (color); note that the glomerulus and convoluted tubules are in the cortex, but the loop of Henle runs down into the medulla.

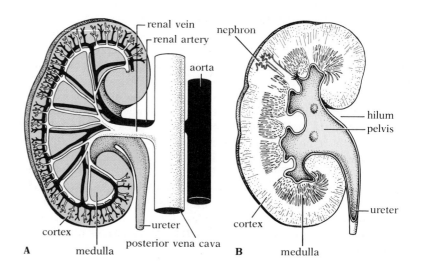

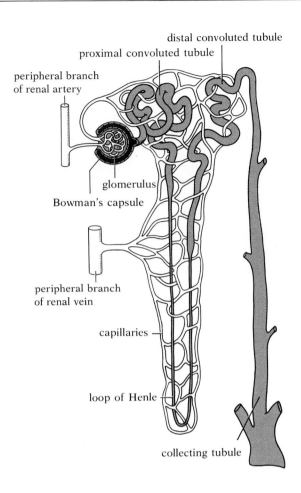

14.17 The human nephron
Each human kidney contains approximately 30–50 km of nephrons. The functioning of the nephron is explained in the text.

the rejoining of the glomerular capillaries. After emerging from the capsule, the arteriole promptly divides again into many small capillaries that form a second dense network around the remaining elements of the nephron. Finally, these capillaries unite once more to form a small vein. The veins from the many nephrons then fuse to form the ***renal vein***, which leads to the posterior vena cava. The kidney is the second of the three places in the mammalian body where blood circuits incorporate two sets of capillaries.

The formation of urine   With the structural relationships in mind, we are now in a position to examine the mechanism of urine formation in the human kidney. We must consider three processes: ultrafiltration, reabsorption, and tubular excretion.

In 1844 the German physiologist Carl Ludwig suggested that a glomerulus acts as a simple mechanical filter—that molecules small enough to pass through the capillary walls and through the thin membranous walls of the capsule filter from the blood into the nephron as a result of the high hydrostatic pressure of the blood in the glomerulus. The advent of modern techniques of microscopy has suggested how such filtration might take

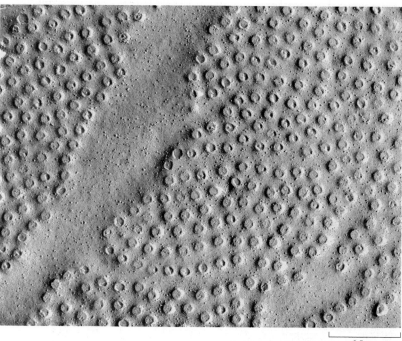

0.5 μm

14.18   Electron micrograph of freeze-etched membrane of a glomerular capillary of a rat
The membrane is perforated by many pores of uniform size. It is presumably through these pores that substances filter out of the glomerulus and into the Bowman's capsule, driven by the high blood pressure built up by the beating of the heart.

place. Pores can be detected in the walls of both the glomerular capillaries (Fig. 14.18) and the Bowman's capsule (Fig. 14.19). The blood pressure presumably forces ions, water, and other small molecules through the pores from the capillaries into the lumen of the capsule.

If the filter explanation is correct, the liquid entering the lumen of the nephron should have basically the same composition of dissolved substances as blood, lacking only the formed elements and the plasma proteins, both of which are too large to filter through the membranes to any appreciable extent. Collection of capsular urine to verify this conclusion is very difficult, and at best can yield only minute quantities for analysis. In spite of the technical difficulties, A. N. Richards of the University of Pennsylvania was able to draw off small samples of glomerular filtrate and to show that the filtrate has essentially the same concentration of dissolved substances (glucose, urea, salts, amino acids, etc.) as blood plasma, just as had been anticipated. Other experiments have shown that if the hydrostatic pressure in the glomerular capillaries is increased, the volume of the filtrate is increased proportionately, and if the hydrostatic pressure is decreased, the filtrate volume declines proportionately. Furthermore, changes in filtrate volume are not accompanied by the changes in the kidney's oxygen consumption that would occur if the kidney were performing work in moving materials from the blood to the capsule. All the evidence, therefore, supports Ludwig's thesis that the cells of the glomerular capillaries and of the Bowman's capsules do not carry out active transport in the movement of materials from the glomeruli into the capsules, and that the work involved is performed by the beating heart as it drives the blood under high hydrostatic pressure into the glomeruli.

But what happens to the filtrate once inside the nephron? If the filtrate in

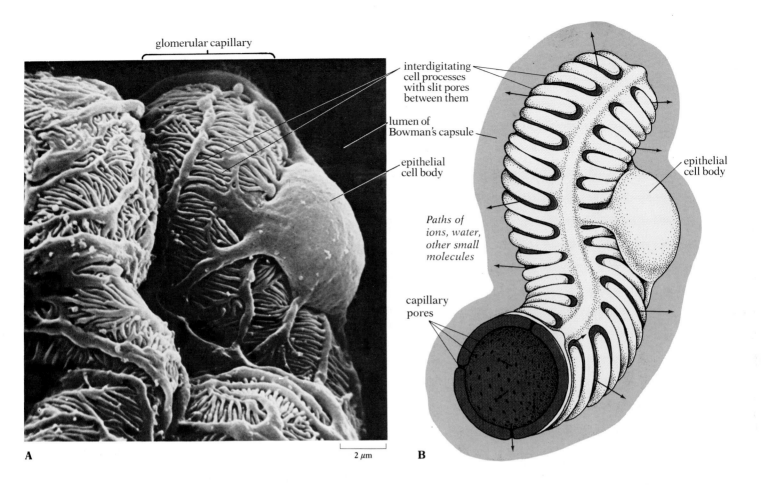

glomerular capillary

interdigitating
cell processes
with slit pores
between them

lumen of
Bowman's capsule

epithelial
cell body

epithelial
cell body

Paths of
ions, water,
other small
molecules

capillary
pores

A

2 μm

B

the nephron were expelled from the body without modification, many essential substances would be lost with it and the process would be extremely wasteful. Consider the drinking that would be necessary to replace the water alone in the 170 liters of filtrate formed every day in the average person's kidneys! Selective reabsorption of most of the water and many of the dissolved materials is one of the functions of the tubules of the nephrons. In humans the filtrate passes first through the ***proximal convoluted tubule***, then through the long ***loop of Henle***, then through the ***distal convoluted tubule***, and finally into the ***collecting tubule*** (Fig. 14.17). The Bowman's capsule and the proximal and distal convoluted tubules are in the kidney cortex layer, while most of the loop of Henle and the collecting tubule are in the kidney medulla. As the filtrate moves through the tubules, as much as 99 percent of the water may be reabsorbed by the cells of the tubule walls and returned to the blood in the capillary network. In this way the human kidneys (and the kidneys of other mammals and of birds) produce concentrated urine—that is, urine that is hypertonic relative to the blood plasma —even though the initial filtrate was nearly isotonic.

To understand the process by which water is removed and urine becomes concentrated, we must examine the structure of the nephron more closely. There is no evidence that any cells have ever been able to evolve a mechanism for the active transport of water itself. When water must be moved, the usual mechanism is the active pumping of ions, with the water following

14.19   Part of a glomerulus from a rat kidney
(A) Scanning electron micrograph showing inner epithelial cells of a Bowman's capsule enveloping two glomerular capillaries. The individual cells that make up the inside layer of the cup of the capsule are so highly modified that they bear little resemblance to any other epithelial cells. Called podocytes, these cells consist of a cell body (of which only one is shown in this SEM) and numerous interdigitating processes that embrace the glomerular capillaries. Spaces, called slit pores, between the interdigitating processes provide an enormous area for the filtration of the blood. Together, podocytes form the inner wall of the lumen of the capsule, through which the filtrate enters; from the lumen the filtrate moves into the tubule system of the nephron. (B) An interpretive drawing of one glomerular capillary with a podocyte cell body and interdigitating processes attached. Blood pressure in the capillary forces ions, water, and other small molecules through its pores and out through the slit pores of the podocyte into the lumen of the capsule.

371

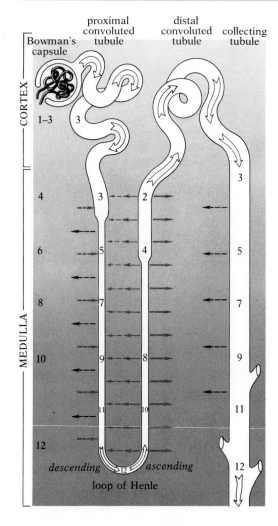

proximal
convoluted
tubule

distal
convoluted
tubule   collecting
tubule

Bowman's
capsule

CORTEX

1–3   3

MEDULLA

4   3   2   3

6   5   4   5

8   7   7

10   9   8   9

11   10   11

12   12

descending   *ascending*   12

loop of Henle

——→ Active transport of salt ions (Na⁺ or Cl⁻,
with the other following electrostatically)

--→ Passive transport of salt ions

- -→ Passive transport of H₂O

⇥ Flow of filtrate, or urine

Figures indicate solute concentrations in hundreds
of milliosmols per liter

passively. In this case the pumps are in the convoluted tubules and ascending limb of the loop of Henle. Two kinds of ions—Na^+ and Cl^- ions, the crucial solutes for controlling the distribution of water in living tissue—are known to cross the walls of the tubules into the surrounding tissue fluid, but at present there is a lively controversy over which ion is actually pumped; whichever it is, the other probably follows it through electrostatic attraction. For years most researchers supposed that Na^+ was the ion being pumped, since vast numbers of Na^+ pumps are found in the membrane of virtually every cell in the body. Now, however, many investigators believe instead that tubule pumps actually transport Cl^- ions, and that Na^+ ions follow electrostatically. Since this controversy continues, we shall describe kidney function in terms of ionic pumps without worrying about which ion moves actively and which just follows along. By the action of the ionic pumps, some 75 percent of the solutes, and the water that follows them, are removed from the filtrate in the proximal convoluted tubule and reabsorbed by the associated capillaries. The remaining filtrate passes into the loop of Henle, where fine-tuning of the filtrate concentration and volume is effected.

In the loop of Henle, once salt ions, Na^+ and Cl^-, are moved out of the ascending limb, some of them, but not all, diffuse passively back into the descending limb, so there is, in effect, a cycling of some of the ions from ascending limb to tissue fluid to descending limb to ascending limb to tissue fluid, and so on. The result is that a salt-ion concentration gradient is maintained in the tissue fluid along the loop, with the concentration lowest in the outer part of the kidney cortex, where the convoluted tubules are located, and highest in the medulla, where the tip of the loop of Henle is located (Fig. 14.20). The wall of the ascending limb of the loop must be impermeable to water, since water does not diffuse out of the tubule as the salt ions are moved out. Consequently the net effect of passage of the fil-

14.20 Fine-tuning of filtrate concentration by the loop of Henle
The hairpin structure of the nephron, the loop of Henle, acts as a countercurrent-multiplier system: it multiplies the effect of active transport, bringing about higher concentration differences than might otherwise be attainable. The action of pumps in the ascending limb of the loop of Henle moves out salt ions (solid color arrows), some of which diffuse back passively into the descending limb (dashed color arrows). This pumping action is relatively weak; at any given point the concentration of the filtrate in the ascending limb and that of the surrounding tissue fluid differ by only two units. Yet because the ionic pumps act upon a filtrate passed along a hairpin loop, they can establish and maintain, in the tissue fluid between the cortex and the inner portion of the medulla, a concentration gradient of roughly nine units. As a result of the gradient, the filtrate passing down the collecting tubule is at every point hypotonic relative to the surrounding tissue fluid. Thus the urine, dilute when it enters the collecting tubule, can steadily lose water (dashed black arrows) by osmosis through the permeable walls of the tubule and become concentrated. The deepening of color from the cortex to the medulla indicates the increase in salt concentration in the tissue fluid.

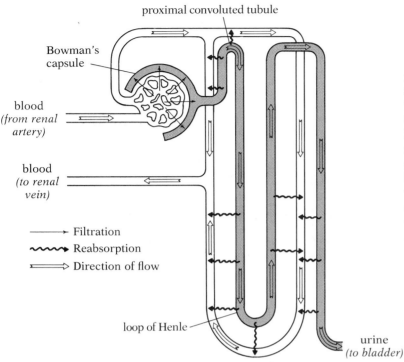

proximal convoluted tubule

Bowman's
capsule

blood
*(from renal
artery)*

blood
*(to renal
vein)*

→ Filtration
〜〜〜➤ Reabsorption
⇒ Direction of flow

loop of Henle

urine
(to bladder)

14.21 Summary of filtration and reabsorption in the nephron
The fluid in the blood arriving from the renal artery is forced through the walls of the capillaries in the Bowman's capsule. The water, ions, and small molecules that pass through this filter are collected and move through tubules leading to the pelvis of the kidney, and from there to the bladder. On the way, water leaving the tubules by osmosis, as well as ions and small molecules actively transported out of the tubules, can be reabsorbed by the capillaries closely associated with the tubules. Most of the reabsorption takes place in the proximal convoluted tubule.

trate through the loop of Henle is the removal from it of some Na^+ and Cl^- but very little water, and the urine in the distal convoluted tubule is actually less, instead of more, concentrated than the filtrate that entered the loop of Henle. But now the urine flows into the collecting tubule, which runs from the cortex through the medulla to the renal pelvis—through a region of increasing salt-ion concentration in the tissue fluid. Since the wall of the collecting tubule is permeable to water, water moves by passive osmosis from the dilute urine in the tubule to the surrounding hypertonic tissue fluid, until the final urine may become essentially isotonic with the highly concentrated tissue fluid in the inner region of the medulla. From here the water is reabsorbed by the nearby capillary bed (Fig. 14.17). Whether the collecting tubule finally releases dilute or concentrated urine depends on whether there is a deficiency or an excess of water in the body at the moment: detectors in the brain measure the osmotic concentration of the blood and then regulate the permeability of the collecting tubules to water.

In the nephron, water is not the only substance reabsorbed into the adjacent capillaries. In a normal healthy person all the glucose, almost all of the amino acids, and most of the salt ions are also reabsorbed and returned to the blood. Much of this reabsorption involves active transport, and thus energy expenditure by the tubule cells. In general, the kidney functions by forcing out of the blood in the glomerulus most molecules small enough to pass through the pores, and then reabsorbing into the capillaries surrounding the convoluted tubules and the loop of Henle only what is to be saved (Fig. 14.21). Hence the kidney effectively and automatically removes many toxins and chemicals by simply not transporting them back into the blood. Obviously this system is far safer than one requiring a special pump for every undesirable element that might turn up in the blood.

In spite of the extensive reabsorption that may take place, urine is more than a concentrated solution of urea and other "waste" substances. In fact, most substances have what is called a kidney threshold level. If the concentration of such a substance in the blood exceeds its kidney threshold level, the excess is not reabsorbed from the filtrate by the capillaries but instead appears in the urine. Glucose is an example of a substance with a high threshold level; ordinarily all glucose in the filtrate is reabsorbed, because the threshold level for glucose is higher than the normal blood glucose level. If, however, the blood-sugar level is abnormally high, as it is in diabetics, sugar appears in the urine. This elimination of excess sugar by the kidneys points up once again that excretory organs do far more than just remove nitrogenous wastes; they play a critical role in maintaining the relatively nonfluctuating internal fluid environment of the organism. In this case, when the liver and/or the peripheral tissues are not functioning properly and the blood-sugar level rises, the kidneys act as a second line of defense. The kidneys likewise help regulate the composition of the blood by keeping the relative concentrations of such inorganic ions as sodium, potassium, and chloride in the blood plasma at a nearly constant level. Whenever the concentration of an ion in the blood, and hence in the glomerular filtrate, exceeds its kidney threshold value, the excess in the filtrate is not reabsorbed but is released in the urine. The remarkably steady level of ionic concentration in the blood and the considerable variation noticeable in the urine suggest the extent of the regulation.

Not all materials are subjected to the filtering/selective-reabsorption strategy we have described. Some larger molecules, such as penicillin, bypass the pores of the glomerulus and are actively removed from the blood in the second bed of capillaries adjacent to the tubules. This tubular secretion[3] supplements glomerular excretion and increases the efficiency of the overall excretory regulation of blood composition.

The glomerular kidney, which probably arose first in ancestral marine vertebrates, doubtless played an important role in enabling the ancestors of modern bony fish to enter fresh water, a hypotonic medium relative to their body fluids. We have seen that freshwater fish are constantly being flooded by water. The glomerular kidney is particularly well suited to pumping out excess water. Modern marine fish, on the other hand, have a problem of water conservation; it is not surprising, therefore, that the development and the activity of their glomeruli have declined. Terrestrial vertebrates have no need to excrete large quantities of water, either. Hence in reptiles and birds, as in the marine fish, the glomeruli have declined, and much of the excretion of uric acid is tubular rather than glomerular. In mammals, by contrast, evolution has tended, not to reduced glomeruli, but to longer convoluted tubules and loops of Henle as well as more efficient water reabsorption; as might be expected, the longest tubules and loops occur in species that inhabit very dry environments.

---

[3] "Tubular secretion" is actually a misnomer, since it is the second bed of capillaries, not the tubules, that does the secreting.

Special excretory adaptations of vertebrates   The kidneys of vertebrates vary considerably in their capacity to produce concentrated urine. You will recall that marine fish are incapable of producing urine more concentrated than their blood and must therefore excrete excess salt by another method—special cells in the gills. Similarly, lacking very efficient kidneys, sea turtles and marine birds (albatrosses and penguins, for example) must excrete the excess salt in the water they drink by some other mechanism; these animals have special glands in their head, near the nose, that are capable of excreting salt in very concentrated solution.

Seals and some whales seldom drink; they get their water from the body fluids of the fish they eat, and thus benefit from the fish's ability to excrete salt through the gills. A fish diet means much protein, however, and much urea to excrete. These animals have kidneys capable of excreting urine with a high urea concentration. Some whales eat marine invertebrates instead of fish, and accordingly take in much excess salt; apparently these species can produce urine with a high salt concentration.

Kangaroo rats living in deserts almost never drink; nor do they get water by eating succulent food. Most of their water is metabolic water obtained during the respiratory breakdown of the dry grains they eat. They must, of course, conserve water extremely well: they are not active during the heat of the day; they do not sweat; they eliminate very dry feces; and they have extraordinarily efficient kidneys capable of producing extremely concentrated urine.

The human kidney is incapable of producing urine with a very high concentration of either salt or urea. And humans have no alternative excretory mechanisms like those of marine fish, turtles, and birds. Adrift at sea, humans are in serious danger indeed. Drinking seawater aggravates their condition because in the process of removing salt from their bodies they excrete more water than they drink. If they try to get their water by eating fish, as seals do, they excrete much water in the process of removing urea. Human kidneys are simply not adapted to life at sea or to life in very dry habitats.

THE CELLULAR BASIS OF ACTIVE TRANSPORT OF IONS

In our discussion of excretion and osmoregulation so far, we have paid little attention to events at the cellular level. We have indicated that active transport of some substances, particularly salt ions, is carried out by the cells in the walls of the kidney tubules and Malpighian tubules, by the osmoregulatory cells in the nasal glands of marine birds and turtles, and by the salt-secreting cells of the gills of bony fish and the rectal glands of elasmobranchs. What is known about the active-transport process?

The sodium-potassium pump   Our repeated assertion that animals tend to maintain a nonfluctuating fluid environment for their cells, and that this fluid has approximately the same osmotic concentration as the cells, should not be taken to imply that the extracellular and intracellular fluids have the same ionic composition. On the contrary, their compositions are very different. All cells, plant and animal, tend to accumulate certain ions in much higher concentrations than are found in the surrounding fluids and to stabilize intracellular concentrations of other ions at levels far below those in the extracellular fluids. For example, the vast majority of cells maintain an internal concentration of sodium ions (Na^+) far below that in the fluids bathing them, while at the same time accumulating potassium ions (K^+) to a concentration many times that in the extracellular fluid.

As we saw in Chapter 4, specific pumps in cell membranes use energy to move substances across the membranes against osmotic and/or electrical gradients. The most important of these is the sodium-potassium pump, a transmembrane protein that changes conformation when phosphorylated by ATP (see Fig. 4.20, p. 103). In one conformation it is open to the outside of the cell, and has a high affinity for K^+ but a low affinity for Na^+. In the other conformation it is open to the inside, and has the opposite ion-binding properties—a low affinity for K^+ and a high affinity for Na^+. The result is that K^+ from outside the cell is bound and released inside, while Na^+ from inside the cell is bound and released outside. This ion pumping creates an electrochemical Na^+ gradient, which is then used to provide the energy needed to transport glucose into the cell (see Fig. 4.18B, p. 101). A similar strategy is used to move amino acids into cells. The sodium-potassium pump is also responsible for maintaining the electrical activity of nerves and muscles, and for regulating the volume of cells. Roughly 30 percent of the ATP consumed by the body at rest is used to fuel the sodium-potassium pump system.

But how do these details explain active transport in the epithelial cells of excretory and osmoregulatory organs? These cells apparently use the same basic pump strategy, but the task they perform is a particularly complex one. Consider the osmoregulatory cells in the gills of a marine fish: these cells must remove salt—Na^+ and Cl^- ions—from the blood and tissue fluid and actively secrete one type of ion into the surrounding seawater (the other will follow along to balance the electric charge). Or consider the cells in the walls of the ascending limb of the loop of Henle in a mammalian kidney: these cells too remove Na^+ and Cl^- ions from the urine, actively secreting one of them into the tissue fluid surrounding the nephron while the other follows. In both instances the cells are doing more than simply expelling salt ions from their cytoplasm: they are picking up ions on one side and expelling them on the other side. In other words, ions are being moved completely across the cell barrier that separates the tissue fluids of the fish from the seawater or that separates the contents of the mammalian nephron from the tissue fluid. Apparently the membranes on the two sides of the cell function differently.

It has been hypothesized that salt ions diffuse passively into the cell on one side and are then actively expelled from the cell on the other side. How would this hypothesis apply to the osmoregulatory cells in the gill of a freshwater fish? These cells must take in salt from the surrounding water and secrete it into the blood-derived tissue fluid. The pump would be active only at the tissue-fluid side of the cell, pumping Na^+ or Cl^- ions from the cell contents into the tissue fluid. This removal of ions would lower the salt-ion concentration in the cell to a point below that in the environmental water, and if the membrane on the environmental side of the cell were permeable to Na^+ and Cl^- these ions would tend to diffuse passively into the cell from the environmental water. In other words, some ions that diffused passively into the cell from the medium on one side would then be actively expelled by the pump from the cell into the tissue fluid on the other side. In a marine fish the situation would be reversed: the pump would be in the membrane on the side of the cell exposed to the seawater, and the ions would diffuse passively into the cell on the tissue-fluid side.

Chapter **15**

CHEMICAL CONTROL IN PLANTS

Most organisms respond to change in their environment: bacteria swim up a food gradient, plants bend their leaves toward the sun, animals generally move toward food and away from predators. Whatever the organism, three steps are involved in the flow of information: reception of the relevant stimuli by the organism, communication of the information from the receptor site to the area where responses are generated, and the response itself. In unicellular organisms, all three functions are carried out in the same cell, but most multicellular organisms have tissues specialized for each function. In animals these systems are so elaborate that we must devote seven chapters to the flow of information through them: two on chemical communication, one each on nervous communication and sensory reception, and three on responses—the physiology behind the muscle movements and the behavior they serve. In plants, however, the ranges of relevant stimuli and possible responses are drastically limited, so a single chapter can cover the entire information pathway.

Plants pass information between cells almost exclusively by chemical means. Though every plant cell, like every animal cell, has an electrochemical potential across its membrane, only animals have evolved a system of fibrous nerves that can transmit electrical messages over long distances to precise targets. (As we have seen, some plants, such as the Venus fly trap, can use electrical signals even though they lack nerves.) The chemicals by which plants transmit signals are generally produced by the cells receiving the relevant environmental stimuli, and are carried by internal-transport systems to target cells, where they communicate their messages by directly or indirectly altering specific chemical reactions. The target cells then initiate the appropriate responses. Control chemicals of this sort, secreted by tissues specialized for their production, are usually referred to as **hormones**. Hormones act as intercellular messengers.

You are probably already aware of how some hormones function in animals. You may know, for instance, that the release of the vertebrate hormone adrenalin into the bloodstream can make an animal suddenly more alert. It signals muscles not essential for immediate life-sustaining action (like those in the digestive tract) to stop working, and redirects the blood that had been involved in noncritical processes to the muscles used for movement. Adrenalin also speeds up the heart rate and breathing, and has a number of other effects that help an animal to survive against odds.

Like most hormones, adrenalin is secreted by specialized tissues, is carried by an internal-transport system to a variety of target cells, and acts on these targets by altering one or more chemical reactions in each.

For communication to take place between cells, each hormone must deliver its message to its special target, and must influence a specific chemical reaction. If the hormone is small and lipid-soluble, it may be able to cross the membrane of the target cell directly, as many steroid hormones do, or it may move into the cell by means of a particular channel or transport system. In the latter case specific proteins, or permeases, in the membrane must recognize the chemical (see pp. 100–103). A large hormone molecule may never actually enter the target cell. Instead, it may pass on its message by binding to a receptor on the outside of the cell membrane, thus bringing about the entrance of ions through a channel, or the release of another chemical messenger inside the cell. These second messengers then affect cellular metabolism in a particular way. The second-messenger mechanism, though very common in animals, has not yet been reported for plants. We examined many of these strategies in Chapter 4 in our discussion of membrane gates and channels (see Fig. 4.18, p. 101).

Once they have gained entrance to the cells, the various hormonal messengers have different ways of acting to bring about particular alterations in cellular metabolism. For instance, hormones can change specific structures (as, we shall see, plant auxins alter the cell wall), modify biochemical pathways in the cytosol, influence the production of enzymes, or interact directly with the chromosomes to turn particular genes on or off (the likely mode of action of plant gibberellins). In this chapter and the next, we shall be concerned with the details of how particular hormones function, and with the role of each in integrating the activities of the many specialized tissues and organs in complex multicellular plants and animals.

HORMONES AND PLANT GROWTH

Most plants have specialized tissues—roots, stems, leaves, and reproductive organs—whose activity must be coordinated if the plant is to grow and reproduce. Chemical control can play a role in orienting and regulating the growth of stems and roots, in timing both reproduction and shedding of leaves, in initiating the germination of seeds, and in regulating many other functions.

Plant hormones, at least the known ones, are produced most abundantly in the actively growing parts of the plant body, such as the apical meristems of the shoot and the root, young growing leaves, or developing seeds or

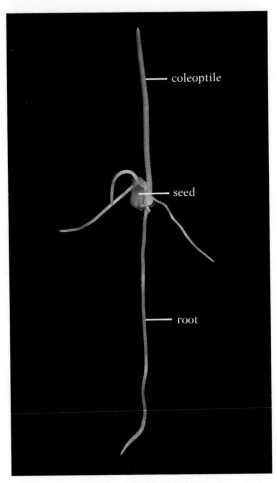

15.1 Photograph of corn coleoptile
The first leaves are rolled up inside the protective coleoptile sheath.

fruits. The tissues in which these hormones are produced, frequently the meristematic tissues themselves, are specialized for hormone production, but they are not so highly specialized as to be concerned with little else, as is often the case with the most highly specialized hormone-producing tissues in animals. There are no separate hormone-producing organs in plants analogous to the endocrine glands of higher animals. Furthermore, plant hormones are predominantly involved in regulating growth and development (they are often called growth regulators), while animal hormones mediate a great variety of functions in addition to growth.

AUXINS

Of the most thoroughly researched plant hormones, one group, collectively known as auxins, displays an amazing variety of effects on the growth of different plant tissues. Plants grow both by cell division, which takes place principally in the apical-meristem areas, and by the elongation and enlargement of cells already present, particularly in stem tissue. It is on the latter sort of growth—cell elongation—that auxins have their greatest effect.

Auxins and phototropism of shoots Everyone is familiar with the strong tendency of many plants to turn toward the light. A potted plant in the living room bends toward a window; you turn the plant so that it will look nicer to people in the room, but discover that in a disconcertingly short time the shoot is again oriented toward the light of the window. This phenomenon of responding to light by turning is called phototropism, from the Greek words for "light" and "turning." (Other tropisms involve turning in responses to other stimuli. Geotropism is a turning response to gravity, hydrotropism a turning response to water.) In plant shoots the phototropism is positive, a turning toward the stimulus; roots, on the other hand, exhibit negative phototropism, a turning away from a light stimulus.

15.2 The Darwins' experiments on phototropism
(A) A coleoptile of canary grass bends toward the light. (B–C) The coleoptile does not bend if its tip is removed or is covered by an opaque cap. (D) The coleoptile does bend if its tip is covered by a transparent cap. (E) It also bends if its base is covered by an opaque tube.

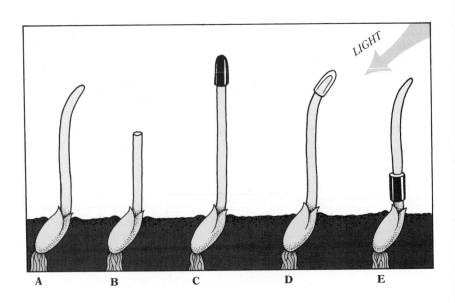

Among the first to investigate the phototropism of plants was the wide-ranging Charles Darwin, who worked on the problem with his son Francis, about 1880. They, like many who followed them, performed their experiments on the cylindrical sheath that encloses the first leaves of seedlings of grasses and their relatives (Fig. 15.1). This sheath, called the **coleoptile**, grows principally by cell elongation, and it exhibits a very strong positive phototropic response. The Darwins showed that if the tip of the coleoptile is covered by a tiny black cap, the plant fails to bend toward light, while control coleoptiles with their tips exposed or covered with transparent caps bend, as expected, toward the light (Fig. 15.2). They observed that a black tube placed over the base of the coleoptile, but not covering the tip, fails to prevent bending. It seemed to be the tip of the coleoptile, therefore, that plays the key role in sensing the light; the tip must then communicate this information to the growth zone (lying just below the tip), where the phototropic response—bending—is accomplished. This was confirmed by experiments in which the Darwins cut off the tip and found that the coleoptile failed to bend, even though control coleoptiles damaged in other ways, but with their tips intact, bent normally. Clearly, it was the absence of the tip and not a reaction to wounding that blocked the phototropic response. The Darwins concluded that light is detected by the tip of the coleoptile and that "some influence is transmitted from the upper to the lower part, causing the latter to bend."

About thirty years later P. Boysen-Jensen in Denmark obtained the first clear evidence that the "influence" postulated by the Darwins was probably chemical rather than electrical or nervous. He removed the tips of oat coleoptiles (thus making the coleoptiles stop growing), placed a thin layer of gelatin on the cut end of each stump, and then placed the tip in the gelatin. Thus the tip was separated from the rest of the coleoptile by a thin layer of gelatin (Fig. 15.3). The coleoptiles resumed growing. If he then illumined the tip from the side, the coleoptile base bent toward the light. The tip had received the light stimulus, and a message from the tip had moved across the gelatin barrier and induced bending in the base. Though this experiment did not completely rule out an electrical or nervous message, it made such a possibility appear highly unlikely and strongly indicated that a diffusible chemical was involved.

That the tip could cause the base of the coleoptile to bend even in the dark was demonstrated by A. Paál in Hungary in 1918. He cut off the tip and then replaced it off center on the stump (Fig. 15.4). If he put the tip on the right side of the stump in the dark, the coleoptile bent to the left; if he put the tip on the left side, the coleoptile bent to the right. Apparently that part of the coleoptile directly under the replaced tip grew much faster than the

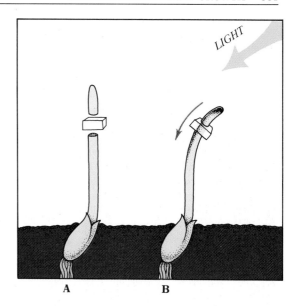

15.3 Boysen-Jensen's experiment
When the tip of an oat coleoptile is cut off, a layer of gelatin is put on the end of the stump, and the tip is replaced (A), the coleoptile will grow and turn toward the light (B). Presumably a chemical (red) moves from the tip, through the gelatin, into the base, and stimulates the plant to turn.

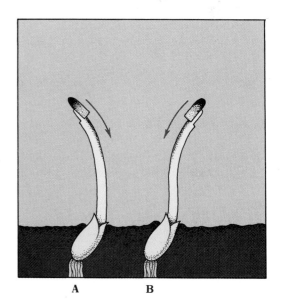

15.4 Paál's experiment
This experiment is performed in the dark. If the tip of a coleoptile is cut off and then replaced right of center, the coleoptile will bend to the left (A); if the tip is placed left of center, it bends to the right (B). (Note that only the tip of the coleoptile is cut off; the rolled-up leaf inside is left intact.)

15.5 Went's experiment

When the tips of coleoptiles are cut off and placed on blocks of agar for about an hour (A), and one of the blocks alone is then put on a stump (B), the stump will resume growing even in the dark. If a block is placed off-center on a stump in the dark (C), the stump will grow and will bend away from the side on which the block rests. Apparently a hormone has diffused from the tips into the blocks, and this hormone can then diffuse from the blocks into the stumps.

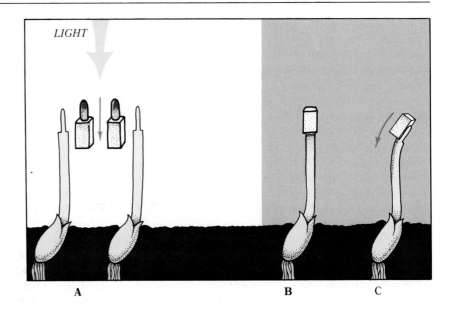

part not under the tip. This asymmetric elongation of the coleoptile caused it to bend away from the side undergoing the greater elongation. A reasonable interpretation of this result might be that the tip continuously produces a substance that moves down the coleoptile stem and causes cells directly below it to grow. Hence, even though the region of growth is below the tip, a coleoptile lacking its tip cannot grow. Viewed in terms of this interpretation, Boysen-Jensen's work would suggest that light coming from the side alters the relative amount of this hypothetical substance to move down each side of the stem, thus causing more growth on one side than on the other. As a result, the stem bends.

Experiments conclusively demonstrating that the growth stimulus moving downward from the tip is a chemical were reported in 1926 by Frits Went, then in Holland. He removed the tips from coleoptiles and placed these isolated tips, base down, on blocks of agar for about an hour (Fig. 15.5). (Agar is a gelatinlike material, made from seaweeds, often used as the base for laboratory culture media.) He then put the blocks of agar, minus the tips, on the cut ends of the coleoptile stumps. The stumps resumed growth for a time. If the agar blocks were put on off-center, the stumps could be made to bend, even in darkness. Plain agar blocks used as controls produced none of these effects. Apparently a growth-stimulating substance had been synthesized in the tips and had diffused into the blocks of agar while the tips were sitting on them. When these blocks were then placed on the stumps, the chemical moved down from them into the stumps and stimulated elongation. This experiment ruled out the possibility that the stimulus was electrical or nervous, for stimuli of these types cannot be stored in agar blocks. Went called this hypothetical diffusible hormone *auxin* (the name is from a Greek word meaning "to grow"). To this day, the identification of auxins is based on Went's experiment: if an agar block containing the substance in question causes a decapitated oat coleoptile to bend in the dark when the block is placed on one side of the cut end, the substance is classed as an auxin.

Many chemicals, some of them found naturally in plants and some synthesized only in the laboratory, have passed Went's test and are commonly called auxins. The one most thoroughly investigated is ***indoleacetic acid*** (Fig. 15.6), which has been isolated from numerous natural sources. There is evidence that indoleacetic acid is the principal (perhaps the only) directly active natural auxin. The observed auxin activity of other natural compounds may result from their conversion into indoleacetic acid by the plant.

The experiments we have discussed have shown that the tip of the coleoptile releases auxin, which moves downward and stimulates cell elongation in the coleoptile. As the results obtained by Paál and Went suggest, there is normally little lateral movement of the auxin after it has been released from the tip; the hormone reaches and stimulates only those cells directly under the point of release. The experiments by Boysen-Jensen, Paál, and Went revealed much about the more general problem of hormonal control of growth, but what do they say about the more specific problem of phototropism, the phenomenon that originally inspired Darwin's experiments? A superficial explanation of the phototropic response is that when light strikes the tip of the plant from one side, it reduces the auxin supply on that side. Consequently the illuminated side of the plant grows more slowly than the shaded side, and this asymmetric growth produces bending toward the slower-growing illuminated side (Fig. 15.7).

We call this explanation superficial because plant physiologists are still not certain how the tip detects the light; some think that the detection pigment is β-carotene (one of the carotenoids), others that it is riboflavin, others that it is a molecule, known as phytochrome, and still others that a combination of pigments is involved. Nor do they know how detection of

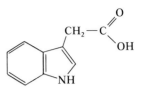

15.6   Indoleacetic acid

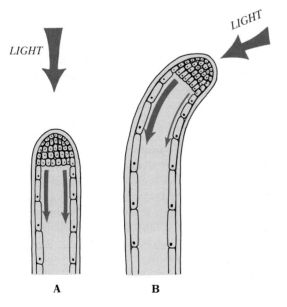

A   B

15.7   Auxin-mediated response to a light source in a young shoot
Light receptors in the growing tip of the plant are thought to respond to light by redistributing auxin, causing it to concentrate on the side of the plant away from the light. In A, with light from directly above, auxin is equally distributed to the growing tissues on all sides. In B, light from the right causes an increase in the supply of auxin to the left side of the plant, which in turn causes an elongation of cells on that side, and a bending toward the light.

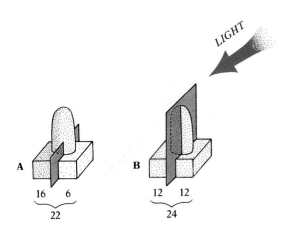

15.8   An experiment demonstrating lateral movement of auxin

The tips of two coleoptiles were placed on agar blocks partitioned by an auxin-impermeable barrier. (A) If the barrier extended only slightly into the base of the coleoptile tip, more auxin was later found in the side of the block away from the light (16) than in the side toward the light (6). (The numbers refer to the degrees of bending the block segments induce in decapitated coleoptiles in the dark; the amount of bending indicates the amount of auxin in the segment.) (B) Approximately the same total amount of auxin was found in the agar block under a coleoptile tip that was completely partitioned, but the amounts in the two sides of the block were the same. This experiment showed that light produces asymmetric auxin distribution by causing lateral movement of auxin, not by destroying auxin on the lighted side.

light is coupled to asymmetric auxin distribution. The evidence indicates that there is active lateral transport of auxin in the tip from the illuminated to the shaded side (Fig. 15.8), but by what mechanism this transport is effected remains unclear.

Cellular basis of auxin action   Once the phenomenon of shoot bending was understood to depend on auxin, work began to focus on the three stages in information flow: how light is detected, how the movement of auxin is controlled, and how auxin actually causes the bending. Though there is still much to be understood about how auxin accomplishes each one of its several functions, progress has been rapid.

Since auxin is produced continually in growing tips, it seems clear that light receptors—whatever chemicals they may turn out to be—respond to the light by simply redistributing the hormone. In response to light, the receptor molecules apparently alter membrane permeability to auxin, thereby directing the lateral movement of the hormone to the side of the shoot away from the light. Horizontal movement and vertical movement of auxin probably involve identical processes. Neither depends on the vascular system, and movement from cell to cell takes place so quickly that active transport must be involved. This conclusion is reinforced by the observation that auxin can move against its concentration gradient, but cannot be transported in the presence of chemicals that inhibit metabolic processes.

Once the auxin reaches the growing cells in the shoot, it binds to receptors that control the flow of H^+ ions across the cell membrane. In turn, these receptors probably activate pumps that transport the ions from the cytoplasm to the cell wall, just outside the membrane. Here the lowered pH seems to activate enzymes that cut the cross-linkages between the cellulose fibrils in the cell wall. As a result the walls soften and the cells can elongate. When there is more auxin on the side of the shoot away from the light, cells there elongate more than their counterparts on the lighted side, and the shoot bends. Cells outside the growing region appear to lack auxin receptors.

Recent evidence suggests that in addition to triggering the pumping of H^+ ions in meristematic tissues, auxin also acts indirectly (through the genetic machinery of the cell) to increase the concentration of enzymes or structural proteins already present or to manufacture entirely new proteins. The identity and role of these proteins is unknown, as is the mechanism (probably also indirect) by which auxin influences their synthesis and reactivity.

Auxins and the geotropism of shoots   Plants respond not only to light but to gravity as well. If you lay a potted plant on its side in the dark and leave it for a few hours, you will find that the shoot has begun to bend upward (Fig. 15.9). This is a negative geotropic response: the shoot turns away from the pull of gravity. It seems reasonable to suppose that auxin might control this kind of plant growth response.

Herman Dolk in Holland showed that in a horizontally placed shoot, the concentration of auxin in the lower side increases while the concentration in the upper side decreases. This unequal distribution of auxin stimulates the cells in the lower side to elongate faster than the cells in the upper side,

15.9   The geotropism of shoot and root
When a growing plant is left lying on its side, the shoot will bend upward and the roots will bend downward.

and the shoot therefore turns upward as it grows. Again, the external stimulus—in this case gravity—is apparently detected in the meristematic tissue in the shoot tip. The meristematic cells probably sense the pull of gravity by its effect on the distribution of the specialized starch plastids called amyloplasts (Fig. 15.10). As the amyloplasts, which are denser than the cytosol, respond to the pull of gravity by settling to the bottom of the cells, they must by some unknown chemical means (perhaps analogous to the way in which phytochrome is thought to control lateral movement of auxin in response to light) create an asymmetric auxin distribution. As we shall see, a different hormone is responsible for the positive geotropism of roots.

Auxins and inhibition of lateral buds   A hormone can have one effect on one type of cell and quite a different effect on another. For instance, the auxin concentration that stimulates stem elongation inhibits growth in the more sensitive buds. Hence, auxin produced in the terminal bud can move downward in the shoot and inhibit development of the lateral buds, while at the same time stimulating elongation of the plant's main stem. The terminal bud thus exerts *apical dominance* over the rest of the shoot, ensuring that the plant's energy for growth will be funneled into the main stem and produce a tall plant with relatively short lateral branches. Longer branches

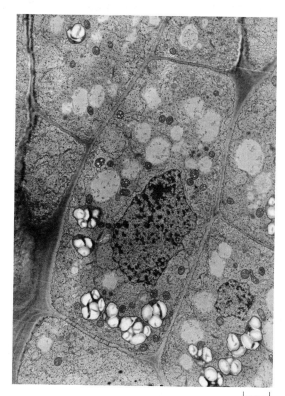

2 μm

15.10   Gravity-sensitive amyloplasts in a root-cap cell
In response to gravity, amyloplasts accumulate at the bottoms of cells. Since the starch they contain is denser than the surrounding cytoplasm, amyloplasts will move through the cytoplasm and settle within minutes at the bottom of a repositioned cell. Organelles of the same density as cytoplasm remain spread throughout the cell.

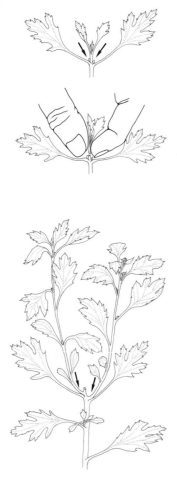

15.11   Inhibition of lateral buds by the terminal bud in chrysanthemum
As long as the terminal bud is intact, the lateral buds marked by arrows will grow very little, if at all. But if the terminal bud is removed, those buds are released from inhibition and grow rapidly, forming the new leaders of the plant.

usually develop only from buds far enough below the terminal bud to be partly free of the apical dominance. If the terminal bud is removed, however, apical dominance is temporarily destroyed, and several of the upper lateral buds will begin to grow, producing branches whose terminal buds soon exert dominance over any buds below them (Fig. 15.11). Flower and shrub growers frequently pinch out the terminal buds of their plants one or more times each season in order to produce bushy well-branched plants with many flowering points instead of tall spindly ones with sparse flowers. Pinching buds will not work, however, for plants in which the young leaves, not the terminal buds, exert control over the lateral buds.

Once two or more branches have begun to develop, neither inhibits the other. Auxin secreted by the terminal bud of one branch does not reach the terminal bud of the other branch in any significant quantity, because it moves mostly downward in the stem (Fig. 15.12). This movement in one direction indicates that the cells involved in transport, like many (perhaps all) other cells, are physiologically polarized—that is, their ends differ from one another in some important way—and therefore transport of certain chemicals occurs in only one direction. The most likely difference in this case is in the orientation of the transport proteins in the top and the bottom membranes.

The action of auxin in inhibiting the growth of lateral buds has sometimes been utilized commercially to prevent the sprouting of stored potatoes (which are stems, not roots). Unless prevented, the buds (commonly called eyes) of potatoes begin to grow during storage, and may produce numerous long sprouts. The sprouting drains nutrients out of the potato tuber itself, often leaving only a shriveled remnant. Treatment with auxin makes it possible to store potatoes for long periods, sometimes as long as three years, with relatively little loss. Recently the use of auxins has declined in favor of irradiation and of treatment with compounds that, like radiation, completely block cell division and thus stop growth altogether.

Auxins and fruit development   An organ whose normal development depends on the stimulatory effect of auxins is the fruit, which develops from the ovary or from the flower receptacle of the plant. In the absence of fertilization, fruit usually does not develop; instead, a weak layer of thin-walled cells, called an *abscission layer*, forms at the base of the flower stalk (Fig. 15.13). This layer soon breaks, under any slight strain, and the withered flower with its ovary falls to the ground. If, on the other hand, fertilization does occur, no abscission layer forms, and the ovary begins to grow rapidly. This period of rapid growth by the ovary (or by the receptacle in some plants), during which it develops into the fruit, is initiated by auxin released by the same pollen grains that bring about fertilization. The continued growth and development of the fruit depends on stimulation by auxins produced by the seeds contained within it.

Once the role of auxins in stimulating fruit development and in inhibiting abscission-layer formation was recognized, production of seedless fruit became a possibility. A few plants sometimes produce seedless fruit naturally, and it seemed likely that in these cases tissues of the ovary (or associated structures) themselves produce so much auxin that fertilization and

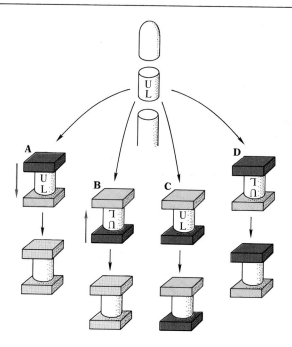

15.12   Experiment demonstrating the polarity of auxin movement

A segment is cut from a coleoptile (top). (A) An agar block containing auxin (dark color) is placed on the upper end of the segment and a block without auxin (gray) is placed on the lower end. Some auxin moves from the one block, through the coleoptile segment, into the other block. (B) The same thing happens even if the whole preparation is inverted, an indication that the movement is not a response to gravity. (C) When the agar block containing auxin is put on the lower end of the coleoptile segment and the block without auxin is put on the upper end, virtually no movement of auxin occurs, and inverting the group (D) makes no difference. Hence auxin must move primarily in only one direction through a coleoptile, and that direction is determined by properties of the cells of the coleoptile itself, not by the pull of gravity.

the resulting development of seeds is unnecessary for fruit growth. If fertilization of plants that normally produce seeds, such as tomatoes, cucumbers, squash, and figs, could be prevented, and if auxins could be artificially supplied to take the place of those normally produced by the seeds, seedless fruits should develop. Experiments along these lines were tried, and in 1934 S. Yasuda in Japan produced seedless cucumbers and in 1936 F. G. Gustafson at the University of Michigan produced seedless tomatoes. Since that time seedless fruits of many other plants have been produced. In some plants, however—notably most single-seeded fruits, such as plums, cherries, and peaches—all attempts to produce seedless fruits have failed, for reasons as yet unknown. In other plants, such as strawberries and blackberries, seedless fruits have been produced, but the hard cases (derived from the ovaries) that normally cover the seeds have remained, so a person eating the berry can't tell whether it is seedless or not.

Treatment of fruit crops with auxins has other commercial applications. Often used to supplement normal pollination in the setting of fruit,[1] it ensures a larger crop. In some cases the size of the individual fruits can also be increased by auxin sprays. Since ripe-fruit drop too is a result of abscission-layer formation, as auxin production by the mature seeds declines, it has become common practice with many fruits to apply auxin sprays to orchards as the time of ripening approaches. By inhibiting abscission-layer formation these sprays reduce preharvest fruit drop. (Fruit that drops to the ground early is largely useless commercially.)

---

[1] Fruit-set is the development of the ovary of a flower into fruit. It usually begins after pollination of the flower, but as we have seen, may also be initiated by applying auxins.

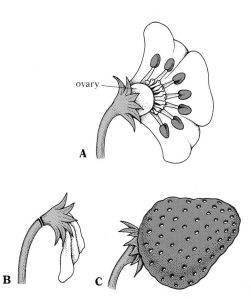

15.13   Formation of fruit

The formation of fruit from the ovary of a flower (A) depends on whether or not fertilization has occurred. If fertilization does not take place, a fragile abscission layer forms at the base of the flower stalk, and the withered flower soon drops off (B). But during fertilization, pollen grains release auxins that initiate the growth and development of the ovary into the fruit (C).

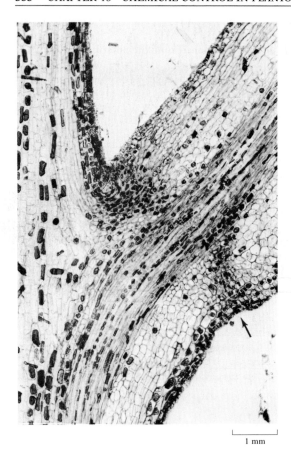

1 mm

15.14 Photograph showing abscission layer at base of petiole of *Coleus* leaf
The arrow indicates the small cells of the abscission layer.

Auxins and leaf abscission   Unfertilized flowers drop off the plant because a special layer of cells, the abscission layer, forms at the base of the flower stalk in the absence of high auxin production in the floral organs, and ripe fruit also drops because of abscission-layer formation. Similarly, the shedding of leaves in autumn (or of diseased leaves at any time of year) by deciduous trees and shrubs usually (though not always) involves abscission-layer formation at the base of the petiole (Fig. 15.14), in part as a result of declining auxin production by the leaf blade. In short, auxin acts as an inhibitor of abscission; another hormone, ethylene (to be discussed below), is the principal promoter of abscission. The actual break in the abscission layer can be initiated by any slight strain like the pressure of a gentle wind, because the cell walls there have been weakened by an increasing concentration of cellulase, an enzyme that digests cellulose; in addition, in some plants, the middle lamella between the cells may become soft and gelatinous, and thus less able to cement the cells firmly together.

Sprays of chemicals that are auxin antagonists are commonly applied to the leaves of cotton just before harvest. The anti-auxins induce formation of abscission layers at the base of the leaves, causing the leaves to fall prematurely and thus making it easier for mechanical pickers to move through the fields and harvest the bolls (Fig. 15.15).

Auxins and cell division   There is good reason to believe that, besides playing a part in cell elongation and abscission-layer formation, auxins are involved in cell division in certain tissues. Apparently it is auxin, moving downward from the buds in early spring, that stimulates renewed activity in the cambium, leading to production of new vascular tissue. As autumn approaches, auxin production by the buds and leaves declines, with the result that cambial activity also declines.

Auxins probably also initiate formation of lateral roots. Such roots usually have their origin in the layer of relatively undifferentiated cells called the pericycle (see Fig. 31.32, p. 815), which is just internal to the endodermis. Most of the time the cells of the pericycle show no meristematic activity. At intervals, however, a small group of cells in the pericycle changes into actively dividing meristematic tissue, giving rise to a new lateral root that bursts through the outer tissues of the main root and enters the soil. There is evidence that the stimulus initiating this meristematic activity in the pericycle comes from auxin. Auxin can, in fact, be applied to the roots of plants to induce lateral branching, but it is uncertain whether the auxin stimulates cell division directly or does so indirectly—by eliciting increased production of another hormone (ethylene) that may then act as the direct stimulant.

It is in the development of adventitious roots from cuttings of such organs as stems or leaves that auxins (generally called rooting hormones in this context) have found a particularly important commercial use. Cuttings from some plants, such as geraniums and willows, will readily root in water or soil without application of hormone, but many plants cannot be propagated in this manner. Application of auxins will often induce formation of roots in these cases, making it possible to propagate vegetatively many valuable strains of plants that might otherwise be lost.

15.15 Effect of auxin antagonists on leaf abscission
Left: A cotton field before treatment. Right: A typical field after treatment with auxin antagonists. Treated cotton plants drop their leaves, allowing mechanical pickers to remove bolls with ease and at the optimum time.

Chemical weed control   Two widely used modern weed killers, or herbicides, are 2,4-dichlorophenoxyacetic acid, usually abbreviated 2,4-D, and 2,4,5-trichlorophenoxyacetic acid, abbreviated 2,4,5-T. Both 2,4-D and 2,4,5-T have many of the properties of auxins, though they do not meet all the auxin tests. These artificial auxinlike chemicals have been used in vast quantities since the 1940s for the control of dandelions and other broad-leaved weeds. Because they are selective in their action and, when used in proper concentrations, will not kill grasses or related monocots, they are of enormous commercial value in combating broad-leaved weeds in lawns, pastures, and fields of corn, wheat, oats, and rice. (Of course, their lack of effect on monocots also means they are useless against crabgrass.) The same compounds were used extensively in Vietnam to defoliate entire forests (exposing the ground, which is normally hidden by the leaf canopy). Though low concentrations of these synthetics have effects similar to those of natural auxin, higher concentrations kill plants by stimulating rapid, uncoordinated, and distorted growth of some body parts while seriously inhibiting the functioning of other parts (Fig. 15.16). The exact manner in which these results are produced is not understood; nor do we understand why broad-leaved plants (dicots) are so much more susceptible than grasses (monocots).

In recent years a number of new herbicides even more selective than 2,4-D or 2,4,5-T have been developed and put into use; most of them are not auxins. The goal, of course, is the eventual discovery of herbicides so selective that any given species of weed can be killed with minimum disturbance to the other plants growing around it. Unfortunately, some of the most promising of these herbicides are proving to be destructive of animal tissue as well, and have been implicated in such disorders as cancer and miscarriage in humans.

15.16 Effect of 2,4-D on a dandelion

GIBBERELLINS

The Japanese have long been familiar with a disease of rice that they call "foolish-seedling disease." Afflicted plants grow unusually tall, but seldom live to maturity. In 1926 a Japanese botanist, E. Kurosawa, found that all such plants are infected with a fungus named *Gibberella fujikuroi*. He showed that when the fungus was moved to healthy seedlings they developed the typical disease symptom of rapid stem elongation. He could also produce symptoms with an extract made from the fungus, and even with an extract made from culture media on which the fungus had grown. Clearly, some chemical was involved.

Several Japanese scientists, working on the problem of foolish-seedling disease during the 1930s, succeeded in isolating and crystallizing a substance from *Gibberella*, now known as **gibberellin**, that produced typical disease symptoms when applied to rice plants. Apparently the gibberellin from the fungus instigates rapid growth of the host plant in the affected area; the fungus then uses the products of its host's elevated metabolism to support its own growth. Since 1950 research on gibberellin has become widespread. More than 50 different substances that can be classed as gibberellins have already been found to occur naturally in fungi and in higher plants. The gibberellin most often used in experimental work is **gibberellic acid** (Fig. 15.17).

The most dramatic effect of gibberellins is their stimulation of rapid stem elongation in dwarf plants and other plants that normally undergo little stem elongation (Fig. 15.18). They have much less effect on most normally tall plants, presumably since gibberellins act by promoting cell elongation rather than cell division, and so cannot much alter cells that are already normally elongated. An attractive hypothesis is that the dwarf varieties are genetically incapable of producing sufficient gibberellin, and administration of extra quantities of the hormone simply makes up for the deficiency and allows the plants to grow more normally.

Though both gibberellin and auxin stimulate stem elongation, they cannot substitute for each other in controlling plant growth. For one thing, the stages of development at which the plant is most sensitive to these two hormones often differ; in wheat coleoptiles, for example, responsiveness to gibberellin appears earlier than responsiveness to auxin. For another thing, gibberellin can move freely in the plant body through both the xylem and phloem, whereas auxin characteristically can move in only one direction; hence gibberellin exerts systemic influences and cannot produce the bending movements that mark auxin-induced responses. The usual tests for gibberellins are based on their ability to stimulate growth in dwarf plants that have a very low natural gibberellin content (dwarf corn and dwarf pea, for example); in such tests auxins would yield negative results, because they have virtually no effect on elongation when applied to intact plants.

Gibberellins play a role in a host of developmental processes besides stem elongation. They can (1) often break seed and bud dormancy; (2) induce the embryos in germinating seeds to produce an enzyme that hydrolyzes starch reserves in the seeds, as described just below; (3) stimulate some biennials to flower during the first year of growth; (4) induce some long-day plants to "bolt" and so produce a flowering stalk when the day

15.17   Gibberellic acid

length is too short for them to flower normally; and (5) stimulate fruit-set in some species.

Gibberellins are derived from the same biosynthetic pathway as vertebrate steroid hormones (like estrogen and testosterone) and are lipid-soluble; as a result they can cross cell membranes. They function by turning particular genes on and off. In many seeds, for example, gibberellin produced in one part of a developing plant embryo activates genes in another part that specify the assembly of the enzyme α-amylase. This enzyme helps convert starch—the seed's supply of stored energy—into the readily usable, high-energy sugar glucose. The effects of gibberellins on molecular events in other stages of the life cycle and in different parts of growing plants are not yet well understood.

Possible practical applications for gibberellins in agriculture are being extensively investigated. An obvious application would be to induce growth to greater height in crops like hay. Unfortunately the tallness induced by gibberellins is usually offset by poorer leaf formation and overall spindliness, so the total weight yield of the crop is increased little, if at all. Gibberellins do, however, have a beneficial effect on celery by inducing rapid growth, which produces tenderer stalks. They have also proved useful in accelerating seed germination in some plants in spring and in producing larger clusters of seedless grapes. Other important uses for these hormones will undoubtedly be found in the future.

CYTOKININS

The technique of *tissue culture*—the growing of cells or bits of tissue on sterile nutrient media in the laboratory—has greatly facilitated research in both plant and animal developmental biology. It was thanks to this technique that a new class of hormones was discovered in the 1950s by Folke Skoog, Carlos O. Miller, and their associates, at the University of Wisconsin.

These botanists developed methods of growing parenchyma tissue from tobacco plants on tissue-culture media. The cells formed a tumorlike mass of tissue called a callus, in which the constituent cells often grew to huge size. They did not, however, undergo complete cell division—sometimes the nuclei divided, but new cell walls did not form. Skoog and his colleagues found that extracts made from old nucleic acids would cause the cells in the callus to divide, even though these cells possessed fully developed vacuoles—an indication that they were mature differentiated cells

15.18 Effect of gibberellic acid on cabbage
The plant at left is normal. The one at right was treated with gibberellic acid. You may recall from Chapter 1 that cabbage is one of several domesticated forms of a tall, spindly ancestor (see Fig. 1.21, p. 17). The loss of normal levels of gibberellic acid, with the consequent development of the desirable compact growth form of cabbage, was the result of intense selective breeding.

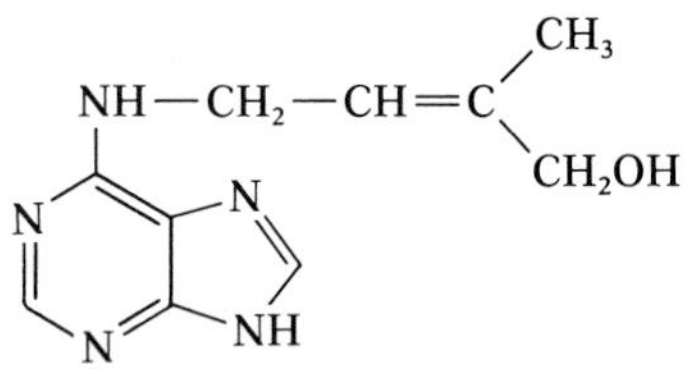

15.19 Cytokinin structure
The most common (and perhaps the only natural) cytokinin is zeatin, shown here.

bearing little resemblance to normal meristematic cells. The compound responsible for providing the stimulus was eventually isolated; it is a degradation product of nucleic acids and can easily be produced in the laboratory.

The substance isolated by Skoog and his associates was not a naturally occurring compound, but in 1964 D. S. Letham and his associates in New Zealand isolated from corn seeds a compound called *zeatin* (Fig. 15.19), which is the most active known naturally occurring representative of a class of hormones promoting cell division: the *cytokinins*.

The action of cytokinin on a callus in tissue culture depends on the presence of auxins. Indeed, the ratio of cytokinin to auxin appears to be of fundamental importance in determining the differentiation of new cells. When there is more auxin than cytokinin, root growth is initiated; when there is more cytokinin than auxin, stems and leaves are generated. In short, the net effect of these contradictory hormones determines what sort of tissue will develop, and so is responsible for controlling plant morphology. In the normal growing plant, cytokinin and auxin act synergistically in some situations and antagonistically in others—for example, they act synergistically in promoting cell division, but antagonistically in influencing the growth of lateral buds. Both hormones influence cell growth, but auxin primarily stimulates elongation, whereas cytokinin promotes cell division. This sort of hormonal interaction is a recurrent theme in vertebrates as well.

Among their numerous other functions, cytokinins (1) stimulate conversion of proplastids into functional chloroplasts; (2) help break dormancy in some seeds; (3) enhance flowering in some plants; (4) promote fruit development in some species; and (5) help retard the onset of senescence (aging), especially in leaves, by maintaining protein and nucleic acid synthesis and helping preserve membrane integrity. As this list, and the discussions of auxins and gibberellins suggest, all the major plant hormones seem to participate in some fashion in nearly all aspects of plant growth and development.

INHIBITORS

Relatively little is known about growth inhibitors, which have effects opposite to those of auxins, gibberellins, and cytokinins. A few have been isolated and identified, but the existence of many others has simply been inferred.

The role of inhibitors in maintaining dormancy has attracted particular interest. It is believed that inhibitors block the activity of some buds and seeds in autumn, thus making sure that they will not begin to grow during a few warm days, only to be killed by the rigors of winter. Dormancy is broken—and the buds and seeds are set free to become active in the next growing season—when gradual breakdown over time, prolonged exposure to cold, or the leaching action of water has helped destroy the inhibitors. In addition, there may be a rise in another hormone (usually gibberellin) that opposes the inhibitors and helps break dormancy.

Inhibitors that must be leached out by water before the seeds can germinate constitute an important evolutionary adaptation in some desert plants. The seeds that fall to the ground will germinate only after long hard rains. Light showers, which might provide enough moisture for germination by seeds not adapted for life in the desert, do not leach out enough inhibitor to allow germination to begin in the desert-adapted seeds; hence no tender young seedlings remain to be killed by the dry conditions that soon follow.

The most important known inhibitor is the hormone ***abscisic acid*** (Fig. 15.20), a chemical synthesized primarily by chloroplasts. It not only helps induce dormancy in buds and seeds but, when applied to an actively growing twig, induces a complex of other changes (including reduced cell division, production of protective scales instead of foliage leaves, deposition of waterproofing substances, and so on) that prepare the plant for the winter. As its name implies, abscisic acid promotes leaf abscission in some plants. It also participates in the control of flowering in some species. In addition to these long-term effects, abscisic acid also has some short-term effects; for example, it plays a role in controlling the stomata, the openings through which air enters and circulates in the leaves. The stomatal guard cells close when the plant begins to lose too much water, and it is abscisic acid that carries this message and thus causes closing. Abscisic acid is also involved in the positive geotropism of roots. As roots grow, the tips synthesize abscisic acid, which is transported to the region of growth. Within the root cells, the amyloplasts, being denser than the cytosol, sink to the bottom, where they interact with the membrane in some way to increase the effective concentration of abscisic acid on the lower side of the root. As a result, growth of the cells on the lower side is inhibited, and the root turns until it is growing downward.

Most of the other known growth inhibitors are not true hormones. These so-called ***secondary plant substances*** include quinones, phenolic acid and its derivatives, and a variety of other compounds. They are usually effective only in very high concentrations and do not appear to play any direct role in metabolism. Being often toxic to insects and other herbivorous animals, they may be of special importance to plants as a means of chemical defense.

ETHYLENE

"One rotten apple spoils the lot." That piece of folk wisdom rests on a familiar fact: when one apple in a barrel goes bad, most of the other apples in that barrel soon go bad too. We now know that the bad apple affects other fruit by the production of the highly volatile compound ethylene (Fig. 15.21). Only in recent years, however, have plant physiologists realized that ethylene is, in fact, a regular plant hormone that plays a variety of roles in the life of plants. Since its production is normally triggered by other hormones—auxin and abscisic acid, primarily—it can be considered in some sense a secondary hormone.

One of the best-studied effects of ethylene is the stimulation of fruit ripening. Once fruit has attained its maximum size, a host of chemical changes begin that cause it to ripen. The ripening process starts with a sudden sharp increase in carbon dioxide output, followed quickly by a sharp decline. This burst of metabolic activity, called the ***climacteric***, is triggered by a hundredfold increase in the concentration of ethylene. Inhibition of ethylene

15.20   Abscisic acid

15.21   Ethylene

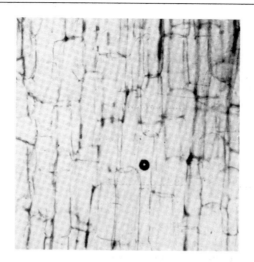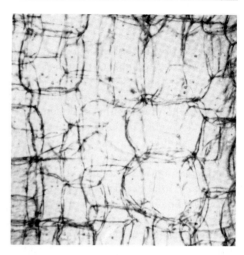

15.22 Effect of ethylene on radial growth in a pea seedling

Left: Untreated cells in a longitudinal section. Right: The same cells after twenty-four hours in air containing 0.5 parts of ethylene per million of air. As this photograph shows, even minute quantities of ethylene have a powerful effect on radial growth.

production, or removal of the ethylene as fast as the fruit produces it, prevents the climacteric, and no ripening occurs. Many commercial fruits are now picked and transported while they are still green and firm, and therefore resistant to damage. They can then be ripened with ethylene gas when ready for sale.

Ethylene also contributes to leaf abscission (apparently by stimulating the production of cellulase) and to various other changes that characterize senescence in a plant or parts of a plant. In addition, it can sometimes stimulate radial growth of stems and roots (Fig. 15.22); it can aid in breaking dormancy in the buds and seeds of some species; and it can help initiate flowering in some plants, such as the pineapple. Moreover, some effects usually attributed to auxin, such as lateral-bud inhibition, probably result in some cases from an auxin-induced increase in ethylene.

Let us summarize the conclusions that can be drawn from our discussion of chemical control of plant growth so far:

1. Cell division, the first phase of growth, is stimulated by cytokinins and auxins, and by other factors that enhance their activity.
2. Cell division is inhibited by a variety of substances, which are but poorly known.
3. The balance between the cytokinins and the inhibitors determines whether a cell will divide or not.
4. Control of cell enlargement, the second phase of growth, involves substances such as auxins and gibberellins, which promote elongation.
5. The various plant hormones, by their mutual interactions and their differential effects on various parts of the plant body, help integrate and coordinate the development of form and function.
6. The aging process, leading to death, is brought on by various senescence-inducing substances, of which ethylene is one of the most important.

CONTROL OF FLOWERING

Flowering is not a random process. Some plants flower early in the spring, others flower in midsummer, and still others, like chrysanthemums, flower in the fall. These simple facts have been known for centuries. But only since 1920 has anything been known of the control mechanisms involved, and some aspects are still not well understood.

PHOTOPERIODISM AND FLOWERING

The intense modern interest in the flowering process dates from the investigations of W. W. Garner and H. A. Allard of the U.S. Department of Agriculture. Working in Beltsville, Maryland, on the tobacco plant, they discovered that a new mutant variety, called Maryland mammoth, grew unusually large (as much as 3 meters tall), but would not flower. They propagated the new variety by cuttings and discovered that it *would* flower in the greenhouse in winter. Though flowering was not the subject Garner and Allard were originally investigating, they became interested in the question of why Maryland mammoth would flower in the greenhouse in winter, but not in the fields in summer. Accordingly they began a series of experiments that opened the way to a whole new area of botanical research. Inventive scientists will not overlook a lead, but rather recognize its importance and pursue it, even if it changes the entire course of their research.

Garner and Allard realized that winter greenhouses and summer fields differ in temperature, moisture, light intensity, day length, and so on. They began experiments that painstakingly eliminated one after another of these environmental factors, until only one was left as the probable controlling factor in flowering—day length. They concluded that the short days of late autumn and early winter induced flowering in Maryland mammoth tobacco. They could get the plants to bloom in summer if they shielded them from the light for a part of each day. Conversely, they could prevent blooming in the greenhouse in winter by extending the day length with electric lights.

Garner and Allard also experimented with Biloxi soybeans. They planted soybeans at two-week intervals from early May through July and found that all the plants flowered at the same time in September, even though their growing periods had differed by as much as 60 days. It was as though they were waiting for some signal from the environment. Garner and Allard were sure that the signal was short days.

Experiments with other species revealed that most plants can be placed in one of three groups: (1) short-day plants, which, like these unique strains of tobacco and soybeans, flower when the day length is below some critical value, usually in spring or fall (examples include chrysanthemum, poinsettia, dahlia, aster, cocklebur, goldenrod, and ragweed); (2) long-day plants, which bloom when the day length exceeds some critical value, usually in summer (beet, clover, petunia, larkspur, black-eyed Susan); and (3) day-neutral plants, which are independent of day length and can bloom whether the days are long or short (dandelion, sunflower, carnation, pansy, tomato, corn, string bean). Garner and Allard called the response by an organism to the duration and timing of light and darkness ***photoperiodism***.

If the critical element in the photoperiodism of flowering is day length, as the terms "long day" and "short day" imply, then we should be able to prevent a long-day plant from flowering at the proper season by shielding it from light for an hour or so during the middle of the day. But if this is done, nothing happens; the plant flowers normally. If, however, a short-day plant is illuminated by a bright light for a few minutes, or even seconds, in the middle of the night during the normal flowering season, it will not bloom (Fig. 15.23H). The same sort of experiment will induce flowering at the

15.23 Comparison of long-day and short-day plants
White bars indicate days and gray bars nights. The hypothetical long-day (short-night) plant shown here has a rather long critical night length of 13 hours, and the hypothetical short-day (long-night) plant has a rather short critical night length of 8½ hours. In other words, in this example the critical night length for the short-night plant is actually longer than that for the long-night plant. The difference is that the critical night length is a *maximum* value for the short-night plant and a *minimum* value for the long-night plant. The short-night plant will flower when the night length is slightly *below* the critical value (A) or when it is much below the critical value (B), but will not flower when it is above the critical value (C); the plant will flower, however, if a long night is interrupted by a flash of light that reduces the period of continuous dark below the critical value (D). Conversely, the long-night plant will flower when the night length is slightly *above* the critical value (E) or when it is much above the critical value (F), but will not flower when it is below the critical value (G); the plant will not flower if a long night is interrupted by a flash of light that reduces the period of continuous dark below the critical value (H).

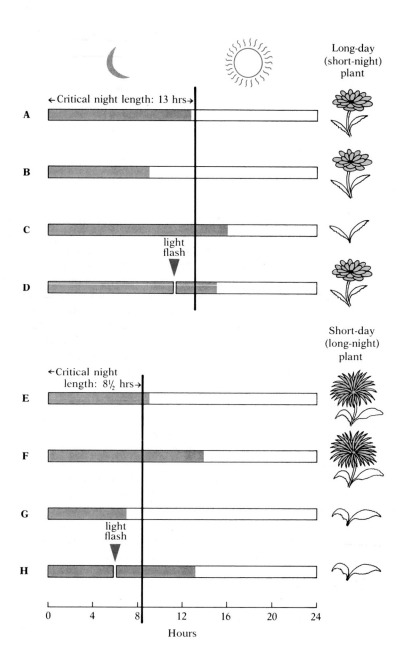

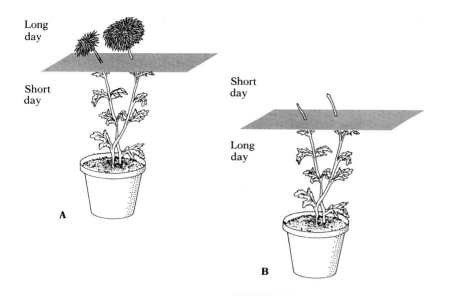

15.24 Chailakhian's experiment
(A) Chailakhian removed the leaves from the top half of a chrysanthemum (a short-day plant) and then exposed the top half of the plant to long days and the bottom half to short days. The plant flowered.
(B) When he did the reverse experiment, the plant did not flower.

wrong season by a long-day plant (Fig. 15.23D). Clearly, then, the critical element of the photoperiod is actually the length of the night, not the length of the day. Instead of speaking of long-day and short-day plants, we would be more accurate to speak of short-night and long-night plants.

The difference between long-day and short-day plants does not depend upon the precise duration of darkness at the time of flowering. Rather, the basic distinction is that long-day (short-night) plants will flower only when the night is *shorter* than a critical value, whereas short-day (long-night) plants will flower only when the night is *longer* than a critical value (Fig. 15.23). The critical night length is thus a maximum value for flowering by long-day plants and a minimum value for flowering by short-day plants.

IS THERE A FLOWERING HORMONE?

Now that we have seen what determines flowering in different plants, we are ready to explore how the photoperiod exerts its influence. Beginning with the 1936 experiments of M. H. Chailakhian in Russia, the evidence has supported hormonal control of flowering. Chailakhian removed the leaves from the upper half of chrysanthemums (which are short-day plants), but left the leaves on the lower half (Fig. 15.24). He then exposed the lower half to short days while simultaneously exposing the defoliated upper half to long days; the plants flowered. Next, he reversed the procedure, exposing the lower half to long days and the defoliated upper half to short days; the plants did not flower. He concluded that day length does not exert its effect directly on the flower buds, but causes the leaves to manufacture a hormone that moves from the leaves to the buds and induces flowering. This hypothetical hormone was named *florigen*.

Further evidence for the existence of a moving stimulus, probably a hor-

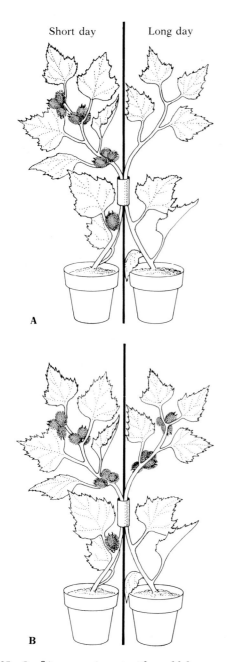

15.25 Grafting experiment with cockleburs
The two plants are separated by a light-tight barrier, but are connected by a graft. The plant exposed to an inducing photoperiod (short days) flowers (A), and shortly thereafter the other plant begins to flower (B).

mone, comes from grafting experiments with cockleburs (which are short-day plants) (Fig. 15.25). If one plant is grafted onto another through a light-tight partition, and if the first plant is exposed to an inducing photoperiod (short days/long nights) while the other is exposed to a noninducing photoperiod (long days/short nights), the plant exposed to short days will flower, and soon thereafter the plant exposed to long days will also flower. Presumably a stimulus from the first plant moves through the graft and induces flowering in the second plant, even though the second plant is exposed to the wrong photoperiod. The same results are obtained if only a single leaf is left on the plant exposed to the inducing photoperiod; apparently the one leaf can produce enough of the stimulus to cause flowering in both plants.

There is evidence that the flower-inducing factor is the same in both short-day and long-day plants: in many cases, if a long-day and a short-day plant are grafted together and then exposed to short days, both will flower. Apparently some substance produced by the induced short-day plant can move through the graft and make the noninduced long-day plant flower too. Cross induction of flowering can also be obtained in grafts between long-day and day-neutral plants or between short-day and day-neutral plants. Ringing experiments show that the stimulus is transported in the phloem.

In some plants there is an added complication, namely that leaves exposed to a noninducing photoperiod actively inhibit flowering. If a light-tight barrier is placed across a cocklebur leaf and the base of the leaf is exposed to a flower-inducing photoperiod while the tip is exposed to a noninducing photoperiod, a nearby bud will flower (Fig. 15.26A). If the reverse experiment is run, with the base of the leaf exposed to a noninducing photoperiod and the tip to an inducing photoperiod, the bud will either not flower or flower only weakly (Fig. 15.26B); perhaps any hormone produced under inducing conditions in the tip of the leaf is destroyed as it passes through the noninduced base. The inhibiting effect can also be seen if one leaf is exposed to an inducing photoperiod, but only if the noninduced leaf is located between the induced leaf and the bud (Fig. 15.26C–D). In cockleburs and other species the inhibition may be local and not transmissible, but in some species, like strawberries, there is evidence for a transmissible inhibitor.

The most obvious hypothesis to explain flowering that emerges from these various experiments is that an inducing photoperiod causes the leaves to increase the production of a hormone that then moves in the phloem to the buds and stimulates development of the flower. A noninducing photoperiod causes the leaves of many (but not all) plants partially to inhibit the production of this hormone. Under natural conditions, then, flower production is triggered when, as the seasons change, the photoperiod passes a critical value and production of the hormone by the leaves exceeds some inhibitory threshold.

Given the extraordinarily powerful biochemical techniques now available, the continued failure to isolate florigen is discouraging. The majority of biochemists working in this area are now inclined to doubt that a separate flowering hormone exists. Instead, it seems more likely that flowering, like so many other plant functions, is actually controlled by the ratio of two

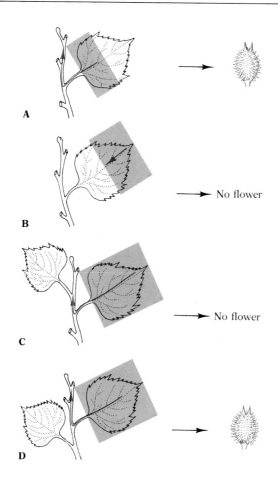

15.26 Some experiments illustrating the inhibitory effect of long days on flowering in the cocklebur
(A) When the tip of the uppermost leaf is exposed to continuous illumination, but the base is covered with black paper to give it a short day, the plant flowers, presumably because some inducing substance synthesized in the shaded part of the leaf has moved (color arrow) to the flower bud. (B) The reverse procedure, shading the leaf tip and illuminating the leaf base, does not result in flowering, presumably because the inducing substance is destroyed as it passes through the illuminated part of the leaf. (C) When one leaf is shaded, but a leaf above it on the stem is illuminated, the plant does not flower; perhaps something at the base of the illuminated leaf destroys the inducing substance. (D) However, if the shaded leaf is located above the illuminated leaf, no such destruction can occur and the plant flowers.

or more other hormones. The best candidates at the moment are gibberellin and auxin, both of which can independently affect flowering in many species. If the hormone-ratio hypothesis is correct, it is probably also the case that both the ratio and the sensitivity to this ratio can be changed by metabolic or environmental factors, like photoperiod.

DETECTION OF THE PHOTOPERIOD

If flowering is set in motion by exposure to the appropriate photoperiod, we must now ask how the photoperiod is detected and measured in the first place. Since interrupting the dark period prevents flowering in a short-day (long-night) plant and induces flowering in a long-day (short-night) plant, the light itself must be detected by the plant. What wavelengths of light are involved? H. A. Borthwick, S. B. Hendricks, and their associates, of the U.S. Department of Agriculture, Beltsville, Maryland, began investigating this question in 1944. They exposed Biloxi soybeans to light of different wavelengths, and found that red light (wavelength about 660 nm) is by far the most effective in inhibiting flowering in these short-day plants; the same red light is very effective in inducing flowering in long-day plants. Later, it was found that far-red light (wavelength about 730 nm), which is invisible to the human eye, has effects exactly contrary to those of red light; it induces flowering in short-day plants and inhibits flowering in long-day plants. Not only do red and far-red light have opposite effects, but each reverses the effect of prior exposure to the other (Fig. 15.27). A short-day (long-night) plant will not flower if its long night is interrupted by a bright flash of red light; if, however, the red flash is followed immediately by a far-red flash, the plant flowers normally. Almost any number of successive flashes can be used, the final effect depending solely on whether the last flash was red or far-red.

The discovery that red light and far-red light can reverse each other's effects led Borthwick and Hendricks to conclude that a single receptor pigment, which they called ***phytochrome***, is involved, and that this pigment exists in two forms: one that absorbs red light (P_r) and one that absorbs far-red light (P_{fr}). It has since been identified as a protein with a prosthetic group that gives it the properties of a pigment—notably the ability to absorb light of certain wavelengths (Fig. 15.28). When P_r absorbs red light, it is rapidly converted into P_{fr}. Conversely, absorption of far-red light by P_{fr} rapidly

15.27 Reactions of long-day and short-day plants to a variety of light regimes
White bars indicate days and gray bars nights. Long-day (short-night) plants flower when the night is shorter than the critical value, or when a longer night is interrupted by an intense red or white flash or by a series of flashes of which the last is red or white. Short-day (long-night) plants give the reverse responses.

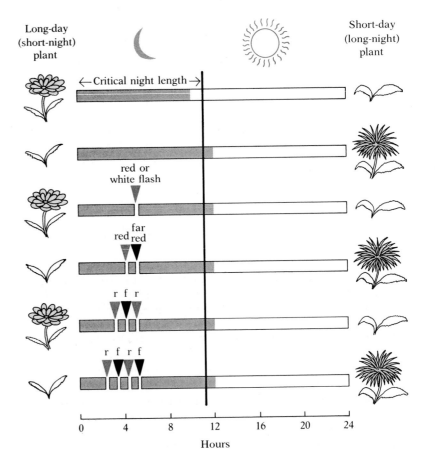

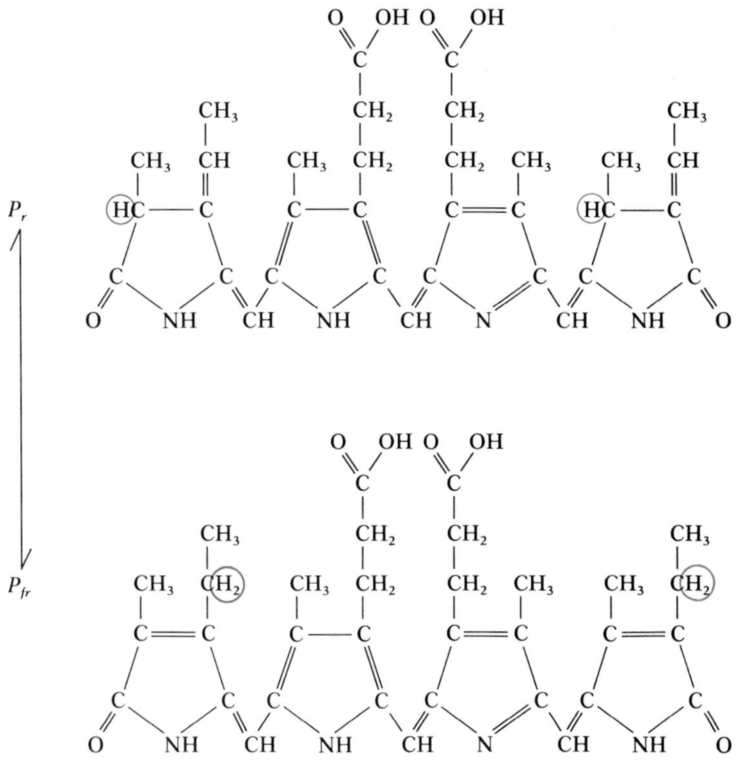

15.28 Phytochrome
Notice that the difference between P_r and P_{fr} lies in the position of two hydrogen atoms (color circles).

converts it into P_r. The P_r form is apparently the more stable of the two; in darkness P_{fr} may revert to P_r over the course of several hours in some plants. In addition, P_{fr} may be enzymatically destroyed. We can summarize these conversions as follows:

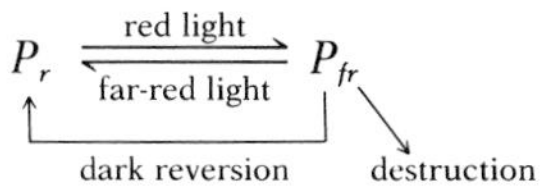

When phytochrome is exposed to red and far-red light simultaneously, the red light dominates and most of the pigment is converted into P_{fr}. Sunlight or light from ordinary electric lamps contains both red and far-red wavelengths; hence, during the day, the phytochrome exists predominantly as P_{fr}. During the night the P_{fr} supply dwindles as a result of both reversion and destruction. The pigment thus gives the plant a way of detecting whether it is day or night.

We began this part of our discussion with the question, How does the plant detect and measure the dark period? We have answered the first half of the question; the plant possesses a sensitive pigment, phytochrome, that responds to presence or absence of light. But what about the crucial second half of the question? It is, after all, the measuring of the dark period that is fundamental to control of flowering.

The most obvious hypothesis would be that the metabolic conversion of P_{fr} into P_r in the dark proceeds so slowly that the amount of conversion occurring between two light periods provides a measure of the length of the intervening dark period. In other words, the system would work like an hourglass. Light would convert all the pigment into P_{fr}. Then during the following dark period the amount of P_{fr} converted into P_r before the next light period would tell the plant how long the dark period had lasted. The evidence is unfortunately against this hypothesis. Though the rate of conversion is strongly temperature-dependent, the plant's measure of time is not significantly influenced by temperature. Moreover, it appears that the supply of P_{fr} declines much too fast to be the basis for time measurement.

The mechanism by which the plant measures the length of the dark period is apparently tied to a phenomenon—now believed to occur in all living cells—involving persistent and regular rhythms in function, rhythms that must be dependent on some internal time-measuring system, or "internal clock." Biological time measurement is an exciting area of modern research, and we shall discuss it at greater length in a later chapter. For the moment, let us say just this: the phytochrome enables the plant to sense whether it is in light or in darkness, but the actual measuring of the time lapse between the moment the plant senses onset of darkness and the moment it senses the next exposure to light must depend on an internal clock.

Once the phytochrome mechanism and the internal-clock mechanism have together indicated to the plant that the photoperiod is appropriate for flowering, the leaves must initiate the next step—altering hormone production or sensitivity to hormone ratios as necessary to bring about actual flowering.

The phytochrome molecules, which are located in the cell membrane and occur in only very tiny concentrations, can powerfully influence a variety of cell activities besides those associated with flowering. For example, germination of some types of seeds requires exposure to red light, which is sensed by phytochrome. Gibberellin-controlled cell elongation, expansion of new leaves, breaking of dormancy in spring, and formation of plastids in cells also involve this pigment. And as we saw earlier, phytochrome is thought by some to be the substance that controls the movement of auxin in stems.

Recent evidence suggests that one of the major functions of phytochrome in plants growing under natural conditions lies in detecting the extent of shading. Because foliage tends to absorb or reflect wavelengths below 700 nm, but to transmit wavelengths in the 700–800 nm (far-red) range, the ratio of red to far-red light in the sunlight reaching a leaf is an indication of the amount of shading by other leaves. This is important information: it would be wasteful for a plant requiring direct sunlight to grow new leaves (or sometimes even to maintain existing ones) in the shade of other leaves, or for a plant pollinated by insects to grow flowers where insects would be unlikely to see them. Clearly, branches should grow out toward the light before producing leaves or flowers. And plants do indeed respond to shading by modifying their normal patterns of stem elongation, amount of branching, leaf pigmentation, and flowering.

Chapter 16

CHEMICAL CONTROL IN ANIMALS

Plant hormones are usually produced by the actively growing parts of plants, and each serves to control growth, development, or both in the tissues it affects. The same hormone often exerts its effects by very different means in different plant tissues. The general pattern in animal hormones stands in sharp contrast to these characteristics. Animal hormones are usually secreted by organs specialized for their production, and after transport through the circulatory system they bind to very specific target tissue. Moreover, though animal hormones help guide growth and development, they also play a large role in regulating metabolism and in maintaining general homeostasis. In fact, as we shall see, the endocrine tissues that produce animal hormones are intimately involved with the nervous system. The evolutionary relationship between nerves and endocrine tissue is important enough to require separate treatment, in Chapter 18.

HORMONES IN INVERTEBRATES

Much interest has developed in the hormones of invertebrate animals in recent years. Hormonal mechanisms have been found in a variety of invertebrates, including arthropods, annelid worms, molluscs, and echinoderms. It seems likely that hormonal control is a general phenomenon in both plants and animals, and that the list of animals in which such control is demonstrated will become steadily longer. At present, however, knowledge of the hormones of most invertebrates is extremely rudimentary. The arthropods, particularly the insects, have been most extensively studied, but even in insects, only a very limited group of hormones involved in growth and development are understood in any detail. Indeed, the first hormone in invertebrates not related to growth control—a chemical in cockroaches that regulates blood-sugar level—was discovered only in 1963. Similar regulatory hormones for salt and water balance and protein metabolism have since been identified.

Much of the early work on insect growth hormones was done in the 1930s in England by V. B. Wigglesworth, who was studying the metamorphosis of insects. Insects show a pattern of growth very different from that of vertebrates. Their body is encased in a hard outer covering, or exoskeleton, that severely limits their size. The insect's tissues grow until they exert considerable pressure against the inner surface of the exoskeleton; further growth is impossible unless the exoskeleton is shed. This is exactly what happens. The insect periodically molts its old exoskeleton and develops a new larger one in its place. Wigglesworth was interested in the mechanisms that control this molting.

Most of his experiments were performed on a bloodsucking bug from South America named *Rhodnius*. This bug goes through five immature or nymphal stages, each separated by a molt, before it becomes an adult. During each nymphal stage it must obtain a blood meal, which engorges and stretches the abdomen. As Wigglesworth demonstrated, this repletion apparently stimulates release of hormones that cause molting at the end of a definite time interval following the meal.

Ordinarily, the last molt (from the fifth nymphal stage to adult) occurs about 28 days after the blood meal. Wigglesworth showed that if *Rhodnius* is decapitated during the first few days after this meal, molting does not occur, even though the animal may continue to live for several months. Decapitation more than eight days after the blood meal does not interfere with molting; a headless adult is produced. Furthermore, if the circulatory system of a bug decapitated eight days after a blood meal is joined to that of a bug decapitated soon after the meal, both bugs molt into adults. Clearly, some stimulus passes via the blood from one insect to the other and induces molting. That stimulus must be a hormone, whose secretion by the head begins about eight days after the blood meal.

It turns out that this **brain hormone**, a polypeptide, stimulates glands in the prothorax (the part of the body immediately behind the head, to which the first pair of legs is attached). The prothoracic glands, in turn, secrete a second hormone, **ecdysone**, which induces molting (Fig. 16.1). Ecdysone, a lipid-soluble steroid, acts on the genes of several types of cells, stimulating the cells to grow and divide.

Wigglesworth became interested in the factors that determine whether a molt will result in an adult or in another immature stage. This is a particularly important question in insects like flies, beetles, and moths, which undergo a radical change, a complete metamorphosis, from immature to adult form—from grub to fly or from caterpillar to moth. Wigglesworth found that a third hormone is involved. This hormone, called **juvenile hormone** or JH, is produced by a pair of glands (corpora allata) located just behind the brain and closely associated with it. When JH is present in high concentration at the time of molting, another immature stage follows the molt. The pupal stage—the changeover stage between the last larval stage and the adult in insects like flies and moths—results from a low concentration of JH. The hormone is absent in the pupa, and when it molts, an adult

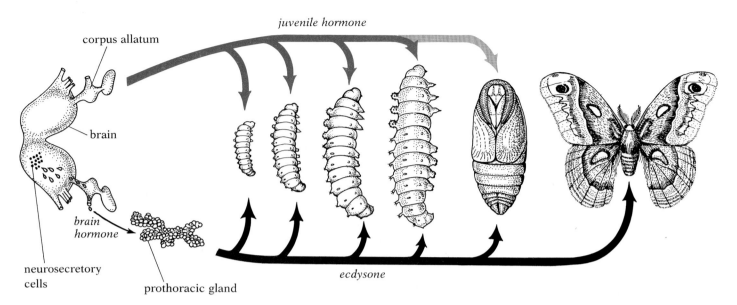

corpus allatum

juvenile hormone

brain

brain hormone

neurosecretory cells

prothoracic gland

ecdysone

16.1 Interactions of juvenile hormone, brain hormone, and molting hormone (ecdysone) in the cecropia silkworm (*Hyalophora cecropia*)

If much JH is present when the insect molts, it will molt into another larval stage. If a low concentration of JH is present, the larva will molt into a pupa. If no JH is present, the pupa will molt into an adult.

results. Removal of the corpora allata from insects in the first or second immature stage results in pupation at the next molt, followed by a molt that results in a midget adult. Conversely, implantation of active corpora allata into insects about to undergo their final molt results in another immature stage instead of an adult; in this way several extra immature growth stages can be inserted into the insect's developmental sequence. These can be followed by pupation and a molt producing an unusually large adult when JH is finally eliminated. Certain trees, notably balsam fir and hemlock, defend themselves against insect pests by synthesizing artificial JH. As a result, insects that attempt to feed on these trees do not ever mature into egg-laying adults.

The hormone-induced morphological changes from larva to adult are accompanied in most insects by profound behavioral changes: adult moths, for instance, fly and walk, drink liquids, and so on, whereas when they were caterpillars they moved in a very different way and chewed solids. Hormones are responsible for initiating many such behavioral changes. The seminal fluid of male mosquitoes, for example, contains a hormone that turns the females' sexual behavior off and induces them to lay their now-fertilized eggs.

Like plant hormones, many insect hormones have different functions in different periods of an organism's life. For example, though the molt from pupa to adult will proceed only if the quantity of juvenile hormone has been substantially reduced from previous stages, secretion of JH usually resumes after metamorphosis is complete. This is because a high concentration of JH is required in many species for females to deposit yolk in their eggs and for males to form mature sperm.

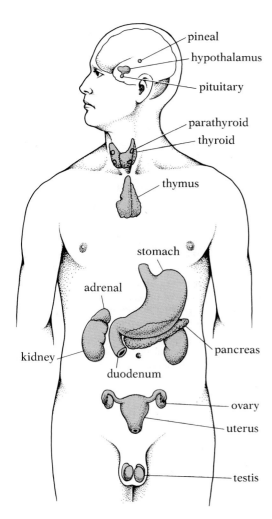

16.2   The major human endocrine organs

HORMONES IN VERTEBRATES

The tissues that produce and release hormones are called ***endocrine*** tissues. The term "endocrine"—from the Greek *endon*, "within," and *krinein*, "to separate" or "secrete"—is meant to convey that the hormones are secreted internally directly into the blood in the capillaries supplying the endocrine tissues, with no special ducts or tubes involved. In fact, endocrine glands are often called the ductless glands. Most vertebrate hormones form weak bonds with plasma proteins and are transported in the blood in a bound form.

Since hormones in mammals are better understood than those of any other group, we shall emphasize the mammalian hormonal system, and especially the human system, in this chapter (Fig. 16.2; Table 16.1).

HORMONAL CONTROL OF DIGESTION

On first consideration, digestion seems too simple to require an elaborate control system: food enters the mouth, moves on to the stomach and the duodenum, and is digested by the enzymes present along the way. But digestion is in fact a complicated process, and is precisely controlled by a series of neural and hormonal mechanisms.

A major step toward understanding digestion came from a series of now-classic experiments performed by the famous Russian physiologist Ivan Pavlov (1849–1936). These showed that both the secretion of saliva and the first phase of gastric secretion (secretion of gastric juice by the stomach) are under nervous control, and that the first phase of gastric secretion actually begins before food reaches the stomach (when the animal tastes or even just sees or smells food). In the second phase of gastric secretion the partially digested food in the stomach leads somehow to the release of more gastric juice. The question Pavlov next sought to answer was whether the mere physical presence of food was sufficient to stimulate this second release.

If Pavlov inserted a piece of meat directly into a dog's stomach without allowing the animal to sense the food, no secretion of gastric juice occurred. Furthermore, stimulation of the stomach wall with a glass rod or with other mechanical devices resulted in very little secretion. If, however, a piece of partly digested meat was inserted directly into the stomach, secretion promptly began. Apparently compounds released from the partly digested meat triggered the secretion. Pavlov concluded that under normal conditions, partial digestion during the first phase of gastric secretion, which is under nervous control, releases compounds that trigger the second phase of gastric secretion, thus making possible the continuation of gastric digestion after nervous stimulation had ceased.

How did the substances from partly digested meat stimulate release of gastric juice? An obvious possibility was that they might act by direct stimulation of the gastric glands. This was ruled out, however, when Pavlov divided the stomach surgically into two chambers and showed that partly digested meat in one chamber stimulated release of gastric juice in both chambers. Direct contact between the meat substances and the glands was thus not necessary to induce secretion. The possibility that gastric secretion was effected by the nervous system was also ruled out, when the same experiment was performed with the stomach completely isolated from nervous control. Perhaps the substances from the partly digested meat were absorbed and carried to the gastric glands by the blood. This, too, was ruled out: injection of the meat substances directly into the blood caused relatively little secretory response, and little or no absorption occurred in the stomach.

At least one other promising possibility remained. Perhaps the meat substances triggered release of a hormone into the blood, and this hormone, in turn, stimulated the gastric glands to begin secreting. In 1905 J. S. Edkins at St. Bartholomew's Hospital, London, combined meat with tissue from the pyloric region of the stomach wall (the region near the opening to the small intestine) and showed that injection of the resulting extract stimulated gastric secretion. Edkins concluded that the meat substances stimulated the mucosa of the pyloric region of the stomach to release a hormone, which he called *gastrin*. He believed that the gastrin was carried by the blood to the gastric glands and stimulated them to secrete gastric juice. Numerous later experiments brought confirmation of these conclusions. For example, when the circulatory systems of two dogs were interconnected, and partly digested food was placed in the stomach of one of the dogs, the gastric glands of both dogs began secreting gastric juice; presumably gastrin from one dog was carried by the blood to the other dog and stimulated its gastric glands.

W. M. Bayliss and E. H. Starling of University College, London, showed in 1902 that secretion of pancreatic enzymes is under the control of a hormone. This hormone, called *secretin*, is released by the mucosal cells of the duodenum when they are stimulated by the acidity of food coming from the stomach. Another hormone, *cholecystokinin*, released by the duodenum under stimulation by acids and fats, stimulates release of bile from the gallbladder.

THE PANCREAS AS AN ENDOCRINE ORGAN

Diabetes mellitus, a disease in which much sugar is excreted in the urine, has been known for centuries, but its causes only began to be understood in the latter half of the nineteenth century. In 1889 two German physicians, Johann von Mering and Oscar Minkowski, who were interested in the role of the pancreas as a producer of digestive enzymes, surgically removed the pancreas from a dog. A short time later they noticed that the dog's urine was attracting an unusual number of ants. Analysis showed that the urine contained a high concentration of sugar. Furthermore, the dog soon developed other symptoms strikingly like those of human diabetes. Von Mering and Minkowski removed the pancreas from other dogs, and diabetes invariably developed. To eliminate the possibility that the extensive damage resulting from so severe an operation might be the causal factor, they performed operations of similar severity in which the pancreas was not actually removed. Since the dogs undergoing this control procedure did not develop symptoms of diabetes, its onset was clearly correlated with the destruction of the pancreas. But operations in which the pancreatic duct

TABLE 16.1 *Important mammalian hormones*

Source	Hormone	Principal effects
Pyloric mucosa of stomach	Gastrin	Stimulates secretion of gastric juice and pancreatic enzymes
Mucosa of duodenum	Secretin	Stimulates flow of pancreatic enzymes; inhibits gastrointestinal motility and gastric acid secretion
	Cholecystokinin	Stimulates release of bile from gallbladder
Pancreas	Insulin	Stimulates glycogen formation and storage, glucose oxidation, and synthesis of protein and fat; inhibits formation of new glucose
	Glucagon	Stimulates conversion of glycogen into glucose
Adrenal medulla	Adrenalin	Stimulates elevation of blood-glucose concentration and other fight-or-flight reactions
	Noradrenalin	Stimulates reactions similar to those produced by adrenalin
Adrenal cortex	Glucocorticoids (corticosterone, cortisol, cortisone, etc.)	Stimulate formation of carbohydrate from protein and fat, thus helping maintain normal blood-sugar levels
	Mineralocorticoids (aldosterone, deoxycorticosterone, etc.)	Stimulate kidney tubules to reabsorb more sodium, chloride, and water, and less potassium
	Cortical sex hormones	Stimulate development of secondary sexual characteristics, particularly those of the male
Thyroid	Thyroxin, triiodothyronine (together called TH)	Stimulate oxidative metabolism; help regulate growth and development
	Calcitonin	Prevents excessive rise in blood calcium
Parathyroids	Parathyroid hormone (PTH)	Regulates calcium-phosphate balance
Thymus	Thymosin	Stimulates immunologic competence in lymphoid tissues
Hypothalamus	Releasing hormones	Regulate hormone secretion by anterior pituitary

was destroyed and the disease did not result made it clear also that diabetes was not correlated with absence of the pancreatic digestive enzymes. The unavoidable inference was that the pancreas participated in body functions other than the digestion.

Mounting evidence pointed to secretion by the pancreas of some substance preventing diabetes in the normal animal, but all attempts at proof failed. Feeding bits of pancreas to diabetic dogs had no effect; if the pancreas contained a control chemical—a hormone—it was destroyed by digestive enzymes. Repeated efforts by numerous investigators to show that injection of an extract made from the pancreas would alter diabetic symptoms also failed. Grinding pancreatic tissue to produce the extracts evidently mixed the hormone with the pancreatic digestive enzymes, which destroyed the hormone. But how could this be avoided?

TABLE 16.1 *Important mammalian hormones (cont.)*

Source	Hormone	Principal effects
Posterior pituitary (storage of hypothalamic hormones)	Oxytocin	Stimulates contraction of uterine muscles; stimulates release of milk by mammary glands
	Vasopressin	Stimulates increased water reabsorption by kidneys; stimulates constriction of blood vessels (and other smooth muscle)
Anterior pituitary	Growth hormone (STH)	Stimulates growth; stimulates protein synthesis, hydrolysis of fats, and increased blood-sugar concentrations
	Prolactin (PRL)	Stimulates milk secretion by mammary glands; participates in control of reproduction, osmoregulation, growth, and metabolism
	Melanophore-stimulating hormone (MSH)	Probably helps regulate salt and water balance; may influence certain types of behavior; controls cutaneous pigmentation in ectotherms
	Thyrotropic hormone (TSH)	Stimulates the thyroid
	Adrenocorticotropic hormone (ACTH)	Stimulates the adrenal cortex
	Follicle-stimulating hormone (FSH)	Stimulates growth of ovarian follicles and of seminiferous tubules of the testes
	Luteinizing hormone (LH)	Triggers ovulation; stimulates conversion of follicles into corpora lutea; stimulates secretion of sex hormones by ovaries and testes
Pineal	Melatonin	Helps regulate production of gonadotropins by anterior pituitary, perhaps by regulating hypothalamic releasing centers
Testes	Testosterone	Stimulates development and maintenance of male accessory reproductive structures, secondary sexual characteristics, and behavior; stimulates spermatogenesis
Ovaries	Estrogen	Stimulates development and maintenance of female accessory reproductive structures, secondary sexual characteristics, and behavior; stimulates growth of the uterine lining
	Progesterone	Prepares uterus for embryo implantation and helps maintain pregnancy

It was known that the pancreas is a compound organ, one containing several types of cells, and these were thought to function independently. There are the cells involved in production and release of digestive enzymes, and there are other quite different cells, called *islet cells* or islets of Langerhans (Fig. 16.3). It seemed likely that the hormone so many people were searching for was produced by the islet cells (more specifically, by β islet cells). In a critical experiment that supported this hypothesis and opened the way for isolation of the hormone, the pancreatic duct was tied off; the result was atrophy of most of the pancreas, but not the development of diabetes. Examination of the atrophied pancreas revealed that the enzyme-producing portion had atrophied, while the islet cells had remained essentially intact. The hormone that prevents diabetes must have come from this portion of the pancreas.

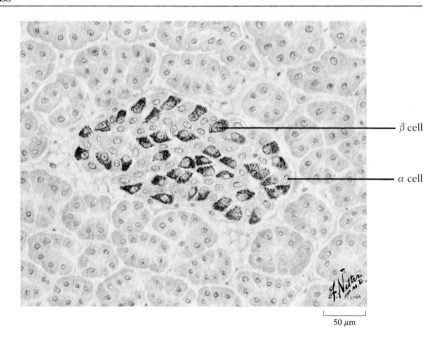

β cell

α cell

50 μm

16.3 Drawing of an islet of Langerhans
The endocrine cells of the pancreas form an islet,
clearly distinct from the surrounding cells, whose
function is secretion of digestive enzymes. The α islet
cells secrete the hormone glucagon; the β islet cells,
insulin.

The hormone, ***insulin***, was finally isolated in 1922 by F. G. Banting and
C. H. Best, working in the laboratory of J. J. R. MacLeod at the University of
Toronto. They tied off the pancreatic ducts of a number of dogs, waited until
the enzyme-producing tissue had atrophied, removed the degenerated pancreas and froze and macerated it in an isotonic medium (freezing prevents
any remaining digestive enzymes from acting), filtered the solution, and
quickly injected the filtered material into diabetic dogs. The dogs showed
marked improvement. Banting and Best also obtained good results with
extracts prepared from the pancreases of embryonic animals; since the
islet cells develop in the embryo before the enzyme-producing cells, there
are no enzymes to destroy the insulin during the extraction procedure.
Banting and MacLeod received the Nobel Prize in 1923 for this important
work.

In most respects Banting and Best followed a procedure considered
standard for demonstrating that a particular organ or tissue has an endocrine function. Let us outline the essential criteria of that procedure:

1. Removal or destruction of the organ in question should result in
   predictable symptoms presumed to be associated with absence of
   the hormone.
2. Administration of material prepared from the organ in question
   should relieve the symptoms.
3. The hormone should be present in both the organ and the blood, and
   the hormone should be extractable from each.

Fortunately, preparing extracts of suspected organs has not always been
as difficult as it was with the pancreas before Banting and Best solved
the problem.

Insulin was crystallized by J. J. Abel of Johns Hopkins University in 1926. As we have already seen, it was the first protein for which the complete amino acid sequence was determined (by Frederick Sanger in 1954, as described on p. 61; despite the familiarity of insulin, however, you should keep in mind that most hormones are not proteins). We now know that the β islet cells of the pancreas first synthesize a much longer polypeptide zymogen—an inactive enzyme precursor—called proinsulin. Insulin becomes active when 35 amino acids are removed from the middle of its precursor (Fig. 16.4).

A high concentration of sugar in the urine is a major symptom of diabetes. How is insulin related to this symptom? Before we can attempt an answer, we must examine the symptom further. The presence of sugar in the urine of a diabetic means, not that the kidneys are functioning improperly, but that the blood-sugar concentration is higher than normal and the kidneys are removing part of the excess. You will recall that the liver plays a critical role in regulating blood-sugar levels. When blood coming to the liver via the portal vein from the intestines contains a higher than normal concentration of sugar, the liver removes much of the excess and stores it as glycogen. Conversely, when blood coming to the liver is low in sugar, the liver converts some of its stored glycogen into glucose, which it adds to the blood. Other parts of the body, particularly the muscles and adipose (fatty) tissue, are also important elements in this regulatory system. When the blood-sugar concentration rises after a carbohydrate meal, part of the excess glucose is stored as glycogen in the muscles, and part is converted into fat and stored by adipose tissues; the rate of oxidation of glucose in the liver and muscles may also increase under these conditions.

This brief outline of the interplay between liver, muscles, and adipose tissues—all three target tissues of insulin—helps explain the following known actions of insulin, which have the net effect of reducing the concentration of glucose in the blood.

1. Insulin stimulates (probably by altering membrane permeabilities) absorption of more glucose from the blood by muscle cells and adipose cells, but it does not influence uptake of glucose by liver cells.
2. It promotes both oxidation of glucose and incorporation of glucose into glycogen in liver and muscle cells—actions whose effect is to reduce the supply of free glucose.
3. It inhibits metabolic breakdown of stored glycogen in liver and muscle cells.
4. It promotes synthesis of fats from glucose by adipose cells and also inhibits metabolic breakdown of fat.
5. It promotes uptake of amino acids by liver and muscle cells, and favors protein synthesis while inhibiting protein breakdown.

Notice that though the first three actions concern carbohydrate metabolism, the last two deal more directly with the metabolism of fats and proteins. But the promotion of fat and protein synthesis and the slowing of their catabolism force the cells to rely more heavily on glucose as a source of metabolic energy; and the result is a reduction in the supply of free glucose. Thus we see once again that the metabolic pathways for all classes of

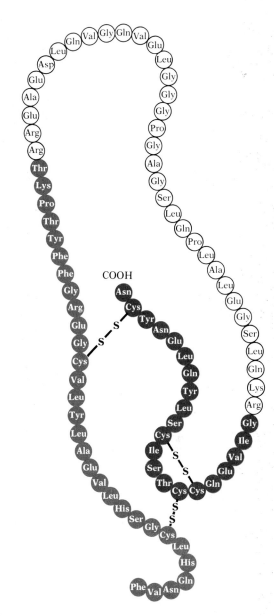

16.4 Activation of human insulin
The protein is synthesized initially as a single polypeptide zymogen—an inactive precursor—with three internal disulfide bonds. Removal of a long section (white circles) by an activating enzyme leaves the active hormone as two separate short polypeptide chains (color) held together by disulfide bonds.

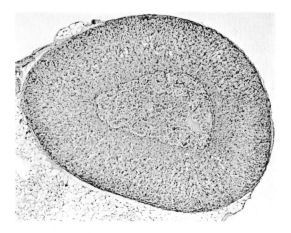

16.5   Cross section of the adrenal gland of a mouse
The medulla (central region) is entirely surrounded by cortex.

nutrients form an interlocking system; alteration of one pathway unavoidably influences the others.

Too much insulin in the system, whether from an overactive pancreas or from administration of too large a dose to a diabetic, can produce a severe reaction called insulin shock. The blood-sugar level falls so low that the brain, which has few stored food reserves of its own, becomes overirritable; convulsions may result, followed by unconsciousness and often death. A naturally occurring excess of insulin is, however, extremely rare. Far more common is a deficiency of insulin (or an insensitivity of the tissues to insulin), and it is this that is called diabetes. The liver and muscles do not convert enough glucose into glycogen, the liver produces too much new glucose, and utilization of carbohydrate in cellular respiration is impaired. The blood-sugar level rises above normal, and part of the excess glucose begins to appear in the urine. More water must be excreted as a vehicle for this glucose, and the diabetic therefore tends to become dehydrated. The glycogen reserves become depleted as more and more glucose is poured into the blood and lost in the urine; yet the body still lacks sufficient energy, because of the impairment of carbohydrate metabolism. As the body begins to metabolize its reserves of proteins and fats—particularly the latter—the diabetic becomes emaciated, weak, and easily subject to infections. As if this were not enough, the excessive but incomplete metabolism of fats releases toxic substances that, seriously disturbing the delicately balanced pH of the body, often have a major part in the eventually fatal outcome of the untreated disease.

The pancreas secretes another polypeptide hormone besides insulin. This hormone, called **glucagon**, is produced in the α islet cells. Glucagon has effects opposite to those of insulin; it causes an increase in blood-glucose concentrations. Fat cells have both insulin and glucagon receptors, and the balance between these contradictory hormones controls the metabolism of the cells. Again we see that the normal functioning of an organism depends on a delicate balance between opposing control systems; if one of these systems is disturbed and the proper balance destroyed, abnormalities result, and in severe cases the abnormalities may lead to disease or even death.

THE ADRENALS

The endocrine organs examined so far—stomach, duodenum, and pancreas—and the gonads (sex organs), which will be discussed in the next chapter, all have other major functions in addition to hormone secretion. They are multipurpose organs; each is an important component in both the endocrine system and some other system. By contrast, the adrenals (and also the thyroid, the parathyroids, and the pituitary) are endocrine glands whose only known function is hormone secretion.

The two adrenal glands, as their name implies (ad- means "at" and renal refers to the kidney), lie very near the kidneys (Fig. 16.2). In mammals each adrenal is actually a double gland, composed of an inner corelike **medulla** and an outer barklike **cortex** (Fig. 16.5). The medulla and cortex arise in the embryo from different tissues, and their mature functions are unrelated. In fact, they remain as two separate pairs of glands in adult fish and amphibians. In reptiles, birds, and mammals the two sets of glands have evolved a close spatial relationship (in reptiles and birds the tissues of the two are actually intermingled, and do not form a distinct cortex and medulla, as in most mammals), but they remain functionally distinct.

The adrenal medulla   The adrenal medulla secretes two hormones, *adrenalin* (also known as epinephrine) and *noradrenalin* (norepinephrine), whose functions are similar but not identical. Both hormones have been isolated, identified, and synthesized in the laboratory (Fig. 16.6). They can be shown to produce a great variety of effects on the body. For example, adrenalin causes a rise in blood pressure, acceleration of heartbeat, increased conversion of glycogen into glucose and release of glucose into the blood by the liver, increased oxygen consumption, release of reserve erythrocytes into the blood from the spleen, vasodilation and increased blood flow in skeletal and heart muscle, vasoconstriction and decreased blood flow in the skin and in the smooth muscle of the digestive tract, inhibition of intestinal peristalsis, erection of hairs, production of "gooseflesh," and dilation of the pupils.

At first glance, this list may seem to include a curious assortment of seemingly unrelated effects, but a more careful examination shows that these reactions occur together in response to intense physical exertion, pain, fear, anger, or other heightened emotional states; they have sometimes been called fight-or-flight reactions. They are in fact more often generated initially by the nervous system, but after any rapid neural signaling, adrenalin (and to a lesser extent noradrenalin) help maintain them. These hormones aid in mobilizing the resources of the body in response to emergencies by stimulating reactions that increase the supply of glucose and oxygen carried by the blood to the skeletal and heart muscles, and that help inhibit functions, such as digestion, which are not immediately important during the emergency and might otherwise compete with the skeletal muscles for oxygen.

It is probably as an antagonist to insulin that adrenalin fulfills its most important normal function. It elevates blood sugar by stimulating the liver to produce glucose from its glycogen reserves, and it acts on muscles to transform their glycogen stores into lactic acid, which is converted into glucose after being transported to the liver by the blood.

16.6   Structural formulas of adrenalin, noradrenalin, and tyrosine
The two hormones are derived from the amino acid tyrosine.

The adrenal cortex   The adrenal cortices are essential for life; their removal is fatal. Death is preceded by a severe disruption of ionic balance in the body fluids, lowered blood pressure, impairment of kidney function, impairment of carbohydrate metabolism with a marked decrease in both blood-glucose concentration and stored glycogen, loss of weight, general muscular weakness, and a peculiar browning of the skin. These symptoms are also seen in varying degrees in individuals whose adrenal cortices are insufficiently active, a condition known as Addison's disease.

The numerous symptoms of adrenal cortical insufficiency listed here are not related to a single hormone. The adrenal cortex is, in fact, an amazing endocrine factory, producing so many different hormones that scientists still have no idea of the total number. All the cortical hormones are chemically similar: all are steroids manufactured through modifications of the

Aldosterone

Cortisone

Cortisol

common membrane component cholesterol, and many differ from each other by only one or two atoms (Fig. 16.7). Yet because of these apparently minor differences, the various hormones have strikingly different functions: they bind to different receptors in target cells, and affect different sets of chemical reactions. Note that these hormones are chemically unlike the other hormones we have discussed. In mammals, only the hormones of the adrenal cortex and those of the gonads and other reproductive structures are steroids. The hormones of the other endocrine organs, as far as is known, are amino acids (or, like adrenalin and noradrenalin, compounds derived from amino acids), short polypeptide chains, or full-sized proteins (like insulin). Cortical steroids, which have enormous importance in vertebrates, may be grouped into three categories on the basis of their functions: (1) those that act primarily in regulating carbohydrate and protein metabolism, called glucocorticoids; (2) those that act primarily in regulating salt and water balance, called mineralocorticoids; and (3) those that function primarily as sex hormones.

Hormones in the first category, **glucocorticoids** (like cortisol, corticosterone, cortisone), cause a rise in blood sugar and an increase in liver glycogen; both effects probably result from an increased rate of conversion of protein into carbohydrate. The hormones also inhibit oxidation of glucose while promoting mobilization of fat reserves. In short, the glucocorticoids tend to elevate blood-sugar levels by stimulating the body to draw on its noncarbohydrate energy sources. When administered to a person with Addison's disease, they restore the blood-sugar level to normal. Thus they act as antagonists to insulin.

Hormones in the second category, **mineralocorticoids** (especially aldosterone), stimulate the cells of the convoluted tubules of the kidneys to decrease reabsorption of potassium and increase reabsorption of sodium and chloride, actions that lead also to increased reabsorption of water. The reabsorption of these substances, in turn, causes a rise in blood volume and blood pressure. Animals deprived of mineralocorticoid soon begin excreting large quantities of urine containing high concentrations of sodium; their blood volume decreases and their blood pressure falls. If not given hormone-replacement therapy, the animals quickly die.

Hormones in the third category are very similar both chemically and functionally to the sex hormones produced by the gonads. Their normal role is not yet fully understood, but we do know that tumors or other dis-

Cholesterol

16.7 Some steroids secreted by the adrenal cortex, and the steroid—cholesterol—from which all are synthesized
Very slight differences in side chains can result in markedly different properties.

turbances of the adrenal cortex may cause excessive secretion of these hormones, especially of male hormones, resulting in masculinizing effects on females and early sexual development in males.

By and large, cortical steroids of all three classes pass through the cell membrane (some of them operating in conjunction with intercellular or intracellular receptors) and then act directly on the DNA, as we shall show in more detail in a later section.

The medical use of some of the cortical hormones has an interesting history. Soon after cortisone was isolated in 1935, it was found to have a remarkable ability to increase a test animal's resistance to exposure, cold, poisons, and other physiological stresses. The first test of the hormone as a therapeutic agent for human beings was performed at the Mayo Clinic on a young women suffering from severe rheumatoid arthritis. All previous attempts to relieve her symptoms had failed, and she could hardly move. By the third day after injections of cortisone were begun she could move easily, and by the eighth day all her symptoms were essentially gone. If the injections were stopped, however, the symptoms promptly returned. Trials of cortisone on other arthritic patients yielded similar results.

Next, various investigators began trying cortisone on a host of other diseases, many of them unrelated. And most of these investigators reported dramatic relief of symptoms, though seldom any actual cures. For example, all pneumonia symptoms of a boy with severe lobar pneumonia disappeared within 24 hours after the administration of hormone (in this case the hormone used was not actually cortisone but a pituitary hormone, ACTH, that stimulates release of cortisone by the adrenal cortex). But the bacteria that cause pneumonia remained in large numbers in his body. Similarly, all disease symptoms of patients with tuberculosis disappeared several days after treatment with hormone was begun, but tuberculosis bacteria still swarmed in their bodies. In each case disease symptoms immediately returned if treatment was suspended.

You can imagine the excitement such results stirred among doctors (and in the newspapers). It was hoped that ways would be found to produce actual cures with cortisone and related hormones, but even if that proved impossible, relief of symptoms for everything from rheumatism to cancer with a single drug seemed to signal the beginning of a new era in medicine. It was thought that stress opens the body to disease symptoms and that cortisone helps the body withstand stress. Hence administration of extra cortisone, by giving the body greater ability to withstand stress, would minimize its susceptibility to the damaging disease symptoms normally elicited by infections and by metabolic disturbances.

Cortisone and related hormones may to some extent actually function in this manner. They may indeed be involved in the body's reaction to stress, and stress may well play a fundamental role in disease. But, unfortunately, they have not proved to be the panacea first envisioned. When administered over a long period of time, they often cause side effects as bad as or worse than the condition being treated; among them are high blood pressure, excessive growth of hair, mental aberrations, lowered resistance to certain infections such as poliomyelitis and tuberculosis, peptic ulcers, and brittle bones that are easily fractured. That these cortical hormones were not found to be a universal panacea is, in fact, not so surprising after all. Being

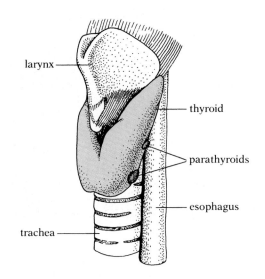

16.8 The thyroid and parathyroid glands

larynx

thyroid

parathyroids

esophagus

trachea

well and active is adaptive, so if the body already made a hormone that relieved illness without generating harmful side effects, the ability to secrete that hormone in response to illness would most likely have been reinforced through natural selection. No such hormone has yet been found.

Cortisone and its chemical relatives are now most frequently used to facilitate healing where administration need be repeated only a few times, or to give partial relief from the symptoms of arthritis and other diseases of connective tissues (where they apparently cause changes in the collagen fibers of such tissues). They are also sometimes used to treat severe allergic diseases, particularly asthma, and some types of lymphatic diseases. And they are used for temporary relief of severe symptoms in emergency situations. Because physicians must weigh the possible harmful side effects against the hoped-for symptomatic relief, they seldom prescribe enough to give full symptomatic alleviation of a chronic condition, preferring to have the patient endure mild symptoms in order to minimize the harmful side effects.

The dilemma presented by the cortical hormones serves as an example of a general problem faced by physicians every day. Most drugs—and other treatments, for that matter—have potential harmful side effects. Physicians must therefore always balance possible good against possible harm, and they must remember that even the safest drugs are dangerous when used in excessive quantity or at the wrong time. The body is, after all, a finely tuned machine, with interactions between its parts so intricate that they still largely defy analysis. There is a risk of damage to the machine when it is subjected to treatment with chemicals that almost always affect more different functions than can be predicted. Medicine has a long way to go before it can eliminate a very large element of guesswork from its practice, but it has come a long way, too, and the future looks encouraging as the increasing tempo of biological discovery provides the basis for new medical applications.

THE THYROID

Most vertebrates have two thyroid glands, located in the neck; in humans the two have fused to form a single gland (Fig. 16.8). There is convincing evidence that the vertebrate thyroids evolved from ventral pouches of the pharynx. These pouches probably functioned originally as channels in which food particles were strained from water currents flowing into the mouth and out through the gill slits. In the vertebrates, however, the developing pouches soon lost all connection to the pharynx and became both structurally and functionally independent of the digestive system. This is but one example of a common evolutionary occurrence—the development of a new structure from an ancestral structure with an apparently unrelated function.

Years ago a condition known as goiter, in which the thyroid may become so enlarged that the whole neck looks swollen and deformed (Fig. 16.9), was very common in some areas of the world, such as the Swiss Alps and the Great Lakes region of the United States. Goiter is often associated with a group of other symptoms, including dry and puffy skin, loss of hair, obesity, a slower than normal heartbeat, physical lethargy, and mental dullness. No cause for this condition was known. Then in 1883 a Swiss surgeon, who believed that the thyroid had no important function, removed the gland from a number of his patients. Most of these patients developed all the symptoms usually associated with goiter, except the swelling of the neck. The result suggested that the normal thyroid must secrete some chemical that prevents these symptoms. The curious fact that patients with no thyroid and patients with the excessively large thyroid of a goiter showed the same complex of symptoms could be explained if the malformed gland of the goiter was, despite its large size, secreting too little hormone. By the 1890s patients with goiters or the other symptoms of hypothyroidism[1] were being successfully treated with injections of thyroid extract or with bits of sheep thyroid in their diets. But nothing more specific was known as yet about the hypothetical thyroid hormone itself. True, a German chemist, E. Baumann, discovered in 1896 that the thyroid contains iodine, an element previously unknown in the body. But little attention was paid to his discovery.

In 1905 David Marine of Western Reserve University noticed that many people in Cleveland had goiters. A high percentage of the dogs also had goiters. Even many of the trout in the streams had goiters. Marine wondered if the goiters might be caused by an insufficiency of iodine in the food and water. When he administered tiny traces of iodine in water to his experimental animals, their goiters and other symptoms disappeared. In 1916 Marine tried his treatment on approximately 2,500 schoolchildren in Akron, Ohio. He fed these children iodized salt. Another 2,500 children, used as controls, were fed uniodized salt. At the end of a specified period he found only two cases of goiter among the children who had eaten iodized salt, but 250 cases among the controls. Though it took years to convince a skeptical public, the use of iodized salt finally became widespread, and hypothyroidism caused by insufficient iodine in the soil and water now seldom occurs in the United States or Europe.

The requirement for iodine became more understandable when a thyroid hormone, now known as ***thyroxin*** or T_4, was isolated in 1914 and synthesized in the laboratory in 1927. It proved to be an amino acid containing four atoms of iodine (Fig. 16.10). Later, another thyroid compound, identical to thyroxin except that it contains only three atoms of iodine, was found. This substance, called ***triiodothyronine*** or T_3, is three to five times more active than thyroxin, but is secreted in smaller amounts. Because T_4 and T_3 have virtually identical effects on target cells, they are usually considered together, under the designation "thyroid hormone" or TH.

The most characteristic effect of TH is stimulation of increased oxidative metabolism of most tissues of the body. The hormone, which is lipid-soluble, is thought to cross the cell membrane directly and act to alter gene expression. As a consequence, TH stimulates increased synthesis of certain enzymes, including mitochondrial respiratory enzymes, that facilitate an elevated basal metabolic rate (BMR). Hyperthyroidism—excessive secretion of TH—produces many symptoms that you might predict: a higher than normal body temperature, profuse perspiration, high blood pressure, loss of weight, irritability, and muscular weakness. It also produces one

---

[1] The prefix *hypo-* means "less than normal," while the prefix *hyper-* means "more than normal." Hence hypothyroidism means less than normal thyroid activity, and hyperthyroidism means more than normal thyroid activity.

16.9 A young Tanzanian woman with goiter
The scars on the woman's neck are the result of folk-medicine treatment.

16.10 Structural formula of thyroxin
Triiodothyronine has the same formula, except that the iodine atom shown at upper left is replaced by a hydrogen.

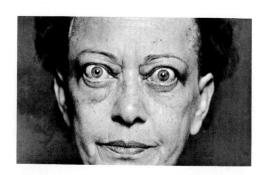

16.11   Exophthalmia

very characteristic symptom that you might not predict, because it lacks an obvious connection to a higher than normal metabolic rate: exophthalmia, a startling protrusion of the eyeballs (Fig. 16.11). Though hyperthyroidism can sometimes be controlled with antithyroid drugs, it is more often treated by surgical removal of part of the gland or by partial destruction of the gland with radioactive iodine.

When hypothyroidism—the opposite of hyperthyroidism—is caused by malfunction of the thyroid gland itself rather than by dietary iodine insufficiency, it is treated by administration of thyroid hormone. The untreated condition is particularly serious when it occurs in the newborn. These victims, grotesque in their development, are called cretins. They are dwarflike, and they never mature sexually. They have very low intelligence, seldom achieving a mental age of more than four or five years. Prevention of cretinism by early administration of hormone to babies showing deficiency symptoms is surely one of the triumphs of modern medicine.

Though many of the symptoms of hypothyroidism—such as slow heartbeat, obesity, physical lethargy, and mental dullness—may well be consequences of a low BMR, other symptoms, especially some seen in cretinism, cannot easily be explained in this way. An example is the highly abnormal distribution of protein in the bodies of cretins; they have an excessive amount of glycoprotein in the skin—the cause of their puffy appearance—and an unusually high concentration of protein in the blood plasma, but their kidneys and liver are severely deficient in protein and therefore markedly underdeveloped. Administration of TH relieves all these symptoms. It seems clear, then, that TH plays an important role in regulating the synthesis and distribution of protein.

The effect of TH on protein metabolism is but one manifestation of its more general role in the regulation of many aspects of development. Most vertebrates cannot develop normal adult form and function without the hormone. Not only is TH necessary for the protein synthesis required for proper growth; it is also essential for functional maturation of the testes and ovaries, and it acts synergistically with growth hormone from the pituitary gland in promoting skeletal development. In many lower vertebrates TH is necessary for metamorphosis and for molting.

In 1961 another thyroid hormone, called *calcitonin*, was discovered. It is functionally unrelated to TH, its chief effect being the prevention of an excessive concentration of calcium in the blood. In lower vertebrates calcitonin is produced by separate glands (the ultimobranchial glands); the corresponding tissue in mammals becomes incorporated into the thyroid during embryonic development.

THE PARATHYROIDS

The parathyroid glands in humans are small pealike organs, usually four in number, located on the surface of the thyroid (Fig. 16.8). They were long thought to be part of the thyroid or to be functionally associated with it. Now, however, we know that their close proximity to the thyroid is misleading; both developmentally and functionally, they are totally separate.

The parathyroid hormone, usually designated PTH, helps to regulate the calcium-phosphate balance between the blood and the other tissues. Consequently it is an important element in maintaining the relative constancy of the internal fluid environment of the body, a subject discussed at length in Chapter 14. We have already seen that such hormones as insulin, glucagon, and calcitonin are also important in this regard. PTH increases the concentration of calcium in the blood (thus functioning as a calcitonin antagonist), and decreases the concentration of phosphate, by acting on at least three organs: the kidneys, the intestines, and the bones. It inhibits excretion of calcium by the kidneys and intestines, and it stimulates release of calcium into the blood from the bones (which contain more than 98 percent of the body's calcium and 66 percent of its phosphate). But calcium in bone is bonded with phosphate, and breakdown of bone releases phosphate as well as calcium. PTH compensates for this release of phosphate into the blood by stimulating excretion of this material by the kidneys. Actually, it overcompensates, causing more phosphate to be excreted than is added to the blood from bone; the result is that the concentration of phosphate in the blood drops as the secretion of PTH increases.

Naturally occurring hypoparathyroidism is very rare, but the parathyroids are sometimes accidentally removed during surgery on the thyroid. The result is a rise in the phosphate concentration in the blood and a drop in the calcium concentration (as more calcium is excreted by the kidneys and intestines and more is incorporated into bone). This change in the fluid environment of the cells produces serious disturbances, particularly of muscles and nerves. These tissues become very irritable, responding even to very minor stimuli with tremors, cramps, and convulsions. Complete absence of PTH is usually soon fatal unless very large quantities of calcium are included in the diet. Injections of PTH are effective in preventing the symptoms.

Hyperparathyroidism sometimes occurs naturally when the glands become enlarged or develop tumors. PTH is then produced in such quantity that the opposing action of calcitonin is no longer able to maintain a proper balance. The most obvious symptom of this condition is bones that are weak and easily bent or fractured, because of the excessive withdrawal of calcium from them.

THE THYMUS

One of the functions of the thymus, a gland in the neck region particularly prominent in young animals, is production of a hormone important in stimulating immunologic competence in the plasma cells of the spleen, lymph nodes, and other lymphoid tissues. This hormone, called ***thymosin***, has been the subject of intensive research in the last few years. The thymus and thymosin will receive further attention in our discussion of the immune system in Chapter 30.

THE PITUITARY AND THE HYPOTHALAMUS

The pituitary (also called the hypophysis) is a small gland lying just below the brain. Like the adrenals, the pituitary is a double gland (Fig. 16.12). It consists of an anterior lobe, which develops in the embryo as an outgrowth from the roof of the mouth, and a posterior lobe, which develops as an out-

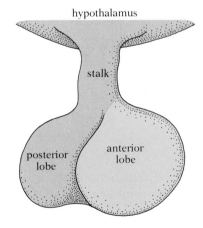

16.12   The pituitary gland

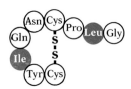

Oxytocin

Vasopressin

16.13 Oxytocin and vasopressin
The two hormones differ by only two amino acids,
but this difference accounts for their distinctive
activities.

growth from the lower part of the brain. The two lobes eventually contact each other as they grow, and the anterior lobe partly wraps itself around the posterior lobe. In time, the anterior lobe loses its original connection with the mouth, but the posterior lobe retains its stalklike connection with a part of the brain called the **hypothalamus**. Despite their intimate spatial relationship, the two lobes remain fully distinct functionally, and we shall consider them separately.

The posterior pituitary   Two hormones, *oxytocin* and *vasopressin*, are released by the posterior pituitary. As Figure 16.13 shows, they are chemically very similar. Each contains nine amino acids, seven of which occur in both compounds. Only two are different, yet they suffice to give the two hormones very different properties.

Oxytocin acts mainly on the muscles of the uterus, causing them to contract; hence a release of oxytocin generates labor in a pregnant woman. What causes this release is not yet understood. Injection of oxytocin can induce labor artificially.

Vasopressin causes constriction of the arterioles, and a consequent marked rise in blood pressure. It also stimulates the kidney tubules to reabsorb more water. This function—the source of its alternative name, "antidiuretic hormone"—is the more important under normal conditions. A human totally lacking vasopressin would have to excrete more than 20 liters of urine daily. The well-known diuretic effect of ethyl alcohol results from its tendency to suppress vasopressin release.

We said earlier that the posterior pituitary originates as an outgrowth of the hypothalamus of the brain. Even in the adult it retains a stalklike connection with the hypothalamus (Fig. 16.12). Oxytocin and vasopressin do not originate in the posterior pituitary, but are produced by nerve cells in the hypothalamus and flow along their axons through the stalk to the posterior pituitary, where they are stored. The storage organ releases the hormones on stimulation by electrical signals from the hypothalamus.

The anterior pituitary   The anterior pituitary (also called the adenohypophysis) is an immensely important organ that produces hormones with widely varying and far-reaching effects. At least seven hormones are secreted by the anterior pituitary in humans.

Prolactin (PRL, also called lactogenic hormone) is the most versatile of all the pituitary hormones. It stimulates milk production by the female mammary glands shortly after the birth of a baby; its continued production depends on the mechanical stimulation provided by a suckling infant, and its absence causes milk production to cease. Prolactin also plays a variety of roles in reproduction, osmoregulation, growth, and the metabolism of carbohydrates and fats.

Another versatile pituitary hormone, *growth hormone* (also called somatotropic hormone or STH), plays a critical role in promoting normal growth, especially in combination with thyroid hormone. It is a very powerful inducer of protein anabolism, favoring both cellular uptake of amino acids and their incorporation into protein; in this role it acts as an antagonist to the glucocorticoids, inhibiting their protein-catabolizing influence. In addition, it is an insulin antagonist, acting to elevate blood-sugar concentration. Growth hormone also stimulates hydrolysis of fats in adipose tissues, thereby increasing fatty-acid concentrations in the blood.

If the supply of growth hormone is seriously deficient in a child, growth will be stunted and the child will be a midget. Oversupply of the hormone in a child results in a giant. (The tallest pituitary giant on record reached a height of 8 feet 11.1 inches.) Both pituitary midgets and pituitary giants have relatively normal body proportions, and their appearance is not unattractive. However, if oversecretion of growth hormone begins during adult life, only certain bones, such as those of the face, fingers, and toes, can resume growth. The result is a condition known as acromegaly, characterized by disproportionately large hands and feet and distorted features—a greatly enlarged and protruding jaw, enlarged cheekbones and eyebrow ridges, and a thickened nose.

The anterior pituitary also secretes a number of very important hormones that exert controlling action on other endocrine organs. These hormones include *thyrotropic hormone* (also called thyroid-stimulating hormone or TSH), which stimulates the thyroid; *adrenocorticotropic hormone* (ACTH), which stimulates the adrenal cortex; and at least two *gonadotropic hormones*, or gonadotropins (follicle-stimulating hormone or FSH, and luteinizing hormone or LH), which act on the gonads. Proper growth and development of these endocrine glands depend on adequate secretion of the appropriate tropic (stimulatory) hormone from the pituitary; if the pituitary is removed or becomes inactive, these organs atrophy and function at very low levels. It is easy to understand why the pituitary is often called the master gland of the endocrine system.

The interaction between the anterior pituitary and the other endocrine glands over which it exerts control is an example of negative feedback, a strategy very common in living systems. Just as the familiar household thermostat senses when the temperature falls below a desired level—specified by its setting—and takes appropriate action by switching on the furnace, so an organic feedback mechanism senses variation from a desired norm and takes appropriate action in response. And just as the thermostat senses when the required temperature has been attained and switches off the furnace, so does the organic feedback mechanism sense when the desired norm has been attained and switch off the action. Negative feedback acts to compensate for change, restoring a lost homeostasis, while positive feedback acts to amplify change.

Thyrotropic hormone provides an example of how the body uses negative feedback. This two-chain glycoprotein is released by the pituitary when the concentration of thyroxin in the blood is low and stimulates increased production of thyroxin by the thyroid; but the resulting rise in concentration of thyroxin in the blood inhibits secretion of more thyrotropic hormone by the pituitary. In other words, the pituitary responds to a low thyroxin level in the blood by sending a chemical messenger that stimulates increased activity by the thyroid. But once the thyroid becomes more active, the increased amount of thyroxin produced tells the pituitary that release of thyrotropic hormone can now be reduced. There is thus a feedback of information from the thyroid to the pituitary. The pituitary exerts control over the thyroid, and the thyroid, in turn, exerts some control over the pituitary. Each sends chemical messengers to the other. But note that

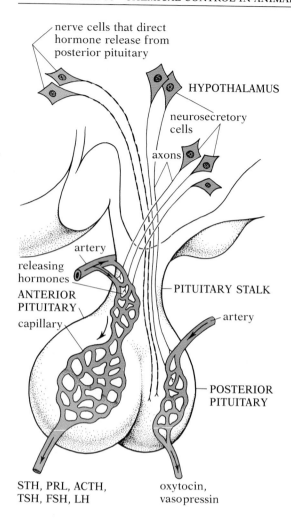

nerve cells that direct
hormone release from
posterior pituitary

HYPOTHALAMUS

neurosecretory
cells

axons

artery

releasing
hormones

ANTERIOR
PITUITARY

capillary

PITUITARY STALK

artery

POSTERIOR
PITUITARY

STH, PRL, ACTH,
TSH, FSH, LH

oxytocin,
vasopressin

16.14   Connections between the pituitary and the hypothalamus

The pituitary and the hypothalamus are intimately associated. Certain nerve cells in the hypothalamus secrete hormones directly into a capillary bed which then empties into another capillary bed in the anterior pituitary. The hormones transported in this manner regulate the release of the hormones synthesized in the anterior pituitary. Other nerve cells in the hypothalamus produce hormones that move down their axons to the posterior pituitary; the hormones are released directly into the capillaries there. The cells of the posterior pituitary capture these hormones from the blood and store them until their release into the bloodstream is triggered by electrical signals from another set of nerve cells in the hypothalamus.

the message from the pituitary to the thyroid is a stimulatory one, while the return message from the thyroid to the pituitary is an inhibitory one; the feedback is negative. The pituitary tends to speed up the system, and the thyroid tends to slow it down. The interaction between the two opposing forces—the antagonist strategy we have seen so frequently in this chapter—produces a delicately balanced control system.

The interaction of the pituitary with the adrenal cortex and with the gonads is similar to its interaction with the thyroid. The pituitary responds to low levels of cortical hormones by secreting more ACTH and to low levels of sex hormones by secreting more gonadotropic hormone. The resulting rise in concentration of cortical hormones or of sex hormones inhibits further secretion by the pituitary.

The anterior pituitary, important as it is as a regulator of other endocrine glands, does not, so far as is known, participate in the control of the pancreas, the adrenal medullae, or the parathyroids. Both the pancreas and the adrenal medullae are controlled in part by the nervous system (the pancreas is also influenced by adrenalin), and the parathyroids are thought to be regulated primarily by the concentration of calcium ions in the blood.

The control function of the hypothalamus   The delicately balanced feedback interaction between the anterior pituitary and other endocrine glands is not the only factor that regulates the regulator. The desired norm must be changed from time to time. For example, many animals, like the plants discussed in the preceding chapter, have annual reproductive cycles. And as in plants, day length is frequently the critical cue. Suppose an animal detects the lengthening of days in spring, and its gonads secrete more sex hormone. How is its perception of the stimulus—increasing day length—able to affect its endocrine glands? Perception of stimuli involves the nervous system, and there are no nervous connections either to the anterior pituitary or to endocrine glands such as the gonads. But nervous tissue is capable of producing hormones, as we saw in the interaction of the hypothalamus and the posterior pituitary. And it is the hypothalamus that acts as the main control center for gathering and integrating neural information bearing on the appropriate "setting" of the chemical thermostat.

As Figure 16.14 indicates, hormones from the hypothalamus regulate hormone secretion by the anterior pituitary, though there is no direct physical connection between the hypothalamus and the anterior pituitary. Neurosecretory cells in the hypothalamus synthesize hormone, and transport it down their axons to capillary beds in the stalk. There is an unusual connection between the blood supplies of the stalk and the anterior pituitary. Arteries to the stalk break up into capillaries, and these capillaries eventually join to form several veins leading away from the hypothalamus. But unlike most veins, these do not run directly into a larger branch of the venous system; instead, they pass downward into the anterior pituitary and there break up into a second capillary bed. Thus hormones secreted into the stalk capillaries travel to the capillaries in the anterior pituitary (Fig. 16.14).

We have encountered two other places in the body where the circulation depends on two beds of capillaries arranged in sequence: the kidney nephrons, where one bed forms the glomerulus and the other envelops the tubules, and the hepatic portal system, where one bed is in the wall of the intestine and the other is in the liver. In both places the special type of circulation provides an important functional arrangement. In the hepatic portal system, for example, many substances picked up by the blood in the first capillary bed are removed from the blood in the second capillary bed. The portal system linking the hypothalamus and the anterior pituitary seems to function in a similar fashion. Thus the relationship between the body's two principal control systems—nervous and endocrine—becomes much easier to understand.

The hypothalamus is now known to produce a variety of special peptide hormones called *releasing hormones*, which, carried directly to the anterior pituitary by the portal vessels, regulate the secretory activity of the anterior pituitary. Thus TSH-releasing hormone (TRH)—a peptide only three amino acids long—stimulates the release of TSH; LH-releasing hormone (ten amino acids long) controls LH levels; corticotropic releasing hormone stimulates ACTH release; growth hormone–releasing hormone stimulates release of growth hormone; and so on. A few of the hypothalamic hormones are inhibitory rather than stimulatory: prolactin release–inhibiting hormone, for instance, inhibits release of prolactin by the anterior pituitary. As we shall see in more detail in Chapter 18, nervous and hormonal control are parts of an integrated control system.

The sequence of responses triggered in our hypothetical animal by the perception of changing day length can now be reconstructed. The animal detects the light with its sense organs; nervous impulses are transmitted from the sense organs to the hypothalamus of the brain; the hypothalamus secretes gonadotropic releasing hormone (GnRH) into the blood; the releasing hormone stimulates the anterior pituitary to increase its secretion of gonadotropic hormones; the gonadotropic hormones stimulate the gonads to secrete more sex hormone; and the sex hormone helps prepare the animal physiologically to begin the activities of the breeding season.

The hypothalamus is not only the point of entry for information transmitted to the endocrine system from the nervous system, it is also one of the major points for the transmission of feedback information from the endocrine system. Let us take the previously discussed negative-feedback effect of a rise in thyroxin on the anterior pituitary as an example. To some extent, probably, the thyroxin exerts negative feedback on the pituitary directly, by inhibiting its secretion of TSH, but to a large extent it does so indirectly, by inhibiting the hypothalamus from secreting TSH-releasing hormone (Fig. 16.15). Similarly, much of the negative feedback action of other hormones is via the hypothalamus.

THE PINEAL

The glandular appearance of the pineal, a lobe in the rear portion of the forebrain, has long intrigued investigators. Aristotle thought it must be the site of the soul, since, unlike the rest of the brain, in which all structures exist as pairs (the right and left lobe of the hypothalamus, for example), the pineal is unpaired. Only recently, however, has the endocrine function of the pineal been demonstrated.

In some lower vertebrates the pineal is eyelike and responds to light both by generating nervous impulses and by secreting a hormone called *mela-*

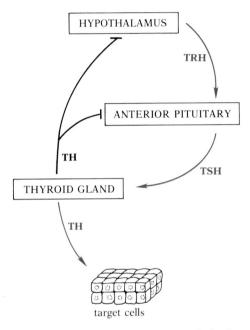

16.15 Feedback control of the thyroid gland Thyrotropic releasing hormone (TRH) from the hypothalamus stimulates the anterior pituitary to release thyrotropic hormone (TSH), which stimulates the thyroid gland to secrete thyroxin (TH). The thyroxin stimulates target cells throughout the body, but it also inhibits both the hypothalamus and the anterior pituitary. (Color arrows indicate stimulatory influences, barred black lines inhibitory influences.)

tonin, which lightens the skin by concentrating the pigment granules in melanophores (pigment-containing cells). It has been shown that in these animals the pineal is intimately involved (presumably through melatonin) in the control of circadian rhythms—cycles of activity repeated approximately every 24 hours. In salamanders, the light sensitivity of the pineal may even provide a means of celestial navigation.

The mammalian pineal, too, secretes melatonin, but it has no light-sensitive cells, and mammals have no melanophores. Though the pineal is a part of the brain, its principal (perhaps only) innervation originates outside the skull cavity in the neck. Hence the mammalian pineal and its hormone have long been a puzzle.

Evidence accumulated in the last few years suggests that the pineal may function as a neurosecretory transducer, converting neural information about light conditions into hormone output. If this view of the process is accurate, the conversion would take place as follows: Information about the light-dark cycle received by the eyes would go first (via the brain) to the nerve cells in the neck, and would then be conveyed to the pineal; the pineal would respond by secreting melatonin in inverse proportion to the amount of light; the melatonin, in turn, would influence the secretion of gonadotropic hormones by the anterior pituitary. It might do this directly, or by causing the secretion of a peptide in the hypothalamus that would then carry the message to the pituitary. The usual effect of melatonin is apparently an inhibitory one; in autumn, for example, as the days get shorter, increased production of melatonin tends to turn off gonadotropic secretion, with the result that the gonads regress and the animal enters a nonreproductive phase.

The sequence of events is different in animals that breed in autumn. Moreover, melatonin may sometimes stimulate the testes instead of inhibiting them; whether the hormone is gonadotropic or anti-gonadotropic probably depends on when in the circadian cycle it is released.

Melatonin may affect parts of the brain other than the hypothalamus, particularly those concerned with locomotor rhythms, feeding rhythms, and other biological rhythms. But until more research is done, all proposals concerning the functioning and effects of the pineal and its hormone remain highly speculative.

MECHANISMS OF HORMONAL ACTION

The number of mammalian hormones and the variety of their functions is enormous. But despite this variety, only a limited number of mechanisms for intercellular communication are at work. Some of these strategies for the communication of somatic control information are very simple, some are familiar from our discussion of the cell membrane (Chapter 4), and some are unique to hormonal control.

As we mentioned in the last chapter, hormones gain entrance to target cells in various ways. Some, like the steroid hormones, cross the cell membrane either directly or bound to a receptor (Fig. 16.16A). A few can pass into the cell through special channels (Fig. 16.16B; this is a rare strategy), and some enter by being transported actively (Fig. 16.16C). Finally, most hormones do not enter the cell at all, but instead bind to extracellular

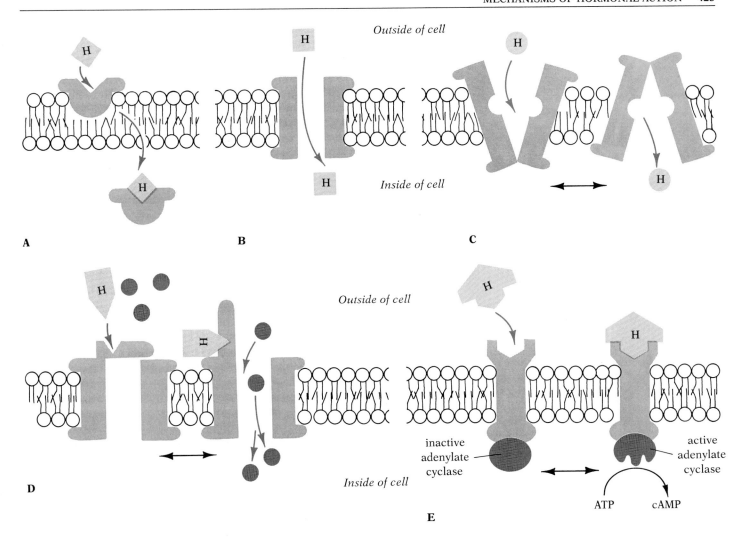

16.16 How hormones transmit messages
A small hormone can enter its target cell by simply diffusing through the membrane or binding to an intercellular receptor molecule (not shown). Other small hormones bind to a membrane receptor and enter the cytoplasm with it as a conjugated pair (A). Rarely, a small hormone may enter by passing through a specific membrane channel (B). More often, it may bind to a receptor which then transports it into the cell, either with or without the expenditure of ATP (C). Once inside, each type of hormone transmits its message by directly or indirectly altering a specific chemical reaction.

A larger hormone may transmit its message by facilitating the entrance into the target cell of a second substance, sometimes an ion, as in the case of the chemically gated ion channels by which most nerve cells communicate (D). Or a larger hormone may bind to and activate a membrane protein (E), which then causes another substance to be generated or activated on the inside of the membrane. In the version of this strategy shown here (and discussed further in the text) the binding of the hormone activates adenylate cyclase, which in turn catalyzes conversion of ATP into a so-called second messenger, cAMP.

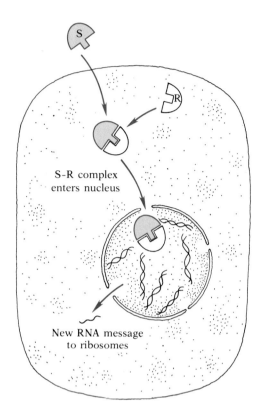

16.17   Model for the mode of action of steroid hormones

The hormone (S), being lipid-soluble, can penetrate the plasma membrane and enter the cytoplasm. There it binds with a receptor protein (R), and the complex then enters the nucleus, where it influences synthesis of RNA by the genes. A new message (in the form of new RNA) is thus sent to the ribosomes, which begin synthesizing the protein coded by the new RNA. The new protein, perhaps an enzyme, will then influence the chemical activity of the cell.

receptors. Extracellular binding causes an intracellular effect, either by opening an ion channel in the membrane (Figure 16.16D) or by activating a second messenger inside the cell (Fig. 16.16E). In the next section we shall consider in more detail how these various mechanisms work to control cell metabolism.

Within the cell, too, methods of hormonal control vary. Some, like testosterone, bind to the DNA and directly affect gene expression and, as a result, the production of specific enzymes. Others, like adrenalin, control the activity of enzymes already synthesized or exert their effects by altering structural proteins. Another set of strategies involves the interaction of different hormones, like insulin and glucagon, with their opposite effects on glucose metabolism. In the next sections we shall consider some of these methods in more detail.

REGULATION OF GENE EXPRESSION

Steroid hormones, such as the ecdysone of insects and the hormones produced in vertebrates by the adrenal cortex and the gonads, can easily move through membranes and penetrate into the cytoplasm of the target cell. There the steroid (S) binds to a specific cytoplasmic receptor protein (R). The complex (S-R) then moves into the nucleus, where it helps regulate the activity of specific genes, influencing what RNA messages are transcribed from the DNA and exported from the nucleus to the cytoplasm (Fig. 16.17). In other words, by interacting with the genetic material of the target cells, steroid hormones help determine what instructions for protein synthesis (especially enzyme synthesis) are sent from the nucleus to the ribosomes.

Thyroxin is another hormone which, it seems likely, can easily pass through membranes and enter cells, to alter the activity of genes in the nucleus or influence mitochondrial enzyme activity, or both.

SECOND MESSENGERS

The role of cyclic AMP   The 1960s saw some exciting progress in the attempt to learn how hormones act on their target cells. A major contribution was that of E. W. Sutherland and T. W. Rall of Western Reserve University, who were investigating the mechanism by which glucagon and adrenalin stimulate liver cells to release more glucose into the blood. They discovered that these hormones stimulate an increase in the intracellular concentration of a compound called cyclic adenosine monophosphate—*cyclic AMP*, or cAMP for short (Fig. 16.18). They found that this compound leads, in turn, to activation of an enzyme necessary for breakdown of glycogen to glucose. Subsequent research has demonstrated that, in addition to glucagon and adrenalin, a large number of other hormones act on their target cells either to increase or to decrease the concentration of cAMP. The far-reaching implication of this work earned Sutherland, now at Vanderbilt University, a Nobel Prize in 1971.

As you have doubtless deduced from its name, cAMP is a compound related to ATP (adenosine triphosphate) (see Fig. 7.2, p. 171). Very widely distributed in nature, it has been found in almost all animal tissues studied (vertebrate and invertebrate) and in bacteria. It is synthesized from ATP in living cells by a reaction catalyzed by an enzyme called ***adenylate cyclase***, which appears to be built into the cell membrane.

As evidence accumulated that many hormones, among them glucagon and adrenalin, did not actually enter their target cells, but rather form weak bonds with receptor sites on the cell membrane, the ***second-messenger model*** of hormonal control was proposed. According to this model, an extracellular first messenger, which is the hormone itself, goes from an endocrine gland to the target cell and there stimulates production of an intracellular second messenger, which is often cAMP. More particularly, the binding of the hormone to a highly specific receptor site on the outer surface of the membrane of the target cell activates adenylate cyclase, which catalyzes production of more cAMP on the inner surface of the membrane (Fig. 16.19). The increased amounts of cAMP then interact with cytoplasmic enzyme systems and thus initiate the cell's characteristic responses to the hormonal stimulation. In other words, the initial extracellular signal (the hormone, or first messenger) is converted into an intracellular signal (cAMP, or second messenger) that the chemical machinery of the cell can more readily understand.

Let us return to the stimulation of glucose production in the liver by glucagon or adrenalin as an example of the mechanism by which the increased amount of cAMP influences the cell. In this case the cAMP activates an enzyme of a class called protein kinase. The activated protein kinase, in turn, activates a second enzyme, which activates a third enzyme, which catalyzes the first reaction in the breakdown of glycogen to glucose. We can diagram

16.18   Cyclic AMP

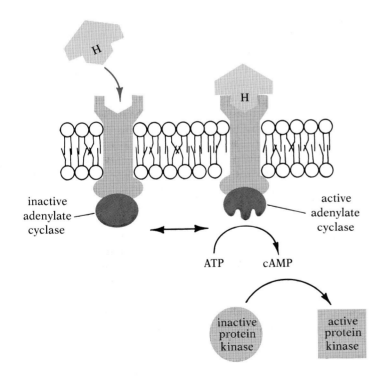

16.19   The second-messenger model
In this version of the strategy, the binding of hormone to a receptor on the membrane results in the intracellular activation of the enzyme adenylate cyclase, which is attached to the membrane receptor. The adenylate cyclase catalyzes conversion of ATP into cAMP, which in turn activates another enzyme, a protein kinase. The protein kinase, in its turn, activates certain enzymes and inactivates others by phosphorylating them.

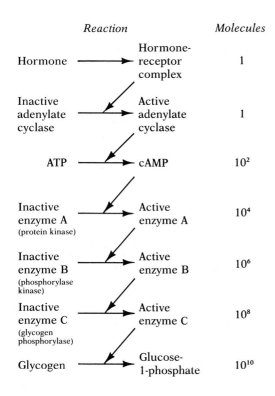

Reaction		Molecules
Hormone $\longrightarrow$	Hormone-receptor complex	1
Inactive adenylate cyclase $\longrightarrow$	Active adenylate cyclase	1
ATP $\longrightarrow$	cAMP	10^2
Inactive enzyme A (protein kinase) $\longrightarrow$	Active enzyme A	10^4
Inactive enzyme B (phosphorylase kinase) $\longrightarrow$	Active enzyme B	10^6
Inactive enzyme C (glycogen phosphorylase) $\longrightarrow$	Active enzyme C	10^8
Glycogen $\longrightarrow$	Glucose-1-phosphate	10^{10}

16.20   Hormonal stimulation of glucose production

this chain of events, beginning with formation of the hormone-receptor complex, as shown in Figure 16.20.

This cascade of enzyme-catalyzed reactions, an example of a *cascade reaction*, makes possible a very great ***amplification*** of the effects of the original hormone-binding event. Because an enzyme can be used over and over again, a single molecule of active adenylate cyclase may catalyze production of about 100 molecules of cAMP. Each molecule of cAMP may, in turn, catalyze production of roughly 100 molecules of active enzyme A, and so on. The net result is that a single molecule of glucagon or adrenalin may lead to release of as many as 10^{10} (10 billion) molecules of glucose, within only one or two minutes. We see, then, why communication requires only very small quantities of hormone.

The above sequence of reactions is not the only result of stimulation of a target cell by glucagon or adrenalin. You will recall from our earlier discussion of those hormones that release of glucose is just one of a variety of ways in which they bring about a rise in blood-sugar levels. Another of their important actions, for example, is to inhibit synthesis of glycogen. This action, like the one already discussed, depends on protein kinase activated by cAMP. The protein kinase (enzyme A), which we previously saw activating an enzyme necessary for glucose production, is here inactivating an enzyme and hence blocking the conversion of glucose into glycogen (Fig. 16.21). There is convincing evidence that protein kinase is an essential link in all hormonal actions mediated by the adenylate cyclase system.

We have seen that both glucagon and adrenalin can initiate the processes we have examined. These two hormones are chemically very different, however. How can they have the same effect on adenylate cyclase? The answer is that though they bind to different receptor molecules on the membrane, both types of receptors are capable of activating adenylate cyclase. Thus the two hormones have additive effects. But this is true only for the liver; the two hormones do not have the same effect on muscle. Acting via adenylate cyclase, adrenalin can stimulate production of glucose from glycogen by muscle cells, but glucagon cannot do so, apparently because the membranes of muscle cells lack receptors for glucagon.

From this example of glucose regulation, we can draw several important conclusions. First, it is the presence or absence of receptors on the cell membrane that determines whether or not a particular cell type is influenced by a given hormone—that is, whether or not it is a target of that hormone. Second, a cell may respond in the same way to two or more different hormones if it has cyclase-activating receptors for each of those hormones; the hormones themselves are specific only for the receptors, not for the reactions that their binding triggers via adenylate cyclase activation.

These rules of hormone action help explain how the many different hormones that depend on the adenylate cyclase system (including—in addition to glucagon and adrenalin—gastrin, secretin, parathyroid hormone, calcitonin, the tropic hormones of the anterior pituitary, and at least some of the hypothalamic releasing hormones) can have their own distinct sets of target cells, even though their immediate effect in every case is activation of adenylate cyclase. It need only be assumed that different types of cells have different hormone-specific receptor sites. Thus cells in the thyroid might have receptors on which only thyrotropic hormone could react, while cells in the adrenal cortex might have receptors on which only ACTH could react. Such an arrangement would ensure that the thyroid would be stimulated by thyrotropic hormone and not by ACTH, and that the adrenal cortex would be stimulated by ACTH and not by thyrotropic hormone.

Much more difficult to answer is the question of why different cells respond so differently to changes in their cAMP content. It is by no means clear why an increase in cAMP should cause increased thyroxin production by thyroid cells, increased cortisone production by adrenal cortex cells, increased glucose production by liver cells, and increased protein synthesis by uterine cells. Two assumptions have to be made. The first is that a vast array of different chemical processes in cells can be regulated by cAMP, and there is abundant evidence that this assumption is valid. The second is that the response of a given cell to a change in cAMP concentration depends on its own chemical makeup. The thyroid, adrenal, liver, and uterine cells have different chemical profiles. Hence a hormonally induced rise in cAMP will take place against vastly different enzymatic backgrounds in each of these four cell types, and the effects of the rise may be correspondingly different.

Various chemicals other than hormones also influence cells by their effects on cAMP levels. Some of these, like histamine, work through adenylate cyclase, but others exert their effect by acting on the enzyme ***phosphodiesterase***, which breaks down cAMP. In a normal cell a certain amount of this enzyme, which keeps the concentration of cAMP in check, is always present. Thus, if adrenalin induces a rise in the cAMP concentration, phosphodiesterase causes a return to more normal levels as soon as the hormone is no longer present. It is clear that a chemical capable of inhibiting this enzyme, like caffeine, or of activating it, like nicotine, can markedly influence the cAMP levels in the cell.

The role of cAMP as a second messenger so impressed some researchers when it was first discovered that they envisioned it as a possible universal mediator of the action of hormones and other hormonelike compounds. This idea was soon shown to be incorrect. For example, recent evidence suggests that the principal effect of insulin (whose action is antagonistic to that of glucagon and adrenalin) is not an alteration of cAMP concentration, but a rise in the concentration of a related substance, cGMP (cyclic guanine monophosphate) that has effects opposite to those of cAMP. Several other control chemicals (especially acetylcholine in the nervous system) are thought to act via cGMP, but the details of the cGMP system, and its relationship to the adenylate cyclase system, are not yet well understood.

Ions as second messengers   We said earlier that hormones that exert their effects on cell metabolism without entering the cell do so either by activating a second messenger or by opening an ion channel. In fact, the cAMP system and the ion-channel strategy are so intimately connected that it probably makes more sense to think of the ions as second messengers as well.

By far the most common second-messenger ion is Ca^{++}. Calcium ions can serve this function because the concentration of Ca^{++} in the cytosol is normally kept low: Ca^{++} ions are actively pumped both into the ER and out of the cell; they are also taken up and stored by the mitochondria, and are

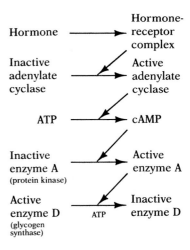

16.21 Hormonal inhibition of glucose synthesis

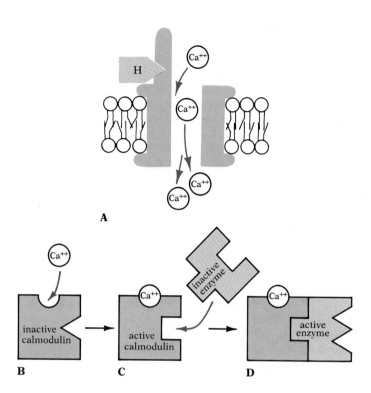

16.22 Calmodulin activation

The entrance of Ca^{++} ions in response to the binding of a hormone to its receptor (A) activates calmodulin (B, C), which in turn binds to and activates one or more specific enzymes in the cell (D). (Calmodulin actually has four Ca^{++} binding sites, only one of which is shown.)

captured by various binding molecules free in the cytosol. As a result, when a hormone binds to its specialized receptor and thereby opens a Ca^{++} channel in the membrane, the strongly favorable electrochemical gradient causes Ca^{++} ions to rush in. These ions then bind to and activate particular intracellular enzymes, the most common and best known of which is **calmodulin**, a protein consisting of 148 amino acids. The Ca^{++}-calmodulin complex then activates enzymes of the particular system being controlled (Fig. 16.22). Which enzymes are regulated in a cell depends on the kind of cell it is—that is, on the particular genes active in the cell. Like cAMP, calmodulin controls different sets of enzymes in different sorts of cells.

The operation of the ion channel and the Ca^{++}-calmodulin complex can be seen in muscle cells. In preparing skeletal muscles for high activity by stimulating the breakdown of glycogen to glucose, adrenalin uses not only cAMP but also Ca^{++} as a second messenger. The binding of adrenalin to its specialized receptor opens a channel to Ca^{++}, and the resulting calmodulin complex binds to phosphorylase kinase which in turn activates glycogen phosphorylase, the enzyme that converts glycogen into glucose-1-phosphate and sends it into the glycolytic pathway. Thus in muscles both cAMP and activated calmodulin are necessary to stimulate the production of glucose; the binding of the activated calmodulin to phosphorylase kinase is part of the kinase's activation. This interaction of cAMP, activated calmodulin, and the several enzymes gives us a glimpse of the elaborate interconnections underlying the chemistry of complex organisms.

INTERACTION OF SEVERAL HORMONES

We have seen many cases in which one hormone triggers another, or in which two hormones act either synergistically or antagonistically to exert control. The antagonistic strategy, of which there are several examples in the next chapter, is relatively common: one aspect of metabolism (concentrations of particular substances in the blood, for instance) will be increased by one hormone and decreased by another. This type of control mechanism has at least two advantages over the use of a single hormone. The first is speed: secreting an antagonist to stop a process is faster than merely waiting for the stimulatory hormone to disappear. The second advantage is precision: a system that depends on the *ratio* of one hormone to another can be very precise in its response, even in regions of the body in which the absolute concentration of the hormones may be low as a result of poor circulation, depletion by receptors upstream, and the like.

LOCAL CHEMICAL MEDIATORS

According to our working definition, animal hormones are substances produced by organs specialized for hormone synthesis; they are transported through the circulatory system; and they exert highly specific effects on target tissue. We have excluded neurotransmitters (chemicals, like acetylcholine, that convey messages from one nerve cell to another or to a muscle cell) because the cell that produces the communication chemical releases it directly onto the target cell, and the effect is to alter the electrical rather than the metabolic activity of that target. Somewhere between the long-distance chemical effects of the hormones of the endocrine system and the direct cell-to-cell electrical interactions of neurotransmitters are the effects of local chemical mediators. These substances are secreted by cells that are not part of specialized chemical-control organs, and in some cases are so rapidly destroyed that they affect only cells in their immediate neighborhood. In other cases they can be transported by the bloodstream to distant targets.

Histamine   Perhaps the best-known local chemical mediator is histamine, a derivative of the amino acid histidine (Fig. 16.23). Histamine is produced primarily by the mast cells in connective tissue. These cells store histamine in large vesicles, and release it when they detect injury or a toxic substance in the vicinity. Histamine signals nearby capillaries to dilate and leak, thus allowing more blood to reach the site and enabling cells of the immune and cell-repair systems to leave the capillaries and reach the tissue in need of help. (Of course, fluid also leaks out, causing swelling.) The histamine also attracts these repair cells, and they in turn secrete compounds that inactivate histamine.

One of the most obvious effects of a cold—the generalized swelling of nasal tissues—results from histamine release, and for this reason the antihistamines in many cold preparations provide some symptomatic relief. Most allergic reactions (among them the rare reaction to wasp or honeybee venom) are a consequence of uncontrolled histamine release after a

16.23   Histamine and its precursor, histidine

16.24   Prostaglandin E, one of the prostaglandins

16.25   Morphine
The ring at the upper center is at right angles to the rest of the molecule.

malfunction of the immune system. The loss of blood fluids to the tissues causes blood pressure to fall, and the swelling of the tissues sometimes blocks the flow of air to the lungs. A severe allergic reaction can be fatal unless treated quickly by an injection of adrenalin, and even mild reactions such as hay fever can be debilitating. We shall examine how the immune system works and what underlies its occasional malfunctions in a later chapter.

Prostaglandins   Since 1957 much attention has been focused on a group of substances called prostaglandins, which, though named for the prostate gland, where they were originally discovered, are now known to be secreted by most animal tissues. Prostaglandins exhibit a bewildering array of actions, including stimulation or relaxation of smooth muscle, dilation or constriction of blood vessels, stimulation of intestinal motility, modulation of synaptic transmission in the nervous system, stimulation of inflammation responses, and enhancement of the perception of pain. The effectiveness of aspirin in combating inflammation and pain is a result, at least in part, of its inhibition of prostaglandin synthesis.

Prostaglandins (Fig. 16.24) are continually being synthesized from phospholipids in the cell membrane and released into the bloodstream or surrounding fluids. They exert control through changes in their rate of synthesis, which in turn is controlled by a great variety of stimuli, including hormones, nervous stimuli, mechanical stimuli, oxygen deprivation, and inflammation. Interestingly enough, prostaglandins can affect the cells that manufacture them, as well as their neighbors. In some cases they circulate in the blood like hormones and exert their effects at distant locations. In others, they reach distant target sites without transport through the bloodstream; the prostaglandins in the semen secreted by male seminal vesicles, for instance, cause contractions of the uterine muscles in the female.

Prostaglandins bind to receptors on target cells. Studies of their effects on a variety of these cells have revealed that they often mimic the effects of the corresponding stimulatory hormones, apparently by utilizing a variant of the cAMP second-messenger mechanism. Like histamines, they are continually being destroyed by enzymes in the intercellular fluid and cell membranes.

Endorphins   Closest to the conceptual line separating hormones and neurotransmitters are the endorphins. Like hormones, they may act on cells at some distance, being carried by the bloodstream, but their targets are nerve cells and they affect electrical rather than metabolic activity. The discovery of this group of compounds resulted from research on opiates, a group of highly addictive pain-killing drugs. It eventually became evident that opiates like morphine (Fig. 16.25) bind to specific receptors on nerve cells, and since investigators had learned that receptors characteristically bind naturally occurring chemicals, the search was on for such chemicals, which would be internally synthesized pain-killing opiates.

The search was successful. The body is now known to synthesize more than half a dozen polypeptides that bind to the opiate receptors. Of these, two short-chain polypeptides are produced by nerve cells throughout the nervous system, while the others (Fig. 16.26), which have the same terminal sequence as the first two, are produced in the anterior pituitary. The common terminal sequence probably explains why they all bind to the same receptors, while the differences in chain length give rise to the variations in their activity. Why there are so many different endorphins, and how they interact with their target cells, is not yet understood.

THE EVOLUTION OF HORMONES

One of the most pervasive trends in the organization of multicellular organisms is the specialization of particular cells and groups of cells. Two points about the evolution of such specialization are worth bearing in mind. First, many simple organisms face the same problems as more complex plants and animals with respect to transport, water balance, control of growth, and so on. Each cell in these unspecialized organisms, therefore, must perform many different duties. Second, evolution usually proceeds by modifying or building on systems and structures already present. Accordingly, it seems likely that some of the problems we share with less specialized organisms may have been solved by them in simpler ways, and that these solutions may in some cases have provided the raw materials for the elaborate mechanisms we now employ.

The evolution of hormonal control seems to underscore both these points. Even animals with specialized endocrine organs also have unspecialized communication networks of the sort that would be expected in their simpler ancestors. We have already seen how various nonglandular cells can communicate through histamine, for instance. A more complex example involves insulin, which plays some roles in the body's information-processing system that we have not yet discussed. Insulin in the bloodstream is largely prevented from entering the brain by a thick membrane—the blood/brain barrier. But the brain is nevertheless full of cells with insulin receptors, for insulin is secreted by the brain cells themselves, which thus turn out to be less specialized than the pancreatic cells devoted exclusively to insulin production. This locally produced insulin serves a regulatory function in the brain quite independent of sugar metabolism in the body.

That this unspecialized insulin communication route is indeed a heritage from simpler forebears is suggested by the discovery of the hormone in insects, protozoans, fungi, and even procaryotes like *E. coli*. The insulin of *E. coli* is so similar to human insulin that it promotes glucose oxidation in human fat cells, thus passing the standard test for insulin activity. In all likelihood, then, insulin originated as a nonglandular messenger before the vertebrate pancreas evolved, and its extended range of activities in complex organisms is an elaboration of its role in simpler organisms.

Nor is insulin the only hormone with a long pedigree: protozoans have biologically indistinguishable versions of *β*-endorphin and ACTH. As for receptors, adrenalin activates the adenylate cyclase system in protozoans just as it does in us, and is blocked by the same chemicals. Just why microorganisms make and respond to what are for us specialized hormones remains to be understood, but the possibility that the specialized hormones and receptor systems in higher plants and animals evolved from their intracellular equivalents in unicellular organisms helps explain why there is no distinct functional line dividing the "official" hormones from the many other chemicals made by cells for controlling metabolism and development.

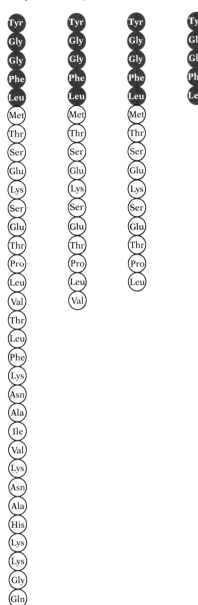

16.26 Endorphins
All four of these endorphins have the same sequence of amino acids at one end. Indeed, the sequences of all four are identical as far as they go. The differences among them result from differences in length.

Chapter 17

HORMONES AND VERTEBRATE REPRODUCTION

From an evolutionary standpoint, reproduction is the ultimate goal of life. All the other aspects of living discussed in this book—nutrient procurement, gas exchange, internal transport, waste excretion, osmoregulation, growth, hormonal and nervous control, and behavior—can be viewed, in a sense, as processes that enable organisms to survive to reproduce. If "the hen is the egg's way of producing another egg," the idea is equally applicable to human beings; we are, in a physiological sense, elaborate devices for producing eggs and sperm, for bringing them together in the process of fertilization, and for giving birth to young. The behavioral, ecological, and evolutionary bases of reproduction, as well as genetics and growth, will be discussed in later chapters. Here we consider the physiology of (primarily) mammalian reproduction as an example of the complex interplay between hormonal and other control mechanisms.

THE PROCESS OF SEXUAL REPRODUCTION

Sexual reproduction in higher animals involves the union of two parental cells called *gametes*, one an egg cell (ovum) and the other a sperm cell (spermatozoon). Occasionally the same individual produces both gametes, which unite in a process known as self-fertilization. This process is common among internal parasites such as tapeworms, whose chances of locating another individual for cross-fertilization are often poor. However, most animals use cross-fertilization, even when, as in earthworms, each individual is hermaphroditic—that is, each possesses both male and female sexual

17.1 Two earthworms mating
The worms are hermaphroditic, each possessing both male and female reproductive organs. They are coupled at two mating points; at one the upper worm is acting as male and the lower worm as female, while at the other they reverse these roles. Some white fluid can be seen at each mating point.

organs (Fig. 17.1). Among vertebrates, sexual reproduction always involves cross-fertilization; even in hermaphroditic species of vertebrates, such as some species of fish, individuals alternate between producing eggs and sperm, but never fertilize themselves. Sexual reproduction, as we have said, depends on bringing together an egg cell and a sperm cell, which then unite in the process of fertilization to form the first cell of the new individual. There are two basic ways in which egg cells and sperm cells are brought together: external fertilization, in which both types of gametes are shed into the surrounding medium and the sperm swim or are carried by water currents to the eggs; and internal fertilization, in which the eggs are retained within the reproductive tract of the female until after they have been fertilized by sperm inserted into the female by the male.

External fertilization is limited essentially to animals living in aquatic environments, both because the flagellated sperm must have fluid in which to swim, and because the eggs, in the absence of a protective coat or shell (which would prevent the sperm from penetrating and fertilizing them), would become desiccated on land. Almost all aquatic invertebrates, most fish (but not sharks), and many amphibians use external fertilization. As you would expect, shedding eggs and sperm into the water of a lake or stream is an uncertain method of fertilization; many of the sperm never locate an egg, and many eggs are never fertilized, even if both types of gametes are shed at the same time and in the same place, as is usually the case. Consequently animals using external fertilization generally release vast numbers of eggs and sperm at one time (Fig. 17.2). And they often go through elaborate behavioral sequences (in which hormonal control is very important) that ensure concurrence in both time and space in the release of gametes by the two sexes.

17.2 Frogs spawning
The smaller male clasps the female in an embrace called amplexus and sprays semen over the eggs she releases. Many hundreds of eggs are produced.

17.3 Internal fertilization by American coppers (*Lycaena phlaesa americana*)
The male butterfly inserts sperm into the reproductive tract of the female. Fertilization occurs within the body of the female, and the eggs are invested with a protective shell before being laid.

Most land animals, both invertebrate and vertebrate, use internal fertilization (Fig. 17.3). In effect, the sperm cells are provided with the sort of fluid environment that is no longer available to them outside the animals' bodies. The sperm can therefore remain aquatic, swimming through the film of fluid always present on the walls of the female reproductive tract. Once fertilized, the egg is either enclosed in a protective shell and released by the female or held within the female's body until the embryonic stages of development have been completed. Internal fertilization requires, of course, very close physiological and behavioral synchronization of the sexes, and this synchronization involves extensive hormonal control.

Let us summarize briefly the characteristic reproductive methods employed by the major classes of vertebrates. Fish, being aquatic, almost always use external fertilization and hence lay eggs with no shell. Though they usually go through elaborate behavioral rituals that help synchronize the release of gametes, huge numbers of eggs and sperm are released at each mating and the waste of gametes is enormous. Amphibians (frogs, salamanders, and the like) evolved from fish, and they too generally use external fertilization; they must therefore return to the water or to a very moist place on land to lay their eggs. Some salamanders have evolved a behavioral sequence in which the male releases a membranous packet (spermatophore) containing sperm that the female picks up with her cloaca. These amphibians have thus evolved a primitive type of internal fertilization, and some of them can mate on land, but their eggs must still be laid in very moist places.

The reptiles, modern representatives of which are snakes, lizards, and turtles, evolved from ancestral amphibians. They were the first vertebrates to be fully emancipated from the ancestral dependence on the aquatic environment for reproduction. As would be expected, they use internal fertilization, and they lay eggs enclosed in tough membranes and shells. Since internal fertilization entails much less waste of egg cells than external fertilization, only a few egg cells are released during each reproductive season. Birds evolved from one group of ancient reptiles, and they too employ internal fertilization and lay eggs with shells.

The eggs of land vertebrates such as reptiles and birds have four different membranes in addition to the shell. These are the amnion, the allantois, the yolk sac, and the chorion (Fig. 17.4). The *amnion* encloses a fluid-filled chamber housing the embryo, which can thus develop in an aquatic medium even though the egg as a whole may be laid on dry land. The *allantois* functions as a receptacle for the urinary wastes of the developing embryo, and its blood vessels, which lie near the shell, function in gas exchange. The *yolk sac*, as its name indicates, encloses the yolk, which is food material used by the developing embryo. The *chorion* is an outer membrane surrounding the embryo and the other membranes.

Like reptiles and birds, mammals use internal fertilization, but (with a few rare exceptions) no shell is deposited around the fertilized egg and it is not laid. Instead, the early embryo with its membranes becomes implanted in a specialized chamber of the female genital tract, and there embryonic development is completed. The young animal is then born alive. The remainder of our discussion here will be concerned with mammalian, and in particular human, reproduction.

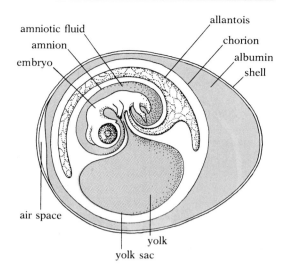

17.4 The embryonic membranes in a bird's egg
Though everything shown here is part of the "egg" in common parlance, the chorion is the outer boundary of structures derived from the true egg cell. The thick layer of albumin, a protein, is outside the cell.

THE HUMAN REPRODUCTIVE SYSTEM

THE REPRODUCTIVE SYSTEM OF THE HUMAN MALE

The male gonads, or sex organs, are the *testes*—oval glandular structures that form in the dorsal portion of the abdominal cavity from the same embryonic tissue that gives rise to the ovaries in females. In the human male the testes descend about the time of birth from their points of origin into the *scrotal sac* (scrotum), a pouch whose cavity is initially continuous with the abdominal cavity via a passageway called the *inguinal canal*. After the testes have descended through the inguinal canal into the scrotum, the canal is slowly plugged by growth of connective tissue, so that the scrotal and abdominal cavities are no longer continuous.

Sometimes the inguinal canal fails to close properly; and even when it does, it remains a point of weakness and is easily broken open again when subjected to excessive strain, as when a man lifts a heavy object. The opening resulting from insufficient closure or from later rupture is known as an inguinal hernia, or rupture; it is the most common type of hernia in human males. If the hernia is large, it must be repaired surgically to prevent a loop of the intestine from slipping through the opening into the scrotal sac, where the intestine may become caught so tightly that its blood supply is cut off and gangrene results. Inguinal hernia is largely a human hazard, attributable to our two-legged stance, which places much strain on the lower abdomen; such hernias are very infrequent in mammals that walk on four legs. In fact, in some mammals the inguinal canal remains partly open, and the testes move back into the abdominal cavity during the nonreproductive season.

Each testis has two functional components: the ***seminiferous tubules***, in which the sperm cells are produced, and the ***interstitial cells***, which secrete male sex hormone. The seminiferous tubules of the human are not functional at the temperatures characteristic of the abdominal cavity; if the testes fail to descend, the germinal epithelium of the tubules eventually degenerates. If, however, the testes descend normally into the scrotal sac, where the temperature is approximately 1.5°C cooler, the germinal epithelium of the seminiferous tubules becomes functional at the time of puberty. Mature sperm cells pass from the seminiferous tubules via many tiny ducts into a much-coiled tube, the ***epididymis***, which lies on the surface of each testis (Figs. 17.5 and 17.6). The sperm are stored in this organ until they are activated by secretions produced by it; they are then released during copulation. (Sperm that have not passed through the epididymis are nonmotile.)

A long sperm duct, the ***vas deferens***, runs from each epididymis through the inguinal canal and into the abdominal cavity, where it loops over the bladder and joins with the ***urethra*** just beyond the point where the urethra arises from the bladder. The urethra, in turn, passes through the ***penis*** and empties to the outside. Notice, then, that the urethra in the mammalian male is a common passageway used by both the excretory and reproductive systems; urine passes through it during excretion and semen passes through it during sexual activity. In more primitive vertebrates the relationship between the excretory and reproductive systems is even closer. In frogs, for example, sperm cells pass from the testes into the kidneys and down the excretory ducts to the cloaca; the reproductive system has no separate vasa deferentia. In the vertebrates there has been an evolutionary trend toward increasing liberation of the reproductive system from its ancestral dependence on the excretory system. There is far more separation in mammalian males than in fish or frogs, but the two systems still share the

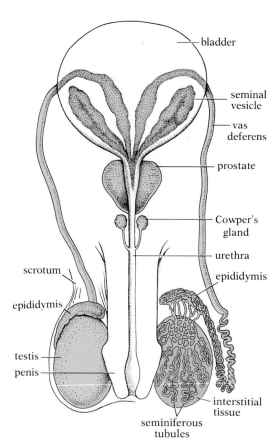

17.5 Reproductive tract of the human male: frontal view

17.6 Reproductive tract of the human male: lateral view

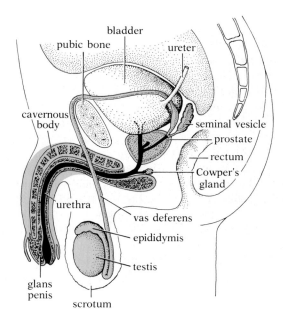

urethra and thus do not have separate openings to the outside. As we shall see, only in mammalian females has complete separation arisen.

As sperm pass through the vasa deferentia and urethra, seminal fluid is added to them to form *semen*, which is a mixture of seminal fluid and sperm cells. The seminal fluid is secreted by three sets of glands: the *seminal vesicles*, which empty into the vasa deferentia just before these join with the urethra; the *prostate*, which empties into the urethra near its junction with the vasa deferentia; and the *Cowper's glands*, which empty into the urethra at the base of the penis (Figs. 17.5 and 17.6). Seminal fluid has a variety of functions: (1) It serves as a vehicle for transport of sperm. (2) It lubricates the passages through which the sperm must travel. (3) As an effectively buffered fluid, it helps protect the sperm from the harmful effects of the acids in the female genital tract. (4) It contains much sugar (mostly fructose), which the active sperm can use as a source of energy. The tiny sperm cells can store very little food themselves; they depend on an external source of nutrients for the respiratory production of the ATP necessary to keep their flagella active. The base of a sperm flagellum is an amazing power plant, packed with mitochondria, in which energy can be extracted from the sugar absorbed from the seminal fluid (Fig. 17.7).

During sexual excitement the arteries leading into the penis dilate, and the veins from the penis constrict, in response to stimulation by nerves of the autonomic system. Much blood is pumped under considerable pressure through the arteries into the spaces in the *cavernous body*, the spongy tissue of which the penis is largely composed (Fig. 17.6). The engorgement of the penis by blood under high arterial pressure causes it to increase greatly in size and to become hard and erect, thus preparing it for insertion into the female vagina during copulation (also called coitus). Note that erection of the penis does not involve activity of skeletal muscles, but is entirely a vascular phenomenon.

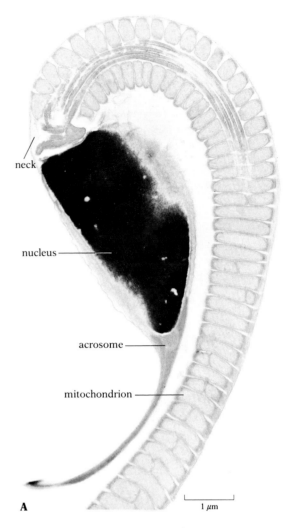

A

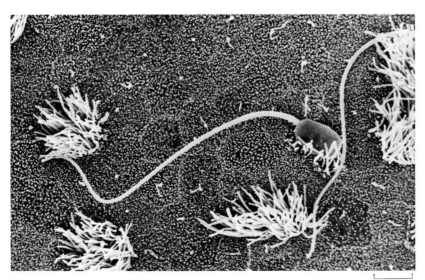

B

17.7 Mammalian spermatozoa
(A) Electron micrograph of longitudinal section of a sperm cell from a kangaroo rat. The sperm head, which contains the acrosome (an enzyme-filled organelle) and the nucleus, is connected by a short neck to a portion of the flagellum tightly packed with spirally arranged mitochondria. There are no mitochondria in the long distal portion of the flagellum, which is not shown here. (B) Scanning electron micrograph of a rabbit sperm cell in the uterus of a female rabbit some hours after copulation.

17.8 The structural formulas of male sex hormone (testosterone) and female sex hormone (progesterone)
Differing in only one side group, these two steroids have amazingly different effects on the body.

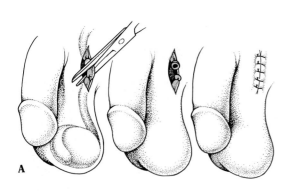

Testosterone   Progesterone

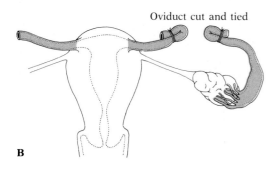

A

Oviduct cut and tied

B

17.9 Surgical birth-control procedures
(A) Vasectomy. A small incision is made in the wall of the scrotum (on each side); the vas deferens (color) is tied and cut, and the incision is sutured. (B) Tubal ligation. Each of the oviducts is cut and tied, so that eggs cannot descend and no fertilization can occur.

When the penis is sufficiently stimulated by friction during copulation, nervous reflexes involving pathways of the autonomic system cause waves of contraction in the smooth muscles of the walls of the epididymides, vasa deferentia, seminal glands, and urethra. These contractions move sperm from the epididymides through the vasa deferentia, combine seminal fluid from the various glands with the sperm, and expel the semen from the urethra. An average of about 100 million sperm cells in about 3.5 ml of semen are released during one human ejaculation.

HORMONAL CONTROL OF SEXUAL DEVELOPMENT IN THE MALE

The testes begin secreting small amounts of the male sex hormone, ***testosterone*** (Fig. 17.8), during embryonic development. The presence of this hormone in the embryo is crucial to the differentiation of male structures. The level of production remains low until the time of puberty, and no sperm cells are produced.

The factors governing onset of puberty are not well understood. One possibility is that the hypothalamus—which during childhood is sufficiently sensitive to testosterone to be inhibited by even low levels of the hormone—becomes less sensitive and begins sending more gonadotropic releasing hormone (GnRH) to the anterior pituitary, stimulating it to increase its secretion of the two gonadotropic hormones, luteinizing hormone (LH) and follicle-stimulating hormone (FSH). The LH induces the interstitial cells of the testes to produce more testosterone. The testosterone, together with FSH, induces maturation of the seminiferous tubules and stimulates them to begin sperm production (spermatogenesis). Maintenance of spermatogenesis over long periods requires the continued presence of testosterone (and the LH that induces it) and of FSH.

Once testosterone appears in appreciable quantity in the system, it stimulates maturation of the accessory reproductive structures and also triggers the complex of changes in the secondary sexual characteristics normally associated with puberty: growth of the beard, growth of pubic hair, deepening of the voice, development of larger and stronger muscles, and the like. If the testes are removed before puberty by castration, these changes in the secondary sexual characteristics never occur. If castration is performed after puberty, there is some retrogression of the adult sexual characteris-

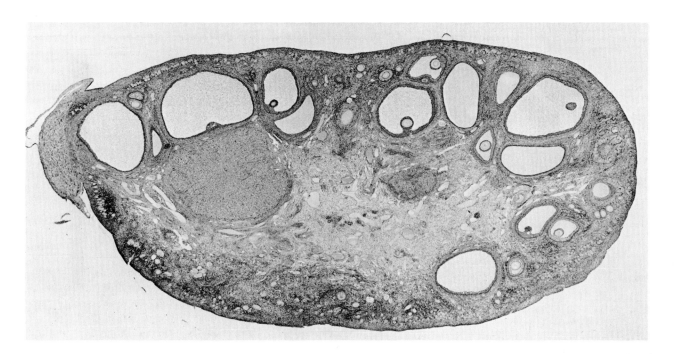

tics, but they do not disappear entirely. Castration after puberty abolishes the sex urge in many animals, but not in man, where psychological factors are of much greater importance than in other animals. Vasectomy, a surgical procedure sometimes confused with castration, is often used as a birth-control method. This minor operation consists of cutting the vasa deferentia to prevent movement of sperm into the urethra (Fig. 17.9A). Vasectomy causes no retrogression of sexual characteristics, because there is no alteration of hormone levels. However, recent evidence indicates that vasectomy sometimes triggers an immune reaction that can cause damage to the circulatory system; hence the safety of this operation is considered questionable by some physicians.

THE REPRODUCTIVE SYSTEM OF THE HUMAN FEMALE

The female gonads, the *ovaries*, are located in the lower part of the abdominal cavity, where they are held in place by large ligaments. Like the testes, the ovaries have the two main functions of producing gametes (in this case egg cells) and secreting sex hormones. At the time of birth, a girl's ovaries already contain a huge number of primordial egg cells, or *oocytes*; estimates range from 100,000 to 1,000,000. During the approximately 30 years of her reproductive life, a woman ovulates about 13 times per year, producing one mature ovum each time, so usually fewer than 400 oocytes ever mature and leave the ovaries. The rest eventually degenerate, and ordinarily none can be found in the ovaries of women past the age of about 50.

Each oocyte is enclosed within a cellular jacket called a *follicle*. The oocyte fills most of the space in the small immature follicle. In the process of maturation, however, the follicle grows bigger relative to the oocyte and develops a large fluid-filled cavity (Fig. 17.10); the oocyte, embedded in a

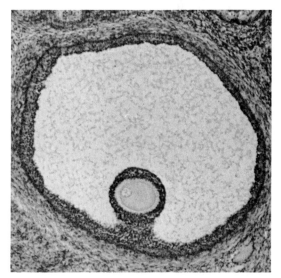

17.10 Photographs of sections of cat ovary
Top: Follicles of different sizes are shown in a section of the entire ovary. The more mature follicles have a large cavity, with the egg cell embedded in a pedestal of epithelial cells that projects into the cavity.
Bottom: Enlarged view of a nearly mature follicle with its egg cell.

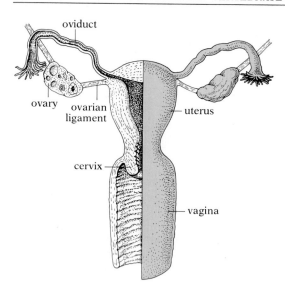

17.11 Reproductive tract of the human female
The wall of one side has been dissected away to reveal the interior structure.

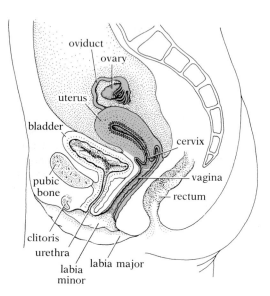

17.12 Reproductive tract of the human female: lateral view

mass of follicular epithelial cells, protrudes into the cavity. A ripe follicle bulges from the surface of the ovary; when *ovulation* occurs, its outer wall ruptures and both the fluid and the detached oocyte are expelled. In the human, only one oocyte is normally released at each ovulation. There is no apparent pattern determining which ovary will ovulate in any given period.

Ovulation releases the oocyte from the ovary into the abdominal cavity. From here, it is usually promptly drawn into the large funnel-shaped end of one of the *oviducts* (Fallopian tubes), which partly surround the ovaries but are not continuous with them (Fig. 17.11). Cilia lining the funnel of the oviduct produce currents that help move the oocyte into the oviduct. If sperm are present to meet the oocyte while it is still in the upper third of the oviduct, the penetration of a sperm through the membrane of the oocyte stimulates it to complete its maturation into a true egg cell (ovum), and almost immediately thereafter the nuclei of the sperm and ovum fuse in the process of fertilization. A method of permanent sterilization of female, sometimes used as a birth-control measure, is tubal ligation. In this operation the oviducts are cut and tied, so that sperm cannot reach the oocyte and no ovum can move down the oviduct into the uterus (Fig. 17.9B). Like vasectomy of the male, tubal ligation causes no change in hormone production.

Each oviduct empties directly into the upper end of the *uterus* (womb). This organ, which is about the size of a fist, lies in the lower portion of the abdominal cavity just behind the bladder (Fig. 17.12). It has very thick muscular walls and a mucous lining containing many blood vessels. If an egg is fertilized as it moves down the oviduct, it becomes implanted in the wall of the uterus, and there the embryo develops until the time of birth. Another method of birth control involves insertion of a plastic ring or spiral into the uterus (Fig. 17.13A). Such intrauterine devices (IUDs) seem to be very effective in preventing pregnancy, probably by preventing implantation in the uterus. However, these devices sometimes cause irritation and/or bleeding in the uterus; hence some women cannot tolerate them, and their safety for prolonged use is questionable.

At its lower end the uterus connects with a muscular tube, the *vagina*, which leads to the outside. The vagina acts as the receptacle for the male penis during the copulation. The great elasticity of its walls makes possible not only the reception of the penis, but the passage of the baby during childbirth.

The uterus and vagina do not lie in a straight line, as Figure 17.11 may seem to indicate. Instead, the uterus projects forward nearly at a right angle to the vagina, as shown in Figure 17.12. The *cervix*, a muscular ring of tissue at the mouth of the uterus, protrudes into the vagina. Devices that block the mouth of the uterus by covering the cervix are widely used in birth control. One such device, the diaphragm, is a shallow rubber cup with a spring around its rim. It is inserted into the vagina and positioned so that it covers the entire cervical region (Fig. 17.13B). It is very effective in preventing sperm from entering the uterus, particularly if it is used in conjunction with a spermicidal jelly or cream.

The opening of the vagina in young human females is partly closed by a thin membrane called the *hymen*. Traditionally the hymen has been regarded as the symbol of virginity, to be destroyed the first time sexual intercourse takes place. Frequently, however, the membrane is ruptured during childhood, by disease or by a fall or as a result of strenuous physical exercise.

The external female genitalia are collectively termed the **vulva**. The vulvar region is bounded by two folds of skin, the labia minor and the labia major, which enclose the space known as the vestibule. The vagina opens into the rear portion of the vestibule, and the urethra opens into the midportion of the vestibule. Note, then, that in the adult mammalian female there is no interconnection between the excretory and reproductive systems, and that the urethra carries only excretory materials. During embryonic development the vagina and urethra share a common opening, but as development proceeds, this opening becomes divided, so that the vagina and urethra have separate openings to the exterior.

In the anterior portion of the vestibule, in front of the opening of the urethra, is a small erectile organ, the **clitoris**, which forms from the same embryonic tissue that gives rise to the penis in the male (see Fig. 32.13, p. 827). Like the penis, it becomes engorged with blood during sexual excitement and is a major site of stimulation during copulation.

HORMONAL CONTROL OF THE FEMALE REPRODUCTIVE CYCLE

As in the male, puberty in the female is thought to begin when the hypothalamus loses its sensitivity to the low levels of sex hormone in childhood and starts secreting more GnRH, which stimulates the anterior pituitary to release increased amounts of FSH and LH. These gonadotropic hormones cause maturation of the ovaries, which then begin secreting the female sex hormones, **estrogen** and **progesterone**. The estrogen stimulates maturation of the accessory reproductive structures (increase in the size of the uterus and vagina, for example) and development of the female secondary sexual characteristics: growth of pubic hair, broadening of the pelvis, development of the breasts, change in the distribution of body fat, and some change in voice quality. The changing hormonal balance also triggers the onset of the **menstrual cycle**. The menstrual cycle is a good example of complex interplay between several hormones and the nervous system.

Rhythmic variation in the secretion of gonadotropic hormones in the females of most species of mammals leads to what is known as the **estrous cycle**—rhythmic variation in the condition of the reproductive tract and in sexual readiness. The females of most species will accept the male in copulation only during those brief periods of the cycle near the time of ovulation when the uterine lining is thickest and sexual readiness is at its height. During such periods the female is said to be "in heat," or in estrus. Many mammals have only one or a few estrous periods each year, but some, like rats, mice, and their relatives, may be in estrus as often as every four days. If fertilization does not occur, the thickened lining of the uterus is gradually reabsorbed by the female's body; ordinarily no bleeding is associated with this process.

The reproductive cycle in humans and some other higher primates differs in several important ways from that of other mammals. There is no distinct heat period, the female being to some degree receptive to the male throughout the cycle. And the thickened lining of the uterus is not completely reabsorbed if no fertilization occurs; instead, part of the lining is

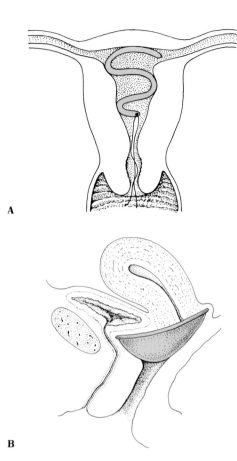

17.13 Two birth-control devices
(A) An IUD in place in the uterus. The strings that run through the cervix permit the woman to make sure the IUD has not been expelled. (B) Diaphragm in position in the vagina. The device covers the mouth of the cervix. It is very effective in preventing sperm from entering the uterus when used with spermicidal jelly or cream.

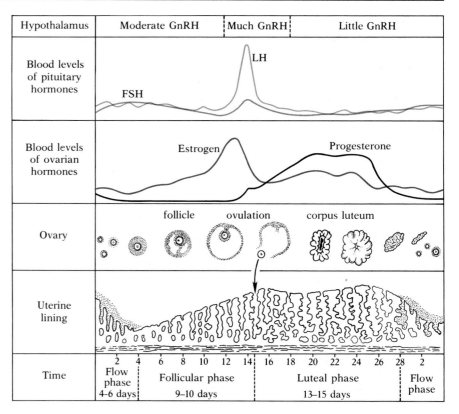

Hypothalamus	Moderate GnRH	Much GnRH	Little GnRH
Blood levels of pituitary hormones		LH	
	FSH		
Blood levels of ovarian hormones	Estrogen		Progesterone
Ovary	follicle	ovulation	corpus luteum
Uterine lining			
Time	2 4 6 8 10 12 14	16 18 20 22 24 26 28	2
	Flow phase 4-6 days	Follicular phase 9–10 days	Luteal phase 13–15 days	Flow phase

17.14   The sequence of events in the human menstrual cycle

The hormone levels shown are only approximate: they are based on recent radioimmunoassay determinations, the results of which vary considerably.

sloughed off during a period of bleeding known as menstruation. The human menstrual cycle averages about 28 days; there are consequently about 13 each year. This is an extremely rough average, however; extensive variation occurs from person to person and from cycle to cycle in the same person.

Let us trace the sequence of events in the menstrual cycle, assuming a period of 28 days. It is customary in medical practice to consider the first day of menstruation as the first day of the cycle, as shown in Figure 17.14. From a biological point of view, however, it is more appropriate to regard the end of the period of bleeding as the beginning of the new cycle. At this point (day 4 in Figure 17.14), the uterine lining is thin and there are no ripe follicles in the ovaries. Under the influence of FSH—follicle-stimulating hormone—from the anterior pituitary, several follicles in the ovaries begin growing, and, influenced by the synergistic action of FSH and LH, they begin secreting the first of the two female sex hormones, estrogen. One of the follicles soon gains ascendancy: it continues to grow and secrete estrogen, while the others cease growing. The estrogen, in turn, stimulates the lining of the uterus to thicken. This follicular phase (growth phase) of the cycle lasts, on the average, about nine to ten days after cessation of the previous menstrual flow.

As the follicle grows it produces more and more estrogen (Fig. 17.14).

The eventually high level of estrogen (or, according to some investigators, the slowing of the rate of estrogen production as it reaches its peak) apparently stimulates, probably via GnRH from the hypothalamus, an abrupt surge of LH from the pituitary. This LH surge triggers ovulation by the ascendant follicle, which by this time is mature and bulging from the surface of the ovary.[1] The mechanism of ovulation is as yet unknown. It is often said that the pressure of the follicular fluid causes the wall to burst, but there is evidence that, on the contrary, this pressure may actually decline slightly just before ovulation. It does seem clear that hormonally induced thinning of the follicular wall is one important factor in ovulation. Whatever the mechanism, ovulation marks the end of the follicular, or growth, phase of the menstrual cycle.

Following ovulation, LH induces changes in the follicular cells that convert the old follicle into a yellowish mass of cells rich in blood vessels. The new structure formed from the ruptured follicle under the influence of LH is called *corpus luteum* (Latin for "yellow body"). From its location on the ovary, the corpus luteum continues secreting estrogen, though not as much as was secreted by the follicle just prior to ovulation.[2] But the corpus luteum also secretes the second female sex hormone, progesterone (see Fig. 17.8).

Progesterone functions in preparing the uterus to receive the embryo. The uterine lining has already thickened substantially under the stimulation of estrogen during the follicular phase; now the progesterone causes maturation of the complex system of glands in the lining. The luteal phase of the menstrual cycle is, in fact, sometimes called the secretory phase, though this name is slightly misleading, since there is also some glandular activity in the uterine lining during the latter part of the follicular phase. Repeated experiments have shown that implantation of a fertilized ovum in the uterus cannot occur in the absence of the changes in the uterine lining produced by progesterone. Progesterone is truly the hormone of pregnancy.

In addition, high levels of progesterone inhibit initiation of the next cycle, though the precise mechanism of this inhibition is not entirely clear. The progesterone probably acts on the hypothalamus to suppress GnRH release, thereby limiting FSH and LH secretion by the pituitary and preventing the LH surge necessary for ovulation. But the progesterone may also act directly on the immature follicles in the ovary, inhibiting their growth and estrogen secretion. As long as progesterone is present in quantity in the system, then, there is little follicular growth. This inhibiting action of progesterone is, of course, an important element in regulating the duration of the menstrual cycle. It is also the basis for the action of birth-control pills. These pills contain synthetic compounds similar to progesterone and estrogen. Taken regularly, they inhibit secretion of FSH and LH

---

[1] Some investigators think that a slight rise in progesterone just prior to ovulation acts synergistically with LH to trigger ovulation.

[2] In rats maintenance of the corpus luteum requires prolactin in addition to LH, but current evidence strongly suggests that in most mammals, including humans, LH alone suffices for both the formation and the secretory activity of the corpus luteum.

(probably by suppressing GnRH release by the hypothalamus) and thus prevent follicular growth and ovulation—and consequently conception. Like most contraceptive techniques, birth-control pills sometimes have undesirable side effects. These include migraine headaches, vascular disturbances, depression, reduced sexual drive, changes in skin pigmentation, and possibly increased risk of certain kinds of cancer (though there is some evidence that they can also reduce the risk of certain other kinds of cancer).

If no fertilization occurs during a normal cycle, the corpus luteum begins to atrophy about eleven days after ovulation, and its secretion of progesterone falls. When this happens, the thickened lining of the uterus can no longer be maintained, and reabsorption of part of the lining begins. Unlike most mammals, humans and some other higher primates cannot reabsorb all the extra tissue laid down during the follicular and luteal phrases of the cycle; part of it must be sloughed off during menstruation, which lasts about four or five days.

The fall of progesterone levels, resulting from atrophy of the corpus luteum frees the hypothalamus from inhibition and allows it to stimulate the pituitary to increase secretion of FSH. The immature follicles are also freed from inhibition and, under the influence of the FSH, begin growing, and a new cycle begins. The sequence of events in a normal menstrual cycle is depicted diagrammatically in Figure 17.14.

From this account it is apparent that the critical event in resetting the system is probably the fall in progesterone levels as a result of atrophy of the corpus luteum. But what causes this atrophy? No sure answer can yet be given, which means that a fundamental aspect of the timing of menstrual cycles remains in doubt. Before sensitive methods for measuring hormone concentrations in the blood were available, it was thought that LH secretion declined gradually after ovulation, and that when levels fell low enough the corpus luteum could no longer be maintained; but it is now known, as shown in Figure 17.14, that LH levels fall precipitously immediately after ovulation, long before the corpus luteum declines. Convincing recent evidence indicates that in cows certain prostaglandins, produced by the nonpregnant uterus and carried to the ovary by a special system of portal blood vessels, act to cut off progesterone secretion. But there is no such portal system in human beings; nor is there solid evidence that the human uterus sends any inhibiting substance to the ovaries. It is possible that the ovary itself produces prostaglandins that turn off progesterone secretion, but the evidence here is still scant.

During the flow phase, particularly in the first few days, secretion of progesterone and estrogen by the old corpus luteum is at a low level, and the follicles of the new cycle have not yet begun producing significant amounts of estrogen. Since a woman's body is accustomed from puberty to functioning in the presence of sex hormones, their withdrawal at the end of the luteal phase of each menstrual cycle is often accompanied by physiological and psychological disturbance including irritability, depression, and sometimes nausea; abdominal cramps, caused by contractions of the uterus, are also common.

Emotional stress sometimes also accompanies the ***menopause***, a period lasting a year or two at the end of a woman's reproductive life. The meno-

tation are brown bears (about five months), pine martens (six months), American badgers (two months), and armadillos (14 weeks).

After implantation of the embryo in the uterine lining, the embryonic membranes form the ***umbilical cord***, through which blood vessels contributed by the allantois run to a large structure, the ***placenta***, formed from the embryonic membranes (primarily the chorion) and from the adjacent uterine tissue (Fig. 17.15). Within the placenta the blood vessels of the embryo and those of the mother lie very close together, but they are not joined and there is no mixing of maternal and fetal blood. Exchange of materials takes place in the placenta by diffusion between the blood of the mother and that of the embryo; nutritive substances and oxygen move from the mother to the embryo, and urinary wastes and carbon dioxide move from the embryo to the mother (see Fig. 13.6B, p. 324).

We saw earlier that progesterone is essential for maintenance of the uterine lining during implantation and pregnancy. But we saw also that in a normal menstrual cycle, when there is no fertilization, the corpus luteum soon begins to atrophy and cuts off the supply of progesterone, with the result that menstruation occurs. Clearly, this sequence of events cannot be allowed to take place after conception, or the uterine lining with the implanted embryo would be sloughed off and lost. And indeed, when conception occurs, the corpus luteum does not atrophy, but lasts through most of the pregnancy. How is this possible? Apparently the chorionic portion of the placenta soon begins secreting a gonadotropic hormone (human chorionic gonadotropin, or HCG). This hormone preserves the corpus luteum, which continues to secrete progesterone and thus sustains the pregnancy. So much HCG is produced in a pregnant woman that much of it is excreted in the urine. Many commonly used tests for pregnancy, among them some available over the counter in drugstores, are based on this phenomenon, and can yield reliable results in less than an hour.

Though the corpus luteum is essential during early pregnancy, it is no longer necessary in humans after the first two months or so. Removal of the ovaries after this time does not terminate the pregnancy, evidently because both estrogen and progesterone, which can still be detected in the urine after excision of the ovaries, are produced by another organ. Apparently the placenta begins to secrete these hormones early in pregnancy, and once this secretion has reached a sufficiently high level the placenta itself can maintain the pregnancy in the absence of progesterone from the corpus luteum. This is not true of the rabbit; removal of a rabbit's ovaries just a few days before the end of pregnancy invariably results in abortion unless progesterone is artificially administered.

HORMONAL CONTROL OF PARTURITION AND LACTATION

Much research is still needed to clarify the complex interactions of hormones that control the birth process (parturition). The best evidence to date indicates that a rise in prostaglandin production by the placenta plays a central role in initiating parturition. The major stimulus for this rise in late pregnancy appears to be an increase in estrogen secretion (also by the placenta), which is itself a result, at least in part, of increased secretion of estrogen precursors by the fetal adrenal cortex. In some mammals, but not

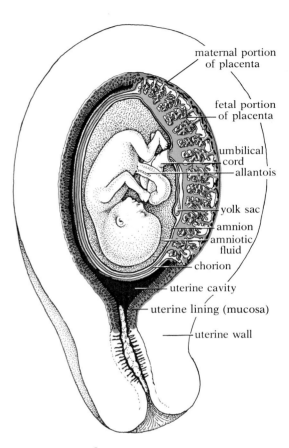

17.15 Uterus of a pregnant woman
The placenta consists of both maternal and fetal components, the latter derived from the chorion. Maternal and fetal capillaries lie side by side, allowing the diffusion of nutrients into the fetal circulation. Nutrients absorbed by the fetal capillaries are carried by the umbilical cord to the embryo, which lies in the amniotic sac, bathed by amniotic fluid.

in primates, the placenta in late pregnancy begins converting progesterone, which severely inhibits muscular contraction, into estrogen; the resulting fall in plasma progesterone levels may make the uterine muscles more susceptible to stimulation. Since oxytocin, released by the posterior pituitary (and perhaps also secreted by the placenta), is known to have a powerful stimulatory effect on uterine contractility, it seems likely that this hormone, too, contributes to the induction of labor.

Another hormone important in parturition is *relaxin*, which is secreted during pregnancy by the ovaries and the placenta. This hormone has the effect of loosening the connections between the bones of the pelvis and thereby enlarging the birth canal and facilitating parturition. Relaxin also aids in softening and dilating the cervix. The activity of relaxin is enhanced by estrogens.

Like the hormonal control of parturition, that of milk secretion is complicated and not completely understood. Growth and development of the mammary glands in humans seem to be controlled by a complex interaction between estrogen, progesterone, thyroxin, insulin, growth hormone, prolactin, glucocorticoids, and human placental lactogen (a potent prolactinlike hormone produced by the placenta). Initiation and maintenance of lactation by mature mammary glands following parturition seem to be controlled primarily by prolactin and glucocorticoids. These hormones apparently become effective in inducing lactation when the high levels of sex hormones, which inhibit lactation, disappear at the time of parturition.

The actual release of milk from the mammary glands involves both neural and hormonal mechanisms. The stimulus of suckling causes nervous stimulation of that part of the hypothalamus that stimulates release of oxytocin stored in the posterior pituitary (in cows, seeing the calf or hearing rattling milk pails has the same effect). The oxytocin, in turn, induces constriction of the many tiny chambers in which the milk is stored in the mammary glands. The constriction forces the milk into ducts that lead to the nipple. Adrenalin inhibits this milk-ejection process.

Chapter 18

NERVOUS CONTROL

As we have seen, the endocrine system provides communication between different cells and tissues in multicellular organisms, providing a critical link in the flow of information that ends with a response in target cells. Chemical communication also makes possible centralized control of the entire organism. This hormonal control is accomplished by a variety of molecular messengers that are able to alter the precise chemistry of their target cells. But for virtually all animals, for many monerans and protists, and even for some plants, hormonal communication and control is simply too slow for the essential activities of integrating sensory information from the environment and coordinating movement. In these organisms, electrical communication has come to fill this need, using as its raw material the sodium-potassium pump, the kinds of membrane receptors involved in chemical control, and (in animals) even exocytosis. We shall look first at how certain monerans and protists solve the problem of rapid communication and coordination, and then at the strategy that evolved in animals.

ORGANIZATION AND EVOLUTION OF NERVOUS SYSTEMS

Almost every response to a stimulus, whether electrical or chemical, involves four stages: (1) detection of the stimulus, (2) conduction of a signal, (3) "processing" of the signal, and (4) response. These stages are evident in all organisms, from relatively simple procaryotes to complex multicellular organisms like ourselves.

MONERANS AND PROTISTS

Eucaryotic nerve cells may have evolved from similar mechanisms that guide the feeding behavior of some procaryotes, notably bacteria. *E. coli,*

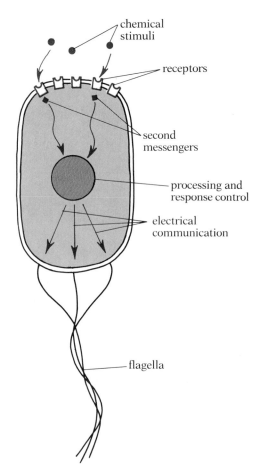

chemical stimuli

receptors

second messengers

processing and response control

electrical communication

flagella

18.1 Model of bacterial chemotaxis
Chemotaxis in *E. coli* illustrates the four stages of information flow in organisms. (1) *Detection* of the stimulus is accomplished by receptor molecules in the membrane and cell wall. (2) *Conduction* is by means of second messengers released in the cell.
(3) *Processing* consists of measuring the concentration of various second messengers to determine whether the environment is getting better or worse.
(4) *Transmission* of the signal eliciting the *response* again involves conduction (this time probably by electrical means). The conduction and processing stages are more complex than indicated here.

for instance, detects the presence of various chemicals in its environment and orients its movements with respect to them, moving predictably up the concentration gradients of sugars, amino acids, and other chemicals it needs, and down the gradients of damaging chemicals such as phenol, inorganic acids, and the like. This process is known as **chemotaxis.** (A taxis is an oriented movement; hence chemotaxis is movement oriented to chemicals, phototaxis is movement oriented toward light, and so on.) Julius Adler of the University of Wisconsin and Donald Koshland of the University of California, Berkeley, have demonstrated by elegant genetic and behavioral tests that *E. coli* does this by sampling its environment with some two dozen kinds of specific receptor molecules, which recognize particular molecules in the same way enzymes recognize their substrates. But these chemical receptor molecules, hundreds of thousands in the membrane and cell wall of each *E. coli*, would be useless without some mechanism for conducting information about what is being sensed to an information-processing system. The signal is chemical—as so often in the endocrine system, each receptor contributes a second messenger—and "conduction" of the second messengers is by diffusion (Fig. 18.1).

Processing in a bacterium may be thought of as a sort of molecular polling: different types of receptors contribute various second messengers to the pool, and the positive "votes" cast by receptors in response, say, to increasing concentrations of a sugar are matched against the negative votes produced by increasing concentrations of noxious compounds. On the basis of this continuous voting process, the bacterium "knows" whether the environment, on the whole, is getting better or worse. The results of this analysis appear to be communicated by electrical signals to the response centers.

The final stage, the response, consists of a brief change in the direction of rotation of the several stiff, helical flagella that propel the bacterium. The result is that the bacterium founders briefly and then strikes out in a new direction, once again sampling to see whether the environment is improving or deteriorating.

Some eucaryotes too accomplish all four stages of information flow within a single cell. In *Paramecium*, for instance, specialized membrane proteins are again responsible for detecting most stimuli, and the response is a brief reversal of the cilia that propel the organism. (*Paramecium* actually backs up before setting off in a new direction.) *Paramecium*'s processing and communication are, however, almost exclusively electrical: the organism is itself in many ways a mobile nerve cell. It is from this sort of mechanism for intracellular communication in early eucaryotes that specialized nerve cells evolved.

ANATOMY OF NERVE CELLS

The anatomy of specialized nerve cells, or **neurons,** often reflects their particular roles in detection, conduction, processing, and response control. Regardless of appearance, however, all neurons have the same functional organization, which enables them to collect information—from the environment directly (as sensory cells), or from other neurons, or from both—and to transmit it to target cells such as other neurons, muscle cells, and

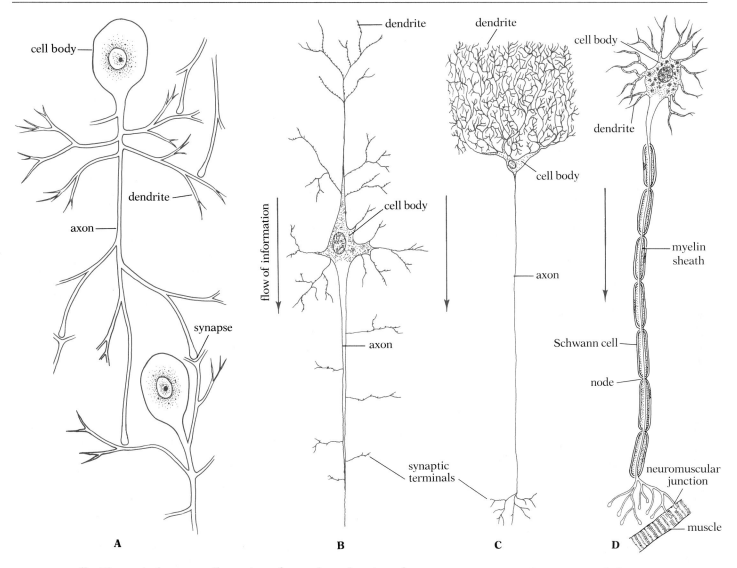

A   **B**   **C**   **D**

secretory cells. The typical nerve cell consists of an enlarged region, the cell body, which contains the nucleus, and one or more long processes, or nerve fibers. The fibers on which information is generally received are the ***dendrites***; those on which it is generally transmitted to other cells are the ***axons*** (Fig. 18.2A–D). Specialized terminals on the ends of axons convey the signals to the dendrites of target cells. In vertebrates the dendrites usually feed into the cell body (Fig. 18.2B–D), whereas in invertebrates the axon is most often connected directly to the dendrites, so that the cell body is out of the path of information flow (Fig. 18.2A). As we shall see, it is on the finger-like dendrites and (in vertebrates) the cell body that much of neural processing takes place.

Dendrites are usually rather short, and neurons characteristically have many of them, receiving input from perhaps several thousand other cells. Most dendrites are profusely branched and have a spiny appearance. When

18.2 A variety of neuron morphologies
Information is normally collected by dendrites, conducted along an axon, and transmitted to other cells by terminals, at information links called synapses. In invertebrates, the cell body of the neuron usually lies out of the information pathway, and the axon is never myelinated (A). In vertebrates (B, C, D), the cell body usually lies between the dendrites and the axon, and longer axons are frequently insulated by a series of myelin sheaths.

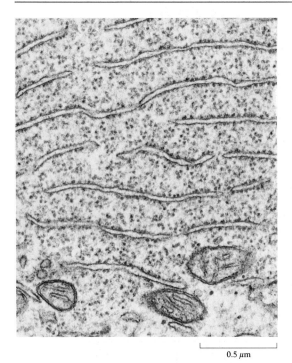

stained and examined under an ordinary light microscope, their cytoplasm appears to contain a great deal of dark, granular material (Nissl substance), which the electron microscope reveals as extensive arrays of flattened cisternae of the endoplasmic reticulum, with a vast number of associated ribosomes (Fig. 18.3).

In contrast to the multiplicity of dendrites, there is usually only one axon per neuron, and it is usually longer and thicker than the dendrites; it may branch extensively, but it does not have a spiny appearance and does not contain Nissl substance. These histological differences reflect the basic functional distinction between dendrites, which receive information from other cells, and axons, which transmit to other cells. Recently, however, it has become clear that the situation is occasionally more complex: dendrites sometimes form *synapses*—information links—to other dendrites, while axons sometimes receive synapses, especially near their bases and their distal terminals. As we shall see, these synapses are of great importance to the processing of information in the nervous system.

Within the central nervous system, or CNS, of vertebrates, neurons are intimately associated with vast numbers of satellite cells called *neuroglia,* or simply glia for short. Some glia provide the neurons with nutrients, and may help maintain a homogeneous environment by absorbing substances secreted by the neurons. Much of this absorbed material is then cycled back to the neurons for reuse. In at least some areas of the CNS, glia provide a framework during development along which neurons migrate and axons grow to reach their targets. One class of glia found in vertebrates, the myelin, is particularly well understood. The membranes of these specialized glial cells wrap around and around the axons of many neurons in the CNS to form a heavily lipid *myelin sheath* (Fig. 18.4), which insulates against "cross talk" between adjacent axons. Many vertebrate axons peripheral to the CNS are enveloped by satellite cells called *Schwann cells* (Fig. 18.5A), which often give rise to myelin sheaths in much the same way as the glial cells within the CNS (Fig. 18.5B). Myelin sheaths serve to speed up the conduction of impulses in axons that have them; they are interrupted at regular intervals called *nodes* (Fig. 18.2D)—points where one glial or Schwann cell ends and another begins.

18.3 Electron micrograph of Nissl substance
At a magnification of × 42,000 the granular material known as Nissl substance, which is visible under the light microscope, clearly consists of flattened cisternae of the endoplasmic reticulum plus numerous ribosomes. Many of the ribosomes are not directly attached to the cisternal membranes.

18.4 Development of the myelin sheath
(A) Initially the unmyelinated axon lies in an inpocketed area of the glial cell. (B) The inpocketed area begins to coil. (C) The glial-cell membrane is wound ever more tightly around the axon, forming what is known as a myelin sheath.

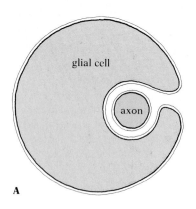

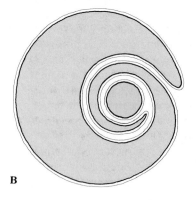

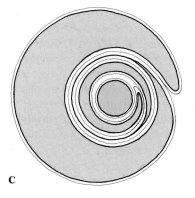

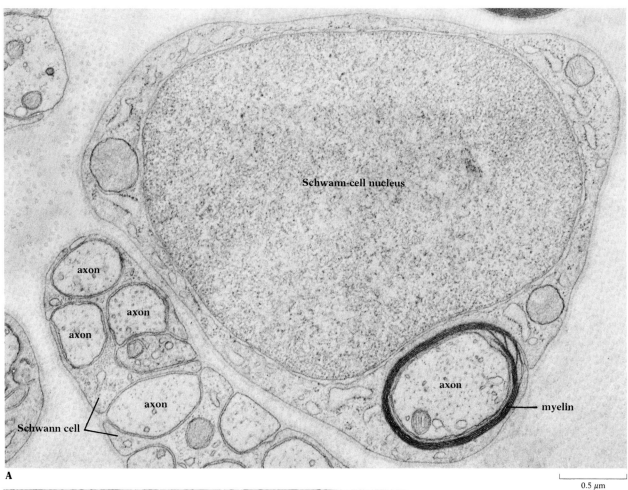

A

0.5 μm

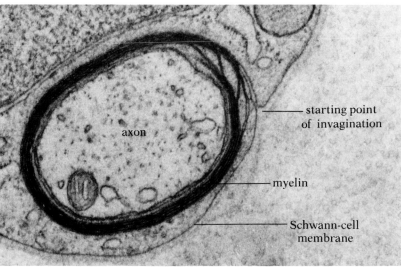

B

0.5 μm

18.5 Electron micrograph of cross section of part of a nerve from a guinea pig

(A) The axons of the neurons are enveloped by Schwann cells; at lower left a single Schwann cell envelops several unmyelinated axons. The axon at lower right (shown also in B) has a myelin sheath, formed from an invaginated coiled portion of the Schwann-cell membrane; note the large nucleus of the Schwann cell. The numerous small circular structures in the spaces between the Schwann cells (as in extreme lower right corner) are cross sections of collagen fibrils. (B) Enlarged view of the myelinated axon.

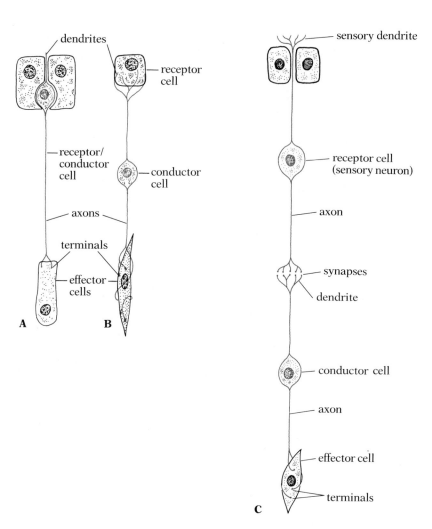

18.6   Simple nervous pathways
(A) In the simplest pathway—the one generating tentacle withdrawal in coelenterates, for example—only two cells are necessary. A single neuron acts as both sensory receptor and conductor of information; the other cell is the effector, and no processing stage is required. Such pathways are called reflex pathways. (B, C) Slightly more elaborate pathways may involve a specialized sensory receptor cell, itself sometimes a neuron, and a separate conductor cell. Again, however, there is no processing. The conductor cells shown here display typical neuron anatomy: a large cell body containing the nucleus, and two narrow projections—the dendrite (sometimes branched) for collecting information, and the axon for carrying it to other cells.

SIMPLE NERVOUS PATHWAYS

All animal groups above the level of the sponges have some form of nervous system, though in some groups it is very primitive. In the tentacles of many coelenterates we see the simplest possible type of nervous pathway—one composed of only two specialized cells: a sensory receptor/conductor neuron, and an effector (response) cell (Fig. 18.6A). Such a pathway generates strictly automatic behavior—tentacle withdrawal, for instance—because the processing stage is essentially missing. We refer to such automatic behavior as a pure *reflex*: there is only one input; there are no alternative pathways for the information to take; the behavioral response involves no coordination of effectors; and since there is no interconnection between the various circuits of this sort, there is no possibility of central neural control.

And yet, simple as such a circuit looks, it does have one kind of flexibility: as a result of repeated stimulation the sensory neuron can become less sen-

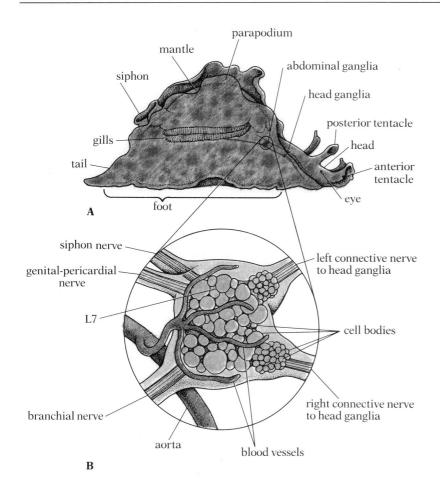

A

B

18.7   The sea slug, *Aplysia*
(A) *Aplysia* is basically a large, shell-less aquatic snail. The head has eyes and chemosensory tentacles. Under the mantle, which is the remnant of a shell, are the delicate gills, which extract oxygen from the water, and the siphon, which moves water across the gills. The mantle cavity also contains the so-called purple gland from which the animal discharges a concealing ink when disturbed. The foot is the organ of movement. The bilaterally symmetrical nervous system (red) is organized as a series of paired ganglia. (B) The ganglia contain the cell bodies of the neurons. The long axons over which they transmit information are organized into cables called nerves. The abdominal ganglia, shown here, actually contain about 2,000 neurons; they control circulation, respiration, and reproduction. The cell known as L7 is involved in defensive behavior, respiration, and circulation.

sitive, through a process known as ***adaptation.*** In coelenterates, for example, adaptation allows the tentacles to adjust to the constant background level of stimulation produced by water currents, so that the defensive reflex is triggered only by an extraordinary stimulus. Sensory adaptation, as we shall see, is a widespread phenomenon.

The gill-withdrawal pathway of *Aplysia*   Most nervous pathways, even in coelenterates, are in fact comprised of at least three separate cells: a sensory receptor cell or neuron, a conductor neuron, and an effector cell (Fig. 18.6B–C). (Sensory receptors respond to appropriate stimulation with an electrical change; when the resulting signal is conducted through an axon to another cell, the receptor is said to be a ***sensory neuron;*** when an axon is lacking and the information is collected by dendrites of a neuron directly from the receptor-cell body, the receptor is referred to simply as a sensory cell.) Neural organization involving separate receptors, conductors, and effectors is particularly well understood in the marine mollusc *Aplysia,* often called a sea slug (Fig. 18.7). Eric Kandel, who with his co-workers at

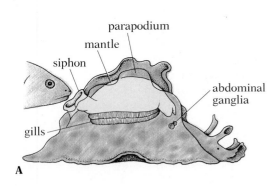

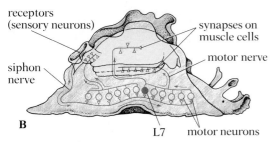

18.8   Gill-withdrawal behavior

Normally, *Aplysia* pumps water across its exposed gills to obtain oxygen (A). When the siphon is disturbed, however, the touch-sensitive receptors send signals through the siphon nerve to the abdominal ganglia; there the sensory information is communicated to a group of motor neurons which, when sufficiently stimulated, signal the muscles of the mantle, gill, and siphon (B). As a result, a disturbed *Aplysia* withdraws these sensitive structures (C). The motor neurons, all of which are actually found in the abdominal ganglia, as well as the sensory and motor nerves, have been rearranged and enlarged here for clarity.

Columbia University has contributed enormously in recent years to our understanding of neural organization, selected this large shell-less snail for study because of the accessibility of its nervous system. The cell bodies of *Aplysia*'s neurons are very large and are found, conveniently, in groups called **ganglia,** where they can be recognized individually and their electrical activity monitored. As in many other animals, the axons along which neurons transmit information are grouped into discrete cablelike bundles —the **nerves** (Fig. 18.7B).

The gill-withdrawal circuit of *Aplysia* begins with a group of touch-sensitive sensory neurons in the siphon that send information to a set of **motor neurons,** which control muscle cells (Fig. 18.8A–B). When touched, the animal quickly withdraws its siphon and gill (Fig. 18.8C). The neural pathway that generates this critical behavior has a built-in potential for control: each motor neuron automatically averages the inputs from all the sensory neurons, and so is not stimulated to conduct by the activity of one or two cells in contact with a stray grain of sand or blade of aquatic vegetation. At first glance it might appear that just one motor neuron would suffice to generate the reflex behavior of gill withdrawal. But each motor neuron controls the muscles in a separate part of the vulnerable gill; to maintain the processes of respiration and circulation in *Aplysia*, the muscles in the various parts of the gill (and of the siphon) must perform carefully timed coordinated movements. Hence, there must be several independent motor neurons.

Interneurons   As more cells are added to a pathway, more processing of information becomes possible. Cells specialized for this crucial role of processing—the middlemen of the nervous system—are known as **interneurons.** The interneuron typically collects and processes input from many cells (often thousands), and passes on the resulting information to its target cells (Fig. 18.9). As we shall see later, it is found even in the gill-withdrawal circuits of *Aplysia*, which includes an alternative pathway: sensory neurons to an interneuron to motor neurons to muscle cells. The signals an interneuron processes can be either excitatory or inhibitory. Indeed, **inhibition,** the ability to counteract excitatory input from other cells, is essential to this information processing, since the nervous system usually operates in a manner resembling the antagonist strategy of the endocrine system: contradictory signals are sent, and the ratio between them determines the cell's response. The importance of interneurons in this processing of contradictory signals would be hard to overemphasize: the brains of animals consist almost entirely of interneurons arranged in complex, highly specialized networks.

NERVE NETS AND RADIAL SYSTEMS

The simple neural pathways we have been discussing join together in most animals to form organized nervous systems. In coelenterates like hydra, the nervous system consists of separate receptor, conductor, and effector cells (18.6B). The conductor cells, however, do not form definite pathways, but

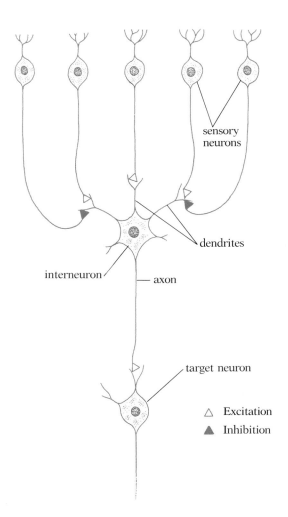

sensory
neurons

dendrites

interneuron

axon

target neuron

△ Excitation

▲ Inhibition

18.9   An interneuron
Interneurons combine information from many cells
to produce a single output. In this example five
sensory neurons communicate with the interneuron,
three that excite it and two that inhibit it. The relative
strength of the various incoming signals determines
whether excitation or inhibition predominates. In the
brain, the input is usually from other interneurons.

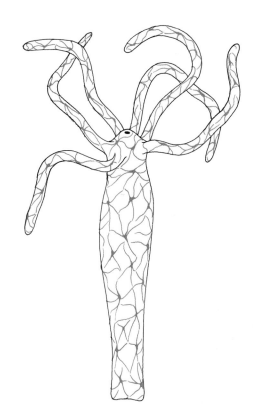

18.10   Nerve-net system of hydra
Conductor cells in organisms with nerve nets are not
organized into specialized pathways. As a result, there
can be no centralized control; only localized
responses to stimuli are possible.

rather interlace to form a diffuse ***nerve net*** running throughout the body
(Fig. 18.10). There is no central control: impulses simply spread slowly
from the region of initial stimulation to adjacent regions, and they can
move in either direction along most fibers, though a few fibers are strictly
unidirectional. This sort of organization suits a sessile organism whose only
means of escape involves moving the tentacle and the side of the body that
has been touched. Nerve-net reactions generally take the form of local
movements and the discharge of stinging cells (nematocysts) into potential
prey. Such a system, lacking the potential for central coordination of complex reactions, can produce only a limited behavioral repertoire.

Other coelenterates display a degree of centralization. Jellyfish, for in-

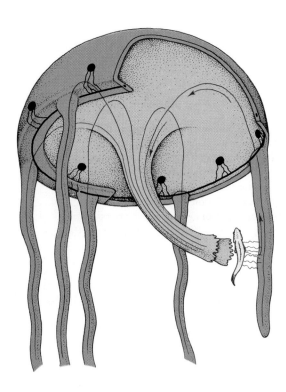

18.11 Nervous system of jellyfish
Jellyfish represent a degree of centralization among radially symmetrical animals. Here neurons are organized into a primitive ring system that serves to synchronize the contractions of the swimming muscles of the bell. In addition, sensory neurons on each of the peripheral tentacles send axons to muscles on the central stalk. When the firing of a nematocyst stimulates certain of these cells, the resulting signals direct the creature's mouth (which is at the base of the stalk) to the affected part of the tentacle for a possible meal.

stance, have a nerve ring in the "bell" portion of the body (Fig. 18.11). The other neurons tend to funnel into the ring, and conduction from one side of the animal to the other is thus more rapid and organized than would be possible with the randomly oriented pathways of a simple nerve net. This centralization is reflected in the swimming movements of jellyfish, which consist of rhythmic, coordinated contractions of the whole bell.

BILATERAL NERVOUS SYSTEMS

The evolution of nervous systems in bilaterally symmetrical animals is evident even in the lowly flatworms. The developmental progression of the bilateral nervous system can be summarized by the following evolutionary trends:

1. The nervous system becomes increasingly centralized by formation of major longitudinal nerve cords (the central nervous system) through which most pathways between receptors and effectors have to pass, and in or near which the cell bodies of most neurons lie.
2. Conduction along nervous pathways becomes restricted to one direction only. A natural distinction thus develops between sensory fibers leading toward the CNS (afferent fibers) and motor fibers leading away from the CNS (efferent fibers).
3. Nervous pathways within the CNS become increasingly complex as large numbers of interneurons are included, a development which permits increased flexibility of response.
4. Cells performing different functions become increasingly segregated within the nervous system, so that distinct functional areas and structures become obvious.
5. An increasing ascendancy of the front end of the longitudinal cords leads to the formation of a *brain,* which becomes more and more dominant—a process known as *cephalization.*
6. The number and complexity of sense organs increases.

These trends are not yet distinct in the most primitive flatworms (those thought to be most like the ancient ancestral forms); such flatworms have only a nerve net much like that of hydra. Some slightly more advanced flatworms show the beginnings of major cords, formed by a grouping of neurons in the nerve nets and running longitudinally, from front to back (Fig. 18.12A). There are often as many as eight of these cords, located ventrally, dorsally, and laterally (Fig. 18.12B). Still more advanced flatworms show a reduction in the number of longitudinal cords, the most advanced having only two, both located ventrally (Fig. 18.12C–E).

Flatworms with the least advanced development of longitudinal cords show very little evidence of any special structure at the anterior end that could be called a brain (Fig. 18.12A), but the tiny swellings present there have nevertheless been charitably labeled "brains." Flatworms at the more advanced end of the spectrum have a much better developed brain (Fig. 18.12E), though even here the brain exerts only limited dominance over the rest of the CNS.

The nearly universal location of an animal's brain in its head (at the anterior end of its body) is not accidental. The evolutionary and functional explanation for the brain's familiar location, which seems so natural, is almost certainly to be found in the animal's direction of movement. The anterior end is usually the part of a bilateral animal that first encounters new stimuli as the animal moves. Natural selection therefore favored development of a particularly high concentration of sense organs in this region, which, in turn, led to an enlargement of the anterior end of the longitudinal nerve cord.

The most primitive version of the brain was probably almost exclusively concerned with funneling impulses from the sense organs into the nerve cords that carried the information to the appropriate motor neurons. The selective advantage of comparing the various sensory inputs—processing the incoming information before passing it on to the muscles—must have led to an increase in the number of interneurons at the front, where the sensory receptors were already concentrated. Thus the brain became an area for analysis rather than simply a collection of sensory cell bodies through which raw sensory data were funneled to the effectors farther back, as can be seen in an organism like the nematode (Fig. 18.13). With this increasing specialization for analysis, the brain became a coordination center as well, developing more and more dominance over the rest of the CNS.

The evolutionary trends whose beginnings can be seen so clearly in flatworms, where intermediate stages between a nerve-net system and a centralized system can be studied in animals still extant, are most fully developed in the vertebrates—particularly the mammals—and in the higher invertebrates, including the annelids, molluscs, cephalopods, and arthropods. In all of these animals there is a high degree of centralization, and the older nerve-net strategy is represented by only a few vestiges in parts of the body where sluggish movements, such as the peristaltic contractions of the mammalian intestine, are controlled by slow, diffuse conduction.

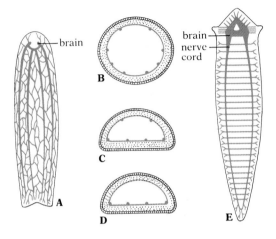

18.12   Nervous systems in flatworms

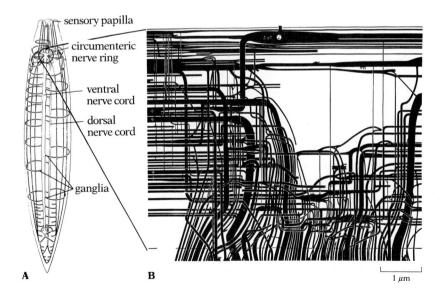

18.13   Nervous system of a nematode
(A) Though the nematode has fewer than 300 neurons, it displays many of the organizational features of advanced nervous systems. Most of the sensory neurons are located at the anterior end in a group of sensory papillae, from which they send information to a centralized processing area—a sort of brain—called the circumenteric nerve ring. From there neurons communicate with the muscles through the ventral nerve cord. (B) A portion of the switchboard-like anatomy of the nematode "brain" has been enlarged to show its formidable organization. The processing of sensory information that goes on in these interneurons determines what actions will be performed, and where they will be directed.

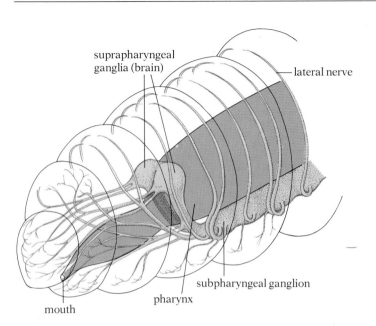

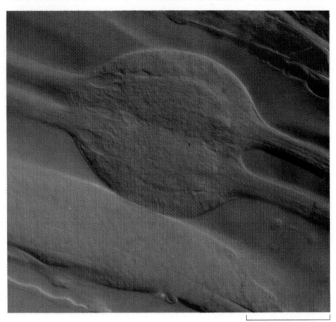

0.1 mm

18.14   Earthworm and insect neural organization
Left: As in nematodes, many of the sensory receptors
in the earthworm are located in the head, and carry
information to the brain (the suprapharyngeal
ganglia) for processing. From there information is
conducted along the ventral cord and ultimately to
the muscles. The earthworm brain serves as a device
for deciding what action is to be taken and where to
direct it. The earthworm brain is far more complex
than that of nematodes. Right: Double ganglion and
paired ventral nerve cord of a living fly larva, viewed
with Nomarski optics.

In the higher invertebrates, the central nervous system is a pair of ventrally located longitudinal cords in which the cell bodies of the neurons
form ganglia, and the fibers are gathered into nerves that act as communication pathways between ganglia (Fig. 18.14). Even primitive annelids have
prominent ganglionic masses, a pair in each body segment, which are connected by nerves running between the segments; almost all the cell bodies
are located in ganglia. The brain is simply another ganglion that happens to
be in the animal's head. Because the brain of invertebrates is normally
formed by the fusion of the four most anterior ganglia, it is larger than the
segmental ganglia, and its mix of cells has a higher proportion of sensory
than of motor neurons. Its dominance over the other ganglia is noticeable,
but limited in comparison to that of the vertebrate brain.

More advanced arthropods, particularly some of the insects, show far
more concentration of coordination in the front end. Moreover, many of
the other segmental ganglia have fused, the result apparently being better
integration of control between the segments. The brain, however, has remained relatively small, and the thoracic ganglia (with which some of the
abdominal ganglia have fused in many organisms) perform many vital coordinating functions.

The persistence of thoracic ganglia in insects is probably correlated with
at least two important functional arrangements of their bodies: the legs and
wings are attached to the thorax, so a concentration of motor coordinating
centers in the thorax is advantageous; and many of the sense organs are located on the legs or the thorax rather than on the head (flies have taste receptors on their feet, for instance—the better to know what they are
walking on—while many insects have ears on the thorax or in their legs).
The original functional autonomy of the various ganglia is still evident in

many insects. The neural "instructions" for many of the so-called behavioral programs that control walking, flying, courting, mating, and stinging, for instance, are stored in the thoracic and abdominal ganglia, and the brain serves primarily to aim the behavior and to turn it on and off. Even headless flies and roaches can learn.

VERTEBRATE NERVOUS SYSTEMS

The central nervous systems of vertebrates (the spinal cord and brain) differ in several important ways from those of annelids and arthropods:

1. The vertebrate spinal cord is single; it is located dorsally; and it forms in the embryo as a tube with a hollow central canal, a remnant of which survives in the adult (see Fig. 18.32). The cords of annelids and arthropods, on the other hand, are double (two cords lying side by side, though often partially fused); they are located ventrally; and they are always solid.
2. The vertebrate spinal cord is not so obviously organized into a series of ganglia and connecting tracts.
3. Though many simple coordinating functions in vertebrates are still performed in the spinal cord, there has been extensive development of the brain, which exerts far more dominance over the entire nervous system than the brain of any annelid or arthropod. Neural pathways that control complex behavior patterns are located exclusively in the brain. In short, the vertebrate brain is the master control center for almost all bodily functions. We shall discuss neural pathways in vertebrates in more detail later in this chapter and in the next chapter.

HOW NEURONS WORK

CONDUCTION ALONG NEURONS

General features of the nerve impulse   Neurons respond to a great variety of stimuli, such as mild electric shock, a pinch, or an abrupt change of pH. Various types of sensory neurons and sensory cells are, as we know, specialized to respond to light, odors, movement, and so on. However, mild electrical stimuli are most often used in research because the intensity and duration of such stimuli can be precisely controlled, and because they do little or no damage to the nerve cell. Consider the following experiment.

Suppose we are working with an isolated neuron. We place two electrodes several centimeters apart on the surface of the axon (Fig. 18.15A). These electrodes are connected to recording equipment, so we can detect any electrical changes at the points they touch on the axon. Now we apply an extremely mild electrical stimulus to the cell body. Nothing happens; our recording equipment shows no change (Fig. 18.15B). We increase the intensity of the stimulus and try again. This time our equipment tells us that an electrical change has occurred at the point of contact with the first electrode, and that a fraction of a second later a similar electrical change has taken place at the second electrode (Fig. 18.15C). We have succeeded in

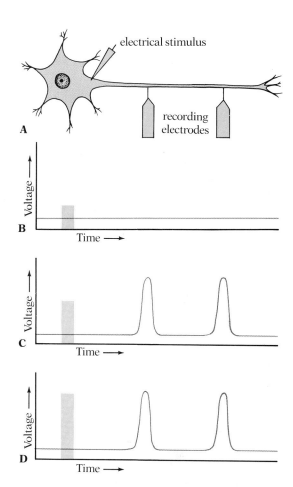

18.15 Initiation and propagation of an impulse
(A) By inserting an electrode into the cell body of a neuron to deliver precise amounts of current, and then monitoring the axon with recording electrodes, we can study the initiation and propagation of an impulse. In B–D the magnitude of the stimulus is shown in gray, at left. (B) If the stimulus is below the cell's threshold for triggering an impulse, neither recording electrode registers any change. (C) If, however, the stimulus is adequate, an impulse is registered as a spike (red) first at electrode 1 and then at electrode 2 as it moves along the axon. Note that the spike has the same magnitude at both places. (D) When a still larger stimulus is used, the speed of movement and the intensity of the impulse are not affected: the impulse is an all-or-none response.

stimulating the axon, and a wave of electrical activity has moved down the axon from the point of stimulation, passing first one electrode and then the other at a rate of 30–90 m/sec. Next we apply a still more intense stimulus, and again we record a wave of electrical change moving down the fiber (Fig. 18.15D), but the intensity and speed of this electrical activity are the same as those recorded from the previous milder stimulation.

We can draw several important conclusions from this experiment: (1) A nerve impulse can be detected as a wave of electrical activity moving along an axon. (2) A potential stimulus must be above a critical intensity or *threshold* if it is actually to stimulate an axon; there is also, it turns out, a minimum duration. (3) Increasing the intensity of the stimulus above the threshold does not alter the intensity or speed of the impulse produced; the axon fires maximally or not at all, in a type of reaction commonly referred to as an ***all-or-none response.***

Immediately an important question comes to mind: if an axon exhibits the all-or-none property with respect to the intensity and speed of an impulse, how do animals normally detect the intensity of a stimulus? There are two ways in which this information is commonly coded: First, as the intensity of the stimulus increases, the number of impulses produced per unit time—that is, the frequency of the impulse—goes up. Second, neighboring cells may have different thresholds, so as the intensity of the stimulus increases, the thresholds of more and more cells are exceeded and more neurons fire. The brain can interpret both the higher firing rate of individual neurons and the greater number of active neurons as indicating a more intense stimulus.

The nature of the impulse   When it was discovered more than a century ago that a nerve impulse involves electrical changes, scientists assumed that the impulse was a simple electric current flowing through a neuron much as other currents flow through wires. It soon became clear, however, that the speed of a nerve impulse is far slower than that of electricity in a wire. Moreover, the cytoplasmic core of an axon offers so much resistance to simple electric currents that they die out after moving only a few millimeters. In fact, any resistance, however low, would cause a simple electric current to diminish in strength as it moved. Yet if we measure a nerve impulse at various points along the axon of a neuron, we find that it remains the same; its strength does not decrease with distance. On the other hand, crushing or poisoning an axon may destroy its ability to conduct impulses even though its electrical conductivity has not been altered. In short, impulse conduction depends on activity in the living cell, activity almost certainly involving chemical processes. The impulse, then, is not a simple electric current, but rather an electrochemical change propagated along the neuron.

The basic outlines of the modern theory of nerve action were proposed in 1902 by Julius Bernstein of the University of Halle in Germany. He knew that the concentrations of certain ions inside the nerve cell and in the surrounding fluids are very different: the concentration of sodium (Na^+) is very low inside the cell, while the concentration of potassium (K^+) and negative ions is very high. (As we saw in Chapter 14, this ionic differential is seen to a lesser extent in most cells.) He also knew that the unequal distri-

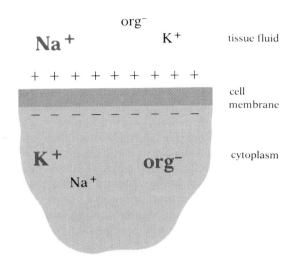

18.16 The polarization of the nerve-cell membrane
The concentration of sodium ions (Na⁺) is much greater in the tissue fluid outside the cell, and the concentration of potassium ions (K⁺) is much greater inside the cell. There is also a much greater concentration of negative organic ions (org⁻) inside the cell. The net effect of these unequal distributions of ions is that the inside of the cell is negative (about −70 millivolts) relative to the outside.

bution of ions results in an electrostatic gradient across the membrane. As a result the inside of the cell has a charge of about −70 millivolts relative to the outside (Fig. 18.16). Though an understanding of membrane function was far from complete in his day, Bernstein suggested that the permeability of the neural membrane varies for different ions, and that it is the great selectivity of the membrane that maintains the separation of ions and the resulting electrostatic gradient. Bernstein further proposed that during the passage of an impulse the selectivity of the membrane is momentarily destroyed. The ions can therefore move freely, and the electrostatic gradient across the membrane falls to zero; in other words, the membrane is momentarily unable to maintain a separation of charge. In addition, this depolarization in some way triggers the depolarization of adjacent patches. To Bernstein the nerve impulse was a wave of depolarization moving along the membrane.

Though Bernstein's hypothesis was widely accepted, there was little experimental evidence either for or against it for many years. Nerve fibers were simply too small for accurate measurements of the changes taking place inside them during impulse conduction. In 1933, however, J. Z. Young at Oxford discovered that squid possess several giant nerve fibers, which may be as much as a millimeter in diameter. These fibers mediate the animal's escape response. They run along the body wall of the squid (Fig. 18.17) and innervate the muscles that enable the squid to propel itself backward at very high speed by explosively expelling water from the mantle cavity through a funnel near the head.

The great size of these squid nerve fibers allows them to conduct nerve impulses very rapidly. In general, the greater the diameter of a nerve fiber, the faster it conducts. Fibers larger than normal, though rarely as large as those of the squid, are also found in many other invertebrates in neural circuits in which very rapid conduction is important. The fibers in squid are especially large because squid are large, so the messages have a far greater distance to travel than in most invertebrates. (As we have seen, vertebrates

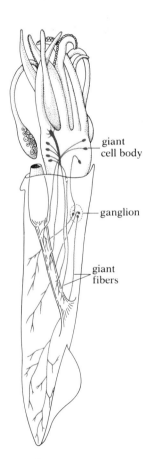

18.17 Giant nerve fibers of a squid

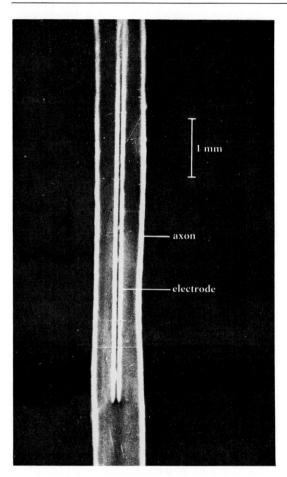

18.18   Photograph of a giant axon of a squid with a glass-tube microelectrode inside it

evolved myelinated fibers as an alternative adaptation for increasing conduction speed.) The discovery of the giant nerve fibers of the squid finally made it possible for biologists to study in detail the events that occur during conduction, and opened a whole new era in neurophysiology.

In 1939 H. J. Curtis and K. S. Cole at Woods Hole, Massachusetts, and A. L. Hodgkin and A. F. Huxley in Plymouth, England, developed a technique for inserting a very thin microelectrode into a squid giant nerve fiber. The microelectrode, which consisted of a glass tube filled with either salt solution or metal, was inserted into one end of the fiber and pushed down it a distance of 1–3 cm (Fig. 18.18). With the microelectrode in place, it was possible to show that the electrostatic gradient across the membrane did indeed change during the passage of a nerve impulse, as Bernstein had predicted. But one thing was wrong: the gradient changed too much. For an instant, the inside of the fiber actually became positive relative to the outside. The membrane, then, did not simply become depolarized; it actually reversed its polarization momentarily. Stimulation does not just destroy the selectivity of the membrane, as Bernstein had thought, for had it done so the membrane potential would have simply gone to zero. Instead, it radically alters this selectivity. The stimulated point on the membrane remains selectively permeable to ions, but the nature of its selectivity changes dramatically. Hodgkin and Bernard Katz proposed in 1947 that the membrane must initially allow far more Na$^+$ ions to enter the cell than it allows other ions to leave. In a long series of now-classic experiments, Hodgkin and Huxley went on to work out a detailed quantitative description of the changes in ionic conductance (including those of both Na$^+$ and K$^+$) during impulse conduction in the squid. Their findings have been found applicable to virtually all nerve cells.

According to the Hodgkin-Huxley model, the membrane of the resting neuron is polarized, with the inside negative relative to the outside. The concentration of Na$^+$ ions is much higher outside, and the concentration of K$^+$ ions is much higher inside. Stimulation causes the membrane to undergo a large but short-lived increase in permeability to Na$^+$ ions. These rush across the membrane into the cell, both because of their natural tendency to diffuse from regions of higher concentration to regions of lower concentration and because they are attracted by the negative charge inside the cell. The inward flux of Na$^+$ is so great, in fact, that for a moment the inside actually becomes positively charged relative to the outside. (This happens because the osmotic concentration of Na$^+$ relative to the inside remains high even when the electrostatic gradient across the membrane becomes zero, or even slightly unfavorable.) A fraction of a second later, the permeability of the membrane to Na$^+$ has returned to normal, while its permeability to K$^+$ has increased greatly. The K$^+$ ions now rush out of the cell because their concentration is higher inside than out and because they are repelled by the momentarily high positive charge inside the cell. This exit of positively charged K$^+$ ions restores the charge inside the cell to its original negative state. In short, the inside surface of the membrane is initially negative; it becomes positive when Na$^+$ ions flood in, and then negative again when K$^+$ ions rush out. The impulse is propagated along the neuron because the cycle of changes at each point alters the permeability of the membrane at the adjacent point and initiates a similar cycle there (Fig. 18.19);

this in turn starts the cycle farther along the axon, and so on, like a chain of falling dominoes.

The Hodgkin-Huxley model of impulse conduction explains why the nerve impulse, unlike a simple electric current moving through a wire, does not decrease in strength as it moves along the fiber. The impulse is constantly being regenerated; the electrical change, or *action potential,* at each successive point is a new event, equal in magnitude to the electrical events at the preceding points. An electric current generated by a battery or dynamo is simply carried passively by a wire, but a nerve impulse derives its energy from the path along which it travels, being generated anew at each successive point along the fiber. (Myelinated fibers conduct impulses faster than nonmyelinated fibers of the same diameter because the points at which the action potential is regenerated are not adjacent patches of membrane, but rather successive nodes.)

The role of gated channels   The propagation of a nerve impulse, or action potential, is basically a membrane phenomenon. It depends on an initial electrostatic gradient across the membrane, followed by a coordinated

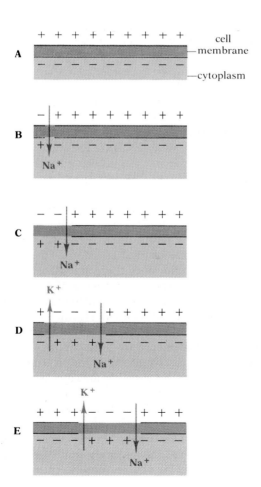

18.19 Model of propagation of a nerve impulse
(A) The interior of a resting nerve fiber is negative relative to the exterior because the ratio of negative to positive ions is higher inside the cell than outside. The interior has a high concentration of potassium ions (K^+), the exterior a high concentration of sodium ions (Na^+). (B) When a fiber is stimulated, the membrane, previously relatively impermeable to Na^+ ions, becomes highly permeable to them at the point of stimulation, and a large number rush into the cell (arrow). The result is a reversal of polarization at that point, the inside of the fiber becoming positive relative to the outside. (C) Meanwhile, because the change in membrane permeability at the point initially stimulated has altered the permeability at adjacent points, Na^+ ions begin rushing inward at those points. (D) An instant later, the membrane at the initial point of stimulation becomes highly permeable to K^+ ions; a large number of these rush out of the cell, and the inside of the fiber once again becomes negative relative to the outside. (E) The cycle of changes at each point alters the permeability of the membrane at adjacent points and initiates the same cycles of changes there; Na^+ ions rush into the cell, and K^+ ions rush out a moment later. The movement of this cycle of changes along the nerve fiber is what we call a nerve impulse.

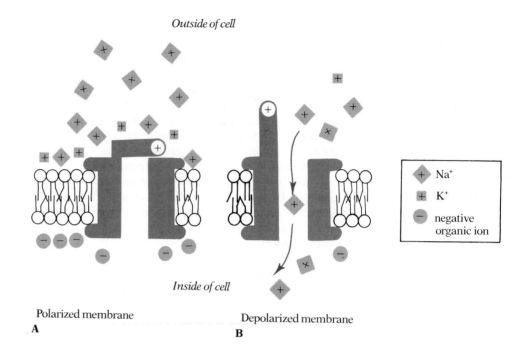

Outside of cell

Polarized membrane
A

Depolarized membrane
B

Inside of cell

◆ Na⁺

⊞ K⁺

● negative organic ion

18.20   Voltage-gated channels

(A) Voltage-gated channels probably have charged gates held closed by the electrostatic repulsion of positive ions outside the cell and the attraction of negative ions inside. (B) When the membrane depolarizes, these gates swing open to expose specific ion channels. As the electrostatic gradient continues to change during the course of the action potential, the gates close.

series of ion-specific changes in permeability. In turn, these permeability changes depend on gated channels; since, as we have seen, both Na^+ and K^+ cross the membrane, there must be channels for each. You may remember that in some of the channels we examined in connection with the endocrine system, a hormone binding to the channel caused an allosteric change that opened a gate and allowed ions to cross through the membrane. The membrane proteins responsible for creating the action potential are ***voltage-gated channels;*** that is, <u>they open and close in response to changes in the electrostatic gradient across the membrane.</u> Current evidence suggests that voltage-gated channels have positively charged gates held shut by both the electrostatic repulsion of like-charged ions outside the membrane and the electrostatic attraction of oppositely charged ions inside. Once a stimulus has caused the membrane to depolarize by a precise amount and the electrostatic interactions are thereby weakened, the gates swing open, exposing specific ion channels (Fig. 18.20). The Na^+ channels open when the stimulus has reached the threshold for firing the neuron in question, while the K^+ channels do not open until enough Na^+ ions have flowed through to depolarize the membrane almost completely. As the electrostatic gradient continues to change as a result of ionic flow, the gates close.

The role of diffusion and electrostatic attraction   If impulse conduction involves an inward flow of Na^+ followed by an outward flow of K^+, how does the neuron reestablish its original ionic balance? In other words, how does it get rid of the extra Na^+ and regain the lost K^+? If the initial ionic distribution were not restored the neuron would eventually lose its ability to conduct impulses, but neurons can continue to conduct impulses indefinitely with only a very brief refractory period (on the order of 0.5–2.0 milliseconds) after each impulse.

Two mechanisms work to keep neurons functioning. In the short run, as an impulse passes and the membrane is depolarized, diffusion and electrostatic attraction seem to restore the electrochemical balance between Na^+ outside and K^+ inside the cell almost instantaneously. To understand how this happens, it is important to realize that virtually all events involving the action potential take place very close to the cell membrane. In a resting nerve fiber, free ions that have been attracted to the cell membrane by the opposite charges on the other side turn it into an area of concentrated charge; the electrostatic gradient across the thin membrane approaches an incredible 10^5 volts/cm. Once the two surfaces of the membrane have been "coated" by ions attracted by the electrostatic force of the oppositely charged ions on the other side, this concentrated layer of charge actually repels the approach of additional ions from the same side (Fig. 18.21).

When an action potential allows ions to cross the membrane, only the ion concentrations near the highly charged inner and outer surfaces of the membrane are affected, simply because the permeability changes are so short-lived that only the closest ions have time to move to and through the membrane. The tiny resulting alteration in charge is rapidly absorbed as the Na^+ ions that crossed into the neuron and the K^+ ions that moved out diffuse into the fluid on either side of the membrane. It is rather like adding a drop of ink to a pond: for a moment there is a dark patch, but this rapidly dissipates as the ink diffuses into the larger volume of water. In nerve cells, diffusion of the added Na^+ inside is aided by the electrostatic attraction of negative ions near the membrane for the oppositely charged ions concentrated on the other side. With the added Na^+ ions thus dissipated within the

18.21 Charging of the membrane
Electrostatic attraction across the cell membrane draws some of the positive ions on the outside and some of the negative ions on the inside to the respective surfaces of the membrane (A–B). As a result, these surfaces become coated with oppositely charged ions. At some point, however, no more ions are attracted: the concentration of positive ions on the outer surface is so high that their electrostatic repulsion of free-floating positive ions balances the attraction of those ions to the negatively charged interior of the cell; on the other side of the membrane an equivalent situation develops with respect to negative ions (C).

Outside of cell

A   B   C

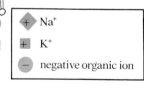

◆	Na^+
▦	K^+
⬤	negative organic ion

Inside of cell

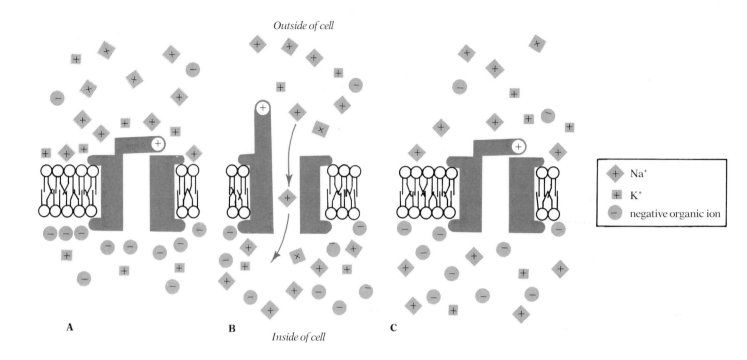

Outside of cell

Na⁺

K⁺

negative organic ion

A   B   C

Inside of cell

18.22 Diffusion after an action potential
The ions that cross the neural membrane during an action potential rapidly diffuse away from the membrane into the abundant fluid inside and outside the cell. (A) Before the Na⁺ channel shown here opens in response to partial membrane depolarization, the membrane is fully charged. (B) When the gate opens, Na⁺ ions rush into the cell, reducing and then reversing the electrostatic gradient across the membrane. (C) After the channel closes, the excess Na⁺ near the inner membrane of the neuron rapidly diffuses into the cytoplasm, and the membrane begins to recharge.

cell, the polarity of the membrane is restored to the resting potential (Fig. 18.22).

Despite the dramatic nature of the events at the neural membrane, only a minute quantity—about 10^{-12} moles—of Na⁺ enters the cell during an action potential. The net effect of a single action potential on the concentration of Na⁺ in the cell as a whole is therefore negligible. The reserve of ions both inside and out is so large that a recently killed nerve cell will continue to conduct action potentials for some time. Eventually, however, the action potentials begin to deteriorate and conduction comes to a halt. If we were to continue polluting our pond with ink, at some point it would begin to change color. Similarly, at some point the internal fluid of the nerve cell can no longer sufficiently dilute the doses of Na⁺ that have been repeatedly allowed to enter. Diffusion and electrostatic attraction are therefore only short-term solutions to the problem of maintaining neuron function.

The sodium-potassium pump What process, then, maintains the long-term ability of active neurons to conduct action potentials? Whatever process preserves the electrochemical gradient across a nerve-cell membrane and accounts for its resting potential (and therefore its ability to conduct impulses) must involve active transport, since to expel the Na⁺ a cell must move it against both concentration and electrostatic gradients, and to regain the lost K⁺ the cell must force it to move against its concentration gradient. We know now that the membrane of the average neuron contains approximately a million sodium-potassium exchange pumps, and that their power is supplied by ATP. Such pumps enable cells actively to extrude Na^+ ions and take up K^+ ions (Fig. 18.23). A single pump moves approximately 200 Na^+ and 135 K^+ ions every second. The development of sodium-potassium pumps, combined with the evolution of ion-specific voltage-gated channels, has been the basis for the evolution of neural transmission.

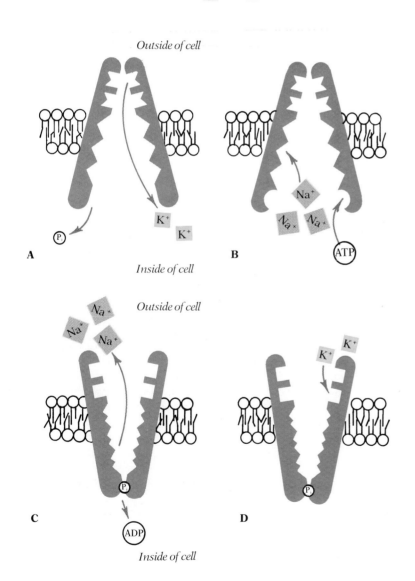

18.23   Model of the sodium-potassium pump
The pump consists of proteins embedded in the cell membrane. When the pump complex is open to the inside of the cell (A, B), the K^+ binding sites no longer bind potassium, so those ions drift free into the cytosol, along with the phosphate group that powered the preceding cycle. The Na^+ sites, on the other hand, become active. Once the Na^+ ions have been bound, ATP binds to its site, thus phosphorylating the complex and causing a conformational change that opens the pump to the outside (C, D); this change severely reduces the protein's affinity for Na^+ (which in consequence drifts free into the extracellular fluid), while it activates the K^+ sites. The binding of K^+ ions causes the complex to return to its former conformation (A) and to release both the phosphate group obtained earlier from the ATP and the K^+ ions. The extent of movement of the pump subunits has been greatly exaggerated for clarity.

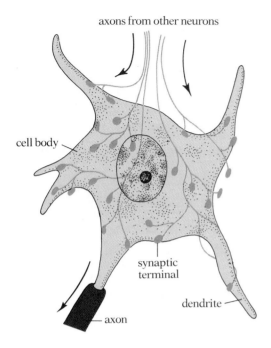

axons from other neurons

cell body

synaptic
terminal

dendrite

axon

18.24   Synapses on a motor neuron
Many axons, each of which branches repeatedly,
synapse on the dendrites and cell body of a single
motor neuron. Each branch of an axon ends in a
swelling called a synaptic terminal.

**18.25   Scanning electron micrograph of synaptic
terminals from *Aplysia***
The synaptic terminals of the numerous axons are in
contact with the cell body of a postsynaptic neuron.
Note that it is the edge of the terminal, not its
flattened end, that characteristically forms the synapse.

TRANSMISSION ACROSS SYNAPSES

The nature of synaptic transmission   Understanding how an impulse
moves down the axon of a nerve cell is the first step in understanding how
neurons communicate with each other and, eventually, how neural control
is accomplished. The axon of one neuron usually synapses with the dendrites or cell body of other neurons. Since the terminal portion of an axon
normally branches repeatedly, a single axon may synapse with many other
neurons; moreover, it usually synapses at many points with each of these
neurons (Fig. 18.24). Each tiny branch on an axon usually ends in a small
swelling called a *synaptic terminal,* or bouton (Fig. 18.25).

In a few cases there is a gap junction between the membrane of the synaptic terminal and the membrane of the adjoining cell (see p. 140). Such a
junction permits direct electrical coupling between the two neurons, so
that an impulse traveling down the axon of the first neuron can pass unhin-
dered to the next neuron. Since electrical synapses minimize the delay in
transmission of impulses, they tend to occur in places in the nervous system
where speed of conduction is of special importance. They also provide a
high degree of certainty that an impulse in the first neuron will give rise to
an impulse in the second neuron.

The vast majority of synapses, however, are not electrical, but chemical.
A space about 20 nm wide, known as the *synaptic cleft,* separates the syn-

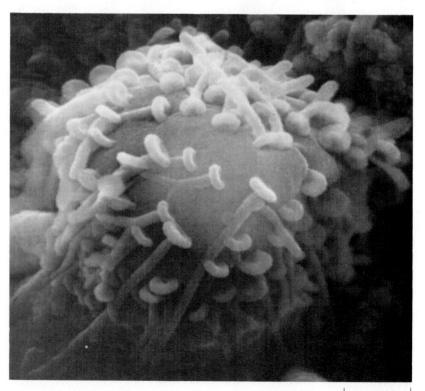

2 μm

aptic terminal of the first (presynaptic) neuron and the membrane of the second (postsynaptic) neuron. Transmission across this cleft is by a diffusible *transmitter* chemical released from tiny *synaptic vesicles* in the terminal (Fig. 18.26); each of the thousands of vesicles may contain as many as 10,000 molecules of transmitter. When an impulse traveling down the axon of the presynaptic neuron reaches the terminal, special voltage-gated calcium channels concentrated at the synapse open and the terminal membrane becomes more permeable to Ca^{++} ions. Because they are 10,000 times more concentrated outside cells, Ca^{++} ions then diffuse into the terminal from the surrounding fluid. The Ca^{++} ions in some way stimulate synaptic vesicles in the terminal to move to the terminal membrane, fuse with it, and then rupture, thereby releasing the transmitter chemical into the synaptic cleft by exocytosis. As we shall see in Chapter 20, Ca^{++} ions play an analogous role in triggering muscle contraction and ciliary action.

The transmitter molecules released into the cleft diffuse across it and bind by weak bonds to highly specific receptors on the *postsynaptic membrane* of the next neuron. The receptors are specific for a neurotransmitter, and function in a manner similar to that of hormone receptors. In the case of the transmitter *acetylcholine* (the chemical by which vertebrate motor neurons communicate with muscle cells), two molecules of transmitter must bind to a receptor to activate it; the contents of a single synaptic vesicle typically activate about 2,000 of these receptors. The binding of transmitter to a receptor opens the gates of a channel, allowing a specific ion to pass through the membrane. This ion movement results in an alteration of the postsynaptic neuron's membrane potential, and a new impulse may be generated by that cell. The channels with receptors for acetylcholine pass mostly Na^+ ions, and a partial depolarization of the cell around the channels results.

Because transmission across a synapse involves a whole series of events —an influx of Ca^{++} ions into the terminal, followed by movement of transmitter vesicles, exocytosis, diffusion of the transmitter across the cleft, and finally the diffusion of ions through the postsynaptic channels—the process is much slower than impulse conduction along a neuron. For this reason, the time a message takes to traverse a neural pathway is always longer than would be expected on the basis of the length of the pathway and the speed of axonal conduction. In general, the more synapses in a neural pathway, the slower the average speed of transmission per unit distance along the pathway.

The story does not end when an impulse has been communicated to a postsynaptic cell by the diffusion of transmitter across the cleft. If the transmitter remained, the postsynaptic receptors would be stimulated indefinitely by the arrival of a single action potential, so there must be a mechanism to destroy the transmitter. For instance, once acetylcholine has diffused across the synaptic cleft and exerted its effect on the postsynaptic membrane of a dendrite or the cell body of the next cell, it is promptly destroyed by an enzyme called *acetylcholinesterase.* By destroying the transmitter, this enzyme makes it possible for the next impulse, with new information, to be transmitted. Many insecticides, such as the organophosphates (also known as nerve gases), are cholinesterase inhibitors. They block destruction of acetylcholine, with the predictable result that an in-

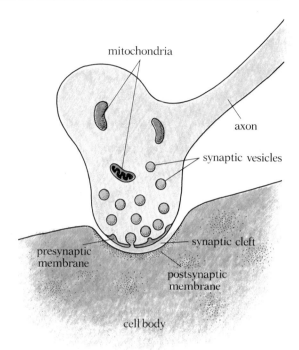

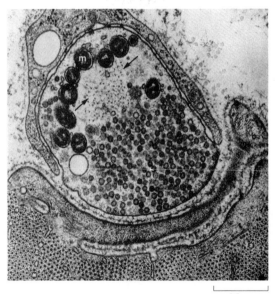

0.5 μm

18.26 The synapse
Each synaptic terminal at the end of an axon encloses numerous synaptic vesicles containing transmitter substance. When vesicles release this substance into the synaptic cleft, the substance diffuses across the cleft and alters the polarization of the postsynaptic membrane of the dendrite or cell body of the next cell.

ADAPTATION, HABITUATION, AND SENSITIZATION

You may recall that many sensory neurons undergo adaptation, the process by which the cell's sensitivity to stimuli is adjusted to match the background level of stimulation. Adaptation explains why we gradually acclimate to the cold water of a swimming pool and to the hot water of a bath, or (in conjunction with changes in pupil size) to bright sunlight after the darkness of a theater or lecture hall. But adaptation, though enormously important, is restricted to individual sensory neurons. Two other neural phenomena, habituation and sensitization, affect the responsiveness of many neurons at once, and so are crucial to behavioral control.

Habituation is superficially similar to adaptation in that after repeated exposure to a stimulus at a given level, an animal comes to ignore that stimulus, but the distinction between the two is as crucial as it is difficult to grasp—some professional neurobiologists still confuse these two very different behavioral mechanisms. The essential difference is that adaptation is a property of *individual* sensory neurons, which basically "tire" after repeated firing; while habituation is a property of an entire behavioral pathway. Adaptation takes place peripherally, in the sensory neurons themselves, while habituation occurs centrally, in a brain or ganglion, in most organisms. (Among the exceptions to this anatomical distinction is *Aplysia*.)

We can see this distinction clearly in the gill-withdrawal behavior we examined earlier. The waning of gill withdrawal under conditions of continual stimulation could be the result of either adaptation or habituation. We know that this behavioral boredom is central habituation, rather than peripheral adaptation, though, because it fulfills several conditions: (1) Repeated stimulation of only a small subset of the sensory neurons will make the animal insensitive to stimulation of the others. If this were adaptation, the previously unstimulated, "untired" cells would still be able to trigger the behavior. (2) The time course of the insensitivity is very long. Habituation lasts hours or even days after sufficient stimulation, whereas sensory adaptation lasts only seconds or minutes after the stimulus is removed.

Finally, (3) the insensitivity can be swept away instantly by some irrelevant stimulus, one that does not directly affect the sensory neurons in the circuit involved. A flash of light or a prod in the tail, for instance, will reawaken the gill-withdrawal circuit in *Aplysia*. This curious but crucial

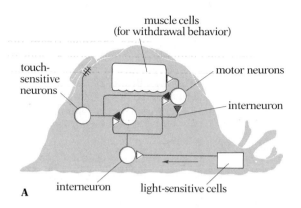

muscle cells
(for withdrawal behavior)

touch-
sensitive
neurons

motor neurons

interneuron

A   interneuron   light-sensitive cells

phenomenon is ***sensitization;*** it involves a message from a higher-level control center that serves to alert an animal and prepare it for action by reawakening it from its habituated state.

Figure A summarizes the relationship between adaptation, habituation, and sensitization, as seen in a behavioral pathway in *Aplysia*. Adaptation can take place in any of the sensory neurons; habituation occurs in the solid-color terminals that are part of the touch-withdrawal pathway. An irrelevant stimulus, shown as coming from the visual system by way of an interneuron (here functioning as the higher-level control center) can sensitize the habituated terminals presynaptically, thereby restoring sensitivity of the pathway to tactile stimuli to (or even beyond) its normal level.

Adaptation, habituation, and sensitization all serve to "tune" an animal's behavior. Paralleling the differing patterns of neural connectivity exemplified by these three phenomena are different membrane events involving the ion-specific channels in the neurons themselves.

Adaptation, for example, involves calcium ions. You may recall that voltage-gated calcium channels are important in synaptic transmission: they admit Ca^{++} ions into the terminal when an action potential arrives, and these ions cause the synaptic vesicles to fuse with the membrane and release transmitter chemical onto the postsynaptic cell. The Ca^{++} ions thus admitted are quickly removed by a variety of intracellular mechanisms, including Ca^{++}-binding enzymes, vesicles, and mitochondrial pumps.

When Ca^{++} ions continue to enter the cell because of repeated firing, however, these systems fail to keep up, and the intracellular Ca^{++} concentration begins to rise. The extra Ca^{++} ions then diffuse into the narrow axon and bind to the gates of special membrane channels (from the inside!); the channels open and allow K^+ ions to pass through. The departure of K^+ ions hyperpolarizes the axonal membrane, so that action potentials coming along the axon cannot reach the terminal to open the voltage-gated calcium channels. As a result, once any Ca^{++} ions remaining in the terminal have been used up, no transmitter can be released (Fig. B). Adaptation is relatively brief, however, since various mechanisms are acting to remove the excess Ca^{++} ions from the axon, while the membrane's sodium-potassium pumps are reintroducing K^+. Thus the neuron's activity is soon restored.

Habituation, which usually results from the repeated firing of an interneuron rather than a sensory neuron, has a different basis. During the first stages of habituation, synaptic vesicles are used up faster than they can be made, and synaptic transmission declines. If stimulation continues in the absence of available vesicles, transmitter release necessarily ceases. At the same time, some of the voltage-gated calcium channels in the terminal have changed structure and now fail to work; without Ca^{++} ions, of course, vesicles cannot fuse with the synaptic membrane and release transmitter, so even after the number of vesicles has returned to normal in the habituated terminal, incoming action potentials from the axon do not cause transmitter release (Fig. B). The exact chemical basis of this change in the Ca^{++} channels is not yet understood.

Sensitization, brought about by an irrelevant stimulus, restores the activity of the Ca^{++} channels, erasing habituation. Sensitization occurs when input from facilitory interneurons monitoring other sensory systems releases serotonin onto the habituated terminals (Fig. B); the binding of serotonin to receptors on the membrane of the terminal causes the enzyme adenylate cyclase to be released into the cytoplasm, where it catalyzes the synthesis of cAMP. At least two additional enzyme intermediates enable cAMP to serve as an intracellular messenger that increases the responsiveness of the Ca^{++} channels to impulses from the axon. With the entry of Ca^{++} restored by this indirect means, the synaptic vesicles can again release their contents and so stimulate the interneurons and motor neurons.

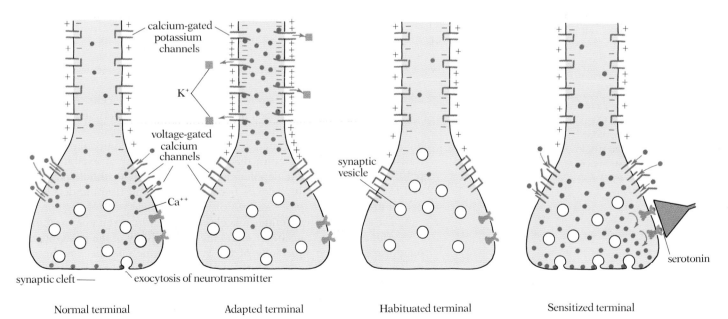

synaptic cleft —— exocytosis of neurotransmitter

Normal terminal   Adapted terminal   Habituated terminal   Sensitized terminal

B

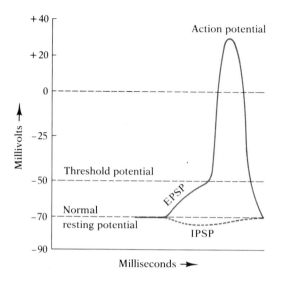

18.27   The effect of transmitter substance on the membrane potential of a neuron

The normal resting potential of a typical neuron is about −70 millivolts. An excitatory transmitter substance slightly reduces that polarization—that is, makes the inner surface of the membrane less strongly negative—thereby creating an excitatory postsynaptic potential (EPSP). If the EPSP reaches the threshold level, which is usually about −50 mv, an impulse is triggered; a sudden inrush of Na$^+$ ions causes the inside of the cell to become positive. A fraction of a second later, K$^+$ ions rush out of the cell, and the inside of the cell again becomes negative. This cycle of electrical changes constitutes the action potential.

If the transmitter substance had been inhibitory, the membrane would have become hyperpolarized (to perhaps −75 mv), a condition called an inhibitory postsynaptic potential (IPSP) (dashed curve), and no action potential would have resulted; the neuron would slowly have returned to its resting potential after release of the transmitter had ceased.

sect exposed to them suffers from uncontrollable spasms as vast numbers of synapses become permanently active. Given in high enough doses, cholinesterase inhibitors affect major physiological processes, and the animal dies.

Besides serving as a transmitter outside the central nervous system, acetylcholine is one of a growing list of transmitters also found in the CNS. In vertebrates, these include noradrenalin (which is also produced in the adrenal medulla as a hormone), serotonin, dopamine, and gamma-aminobutyric acid (GABA). As we shall see, certain disorders such as schizophrenia and severe depression, once blamed on vague emotional disturbances, are now known to be triggered by biochemical malfunctions of CNS transmitters, receptors, and previously unknown neural hormones. These welcome discoveries are beginning to open the way to relatively precise physiological treatments for certain emotional disorders.

The action of transmitter chemicals   We said earlier that neurons—particularly interneurons—collect and average information from many cells. We are now in a position to understand just how this processing is accomplished on the membrane. In the case of transmitters like acetylcholine, the inward flow of Na$^+$ ions slightly decreases the polarization of the neuron so that the inside becomes less negative relative to the outside, creating what is called an excitatory postsynaptic potential, or **EPSP** (Fig. 18.27). If the EPSP is sufficiently large, it may spread to the base of the cell's axon (called the axon hillock), depolarize the membrane past its threshold, and trigger an impulse that will move down the axon to the next synapse.

The transmitter chemicals released by the terminals of some neurons have the opposite effect: they *increase* the polarization of the postsynaptic membrane and make the neuron harder to fire. These transmitters produce their inhibitory effects by binding to and opening channels in the postsynaptic membrane that pass chloride ions (Cl$^-$) to the inside (or, in some cases, K$^+$ to the outside). The Cl$^-$ ions can enter even though the cells are negatively charged with respect to the surrounding fluids because the unfavorable electrostatic gradient is overcome by the strong concentration gradient: Cl$^-$ ions are five times more concentrated outside cells. As Cl$^-$ ions enter (or K$^+$ ions leave), the cell membrane becomes hyperpolarized—the inside of the cell becomes more negative relative to the outside—and an inhibitory postsynaptic potential, or **IPSP**, is produced. Additional excitatory transmitters, and hence more than the usual number of excitatory impulses, are then needed to depolarize the membrane to threshold potential and fire the neuron. The balance between EPSPs and IPSPs underlies all neural processing.

At one time every transmitter was classified as either excitatory or inhibitory. But since the transmitter alters the permeability of the postsynaptic membrane, the ion specificity of the gated channels in the membrane, rather than the character of the transmitter itself, should determine whether it is inhibitory or excitatory. Indeed, we now know that while acetylcholine is excitatory at most neuromuscular junctions, it is inhibitory on heart muscle. This suggests that the channels associated with the acetylcholine receptors of heart muscle must be different from those of other muscles, and that in the same way transmitters may open different sorts of channels on different target tissues.

There was also a time when all synaptic transmission was thought to cause a direct alteration of membrane potential. But given the functional similarities between the postsynaptic membrane and the target cells of the endocrine system, it is reasonable to expect that synapses might also use a second-messenger strategy to alter the chemistry of the nerve cell. Indeed, a variety of transmitters—neuropeptides such as serotonin and noradrenalin—activate the adenylate cyclase system of nerve cells, thus providing additional support for the idea that the nervous and endocrine systems share a common evolutionary origin. In general, the second-messenger neurotransmitter system causes long-term changes in the excitability of the postsynaptic cell—changes of the sort that are thought to underlie learning and memory, for example. Transmitters that act directly on ion channels are responsible for short-term electrical events: EPSPs and IPSPs.

Both the excitatory and inhibitory synapses we have discussed so far determine whether or not a postsynaptic cell will fire an impulse. There is now abundant evidence that synapses are also subject to **presynaptic inhibition** and **presynaptic facilitation.** In such cases the terminals of inhibitory or excitatory interneurons synapse on the terminals of the presynaptic cell (Fig. 18.28) and act to alter the number of synaptic vesicles that release transmitter. This form of inhibition or facilitation has the decided advantage of modifying excitatory input to the postsynaptic cell from one information source without altering the responsiveness of the postsynaptic cell to other inputs.

The integrative function of neurons   Chemical synapses are points of resistance in neural circuits. An impulse may travel to the end of the axon of one neuron but die there because not enough excitatory transmitter is released to initiate the all-or-none response in the next neuron in the pathway. Indeed, it is rare that a single impulse from a single neuron can fire the next cell in the circuit. Ordinarily, excitatory transmitters from many different terminals must be released more or less simultaneously to fire a target cell. The individual EPSPs then combine to produce a large enough resultant EPSP to exceed the threshold of the target cell and so trigger an impulse. This additive phenomenon is called **summation.** It can be spatial, as when several adjacent terminals from different neurons fire simultaneously; or it can be temporal, as when one cell fires at such high rate that individual EPSPs overlap in time; or it can be a combination of the two.

Summation on the postsynaptic membrane is algebraic: if both excitatory and inhibitory transmitters are released at the same time, the resulting EPSPs and IPSPs add according to sign. The result of this phenomenon, which is similar to the molecular "voting" of the receptors in bacterial chemotaxis that we discussed earlier, is a postsynaptic potential that reflects the overall pattern of incoming information. This kind of neural processing, evident in most species of animals, is known as **integration.** The cell integrates all the signals that converge on it (signals which, in the case of an interneuron or a motor neuron, may be coming from thousands of different interneurons or sensory neurons) and either fires an impulse or re-

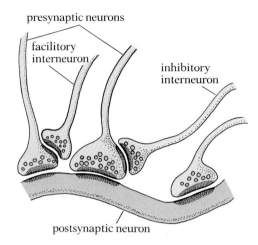

presynaptic neurons

facilitory
interneuron

inhibitory
interneuron

postsynaptic neuron

18.28 Presynaptic inhibition and facilitation
In presynaptic inhibition, an inhibitory interneuron synapses on the terminal of the presynaptic neuron. When the inhibitory cell releases its transmitter (often GABA), fewer vesicles in the terminal of the presynaptic neuron will release their own transmitter substance, with the result that the postsynaptic neuron receives less stimulation from this particular pathway. Similarly, in presynaptic facilitation, the facilitory interneuron releases its transmitter (often serotonin) onto the terminal of the presynaptic neuron, with the result that more vesicles in the terminal of the presynaptic neuron will release transmitter, and the postsynaptic neuron will receive increased stimulation from that pathway. Unlike postsynaptic inhibition, which reduces the responsiveness of the postsynaptic neuron to all excitatory inputs, presynaptic inhibition and facilitation leave unchanged the responsiveness of the postsynaptic neuron to inputs from other sources.

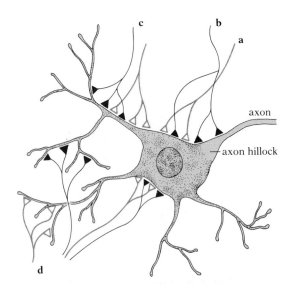

18.29 The integrative function of a neuron
The neuron receives both excitatory synapses (open triangles) and inhibitory synapses (solid triangles) from many different sources. The synapses vary in their distance from the base of the axon (the axon hillock), where impulses are generated. Whether or not the neuron will fire an impulse is determined at any given moment by the algebraic sum of all the individual EPSPs and IPSPs arriving at the axon hillock from the various synapses. But unlike an axonic impulse, which shows no decrement with distance, depolarizations (EPSPs) or hyperpolarizations (IPSPs) decrease in magnitude as they spread along a dendrite or the cell body. Hence impinging interneurons such as **a** and **b,** which synapse near the axon hillock, can more easily influence the neuron's firing than interneurons such as **c** and **d,** which synapse on the cell at a distance from the hillock; only if these latter interneurons fire at a very high rate are they likely to have a major effect. In short, the geometry of the synapses on a neuron biases the integration process; inputs from some interneurons are given greater weight than inputs from other interneurons.

mains silent (Fig 18.29). It is by combining many such simple yes-no, on-off decisions from a host of different cells that the nervous system processes the sensory information it receives.

Synaptic malfunctions, particularly those involving transmitter irregularities, have been implicated in a variety of neurological diseases. Because synapses are responsible for information processing in the nervous system, and because their proper function depends upon a very delicate balance among a variety of presynaptic enzymes, ion channels, transmitters, deactivating enzymes, and postsynaptic receptors and ion channels, it is not surprising that synaptic malfunctions can be devastating. Used with caution and informed medical supervision, some drugs that exert their effects at synapses can give relief from anxiety, severe neurological pain, or diseases involving biochemical disorders of the synapses. But used improperly, the same agents can induce symptoms strikingly similar to those seen in certain mental disorders, and in some cases the symptoms may be long-lasting or even permanent.

Neurological drugs alter synaptic function in a variety of ways. They can turn off certain synapses by (1) interfering with synthesis of the appropriate transmitter; (2) blocking uptake of the transmitter into synaptic vesicles; (3) preventing release of transmitter from the vesicles into the cleft; or (4) blocking the receptor sites on the postsynaptic membrane, so that the transmitter has no effect even when released. By contrast, other drugs can induce excessive and uncontrolled firing of postsynaptic cells by (1) stimulating massive releases of transmitter; (2) mimicking the effect of transmitter; or (3) inhibiting the destruction of the transmitter once it has done its job, as in the case of the cholinesterase inhibitors mentioned earlier.

The physiological mode of action of a drug may help explain the behavioral symptoms the drug induces. Amphetamine, for example, acts as a stimulant because it increases the release of noradrenalin in the brain. Reserpine, on the other hand, acts as a tranquilizer because it blocks the uptake of noradrenalin into synaptic vesicles and hence prevents its release. Thus the contrast in the behavioral symptoms produced by these drugs is explained by their opposite effects on the same synapses.

Recent research has provided partial explanations for the action of some other important neurological drugs. Nicotine acts as a stimulant because it mimics the effect of acetylcholine. Chlorpromazine, a commonly used tranquilizer, inhibits transmission of impulses at both acetylcholine- and noradrenalin-mediated synapses by combining with receptor sites on the postsynaptic membranes and thereby blocking the transmitter chemicals. Local anesthetics—cocaine, for instance—act by binding inside sodium channels and thereby blocking them. (Cocaine also inhibits the uptake of noradrenalin, thereby affecting the nervous system in other ways.) LSD (lysergic acid diethylamide) produces its characteristic derangement of sensory experience and other mental functions by combining indiscriminately with receptor sites for serotonin. The benzodiazepines, of which Valium is the most widely prescribed, interact synergistically with the inhibitory transmitter GABA to open chloride channels and thus inhibit synaptic transmission. The mode of action of the active ingredient of marijuana (tetrahydrocannabinol) is not yet known.

Several previously misunderstood neurological disorders have now been

10 μm

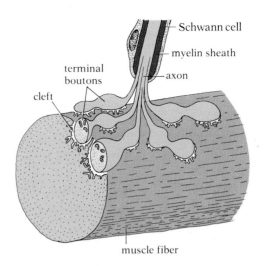

18.30 Neuromuscular junctions
Left: Toward its end, an axon supplying a muscle in a snake branches extensively and forms neuromuscular junctions on individual muscle fibers. Right: A close-up sketch of one neuromuscular junction; the junction is formed by branches of the axon, with their terminals, and the specialized adjacent portion of the muscle fiber. As in a synapse between two neurons, there is a cleft between muscle and nerve cells.

traced with more or less certainty to transmitter problems. Severe depression has been firmly linked to a defect in the serotonin transport system: the brains of many suicide victims have about half of the normal serotonin level and only two-thirds of the usual number of binding sites. Abnormal dopamine levels are now tied to one form of schizophrenia, and the chemical basis of the manic-depressive syndrome is sufficiently well understood to be treated chemically (with lithium) rather than through psychoanalysis. Recent evidence suggests that even severe autism, a tragic disability that affects a person's power to communicate and to relate to others, may prove to result from a biochemical defect.

TRANSMISSION FROM NERVE TO MUSCLE

Just as there is a gap at the synapses between successive neurons in a neural pathway, there is also a gap between the terminals of an axon and the effector it innervates. When the effector is skeletal muscle, the gap is usually located within a specialized structure, the *neuromuscular junction* (or motor end plate), formed from the end of the axon and the adjacent portion of the muscle surface (Figs. 18.30 and 18.31). Transmission across this gap is by transmitter chemicals, just as it is in most synapses between neurons, and

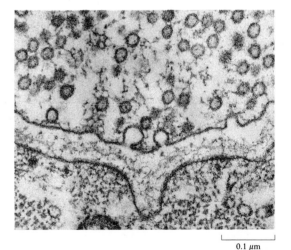

0.1 μm

18.31 Electron micrograph of portion of a neuromuscular junction of a frog
The upper half of the micrograph shows part of the terminal of an axon containing numerous synaptic vesicles, some of them releasing neurotransmitter. The lower half shows part of a muscle cell. There is a distinct cleft between the two cells.

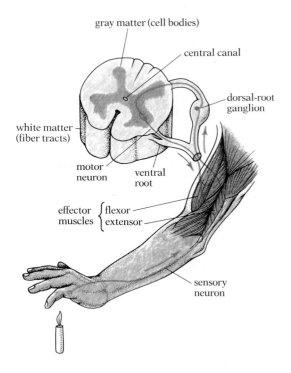

gray matter (cell bodies)

central canal

dorsal-root ganglion

white matter (fiber tracts)

motor neuron

ventral root

effector muscles { flexor
                   extensor

sensory neuron

18.32   Hand-withdrawal circuit
When the hand touches something hot, signals from pain receptors reach motor neurons in the spinal cord, which in turn quickly retract the arm. These motor neurons both activate the flexor muscles and inhibit the extensors. (The central canal through the gray matter is a remnant of the hollow nerve cord evident early in development.)

the biochemical processes underlying neuromuscular transmission are very similar to those involved in ordinary chemical synapses.

Several kinds of neural networks in animals allow precise control of muscles. Typical of most invertebrates is a system that involves three classes of motor neurons: a "fast" fiber that produces large, rapid contractions; a "slow" fiber that generates slow, more finely controlled movement; and an inhibitory fiber that serves to block contraction. In higher invertebrates the muscles act like interneurons, integrating inhibitory and excitatory inputs. Many invertebrate muscle fibers display a graded response: they contract in proportion to the rate of stimulation. In vertebrates the movement of skeletal muscles involves quite a different strategy. First, there are many more fibers per muscle, but each responds in an all-or-none fashion. Second, there are no inhibitory motor neurons; instead, muscles with opposite effects are paired—one that extends a finger, for instance, with one that retracts it. While in higher invertebrates muscle fibers integrate their various inputs themselves, in vertebrates this processing is accomplished by the central nervous system. There are far more motor-neuron fibers at the muscles, each with correspondingly less potency, and the CNS exerts control (1) by regulating the number that are active and their rate of firing, and (2) by balancing the signals to paired muscles, so that they pull against one another.

The transmitter at the neuromuscular junctions of vertebrate skeletal muscle is acetylcholine. A variety of drugs that cause paralysis—the poison produced by the bacterium responsible for botulism, for instance, and the famous neuromuscular blocker used on the poison arrows of South American Indians, curare—do so by blocking transmission between the motor neurons and the muscles. Curare is now known to act as an acetylcholine mimic: it binds to the receptors but fails to open the gates. To make matters worse, acetylcholinesterase cannot inactivate it. The receptor sites are permanently blocked, and neuromuscular transmission ceases. The action of botulin toxin, on the other hand, is not well understood. It first works presynaptically to cause terminals at neuromuscular junctions to be overactive, which results in muscular tremors and paralysis. But then it silences these same terminals, and death results. The neuromuscular junction is also the site of at least one deadly viral disease: rabies. The extreme virulence of rabies virus and the wide phylogenetic range of its hosts seem to be the consequence of the virus's specific affinity for acetylcholine receptors.

NEURAL PATHWAYS IN VERTEBRATES

Our developing understanding of neural control demonstrates it to be a complex phenomenon involving both physiological events at the level of the individual neuron and integrative process between systems of neural cells. In vertebrates the simplest form of neural control is accomplished through the **reflex arc.**

REFLEX ARCS

A sensory neuron connected to a motor neuron forms the simplest circuit in the nervous system. Circuits of this sort are seen in reflex arcs controlling behavioral responses that must occur quickly, such as emergency reactions and the automatic maintenance of some kind of equilibrium. In this section we shall examine the organization of reflex arcs in vertebrates.

A good example of a familiar emergency reaction is the withdrawal reflex. When we touch something hot, our hand jerks back automatically. The sensory neurons involved in this response run from the hand to the spinal cord; their cell bodies are located in a ***dorsal-root ganglion*** (or spinal ganglion) that lies just outside the spinal cord near its dorsal (posterior) side (Fig. 18.32). The axons enter the cord dorsally and synapse with the dendrites or cell bodies of motor neurons within the gray matter of the spinal cord. The axons of the motor neurons exit the cord ventrally and run to the muscles. In this reflex, a strong signal from the appropriate sensory cells both fires the flexor muscles and inhibits (relaxes) the extensor muscles, so this crucial motor response is well under way before the signals responsible for the conscious sensation of pain (which exit the reflex pathway in the spinal cord) ever reach the brain for analysis.

The kind of circuit that automatically maintains equilibrium is exemplified by the well-known knee-jerk reflex. Doctors regularly test for this response by tapping a patient's knee with a special rubber hammer (Fig. 18.33). The sensory elements are stretch receptors that measure the degree to which a particular muscle is stretched. As the force against which the muscle must act—the amount of weight on one leg in this case—increases, the muscle is stretched, and the receptors signal this fact through sensory neurons to the spinal cord. As in the previous example, the information is sent both to the brain for analysis and to motor neurons for immediate interim action. In this particular circuit, the motor neurons are those controlling the very muscles being monitored by the stretch receptors that have been activated. The arrival of signals from the receptor increases the firing rate of the motor neurons, and the muscles—extensors in this case —tighten to accommodate the added load.

This sort of automatic compensation provides an example of negative feedback; the analogy of the household thermostat we used in reference to hormonal action is applicable here as well, for a change in sensory input produces a self-correcting response that tends to cancel the change and return the system to its normal state. In this instance the tightening of the extensor in response to the stretch-receptor signal serves to shorten the muscle, thereby eliminating the stretch felt by the receptor, and so turning off the sensory signal. Thus the knee-jerk reflex automatically, without the organism's awareness, tunes posture.

Using these very simple reflexes as a model, we can make several generalizations about the spinal reflex arcs of vertebrates:

1. For a particular reflex arc there is never more than one neuron, however long it must be, to carry the sensory information to the spinal cord (there may, of course, be many such neurons running side by side serving the same function).
2. The cell body of the sensory neuron is always outside the spinal cord in a dorsal-root ganglion.
3. The axons of sensory neurons always enter the spinal cord dorsally.
4. The axons of motor neurons always leave the spinal cord ventrally.

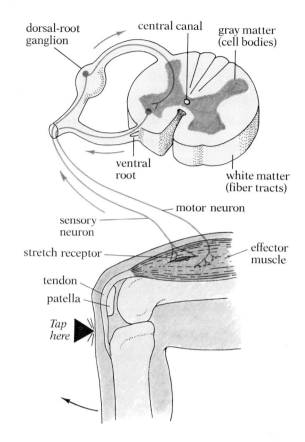

18.33   The knee-jerk reflex
This reflex, which is part of the automatic postural-control system, measures the load on the muscle (as reported by stretch receptors) and then adjusts the firing rate of the motor neurons to that muscle to compensate for any changes. The knee jerk itself results from the brief overloading of the muscle and its receptor when the tendon is tapped sharply.

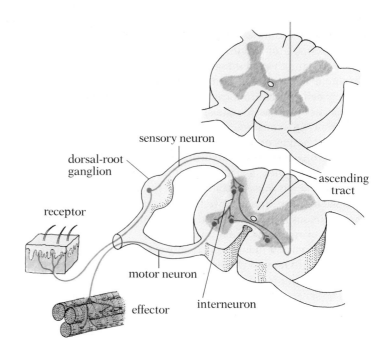

18.34 Diagram of a reflex arc with interneurons
The sensory neuron synapses with several
interneurons in the gray matter of the cord. Some of
these interneurons may synapse directly with motor
neurons on the same side, but some cross to the
other side of the cord and there synapse with other
motor neurons and with additional interneurons that
run in ascending tracts through the cord to the brain.

The sensory and motor neurons of the knee-jerk reflex run through the
same nerve, even though they carry impulses in opposite directions. A
nerve containing both sensory and motor fibers is called a mixed nerve. All
the nerves connected to the spinal cord are mixed. There are 31 pairs in our
species, all of which branch repeatedly after leaving the spinal cord, giving
rise to smaller nerves that innervate most parts of the body below the head.
Some nerves, on the other hand, connect directly to the brain rather than to
the spinal cord. In humans there are 12 pairs of these cranial nerves, some
purely motor, some purely sensory, and some mixed.

So far we have seen reflexes in their simplest isolated form. Very few reflex pathways, however, involve only two neurons in series. At least one interneuron is usually interposed between the sensory neuron and the motor
neuron (Fig. 18.34), and many interneurons may be involved, their number
depending on the balance between how quickly a reaction is needed and
how much sensory integration and muscle coordination is desirable.

It is important to keep in mind that a reflex arc, whether it includes few
cells or many, makes two sorts of connections with other neural pathways.
First, it almost always sends information to the brain, where instructions to
counteract or augment the behavioral reaction can be issued. If you know
that the doctor is going to strike your knee, for instance, and so have sufficient time to issue neural commands to modify the reaction, you can consciously either inhibit or exaggerate the response.

The second sort of connection is with other reflex arcs. Consider for a

Functions of reflex arc

moment the withdrawal reflex for a foot rather than a hand. You are walking barefoot and your right foot steps on a thorn. Impulses from pain receptors in the foot immediately ascend along sensory neurons to the spinal cord, activate the appropriate motor neurons, and descend to the muscles of the leg to excite the flexors and inhibit the extensors. At the same time, the interneurons send messages across the spinal cord to certain of the circuits for the other leg. These result in a reflex extension of the left leg in anticipation of the increased load it is about to bear. As the right foot is lifted from the ground, the stretch receptors in the left leg take over control of their circuits to accommodate more precisely the extra weight. But now a new problem arises—balancing on one leg. Additional reflex arcs, involving the muscles of the ankle, must begin to operate, and so a cascade of independent circuits are recruited to handle the situation.

Meanwhile, of course, some of the interneurons have sent information to the brain. Part of the information goes to portions of the brain concerned with our general level of awareness, and serve to make the victim of the thorn more alert and "wide awake": the process of sensitization. Still other messages have been sent to the part of the brain known as the cerebellum, where conscious motor control is coordinated. If the reflex arcs have proved inadequate for the situation or other behavior seems called for, the cerebellum can issue commands to the muscles to take the necessary action.

Reflex circuitry is able to control and coordinate a variety of simple responses and automatically fine-tune behavior such as walking, whose details must constantly be adjusted as body weight rhythmically shifts. Impressive as these interacting circuits are, however, we shall see that they do not provide the basis of the more complex behavior patterns of most animals.

THE AUTONOMIC NERVOUS SYSTEM

The central nervous system of vertebrates serves as a coordinating system for two kinds of pathways: somatic and autonomic. **_Somatic_** pathways, exemplified by reflex arcs and by more complex behavioral circuits to be discussed later, innervate skeletal muscle and include sensory and motor neurons lying largely outside the CNS. They involve, potentially, some conscious control of behavior, or at least an awareness that the behavior has occurred. **_Autonomic_** pathways, in contrast, are basically unconscious internal reflexes; they are not ordinarily under the control of the will and usually function without our being aware of them. They innervate the heart, some glands, and the smooth muscle in the walls of the digestive tract, respiratory system, excretory system, reproductive system, and blood vessels. Autonomic pathways are controlled largely by the hypothalamus, which you may recall is the same part of the brain that orchestrates the many endocrine functions of the pituitary. The pathways of the autonomic nervous system differ structurally from somatic pathways in having two motor neurons instead of one. Somatic processing takes place largely in the CNS, while autonomic processing occurs in ganglia outside the CNS.

The autonomic nervous system (ANS) is composed of two parts, the sym-

18.35   The autonomic nervous system

Of the 12 cranial and 31 spinal nerves, four cranial nerves (see gray area of brain at upper right, where they emerge) and about half the spinal nerves (see colored and gray segments, where they emerge from the cord) contribute neurons to the autonomic nervous system, which innervates internal organs.

The ANS is customarily divided into two parts: the sympathetic and the parasympathetic systems. The pathways of both usually have two motor (efferent) neurons; a first (presynaptic) neuron exits from the central nervous system and synapses with a second (postsynaptic) neuron that innervates the target organ.

The presynaptic neurons of the sympathetic system exit from the thoracic and upper lumbar regions of the spinal cord, and synapse with the postsynaptic neurons in a series of small ganglia (circles) lying near the cord or in larger ganglia in the abdominal cavity; the postsynaptic neurons then run from the ganglia to the target organs. The presynaptic neurons of the parasympathetic system exit from the medulla of the brain and from the sacral region of the spinal cord. These are very long neurons that run all the way to the target organ, where they synapse with short postsynaptic neurons.

Most but not all internal organs are innervated by both the sympathetic and the parasympathetic system.

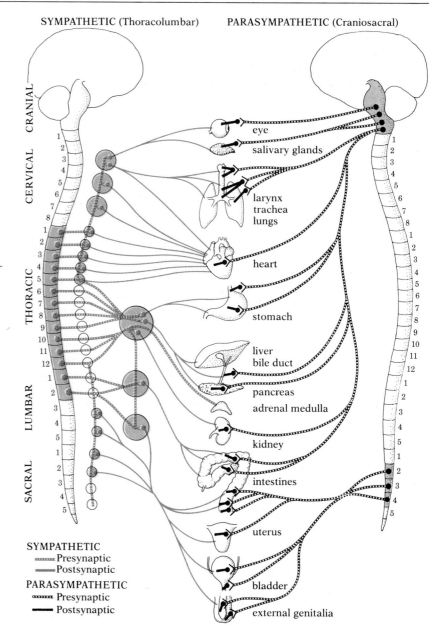

SYMPATHETIC (Thoracolumbar)   PARASYMPATHETIC (Craniosacral)

SYMPATHETIC
⚬⚬⚬ Presynaptic
— Postsynaptic
PARASYMPATHETIC
▪▪▪ Presynaptic
— Postsynaptic

pathetic and the parasympathetic systems (Fig. 18.35), which differ both structurally and functionally. In the **sympathetic system**, the cell bodies of the first motor neurons lie *inside* the thoracic and lumbar portions of the spinal cord. The axons of these neurons exit ventrally from the cord and run to ganglia lying near the cord; there they synapse with the second motor neurons, whose cell bodies lie in the ganglia. Because the synapse between the first and second motor neurons occurs in a ganglion that is at a distance from the target organ, the axon of the second motor neuron is often quite long.

Two principal structural differences distinguish the ***parasympathetic system*** from the sympathetic system. First, the cell bodies of the first motor neurons in the parasympathetic system lie in the brain and in the sacral region of the spinal cord. Second, the synapses between the first and second motor neurons of the parasympathetic system occur in the immediate vicinity of the target organs, or even inside the walls of those organs. As a result, the axons of the second motor neurons are relatively short.

Most internal organs are innervated by both sympathetic and parasympathetic fibers, with the two systems functioning largely in opposition to each other. Together they fine-tune the balance between what we might call active and passive behavior. At one extreme, the sympathetic system prepares an animal for emergency action during a crisis: it shuts down oxygen-consuming processes, such as digestion, that are not immediately essential, and prepares the machinery that may be necessary for defense, for fighting or fleeing. At the other extreme, the parasympathetic system restores order or passivity after a crisis has ended, restarting the important but not immediately critical processes that have been shut down, and freeing the system of anxiety-producing chemicals. Normally, of course, animals function somewhere along a continuum between the highly active, emergency behavioral state produced by the sympathetic system and the passive, almost vegetative behavior produced by the parasympathetic system. In general, the sympathetic system gives rise to the same effects as the hormones of the adrenal medulla—the fight-or-flight reactions—but does so more rapidly. Present evidence suggests that this nervous mechanism is far more important than the endocrine system in preparing an animal for emergency situations, as indeed it should be considering how much faster a pathway it is. The reason the sympathetic nervous system and the hormones of the adrenal medulla produce similar effects seems clear. While the transmitter at the neuromuscular junctions of both somatic and parasympathetic pathways is acetylcholine, the transmitter at the sympathetic terminals is noradrenalin (or, in a few cases, adrenalin). The effect on the target organs is identical because the transmitters of the sympathetic system and the hormones of the adrenal medulla are identical.

The reason the adrenal medulla releases the same substances as the sympathetic system apparently lies in a fascinating functional and evolutionary relationship between them. As we have seen, autonomic pathways normally include two motor neurons. There is a single exception: the sympathetic pathway to the adrenal medulla has only one motor neuron (Fig. 18.35). It appears that the adrenal medulla forms from presumptive nervous tissue (tissue that is destined to become neurons) in the embryo, and is itself actually the highly modified second motor neuron in this sympathetic pathway, specialized for neurosecretion rather than conduction. In other words, evolution has converted what was probably once a motor neuron of the sympathetic nervous system into an endocrine gland specialized for high-quantity secretion of the very substances that nearly all second motor neurons in this system produce.

Here, then, is another example of the close interrelationship between the

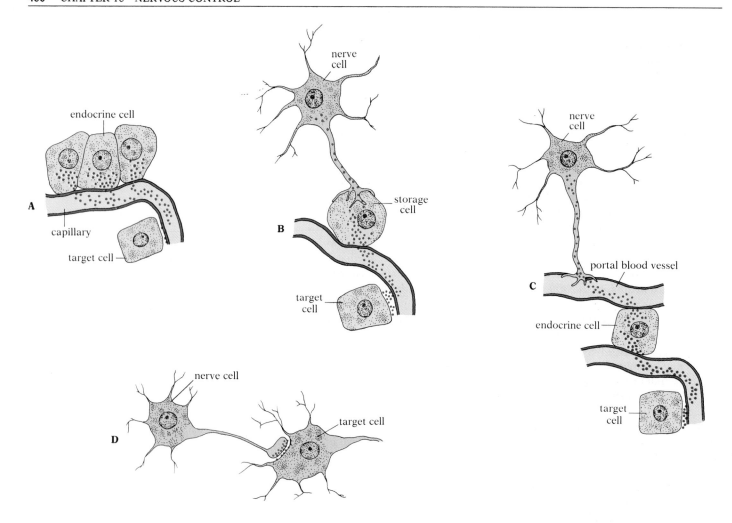

18.36   Four arrangements for chemical control compared
(A) A typical endocrine cell secretes its hormone directly into the blood, which carries the hormone to the target cells. (B) Some hormones secreted by nerve cells in the hypothalamus are stored in cells of the posterior pituitary, and later released into the blood. (C) Releasing hormones secreted by nerve cells in the hypothalamus are carried by the blood in a special portal system to endocrine cells of the anterior pituitary, which respond by secreting their own hormones into the general circulation. (D) A more typical nerve cell secretes its transmitter substance directly onto the target cell; no transport by the blood is involved.

nervous system and the endocrine system, and of the similarity in their mechanisms of action (Fig. 18.36). Neurosecretory activity like that of the hypothalamus in its interaction with the pituitary, and like that of the insect brain, which produces the hormone that regulates the prothoracic gland, is not an unusual or isolated phenomenon. Neurosecretion is fundamental to nerve action. Impulse transmission across gaps in neural pathways depends on it. It is likely, therefore, that natural selection favored a nervous system that functions as a high-speed elaboration of the endocrine system. This would explain why nervous tissue has many endocrine functions, and many of the chemicals secreted by nervous tissue are basically hormones.

Autonomic control of heartbeat   The nervous control of heart rate provides a good example of autonomic control, in which the effector—heart muscle in this case—is not under conscious control (although we may consciously alter its rate of contraction indirectly, by deliberately working into an emotional or excited state, or by deliberately relaxing).

As we saw in Chapter 13, nervous impulses are not necessary to make the heart beat, since beats are initiated directly by the S-A node in the wall of the heart itself. But the rhythm of the S-A node can be modified by impulses coming to it from two sets of nerves: sympathetic, which excite the S-A node, and parasympathetic, which inhibit it. Though the sympathetic nerves to the heart issue from the CNS in the thoracic region of the spinal cord, the impulses they carry originate in a cardiac-accelerating center in the medulla of the brain. Similarly, the impulses carried by the parasympathetic nerves to the heart originate in a cardiac-decelerating center in the medulla.

If blood pressure is high, pressure receptors in the carotid artery of the neck and in the arch of the aorta (and to a lesser extent in other arteries) are stimulated. Impulses travel from them to the medulla, where the sympathetic pathways from the cardiac-accelerating center are inhibited, while the parasympathetic pathways from the cardiac-decelerating center are activated. The result is that heart rate is slowed. As blood pressure falls, however, there is less stimulation of the pressure receptors, which consequently send fewer impulses to the medulla. The sympathetic pathways, freed of inhibition, begin carrying more impulses from the accelerating center to the S-A node, while the parasympathetic carry fewer. This pattern of automatic alternation of firing rate may be enhanced through stimulation of the accelerating center by impulses from chemoreceptors in the carotid artery and aorta, which respond to lowered O_2 and elevated CO_2 in the blood. It may also be enhanced through stimulation of the accelerating center by impulses from receptors in the wall of the right atrium, which respond to the filling of the atrium with blood. The result is an acceleration of heart rate.

The actual rate of heartbeat thus depends partly on the relative activity of the accelerating and decelerating centers in the medulla, while the activity of these centers in turn depends partly on the amount of excitation they receive from the stretch receptors and the chemoreceptors in the arteries. In this respect, these autonomic reflex circuits serve to fine-tune heart rate by means of a negative-feedback loop. The two centers in the medulla are also significantly influenced by signals from other parts of the brain. For example, when a person suddenly sees something frightening, the response, a quickening of the pulse, is prompted by impulses from processing centers in the brain that send signals to the medulla. These two classes of control, one a feedback-regulated system that maintains bodily processes on an even keel, and the other an emergency system to take over in life-threatening situations, is an interesting analogue to the parasympathetic/sympathetic systems of the CNS. It also reminds us of the two kinds of reflexes discussed earlier: those, such as the knee-jerk reflex, that maintain equilibrium and those, such as the withdrawal reflex, that handle emergencies.

THE NEURAL CONTROL OF MORE COMPLEX BEHAVIOR

Not long ago, most scientists supposed that complex behavior—behavior that integrates various sensory inputs and is expressed through coordinated movements of several muscles in sequence—was organized through interacting reflex arcs. Each reflex arc presumably monitored a single sensor and controlled a single muscle (or a pair of antagonistic muscles, as in the

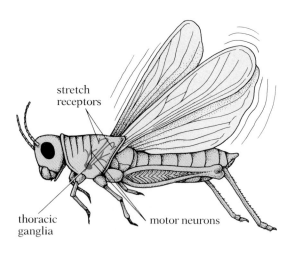

stretch
receptors

thoracic
ganglia

motor neurons

18.37   Flight circuitry in a locust
The motor neurons for the flight muscles are located
in the thoracic ganglia, and the stretch receptors for
the flight muscles send the information they collect
about muscle position to the same ganglia. When
flight is triggered by outside cues (loss of foot contact
with the ground, for instance), a circuit in the
thoracic ganglia is activated that generates
instructions for all the flight muscles. The information
from the stretch receptors is used to fine-tune flight
behavior.

hand-withdrawal reflex). Learned behavior—maze running in rats, for instance—was supposed to come about through the linking of such reflex
circuits into chains. This plausible model of the organization of behavior
dominated the thinking of many researchers through most of this century.

Insect flight   The idea that complex behavior is built out of chains of reflexes was effectively disproven about 1960 by the research of Donald Wilson of Stanford University. Wilson was studying the circuitry underlying
locust flight. Here was an obvious candidate to be understood as chains of
reflexes: one set of flight muscles would raise the wings until the stretch receptors in the other set were sufficiently active to trigger the contractions
that brought the wings down. As the wings came down, the stretch receptors in the "up" set of muscles would signal the increasing strain until what
might be called the "up reflex" was triggered; the wings would then be
pulled up until the receptors in the "down" muscles triggered the "down
reflex"; and so on *ad infinitum*.

By tracing out the circuitry, Wilson discovered the requisite stretch receptors, and he cut the nerves running from the receptors to the thoracic
ganglia. Wilson expected that once the triggering elements had been removed from the circuit, the behavior would collapse. To his surprise, the
locusts continued to fly almost normally. The surgery did, to be sure, have
some effect: the wing beat rate dropped slightly and coordination suffered
somewhat. But aside from the fine-tuning, the behavior remained intact.
Subsequent investigation revealed that the source of the locusts' continuing
ability to fly without their stretch receptors lies in another circuit—a *motor
program*—in the thoracic ganglia that generates all the commands necessary to produce flight without any need for external input from stretch receptors (Fig. 18.37). Normally, of course, the circuit is fine-tuned by the
stretch receptors.

Since Wilson's discovery, many of the complex rhythmic or sequential
behaviors that have been investigated in detail in vertebrates and invertebrates have been found to rely on specialized self-contained circuits that
orchestrate muscle movement in the proper order and with the correct
timing; no complex behavior based on reflex chains is known. The list of
behaviors controlled by motor programs includes feeding, walking, swimming, scratching, vocalizing, building, attack, courtship and mating, and so
on. We shall examine the mechanisms underlying the organization of behavior in Chapter 21.

Feeding in *Aplysia*   As research techniques have improved, increasing attention has been directed toward neurophysiological mapping, cell by cell,
of the circuits responsible for complex behavior. Most of the motor-program networks that are well understood at present involve rhythmic and repetitive movement such as we see in locomotion. Out of practical necessity,
the research focuses on types of behavior that proceed uniformly for minutes or even hours, so that scientists have time to find two cells in the same
circuit, establish the nature of their connections, recording their activity
simultaneously, and finally mark them (usually by means of dye injection).
Needless to say, the more cells in a circuit, and the more elaborate their in-

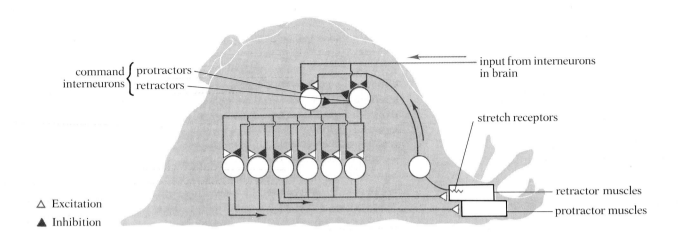

command { protractors
interneurons { retractors

input from interneurons
in brain

stretch receptors

△ Excitation
▲ Inhibition

retractor muscles
protractor muscles

terconnections, the more time-consuming the task of mapping and understanding the circuit will be.

A good example of an almost fully mapped behavioral pathway is the circuit controlling feeding in *Aplysia*. It illustrates the four steps in information flow—detection, conduction, processing, and response—in the context of relatively complex behavior. *Aplysia*, which feeds on seaweed, has chemosensitive tentacles on its head that detect potential food. Signals from these receptors go to the *Aplysia* brain for analysis. If the olfactory processing network determines that the animal is sensing suitable food (and molluscs are able to learn food odors), it sends signals to the interneurons that control feeding behavior. But these interneurons also receive input from other parts of the brain that may *prevent* feeding. Escape, for example, takes precedence over feeding, so when the animal senses danger, low-priority activities like feeding are instantly terminated.

When feeding is to proceed, a motor program is activated (Fig. 18.38); as in many invertebrate behaviors, the CNS simply stops inhibiting the appropriate circuit in a ganglion. In feeding, 22 muscles are controlled by six motor neurons that are themselves coordinated by two antagonistic groups of "command" interneurons, the protractors and the retractors. The protractor and retractor muscles are analogous to extensors and flexors, respectively. The protractor interneurons spontaneously produce rhythmic bursts of impulses every three to four seconds. Signals from these cells activate in a particular order the muscles that open the mouth and bring its rasplike radular "teeth" down into position to bite its food. At the same time, these cells also inhibit the retractor motor neurons and their command interneurons.

When the rhythmic cycle of the protractor interneurons brings them into their quiet phase, they no longer inhibit the retractor interneurons. These cells, which fire continuously in the absence of inhibition, now activate the retractor motor neurons and inhibit those controlling the pro-

18.38 The *Aplysia* feeding circuit
Aplysia feeds by opening its mouth with protractor muscles, and then closing it with retractor muscles. The protractor interneurons stimulate protractor motor neurons and simultaneously inhibit retractor motor neurons, while the retractor interneurons have the opposite effect. The two sets of interneurons coordinate their activity by taking turns (see text). The self-generated pattern of opening and closing can be modified by information arriving from stretch receptors in the retractor muscles. When the load on these muscles is unusually high, indicating that the piece of seaweed is too large or tough for the normal bite pattern, the stretch receptors cause the next opening phase to be extended. Input from interneurons in the brain keeps the circuit turned off except when it is needed.

tractor muscles, thereby directing the sequence of muscle movements necessary to produce a biting movement and closing of the mouth.

The feeding circuit in *Aplysia* illustrates the three basic characteristics of a typical motor program: (1) The program is run by a self-contained circuit that coordinates the movement of several muscles. (2) Though self-contained, it takes advantage of sensory feedback to fine-tune the behavior automatically, in this instance by extending and strengthening the mouth-opening phase when appropriate. (3) It is under direct central control from the brain. In *Aplysia* the brain integrates information from chemosensory cells and "hunger" sensors, and then allows this motor program to be active when it is needed. Similarly, when other behavior becomes necessary —escape from a predatory starfish, for instance—the brain can switch off the feeding behavior. The brain can also order major changes in the rate of the behavior, lengthening or shortening the cycle time as appropriate.

The strategy of using two interacting sets of command cells, like *Aplysia*'s protractors and retractors, is known as ***reciprocal inhibition.*** It is thought to be the means by which virtually all rhythmic behavior is organized. Motor programs appear also to underlie nonrhythmic behavior such as swallowing, smiling, and so on. As we shall see in Chapter 21, the strategy that has evolved for orchestrating even more complex behavior—the building of a bird nest, for instance, or the sequence by which a cat stalks, chases, catches, kills, disembowels, and eats a mouse—has been to construct many separate motor programs and then to devise a higher-level circuit to coordinate the interactions of these individual behavioral elements.

The realization that so much of an organism's behavior is accomplished by interacting groups of neural circuits very similar to those of *Aplysia* is a major achievement for modern neurophysiology.

Chapter 19

SENSORY RECEPTION AND PROCESSING

SENSORY RECEPTION

Even simple procaryotes like *E. coli* derive information about the world from a complex series of chemical reactions between the cell membrane and substances in the environment, and "process" that information in simple ways to produce a behavioral reaction. In animals from *Aplysia* to humans, specialized receptor cells detect physical changes in their surroundings, to which the animals may then respond in adaptive ways. In this chapter we shall explore various mechanisms for detecting mechanical and chemical changes in the environment and translating them into neural stimuli, a process known as ***sensory transduction.*** We shall also examine what is currently known about how nervous systems process the impulses produced by environmental stimuli, and how this processing results in the mental experiences we call sensations. Finally, we shall turn our attention to the question of how the vertebrate brain is organized, and how that organization reflects the evolution of increasingly sophisticated levels of neural processing.

MECHANISMS OF RECEPTOR FUNCTION

How environmental changes produce neural impulses Most receptor cells respond to a variety of stimuli, provided the stimuli are unusually strong; for example, a physical shock causes our photoreceptors to "see stars." But sensory receptors are useful to organisms only because each class of cells is maximally responsive to just one kind of stimulus. The range of sensory specializations includes pressure, heat and cold, concentrations of particular chemical compounds, vibrations, light, electric and magnetic fields, and so on. But regardless of its specialization, each type of receptor translates the stimulus into a change in membrane polarization—the common currency of the nervous system.

How do environmental changes cause receptor cells to depolarize or hyperpolarize? You can probably guess that in most sensory cells the stimulus opens or closes gated channels. For instance, when light is absorbed by visual pigments, the activated pigments close sodium channels in the membrane, thereby hyperpolarizing the sensory cells. Sodium channels in stretch-receptor cells are thought to open (and therefore to depolarize the membrane) simply as a result of the physical distortion of the membrane caused by stretching.

The receptor proteins in the membranes of sensory cells of the nose and tongue open Na^+ channels when the appropriate chemical substrates bind to them. When enough sugar molecules, for example, bind to receptor proteins on a sugar-sensitive cell, so many sodium channels open that the sensory cell depolarizes past threshold and fires. An adjacent cell that has receptor proteins for bitter chemicals like quinine will not respond, since its ion channels are not affected. In the presence of sufficient quinine, however, enough of its channels will open to depolarize *it* past threshold. The sensations of sweet and bitter are thus conveyed by identical means—the opening of Na^+ channels, followed by depolarization, followed by the generation of action potentials. They are distinguished because the different kinds of sensory cells are connected to different targets in the brain. Specific sensations are the brain's interpretation of incoming stimuli. What an animal senses, then, is shaped and constrained by the nature of its sensory receptors and by the way in which its nerve cells are wired together.

Varieties of receptor response We have seen that many sensory cells operate when specific environmental stimuli in one way or another open or close ion channels in the membrane. But how do sensory cells encode the strength and (in many cases) the temporal pattern of the stimuli? Perhaps the simplest and best-understood example is provided by the stretch receptor.

In 1950, Bernard Katz of University College, London, demonstrated that the stretching of a neuromuscular spindle produces a local depolarization of the receptor-cell membrane. When this depolarization, known as the *generator potential,* reaches threshold level, it triggers an action potential in the nerve fiber (Fig. 19.1). An increase in the intensity of the stimulus (increased stretching in this case) causes a proportional rise in the generator potential. This rise, in turn, increases the frequency of the triggered impulses. Thus the frequency of output from the receptor is a direct measure of the strength of the stimulus.

But stretch-receptor cells differ in their behavior from most sensory cells. For one thing, they do not fire unless stimulated, whereas most sensory cells fire, at a slow *basal rate,* even when unstimulated. Moreover stretch-receptor cells continue to fire more or less indefinitely at a rate proportional to the stimulus intensity, whereas most sensory cells undergo adaptation. For example, when a simple heat receptor in the skin is stimulated by warm water, it begins to fire at a higher rate. But as time goes on,

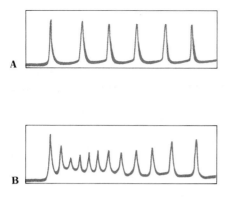

19.1 Relation of impulse pattern to intensity of stimulation of a stretch-receptor cell
(A) A neuromuscular spindle is stretched slightly. The stimulus thus applied produces a generator potential (shown here as an upward shift in the base line) that triggers a series of impulses (action potentials).
(B) The same neuromuscular spindle is stretched more. The generator potential (base line) rises higher, and the frequency of the impulses increases. Thus the frequency of impulses is a function of the intensity of the generator potential, which in turn is a function of the strength of the stimulus.

the sensory cell becomes less sensitive to temperature and the firing rate declines. Eventually the firing rate drops to the basal level, and we no longer sense that the water is warm.

This slow adaptation strategy makes good sense for most sorts of receptor cells because it allows them to operate over a far larger range of stimulus intensities than would otherwise be possible. For example, we are sensitive to temperature variations over a range of at least 50°C. If a cell responded to these changes by varying its output frequency from, say, zero to 100 impulses per second, its ability to sense accurately the sorts of small temperature changes we normally experience and react to would be very poor: these small changes would produce only small changes in the cell's output frequency. Instead, however, the receptor cell responds accurately only over a range of 10°C, but through adaptation its sensitivity can be centered around the current temperature. Hence it can report with high accuracy on changes in temperature when the differences are small, as they usually are. The receptor cells are also able to accommodate slow changes in the current temperature range, but must readapt to any dramatic change.

Cells that adapt slowly are intermediate between the two extremes of receptor response. Cells at one end of this spectrum of adaptation rates—cells that adapt almost instantly—are called **phasic.** Those at the other end—cells that essentially never adapt—are called **tonic** (Fig. 19.2). A stretch-receptor cell is a good example of a tonic cell: it continuously provides an accurate, unadapted measure of muscle loading (the amount of

19.2 Varieties of receptor response
The response of most sensory receptor cells to a stimulus usually falls somewhere between the two extremes of phasic and tonic response. A highly tonic receptor fires at a continuous rate proportional to the strength of the stimulus (color) and immediately returns to its unstimulated state once the stimulus is removed. By contrast, a highly phasic receptor returns to its basal firing rate almost immediately after the onset of the stimulus, even while the stimulus continues to be applied. Removal of the stimulus on the highly phasic cell causes its firing rate to drop slightly below its basal level, until adaptation to the removal has taken place.

Since tonic cells respond efficiently only over a relatively narrow range of stimulus frequency, a series of receptors with differing ranges is required to monitor the strength of the stimulus effectively. Phasic cells provide little information about the magnitude of a stimulus; instead, they provide essential information about the onset and end of stimulation.

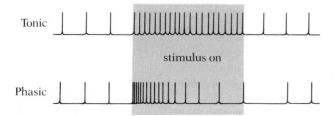

Tonic

stimulus on

Phasic

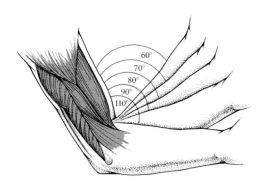

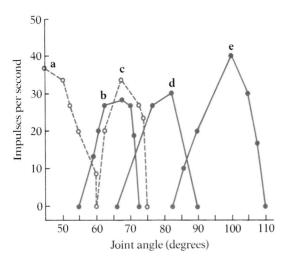

19.3 How stretch receptors measure joint angles
Each receptor cell measures with great precision the degree to which a muscle stretches (which in this case serves to define the angle between the upper and the lower arm), but the limited maximum firing rate of each cell restricts the range over which its measurements can be made. As a result, many different receptor cells, with overlapping sensitivity ranges, come into play. Here five cells (out of a much larger population) with overlapping ranges encode the joint angle over a wide range. Notice how the pattern of activity shifts over a range of just 10 degrees: at 60°, for example, spindle **b** responds with 18 impulses per second, while **c** is silent; at 65°, **b** responds with 28 per second and **c** with 31 per second; at 70°, **b** responds with 19 per second and **c** with 28 per second. The relative firing rates are unique for each angle.

stress "loaded" onto the receptor). Accuracy in this task, of course, means that an individual sensory element can function only over a rather narrow range, but this problem is circumvented by the presence of a series of receptor cells with different but overlapping ranges (Fig. 19.3).

Phasic cells, on the other hand, are specialized to detect change. Invertebrate muscles, for instance, have two sorts of stretch-receptor cells: a nonadapting one virtually identical to the tonic cells seen in vertebrates, and a "fast" receptor cell, a phasic cell that is sensitive to changes in muscle loading. The phasic cell produces a burst of impulses at the slightest alteration in muscle stretch, thus providing very precise information about the time of a change, though very little about the actual *amount* of loading. Such phasic cells are quite common in the vertebrate nervous system, but most sensory cells fall somewhere between the purely tonic and the purely phasic.

SENSORY RECEPTORS OF THE SKIN

There are several types of sensory receptors in vertebrate skin (see Fig. 6.16, p. 153), and each has an analogue in invertebrates. The vertebrate receptors are concerned with at least five different senses: touch, pressure, heat, cold, and pain. Some of the skin receptors, particularly those concerned with pain, are simply the unmyelinated terminal branches of neurons (Fig. 19.4A). Others are nets of nerve fibers surrounding, in mammals, the bases of hairs (Fig. 19.4B); these fibers, which are particularly important to the sense of touch, are stimulated by the slightest displacement of the tiny hairs present on most parts of the body. Other skin receptors are more complex, consisting of nerve endings surrounded by a capsule of specialized connective-tissue cells (Fig. 19.4C).

The relative abundance of the various types of receptors differs greatly in humans: pain receptors are about 30 times as abundant as cold receptors, while cold receptors are about 10 times as abundant as heat receptors. Nor are the receptors distributed evenly over the body: touch receptors, for instance are much more densely packed in the lips than on the back, while the fingertips have more touch receptors than the back, but fewer than the lips. The differences in density correlate well with the different functions of the body parts. You can ascertain differences in touch-receptor density by prodding your skin gently with a pair of needles. By moving the needles slightly farther apart after each test until you feel two distinct points of stimulation, you can judge just how tightly or thinly packed the sensors are at various locations.

THE PROPRIOCEPTIVE AND VISCERAL SENSES

Unlike the receptors in the skin, which receive information from the outside environment, some other widely dispersed receptors in the body function primarily in collecting data about the condition of the body itself. Among these immensely important monitors are the stretch receptors (proprioceptors) in the muscles and tendons (Fig. 19.5). They are sensitive to the changing tensions of muscles, and send impulses to the central nervous system informing it of the position and movement of the various parts of the body. They give rise to the knee-jerk reflex (see Fig. 18.33, p. 481).

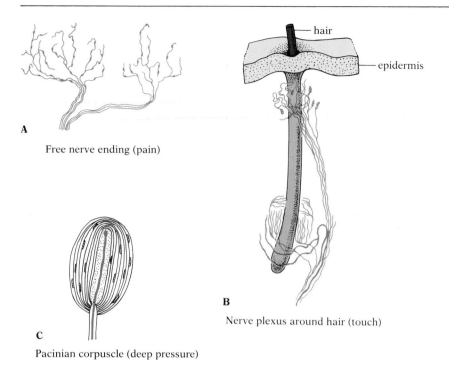

A

Free nerve ending (pain)

hair

epidermis

C

Pacinian corpuscle (deep pressure)

B

Nerve plexus around hair (touch)

19.4 Three receptors of the skin and the senses they mediate

Other internally dispersed receptors include those of the so-called visceral senses, located in the internal organs. Among those mentioned earlier are two groups of receptors in the carotid artery, sensitive respectively to elevated CO_2 concentration in the blood, and to high blood pressure. They are important in the automatic control of heartbeat. We seldom perceive the activity of such visceral receptors consciously; responses to their messages are usually mediated by the autonomic system. Certain of the visceral receptors, though, can contribute to the conscious sensations of hunger, thirst, and nausea.

THE SENSES OF TASTE AND SMELL

The receptors for taste and smell are chemoreceptors—they are sensitive to solutions of certain types of chemicals that bind to them by weak bonds. These two senses can interact, so when we speak of a taste, we are more often than not referring to a compound sensation produced simultaneously by taste and olfactory (smell) receptors. One reason that hot foods have more "taste" than cold ones is that they vaporize more chemicals, which then reach the olfactory receptors, either from the outside through the nose, or from the mouth upward through the nasal passages. The reason we may find food tasteless when we have a cold is that the mucus coating our inflamed nasal passages blocks the olfactory receptors.

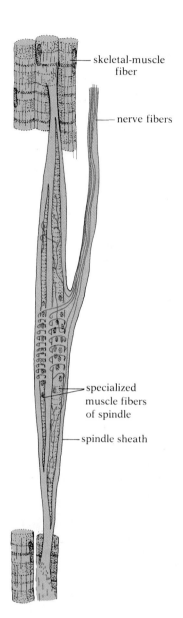

skeletal-muscle fiber

nerve fibers

specialized muscle fibers of spindle

spindle sheath

19.5 A stretch receptor in skeletal muscle
The terminal branches of sensory nerve fibers are intimately associated with several specialized muscle fibers that form the apparatus called a neuromuscular spindle.

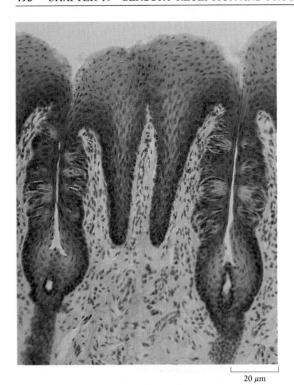

20 μm

19.6   Photograph of section of mammalian tongue
The taste buds are located in the walls of the deep
narrow pits.

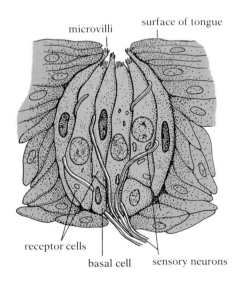

microvilli   surface of tongue

receptor cells

basal cell   sensory neurons

19.7   The structure of a taste bud
Each taste bud contains specialized receptor cells
bearing sensory microvilli that are exposed in pits on
the tongue surface. The ends of sensory neurons
(color) are closely associated with the receptor cells.

Taste   The receptor cells for taste in our species are located in *taste buds*
(Fig. 19.6) on the upper surface of the tongue (and, to a lesser extent, on the
surface of the pharynx and larynx). The receptor cells themselves are not
neurons, but specialized cells with microvilli on their outer ends (Fig. 19.7).
The ends of nerve fibers lie very close to these receptor cells, and when a
receptor is stimulated it generates impulses in the fibers.

Subjective experience has long told researchers that there are four basic
tastes: sweet, sour, salt, and bitter. The purest stimuli for these sensations
are sucrose, H^+, NaCl, and quinine, respectively. In humans, receptors that
respond maximally to the four basic taste stimuli are not evenly distributed:
sweet and salt are most easily sensed at the front of the tongue, bitter at the
back, and sour at the sides. Unlike the four "pure" stimuli, though, most
chemicals stimulate sensations of more than one category. The differential
blending of these four sensations is thought to carry the information by
which we distinguish tastes.

Some amount of blending is, in fact, inevitable. When neurophysiologists
first succeeded in recording the activity of taste receptors in insects, where
they are conveniently located near sensory hairs on the feet as well as on
the mouthparts (Fig. 19.8), they usually found four or five taste-receptor
cells per sensory hair. Each cell was neatly tuned to a particular taste class
(though in many species one or another class was absent): one salt cell, one
sugar cell, one cell for sour, one for bitter, and frequently one for water. In
recordings of the activity of vertebrate taste buds, however, this sharp distinction is absent: a cell, primarily responsive to salt, for instance, also responds (albeit less vigorously) to sweet, bitter, and sour. The advantage, if
any, of this lack of absolute discrimination is not known.

The subjective nature of sensations in general is illustrated by the large
individual differences in taste perception. For example, most people find
saccharin sweet, but to a substantial minority this artificial sweetener
seems bitter. Individuals also differ greatly in their tolerance of sugary,
salty, sour, and bitter tastes: lemon candy may be too sweet for one person,
too sour for another, and just right for a third. Such differences in taste
probably have a biological component; to some extent, the nervous system
is as responsible for our sensations as the stimuli themselves.

Smell   The receptor cells for the sense of smell are located on the antennae of some insects and, for the most part, in the noses of terrestrial vertebrates. In humans, for instance, the olfactory receptors are located in two
clefts in the upper part of the nasal passages (Fig. 19.9A). Unlike the receptor cells of taste, the olfactory receptors are true neurons. The cell bodies
of most of these neurons lie embedded in the epithelial layer of the walls of
the olfactory areas of the nasal chamber (Fig. 19.9B). Dendrites run from
the cell bodies to the surface of the epithelium, where they bear a cluster of
modified cilia, which apparently hold the receptor sites.

Many attempts have been made to identify a group of primary odors from
which all the more complicated odors can be derived, but with little success. In part this may be a result of our species' relative insensitivity to
odors as compared to most other animal species. A more important problem is that the chemical analysis of substances that smell very much alike
quite often reveals that they are not at all similar chemically. R. W. Mon-

A

0.1 mm

B

0.1 mm

19.8   Taste receptors in a blowfly (*Calliphora vicina*)
(A) Most of the sensory hairs surrounding the labellum in this scanning electron micrograph of a proboscis of a blowfly contain taste receptors. (B) The two small, upward-curving hairs, one above each of the large claws on the lower part of the fly's foot, also contain taste receptors. Some of the forward-projecting hairs are probably touch receptors.

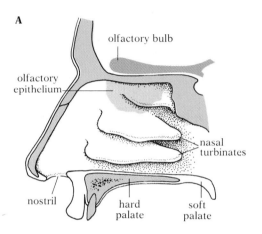

A

olfactory bulb

olfactory
epithelium

nasal
turbinates

nostril

hard
palate

soft
palate

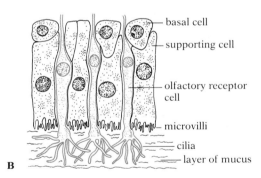

basal cell

supporting cell

olfactory receptor
cell

microvilli

cilia

layer of mucus

B

19.9   Human olfactory receptors
(A) The receptor cells are located in the olfactory epithelium, whose distribution on the wall of the upper portion of the nasal cavity is indicated in light pink. The olfactory bulb of the brain receives information from the receptor cells. (B) The cell bodies of most of the receptor cells are in the olfactory epithelium; their sensory cilia protrude into a layer of mucus on the surface of the epithelium.

crieff of Scotland suggested in 1949 that the spatial configuration of molecules, which is crucial in enzyme-substrate reactions as well as in the binding of transmitter to receptor and of antibody to antigen, might be the basis of smell. This seems highly likely, and after examining the size and shape of some 600 organic compounds whose odors had been well described, John E. Amoore at Oxford University formalized this suggestion into a stereochemical hypothesis of smell, published in 1952.

19.10   Antennae bearing olfactory receptors of an adult male polyphemus moth (*Antheraea polyphemus*) The thousands of sensory hairs extending perpendicularly from the combs of the male moth's antennae bear specialized cells that respond to minute amounts of the female's pheromone. While millions of molecules may be required to activate a human olfactory neuron, just a single molecule of pheromone at a receptor cell on the surface of a sensory hair triggers the moth's cell.

Amoore's scheme proposed the existence of seven primary odors: camphoraceous, musky, floral, pepperminty, ethereal (etherlike), pungent, and putrid. He specified the size and shape a substance must have to stimulate the receptors for each of these basic odors. Thus, according to Amoore, camphoraceous-smelling molecules would be roughly spherical, with a diameter of 0.7 nm; musky-smelling molecules, disc-shaped and about 1 nm in diameter; floral-smelling molecules would have the form of a disc with a flexible, elongate side group attached; ethereal-smelling molecules would be thin and rod-shaped; and so on. The receptor sites on the sensory cells, according to Amoore's hypothesis, would have shapes into which portions of such molecules could fit more or less perfectly. As with taste, the relative activity of these basic receptor types generated by a chemical would determine its classification. The available evidence, however, is ambiguous, and Amoore's provocative hypothesis remains controversial. Recent results from work on invertebrates, for instance, places the number of discrete receptor classes at about three dozen.

Though the stereochemical hypothesis, with its classes of receptor types, may prove to be an oversimplification, it seems clear that highly specific binding proteins *are* the basis of an important class of olfactory receptors. Many animals are equipped to recognize at very low concentrations odors of special importance to their species. For example, many animals recognize predators chemically: the escape response of *Aplysia* is triggered by the species-specific odor of starfish, and many fish flee when they detect the odor produced by the damaged skin of conspecifics (members of the same species). Male moths ready to mate can detect the unique odor of the female of their species from enormous distances (Fig. 19.10). Organic chemists have been able to synthesize these so-called pheromones to attract and trap male gypsy moths, and thus to combat their locustlike destructiveness. The olfactory vocabularies of other species include precisely defined chemical messages for "attack," "flee," "come here," and the like. In each such case the olfactory receptor employs a highly specific binding protein to sample the air (or water) for information.

THE SENSE OF VISION

Light-sensitive structures have evolved independently in a vast number of plants and animals. Though electromagnetic radiation takes many forms, ranging from low-energy radio waves to high-energy gamma rays, only the middle wavelengths have enough energy for vision, photosynthesis, and phytochrome conversion, but not so much that they can damage tissue. It is this relatively narrow band of wavelengths that so many unrelated kinds of organisms use to gather information about their environments, and it is this portion of the spectrum of electromagnetic radiation that we call light (see Fig. 8.3, p. 199).

Light receptors of animals   Almost all animals respond to light. Even some Protozoa react quickly to changes in light intensity, often moving toward or away from brightly lit areas. These organisms have a specialized region containing a pigment that undergoes chemical changes when exposed to light. The light-sensitive pigment common to many animals is a protein to which a portion of a carotenoid molecule is attached. Carotenoids, you may recall, are yellow or orange light-sensitive pigments in plants. Animals must obtain carotenoid in their diet as a vitamin (vitamin A), since it can be synthesized only by plants.

The light receptors of many invertebrates, like those of Protozoa, do not function as eyes in the usual sense of the word. Some of these very simple receptors do nothing more than indicate the general light intensity; they do not form images (Fig. 19.11). In planaria (see Fig. 6.33, p. 163), light receptors are arranged within a cup-shaped organ called an *eye cup;* the shadow cast by the opaque edge of the cup defines the precise direction and elevation of the source of light (Fig. 19.12). Eye cups, however, are incapable of forming images. They merely register the presence or absence of light from a certain direction.

Compound eyes   From this modest beginning, two considerably more complex strategies have evolved that enable most animals to see images of the world about them. These are the compound eye and the camera eye. The *compound eye* uses what is basically an array of tiny eye cups, each modified into a tube, called an *ommatidium,* that points out at the world in

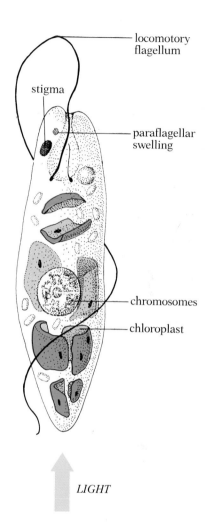

19.11   *Euglena*
During part of the day the photosynthetic protist *Euglena* swims toward light; at other times it swims away. It points itself away from light, as shown here, by turning until its opaque stigma casts a shadow on the photosensitive paraflagellar swelling. To swim toward light, the organism turns until there is no shadow. The paraflagellar swelling, as its name implies, is involved in controlling flagellar movement.

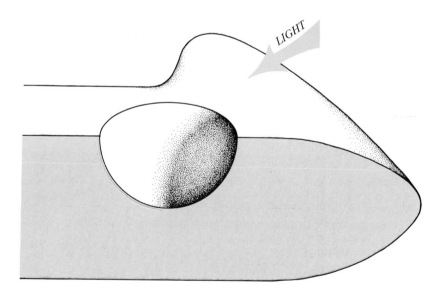

19.12   A planarian eye cup
The planarian eye cup provides directional information that the organism can use to maneuver in the environment; it does not form images. The direction of a light source is indicated by the location of the shadow cast by the cup's opaque edge onto receptors within the organ, as shown in this cross section.

a slightly different direction (Figs. 19.13 and 19.14). Light entering through the lens and crystalline cone of each ommatidium is focused onto an array of seven to nine elongated receptor cells, known as reticular cells. A specialized area—the ***rhabdomere***—runs the entire length of each of these cells. The rhabdomeres, all located in the central axis of the ommatidium, where they jointly form the rhabdom, have layer upon layer of microvilli on which are spread light-sensitive pigments. With so many layers for light to pass through, compound eyes are more efficient at absorbing photons than the camera eyes we shall examine next.

In many species with compound eyes the individual rhabdomeres have pigments that absorb maximally (and are therefore most sensitive to) light from particular parts of the spectrum. In honey bees, for instance, there are three pigments in the cells of each ommatidium: two cells have a pigment that absorbs green light most effectively, two cells absorb primarily blue light, three respond best to ultraviolet (UV), and the last two absorb either green (if the ommatidium is in the ventral half of the eye, where it will most often see vegetation) or UV (if in the dorsal half, where it will normally view the sky). The differing sensitivities of these three pigments, tuned respectively to green, blue, and UV light, form the basis of color vision in most insects.

The picture produced by an insect brain from the information it receives from a compound eye is probably a grainy mosaic of the world rather like needlepoint, with far less precise delineation of objects in the visual field than we experience (Fig. 19.15). Several advantages of compound eyes, however, compensate for this low spatial resolution. Besides being very efficient at absorbing light, they are very small and lightweight, which is important for a flying insect. In addition, most compound eyes permit arthropods to see details of movements that are far too rapid for our eyes.

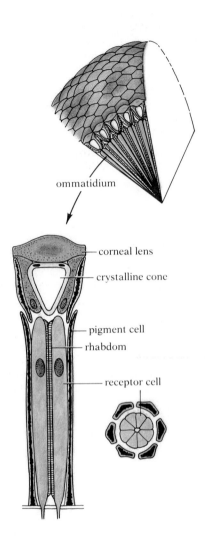

19.13 Ommatidia from the compound eye of a typical diurnal insect
Top: Section of compound eye. Bottom: Longitudinal and cross sections of one ommatidium. The lens and crystalline cone focus incoming light rays into the rhabdom, a translucent cylinder formed by the highly specialized microvilli in the rhabdomeres of the eight receptor cells; photosensitive pigment is located in the microvilli. The pigment cells surrounding the ommatidium contain a dark pigment that prevents passage of light from one ommatidium to another.

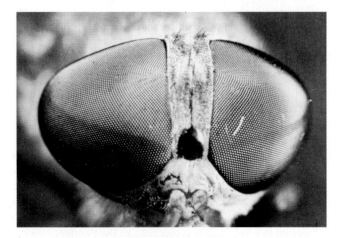

19.14 The compound eyes of a horse fly
Each eye is composed of a huge number of ommatidia.

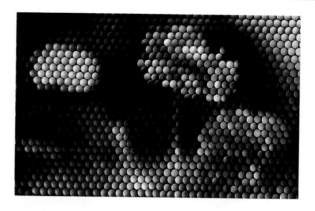

19.15 View through a compound eye
The visual world of animals with compound eyes is thought to be broken up into a mosaic of spots. Here dahlias are shown as seen by human beings (left) and as a honey bee might see them from about 10 cm away (right). This picture, taken through an optical device, is not a perfect representation: bees see ultraviolet light and not red, while this photograph does not reproduce ultraviolet and includes red. Moreover, the circles should be vertically elongated ellipses, to reflect the peculiar anatomy of the bee eye.

19.16 Cephalopods, with pinhole eyes
Cephalopods such as these juvenile chambered nautili, here seen feeding on fish, are the only organisms with pinhole eyes.

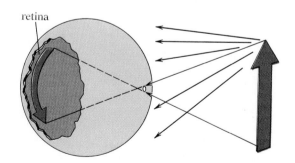

19.17 How a pinhole eye works
Light reflected from objects in the environment passes through the pinhole and projects a precise (though inverted) image on the retina. The main disadvantage of this strategy is that the image is relatively dim, since very little light is admitted.

Camera eyes There are two versions of the *camera eye,* the strategy employed by vertebrates and some molluscs for forming a visual image of the world. By far the rarer of the two is the *pinhole eye* of organisms such as the chambered nautilus (Fig. 19.16). The pinhole eye is simply a covered eye cup with a tiny opening in its surface. Light from the world outside passes through the hole and is projected onto the array of receptors (the *retina*) at the back (Fig. 19.17). A great disadvantage is that very little light can enter; if the opening were widened to admit more light, the image would become blurred.

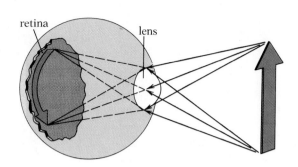

19.18   How a lens eye works
Light reflected from objects in the environment
arrives at a lens and is focused on the retina. The
lens eye admits much more light than the pinhole eye,
but has great constraints on its focusing ability.

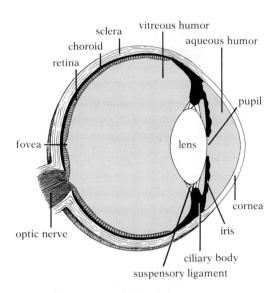

19.19   The human eye
The drawing shows a diagrammatic section through
the eye.

The **lens eye,** on the other hand, can have a far larger opening for light,
since the lens focuses images on the retina (Fig. 19.18). This more elaborate
approach, however, is not without its problems. Unlike a pinhole eye, a lens
eye can focus on objects at only one distance at any one time. Mammals,
birds, and some reptiles have muscles to change the shape (and thus the
focus) of their lenses, while fish, amphibians, and other reptiles move the
whole lens toward or away from the retina in the manner of a true camera.
Moreover, the retina and lens must have just the right spatial relationship: if
the retina is too close to the lens, the animal will have difficulty seeing
nearby objects (that is, it will be farsighted), whereas if the retina is too far
away, distant objects cannot be brought into focus (the animal is near-
sighted). In addition, irregularities in the lens itself may give rise to astig-
matism, in which images of objects at the same distance from the lens but in
different parts of the visual field come into focus at different distances. An
insuperable disadvantage of the camera eye for many creatures is its bulk: a
camera eye large enough to provide the same visual resolution as the compound eye of a honey bee would weigh more than the bee itself.

The human eye   The adult human eye is a globe-shaped lens eye with a diameter of about 2.5 cm (Fig. 19.19). It is encased in a tough but elastic coat
of connective tissue, the **sclera.** The anterior portion of the sclera, the **cornea,** is transparent and more strongly curved; it functions as the first element in the light-focusing system of the eye. Just inside the sclera is a layer
of darkly pigmented tissue, the **choroid,** through which many blood vessels
run. The choroid is important both as the structure providing the blood
supply to the rest of the eye and as a light-absorbing layer that, like the black
inner surface of a camera, helps prevent internally reflected light (and light
from outside the eye that has not entered through the lens) from blurring
the image. In nocturnal animals, by contrast, the choroid layer is usually
highly reflective; though it does reduce resolution, it increases sensitivity
by sending unabsorbed light back through the receptor layer for another
try. This mirrorlike layer accounts for the way a cat's eyes seem to glow in
the dark.

Just behind the junction between the main part of the sclera and the cornea, the choroid becomes thicker and has smooth muscles embedded in it;
this portion of the choroid is called the **ciliary body.** Anterior to the ciliary

body, the choroid leaves the surface of the eyeball and extends into the cavity of the eye as a ring of pigmented tissue, the **iris.** The iris contains smooth muscle fibers arranged both circularly and radially. When the circular muscle fibers contract, the opening in the center of the iris (known as the **pupil**) is reduced; when the radial muscles contract, the pupil is dilated. The iris thus regulates the amount of light admitted to the eye in much the same way that the diaphragm of a camera controls the lens aperture.

The **lens,** which functions as the second element in the light-focusing system, is suspended just behind the pupil by a **suspensory ligament** attached to the ciliary body. The exact shape of the lens is controlled by the tension applied by an array of tiny muscles mounted here. The lens and its suspensory ligament divide the cavity of the eyeball into two chambers. The chamber between the cornea and lens is filled with a watery fluid, the aqueous humor. The chamber behind the lens is filled with a gelatinous material, the vitreous humor.

The **retina,** which contains the receptor cells, is a thin tissue covering the inner surface of the choroid. It is composed of several layers of cells: the receptors, sensory neurons, and interneurons. The receptors are of two types, rods and cones (Fig. 19.20). The **rod cells** are more abundant toward the periphery of the retina, and are exceedingly sensitive to light; they allow us to see in dim light, but produce colorless, poorly defined images. The **cone cells,** which are specialized for color vision in bright light, are especially abundant in the central portion of the retina, an area known as the **fovea.** Because of the high density of receptors in the fovea, we are able to see the small area in the center of the visual field in fine detail.

The rods and cones synapse in the retina with short sensory neurons (bipolar cells), which themselves synapse with the retinal ganglion cells,

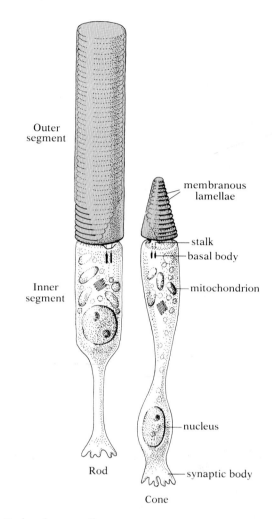

19.20 Rod and cone cells

Above: Cells from a human retina. The outer segment and the stalk connecting it to the inner segment develop as a highly specialized cilium. Electron microscopy reveals that a basal body is located at the inner end of the stalk, and that the nine peripheral microtubules characteristic of all cilia run from it through the stalk into the outer segment; as is true of most cilia that have lost their motile properties, the two central microtubules of motile cilia are absent. The visual pigment is located in the numerous membranous lamellae of the outer segment. The pigments of rods and cones are different; those of cones are responsible for color vision. Left: Scanning electron micrograph of rod and cone cells from a pig.

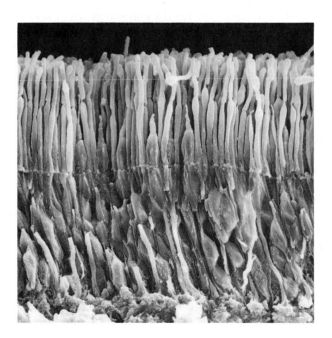

whose axons, bundled together as the optic nerve, run to the visual centers of the brain (Fig. 19.21). As we shall see, the interconnection of neurons in the retina enables the eye to modify extensively the information transmitted from the hundred million or so receptor cells through the few million axons of the optic nerve to the brain. Oddly enough, in most vertebrates these processing cells lie *between* the lens and the receptors, a location that requires a sizable hole in the retina (known as the blind spot), for the optic nerve fibers to pass through. In squid, octopus, and snakes, the processing cells are behind the receptors, an arrangement that creates no blind spot in their vision.

The light sensitivity of rods and cones   Both rods and cones contain light-sensitive pigments. In rods the pigment, which is built into the membranes of the flattened vesicles in the outer segment (Fig. 19.20), is called **_rhodopsin._** It consists of a protein (opsin) bonded to a prosthetic group called retinal, which is a derivative of vitamin A. When a molecule of rhodopsin is struck by a photon of light, the retinal is converted into a slightly different isomer (Fig. 19.22). This conversion is entirely light-driven; no

19.21   The cells of the human retina
The receptor cells (rods and cones) synapse at their base with bipolar cells. The bipolar cells, in turn, synapse with ganglion cells; the axons of the ganglion cells form the optic nerve, which runs from the eye to the brain. Hence information that follows the most direct route to the brain moves from receptor cell to bipolar cell to ganglion cell to brain. Processing of information can occur within the retina because often several receptor cells synapse with a single bipolar cell and several bipolar cells synapse with a single ganglion cell. Besides convergence of information, there is lateral transfer of information from pathway to pathway via horizontal cells (each of which receives synapses from many receptor cells and synapses on many bipolar cells and on other horizontal cells) and via amacrine cells (which both receive synapses from and synapse on bipolar cells, and also synapse on many ganglion cells). Note that the retina is arranged anatomically in reverse order from what might be expected; the receptor cells are in the back of the retina, and light must pass through the nerve cells to reach them.

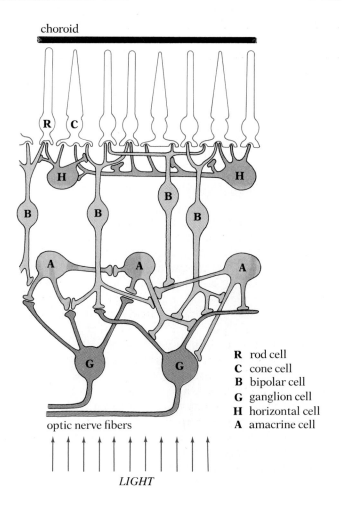

R   rod cell
C   cone cell
B   bipolar cell
G   ganglion cell
H   horizontal cell
A   amacrine cell

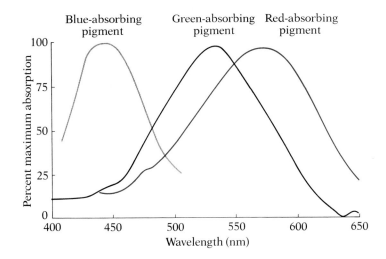

19.22 The light-driven change of retinal from one isomer into another
With the absorption of light energy, the shape of the side chain is altered.

Blue-absorbing pigment   Green-absorbing pigment   Red-absorbing pigment

19.23 Absorption spectra of the three cone pigments of primates
The red-absorbing pigment actually has its maximum sensitivity in the yellow region of the spectrum. The relative narrowness of the absorption spectrum of the blue-absorbing cone is a result of faint yellow pigments in the lens and cornea, which absorb ultraviolet light.

enzyme is required. The isomerization of the retinal leads, in turn, to conformational changes in the protein and, in vertebrate photoreceptors, to hydrolysis of the bond between the retinal and the protein. The result is a change in the polarity of the plasma membrane of the rod cell (probably through the release of Ca^{++} ions from the flattened vesicles), which alters transmitter release at synapses between the rod cell and the bipolar cells (or the horizontal cells). In the dark, retinal is reconverted to the original isomer and rebonded to opsin to form functional rhodopsin.

The mechanism of cone vision is much more complex. There are three classes of cones, each containing a different pigment. All three human pigments have retinal as their prosthetic group, and all three absorb light over a wide range of wavelengths. However, their protein components differ slightly, and as a result, the range of wavelengths for each pigment centers on a different part of the spectrum (Fig. 19.23).

It may seem odd that we can distinguish colors so well when the absorption curves of the three pigments, particularly those of the green-absorbing and red-absorbing pigments, overlap so much: green-absorbing cones, for instance, may respond to orange, yellow, green, blue-green, or even blue light, and the "red" curve actually peaks in the yellow. Current theory suggests that in processing color vision the nervous system compares and contrasts the output of the cones in pairs, to obtain the fine color distinctions we consciously perceive.

We are so accustomed to the human version of color vision that we tend to assume that the colors we see are somehow "true," and that other animals must see the world in the same way we do. Humans and other primates, however, are rather unusual among the mammals in possessing well-developed color vision. Most mammalian species see the world in varying shades of gray. By contrast, most birds and many fish and reptiles appear to have color vision, and some even detect light in the ultraviolet range (UV). Some of these color-sensitive vertebrates, however, have only two types of cones—certain species of fish, for instance, have only green and blue pigments. Often in birds and reptiles color sensitivity is further complicated by filtering: a colored oil droplet lies between the lens and the pigments so that only one part of the spectrum filters through to the receptor.

As we saw earlier, arthropods often have color vision. Most, however, do not have the same range of color vision as our species. Honey bees, for ex-

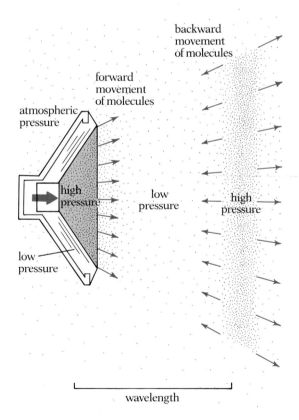

19.24 Sound generation by a loudspeaker
By vibrating and setting air molecules into motion, a loudspeaker creates waves of high and low pressure; these are the physical bases of sound. Various kinds of ears have evolved, which respond either to particle movements or to waves of pressure. What we hear is the result of the processing of sensory input in the nervous system. In this diagram, air pressure is indicated by the density of dots, while the directions of particle movement are shown by arrows. The distance between one band of high pressure and the next is one wavelength.

A

B

19.25 Two sound detectors in insects
(A) The antennal hairs on the head of this male mosquito can resonate, and are therefore very sensitive to sounds of the proper frequencies. (B) The membrane-covered organs on the forelegs of a katydid, like the antennal hairs of mosquitoes, detect the displacement of air particles.

ample, are blind to red but sensitive to UV, a range of the spectrum they use for navigation. Thus animals have widely varied and very selective color perceptions; each species sees the colors that are useful in its world.

THE SENSE OF HEARING

Like vision, hearing is an important sensory ability among animals. For many species, sounds in the environment provide crucial information about predators, prey, and conspecifics. A variety of strategies for detecting them have therefore evolved.

Sound is produced when an object vibrates, setting in motion the particles of the medium (usually air) and thus generating alternating bands of high and low pressure (Fig. 19.24). The number of vibrations per second is the sound's frequency, and is given in hertz (Hz). Many insects have long hairs that are moved back and forth by the air movement (Fig. 19.25A). Like tuning forks, these hairs have strong resonance, so that they respond best to vibrations within a narrow range—typically the frequencies emitted by other members of the species. Most vertebrates, on the other hand, are sensitive to the pressure changes associated with sound waves. We shall only discuss the human ear—a pressure-sensitive ear—in detail.

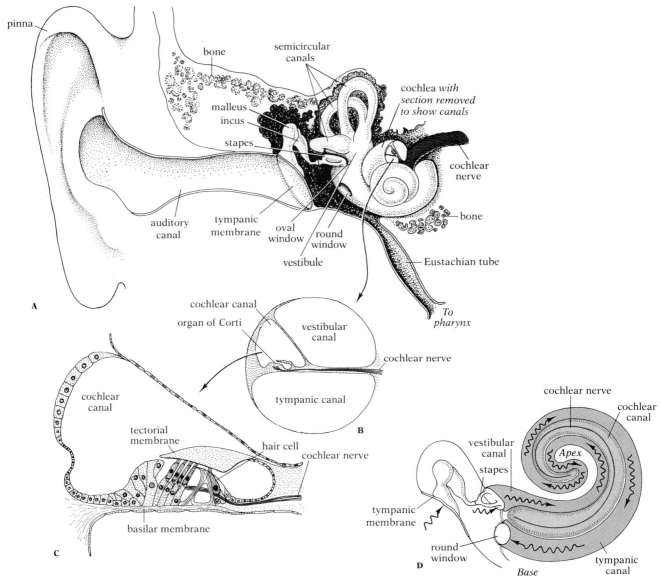

19.26 The human ear

(A) The major parts of the outer, middle, and inner ear (see text for description). (B) Enlarged cross section through one unit of the coil of the cochlea, showing the relationship between the vestibular, cochlear, and tympanic canals and the location of the organ of Corti. (C) Enlarged diagram of the organ of Corti, which rests on the basilar membrane separating the cochlear and tympanic canals. When the basilar membrane vibrates and moves the sensory hair cells up and down, the hairs rub against the tectorial membrane overhanging them. The resulting deformation of the hairs produces a generator potential in the hair cells, which triggers impulses in sensory neurons running from the organ of Corti to the brain. (D) Diagram of the relationship between the middle ear and the cochlea, here pictured partially uncoiled to show its canal system more clearly. When the stapes moves against the oval window, the fluid in the vestibular and tympanic canals oscillates (wavy arrows); the oscillation is made possible by the flexible round window, which permits relief of pressure. High-frequency sounds stimulate hair cells near the base of the cochlea, and low-frequency sounds stimulate hair cells near the apex.

Structure of the human ear   The human ear is divided into three parts: the *outer ear,* the *middle ear,* and the *inner ear* (Fig. 19.26A). The outer ear consists of the ear flap or pinna, and the auditory canal. At the inner end of the auditory canal is the *tympanic membrane,* more commonly known as the eardrum.

On the other side of the tympanic membrane is the chamber of the middle ear. The air in this chamber is normally at the same pressure as that in the surrounding atmosphere. When the high-pressure part of a sound wave enters the ear, the pressure forces the eardrum to bulge slightly into the middle ear; when the low-pressure part of a wave enters, the higher-pressure air in the middle ear forces the eardrum to bulge into the outer ear. As a result, the eardrum vibrates, faithfully reproducing the pattern of pressure variation, and transforming it into mechanical movement.

The middle ear is connected to the pharynx via the *Eustachian tube,* which opens when we swallow or yawn, allowing the pressure in the middle ear to equalize with the pressure outside. Periodic equalization is necessary because the external air pressure can change noticeably as the barometer rises or falls, and as we change altitude. During the course of the all-too-common cold, clogging of the Eustachian tube with mucus may destroy our ability to accommodate pressure changes.

Three small interconnected bones, the malleus, the incus, and the stapes (commonly known as the hammer, the anvil, and the stirrup), extend across the chamber of the middle ear from the tympanic membrane to a membrane known as the *oval window.* These bones amplify the sound waves received by the tympanic membrane. Another membrane, the *round window,* lies just below the oval window.

On the inner side of the oval and round windows is the inner ear, a complicated labyrinth of interconnected fluid-filled chambers and canals. The upper group of chambers and canals is concerned with the sense of equilibrium. The lower portion of the inner ear, to which the upper chambers are connected, consists of a long tube coiled like a snail shell. This is the *cochlea* (the term is from the Latin for "snail shell"), which is the organ of hearing. Inside the cochlea are three canals (Fig. 19.26B–D): the vestibular canal, which begins at the oval window; the tympanic canal, which begins at the round window and connects with the vestibular canal; and the cochlear canal, which lies between the other two. All three canals are filled with fluid.

The sensory portion of the cochlea, called the *organ of Corti* (Fig. 19.26C), projects into the cochlear canal from the *basilar membrane,* which forms the lower boundary of the cochlear canal. The organ of Corti consists of a layer of epithelium on which lie rows of specialized receptor cells bearing sensory hairs at their apices. Dendrites of sensory neurons terminate on the surfaces of the hair cells. Overhanging the hair cells is a gelatinous structure, the *tectorial membrane,* into which the hairs project. When vibrations of the basilar membrane cause the sensory hairs to move up and down against the less mobile tectorial membrane, deforming the hairs, a generator potential is produced in the hair cells, which in turn stimulate the sensory neurons.

19.27   The mechanism of frequency detection in the human cochlea

The cochlea, here shown uncoiled, includes a basilar membrane that becomes progressively wider and thicker as it extends away from the oval window. Sound waves of different frequencies give rise to pressure waves of corresponding frequencies in the cochlear fluid. These in turn cause the basilar membrane to vibrate, at different points for different frequencies, thus stimulating surrounding sensory hairs to activate the sensory neurons that conduct impulses to the auditory centers of the brain. Processing of this auditory information in the brain results in discrimination between different frequencies, or pitches.

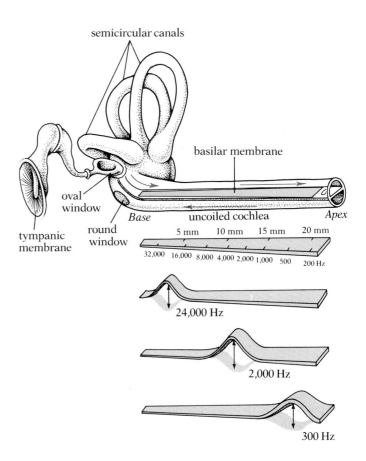

19.28   Otoliths in the inner ear

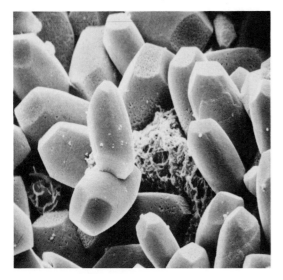

The crystals of calcium carbonate, called otoliths, shown in this scanning electron micrograph are surrounded by a gelatinous film in which hairs are embedded. When the head is tilted, the altered pull of the otoliths on the hairs generates signals to the brain.

How the ear works   We can now trace the steps in the reception of sound waves by the ear. Vibrations in the air pass down the auditory canal of the outer ear and cause the tympanic membrane to vibrate. These vibrations are transmitted across the cavity of the middle ear to the oval window by the chain of middle-ear bones. The leverlike arrangement of these bones, and the relatively small surface area of the oval window (about 1/30 that of the tympanic membrane), increase the force of the transmitted vibrations. The result is a great enhancement of the system's capacity to detect faint sounds.

The movements of the oval window result in movements of the fluid in the canals of the cochlea: an inward push by the middle-ear bones sends some of the fluid in the vestibular canal into the tympanic canal, causing the round window to bulge outward. The process reverses each time the middle-ear bones draw back from the oval window. These movements of the cochlear fluids are at exactly the same frequency as the vibrations entering the outer ear. The pressure waves in the fluid of the cochlea cause the basilar membrane to move up and down, rubbing the hairs of the hair cells against the tectorial membrane. Thus stimulated, the hair cells activate the sensory neurons that carry impulses to the auditory centers of the brain.

We are able to distinguish pitch (the frequency of sound waves) because the basilar membrane becomes regularly wider and thicker as its distance from the oval window increases. Different parts of the membrane are stimulated to maximum movement by sound waves of different frequencies (Fig. 19.27). Just as thicker violin and guitar strings resonate to lower-frequency vibrations, so too the thicker parts of the basilar membrane vibrate when low-frequency vibrations stimulate them, giving rise to the sensation of low pitch. Similarly, the thinner parts of the membrane resonate in the presence of high-frequency vibrations, giving rise to the sensation of high pitch. At very low frequencies—a few hundred hertz (cycles per second) and below—the firing of the hair cells keeps pace with the vibrations directly.

THE SENSES OF EQUILIBRIUM AND ACCELERATION

The upper portion of the labyrinth of the inner ear is composed of three **semicircular canals** and a large vestibule that connects them to the cochlea (Fig. 19.26). Inside the vestibule are two chambers, the **utriculus** and the **sacculus,** oriented perpendicularly to one another. Each contains sensory hairs embedded in a gelatinous layer that surrounds crystals of calcium carbonate called **otoliths** (Fig. 19.28). Any change in the position of the head causes the otoliths to exert more pull on some hairs than on others. The relative strength of the pulls tells the cerebellum the position of the head relative to gravity.

Similar sensory devices are found in invertebrates (Fig. 19.29.) For example, crayfish and lobsters have organs of equilibrium called **statoliths,** which consist of sand grains resting on beds of sensory hair cells. When the animal molts it loses the lining of the statolith and with it the sand grains, but the animal normally shovels in new sand grains when it has finished molting. If a crayfish is kept in a tank containing iron filings instead of sand, it will replace the sand with filings after molting. If we then hold a strong magnet near the crayfish, its iron otoliths will be pulled toward the magnet and the animal will orient itself as if the magnet were the pull of gravity. By properly positioning the magnet, researchers have even made such disoriented crayfish swim upside down.

In vertebrates, the design of the semicircular canals makes it possible to sense acceleration (change in the speed or direction of motion). Like the axes of a three-dimensional grid, each of the three canals is oriented at right angles to the other two (Fig. 19.27). At the base of each canal is a small chamber containing a tuft of sensory hair cells. When the head is moved or rotated in any direction, the fluid in the canals lags behind because of its viscosity. This results in increased pressure on the hair cells, which then send signals to the cerebellum. The brain, by integrating the different amounts of stimulation it receives from each of the three canals, can then determine very precisely the direction and rate of acceleration. The canals are not, however, perfect sensory organs. If the head is rotated for an extended period and then stopped abruptly, the fluid will continue to circulate. The result is the kind of dizziness we feel after being spun in a chair.

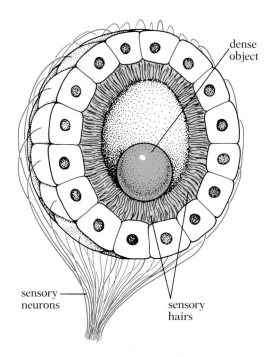

19.29 Diagram of the interior of a statolith
Statoliths are the equilibrium organs of invertebrates; each consists of a chamber lined with sensory hairs enclosing one or more dense objects, such as sand grains. Since gravity pulls dense objects downward, the location of the sensory hairs that are stimulated by the weight of the object indicates the direction of gravitational pull.

19.30 The infrared "eye" of a pit viper
Pit vipers like the western rattlesnake, shown here, are known for their ability to see in daylight, and to detect warm-blooded prey in the dark by means of a pit organ. The organ (arrow) is sensitive to the infrared radiation of warm-blooded animals; it actually forms crude images.

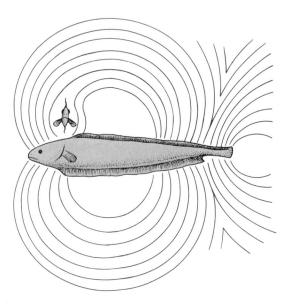

19.31 Electrolocation by a weakly electric fish
Nearby objects distort the field. Objects that are good conductors concentrate the lines of electric force on a small area of the fish's body, whereas bad conductors (here exemplified by a potential prey fish) spread the lines.

SENSORY SPECIALIZATIONS

We have already noted that bees, fish, and humans see different sets of colors; their visual ranges are well suited to their various needs. The same pattern is evident in hearing: bats, for instance, broadcast ultrasonic sound waves and use the echoes to locate obstacles and prey. These predators can hear sounds as high as 75,000 Hz. Many insects like moths and locusts that are hunted by bats also hear ultrasonic sounds, and begin evasive maneuvers when they hear the bats' sonar pulses. Other animals, such as rats, use the ultrasonic range for communication. Pigeons, on the other hand, seem to be sensitive to sounds of extremely low frequencies.

In addition to vision and hearing, many animals have senses that allow them access to information from other sources, among them such varied aspects of the environment as heat, electricity, and magnetism. We shall examine a few of these senses briefly. There is no reason to suppose that all or even most specialized sensory abilities have yet been discovered.

Infrared vision   We sense strong radiation in the infrared (IR) frequency range as heat. Some animals have specialized sensory structures that allow them to locate very faint IR sources. The most advanced of these structures are the infrared "eyes" of rattlesnakes and other pit vipers (Fig. 19.30). Intermediate between eye cups and pinhole eyes, these pit organs form crude images on a highly heat-sensitive retina, enabling the snakes to hunt warm-blooded prey in the dark, while conventional light-sensitive eyes enable them to see in daylight. The pit organs have evidently evolved from the heat sensors of the skin, rather than from any light-sensitive structure.

The electric sense   Three classes of fish use electric fields. Members of the first class, the strongly electric fish, do not actually sense electric fields; instead they use highly modified muscle cells like a series of batteries to produce a strong electric charge. Electric eels and some rays stun or kill both prey and potential predators with this charge.

Members of the second class, the weakly electric fish, are either nocturnal or live in murky water, where vision is almost useless. They compensate for the lack of visual information by producing an electric field around themselves, which they monitor for signs of objects that disturb it (Fig. 19.31). They can also perceive the electrical signals broadcast by other members of their own species, an ability that is important in intraspecific communication and mating. The sensory organs of a weakly electric fish are at the base of long, low-resistance, jelly-filled canals that radiate through the body from the head and monitor the electrical field at points all over the body. The receptors are simply modified nerve cells that are extremely sensitive to changes in the flow of current—in this case, the weak current generated by the fish itself and then flowing through the water and down the canals to the receptors.

Members of the third class, the passive electric fish, simply use their electrosensory apparatus to monitor their surroundings. These animals, which include the sharks, can sense the minute electrical fields generated by the neuromuscular activity of their prey. Alerted by the odor of prey, a shark can use its electric sense to detect a buried flounder, for example, from a distance of a meter or more.

The magnetic sense   Various organisms are sensitive to the magnetic field that envelops the earth. This field, which is generated by the flow of molten material in the earth's core and, to a lesser extent, by the flow of ions in the atmosphere, varies more or less regularly from place to place, and irregularly when solar storms shower bursts of charged particles on the earth.

Magnetotactic bacteria (Fig. 19.32) build within themselves chains of magnetite (lodestone) that rotate them into alignment with the lines of the earth's magnetic field. These lines point north and down in the Northern Hemisphere, and so guide the many species of mud-dwelling, magnetite-bearing bacteria down to the bottom of the stagnant ponds and marshes they inhabit. In the Southern Hemisphere, where the lines of the magnetic field point south and down, the polarity of the chain is reversed. At least one species of alga also uses magnetite. This kind of magnetic orientation probably evolved from more conventional weighting: most mud-dwelling bacteria have dense crystals at their front ends. The weight forces the front end down, and so aims a swimming bacterium toward the muddy bottom. Magnetite, the densest substance known to be synthesized by living organisms, would have made an excellent weight, and the subsequent development of a chain of these crystals would have provided an even more effective way to orient the bacteria.

Behavioral studies have shown that many other organisms sense the earth's magnetic field. Sharks and most rays use their electrical sensitivity to detect the direction of this field. Among the other animals able to detect it are honey bees, homing pigeons, various migratory birds, tuna, and salmon. The basis of this ability in such animals is not yet known, but localized deposits of magnetite have been found in association with the nervous system in each of them. From what we know of the detection and processing of sensory information, we can see that the physical twisting of a magnetic crystal as it seeks to align itself with the earth's field, or the bending of a hair bearing a chain of magnetic crystals, could conceivably alter membrane permeability and so provide a useful signal. Despite reports of magnetic sensitivity in our species from R. R. Baker of the University of Manchester, many careful tests by Kraig Adler and W. T. Keeton at Cornell, Kenneth Able of the State University of New York, Albany, J. L. Gould of Princeton, and J. L. Kirschvink of the California Institute of Technology have failed to substantiate this view. This may not be too surprising: we have seen repeatedly that sensory abilities tend to mirror an animal's needs. Since our species is unlikely to have evolved as a regular long-distance migrant, there may never have been a selective advantage for the evolution of a compass sense.

SENSORY PROCESSING AND THE BRAIN

When a sensory receptor is activated, the information it receives must be carried to the part of the central nervous system that is specialized to deal with it, and the information must then be processed. The network of recep-

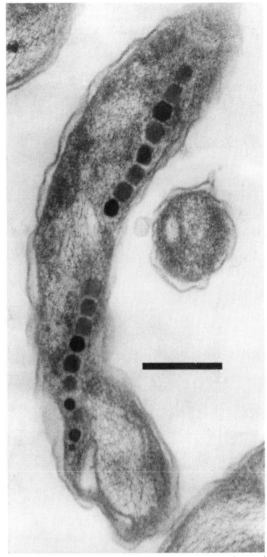

0.25 µm

19.32 A magnetotactic bacterium
The chain of dark magnetite crystals within this mud-dwelling organism causes it to rotate into alignment with the lines of the earth's magnetic field. When it divides, each daughter cell gets one chain.

19.33 A miswired frog

In an experiment conducted by Marcus Jacobson of the University of Utah and his colleagues, a patch of skin from the back of a tadpole was exchanged with a patch from its abdomen. The two patches continued to develop their original color patterns, and sent their sensory axons to their original targets in the CNS. As a result, when the transplanted patch on the back of the adult was tickled, the frog automatically scratched the place on its abdomen where the patch had originally begun to develop.

A

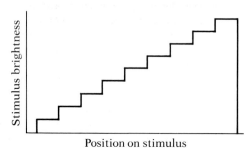

B

C

19.34 Mach bands

Lateral inhibition gives rise to a phenomenon known as Mach bands. For this series of grays (A), the actual brightness of each strip in relation to the others is shown (B), along with the perceived brightness (C). The optical illusion is evident as an exaggerated contrast between light and dark at the boundary between each two strips.

tors, sensory neurons, interneurons, and brain cells is intricate and precisely ordered. Understanding this network and how it forms during the development of an organism is an active and interesting area of study.

The means by which the first axons destined to carry specific sorts of information "find" their proper targets during development are not yet well understood. During development, axons from sensory neurons seem to "know" where they ought to terminate. In fact, experiments indicate that receptors transplanted from one location to another during development often send their axons to the part of the brain that would have been their destination if no transplantation had occurred. Moreover, the brain interprets input from these receptors as coming from the original location. For example, if a patch of skin with touch receptors from the back of a sufficiently developed tadpole is exchanged with a patch from its abdomen, the adult frog will scratch its abdomen when the skin on its back is irritated (Fig. 19.33). This experiment emphasizes the dependence of sensation on the structure of the central nervous system: the interpretation of incoming action potentials is based on the destination in the brain of the axons carrying them, not on the actual external stimulus or on some special quality of the impulses themselves.

SENSORY PROCESSING

Though the question of how the hundreds of millions of neural wires come to be correctly connected to the brain remains to be answered, an understanding of how information is processed once it reaches the CNS is growing steadily. A significant development came from studies of a primitive invertebrate—the horseshoe crab, *Limulus*. The principles discovered there have provided the key to understanding information processing in virtually every other organism investigated.

Lateral inhibition While studying how the compound eye of *Limulus* receives and transmits a picture of its world to the animal's brain, H. K. Hart-

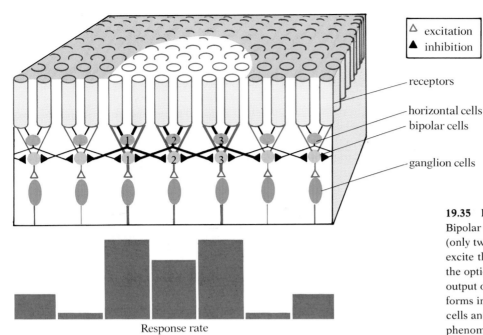

excitation △
inhibition ▲

receptors

horizontal cells

bipolar cells

ganglion cells

Response rate

19.35 Lateral inhibition in the vertebrate retina
Bipolar cells add the output of several receptors (only two are shown for each bipolar cell here) and excite the ganglion cells, which join together to form the optic nerve. Each horizontal cell also adds the output of several receptors and sends out an axon that forms inhibitory synapses with neighboring bipolar cells and prevents them from firing maximally, in the phenomenon called lateral inhibition. For simplicity, amacrine cells have been omitted from this highly schematic representation of a section of vertebrate retina.

The operation of lateral inhibition can be seen in this diagram in the processing of a spot of light falling on six receptors. The activity of the various nerve fibers is indicated in the diagram by their relative width, with the greatest width showing the highest level of activity. Stimulated by input from receptors, bipolar cell 1 fires a ganglion cell, but horizontal cell 1 inhibits the bipolar cell to its left, causing its firing to drop below the basal rate, and also inhibits the cell to its right, bipolar cell 2. Being inhibited by only one adjacent cell, bipolar cell 1 fires more strongly than bipolar cell 2, which is also stimulated by input from receptors, but is inhibited by horizontal cells 1 and 3. Bipolar cell 3, like bipolar cell 1, is inhibited by only one adjacent horizontal cell, so it fires more strongly than bipolar cell 2. This pattern of lateral inhibition, which is reflected in the graph of the response rate of ganglion cells at the bottom of the diagram, explains why the operation of the vertebrate retina accentuates the difference between light and dark along an edge. The cells just outside the illuminated area, receiving inhibition but no excitation, fire below their basal rate, thereby reporting their part of the visual field to be relatively darker than it is. Cells just inside the illuminated area, being inhibited by fewer adjacent cells than those toward the center, report their part of the visual field to be lighter than it is toward the center. This strategy of lateral inhibition has been found to work in decoding a variety of features of the visual world in both invertebrates and vertebrates.

line and his colleagues at the Rockefeller Institute discovered that the activity of receptors in one ommatidium has a dramatic effect on the behavior of receptors in other ommatidia. In the dark the receptors fire at a slow but regular basal rate. When a single ommatidium is illuminated, its receptors, not surprisingly, begin to fire at a more rapid rate; at the same time, however, the receptors in the adjacent ommatidia fall silent, inhibited from firing by their neighbors' excitation. This phenomenon is known as *lateral inhibition*. The net result is that in the picture transmitted the contrast between light and dark is enhanced at an edge: all the cells next to the stimulated receptors have ceased to fire even at their low basal rate.

Hartline and his associates realized immediately that this strategy of lateral inhibition, or its equivalent, is probably at work in many animals, including humans. Subsequent investigations have shown their guess to be correct; this kind of neural comparison-contrast strategy has been found in the visual wiring of every creature studied (though in vertebrates the receptor cells do not interact directly). It explains the familiar but previously puzzling optical illusion known as Mach bands: the contrast between two areas of differing shades of gray is enhanced at the border, with the edge of the lighter area appearing paler than the rest of that area, while the adjoining edge of the darker area appears even darker than the rest (Fig. 19.34). The effect of lateral inhibition is not solely that of generating optical illusions; it serves to pick out and enhance subtle features in the environment that might otherwise be missed.

In the vertebrate retina, horizontal cells play an active role in lateral inhibition.[1] Horizontal cells synapse on bipolar cells, which in turn synapse on ganglion cells. Ganglion cells transmit visual information to the brain (Fig. 19.35). Since, in the simplest case, a ganglion cell is stimulated (via bi-

---

[1] In fact, lateral inhibition in vertebrates also involves amacrine cells. For simplicity, this discussion focuses only on the activity of horizontal cells.

polar cells) by light falling on a small group of receptors, and is inhibited by light falling on the surrounding receptors, the ganglion cell responds best to a small spot of light on the activating receptors, surrounded by darkness. The ganglion cell, with its associated horizontal and bipolar (and amacrine) cells, is often called a ***spot detector*** (Fig. 19.36).

A second class of spot detector is wired to respond to the opposite situation, a small dark spot on a bright background. As a result, the picture falling on the retina is abstracted and transmitted to the brain as a pattern of light and dark spots.

Other things are going on in the retina as well. Similar circuits are encoding color information by comparing the output of adjacent cone cells. This sort of comparison is necessary because the cones are very broadly tuned (Fig. 19.23); a small hyperpolarization induced in a "blue" cone, for instance, may represent a dim blue light or a bright green one. The ambiguity is resolved by comparing the response of the blue cone with that of an adjacent green cone: if the green cone is unstimulated, the light cannot be green. Taken together, the relative responses of the three kinds of cone cells uniquely specify the wavelength of incoming light, thus providing the basis of our sensation of color. Comparing the output of two broadly tuned receptors in this way is called opponent processing, and the mechanism of color identification just described is known as color-opponent processing. Opponent processing is not confined to the retina. Its comparison-contrast strategy is almost certainly used, for example, when the nervous system measures joint angles (Fig. 19.3) or pitch (Fig. 19.27). Indeed, it is the means used for virtually every kind of sensory discrimination in vertebrates and invertebrates alike.

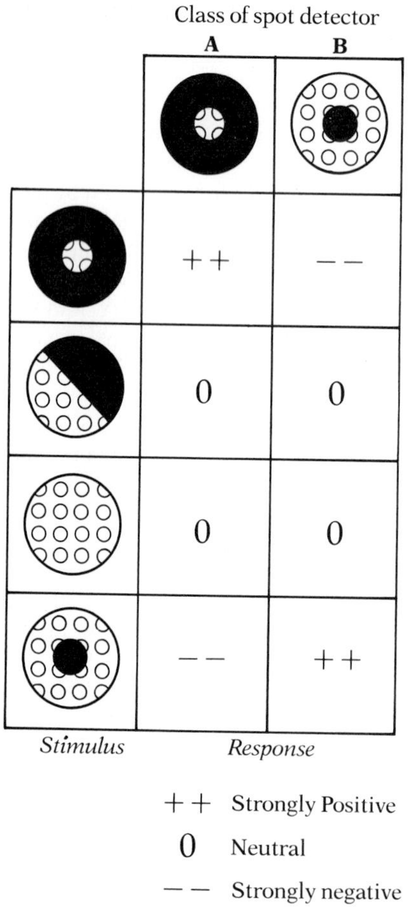

19.36 Two kinds of spot detectors
The ganglion cells ultimately responsive to light falling on receptors in the retina belong to several classes; two particular classes of spot detectors (each of which consists in this diagram of sixteen receptors, as well as bipolar, horizontal, and amacrine cells, and a single ganglion cell, not shown) respond to spots of light in the receptive field. The first class of spot detector (A) is most strongly excited when a spot of light falls in the center of its receptive field, and is inhibited when light falls on receptors to the outside. This can be seen in the spectrum of responses in the table, ranging from the strongly positive response (+ +) to a focused spot of light through the relatively neutral responses in between (0) to the strongly negative response (− −) to darkness focused on the receptor cells at the center. The second class (B) responds in exactly the opposite way.

EVOLUTION OF THE VERTEBRATE BRAIN

Information from vertebrate sensory organs like the retina frequently undergoes some processing before it reaches the brain. In addition, once in the brain, the information usually undergoes processing that yields increasingly abstract versions of the original sensory input. The brain, then, is the organ of massive sensory processing as well as of motor coordination and control. How is this impressive neural computer organized, and how has it evolved?

A comparison of the partly developed brains of vertebrate embryos, from primitive fish to humans, reveals the structural similarities among vertebrate brains: all begin as three irregular swellings at the anterior end of the longitudinal nerve cord. In more advanced vertebrates these three regions undergo much modification in the course of development, displaying discretely thickened areas in their walls and distinctive outgrowths in other places. Despite these changes, however, the original three divisions of the brain can still be recognized even in the most advanced vertebrates, humans included. The three divisions are the *forebrain,* the *midbrain,* and the *hindbrain* (Fig. 19.37).

The vertebrate brain contains a series of hollow compartments, known as ventricles, that connect to the central canal of the spinal cord. The canal and the ventricles contain cerebrospinal fluid, which is kept circulating by the beating of cilia on the epithelial cells that line the ventricles. Both the brain and the cord are wrapped in three protective membranes, the *meninges.* These are the pia, the innermost membrane, lying on the surface of the brain and spinal cord; the fragile arachnoid, just above the pia; and the tough dura, separating the other two from the inner surface of the skull and vertebrae. The spaces between the three meninges are filled with cerebrospinal fluid, which cushions the nervous tissue against damage.

Very early in its evolution, the vertebrate brain underwent modifications that set the stage for later evolutionary trends. Briefly, the modifications were these:

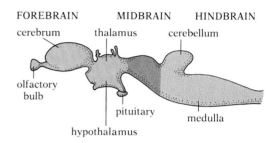

19.37 Diagram of the principal divisions of the vertebrate brain

1. The ventral portion of the hindbrain, the *medulla oblongata,* became specialized as a control center for some autonomic and somatic pathways concerned with visceral functions—breathing and heart rate, for instance—and as a connecting tract between the spinal cord and more anterior parts of the brain. At the same time, the anterior dorsal portion of the hindbrain became much enlarged as the *cerebellum,* a structure concerned with balance, equilibrium, and muscular coordination (see p. 483).
2. The dorsal part of the midbrain became specialized as the *optic lobes*—the visual centers associated with the optic nerves.
3. The forebrain became divided into an anterior portion consisting of the *cerebrum,* with its prominent olfactory bulbs, and a posterior portion consisting of the *thalamus* and *hypothalamus.*

Continual evolution has made few changes in the hindbrain, though the cerebellum has become larger and more complex in many animals, particularly those with large bodies requiring fine muscle control. The most obvi-

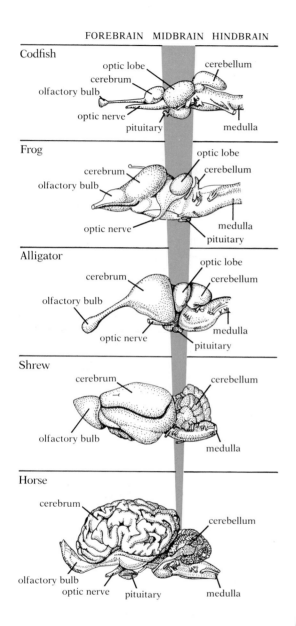

FOREBRAIN   MIDBRAIN   HINDBRAIN

Codfish
optic lobe
cerebrum
olfactory bulb
cerebellum
optic nerve
pituitary
medulla

Frog
optic lobe
cerebrum
cerebellum
olfactory bulb
optic nerve
medulla
pituitary

Alligator
optic lobe
cerebrum
cerebellum
olfactory bulb
optic nerve
medulla
pituitary

Shrew
cerebrum
cerebellum
olfactory bulb
medulla

Horse
cerebrum
cerebellum
olfactory bulb
optic nerve
pituitary
medulla

19.38 Evolutionary change in relative size of midbrain and forebrain in vertebrates
In this evolutionary sequence the relative size of the midbrain shows a marked decrease, and that of the forebrain a very considerable increase.

ous evolutionary change has been the steady increase in the size and importance of the cerebrum, with a corresponding decrease in the relative size and importance of the midbrain (Fig. 19.38).

The evolution of the vertebrate brain can be reconstructed through comparison of the brains of various species: structures that are shared probably represent a common evolutionary heritage. When we compare fish, amphibians, reptiles, and mammals, a fairly coherent picture emerges.

It appears, first of all, that the ancestral cerebrum was only a pair of small smooth swellings concerned chiefly with the sense of smell. As in the spinal cord, the gray matter (cell bodies and synapses) was mostly internal. The synapses functioned predominantly as relays between the olfactory bulbs and more posterior parts of the brain; little, if any, processing of sensory information occurred in the cerebrum. The cerebrums of many modern fish are still little more than relay stations, though the areas of gray matter are more massive. In amphibians, which evolved from ancestral fish, there was an expansion of the gray matter and a multiplication of synapses between neurons. No longer was the cerebrum only a relay station; it now functioned as a processing center for impulses coming to it from various sensory areas of the brain. Slowly, much of the gray matter moved outward from its initially internal position, until it came to lie on the surface of the cerebrum. This surface layer is known as the **cerebral cortex** ("cortex" means bark).

In certain advanced reptiles a new component of the cortex, called the **neocortex** (neopallium), arose at a point on the anterior surface of the cerebrum. Mammals, which evolved from reptiles of this type, show the greatest development of the neocortex. Even in primitive mammals the neocortex has expanded to form a surface layer covering most of the forebrain. This does not mean that the old cortex of the ancestral brain has been reduced; it has simply been pushed to an internal position by the immense increase in relative size of the neocortex.

As the neocortex, a major coordinating center for sensory and motor functions involving all senses and all parts of the body, continued to expand, both by relative increase in its total size and by folding, which increased its surface area, it became more and more dominant over the other parts of the brain. The midbrain had been the chief control center in the earliest vertebrates. Then the thalamus portion of the forebrain became a major coordinating center, first sharing this function with the midbrain, then becoming dominant. Finally, with the rise of the neocortex and its preempting of many control functions from both the midbrain and the thalamus, the midbrain was left as a small connecting link between the hindbrain and the forebrain; it remains a control center for many unconscious mechanisms and some of the simpler visual functions; it also continues to play a major role in control of emotions.

This increase in brain size and complexity from fish—the vertebrates with the simplest brains and smallest cerebrums—through amphibians and reptiles to mammals, though it does suggest the likely evolution of the vertebrate brain, should not be taken to imply that the brain of each type of organism has now ceased to evolve. On the contrary, the fish brain has continued to evolve since the rise of amphibians, and the amphibian central nervous system has likewise continued to evolve since reptiles diverged into their own evolutionary line. Though the most primitive vertebrate brains are by and large found in fish, the brains of some species of modern fish—the weakly electric fish in particular—are relatively large and complex. The size and complexity of the brains of present-day vertebrate species is determined by the complexity of each species' way of living, rather than by its phylogenetic status.

The evolution of the vertebrate brain in response to more complex behavioral challenges has taken several forms. The most obvious changes have been in the areas of the brain that handle specialized behavior. In weakly electric fish, for example, the area responsible for analyzing the input from the electric field the animal surrounds itself with—that is, the area used in identifying and locating prey and conspecifics—is greatly enlarged and more conspicuously compartmentalized. The visual area of the brain of weakly electric fish, by contrast, is not enlarged.

As specific areas of the vertebrate brain have increased in size, their internal organization has also become more complex. This evolutionary progression begins with brain areas that are unstructured: the neurons in these unmodified areas are scattered throughout. In brain areas of modest specialization, neurons are grouped into **nuclei** with cell bodies and dendrites at the center and axons on the periphery. These axons bring information in and carry it out, and are channeled from one nucleus to another in nerve tracts. In even more highly evolved areas these nuclei subdivide, and ultimately form regions, called **laminations,** in which loose layers of cell bodies and axons alternate. The most specialized brain nuclei have a tightly laminated structure with extremely regular patterns of connections both within and between layers. As we shall see when we examine the highly laminated visual cortex of mammals, this structural strategy makes possible complex synaptic organization and information processing.

In mammals, the increase in cortical area has resulted in a reduction in the relative amount of cortex devoted to strictly sensory and motor functions. The emphasis has shifted instead to the addition of so-called association areas—areas in which information from different sensory systems converges and memory formation and storage occur. These make more flexible and complex behavior possible. The brain of our species represents the most extreme manifestation of this trend (Fig. 19.39).

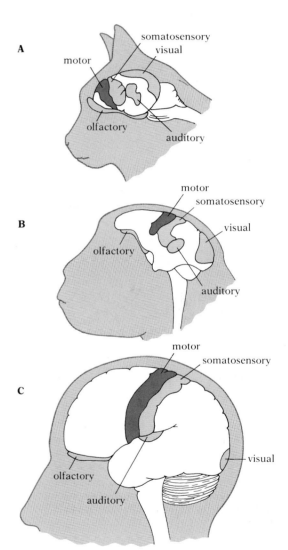

19.39 Proportion of cerebral cortex devoted to sensory and motor functions in three mammals
(A) In the cat sensory (color) and motor (dark gray) areas constitute a major portion of the cortex. (B) In the monkey the proportion of cortex devoted to association areas (white) is much greater than in the cat. (C) In humans the sensory and motor areas occupy a relatively small percentage of the cortex, most of the cortical area being devoted to association.

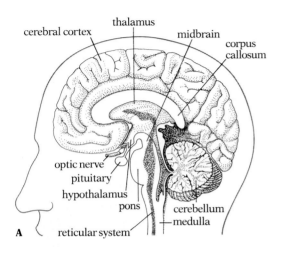

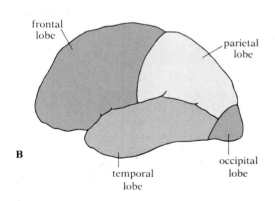

19.40 The human brain
(A) Sagittal section (longitudinal section through the
midline) showing major parts of the brain.
(B) Diagram of left side of the cerebral cortex
showing its four main lobes. These lobes are
duplicated in the right hemisphere of the brain.

THE MAMMALIAN FOREBRAIN

Much of our knowledge of vertebrate brain function is derived from research on rats, cats, monkeys, and chimpanzees. A considerable amount of information has accumulated about the human brain as well, most of it derived from electrical stimulation during brain surgery and from observations of the effects of tumors and accidental damage. The mammalian forebrain, as we have seen, is made up of the thalamus, the hypothalamus, and the cerebral cortex.

The thalamus   The thalamus (Fig. 19.40A) is the major sensory-integration center in lower vertebrates; in higher vertebrates it has become in large part a sensory relay station on the way to the cerebrum, where complex integration takes place. Nevertheless, the thalamus continues to play an integrative role even in humans, and parts of it are intimately involved in memory.

The thalamus also contains part of an extremely important neural formation known as the ***reticular system*** (Fig. 19.40A), a densely interconnected network of neurons that runs through the brainstem of the medulla and midbrain as well as the thalamus. Every sensory pathway running to higher centers of the brain sends side branches to the reticular system, as does every descending motor pathway. In this way, the reticular system is able to "listen in" on whatever is coming into or leaving the brain. It also sends a great many fibers of its own to areas of the cortex, brainstem, and spinal cord.

Microelectrode recordings from reticular cells reveal that many of the system's neurons are relatively unspecific: a single neuron may respond to stimulation of pain receptors in the foot, touch receptors on the hand, sound receptors in the ear, and light receptors in the retina. A major function of this curious area of the brain seems to be sensitization—the activation of other parts of the brain upon receipt of appropriate stimuli from sensory receptors, of whatever type or location. It acts as the brain's arousal system, awakening and focusing an animal's attention on some change in the world around it. This adaptive response ensures that an animal will be alerted to anything that may be a predator, a potential mate, or food. One reason falling asleep is easiest in a dark, quiet bedroom is that fewer signals from the sensory receptors reach the reticular system, so it does less to arouse the brain. Barbiturates, which were once commonly used in sleeping pills, block the reticular activating system and thus facilitate deep sleep. Destruction of the reticular system suspends the brain's ability to respond to sensory stimuli, and results in a permanent comatose state.

The reticular system does not, however, arouse animals indiscriminately. It selectively enhances or suppresses incoming sensory information on the way to the cortex to be processed. In a very real sense, the reticular system "decides" what an animal will be most aware of—when it should pay more attention to sounds, say, than to its touch receptors, and vice versa. Such filtering is essential, since hundreds of millions of sensory receptors continually flood the brains of most mammals with irrelevant information. The brain simply does not have the capacity to deal with even a tiny fraction of this material at any given time. Some of the filtering is actually done peripherally, with the reticular system sending out inhibitory impulses to block incoming signals before they ever reach the central nervous system. The system also appears able to modulate motor commands issued by the cortex, amplifying some and attenuating others.

The hypothalamus The hypothalamus is the part of the brainstem just ventral to the thalamus (Fig. 19.40A). As we saw in Chapter 16, its major functions include the synthesis of the hormones stored in the posterior pituitary and the secretion of the releasing hormones that help regulate the anterior pituitary. The hypothalamus is thus a crucial link between the neural and endocrine systems.

The hypothalamus is also the most important control center for the visceral and emotional responses of the body. By stimulating various parts of the hypothalamus with microelectrodes, researchers have been able to identify centers in the hypothalamus that control hunger, thirst, body temperature, water balance, and blood pressure, as well as sexual desire, pleasure, pain, hostility, and so on. It is possible to induce behavior appropriate to each of these states by inserting microelectrodes surgically into the control centers of experimental animals and then stimulating the centers electrically. Rats with electrodes in their pleasure centers will spend virtually all their time pressing levers that turn on a tiny current in those centers, ignoring food and water almost to the point of starvation. Animals with electrodes in appropriate parts of the hypothalamus can be made to feel sated one moment and hungry the next, cold and then hot, angry and then calm. Cats wired in this way may be friendly one moment and in a rage the next, with their fur erect, eyes wide, and claws out. They can be made to break off an attack on a mouse and cower in a corner in fear of the very creature they were attacking only a second earlier. The centers controlling such conflicting responses may be separated by only a fraction of a millimeter in the hypothalamus.

Research of this sort has made it clear that the hypothalamus is the major integrating center for both visceral and emotional responses. It is not the only region controlling emotional responses, however, but rather one of a functionally related set of structures known as the ***limbic system.*** These areas, which ring the anterior end of the brainstem (Fig. 19.41), develop mostly from the forebrain, though some originate in the midbrain.

The cerebral cortex In higher vertebrates, the cortex performs most of the complex processing by which the sensory information relayed by the hypothalamus is analyzed and used. Because the cortical cells are situated at the surface of the brain, electrical stimulation and recording are relatively convenient. As a result, we know a good deal about where and how sensory information is processed in the cortex.

Neurobiologists quickly realized that, unlike the reticular system, where the neurons seem to be relatively unspecialized, the cortex contains discrete areas activated by different sensory systems (Fig. 19.39). Moreover, these areas display a logical internal organization. In the somatosensory area, for instance, each part of the body has its own representation (Fig.

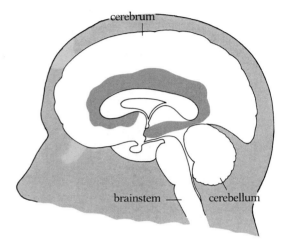

19.41 The limbic system of the human brain
The limbic system (color) is not an anatomically distinct structure, but rather a group of brain areas that are related functionally in giving rise to feelings and emotions.

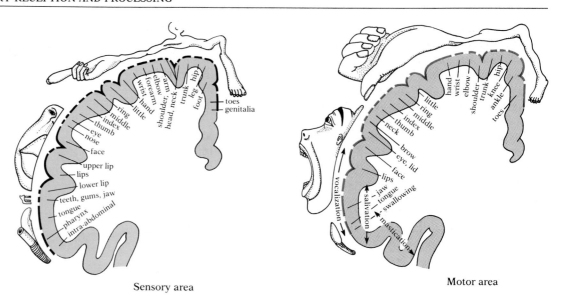

Sensory area

Motor area

19.42   Functional map of the sensory and motor cortex

The somatosensory strip of the human cerebral cortex (left) receives sensory input from all parts of the body. On this strip adjacent parts of the body are usually represented next to each other. Parts of the body with large numbers of sensory receptors, such as the lips, tongue, and fingers, are represented by correspondingly large areas on the cortex. A strip from the adjoining motor-control area of the cortex (right) shows similar organization: mapping proceeds from toe to head; adjacent parts of the body are usually represented next to each other; and those with many small muscles for fine motor control occupy especially large amounts of space.

19.42). To be sure, it is a distorted map—the 4 square cm of lip surface are represented by a greater area of cortex than the 6,000 square cm of neck, trunk, hips, and upper legs (reflecting, of course, the relative density of receptors in these regions)—but the elements are arranged sequentially in an anatomically logical order, with contiguous regions on the body usually mapped on contiguous areas of the cortex. At an even higher level of detail within this strip of cortex, these positional relationships persist: an array of bristles on the face of a mouse is matched by an arrangement of cell groups on the cortex. The motor area, which lies alongside the somatosensory area, is mapped almost precisely in parallel.

Logical internal organization is seen also in the visual system. The optic nerve leaving one retina meets its counterpart from the other eye at the *optic chiasm.* In animals with binocular vision (in which the fields of view of the two eyes overlap), the axons of each optic nerve separate into two parts there. Those from the cells in the left half of the retina of each eye (which look out at the same right half of the visual world) join to travel together to the left side of the thalamus. Those from cells in the right half of each retina go to the right side of the thalamus (Fig. 19.43). Hence the neurons from the left halves of both retinas synapse in the left lateral geniculate nucleus (LGN) of the thalamus, while those from the right halves of both retinas synapse in the right LGN, and the fibers from the two eyes are thus brought together. Fibers then leave the LGN and synapse in the left or right *primary visual cortex* consecutively: that is, fibers from the uppermost, farthest left patch of the left-eye retina are mapped on the left visual cortex next to those from the uppermost, farthest left patch of the right-eye retina, and so on. Because of the density of receptors in the fovea of the retina, the area devoted to the center of the visual field is greatly enlarged in the visual cortex. In birds and many lower vertebrates visual information is mapped on an analogous but independently evolved area known as the optic tectum.

Sensory processing in the cerebral cortex   The work of David Hubel and Torsten Wiesel at Harvard has revealed a great deal about what happens to visual information in the cortex. As we have seen, lateral inhibition in the retina encodes the visual world into a series of spots of two classes: bright-center/dark-surround and dark-center/bright-surround. In lower vertebrates the retina, thalamus, and tectum add three additional classes of feature-detector circuits. (These classes are also found in other vertebrates and in invertebrates as well.) Feature detectors of the first class are spot-motion detectors. Each of these responds only to spots of a particular size moving in a specific direction at a certain rate (Fig. 19.44B). Different spot-motion detectors have different preferences, and virtually all contingencies appear to be represented in every animal. In frogs and toads such detectors are wired into circuits that aim and trigger the bug-catching behavioral responses. Spot-motion detector circuits work by comparing the output of a series of simple retinal spot-detector circuits (like those

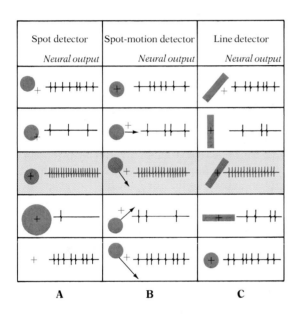

	Spot detector	Spot-motion detector	Line detector
	Neural output	*Neural output*	*Neural output*

A   **B**   **C**

19.44 Feature detectors in the visual system
The response characteristics of three simple sorts of feature detectors are represented. In each case, the center of the receptive field is indicated by +. The spot detector shown here (A) is most responsive to a spot of light that is small and centered in the receptive field. Spot-motion detectors respond to specific patterns of movement. The

one shown here (B) is stimulated by a small spot moving 135° to the right of vertical. Moreover, as the middle and bottom examples indicate, only a specific rate of movement stimulates a particular spot-motion detector maximally. Line detectors respond to lines of specific width and direction. The one shown here (C) requires a narrow bar oriented 45° to the right of vertical for its stimulus.

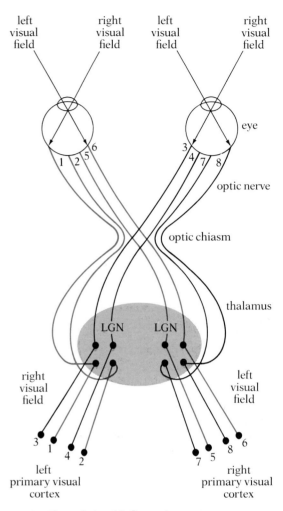

19.43 Flow of visual information
Axons from the ganglion cells in the retina run in the optic nerve to the optic chiasm. There, in creatures with binocular vision, the fibers from receptors looking out on the left half of the visual field in the left eye join those representing the left half in the right eye and travel to the right lateral geniculate nucleus (LGN) of the thalamus. Similarly, information from the right visual field of each eye projects to the left LGN. From there each nucleus sends axons to the primary visual cortex, where the two images of its half of the world, one from each eye, are integrated. Though the inputs from the left and right visual fields are not initially integrated anatomically in the brain, we see no division between them in our conscious experience.

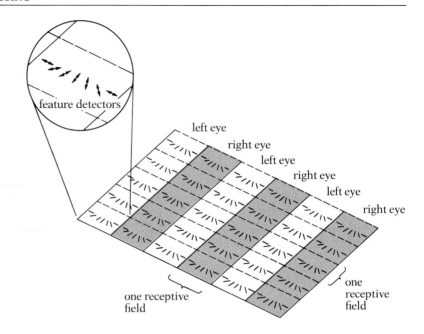

19.45   Diagram of the organization of a layer of visual cortex
Visual information arriving in the LGN is transmitted to specific cells that are arranged, several abreast, in narrow stripes. Each segment within a stripe contains cells responding to the receptive field of one group of receptors in one eye. Adjacent segments in the same stripe and in the next parallel stripe for the same eye code for contiguous receptors in the retina. Within these stripes are ordered sequences of line detectors, each maximally responsive to one angle.

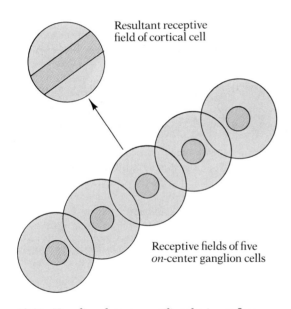

19.46   How line detectors utilize the input from spot detectors
Line detectors compare the signals coming from a series of spot detectors. Depending on which and how many spot detectors converge on a cell, a feature detector of this type can code for a line of any width, length, and orientation. Similarly, feature detectors have been found that respond to "hypercomplex" conformations like points or curved edges or even certain shapes.

shown in Figure 19.36) over time: the motion-detecting feature detector is inhibited for all but one pattern of signals.

Feature detectors of another class respond only to lines of a particular orientation (Fig. 19.44C). Detectors for all possible orientations of lines are found in every vertebrate brain, but vertical and horizontal detectors seem to be especially numerous. In the cortex of higher vertebrates, the angle to which these line detectors respond best is continuous across each little bit of the cortex, and each unit of the grid thus established corresponds to the receptive field of one spot detector in one eye (Fig. 19.45). A line detector works by comparing signals coming (via the LGN) from arrays of spot detectors in the retina. It is active only if the feature it recognizes is present (Fig. 19.46).

Feature detectors of still another class sort for rates and directions of motion in these lines. Each such detector has a well-defined specificity, and fires only if it detects motion at a particular speed *and* in a particular direction. Other feature detectors combine the input from the two eyes and, by comparing what the same parts of the two retinas are seeing, are able to encode distance. Vertebrates also have many other feature detectors, whose functions are yet to be fathomed. Some of these may correspond to the so-called hypercomplex feature detectors for corners, curvature, specific rates of approach, and the like, found in higher invertebrates. In any case, incoming information is transformed in the retina into spots, and the many other layers in this highly laminated area of cortex process this input.

The strategy of the vertebrate visual system, then, is not to transmit a faithful picture of the world passively to some sort of neural TV screen at the back of the head, but rather to sort through the data, first breaking it down into spots, and then rebuilding it in terms of a series of increasingly complex, abstract, but discrete features. The auditory system seems to be structured in much the same way. The logic behind this highly organized processing is not yet clear, though some indications of its possible function in directing and triggering behavior in various species will be discussed in a later chapter.

In primates, at least, the much-edited picture of the world in the left or right primary visual cortex is then passed to several nearby places in the cortex where further transformations are made (Fig. 19.47). Some of these areas may be involved in visual memory and visual association. One in particular—the posterior parietal cortex—seems to decide what in the visual field ought to be subject to the high-resolution scrutiny of the fovea, and then to issue instructions to aim the eyes appropriately. Others appear to bring the visual world at least roughly into register with other senses—touch and hearing most commonly—so that stimuli from a particular direction from any sensory modality map to the same cells on a single array.

Complex but regular organization like that of the visual system appears to be the rule rather than the exception in the brain. Research is showing that from insects to primates, brains are specialized into nuclei consisting of relatively precise three-dimensional matrices of neurons. These nuclei receive input from particular sensory receptors and from other nuclei, and send information out to other nuclei, as well as to the appropriate motor areas.

But as the computerlike organization of the brain has become apparent, and with it the almost unlimited possibilities for memory and sensory association, the question of how our conscious experience arises has become more difficult. We have no awareness of the spot and line detectors that produce our view of the world, for instance; to our conscious minds the visual world is a well-knit and seamless whole, rather than the miscellaneous collection of discrete bits of information that the retina receives. Nor is it at all obvious how thought as such can emerge from even a vast well-ordered array of neurons. Though we are finally in a position to understand how some of the specialized nuclei work, we have not yet discovered how they interact with each other to produce creative thought, aesthetic judgment, symbolic language, and the other traits usually thought to be uniquely human. Some of these questions will become less mysterious after we look at the mechanisms and evolution of animal behavior, but no one knows yet whether the human mind will prove capable of understanding itself.

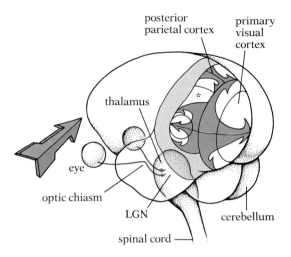

19.47 Projection of visual images in the cortex
The right half of the visual field is shown here as it is transmitted from the retinas, via the LGN, to the cells in the left primary visual cortex of a squirrel monkey. (Part of this cortex is hidden between the two hemispheres in this view.) The visual world we (and presumably the monkey) perceive is a spatially faithful picture except that the part seen by the fovea (the middle of the arrow) is magnified to allow high-resolution processing of the image in the center of the visual field. The picture is transmitted from the primary visual cortex to several other areas. In many of these the top and bottom halves of the visual field are separated. The organization of the visual area marked with an asterisk has not yet been traced.

Chapter 20

EFFECTORS

Effectors are the parts of the organism that do things, carrying out the organism's response to stimuli. Their actions are as various as secretion by glands, production of light by fireflies, phototropic and geotropic responses in plants, cytoplasmic streaming in both plant and animal cells, and, most familiarly, muscle movements of animals.

As we saw in a previous chapter, plants are capable of slow movements in response to light and gravity, movements produced by differential growth rates controlled by hormones. Many plants are also capable of some types of rapid movement. For example, some leaves droop or fold at night and expand again in the morning. The flowers of many plants open and close in a regular fashion at different times of day. The leaves of sensitive plants *(Mimosa)* fold and droop within a few seconds after being touched (Fig. 20.1). The leaves of the Venus fly trap close rapidly around insects that have landed on them (see Fig. 9.13, p. 236). The seed pods of some plants snap open at maturity, vigorously expelling their seeds.

All these movements are far too rapid to depend on differential-growth changes. Another mechanism, turgor-pressure change, is involved. Leaves droop when certain of their cells lose so much water that they are no longer turgid enough to give rigidity to the leaf. Flowers fold when specially sensitive cells arranged in rows along the petals lose their turgidity, and they open again when these cells regain their turgidity. Rapid changes in turgidity in special effector cells located along the hinge of the leaf of the Venus fly trap are responsible for that plant's curious behavior. Similarly, rapid turgidity changes in specialized effector cells at the bases of the leaflets and petioles of the sensitive plant are responsible for that plant's response to a touch or other mechanical stimulus.

It is clear, then, that active movement is not exclusively a characteristic of animals. Nonetheless, it remains true that the most elaborate mechanism for producing locomotion are found in the animal kingdom, and it is on these that we shall concentrate in this chapter (disregarding the many effector actions, such as glandular secretion, that occur without gross movement of the animal).

Effectors are often controlled by the nervous system; unlike sensory receptors and conductor cells, however, effector cells are not themselves part of the nervous system, even though they are the last components of reflex arcs. Among the numerous effector systems not under nervous control are the nematocysts of coelenterates, cilia and flagella, and some smooth muscles of vertebrates.

As was suggested in Chapter 5, the underlying mechanism of almost all the various kinds of effector action—from cytoplasmic streaming to muscular movement—depends on either microfilaments or microtubules.

THE STRUCTURAL ARRANGEMENTS OF MUSCLES AND SKELETONS

The most prominent effectors in all multicellular animals except the sponges are the muscles—tissues composed of specialized contractile cells. All cells possess some ability to contract. But as multicellular animals became larger and more complex, and as division of labor among their cells and tissues increased, they evolved elongate cells specialized for contraction, which became the principal effectors of movement in higher animals.

ANIMALS WITH HYDROSTATIC SKELETONS

The first multicellular animals (disregarding the sponges) must have been small, perhaps on the order of one millimeter in length. They probably swam by means of cilia. Even today the smallest flatworms and the tiny larvae of many coelenterates depend primarily on cilia as their locomotory effectors. But cilia are practical only in very small organisms. As animals evolved larger size, they evolved contractile tissues that first supplemented and then supplanted the cilia as the chief effectors of locomotion. Though coelenterates have only very primitive contractile fibers, which do not constitute distinct tissues, rhythmic contractions of such fibers in the bell of a jellyfish enable it to swim weakly (see Fig. 18.11, p. 460), and contractions of other fibers enable a jellyfish or a hydra to move its tentacles. The hydra is even able to move by turning somersaults (Fig. 20.2), a rather surprising type of movement in an animal with such primitive nerve and muscle cells.

20.1   An example of rapid movement in plants
Within a few seconds of being touched, the sensitive plant (*Mimosa pudica*) can respond by folding its leaves. Rapid decreases in turgor pressure affect the rigidity of the leaves, causing them to droop or close.

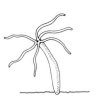

20.2   The somersaulting locomotion of hydra

A

0.5 mm

Some animals, such as flatworms and snails, glide forward slowly as waves of contraction pass along their longitudinal muscles. With each wave, points on the lower surface of their bodies may advance a fraction of a millimeter; these points may then grip the substrate and, as the wave passes backward, act as anchors toward which more posterior parts of the lower surface are drawn. In this type of locomotion, little use is made of the circular muscles of the body or of the hydrostatic properties of the body contents.

In many animals the muscle fibers of the body wall are arranged in prominent longitudinal and circular layers. The fibers in these two layers are antagonistic to each other—that is, they produce opposite actions. Contraction of the longitudinal muscles shortens the animal; contraction of the circular muscles lengthens it. Because the semifluid body contents resist compression, and thus function as a **hydrostatic skeleton,** the body volume remains constant and the shortening is accompanied by a compensating increase in diameter.

The most complete exploitation of the potentialities of hydrostatic skeletons is seen in certain annelid worms, such as earthworms. Here the body cavity is partitioned into a series of separate fluid-filled chambers (see Fig. 14.13, p. 366). Correlated with this **segmentation** of the body cavity is a similar segmentation of the musculature; each segment of the body has its own circular and longitudinal muscles. It is thus possible for the animal to elongate one part of the body while simultaneously shortening another part. The result is peristalsis, a series of alternating waves of contraction, activating first longitudinal and then circular muscles. During the longitudinal contraction of a segment, hard bristles known as setae are protruded to provide traction (Fig. 20.3).

For a worm with an unsegmented body cavity, it would not be so easy to perform a variety of localized movements, because changes in the fluid pressure would be freely transmitted to all parts of the body. Segmentation

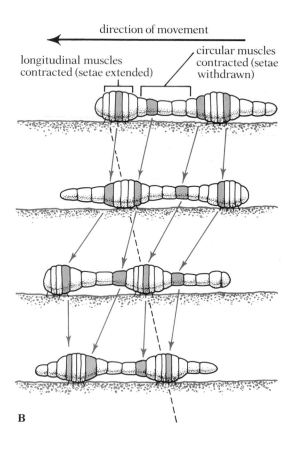

direction of movement

longitudinal muscles contracted (setae extended)

circular muscles contracted (setae withdrawn)

B

20.3 The use of setae, together with peristalsis, to create movement in earthworms
(A) This scanning EM of an earthworm shows the bristlelike setae of each segment. (B) The earthworm, represented here with 20 segments, uses its hydrostatic skeleton to generate movement. Some segments are shown in color and connected by arrows for easier identification. As the longitudinal muscles of a segment contract, the segment becomes short and thick, and its setae are extended to anchor the worm to the substrate. As the circular muscles of a segment contract, the opposite occurs; the segment becomes long and thin, extending forward as it loses contact with the substrate. Peristalsis—alternating waves of contraction of circular and longitudinal muscles—thus enables the earthworm to move. The progress of one wave of peristaltic contractions is indicated by the dashed line.

carries with it, of course, a necessity for some degree of segmental organization of the nervous system and for serial repetition of other organs such as the nephridia.

Present evidence indicates that segmentation of the annelid type evolved as an adaptation for burrowing. The compartmented hydrostatic skeleton aids movement by peristaltic waves, which can develop considerable thrust against the substrate and push aside the soil particles. Though there are many unsegmented worms that burrow by peristaltic wave motions, none can develop as much thrust or burrow so continuously and effectively as segmented worms.

Many marine annelids possess paired lateral flaps on each segment called parapodia (Fig. 20.4), which may have evolved first as devices for producing respiratory and feeding currents inside the tubes in which many rather sedentary annelids live; they often still function in this way and may also serve as primitive gills and as locomotory appendages. Those annelids

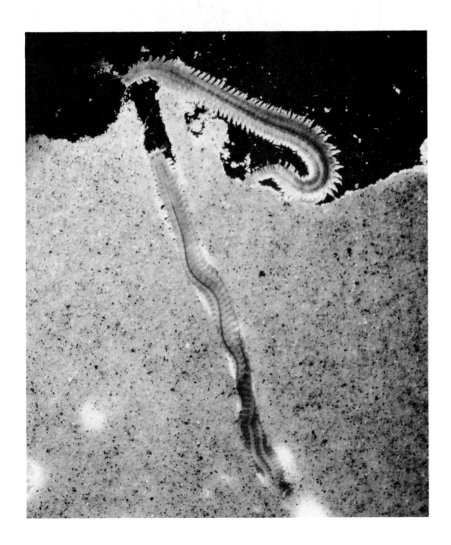

20.4 Marine worms (*Nereis diversicolor*) with parapodia
The worms get tiny particles of food and an adequate supply of oxygen as the waving parapodia produce water currents in the tubelike burrows the animals inhabit. Richly supplied with blood vessels, the parapodia also function as gills (see Fig. 11.10, p. 281).

20.5 A marine annelid

This cross section through a segment of the marine annelid *Nereis* shows the circular and longitudinal muscles and the tough outer cuticle. The evolutionary loss of the strong cuticle made peristaltic movement in soft-bodied annelids like earthworms possible. In marine annelids the muscles serve instead to create an internal peristalsis that both pumps blood and turns the animal's body in rhythmic snakelike undulations that help it burrow into the ocean floor. The parapodia are moved from inside like oars by muscles that run from the cuticle to the inner ends of the hard setae.

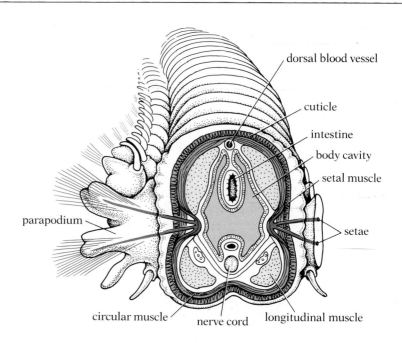

dorsal blood vessel

cuticle

intestine

body cavity

setal muscle

setae

parapodium

circular muscle

nerve cord

longitudinal muscle

whose parapodia are particularly well adapted for locomotion have, in addition to muscles arranged in longitudinal and circular layers, many large muscles running at odd angles. In part these muscles pull against other muscles that act against the hydrostatic skeleton, but in part they also pull against the body wall, which in these annelids has a cuticle much tougher and less pliable than that of earthworms. The cuticle is so tough, in fact, that these worms, because their girth is fixed and there can be little alternating swelling and constriction, generally do not move by peristaltic contractions (Fig. 20.5). Many of them lack the high degree of internal segmentation of earthworms and their relatives, though they do have segmentation of the musculature of the body wall. It was probably from ancestral annelid worms of this general type that the soft-bodied annelids (like earthworms) and the arthropods evolved.

ANIMALS WITH HARD JOINTED SKELETONS

Exoskeletons vs. endoskeletons   The arthropods and the vertebrates are much the most mobile of the multicellular animals. Both groups possess paired locomotory appendages—legs and sometimes wings. Neither depends on a hydrostatic skeleton for the mechanical resistance against which their muscles act; each has evolved, instead, a hard jointed skeleton, with most of the skeletal muscles so arranged that one end is attached to one section of the skeleton and the other end to a different section (Fig. 20.6). Hence, when the muscle contracts, it causes the skeletal joint between its two points of attachment to bend. In many ways, then, the skeletal and muscular systems of arthropods and vertebrates show striking functional similarities. But these two great groups of animals evolved (if we read the evidence correctly) from entirely different ancestral stocks; they represent the highly successful products of two very different evolutionary lines. It is not surprising, therefore, that along with their striking similarities the skeletons and muscles of these two groups display significant differences. The two groups of animals have evolved many similar adaptations to the same functional problems, but they have arrived at those adaptations in entirely different ways.

The most obvious difference between the skeletal systems of arthropods and vertebrates is that arthropods have an *exoskeleton*—a hard body covering with all muscles and organs located inside it—whereas vertebrates have an *endoskeleton*—a framework with muscles attached, embedded within the organism. Besides functioning as structures against which muscles can pull, both types of skeleton are important in providing shape and structural support for animals, particularly animals living on land, where the buoyancy of water is not available for support; in this respect they are analogous to the rigid xylem, which is a critical factor in enabling land plants to attain large size. Exoskeletons, which are composed of noncellular material secreted by the epidermis, function also as a protective armor for the softer body parts and as a waxy barrier preventing excessive water loss by terrestrial arthropods.

Exoskeletons obviously impose difficulties in overall growth, however, and periodic molting of the exoskeleton and deposition of a new one are necessary to permit increase in size. Further, the mechanics of any exo-

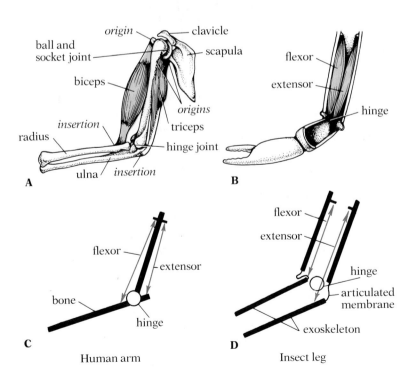

20.6 Mechanical arrangement of muscle and skeleton in a human arm and an insect leg
(A) When the biceps of the human arm contracts, the arm is flexed (bent) at the elbow. The triceps has the opposite action; when it contracts, the lower arm is extended. (B) The comparable flexor and extensor muscles in the insect leg have the same action, even though the muscles are inside the skeleton. The joints in this example are so arranged that each segment, when bent, is perpendicular to its neighbors. Notice that only one of the two "fingers" of the claw can move. (C–D) The similarities between these two simple lever-joint strategies can be seen in these diagrams.

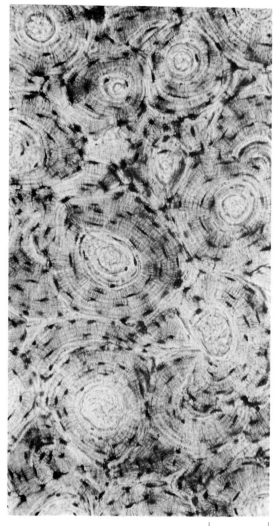

0.1 mm

20.7 Photograph of cross section of bone, showing Haversian systems

Each Haversian system is seen as a nearly round area. The light circular core of each system is the Haversian canal, through which blood vessels pass. Around the Haversian canal is a series of concentrically arranged hard lamellae. The elongate dark areas between the lamellae are cavities, called lacunae, in which the bone cells are located. The numerous very thin dark lines running radially from the central canal across the lamellae to the lacunae are canaliculi through which tissue fluid can diffuse.

skeletal system impose limitations on the possible size of the animal. Since the weight of an animal is a function of its volume (length × width × height), a doubling of an animal's linear dimensions increases its weight by a factor of 8 (that is, 2^3). To support this added weight, the cylindrical exoskeletons of arthropods, for example, must become disproportionately thicker and heavier with increasing length. As a result, all the large arthropods, such as lobsters, are aquatic: the buoyancy of the water they live in provides much of the support their weight requires.

The same limitations apply in a somewhat less drastic way to vertebrates. The endoskeletons that provide support and protection for an animal's delicate internal organs must be strong enough for the disproportionate increase of weight in larger animals. Elephants have thicker bones than antelope, and the largest vertebrates, the whales, like lobsters, are aquatic. But endoskeletons have two advantages. They are made of stronger materials than exoskeletons, and they are located inside the attached muscles instead of enclosing them. Unconfined by the walls of an exoskeleton (Fig. 20.6C–D), the muscles of vertebrates can be of sufficient size to support a body of relatively large volume. In small animals, exoskeletons and endoskeletons are about equally effective, and in very small ones exoskeletons are probably superior.

The vertebrate skeletal and muscular systems Vertebrate skeletons are composed primarily of bone and/or cartilage, two types of connective tissue mentioned earlier (see pp. 150–51).

Cartilage is firm, but not as hard or as brittle as bone. In all vertebrate embryos it is the primary component of the skeleton, and in some adult vertebrates—notably the sharks, skates, and rays—a cartilaginous skeleton persists throughout life. But in most vertebrates bone progressively replaces the cartilage as development proceeds; some cartilage is usually retained, however, where firmness combined with flexibility is needed, as at the ends of ribs, on the articulating surfaces in skeletal joints, in the walls of the larynx and trachea, in the external ear, and in the nose.

Some bones are partly "spongy," consisting of a network of hardened bars with the spaces between them filled with marrow. Other bones are more compact, their hard parts appearing as an almost continuous mass with only microscopic cavities in them. The shafts of typical long bones, like those of the upper arm and thigh, consist of compact bone surrounding a large central marrow cavity. In adults the marrow in the cavities of the shafts of long bones is primarily of the yellow fatty variety, while the marrow in the flat bones of the ribs and skull and in the ends of long bones is primarily of the red variety and is active in the production of blood cells. There is no sharp distinction between the two types of marrow, however, and they may grade into each other. Even the most characteristic red marrow contains about 70 percent fat.

Compact bone is composed of structural units called **_Haversian systems_** (Fig. 20.7). Each such unit is irregularly cylindrical and is composed of concentrically arranged layers of hard inorganic matrix surrounding a microscopic central Haversian canal. Blood vessels and nerves pass through this canal. The scattered irregularly shaped bone cells lie in small cavities located along the interfaces between adjoining concentric layers of the hard matrix. Exchange of materials between the bone cells and the blood vessels in the Haversian canals is by way of radiating canaliculi that penetrate and cross the layers of hard matrix.

Vertebrate skeletons are customarily divided into two components: (1) the axial skeleton, which is the main longitudinal portion, composed of the skull and the vertebral column with its associated rib cage; and (2) the appendicular skeleton, which includes the bones of the paired appendages (fins, legs, wings) and their associated pectoral and pelvic girdles (Fig. 20.8). Some bones are joined together by immovable joints or sutures, as in the case of the numerous small bones that together constitute the skull. But many others are held together at movable joints by **ligaments.** Skeletal muscles, attached to the bones by means of **tendons,** produce their effects by bending the skeleton at these movable joints. The force causing the bending is always exerted as a pull by contracting muscles; muscles cannot actively push. Straightening or reversal of the direction in which a joint is bent must be accomplished by contraction of a different set of muscles.

If a given muscle is attached to two bones with one or more joints between them, contraction of the muscle generally causes movement of only one of the two bones, the other being held relatively rigid by other muscles. The end of muscle attached to the essentially stationary bone—generally the proximal end in limb muscles—is called the **origin,** and the end of the muscle attached to the bone that moves—generally the distal end in limb muscles—is called the **insertion** (Fig. 20.6A). The movable bones behave like a lever system with the fulcrum at the joint. A single muscle sometimes has multiple origins and/or insertions, which may be on the same or on different bones. The action resulting from contraction of any specific muscle depends primarily on the exact positions of its origins and insertions and on the type of joint between them.

Actually, under normal circumstances, muscles do not contract in isolation. The nervous system sends impulses to entire sets of muscles that operate in antagonistic groups—when one group of muscles is strongly contracted, an antagonistic group exerts a weaker opposing pull, to fine-tune the movement. Moreover, this familiar principle of control permits the antagonistic muscles to reverse the direction of movement instantly should that be necessary. In addition, other muscles (synergists) guide and limit the movement even further. To understand the action of a muscle, therefore, one must know, in addition to the muscle's own origins and insertions, the positions, actions, and relations of its antagonists and synergists. Since even the simplest action—taking a step, for instance—involves a complicated, coordinated pattern of activity by a large number of individual muscles, understanding the physiology of an isolated muscle has only a limited value. The various types of skeletal joints and the numerous muscle-joint arrangements employ the mechanical principles of pulleys, levers, and braces, and make a fascinating subject of study for those interested in biological engineering and mechanics.

The types of muscle Three types of muscle tissue are recognized in vertebrates: skeletal muscle, smooth muscle, and heart or cardiac muscle.

Skeletal muscle (also called voluntary or striated muscle) produces the movements of the limbs, trunk, face, jaws, eyeballs, etc. It is by far the most

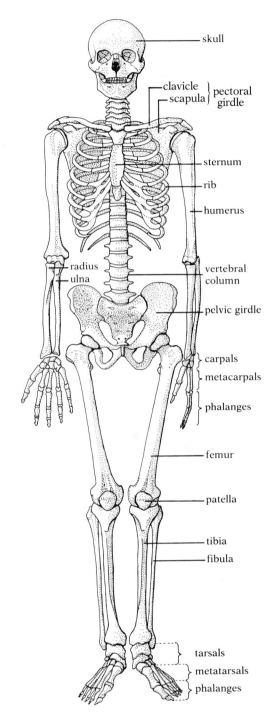

20.8 Human skeleton

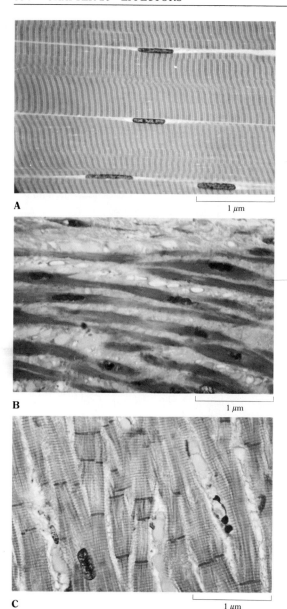

A   1 μm

B   1 μm

C   1 μm

20.9   Photographs of three kinds of vertebrate muscle fiber
(A) Portions of four skeletal-muscle fibers from a monkey. Each fiber has several nuclei, located on its outer sheath, and is crossed by alternating light and dark bands, or striations. (B) Spindle-shaped smooth-muscle fibers from a human blood vessel. (C) Human cardiac muscle. The thick dark lines are places, called intercalated discs, where one cell ends and another begins. Cardiac muscle cells often bifurcate, producing a complex three-dimensional network.

abundant tissue in the vertebrate body. Most of what we commonly call meat is skeletal muscle. Each skeletal-muscle cell—or fiber, as it is usually called—is roughly cylindrical, contains many nuclei, and is crossed by alternating light and dark bands called **striations** (Fig. 20.9A). The fibers are usually bound together by connective tissue into bundles, which in turn are bound together by more connective tissue to form muscles. A muscle, then, is a composite structure made up of many bundles of muscle fibers, just as a nerve is composed of many nerve fibers bound together. Skeletal muscle is innervated by the somatic nervous system.

There are two types of skeletal muscle: red muscle (or slow-twitch muscle) and white muscle (or fast-twitch muscle). **Red muscle** has a rich blood supply, numerous mitochondria, and much **myoglobin,** a compound similar to hemoglobin that forms a loose combination with oxygen and stores it in the muscle. Red muscle oxidizes fatty acids as its primary source of energy. It contracts rather slowly, and is specialized for long-term activity without appreciable fatigue. By contrast, **white muscle** has a more limited blood supply, few mitochondria, and a low myoglobin content. It depends almost entirely on anaerobic breakdown of glycogen for its energy supply. It is specialized for very fast contractions and can develop great tension, but only for a short period, because it fatigues rapidly. The light and dark meat of chicken provide a familiar example of these two types of muscle. The dark meat of the thigh is composed primarily of the slower, high-endurance red muscle needed for continuous walking about, while the white meat of the breast is largely made up of the fast, low-endurance white muscle needed for an occasional escape from danger. By contrast, the breast and wing muscles of birds that fly a great deal consist of red fibers rather than white, while the leg muscles of relatively sedentary animals like rabbits, which use their legs primarily for rapid escape, consist predominantly of white fibers. In humans, the fibers of leg muscles, which must be able to support the weight of the body for extended periods, are mostly red, while the fibers of arm muscles are white.

Smooth muscle (also called visceral muscle) forms the muscle layers in the walls of the digestive tract, bladder, various ducts, and other internal organs. It is also the muscle present in the walls of arteries and veins. The individual smooth-muscle cells are thin, elongate, and usually pointed at their ends (Fig. 20.9B). Each has a single nucleus. The fibers are not striated. They interlace to form sheets of muscle tissue rather than bundles. Smooth muscle is innervated by the autonomic nervous system. In some parts of the body, like the intestinal wall and the uterus, adjacent smooth-muscle cells are interconnected by gap junctions (described on p. 140). The result is electrical coupling, which ensures contraction of the whole muscle as a single unit.

The functional differences between vertebrate skeletal muscle, which is primarily concerned with effecting adjustments to the organism's external environment, and vertebrate smooth muscle, which brings about movements in response to internal changes, are reflected in differences in their physiological characteristics. Cells of skeletal muscle are innervated by only one nerve fiber; they contract when stimulated by nerve impulses and relax when no such impulses are reaching them. Smooth-muscle cells, by contrast, are usually innervated by two nerve fibers, one from the sympathetic system and one from the parasympathetic system; they contract in response to impulses from one of the fibers and are inhibited from contracting by impulses from the other. Skeletal muscles cannot function normally in the absence of nervous connections and actually degenerate when deprived of their innervation, but smooth muscle (like cardiac muscle) can often contract without any nervous stimulation, as is commonly the case in peristaltic contractions of the intestine.

There are other contrasts: The action of skeletal muscle is more rapid, but smooth muscle can remain contracted longer. Skeletal muscle is more sensitive to electrical stimuli than smooth muscle, while the latter is more sensitive to chemical stimuli. Skeletal muscle has a definite resting length, whereas smooth muscle does not; yet smooth muscle contracts more readily in response to stretching. Finally, and most important, smooth muscle is slow, but energy-efficient: it uses only about 10 percent of the ATP required by skeletal muscle to produce the same strength of contraction.

Cardiac muscle, the tissue of which the heart is composed, shows some characteristics of skeletal muscle and some characteristics of smooth muscle. Its fibers, like those of skeletal muscle, are striated. But like smooth muscle, it is innervated by the autonomic nervous system, and its activity is more like that of smooth muscle. Where two separate cardiac-muscle fibers (cells) meet, their adjacent membranes are so tightly appressed and complexly interdigitated, and have so many desmosomes and other fibrous reinforcements, that for many years these areas were not recognized as cellular junctions. The sites of these junctions are visible under light microscopes as dark-colored discs called intercalated discs (Fig. 20.9C).

These descriptions of muscle types, however, do not apply in all respects to the muscles of invertebrate animals. All the muscles of insects are striated, even those in the walls of their internal organs; many other invertebrates possess only smooth muscles. Some invertebrates have evolved arrangements enabling them to perform the same action in two different ways. Thus scallops, which swim by opening and closing their shells in a flapping motion, have two sets of shell-closing muscle fibers: a striated set, whose fast short-term action is used in swimming, and a smooth set, whose slower but longer-lasting action is used for holding the shells tightly closed when the scallop is at rest or is attacked by a predator.

THE PHYSIOLOGY OF MUSCLE ACTIVITY

THE GENERAL FEATURES OF SKELETAL-MUSCLE CONTRACTION

Individual muscle fibers resemble individual nerve cells in firing only if an impinging stimulus is of threshold intensity, duration, and rate. Like nerve cells, relaxed vertebrate muscle fibers seem to exhibit the all-or-none property. If we were to administer a stimulus above the threshold value to an excised vertebrate muscle fiber, we would obtain the same degree of contraction whatever the value of the stimulus, provided it was not so strong as to damage the cell.

But this description of the response of an isolated, relaxed muscle fiber leaves several questions unanswered. In particular, while a relaxed isolated

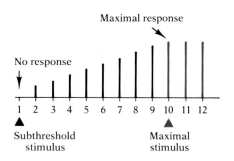

20.10 Response of a muscle to stimuli of various intensities

The numbers indicate the intensities at which stimuli are administered, and the height of the bars shows the strength of muscle response. Stimulus 1 is very weak and elicits no response; it is subthreshold. Stimulus 2 is somewhat stronger and proves to be above threshold, for the muscle contracts. Each stimulus from 3 to 10 is slightly stronger than the preceding, and each elicits a correspondingly stronger muscle contraction. Stimuli 11 and 12 are stronger than 10, but the muscle gives no greater response, indicating that 10 elicits a maximal response.

fiber may react in an all-or-none fashion, intact muscles do not. We may use the same muscles to lift a pencil that we use to hoist a 10-kg weight, and it is easy to demonstrate in the laboratory that an individual muscle is capable of giving graded responses, depending on the strength of the stimulation. Suppose we measure the contraction of a frog leg muscle when it is stimulated. If we administer a stimulus barely above the threshold intensity, the muscle gives a very weak twitch. If we apply a slightly stronger stimulus after a few seconds' delay, the muscle gives a slightly stronger twitch. Increasing the strength of the stimulus elicits an ever-stronger contraction from the muscle, until we reach the point at which further increases in the stimulus do not increase the strength of the response. The muscle has reached its maximal response (Fig. 20.10).

How can these results be explained if muscle fibers give all-or-none responses? One kind of explanation invokes the interaction of the many different fibers in each muscle. For instance, since the threshold values of these fibers are not all the same, and since different muscle fibers may be innervated by different nerve fibers and these may not all fire at the same time, an increase in the strength of the stimulus above the threshold level may elicit a greater response from the whole muscle by stimulating more muscle fibers. Ultimately, however, all the fibers will be stimulated to respond, and the muscle will thus have reached a maximal response; increasing the intensity of the stimulus further will not increase the response, since there are no more fibers to stimulate. (This description applies only to vertebrate skeletal muscles: the striated muscle fibers of invertebrates seldom exhibit the all-or-none response, but instead contract in proportion to the frequency of stimulation.)

Though the interaction of the various muscle fibers in a muscle explains some of the grading in responses generated by relaxed all-or-none fibers, it does not account for everything, particularly if the fiber is not fully relaxed. The experiment with the frog muscle requires an appreciable time delay between stimuli; each response is induced by a single brief stimulus, and

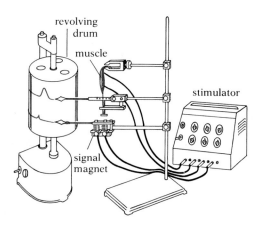

20.11 Kymograph apparatus for studying muscle contraction

The drum of the kymograph, which is covered with paper, revolves at a constant speed. The muscle is mounted in such a way that when it contracts it raises a stylus that writes on the revolving drum. Wires lead from a stimulator to the muscle and also to a signal magnet that can deflect a second stylus. At the moment when a stimulus is sent to the muscle, the signal stylus is deflected, producing a blip in the trace it is drawing on the revolving drum; one such blip is shown on the trace in the picture. This blip indicates exactly when the stimulus was administered. The stimulus causes the muscle to contract, raising the stylus to which it is attached and producing a corresponding rise in the trace the stylus is drawing on the revolving drum. As the muscle relaxes, the stylus is lowered and the trace it is drawing falls. Thus the trace drawn on the kymograph drum gives us a record of the contraction pattern of the muscle.

the muscle must have time to relax fully before it is stimulated again. Such a muscular response is called a *simple twitch.* Suppose that an isolated frog muscle has been attached to a kymograph apparatus (Fig. 20.11) in such a manner that every time the muscle is stimulated by an electric shock it moves a lever that writes on the revolving drum of the kymograph. The stronger the contraction of the muscle, the higher the lever will be pulled. If a single adequate stimulus is administered to this muscle, there is a brief interval during which no contraction occurs. This is the *latent period*—the interval between stimulation of the muscle and the commencement of the shortening process—which usually varies between 0.0025 and 0.004 seconds (Fig. 20.12). The latent period is followed by the *contraction period,* and this phase is followed immediately by a *relaxation period.* These three phases comprise a single simple twitch of the muscle.

Now suppose that we apply a series of rapid stimuli to a relaxed excised muscle. In this case the muscle will not have completely relaxed after contracting in response to one stimulus when the next stimulus arrives. When this happens, a contraction results that is greater than either stimulus alone would produce (Fig. 20.13). There has been a *summation* of contractions, the second adding to the first. If the initial stimulus was submaximal, the summation may result in part from recruitment of additional muscle fibers by the second stimulus. But this is not the whole story. Summation can occur even if the individual stimuli are of maximal intensity—that is, even if all fibers in the muscle are activated by each stimulus. Some change in the physiological condition of the muscle fibers during activity must account for the increased strength of contraction. It is not strictly true, then, that individual muscle fibers give an all-or-none response. Each fiber will respond with a certain strength of contraction to a single isolated stimulus if that stimulus is above the threshold level and the fiber is relaxed, but the fiber will respond a little more strongly if it is given so rapid a series of stimuli that it cannot relax between impulses. The initial contraction produces chemical changes that make the fiber momentarily more irritable. These changes may be a consequence of the increase in temperature that results from the waste heat of contraction, or may be some other secondary consequence of the chemistry of contraction and its control discussed below.

When stimuli arrive with extreme rapidity, a muscle can't relax at all between successive stimuli. In this case, individual contractions become indistinguishable, and fuse into a single sustained contraction known as *tetanus.* For the same reasons that apply to summation, of which tetanus is a form, a tetanic contraction is greater than a maximal simple twitch of the same muscle. Normally, a high percentage of our actions involve tetanic contractions rather than simple twitches, because a volley of nerve impulses is sent to the muscle. If, however, a tetanic contraction is maintained too long, the muscle will begin to fatigue, and the strength of its contraction will fall, even though the stimuli continue at the same intensity. Fatigue probably results from an accumulation of lactic acid, a depletion of stored energy reserves, and other chemical changes.

Some muscles are never completely relaxed, but are kept in a state of partial contraction called muscle tone or *tonus.* Tonus is maintained by alternate contraction of different groups of muscle fibers, so that no single fiber has a chance to fatigue.

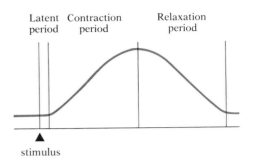

20.12 Kymograph record of a simple twitch
The duration of the latent period, which is too short to be measured even by the most sensitive kymograph, is much exaggerated here.

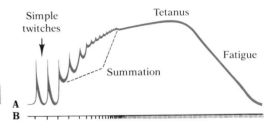

20.13 Summation and tetanus in muscle response
When the stimuli (line B) are widely spaced, the muscle has time to relax fully before the next stimulus arrives, and simple twitches result. (Because the drum is revolving much more slowly than in Figure 20.12, each simple twitch is recorded as a sharp spike on the trace.) As the frequency of the stimuli increases, the muscle does not have time to relax fully from one contraction before the next stimulus arrives and causes it to contract again. The result is summation—contractions that are stronger (and hence produce taller spikes in the trace) than any single simple twitch. If the stimuli are very frequent, the muscle may not relax at all between successive stimulations; the resulting strong sustained contraction is called tetanus. If the very frequent stimulation continues, however, the muscle may fatigue and be unable to maintain the contraction.

THE MOLECULAR BASIS OF CONTRACTION

As you would expect, the energy for muscle contraction comes from ATP. But so little ATP is actually stored in the muscles that a few muscular twitches might quickly exhaust the supply, were it not for another high-energy phosphate stored in the muscles. In vertebrates and some invertebrates, particularly echinoderms, this compound is *creatine phosphate,* formed by linkage of a high-energy phosphate group to creatine. Many invertebrates use a similar compound, arginine phosphate. These two compounds—creatine phosphate and arginine phosphate—are called *phosphagens.* The phosphagens cannot supply energy directly to the contraction mechanism of muscle, but they can pass their high-energy phosphate groups to ADP to form ATP, and the ATP can then act as the direct energy source for contraction. Enough high-energy phosphate is stored in the muscle to enable it to contract strongly during the several seconds' delay before the machinery of glycolysis and cellular respiration can be speeded up.

If the demands on the muscles are not great, much of the energy used to replenish the supply of phosphagens and ATP may come from the complete oxidation of nutrients to carbon dioxide and water. During the unavoidable delay before adjustments of the respiratory and circulatory systems increase the oxygen supply to the active muscles, some of the O_2 for oxidative phosphorylation in red muscles may come from oxygenated myoglobin. Myoglobin, you may recall (p. 534), is a special oxygen-storage protein, unique to muscle cells, that forms a loose combination with O_2 while the O_2 supply is plentiful and stores it until the demand increases.

But during violent muscular activity, such as strenuous exercise or the lifting of a very heavy object, the energy demands of the muscles (especially white muscles) may be greater than can be met by complete respiration unaided, because sufficient oxygen cannot get to the tissues fast enough. Under these circumstances lactic acid fermentation occurs; the muscles obtain the extra energy they need from the inefficient anaerobic processes of glycolysis alone, and thus incur what physiologists call an *oxygen debt.* Some of the lactic acid accumulates in the muscles, but much of it is promptly transported by the blood to the liver. When the violent activity is over, a period of hard breathing or panting helps supply the liver with the large quantities of O_2 it requires as it reconverts the lactic acid into pyruvic acid, some of which it oxidizes, using the energy thus obtained to resynthesize glycogen from the rest of the lactic acid. In this manner the oxygen debt is paid off.

If ATP is the immediate source of energy for muscle contraction, how is the ATP coupled to the contraction process and what is that process? Analysis of muscle shows that the major components of its contractile parts are two proteins, *actin* and *myosin.* V. A. Engelhardt and M. N. Ljubimova of the Academy of Sciences in Moscow demonstrated in 1939 that myosin can function as an enzyme that catalyzes removal of the terminal phosphate group from ATP. This is, of course, precisely the energy-liberating reaction we would expect to find coupled to the contraction process. Might myosin function in the intact muscle both as one of the contractile elements and as the enzyme that makes energy available for contraction? Proteins often ex-

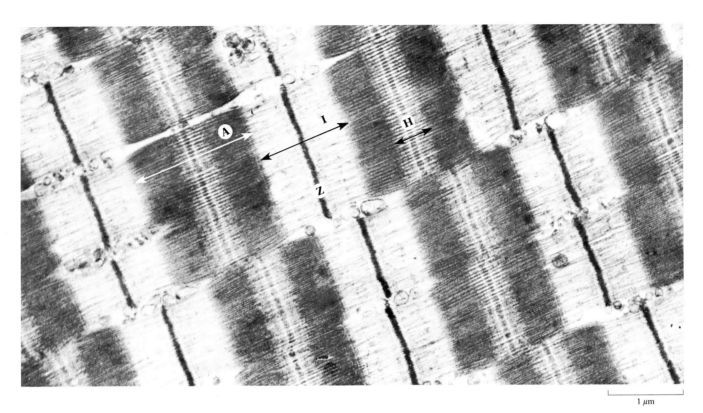

20.14 Electron micrograph of skeletal muscle from a rabbit
The myofibrils run diagonally across the micrograph from lower left to upper right; each looks like a ribbon crossed by alternating light and dark bands. The wide light bands are I bands; there is a narrow Z line in the middle of each. The wide dark bands are A bands, each with a lighter H zone across the middle.

hibit dual functions, acting simultaneously as structural elements and as enzymes.

Evidence for the dual function of myosin in muscle contractions comes from experiments performed by Albert Szent-Györgyi in Woods Hole, Massachusetts. Szent-Györgyi showed that if actin and myosin are separately extracted from muscle, purified, and put into solution together, they will combine spontaneously to form a loose complex known as ***actomyosin.*** If the actomyosin complex is then precipitated and artificial fibers are prepared from it, the fibers will contract when exposed to ATP. Neither actin nor myosin alone contracts. Clearly, then, the actomyosin complex is a contractile material. And, clearly, the complex (actually its myosin component) can itself liberate from ATP the energy necessary for its own contraction, since in this experiment no other possible enzyme is present.

If actomyosin really is the contractile component of muscle, by what process does it contract? One way for a protein complex of this sort to shorten would be to fold more tightly. This hypothesis, once favored by many biologists, has long since been abandoned, on both anatomical and physiological grounds. Let us examine the anatomical evidence.

We have seen that a skeletal muscle is characterized by striations of light and dark bands and that it is composed of numerous muscle fibers (cells) bound together by connective tissue. Examination of these fibers under very high magnification reveals that they, in turn, are filled with numerous long thin myofibrils, each about 1–2 µm in diameter, with mitochondria in the cytoplasm between them. The myofibrils show the same pattern of cross striations as the fibers of which they are a part (Fig. 20.14). There is an

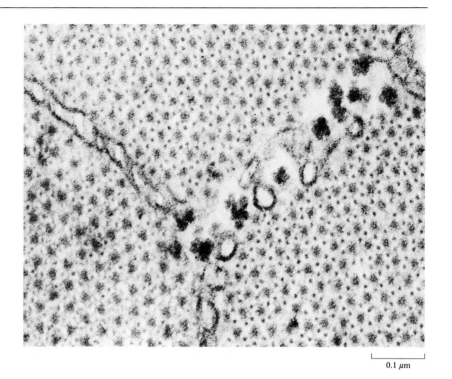

0.1 μm

20.15 Electron micrograph of cross section of frog skeletal muscle
The thin actin filaments are arranged in a hexagonal pattern around the thick myosin filaments. Parts of three myofibrils are seen here in cross section, separated by a structure associated with the sarcomeres called the sarcoplasmic reticulum.

alternation of fairly wide light and dark bands, which have been called *I bands* and *A bands* respectively. In the middle of each dark A band is a region lighter than the rest of the A band, but darker than the I bands—the *H zone.* In the middle of the light I band is a very dark thin line called the *Z line.* The entire region of a myofibril from one Z line to the next is called a *sarcomere.* The sarcomeres are the functional units of muscular contraction.

Might the striations so characteristic of skeletal muscle be a structural reflection of the functional contractile units? According to a variety of evidence, yes. Chemical analysis shows that myosin is concentrated in the A bands and actin is concentrated in the I bands. Furthermore, A. F. Huxley of Cambridge University showed in 1954 that the relative widths of the bands change as the fiber contracts; the I bands and H zones become narrower, but the A bands change very little, with the result that the A bands are moved closer together.

At about the time A. F. Huxley was performing his experiments under a high-power light microscope, H. E. Huxley (no kin to A. F.), then at University College, London, began studying muscle with the electron microscope, and found that within each myofibril there are two types of filaments, thick ones and thin ones, arranged in a very precise pattern (Fig. 20.15). The two are interdigitated, with the thick ones located exclusively in the A bands and the thin ones primarily in the I bands, but extending some distance into the A bands. This distribution explains the different appearances of the A bands, I bands, and H zones. Each dark A band is precisely the length of one region of thick filaments; it is darkest near its borders, where the thick and thin filaments overlap, and lighter in its mid-region, or H zone, where only

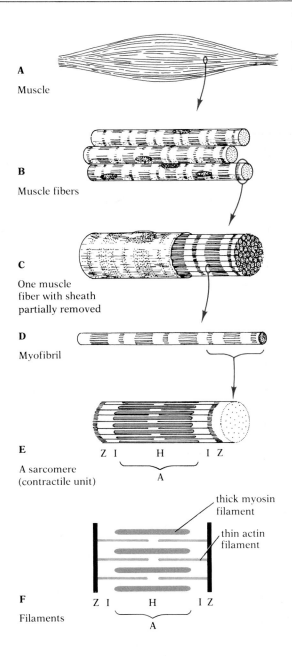

A

Muscle

B

Muscle fibers

C

One muscle
fiber with sheath
partially removed

D

Myofibril

E

A sarcomere
(contractile unit)

Z I   H   I Z

A

F

Filaments

thick myosin
filament

thin actin
filament

Z I   H   I Z

A

20.16   The component parts of skeletal muscle
The pattern of light and dark bands visible in
myofibrils under high magnification (D; see also Fig.
20.14) results from the interdigitation of actin and
myosin filaments in each myofibril. As shown in F,
the A band corresponds to the length of the thick
myosin filaments (color); the lighter H zone is the
region where only the thick filaments occur, while the
darker ends of the A band are regions where thick
and thin filaments overlap. The I band corresponds
to regions where only thin actin filaments occur. The
Z line is a structure to which the thin filaments are
fastened at their midpoints.

the thick filaments are present (Fig. 20.16). Each light I band corresponds
to a region where only the thin filaments are present. The Z line is a structure to which the thin filaments are anchored at their midpoints and against
which they exert their pull during contraction; it also functions to hold the
filaments in proper register. The protein that anchors the actin filaments to
the Z line is called α-actinin, while another, known as desmin, keeps the
actin in proper position.

Relaxed

Moderately contracted

Strongly contracted

20.17 Arrangement of actin (gray) and myosin (color) filaments in a sarcomere in relaxed and contracted states

The observations of the two Huxleys led each of them independently to propose a new theory of muscle contraction—that, instead of folding, the filaments telescope together by sliding past each other. If the filaments slide together, the zone of overlap between thick and thin filaments will increase until the thin filaments from the I bands on the two sides of an A band actually meet and overlap slightly; this sliding together will reduce the width of the H zone and even obliterate it entirely if the thin filaments meet (Fig. 20.17). The sliding together will also pull the Z lines closer together and greatly reduce the width of the I bands. But the width of the A bands will be minimally altered, since these correspond to the full length of the thick filaments, which remains the same (except, perhaps, for a slight crumpling from contact with the Z lines under conditions of extreme contraction). Thus the sliding theory accounts for the changes observed in sarcomeres. But how the sliding is brought about remained to be discovered.

Analysis shows that the thick filaments are composed of myosin and the thin filaments primarily of actin. We have already seen that myosin and actin must unite to form the actomyosin complex before acquiring contractile properties. Some sort of connection must exist, then, between the thick myosin filaments and the thin actin filaments. And indeed, electron micrographs show what appear to be small cross bridges between the filaments (Fig. 20.18). The evidence suggests that they are portions of the thick filaments (Fig. 20.19A). A single thick filament is a bundle of myosin molecules, each of which is composed of an elongated tail portion and a pair of globular heads (Fig. 20.19B, C). When a myosin molecule is fragmented by treatment with proteolytic enzymes, it is the head regions that bind actin and hydrolyze ATP—a strong indication that they are the cross bridges.

According to the Huxley sliding-filament theory, the cross bridges act as hooks or levers that enable the myosin filaments to pull the actin filaments (Fig. 20.20). The cross bridges bend toward the actin, hook onto it at spe-

20.18 Electron micrograph of insect flight muscle, showing cross bridges between filaments The thick (myosin) and thin (actin) filaments are connected by cross bridges composed of myosin heads. Essentially the same structural pattern is seen in the myofibrils of skeletal muscle.

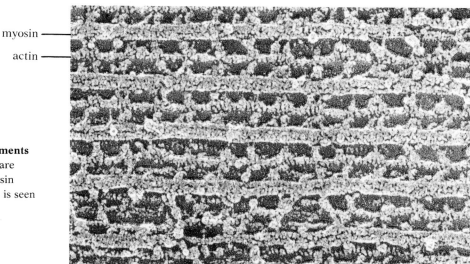

myosin

actin

0.05 μm

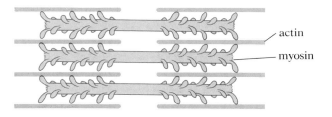

actin

myosin

A

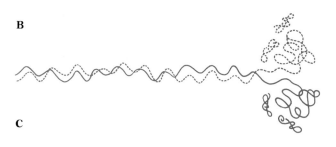

B

C

20.19 Molecular structure of myosin filaments
(A) Each myosin filament (color) is linked to the adjacent thin filaments (gray) by numerous cross bridges. (B) The myosin filament is composed of a bundle of elongate molecules, each with a double club-shaped head, which acts as the cross bridge. Each thick filament has about 500 heads. (C) The tail of a myosin molecule is composed of two intertwined polypeptide chains. Each of its heads is formed by the coiled free end of one of the tail chains, plus two smaller polypeptides.

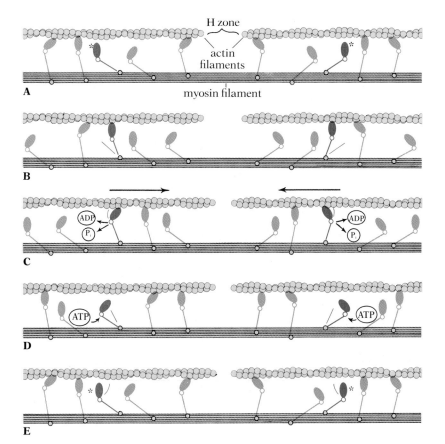

20.20 Action of cross bridges during contraction
Each of the 500 or so myosin heads of a thick filament acts independently. This drawing traces the action of two heads through a cycle of movement on both sides of the H zone in one sarcomere. (A) On each side, an ATP-activated head is ready to bind to one of the spherical subunits in the actin filament. (B) Binding takes place. (C) The spent ATP is released in the form of ADP and phosphate, and this release triggers an allosteric change that enables the myosin cross bridges to bend and pull the actin filaments toward the H zone—to the left on the right side, and to the right on the left side. (D) As a result of the bending, the myosin heads lose their affinity for the actin and drift free. At the same time each becomes able to bind another ATP. (E) ATP has activated the heads by inducing an allosteric change that "cocks" them in preparation for another power stroke. The activated state is indicated by the asterisks. Research indicates that each myosin head goes through 5–10 of these cycles every second during contraction.

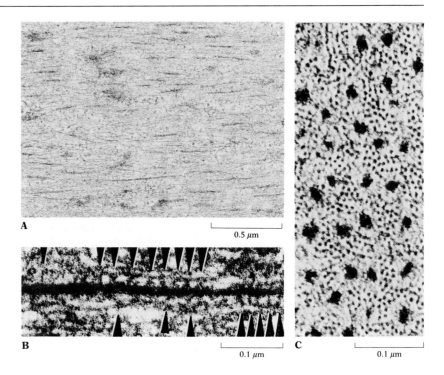

20.21 Electron micrographs of smooth muscle from a vein of a rabbit
(A) Longitudinal section. Filaments can be seen, but their arrangement is not as regular as in skeletal muscle, and they are not partitioned into sarcomeres. (B) Under very high magnification, cross bridges can be made out (the locations of some are indicated by the dark wedges). (C) Cross section. The thick myosin filaments (ca. 15 nm) are surrounded by numerous thin actin filaments (ca. 5–8 nm). The ratio of about 15 to 1 of thin to thick filaments contrasts with the 10-to-1 ratio typical of striated muscle. The disposition of the thin filaments around the thick filaments does not appear as orderly as in striated muscle.

cialized receptor sites, and then bend in the other direction, pulling the actin with them; they then let go, bend back in the original direction, hook onto the actin at a new active site, and pull again. In other words, the sliding together of the filaments is created by a ratchet mechanism. The necessary energy comes from the hydrolysis of ATP by the myosin.

Though the remarkably precise arrangement of filaments found in striated-muscle cells has not been found in smooth-muscle cells, the evidence suggests that such cells contract by essentially the same mechanism (Fig. 20.21).

THE ELECTROCHEMICAL CONTROL OF CONTRACTION

Now that we have seen how ATP fuels the ratchet action of myosin filaments in muscles, you may well wonder how the muscles know when to "burn" this fuel, and why they don't remain perpetually contracted. As we have already seen in the nervous and endocrine systems, chemical control can be exerted at various levels. In muscles, the critical step is the binding of the myosin heads to the actin filaments. How, then, does a nervous impulse at a neuromuscular junction signal and control the molecular binding of myosin heads that results in contraction?

Like the membrane of a resting neuron, the membrane of a resting muscle fiber is polarized, the outer surface being positively charged in relation to the inner one. Stimulatory transmitter substance released by a nerve axon at a neuromuscular junction (see Fig. 18.30, p. 479) causes a momentary depolarization of the muscle membrane. If the depolarization reaches the threshold level, an impulse, or action potential, is triggered and propagated over the surface of the fiber by the same combination of voltage-gated Na^+ and K^+ channels that is seen in the membrane of nerve cells. The action potential activates the contraction process indirectly, by means of Ca^{++} ions.

The first clue to the ionic control of contraction in muscles came in 1949, when L. V. Heilbrunn and Floyd J. Wiercinski of the University of Pennsylvania injected a wide variety of substances into muscle fibers and found that the only one that caused contraction was a salt of calcium. It was reasonable to suppose, therefore, that calcium ions (Ca^{++})—of which some normally flow into the fibers when they are stimulated—cause the cross bridges to become active. When this idea was first put forward, however, at least two major objections were raised against it. First, contraction of a vertebrate muscle fiber requires the essentially simultaneous shortening of all its many myofibrils, but the myofibrils in the center of a fiber are so far from the surface that Ca^{++} ions from outside could not possibly diffuse fast enough to reach them in the short interval between stimulation of the fiber and its contraction. Second, there is evidence that not enough Ca^{++} enters the cell to account for the sustained contraction resulting from rapid volleys of nerve impulses.

It was the rediscovery in 1955 of an extensive network of tubules in muscle fibers that opened the way to a solution of the problem of the coupling of excitation and contraction. We say "rediscovery," because the tubules had been well known to histologists before the First World War, but had been forgotten until Stanley Bennett and Keith R. Porter, working at the Rockefeller Institute, detected them again by the then new technique of electron microscopy. The tubules were soon found to comprise two separate but functionally related systems: the ***sarcoplasmic reticulum,*** which does not open to the exterior, and the ***T system,*** or transverse tubule system, which is part of the plasma membrane surrounding the fiber.

The sarcoplasmic reticulum is the muscle cell's highly specialized version of the ubiquitous endoplasmic reticulum. Its membranous canals form a cufflike network around each of the sarcomeres of the myofibrils (Fig. 20.22). The sarcoplasmic reticulum at the distal end of one sarcomere and that at the proximal end of the next sarcomere beyond it are very close together, but lying between them, at the level of the Z line, is usually a tubule of the T system (two tubules in mammals). Though the sarcoplasmic reticulum and the T tubules are in direct contact, there is no interconnection between their lumina (cavities), and hence no mixing of their contents.

When an action potential is propagated across the surface of the muscle cell, it also penetrates into the interior of the fiber via the membranes of the T tubules. The action potential moves much faster than diffusing ions, fast enough so that the stimulus for contraction can reach all the myofibrils at nearly the same instant and the myofibrils near the surface and those in the center of the fiber can contract together.

The intimate association between the T tubules and the sarcoplasmic reticulum allows action potentials moving along the membrane of a T tubule to trigger the reticulum. The reticulum contains large quantities of Ca^{++} ions, which, as Heilbrunn and Wiercinski found, cause myofibrils to contract. The action potential induces a sharp, very marked increase in the per-

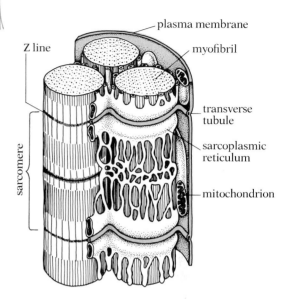

20.22 Sarcomeres with associated sarcoplasmic reticulum

Shown here are sarcomeres typical of amphibian skeletal muscle. The sarcomeres of mammalian skeletal muscle differ in having a separate transverse tubule (and associated sarcoplasmic reticulum) on each side of each Z line, rather than one tubule per line.

20.23 The role of calcium in stimulation of muscle contraction
(A) A sarcomere in the resting (relaxed) condition. Ca^{++} ions (color) are stored in high concentration in the sarcoplasmic reticulum. (B) The polarization of the membranes of the T tubules is momentarily reversed during an action potential (impulse), and this reversal of polarization induces release of the Ca^{++} ions, which spread over the sarcomere and stimulate contraction.

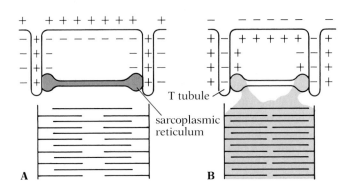

T tubule

sarcoplasmic reticulum

A   B

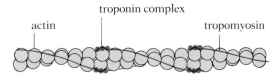

actin   troponin complex   tropomyosin

20.24 Molecular structure of a thin filament
Globular subunits of actin form two helically coiled rows. Molecules (likewise globular) of the regulatory protein complex troponin are evenly spaced along the rows of actin, and the long thin molecules of another regulatory protein, tropomyosin, run along the length of the rows.

meability of the reticular membranes to Ca^{++} ions, allowing these to escape in large numbers (Fig. 20.23). It is this suddenly released intracellular Ca^{++} that is the direct stimulant for contraction.

The question now becomes, How does the calcium trigger contraction of the muscle fibers? To answer, we must look more closely at the structure of the thin filaments. As we have already seen, the main protein in the thin filaments is actin; in addition, these filaments contain important regulatory proteins, tropomyosin and the troponin complex.

The subunits of the actin molecule are globular and form two helically intertwined rows along which run the long thin molecules of the first regulatory protein, **tropomyosin** (Fig. 20.24). In the resting muscle the tropomyosin prevents the actin from binding to the cross bridges from the thick myosin filaments—probably by masking its binding sites for myosin. The molecules of the other regulatory proteins, the **troponin complex,** are also globular and occur in triplets near every seventh pair of actin units; each complex has three binding sites: one for actin, one for tropomyosin, and one for Ca^{++} ions.

When Ca^{++} ions are released from the sarcoplasmic reticulum, they are picked up by the calcium-binding sites of the troponin complex; troponin reacts to this binding with a conformational change. The conformational change effects a shift in the position of tropomyosin, as a result of which tropomyosin ceases its inhibition of actin. The actin thus becomes free to bind with cross bridges from the myosin, and the contraction process is initiated.

We now have all the elements necessary to understand the current model of cross-bridge action and its stimulation. In a resting muscle the cross bridges—the globular myosin heads of the thick filaments—have been "cocked" (activated) by ATP, but they cannot bind to the thin actin filaments because tropomyosin is inhibiting the binding sites on the actin molecules (Fig. 20.25A). When stimulation from a motor nerve triggers an action potential and the action potential, transmitted along the T tubules, penetrates into the interior of the muscle fibers, the sarcoplasmic reticulum releases Ca^{++} ions. Some of these ions bind to the troponin complex, which thereupon undergoes a conformational change displacing the tropomyosin and exposing the myosin-binding sites of the actin to the cross bridges (Fig. 20.25B). The binding of the myosin causes the cross bridges to bend, thus initiating the power stroke that forces the filaments to slide along each other (Fig. 20.25C). When a new ATP molecule binds to a myosin cross bridge after the attachment to actin has been broken, the head is forced back to its original "cocked" conformation (Fig. 20.25D–E).

As long as free Ca^{++} ions (and ATP) remain available, as when the nerve continues to stimulate the muscle, the cycle of cross-bridge binding, power stroke, and recovery flip can occur over and over again, as the muscle continues to contract. But if nervous stimulation ceases, the muscle relaxes, because a calcium pump in the membrane of the sarcoplasmic reticulum quickly moves the Ca^{++} back into the reticulum; the troponin-tropomyosin system can then resume its inhibition of the myosin-binding sites on the actin. ATP, then, is not "burned" by a muscle fiber unless release of Ca^{++} has set in motion the steps leading to a new contraction. As a result, there need be no cost to maintaining a contraction—a point well illustrated by the phenomenon of rigor mortis: after death, muscles initially relax as nervous stimulation ceases, but several hours later they contract as the sarcoplasmic reticulum breaks down and releases calcium, leaving the body rigid.

This description applies only to vertebrate skeletal muscle. Smooth muscle, though similar in many ways, displays some interesting differences. In skeletal muscle ATP activates the myosin heads and Ca^{++} ions trigger movement by binding to the troponin complex of the actin filaments. In smooth muscle the Ca^{++} ions activate the myosin, through two intermediate enzymes, before the ATP becomes involved. This helps explain why smooth muscle acts so slowly. It also accounts for the capacity of smooth muscles to be activated by hormones: in our examination of vertebrate hormones in Chapter 16, we saw that hormones frequently work by opening membrane channels specific for Ca^{++} ions, with these ions, in their

20.25 Model for the stimulation of muscle contraction

(A) In a resting muscle the myosin cross bridges which ATP has already activated (as indicated by the asterisk) cannot bind to the actin in the thin filament, because the binding sites are masked by tropomyosin. (B) The binding of Ca^{++} ions to the troponin complex causes a conformational change that slightly displaces the tropomyosin. The active sites of the actin are thus exposed, and the cross bridges bind to the actin. (C) The binding of each myosin cross bridge to actin, with the concomitant release of ADP and phosphate, initiates a conformational change in the cross bridge, whose bending—the power stroke—forces the filaments to slide along each other. (D) The myosin head then dissociates from the actin, and its ATP-binding site becomes available once more. (E) ATP binds to the myosin head, which is thus "cocked" in preparation for a new stroke.

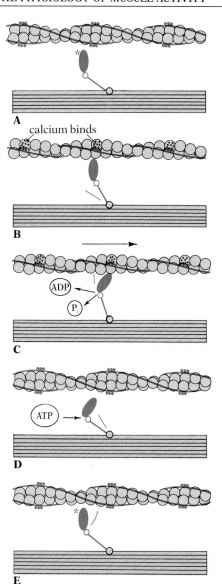

20.26 Tropomyosin in a nonmuscle cell

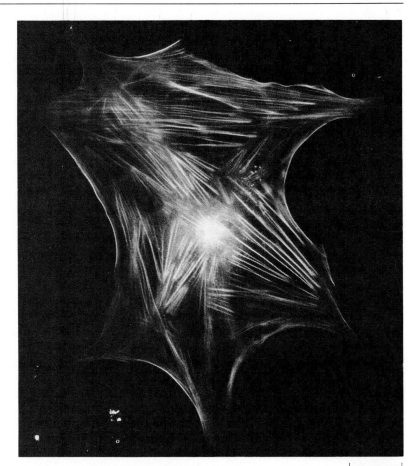

20.27 α-actinin (left) and actin (right) in a rat fibroblast cell

Left: The α-actinin appears to be located at discrete points on the plasma membrane of the cell. Since in striated muscle α-actinin is the principal protein in the Z lines, to which the actin filaments are attached, the α-actinin in the plasma membrane probably provides the anchor points for actin in nonmuscle cells. Right: Actin filaments appear to run between distinct points on the membrane, lending strong support to this idea.

0.05 mm

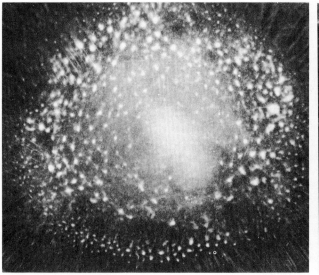

0.02 mm

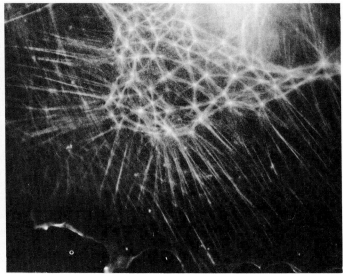

0.02 mm

role as second messengers, then binding to and activating an enzyme complex in the cytosol, such as calmodulin. This is precisely what happens in smooth muscle, where calmodulin takes the role played by troponin in skeletal muscle. In fact, one of the three subunits of the troponin complex in skeletal muscle is a modified calmodulin molecule. There is yet another connection between the endocrine system and the smooth-muscle strategy of vertebrates: cyclic AMP acts as a second messenger to activate the myosin. The critical evolutionary steps from the utilitarian but less elaborate slow-contraction system of smooth muscle to the highly organized fast-contraction system of skeletal muscle must have been the evolution of the Z line dividing contraction units into discrete sarcomeres, the minor modification of calmodulin into one of the three subunits of troponin, and the evolution of the endoplasmic reticulum into a sarcoplasmic reticulum specialized for Ca^{++} transport to the myofibrils.

EFFECTORS OF NONMUSCULAR MOVEMENT

MOVEMENTS PRODUCED BY MICROFILAMENTS

We have seen that muscular contractions depend on microfilaments—the thick myosin filaments and the thin actin filaments. Do other types of cellular movement depend on microfilaments as well? In many cases the answer is an emphatic yes. Not only do many eucaryotic cells—nearly all, perhaps —contain microfilaments, but the most common of these are composed of actin, which often constitutes 2–4 percent (sometimes as much as 15 percent) of the total cellular protein. New staining techniques have made it possible for us to see the filaments themselves. The pattern of sarcomeres found in striated muscle is not seen in nonmuscle cells, but the arrangements are scarcely less orderly than those in smooth muscle. Figure 20.26 shows how tropomyosin appears in a nonmuscle cell; it is probably associated with the actin filaments just as it is in muscle.

In addition to actin, nonmuscle cells usually contain myosin, though the quantity is about one-tenth as great (relative to actin) as the quantity found in striated muscle; in this respect, again, there is a resemblance to smooth muscle. As in muscle, the complexing of actin and myosin allows the myosin to extract energy from ATP, breaking it down into ADP and inorganic phosphate, and contractions are stimulated by an increase in free intracellular calcium. You will recall that Z lines, the structures that actin filaments pull against in striated muscle, are composed largely of α-actinin. This protein is regularly found in the plasma membrane of nonmuscle cells, where it apparently provides points of attachment for the actin microfilaments (Fig. 20.27). Evidently, the same fundamental mechanism that is responsible for muscular movement is at work in many other kinds of movement as well, a prime example of the unity in diversity so typical of the living world; only in procaryotic cells are actin and myosin absent.

A striking example of actomyosin-mediated nonmuscular movement in eucaryotes is found in the microvilli. As shown in Figure 20.28, a bundle of microfilaments runs through the core of the microvilli of intestinal epithelial cells. Anchored by α-actinin at the tip of a microvillus, these microfilaments can cause the microvillus to move back and forth in a gentle waving

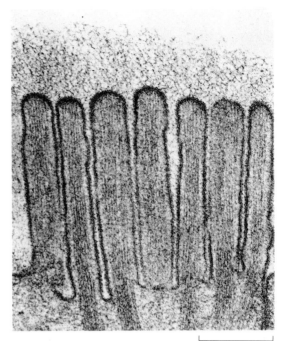

0.2 μm

20.28 Microfilaments in the microvilli
The actin microfilaments in the microvilli provide rigidity; they are, in turn, anchored to the belt desmosomes—also partly composed of actin microfilaments—which provide rigidity for the cell (see Fig. 6.1, p. 139).

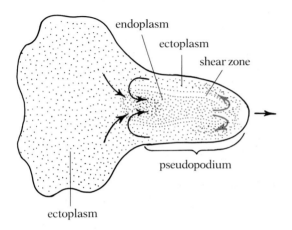

20.29 Amoeboid movement
The endoplasmic core slides forward in a pseudopodium between the peripheral layers of stationary ectoplasm. As the core moves forward, ectoplasm at the rear of the pseudopodium and from the rest of the cell is converted into endoplasm, while endoplasm at the front of the pseudopodium is converted into ectoplasm.

motion. There is now evidence that transient microvilli, which probably function as sensory devices, form in many other types of cells too; they can suddenly pop out from the cell surface and be retracted with equal rapidity, probably through the agency of actomyosin microfilaments. Less dramatic, but of great importance in the lives of cells, are the numerous actomyosin-mediated changes of cell shape that help mold the tissues during embryological development.

A particularly important form of movement apparently mediated by an actomyosin system is cytoplasmic streaming. Whether this process occurs in plant cells (see Fig. 12.1, p. 296), in the movement of vertebrate white blood cells, or in the formation of pseudopodia by an amoeba (see Fig. 4.22, p. 104), it appears to depend on actin microfilaments. In each case treatment of the cell with cytochalasin B, a reagent that blocks the normal activity of actin microfilaments, results in a cessation of streaming. The exact mechanism of movement is still unclear, however.

Let us look in more detail at amoeboid movement. As in all cases of cytoplasmic streaming, movement is restricted to a central core of the cytoplasm, the endoplasm (Fig. 20.29). When the endoplasm reaches the advancing end of the pseudopodium, it spreads peripherally, forming a stiffer gel-like nonmoving layer, the ectoplasm. There is continuing conversion of ectoplasm into endoplasm at the rear of the pseudopodium and from the rest of the advancing cell, and conversion of endoplasm into ectoplasm at the front. Endoplasm slides forward over the lower layer of stationary ectoplasm, which rests on the substratum, and then becomes converted into ectoplasm, while the rearmost ectoplasm is simultaneously being converted into endoplasm and becoming mobile once more. The resulting movement of the cell is much like that of a bulldozer on its caterpillar track.

As to the motive force for the endoplasm, it is not clear whether the endoplasm is squeezed forward by the contractions of microfilaments near the rear of the pseudopodium or whether, on the contrary, it is pulled by contractions near the advancing end of the pseudopodium. Possibly both mechanisms operate. Since the actin microfilaments are much more abundant in the rear region than in the advancing portion of the pseudopodium, the push-from-behind model seems somewhat more likely.

MOVEMENTS PRODUCED BY MICROTUBULES

In addition to microfilament-mediated movements, there is a class of microtubule-mediated movements in eucaryotic cells. We saw in Chapter 5

20.30 The stroke of a cilium
In the power stroke the stalk is extended fairly rigidly and swept back by bending at its base. The recovery stroke brings the cilium forward again, as a wave of bending moves along the stalk from its base; at no time during the recovery stroke is much surface opposed to the water in the direction of movement.

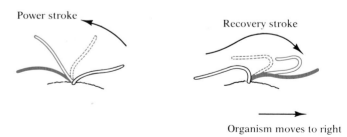

Power stroke

Recovery stroke

Organism moves to right

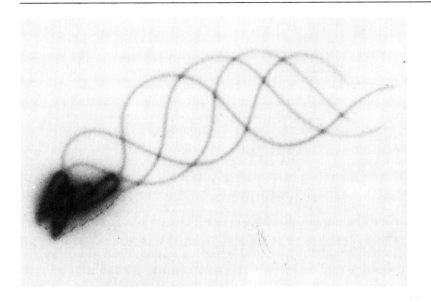

20.31 Multiple-exposure photograph of flagellum movement in a sea-urchin spermatozoon
Successive waves of bending that move along the flagellum, from its base toward its tip, push against the water and propel the sperm forward. This photograph was taken using four light flashes 10 milliseconds apart.

that microtubules are composed of many molecules of a protein called tubulin (see Fig. 5.19, p. 126). Microtubules are primarily eucaryotic organelles; they occur only in one major group of procaryotic cells.

The beating of cilia and flagella is the most obvious example of microtubule-mediated movement (Figs. 20.30 and 20.31). As we have already seen, the stalks of all eucaryotic cilia and flagella are alike in exhibiting the 9 + 2 arrangement of internal microtubules, in which nine, usually double, microtubules near the periphery of the stalk surround two single ones located near the center (see Fig. 5.25, p. 131). Recent electron-microscope studies have revealed many more details about the relationship among the nine microtubules. The two single microtubules are surrounded by a sheath, to which each of the outer doublets is connected by a radial spoke; the outer doublets are linked, and each has two clawlike arms (Fig. 20.32).[1]

In a series of definitive experiments, Ian Gibbons of Harvard University showed that a protein called **dynein,** found in cilia, can extract energy from ATP and, further, that when all the dynein is extracted from cilia the doublets are left armless, and when the dynein is restored the arms reappear—a clear demonstration that the arms are the site of ATP hydrolysis for the cilia. The resemblance of this arrangement to the localization of ATP hydrolysis in the cross bridges of muscle is obvious. The dynein arms may well function as cross bridges between adjacent doublet microtubules. Current models of ciliary movement postulate that the arms provide the basis for a ratchetlike mechanism that enables ciliary microtubules to "walk along" or slide over one another. The models postulate, further, that because of the shear resistance within the cilium, due in large part to the radial spokes that

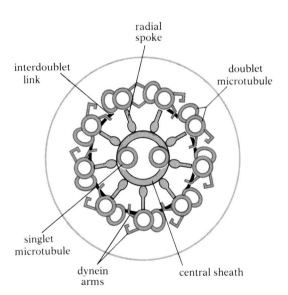

20.32 Internal structure of a cilium seen in cross section
See text for description.

---

[1] The flagella of most bacteria—*E. coli,* for example—lack microtubules and, consequently, the internal structure described here. These flagella are in the form of relatively stiff helices, which rotate like propellers to push the organism along.

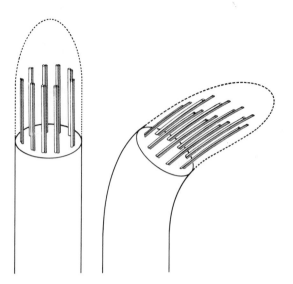

20.33 Bending of a cilium produced by sliding of microtubules past each other
The microtubules on the concave side of the bend (color, right) have slid tipward. Because the tubules are all interconnected, their changes in relative position can be accommodated only if the stalk of the cilium bends.

bind all the doublets to the central sheath, the sliding of the doublets past one another brings about a bending of the ciliary stalk (Fig. 20.33).

The remarkable resemblance in mode of action between the tubulin-dynein system of cilia and flagella and the actin-myosin system of muscle might suggest that the two are evolutionarily related, and that one is derived from the other. But there is no evidence to support this view. The amino acid sequences in tubulin and dynein show no similarities to those of actin and myosin. These two systems evolved their similar mechanisms for producing ratchet-driven sliding motion independently; this is a truly impressive example of what is called convergent evolution—the independent evolution of two very similar solutions to the same problem, as exemplified by the camera eyes of cephalopods and vertebrates (see Chapter 19) or the wings of birds and bats.

Microtubule-mediated movement is not limited to cilia and flagella. In the process of nuclear division (mitosis), to be discussed in Chapter 23, a spindle of microtubules functions in moving chromosomes as new nuclei are organized (see Fig. 23.12D, p. 624). And the transport of mitochondria and membranous vesicles through the axon of a neuron, from the cell body to the axonal tip (a process known as rapid axoplasmic flow), also depends on microtubules; treatment of the axon with colchicine, a reagent that causes breakdown of microtubules and blocks assembly of new ones, inhibits the transport. According to current hypotheses, the organelles being transported are passed along the microtubules by arms composed of protein that can hydrolyze ATP, in a manner strongly reminiscent of the action of the dynein arms in cilia and flagella.

Chapter 21

ANIMAL BEHAVIOR

In the last six chapters we have followed the flow of information through organisms. We have seen the various ways environmental stimuli are "processed" and transmitted to effectors to produce appropriate responses. In this chapter we shall examine the *mechanisms* behind the behavior of animals: how the behavioral repertoires of animals are organized through an elaborate coordination of sensory organs, neural pathways and processing networks, the endocrine system, and effector organs. In the next chapter we shall concentrate on the *evolution* of behavior: how selection for variations in behavior that confer an advantage cause behavior to evolve to suit the lifestyle of each species. The study of the mechanisms and evolution of animal behavior is known as ***ethology.***

The idea that something as intangible as behavior could be inherited has always been controversial. But by the eighteenth and nineteenth centuries, naturalists had catalogued thousands of observations, and Darwin could point to many examples of adaptive behavior that was clearly inherited. Perhaps the example most compelling to Darwin was the life history of the European cuckoo, which lays its eggs in the nests of other birds (Fig. 21.1). The cuckoo chick hatches in the nest of one of several possible host species, and even before its eyes are open, it ejects the hosts' own eggs and chicks (Fig. 21.2A). The foster parents, unable to recognize the cuckoo as an intruder, feed and care for the chick until it is fledged, by which time it is considerably larger than they are (Fig. 21.2B). Even though it has probably never seen or heard another cuckoo of either sex, the young bird is able to find and recognize a suitable mate the following spring, to court, and to copulate, all in a manner typical of its species.

How, Darwin wondered, is a cuckoo able to do precisely the right thing at

21.1 Egg removal by a cuckoo
A cuckoo about to lay an egg in a reed warbler's nest first removes one of the warbler's eggs. The warbler apparently doesn't distinguish size differences among the eggs, but the proper count is crucial. If she found one egg too many, she would abandon the nest, and both her own eggs and the cuckoo's along with it.

A

B

21.2 Egg ejection by a fledgling cuckoo
(A) A newly hatched cuckoo rolls the eggs of its host
out of the nest. As a result the young bird does not
have to share any of the food its foster parents
collect. (B) The unwitting foster parent (a hedge
sparrow, left) continues to feed the cuckoo even when
it has grown to several times the parents' size.

the right time having had little or no opportunity to learn; and when something must be learned—the appearance of the host species, for example,
which the female cuckoo must recognize in order to locate the proper nest
for her egg—how does the bird "know" to ignore a world full of distracting
information and focus on exactly what must be memorized? Amazing as it
seems, the baby cuckoo must inherit the essential "instructions" in its
genes—genes that direct the wiring of its nervous system during development. The underlying instructions that direct learning and behavior are
known popularly as instinct; behavior like that of the cuckoo, which depends largely on inherited mechanisms, is usually referred to as innate or
instinctive.

As we shall see, it is extremely unlikely that any behavior can be classified
as strictly innate or strictly learned: even the most rigidly automatic behavior depends on the environmental conditions for which it evolved, while
most learning, flexible as it seems, appears to be guided by innate mechanisms. *Instinct,* then, can be defined as the heritable, genetically specified
neural circuitry that organizes and guides behavior. The behavior that is
thereby produced can reasonably be said to be at least partially innate.

THE FUNDAMENTAL COMPONENTS OF BEHAVIOR

SENSORY WORLDS

A major turning point in the study of behavior came around 1915, when the
great Austrian zoologist Karl von Frisch discovered that honey bees had a

range of sensory experience largely outside our own. Von Frisch had begun by wondering why flowers are colorful. The notion that insects might have color vision, and that colorful flowers might therefore appear attractive to their pollinators, was hardly taken seriously at the turn of the century. Nevertheless, von Frisch put the issue to a test; choosing neither physiological nor biochemical methods, he began a new tradition by designing behavioral tests. First he trained honey bees to collect sugar solution from a dish placed on a blue card. Later he removed the card and food, and set out another blue card among cards of varying shades of gray, with an empty dish on each (Fig. 21.3). He reasoned that if bees had only black-and-white vision, they would confuse at least one shade of gray with blue. In fact, however, the bees had no difficulty whatever in distinguishing the card of the color they had been trained to recognize.

Subsequent work revealed that bees cannot distinguish red from dark gray—that is, they are blind to red—but their vision extends well into the ultraviolet (UV) range. Von Frisch therefore examined the world through UV filters, to see what bees were seeing, and he discovered that bee-pollinated flowers show a distinctive bull's-eye pattern, with a dark center (Fig. 21.4). Later he and his students found that bees have several capabilities unavailable to humans: they see polarized light, hear sounds inaudible to humans, smell carbon dioxide and humidity, as well as certain odors too faint for our noses, and sense the earth's magnetic field. The work of von Frisch and those he inspired made it clear that the sense organs of various animals have evolved to provide them with the range of sensory experience that suits their particular needs.

21.3 A test for color vision
After being trained to go to a feeding dish on a blue card, honey bees were presented a varied array of cards, each under an empty food dish. The bees demonstrated their color vision by searching for food only on the blue card, where they had been trained to expect it.

A

B

21.4 Flowers in visible and UV light
To us, the primrose willow appears to be a nearly uniform yellow (A), but to bees, which can see UV light, the center is marked by a dark bull's-eye (B).

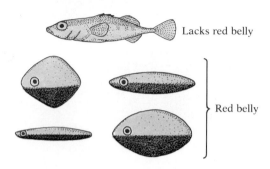

21.5 Models of male stickleback
The realistically shaped model lacking the red belly was attacked by male sticklebacks much less often than the oddly shaped models with red bellies.

SIGN STIMULI AND RELEASERS

The discoveries of von Frisch and others led a Bavarian naturalist, Konrad Lorenz, to realize that though the private sensory worlds of many animals overlap, so that they share the same perceptions, their brains may interpret these perceptions very differently. This discovery came in the early 1930s, when Lorenz noticed that he was attacked by his pet jackdaws whenever he carried something black hanging from his hand. The birds, which are themselves black, seemed to interpret any dangling black object as a fellow jackdaw in distress. It seemed to Lorenz that the birds, though evidently capable of recognizing each other as individuals in other situations, were ignoring most of what they could see and focusing instead on a small (and in this case misleading) subset of cues in this situation.

Lorenz and the ethologist Niko Tinbergen examined this phenomenon further, and found that many animals are highly responsive to specific stimuli. For example, both males and females of the common, minnowlike stickleback recognize breeding territorial males by the red stripe on their ventral surfaces. Sticklebacks are so thoroughly attuned to the red stripe that they are oblivious to additional cues that might otherwise be useful, such as the size and shape of the object displaying the color (Fig. 21.5). Tinbergen reported, in fact, that his territorial males would attempt to attack passing red British mail trucks that were visible through the sides of their aquariums. In behavioral terms a *sign stimulus* is any simple signal, such as the red stripe, that elicits a specific behavioral response. The specificity of a sign stimulus for a particular behavioral reaction is illustrated by Lorenz's observation that a mother hen can recognize a chick in trouble only by its special distress call, even when the chick is clearly visible (Fig. 21.6).

Sign stimuli seem to organize behavior throughout the animal world. Ground-nesting birds such as gulls and geese instinctively rotate their eggs —a behavior that prevents the embryos from sticking to their shells—and then carefully roll any eggs that escaped during this procedure back into their nests. But these birds will also roll flashlight batteries, golf balls, beer bottles, and a variety of other rounded objects into their nests, apparently mistaking them for eggs. Similarly, male robins will attack anything red on their territories during the breeding season. The response of these birds to seemingly inappropriate cues does not indicate inability to see clearly. Rather, experiments with these and other examples of the same sort of rote behavior have shown that certain stimuli trigger a particular behavior because the organism possesses an exaggerated neural sensitivity to these stimuli: because sign stimuli are said to "release" specific behaviors, they are frequently called *releasers.*

Depending on the sensory world of the organism in question, behavior may be released by cues from many sensory modalities. *Pheromones,* odors to which specialized receptor cells may be attuned, often serve as releasers. The odor of starfish that triggers an escape response in *Aplysia,* as described in Chapter 19, is a pheromone. Sounds, too, may act as releasers. For example, the high-frequency sounds produced by the wingbeat of a female mosquito attract males of the same species for mating (Fig. 21.7). The cries of bats (or the similar frequencies produced by jingling keys) release evasive maneuvering in certain moths. The bull's-eye pattern of many

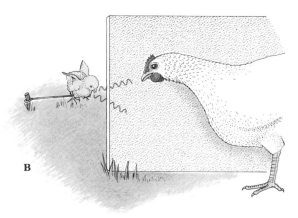

A

B

flowers in UV light, which attracts bees, is a visual releaser. A sensitivity to the releasers their species employs is present in most animals at birth, and generates a variety of species-specific behaviors. The young of most species are also born with the releasers necessary to direct and trigger parental care. The success of the cuckoo chick, for instance, is assured because it possesses at birth an orange throat patch and specialized peeping, the releasers that cause its surrogate parents to recognize and feed their offspring.

Releasers have a great advantage: they can initiate certain critical behavioral responses automatically, thus bypassing the time-consuming and error-prone process of learning. But releasers have at least one major disadvantage as well, in that they may be triggered by crude and inappropriate stimuli. As we shall see, a combination of learning and releasers frequently provides animals with the best features of both.

RELEASER RECOGNITION

As releasers have come to be better understood, it has become clear that many of them—perhaps all—depend on the feature-detector circuits described in Chapter 19. Working independently, Niko Tinbergen at Oxford and Jack Hailman, now at the University of Wisconsin, conducted extensive experiments with models to determine which characteristics of the parent

21.6   Difference in response by a hen to her chick's visual and vocal distress signals
(A) The hen ignores the chick if she cannot hear its calls, even though its actions are clearly visible.
(B) Distress calls elicit vigorous reaction from the hen even when she cannot see the chick.

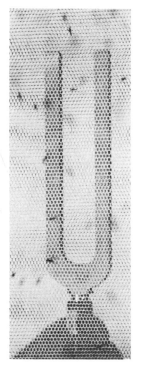

21.7   Photographs showing response of male mosquitoes to a tuning fork
Left: The fork is silent. Right: The fork is vibrating at about the same frequency as the female mosquito's wingbeat, and its sound is attracting males.

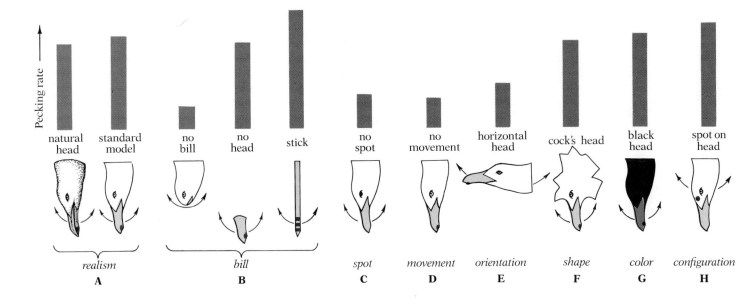

21.8 Releasers and pecking rate in herring gull chicks

To determine which characteristics of a herring gull's head release pecking in the chicks, Tinbergen and Hailman offered the chicks a series of models. Except as noted, all were held vertically and moved back and forth horizontally, as indicated by the arrows. It turned out that the standard flat cardboard model is slightly more effective than the head of a real bird (A), and a disembodied bill (B) is almost as good. The spot (C), movement (D), and bill orientation (E) are crucial, but head shape (F) and color (G) are not. The color of the bill and spot (G) are likewise of little consequence, as long as they contrast well. The model with the misplaced spot (H) was moved so that the spot traversed the same arc as the bill spot on the other models. The two releasers for pecking, then, are a vertical bar moving horizontally, and a moving spot that contrasts with its background. A model emphasizing these features—a narrow stick with three spots that is waved back and forth (B)—is a supernormal stimulus.

herring gull incite the chicks to peck at the adult's beak for food. The experiments indicated that the baby birds respond instinctively to the red spot on the parent's downward-pointing beak, as it is waved slowly back and forth. However, they lack any internal "picture" of the parent gull. A model with no head, or with a misplaced spot, is effective (Fig. 21.8B, H); indeed, an unrealistic model that emphasizes spots and vertical bars is actually more effective than a normal gull head (Figs. 21.8B and 21.9). Exaggerated features of this sort, that are superior to the natural stimulus in releasing a response, are known as ***supernormal stimuli.***

The two releasers operative in this instance—a spot moving horizontally, its color relatively unimportant as long as it contrasts with the background, and a vertical bar moving horizontally—combine in such a way that the releasing values of two stimuli add together in the chick's central nervous system to produce an increased probability of response. This type of combination, common throughout nature, is called ***heterogeneous summation;*** it increases the specificity of the recognition system.

As Hailman later pointed out, the two releasers that enable chicks to recognize their parents correspond to two classes of feature detectors: a horizontal-motion spot detector, and a horizontal-motion vertical-line detector. An equivalent pattern has been confirmed by studies of how the common European toad recognizes prey. Over the past two decades Jörg-Peter Ewert of the University of Kassel in Germany has defined the stimuli that release a prey-capturing sequence in which the toad turns toward an object and fires its long sticky tongue. The single most powerful releaser of this behavior seems to be a bar moving along its long axis—a cue naturally provided by the worms and centipeds the toad favors (Fig. 21.10). This sign stimulus is even more effective in the presence of the odor of prey, which the toad recognizes innately.

21.9 Pecking by a herring gull chick
Young herring gull chicks peck at the red spots on their parents'
bills from birth, and will peck at the cardboard model on the
right in this picture as frequently as they will at a real bird. But
the chick's pecking behavior is so completely dictated by a set of
releasers—one or more red spots on a vertical bar moved
horizontally—that it will ignore a full representation of a herring
gull to peck at the stick painted with three contrasting spots.

Recording from the visual areas of the toad thalamus and tectum, Ewert
found a sensory "map" of the toad's visual world, one layer of which contained feature-detector circuits tuned specifically to prey shape and movement in the lower half of the toad's visual field—the usual location of its
terrestrial prey. Using microelectrodes, Ewert could sometimes stimulate
an individual cell and thus cause the toad to turn to the approximate direction of the corresponding spot in the visual field and launch its tongue at an
imaginary target. In the course of his research Ewert also found two other
classes of feature detectors in toads—one for recognizing moving spots,

A

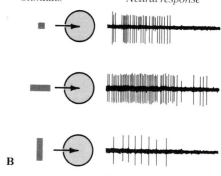

B

Stimulus   *Neural response*

21.10 Prey-capture behavior by a toad
(A) A toad responds to an elongated stimulus moving lengthwise
by turning toward it and striking at it with the tongue. The toad
then "swallows" (with eyes shut) and wipes its mouth, even
though it has not actually caught anything. (B) This behavior is
controlled by feature detectors in the toad brain that respond
preferentially to a wormlike stimulus (horizontal bar, middle row)
moving into their receptive fields.

21.11 Egg rolling by a goose
When a goose sees an egg outside her nest she rises, touches the egg with her beak, and then rolls it back in. She completes the same recovery behavior when the object she sees is a beer bottle, or when the egg is removed after she has begun to reach for it.

presumably flying insects, and another for recognizing potential predators (essentially any large, moving, and nearby object), to which the toad responds with crouching or flight.

Visual, auditory, and olfactory feature detectors also underlie much of the instinctive recognition of conspecifics so necessary in most animal communication.

MOTOR PROGRAMS

FIXED-ACTION PATTERNS

Lorenz and Tinbergen noticed in the 1930s that some behaviors appear to be all-or-none responses to releasers: they tend, once begun, to run to completion regardless of the situation. A dramatic example is the egg-rolling response of geese. If a goose notices an egg outside her nest, she rises, extends her neck until the egg is touching the underside of her bill, rolls it gently back into the nest, and then settles down to continue brooding (Fig. 21.11). To the casual observer this behavior appears to show thought on the goose's part: she has recognized a problem and solved it. But as noted earlier in this chapter, despite the experience of continually turning the eggs in the nest, a goose will recover and brood a variety of objects that are clearly not eggs. She apparently has only the vaguest idea of what an egg is. If an egg is removed while she is reaching for it, the goose will go on as if nothing had happened, rolling the nonexistent egg carefully into the nest. In short, egg rolling is an independent behavioral unit which, once initiated, proceeds to completion with little or no need for further feedback. Lorenz and Tinbergen called such units *fixed-action patterns.*

Examples of fixed-action patterns are common. Among those we have discussed in this and previous chapters are the sequence by which the baby cuckoo disposes of its hosts' eggs and chicks, the prey-capture behavior of toads, the feeding, gill-withdrawal, and escape responses of *Aplysia,* the flight of locusts, and swallowing in humans.

The fixed-action patterns that Lorenz and Tinbergen identified are what we now call motor programs (see p. 488). We shall use this more modern term in the discussions that follow, but many researchers still employ the older term to distinguish behavioral units like egg rolling that are almost completely independent of sensory feedback, from the feedback-dependent, and often learned, motor programs (like scratching, shoe tying, or piano playing) that underlie much of the behavior of higher vertebrates.

MATURATION AND MOTOR LEARNING

It is often difficult to be sure whether or not a motor behavior has been learned if it is not actually exhibited at birth. In some cases, though, particularly among the invertebrates, opportunities for learning are so slight that even many behaviors seen only in adults must be regarded as innate. A wasp that specializes in capturing honey bees, for instance, must be equipped from birth to spin a cocoon, emerge and dig out of its particular kind of burrow or chamber, groom itself, fly, court and mate, pounce on bees, sting them in an unarmored patch under the neck without being itself stung,

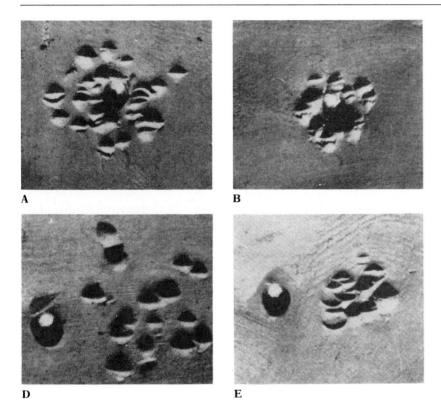

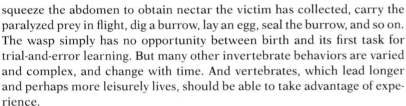

squeeze the abdomen to obtain nectar the victim has collected, carry the paralyzed prey in flight, dig a burrow, lay an egg, seal the burrow, and so on. The wasp simply has no opportunity between birth and its first task for trial-and-error learning. But many other invertebrate behaviors are varied and complex, and change with time. And vertebrates, which lead longer and perhaps more leisurely lives, should be able to take advantage of experience.

Nevertheless, even in vertebrates many behaviors that appear to be learned are actually innate. A classic example was provided by Eckhard Hess of the University of Chicago in an elegant experiment that demonstrated the maturation of a motor program. He fitted newborn chicks with tiny goggles that deflected their vision seven degrees to the right, and recorded the accuracy of their pecks by providing a target (a nailhead that, like seed, acted as a sign stimulus for pecking) set in soft clay. The pecks of both the normal chicks and those with the goggles were scattered, but the marks of the chicks with goggles were well to the side of the target (Fig. 21.12A, D). A few days later, chicks of both groups were able to produce a tight cluster of pecks, but those with the goggles were still missing the target as much as before (Fig. 21.12B, E). Apparently the chick's circuitry for aiming and pecking is already wired in at birth, and the normal improvement in accuracy is a simple consequence of increased nerve and muscle coordination rather than of learning.

By contrast, many behaviors we know to be learned take on the appearance of fully innate motor programs. Walking, for instance, which is innate

21.12 Maturation of pecking behavior in chicks
Newborn chicks peck at a target with fair accuracy (A), but their aim improves with age until at four days the pecks are tightly clustered (B). This improvement could be the result of some sort of maturation—better vision, perhaps, or strengthened neck muscles—or of learning, by which the chick recognizes and corrects its errors. Eckhard Hess pitted these alternatives against one another by raising chicks with goggles that deflected their vision to the right (C). As newborns, such birds produce the usual set of scattered pecks, but the pecking is well to the right of the target (D). By the fourth day, the pecks are tightly clustered but still misdirected (E), indicating that chicks are unable to learn to adjust their aim. The coordination of beak and eye involved in pecking must therefore be a wholly innate behavior, which matures without benefit of learning.

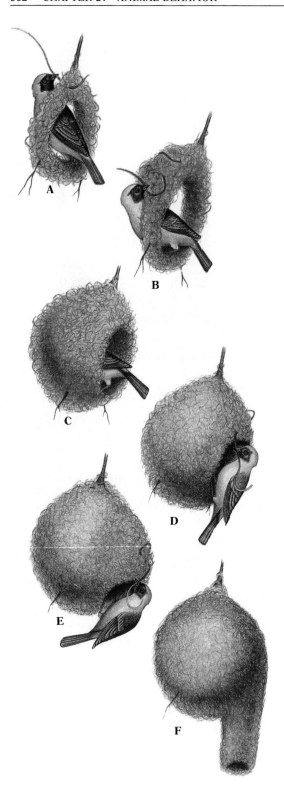

in most species, must be learned initially in ours. Though the alternation of the legs and the interacting reflex arcs responsible for walking come pre-wired at birth—a properly supported human infant will perform walking motions on the delivery table—humans must learn to balance, once they have matured enough to support their own weight. Yet after the difficult process of learning has been completed, simple walking becomes automatic. Swimming and bicycle riding present the same story, and as everyone knows, once painstakingly learned, neither is ever completely forgotten. The same pattern may be seen in other animals: learned behavior can become stereotyped and largely automatic—that is, take on the characteristics of an innate motor program. This apparently occurs as new motor-program circuits are wired up in the brain. In fact, studies of the effects of localized brain damage in humans reveal that special areas of the cortex and cerebellum are reserved for learned motor programs. These areas are localized to a high degree, with one region of the cortex, for example, constructing and storing motor programs requiring fine finger control—typing, writing, weaving, playing the piano, tying shoes, and the like—while another is devoted to programs involving the kind of limb movements necessary for swimming and kicking a ball. This freeing of learned behavior from detailed conscious control allows conscious attention to focus on new problems—an advantage for humans and for other animals too: a bird that has learned to shell seeds automatically can devote its attention to watching for predators while it eats.

ORCHESTRATION OF BEHAVIOR

Before we consider the major role learning frequently plays in complex motor behavior, we must look at the degree of complexity and flexibility that is possible without learning. Observation of animals, and experimentation with them, have made it clear that exceedingly complex behavior can be choreographed by instinct. For instance, all birds, so far as is known, follow innate instructions in building their nests, instinctively recognizing appropriate material and suitable locations. Bird nests vary in complexity from a simple hollow scraped in the ground to elaborate multilayer structures with a different material for each layer—a robin will first use sticks, then twigs, then mud, and finally grass. The substances used to join these materials and to attach the nest to its support are as species-specific as the nests themselves, varying from dung to spider webs to special adhesive saliva.

21.13 Construction of a village weaverbird nest
A male village weaverbird begins with a suitable forked branch as a support and weaves around himself until he has a circular perch (A). Then he stands facing out of the circle and weaves around himself (B) until he has completed a roof and nest cup (C). Then he weaves backward to create a doorway (D–E). Finally, if he is successful in obtaining a mate, he weaves an entrance tube, which helps keep out nest predators (F).

The means by which the many steps of nest construction are orchestrated are nowhere clearer than in the nest-building *tour de force* of weaverbirds, as observed by N. E. Collias and E. C. Collias of the University of California, Los Angeles. A male weaverbird begins by selecting a suitable branch—usually in the shape of an inverted Y—to support the nest (Fig. 21.13A). The male then begins collecting green material, normally vegetation, which he tears into long strips suitable for weaving. He next initiates a "do loop"—a series of steps repeated over and over until a preset goal or criterion is reached—that involves weaving the strips one by one over and around himself in an arc: he puts the tip of the strip through a narrow opening, vibrates his bill until the tip comes out the other side, grasps the strip and pulls it through, and then begins again, on and on until the strip is entirely woven in. When all but the entrance to the nest is complete (Fig. 21.13E), the male breaks off building and begins an instinctive courtship display, in which he hangs upside down from the nest, vibrating his outspread wings and calling. If a female accepts the male, she lines the nest cup with a layer of soft grass and then a layer of feathers, while the male builds an entrance tube (Fig. 21.13F) that serves to keep out snakes and other nest predators.

The building of this complex structure is organized into relatively simple subroutines, each with its own motor program and clearly defined criteria for terminating the do loop. For example, since Collias could end the thatching phase (during which wide strips are woven into the roof to make the nest waterproof) by covering the top of the nest with an opaque cloth, the bird is evidently programed to stop thatching when an opaque layer is complete. Once the bird had begun his next subroutine, Collias could remove the cloth and the bird would not resume thatching. Most complex innate behavior is probably structured along these lines: specific behavioral subroutines, cued by signals from the environment, start and stop in a preset ordered sequence.

MOTIVATION AND DRIVE

Different species behave differently—for instance, weaverbirds build nests quite different from those of robins or geese. In addition, the same animal may behave differently at different times—birds are not always building nests, flying south, or courting potential mates. The forces from within an animal that motivate it to do one thing now and another later are known as *drives.* A drive can have two basic effects, neither of which is well understood: it can alter an animal's threshold to stimuli, thus making a particular behavior more or less likely, and it can substitute entirely new programs for old ones. Present evidence strongly suggests that drives are shaped by the sorts of hormonal control, proprioceptive monitoring, and habituation we have examined in previous chapters.

MODULATION OF BEHAVORIAL PRIORITIES

In response to simultaneously active drives, an animal must choose among several behaviors, such as searching for food, searching for water, attempting to attract a mate, grooming itself, patrolling its territorial boundaries to

21.13G The nest of a thick-billed weaverbird
Unlike the village weaver, *P. cucullatus* (shown in A through F), the thick-billed weaver, *A. albifrons* (shown here) builds its nest in the marshlands and begins nest building by constructing a bridge between the upright supports provided by cattails and other marsh stems. The strategy of nest building employed by thick-billed weavers is intermediate between the typical technique of other birds and the weaving strategy of the more advanced weaverbirds. Both village and thick-billed weavers are indigenous to sub-Saharan Africa.

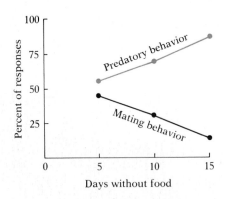

21.14 Change in response of a male jumping spider to a model as a function of the number of days he has gone without food
Among the prey of male spiders are some insects that closely resemble female spiders. It is therefore possible to construct models that can release either prey capture or mating, depending on the motivational state of the spider. The longer he has gone without food, the more likely he is to exhibit predatory behavior instead of mating behavior.

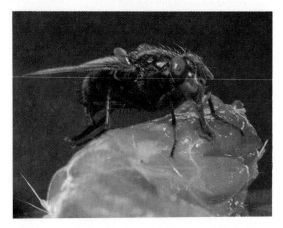

21.15 A blowfly, with its inflatable proboscis extended

guard against intruders, repairing its nest or burrow, or even playing. At any given moment, some of these behaviors will be more important than others. Thus escape behavior almost always takes precedence. Current priorities, established by the shifting thresholds of response of various drives, can help the animal select among the many behavioral possibilities that may be competing for its attention, a process often referred to as "time-sharing." These priorities determine, for instance, whether a spider will feed or mate (Fig. 21.14). Performing the behavior with the highest priority will lower the urgency of the drive that motivated it, and behavior that was formerly less important will then become the animal's highest priority. What behavior an animal chooses depends not only on the relative urgency of various drives, but on opportunity as well: an animal only mildly hungry but very thirsty will prefer water to food if both are present, but will probably eat if no water is available.

One type of mechanism by which drives are modulated is exemplified in hunger and thirst. In mammals the sensation of thirst is controlled by a small group of cells in the hypothalamus that measure the salt content of the blood. Injecting even a minute quantity of salt near these cells makes an animal extremely thirsty, radically altering its priorities and hence its subsequent behavior: the animal's drive directs it to ignore all other concerns (sometimes even predators) and search desperately for water. A mild increase in the salt level of the blood simply moves an organism's need to find water higher on its list of priorities.

This kind of interaction between internal monitors and behavior can be observed in many animals. A hungry blowfly, for instance, is motivated to fly around in search of food. Once it smells the odor of fermenting organic matter (its sign stimulus for food), it turns and tracks the scent upwind. After landing, the fly walks about until it actually steps into the food, thereby initiating a series of linked motor programs: Taste cells on the fly's feet fire, directing the fly to turn toward the food and extend its inflatable, strawlike proboscis (Fig. 21.15). Contact with the substrate causes sensory hairs to direct the lips of the proboscis to open, exposing yet another set of taste cells. If these too report the presence of suitable food, the fly begins to pump it into its crop. When stretch receptors in the crop report that organ to be full, the pumping stops until the contents of the crop are emptied into the gut; once the inhibition from the crop stretch receptors is removed, pumping begins again. This cycle continues until stretch receptors in the foregut report that the digestive system is full. At this point the fly retracts its proboscis, grooms, and sets off, now most strongly motivated by a drive other than hunger, perhaps the need for a mate.

This network of sign stimuli, motor programs, and feedback control was sorted out by Alan Gelperin and Vincent Dethier at Princeton University. They then confirmed the role of the stretch receptors in the crop by cutting the nerve that carries their output to the central nervous system. Flies that had undergone this operation ate indefinitely, some literally exploding from their inability to restrain themselves.

Hunger and thirst, then, act as internal stimuli that can redirect an animal's behavior so that its focus is on eliminating these needs. They are part of a larger feedback mechanism for maintaining homeostatic blood levels. But most drives do not conform to this simple "appetitive" model (so called

A

B

C

D

because animals seem to seek to perform certain behaviors to satisfy some sort of appetite). Geese, for instance, do not actively seek out eggs to roll after a period in which misplaced eggs are lacking, nor do they appear "starved" for predators after a period of calm. The physiological bases of nonappetitive drives are not understood.

CHANGING BEHAVIORAL PROGRAMS

In addition to altering thresholds to stimuli, drives can act by bringing in or retiring entire behavioral programs. A migratory bird, for example, must dramatically alter its dietary intake and metabolism to accumulate the fat reserves necessary for its annual journey, and then must set off in the correct direction at the appropriate time of year. How does the bird come to switch its behavior in anticipation of the need, even when it has never migrated before? And why do geese retrieve eggs outside their nests only in the period from about a week before laying until a week after hatching, while at other times the same cue is ignored? Though answers to such questions are currently incomplete, we do know that animals often behave as though some sort of timer is repeatedly switching their behavior patterns from the set appropriate for one stage of their life cycle to the set appropriate for the next stage.

The brooding behavior of ring doves is a particularly well understood example of such automatic switching of behavior patterns (Fig. 21.16). In the normal course of events ring doves court and pair, build a nest, lay two eggs in it (one day apart, both in the late afternoon), and begin to brood or incubate them. About 16 days later the eggs hatch and the adults feed the young a liquid secretion known as crop milk, produced by special glands in their throats. Erich Klinghammer, now at Purdue University, and Eckhard Hess have elucidated some of the mechanisms behind the changing behavior of

21.16 Behavior of a pair of ring doves
A pair of laboratory-reared doves court (A) and mate (B). They go on to build their nest in a ceramic dish (C) and rear offspring (D) in this artificial environment. Their remarkable willingness to continue this behavior outside of their natural environment has made doves one of the standard laboratory animals.

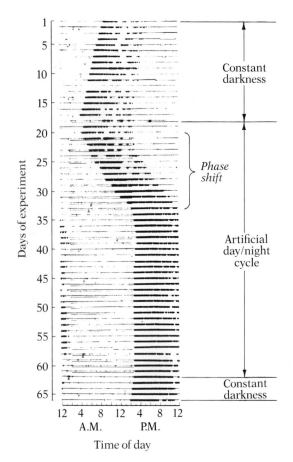

21.17   Record of a circadian-rhythm experiment
In constant darkness (days 1–17) a flying squirrel
(*Glaucomys*) continues to spend about 10 hours in
foraging activity and then 14 hours in resting.
Because the lengths of these periods are only
approximate, the time of foraging drifts slowly with
respect to a 24-hour day. When an artificial day/night
cycle is imposed (days 18–61), the activity rhythm
gradually shifts into phase with it. When constant
darkness is resumed (day 62), the period of foraging
again begins to drift.

these birds. An important discovery came when they provided nesting ring
doves with foreign eggs, in preparation for tests in which the young of one
species were to be reared by adults of another. The ring doves produced
crop milk 16 days after eggs appeared in the nest, regardless of whose eggs
they were, when the pair's own eggs had been laid, or when the chicks
hatched. Evidently, the *sight of the eggs* initiates the ticking of a 16-day behavioral timer in ring doves that controls the metabolic and physiological
preparations necessary for feeding the young. Seeing the eggs, it turns out,
also triggers the drive to perform the appropriate brooding and feeding behavior. In the ring dove, this particular timer controls the level of the hormone prolactin in the parents. Prolactin binds to specific neurons in the
brain and activates the neural circuits involved in parental behavior and in
the production of crop milk. The exact level of prolactin is critical: incubation is triggered by a low concentration, while the production of crop milk
requires a higher level. Similar behavioral timers are most likely involved
in egg rolling and countless other motor behaviors, though they may not all
use specific hormone levels to switch behavioral circuits on and off.

CIRCADIAN RHYTHMS

The most obvious of the internal timers that switch behavioral routines on
and off and modulate drives is the nearly universal one that controls daily
behavioral cycles—the cycle of sleeping and waking, for example. We
know that this "clock" is independent of outside influences, since experiments show that the animals' cycles persist even in the absence of outside
cues. A nocturnal animal like the flying squirrel, for instance, will continue
to alternate about ten hours of foraging with 14 hours of resting even in
continuous darkness (Fig. 21.17). The internal nature of timers is also demonstrated by an experiment in which Karl C. Hamner of the University of
California, Los Angeles, took hamsters, fruit flies, cockroaches, cockleburs,
soybean plants, and bread molds to the South Pole and placed them on a
turntable set to rotate at exactly the same speed as the earth but in the opposite direction. In other words, he exposed these organisms to conditions
where no earthly indications of daily time existed. Yet the regular 24-hour
cyclic activities of these organisms continued.

Because the period of animal clocks is approximately (though not exactly) 24 hours, the cycles they control are called ***circadian rhythms*** (from
the Latin *circa*, "about," and *dies*, "day"). But under experimental conditions, activity rhythms drift with respect to the 24-hour day (Fig. 21.17). The
rhythms of organisms can be reset by a species-specific hierarchy of cues,
the most important of which is usually light. The clock's response to resetting (also known as "phase shifting" or, more commonly, "clock shifting")
has a time schedule of its own. A flash of light seems to be taken by many
animals as a sign stimulus for dawn; if experienced in the six hours before
dawn is expected it will advance the rhythm, and if experienced in the dark
in the first few hours after dawn *ought* to have occurred, it will delay it. The
process of resynchronizing the circadian clock to fit external cues may take
up to several days, depending on the extent of the difference between internal and external time (Fig. 21.17). A difference of this sort is, of course, the
basis of "jet lag."

Manifestations of circadian clocks are not restricted to the sorts of gross reactions of whole organisms mentioned so far. There is ample evidence that clock phenomena are characteristic of virtually all living things, whether individual cells or whole multicellular plants or animals. Aspects of cellular function that vary in approximately 24-hour cycles include enzyme activity, osmotic pressure, respiration rate, growth rate, membrane permeability, bioluminescence, sensitivity to light and temperature, and reactions to various drugs. Physicians are becoming increasingly aware that the proper dosage of a drug may be very different at different times of day; in some cases, what constitutes a beneficial dose at one time may actually be lethal at another.

Circadian clocks are by no means the only short-term behavioral timers known. Animals living near the seashore have a 13-hour tidal clock and a 27-day lunar rhythm that together enable them to anticipate the time and day of peak tides. The precision of this combination is obvious in a species of intertidal midge that emerges from the sand, mates, and lays eggs in the period between 2 and 5 A.M. during the month's lowest low tide. Such accuracy is essential since this organism has only a two-hour life-span as an adult.

Although researchers have uncovered countless behavioral timers that enable animals to prepare for various regularly occurring situations, the physiological basis of timers is not yet well understood. The internal rhythms of unicellular organisms, however, provide some intriguing clues. For instance, the rhythms of the green alga *Acetabularia* appear to be set by the cell's nucleus, but its cytoplasm then appears to be perfectly capable of maintaining the rhythm and of carrying out phase shifting when appropriate. Analogously, various cells and organs in multicellular animals are able to maintain their own cycles, though they are normally all kept in synchrony by a master clock in the brain.

The means by which synchrony between the master clock and the numerous local timers is maintained appears to be chemical in some organisms and neural in others. James Truman of the University of Washington has shown, for instance, that the emergence of moths from their pupal cases is controlled by a circulating hormone rather than by nervous impulses. These insects all begin to emerge at a particular time of day, their clocks having been set days earlier by the first light of dawn. Truman exposed one group of moths to an artificial dawn, and then inserted their brains into the abdomens of a second group of moths whose brains had previously been removed and who had been kept many hours out of phase with the first group. By releasing a hormone into the body fluid of their new hosts, the transplanted brains caused those moths to emerge at the time appropriate for the donors. Subsequent experiments with birds have shown that isolated pineals implanted in pineal-less birds produce similar results, which suggests that the pineal acts as the master clock in birds. The question of how temporal information is communicated chemically, and of how chemical processes, which typically take place during microseconds, can time periods of hours, days, and years, is yet to be answered.

LEARNING AND BEHAVIOR

So far we have focused on the sorts of behavior that are controlled by genetically specified hormones and neural circuits. Such behavior is modulated

21.18 Pavlov's experiment showing classical conditioning
The device on the dog's cheek measures salivation, the unconditioned response; and the dish at left contains meat powder, the unconditioned stimulus. The conditioned stimulus is the light.

adaptively by drives, triggered and guided by external cues. It is subject only to the modifying influence of specific sensory feedback, and—where it is under neural control—of the cellular phenomena of adaptation, habituation, and sensitization, as discussed in Chapter 18. It can, as in nest building, be amazingly complex. But purely instinctive behavior frequently lacks adaptive flexibility. The changing environmental and social conditions to which many species must adapt require the behavioral flexibility provided by learning.

CONDITIONING

Many early psychologists believed that all behavior, whether simple or complex, is the result of conditioning—that is, of the organism's learning from its experiences in the environment what to do in order to survive. Though this view has turned out to be incorrect, conditioning studies have revealed much about how animals learn.

Classical conditioning Psychologists recognize two general forms of conditioning. The first, discovered by the Russian physiologist Ivan Pavlov (1849–1936), is known as Pavlovian conditioning or *classical conditioning.* Pavlov encountered this form of learning during his pioneering studies of the physiology of digestion. While measuring the quantity of saliva produced by dogs when they see food, Pavlov noticed that if an irrelevant stimulus—a light or a ringing bell, for instance—appeared just before the dogs saw the food, the dogs came in time to associate the stimulus so closely with the expected food that the light or bell alone would trigger salivation (Fig. 21.18). They had been conditioned.

Classical conditioning, then, begins with an unconditioned (innate) response (UR) that inevitably follows an unconditioned (instinctively recognized) stimulus (US). A novel, conditioned stimulus (CS) is then paired with the US, and after this process has been repeated sufficiently, the CS alone will elicit the response (which at this point may be referred to as the CR). In short, the animal comes wired with the sequence US → UR; the environment (or the researcher) provides the relationship CS + US → UR; and the animal generalizes from a series of individual experiences, so that CS → UR. Animals as diverse as *Aplysia* and humans can be trained through classical conditioning.

Multitudes of experiments similar to Pavlov's have shown that classical conditioning is an instinctive process by which animals free themselves from exclusive dependence on simple releasers, or sign stimuli, for the recognition of important objects and individuals in their environments. Of course, species differ dramatically in the range of acceptable conditioned stimuli and in their ability to spot subtle CS–US connections.

Trial-and-error learning Early psychologists believed that chains of linked conditioned stimuli and unconditioned responses could explain complex behavior. B. F. Skinner of Harvard University, however, recognized that classical conditioning cannot plausibly account for such behavior as the ability of a rat (or even an ant) to learn a maze with 30 choice points. He pointed out that many animals seem instead to learn by doing—by experiencing the consequences of their behavior and altering it accordingly. This second learning strategy is called *operant conditioning*, or *trial-and-error learning.* In the laboratory, experimenters train animals to perform a behavior (Fig. 21.19); in nature, animals train themselves.

The capacity for trial-and-error learning confers a considerable advantage on animals, since it allows them to acquire motor behaviors that are not instinctive. A seed-eating bird, for instance, does not have innate motor programs for picking up the various kinds of seeds in its environment, cracking them open, and separating the kernels from the shells. Instead, the bird has an innate ability to recognize seedlike objects, along with a drive to experiment with anything that looks like a seed. Getting into the first sunflower seed may take a finch several minutes of manipulation with its beak and tongue, but the kernel the bird ultimately harvests provides the reward that motivates it to try another. By trial and error, the finch discovers which muscle movements help get at the seed and which are irrelevant; experience in opening a succession of seeds slowly shapes the bird's harvesting behavior into a quick and efficient series of movements that finally become automatic—a learned motor program.

Conditioning biases Most researchers originally thought that animals could be taught to associate any CS with any US, and that any motor behavior an animal was physically able to perform could be taught through trial and error. Most animals, however, have been found to have strong species-specific "biases" that channel their learning along paths that are adaptive for them. Chicks, for instance, can learn to associate a particular sound with an impending shock and perform certain sorts of behavior (running or wing flapping, but not pecking) to avoid it. They also readily learn to associate color cues with a food reward and to peck (but not flap) in response. But they are virtually unable to associate sound with food or color with an

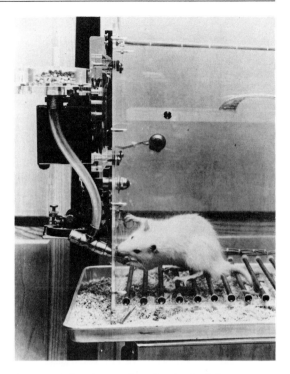

21.19 Trial-and-error learning in the laboratory
A pellet of food is dispensed from the apparatus when the rat presses the bar in response to the correct stimulus. This behavior is shaped by rewarding ever-closer approximations of the desired performance. The rat may be fed at first for being in the correct end of the box, and then only for accidentally touching the bar. Then the reward threshold may be raised to require actually pressing it. Finally the task of pressing in response to a particular stimulus is added.

A

B

21.20 Imprinted goslings
(A) Young goslings stay close to their own parent, whom they readily distinguish from the other geese. (B) Having been imprinted on Lorenz during their first day of life, these goslings follow him as if he were their parent.

impending shock. For chicks, color is a better predictor of food than sound, and pecking a more appropriate behavioral "experiment" than flying or running away; sound is a better predictor of the threat of predation, and flight is the appropriate response.

Animals, then, can be (and usually are) instinctively predisposed to recognize specific stimuli and to try specific sorts of behaviors in specific contexts. Clearly these biases reflect the cues and behaviors most likely to prove useful under natural conditions. Instinct thus helps focus learning, so that animals can modify their behavior quickly and adaptively. Such examples demonstrate the oversimplification involved in dividing behavior into "instinctive" and "learned."

SELECTIVE LEARNING

From what we know of adaptive innate biases in classical conditioning and trial-and-error learning, we can understand how, through natural selection, highly specialized learning programs have evolved. Though many species exhibit learning programs of this sort, some of the most interesting data come from birds.

Parental imprinting in birds Konrad Lorenz was probably first prompted to study animal behavior from a biological perspective by the phenomenon he subsequently labeled imprinting. Precocial young birds—those able to walk from the moment of birth—must follow their parents if they are to survive, and must therefore be able to recognize them. Imagine a newly hatched gosling faced with the task of correctly identifying its parents. From birth, its visual field is full of an enormous variety of objects in the environment, and yet it does identify its parents, follows them, and, when offered a choice later, is able to distinguish its parents from all other geese (Fig. 21.20A). Lorenz found that he could induce goslings to follow him instead if he removed the parents from view during the first day after hatching, and then walked away from the young birds while producing the appropriate species-specific call (Fig. 21.20B). The same trick, however, would not work if he waited until the third day. The process by which the goslings follow their parents and memorize enough about them to ensure future recognition is now known as *parental imprinting.* From his observations, Lorenz drew several conclusions about this process: it involves a *critical period,* also known as the sensitive phase, during which the learning must take place (Fig. 21.21). It requires releasers—the call and movement away (the unconditioned stimuli)—to trigger and direct both following and learning (the unconditioned responses); indeed, the young will follow the first moving object they see, particularly if it makes the correct sound. It involves neither reward nor punishment. It is irreversible. And it normally establishes an ability to recognize the parent (the conditioned stimulus).

We now know that parental imprinting is seen in mammals as well as birds. The set of cues for parental recognition seems to be species-specific: auditory, visual, or (as is most often the case in mammals) olfactory. The critical period is normally early and brief, and the learning is generally not reversible.

Sexual imprinting in birds   Parental imprinting has since been shown to be part of a broader phenomenon. The other sort of imprinting Lorenz explored is *sexual imprinting,* the process by which many animals learn to recognize their species and, in many cases, their close relatives. Though most animals can identify reproductively ready members of the opposite sex of their own species by means of innately recognized cues, there are circumstances, particularly among birds and mammals, in which the animals require more detailed information than is provided by instinct alone. In the North Atlantic, for example, four closely related and morphologically similar species of gull often nest together, but rarely interbreed. The only reliable morphological difference between them seems to be the color of the iris and of the fleshy ring that encircles the eye. By exchanging eggs of some birds between species and by painting the eye rings of others, N. G. Smith showed that the offspring use the eye color of their parents as a behavioral cue: when they have become adults, females choose mates whose eye ring and iris colors match those of the birds that raised them, whether natural or foster parents, while the males will only copulate with females that have the color combination they saw as young birds. The chicks, then, imprint on the parental eye color and later use it, in addition to other cues such as calls and postures common to gulls in general, in selecting a mate.

Other imprinting   Many parasitic birds utilize imprinting. Hatched in strange nests by parents of another species, the European cuckoos we discussed earlier, for instance, memorize the songs of their host species and later employ this knowledge to locate suitable hosts for their own young. Identifying the correct host is essential, for each cuckoo female lays eggs that pass for those of only one host species. A mismatch means her offspring may not survive, since many hosts eject eggs that do not resemble their own or, failing that, abandon the nest.

The host birds' behavior probably results from imprinting on the eggs themselves. Many species that live in close quarters, such as guillemots, memorize their own eggs so well that after a few hours even those of other birds of the same species are rejected. Similarly orioles, which are parasitized by cowbirds, will accept and imprint on a cowbird egg introduced *before* the host finishes laying her own. After her own eggs are laid, the oriole's imprinting program ceases, and she rejects any interlopers.

Another well-studied example of innately guided learning involves the way birds learn their species' songs. As Peter Marler of Rockefeller University has shown, most songbirds instinctively recognize only certain elements of their species' song; apparently their auditory system is structured to detect these particular features. The elements they recognize function as acoustic releasers to trigger a detailed memorization program of the father bird's song during a critical period. As a result, a juvenile white-crowned sparrow, for example, memorizes only a white-crowned sparrow song (Fig.

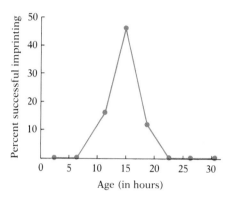

21.21   The sensitive phase for imprinting the following response of ducklings
Individual ducklings of various ages (in hours) were exposed to a moving decoy for one hour. They were then tested to see whether they had become imprinted on that object. The results showed that imprintability was at its peak when the birds were 15 hours old. Exposure to a suitable object too early or too late was ineffective.

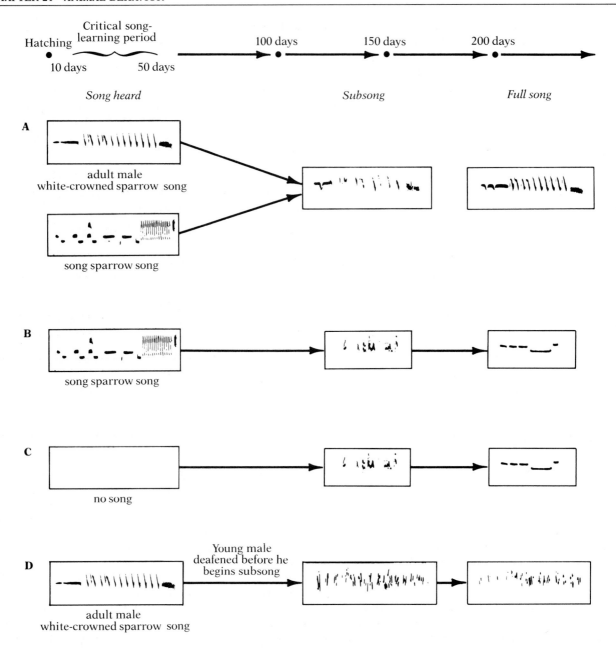

21.22 Song learning
Juvenile white-crowned sparrows learn their species' song during a critical period that runs from about 10 to 50 days after hatching. By 150 days they begin to practice, making syllable sounds. By 200 days they have developed a stable song closely matching the one heard during the critical period. Their learning is selective; birds exposed to several different songs learn only that of their own species (A). Birds offered just the song of another species (B), or no song at all (C), learn nothing, and at maturity sing only a simple tune. Birds deafened before beginning to practice (D) never sing anything melodic. However, if deafening occurs after practice has crystallized the song, it subsequently remains unchanged; evidently the perfected song is stored as an automatic motor program, a neural circuit in the brain. (These visual representations of songs were made with a sound spectrograph, a device that represents frequency on the vertical axis, time on the horizontal axis, and intensity by the darkness of the trace.)

21.22A). A bird that hears no song during the critical period, or only the songs of other species, learns nothing, and later produces a song that contains only the basic innate elements (Fig. 21.22B–C).

But even the production of the innate song requires learning: it is not stored as a motor program, so the young bird must experiment and learn how to sing it. The need for trial-and-error motor learning is illustrated by experiments in which young birds were deafened after the critical memorization period but before they had the opportunity to practice. Even birds that had never heard their species' song were never able to produce anything melodic (Fig. 21.22D). Apparently both the innate and the memorized songs are stored as acoustic images, and by trial and error the young birds attempt to produce a match. Since the deafened birds were unable to hear themselves practicing, they were unable to master the production of their song even though they "knew" how it should sound.

Various other types of imprinting are now known. Female ducks, for instance, imprint on nest height on their second day of life, and subsequently build their own nests high or low as a consequence. Mice imprint on certain features of their birthplaces, and when they later disperse and choose their own home ranges they pick those with similar features. This behavior is adaptive because it biases the animal toward areas that resemble the one that allowed it to survive and are therefore likely to allow its progeny to survive. Salmon imprint on the odor of their home stream on the day they begin their journey to the sea, and use that memory years later on their way upstream from the ocean to track the tiny tributary in which they were born (Fig. 21.23). Homing pigeons imprint on the location of their home loft as fledglings and will return to it even after years of life in a cage hundreds of kilometers away. Such examples are virtually endless. If the conditions are stable—that is, if the context, timing, and general sorts of cues that will be useful are available in advance of need—then innately specified guidance that channels learning into particular paths will be adaptive, and will be very likely to evolve.

CULTURAL LEARNING

What makes our species unique is in large part our ability to pass on information from individual to individual and from generation to generation, thereby saving others the risky and time-consuming exercise of rediscovering by trial and error the lessons of the past. Our culture, then, is cumulative; each generation stands on the shoulders of the past.

Among other animals, too, there has evolved the ability to pass novel and useful information from generation to generation. Some of this new data is initially acquired through classical conditioning, some through trial-and-error learning, and some through a combination of both.

Food learning   Most specialist feeders (animals with diets confined to a limited number of substances) have innate mechanisms for recognizing their particular foods, but generalists, with more catholic tastes, are most often equipped with a mixture of general innate guidance and a capacity to learn which foods are edible, and how best to handle them. Probably the most thoroughly understood examples come from the experimentally con-

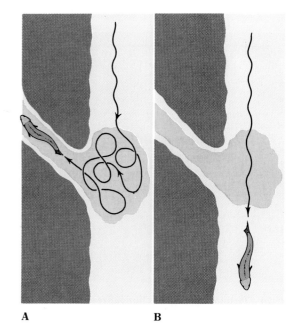

A   B

21.23   Tracks of salmon when they encounter morpholine
In one series of experiments (not discussed in the text), Arthur D. Hasler and his associates at the University of Wisconsin imprinted young salmon on the odor of a chemical called morpholine. Later, they used ultrasonic tracking to follow the adult salmon as they swam southward along the shore of Lake Michigan in search of a spawning stream. Morpholine was released in the area indicated in color. Fish previously imprinted on morpholine stopped their southward migration there, began to circle, and swam up the morpholine-scented stream (A). Those not previously imprinted on the chemical typically swam through the morpholine-scented area without pausing (B).

21.24 Termite hunting in chimpanzees
Tool using and social learning are crucial elements in the chimpanzee social system. Here an adult has selected and pruned a stick, which it uses to "fish" for ants or termites. Young chimpanzees observe this process, thus gaining cultural information on how to obtain food.

venient ground-dwelling birds, such as jungle fowl, grouse, and domestic chickens. Work by Eckhard Hess has shown that chicks are born with strong innate preferences with regard to the size, shape, and color of food. In addition, however, under natural conditions, the mother hen influences them by pecking at food, picking it up, dropping it, uttering an innately recognized food call, and so attracting their attention and causing them to peck at it. The chicks soon learn from experience what is and is not edible. The strength of the bias introduced by the parent has been shown by experiments in which a mixture of grain dyed orange and green was offered the chicks, while at the same time they saw through a transparent barrier a hen trained to avoid one of the colors, or a crude model hen that pecked at only one color. Though the grain of each color was equally edible, the chicks developed a strong preference for the color selected by the hen.

Such experiments indicate that along with their innate preferences for certain foods, chicks have an innate predisposition to attend to and copy the food-selection behavior of the parent, with the result that knowledge is passed from generation to generation. Inborn predispositions and learning through experience act in parallel in the chicks for the transfer of cultural information, just as they do when a bird's innate song is modified and elaborated by learning. Virtually the same pattern is seen in a variety of birds and mammals, and is of special importance among primates (Fig. 21.24). It is a strategy that allows novel food to be discovered and added to the diets of the discoverer and its associates.

Enemy recognition In recognizing their potential enemies, many animals display knowledge obtained through learning and cultural transmission. A careful look at this phenomenon reveals that at least in some instances the learning is innately guided. Birds, for example, generally face two classes of threat: nest predators, such as owls, crows, and snakes, which attack eggs or chicks; and adult predators, such as hawks and cats, which capture and kill grown birds. The behavior in the face of these two types of enemies is quite different: all the nesting birds in an area mob (attack *en masse*) a nest predator, but the same birds hide when a potential adult predator is seen. Birds appear to recognize a few animals of each class innately, but for the most part must learn who their enemies are. This is accomplished without the young's having to experience life-threatening attacks directly.

By means of a clever series of experiments on European blackbirds, Eberhard Curio of the University of the Ruhr in Germany discovered the programming that underlies enemy recognition. Curio placed two cages of blackbirds on opposite sides of a hallway, in sight of each other. Between the two cages he installed a four-compartment box that allowed the occupants of each cage to see an object on their side, but not on the other side. Curio then presented a stuffed owl to the birds in one cage and a harmless and unfamiliar bird, an Australian honey guide, to those in the other. The birds that saw the owl began at once to deliver the mobbing call and attempted to attack the stuffed figure through the cage. The birds on the other side, seeing only the honey guide (which unconditioned blackbirds ignore), and seeing and hearing the mobbing birds, began attempting to mob the honey guide. These birds then passed on the practice of mobbing honey guides to other birds, and these passed it on to still others. Curio saw this mindless aversion to a creature that had never harmed a single blackbird transmitted through six generations in the laboratory. He was able to repeat this same piece of enculturation—blind, but adaptive in the wild—even with a plastic bottle as the object of official hatred. It seems clear that the innately recognized mobbing call is a sign stimulus for classical conditioning: when a bird hears the mobbing call, it automatically identifies the object of the call as an enemy. The warning signals of various other animals—the trumpeting of elephants, for instance—serve the same function, alerting the experienced and teaching the young about danger before they come to harm.

AVIAN NAVIGATION

Of all the astonishing feats resulting from the interplay of sign stimuli, motor programs, drives, and innately guided learning, perhaps none is more impressive than the performance of many creatures in finding their way over great distances through unfamiliar territory. Animals as diverse as butterflies, sea turtles, and hummingbirds migrate thousands of miles to places they may never before have visited. Their almost incredible navigational abilities arise from a variety of complex neurological and physiological systems that are still to be worked out in detail, but a few animals are beginning to give up their secrets.

Among birds, two separate navigational strategies are evident. In one, the creatures are preprogramed to fly a certain course. Wolfgang Wiltschko of the University of Frankfurt has shown that in the fall garden warblers from northern Germany will fly (or, in orientation cages, attempt to fly) southwest for several weeks, and then southeast for several more—following a course that, in the wild, carries them down through Spain, across Gibraltar, and into their winter ranges in Africa (Fig. 21.25). Timothy and Janet Williams of Swarthmore College discovered that many small birds in North America migrate to the East Coast, wait for a low-pressure front, and then fly southeast. In general this course results in their catching winds that carry them to South America, though if the winds fail these birds perish at sea by the millions.

This compass-and-timer strategy is good enough when the target is a continent, but will hardly serve when the goal is small. Many migrating animals need to know precisely where they are even when in an area for the first time. This need is filled by a mysterious but very real ability known as *map sense,* which represents something quite different from a mental map of a familiar area. An animal with a map sense behaves as though always aware of longitude and latitude. The nature of this map sense is one of the most intriguing mysteries in modern biology.

Though many migrating birds have a map sense, their journeys twice a year leave much to be desired when it comes to experimentation. The animal many researchers prefer to work with, therefore, is the homing pigeon. A good homer can be taken from its loft and transported hundreds or even thousands of kilometers in total darkness, and when released it will circle

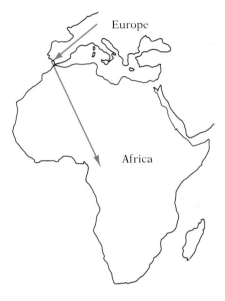

21.25   Warbler migration
Some European garden warblers reach their winter grounds after a two-leg journey. They know at birth the two flight bearings they need, and how long to fly in each direction.

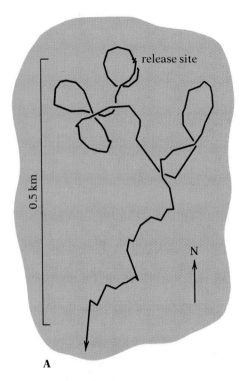

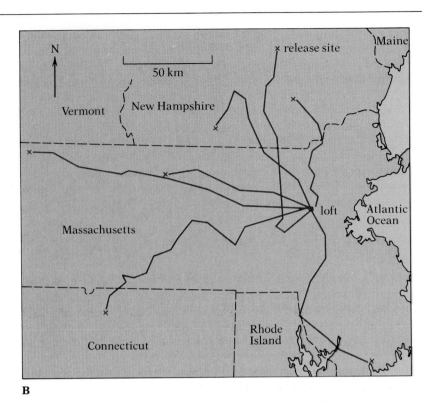

21.26 Pigeon homing
(A) Pigeons usually begin their journey home by circling the release site, but quickly set off along an irregular course for home. (B) The actual routes are rarely straight and direct, and indicate that new map measurements must be taken from time to time to make mid-course corrections.

briefly and then fly off in roughly the direction of home (Fig. 21.26). Homing pigeons have both a compass sense and a map sense. To understand why they need both, imagine yourself kidnapped, taken 100 km away (to the south, say) in a windowless vehicle, and released. If you had a compass you would know which direction was north, but unless you knew in which direction you had been taken, that information would be useless. On the other hand, if you had a map sense, you would know that you were south of home, but without a compass you would be at a loss to put that information to use.

There are, in fact, ways to do without one or the other: if you had just a compass and could watch it on the trip out, you could, as do animals like honey bees, reconstruct the outward journey and then retrace the route; if you had just a map sense, you could move in one direction for a few minutes, take a new map reading, see whether you were getting closer or farther away, and so by means of a "getting warmer" system find your way home. Pigeons can be transported in the absence of all visual and magnetic cues to direction—even anesthetized—and still home perfectly well, which indicates that cues perceived on the outward journey are not necessary to their successful orientation. That they do not rely solely on successive map readings has been proven by clock-shift experiments, described later in this section, which indicate that they have a compass sense as well.

The workings of the compass sense in pigeons and migratory birds are now fairly well understood. Like many insects, pigeons and other diurnal

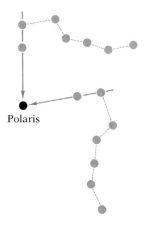

Polaris

21.27 An example of how north can be located by star patterns
An arrow through the two end stars in the cup of the Big Dipper (Ursa Major) points toward Polaris, the North Star. Though the position of the constellation changes during the night, the same stars always determine an arrow pointing toward Polaris; hence directions can be determined without need of time compensation. Many different star patterns could be used for finding direction in this way.

birds use the position of the sun as their standard cue. Of course, the sun's position depends on the time of day, and birds appear to have an internal time sense enabling them to allow for the westward movement of the sun from morning to night. Similarly, nocturnal migrants use a learned picture of the stars to set their course. The roles of the sun and stars in avian navigation are demonstrated by the behavior of caged birds under artificial skies: during migration season, the animals display an intense desire to escape in the direction their wild conspecifics are flying. A sudden shift of the artificial sun or pattern of stars results in an immediate compensatory change in the direction in which the caged birds are struggling to go.

Experiments by Steven T. Emlen at Cornell University have made it clear that nocturnal migrants memorize the constellations while they are still nestlings, using the North Star, around which all other stars in the night sky appear to rotate, as their point of reference. As a result, they are able to infer north from even a small patch of sky (Fig. 21.27). Such birds can be raised under an artificial sky, with an arbitrary pattern of stars rotating about a pole at any chosen compass point. When the time to migrate arrives they then attempt to set off in the appropriate direction relative to the star patterns they observed during the critical period.

Homing pigeons demonstrate their use of the sun compass in an equally dramatic way. Correctly interpreting the sun's direction depends on an internal timer, which is sensitive to manipulations of the day/night cycle. For instance, a pigeon kept in a room whose lights go on, and later off, six hours early—on at midnight and off at noon—will misinterpret the sun's position accordingly. When such a bird is released at true noon, its internal clock reads 6 P.M. It sees the sun in the south, but because it has been clock-shifted, interprets the sun's position as indicating west. Therefore, if its home is to the south, it will fly 90 degrees to the left of the sun; attempting to fly south, it heads east (Fig. 21.28).

But pigeons can also home under an overcast sky. If they use the sun as their compass, what guides them when it is invisible? William T. Keeton of Cornell University attacked this question by releasing both normal and

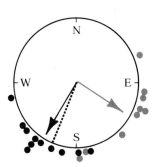

21.28 Effect of a six-hour-fast clock shift on the initial bearings chosen by homing pigeons on a sunny day
Each dot indicates the bearing chosen by one bird; black dots represent control birds, color dots experimental birds. The dashed line marks the proper homeward direction. The arrows show the mean bearing (average direction) for each group. The length of each arrow is proportional to the degree of clustering of the dots: if all the birds flew off in the same direction, the resulting arrow would touch the circle; if the departure directions were widely scattered, the arrow would be very short. In this experiment the mean bearing of the clock-shifted birds (color arrow) is about 90 degrees to the left of the mean bearing of the control birds (black arrow). The experimental birds have been clock-shifted a quarter of a day, and they have made an error of a quarter of a circle in reading the sun compass.

21.29 Effect of a six-hour-fast clock shift on the initial bearings chosen by homing pigeons on a totally overcast day
When the sun is not visible, the experimental birds choose bearings (color dots) not significantly different from those of control birds (black dots); the 90-degree deflection of their bearings seen on sunny days (compare Fig. 21.28) is not evident. In the absence of the sun compass, the birds appear to orient by some other system, which does not require time compensation.

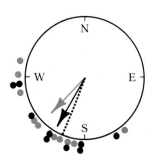

21.30 Effect of magnets on pigeons
When pigeons with magnets on their heads (color dots) are released on sunny days, the birds, relying on the sun, are not much affected. On cloudy days, however, magnets have a dramatic effect.

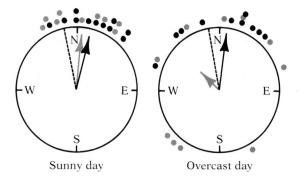

Sunny day   Overcast day

clock-shifted birds on cloudy days (Fig. 21.29). The results are clear and dramatic: pigeons are able to home on overcast days, and they are then using cues that are not time-dependent, for the departure bearings of clock-shifted birds are not rotated. Obviously pigeons have a backup system. Keeton guessed that the backup compass might be magnetic, and to test this possibility released pigeons carrying either magnets or brass bars on both sunny and overcast days. As might be expected, there was little effect when the sun was visible, since celestial cues take precedence over all others when they are available. On cloudy days, however, the birds carrying magnets were disoriented (Fig. 21.30).

Though research on the compass senses of animals has progressed steadily, understanding of the map sense is still limited. At present we know only what cues are *not* involved, and what general accuracy is achieved. The intriguing data on the accuracy of the map sense comes from experiments by Klaus Schmidt-Koenig of the University of Göttingen and Charles Walcott, now at Cornell University. They released pigeons wearing translucent lenses that prevented them from perceiving shapes. That these birds were able nevertheless to navigate to within a few kilometers of home (Fig. 21.31) indicates that they possess a very precise map sense.

At present there are two major hypotheses about how the map sense might work. According to the first, developed by Floriano Papi of the University of Pisa, the birds, while at home, memorize the odors carried by the winds from various directions and construct an olfactory map. Thus a bird might learn that at its home site a westerly wind always smells of the sea, while a breeze from the north carries the fragrance of a pine forest. Taken to a pine woods, it would then associate the local odor with the scent carried to its home by northerly winds, and fly south. But this bird could be carried to one of dozens of pine forests located in various directions with respect to its loft, so how it could possibly know which way to fly home remains difficult to imagine.

The other hypothesis, advanced by Walcott and J. L. Gould of Princeton University, involves the earth's magnetic field. The exact strength of the field varies with latitude, roughly doubling between the equator and each pole. An ability to measure the field strength precisely could provide information at least about latitude. Several lines of evidence support the notion that small changes in field strength affect a pigeon's judgment of where it is. For instance, during magnetic storms (which are created by sunspots that dump enormous numbers of ions into the upper atmosphere) pigeons depart increasingly off to one side of the homeward bearing—up to 40 degrees in some cases, even though the effect of such a storm on a compass is about 1 degree and the primary reference, the sun, may be clearly visible. Moreover, these birds home significantly more slowly than usual. Similarly, pigeons released at locations where the earth's magnetic-field strength is anomalously high because of local conditions tend to be disoriented even

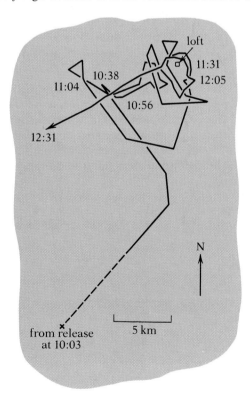

21.31 Flight of a pigeon that cannot see shapes
Pigeons that have arrived in the vicinity of home use visual landmarks to locate the loft. Those wearing frosted contact lenses cannot see shapes, but nevertheless know when they are in the vicinity of home and fly wide circles nearby. The track of such a pigeon is shown (somewhat abridged) in a representative example here; it indicates a map sense accurate to within a few kilometers.

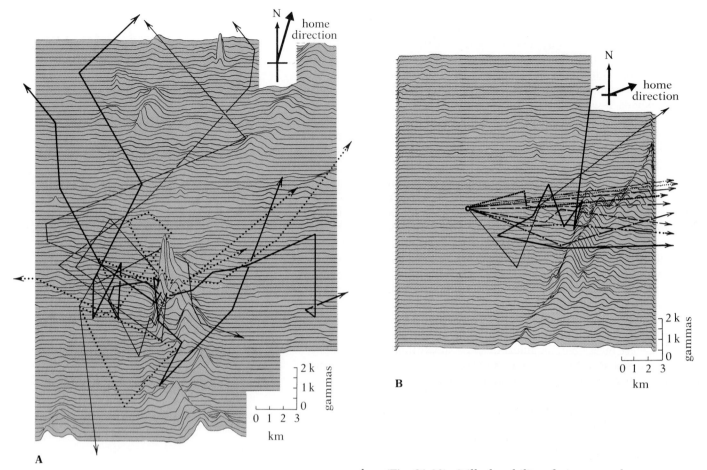

A

21.32   Effect of a magnetic-field anomaly on homing

The strength of the earth's magnetic field increases about 6 gammas per kilometer in the northeastern United States. Shown here in a three-dimensional plot is an anomaly in field strength at Iron Mine Hill, Rhode Island. (A) Pigeons released at Iron Mine Hill are almost completely disoriented even on sunny days. Their confusion persists until they have been out of the anomalous area for some time. (B) Released at a normal site—Worcester, Massachusetts, in this example—pigeons depart in one general direction. (The total intensity of the earth's field is about 50,000 gammas in this area.)

on sunny days (Fig. 21.32). Still, the ability of pigeons to home on sunny days while wearing strong magnets is difficult to reconcile with the magnetic-map hypothesis. There remains, of course, the intriguing possibility that the map sense involves cues of which we are not yet aware, or neural processing more elaborate than any yet discovered. It is this sort of mystery that helps make animal behavior one of the most interesting and exciting fields in modern biology.

Chapter 22

THE EVOLUTION OF BEHAVIOR

Most animal behavior is a product of the interplay of a small set of neural phenomena—releasers, motor programs, drives, and learning. These mechanisms and processes define in each species the range of behavioral possibilities that *can* evolve, but they tell us nothing about exactly what behavior *will* evolve. Natural selection can effect changes in a species' behavior through differential survival: individuals whose behavioral traits improve their chances for survival and reproduction under existing circumstances will increase in the population. The set of traits that promote survival will therefore come to predominate in the species as long as the circumstances do not change. Just which traits these will be for the individual animal will depend on its environment, its food sources, and the nature of its interactions with other members of its species. The branch of ethology that seeks to understand how the dynamic interaction of animals with their environments and conspecifics brings about the evolution of behavior is called ***behavioral ecology.***

The first section of this chapter will examine some of the means that have evolved for animals to interact with others of their species—to communicate. The second section will explore the evolution of behavior, with particular emphasis on the evolution of sociality in animals. In the final section, we shall apply what is currently known about the mechanisms and evolution of social behavior to the study of particular social species, including our own.

COMMUNICATION

Animals are frequently categorized as either solitary or social—living by themselves or living with conspecifics in pairs or groups. Those that are sol-

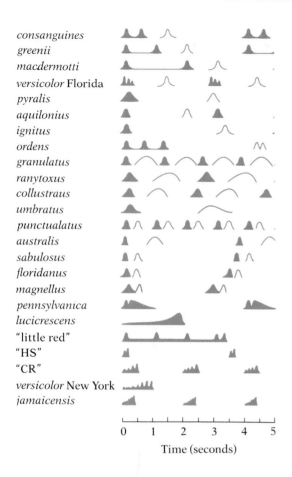

consanguines	
greenii	
macdermotti	
versicolor Florida	
pyralis	
aquilonius	
ignitus	
ordens	
granulatus	
ranytoxus	
collustraus	
umbratus	
punctualatus	
australis	
sabulosus	
floridanus	
magnellus	
pennsylvanica	
lucicrescens	
"little red"	
"HS"	
"CR"	
versicolor New York	
jamaicensis	

0   1   2   3   4   5

Time (seconds)

22.1   Firefly codes
The signals of various flying male fireflies (*Photuris*)
are shown in solid color, while the female response is
indicated in outline. Each female response occurs at
a species-specific interval after the male flash. For
some species the response is not yet known.

itary come together with other members of their species only to mate. At
this crucial point—crucial to the perpetuation of their genes—each animal
must find a reproductively ready member of the opposite sex of its own
species. Communication probably first evolved to accomplish this goal, and
from the mechanisms of sexual communication evolved the wide variety of
signals that now communicate mood, intention, and other information necessary to maintain order and stability in social groups. We shall begin by
exploring sexual communication, and then examine the more complex
phenomena of social communication.

SEXUAL COMMUNICATION

The goal of most animal communication, like the ultimate goal of behavior
in general, is survival and reproduction. Most animals, from planaria to
primates, are solitary for all or part of their lives, and so must actively seek
out their own kind for mating. Since such animals have little or no opportunity to learn where to look for a suitable mate, how to recognize one, or
what to do when one is found, much of the behavior associated with mating
must be innate.

Sensory channels   The most primitive and widespread channel of communication is chemical. Many species of unicellular organisms depend on
chemoreceptors to recognize when they have bumped into another individual of their species, while others, like certain slime molds that aggregate periodically to reproduce, locate each other through pheromone (odor) trails.
Only slightly more elaborate is the mating system of most moths: females
release a species-specific pheromone, and males follow the odor upwind to
its source. Pheromones also play an important role in groups as diverse as
beetles, aquatic invertebrates, and mammals. In each case the role is basically the same: the odor informs potential mates of the species, sex, reproductive readiness, and location of an appropriate mate.[1]

A number of species employ other sensory modalities. Fireflies, for instance, produce pulsed signals that can be seen by other fireflies at great
distances. Males fly about flashing according to a species-specific code
while females wait on vegetation, flashing in answer (Fig. 22.1). The releaser in firefly communication is, in most cases, the interval between
pulses. Both sexes are thoroughly tuned to a specific set of intervals; with a
bit of experimentation you can lure males in with a penlight.

Many other species employ precisely timed auditory versions of the firefly system. Both crickets and frogs generate species-specific calling songs
whose temporal characteristics assure that any possible ambiguity of species or sex is minimized (Fig. 22.2). To our eyes and ears, the rhythmic pattern of pulses appears to be the most useful characteristic, but the feature
detectors of the females seem to be tuned instead to the intervals between
pulses. Hence a scrambled song that sounds totally different to our ears but
faithfully preserves the intervals is as acceptable to females as the normal
song.

[1] Pheromones are detected by olfactory receptors. As we saw in Chapter 19, taste receptors
can detect only sweet, sour, bitter, and salt. More refined discrimination requires olfactory reception.

In species such as birds and mammals for whom fine frequency (pitch) discrimination is possible, different frequencies can be used to convey different messages. Among birds, species recognition is based not only on temporal intervals and sound frequency, but on the rate of change of frequency with time. Even in species whose song is learned, feature detectors responsive to a specific combination of these three characteristics (and perhaps others) sensitize the animal to what it is to learn—what to sing, in males, and what to listen for, in females.

Specificity through multiple signals   Much of the species-specificity of animal communication depends upon the *simultaneous* presence of several cues at once—rather in the manner of a fraternity handshake—to exclude all creatures that do not "belong." Obviously the multiplicity of cues reduces the potential for mistakes.

In a second and even more effective technique, the several cues must also appear in a particular order. The courtship sequence of queen butterflies provides a good illustration of this more elaborate form of communi-

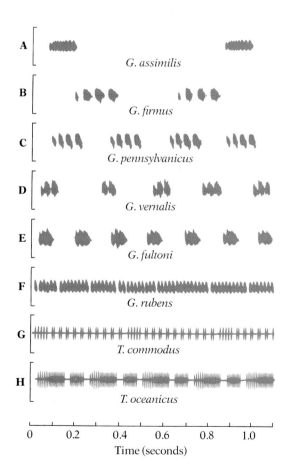

22.2   Cricket calling
Male crickets produce a pulsed calling song that attracts females of their species. The songs of the six species of *Gryllus* (A–F) are less complex than those of the two species of *Teleogryllus* (G, H). When a female approaches, the males switch to a courtship song.

22.3 Butterfly courtship

Courtship of the queen butterfly involves a series of signals from male to female and from female to male, presented in a particular order. Since the failure of either individual to produce the right signal at the right time breaks off the courtship, only reproductively ready members of opposite sexes of the same species mate.

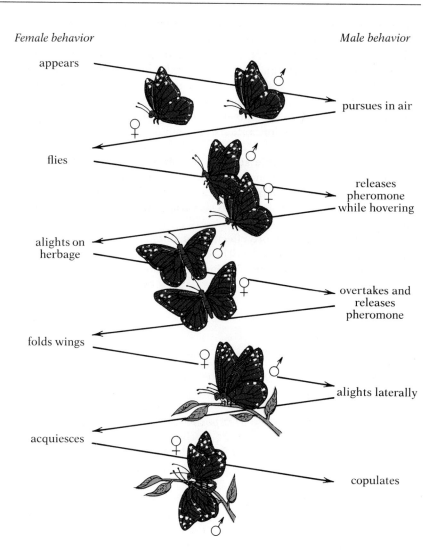

Female behavior   *Male behavior*

appears — pursues in air

flies — releases pheromone while hovering

alights on herbage — overtakes and releases pheromone

folds wings — alights laterally

acquiesces — copulates

22.4 Deception among fireflies

After mating with a male of her own species, the female began to respond correctly to the signals of males of another species. When a male landed she captured and began to eat him, as shown here.

cation (Fig. 22.3). A male will chase any rapidly flapping object—including another butterfly in flight—until he overtakes it. The flashing of the fluttering wings is the releaser that triggers the male's courtship behavior, causing him to extrude a pair of brushlike "hairpencils" from his abdomen that emit a species-specific pheromone as he hovers over the object of his pursuit. If the latter is of the correct species and sex, and is reproductively ready, she then alights on some nearby vegetation. (So, of course, might a falling leaf, which males often pursue.) The male hovers above the female, sweeping his hairpencils over her antennae. If all goes well the female then closes her wings, thus signaling for the male to alight and begin mating. Each step of the courtship sequence must be performed correctly for mating to take place. This type of strategy, involving a specific series of releasers, is almost universal.

Deception   Having to advertise for a mate carries the obvious risk of attracting predators; less apparent, perhaps, is the potential it provides for deception. Various predatory organisms specialize in luring other animals with the victims' own attractants. After mating, for instance, the females of at least one species of firefly switch over to answering the flashes of several other species. Males that are deceived are unceremoniously eaten (Fig. 22.4).

More interesting in their implications are cases in which members of one species attempt to deceive others of the same species. The hanging-fly male characteristically hunts for food—most often other, smaller species of fly —and offers it as a nuptial gift to the female he has attracted with a pheromone (Fig. 22.5). The nutrients provided by the offering enable the female to produce eggs. While a female eats the prey the male has captured, he mates with her.

Some males, however, have a less straightforward method of winning mates. Instead of spending time and energy looking for prey, they fly upwind to a signaling male, pretend to be female, take the proffered prey, and attempt to fly away with it. The potential benefits to an individual able to deceive others are often so great that many animal societies have evolved elaborate safeguards to exclude cheaters.

SOCIAL COMMUNICATION

Sociality is a matter of degree: crickets and fireflies, for instance, are solitary except for a few seconds of mating; at the other extreme, ants and honey bees live in large colonies, and are so dependent on community life that in isolation they die. Most social animals fall somewhere in between. Many birds pair with a member of the opposite sex for a few weeks to rear offspring, and may join winter feeding flocks for protection. The nature and degree of a species' sociality largely determines the kinds of social signals it will need.

Social communication may utilize several behavioral channels, or may involve only one. When we talk on the telephone we are communicating acoustically, but when we talk to each other face to face we use gestures, facial expressions, and other sorts of body language to elaborate and clarify the vocal messages we are sending. The roles played in nonhuman communication systems by gestures, postures, and acoustical signals are often difficult to sort out.

Several examples of social communication have already been discussed; such signals as begging calls and pecks, food calls, and alarm calls serve this purpose. Virtually all these messages are delivered and recognized instinctively; neither the young nor first-time parents have had an opportunity to learn how to use them. But among highly social animals new levels of complexity in communication (and sometimes new behavioral mechanisms) begin to emerge as individuals need signals to show mood or intention, to coordinate hunting or escape, to indicate social status, and so on.

Dance communication in honey bees   For pure elegance and complexity, few communication systems can equal the dance language of honey bees. The ability of the forager bee (which scouts for food) to inform her

22.5   Courtship among hanging flies
Males attract females with a pheromone, offer a gift of food (a dead fly in this case), and mate with the female while she is consuming it. Some males acquire the gift by hunting flies, while others steal it from a courting male by mimicking female behavior. Still other males suck out all the nutritive juices of their prey before offering it as a gift, but females that discover the deception will refuse to mate with the male.

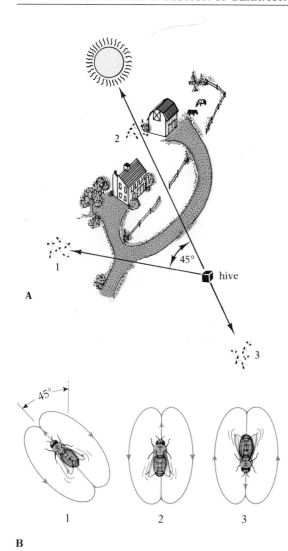

22.6   An example of bee language
(A) Three different sources of food were located (1) 45 degrees to the left of the sun as seen from the hive; (2) straight toward the sun; and (3) straight away from the sun. (B) The dances performed on the vertical comb in the darkened hive by forager bees from these food sources were oriented 45 degrees to the left of vertical (1), straight up (2), and straight down (3). In short, the vertical direction on the comb symbolizes the sun. The dancing bee in the photograph, her legs bearing pollen, is closely surrounded by attenders.

hivemates of the distance and direction of a good source through a symbolic system of communication seems hardly less amazing now than it did when first discovered by Karl von Frisch in 1945. It had been known for decades that returning foragers often perform dances, and that subsequently other bees visit the food sources these scouts have found, but not until 1943 were foragers trained to visit food sources so far from the hive that it was clear that subsequent visitors were indeed not searching randomly for the food. With this realization, von Frisch began to observe the dances (which normally occur in the darkness of the hive on vertical sheets of comb) in special observation hives. He divided the foragers into two groups, marking the members of each with a different color of paint, and trained each group to visit a food source at a different location. Von Frisch was then able to decode the dance by observing how the differences between the two groups' dance movements related to differences in the messages being conveyed.

In the significant central portion of the dance, the forager vibrates, or "waggles," her body from side to side and simultaneously produces sound bursts by vibrating her folded wings. All the information that recruits need is conveyed in these acoustically emphasized waggle runs. "Up" on the comb is always the direction of the sun, and the angle the dance runs with respect to vertical corresponds to the direction of the food with respect to the sun (Fig. 22.6). For example, if the food is 45 degrees to the left of the sun, as seen from the hive, the dance will point 45 degrees to the left of vertical. Even in the darkness of the hive the dance attenders are able in some way to perceive the angle of the dance.

Distance can be determined from the duration of the waggle run or the number of waggles in each run (Fig. 22.7). Each waggle specifies a particular increment of distance—about 40 m in the case of von Frisch's honey bees. The dance communication system is called a language because it refers to objects distant in both space and time (that is, the animal is not simply pointing and grunting) and because it is symbolic ("up," for instance, is an arbitrary symbol for the sun's direction; "down" or any other direction could have been used just as well). That the waggle as an indication of distance is a relatively arbitrary symbol is emphasized by the discovery that different races of bees have different distance dialects (Fig. 22.7).

These dialects are entirely instinctive: cross-fostered bees misread the dances of their adopted hive.

After von Frisch's discovery, Adrian Wenner at the University of California, Santa Barbara, and his colleagues mounted a strong challenge to the idea that insects could have a symbolic language. Beginning in the late 1960s they performed a series of experiments that suggested that foragers indicated the location of food by odor cues alone, and that the correlations between location and the dance were accidental. Though their evidence was plausible, and none of von Frisch's experiments seemed conclusive under careful, skeptical scrutiny, we now know that the dance really is a language. James L. Gould of Princeton University took advantage of two relatively obscure facts about bee behavior to demonstrate conclusively that the dance communicates specific information. The first is that if the sun or a sufficiently bright light is visible within the hive, the bees will orient their dances to it instead of to the vertical. While a food source in the direction of the sun would normally elicit a dance aimed directly up on the comb, a bright light 90 degrees to the right of the comb causes the dances to be rotated 90 degrees to the right. Since both dancers and dance attenders see this light, both continue to employ the same point of reference, and no confusion ensues.

The second fact is that when the ocelli of bees—three tiny, single-facet eyes located between their two large compound eyes—are covered over, the bees become about 10 times less sensitive to light. They can still see in bright daylight, but tend not to venture out near dawn or dusk, and do not go out at all on heavily overcast days. By painting over the ocelli of foraging bees on sunny days and then offering a bright light within the hive, researchers can eliminate the bees' unanimity about their reference point. Normal, unpainted foragers perform dances rotated by the light, but the ocelli-painted bees, apparently unable to see the artificial sun, dance as if it were not there. If the dance language actually communicates information symbolically, the attenders, able to see the light, should misinterpret the dances of these foragers in a regular way and fly to a predictable location well away from the actual food. If odor were the sole mechanism, the recruits should not be fooled by the forager's "lie." In point of fact, most recruits do fly to the station indicated by the dance.

Learning and social communication   As far as we know, the honey-bee dance language is second only to human speech in its ability to convey complex information. Though used to relay information about water, nectar, pollen (which bees collect for its protein), propolis (tree sap, used by the bees to seal openings and entomb unwanted objects too large or awkward to remove from the hive), and new hive sites, it is basically a closed system under instinctive control: bees can perform or understand dances with no previous experience, but can use them only to specify the distance, the direction, and (in ways we have not discussed) the desirability of a location. In other highly social animals, however, communication may be more flexible, and may involve neural mechanisms beyond those we have so far discussed.

Postural and facial cues play a role in many bird and mammalian communication systems. Dogs, for instance, solicit play with a half-crouch, tail

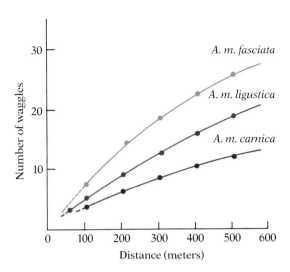

22.7   Distance codes in the bee dance
Different races of honey bee use different dialects to indicate distance. For the German honey bee (*Apis mellifera carnica*), each waggle corresponds to about 40 m; for the Italian honey bee (*A. m. ligustica*), a waggle corresponds to 25 m; for the Egyptian honey bee (*A. m. fasciata*), a waggle represents about 15 m.

Increasing anger ⟶

I
n
c
r
e
a
s
i
n
g

f
e
a
r

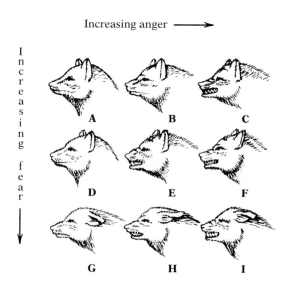

22.8 Facial signals of dogs
Dogs are among the many species of mammals and birds that use body postures along with other signals to communicate mood. Shown here is the simultaneous expression of varying degrees of anger and fear by dogs. Since these two states can vary independently a dog can be purely angry (C), purely frightened (G), or both angry and frightened (I). Tail position also signals degrees of fear, as in the classic tail-between-the-legs posture.

up and wagging. This signal is species-specific; cats, for instance, often misinterpret it, since their "vocabulary" includes no corresponding posture. Fear and aggressiveness are shown by facial cues easily read by other canids (Fig. 22.8). These may or may not be accompanied by acoustical signals such as short barks or growls. How do animals recognize and decode these visual messages? Unlike the acoustical and chemical signals we have discussed, complex visual information cannot readily be encoded by simple releasers in the brain, and the neural mechanisms remain unknown.

BEHAVIORAL ECOLOGY

As we saw at the outset of this chapter, the behavior of animals is modified and adapted through natural selection to suit species' particular lifestyles. The dances of tropical honey bees, for example, take place in the open (Fig. 22.9) and are practically silent; to obtain information about food sources, the dance attenders watch the lifted waggling abdomens of the dancers. When the habit of nesting in tree cavities evolved (probably as a way of avoiding predation), natural selection must have favored colonies whose foragers buzzed during the waggle run, helping other bees to obtain information from the dances in the total darkness of the nest.

Selection also seems to have favored colonies with dialects appropriate to their foraging range. Tropical bee species, and tropical races of the familiar honey bee *Apis mellifera*, usually find food close by. In their dialects each waggle corresponds to only a few meters. Races in cold areas like northern Europe, however, must forage over large areas. In their dialects each waggle corresponds to as much as 50 meters.

Another effect of natural selection on animal behavior has been the substitution of ritual, especially in courtship, for bright visual releasers, which for many species became increasingly costly as they attracted not only potential mates but predators. Some species of tropical bower bird, for instance, have evolved elaborate displays (Fig. 22.10) that take the place of the showy female-attracting—and predator-attracting—plumage of more primitive species. Now possessing only token patches of colored feathers, the males build structures called bowers, deck them with objects of their species-specific color, and wave these objects at females, just as males of species still bearing color patches display them in courtship.

In the last chapter we saw too how, through natural selection, the learning of animals of particular species becomes "tuned" to that species' needs. Now we shall look more closely at the ecological contingencies that make different behavior patterns appropriate for different species, and thus in large measure drive behavioral evolution. We shall pay particular attention to the factors favoring some degree of social organization.

NICHE AND HABITAT

One unavoidable question Darwin faced was how the world came to support so many species in relative harmony and stability. Darwin concluded that for a species to survive and remain stable it must fulfill three important requirements: (1) The species must avoid mating with other species (species-specific sexual communication, discussed earlier, serves this pur-

A

B

22.9 Nest of a tropical honey bee
The dwarf honey bee (*Apis florea*) builds a single sheet of comb that hangs unsheltered from a tree limb (A). Communication dances are performed on the top of the bee-covered comb (B); the bees have been removed to reveal the dance platform.

pose). (2) The species must not compete with another species for exactly the same food, since the victory of one of the competing species would mean the inevitable extinction of the other. (3) The species must not consume its entire food supply, whether animal or vegetable, thereby eliminating itself along with those species that nourished it.

The second of these requirements for stability, stating that no two species can long coexist if they eat exactly the same thing in the same place at the same time, is known as the ***Niche Rule.*** Each species in an area occupies its own niche. Niches, contrary to common usage, are not preordained "slots" into which species fit; rather, they may be thought of as professions, defined by the way members of the individual species actually make their living. A species' ***habitat*** would then be its residence—the physical place that the given species occupies. A single lake may represent the entire habitat of a species of fish, while for geese, which also live on lakes but fly great distances, lakes of an entire continent may constitute the habitat. In a sense, the biological world is organized like an enormous Rotary Club, with only one member (species) from each profession (niche) allowed per town (habitat). Different species can occupy the same sort of niche in different areas —in Australia marsupials fill niches similar to those that conventional mammals occupy on other continents. And as the environment of an area changes over time, the exact species composition in the area, and the precise niche of each species, inevitably change.

As a result of natural selection, species occupy niches to which their behavior—in hunting or food gathering, most obviously—is well suited; similarly, their behavior is well adapted to the nature of the habitat and its predators. Niche and habitat are also major determinants of whether sociality evolves, and if so, what kind. Social systems evolve primarily when groups of animals are better able than individuals to control resources like food and suitable habitats, and perhaps to defend against predators. Animals like mosquitoes, which do not compete with one another for food (mammalian blood, in this instance), and could not, as a group, better defend themselves against predators like birds, do not evolve group organiza-

22.10 Decorative display in tropical bower birds
In an effort to attract a mate, males of various species of tropical bower bird will decorate their bowers with colorful objects to which their mates are genetically attuned. When used in elaborate display rituals, these objects take the place of the colorful plumage the birds lack. This male satin bower bird has a particular fascination for the color blue.

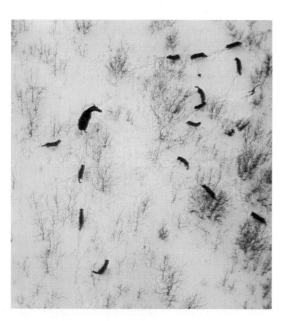

22.11 Population control by predation
A pack of wolves on Isle Royale corners an adult moose. Despite the long chase, however, this individual was too strong and healthy to be worth the risk of attacking at close quarters. The wolves eventually left to find a more likely target.

tion. For them there is no selective advantage to sociality. In many species of birds and mammals, on the other hand, individuals must compete for limited resources. For them, living in groups has possible advantages, as we shall see in a later section.

POPULATION CONTROL

Of particular importance in the evolution of sociality is the problem of population control. As Malthus pointed out (see Chapter 1), humans tend to reproduce beyond the maximum that the environment can support over an extended period—a point now known as the *carrying capacity*—and the population is held in check by starvation, disease, and the like. Most individuals in Mathus's time lived at the edge of starvation, and the hazards of childbirth resulted in a high death rate for both infants and mothers.

As Darwin recognized, these problems are not unique to the human species. But while population control may be good for a *species* as a whole, it is difficult from an evolutionary perspective to imagine how natural selection could favor *individuals* that would, for example, avoid eating all the available food, and consequently produce fewer offspring than would otherwise be born to them. How can an individual of a species show sufficient restraint for the species' long-term good if its own immediate evolutionary advantage lies in consuming as much as possible in order to reproduce? And yet in the long run most species do not eat everything available; mechanisms other than simple starvation keep populations below the carrying capacity of their environment. How can such apparent self-restraint evolve?

There are at least four means by which the niches, habitats, and behavior of various species help prevent wholesale starvation: "opportunistic" life history, programed dispersal, regulation by predators and disease, and, most interesting, territoriality, an aspect of social organization.

For opportunistic species—those, like mosquitoes, adapted to rapid reproduction during brief periods when resources are abundant—restraint is normally unnecessary. Mosquitoes, for example, are rarely so numerous that they have to fight with each other for access to the last remaining mammal in the habitat not fully drained of its blood. Environmental factors (namely the coldness of winter and the availability of pools of water) rather than their own density limit population growth. The second method by which some animals avoid starvation is evident in the behavior of aphids and lemmings, which are programed to disperse and seek new habitats when overcrowding threatens.

Control by predators   Unlikely as it may seem, predators and disease often function as a third method of maintaining the ecological balance. This ironic strategy is dramatically illustrated by a well-known case study involving the moose population of Isle Royale, an island in Lake Superior.

In 1908, a small group of moose chanced to walk the 25 km from Canada across the frozen surface of Lake Superior to Isle Royale. With no predators to limit their numbers, the moose population on this 544-square-kilometer island grew steadily to about 3,000 in 1935. Since the vegetation could not support such numbers, the result was mass starvation: about 90 percent of the moose died. As the vegetation recovered, the moose population began slowly to increase again, until in 1948 it again peaked at 3,000 and crashed. In 1949 a group of wolves inexplicably also crossed the ice to Isle Royale. The wolves began to prey on the moose, for the most part attacking the sick, the old, and the unlucky young, constantly testing the herd for signs for weakness (Fig. 22.11). A stable balance of two dozen wolves, 800 moose, and a healthy crop of grass was achieved. As this example demonstrates, prey-predator relationships can lead toward a dynamic balance, with predation an important factor in keeping the population below the carrying capacity of the habitat.

Territories   A fourth adaptation that enables a species to avoid starvation involves social organization. Population size and density will be limited if individuals must compete to gain control of an area containing enough critical resources to permit bearing and raising young. Such a system of strictly defended areas limits population density by denying some members of the population access to certain resources. This, the single most common method of dividing the spoils, is known as *territoriality.* Each territory, or at least each hotly contested one, has a local abundance of a critical resource —food, mates, nesting sites, or whatever.

When individuals compete for a territory, or two territory holders encounter each other along a mutual boundary, they employ a repertoire of species-specific social signals. The behavior of many songbirds provides a familiar example of this strategy: males arrive early in the season and contest for space, singing to warn off neighbors and landless males as well as to attract females. Experiments have shown that males are definitely being excluded under this system: when males with territories are systematically removed, "floater" males, in seemingly endless supply, step in to occupy the vacated territories.

The most frequent pattern of territoriality among birds is a matrix of territories, each occupied by a pair that feeds, nests, and rears young within its confines. Tawny owls, for instance, live and hunt at night on well-defined territories that they occupy year round (Fig. 22.12). These birds can lay up to four eggs per season, so if each pair reproduced without restraint and all of the offspring survived to do the same, at the end of a decade some three million birds would occupy a patch of woods that supports only 50.

But despite this reproductive potential, the owls do not invest their time and energy in filling the world with their kind. They lead a long reproductive life, so it is perhaps not surprising that selection has favored mechanisms by which each pair undertakes the exhausting and life-shortening enterprise of childbearing only when conditions (as indicated by the relative abundance of mice) are ideal, giving the offspring the best chance to survive to reproductive age. For example, H. N. Southern at Oxford University showed that in a typical year in one habitat, 8 of the 25 resident pairs do not even breed, and another 9 cut their losses early by refusing to incubate the eggs they have laid. Two more pairs allow their chicks to starve, and the remaining 6 pairs lay an average of three eggs each rather than four, so that in the end only 18 of the 100 potential offspring are fledged. Since in an average year only 11 adults die, creating 11 vacancies in the habitat (Fig.

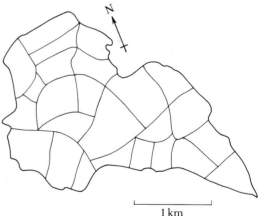

22.12 Owl territories
Pairs of tawny owls have divided this patch of habitat into a matrix of well-defined territories whose number and boundaries are relatively stable from year to year. The habitat supports 50 individuals.

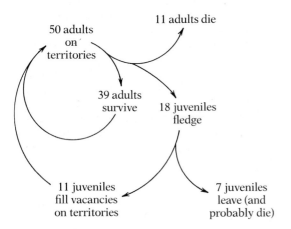

22.13 Typical yearly population flow in a group of tawny owls
A typical year for a tawny-owl population is shown here. On average, 18 juveniles are fledged (though in theory as many as 100 eggs could have been laid) to fill 11 vacancies left in the matrix of territories by the deaths of adults. Seven juveniles are therefore left to seek out vacancies in other habitats, but most do not survive.

22.13), even this number of surviving young is too high. Seven juveniles must leave to seek their fortunes elsewhere, and careful banding studies indicate that their prospects are poor.

Mammals frequently follow similar patterns. Males of many species of African antelope divide up the rich grasslands into a territorial matrix. Excluded males—typically the young and the weak—have the choice of risking injury or death by fighting older, stronger, more experienced males, or of defending nearly worthless spots on the periphery, where the risk of predation is high (Fig. 22.14), or of joining the "bachelor herd" of disenfranchised males. Each must assume one of these behavioral roles, with the individual's choice probably depending on his physical condition as reflected by hormone level. Females wander over the array of defended grasslands eating where they please, and the resident male mates with any who chance to come into estrus on his territory. Naturally, the females remain longest on the richest territories, so the males that possess the best territories have the best chance of siring offspring.

GROUP LIVING

We have seen how selection for one particular mechanism of population control—territoriality—leads to a modest degree of social behavior. It is the distribution of critical resources like food that in large part determines whether territoriality is adaptive. Tawny owls and many antelopes, for example, can find virtually everything they need in an area small enough to defend. And when resources are too dispersed or unpredictable for a single individual to defend, sometimes a group can cooperate to hold an area large enough for them all. But even when territoriality is of no real advantage in assuring the availability of resources, the behavior of predators can select for social living.

22.14 Peripheral territory in a wildebeest habitat
The risk this low-ranking male takes by defending his relatively poor territory is probably balanced by two possible benefits: there is always a chance that a female grazing on his territory may come into estrus while there; and as a territory-holding male he is in a position to judge the fitness of males on adjacent territories, and can attempt to take over a better area if its owner begins to look vulnerable.

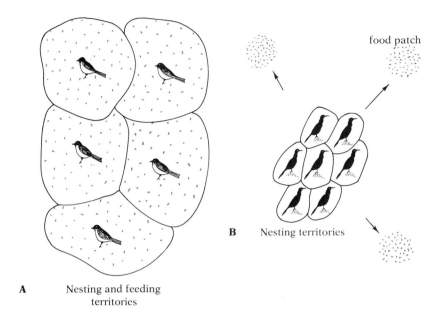

A   Nesting and feeding
territories

B   Nesting territories

22.15   Territory used for both feeding and nesting compared with territory used only for nesting
(A) When food is distributed more or less evenly throughout the area, it may be energetically efficient for each individual to occupy and defend its own territory for both feeding and breeding. (B) If food occurs in unpredictable patches, often far from good nesting sites, it may be more efficient for the individuals to occupy only small nesting territories in a colony and to forage as a group. This type of organization leads to more intimate and structured social systems.

Consider two closely related species of blackbirds, the familiar redwing and the less common tricolor. Their habitats overlap in the western United States, but their social systems are quite different. Redwing blackbirds live on territorial matrices in marshes, and harvest the insects that emerge within their territories; their feeding grounds and their breeding grounds are the same. Tricolors, on the other hand, nest in tight clumps and fly large distances to gather ripening seeds (Fig. 22.15). Because of the dispersed, unpredictable, and transient nature of the tricolors' food, it is not possible for individuals or even groups to defend stable feeding territories. Nevertheless, instead of nesting and foraging as individuals, they do both as a flock. Their nest sites represent breeding territories, and are grouped together in locations relatively safe from nest predators.

Why, we might ask, do tricolor blackbirds forage as a flock rather than as individuals? Flocking behavior, like any other characteristic, has its costs and its benefits; if the benefits of flocking for the individual outweigh the costs, then the behavior is adaptive and will be selected for. One obvious cost of flocking is that the individual must compete for food with other birds scouring the same ground for grain. In tricolors, and in many other species, two benefits outweigh such costs: (1) The birds can indirectly pool information about the location of food, so that a far larger area can be monitored than would otherwise be possible. (2) The birds are safer from predators. Field measurements have shown that the many watchful eyes of the flock make it aware of predators when they are still so far away that an individual feeding alone would not notice them. As a result, individuals can spend more time feeding. Furthermore, predators seem to have more difficulty capturing a bird when it is part of a flock. A falcon or other birds of prey cannot plunge through an airborne flock without risking a potentially damaging collision (Fig. 22.16). Even when it is attacking birds that are just

22.16   Flocking defense against a falcon by flying starlings
Starlings usually fly in a loose formation, and continue to do so even in the presence of a falcon if they are above the falcon. But if they are below, they form a tight flock. The falcon cannot swoop at a bird in such a flock without risking serious damage to itself by crashing into other birds. The falcon will swoop only if an individual starling becomes separated from the flock.

beginning to scatter from the ground, the confusion created by dozens of pairs of wings and potential targets crisscrossing one another distracts a predator. Similarly, bachelor males among the African antelope are innately impelled to form their own herds primarily because herds confer the protection of group-generated confusion, and the females likewise graze across the territorial matrix in groups: during a chase, lions are likely to be distracted from one potential victim and turn to pursue another. Such time-consuming indecision usually reduces the predator's chance of catching either one.

Among animals that, like the tricolor blackbird, have territories devoted exclusively to breeding, suitable areas for such territories may be so scarce that they do not accommodate all reproductively ready individuals. Some animals may then be excluded from the breeding population entirely. Seals, for instance, typically bear their young in the protection of secluded beaches surrounded by cliffs. On shore, newborn seals are safe from killer whales and other aquatic predators, while the cliffs isolate the beach from terrestrial predators as well. Since such ideal breeding grounds are rare, males fight (and frequently die) over them. The winner enjoys the right to mate with arriving females, and so collects a harem of individuals that are actively searching for a safe spot to mate, give birth, and rear their young.

The social "nesting" of seals is a consequence of a shortage of safe breeding grounds. But some other animals—gannets, for example—ignore large areas that are suitable for nesting and crowd together in apparently arbitrary spots (Fig. 22.17). Here again, the behavior is usually shaped by the threat of predation. Nest predators are seen sooner and attacked more effectively by a group than by an individual. For predators, too, the advantages of group living may outweigh the disadvantages. A group of cooperating lions, for instance, can encircle the antelope more effectively and subdue larger prey than can an individual.

Group living is not entirely free of disadvantages, however. Dense colonies can aggravate problems of disease (which spreads more rapidly in groups) and of intragroup aggression—gulls often steal each other's bits of nest material or eat each other's eggs.

22.17   A gannet colony

DOMINANCE

Many animal social systems revolve around contests that establish the likely winner of a fight with a minimum of risk. Mountain sheep, for example, engage in highly coordinated and stylized duels in which two males crash into each other's well-armored heads and back-curved horns (Fig. 22.18). The physics of the collision make the stronger male the likely winner, while the specially shaped horns act as bumpers, rather than weapons, so this test of strength is relatively safe for both opponents. The specialized weaponry developed for protection from or predation on other species is rarely used on conspecifics in any effective way: poisonous snakes wrestle without striking; fish lock jaws but do not bite; antelope push and fence with their horns but will not stab. Considering the risk of unconstrained fighting even to the probable winner, the selective advantage of this innate restraint (known as *ritualization*) is clear: what would be the use of winning an all-out contest if even the winner was left exhausted, injured, or easy prey to watchful predators? However, when two contestants are so evenly matched that there is no clear victor, a ritual contest can sometimes turn into a brutal fight.

The genetic orchestration of aggression through behavioral and hormonally mediated strategies has enormous advantages. In the working out of territories, for example, ritualized contests usually reduce the risk of injury. In more social species, they actually lessen the frequency of serious fighting and stabilize the group through the formation of a stable ***dominance hierarchy***, or pecking order, in which every member knows which individuals it can defeat, and which can defeat it. The formation of such a hierarchy causes fighting practically to disappear—at least until some outsider tries to break in. Every individual keeps its place, since it is aware of the probable outcome of a challenge to animals higher on the scale. Even on a territorial matrix, this principle holds. Locations on the matrix are seldom equally advantageous, and territories with superior grazing or hunting are hotly contested at the beginning of each season: the eventual occupants of better territories hold higher positions in the dominance hierarchy. Curiously enough, an individual that has won and held a spot in the matrix is almost certain to win an encounter on its own territory even against a somewhat physically superior neighbor. The programed willingness to sacrifice more to defend one's own territory evidently swings the balance toward the incumbent.

As a result of this sort of social machinery, the day-to-day workings of societies—particularly societies of mammals, in which memory, the ability to recognize one's neighbors, and the capacity for subtle signaling among group members reach their height—go on smoothly. Overt aggression in such species is generally reserved for mutual defense against rival groups and, to all appearances, helps to stabilize the social bonds within the group. In consequence, intragroup aggression usually lacks that aura of violence and disruption that we tend to associate with fighting. Instead of being disruptive, aggression seems to be a programed behavioral trait that provides the cement for social order, and enhances the fitness of each individual.

22.18 A dominance ritual in mountain sheep Male mountain sheep work out a dominance hierarchy through a ritualized duel in which a run on hind legs culminates in a loud collision. The clash is followed by a head display. In many group-living species the results of such encounters determine the way individuals will interact.

ALTRUISM

Another phenomenon that occurs in social animals, and one that posed serious problems for Darwin as he endeavored to explain how it is that only the fit survive, is the apparently altruistic behavior of individuals in many species. ***Altruism*** is one individual's sacrifice of itself for another, or more precisely, an individual's willingness to lower its own fitness—its personal reproductive potential—thereby raising another's.

The concept of biological altruism has, of course, nothing in common with the uniquely (as far as we know) human moral concept of self-sacrifice for others who may be totally unrelated or even unknown to us. Biological altruism is much more pragmatic: the question here is not so much what motivates self-sacrifice, as how it can persist in the natural world at all. Parents, of course, sacrifice themselves to feed and protect their offspring, but since the number of offspring successfully reared to adulthood is the usual measure of natural fitness, this behavior exemplifies not altruism, but self-interest. The evolutionary problem arises when an animal's sacrifice benefits individuals that are not its offspring.

The classic example of altruistic behavior both in Darwin's day and in ours is the caste system of social insects. The tens of thousands of female honey bees in a hive do not reproduce; instead they devote their energies to rearing the offspring of their queen. Insects like honey bees are thus ample illustration that altruism can evolve and persist in a world shaped by natural selection.

Group selection   The best early attempt to explain altruism on a broad scale was made in the 1950s by V. C. Wynne-Edwards of the University of Edinburgh. Wynne-Edwards offered a detailed theory based on the view then popular that keeping the population below the carrying capacity of the habitat served the good of the group or species; he also argued that species were favored by natural selection if they were so organized that only the fittest—the largest, strongest, or ablest—reproduced. Hence, he suggested, territoriality and ritualized aggression have evolved to provide a means of identifying the less fit, which then step aside for the greater good of the group. In his view, for example, altruism was displayed by the less fit mountain sheep when it did not fight to the death, but rather withdrew so that a more fit male could reproduce. Races or species lacking such a mechanism would, he argued, become weak and overcrowded, and soon die out. For decades biologists had supposed that some explanation along these lines was possible, but with Wynne-Edwards's detailed exploration and application of the idea to a variety of specific cases it became obvious that a mechanism of *group selection*—selection operating on entire groups, rather than on individuals—simply could not work. The problem is that genes coding for altruism would spread more slowly than genes coding for selfishness, since the individuals programed for self-sacrifice would leave fewer offspring—or none at all. In short, selfishness would spread in a group because of individual selection: with altruism less adaptive in the short run than selfishness, the less altruistic strategies would inevitably take over.

Kin selection   Soon after Wynne-Edwards's work focused attention on the problem of altruism, a major conceptual advance was made by W. D. Hamilton at the London School of Economics. Hamilton realized that for the altruism that is evident in many social systems to evolve, genes that code for neural circuits leading to altruism must have succeeded better even in the short run than genes that result in selfishness. He also noticed, however, that the true measure of a gene's success, or fitness, is not whether a particular *individual* possessing it reproduces, but rather whether that gene is found in more individuals in the next generation. Genes that cause one animal to be altruistic and forgo reproduction will still survive in the population if the resulting altruistic behavior sufficiently enhances the fitness of other individuals that carry the same genes. Of course, the easiest way to judge which other animals share genes is by determining how closely individuals are related genetically: siblings, for instance, normally have half their genes in common, so there is a 50 percent chance that any particular gene carried by one is possessed by the other. Similarly, each parent shares half its genes with its offspring, but only a quarter of its genes with each

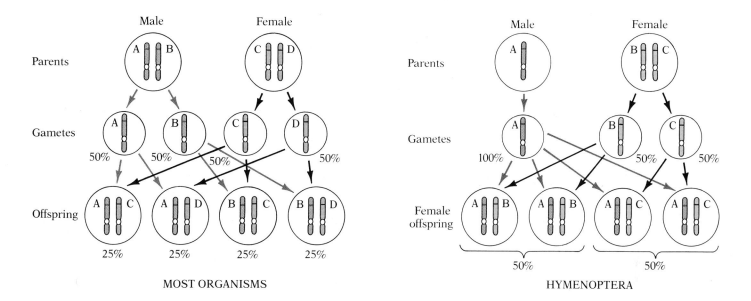

MOST ORGANISMS

HYMENOPTERA

grandchild. In short, for genetically programed altruism to evolve, Hamilton concluded, the donor and recipient must be close kin, and the benefit must be much larger than the sacrifice: the *advantage* received by one individual from an altruistic act performed by its sibling, for example, must be at least double the *loss* in fitness incurred by the altruist.

Hamilton applied this idea of **kin selection**—altruism that increases the fitness of kin—to the social insects. He pointed out that the Hymenoptera (ants, bees, and wasps) have an unusual genetic characteristic: most organisms have two copies of each chromosome, but male Hymenoptera develop from unfertilized eggs and have only one. Under normal circumstances, when two organisms other than Hymenoptera mate, each contributes one copy of its chromosomes to each of its offspring—otherwise the species' chromosome number would double with each generation. Male Hymenoptera, however, pass *all* their chromosomes to each individual they sire. The result is that all a male's daughters have exactly the same set of paternal genes, and a randomly chosen set of half their mother's. In consequence, on average two sisters have three-quarters of their genes in common, rather than the normal one-half (Fig. 22.19). Thus in the Hymenoptera, sisters share more genes with each other and with any new offspring produced by their mother and father than they would share with their own daughters if they were to reproduce. Hence, Hamilton concluded, genes for altruism directed toward the viability of sisters—impelling them, for example, to nurture their younger siblings instead of having offspring of their own—could increase in frequency in these insects.

Even among genetically more conventional species, Hamilton theorized, circumstances might cause selection for altruistic behavior toward kin. Subsequent research has turned up many such cases, perhaps the best understood of which involves burrow-dwelling ground squirrels. These crea-

22.19 Inheritance in the Hymenoptera
In the Hymenoptera (ants, bees, and wasps) males have only half the normal number of chromosomes. All female offspring therefore receive exactly the same set of chromosomes from their father, while the chance of two daughters having a particular maternal chromosome in common is only 50 percent. As a result, sisters on average are more genetically similar to each other than to their mother.

22.20 A ground squirrel giving an alarm call
Alarm calls are sounded at the approach of airborne or terrestrial predators; there is a different call for each. The calls, which involve risk in that they draw attention to the individual giving the alarm, are produced primarily by females with close relatives living nearby.

tures hibernate for nearly two-thirds of the year, becoming fully active in their mountain habitats only during the short summer, when they eat grass and seed and reproduce. In these months females defend territories against ground squirrels they consider intruders. In addition, when predators approach, certain females consistently give alarm calls (Fig. 22.20)—thus increasing their chance of being attacked—while others never call. Males fight over estrous females, and a few older males are usually most successful. Finally, juvenile females overcome their territoriality and clump together in small groups to hibernate near their birthplace, while males disperse and pass the winters alone.

Without the model provided by the hypothesis of kin selection, none of the ground squirrels' behavioral jumble makes much sense. But by marking hundreds of ground squirrels and their young year after year, Paul Sherman of Cornell University has been able to keep track of who is related to whom, and thereby to sort out the ground-squirrel social organization.

The major source of mortality, he found, is the unpredictably long and cold winter of their mountain habitat. In addition, predation by coyotes, weasels, badgers, bears, and hawks takes about 5 to 10 percent of the population each year. Fighting kills another 5 to 10 percent of the adults, and adults kill roughly 10 percent of the young annually.

Surviving the first winter, controlling sufficient food resources, guarding against predators and aggressive conspecifics, and reproducing early are crucial to fitness in ground squirrels. An estrous female accepts whichever male is able to fight off the other males long enough to mate with her. In general these males are larger and older, so we may suppose that the winning male's genes have sufficiently proved their fitness to justify the females' passive acquiescence in obtaining half the genetic information her offspring will preserve.

The squirrels probably learn to recognize their relatives in their first day or two above ground; any squirrel encountered by a youngster regularly on its mother's territory is almost certainly a relation. According to Hamilton's theory of kin selection, the squirrels should favor low-risk altruism directed toward related individuals.

A careful examination of the kin records derived from marking studies and blood tests shows that, in fact, the females take small risks to aid their kin. Their social organization serves to help kin in facing the challenges of winter survival, food scarcity, predation, and infanticide. Group hibernation is clearly an efficient, low-risk way to survive winter by sharing warmth, and the programed toleration of kin means that only females that are related will cooperate in this way. Only relatives—sisters, mothers, daughters, and their offspring—are allowed to share food during the warm season. All others are attacked. Thus the curious variability in the behavior of female ground squirrels—their sharing of territories and food with some squirrels while they chase out others—turns out to be based on kinship. Warning calls are produced much more frequently by older females with living kin than by males or childless females, and the advantage of such a warning to a female's nearby relations outweighs the risk she herself incurs. Finally, infants are never killed by relatives; rather, they are killed by homeless squirrels that would benefit from a reduction in competition. Hence, related females cooperate in driving off unrelated animals.

Reciprocal altruism   Though kin selection appears to be widespread and helps explain much of the behavior of social animals, careful studies inspired by Hamilton's hypothesis have turned up at least as many cases in which kinship is not a factor in altruistic behavior. The most common motive force behind altruism after kin selection is probably reciprocity. The behavior in which an animal confers a favor on another animal, not necessarily a relative, in the expectation of eventual repayment, is called ***reciprocal altruism.*** Two social animals may cooperate to groom places on each other that neither can reach, for example (Fig. 22.21). In most cases of reciprocal altruism, animals must be able to recognize and discriminate against cheaters, those chronically unwilling to repay favors. Failure to punish cheaters usually means that reciprocal altruism cannot persist in a population, since genes for the purely selfish strategy of nonreciprocation are then inevitably more fit. The necessity for keeping track of cheaters suggests that reciprocal altruism should most often be found in small groups of animals capable of recognizing each other individually, which seems in fact to be the case.

The realization that much of what we recognize as altruistic behavior is ultimately selfish underscores the point that innate behavior, like all other specializations, whether morphological, physiological, or behavioral, enhances the fitness of the genes coding for it.

MATING STRATEGIES

We have seen how ecological factors affect selection for solitary, territorial, or group living, and for various sorts of altruism. Another set of behavioral alternatives subject to strong selection involves mate choice. The reason is clear: in diploid animals (those with two sets of chromosomes), half the genes of any individual's offspring come from another individual. Any behavior, then, that serves to secure genes of greater quality will be favored by natural selection. As in the case of the other behavioral alternatives we've examined, the mate-choice method differs from species to species. Indeed, individuals in many species seem to make no choices at all (female mosquitoes, for example, copulate with practically any male when the time is right), while in other species choosing involves elaborate rituals.

One goal of behavioral ecology is to understand how environmental variables help determine whether the individuals of a species will pair for life (***monogamy***) or mate with many individuals (***polygamy***), whether males will court females or vice versa, and whether courted individuals will pay more attention to a suitor's morphology, to the vigor of the suitor's display, or to the quality of the resources the suitor controls—territory or a nuptial gift, for instance.

Sexual selection and contests between males   Darwin postulated that a special sort of selection is involved in mate choice. Most selection pressures, he reasoned, are shared by the two sexes of a species, and basically involve survival. But ***sexual selection*** is selection for characteristics necessary to attract and keep mates, and the selection pressure may be confined to just one sex. Individual fitness is best served if those potential mates with generally high genetic quality are preferred. Darwin divided the phenom-

22.21   Mutual grooming in penguins
Each animal grooms an area the other cannot reach.

ena of sexual selection into two categories—one involving contests, the other female choice.

The contest, or competition, form of sexual selection is very widespread. Individuals of one sex, most often the males, compete for territories or position in a dominance hierarchy. We have already seen examples of this form of sexual selection. Male wildebeest, for instance, compete for the grassiest territories; females, by the simple act of grazing where the grass is most plentiful, increase their chances of mating with a physiologically (and probably genetically) superior male.

In some animals the contests are not linked to territoriality. The logic of the fight-it-out strategy, as seen in ground squirrels, for instance, appears to be that a male's ability to survive several seasons and defeat the other males in the neighborhood is a good indication of the quality of his genes (or at least those for physical endurance). Similarly, male mountain sheep work out a hierarchy based on strength and age that seems to reflect overall genetic quality. Researchers disagree over whether females in these social systems have the victors of these contests forcibly imposed on them or whether they simply respond to the stronger male.

Female-choice sexual selection In other species, males do not engage in contests; instead, they perform ritualized displays, and females appear to choose among them. We say "appear" because some researchers deny the existence of female choice. But consider male widow birds, which have remarkably long showy tail feathers, which they display in specialized acrobatic performances (Fig. 22.22). Though it has long seemed likely to most observers that these feathers are attractive to females, the possibility existed that they might instead confer fitness on the males in contests with other males, or in gathering food, establishing territories, and so on. By cutting off part of the tail feathers of some males and gluing those pieces to the tail feathers of other males after their territories had been established, Malte Andersson of the University of Gothenburg in Sweden showed that females select a male solely on the basis of his tail length. Similarly, R. J. Bischoff, J. L. Gould, and D. I. Rubenstein at Princeton have shown that female guppies choose males in large part on the basis of tail area (see Fig. 33.4, p. 854). In these cases at least, sexual selection is clearly the force that has led to exaggerated tails. Structures like these tails that differ between the sexes of a species are known as *sexual dimorphisms*. It seems likely that many sexual dimorphisms that have no obvious role in combat between males do have a role in female-choice sexual selection.

Since female-choice sexual selection evidently exists in at least some species, you may wonder how such a system could have evolved. Here again there is considerable controversy. If the sexual dimorphism of a male accurately reflects his genetic fitness, then selection will favor those females that are attracted to the most conspicuous dimorphisms. Genes that code for neural circuits that respond to ever-greater stimuli, then, will increase in the population. But how can a large tail indicate fitness? If anything, exaggerated sexual dimorphisms are burdens that not only tax the metabolic resources of males, making movement and feeding more difficult, but also make them more conspicuous to predators and often slower to escape. In

22.22 A male widow bird in flight

fact, a male's ability to survive to reproductive maturity despite the handicap imposed by his size or shape may be an indirect indication of his fitness. A female programed to choose the male with the largest handicap, then, might enjoy the dual benefit of producing daughters that are physically superior and sons that are both superior and likely to be attractive to females.

If the females of some species recognize males only by means of a releaser, however, male dimorphisms could also evolve without any correlation with fitness. The possession of the proper stimuli alone would be sufficient to recruit females. Such species would be caught in a genetic trap: only females that recognized the male releaser (the dimorphism) could mate with their species reliably, and males with the most supernormal stimuli would pass their genes on disproportionately in the population, no matter how maladaptive the trait might prove to be.

PARENTAL INVESTMENT

The amount of time adult animals spend raising offspring varies radically from species to species. At one extreme are animals adapted to produce as many young as possible with minimal investment of time and energy; at the other are animals that invest a great deal of both to enhance the fitness of only a few offspring. Most fall somewhere in between.

The relative advantages of these investment alternatives differ not only among species, but between the sexes. In most mammals, for example, the males invest much less time than do the females in caring for the young. Once fertilization has occurred, the developing fetus is tucked safely away inside the mobile female, and once born, it is suckled by its mother. In short, a male mammal usually can do little to enhance his offspring's (or his own) fitness by remaining with his mate. As a result, males form long-term bonds with females and help with rearing the young in only 10 percent of mammalian species. Of course, the female too can be programed to help or to abandon the young, as considerations of fitness dictate, and in some species, such as the seahorse and the jaçana (a tropical bird), the male is left to care for the progeny.

By contrast, for some species it has proven advantageous to have both parents care for the young. The male robin, for instance, could have been programed to desert his nest once the eggs were laid, and to attempt to establish another territory and attract another mate. But natural selection has favored male robins that stay at home, and the reason is clear: by remaining at the nest, he is able to enhance his fitness. He can guard the nest while his mate feeds, and when the eggs hatch he can help gather food for the voracious young. In fact, 93 percent of bird species form monogamous pair bonds and divide the labor more or less equally between the sexes.

SOCIAL ANIMALS

Now that we have looked at various strategies that shape the social systems of different species, we shall examine the behavioral ecology of a representative social insect and a group of social mammals.

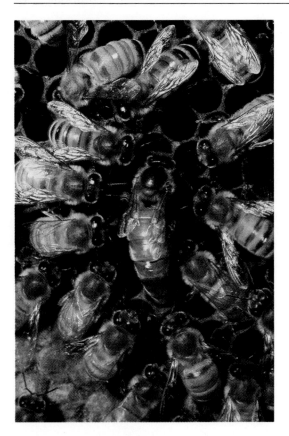

22.23   A queen honey bee
Each colony normally has only one queen (center),
who is constantly attended by young worker bees.
The queen, whose only function is to lay eggs,
produces pheromones that both attract workers and
repress their ovaries. The ovary-repressor pheromone
is unusual in that it is not transmitted through the
air, but instead is consumed by the attendants when
they groom the queen, and passed to other workers as
part of the normal process of food exchange.

HONEY BEES

Except for the primates, the social insects, particularly the Hymenoptera,
have the most complex and best-understood social systems known. We shall
look at honey bees, not because their societies are more complex than those
of highly social ants, but because much more is known about how bee societies are organized.

Honey bees evolved tens of millions of years ago in the tropics, where
three primitive species are still to be found. The colony, consisting of thousands of individuals, persists for years, raising a new queen and dividing
into two when it has reached a certain size. As we saw earlier, tropical
honey bees build sheets of comb that hang exposed from tree limbs (see Fig.
22.9). The individual cells of which the comb is made are used to raise new
bees and to store honey and pollen. The adaptation that allowed the familiar species *Apis mellifera* to survive the winter in temperate zones is its habit
of building the comb inside a hollow tree, where the waste heat from the
bees' metabolism can provide warmth during cold weather. This behavioral preference for cavities of some sort also protects the colony from predation.

Honey bees have three castes: a single queen (Fig. 22.23), whose only
function is to lay eggs; tens of thousands of nonreproductive female
workers, who build and maintain the hive, rear young, and gather food; and
the males, or drones, of whom there may be several thousand in the spring,
when the colony is producing new queens. The life cycle of the colony
starts in the spring, as the workers begin to forage. Since honey bees, unlike
other bees and wasps, overwinter as a group rather than as isolated individuals, they vastly outnumber the other insects and so have a supply of spring
food almost entirely to themselves. During this period of abundant food, the
colony invests heavily in raising new bees: the queen may lay up to three
thousand eggs a day. As the colony gets large, the workers begin to prepare
for swarming—the process by which the colony divides. The queen is fed
less and less, so that she becomes light enough to fly, and new queens are
reared.

The mechanism of social control during this and most other stages of the
life cycle of the honey bee is chemical. The queen continually produces
pheromones that not only inhibit the ovaries of the worker bees but also attract the workers, and induce them to feed her. Furthermore, the same
pheromones inhibit the construction of the special cells used to rear new
queens. When the population of the colony becomes so large that the quantity of pheromone produced by the queen is insufficient to repress the
workers, their behavior changes: queen cells are built, new queens are
reared, and the old queen is no longer fed. Swarming takes place about two
weeks later.

Swarming is initiated by acoustical messages. As the new queens mature
in their cells, the old queen occasionally produces a series of sound pulses.
When a new queen is ready to emerge she answers with another sound pattern. This signals the colony that swarming is now possible, but swarming is
actually initiated by a different acoustical signal, produced by worker bees
when the time of day and the weather are just right.

The swarm, which consists of about half the bees and the old queen,

22.24 A honey-bee swarm
The bees form a huge ball in the fork of a tree. Scout bees fly out from the swarm in search of possible new nesting sites.

leaves the hive and settles on the branch of a nearby tree (Fig. 22.24). From here scouts explore the habitat for a new cavity. They communicate their finds to other bees by means of waggle dances on the side and top of the swarm. For two or three days the scouts visit each other's finds, and dance more and more vigorously for the cavity they consider the best. When a consensus has been reached among the scouts, the swarm departs to the new location and begins building comb.

Meanwhile, in the old hive one of the queens will have hatched and killed any other developing queens. Later she will fly out to a special mating area, where she will mate with several drones, obtaining and storing the sperm that will fertilize the tens or hundreds of thousands of eggs she will lay during her reproductive life. Once back in the hive, the queen becomes an efficient egg-laying machine.

The workers, the labor force in the insect caste system, pass through a series of age-dependent tasks, from cleaning cells, to nursing larvae and the queen, to building comb, to guarding the hive from predators, until at about three weeks of age they begin to forage. Those bees that escape such common mishaps as getting lost or being captured by hunting wasps or the ambush bugs and crab spiders that lurk on flower blossoms (see Fig. 33.29, p. 873) end by simply wearing out—falling unnoticed to the ground on some foraging trip after having flown perhaps 2,000 km during a six-week lifespan, to collect less than 30 grams of nectar. The worker's life is finely regulated by communication—specific acoustical messages are involved in swarming and add information to dance communication, while special odors guide her to the hive and to rich food sources, elicit attack or alarm, and so on. A worker who dies in the hive produces one final odor that plays a role in social communication: a dead bee exudes a pheromone that instructs her sisters to remove her body from the hive. So powerful is this particular message that painting a drop of the substance on even a live queen will cause the workers to discard her.

The coordination engendered by the powerful force of kin selection and maintained by highly specialized behavioral programing has made the honey-bee colony a smoothly operating corporation of individuals that fully exploits the many advantages of social living. Sociality makes possible the maintenance of the nest at a relatively precise temperature and humidity, and in honey bees, division of labor creates several subpopulations of efficient specialists. The protective cavity and the large colony size permit effective group defense against even the largest of predators, and the colony's numbers, combined with the dance communication system, facilitate the rapid recruitment of foragers, a process that enables bees to exploit new resources before other species (or colonies) locate them.

But sociality has its costs. Specialized predators such as the oriental hornet and one species of digger wasp feed on foragers as they come and go. The birds aptly known as honey guides lead honey badgers to the hive and share in the spoils after a badger tears the nest open. Wax moths, which can destroy the framework of the colony, can be devastating. Moreover, bees suffer from various "social diseases," including foulbrood (in which a microorganism kills and consumes larvae and pupae) and infestation by a variety of viruses and a motley collection of bee mites and bee lice, all of which depend on the social interactions of bees to move them from one host to another. Still, the advantages of sociality far outweigh these costs, and the honey-bee colony remains a model of efficiency and order.

PRIMATES

Primate social organization varies from one species to another, encompassing the solitary lifestyle of the orangutan as well as the highly social and seemingly promiscuous behavior of the chimpanzee, but some general ecological and behavioral trends are evident. With only a few exceptions, primates are arboreal herbivores and insectivores; they live in trees (out of the range of most predators) on generalist diets. Though human DNA is most similar to the DNA of chimpanzees and gorillas, our behavioral evolution may have more in common with the evolution of baboons and macaques.

In the distant past, baboons and our early not-yet-human ancestors evidently occupied the same niche; about 6 million years ago, after the African savannas, or grasslands, came into being, baboons and human forerunners alike descended from the trees and began to employ a very precise hand–eye coordination to harvest the seeds, corms (underground stems), and berries of the savannas. Baboons still occupy this niche, though their diet today includes insects and even young vervet monkeys in addition to berries and seeds. Some present-day species, like the olive baboon, forage in parties of 40–80 individuals in the dry woodlands and grasslands, retreating to the trees for protection at night. Their large troop size effectively discourages predation by leopards and hunting eagles.

Adult females form the stable core of baboon society. They range for many years over vast areas of savanna and woodland, remaining together in relative stability. This social stability is fostered by a hierarchy, established by relationship, within the female group; as a juvenile female matures, she "inherits" a place in the hierarchy just below that of her mother. Interestingly, dominant females tend to have more female offspring, while subordinate females have more sons. Since a daughter inherits rank, while a son soon leaves the troop to establish rank elsewhere, this curious conspiracy of biological factors seems to make sense.

Infants with mothers at the top of the hierarchy have a very different childhood from their counterparts near the bottom. The offspring of dominant females are rejected sooner and more forcefully by their mothers (perhaps thereby allowing the mothers to rear more offspring); these young become more independent and interact more with other females and males. The infants of low-ranking females, on the other hand, are watched longer and are more vigorously guarded by their mothers.

Adult male baboons in the troop also form a hierarchy among themselves. Unlike rank in the female hierarchy, male rank is established and maintained through fighting skill and social alliances with other males. In fact, male cliques form, each with its own internal hierarchy; one clique becomes dominant, and each member will rush to the aid of other members to maintain that dominance. The result of such mutual aid is a high degree of stability: a particular male from outside the group may be capable of defeating any individual in it, but he has no chance against the clique as a whole (Fig. 22.25).

Males may also play a protective role for females and infants in the troop: they feed young, carry them, and shield them from danger and dominant females, and ranking males may provide protection for subordinate females. Indeed, if a subordinate female enjoys the tolerance of a dominant male, she may threaten dominant females with no fear of retaliation. A mother protected by a ranking male is generally better able to protect her young as well as herself. Male protection depends on female response; the male can offer his protection but cannot impose it unless the recipient is willing.

Other species of primate display equally elaborate hierarchal systems. Social systems based on individual recognition and mutually accepted rank benefit both high- and low-ranking individuals and confer great stability on a group. Everyone has a place and knows it. As a result, the society appears calm and well ordered, at least on the surface. An acute observer, however, sees a constant tension, an intricate web of veiled threats, passive displacements from position, and subtle requests for reassurance that underlie this stability. To what extent the great variety of primate behavior can hold a message for our species, with its apparently unique evolutionary trajectory, is a subject of stimulating debate.

22.25   A primate alliance
The two baboons at the right are cooperating to threaten the baboon on the left, who could defeat either alone, but is unwilling to fight the two together. Such alliances permit small groups to control a sizable community.

THE EVOLUTION OF HUMAN BEHAVIOR

The weight of fossil evidence suggests that our species evolved as a ground-dwelling hunter of animals and gatherer of vegetation, living in small groups in the bush and plains of Africa. As early as 7 million years ago a

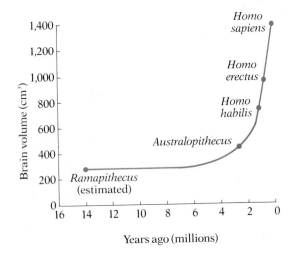

22.26 Evolution of brain size in humans
The brain capacity of the evolutionary line that led to humans increased at an extraordinary rate over the last few million years. Since brain volume is a rough measure of brain specialization when variables like body size can be excluded, it seems likely that selection strongly favored our species' many specializations, which form much of what we call intelligence, during a brief, relatively recent period.

hominid ancestor was clubbing to death and eating small game—lizards, tortoises, snakes, porcupines, savanna baboons, and the like—as well as consuming readily available seeds and berries. About 1.5 million years ago, after 2 million years of rapid brain growth (Fig. 22.26) that probably reflects a strong selection pressure for intelligence, our species' immediate ancestor, *Homo erectus*, emerged to become the most dangerous animal on earth, hunting in groups with stone weapons to kill large animals like giraffe that had previously known little predation.

Our own species, *Homo sapiens*, also originally lived in hunter-gatherer societies, but the domestication first of animals and then of plants, only 10,000–20,000 years ago, made possible great changes in social structure. With domesticated herds and cultivated grain providing reliable sources of food, groups no longer had to be nomadic and small to follow game. The result was the beginnings of civilization as we see it now across most of the world, characterized by cities, division of labor, private wealth, and large-scale central authority.

Modern civilization has created a new environment for our species, to which we must adapt as best we can. It is logical to suppose that our societies represent a compromise between the behaviors that evolved under the pressure of selection for a hunter-gatherer niche, and the new behaviors required by the technology on which we now so completely depend. Conflicting demands from the evolutionary past and the cultural present may create many of the stresses we associate with modern civilization. But the fossil evidence of human evolution can tell us little about the more interesting questions of early human behavior. Though there is evidence that our ancestors must have been social, we have no idea what their social system may have been. And though their diet is well understood, we still do not know whether our evolution was shaped by a relative scarcity of animal protein, or carbohydrate, or salt—or indeed whether the critical determinants involve food at all. We shall return to the evidence of the human fossil record in the last chapter of this book.

Hunter-gatherer cultures Possible insight into the cultural evolution of human behavior may be derived from the study of the rare groups of hunter-gatherers. Until relatively recently, these groups have been untainted by the technology and social adjustments necessary for crop cultivation and the domestication of animals. In the past these peoples often exhibited what are, according to E. O. Wilson of Harvard, universal human behavioral characteristics: property rights, body adornment, incest taboos, sexual roles, rites of passage, intraspecific war, and belief in the supernatural. The best understood of the primitive tribes are the !Kung bush people of the Kalahari Desert in southern Africa. (The ! represents a tongue click not present in our alphabet.) Over the past few decades the number of !Kung groups depending exclusively on hunting and gathering has dwindled to zero, and so this account of the hunter-gatherer culture does not describe the vast majority of present-day !Kung groups. It is also worth keeping in mind that the presumed cultural universal of intraspecific war was absent from traditional !Kung culture. The following description of !Kung culture is based on information gathered over several decades and most recently synthesized by Richard B. Lee of the University of Toronto.

A   B

22.27 Hunter-gatherers
!Kung groups depended on both hunting (A) and
gathering (B).

In the past !Kung adults, like the members of many nonhuman species, generally formed monogamous pair bonds and invested heavily in their off-spring, but unlike animals of the other predominantly monogamous species, which tend to live as isolated pairs, they lived in groups of 20 to 30. These groups occupied traditionally undefended areas of about 500 square kilometers, centered on sources of water. Group interchange was frequent, and individuals often visited other bands for days or even weeks. The !Kung divided the labor of obtaining food, with women gathering fruits and vegetables and men hunting game in small groups (Fig. 22.27). Since gathering is more efficient than hunting, women harvested about 60 percent of the protein and carbohydrates, yet worked fewer hours than the men. Though less efficient in terms of quantities obtained, hunting does provide essential amino acids, vitamins, and minerals. Since food from both sources was seasonal, the rhythm of the seasons controlled !Kung social organization. The limiting factor was water. Since about 95 percent of the rain falls during a six-month period, there is an extended drought each year. During the rainy season a group would settle in a nut forest and eat their way out of it, collecting vegetables, fruit, and nuts over an ever-widening circle until the round trip for foraging reached 15–20 km. With the approach of the dry season, !Kung groups retreated to the most dependable water holes and made do with whatever food they could locate until the arrival of the next rainy season.

The controlling element in the behavioral ecology of the !Kung, as in most animals, was the variability and density of the food supply. Since gathering was done from a base camp, at ever-greater distances, it is clear that the larger the group, the larger the area that had to be harvested to support the population—and the farther individuals had to walk. Hence, if gathering had been the only activity, groups would have been very small. Hunting, however, tended to increase group size. The men normally hunted in pairs, and these teams frequently failed to find or to kill prey. The greater the number of pairs on the move, the more likely it was that at least one of them

22.28   A !Kung hunter preparing poisoned arrows

would be successful. Since a single wildebeest provides about 200 kg of meat, more than enough to carry a group of 30 through two weeks, an increase in the number of hunting pairs could dramatically improve the diet of the group as a whole.

!Kung social organization traditionally depended on reciprocal altruism within and between groups. Within a group, hunters and gatherers would share food daily, and the spottiness of rain meant that a group would have to borrow water from a neighboring group in some years and lend its own in others. The cement for this altruism appears in part to have been kinship. The groups were complex kin associations, and the choice of leaders was determined largely by kinship. Requests to use part of another group's territory or water were inevitably made through lines of kinship, and visits to neighboring groups always took the form of kin visiting kin. Strong cultural traditions were also of great importance in group cohesion. Hunting and gathering both depended on a fairly sophisticated, culturally transmitted technology of digging sticks, woven nets, canteens, pouches, snares, spears, bows, poisoned arrows (Fig. 22.28), knives, fire, and so on.

Though the !Kung may be considered to have represented a relatively early stage in the cultural evolution of human behavior, it is difficult to separate reliably the innate from the cultural even among these so-called primitive peoples. Their technology was already impressive, and culture played an essential part in their lives. Moreover, the !Kung and other hunter-gatherers survived until relatively recently because they occupied habitats unsuitable for farming or grazing; the selective forces shaping behavior may have been different in the richer habitats where the majority of early humans lived.

Infant behavior   For unenculturated humans we must turn to newborns. Though we can hardly expect our observation of infants to tell us much about social organization or behavior that matures later in life, it does yield unambiguous evidence that our species is provided with at least some degree of innate guidance. Infants enter the world with a remarkable repertoire of innate behaviors and reactions. They are able to cling tenaciously, walk in a rhythmic and coordinated way if supported, swim, locate both sounds and odors, search for a nipple and suck rhythmically, distinguish human speech from other sounds, and, of course, perform that well-coordinated unit of communicative motor behavior, crying. When more demanding testing becomes practicable within a week or two, we find that infants presented with a series of expanding silhouettes that suggest an approaching object will interpret them as dangerous, and perform a stereotyped set of defensive maneuvers. Coordinated reaching also begins at this age; the innate character of this behavior seems indicated by its relative immunity to feedback: an infant continues the reaching movement for the same length of time even if the presented object is shifted or made to disappear, and follows with a stereotyped grasping movement regardless of the shape of the object or even of whether or not it is present.

Four-week-old infants begin to focus on faces and smile, a behavior that is as mindless as it is endearing: blind children evidence the same fixation of the eyes (directed at the source of sound) and produce an equally winning smile though they have had no opportunity to observe one (Fig. 22.29).

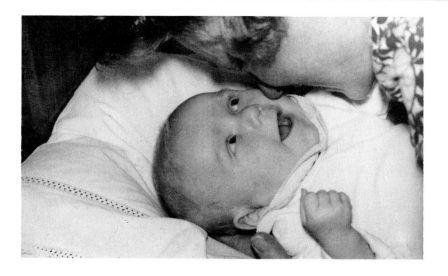

22.29 Smiling
Smiling appears spontaneously in human infants at about four weeks of age. The innate nature of this motor program is illustrated by the smile of this eleven-week-old congenitally blind girl. Her eyes have fixated on the source of her mother's voice, a complex behavior that is also innate. Smiling helps cement a strong emotional attachment between parent and child.

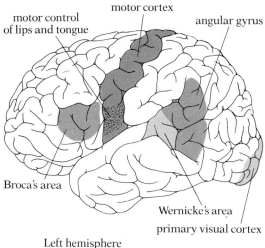

22.30 Anatomy of language
The human brain contains many specialized centers concerned with language, particularly in the left cerebral hemisphere (shown here). Sounds are detected by the ears and processed in the midbrain (not shown). Spoken language is then sent to Wernicke's area. Written language is somehow abstracted from the other material processed in the primary visual cortex and sent to the angular gyrus, where it is translated into sound. It too is next sent to Wernicke's area, which extracts meaning from incoming language, whatever the source. Language production too begins in Wernicke's area. Thoughts are encoded there into crude linguistic outlines, and these are sent to Broca's area, which refines them into grammatical sentences. Finally, Broca's area transmits directions to the adjacent motor cortex, which controls the organs of speech, to produce utterances.

Laughing, pouting, frowning, and the whole range of facial expressions typical of responses to sweet, sour, and bitter tastes, are also seen in congenitally blind infants. And, as the Swiss psychologist Jean Piaget demonstrated, children the world over pass through essentially the same series of developmental stages, virtually immune to instruction in a particular area until some critical time, when even the slightest urging may trigger the behavior. For instance, the concept that the mass of a given amount of clay stays the same regardless of whether the clay is rolled into a ball or into a cylinder cannot be taught: children believe the cylinder to be heavier (and larger), and indicate their opinion both verbally and in the degree to which they tense their arm muscles when the test object is handed to them. And yet, though coaching is ineffectual, a day comes at about the same age whether or not the child has received previous instruction, when the idea of conservation of mass is suddenly "learned."

By far the most impressive evidence for the role of innate programing in infants, however, comes from studies of language acquisition. Human culture depends almost exclusively on language—on our ability to transfer information from one person to another about things and events distant in space and time—and we normally think of language as a purely intellectual achievement. Yet when examined in more detail, language learning reveals patterns reminiscent of the elegant but innately directed learning of birdsong.

Language acquisition Much remains to be discovered about language processing, but data now in hand indicate that the human brain is already wired at birth with the neural circuitry necessary for language. Linguistic meaning seems to be processed on one side (usually the left), in discrete, well-defined areas (Fig. 22.30). For written language the processing begins in the primary visual cortex, where what we see is analyzed before being sent to the so-called *angular gyrus;* there written words are translated into sounds, which are then passed to **Wernicke's area.** Spoken language is sorted from other sounds in the midbrain, which sends the lowest frequencies of speech to Wernicke's area. Wernicke's area, then, is the destination for both written and spoken language. Higher frequencies go to the right side of the brain, where the emotional overtones of the speech are ascertained. So segregated are the intellectual and the emotional functions of the brain that many people with right-hemisphere damage can understand the *meaning* of a spoken sentence, but cannot say whether the speaker was happy or sad, angry or ironic. Conversely, people with left-hemisphere lesions (any damage from injury or disease that interrupts the flow of information) can often judge the mood and intention of a speaker, and yet have no idea of what has been said.

Wernicke's area is also the processing center most involved in the individual's own spoken and written expression. It is here that thoughts are formulated into crude linguistic structures before being sent to **Broca's area** for grammatical refinement. Lesions in Broca's area often leave a patient knowing what he wants to say, but unable to express it according to the accepted rules of tense, declension, number, gender, and so on; they may rob the speaker of such linguistic signposts as pronouns, conjunctions, and prepositions. Lesions in Wernicke's area, on the other hand, leave the patient talking perfectly grammatical nonsense.

The human brain is able to distinguish among consonants at birth: all infants, regardless of the linguistic heritage into which they are born, distinguish the same set of 40 consonants. (Each of the thousands of human language systems use some subset of these consonants—about two dozen in standard English—but babies recognize them all whether they have had an opportunity to hear them or not.) Like birds, human infants have a babbling phase; the fact that it begins at a characteristic age, even in deaf children, suggests that language acquisition is under genetic control. Human infants who are able to hear themselves learn to manipulate their vocal apparatus to make language sounds and then to talk, and children who become deaf after they have learned to talk retain the ability to speak. Apparently the motor programs involved in the production of speech, like those of birdsong, are stored as neural circuits in the brain and then require little feedback.

Even the *structure* of linguistic utterances may be inborn. Lila Gleitman of the University of Pennsylvania has observed that deaf children, left to themselves, create mutual communication systems that share many of the grammatical features of spoken language.

All of the present evidence, then, suggests that language is the product of genetically specified neural circuitry in combination with the cultural environment, and that the drive to acquire language is under strong genetic control. Given further analysis, the intellectual development of infants that Piaget charted—particularly the acquisition of logical and physical concepts—may be found to fall into a similar pattern. The organizing principles of the study of animal behavior can be applied to many aspects of human behavior. A biological consideration of human intellectual development may offer real hope for constructive treatment of the devastating linguistic and cognitive abnormalities so prevalent among our species.

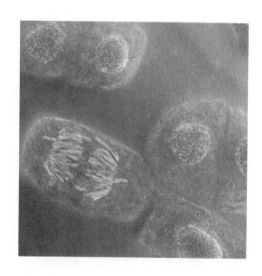

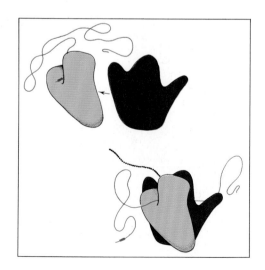

PART **III**

THE PERPETUATION
OF LIFE

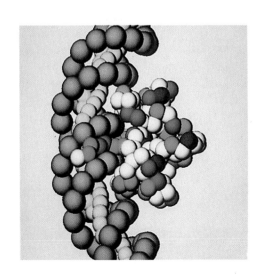

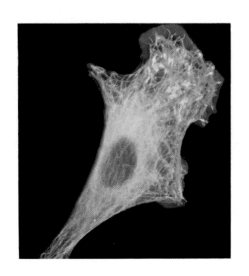

Chapter 23

CELLULAR REPRODUCTION

The processes of life are guided by an elaborate series of information transfers. The genes of an organism's DNA contain all the information necessary to build and operate the organism—the information capable of orchestrating its development from a single cell into a complex corporation of tissues and organs, and of directing events ranging from cellular chemistry to the organism's behavior.

The transfer of this information occurs in two directions: first, the instructions in the DNA are used, through the production of RNA and enzymes, to direct biochemical events; second, and equally important, the instruction set is duplicated and transferred to each new cell, whether it be a cell added as the organism grows, or one of the gametes that unite to form an entirely new organism. In later chapters of this part we shall see how the information stored in a sequence of nucleotides in the DNA of chromosomes is transcribed into a complementary messenger RNA copy, and how the sequence of bases in the RNA is translated into a series of amino acids to create a structural protein or one of the enzymes that control cellular chemistry (Fig. 23.1). But we must begin by examining how instructions are transmitted from cell to cell and from parents to offspring. In this chapter we shall see how genetic information is transferred when new cells are formed.

THE TRANSFER OF GENETIC INFORMATION

Most of the DNA in a cell is contained in the nucleus; smaller quantities are found in mitochondria and chloroplasts. The importance of the nucleus in the transmission of hereditary information has been shown in a variety of ways. Joachim Hämmerling of the Max Planck Institute for Marine Biology in Berlin carried out his demonstrations with the unicellular green alga

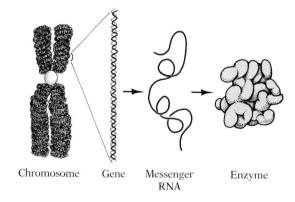

| Chromosome | Gene | Messenger RNA | Enzyme |

23.1 Flow of information in a cell
Genes are located in the chromosomes; a typical eucaryotic nuclear chromosome is depicted here. The sequence of nucleotides in a gene determines the corresponding sequence in a messenger RNA, which in turn is used to assemble amino acids in the sequence unique to a particular structural protein or enzyme.

23.2 The green alga *Acetabularia mediterranea*

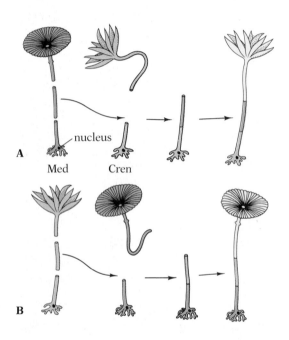

23.3 Hämmerling's experiments on *Acetabularia mediterranea* (med) and *A. crenulata* (cren)
(A) A piece of med stalk (dark green) grafted onto a cren base (beige) regenerates a cren cap (after first regenerating a cap of intermediate type, not shown here). (B) A piece of cren stalk (beige) grafted onto a med base (dark green) regenerates a med cap. In each case the characteristics of the regenerated cap (light green) are determined by the part containing the nucleus.

Acetabularia. This curious plant, though it is composed of only one cell, consists of a branching rootlike base, a stalk, and a cap. The nucleus is in the base. Hämmerling worked with two closely related species, *Acetabularia mediterranea* and *Acetabularia crenulata*—med and cren for short. Med's cap is disc-shaped, with scalloped edges (Fig. 23.2); cren's is deeply dissected into nearly separate lobes.

Hämmerling showed that if he amputated the stalk and cap of a cren individual and grafted a med stalk without a cap onto the cren base, a new cap was eventually regenerated, intermediate in appearance between med and cren. If this cap was amputated, the next cap regenerated was a typical cren one (Fig. 23.3A). In other words, this second cap did not take its characteristics from the med stalk, on which it grew, but from the cren base, which contained the nucleus. The reciprocal experiment yielded similar results: a cell consisting of a med base and a cren stalk regenerated a med cap (Fig. 23.3B). The nucleus was obviously the controlling factor. The intermediate appearance of the first cap regenerated resulted from the presence of mRNA and some proteins in the cut section of stalk which were retained from the original nucleus. By the time the second cap formed, their influence had given way to that of the nucleus in the new base.

Hämmerling's experiments showed that the information or blueprint that determines the development of the individual organism comes from the nucleus. It follows, therefore, that when a cell divides, the nuclear information must be transmitted in orderly fashion to both of the new cells. The division cannot be a simple splitting of the information into two halves, because this would give neither of the new cells a satisfactory blueprint. Just as you cannot have two buildings erected by cutting one blueprint in two and giving half to each of two contractors, so two new cells cannot develop if each receives only half the necessary information from the parental

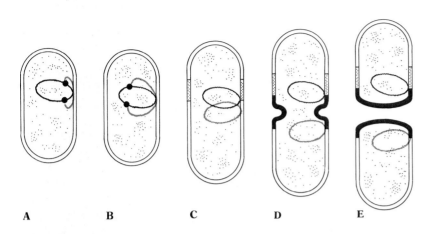

A   B   C   D   E

23.4 Binary fission of a procaryotic cell
(A) The circular chromosome of a cell is attached to the plasma membrane near one end; it has already begun replication (the partially formed second chromosome is shown in red). (B) Replication is about 80 percent complete. (C) Chromosomal replication is complete, and the second chromosome now has an independent point of attachment to the membrane. During replication additional membrane and wall (stippled) was formed. (D) More new membrane and wall (black) has formed between the points of attachment of the two chromosomes. Part of this growth forms invaginations that will give rise to a septum cutting the cell in two. (E) Fission is complete, and two daughter cells have formed. The chromosomes, which are actually so long that they must be looped and tangled to fit into a cell, are depicted here as small open circles. At the scale of this drawing, the actual circumference of each chromosome would be about 30 m.

cell. If you wanted two contractors to erect identical buildings simultaneously, you would duplicate the necessary blueprint and give a complete copy to each. The same applies to the cell. If a parental cell is to divide and produce two viable new cells, it must first make a complete copy of the genetic information in its nucleus and then, as it divides, give one complete copy to each daughter cell. In other words, division of the nucleus is not simply a process of halving; it is a process of duplicating genetic information and distributing the duplicates.

Procaryotic chromosomes Cell division appears to be much less complex in procaryotes, such as bacteria, than in the eucaryotic cells of true plants and animals. After the procaryotic cell has elongated sufficiently to form two independent daughter cells, the single circular chromosome replicates and the resulting second chromosome attaches to a different point from the first one on the expanding plasma membrane (Fig. 23.4). Next, new plasma membrane and wall material form near the midpoint of the parental cell and grow slowly inward, cutting through both the cytoplasm and the nucleoid (Fig. 23.5). Thus each new daughter cell receives a com-

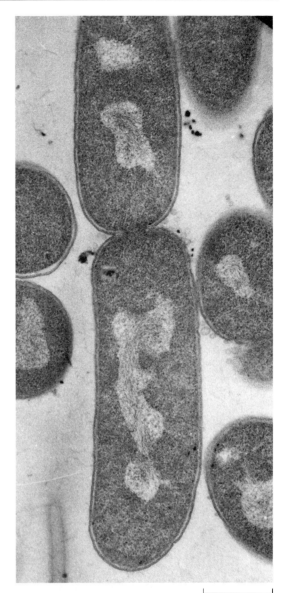

0.5 μm

23.5 Electron micrograph of a dividing bacterial cell
New wall is growing inward, cutting this cell of *E. coli* in two. The nucleoid (white areas inside cells) has already divided.

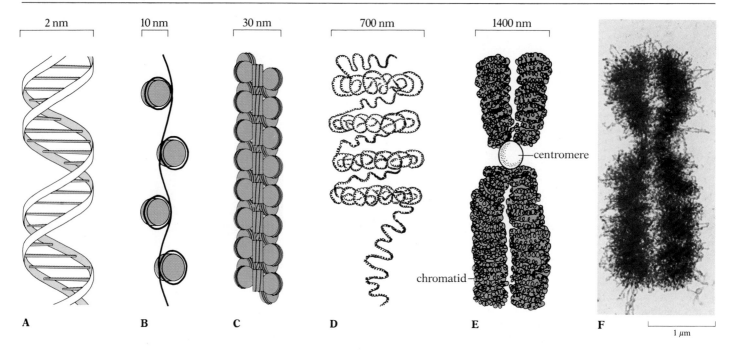

| 2 nm | 10 nm | 30 nm | 700 nm | 1400 nm | |

centromere

chromatid

A   B   C   D   E   F

1 μm

23.6 Packaging of DNA in a eucaryotic nuclear chromosome

The double helical strand of DNA (A) is wound around nucleosome cores about 10 nm in diameter (B). This chain of spools is itself coiled in some way to produce a thick strand about 30 nm in diameter (C). The arrangement shown here is hypothetical. Early in cell division the thick strands are collected into a long series of loops, which are wound into a helix (D). The result is the ragged appearance of the chromosome, as seen in the drawing (E) and in the electron micrograph of a human chromosome (F).

plete chromosome, a full complement of genetic information encoded by perhaps a million base pairs. This process is known as transverse fission, or **binary fission.**

Eucaryotic chromosomes   The chromosomes of eucaryotes differ from those of procaryotes in several ways. First, except in mitochondria and other organelles, eucaryotic chromosomes are linear rather than circular. Second, eucaryotic chromosomal DNA in the nucleus is wound on spool-like structures called **nucleosome cores** (Fig. 23.6B). Each nucleosome core is formed from two copies of each of four kinds of **histone** protein. About 200 base pairs of DNA fit on each core to form a nucleosome. The nucleosomes are normally packed into thick strands (Fig. 23.6C). During cell division the chromosomes condense further as the strands are organized into loops (Fig. 23.6D), creating the structures visible during the division process (Fig. 23.6E).

The third major difference between eucaryotic and procaryotic chromosomes arises because most eucaryotes receive chromosomes from two parental cells—from a sperm cell and an egg cell, for instance—rather than from just one. Chromosomes of eucaryotes therefore occur in homologous pairs, each consisting of one chromosome from each parent (Fig. 23.7), whereas the procaryotic chromosome is unpaired. The homologous chromosomes of a pair have similar distinctive shapes, and the genes they contain are in the same order and code for the same products during protein synthesis. Though chromosomal number may vary enormously between species—cats, for instance, have 19 pairs, fruit flies *(Drosophila melanogaster)* 4 pairs, onions 8 pairs—the chromosomal number within a species

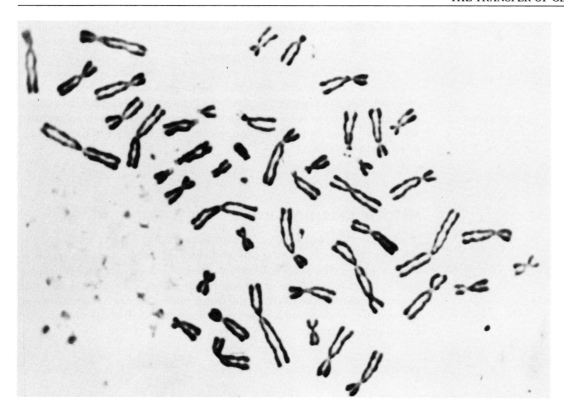

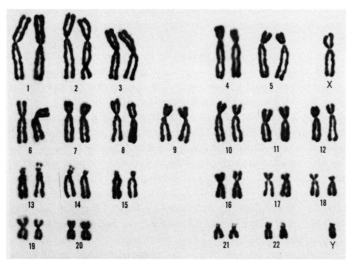

23.7   Photograph of chromosomes of a human male
At bottom, the chromosomes have been arranged as homologous pairs and numbered according to accepted convention. A human somatic cell (any cell except the egg or sperm cells) contains 23 pairs of chromosomes, including a pair of sex chromosomes (for a male, X and Y).

is normally uniform. The 23 pairs of human chromosomes have been studied extensively, and their distinctive "personalities" are beginning to be familiar to researchers interested in the bases of human genetic diseases and developmental abnormalities.

When eucaryotic chromosomes condense and become visible early in

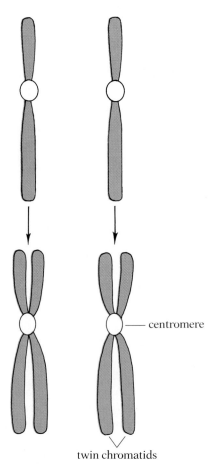

— centromere

twin chromatids

Homologous chromosomes

23.8 Production of chromatids through replication
At some time prior to cell division the genetic
material of each chromosome duplicates itself. As a
result, twin chromatids take form upon condensation,
joined to each other by a centromere. For clarity, the
unduplicated chromosomes are shown in condensed
form, even though condensation normally takes place
only after duplication.

cell division, each chromosome has already been duplicated to produce
compact twins, connected by a *centromere* (Fig. 23.8). Each member of this
bound pair is referred to as a *chromatid.*

We referred earlier to two kinds of cell division; these are division as part
of the growth of an organism, also known as somatic or mitotic cell division, and division to produce the gametes that give rise to new individuals,
also known as meiotic cell division. We shall describe each of these in turn,
in separate sections.

MITOTIC CELL DIVISION

Cell division in eucaryotic cells involves two fairly distinct processes that
often, but not always, occur together: division of the nucleus and division of
the cytoplasm. The process by which the nucleus divides to produce two
new nuclei, each with the same number of chromosomes as the parental
nucleus, is called *mitosis* (from the Greek *mitos,* "thread"). The process of
division of the cytoplasm is called *cytokinesis.* We shall discuss mitosis and
cytokinesis separately.

MITOSIS

For convenience, it is customary to separate each mitotic cycle, from one
cell division to the next, into a series of stages, each designated by a special
name. Though each stage will be discussed separately here, the entire process is a continuum rather than a series of discrete occurrences.

Interphase   The nondividing cell is said to be in the interphase state. The
nucleus is clearly visible as a distinct membrane-bounded organelle, and
one or more nucleoli are usually prominent. But chromosomes, as ordinarily pictured, are not visible in the nucleus; there are none of the distinct
rodlike bodies that microscopes and sophisticated staining techniques
allow us to see in a dividing cell (Fig. 23.9). Interphase chromosomes are so
thin and tangled that they cannot be recognized as separate entities. They
appear only as an irregular granular-looking mass of chromatin material.

In interphase animal cells there is a special region of cytoplasm just outside the nucleus that contains two small cylindrical bodies oriented at right
angles to each other. These are the *centrioles* (see Fig. 5.22, p. 128, and Fig.
5.23B, p. 129), which will move apart and become associated with the poles
of the mitotic apparatus of the dividing cell. In many animal cells this separation of the centrioles occurs just before the onset of mitosis, but in some
cells it occurs during interphase, long before mitosis begins. No centrioles
have yet been detected in the cells of most seed plants, but they do occur in
some algae, fungi, bryophytes (mosses, liverworts, etc.), and ferns in association with the production of motile sperm. Despite their high visibility,
however, the centrioles are not essential for cell division even in animal
cells; evidently they are followers rather than organizing centers during
mitosis.

In the past, interphase cells were commonly called resting cells, but that terminology has been rejected as grossly inappropriate. An interphase cell is definitely not resting; it is carrying out all the innumerable activities of a living, functioning cell—respiration, protein synthesis, growth, differentiation, and so forth. During interphase, furthermore, the genetic material is replicated and all the cellular machinery is duplicated in preparation for the next division sequence.

The replication of the genetic material does not, however, begin immediately after completion of the last division sequence. There is a gap in time, designated the G_1 stage, before genetic replication. This is when ribosomes and organelles begin to be duplicated. Next comes the S stage, during which the synthesis of new DNA takes place, along with the further duplication of organelles. Another period, designated G_2, separates the end of replication from the onset of mitosis proper. During this time the cell prepares for mitosis. These three subdivisions of interphase, together with mitosis (the M stage), constitute what is called the *cell cycle* (Fig. 23.10). Cells in tissue culture show striking morphological changes as they pass through the cell cycle, but the significance of these changes, especially in the intact organism, is not well understood.

The duration of the complete cell cycle can vary greatly. Though usually lasting 10–30 hours in plants and 18–24 hours in animals, it may be as short as 20 minutes in some organisms or as long as several days or even weeks. All the stages can vary in duration to some degree, but by far the greatest variation occurs in the G_1 stage. At one extreme very rapidly dividing embryonic cells may pass so quickly through the G_1 stage that it can hardly be said to exist at all, whereas at the other extreme some cell types become arrested in the G_1 stage and may never divide again; differentiated skeletal-muscle cells and nerve cells, for example, are arrested in the G_1 stage and normally do not divide. There are a few cases in which nondividing cells are arrested in the G_2 stage: they have replicated their DNA, but do not divide. The heart-muscle cells of human adults are an example of cells in G_2 arrest.

Both G_1 and G_2 arrest appear to result from a failure to produce some essential control chemical. If the nucleus from a cell in G_1 arrest is transplanted into a cell that is just entering the S stage, the transplanted nucleus will promptly be activated and itself enter the S stage, apparently because it has been stimulated by a control substance present in the cytoplasm of the host cell. Similarly, when a cell in G_2 arrest is fused with a mitotic cell, its chromosomes soon begin to condense and the cell enters mitosis. From such experiments we can deduce that the missing control substance in G_1 arrest is necessary for initiation of DNA synthesis, and that the missing control substance in G_2 arrest acts as a chromosomal condensing factor.

In nearly all cultured animal cells, production of these control substances depends on stimulation by *growth factors* (also called mitogenic factors or mitogens) from blood serum. These factors, most of which are peptides or small proteins, often have very specific target tissues. Thus there is a nerve growth factor (NGF) essential for mitosis of embryonic sympathetic nerve cells, an epidermal growth factor (EGF) that causes proliferation of epithelial cells, a fibroblast growth factor (FGF) that stimulates division of fibroblasts and of the cells lining small blood vessels, and so forth. Tissue-specific control of growth is an essential part of the develop-

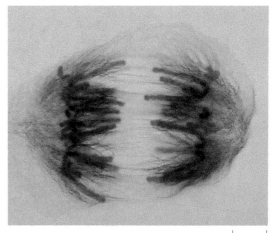

10 μm

23.9 Chromosomes in a dividing cell of the African blood lily
The separating chromosomes for each of the new nuclei are easily distinguishable.

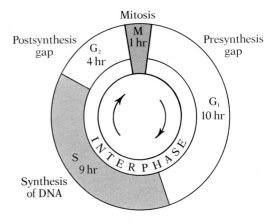

23.10 The cell cycle
This particular cycle assumes a period of 24 hours, but some cells complete the cycle in less than an hour and others take many days. Similarly, the ratios of the four stages of the cycle vary, with G_1 exhibiting the most variation.

2. EARLY PROPHASE
Centrioles begin to move apart.
Chromosomes appear as long thin threads.
Nucleolus becomes less distinct.

3. MIDDLE PROPHASE
Centrioles move farther apart.
Asters begin to form.
Twin chromatids become visible.

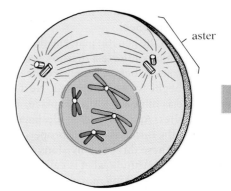

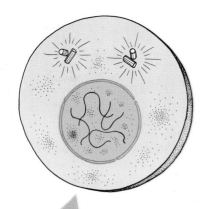

aster

nucleolus

centrioles

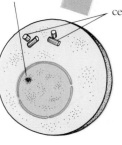

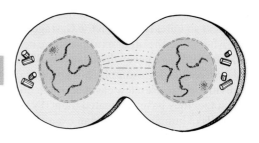

1/9. INTERPHASE
Cytokinesis is complete.
Nuclear membranes are complete.
Nucleolus is visible in each cell.
Chromosomes are not seen as distinct
  structures.
Replication of genetic material occurs
  before the end of this phase.

8. TELOPHASE
New nuclear membranes begin to form.
Chromosomes become longer, thinner,
  and less distinct.
Nucleolus reappears.
Centrioles are replicated.
Cytokinesis is nearly complete.

23.11 Mitosis and cytokinesis in an animal cell
In these drawings, color and gray are used to
distinguish the two pairs of homologous chromosomes.

ment of complex multicellular organisms. We shall look more closely at the
phenomena surrounding it in Chapters 31 and 32.

Let's assume now that the cell we are examining has passed through the
G_1, S, and G_2 stages of interphase and is entering mitosis proper—itself a
process so complex that it is customarily divided into four stages.

Prophase During prophase, the cell readies the nucleus for the crucial
separation of two complete sets of chromosomes into two daughter nuclei.
As the two centrioles of an animal cell move toward opposite sides of the

4. LATE PROPHASE
Centrioles nearly reach opposite sides
of nucleus.
Spindle begins to form and kinetochore
microtubules project from centromeres
toward spindle poles.
Nuclear membrane is disappearing.
Nucleolus is no longer visible.

5. METAPHASE
Nuclear membrane has disappeared.
Kinetochore microtubules from centromere
of each twin-chromatid chromosome link
up with spindle microtubules; other
spindle microtubules interact with spindle
tubules from opposite pole.

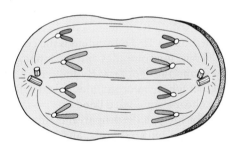

polar microtubule

kinetochore microtubule

spindle

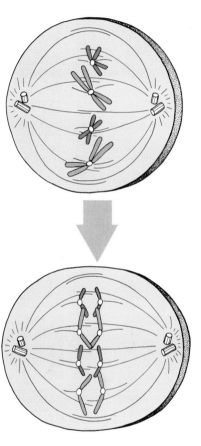

7. LATE ANAPHASE
The two sets of new single-chromatid
chromosomes are nearing their
respective poles.
The poles are being pushed apart.
Cytokinesis begins.

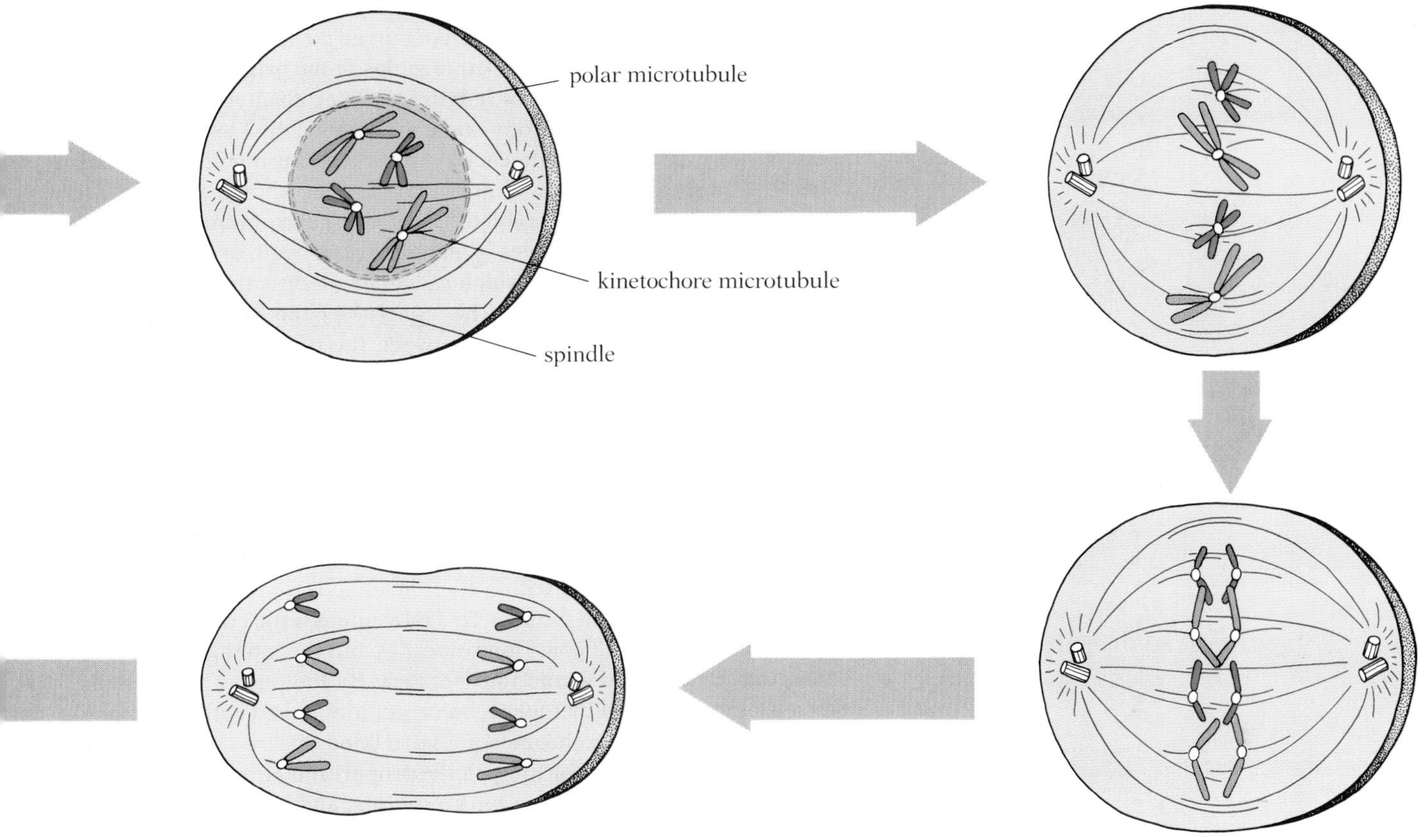

6. EARLY ANAPHASE
Centromeres have split and begun moving
toward opposite poles of spindle.
Spindle microtubules from opposite poles
force the poles apart.

nucleus, the initially indistinct chromosomes begin to condense into visible
threads, which become progressively shorter and thicker and more easily
stainable with dyes. When the chromosomes first become visible during
early prophase, they appear as long, thin, intertwined filaments, but by late
prophase the individual chromosomes can be clearly discerned as much
shorter rodlike structures. As the chromosomes become more distinct, the
nucleoli become less distinct, often disappearing altogether by the end of
prophase (Fig. 23.11).

The shortening of the chromosomes during cell division has one obvious

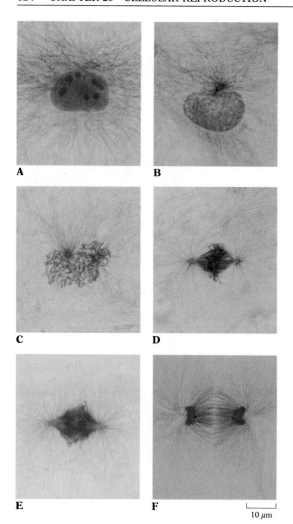

23.12 Mitosis in a cell from the heart endothelium of a frog

(A) Interphase: a diffuse network of microtubules (stained red) is visible. (B) Early prophase: the chromosomes begin to condense; the centrioles have not yet separated. (C) Late prophase: prophase ends as the nuclear membrane disappears; the centrioles have moved apart, and the two asters are prominent. (D) Metaphase: the chromosomes are lined up along the equator of the spindle, midway between the two asters. (E) Middle anaphase: the two groups of chromosomes have begun to move apart from each other, toward their respective poles of the spindle. (F) Early telophase: the two groups of chromosomes are being organized into new nuclei; cytokinesis has begun, and the line dividing the two daughter cells is clearly visible.

advantage. In their shorter form, they can be moved about freely without becoming hopelessly tangled. But only in their long uncoiled form can the chromosomes perform the critical task of RNA synthesis during interphase.

Examined under very high magnification, an individual chromosome from a late-prophase nucleus can be seen to consist of separate twin chromatids (Fig. 23.8). The replication that occurred during interphase resulted in two identical copies of the DNA molecules of the original chromosome, by a process we shall discuss at length in a later chapter. Hence the two chromatids of a prophase chromosome are genetically identical.

Somewhere along its length, each pair of twin chromatids has a centromere. By late prophase a few short microtubules, known as *kinetochore microtubules,* have begun to radiate from each centromere (Fig. 23.11:4). But earlier, as the two centrioles of an animal cell begin to move apart, another system of microtubules, which radiate in all directions, appears near each centriole (Fig. 23.11:2–3). This consists of the *polar microtubules,* which form a basketlike structure between the two centrioles as mitosis continues, and the blindly ending astral microtubules, collectively called an *aster,* which radiate in other directions from each centriole (Fig. 23.12C–D). Together, the various microtubules[1] form the apparatus known as the mitotic *spindle.*

Metaphase The stage of metaphase proper is preceded by a short period known as the prometaphase. The chromosomes, which at first were distributed essentially at random within the nucleus, begin to move toward the middle, or equator, of the spindle. This movement is the consequence of an interaction between the polar microtubules and the kinetochore microtubules, which have now come into contact. By the end of prometaphase each chromosome has kinetochore microtubules linked with microtubules from each pole, and the chromosomes are lined up.

During the brief stage of metaphase proper, the chromosomes are arranged on the equatorial plane of the spindle, and in side view appear to form a line across the middle of the spindle (Fig. 23.12D). Metaphase ends when the centromere of each pair of twin chromatids splits. Each chromatid then becomes an independent chromosome (that is, one having its own centromere); the number of independent chromosomes in the nucleus therefore doubles.

Anaphase The preceding stages of mitosis prepare the nucleus for the critical event that occurs in anaphase: the separation of two complete sets of chromosomes. At the beginning of anaphase the centromeres that hitherto held the twin chromatids together have just broken apart. The two new chromosomes now begin to move away from each other, one going toward one pole of the spindle and the other going toward the opposite pole. This movement toward the respective poles is accomplished in two ways. The kinetochore microtubules move poleward, shortening as they go, and pull the attached chromosomes with them. Meanwhile, the polar microtubules

---

[1]Until very recently, the hollow microtubules were thought to be solid fiber. Hence the references to "spindle fibers" and "astral fibers" in the literature.

from opposite ends of the dividing cell form cross bridges between themselves (similar to the dynein arms that are thought to enable cilia to move) and push the poles apart (Figs. 23.11:6 and 23.12E), becoming longer (by the addition of molecular subunits to their ends) as they do so. By late anaphase the cell contains two groups of chromosomes that are widely separated, the two clusters having almost reached their respective poles of the spindle (Fig. 23.11:7). Cytokinesis often begins during late anaphase.

Telophase Telophase (Fig. 23.12F) is essentially a reversal of prophase. The two sets of chromosomes, having reached their respective poles, become enclosed in new nuclear membranes as the spindle disappears. Then the chromosomes begin to uncoil and to resume their interphase form, while the nucleoli slowly reappear (Fig. 23.11:8). Cytokinesis is often completed during telophase. Telophase ends when the new nuclei have fully assumed the characteristics of interphase, thus bringing to a close the complete mitotic process. What was a single nucleus containing one set of twin-chromatid chromosomes in prophase is now two nuclei, each with one set of single-chromatid chromosomes.

NUCLEAR DIVISION DURING MITOSIS

The preceding description of mitotic division fits most but not all eucaryotes. Most of the deviations from the pattern illustrated in Figure 23.11 are seen in primitive eucaryotes, and they may indicate how the typical eucaryotic pattern of nuclear division evolved from the sequence of cell division in procaryotes. Some primitive eucaryotic dinoflagellates, for example, even though they have numerous chromosomes packaged in a membrane-bounded nucleus, still divide by a process remarkably like the binary fission of procaryotes. The chromosomes, which are attached to the inner surface of the nuclear membrane by very short kinetochore microtubules running from their centromeres, are first replicated, and then separated into two groups by growth of the membrane between the points of attachment of the original chromosomes and their replicates. Thus two new nuclei are organized.

These dinoflagellates rely on microtubules in their cytoplasm to determine the direction of nuclear division. Bundles of these tubules run in parallel from each end of the cell toward the other, meeting in "tunnels" in the nucleus (Fig. 23.13A), and the two new nuclei move away from each other along this scaffolding. There is, however, no physical connection between the microtubules and the nuclear membrane, and the means by which the daughter nuclei are pulled apart is not known.

Advanced dinoflagellates and some fungi and protozoans illustrate a possible further evolutionary progression in their method of cell division, while still relying on cytoplasmic microtubules to orient the process. As in the primitive dinoflagellates, their chromosomes are attached to the nuclear membrane by very short kinetochore microtubules from the centromeres, and bundles of long spindle microtubules that run through a nuclear tunnel determine the direction of nuclear division. In these organisms, however, there is just a single tunnel. Additional spindle microtu-

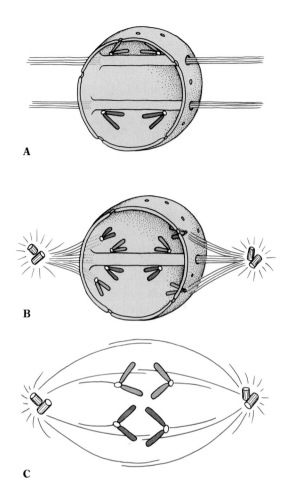

23.13 Evolution of nuclear division
(A) In primitive dinoflagellates, parallel bundles of microtubules run from each end toward the other, meeting and interacting in tunnels in the nucleus. The chromosomes are attached to the nuclear membrane and separate as the nucleus divides, and the daughter nuclei slide along the microtubules. (B) In advanced dinoflagellates there is only a single microtubule tunnel through the nucleus. Additional microtubules run from the poles to the nuclear membrane and pull the daughter nuclei apart. (C) In higher eucaryotes the nuclear membrane disappears during division and certain of the polar microtubules interact with the kinetochore microtubules to pull the individual chromosomes toward the poles.

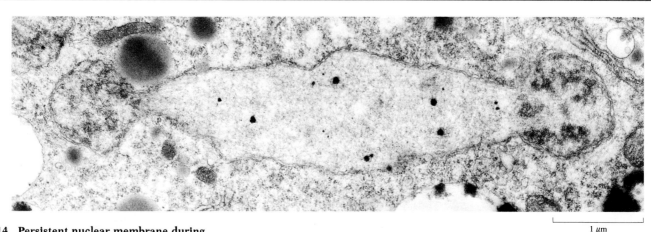

1 μm

23.14 Persistent nuclear membrane during mitosis in the fungus *Catenaria*
In this fungus, part of the spindle runs through the nucleus. Here two new telophase nuclei, at opposite ends of the photograph, are being pinched off by constrictions of the nuclear envelope. The middle part of the envelope, between the new nuclei, will eventually disintegrate.

bules, originating at the same poles as the transnuclear bundles, run to the nuclear membrane and physically pull the two nuclei apart (Figs. 23.13B and 23.14).

As we have seen, in higher eucaryotes the nuclear membrane disappears before division, and there is no apparent bundling of microtubules. Instead of securing the chromosomes to the nuclear membrane, as they do in some dinoflagellates, kinetochore microtubules connect directly to polar microtubules. These tubule complexes exert a direct pull on the chromosomes, rather than on the daughter nuclei (Fig. 23.13C).

The cell-division cycle of higher eucaryotes requires the intricate nuclear membrane to be disassembled and reassembled. Partial disassembly is also necessary in some eucaryotes with persistent nuclear membranes, as seen in Figure 23.14. The disassembly process is not fully understood. Phosphorylation of one of the three proteins that form a thin lamina just inside the nucleus appears to be the cue that triggers disintegration of the membrane. One of these three proteins remains embedded in the fragments of nuclear membrane as they are incorporated temporarily into the endoplasmic reticulum of the cytoplasm, perhaps serving to mark these specific bits of membrane for recovery at the end of telophase. The fate of the proteinaceous pores of the nucleus is unknown, though some evidence suggests that these specialized channels may become temporarily associated with the chromosomes.

CYTOKINESIS

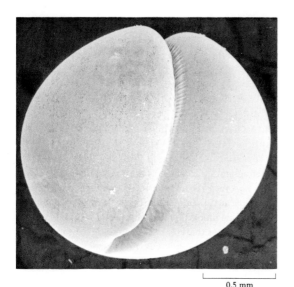

0.5 mm

23.15 Scanning electron micrograph of a dividing frog egg
The cleavage furrow is not yet complete (see bottom of cell). Note the puckered stress lines in the furrow.

We said earlier that division of the cytoplasm often accompanies division of the nucleus and that it often begins in late anaphase, reaching completion during telophase. But this is not always the case. Mitosis without cytokinesis is common in some algae and fungi, producing coenocytic plant bodies (bodies with many nuclei, but no, or few, cellular partitions). It regularly occurs during certain phases of reproduction in seed plants and certain other vascular plants. It is also common in a few lower invertebrate ani-

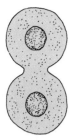 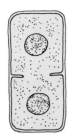 

Animal cell   Algal cell   Higher-plant cell

23.16   Three mechanisms of cytokinesis
(A) Cytokinesis in animal cells typically occurs by a pinching-in of the plasma membrane. (B) In many algal cells cytokinesis occurs by an inward growth of new wall and membrane. (C) In higher plants cytokinesis typically begins in the middle and proceeds toward the periphery, as membranous vesicles fuse to form the cell plate.

mals with coenocytic bodies. During the early development of insect eggs, mitosis without cytokinesis produces hundreds of nuclei in a limited amount of cytoplasm; later, cytokinesis cuts up this cytoplasm to produce many new cells in a very short time.

Cytokinesis in animal cells   Division of an animal cell normally begins with the formation of a *cleavage furrow* running around the cell (Fig. 23.15). When cytokinesis occurs during mitosis, the location of the furrow is ordinarily determined by the orientation of the spindle, in whose equatorial region the furrow forms (Fig. 23.11:7). The furrow becomes progressively deeper, until it cuts completely through the cell (and its spindle), producing two new cells.

Very little is known about the mechanism of formation of the cleavage furrow. Since the location of the furrow is usually related to the position of the centrioles and spindle, an early hypothesis was that some of the astral microtubules of the mitotic apparatus were attached to the cell surface and pulled the surface inward to form the cleavage furrow. But cytokinesis is known to occur in many instances long after mitosis is complete and the spindle has disappeared; indeed, removal of the entire mitotic apparatus from a sea-urchin egg by micromanipulation does not inhibit furrow formation. Recent research suggests that a dense belt of actin and myosin microfilaments at the site of the cleavage furrow is probably responsible. This view is supported by the finding that drugs like cytochalasin, which block the activity of microfilaments, stop cytokinesis. Since agents that bind actin and myosin also stop the cleavage process, the sort of actin-myosin interaction known to be involved in cell movement, as discussed in Chapter 5, is probably responsible for cytokinesis as well.

Cytokinesis in plant cells   Since plant cells possess relatively rigid cell walls, which cannot develop cleavage furrows, it is not surprising that cytokinesis is different in plant cells. In many fungi and algae, new plasma membrane and wall grow inward around the wall midline until the growing edges meet and completely separate the daughter cells (Fig. 23.16). In higher plants a special membrane, called the *cell plate,* forms halfway between the two nuclei (at the equator of the spindle if cytokinesis accompa-

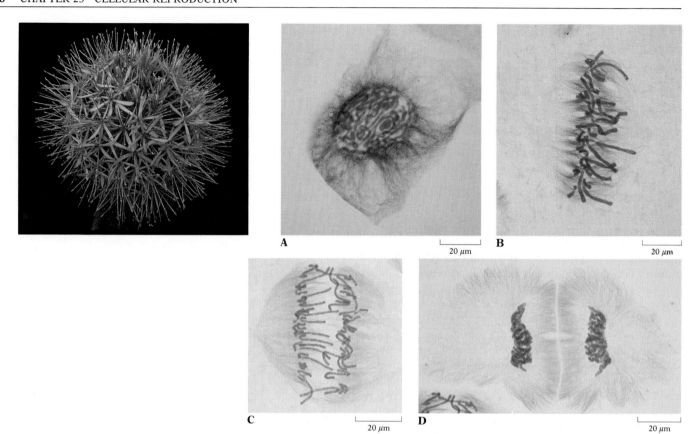

23.17   Cell division in a plant, the African blood lily
(A) Prophase: the chromosomes have condensed; microtubules are visible. (B) Metaphase.
(C) Anaphase: the two groups of chromosomes are moving to opposite poles of the cell. (D) Telophase: a cell plate has begun to form.

nies mitosis; Fig. 23.17D). The cell plate begins to form in the center of the cytoplasm and slowly becomes larger until its edges reach the outer surface of the cell and the cell's contents are cut in two. We see, then, that cytokinesis of higher-plant cells progresses from the middle to the periphery, whereas cytokinesis in animal cells progresses from the periphery to the middle.

The cell plate forms from membranous vesicles that are carried to the site of plate formation by the microtubules that remain after mitosis. There they first line up and then unite (Fig. 23.18). The vesicles are derived mainly from the Golgi and, to a lesser extent, from the ER. As the vesicle membranes fuse with one another and, peripherally, with the old plasma membrane, they constitute the partitioning membranes of the two newly formed daughter cells. The contents of the vesicles, trapped between the daughter cells, give rise to the middle lamella and to the beginnings of primary cell walls.

MEIOTIC CELL DIVISION

As we have seen, mitosis serves to maintain a constant number of chromosomes in somatic cells. What would be its effect in reproductive cells? As you know, in sexual reproduction two gametes (the egg and the sperm) unite to form the first cell (zygote) of the new individual. If those two gam-

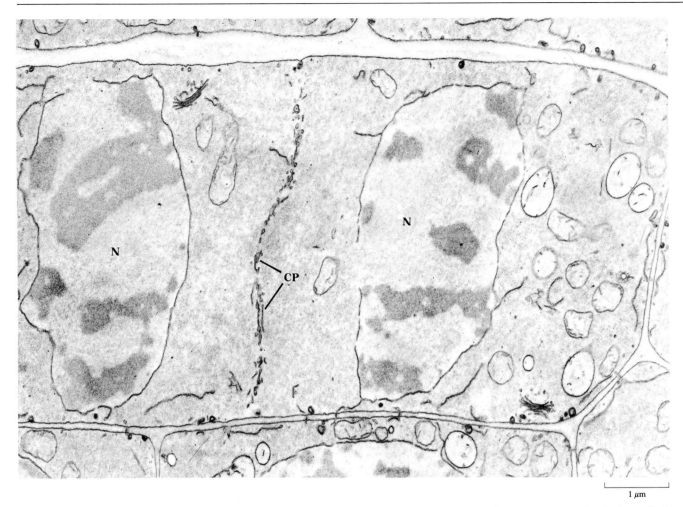

1 μm

23.18 Electron micrograph of a late-telophase cell in corn root, showing formation of cell plate
Mitosis has been completed, and the two new nuclei (N) are being formed; the chromosomes (dark areas in the nuclei) are no longer visible as distinct structures, but the nuclear membranes are not yet complete. A cell plate (CP) is being assembled from numerous small vesicular structures. At the lower end of the nucleus on the right, a length of endoplasmic reticulum can be seen that appears to run from the nuclear membrane to the cell wall, through the wall, and into the adjacent cell.

etes were produced by normal mitosis in human beings or in the hypothetical four-chromosome organism of Figure 23.11, the zygote produced by their union would have double the normal number of chromosomes, and at each successive generation the number would again double, until the total chromosome number per cell approached infinity. This does not happen: the chromosome number normally remains constant within a species. At some point, therefore, a different kind of cell division must take place, a division that reduces the number of chromosomes by half, so that when the egg and sperm unite in fertilization the normal diploid number is restored. This special process of reduction division is called *meiosis* (from the Greek word for "diminution"). In all multicellular animals meiosis occurs at the time of gamete production. Consequently each gamete possesses only half the species-typical number of chromosomes. It is important to note that in the reduction division of meiosis the chromosomes of the parental cell are not simply separated into two random halves; the diploid nucleus contains two of each type of chromosome, and meiosis partitions these chromosome pairs so that each gamete contains one of the two homologues. Such a cell,

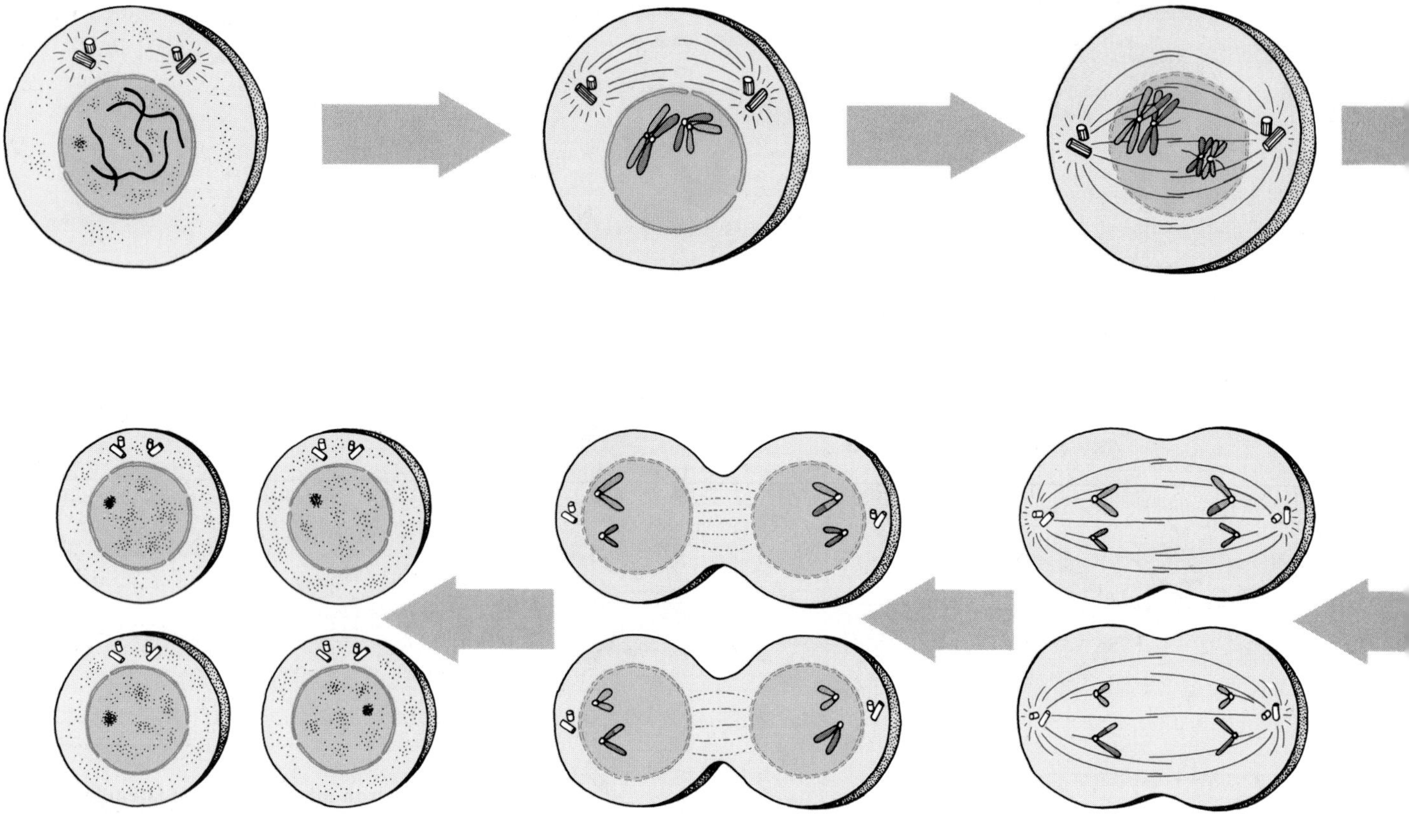

1. EARLY PROPHASE
Chromosomes become visible as
long, well-separated filaments;
replication has already occurred.

2. MIDDLE PROPHASE I
Homologous chromosomes become
shorter and thicker, and synapse;
crossing over takes place.

3. LATE PROPHASE I
The tetrad structure of the synapsed
chromosomes, and the chiasmata created
by crossing over, become visible.
Nuclear membrane begins to disappear.
Kinetochore microtubules arise from
the centromeres.

12. INTERPHASE

11. TELOPHASE II

10. ANAPHASE II

23.19 Meiosis in an animal cell
In these diagrams, the members of each pair of
homologous chromosomes are shown in different
colors to aid in visualizing the results of crossing over.

with only one of each type of chromosome, is said to be **haploid.** When two
haploid gametes unite in fertilization, the resulting zygote is diploid, having
received one of each chromosome type from the sperm of the male parent
and one of each type from the egg of the female parent.

THE PROCESS OF MEIOSIS

Complete meiosis (Fig. 23.19) involves two successive division sequences,
which result in four new haploid cells. The first division sequence accomplishes the reduction in the number of chromosomes; the second separates
chromatids. The same four stages as in mitosis—namely prophase, metaphase, anaphase, and telophase—are recognized in each division sequence.

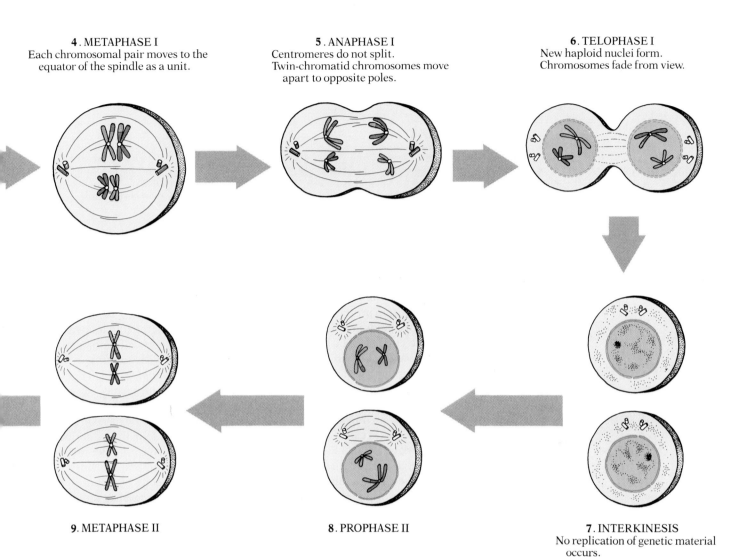

4. METAPHASE I
Each chromosomal pair moves to the equator of the spindle as a unit.

5. ANAPHASE I
Centromeres do not split. Twin-chromatid chromosomes move apart to opposite poles.

6. TELOPHASE I
New haploid nuclei form. Chromosomes fade from view.

9. METAPHASE II

8. PROPHASE II

7. INTERKINESIS
No replication of genetic material occurs.

First prophase   Many of the events in prophase I of meiosis superficially resemble those in the prophase of mitosis. The individual chromosomes come slowly into view as they coil and become shorter, thicker, and more easily stainable. The nucleoli slowly fade from view, and, finally, the nuclear membrane disappears and the spindle is organized. Radioactive-tracer studies show that the replication of the genetic material occurs during the interphase that precedes prophase I, as in mitosis. However, there are important differences between the prophase of meiosis and that of mitosis.

The chief of these differences is that in meiosis the members of each pair of homologous chromosomes move together and lie side by side (Fig. 23.19:2). Also, instead of being fastened only by a centromere, as in mitosis,

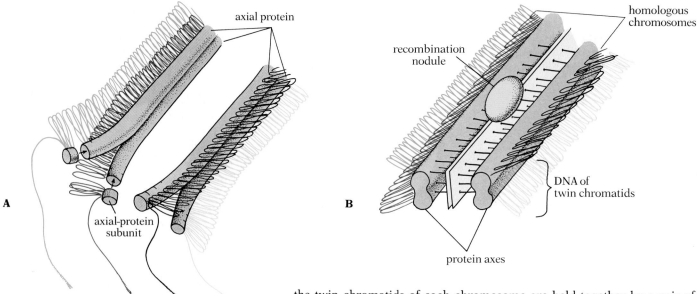

A

axial protein

axial-protein subunit

B

recombination nodule

homologous chromosomes

DNA of twin chromatids

protein axes

DNA of twin chromatids   protein axes   DNA of twin chromatids

C

0.2 μm

23.20   Formation of the chromosomal synapse
(A) The axial proteins gather the replicated DNA of each chromosome—that is, the twin chromatids—into a long series of paired loops. The loops are longer than shown here, and intermingle. Some models assume just a single protein axis for each pair of twin chromatids. (B) The protein axes then bring the homologous chromosomes into register, and cross bridges form. Protein complexes (recombination nodules) appear, which mediate the process known as crossing over. (C) Longitudinal section through a synaptonemal complex from a meiotic cell of the ascomycete *Neottiella*.

the twin chromatids of each chromosome are held together by a pair of long, thin protein axes that run their entire length (Fig. 23.20A), and the two chromatids are indistinguishable. The DNA is gathered into loops as part of the condensation process. The protein axes of the two homologous chromosomes now join by means of protein cross bridges to form an intricate compound structure known as a ***synaptonemal complex***, which lines up the four chromatids (often referred to as a tetrad) in perfect register (Fig. 23.20B). This process is known as ***synapsis***.

Now an important process known as ***crossing over*** begins. Large protein complexes called recombination nodules appear along the ladderlike cross bridges (Fig. 23.20B). These nodules probably determine the locations at which genetic material will be exchanged between homologous chromosomes. The number of nodules depends on the species of organism and the length of the chromosome; each human chromosome has an average of three. Next, each nodule begins a process by which two of the chromatids, one from each of the homologous chromosomes, are clipped open at precisely the same place, and the resulting fragments are spliced to each other (Fig. 23.21). The all-important result of this splicing is that the chromatids of the tetrad no longer form two sets of twins; the recombined chromatids are now hybrids, containing genetic material descended from both the mother's and the father's homologous chromosomes.

When the synaptonemal complex begins to break up and the two homologous chromosomes drift slightly apart in late prophase, the points at which crossing over has taken place begin to become visible (Fig. 23.22). The hybrid chromatids produced by crossing over link the two homologous chromosomes at these points, called ***chiasmata***. Each chiasma represents one crossover event. Each crossover event can involve a different pair of chromatids. The process is summarized in Figure 23.23. Crossing over is not rare or accidental: it is a frequent and highly organized mechanism for generating hybrid chromatids, and has significant adaptive value, as we shall see presently.

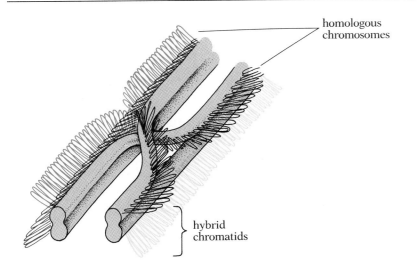

homologous
chromosomes

hybrid
chromatids

23.21   Crossing over

When two homologous chromosomes have synapsed, chromatids from one chromosome exchange fragments with chromatids from the other chromosome to create hybrid chromatids. One such exchange is shown here.

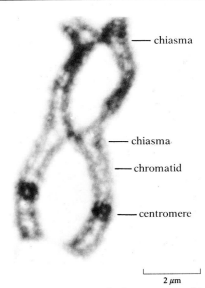

chiasma

chiasma

chromatid

centromere

2 μm

23.22   Photomicrograph showing two chiasmata between homologous chromosomes

Each chromosome is clearly recognizable as a pair of chromatids. The centromeres, too, are clearly visible. Crossing over has taken place at two points—the chiasmata. Note that the crossing over involves only one chromatid from each chromosome. These chromosomes, of a plethodontid salamander from Costa Rica, are from a spermatocyte in prometaphase of meiosis I.

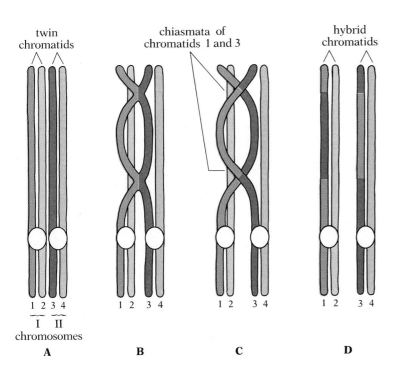

twin
chromatids

chiasmata of
chromatids 1 and 3

hybrid
chromatids

1 2 3 4

I   II
chromosomes

1 2   3 4

1 2   3 4

1 2   3 4

A   B   C   D

23.23   Schematic summary of synapsis and crossing over

(A) Crossing over begins as homologous chromosomes (I and II), each consisting of twin chromatids (1,2 and 3,4), are brought into register and synapse. (The synaptonemal complex has been omitted for clarity.) (B) Parts of separate chromatids are spliced together, while the protein axis (not shown) keeps each segment firmly attached to its twin. (C) When the synaptonemal complex breaks up, the homologous chromosomes begin to drift apart, but the chiasmata of spliced hybrid chromatids prevent them from separating. (D) By the end of prophase, the axial connections between the chromatids have disintegrated and the chromosomes appear double-armed. The homologous chromosomes separate fully during anaphase.

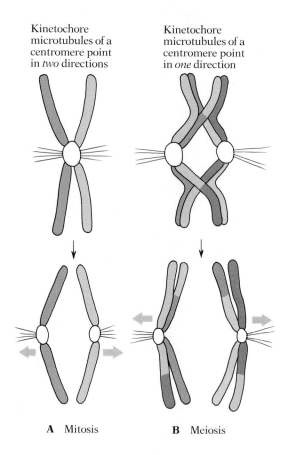

Kinetochore microtubules of a centromere point in *two* directions

Kinetochore microtubules of a centromere point in *one* direction

A Mitosis

B Meiosis

23.24 Kinetochore microtubules in mitosis and meiosis

(A) In mitosis, each centromere sends kinetochore microtubules toward both poles. As a consequence, the two chromatids are pulled apart and wind up at opposite ends of the cell during anaphase. (B) In meiosis I, however, a centromere sends microtubules toward only one of the two poles, with the homologous centromere sending microtubules toward the other pole. The result is that in anaphase I the chromatids remain joined, and homologous two-chromatid chromosomes wind up in separate cells.

The gradual disintegration of the protein axis that has linked the chromatids since the beginning of prophase allows greater separation of the individual chromatids, and as prophase ends, the familiar pair of chromosomes, each made up of two chromatids, again becomes visible. Most chromosomes now consist of two hybrid rather than twin chromatids, since genetic material has been exchanged between the chromatids of homologous chromosomes. Finally, in late prophase, spindle microtubules appear, radiating from the two poles of the cell, and kinetochore microtubules arise from the centromeres. But there is a distinct difference between the organization of the centromeric microtubules in mitosis and in meiosis. In mitosis, the kinetochore microtubules from each centromere project toward both poles (Fig. 23.24A). In meiosis the microtubules from a centromere go to only one pole, a different pole for each of the two centromeres in a homologous pair (Fig. 23.24B).

First metaphase In mitosis each chromosome consists of a pair of twin chromatids, and moves independently to the midline of the cell. In meiosis each chromosome typically consists of two hybrid chromatids, and homologous chromosomes move as a unit to the midline in preparation for anaphase. As a result, in meiosis the number of independent units waiting at the midline for cell division is only half the number of mitosis (Fig. 23.19:4).

First anaphase In mitosis, metaphase ends and anaphase begins when the centromere of each twin chromatid splits, and the two independent single-chromatid chromosomes thus formed move away from each other toward opposite poles of the spindle. But in meiosis I the separation occurs instead between the two chromosomes that have been joined since the middle of prophase (Fig. 23.24). Since each chromosome has its own centromere, there is no splitting of centromeres. With their respective centromeres attached through spindle and kinetochore microtubules to only one pole, the homologous chromosomes move away from each other toward opposite poles during anaphase. Thus in our hypothetical organism, only two chromosomes, each with two hybrid chromatids, move to each pole (Fig. 23.19:5), in contrast to the four single-chromatid chromosomes that move to each pole in mitosis. Because the synaptic pairing was not random, but involved the two homologous chromosomes of each type, the two daughter nuclei get not just any two chromosomes, but rather one of each type.

First telophase Telophase of mitosis and meiosis are essentially the same, except that each of the two new nuclei formed in mitosis has the same number of chromosomes as the parental nucleus, whereas each of the new nuclei formed in meiosis has half the chromosomes present in the parental nucleus (Fig. 23.19:6). At the end of telophase of mitosis, the chromosomes are single when they fade from view; at the end of telophase I of meiosis, the chromosomes are composed of two chromatids each when they fade from view.

Interkinesis Following telophase I of meiosis, there is a short period called interkinesis, which is similar to an interphase between two mitotic division sequences except that no replication of the genetic material occurs

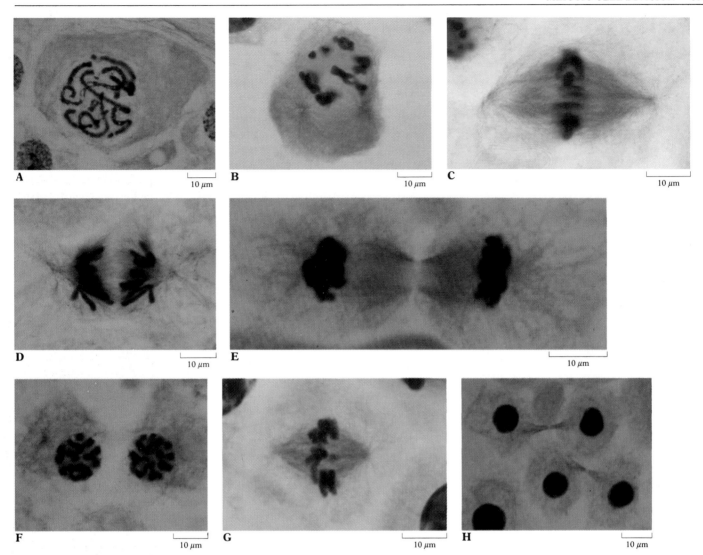

and hence no new chromatids are formed (replication is unnecessary, since each chromosome already has two chromatids when interkinesis begins).

Second division sequence of meiosis The second division sequence of meiosis, which follows interkinesis, is essentially mitotic from the standpoint of mechanics, though the functional result is different (Fig. 23.19:8–11). The chromosomes do not synapse; they cannot, since the cell contains no homologous chromosomes. Each two-chromatid chromosome moves to the midline independently, and its centromere sends kinetochore microtubules toward each pole. At the end of metaphase II each centromere splits, and during anaphase II the single-chromatid chromosomes thus formed move away from each other toward opposite poles of the spindle. The new nuclei formed during telophase II are therefore haploid (Fig. 23.25).

In summary, then, the first meiotic division produces two haploid cells containing double-chromatid chromosomes. Each of these cells divides

23.25 Meiosis in a cell from the grasshopper *Mongolotetix japonicus*

(A) Early prophase I: the chromosomes are seen as long filaments. (B) Middle prophase I: synapsing of homologous chromosomes takes place. (C) Metaphase I: chromosomal pairs line up at the equator. (D) Anaphase I: homologous chromosomes have separated and are moving to opposite poles. (E) Telophase I: division into two haploid nuclei has begun. (F) Prophase II: early in this stage, separate chromosomes are barely distinguishable. (G) Metaphase II: chromosomes in each cell are at the equator, and the spindles are evident. (H) Telophase II: four new haploid nuclei can be seen.

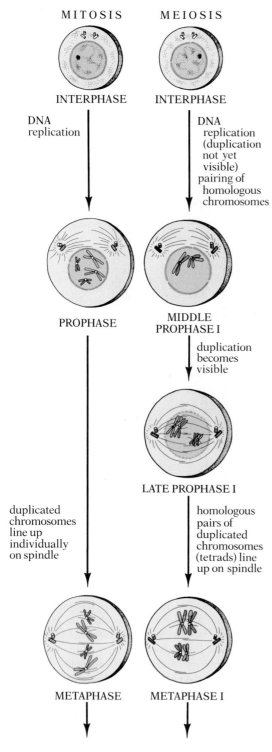

MITOSIS   MEIOSIS

INTERPHASE   INTERPHASE

DNA replication

DNA replication (duplication not yet visible) pairing of homologous chromosomes

PROPHASE   MIDDLE PROPHASE I

duplication becomes visible

LATE PROPHASE I

duplicated chromosomes line up individually on spindle

homologous pairs of duplicated chromosomes (tetrads) line up on spindle

METAPHASE   METAPHASE I

23.26 Comparison of mitosis and meiosis

again in the second meiotic division, producing a total of four new haploid cells containing single-chromatid chromosomes. Mitosis and meiosis are compared in Figure 23.26.

WHY SEXUAL RECOMBINATION?

Meiosis, and the process of sexual reproduction it makes possible, form an extremely elaborate mechanism that allows organisms to pool genetic information through sexual recombination. But why should living things recombine their DNA? Organisms that survive the trials of life long enough to reproduce must have a fairly satisfactory combination of genes. Why tinker with success? Why not reproduce asexually, like haploid bacteria, to create a *clone* of genetically identical offspring? There are several possible answers to this controversial question.

The most obvious consequence of sexual reproduction is that it creates offspring with new combinations of characteristics. Each member of a homologous pair of chromosomes contains genes coding for the same kinds of RNA, structural proteins, and enzymes. But the exact base sequences in two homologous genes are not necessarily identical. Instead, the copy of a gene inherited from an organism's father is often slightly different from the copy inherited from the mother. Different versions of the same gene are called *alleles,* and the enzymes or structural proteins they produce are likely to have slightly (or even very) different activities. For example, blue eyes develop in people who have two homologous eye-pigment genes that code for a defective colorless screening pigment; people with brown eyes have at least one copy of the allele that codes for a functional pigment. Since an individual organism is a mix of alleles from two different parents, its chromosomes are almost certain to contain an ensemble of alleles different from that of either parent, and many aspects of its morphology, physiology, and behavior will be correspondingly different. In the next two chapters we shall examine how different genes, as well as different alleles of the same gene, interact. But right now let us see how the genetic variation created in a population by sexual reproduction might be desirable.

The earth is not static: the weather and other aspects of the environment are in constant flux, and fully 99 percent of all the species that have existed in the earth's history are now extinct. They perished because they were not equipped to deal quickly or effectively enough with changes in their habitats and the competition of other species. The long-term survival of an individual's genes depends, therefore, on their ability to maintain efficient metabolic operation under normal conditions, and yet to accommodate sudden change. A population of *Streptococcus* thriving on an abundance of human metabolic products, for instance, may suddenly be decimated by a plague of penicillin. Only those few members of the population carrying genes for certain variations in the chemistry of their cell walls can survive the toxin and continue to reproduce. Likewise, only a small fraction of a population of mosquitoes—individuals with specific combinations of genes—survive to produce populations resistant to potent pesticides like DDT. Sexual recombination helps create in each generation an almost infinite variety of new combinations of alleles and of different genes, thereby increasing the odds that threatening changes can be accommodated.

Though the novel combinations of genes and other more complicated genetic rearrangements (discussed in later chapters) that result from sexual reproduction can confer enormous benefits when conditions change, they have costs as well. If there were one optimal combination of, say, fur color and body size for a given species in a particular habitat, no other could be as good; the many individuals varying from the most advantageous combination would necessarily be less fit. Since natural selection acts against characteristics that definitely reduce fitness, the range of variation may not be very large: there are no orange ermine, and even a slightly off-white ermine is at a disadvantage in the snow. The more stable an organism's habitat, the more likely it is that selection will have lowered variability. For the ermine, winter always comes, the snow is inevitably white, and all the fur-pigment alleles code for white. Less stable or predictable habitats, however, usually keep genetic variability higher, since circumstances may favor one set of characteristics at one time, and another set at another time. Many species reproduce asexually when the environment is favorable, but resort to sexual recombination when the environment becomes hostile. Aphids, for instance, typically reproduce clones of new wingless aphids on the stems they are sucking in the spring, but can reproduce sexually to create winged offspring, able to seek out new plants or new mates, when the situation demands. Even bacteria can manage to reproduce by their own strange form of sexual recombination when the need for novel genetic combinations arises.

As a general rule, then, we can say that unpredictably variable environments favor species that recombine sexually, whereas predictable, static environments can favor species that are at least temporarily asexual. Despite the many curious exceptions to and variations on the theme of sexual reproduction, it is the strategy for genetic recombination that dominates the living world.

Sources of variation Some novel combinations will be lethal, causing the early death of individuals in whom they occur, but others will be well adapted to live, and still others, which may even confer a disadvantage under normal circumstances, may persist in a small part of the population until a propitious environmental change makes them suddenly highly adaptive. These new combinations are generated in two ways.

First, we've seen that the two gametes that fuse to form a new diploid individual come from different organisms, so the offspring will inevitably represent a new combination of traits rather than a carbon copy of one parent. Even when the same parents produce many offspring, they are (with the infrequent exception of identical twins) each different. To understand why variation at this level arises, let's consider a hypothetical six-chromosome organism. Three chromosomes came originally from the gamete of the male parent (the product of meiosis), and three from the gamete of the female parent, to form the three pairs of homologous chromosomes in this diploid organism. Each time meiosis occurs in an individual producing gametes for reproduction, all the maternal chromosomes *might* go into one gamete and all the paternal ones into another, but many other combinations are equally likely. If an organism has three pairs of chromosomes,

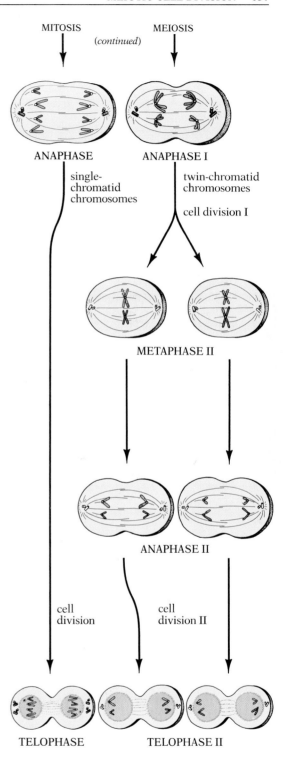

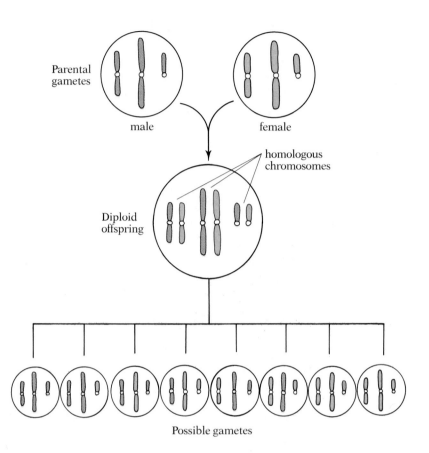

23.27 Variation in gametes
Even without crossing over, a diploid organism produces 2^n different kinds of gametes, where n is the number of pairs of homologous chromosomes. In this hypothetical organism with an n of 3, there are 8 (2^3) possible kinds of gametes.

eight different combinations may be found in the gametes (Fig. 23.27). In the zygote formed by two gametes in such a species, 64 (8×8) different combinations of chromosomes are then possible. For humans, with our unusually large number of chromosomes, the number of different chromosome combinations possible in the offspring of the same two parents is about 7×10^{13}.

Simple sexual recombination, then, produces a substantial amount of variation, and this would be obtained even if the elaborate two-stage process of meiosis were replaced by a much simpler sequence: the single-chromatid homologous chromosomes with which the cell ended the previous telophase of mitosis could in theory segregate into two haploid cells. The replication of DNA before meiosis, and the whole second division sequence, would then be unnecessary. But this shortcut, though economical, would rob the organism of the benefits of crossing over, the second way in which new combinations are generated.

The highly ordered process of crossing over increases the variety of possible gametes astronomically, since crossover points are essentially random, and virtually every hybrid chromatid is spliced together at a unique set of points. If we narrow our focus to an example using only two of a

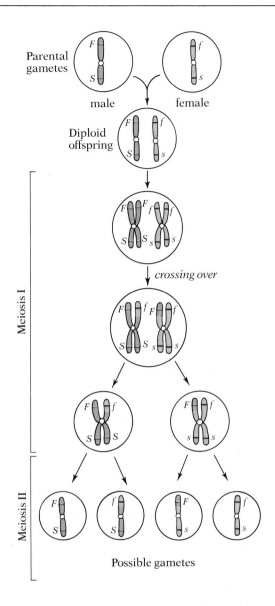

23.28 Variation in gametes from crossing over
In this example, the parents contribute chromosomes with two different alleles of genes for fur color (*F* and *f*) and body size (*S* and *s*); these chromosomes form a homologous pair in the diploid offspring. At meiosis, crossing over breaks up the paternal and the maternal combinations, so some of the resulting gametes have hybrid chromosomes, unlike those of either parent.

chromosome's thousands of genes, we can get an idea of what happens to the genetic information during meiosis. Perhaps one gene at one end of the chromosome controls the animal's fur color, and a gene at the other end codes for body size.

Let's assume that the father contributes a gene coding for black fur (which we can call *F*) and a gene coding for a large body (*S*), while the alleles from the mother are different, coding for gray fur (*f*) and for a small body (*s*) (Fig. 23.28). Though the *F* and *S* of the chromosome contributed by the father remain together (or ***linked***) throughout the *mitotic* life of a cell, as do the *f* and *s* on the maternal chromosome, crossing over in *meiosis* can rearrange things. As a result, the chromosome in the offspring's gamete is

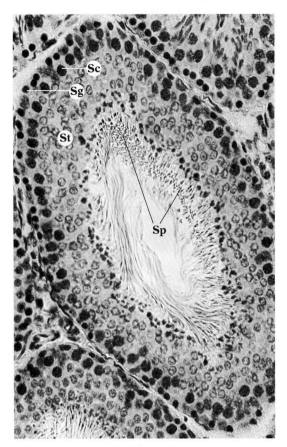

likely to have a new combination—*F* and *s*, or *f* and *S*—rather than one of the parental combinations. In fact, since crossing over usually occurs at several locations and can involve any two homologous chromatids (see Fig. 23.23), the chance of a parental chromosome's surviving meiosis unshuffled is quite a bit lower than the 50 percent of Figure 23.28. Furthermore, crossing over can occur even within a gene, so if the copies of a particular gene are different alleles, two entirely novel alleles can also be created.

THE TIMING OF MEIOSIS IN THE LIFE CYCLE

Meiosis in the life cycle of animals   With rare exceptions, higher animals exist as diploid multicellular organisms through most of their life cycle. At the time of reproduction, meiosis produces haploid gametes, which, when their nuclei unite in fertilization, give rise to the diploid zygote. The zygote then divides mitotically to produce the new diploid multicellular individual. The gametes—sperm and egg cells—are thus the only haploid stage in the animal life cycle (see Fig. 23.32C).

In male animals, sperm cells (spermatozoa) are produced by the germinal epithelium lining the seminiferous tubules of the testes (Fig. 23.29). When one of the epithelial cells undergoes meiosis, the four haploid cells that result are all quite small, but approximately equal in size (Fig. 23.30). All four soon differentiate into sperm cells with long flagella, but with very little cytoplasm in the head, which consists primarily of the nucleus. This process of sperm production is called *spermatogenesis.* Meiosis in human

23.29   Photograph of cross section of rat seminiferous tubule, showing spermatogenesis
The dark-stained outermost cells in the wall of the tubule are spermatogonia (Sg), which divide mitotically, producing new cells that move inward. These cells enlarge and differentiate into primary spermatocytes (Sc), which divide meiotically to produce secondary spermatocytes and then spermatids (St). The spermatids differentiate into mature sperm cells, or spermatozoa (Sp), whose long flagella can be seen in the lumen of the tubule in this photograph.

23.30   Schematic illustration of spermatogenesis and oogenesis in an animal
In some animals the first polar body does not divide.

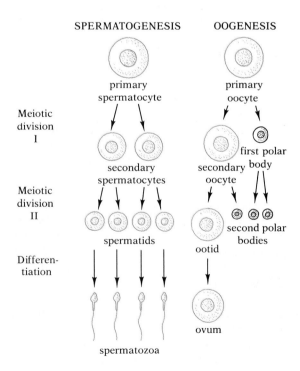

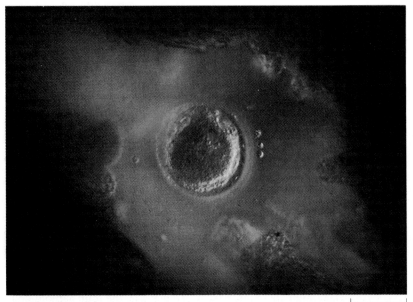

0.5 mm

23.31 Human egg cell with polar bodies
The polar bodies are the three small circular
structures at the right.

males begins at puberty. In female animals the egg cells are produced
within the follicles of the ovaries by a process called *oogenesis.* When a cell
in the ovary undergoes meiosis, the haploid cells that result are very unequal in size. The first meiotic division produces one relatively large cell
and a tiny one called a first *polar body.* The second meiotic division of the
larger of these two cells (secondary oocyte) produces a tiny second polar
body and a large cell that soon differentiates into the egg cell (or ovum).
The first polar body may or may not go through the second meiotic division.
If it does redivide, there are three polar bodies altogether (Fig. 23.31). Thus,
when a diploid cell in the ovary undergoes complete meiosis, only one mature ovum is produced (Fig. 23.30); the polar bodies are essentially nonfunctional. By contrast, a diploid cell undergoing complete meiosis in the
testis gives rise to four functional sperm cells. In human females, the oocytes complete the first meiotic prophase in the fetal ovaries, and then
complete meiosis upon ovulation. The interval between these two events
can exceed forty years.

The advantage of the unequal cytokinesis of oogenesis is obvious. By this
mechanism an unusually large supply of cytoplasm and stored food is allot-
ted to the nonmotile ovum for use by the embryo that will develop from it.
In fact, the ovum provides almost all the cytoplasm and initial food supply
for the embryo. The tiny, highly motile sperm cell contributes, essentially,
only its genetic material.

Meiosis in the life cycles of plants   That meiosis produces gametes in animals does not mean that it must do so in all organisms. There is no inherent
reason why the cells resulting from meiosis must be specialized for sexual
reproduction. And, indeed, they are not specialized in this way in most

A PRIMITIVE PLANT

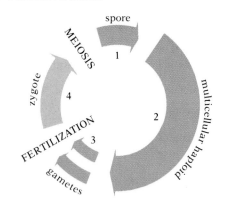

B INTERMEDIATE PLANT

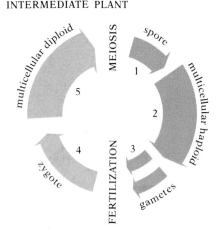

C ANIMAL

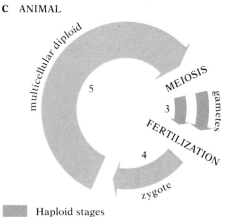

■ Haploid stages

■ Diploid stages

plants. Meiosis in plants usually produces haploid reproductive cells called ***spores,*** which often divide mitotically to develop into haploid multicellular plant bodies.

Let us briefly examine the life cycle of plants like ferns, in which meiosis produces spores rather than gametes (Fig. 23.32B). While the plant is in the diploid portion of its life cycle, certain cells in its reproductive organs divide by meiosis to produce haploid spores (stage 1). These spores divide mitotically and develop into haploid multicellular plants (stage 2). The haploid multicellular plant eventually produces cells specialized as gametes (stage 3). Notice that the gametes are produced by mitosis, not meiosis, because the cells that divide to produce the gametes are already haploid. Two of these gametes unite in fertilization to form the diploid zygote (stage 4), which divides mitotically and develops into a diploid multicellular plant (stage 5). In time, this plant produces spores, and the cycle starts over again.

As we shall see in some detail in a later chapter, the various groups of plants vary greatly in the relative importance of the diploid and haploid phases in their life cycles. In some, like ferns, the two phases are approximately equal, as shown in Figure 23.32B (stages 2 and 5). A very few plants have cycles almost like the animal life cycle shown in Figure 23.32C; stages 1 and 2 are absent, and the haploid phase of the cycle is represented only by the gametes. In the flowering plants, stages 1 and 2 have not been abandoned altogether, but stage 2 has been reduced to a tiny three- to eight-cell entity that is not free-living, and the plant spends most of its life cycle as a multicellular diploid organism (stage 5).

At the opposite extreme are some primitive plants like algae, whose life cycles are almost completely the reverse of those of animals (Fig. 23.32A). There is no stage 5. The diploid zygote (stage 4) formed by the union of gametes promptly undergoes meiosis to produce four haploid spore cells (stage

23.32 Three types of life cycles
(A) In some very primitive plants the diploid phase is represented only by the zygote, which quickly divides by meiosis to produce haploid spores, which may divide mitotically to produce a multicellular haploid plant. (B) In most multicellular plants there are two multicellular stages, one haploid and one diploid (stages 2 and 5); the relative importance of these two stages varies greatly from one plant group to another (the cycle shown here is an intermediate one in which stages 2 and 5 are nearly equal). In flowering plants the multicellular diploid stage (stage 5) is the major one, and the multicellular haploid stage (stage 2) is much reduced, being represented by a tiny organism with very few cells. (C) Animals and a very few plants have a life cycle in which meiosis produces gametes directly—the spore stage (stage 1) and the multicellular haploid stage (stage 2) being absent.

1), each of which divides mitotically and develops into a haploid multicellular stage (stage 2) in which the organism passes most of its life. This multicellular plant eventually produces, by mitosis, cells specialized as gametes (stage 3), which unite to form the diploid zygote and start the cycle over again. In such an organism, then, the haploid phase of the cycle (particularly stage 2) is dominant, and the only diploid stage—the zygote—is very transitory.

Examination and comparison of the three life cycles illustrated in Figure 23.32 allow us to make several important generalizations:

1. Meiosis produces either gametes, specialized for sexual reproduction, or spores, specialized for asexual reproduction.
2. Only diploid cells can divide by meiosis, but both haploid and diploid cells can divide by mitosis.
3. If mitosis produces a multicellular organism after fertilization but before meiosis, that organism is diploid.
4. If mitosis produces a multicellular organism after meiosis but before fertilization, that organism is haploid.
5. The haploid phase of the life cycle of multicellular animals is represented only by the gametes (stages 1 and 2 being absent).
6. Most multicellular plants include all five stages in their life cycles, but the relative importance of these varies greatly.
7. With some major exceptions, the haploid stages are dominant in the more primitive plants, the diploid stages in the more advanced plants.

Chapter 24

MENDELIAN GENETICS

For at least as long as history holds records of the subject, people have known that many traits, whether morphological, physiological, or even behavioral, can be inherited. And for millennia this knowledge has been used in the breeding of more desirable grains, vegetables, fruits, flowers, and domestic animals. But though breeders learned that many desirable characteristics, like the compact growth form of domestic cabbage (see Fig. 1.21, p. 17), may persist through the generations, or breed true, they could not explain why other characteristics tend to disappear for a time or to diminish in frequency in successive generations. Nor could they explain why the so-called blending inheritance, in which children were supposed to embody a recognizable blend of parental traits, did not always occur—why, indeed, instead of combining traits of both parents, a child might even exhibit a trait not present in either one. How could a blue-eyed child be born of brown-eyed parents, for instance? The difficulty of reconciling these seemingly contradictory phenomena remained a major intellectual challenge for scientists even to the middle of the nineteenth century.

The first person known to have made sense of these conflicting phenomena was the Austrian monk Gregor Mendel. Working at first to breed an industrious yet docile honey bee, Mendel began his experiments in inheritance by crossing a hardworking race of German bee, *Apis mellifera carnica*, with a prettier and more gentle Italian race, *A. m. ligustica*. The result of this cross was neither pretty, industrious, nor gentle. Here was a perfect example of the problem.

In 1856, Mendel wisely shifted his attention to garden peas, an organism he found considerably more manageable. The experiments he performed over the next twelve years laid the groundwork for what is now called the chromosomal theory of inheritance, the first element in the modern science of genetics. First published in 1866, Mendel's results are all the

TABLE 24.1 *Mendel's results from crosses involving single character differences*

P characters	F₁	F₂	F₂ ratio
1. Round × wrinkled seeds	All round	5,474 round : 1,850 wrinkled	2.96 : 1
2. Yellow × green seeds	All yellow	6,022 yellow : 2,001 green	3.01 : 1
3. Red × white flowers	All red	705 red : 224 white	3.15 : 1
4. Inflated × constricted pods	All inflated	882 inflated : 299 constricted	2.95 : 1
5. Green × yellow pods	All green	428 green : 152 yellow	2.82 : 1
6. Axial × terminal flowers	All axial	651 axial : 207 terminal	3.14 : 1
7. Long × short stems	All long	787 long : 277 short	2.84 : 1

more remarkable in that they represent the single line of lucid genetic research at the time. A combination of insight and careful methodology—Mendel meticulously counted all the different types of progeny produced in his experiments—led him to conclusions reached in his day by no other researchers on inheritance. Unlike so many other great discoveries, Mendel's seem to have been the achievement of an isolated genius.

MONOHYBRID INHERITANCE

EXPERIMENTS BY MENDEL

Mendel began with several dozen strains of peas, mostly purchased from commercial sources. He raised each variety for several years to discover which strains had recognizable morphological variations that bred true.

Mendel's results In the end, Mendel reported his work on seven of the numerous characteristics of garden peas that he studied. He noticed that each of these seven characteristics occurred in two contrasting forms; the seeds were either round or wrinkled, the flowers were red or white, the pods were green or yellow, and so on (Table 24.1). When Mendel cross-pollinated plants with contrasting forms of just one of these characteristics, he found that all the offspring (usually referred to as the *F₁*, or first filial, generation) were alike and resembled only one of the two parents (the *P*, or parental, generation). When these offspring were crossed among themselves, however, some of their offspring (the *F₂*, or second filial, generation) showed one of the original contrasting traits and some showed the other (Table 24.1). In other words, a trait that had been present in one of the parents, but not in any of their children, reappeared in the next generation, just as blue eyes can reappear after a generation or more of brown eyes.

Mendel's cross of red-flowered with white-flowered plants, for example, produced red flowers in all plants of the F₁ generation (Fig. 24.1). Similarly, when he crossed plants having round seeds with plants having wrinkled

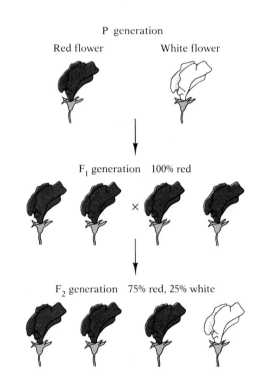

24.1 Results of Mendel's cross of red-flowered and white-flowered peas

seeds, all the offspring had round seeds. Apparently one form of each characteristic had taken precedence over the other: red color had taken precedence over white in the flowers, and round form had taken precedence over wrinkled in the seeds. Mendel referred to the traits that appear in the F_1 offspring of such crosses as **dominant characters,** and the traits that are latent in the F_1 generation (in these examples, white flowers and wrinkled seeds) as **recessive characters.**

When Mendel allowed the F_1 peas from the cross involving flower color, all of which were red, to breed freely among themselves, their offspring (the F_2 generation) were of two types; there were 705 plants with red flowers and 224 with white flowers. The recessive character had reappeared in approximately one-fourth of the F_2 plants. Similarly, when F_1 peas from the cross involving seed form, all of which had round seeds, were allowed to breed freely among themselves, their F_2 offspring were of two types; 5,474 had round seeds and 1,850 had wrinkled seeds. Again, the recessive character had reappeared in approximately one-fourth of the F_2 plants. The same thing was true of crosses involving the other five characters that Mendel reported on in his classic paper (Table 24.1); in each case the recessive character disappeared in the F_1 generation, but reappeared in approximately one-fourth of the plants in the F_2 generation. We can summarize the results of the experiment involving flower color as follows:

$$
\begin{array}{lccc}
\text{P} & \text{red} & \times & \text{white} \\
 & & \downarrow & \\
\text{F}_1 & & \text{all red} & \\
 & & \downarrow & \\
\text{F}_2 & \text{¾ red} & & \text{¼ white}
\end{array}
$$

Mendel's conclusions Experiments of this type led Mendel to formulate and test a series of hypotheses. Finally, he drew the revolutionary conclusion that each pea plant possesses two hereditary "factors" for each character, and that when gametes are formed the two factors segregate and pass into separate gametes. Each gamete, then, possesses only one factor for each character. Each new plant thus receives one factor for each character from its male parent and one for each character from its female parent. That two contrasting parental traits, such as red and white flowers, can both appear in normal form in the F_2 offspring indicates that the hereditary factors must exist as separate entities in the cell; they do not blend or fuse with each other. Thus the cells of an F_1 pea plant from the cross involving flower color contain, according to Mendel, one factor for red color and one factor for white color, the factor for red being dominant and the factor for white being recessive. But their coexistence in the same nucleus does not change them; the red and white do not alter each other. They remain distinct, and segregate unchanged when new gametes are formed. Mendel referred to the principle embodied in this pattern of inheritance, with the genetic factors particulate rather than blending, as the Principle of Segregation.

You will have noticed that Mendel's conclusions are consistent with what we now know about the chromosomes and their behavior in meiosis: the diploid nucleus contains two of each type of chromosome, one homologous chromosome from each parent. Each of the two chromosomes in any given pair bears genes for the same characters; hence the diploid cell contains two copies of each type of gene; these are the two hereditary factors for each character that Mendel described.[1] Since homologous chromosomes segregate during meiosis, gametes contain only one chromosome of each type and hence only one copy of each gene, just as Mendel deduced.

Mendel's theories seem rather obvious in the light of the events of cell division. But he did his work before the details of cell division had been learned—before, in fact, the significance of chromosomes for heredity had been discovered. Mendel arrived at his conclusions purely by reasoning from the patterns of inheritance he detected in his experiments, without any reference to the structural components of the cell or its nucleus.

Oddly enough, Mendel's well-controlled experiments and brilliant deductions had little effect on the scientific world. Many historians argue that the scientific "establishment" was not ready for the radical notion of particulate inheritance, but Mendel himself may have underestimated the value of his conclusions, for reasons that will become apparent in the next chapter. In any case, Mendel abandoned his effort to comprehend the rest of the puzzle in 1868. In 1900, after cell division and chromosome segregation had been observed and described in detail, three biologists—Hugo De Vries in Holland, Carl Correns in Germany, and Erich von Tschermak-Seysenegg in Austria—independently rediscovered the phenomenon of segregation and, almost immediately, Mendel's original paper.

A modern interpretation of Mendel's experiments   Mendel's results make perfect sense in the light of what we now know about chromosome function and anatomy. Let's consider his results on flower color in peas from a modern perspective. In the cells of a pea plant, a gene for flower color exists at the same location, or *locus*, on two homologous chromosomes (Fig. 24.2). The gene can exist in many different forms, or *alleles,* but no individual can have more than two: one on each member of the pair of homologous chromosomes involved. It is customary to designate genes by letters, using capital letters for dominant alleles and small letters for recessive alleles. Flower color results from the presence of pigment molecules, whose structure is determined by the genetic information in the chromosomes; Mendel's red allele (C) encodes a functional pigment that absorbs all wavelengths of light except the red that we see reflected, while the white allele (c) either encodes a nonfunctional pigment molecule that reflects all wavelengths of light, and so appears white, or fails to produce any pigment at all. Since a diploid cell contains two copies of each gene, it may have two copies of the same allele or one copy of one allele and one copy of another allele (Fig. 24.2). Thus cells of a pea plant may contain two copies of the allele for red flowers (C/C), or two copies of the allele for white flowers (c/c), or one copy of the allele for red and one copy of the allele for white (C/c).

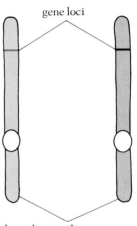

gene loci

homologous chromosomes

24.2   Anatomy of segregation
Somatic cells have pairs of homologous chromosomes, a pair consisting of one chromosome from each parent. Each gene is found in two copies, one on each chromosome of the homologous pair, at corresponding loci. When the genes at the corresponding loci are different, they are called alleles. In meiosis, each gamete receives a copy of only one chromosome from each homologous pair, and hence only one of the two alleles.

---

[1] For the moment, we shall define a gene as the part of a chromosome coding for a single molecule of RNA, whether that RNA in turn is used directly (as a component of a ribosome, for instance) or serves as a messenger to be translated into a polypeptide chain. As we shall see in later chapters, there are many minor exceptions to this generally satisfactory definition.

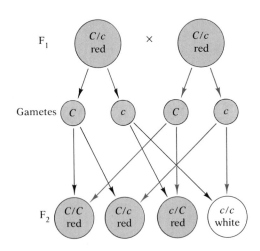

24.3 Gametes formed by F₁ individuals in Mendel's cross for flower color, and their possible combinations in the F₂

Cells with two copies of the same allele (*C/C* or *c/c*) are said to be **homozygous** for that trait. Those with one each of two different alleles (*C/c*) are said to be **heterozygous**. (The slash between letters indicates that the two copies of a gene are on separate chromosomes.)

Unfortunately, we cannot tell by visual inspection whether a given pea plant is homozygous dominant (*C/C*) or heterozygous (*C/c*), because the two types of plants look alike: both have red flowers. In other words, where one allele is dominant over another, the dominant allele takes full precedence over the recessive allele, and a heterozygous organism exhibits the trait determined by that dominant allele; one copy of the dominant allele is as effective as two copies in determining the character trait. For this reason, there is often no one-to-one correspondence between the different possible genetic combinations (**genotypes**) and the possible appearances (**phenotypes**) of the organisms. In the example of flower color in peas discussed here, there are three possible genotypes, *C/C*, *C/c*, and *c/c*, but only two possible phenotypes, red and white. The heterozygote (*C/c*) produces a red phenotype because the red allele is dominant simply because the functional red pigment it specifies masks the product, if any, of the white allele.

The functional/nonfunctional dichotomy is the most common basis of dominant and recessive phenotypes. Many traits result from alleles that code for nonfunctional structural proteins or inactive enzymes; in the presence of an active enzyme, or functional structural protein, the existence of the nonfunctional recessive version is hidden. As we shall see, though, many alleles cannot easily be categorized as dominant or recessive.

Using the modern conception of gene function, we can rewrite the summary of Mendel's pea cross as follows:

P		*C/C*	×	*c/c*
		red		white
			↓	
F₁		*C/c*	×	*C/c*
		red		red
			↓	
F₂	*C/C*	*C/c*	*c/C*	*c/c*
	red	red	red	white

Here we see both the genotypes and the phenotypes of the plants in all three generations. Mendel began with a cross in the parental generation between a plant with a homozygous dominant genotype (red phenotype) and a plant with a homozygous recessive genotype (white phenotype). All of the F₁ progeny had red phenotypes, because all of them were heterozygous, having received a dominant allele for red (*C*) from the homozygous dominant parent and a recessive allele for white (*c*) from the homozygous recessive parent. But when the F₁ individuals were allowed to cross freely among themselves, the F₂ progeny they produced were of three genotypes and two phenotypes; one-fourth were homozygous dominant and showed red phenotypes, two-fourths were heterozygous and showed red phenotypes, and one-fourth were homozygous recessive and showed white phenotypes. Thus the ratio of genotypes in the F₂ was 1:2:1, and the ratio of the phenotypes was 3:1.

How do we figure out the possible genotypic combinations in the F₂? This is an easy matter in a monohybrid cross (a cross involving only one character) such as this. All individuals in the F_1 generation are heterozygous (C/c); that is, they have one of each of the two types of alleles. Each of these two alleles is located on a different one of the two chromosomes of a homologous pair. In meiosis these two chromosomes synapse, move onto the spindle as a unit, and then separate, moving to opposite poles, so that the chromosome bearing the C allele is incorporated into one new haploid nucleus and the chromosome bearing the c allele is incorporated into the other new haploid nucleus. This means that half the gametes produced by such a heterozygous individual will contain the C allele and half the c allele. When two such individuals are crossed (Fig. 24.3), there are four possible combinations of their gametes:

C from male parent, C from female parent
C from male parent, c from female parent
c from male parent, C from female parent
c from male parent, c from female parent

The first of these four possible combinations produces homozygous dominant offspring (red); the second and third produce heterozygous offspring (also red); and the fourth produces homozygous recessive offspring (white). Since each of these four combinations is equally probable, we would expect, if a large number of F_2 progeny are produced, a genotypic ratio close to $1:2:1$ and a phenotypic ratio close to $3:1$, just as Mendel found.

An easy way to figure out the possible genotypes produced in the F_2 is to construct a so-called Punnett square, named after the Cambridge geneticist R. C. Punnett. Along a horizontal line, write all the possible kinds of gametes the male parent can produce; in a vertical column to the left, write all the possible kinds of gametes the female parent can produce; then draw squares for each possible combination of these, as follows:

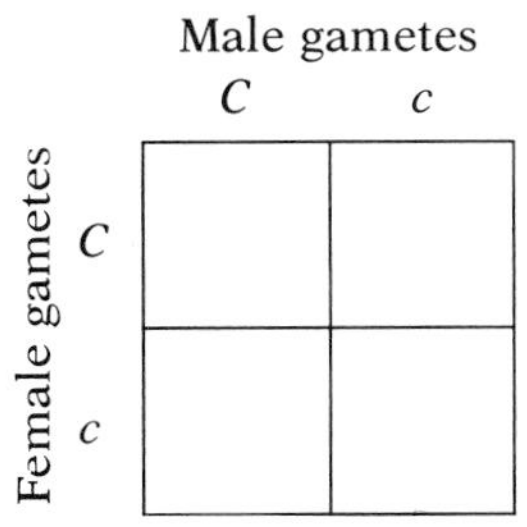

Male gametes
	C	c
C		
c		

Female gametes

Next, write in each box, first, the symbol for the female gamete and, second, the symbol for the male gamete. Each box will then contain the symbols for the genotype of one possible zygote combination from the cross in question. A glance at the completed Punnett square in Figure 24.4 shows that the cross yields the expected $1:2:1$ genotypic ratio and, since dominance is present, the expected $3:1$ phenotypic ratio.

Extensive investigation of a vast array of plant and animal species by thousands of scientists has demonstrated conclusively that the results Mendel obtained from his monohybrid crosses, and the interpretations he

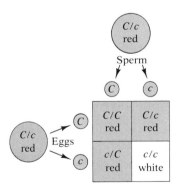

24.4 Punnett-square representation of the information shown in Figure 24.3

placed on them, are not limited to garden peas but are of general validity. Whenever a monohybrid cross is made between two contrasting homozygous individuals, regardless of the character involved, the expected genotypic ratio in the F_2 is $1:2:1$. And whenever there is dominance, the expected phenotypic ratio is $3:1$. If these ratios are not obtained in large samples, some complicating condition must be present.

THE TEST CROSS

We have seen that in a monohybrid cross involving dominance the homozygous dominant progeny and the heterozygous progeny have the same phenotype and cannot be distinguished by inspection. But it is often of practical importance to distinguish between individuals with these two genotypes—in breeding, for example. Homozygous individuals breed true: matings between two such individuals produce offspring all of a single genotype and phenotype, like their parents. But heterozygous individuals do not breed true; as we have seen, a cross between two such individuals may produce offspring of three genotypes and two phenotypes. Consequently the identification of homozygous individuals is of the utmost importance to an animal or plant breeder who is trying to establish true-breeding strains of animals or plants. This is no problem when the breeder is interested in a recessive character, because homozygous recessive organisms can readily be recognized by their phenotypes. But if the breeder is interested in establishing a strain that is true-breeding for a dominant character, he needs a test that will enable him to tell whether a given individual is homozygous dominant or heterozygous.

The test used for this purpose is a cross between the individual of unknown genotype and a homozygous recessive individual, which can be recognized by its phenotype. Suppose, for example, we want to know whether a particular red-flowered pea plant has a genotype of C/C or C/c. The most we can say from simple inspection is that it is red and hence must have at least one C allele; half of its genotype for flower color is known, and half is unknown and can be designated by a dash: $C/-$. We now cross this plant with a white-flowered plant, which can only have the genotype c/c. The results should give us the information we seek. If the plant in question has $C/\underline{C}$ as its genotype, the cross will turn out as follows:

$$
\begin{array}{ccc}
 & C/\underline{C} \;\; \times \;\; c/c & \\
 & \text{red} \qquad\quad \text{white} & \\
 & \downarrow & \\
C/c \quad & C/c \qquad C/c & \quad C/c \\
\text{red} \quad & \text{red} \qquad\; \text{red} & \quad \text{red}
\end{array}
$$

If, however, the plant has the other possible genotype, $C/\underline{c}$, the cross will turn out as follows:

$$
\begin{array}{ccc}
 & C/\underline{c} \;\; \times \;\; c/c & \\
 & \text{red} \qquad\quad \text{white} & \\
 & \downarrow & \\
C/c \quad & C/c \qquad c/c & \quad c/c \\
\text{red} \quad & \text{red} \quad\; \text{white} & \quad \text{white}
\end{array}
$$

The unknown half of the test plant's genotype, then, may be either a C allele or a c allele. If it is a C allele, the full genotype of the plant is C/C; such a plant, when test-crossed with a homozygous recessive (c/c) plant, will produce progeny all of which are heterozygous (C/c) and show a red phenotype (Fig. 24.5A). If, on the other hand, the unknown is a c allele, the full genotype of the plant is C/c; such a plant, when test-crossed with a homozygous recessive plant, will produce progeny of which half are heterozygous and show a red phenotype and half are homozygous recessive and show a white phenotype (Fig. 24.5B). If a large number of progeny are produced from our test cross and all of them show the dominant phenotype (in this case red flowers), the chances are great that the test plant's genotype is C/C. If, however, some of the progeny show the dominant phenotype (red) and some of the recessive phenotype (white), we know that the test plant's genotype is C/c. We expect a 1:1 ratio in this case, but even if our results happened to depart considerably from this expected ratio, we should still feel certain that the test plant's genotype is C/c. The reason is obvious: If any of the progeny show the recessive phenotype, their genotype must be c/c, which means that they received a c allele from each parent; hence the test plant must have a c allele in its genotype. Since we already know that it has at least one C allele, we combine these two pieces of information to write its genotype as C/c.

PARTIAL DOMINANCE

The seven characters of peas that Mendel used in the experiments reported in his paper were all of the kind in which one allele, known as a full dominant, shows complete dominance over the other. Many characteristics in a variety of organisms show this mode of inheritance. But many others do not, including some that Mendel himself studied but did not discuss in his paper. When a heterozygous individual clearly shows effects of both alleles, the alleles are referred to as partial, or incomplete, dominants.

In many cases of partial dominance, heterozygous individuals have a phenotype that is actually intermediate between the phenotype of individuals homozygous for one allele and the phenotype of individuals homozygous for the other allele. For example, crosses between homozygous red snapdragons and homozygous white snapdragons yield pink snapdragons. When these pink plants are crossed among themselves, they yield red, pink, and white offspring in a ratio of 1:2:1, as follows:

$$
\begin{array}{cccccc}
P & & R/R & \times & R'/R' & \\
& & \text{red} & & \text{white} & \\
& & & \downarrow & & \\
F_1 & & R/R' & \times & R/R' & \\
& & \text{pink} & & \text{pink} & \\
& & & \downarrow & & \\
F_2 & R/R & R/R' & & R'/R & R'/R' \\
& \text{red} & \text{pink} & & \text{pink} & \text{white}
\end{array}
$$

Notice that when there is partial dominance both alleles are designated by a capital letter, and one is distinguished from the other by a prime, as here, or by a superscript (C^r for red, C^w for white).

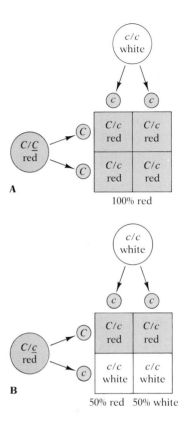

24.5 A test cross
Homozygous dominant ($C/\underline{C}$) and heterozygous ($C/\underline{c}$) red-flowered individuals are to be distinguished by crossing them with a homozygous recessive (c/c) white-flowered individual. (A) If the red-flowered plant is homozygous, all the progeny of the test cross will be heterozygous (C/c) and will have red flowers. (B) If the red-flowered plant is heterozygous, 50 percent of the progeny will be heterozygous (C/c) and will have red flowers, and 50 percent will be homozygous recessive (c/c) and will have white flowers.

The chemical events underlying the production of pink flowers are not fully understood. Heterozygous pea flowers, as we have seen, appear red. One likely explanation is that the red pigment of pea flowers is so intense that it masks the white pigment totally, while the less saturated red pigment of snapdragons does not. It is instructive to see how little pure red pigment must be mixed into a gallon of white base paint to produce a can of bright red paint.

In other cases of partial dominance, the heterozygous phenotype is not so obviously intermediate between the two homozygous phenotypes. For example, in a certain strain of chickens a mating between a black chicken and a so-called splashed white chicken produces offspring all of which have a distinctive appearance called blue Andalusian. A cross between two blue Andalusians produces black, blue Andalusian, and splashed white offspring in a ratio of $1:2:1$, as follows:

P   black $\times$ white
                       $\downarrow$

F_1 blue $\times$ blue
                       $\downarrow$

F_2 black   blue   blue   white

Here the gene products of two alleles, one for black and one for white, are interacting to produce a phenotype somewhat different from the gray that might be expected as the intermediate between black and white.

To summarize, the inheritance pattern of partial dominance differs in the following ways from that in complete dominance: (1) the F_1 offspring of a monohybrid cross between parents each homozygous for a different allele have a phenotype different from both parents, and (2) the F_2 phenotypic ratio is $1:2:1$ (just like the genotypic ratio) rather than $3:1$.

MULTIHYBRID AND MULTIGENIC INHERITANCE

We have limited our discussion so far to crosses involving a single gene (a monogenic trait) controlling a single character (a monohybrid cross). But most characters are controlled by several different genes at once, and so involve multigenic inheritance; in addition, organisms have many different characters, and so most crosses are usually multihybrid crosses. Trihybrid and trigenic crosses mark the practical limits of Mendelian analysis.

THE BASIC DIHYBRID RATIO

Mendel's experiments on garden peas were not limited to single characters, but sometimes involved two or more of the characters listed in Table 24.1. For example, he crossed plants bearing round yellow seeds with plants producing green wrinkled seeds. The resulting F_1 plants all had round yellow seeds. That is, all showed dominant phenotypes, as we have come to expect. When these plants were crossed among themselves, the resulting F_2 progeny showed four different phenotypes:

    315 had round yellow seeds
    101 had wrinkled yellow seeds
    108 had round green seeds
     32 had wrinkled green seeds

These numbers represent a ratio of about $9:3:3:1$ for the four phenotypes.

This experiment produced a new and interesting result: a dihybrid cross can produce new plants phenotypically unlike either of the original parental plants; here the new phenotypes were wrinkled yellow and round green. It demonstrated, in other words, that during meiosis the genes for seed color and the genes for seed form do not necessarily remain paired as they were in the parental generation, but can separate and reassemble in different allelic combinations in the gametes. From what we know of meiosis, this result implies that the genes for seed color are on the chromosomes of one homologous pair while the genes for seed form are on the chromosomes of another pair. Consequently, the genes for the two characters segregate independently during meiosis, sometimes producing new phenotypes.

The $9:3:3:1$ phenotypic ratio is characteristic of the F_2 generation of a dihybrid cross (with dominance) in which the genes for the two characters are ***independent*** (located on nonhomologous chromosomes). Each independent gene behaves in a dihybrid cross exactly as in a monohybrid cross. If we view Mendel's F_2 results as the product of a monohybrid cross for seed color (ignoring seed form), we find that there were 416 yellow seeds $(315 + 101)$ and 140 green seeds $(108 + 32)$, which closely approximates the $3:1$ F_2 ratio expected in a monohybrid cross. Similarly, if we treat the experiment as a monohybrid cross for seed form and ignore seed color, the F_2 results also show a phenotypic ratio of approximately $3:1$. The dihybrid F_2 ratio of $9:3:3:1$ is thus simply the product of two separate and independent $3:1$ ratios.

Let us examine in somewhat more detail a cross of this type, using the symbols R for the allele for round seed and r for the allele for wrinkled seed, and the symbols G for the allele for yellow seed and g for the allele for green seed. In the summary of Mendel's cross below, the dash means that it does not matter phenotypically whether the dominant or the recessive allele occurs in the spot indicated.

P		R/R G/G round yellow	$\times$	r/r g/g wrinkled green	
			$\downarrow$		
F_1		R/r G/g round yellow	$\times$	R/r G/g round yellow	
			$\downarrow$		
F_2	$9\ R/-\ G/-$ round yellow	$3\ r/r\ G/-$ wrinkled yellow		$3\ R/-\ g/g$ round green	$1\ r/r\ g/g$ wrinkled green

The round yellow parent could produce gametes of only one genotype, RG. The wrinkled green parent could produce only rg gametes. When RG gametes from the one parent united with rg gametes from the other parent in the process of fertilization, all the resulting F_1 offspring were heterozygous for both characters (R/r G/g) and showed the phenotype of the dominant parent (round yellow). Each of these F_1 individuals could produce four different types of gametes, RG, Rg, rG, and rg. When two such individuals were crossed, there were 16 possible combinations of gametes

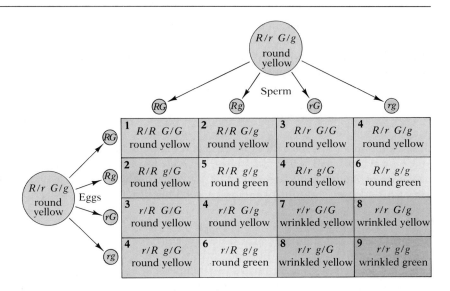

24.6 Punnett-square representation of an F₁ dihybrid cross

When two individuals heterozygous for both seed color and shape were crossed, Mendel obtained roughly nine plants that produced round yellow seeds, three that produced round green seeds, and three that produced wrinkled yellow seeds for every one that produced wrinkled green seeds. Underlying these results were nine different genotypes, as indicated by the numbers in the boxes. Combinations of alleles that represent identical genotypes are shown with the same number; combinations of alleles that give rise to identical phenotypes have the same shading.

(4 × 4). As shown in Figure 24.6, these 16 combinations included nine genotypes, which determined four phenotypes in the ratio of 9 : 3 : 3 : 1.

As Figure 24.6 indicates, one way to determine the genotypic and phenotypic ratios in a dihybrid cross is to construct a Punnett square and then count the number of boxes representing each genotype and each phenotype. This method, though satisfactory in a monohybrid cross and only moderately laborious in a dihybrid cross, becomes prohibitively tedious in a trihybrid cross or a cross involving more than three characters. There is an alternative procedure that is much easier. It is based on the principle that *the chance that a number of independent events will occur together is equal to the product of the chances that each event will occur separately.* This principle is known as the Product Law. It is best explained by an example.

Suppose we want to know how many of the 16 combinations in Mendel's cross will produce the wrinkled yellow phenotype. We know that wrinkled is recessive; hence it is expected in ¼ of the F₂ individuals in a monohybrid cross. We know that yellow is dominant; hence it is expected in ¾ of the F₂ individuals. Multiplying these two separate values (¼ × ¾) gives us ³⁄₁₆; 3 of the 16 possible combinations will produce a wrinkled yellow phenotype. Similarly, if we want to know how many of the combinations will produce a round yellow phenotype, we multiply the separate expectancies for two dominant characters (¾ × ¾) to get ⁹⁄₁₆.

The Product Law applies equally well to more complex examples. Suppose we want to find out, for a trihybrid cross involving the two seed characters and flower color, what fraction of the F₂ individuals (produced by allowing *C/c R/r G/g* individuals to cross among themselves) will exhibit a phenotype combining red flowers, wrinkled seeds, and yellow seeds. The separate probability for red flowers in a monohybrid cross is ¾, that for wrinkled seeds is ¼, and that for yellow seeds is ¾. Multiplying these three values ($\frac{3}{4} \times \frac{1}{4} \times \frac{3}{4}$) gives us $\frac{9}{64}$. This tells us that of the 64 possible combinations in a trihybrid cross, 9 will produce the phenotype here specified. Now try determining the proportion of offspring that will exhibit a phenotype combining white flowers, round seeds, and green seeds.

GENE INTERACTIONS

We have already seen how different alleles of the same gene can interact when both are partially dominant. Separate genes can also interact. Indeed, most phenotypic characteristics, whether they are morphological, like flower color, or more subtle chemical characteristics, like enzyme pathways, are controlled by several genes. You may recall that glycolysis, for instance, is a dozen-step process. Each step requires a different enzyme, and each enzyme is produced by the action of a specific gene. Most of these enzymes depend on one or more others operating earlier in the chemical chain reaction to provide a substrate with which they can interact and so in turn provide a substrate suitable for the next enzyme in the series.

Glycolysis is only one of the many life processes that require the "cooperation" of many genes. But tracking the influence each gene brings to bear in such complicated interactions can be difficult: an individual unable to perform glycolysis at a normal rate, for instance, may be homozygous recessive for any of twelve separate genes—homozygous recessive because a heterozygous individual produces at least some functional enzyme. How can we sort out gene functions when genes depend on each other for expressing their true phenotype? We shall look briefly at ways of solving these problems, and then at how the problems themselves provide useful clues about the operation of cellular chemistry.

Complementary genes   Genes that are mutually dependent are normally detected during crosses between individuals with similar recessive phenotypes. This sort of test can show whether the trait in question is produced by the same gene in both, or by two separate genes. For example, a normal zebra finch is brightly colored, but there are also two distinct forms of albino finch. One form of albino zebra finch, though almost totally white, has a slight brown tinge near its eyes. The other is a classic albino. But if the two forms are crossed, the offspring are normal wild types (the most common naturally occurring form). This result indicates that each form of albinism is produced by a separate gene, at a different locus on the chromosome; the offspring of the cross, being heterozygous in both genes, display the normal phenotype (Fig. 24.7).

A failure of the normal pigmentation system in zebra finches, then, can occur at two known points: the pigment protein can be defective, which is the problem in strain B; or the control mechanisms that turn on pigment production and control the amount made can be defective, which is the problem in strain A. A finch that is homozygous recessive for either element of the pigment system will be an albino. If two white individuals with the *same* defect are crossed, all the progeny will be white, but if two white individuals with different defects are crossed, some or all of the offspring will be normal (Fig. 24.7). This reversion to the normal phenotype in offspring

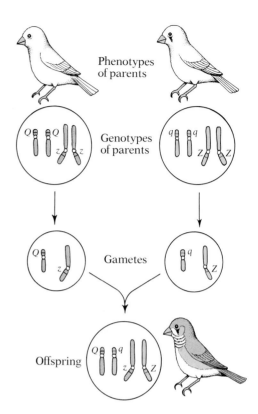

24.7 Complementation test with zebra finches
There are two varieties of albino zebra finch. When these are crossed, the offspring have the slate-gray bodies and colorful markings of normal wild zebra finches. The genetic basis of this phenomenon is shown here. Finches of one strain (right) are homozygous for the recessive allele at the Q locus (genotype q/q), and therefore fail to produce functional pigment. Those of the other strain are homozygous recessive at the Z locus (genotype z/z), and lack normal color because of flaws in the mechanisms controlling gene activity. The offspring of the cross are heterozygous at both loci, and so can produce the normal amount of pigment. In the example shown here, both parents are homozygous for both genes. If one or both of the parents were heterozygous for one gene (with the genotype Q/q z/z or q/q Z/z), we would expect some of the offspring to be normal and some to be albino.

heterozygous at both loci exemplifies the phenomenon known as complementarity, and the two mutually dependent genes are said to **complement** each other.

Complementarity can lead to some unusual phenotypic ratios in crosses. Consider the following dihybrid cross between zebra finches:

P	$Q/Q \ Z/Z$	$\times$	$q/q \ z/z$	
	normal		albino	

$\downarrow$

F_1	$Q/q \ Z/z$	$\times$	$Q/q \ Z/z$	
	normal		normal	

$\downarrow$

F_2	9 $Q/- \ Z/-$	3 $Q/- \ z/z$	3 $q/q \ Z/-$	1 $q/q \ z/z$
	normal	albino	albino	albino

In a normal dihybrid cross the four groups of genotypes shown in the F_2 would produce four phenotypes in a ratio of $9:3:3:1$; here, though, they have produced only two phenotypes in a ratio of $9:7$, which is quite different from any ratio we have previously encountered. The last three phenotypic classes of the normal $9:3:3:1$ ratio have been combined.

Epistasis   As we have just seen, complementary genes find expression in the dominant phenotype if and only if the organism is heterozygous or homozygous dominant at both loci. Each gene in some sense has veto power over the other, because their products must cooperate. In **epistasis,** by contrast, these two genetic "votes" do not have equal power: because of the biochemistry of the interaction of their products, only one of the genes can be vetoed by the other. When one gene has the effect of suppressing the phenotypic expression of another gene but not vice versa, the first gene is said to be epistatic to the second. (The Greek root of *epistasis*, logically enough, means "standing upon.") In guinea pigs, for example, a gene for the production of the skin pigment melanin is epistatic to one for the *deposition* of melanin. The first gene has two alleles: C, which causes pigment to be produced, and c, which causes no pigment production; hence a homozygous recessive individual, c/c, is an albino. The second gene has an allele B that causes deposition of much melanin, which gives the guinea pig a black coat, and an allele b that causes deposition of only a moderate amount of melanin, which gives the guinea pig a brown coat. Neither B nor b can cause deposition of melanin if C is not present to make the melanin. We can summarize a cross involving these two genes as follows:

P	$C/C \ B/B$	$\times$	$c/c \ b/b$	
	black		albino	

$\downarrow$

F_1	$C/c \ B/b$	$\times$	$C/c \ B/b$	
	black		black	

$\downarrow$

F_2	9 $C/- \ B/-$	3 $C/- \ b/b$	3 $c/c \ B/-$	1 $c/c \ b/b$
	black	brown	albino	albino

Instead of an F_2 phenotypic ratio of $9:3:3:1$, this cross has yielded a ratio of $9:3:4$. The last two phenotypic classes of the normal $9:3:3:1$ ratio have been combined.

It is important to distinguish epistasis from dominance, which it superficially resembles. Dominance is the phenotypic expression of one member of a pair of alleles at the expense of the other. Epistasis is the suppression by one gene of the phenotypic effect of another entirely different gene. Dominance refers to interaction between alleles, epistasis to interaction between nonallelic genes.

Collaboration Sometimes two genes influencing the same character interact to produce a novel phenotype that neither gene could produce independently. Such collaborative interaction is seen in the control of the form of the comb in chickens (Fig. 24.8). One gene, R, produces rose comb, while its recessive allele, r, produces single comb. Another gene, P, produces pea comb, while its recessive allele, p, also produces single comb. When R and P occur together, they collaborate to produce walnut comb, a type of comb that neither could produce alone. Rose comb is characteristic of Wyandotte chickens, and pea comb is characteristic of Brahma chickens. A cross between a Wyandotte and a Brahma could be summarized as follows:

P	$R/R \ \ p/p$	$\times$	$r/r \ \ P/P$	
	rose		pea	
		$\downarrow$		
F_1	$R/r \ \ P/p$	$\times$	$R/r \ \ P/p$	
	walnut		walnut	
		$\downarrow$		
F_2	$9 \ R/- \ P/-$	$3 \ R/- \ p/p$	$3 \ r/r \ P/-$	$1 \ r/r \ p/p$
	walnut	rose	pea	single

Notice that a cross of the more usual sort—between a homozygous dominant walnut, $R/R \ P/P$, and a homozygous recessive single, $r/r \ p/p$—would yield the same F_1 and F_2 pattern. Furthermore, the $9:3:3:1$ ratio obtained in either case is the same that we observe in conventional dihybrid crosses—Mendel's cross between peas that produced round yellow seeds and peas that produced wrinkled green seeds, for example. The difference, then, is descriptive: when collaboration occurs, it is not possible to identify the action of each gene on the basis of its phenotypic effects in the clear-cut way Mendel could with a cross involving seed shape and color.

Modifier genes Probably no inherited characteristic is controlled exclusively by one gene pair. Even when only one principal gene is involved, its expression is influenced to some extent by countless other genes with individual effects often so slight that they are very difficult to locate and analyze. An example is eye color in human beings.

Human eye color is largely controlled by one gene with two alleles—a dominant allele, B, for brown eyes, and a recessive allele, b, for blue eyes. Brown-eyed people (B/B or B/b) have branching pigment cells containing melanin in the front layer of the iris. Blue-eyed people (b/b) lack melanin in

Single

Pea

Walnut

Rose

24.8 Comb types in chickens

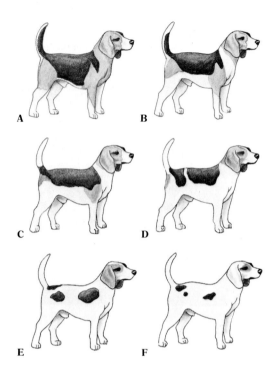

24.9 Variation in spotting of beagle dogs as a result of the action of many modifier genes

the front layer; the blue is an effect of the black pigment on the back of the iris as faintly seen through the semiopaque front layer.

This description of the inheritance of eye color on the basis of a single-gene system assumes only two phenotypes, brown and blue. And it is, in fact, possible to assign most people to one or the other of these two phenotypic classes. But we all know that eyes exhibit endless variations in hue; everyday terminology recognizes eyes as green, gray (both genetically forms of blue), hazel, and black (both forms of brown), to name the most familiar variations. Obviously, then, an explanation of eye color in terms of a single-gene system is an oversimplification. Many modifier genes are also involved, some affecting the amount of pigment in the iris, some the tone of the pigment (which may be light yellow, dark brown, etc.), some its distribution (even over the whole iris, or in scattered spots, or in a ring around the outer edge, etc.). In fact, in rare cases two blue-eyed people can have a brown-eyed child, because one of them, in whom the lack of pigmentation is a consequence of the action of modifier genes, actually carries the genotype B/b instead of b/b.

Another example of the action of modifier genes is seen in the size of the spots of beagle dogs (Fig. 24.9). A great many modifier genes are involved; no one of them by itself produces marked effects, but in combination they can radically alter the dogs' appearance.

Multiple-gene inheritance   So far we have discussed characteristics that have only a limited number of relatively distinct phenotypes. Pea flowers are either red or white, pea seeds either yellow or green, chicken combs either walnut or rose or pea or single. True, modifiers may blur the boundaries of the classes, as in human eye color or spotting in dogs, but a fairly limited number of major phenotypes—blue vs. brown or spotted vs. unspotted—can be meaningfully discussed. Many traits, however, show much greater phenotypic variation, with less distinct boundaries. Human height, skin pigmentation, and IQ are just three of many possible examples. What can be said about the genetic basis of characteristics such as these?

One explanation of these more complex genetic phenomena is that two or more separate genes can affect the same character in the same way, in an additive fashion. The first clear demonstration of this kind of interaction came in 1909, when the Swedish geneticist Herman Nilsson-Ehle showed that the color of wheat kernels, which can vary from white through various shades of pink and red to a very dark red, results from the interactions of three genes. Each gene has two alleles, a partially dominant allele for red and a partially dominant allele for white. Dark-red kernels are homozygous for the red allele in all three genes, while pure-white kernels are homozygous for the white allele in all three genes. All the phenotypes in between result from different heterozygous mixtures of the alleles (Fig. 24.10).

Nilsson-Ehle's method for showing that three genes are involved deserves closer examination. He began, as did Mendel, by crossing homozygotes, in this case a pure-white and a pure dark-red line, to get medium-red F_1 hybrids. He then crossed the hybrids and obtained F_2 offspring; these fell into seven classes of color, ranging from white to dark red, in the ratio 1:6:15:20:15:6:1. Using a Punnett square (Fig. 24.10), we can see that Nilsson-Ehle's results are what we would predict for a cross involving three

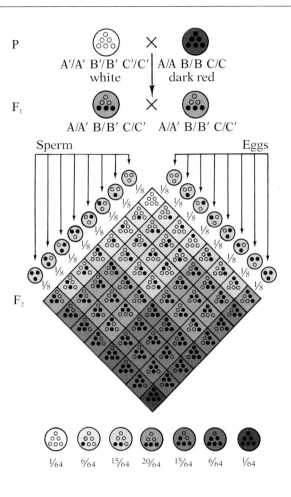

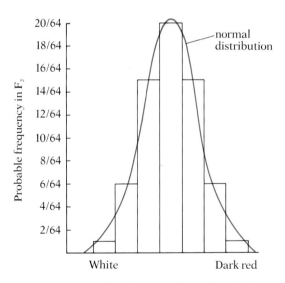

24.10 Nilsson-Ehle's trigenic cross

A line of wheat with pure-white kernels and a line with dark-red kernels were crossed to produce hybrid medium-red F_1 offspring, which were then crossed to produce offspring with the distribution of phenotypes shown here. Nilsson-Ehle concluded from his observations that (1) there are three genes for kernel color in wheat, each with a partially dominant allele for red (A, B, C) and a partially dominant allele for white (A', B', C'); (2) that the genotype of the dark-red line is A/A B/B C/C, while that of the white line is A'/A' B'/B' C'/C'; and (3) that the F_1 hybrids therefore have the genotype A/A' B/B' C/C'. A gradation of phenotypes is seen in the offspring because all the alleles are additive. Eight different kinds of gametes occur because each of the genes for color is on a different chromosome; the exact way this comes about in a three-chromosome species is shown in Figure 23.27 (p. 638). The existence of the seven distinct phenotypes of F_2 offspring in the ratio $1:6:15:20:15:6:1$ is predicted by this Punnett square of a trigenic cross postulating partial dominance in kernel color.

24.11 Frequency distribution of kernel color in wheat

As Nilsson-Ehle showed, a cross of heterozygous wheat kernels results in offspring with seven classes of kernel color, their relative frequencies producing a so-called normal distribution centered around light red.

genes, but not for a cross involving two genes, or four, or any other number. If we graph the probable frequency of colors in the F_2, we obtain a jagged approximation of the bell-shaped curve that represents the so-called normal distribution of most continuously varying traits in a population, traits like skin color, height, IQ, and the like (Fig. 24.11). The greater the number of phenotypes, the less jagged the distribution should appear. In most cases, however, the genes involved in multiple-gene effects do not all contribute equally to the phenotype, and the effects of modifier genes and of the environment also tend to complicate the analysis of the data. As a result, it is usually impossible to determine the number and nature of the genes involved simply by scrutinizing the distribution of phenotypes in the F_2.

PENETRANCE AND EXPRESSIVITY

Because genes interact, an individual can carry a dominant allele that is not expressed phenotypically. Complementary or epistatic genes, for example, or a combination of modifier genes, may prevent the expression of a domi-

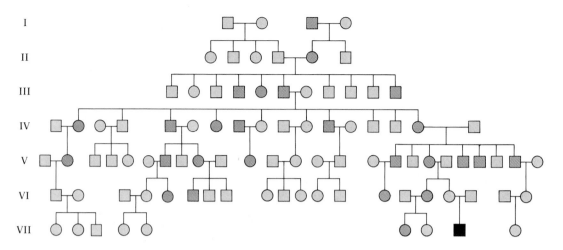

24.12 Pedigree of syndactyly in seven generations of one family

Squares represent males, and circles represent females. Color indicates syndactyly and gray indicates normal phenotype. The character appears to be inherited as a simple autosomal dominant: there is no correlation with sex, and individuals with syndactyly have a parent who also shows this trait. There is one exception: an individual (black square) who has syndactyly even though neither of his parents shows the trait. Presumably the gene was present in his mother without expression.

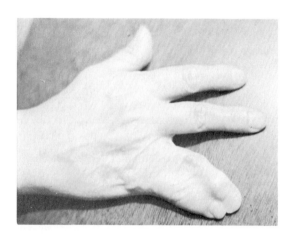

24.13 Photograph showing syndactyly

nant allele. And even when an allele is expressed, the effects of other genes may modify the degree of intensity of expression. We can speak, therefore, of the incomplete penetrance and variable expressivity of certain genes. *Penetrance* refers to the percentage of individuals that, carrying a given gene, actually express that gene's phenotype. *Expressivity* denotes the manner in which the phenotype is expressed.

Incomplete penetrance and variable expressivity are illustrated by a gene in human beings that causes blue sclera, a condition in which the whites of the eyes appear bluish. This gene usually behaves as a simple dominant, and anyone who possesses the gene, whether in heterozygous or homozygous form, might be expected to show the blue-sclera phenotype. In fact, however, only about 9 out of 10 people who have the gene actually show the phenotype. We can say, therefore, that the penetrance of this gene is about 90 percent (or 0.9). Among those that do show the phenotype, the expressivity is variable, with the intensity of the bluish coloration ranging from very pale whitish blue to very dark blackish blue.

Figure 24.12 shows a portion of the pedigree of the Hancock family of Virginia, which for generations has had many members with syndactyly of the ring and little fingers—the two fingers are joined by a web of muscle and skin (Fig. 24.13). This character, like blue sclera, exhibits variable expressivity: a few individuals have three fingers webbed; most have two fingers fully webbed; some have two fingers partially webbed; and a few have a crooked finger with no webbing.

Like blue sclera, too, syndactyly exhibits incomplete penetrance. Of the more than 50 persons in this family known to have had syndactyly, all but one had a parent who also had syndactyly. This pattern of inheritance is very close to the one expected for a dominant character; as a rule, only individuals with a parent showing the character will show the character themselves, because anyone carrying the dominant gene ordinarily shows its phenotype. (By contrast, a recessive character often appears in a person

24.14 Effect of temperature on expression of a gene for coat color in the Himalayan rabbit
(A) Normally, only the feet, tail, ears, and nose are black. (B) Fur is plucked from a patch on the back, and an ice pack is applied to the area. (C) The new fur grown under the artificially low temperatures is black. Himalayan rabbits are normally homozygous for the gene that controls synthesis of the black pigment, but the enzyme encoded by the gene is active only at low temperatures (below about 33°C).

neither of whose parents showed the character, because they were heterozygous.) The pedigree in Figure 24.12 indicates that syndactyly in this family is inherited as a simple dominant, and that the abnormality in the one man having two parents with normal fingers probably indicates incomplete penetrance of the gene for syndactyly: the man's mother must have been carrying the gene, even though she did not exhibit its phenotype.

Both the examples cited make it clear that penetrance and expressivity are aspects of the same phenomenon. Lack of penetrance of the gene for blue sclera produces white color, which is simply one extreme of the expressivity gradient from white through very pale blue to dark blue. Lack of penetrance of the gene for syndactyly is simply one extreme of the expressivity gradient whose other extreme is the full webbing of three fingers.

As we have indicated, incomplete penetrance and variable expressivity may result from peculiarities of the genetic background against which the gene in question must act. This explanation points up a general truth: *The action of any gene can be fully understood only in terms of the overall genetic makeup of the individual organism in which it occurs.*

Penetrance and expressivity are often also affected by the environment. For example, *Drosophila* homozygous for the gene for vestigial wings have wings that are only tiny stumps if they are reared at normal room temperatures (about 20°C), but if they are reared at temperatures as high as 31°C, their wings grow almost as long as normal wings. Himalayan rabbits are normally white with black ears, nose, feet, and tail (Fig. 24.14), but if the fur on a patch on the back is plucked and an ice pack is kept on the patch, the new fur that grows there will be black; the gene for black color can express itself only if the temperature is low, which it normally is only at the body extremities. The same phenomenon underlies the coloring of Siamese cats. It is thought to occur because in the affected animals an enzyme required for pigment synthesis retains its normal active conformation only at low temperatures, and denatures—changes to an inactive conformation—when heated to body temperature. Such a susceptibility to heat might result from an inherited change in the DNA coding for one of the amino acids in the enzyme. The consequent change in the amino acid could reduce the number of internal hydrogen or disulfide bonds in the enzyme, making it unstable and hence inactive at higher temperatures.

We see, then, that the expression of a gene depends both on the other genes present (the genetic environment) and on the physical environment (temperature, sunlight, humidity, diet, etc.). We don't inherit characters.

We inherit only genes, only potentialities; other factors govern whether or not the potentialities are realized. All organisms are inevitable products of both their inheritance and their environment.

THE EVALUATION OF EXPERIMENTAL RESULTS

The meaning of predicted ratios   We have mentioned ratios such as 1:2:1 or 3:1 or 1:1 or 9:3:3:1, expected in the results of various types of crosses. Geneticists frequently use the phenotypic ratios obtained in breeding experiments to deduce the underlying genetic phenomena. And, conversely, geneticists test their hypotheses about gene interactions by predicting the results of novel crosses; if a researcher believes the genotypes for flower color in two lines of peas are *C/c* and *c/c*, he or she can predict that the phenotypic ratio of a cross will be 1:1. Does this mean that exactly equal numbers of red-flowered and white-flowered pea plants must and will be produced by such a cross? No, it simply means that, on the average, the plants produced by this cross ought, if the hypothesis is correct, to have red flowers about as often as white flowers. The same goes for the sex of human children: the chances that any particular new baby will be a boy instead of a girl are roughly fifty-fifty, and we expect that for any large number of young children, such as all those born in a fair-sized city this year, the ratio of boys to girls will closely approximate 1:1. But we are not particularly surprised if any given family has four sons and one daughter or six daughters and one son, any more than if we flip a coin and get heads three times in a row. In short, in any small sample large departures from the predicted results are not unusual or surprising.

Let's examine the basis for the prediction of a 1:1 ratio in the progeny of the presumptive *C/c* × *c/c* cross. As we saw in discussing the test cross, all gametes produced by a homozygous recessive parent will be alike with respect to the gene for flower color; all would bear the *c* allele. It is therefore the heterozygous parent that would determine which of the two possible genotypes (*C/c* or *c/c*) the offspring will have.

The hypothesis is that the diploid cells of the heterozygous parent have one *C* allele and one *c* allele in their nuclei. When these cells undergo meiosis, half the new haploid cells would receive the chromosome bearing the *C* allele and half would receive the homologous chromosome bearing the *c* allele. Therefore we expect that half the gametes produced by this plant carry the *C* allele, and half carry the *c* allele. Which of these two types of gametes would be involved in any given fertilization is, apparently, purely a matter of chance. Consequently the probability that any given gamete from the homozygous recessive parent in this cross would be fertilized by a gamete carrying the *C* allele and produce a red-flowered offspring is 0.5 (which may also be expressed as a percentage or as a fraction: 50 percent or ½). The probability for fertilization by a gamete carrying the *c* allele, which would result in a white-flowered offspring, is likewise 0.5. If 100 fertilizations occur, we would expect, based on our hypothesis concerning the parental genotypes, about 50 red-flowered offspring (0.5 × 100) and about 50 white-flowered offspring. This is the basis for the predicted 1:1 ratio.

The whole matter hinges on the segregation in meiosis of the *C* and *c* alleles of the heterozygous parent and the incorporation of these alleles into gametes that have an equal probability of fertilizing the gametes of the other parent. But since the type of gamete involved in each individual fertilization is purely a matter of chance, it would actually be surprising to get an exact fifty-fifty split in a sample of 100, even if the hypothesis is correct and that is the predicted ratio.

The need for statistical analysis   Now suppose we have made a cross in which the predicted phenotypic results are 1:1. Suppose, further, that our actual results are 45 of one phenotype and 55 of the other. A ratio of 45:55 is fairly close to 1:1, and we might well conclude that our actual results are close enough to the predicted results to justify our assuming that the deviation arose by chance alone. But suppose in another similar cross we get actual results of 5 and 15. This is noticeably further from the predicted 1:1 ratio. Should we conclude that here, too, the deviation can reasonably be attributed to chance, or should we conclude that there may be some genetic explanation for the deviation, and that our original prediction should have been different?

Scientists in all fields of research constantly encounter the same fundamental question—whether the deviations they observe in their experimental results are significant or not. They cannot rely simply on a guess. They cannot just say, "That looks pretty close to what I predicted," or "That looks rather far off, perhaps I overlooked something." To help them arrive at a decision, they can refer to generally agreed upon standards based on the mathematical probability that the observed deviations in their sample could occur by chance alone. This type of mathematical treatment of data is known as statistical analysis.

One point is immediately apparent: in determining statistical significance, the size of the sample is of critical importance. A large deviation from the predicted results is not unusual in small samples, but the same percentage of deviation in a large sample may be very surprising. We said earlier that a family with four sons and one daughter would not be considered unusual, even though the predicted sex ratio is approximately 1:1. But if out of 500 children born in a city in a particular year, 400 were boys and only 100 were girls, we would strongly suspect something amiss. When the predicted ratio is 1:1, an observed ratio of 4:1 is not alarming in a sample of only 5, but in a sample of 500 it is cause for reexamination. Or, to take another example, if we toss a coin 10 times, we can logically predict 5 heads and 5 tails, but we are not greatly surprised if in fact we get 7 heads and 3 tails (a ratio of 7:3 instead of 1:1). But if we toss the coin 100 times and get 70 heads and only 30 tails (again a 7:3 ratio), we suspect that something is wrong either with the coin or with the way it is being tossed. In other words, a ratio of 7:3 seems reasonable enough when the sample is small, but unreasonable when the sample is large. Clearly, then, tests of significance must take into account both the amount of deviation and the sample size. Statisticians have devised many mathematical tests for evaluating experimental or observational data. Though these tests differ in their form and in the sorts of data to which they can validly be applied, all are simply ways of calculating the probability that the deviations of the observed values are due to chance alone.

THE CHI-SQUARE TEST

One test of statistical significance was devised by Karl Pearson of the University of London in 1900; it represented a fundamental breakthrough for evaluating the results of experimental science. Pearson's so-called chi-square (χ^2) test is particularly applicable to many genetic experiments. This test measures whether any deviation from the predicted norm that occurs in experimental results exceeds the deviation that might occur by chance. The formula for chi-square is

$$\chi^2 = \Sigma(d^2/e)$$

where d is the deviation from the expected value, e is the expected value, and Σ means "the sum of."

Let us return to the two hypothetical crosses discussed earlier in which we expected that the phenotypic ratio would be $1:1$. In one cross we actually got values of 45 and 55 instead of 50 and 50, and in the other we got values of 5 and 15 instead of 10 and 10. We want to know in each case whether the deviation of the observed from the expected values can reasonably be attributed to chance or whether we must look for some other explanation. Let us first determine the chi-square value for the $45:55$ experiment:

	First phenotype	Second phenotype
Observed values	45	55
Expected values (e)	50	50
Deviation (d)	−5	+5
Deviation squared (d^2)	25	25
d^2/e	$25/50 = 0.5$	$25/50 = 0.5$
$\chi^2 = \Sigma(d^2/e)$		$= 0.5 + 0.5 = \textbf{1.0}$

Next let us determine the chi-square value for the $5:15$ experiment, following the same procedure:

	First phenotype	Second phenotype
Observed values	5	15
Expected values (e)	10	10
Deviation (d)	−5	+5
Deviation squared (d^2)	25	25
d^2/e	$25/10 = 2.5$	$25/10 = 2.5$
$\chi^2 = \Sigma(d^2/e)$		$= 2.5 + 2.5 = \textbf{5.0}$

Notice that in each of these experiments the absolute deviations of the observed values from the expected values are the same: a deviation of 5 in each phenotype. But notice also that the chi-squares obtained in the two experiments are very different—the one for the experiment based on a sample of 20 being five times as large as the one for the experiment based on a sample of 100. This illustrates well how sensitive chi-square is to sample size: the difference in sample size alone has made the great difference in the two chi-square values.

Each of these experiments involves only two classes, in this case two different phenotypes. Hence their chi-square values were calculated on the basis of only two squared deviations. But suppose we had been analyzing an experiment involving three different phenotypes. Then the chi-square would have been calculated on the basis of three squared deviations, and it is only reasonable to expect that the chi-square value obtained would have been higher than one based on only two. It is clear, then, that in evaluating chi-square values we must also take into account the number of classes on which they are based. By convention, the number of independent classes in a chi-square test is termed the degree of freedom. The number of independent classes is usually one less than the total number of classes in the experiment. Thus, in our experiments involving two phenotypes, there is only one independent class and one degree of freedom, while in an experiment involving three phenotypes there would be two independent classes and two degrees of freedom. A moment's thought will tell you why this is so. In our experiment based on a sample of 100, once we know that 45 offspring show the first phenotype, we automatically know that 55 must show the other phenotype. Since we know the total, the number in one class automatically tells us the number in the other class. In other words, the number in the second class is dependent upon the number in the first class. Therefore, only the first class is an independent class. The same reasoning applies if we perform an experiment involving three different phenotypes, and the total number of observations in our sample is 100; once we know the number showing the first and second phenotypes, we automatically know the number showing the third phenotype, because the number in the third class is dependent upon the number in the first two classes.

We now know the chi-square values (1.0 and 5.0) and the degrees of freedom (one for each experiment) for our two hypothetical experiments. The next step is to consult a table of chi-square values such as the one shown here. This table

Probabilities for certain values of chi-square[a]

Degrees of freedom	$P = 0.20$ (1 in 5)	$P = 0.10$ (1 in 10)	$P = 0.05$ (1 in 20)	$P = 0.01$ (1 in 100)	$P = 0.001$ (1 in 1,000)
1	1.64	2.71	3.84	6.64	10.83
2	3.22	4.60	5.99	9.21	13.82
3	4.64	6.25	7.82	11.34	16.27
4	5.99	7.78	9.49	13.28	18.46
5	7.29	9.24	11.07	15.09	20.52
6	8.56	10.64	12.59	16.81	22.46
7	9.80	12.02	14.07	18.48	24.32
8	11.03	13.36	15.51	20.09	26.12
9	12.24	14.68	16.92	21.67	27.88
10	13.44	15.99	18.31	23.21	29.59
15	19.31	22.31	25.00	30.58	37.70
20	25.04	28.41	31.41	37.57	45.32
30	36.25	40.26	43.77	50.89	59.70

[a] Based on a larger table in R. A. Fisher, *Statistical Methods for Research Workers*, 10th ed., Oliver & Boyd, 1946.

gives five different chi-square values for each of a series of different degrees of freedom, and gives the probability (P) that an amount of deviation as great as or greater than that represented by each chi-square value would occur simply by chance.

Now let us evaluate the results obtained in the first of our hypothetical experiments. Here the deviation of our results from those expected was such as to yield a chi-square value of 1.0. The experiment had one degree of freedom. According to the table, a value as high as or higher than 1.64 has a chance probability of 0.20 (20 percent); that is, deviation from the expected as great as or greater than that represented by 1.64 will occur about once in five trials by chance alone. Our chi-square is less than 1.64; hence the amount of deviation in the experiment can be expected to occur by chance even more often than once in five trials. Most biologists agree that deviations having a chance probability as great as or greater than 0.05 (5 percent, or 1 in 20) will not be considered statistically significant. Since the deviation in our experiment has a chance probability much greater than 5 percent, it is not regarded as statistically significant, and is presumed to be a chance deviation, which can be disregarded.

In our second experiment, the chi-square value representing the deviation from the expected results turned out to be 5.0. Again there was one degree of freedom. Looking at the listings in the table for one degree of freedom, we find that the value of 5.0 is greater than 3.84,

which has a probability of 0.05 (5 percent), but less than 6.64, which has a probability of 0.01 (1 percent). Hence the probability that the deviation in this experiment resulted purely from chance is less than 5 percent but greater than 1 percent.

According to biological convention, then, the deviation from the expected results in our second experiment is significant. A geneticist would suspect that some factor other than chance was involved in producing the disagreement between the results and the predictions, and would begin the search for a reasonable explanation; perhaps the observations were at fault, or perhaps an error was made in carrying out the experiment, or perhaps the assumptions concerning the genetics involved in this cross need modification so that we can make better predictions. In other words, we don't immediately abandon the hypothesis just because results and predictions in this one experiment do not agree, but we do become suspicious of the hypothesis and perform other experiments designed to test it further. In this particular case, one of the first things to do is to perform a similar experiment using a larger sample to minimize chance error. After all, as we saw when we calculated the outcome of a dihybrid cross, the probability of two events happening together is the product of their individual probabilities of happening alone. The probability of this deviation occurring twice by chance is 0.05×0.05, or only 0.25 percent.

GENETICS PROBLEMS

The best way to gain an understanding of genetics is to work with it. The fundamental principles discussed above will become clearer to you, and you will grasp them more surely, if you carefully think through the following problems, which illustrate the various patterns of inheritance treated in this chapter. Additional problems will be found in the next chapter, in the Study Guide accompanying this book, and in the genetics textbooks listed at the end of this book as readings for this and the next chapter.

1. In squash an allele for white color (W) is dominant over the allele for yellow color (w). Give the genotypic and phenotypic ratios for the results of each of the following crosses:

$$W/W \quad \times \quad w/w$$
$$W/w \quad \times \quad w/w$$
$$W/w \quad \times \quad W/w$$

2. A heterozygous white-fruited squash plant is crossed with a yellow-fruited plant, yielding 200 seeds. Of these, 110 produce white-fruited plants, while only 90 produce yellow-fruited plants. Using the chi-square test, would you conclude that this deviation is the result of chance, or that it probably represents some complicating factor? What if there were 2,000 seeds, and 1,100 produced white-fruited plants while 900 produced yellow-fruited individuals?

3. In human beings, brown eyes are dominant over blue eyes. Suppose a blue-eyed man marries a brown-eyed woman whose father was blue-eyed. What proportion of their children would you predict will have blue eyes?

4. If a brown-eyed man marries a blue-eyed woman and they have ten children, all brown-eyed, can you be certain that the man is homozygous? If the eleventh child has brown eyes, what will that show about the father's genotype?

5. A brown-eyed man whose father was brown-eyed and whose mother was blue-eyed married a blue-eyed woman whose father and mother were both brown-eyed. The couple has a blue-eyed son. For which of the individuals mentioned can you be sure of the genotypes? What are their genotypes? What genotypes are possible for the others?

6. The litter resulting from the mating of two short-tailed cats contains three kittens without tails, two with long tails, and six with short tails. What would be the simplest way of explaining the inheritance of tail length in these cats? Show genotypes.

7. When Mexican hairless dogs are crossed with normal-haired dogs, about half the pups are hairless and half have hair. When, however, two Mexican hairless dogs are mated, about a third of the pups produced have hair, about two-thirds are hairless, and some deformed puppies are born dead. Explain these results.

8. In peas an allele for tall plants (T) is dominant over the allele for short plants (t). An allele of another independent gene produces smooth peas (S) and is dominant over the allele for wrinkled peas (s). Calculate both pheno-

typic ratios and genotypic ratios for the results of each of the following crosses:

$$T/t \ \ S/s \quad \times \quad T/t \ \ S/s$$
$$T/t \ \ s/s \quad \times \quad t/t \ \ s/s$$
$$t/t \ \ S/s \quad \times \quad T/t \ \ s/s$$
$$T/T \ \ s/s \quad \times \quad t/t \ \ S/S$$

9. In hogs an allele that produces a white belt around the animal's body is dominant over the allele for a uniformly colored body. An allele of another independent gene produces fusion of the two hoofs on each foot (an instance of syndactyly); it is dominant over the allele that produces normal hoofs. Suppose a uniformly colored hog homozygous for syndactyly is mated with a normal-footed hog homozygous for the belted character. What would be the phenotype of the F_1? If the F_1 individuals are allowed to breed freely among themselves, what genotypic and phenotypic ratios would you predict for the F_2?

10. In watermelons the alleles for green color and for short shape are dominant over the alleles for striped color and for long shape. Suppose a plant with long striped fruit is crossed with a plant heterozygous for both these characters. What phenotypes would this cross produce and in what ratios?

11. In the fruit fly *Drosophila melanogaster*, vestigial wings and hairy body are produced by two recessive alleles located on different chromosomes. The normal alleles, for long wings and hairless body, are dominant. Suppose a vestigial-winged hairy male is crossed with a homozygous normal female. What types of progeny would be expected? If the F_1 from this cross are permitted to mate randomly among themselves, what progeny would be expected in the F_2? Show complete genotypes, phenotypes, and ratios for each generation.

12. Suppose a hairy female *Drosophila* heterozygous for vestigial wing is crossed with a vestigial-winged male heterozygous for the hairy character. What will be the characteristics of the F_1?

13. In some breeds of dogs a dominant allele controls the characteristic of barking while trailing. In these dogs an allele of another independent gene produces erect ears; it is dominant over the allele for drooping ears. Suppose a dog breeder wants to produce a pure-breeding strain of droop-eared barkers, but he knows that the genes for silent trailing and erect ears are present in his kennels. How should he proceed?

14. A dominant allele, A, causes yellow color in rats. The dominant allele of another independent gene, R, produces black coat color. When the two dominants occur together ($A/-\ R/-$), they interact to produce gray. Rats of the genotype $a/a \ \ r/r$ are cream-colored. If a gray male and a yellow female, when mated, produce offspring approximately ⅜ of which are yellow, ⅜ gray, ⅛ cream, and ⅛ black, what are the genotypes of the two parents?

15. What are the genotypes of a yellow male rat and a black female that, when mated, produce 46 gray and 53 yellow offspring? Does the chi-square test bear you out?

16. In Leghorn chickens colored feathers are produced by a dominant allele, C; white feathers are produced by the recessive allele, c. The dominant allele, I, of another independent gene inhibits expression of color in birds with genotypes C/C or C/c. Consequently both $C/-\ I/-$ and $c/c \ -/-$ are white. A colored cock is mated with a white hen and produces many offspring, all colored. Give the genotypes of both parents and offspring.

17. If the dominant allele *K* is necessary for hearing, and the dominant allele *M* of another independent gene results in deafness no matter what other genes are present, what percentage of the offspring produced by the cross *k/k M/m* × *K/k m/m* will be deaf?

18. What fraction of the offspring of parents each with the genotype *K/k L/l M/m* will be *k/k l/l m/m?*

19. Suppose two *D/d E/e F/f G/g H/h* individuals are mated. What would be the predicted frequency of *d/d E/E F/f g/g H/h* offspring from such a mating?

Chapter 25

NON-MENDELIAN PATTERNS OF INHERITANCE

Mendel's exposition of the patterns of inheritance was simple yet revolutionary. In the characters he described, phenotypes are controlled by genes that sort independently during meiosis, and each gene can occur in two forms—dominant and recessive. But we have already seen that in addition to the simple dominance scheme Mendel identified, other kinds of genetic interactions can occur, which modify phenotypic expression. In this chapter we shall examine fundamentally different, non-Mendelian patterns of inheritance; together with Mendel's original insights, these form the modern picture of inheritance.

MULTIPLE ALLELES

As we mentioned earlier, genes may exist in any number of allelic forms. Of course, under normal circumstances, the maximum number of alleles for each gene that any diploid organism can possess is two, because the organism has only two copies of each gene. But many other alleles may be present in the population to which it belongs. Let's look at the effect of multiple alleles on patterns of inheritance.

Eye color in *Drosophila* One of the first examples of multiple alleles was discovered in the tiny fruit fly, *Drosophila melanogaster*—a species whose genetics have been extensively studied. Though normally red-eyed, fruit flies may have eyes of other colors—white, eosin (a brightly fluorescing red), wine, apricot, ivory, or cherry. Each of these eye colors is controlled by a different allele of the same gene; about two dozen such alleles have been discovered, and others may well exist. The allele for the wild-type eye (red) is dominant over all the rest—that is, the normal red pigment masks

TABLE 25.1 *Antigen and antibody content of the blood types of the A-B-O series*

Blood type	Blood contains	
	Cellular antigens	Plasma antibodies
O	None	anti-A and anti-B
A	A	anti-B
B	B	anti-A
AB	A and B	None

TABLE 25.2 *Transfusion relationships of the A-B-O blood groups*

Blood group	Can act as donor to	Can receive blood from
O	O, A, B, AB	O
A	A, AB	O, A
B	B, AB	O, B
AB	AB	O, A, B, AB

the pigment produced by any other allele. When two of the other alleles occur together in a heterozygous fly, however, they produce an intermediate eye-color phenotype.

Human A-B-O blood types   A well-known example of multiple alleles in human beings—and a relatively simple one, since only a few alleles are involved—is that of the A-B-O blood series, in which four blood types are generally recognized: A, B, AB, and O. The erythrocytes in type A blood bear antigen A on their surface; in type B, antigen B; in type O, neither A nor B; in type AB, both A and B.

An antigen, as you may recall, is a chemical capable of triggering an immune reaction by which antibodies are produced that bind to and help destroy that particular antigen and the cell that bears it. We shall examine the workings of the immune system in detail in a later chapter; right now, to understand blood groups, we must know three things: first, every cell has surface proteins, many of which can serve as antigens; second, the surface proteins of different kinds of cells are different, and even the surface proteins of one particular kind of cell usually differ somewhat in unrelated individuals; and finally, very early in life, the immune system of each individual becomes insensitive to the surface proteins of that individual's cells.

Because the immune system soon becomes insensitive to antigens present from birth, an individual whose red corpuscles bear antigen A has no antibodies—called anti-A—to that antigen. Similarly, an individual with antigen B has no anti-B antibodies. The person with type AB blood, therefore, having corpuscles that bear both A and B antigens, will have neither anti-A nor anti-B in the blood plasma; the person with type O blood, on the other hand, having corpuscles that bear neither antigen, will have both anti-A and anti-B in the blood plasma (see Table 25.1). In short, a person's plasma contains antibodies corresponding to any antigens his own corpuscles do not bear. This is one of the few cases in which the body normally synthesizes antibody against an antigen to which it has not actually been exposed.

The presence of these antigens and antibodies in the blood has important implications for blood transfusions. Because the antibodies present in the plasma of blood of one type tend to react with the antigens on the erythrocytes of other blood types and cause clumping, it is always best, when transfusions are to be given, to obtain a donor who has the same blood type as the patient. When such a donor is not available blood of another type may be used, provided that the *plasma of the patient* and the *erythrocytes of the donor* are compatible; in other words, doctors can usually ignore the erythrocytes of the patient and the plasma of the donor. The reason is that, unless the transfusion is to be a massive one or is to be made very rapidly, the donor's plasma is sufficiently diluted during transfusion so that little or no agglutination occurs. This means that type O blood can be given to anyone, because its erythrocytes have no antigens and hence are obviously compatible with the plasma of any patient; type O blood is sometimes called the universal donor. But type O patients can receive transfusions only from type O donors, because their plasma contains both anti-A and anti-B and hence is obviously not compatible with the erythrocytes of any other class of donor. Luckily for those of us with type O blood, it is the most common variety in most parts of the world. Conversely, people with the rare type AB blood, whose plasma contains no anti-A or anti-B antibodies, are universal recipients, but cannot act as donors for any except type AB patients. Table 25.2 summarizes these transfusion relationships.

At first glance, you might suppose that two independent genes are involved in the A-B-O system, one determining whether the A antigen is present and another whether the B antigen is present. But actually the inheritance of the A-B-O groups is for the most part controlled by three alleles of the same gene, here designated I^A, I^B, and i. Both I^A and I^B are dominant over i, but neither I^A nor I^B is dominant over the other. Accordingly the four blood-type phenotypes correspond to the genotypes indicated in Table 25.3. From our discussion of the molecular basis of dominance in the preceding chapter, you can probably guess that I^A and I^B are alleles that code for different functional proteins, while i codes for a nonfunctional protein.

Blood typing is often used as a source of evidence in paternity cases in court. For example, a man with type O blood could not possibly be the father of a child with type A blood whose mother is type B. The child's true father must be either type A or type AB, because the child must have received its I^A allele from its father; an O man has no such allele. Similarly, a man with type AB blood could not possibly be the father of a type O child, because the child must have received an i allele from its father, but an AB man has no such allele. Of course, blood-type analysis can only determine who could *not* be the father.

As indicated in Table 25.4, the frequencies of the various A-B-O blood types vary in populations of different ancestral extraction. Anthropologists have found data on the frequencies of blood types useful in tracing the prehistoric movements and derivations of the various subgroups of the human species. Since the most frequent phenotype in most human populations is type O, which corresponds to the homozygous recessive phenotype, the i allele is more common than the I^A or I^B alleles. Here we have a good illustration of an important fact: whether an allele is dominant or recessive does not determine whether it will be common or rare in the population. Many people have the mistaken impression that dominant alleles are always the common ones and recessive alleles the rarer ones. "Dominant" and "recessive" describe the way the alleles interact when they occur together in a heterozygous individual; they do not indicate which allele determines the more advantageous phenotype. Natural selection tends to increase the frequency of the allele that determines the more adaptive phenotype, whether that allele is dominant or recessive, and it tends to decrease the frequency of the allele that determines the less adaptive phenotype; it is the relative adaptiveness of the phenotype that determines which allele is the more common.

Rh factors The A and B antigens are not the only surface proteins on red blood corpuscles. You have probably heard of Rh factors, the antigens produced by alleles of the Rh gene. Individuals whose two copies of the Rh gene code for nonfunctional products are said to be Rh-negative (Rh⁻); theirs is a situation analogous to that of type O individuals, in whom both copies of the blood-type gene code for nonfunctional surface protein. Indi-

TABLE 25.3 *Genotypes of the A-B-O blood types*

Blood type	Genotype
O	i/i
A	I^A/I^A or I^A/i
B	I^B/I^B or I^B/i
AB	I^A/I^B

TABLE 25.4 *Frequencies of A-B-O blood groups in selected populations*

Population	O	A	B	AB
United States whites	45%	41%	10%	4%
United States blacks	47	28	20	5
African Pygmies	31	30	29	10
African Bushmen	56	34	8	2
Australian aborigines	34	66	0	0
Pure Peruvian Indians	100	0	0	0
Tuamotuans of Polynesia	48	52	0	0

viduals with at least one functional Rh allele (and there are many, including eight common ones) are said to be Rh⁺. Another gene for surface proteins on red corpuscles has two alleles, *M* and *N*, giving rise to the genotypes *M/M*, *M/N*, and *N/N*. Each of these groups of blood antigens can create immunological problems during blood transfusions.

25.1   A creeper hen
Her legs are very short and she cannot walk normally. Such a hen is heterozygous for an allele that is lethal when homozygous.

MUTATIONS AND DELETERIOUS ALLELES

A variety of influences can cause slight changes in the chemical structure of a gene. Such changes are called ***mutations.*** For reasons that will become clear in the next chapter, the rate at which any particular gene undergoes mutation is ordinarily extremely low. But every individual organism has a very large number of different genes, and the total number of genes in all the individuals of a species is vast indeed. Hence mutations are constantly occurring within a species; pure chance determines in which individual any given mutation will occur.

Now, every living organism is the product of billions of years of evolution and is a finely tuned, smoothly running, astoundingly intricate mechanism, in which the function of every part in some way influences the function of every other part. By comparison, the best Swiss watch is simple indeed. If you were to take such a watch, remove its back, and make some random change in its parts, the chances are very great that you would make it run worse rather than better. A random change in any delicate and intricate mechanism is far more likely to damage it than to improve it. Genes are somewhat different in that most mutations do not greatly alter the gene product and therefore have little or no phenotypic effect. However, since mutational changes in genes occur at random, it is easy to understand why the vast majority of the mutations that do have obvious phenotypic effects are deleterious. Only very rarely is a mutation beneficial.

Heterozygous vs. homozygous effects   When a deleterious allele arises by mutation, natural selection can act against it only if it causes some change in the organism's characteristics. Selection acts directly on phenotypes and only indirectly on genotypes. Because dominant deleterious mutations will be expressed phenotypically, they can be eliminated from the population rapidly by natural selection, and most mutations that persist in a population will therefore be recessive to the normal alleles. Recessive mutations are also relatively abundant because new gene products are likely to be less active than normal ones. Since the probability of the same mutation's occurring twice in the same diploid individual is vanishingly slight, most new alleles appear in combination with a normal, usually dominant allele. The individual is then heterozygous for that particular trait, with the new mutant allele generally masked, and its deleterious effects not fully expressed. As a result, natural selection cannot eliminate it from the population very rapidly. Deleterious alleles that are not dominant may be retained in the population in heterozygous condition for a long time. Clearly, organisms that are diploid throughout their lives are much less sensitive to mutation than those that have extended haploid stages.

When a mating occurs between two diploid individuals carrying the same deleterious recessive allele in heterozygous condition, about one-fourth of the progeny will be homozygous for the deleterious allele, and these homozygous offspring will have the harmful phenotype. The phenotype may even kill the organism. An allele whose phenotype, when expressed, results in the death of the organism is called a ***lethal.*** The occurrence of lethals can modify the phenotypic ratios obtained in the progeny of some crosses, as the following example shows:

In chickens one allele of a certain gene, when it occurs in heterozygous condition with the normal allele, causes the chicken to be a "creeper," with short crooked legs (Figure 25.1). When two creeper chickens are crossed, their offspring are of two phenotypes, normal and creeper, in a ratio of approximately 1 : 2. This ratio, which is different from any we have previously encountered, occurs because about one-fourth of the incubated eggs fail to hatch. If the embryos that died before hatching are regarded as a third phenotypic class, the cross can be said to have produced a phenotypic ratio of 1 : 2 : 1, the typical one for the F_2 generation of a monohybrid cross where dominance is lacking. The ratio of 1 : 2 seen in the live chicks is the result of the lethality of the creeper allele when it occurs in homozygous condition.

In numerous instances alleles harmful or even lethal when homozygous are actually beneficial when heterozygous. For example, in England there is a breed of cattle called Dexter, a good beef producer, for which it is impossible to establish a pure-breeding herd because some of its most desirable characteristics are caused by the heterozygous expression of an allele that is lethal when homozygous.

An example in human beings is the allele for ***sickle-cell anemia,*** an ailment we encountered in Chapter 13. In an individual homozygous for the sickle-cell allele, a serious abnormality of the red blood corpuscles results: the corpuscles are curved like a sickle and bear long filamentous processes (Fig. 25.2). These abnormal corpuscles tend to form clumps and to clog the smaller blood vessels. The resulting impairment of the circulation leads to severe pains in the abdomen, back, head, and extremities, and to enlargement of the heart and atrophy of brain cells. In addition, the tendency of the deformed corpuscles to rupture easily brings about severe anemia. As might be expected, victims of sickle-cell anemia usually suffer an early death. Individuals heterozygous for the sickle-cell allele sometimes show mild symptoms of the disease, but the condition is usually not serious.

You might suppose that natural selection would operate against the propagation of any allele so obviously harmful and that such an allele would be held at very low frequency in the population. But the allele is surprisingly common in many parts of Africa, being carried by as much as 20 percent of the black population. What is the explanation? A. C. Allison of Oxford showed that individuals heterozygous for this allele have an unusually high resistance to malaria. Since malaria is very common in many parts of Africa, the sickle-cell allele must be regarded as beneficial when heterozygous. Hence, in Africa there is selection for the allele because of its heterozygous effect on malarial resistance and selection against it because of its homozygous production of sickle-cell anemia. In other parts of the world, however, malaria is infrequent, and the benefits of the sickle-cell allele are outweighed by its costs. The balance between the two opposing selection pressures determines the frequency of the allele in any population, and it

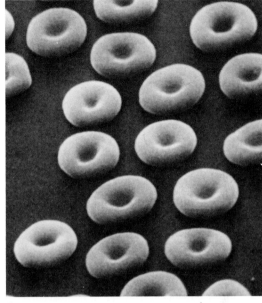

A   5 μm

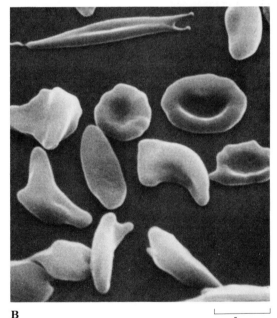

B   5 μm

25.2   Scanning electron micrographs of normal and sickled erythrocytes

Normal erythrocytes (A), which are biconcave discs, look dramatically different from sickled cells (B). Some of the sickled cells seen here bear the filamentous processes which may cause clogging of the body's smaller blood vessels.

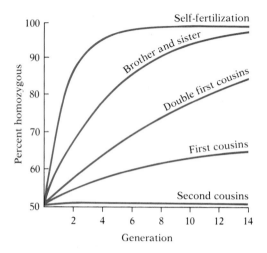

25.3   Graph showing percentage of homozygotes in successive generations under different degrees of inbreeding
It is assumed in this graph that the initial condition is two alleles of equal frequency (hence the 50 percent homozygosity at the start). If a similar graph were drawn for very rare recessive alleles, the rises in homozygosity would be much steeper, particularly in the three curves for cousin-to-cousin matings. Thus matings between first or second cousins, though of little effect in the situation graphed here, may greatly increase the percentage of homozygosity of rare, perhaps deleterious recessive alleles. (Double first cousins result when siblings of one family marry siblings of another family—when two brothers marry two sisters, for example; the offspring of such marriages are first cousins through *both* of their parents rather than through just one of them.)

comes as no surprise that the frequency of the sickle-cell allele in the descendants of African blacks living in the United States has steadily declined.

The case of sickle-cell anemia is a dramatic example of an allele that has more than one effect. Such an allele is said to be **pleiotropic.** Pleiotropy is, in fact, the rule rather than the exception. All genes probably have many effects on the organism. Even when a gene produces only one perceptible phenotypic effect, it doubtless has numerous physiological effects more difficult to detect. For example, the temperature-sensitive coat-color allele of Siamese cats mentioned in the preceding chapter also causes a mysterious misrouting during development of the axons carrying visual information from the eye to the brain. Many Siamese cats are cross-eyed, thus compensating for their visual miswiring.

The effect of inbreeding   The conditions under which genes cause deleterious phenotypes explain the danger of matings between closely related human beings. Everyone probably carries in heterozygous combination many alleles that would cause harmful effects if present in homozygous combination, including some lethals. But because most of these deleterious alleles originated as rare mutations, and are limited to a tiny percentage of the population, the chances are slight that two unrelated persons will be carrying the same deleterious recessive alleles and produce homozygous offspring that show the harmful phenotype. The chances are much greater that two closely related persons will be carrying the same harmful recessives, having received them from common ancestors, and that, if they mate, they will have children homozygous for the deleterious traits. In short, close inbreeding increases the percentage of homozygosity, as Figure 25.3 shows. You can see from the graph that brother-sister matings and matings between double first cousins cause rapid increases in homozygosity, and that matings between first cousins cause slight increases. Many species have evolved specific behavioral or (particularly in plants) physiological mechanisms to prevent close relatives from mating.

Diploidy vs. haploidy   Before leaving this discussion of mutation, harmful alleles, and the implications of homozygosity vs. heterozygosity, we might speculate for a moment on why the diploid stages of the life cycle have such marked dominance over the haploid stages in most higher plants and animals. Why should natural selection have favored diploidy rather than haploidy? One possible reason, mentioned earlier, is that new harmful mutations immediately exert their effect in a haploid organism; they cannot be masked by a dominant allele. But in diploid species an organism can survive a harmful mutation if the new allele is recessive. Moreover, the mutant allele can be carried for generations in the population, perhaps exerting a beneficial effect when heterozygous. Such an allele may also serve as a latent source of variation; the day may come when the environment or the genetic makeup of the organism has changed so much that the allele is no longer deleterious even when homozygous. In a later chapter we shall explore some of the elaborate strategies by which eucaryotes can benefit from this potential for tolerating latent variation.

SEX AND INHERITANCE

SEX DETERMINATION

The sex chromosomes We have said repeatedly that a diploid individual has two of each type of chromosome, identical in size and shape, and hence two copies of each gene in every cell. We must now qualify that statement: in most higher organisms where the sexes are separate (that is, where males and females are separate individuals), the chromosomal endowments of males and females are slightly different. In general, one of the two sexes has one pair of chromosomes in which the members differ markedly from each other in size and shape. These are the *sex chromosomes,* which play a fundamental role in determining the sex of the individual; they exhibit their sex-specific effects primarily in the brain, certain endocrine organs, cells in sexually dimorphic portions of the external and internal anatomy, and the gonads. All other chromosomes are called *autosomes.*

Let's look first at the chromosomes of *Drosophila* and of human beings. In each case the sex chromosomes are of two sorts: one, bearing many genes, is conventionally designated the *X chromosome,* and one of a different shape and bearing only a few genes is designated the *Y chromosome.* In both humans and fruit flies normal females have two X chromosomes, while males have one X and one Y. The diploid number in *Drosophila* is eight (four pairs); a female therefore has three pairs of autosomes and one pair of X chromosomes, but a male has three pairs of autosomes and a pair of sex chromosomes consisting of one X and one Y (Fig. 25.4). The diploid number in human beings is 46 (23 pairs); a female therefore has 22 pairs of autosomes and one pair of X chromosomes, while a male has 22 pairs of autosomes plus one X and one Y (see Fig. 23.7, p. 619).

When a female produces egg cells by meiosis, all the eggs receive one of each type of autosome plus one X chromosome. When a male produces sperm cells by meiosis, half the sperm cells receive one of each type of autosome plus one X chromosome, while the other half receive one of each autosome plus one Y chromosome. In short, all the egg cells are alike in chromosomal content, but the sperm cells are of two types occurring in equal numbers (Fig. 25.4). When fertilization takes place, the chances are approximately equal that the egg will be fertilized by a sperm carrying an X chromosome or by a sperm carrying a Y chromosome. If fertilization is by an X-bearing sperm, the resulting zygote will be XX and will develop into a female. If fertilization is by a Y-bearing sperm, the resulting zygote will be XY and will develop into a male. We see, therefore, that the sex of an individual is normally determined at the moment of fertilization and depends on which of the two types of sperm fertilizes the egg.

This XY-male system, though characteristic of many plants and animals (including all mammals), is by no means universal. Birds, butterflies and moths, and a few other animals have just the opposite system, where XX is male and XY is female. (To distinguish this from the usual XY system, the symbols Z and W are often substituted, ZZ being male and ZW female.) In still other species—many reptiles and marine fish, for example—sex is environmentally determined: the sex of alligators depends on the tempera-

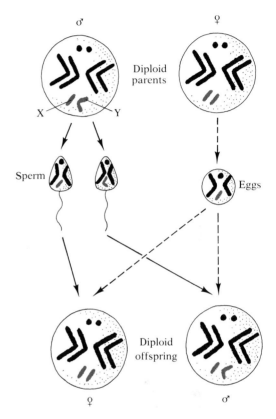

25.4 Chromosomes of male and female *Drosophila melanogaster*
There are three pairs of autosomes and one pair of sex chromosomes. Males (symbolized by ♂) have one X chromosome and one Y chromosome; since these separate at meiosis, half the sperm carry an X and half carry a Y. Since females (♀) have two X chromosomes, all eggs have an X. The sex of the offspring depends on which type of sperm fertilizes the egg.

ture of the eggs during development, while some coral-reef fish reverse sex as adults to take advantage of variations in food and competition. And as we saw in Chapter 22, an entirely different system exists in the Hymenoptera (ants, bees, and wasps), with the males haploid (having developed from unfertilized eggs) and the females diploid. The discussion that follows will focus primarily on the XY-male system of our species.

The role of the Y chromosome in sex determination Occasionally the members of a homologous pair of chromosomes fail to separate properly in meiosis and both move to the same pole. The effect of such "nondisjunction" is that one of the daughter cells receives one too many chromosomes and the other daughter cell receives one too few. If a human gamete carrying an extra chromosome is involved in fertilization, the zygote produced has 47 chromosomes (the normal two of most types plus three of one type) instead of the normal 46. Sometimes it is the sex chromosomes that fail to separate in meiosis, and XXY individuals may result. Since XXY individuals in *Drosophila* are essentially normal females, it seemed reasonable to suppose that only the X chromosomes function in sex determination, with one producing a male and two a female, regardless of the presence or absence of a Y chromosome. And indeed in *Drosophila*, sex is determined by the ratio of certain gene products produced by the X chromosomes to certain products produced by the autosomes. A low ratio results in a male, while a high ratio produces a female. The Y chromosome in *Drosophila* does not determine the sex of the offspring, though it does contain one or more genes essential for normal male fertility.

The idea that the number of X chromosomes is the one crucial variable for determining sex in most animals was also consistent with the observation that grasshoppers and certain other animals lack the Y chromosome entirely—females have two X chromosomes and males a single unpaired X. As a result, females have one more chromosome than males. Such a system of sex determination is known as the XO system, with O designating the absence of a chromosome.

We now know that sex determination of this kind, in which the Y chromosome is virtually irrelevant, is not universal. The human Y chromosome, for example, bears genes with strong male-determining properties, and it is the presence of the Y that determines maleness and its absence that determines femaleness. Whereas an XXY *Drosophila* is a female, an XXY human is a male.

SEX-LINKED CHARACTERS

Many genes occur on the X chromosome and not on the Y chromosome. Such genes are said to be sex-linked. The inheritance patterns for the characteristics controlled by sex-linked genes are completely different from those for characteristics controlled by autosomal genes, for obvious reasons. Females have two copies of each sex-linked gene, one from each parent, but males have only one copy of each sex-linked gene, and that one copy always comes from the mother, since the father contributes a Y chromosome instead of an X to his sons. Hence, in the male, all sex-linked characteristics are inherited from the mother only. And since the male has only one copy of each sex-linked gene, recessive alleles cannot be masked, and recessive sex-linked phenotypes, such as common baldness, occur much more often in males than in females.

Sex linkage was discovered in 1910 by the great American geneticist Thomas Hunt Morgan of Columbia University. It was Morgan who began the systematic use of *Drosophila* in genetic studies. This little fruit fly made it possible to perform in a few months experiments that would have taken Mendel years to perform on peas. *Drosophila* can be easily and economically cultured in large numbers in the laboratory. They can produce a new generation every 10 or 12 days, and are subject to a remarkable number of easily detectable genetic variations. Most of the modern knowledge of eucaryotic genetics derives from work on this tiny insect.

The first sex-linked trait discovered by Morgan was white eye color in *Drosophila*. This mutation arose spontaneously from true-breeding red-eyed stock; it is controlled by a recessive allele *r*. The normal red eye color is controlled by a dominant allele *R*. If homozygous red-eyed females are crossed with white-eyed males, all the F_1 offspring, regardless of sex, have red eyes, since they receive from their mother an X chromosome bearing an allele for red. In addition, the F_1 females receive from their father an X chromosome bearing an allele for white eyes, but the allele for red, being dominant, masks its presence. The F_1 males, like the females, receive from their mother an X chromosome bearing an allele for red eyes. But unlike the females, they receive no gene for eye color from their father, who contributes a Y chromosome instead of an X (in writing the genotype of a male for a sex-linked character, the Y is customarily shown, in order to indicate clearly that no second X chromosome is present and hence there is no second copy of the sex-linked gene). We can summarize this cross as follows (♀ denotes females, ♂ males):

P R/R × r/Y
         red-eyed ♀' white-eyed ♂
             ↓

F_1 R/r × R/Y
        red-eyed ♀   red-eyed ♂
           ↓

F_2 R/R r/R R/Y r/Y
    red-   red-   red-   white-
    eyed ♀   eyed ♀   eyed ♂   eyed ♂

Notice that when the F_1 flies of this cross are allowed to mate among themselves, the F_2 flies show the customary 3:1 phenotypic ratio of a monohybrid cross where dominance is present. But notice also that this 3:1 ratio is rather different from the 3:1 ratio obtained in a cross involving autosomal genes. In an autosomal cross there is no correlation of phenotype with sex, but in this cross all F_2 individuals showing the recessive phenotype are males. In other words, an autosomal cross gives a 3:1 F_2 ratio for both females and males, but this cross yields females of a single phenotype and males with a 1:1 phenotypic ratio.

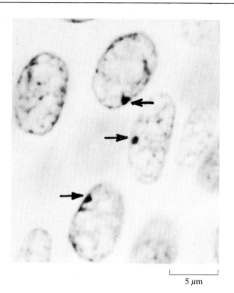

5 μm

25.5   Nuclei from epidermal cells of a human female
The arrows indicate the Barr bodies. Since Barr bodies are present in the cells of female fetuses, the sex of an unborn child can be ascertained by examination of the nuclei of cells sloughed off into the amniotic fluid of the mother's uterus.

Now let us examine the reciprocal cross, where the parental generation consists of homozygous white-eyed females and red-eyed males. We can summarize this cross as follows:

P		r/r	$\times$	R/Y
		white-eyed ♀		red-eyed ♂

↓

F_1		r/R	$\times$	r/Y
		red-eyed ♀		white-eyed ♂

↓

F_2	r/r	R/r	r/Y	R/Y
	white-eyed ♀	red-eyed ♀	white-eyed ♂	red-eyed ♂

Notice that both the F_1 and the F_2 generations differ in phenotypic makeup from a normal autosomal cross and from the reciprocal cross for the same sex-linked trait. In the F_1 all females show the dominant phenotype, all males the recessive phenotype. In the F_2, instead of a 3:1 ratio, there is a 1:1 ratio in each sex. Comparison of the two reciprocal crosses makes it clear that when a sex-linked trait is involved in a cross the results depend on which parent shows the trait (or carries the allele for the trait). By contrast, in crosses involving autosomal genes it does not matter which parent possesses the allele in question; the results of autosomal reciprocal crosses are identical.

Two well-known examples of recessive sex-linked traits in human beings are red-green color blindness and hemophilia. Color blindness occurs in about 8 percent of white males in the United States and in about 4 percent of black males. It occurs in only about 1 percent of white females and about 0.8 percent of black females. It is expected, of course, that more men than women will show such a trait. A man needs only one copy of the allele to show the phenotype, and he can inherit this one copy from a heterozygous mother who is not herself color-blind. But for a woman to be color-blind, she must have two copies of the allele, and so be homozygous; not only must her father be color-blind, but her mother must be either color-blind or a heterozygous carrier of the allele. Since the allele is not very common in the population, it is not likely that two such people will marry; hence the low number of color-blind women.

The statement that females have two copies of each sex-linked gene whereas males have only one, though technically correct, requires qualification. In most interphase somatic cells of females, one of the X chromosomes condenses into a tiny dark object called a **Barr body** (Fig. 25.5). The genes on this condensed X chromosome are inactive. Hence, a normally functioning female cell contains only one active copy of each sex-linked gene. (Even in cells with abnormal numbers of sex chromosomes—XXX or XXXX—only one X chromosome is functional; the others are all condensed into Barr bodies.) Why, then, are sex-linked recessive traits expressed in females only when they are homozygous? And why, if the cells of both males and females contain only one active copy of each sex-linked gene, are the patterns of inheritance in the two sexes markedly different?

The explanation is that it is not always the same X chromosome that con-

25.6   X-chromosome inactivation in cats
One of the most common natural demonstrations of X-chromosome inactivation is the coat color of tortoiseshell (calico) cats. A gene for color found on the cat's X chromosome has two common alleles—black and yellow (or orange). Males can have only one allele, so (in the absence of modifier genes) they are either yellow or black, usually with various white markings. Females, however, can have both alleles, and after inactivation some cells will express the black allele and others the yellow allele. Large patches of cells with one allele or the other develop, and the result is a mosaic of yellow, black, and (usually) white patches. Except for rare XXY individuals, all cats with both yellow and black fur are females.

denses into a Barr body in the different somatic cells of a given individual. Let us call the two X chromosomes X_1 and X_2. Characteristically, about half the cells in any given female show active X_1 chromosomes, the other half active X_2 chromosomes, with no discernible pattern as to which chromosome, X_1 or X_2, is active in which cell. For example, consider the distribution of alleles of the sex-linked gene for the enzyme glucose-6-phosphate dehydrogenase; one of the alleles codes for an active version of the enzyme, the other for a defective version. When the erythrocytes of women heterozygous for these two alleles were examined, roughly half the erythrocytes in each individual revealed normal enzyme activity, the other half a complete lack of it. Apparently female cells differ in their effective genetic makeup as far as sex-linked traits are concerned; women are, in a sense, genetic mosaics for sex-linked traits. For some of these traits, the mosaic pattern finds phenotypic expression; an example is the coat color of tortoiseshell (calico) cats (Fig. 25.6). For others, it does not. It seems, for instance, that as long as half the cells are normal in a woman heterozygous for red-green blindness or for hemophilia, she will not be color-blind or hemophilic.

The phenomenon of X-chromosome inactivation explains why the sex of mammals depends on the presence or absence of the Y chromosome rather than on the ratio of X chromosomes to autosomes, as in *Drosophila:* because of inactivation each mammalian somatic cell has only one functional X chromosome, and so is either XO (female) or XY (male); both sexes have the same number of X-chromosome genes available for transcription. In *Drosophila,* on the other hand, females have twice the number males have. As we have seen, the resulting difference in the amount of certain gene products determines the sex of the individual. But for many other products, various mechanisms have evolved to achieve ***dosage compensation***—to equalize the amount resulting from transcription of the two X chromosomes in females with the amount resulting from transcription of the single

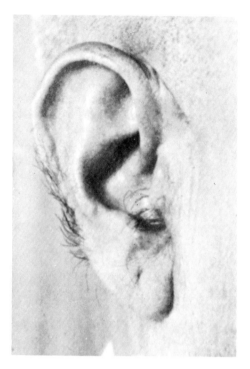

25.7   Hairy pinna
This trait is thought to be determined by a gene on the Y chromosome.

chromosome in males. One of the most common methods of dosage compensation is a feedback system by which the transcription rate is automatically adjusted to match a cell's requirements. On average, then, genes on the single X chromosome of *Drosophila* males do double duty: they are transcribed twice as often as the corresponding genes on any one female X chromosome.

GENES ON THE Y CHROMOSOME

Genes unique to the Y chromosome are termed ***holandric.*** The phenotypic traits they control appear, of course, only in males (Fig. 25.7). There are apparently very few genes on the human Y chromosome; we have mentioned the maleness determiners, which, probably absent in *Drosophila*, are thought to be on the Y chromosome in humans.

In some species a few genes occur on both the X and the Y chromosome; their inheritance patterns are the same as for autosomal genes. No such genes have been conclusively demonstrated in humans.

SEX-INFLUENCED CHARACTERS

As we have seen, sex-linked genes may control characters not customarily regarded as "sexual." Furthermore, not all genes commonly associated with sex are sex-linked; many genes that control "sexual" characters are located on the autosomes. For example, a number of genes that control growth and development of the sexual organs, such as the penis, the vagina, the uterus, and the oviducts, or that control distribution of body hair, size of breasts, pitch of voice, or other secondary sexual characteristics, are autosomal and are present in individuals of both sexes. That their phenotypic expression is different in the two sexes indicates that they are sex-limited, not that they are sex-linked. Apparently the sex chromosomes determine what hormones are to be synthesized in each sex, and these hormones influence the activity of the sex-limited autosomal genes secondarily, either inhibiting or stimulating them.

LINKAGE

As we saw in the last chapter, Mendel's observations led to two generalizations. The first was that each individual carries two copies of every hereditary factor—every gene—and these copies segregate during gamete formation; this Principle of Segregation is often called Mendel's first law. The second generalization was that when several genes are involved in a cross (as in a dihybrid cross), they sort out into the gametes independently of one another; this Principle of Independent Assortment is frequently referred to as Mendel's second law. All seven of the traits Mendel reported on from garden peas did indeed assort independently, and were free of the complications of modifiers, partial dominance, sex linkage, multiple alleles, and so on. But independent assortment describes only the simplest genetic interactions—the behavior of genes on separate chromosomes. Mendel almost certainly knew that some pairs of factors do not assort independently to produce offspring in neat 9:3:3:1 ratios, and that some do not show a simple pattern of recessiveness and dominance. He probably chose to ignore such anomalies, and the theory he published nevertheless represented a tremendous step forward in understanding inheritance. But these nagging problems—particularly the failure of many crosses to exhibit independent assortment—were well known to geneticists. Not until the discovery and description of chromosomes could such major obstacles to Mendel's theory be resolved.

The chromosomal basis of linkage   Cytology, the study of cells with microscopes, was in its infancy in Mendel's day. Though good microscopes existed, chromosomes had not yet been discovered, because appropriate stains had not yet been developed. A hint that something in the nucleus might be special came in 1869, three years after Mendel's paper, when a young Swiss student named Friedrich Miescher discovered in cell nuclei large amounts of a strange acid he called nuclein. The actual carriers of genetic information were not seen until 1882, when Walther Flemming successfully stained mitotic chromosomes. With the observation and description of mitosis, the stage was set for the reevaluation of Mendel's work. In 1900, shortly after the rediscovery of Mendel's paper, W. S. Sutton of Columbia University pointed out the striking accord between Mendel's conclusion that hereditary factors (genes) occur in pairs in somatic cells, and separate in gametogenesis, and the recent cytological evidence that somatic cells contain two of each kind of chromosome and that these chromosomes segregate in meiosis. Sutton interpreted this agreement as powerful evidence that the chromosomes are the bearers of the genes.

Now that Mendel's ideas could be directly related to observations made with the aid of a microscope, scientists could focus on what appeared to be anomalies. In time it became apparent, for example, that traits assort independently, as specified by Mendel's second law, only when their respective genes occur on two different chromosomes. As noted in Chapter 23, genes that occur on the same chromosome do not assort independently during meiosis (unless separated by crossing over), but rather segregate into the same gamete. Such genes are said to be **linked**. One of the first examples of linkage was reported in 1906 by R. C. Punnett and William Bateson of Cambridge University. They crossed sweet peas that had purple flowers and long pollen with ones that had red flowers and round pollen. All the F_1 plants had purple flowers and long pollen, as expected (it was already known that purple was dominant over red and that long was dominant over round). The F_2 plants from this cross did not show the expected 9:3:3:1 ratio, however, but a highly anomalous one. Next, Bateson and Punnett tried a test cross, crossing the F_1 plants back to homozygous recessive plants (with red flowers and round pollen). Their results were equally anomalous. Using the symbols B for purple, b for red, L for long, and l for round, we can summarize them as follows:

$$BbLl \quad \times \quad bbll$$

purple long		red round	

↓

7 *BbLl*	1 *Bbll*	1 *bbLl*	7 *bbll*
purple long	purple round	red long	red round

You will recall that in a test cross, the phenotypic ratio of the offspring depends on the genotype of the parent showing the dominant phenotype, since the recessive parent produces only one kind of gamete. In this example the homozygous recessive red round parent could produce only *bl* gametes. Hence it was the gametes of the heterozygous purple long parent that must have determined the phenotype of the offspring. According to the Principle of Independent Assortment, this parent should have produced four kinds of gametes *(BL, Bl, bL, and bl)* in equal numbers. When united with the *bl* gametes from the homozygous recessive parents, *BL* gametes should have given rise to purple long offspring, *Bl* gametes to purple round, *bL* to red long, and *bl* to red round, and these four phenotypes should have occurred in equal numbers, in a 1:1:1:1 ratio. But the result Bateson and Punnett actually obtained—a ratio of 7:1:1:7—makes it appear that the heterozygous parent produced far more *BL* and *bl* gametes than *Bl* and *bL* gametes.

Only in 1910 did Thomas Hunt Morgan, who had obtained similar results from *Drosophila* crosses, provide the explanation accepted today. He postulated that the anomalous ratios were caused by linkage. Hence we should write the genotypes of the parents in Bateson and Punnett's test cross *BL/bl* and *bl/bl*, to show, by the positions of the slashes, that *B* and *L* are on one chromosome and *b* and *l* on the other (we would have written these genotypes *B/b L/l* and *b/b l/l* if the genes were not linked).

Now, if in Bateson and Punnett's cross the genes for purple and long and the genes for red and round were linked, we might expect the *BL/bl* parent in the test cross to have produced only two kinds of gametes, *BL* and *bl*, and the test cross to have yielded offspring of only two phenotypes, purple long and red round, in equal numbers. Yet the cross also yielded some purple round and red long offspring. How could the *BL/bl* parent have produced *Bl* and *bL* gametes? Morgan suggested that some mechanism occasionally breaks the original linkages between purple and long and between red and round and establishes in a few individuals new linkages between purple and round and between red and long, making possible the production of *Bl* and *bL* gametes. The mechanism of this recombination is, of course, crossing over (see Fig. 23.23, p. 633). Crossing over increases the number of genetic combinations a cross can produce.

Chromosomal mapping If we assume, as did the geneticist Alfred H. Sturtevant—then an undergraduate working in Morgan's lab—that the probability of breakage is approximately equal at any point along the length of a chromosome, then the greater the distance between two linked genes, the greater the frequency with which they will cross over, because there are more points between them at which a break may occur. Or, to be more precise, the frequency of recombination between any two linked genes will be proportional to the distance between them. Sturtevant postulated that the percentage of recombination can therefore serve as a tool for mapping the location of genes on chromosomes. We speak in terms of the percentage of recombination rather than the number of crossover events because most chromosomes cross over at more than one place, so two crossover events can cancel each other and so go undetected (Fig. 25.8).

The percentage of recombination gives us no information about the ab-

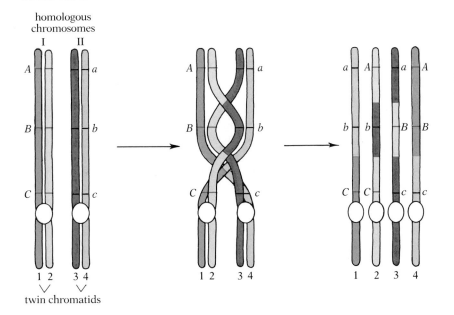

homologous
chromosomes

I   II

A ━━ ━━ a

B ━━ ━━ b

C ━━ ━━ c

1 2   3 4
⋁   ⋁
twin chromatids

25.8 Linkage and recombination

After this typical round of crossing over, the three genes originally linked as *ABC* and *abc* are combined in four ways: *abC, AbC, aBc,* and *ABc.* Notice that the farther apart two genes are, the more likely it is that a crossover event will take place between them. In this example, crossing over between *A* and *C* takes place in every instance, whereas crossing over between *A* and *B* occurs only half as often. Notice also that crossing over between two genes does not always recombine them: because chromatids 2 and 3 have two compensating crossover events, *A* and *C* remain together, as do *a* and *c,* even though segments between the two genes have been interchanged. As a result, recombination frequency is always lower than crossover frequency.

solute distances between genes, but it does give us information about gene order and relative distances. By convention, one unit of map distance on a chromosome is the distance within which recombination occurs 1 percent of the time. In Bateson and Punnett's test cross, 2 out of 16 of the offspring were recombinant products of crossing over. Two is 12.5 percent of 16; hence the genes controlling flower color and pollen shape in the sweet peas of this cross are located 12.5 map units apart.

Suppose we know that linked genes *B* and *L* are 12.5 map units apart. And suppose we find another gene, *A,* linked with these, that crosses over with gene *L* 5 percent of the time. How do we determine the order of the genes? The order could be *B–A–L:*

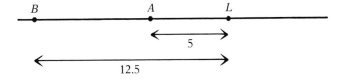

or it could be *B—L—A:*

Obviously, the way to decide between these two alternatives is to determine the frequency of recombination between *A* and *B.* If this frequency is 7.5 percent (12.5 − 5), then we know that the first alternative is correct; if it is

17.5 percent (12.5 + 5), then we know that the second alternative is correct. In this way, by determining the frequency of recombination between each gene and at least two other known genes, it is possible to build up a map showing the arrangement of many different genes on a chromosome (Fig. 25.9).

One important implication of the fact that the frequencies of recombination permit the mapping of genes should be noted: all known frequencies agree with a model of the chromosome in which the genes are sequentially arranged along the chromosome. Therefore one possible definition of the gene is that it is a region (or locus) on a chromosome that controls one or more characteristics of the organism. If two characters are always linked, and recombination by crossing over never occurs, then we can assume that they are controlled by the same point on the chromosome—that is, by the same gene. If, on the other hand, they do recombine, however seldom, then we can say that they must be controlled by different genes. Recombination has thus been the classic test of whether two characters are controlled by one chromosomal locus (gene) or by two separate chromosomal loci (genes). According to this view, the gene is the smallest unit of recombination. We shall examine other possible definitions of the gene later.

Polytene chromosomes Morgan and Sturtevant's propositions—that the genes are bound together in sequence in a limited number of paired linkage groups, that genes belonging to homologous linkage groups can undergo

25.9 Chromosome map of *Drosophila melanogaster*
Only a few of the many known genes are shown. The figures indicate their position in recombination map units from the zero end of the chromosome. Neither the Y chromosome nor the tiny fourth chromosome is shown.

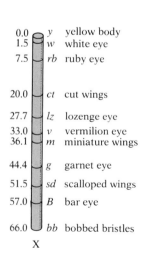

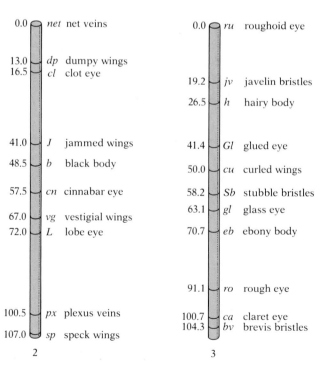

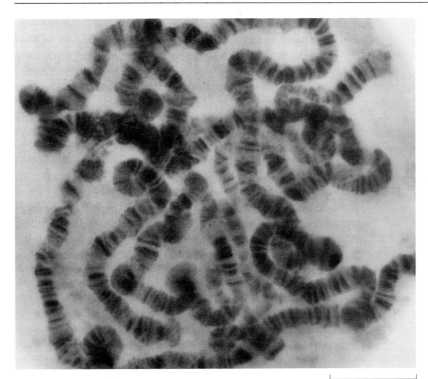

25.10 Photograph of giant chromosomes from salivary gland of *Drosophila melanogaster*
Note the pattern of banding by which different parts of the chromosomes can be identified.

10 μm

orderly recombination by crossing over, and that the frequency of recombination reveals both the linear order and the relative distances apart of the genes in each linkage group—were derived from breeding experiments and did not depend on knowledge of the chromosomes. Morgan and most other geneticists of his day were convinced of the validity of Sutton's theory that the genes are located on the chromosomes, but they recognized that the evidence for Sutton's theory was entirely circumstantial. They knew that genes occur in pairs, and that chromosomes do too. They knew that the members of each pair of genes separate at meiosis, as do the chromosomes. And they knew that the number of linkage groups in each species examined corresponds to the number of pairs of chromosomes (there are four linkage groups and four pairs of chromosomes in *Drosophila*, and there are ten linkage groups and ten pairs of chromosomes in corn). But no one had ever actually demonstrated the presence of a gene on a chromosome.

Eventually, stronger evidence for the chromosomal theory of genes came from the study of the giant chromosomes in the salivary glands of the larvae of many flies, including *Drosophila* (Fig. 25.10). Known as polytene chromosomes (from *poly-*, "many," and *taenia*, "ribbon"), these are more than 200 times larger than normal chromosomes, and can easily be studied in detail through an ordinary microscope. The product of repeated replication of the chromosomal material during interphase without accompanying separation of the strands by mitosis, each polytene chromosome is composed of a very large number of identical strands lying side by side (Fig.

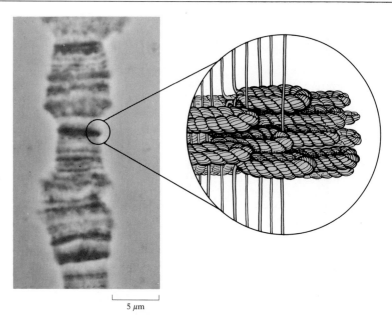

5 μm

25.11 Polytene bands

Polytene chromosomes consist of up to a thousand copies of a chromosome bound together in register. They are found only in certain highly active tissues, and are formed by repeated replication without segregation. The bands are probably regions in which the DNA is looped. These regions appear to be the genes, with the interchromomeric DNA which connects adjacent bands serving as a structural spacer.

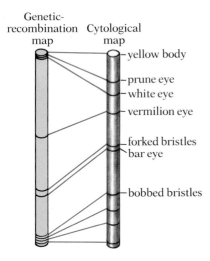

Genetic-recombination map   Cytological map

- yellow body
- prune eye
- white eye
- vermilion eye
- forked bristles
- bar eye
- bobbed bristles

25.12 Comparison of genetic-recombination and cytological maps of a portion of the X chromosome in *Drosophila melanogaster*

The two methods of mapping yield the same sequence for the genes, but the spacing is very different.

25.11). As a result, cells with polytene chromosomes have many copies of each gene available for transcription. This is one of several strategies, known as gene amplification, by which certain cells can greatly increase the transcription of one or more specific genes by making multiple copies of those genes. (Gene amplification will be discussed further in a later chapter.) Polytene chromosomes were first described in 1881, but not until 1930 did geneticists pay serious attention to them and recognize their potential for the study of gene arrangement.

When stained appropriately, polytene chromosomes have a banded appearance; the bands correspond to areas in which the DNA is organized into long clusters of loops (Fig. 25.11). Differences in the widths of the bands and the spacings between them enable researchers to identify with great precision characteristic regions of the chromosome. For instance, detailed comparative studies of chromosomal abnormalities have made it possible to determine the location of individual genes, and these can often be recognized by sight in relation to specific bands. In fact, the bands may well *be* the genes, looped out from the chromosome axis to be accessible for transcription, with intervening regions of DNA as spacers. In any event, cytological studies of polytene chromosomes have provided a second way of mapping the arrangement of the genes on the chromosomes, and such mapping has fully corroborated the sequences of genes deduced from recombination frequencies. However, the two techniques do not always agree about the spacing between genes (Fig. 25.12). Apparently breaks do not occur with equal facility at all points along the chromosomes; some parts of the chromosome are more susceptible to breakage than other parts. However, even though maps based on cytological examination give a more accurate picture of the relative distances between the genes, maps based on recombination frequencies are still constructed because they are more useful than cytological maps in predicting the results of crosses involving linked genes.

CHROMOSOMAL ALTERATIONS

Structural alterations   Besides crossing over, in which corresponding segments are interchanged between homologous chromosomes, there are other kinds of chromosomal rearrangements that occur much less frequently. Some involve interchange of segments between nonhomologous chromosomes; others involve alterations within a single chromosome.

In one form of alteration, called **translocation,** portions of two nonhomologous chromosomes are interchanged, with a consequent modification in linkage groups. Suppose, for example, that a pair of bar-shaped chromosomes in a certain species bear the genes *ABCDEFG* and that a pair of J-shaped chromosomes bear the genes *lmnopqrst.* If the *EFG* end of one bar chromosome and the *st* end of one J chromosome were interchanged, the result would be a shorter bar chromosome bearing only the genes *ABCDst* and a longer J chromosome bearing the genes *lmnopqrEFG.* By changing the linkage relationships of genes, translocations can have important effects on phenotypes.

Sometimes a piece breaks off one chromosome and fuses onto the end of the homologous chromosome. Such an alteration is a kind of **duplication.** An example would be loss of the *ABC* portion from a chromosome bearing the genes *ABCDEFGH* and fusion of this portion onto the homologous chromosome. The chromosome in which the loss occurred would thus bear only the genes *DEFGH,* while the chromosome undergoing the duplication would bear the genes *ABCABCDEFGH.*

Since translocations and duplications do not result in a net loss of genes in a cell, they have little effect on a somatic cell. (We shall look at a major exception to this picture in our discussion of cancer in a later chapter.) But these chromosomal rearrangements create problems in meiosis, when homologous chromosomes first synapse and then separate in the formation of gametes: If one chromosome of a pair has lost genes in a duplication event, half the gametes produced will lack these genes. If one chromosome of a pair has undergone a translocation exchange with a nonhomologous chromosome, synapsis with its homologue may be difficult, and meiosis may be blocked altogether. When meiosis does occur, the two nonhomologous chromosomes that have exchanged genes in translocation (losing some and gaining others) will segregate together only half the time. When they do not, half the resulting gametes will have nonhomologous duplications, while the other half will have two chromosomes with genes missing. A gamete that has lost essential genes may be inviable. Any zygote that does result from fusion involving such a gamete will lack the second copy of the genes in question; hence deleterious recessive alleles carried by the other gamete may be exposed.

In the examples just cited the fragments broken off the original chromosomes were not lost to the somatic cell, because they fused onto other chromosomes. But sometimes the broken-off portions do not fuse onto any chromosome and are lost entirely. This type of chromosomal alteration is called a **deletion.** Any chromosomal fragment that does not have a centromere will fail to move along the spindle during cell division and hence will not be incorporated into either daughter nucleus. One of the harmful ef-

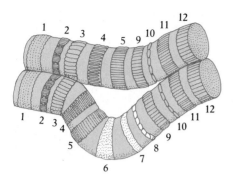

25.13 Synapsis of a deletion chromosome with its normal homologue

The upper chromosome has lost the region containing bands 6, 7, and 8. Consequently, when it synapses with a normal chromosome, that chromosome buckles outward opposite the deletion region. Since this drawing is of a pair of polytene salivary-gland chromosomes, banding patterns are visible, making the location of the deletion obvious. Synapsis occurs between polytene chromosomes even though meiosis does not take place.

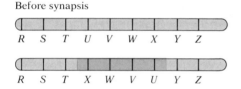

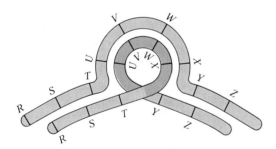

25.14 A hypothetical inversion and its effect on synapsis

Top: The upper chromosome has the normal gene sequence; the lower chromosome has undergone an inversion—the *UVWX* part has broken out, turned around, and fused back in. Bottom: When the normal chromosome and the chromosome with the inversion synapse during meiosis, they form complicated loops that bring corresponding genes into adjacent positions.

fects of intense radioactivity on somatic cells is the large amount of chromosomal breakage it produces, with the consequent loss of important genes on deleted fragments. The occurrence of a deletion can often be detected cytologically in cells undergoing meiosis, because when the aberrant chromosome synapses with its normal homologue the normal chromosome buckles at the point where material is missing from the aberrant chromosome (Fig. 25.13).

Another common form of chromosomal alteration, involving only one chromosome, is called an *inversion.* A portion of a chromosome breaks out, turns around, and fuses back in its original position, but with its ends reversed. Suppose a chromosome bears the genes *RSTUVWXYZ*, in that order. If the *UVWX* segment were to break out and become reattached in reverse order, the result would be a chromosome with the gene sequence *RSTXWVUYZ* (Figs. 25.14 and 25.15). Some such inversions have little phenotypic effect, since the same genes are still all present on the same chromosome. But sometimes there is a phenotypic change, apparently the result of what is known as the *position effect,* whereby the expression of a given gene is influenced by the genes close to it on the chromosome. In the original sequence in the example, gene *U* was located between genes *T* and *V* and was at a distance from gene *Y;* but in the inverted sequence, *U* is between *V* and *Y* and is at a distance from *T*. Any effect *T* may have had on the expression of *U* may be reduced as a result of the inversion, while an effect of *Y* on *U* may be increased.

Changes in chromosome number   As we saw earlier, the separation of chromosomes in cell division does not always proceed normally, and chromosomes that should have moved away from each other to opposite poles of the spindle may move instead to the same pole and become incorporated into the same daughter nucleus. The result of such *nondisjunction* may be production of an organism with an extra chromosome (or sometimes two, three, or more extra chromosomes). The presence of three chromosomes of one type in an otherwise diploid individual is called *trisomy.*

Occasionally cell division may be so aberrant that all the chromosomes move to the same pole, giving rise to a daughter cell with twice the normal number of chromosomes. If this happens during meiosis, the gamete produced is diploid instead of haploid. If such a gamete unites at fertilization with a normal haploid gamete, a triploid zygote results; if it unites with another diploid gamete, also produced by aberrant meiosis, a tetraploid zygote results. Cells or organisms that have more than two complete sets of chromosomes—that are triploid, tetraploid, hexaploid, etc.—are said to be *polyploid.*

Polyploidy has apparently occurred rather often in plants, in which it has sometimes given rise to new species adaptively superior to the original diploid species under certain environmental conditions. Polyploidy can be stimulated in the laboratory by treating plants with certain chemicals that cause nondisjunction during cell division. This procedure has been used in the production of many of the new strains of cultivated plants developed in the last few decades (Fig. 25.16). Polyploidy is very rare in animals and has probably not been an important factor in the origin of new animal species;

it may, however, have played a role in the evolution of the Hymenoptera, in which the various species usually have 4, 8, 12, or 16 chromosomes, and in the evolution of certain groups of fish and perhaps even some frogs, in which similar patterns exist.

EXTRANUCLEAR INHERITANCE

As if all the important complications and exceptions to Mendel's profoundly stimulating model we've discussed so far were not enough, certain subcellular organelles provide yet more exceptions. For instance, the geneticist Carl Correns demonstrated unequivocally in 1909 that in many plants the genes for leaf variegation (the white patterns on the green leaves) come solely from the mother. We now know that these genes are not sex-linked in the classic sense, but rather are located on chromosomes in the chloroplasts and, like those in the mitochondria, are transmitted to progeny through the cytoplasm of the eggs. (You may recall that these organelles are thought to have originated as autonomous symbionts with the full complement of genetic material necessary to sustain and perpetuate themselves.) The male gametes of higher eucaryotes normally do not contribute organelles to the egg, while the cytoplasm of eggs contains a sizable endowment of maternal enzymes and organelles; this maternal "dowry" influences early development of the zygote before the information from the genes of the male gamete takes effect.

25.15 A chromosomal inversion in *Drosophila*
This segment of the third chromosome of *Drosophila* has an inversion similar to the one diagramed in Figure 25.14. Though not immediately obvious in this photograph, the chromosome is one of a pair of polytene salivary-gland chromosomes undergoing synapsis.

25.16 Induced polyploidy in alfalfa
Tetraploidy and octopolyploidy in alfalfa and other commercially valuable crops can be induced experimentally. Tetraploid alfalfa (middle) is the most stress-tolerant, and is the type cultivated by farmers; cultivated alfalfa is a naturally occurring tetraploid. Octopolyploid alfalfa (right) grows well in the greenhouse, but is sensitive to the stress induced by lack of water in the field. (Diploid alfalfa is shown at left.) The drug colchicine, often used to induce polyploidization in the somatic cells of plants, works by arresting cell but not nuclear division. The polyploids shown here, however, were obtained experimentally by a process known as sexual polyploidization, in which gametes with unreduced numbers of chromosomes are produced during meiosis. The union of such gametes results in tetraploid plants; repetition of the process in the next generation yields octopolyploids. Sexual polyploidization results in plants that are superior to the polyploids produced chemically, and is the principal form of polyploidization of plants in nature.

TRISOMY IN HUMANS

In humans most trisomies are lethal. Trisomy-18 (Edwards' syndrome) and trisomy-13 (Patau's syndrome), for example, produce physical malformations, and mental and developmental retardation, so severe that most afflicted infants die within a few weeks after birth. Because trisomies of most other autosomes result in spontaneous abortion, they are not found in live births. Two kinds of trisomy—trisomy-21 (Down's syndrome) and trisomies of the sex chromosomes—are exceptional in that their victims may survive.

Down's syndrome (formerly often called mongolian idiocy), in which three chromosomes of type 21 occur in the individual's cells, was the first clinical condition ever linked to a chromosomal abnormality. It is associated with a variety of characteristic physical features (broad head, rounded face, perceptible epicanthic folds of the eyes, a flattened bridge of the nose, protruding tongue, small irregular teeth, short stature) and also mental retardation (the modal IQ is about 42). The incidence of Down's syndrome is often related to the age of the mother. It occurs in fewer than one out of 1,000 births to women under 20; it is more than seven times more common in births to women 35–39 years old, more than twenty times more common when the mother is 40–44, and more than fifty times more common when she is 45 or older. A similar association with the age of the mother is seen in Edwards' and Patau's syndromes.

Trisomy of the sex chromosomes can take several forms. In one, called Klinefelter's syndrome, the chromosomal makeup is XXY, and the individuals are males. Though the symptoms of the condition are variable, and some of those affected are nearly normal, most show a variety of physical abnormalities and mental retardation; moreover, they often suffer from thyroid dysfunction, chronic pulmonary distress, and diabetes.

Males with a second type of sex-chromosome trisomy, the XYY syndrome, generally show fewer and less severe abnormalities, though they often have poorly developed genitalia and subnormal intelligence. Because the incidence of XYY individuals is often significantly higher in penal institutions than in the general population, some investigators have suggested that men with the XYY condition are predisposed to aggressive behavior, but the evidence for this conclusion is weak.

Women with triple-X syndrome (XXX) usually have underdeveloped sexual characteristics and often subnormal intelligence, but since their abnormalities are not debilitating, most live a relatively normal life.

These trisomic conditions, as well as many other kinds of genetic or chromosomal diseases, can be detected during embryonic development by the process of *amniocentesis,* in which amniotic fluid containing sloughed-off epidermal cells from the fetus is withdrawn from the uterus with a long needle inserted through the mother's abdominal wall. The fetal cells are then cultured and examined for abnormalities. A new technique, in which embryonic tissue cells (from the chorionic villi) are obtained directly through the maternal cervix, promises earlier and safer detection of these and other devastating handicaps.

GENETICS PROBLEMS

1. Suppose that an allele, *b,* of a sex-linked gene is recessive and lethal. A man marries a woman who is heterozygous for this gene. If this couple had many normal children, what would be the predicted sex ratio of these children?

2. Red-green color blindness is inherited as a sex-linked recessive. If a color-blind woman marries a man who has normal vision, what would be the expected phenotypes of their children with reference to this character?

3. A man and his wife both have normal color vision, but a daughter has red-green color blindness, a sex-linked recessive trait. The man sues his

wife for divorce on grounds of infidelity. Can genetics provide evidence supporting his case?

4. Suppose a pigeon breeder finds that about one-fourth of the eggs produced by one of his prize pairs do not hatch. Of the young birds produced by this pair, two-thirds are males. Give a possible explanation for these results. (Remember the mechanism of sex determination in birds.)

5. It is exceedingly difficult to determine the sex of very young chickens, but it is easy to tell, by visual observation, whether or not they are barred. The barred pattern is inherited as a sex-linked dominant. Set up a cross allowing the sex of all chicks to be determined when they hatch.

6. In cats short hair is dominant over long hair; the gene involved is autosomal. An allele, B^1, of another gene, which is sex-linked, produces yellow coat color; the allele B^2 produces black coat color; and the heterozygous combination B^1/B^2 produces tortoiseshell (calico) coat color. If a long-haired black male is mated with a tortoiseshell female homozygous for short hair, what kind of kittens will be produced in the F_1? If the F_1 cats are allowed to interbreed freely, what are the chances of obtaining a long-haired yellow male?

7. The diagram shows three generations of the pedigree of deafness in a family. Black circles indicate deaf individuals. An arrow on a circle indicates a male, a cross below a circle a female. State whether the condition of deafness in this family is inherited as

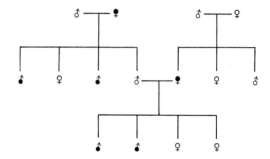

> *a.* a dominant autosomal characteristic
> *b.* a recessive autosomal characteristic
> *c.* a sex-linked dominant characteristic
> *d.* a sex-linked recessive characteristic
> *e.* a holandric characteristic

8. In *Drosophila melanogaster* there is a dominant allele for gray body color and a dominant allele of another gene for normal wings. The recessive alleles of these two genes result in black body color and vestigial wings respectively. Flies homozygous for gray body and normal wings were crossed with flies that had black bodies and vestigial wings. The F_1 progeny were then test-crossed, with the following results:

Gray body, normal wings	236
Black body, vestigial wings	253
Gray body, vestigial wings	50
Black body, normal wings	61

Would you say that these two genes are linked? If so, how many units apart are they on the chromosome?

9. In rabbits the dominant allele of a gene produces spotted body color, and the recessive allele solid body color. The dominant allele of another gene produces short hair, and the recessive allele long hair. Rabbits heterozygous for both characteristics were mated with homozygous recessive rabbits. The results of this cross were as follows:

Spotted, short hair	96
Solid, short hair	14
Spotted, long hair	10
Solid, long hair	80

What evidence for linkage is shown in this cross? Give the percentage of recombination and the map distance between the genes.

10. In *Drosophila melanogaster* the genes for bristle shape and for eye color are known to be about 20 units apart on the same chromosome. Individuals homozygous dominant for these genes were mated with homozygous recessive individuals. The F_1 progeny were then test-crossed. If there were 1,000 offspring from the test cross, how many of the offspring would you predict would show the crossover phenotypes?

11. The recombination frequency between linked genes *A* and *B* is 40 percent; between *B* and *C*, 20 percent; between *C* and *D*, 10 percent; between *C* and *A*, 20 percent; between *D* and *B*, 10 percent. What is the sequence of the genes on the chromosome?

12. Suppose that nondisjunction resulted in the production of new individuals with the following chromosomal abnormalities: XO, XXX, XYY, XXXX, XXXY, XXXXY. Indicate the expected phenotypic sex corresponding to each of these chromosomal combinations if it occurred (*a*) in a human; (*b*) in a *Drosophila*. How many Barr bodies would there be in human cells showing each of these combinations?

13. If a man with blood type B, one of whose parents had blood type O, marries a woman with blood type AB, what will be the theoretical percentage of their children with blood type B?

14. Both Mrs. Smith and Mrs. Jones had babies the same day in the same hospital. Mrs. Smith took home a baby girl, whom she named Shirley. Mrs. Jones took home a baby girl, whom she named Jane. Mrs. Jones began to suspect, however, that her child had been accidentally switched with the Smith baby in the nursery. Blood tests were made: Mr. Smith was type A, Mrs. Smith type B, Mr. Jones type A, Mrs. Jones type A, Shirley type O, and Jane type B. Had a mixup occurred?

Chapter 26

THE STRUCTURE AND REPLICATION OF DNA

It is difficult to appreciate how far the study of genetics has come in the last four decades. By 1950, cytological studies and mapping based on test crosses had shown that the physical units of heredity—the genes—are arranged in linear sequence on the chromosomes. And yet at that time it was still virtually impossible to say anything useful about the structure of genes or how their properties might control life processes. Since then, however, unprecedented progress in uncovering the molecular basis of genetic control has produced a revolution whose effects have yet to be fully measured. Today we can describe in remarkable detail how genes duplicate themselves, direct the synthesis of specific proteins, regulate their own activity, both cause and prevent disease, and even how evolution at the molecular level may take place. Furthermore, new genetic discoveries are still being made at a breathtaking pace.

Modern molecular genetics begins with the study of DNA structure; in this chapter we shall first look at the structure of the genetic material and then go on to examine the process of chromosome duplication, more formally known as ***replication***. This process enables each daughter cell produced during cell division to receive its own copy of each chromosome (and thus a complete set of genes) from the parental cell.

In the next chapter we shall look at how the information in the genes is "read out" and used: in the first step, particular genes (as opposed to entire chromosomes) are selectively copied, encoded as RNA, in a process known as transcription; in the second step, these transcripts are used to direct the synthesis of proteins, in a process called translation (Fig. 26.1, p. 694).

26.1   The flow of genetic information
Replication of a chromosome (A) enables both
daughter cells produced during cell division to
receive identical genetic information from the
parental cell. The process occurs in procaryotes and
eucaryotes alike, though it differs in some details.
Transcription and translation (B) also occur, with
some differences, in both procaryotes and eucaryotes:
an RNA version of a gene is produced by the process
of transcription; most such transcripts bind to a
ribosome, which—guided by the information carried
by the RNA—then assembles amino acids for the
synthesis of a particular protein. In eucaryotes, the
transcript is "processed" before moving to the
ribosome for translation into polypeptides. The details
of transcription and translation are presented in
Chapter 27.

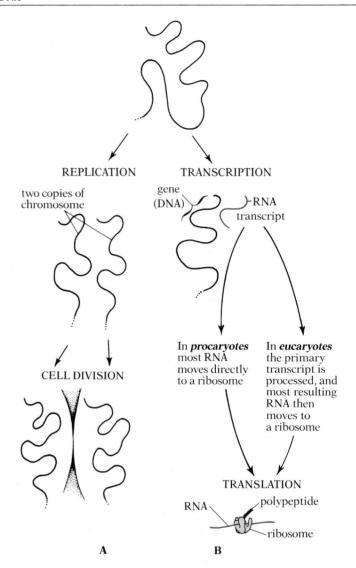

THE CHEMICAL NATURE OF THE GENETIC MATERIAL

THE DISCOVERY OF DNA AND ITS FUNCTION

The composition of chromosomes   The revolution brought about by molecular biology has its roots in the early attempts to ascertain the types of
compounds present in cell nuclei. In 1868 the young Swiss biochemist
Friedrich Miescher showed that when a cell is treated with pepsin to digest
its proteins, the nucleus shrinks but remains essentially intact. He showed
that the same nuclear material that can withstand peptic digestion also behaves totally unlike protein when treated with a variety of other reagents,
and that it contains phosphorus in addition to the carbon, oxygen, hydro-

gen, and nitrogen that would be expected if it were a protein. Taken together, these results meant that the nucleus contains large quantities of both protein and some hitherto unrecognized nonproteinaceous compound. The latter, which Miescher called nuclein, has since been named nucleic acid. Further research has shown that cells contain several sorts of nucleic acids, some not restricted to the nucleus; the type studied by Miescher was DNA (deoxyribonucleic acid).

In 1914 Robert Feulgen, a German chemist, devised a method of selectively staining DNA a brilliant crimson. When, ten years later, Feulgen applied his technique to whole cells, he found that the nuclear DNA is restricted to the chromosomes. Other workers have since used Feulgen staining to measure the DNA content of nuclei from many types of cells. They have shown conclusively that all the somatic (body) cells of a given organism ordinarily contain the same amount of DNA—even though cells from such different tissues as liver, kidney, heart, nerve, and muscle differ drastically in the amounts of other substances they contain—and, further, that egg and sperm cells contain only half as much DNA as the somatic cells. Since biologists had already concluded that mitosis distributes a complete set of genes to every somatic cell, regardless of its eventual role, and that meiosis distributes to every gamete cell exactly half the amount of genetic material found in the somatic cells, the discovery that the amount of DNA is usually constant in all somatic cells within a species, but is halved in the gametes, suggested that DNA rather than protein might be the essential material of the genes. But many workers refused to take this possibility seriously. The chromosomes of most organisms contain both protein and DNA, and most biologists assumed that the protein must be the genetic material, because only protein seemed to have the chemical complexity necessary to encode so much information.

DNA vs. protein   In 1928 Fred Griffith, an English medical bacteriologist, published a paper describing some experiments, now considered classic, on pneumococci, the bacteria that cause pneumonia. Griffith studied the effects on mice of a virulent strain of bacteria (S) and a nonvirulent strain (R). He showed that mice injected with live strain-R bacteria survived, mice injected with live strain-S bacteria soon died, but mice injected with heat-killed strain-S bacteria survived. These results (Table 26.1) were readily understandable. But the results of another of his experiments were thoroughly perplexing: mice injected with a mixture of live strain-R and heat-killed strain-S bacteria also died. How could a mixture of nonvirulent and dead bacteria have killed the mice? Griffith examined the bodies of the dead mice and found that they were full of live strain-S bacteria! Where had they come from? After many careful experiments, he became convinced that somehow the live strain-R bacteria had been transformed into live strain-S bacteria by material from the dead strain-S cells. The transformed bacteria, when cultured, reproduced new strain-S bacteria. Presumably, hereditary material from the dead bacteria had entered the live strain-R cells and changed them into strain-S cells.

By 1931 other workers had shown that the rodent host was not essential for bacterial *transformation*; it could occur just as well in test-tube cultures. Two years later James L. Alloway of the Rockefeller Institute (now

TABLE 26.1   *Griffith's results*

Bacteria injected	Reaction of mice
Live strain R	Survived
Live strain S	Died
Dead strain S	Survived
Live strain R plus dead strain S	Died

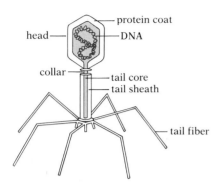

26.2   Interpretive drawing of a bacteriophage

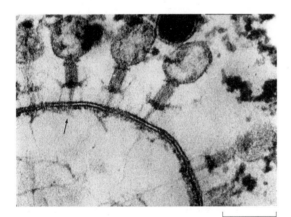

0.1 μm

26.3   Bacteriophage replication
Bacteriophage attach to the bacterial cell wall by their tail fibers and inject their genetic material into the cell. Once inside, the genetic material takes over the metabolic machinery of the cell and puts it to work making new phage.

Rockefeller University) showed that not even whole strain-S cells were necessary; live strain-R cells in a test tube could be transformed into strain-S bacteria by a cell-free extract of S bacteria. Apparently the hereditary characteristics of the virulent strain were transmitted to the cells of the nonvirulent strain by some substance that could withstand both the killing and extracting of the cells in which it had originally been contained.

The work of Griffith, Alloway, and others in the late 1920s and early 1930s was interesting to other biologists, but its full significance was not appreciated at that time. It was not until 1944 that O. T. Avery, Colin MacLeod, and Maclyn McCarty of the Rockefeller Institute demonstrated through purification techniques that the transforming agent was DNA; nothing else was necessary. From our present perspective this seems like strong evidence that DNA rather than protein or a nucleoprotein complex is the essential genetic material, but at that time many scientists remained unconvinced. For all anyone knew, the DNA from the virulent strain might merely activate the protein-based genes in the nonvirulent strain.

During the next ten years, however, the evidence for DNA steadily became stronger. At least 30 different examples of bacterial transformation by purified DNA were described. And strong evidence came from another source: the studies by Alfred D. Hershey and Martha Chase, of the Carnegie Laboratory of Genetics, of a special type of virus that attacks the bacterium *Escherichia coli*, which is abundant in the human digestive tract. This type of bacteria-destroying virus is called **bacteriophage**—phage for short (from the Greek *phagein*, "to eat").

Viruses, tiny parasites that subvert the cellular machinery of host organisms to reproduce their own DNA, are composed primarily of a protein coat and a nucleic acid core. The electron microscope reveals that certain phage viruses are structurally more complex than many other types. Their protein coat is divided into a head region and an elongate tail region made up of a hollow core, a surrounding sheath, and six distal fibers (Fig. 26.2). The DNA is in the head. Electron micrographs show that when such a phage attacks a bacterial cell it becomes attached by the tip of its tail to the wall of the bacterial cell (Fig. 26.3); the tip of the tail contains a protein that binds with a specific component of the bacterial wall. The protein coat of the phage never enters the bacterial cell, but within an hour after the phage becomes attached to the cell wall, new phage appear within the bacterium. At the appropriate time, the phage causes the bacterial cell to produce an enzyme called lysozyme, and other enzymes; as a consequence of this digestive action, the bacterial cell lyses, or ruptures, releasing hundreds of new phage into the surrounding medium. The new phage released from a lysed bacterium are genetically identical with those that initiated the infection. Hereditary material must have been injected into the bacterial cell by the phage attached to its wall, and this hereditary material must have usurped the metabolic machinery of the bacterium and put it to work manufacturing new phage genes and the many proteins necessary to construct the complex protein coat, as well as the several enzymes that shut off much of the host's own protein synthesis and in due time lyse the cell.

Hershey and Chase designed an experiment to determine whether the infecting phage injects into the bacterium only DNA or only protein or some of both. Since DNA contains phosphorus but no sulfur, whereas protein

26.4   Diagram of a nucleotide from DNA
A phosphate group and a nitrogenous base are attached to deoxyribose, a five-carbon sugar. The carbons in deoxyribose are designated 1′–5′, as shown here, though the numbers are not normally included in molecular diagrams. The phosphate group is bound to the 5′ carbon of the sugar, while a hydroxyl group is bound to the 3′ carbon.

contains sulfur (in the disulfide bonds described in Chapter 3) but no phosphorus, they were able to distinguish DNA from protein by using radioactive isotopes of phosphorus and sulfur. They cultured phage on bacteria grown on a medium containing radioactive phosphorus (^{32}P) and radioactive sulfur (^{35}S). The phage incorporated the ^{35}S into their protein and the ^{32}P into their DNA. Hershey and Chase then infected nonradioactive bacteria with the radioactive phage. They allowed sufficient time for the phage to become attached to the walls of the bacteria and inject hereditary material. Then they agitated the bacteria in a blender in order to detach what remained of the phage from their surfaces. Analysis of these remains showed that they contained a substantial amount of ^{35}S but little ^{32}P, an indication that only the empty protein coat had been left outside the bacterial cell. Analysis of the bacteria showed that they contained much ^{32}P but little ^{35}S, an indication that only DNA had been injected into them by the phage. Since new phage were produced in these bacteria, DNA alone had to have been sufficient to transmit to the bacteria all the genetic information necessary for their production. This experiment, reported in 1952, strongly supported the earlier conclusions based on transformation experiments that nucleic acids, not proteins, constitute the genetic material.

THE MOLECULAR STRUCTURE OF DNA

We have already seen that a molecule of DNA is composed of building blocks called nucleotides, each of which is itself composed of a five-carbon sugar bonded to a phosphate group and a nitrogenous base (Fig. 26.4). By convention, the five carbons are designated by numbers, 1′–5′. There are four kinds of nucleotides in DNA, which differ from one another in their nitrogenous bases. Two of the bases, **adenine** and **guanine**, are purines, which are double-ring structures; the other two, **cytosine** and **thymine**, are pyrimidines, which are single-ring structures (see Fig. 3.28, p. 69). In a DNA molecule the nucleotides are arranged in a specific sequence; the sugars are held together by the phosphate groups that link the 3′ carbon of one sugar to the 5′ carbon of the next; the nitrogenous bases are arranged as side groups off the chains (Fig. 26.5). DNA molecules ordinarily exist as

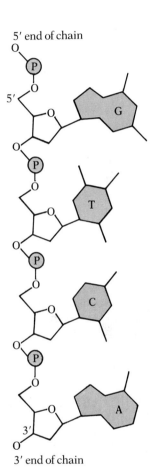

26.5   Portion of a single chain of DNA
Nucleotides are linked together by bonds between their sugar and phosphate groups. The nitrogenous bases (G, guanine; T, thymine; C, cytosine; A, adenine) are side groups. In this diagram P represents the main components of each phosphate group—the phosphorus atom with its hydroxyl and the double-bonded oxygen; only the oxygen atoms in the connecting chain are shown separately.

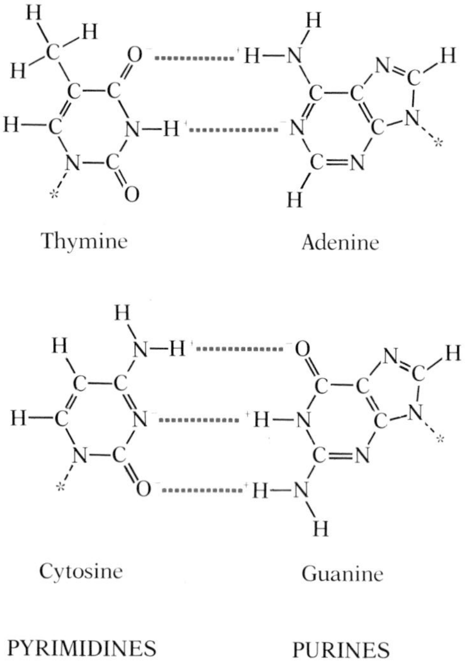

Thymine   Adenine

Cytosine   Guanine

PYRIMIDINES   PURINES

26.6 Bonding of nitrogenous bases in nucleotides
Because of the differing electronegativities of oxygen, hydrogen, nitrogen, and carbon, the nitrogenous bases of DNA have polar segments. The spacing and polarity of these segments permit thymine to form hydrogen bonds with adenine, and cytosine to bond with guanine. (The asterisk marks the point of attachment of each base to a sugar.)

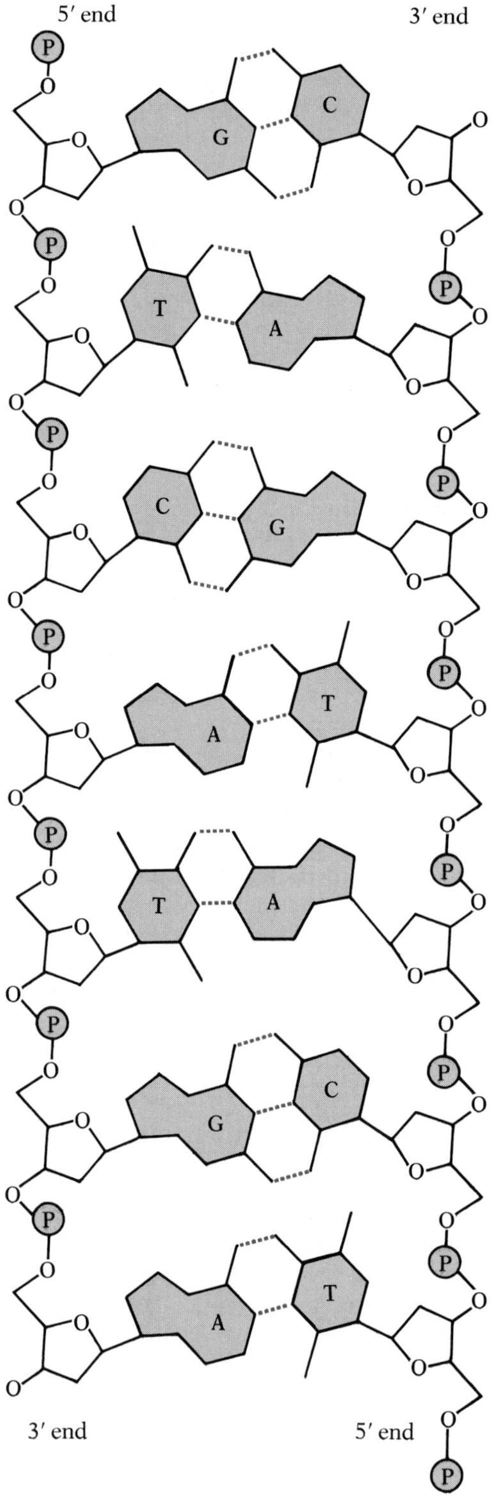

26.7 Portion of a DNA molecule uncoiled
The molecule has a ladderlike structure, with the two uprights composed of alternating sugar and phosphate groups and the cross rungs composed of paired nitrogenous bases. Each cross rung has one purine base (a pentagon attached to a hexagon) and one pyrimidine base (a hexagon). When the purine is guanine (G), the pyrimidine with which it is paired is always cytosine (C); when the purine is adenine (A), the pyrimidine is thymine (T). Adenine and thymine are linked by two hydrogen bonds (striped bands), guanine and cytosine by three. Note that the two chains run in opposite directions: the free phosphate is linked to the 5′ carbon at the upper end of the left chain and at the lower end of the right chain.

double-chain structures, with the two chains, or strands, held together by hydrogen bonds between their nitrogenous bases; such bonding can occur only between cytosine and guanine or between thymine and adenine (Fig. 26.6). Thus the sequence of bases in one strand determines the complementary sequence in the other (Fig. 26.7). Notice that the polarities of the two strands are opposite: one runs from 5′ to 3′, while the other goes from 3′ to 5′.[1] Finally, the ladderlike double-chain molecule is coiled into a double helix (Fig. 26.8), and stabilized by hydrogen bonds between bases within each chain that are close to one another, much as the alpha helix of a protein is stabilized by hydrogen bonds between the amino acids (see Fig. 3.20, p. 64).

Determining the structure of so complicated—and important—a molecule as DNA had become an irresistible challenge to many scientists. In 1950 almost nothing was known about the spatial arrangement of the atoms within the DNA molecule; nor was it known how this molecule could contain within it the necessary information for replicating itself and for controlling cellular function. About this time, several workers began applying the techniques of X-ray diffraction analysis to DNA. Outstanding among them were Rosalind Franklin and Maurice H. F. Wilkins of King's College, London, who succeeded in obtaining much sharper X-ray diffraction patterns than had previously been obtainable. Francis H. C. Crick of Cambridge University had just developed mathematical methods for interpreting X-ray patterns of protein helices, and working from the Franklin and Wilkins photographs, was able to show that crystalline DNA had to be a helix with three major periodicities: one of 0.34 nm, one of 2.0 nm, and one of 3.4 nm.

Now began a collaboration whose outcome would rank as one of the major milestones in the history of biology. James D. Watson and Crick, working at Cambridge University, decided to try to develop a model of the structure of the DNA molecule by combining what was known about the chemical content of DNA with the information gained from Crick's analysis of Franklin and Wilkins's X-ray diffraction studies, as well as with data on the exact distances between bonded atoms in molecules, the angles between bonds, and the sizes of atoms. Watson and Crick built scale models of the component parts of DNA and then attempted to fit them together in a way that would agree with the information from all these separate sources.

They were certain that the 0.34 nm periodicity corresponded to the distance between successive nucleotides in the DNA chain, the 2.0 nm periodicity to the width of the chain, and the 3.4 nm periodicity to the distance between successive turns of the helix. Since 3.4 is exactly ten times the distance between successive nucleotides, each turn of the helix had to be ten nucleotides long.

Having made these essential assumptions about the meaning of the X-ray diffraction data, Watson and Crick tried to correlate them with the infor-

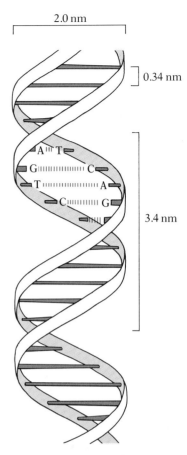

26.8 The Watson-Crick model of DNA
The molecule is composed of two polynucleotide chains held together by hydrogen bonds between their adjacent bases. The double-chained structure is coiled in a helix. The width of the molecule is 2.0 nm; the distance between adjacent nucleotides is 0.34 nm; and the length of one complete coil is 3.4 nm. Interactions between bases within each chain help stabilize the molecule in the helical shape shown here.

[1] This strand polarity has nothing to do with the unequal distribution of charge that gives rise to polar molecules.

mation from other sources. They immediately ran into a discrepancy: they calculated that a single chain of nucleotides coiled in a helix 2.0 nm wide with turns 3.4 nm long would have a density only half as great as the known density of DNA. An obvious inference was that the DNA molecule is composed of two nucleotide chains rather than one. Now they had to determine the relationship between the two chains within the double helix. They tried several arrangements of their scale model and found that the one that best fitted all the data had the two nucleotide chains wound in opposite directions within a hypothetical cylinder of appropriate diameter, with the purine and pyrimidine bases oriented toward the interior of the cylinder (Fig. 26.8). With the bases oriented in this manner, hydrogen bonds between the bases of opposite chains could supply the force to hold the two chains together and to maintain the helical configuration. In other words, the DNA molecule, when unwound, would have a ladderlike structure, with the uprights of the ladder formed by the two long chains of alternating sugar and phosphate groups, and with each of the cross rungs formed by two nitrogenous bases loosely bonded to each other by hydrogen bonds (Fig. 26.7).

Watson and Crick soon realized that each cross rung must be composed of one purine base and one pyrimidine base. Their scale model showed that the available space between the sugar-phosphate uprights was just sufficient to accommodate three ring structures. Hence two purines opposite each other occupied too much space, because each had two rings, for a total of four, and two pyrimidines opposite each other did not come close enough to bond properly, because each had only one ring. This left four possible pairings: A–T, A–C, G–T, and G–C. Further examination revealed that though adenine (A) and cytosine (C) were of the proper size to fit together into the available space, they could not be arranged in a way that would permit hydrogen bonding between them; the same was true of guanine (G) and thymine (T). Therefore neither A–C nor G–T cross rungs could occur in the DNA molecule. This left only A–T and G–C. Both of these base pairs seemed to fulfill all requirements. It did not seem to matter in which order the bases occurred; the essential requirement was that adenine and thymine always be paired with each other and that guanine and cytosine always be paired. This pairing quite unexpectedly explained an earlier finding by Erwin Chargaff and his colleagues at Columbia University that all samples of DNA, regardless of other differences in their composition, are alike in always containing exactly equal amounts of adenine and thymine nucleotides, and exactly equal amounts of guanine and cytosine nucleotides. The amounts of adenine and thymine are always equal because these two bases are always paired, and, similarly, the amounts of guanine and cytosine are always equal because these are always paired.

In summary, then, the Watson-Crick model of the DNA molecule shows a double helix in which the two chains, composed of alternating sugar and phosphate groups, are bonded together by hydrogen bonds between adenine and thymine from opposite chains and between guanine and cytosine from opposite chains. This model, in essentially the same form in which it was first proposed in 1953, has been consistently supported by later research, and it has received general acceptance. Watson, Crick, and Wilkins were awarded the Nobel Prize for their critically important work. (Franklin, who would probably have shared this honor, died before she could do so.)

THE REPLICATION OF DNA

REPLICATION FROM A TEMPLATE

The theory of Watson and Crick   DNA, if it is the genetic substance, must have built into it the information necessary to replicate itself and to control the cell's attributes and functions. One of the most satisfying things about the Watson-Crick model of DNA is that it immediately suggests a way in which the first of these two requirements may be met.

Since the DNA of all organisms is alike in being a polymer composed of only four different nucleotides, the essential distinction between the DNA of one gene and the DNA of another gene must be—aside from the total number of nucleotides—the sequence in which the four possible types of base-pair cross rungs (A–T, T–A, G–C, and C–G) occur. The basic question of genetic replication is, then, assuming that an adequate supply of the four nucleotides has already been synthesized, what tells the cell's biochemical machinery how to put these nucleotide building blocks together in exactly the sequences and quantities characteristic of the DNA already present in the cell?

Watson and Crick pointed out that if the two chains of a DNA molecule are separated by rupturing the hydrogen bonds between the base pairs, each chain provides all the information necessary for synthesizing a new partner identical to its previous partner. Since an adenine nucleotide must always pair with a thymine nucleotide, and since a guanine nucleotide must always pair with a cytosine nucleotide, the sequence of nucleotides in one chain, or strand, precisely specifies what the sequence of nucleotides in its complementary strand must be. In other words, if the cell could separate the two chains in its DNA molecules—much as one might unzip a zipper—it could line up nucleotides for a new chain next to each of the old chains, putting each type of nucleotide opposite its proper partner. As the nucleotides were arranged in the proper sequence, they could be bonded together to form the new chain. Thus, separating the two chains of a DNA molecule and using each chain as a template or mold against which to synthesize a new partner for it would result in two complete double-chained molecules identical to the original molecule (Fig. 26.9).

Experimental support for the theory   Satisfying as it was, this template theory of DNA replication was pure speculation, unsupported by any experimental evidence, when it was first put forward by Watson and Crick in 1953. Since then, convincing evidence has come from the work of a number of investigators.

In 1957 Arthur Kornberg and his associates at Washington University in St. Louis developed a method for achieving DNA synthesis in a test tube. They extracted from cells of *Escherichia coli* enzymes that can catalyze the

26.9   Replication of DNA
As the two polynucleotide chains of the old DNA (black) uncoil, new polynucleotide chains (color) are synthesized on their surfaces. The process produces two complete double-chained molecules, each of which is identical in base sequence to the original double-chained molecule. Replication actually begins not at the end of the molecule, as shown here for simplicity, but at specific internal points.

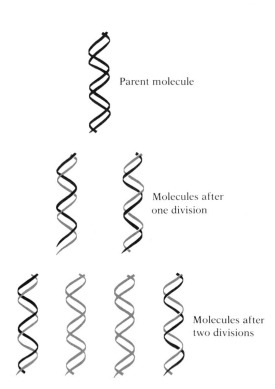

Parent molecule

Molecules after
one division

Molecules after
two divisions

26.10 Results of the Meselson-Stahl experiment
The parent DNA molecule (black chains) contained
only heavy nitrogen. After one division, the DNA had
an intermediate density, an indication that half the
nitrogen in each molecule was heavy and half was
light; the two heavy parental chains had separated,
and each had acted as the template for synthesis of a
complementary light chain (color). Even after
several additional duplications, the two original
heavy parental chains remained intact.

synthesis of DNA—DNA polymerases—and combined them with a plentiful
supply of the four nucleotides as raw material. These nucleotides had been
activated by ATP—that is, a high-energy phosphate group had been attached, to provide the energy for the later reaction steps. The nucleotides
also contained a radioactive isotope of carbon (^{14}C). After the experimenters had added some DNA to serve both as a primer—a starting point
for the expected reactions—and as a potential template, they incubated the
preparation, and they found that new DNA containing ^{14}C appeared; the labeled nucleotides had been built into new DNA chains. This synthesis
would not occur unless DNA was present at the beginning of the reaction.
Kornberg was able to show that the ratio of adenine and thymine to guanine
and cytosine in the new DNA was precisely the same as that in the primer
DNA. These experiments and others indicated that the newly synthesized
DNA was identical to the DNA added to the original mixture—that the
primer DNA had indeed functioned as a template. Kornberg received a
Nobel Prize for this work.

Kornberg's experiment seemed to demonstrate that DNA synthesis cannot take place in the absence of template DNA, which presumably provides
the base-sequence information necessary to guide the synthesis. It also
strongly suggested that new DNA is always a copy of the template DNA. But
it did not actually provide any evidence that the copy mechanism works in
the way proposed by Watson and Crick. More direct support for their model
came from an experiment reported in 1958 by Matthew S. Meselson and
Franklin W. Stahl, then of the California Institute of Technology.

Meselson and Stahl grew *E. coli* for many generations on a medium
whose only nitrogen source was the heavy isotope ^{15}N. Eventually all the
DNA in these bacteria contained the heavy isotope instead of the normal
isotope ^{14}N. Then the nitrogen source was abruptly changed from ^{15}N to ^{14}N.
Cell samples were removed at regular intervals thereafter, and the DNA
was extracted from them and subjected to a complicated procedure designed to separate DNA of different densities.

The experiment showed that when cells containing only heavy DNA
(DNA in which both chains had only ^{15}N in their purine and pyrimidine
bases) were allowed to undergo one division in the ^{14}N medium, the DNA of
the new cells was intermediate in density between heavy DNA and light
DNA. In other words, the nitrogen in the DNA of the new cells was half ^{15}N
and half ^{14}N. This is precisely what would be expected if the two chains of
the heavy parental DNA separated and acted as templates for the synthesis
of new partners from nucleotides containing only ^{14}N (Fig. 26.10): each
new DNA molecule should be composed of one heavy chain from the parent and one light chain newly synthesized, the molecule thus having intermediate density.

To prove that all the ^{15}N really was in one chain of the intermediate-den-
sity DNA and all the ^{14}N in the other chain, Meselson and Stahl subjected the
DNA to a treatment that breaks the hydrogen bonds between the bases and
separates the chains. Sure enough, this procedure produced heavy chains
and light chains; the two isotopes had not been distributed randomly
throughout the DNA molecule, but had been localized each in one of the
chains, just as would be predicted from the Watson-Crick theory.

Mechanisms of DNA replication   The process of replication is complex and anything but haphazard: hydrogen bonds stabilizing the helical shape and linking the two chains of the DNA molecule must be broken, and the chains separated; complementary nucleotides must be paired with the nucleotides of each existing chain; the new nucleotides must be covalently linked to form a chain; and so on. Every step is managed by specific enzymes, and takes place quickly and accurately. In *E. coli*, for instance, a complex of several enzymes, collectively known as a polymerase, adds an average of 500 base pairs each second, and there is only one error in every billion pairs copied.

The basic task of DNA replication is the same in procaryotes and eucaryotes: the process in bacteria and the process in humans display many similarities. For example, though the chromosomes differ in shape—bacterial chromosomes are circular while ours are linear—replication begins in both at particular spots in the DNA and proceeds in both directions away from the initiation site; replication in eucaryotes never begins at an end. In both procaryotes and eucaryotes, the individual chains are copied in only one direction: the replication enzymes move along the parental strands from 3' to 5', generating a complementary strand running from 5' to 3'; this means that one chain can be copied continuously but the other must be replicated "backward," in discontinuous segments (as described in the box). Both groups of organisms have mechanisms that locate and correct errors; both have special mechanisms that prevent the chains from tangling.

But along with these general similarities come differences in detail. For example, because eucaryotic chromosomes are packaged on histone spools to form nucleosomes, replication proceeds at a rate of only 50 base pairs per second; it takes time for the DNA to be unwound from the spools and, after a duplicate set of spools has been synthesized, to be rewound. And because eucaryotic chromosomes are much larger than procaryotic chromosomes, replication is initiated at many independent sites on each chromosome simultaneously; otherwise, it might take literally weeks or even months for a complex eucaryotic cell to divide. While there are many such differences, we shall stress the common features of genetic information flow in later chapters, focusing first on the relatively well-understood mechanisms used by procaryotes, and then on the modifications seen in eucaryotes.

DNA REPAIR

The precise replication of DNA is essential for normal cell function. The chromosomes contain all the instructions for building and running an organism, and random changes in those instructions are far more likely to disrupt a pathway—by destroying the delicate architecture of an essential enzyme, for instance—than to improve it. Since the instructions for the assembly of the thousands of structural proteins, enzymes, regulatory proteins, and so on are each hundreds or thousands of bases long, an error rate as low as one in a thousand bases, though it may seem insignificant, is far too great. Not surprisingly, then, special mechanisms have evolved in both procaryotes and eucaryotes that keep uncorrected replication errors at a

REPLICATION OF THE *E. COLI* CHROMOSOME

The process of replication in *E. coli* illustrates the complex series of steps necessary even in a relatively simple organism to make an accurate copy of a chromosome. Highly specific enzymes underlie each event. Like all enzymes they recognize specific reactants by their complementary shape and pattern of polar charges. And as enzymes often do, they work together to rearrange the bonds in various reactants and produce a final product—in this case, a complete copy of the circular bacterial chromosome (see Fig. 23.4, p. 617). Enzymes in complex pathways work nearly simultaneously, but for clarity we shall discuss the events of this process in sequence.

In *E. coli*, the process begins when a protein, DNA B, recognizes an initiation site on the chromosome (probably by its particular sequence of bases) and binds to it (top figure). Next, molecules of the enzyme DNA gyrase begin to relax the supercoiling of the chromosome on each side of the DNA B protein. As the two DNA gyrase molecules move away from the initiation site, two molecules of "rep" enzyme (also called DNA helicase) unzip the double helix, using energy from ATP to break the hydrogen bonds that hold the bases together. Single-strand binding proteins (SSB) then form a scaffolding, which holds the two strands apart and prevents them from rebinding to each other spontaneously (A, bottom figure). These steps must occur before the actual replication of DNA can begin.

Next, **DNA polymerase III**, a complex consisting of several enzymes bound together, begins to replicate one of the two DNA strands by binding to it and adding complementary bases. A complex is necessary because DNA polymerase must catalyze several different reaction steps and must be able to use four different nucleotides as reactants, depending on what it "reads" from the strand it is copying (the template strand). Presumably there are four active sites, one each for adenine, cytosine, guanine, and thymine. Once the complex has read the template strand and brought the complementary nucleotide into place, it catalyzes the binding of the nucleotide to the growing complementary strand. The nucleo-tides that are added have been activated by the addition of a high-energy phosphate group from ATP, which provides the energy for this step.

As in all enzymes, the active site in each component of DNA polymerase III is very specific; it binds only DNA and a particular nucleotide. Just as carboxypeptidase digests a protein only from its carboxyl ends, so DNA polymerase III can add nucleotides only to the 3′ end of the DNA strand. Two serious problems result: First, there must be a free 3′ end of DNA available to the polymerase. That is, replication must already have begun, so that there is a bit of DNA complementary to the template strand, with a free 3′ end. How replication of this first bit is accomplished is not yet known. Second, since the two strands of the double helix have opposite polarities, with one running from 5′ to 3′ and the other from 3′ to 5′ (Fig. 26.7), they must be copied in opposite directions. One strand can be copied by a DNA polymerase III following along behind the rep enzyme as it unzips the DNA (A, bottom figure), but the other strand must somehow be copied "backward." The DNA formed by the DNA polymerase that follows the rep enzyme is known as the ***leading strand***, while the DNA synthesized backward is known as the ***lagging strand***. The latter is formed bit by bit in a looping "backstitch" pattern. Enzymes known as RNA primases construct primers—short complementary strands about 10 bases long—at random intervals alongside the uncopied template for the lagging strand (A, B). These segments provide the 3′ free end for the DNA polymerase III; from here it works backward, copying the strand, until it reaches the preceding RNA primer segment (B). One strand, then, is copied continuously while the other is copied in sections, known as Okazaki fragments, that are from 1,000 to 2,000 bases long.

Now a series of enzymes must patch the fragments together into a continuous strand. ***DNA polymerase I*** removes the 10-base RNA primer segments and replaces them with DNA (C). Then the fragments are welded together by DNA ligase (D), and the new strand is finished.

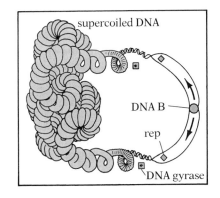

supercoiled DNA

DNA B

rep

DNA gyrase

Symbol	Substance	Function
○	DNA B	finds and marks initiation site
▣	DNA gyrase	relaxes supercoiling
◆	rep	separates DNA strands
⬭	SSB	holds strands apart
○	RNA primase	primes lagging strand for replication
◁	DNA polymerase III	synthesizes complementary DNA
◇	DNA polymerase I	erases primer and replaces with DNA
▽	DNA ligase	welds gaps

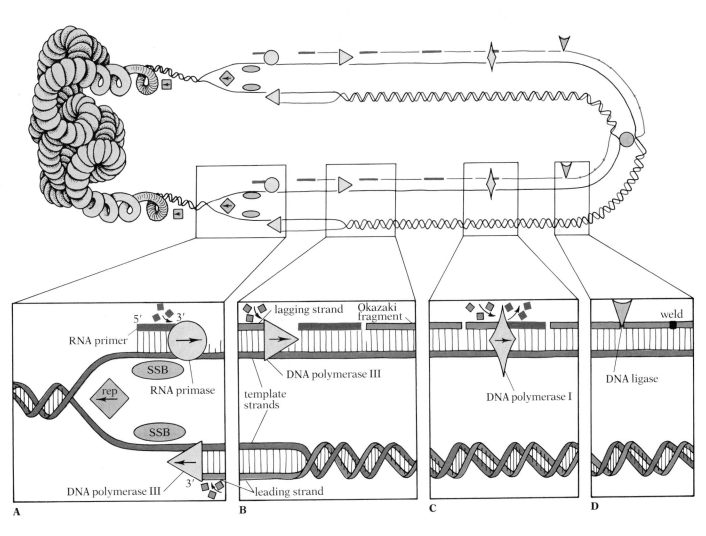

A

5' 3'
RNA primer
SSB
rep
RNA primase
SSB
DNA polymerase III 3'

B

lagging strand Okazaki fragment
DNA polymerase III
template strands
leading strand

C

DNA polymerase I

D

weld
DNA ligase

```
3' ···—G—C—T—T—T—T—T—T—G—G— ··· 5'
5' ···—C—G—A—A—A—A—A—A—C—C— ··· 3'
A
```

```
3' ···—G—C—T—T—T—T—T—T—G—G— ··· 5'
5' ···—C—G—A—A—A—A—A—A—C—C— ··· 3'
B
```

```
            T
3' ···—G—C   T—T—T—T—T—G—G— ··· 5'
5' ···—C—G—A—A—A   A—A—C—C— ··· 3'
                A
C
```

```
repair
enzyme         T
3' ···—G—C   T—T—T—T—T—G—G— ··· 5'
5' ···—C—G—A—A—A   A—A—C—C— ··· 3'
D                    A
```

```
3' ···—G—C—T—T—T—T—T—G—G— ··· 5'
5' ···—C—G—A—A—A—A—A—C—C— ··· 3'
E
```

26.11 Misalignment deletion
A sequence with an extended series of A–T pairs (A) is especially susceptible to deletion. When, by chance, a series of hydrogen bonds breaks simultaneously, the two strands of the helix separate transiently (B). When they pair again, there is a small chance that they will be misaligned (C). If the resulting distortion is detected by DNA repair enzymes, the two unpaired bases will be cut out by the enzymes (D) and the loose ends will be reattached with DNA ligase (E), leaving each strand of DNA one nucleotide short.

very low level. As a further safeguard, there are enzymes that locate and repair most of the damage that occurs to the DNA between bouts of replication. Like the other enzymes we have discussed, repair enzymes bind to particular substrates (in this instance faulty or damaged areas of the DNA) that display the patterns of spacing and polar charges to which their active sites bind; they loosen specific bonds, and catalyze the formation of new ones.

Repair during replication The initial error rate of the replication enzymes of both procaryotes and eucaryotes is about 3 in 10^5 pairs. (This value, like much of what is known about how DNA polymerases work, is obtained from studies of polymerase complexes with one or more inactive component enzymes. From the change in the operation of the complex, the function of the inactive enzyme can often be deduced. In this case, replication proceeds when the enzymes under study are inactive, but most errors are not corrected.) If the errors produced at this rate were left uncorrected, the result would be a mistake in roughly 3 percent of each cell's proteins—perhaps 1,000 changed proteins in every human cell after each replication. Fortunately, the DNA polymerase complex includes one or more enzymes that successively "proofread" each base, clipping out mistakes. Yet other enzymes in the polymerase complex then substitute a rematched base for the excised unit; the complex then moves on without checking a second time. As we saw when we used the Product Law to calculate the probabilities of independent events in Chapter 24, a second pass reduces the chance of an error to the product of the two individual probabilities. In this instance, therefore, the error rate after repair is $(3 \times 10^{-5}) \times (3 \times 10^{-5})$, or roughly 1 in 10^9. For humans, with our 30,000 functional genes, each of which has an average of about 1,500 bases on each of its two strands, this corresponds to an error in a gene somewhere in the genome once every ten replications.

Repair of mutations The integrity of the genetic message is also threatened by alterations in base sequences induced by heat, radiation, and various chemical agents. The rate at which these mutations occur is astoundingly high: thermal energy alone, for instance, breaks the bonds between roughly 5,000 purines (adenine and guanine) and their deoxyribose backbones in each human cell every day. Cytosine is chemically converted to uracil (a nucleotide normally found only in RNA, and misread by DNA replication and transcription enzymes) at a rate of about 100 per cell per day. Ultraviolet radiation from sunlight fuses together adjacent thymines at a high rate in exposed epidermal cells. And yet, because of the continuous operation of repair enzymes, the rate at which mutations accumulate in cells, on average, is even lower than that of uncorrected errors in replication.

The strategy for repairing chromosomes is basically the same for the many sorts of mutations as for replication errors: enzymes locate and bind to the faulty sequences and clip out the flaws, and the intact complementary strand guides repair. A remarkable array of specific enzymes is involved; for instance, the chromosomes are scanned for chemically altered bases by fully 20 different enzymes, each specific for a particular class of problem. Another 5 or so enzymes are specific for faulty covalent bonding

Cytosine   Methylated   Thymine
                    cytosine

26.12   Deamination of methylated cytosine
Cytosine is sometimes methylated by enzymes. When a methylated cytosine undergoes deamination, probably as the result of oxidation by a mutagenic chemical, the result is a normal thymine. This change cannot be corrected reliably because the repair system cannot determine whether it is the thymine or the guanine—the base in the complementary strand to which the cytosine was originally paired—that is incorrect.

between a base and some other chemical, or between adjacent bases on one strand; the fusion of thymines induced by ultraviolet radiation is the most common error of this type. Other enzymes bind at the sites of missing bases, like the purines so easily lost to thermal energy. In all, some 50 enzymes locate and correct errors.

In view of this elaborate repair system, how do mutations survive to generate much-needed genetic variability? Most obviously, no enzymatic system is perfect: some errors are missed, and others are repaired incorrectly. Then, too, when a mutation occurs just before or during replication, there may not be time for detection and repair. Still other mutations are not overlooked, but are actually created by the repair system.

One well-understood class of mutation generated by repair enzymes is deletion. Recent research demonstrates that small deletions (loss of a few base pairs) do not occur at random locations; some base sequences are more susceptible to deletion than others. The reason for this differential susceptibility is the relative weakness of the base pairing between adenine and thymine: they are connected by only two hydrogen bonds, whereas cytosine and guanine are connected by three. The hydrogen bonds are continually breaking and re-forming, and it frequently happens, particularly in regions rich in adenine–thymine pairs, that by chance all the bonds in a small segment of DNA are broken simultaneously. Normally, the bonds re-form spontaneously, but occasionally the rebonding is incorrect. For example, when a region with an extended series of adenine–thymine pairs undergoes a transient separation, there is a small chance that the two strands will be misaligned when they pair again. DNA repair enzymes may then remove the unpaired bases and thereby introduce a misalignment deletion (Fig. 26.11). The "repair" is likely to alter the "meaning" of the gene by causing errors during translation of the gene's mRNA.

Finally, some mutations simply cannot be detected by the repair enzymes. A cytosine that has already been modified by the addition of a methyl group (—CH_3), for example, can then lose its amino group (—NH_2), producing a thymine (Fig. 26.12). The thymine is mismatched with guanine, the partner of its predecessor. But since thymine is a normal base, there is no way to determine whether the incorrect base is the thymine in one chain or the guanine in the other.

The methylated cytosine problem arises in an interesting way, one that involves both costs and benefits. In bacteria, enzymes known as ***endonucleases***, present in the cytoplasm, break up the DNA of invading viruses. (The DNA-chopping ability of certain endonucleases forms the basis of recombinant DNA research.) If the enzymes digest invading DNA, why don't they also cut up the bacterial chromosome? The answer is that in the bacterial DNA the cytosines in the sequences the endonucleases find and cut are methylated, so the endonucleases cannot bind to them. But bacterial DNA protected in this way from endonuclease action is susceptible to mutation by conversion of the cytosines into thymines. Furthermore, viral DNA that escapes the endonucleases long enough to be replicated within the cell may itself be methylated at the appropriate places by the bacterial enzymes. It then becomes undetectable to the endonucleases and can more easily subvert its host. In eucaryotes, methylation is thought to serve primarily as a basis for gene regulation.

TRANSCRIPTION AND TRANSLATION

In the last chapter we saw how a complex set of enzymes makes new copies of the DNA in chromosomes and corrects most errors that arise. But the all-important sequence of bases in the DNA must not only be replicated and repaired, it must be used to produce both the proteins that form cellular structures and the enzymes that direct cellular metabolism. In this chapter we shall look at how information is actually encoded in DNA, how the information is transcribed into ribonucleic acid (RNA), and how some of this RNA is used directly and the rest—messenger RNA—is translated into protein.

TRANSCRIPTION

With only a few exceptions, every cell has a full set of chromosomes—a set replicated and repaired and passed on at every cell division. But at any given moment, only a small proportion of the thousands of genes in the cell are active. Which genes are needed depends not only on what the cell is doing—dividing, growing, resting, moving—but on the cell's environment and, if it is part of a complex multicellular organism, on its specialty. The small fraction of the information in the DNA that is necessary at any particular time is read out by enzymes very selectively.

In this chapter we shall look at how the enzymes recognize the active genes and ignore the rest, and how they know where genes begin and end. (Replication, which is an all-or-none process, does not make any distinction between genes.) In a later chapter we shall take up the question of how genes are activated and inactivated according to the needs of the cell.

HOCH$_2$ O OH

H H
H H
OH **OH**
Ribose

HOCH$_2$ O OH

H H
H H
OH H
Deoxyribose

27.1 Ribose and deoxyribose
The two five-carbon sugars differ only at the site shown in color, where deoxyribose lacks an oxygen atom that is present in ribose.

Messenger RNA The first question to arise when study of the flow of information from DNA to protein began in the 1950s was whether proteins are synthesized directly off the DNA through the intervention of the appropriate enzymes, or indirectly by means of some intermediary substance. Since in eucaryotes nearly all the DNA is found in the nucleus, while the process of protein synthesis was known to take place in the cytoplasm, most researchers suspected that a molecular middleman must exist, but the identity of this substance was unknown.

In the early 1940s several molecular biologists had shown that cells in tissues such as the vertebrate pancreas, where protein synthesis is particularly active, contain large amounts of RNA. Since this nucleic acid is present in only limited quantities in cells that do not produce protein secretions, such as those in muscle and kidney, there seemed to be a strong correlation between protein synthesis and RNA. Moreover, it had long been known that RNA, unlike DNA, occurs in the cytoplasm as well as in the nucleus. Subsequent radioactive-tracer experiments demonstrated that RNA is synthesized in the nucleus and moves from the nucleus into the cytoplasm. All these lines of evidence suggested that RNA might be the chemical messenger between the DNA of the nucleus and the protein-synthesizing cytoplasmic ribosomes.

Though RNA and DNA are very similar compounds, they differ in three important ways, as we saw in Chapter 3: (1) The sugar in RNA is ribose, whereas that in DNA is deoxyribose (Fig. 27.1). (2) RNA has **uracil** where DNA has thymine. (3) RNA is ordinarily single-stranded, whereas DNA is usually double-stranded.

Despite these differences, it was obvious that DNA could easily act as a template for the synthesis of RNA. The synthesis would proceed in essentially the same way as that of new DNA: the two strands of a DNA molecule would uncouple and RNA would be synthesized along one of the DNA strands (Fig. 27.2). For every adenine in the DNA template, a uracil ribonucleotide, rather than a thymine, would be added to the growing RNA strand; for every thymine in the DNA, an adenine ribonucleotide would be added; for every guanine, a cytosine; and for every cytosine, a guanine (Fig. 27.3). In short, geneticists found that the synthesis of RNA, in a process now called **transcription**, could operate exactly like replication; the resulting strand of RNA, which would be complementary to the transcribed DNA, could act as the intermediary.

27.2 Transcription of a gene
As the polymerase complex moves along the DNA, it catalyzes transcription of only one of the two DNA strands.

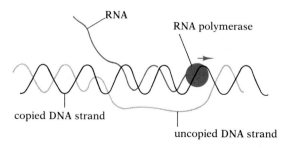

RNA

RNA polymerase

copied DNA strand

uncopied DNA strand

We now know that there are three major types of RNA: ***messenger RNA (mRNA)***—the type just discussed—carries the information necessary to specify the sequence of amino acids in a protein to the ribosomes where proteins are synthesized; ***ribosomal RNA (rRNA)*** forms part of the ribosomes; and ***transfer RNA (tRNA)*** brings amino acids to the ribosomes during protein synthesis. We shall look at how ribosomes work in a later section. In discussing transcription we focus on the synthesis of messenger RNA, but the process is similar for all three types.

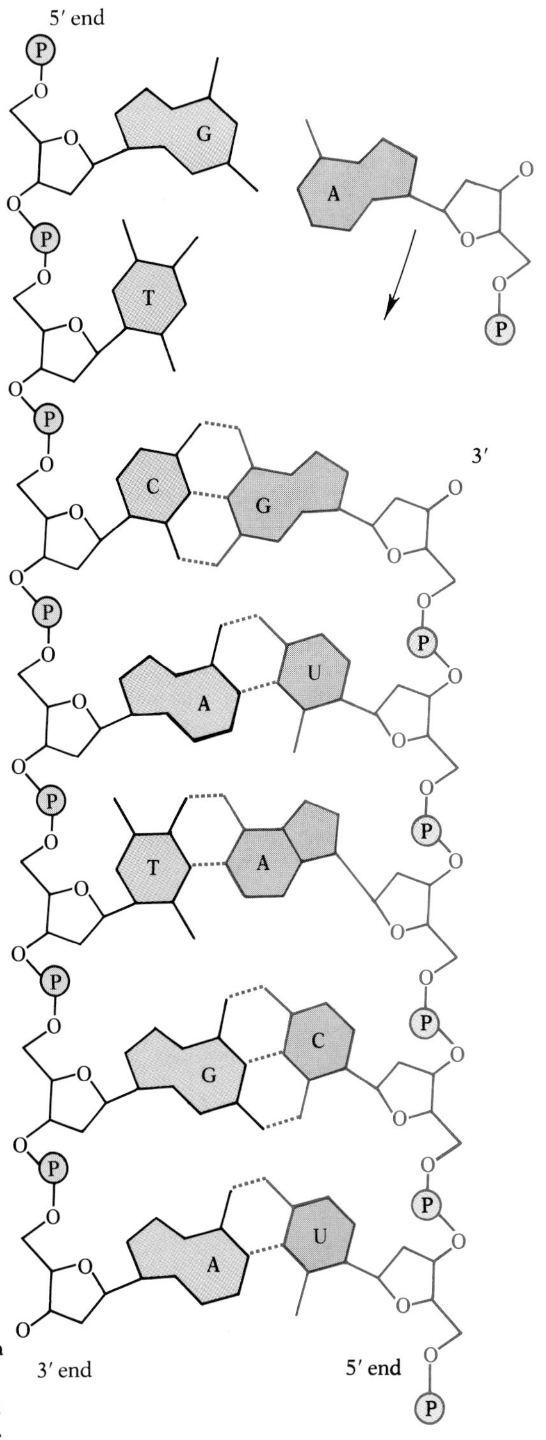

MECHANISMS OF TRANSCRIPTION

Transcription of DNA, with the resulting synthesis of RNA, is accomplished by an enzyme complex—***RNA polymerase***—which both binds to the DNA and opens up the helix. Like DNA polymerase, the transcription complex moves from the 3' end to the 5' end of the DNA, and synthesizes a strand of the opposite polarity (5' to 3'), in this case a strand of RNA, composed of ribonucleotides. In procaryotes, a single RNA polymerase is responsible for all RNA synthesis. In eucaryotes, RNA polymerase II is responsible for the synthesis of mRNA. Two other RNA polymerases synthesize ribosomal RNA and other RNAs that serve structural (and perhaps even enzymatic) roles; they work in much the same way.

Transcription in procaryotes The strategy just described raises an important question: how does RNA polymerase know where to start and stop synthesizing the messenger? The answer is that specific control sequences marking the beginning and end of a gene are embedded in the strands of DNA. The first pair of these is called the ***promoter***; active sites of the RNA polymerase bind specifically to regions with the base sequences of the promoter. The promoter in *E. coli*, reading from 5' to 3' on the complementary strand, usually begins with TTGACA or a very similar sequence.[1] This is followed by a sequence of roughly 17 bases with another function; then comes

---

[1] Though the RNA polymerase copies the template strand of the DNA, the custom is to refer to the sequences in question as they exist on the complementary strand. This method of reference is convenient because the RNA sequences produced in transcription are identical to those of the corresponding regions of the complementary DNA strand—except, of course, that U is substituted for T.

27.3 The synthesis of RNA by transcription of a DNA template
The sugar in RNA (ribose) is slightly different from that in DNA (deoxyribose), and uracil (U) takes the place of the thymine in DNA.

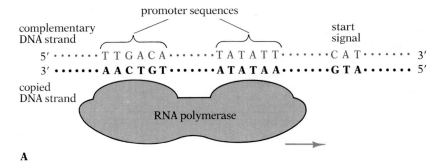

A

27.4   Transcription signal in *E. coli* DNA
(A) RNA polymerase binds to a region of DNA with the sequences shown. Once the polymerase is bound, transcription begins with the GTA signal (corresponding to the CAT sequence on the complementary strand) and proceeds until a termination signal is encountered. (B) The termination signal consists of a "self-complementary" sequence—a sequence that can bind to itself—and four to eight adenines (corresponding to thymines on the complementary strand). At this point, the tail of the mRNA forms a short double-helix hairpin, and the polymerase stops synthesizing. Note that RNA polymerase inserts a uracil wherever DNA polymerase would add a thymine.

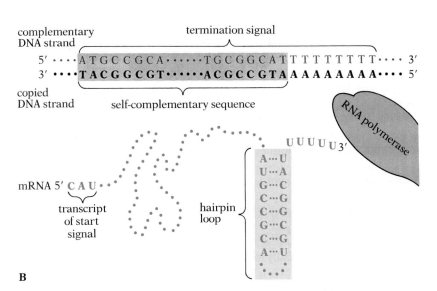

B

TATATT or a very similar sequence (Fig. 27.4A). The polymerase recognizes these two regions in the DNA and is large enough to bind simultaneously to both. The sequences shown in Figure 27.4 are called ***consensus sequences***, since each base shown is the one most commonly found at the location in question in procaryotes. (Much less is known about eucaryotic promoters, but the sequence TATA appears to be important in them as well.) It should be noted that the promoter sequences of most genes differ in one way or another from the consensus sequences; more than a hundred variations have been discovered in *E. coli*, for example. Minor variations in the promoter sequences cause the polymerase to bind less strongly and less often to some of them, so some genes are transcribed less frequently than others. We shall look at more sophisticated mechanisms for controlling rates of transcription in a later chapter.

Once bound, the polymerase does not begin synthesizing mRNA immediately. Instead, synthesis begins at the start signal—often CAT—located about seven bases beyond the binding point, toward the 3′ end of the complementary strand. In most procaryotes this synthesis then continues until the polymerase encounters a termination signal. The termination signal has two components. First, there is a region with a base sequence that permits the corresponding bases in the tail of the mRNA to pair off and form a small loop, known as a *hairpin loop* (Fig. 27.4B). This is followed on the complementary strand by a run of four to eight thymines. When the polymerase has completed transcription of the termination signal, transcription ceases and both the polymerase and the mRNA drift free of the DNA.

Transcription in eucaryotes—messenger RNA processing The mechanism of mRNA synthesis just described is found in procaryotes and their presumptive descendants, mitochrondria and chloroplasts. Transcription in eucaryotic nuclei is more complicated—so much so that its details are just being worked out. One complication is that eucaryotic mRNA is "tagged" on both ends: a "cap" of 7-methylguanosine is added at the front (5′) end, while a "tail" of 100–200 adenines is affixed to the 3′ end. The result—what biologists now refer to as the *primary transcript*—is still not a usable messenger. In some sense it is only a rough draft.

The dramatic difference between a primary transcript and functional mRNA came in 1977 as a total surprise: Phillip A. Sharp of the Massachusetts Institute of Technology discovered, while working with eucaryotic genes parasitized by a virus, that though the primary transcripts are about 6,000 bases long, the mRNA actually transcribed is only about one-third that length. Subsequent work has demonstrated the same pattern in transcripts of normal eucaryotic genes: large specific regions *within* the primary transcript of most eucaryotic messengers must be removed in the nucleus to create a functional mRNA molecule. The regions of the primary RNA transcript that survive this processing and operate during protein synthesis (as well as the parts of the gene that gave rise to these sections) are called *exons*, while the intervening sequences of the primary transcript, which are removed in the nucleus (and the corresponding regions of the gene), are referred to as *introns*. Experiments have shown that despite their early removal, many introns are necessary to the functioning of RNA: most mRNA transcribed from artificially manufactured genes lacking introns fails to get into the cytoplasm, while mRNA from genes with some introns intact is often processed correctly and slips through the nuclear pores to the ribosomes.

Intron removal is a formidable task. Some genes have as many as 50 introns, and a mistake of even a single base in the excision process can render the mRNA useless. The precise beginning and end of introns on the primary transcript must be marked by signals so that they can be recognized and removed. Like the promoter sequence and the start signals of DNA, these consist of specific base sequences, which indicate the boundaries:

intron (10^2–10^4 bases long)

5′ · · · · · AG*GUAAGU* · · · · · · · · *CAG*G · · · · · 3′
end of exon   beginning of exon

The signals are probably recognized by a short bit of RNA found in a curious RNA/protein complex known as the small nuclear ribonucleoprotein par-

27.5 Messenger RNA processing in eucaryotes
RNA polymerase binds to a promoter sequence and moves along the DNA strand, synthesizing a complementary RNA strand (A). A 7-methylguanosine cap is added at the beginning of the transcript by another enzyme. After the RNA polymerase transcribes the termination signal, the RNA between the last exon and the sequence of four to eight adenines is removed; a separate enzyme extends the poly-A tail by 100–200 bases, completing the primary RNA transcript (B). With the probable aid of a small nuclear ribonucleoprotein particle (snRNP), any introns are removed, and the resulting transcript is then spliced (C). Finally, the mature mRNA is ready for export to the cytoplasm (D). (The configuration shown in B and C is greatly simplified for clarity; the actual process probably involves a sequence of steps: binding of the snRNP to the exon I/intron boundary, cutting of the RNA chain at that point, movement of the complex to the intron/exon II boundary, cutting at that point, splicing of exon I to exon II, and finally, release of the intron. The snRNP may require the aid of free nuclear enzymes for some of these steps.)

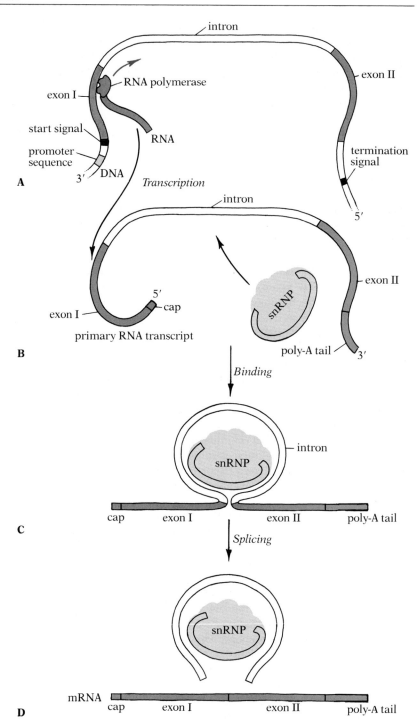

ticle—snRNP (Fig. 27.5). The RNA in this particle is like ribosomal RNA in that it is used directly, and has an enzymatic and/or structural role. It has a base sequence complementary to the boundary to which it binds, and it also binds to the sequence marking the end of the intron. Enzymes catalyze the breakage of the bond between the primary-transcript nucleotides at each intron/exon boundary; the intron drifts away (and is thought to be digested by other enzymes); and other enzymes splice the two exons. Once all the introns have been removed, the mature mRNA is exported to the cytoplasm.

TRANSLATION

Once the instructions in the DNA are transcribed into mRNA (directly in procaryotes, or with an intervening bout of intron removal in eucaryotes), the sequence of bases in the messenger must still be utilized in determining the arrangement of amino acids to form a polypeptide. This process, which occurs on the ribosomes, is called *translation*: information is translated from one molecular "language" to another. We shall look first at the two languages, and then at the process of translation.

THE GENETIC CODE

The codon When Watson and Crick discovered that DNA is essentially a linear array of four nitrogenous bases—adenine, cytosine, guanine, and thymine—it became clear that the unique sequence of amino acids found in a particular protein must be encoded by *groups* of bases. After all, if each DNA base designated a particular amino acid, there would have to be 20 different bases instead of four. If the bases were taken two at a time—AA, AC, AG, AT, CA, CC, and so on—only 16 combinations would be possible. Since 20 amino acids are present in proteins, the information has to be encoded in sets of three or more bases. These coding units are now called **codons**.

Before looking at how the codon's length was worked out, let us speculate, as Crick and his associates did, about constraints on the code. First, it is clear that if a specific series of bases codes for one amino acid, the genetic message can only be translated in one direction. Just as particular letters spell different words when their sequence is reversed—as in "tide" and "edit," for example—so particular bases will code for different amino acids when their sequence is reversed. And indeed it has turned out that a codon read backward usually specifies a different amino acid from that specified by the same codon read forward. Moreover, it seems clear that the code will have to indicate where translation is to start and stop. For the complex of RNA, structural proteins, and enzymes in the ribosome to begin even one base out of phase would alter the grouping of the subsequent bases into codons and make the message perfectly meaningless.

Crick and his associates at Cambridge University established the length of the codon in 1961. Bacteria infected with viruses were treated with compounds called acridines, which insert or delete nucleotides from DNA. Crick reasoned that adding or deleting a single nucleotide would render a message meaningless because it would disrupt the translation process by shifting the apparent starting point in subsequent coding units. Suppose the message is

THE BIG RED ANT ATE ONE FAT BUG

Deletion of the first *E* will make it

THEB IGR EDA NTA TEO NEF ATB UGX

A single deletion redefines all subsequent codons, and if it occurs near the beginning of the message, the sequence of the protein is completely disrupted. Two deletions, Crick reasoned, would have the same effect, but if the number of deletions corresponded to the codon length, and if they all occurred near the beginning, translation would produce a mostly correct protein, an enzyme that would probably retain some activity:

THEBI GRED ANT ATE ONE FAT BUG

Crick and his associates used various concentrations of acridines to make different numbers of nucleotide deletions, and determined for each concentration whether or not active enzymes were produced. Subjecting these results to elaborate statistical analyses, they concluded that codons were three bases long.

Deciphering the code The next problem to be solved in understanding translation was to determine the exact relationships between codons and amino acids. Marshall W. Nirenberg and Heinrich Matthaei of the National Institutes of Health used an enzymatic process developed by Severo Ochoa at New York University to link nucleotides together to create synthetic RNA. When uracil was the only nucleotide available to the enzyme, for instance, a long polyuracil chain of mRNA was synthesized: UUUUUUUUU. Similarly, when adenine alone was supplied, a polyadenine chain was formed.

When polyuracil obtained by this method was used instead of normal mRNA for protein synthesis in an artificial environment, a polypeptide chain composed only of the amino acid *phenylalanine* resulted—a clear indication that the codon UUU codes for phenylalanine. Nirenberg and Matthaei also showed that AAA codes for *lysine*, GGG for *glycine*, and CCC for *proline*.

It was understandably more difficult to interpret codons composed of two or three different nucleotides. If, for example, uracil nucleotides and guanine nucleotides were made available to the enzyme in a ratio of 2 to 1, artificial mRNA in which the codons GUU, UGU, and UUG predominated was synthesized. When this mRNA was used as a template in polypeptide synthesis, the amino acids *cysteine, valine,* and *leucine* were the principal ones incorporated into the polypeptide, but it was impossible to determine by this technique which of the three triplets coded for which of the amino acids.

In 1964 Nirenberg and Philip Leder of the National Institutes of Health developed a technique for getting ribosomes to bind to RNA trinucleotides (three nucleotides bonded together in sequence) of known composition. A trinucleotide, acting as though it were a short piece of mRNA, would then cause the ribosome to bind a specific amino acid—an early step in translation. For example, if they used a ribosome that had bound to a trinucleotide composed of UUU, phenylalanine would couple with it. Since it is relatively easy to synthesize trinucleotides with a particular base sequence, each of the 64 possible triplets could be synthesized and complexed with ribosomes, and the resulting amino acid association would then be determined. A few trinucleotides were not entirely specific in their binding, however, so some ambiguity in codon interpretation remained.

Shortly after the trinucleotide-binding technique became available, procedures for producing synthetic mRNA polymers with known repeating sequences (such as AAGAAGAAG · · ·) were developed by H. G. Khorana at the University of Wisconsin, and these helped resolve the remaining ambiguities in the code. Table 27.1 summarizes the genetic code.

As the genetic dictionary of the table makes clear, all but two amino acids (methionine and tryptophan) are represented by more than one codon. The synonymous codons—those coding for the same amino acid—usually have the same first two bases, but differ in the third; thus CCU, CCC, CCA, and CCG all code for proline. In fact, U and C are always equivalent in the third position, and A and G are equivalent in 14 of 16 cases. Notice also that three codons—UAA, UAG, and UGA—serve as "punctuation," marking the end of a message.[2]

[2]Certain minor inconsistencies in the genetic code have been uncovered. For example, in some ciliates the codons UAA and UAG, which are normally termination signals, instead code for glutamine; in one bacterium, UGA, also normally a termination signal, codes for tryptophan.

TABLE 27.1 *The genetic code (messenger RNA)*

First base in the codon	Second base in the codon				Third base in the codon
	U	C	A	G	
U	Phenylalanine	Serine	Tyrosine	Cysteine	U
	Phenylalanine	Serine	Tyrosine	Cysteine	C
	Leucine	Serine	*Termination*	*Termination*	A
	Leucine	Serine	*Termination*	Tryptophan	G
C	Leucine	Proline	Histidine	Arginine	U
	Leucine	Proline	Histidine	Arginine	C
	Leucine	Proline	Glutamine	Arginine	A
	Leucine	Proline	Glutamine	Arginine	G
A	Isoleucine	Threonine	Asparagine	Serine	U
	Isoleucine	Threonine	Asparagine	Serine	C
	Isoleucine	Threonine	Lysine	Arginine	A
	Methionine	Threonine	Lysine	Arginine	G
G	Valine	Alanine	Aspartic acid	Glycine	U
	Valine	Alanine	Aspartic acid	Glycine	C
	Valine	Alanine	Glutamic acid	Glycine	A
	Valine	Alanine	Glutamic acid	Glycine	G

RIBOSOMES

The process of translation takes place on ribosomes. A ribosome consists of a large and a small subunit, each a complex of ribosomal RNA, enzymes, and structural proteins (Fig. 27.6). As we saw in Chapter 5, the nucleolus is the site of synthesis and preliminary assembly of ribosomes. When not carrying out protein synthesis, the ribosomal subunits exist in the cytoplasm as separate entities. The large subunit of procaryotic ribosomes is composed of two molecules of rRNA and roughly 35 proteins; the small subunit has one molecule of rRNA and approximately 20 proteins. Eucaryotic ribosomes are somewhat larger. In both eucaryotes and procaryotes, the rRNA itself is highly structured: long segments of complementary base pairs are held together in a helical chain by hydrogen bonds, creating a complex pattern of arms and loops (Fig. 27.7). Though the function of this structure is not yet known, it is probably crucial to ribosomal activity.

In procaryotes, the 5' end of the mRNA binds first to the small subunit, which is then able to bind the large subunit (Fig. 27.8). The binding of the mRNA probably involves base pairing between a section of the rRNA and a binding signal—usually AGGAGGU, near the end of the mRNA. This signal binds in the small subunit's cleft (Fig. 27.8B–C), along with the initiation codon—AUG—located just a few bases farther along on the mRNA. When binding of the large subunit is complete, translation can begin. AUG codes for methionine, so methionine is always the first amino acid incorporated during translation.

Because procaryotes have no nuclear membrane, and the process of transcription is not segregated from the ribosomes and other translational

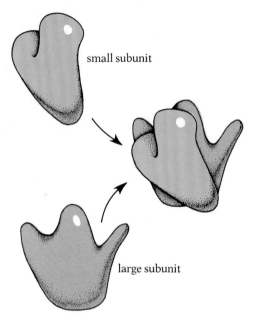

27.6 A ribosome
A functional ribosome is formed when the two kinds of independent subunits join. This joining can take place only after mRNA binds to the small subunit.

small subunit

large subunit

27.7 Ribosomal RNA
Ribosomal RNA has many stretches of complementary base pairs (color), which allow the chain to fold back on itself and form double-stranded helical arms. The open loops represent areas in which the bases are not complementary. This arrangement probably plays an important role in ribosomal function. The rRNA shown here is found in the small ribosomal subunit of an archaebacterium. Its three-dimensional structure is not yet known.

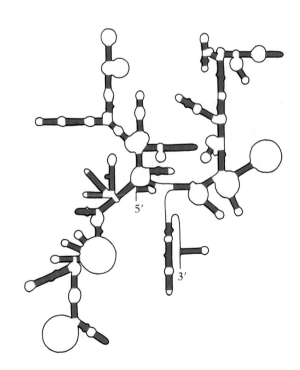

5'

3'

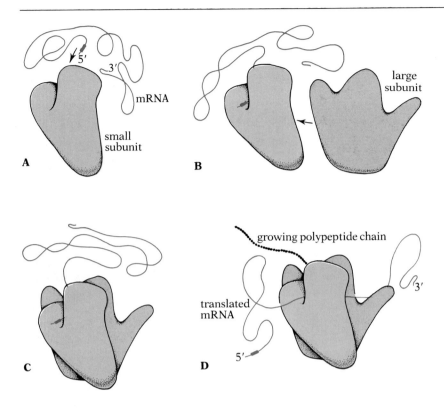

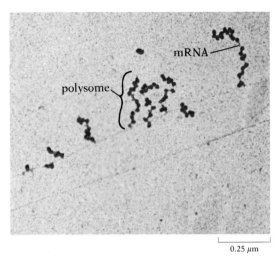

0.25 μm

27.8 Synthesis of a polypeptide chain by a ribosome
Free ribosomes exist as two separate subunits (A–B). A signal
sequence on the mRNA binds to the small subunit, causing a
change that permits it to bind the large subunit (B–C). The
ribosome then translates the mRNA into a polypeptide chain,
producing a protein (D). The mRNA is much longer than shown
here, and additional ribosomes can bind and begin translation
while the first ribosome is still at work.

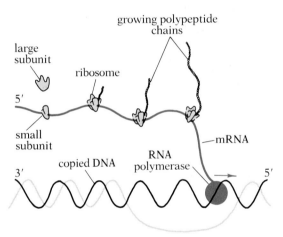

27.9 Simultaneous transcription and translation
In procaryotes and in eucaryotic organelles,
translation can begin as soon as the first part of a
message has been transcribed from the chromosome,
and several ribosomes can be involved
simultaneously. The complex of mRNA and two or
more ribosomes is often called a polysome. In the
micrograph, the two thin horizontal lines are DNA;
the upper strand is being transcribed and,
simultaneously, the RNA transcripts are being
translated by ribosomes (dark spots). Note that the
transcripts become longer from left to right as
transcription of the gene proceeds. The growing
polypeptide chains are not visible.

machinery in the cytoplasm, ribosomes are able to bind one end of an
mRNA molecule and begin translating it into protein while the RNA polymerase is still transcribing the rest of the message from the chromosome.
(The same situation occurs in translation in eucaryotic chloroplasts and
mitochondria.) In both procaryotes and eucaryotes, while the first ribosome to bind is translating later parts of the message, additional ribosomes
can bind and begin translation (Fig. 27.9).

In eucaryotes, four steps intervene between transcription and translation. We have already discussed three of these: modification of the transcript at the 5′ end, modification at the 3′ end, and splicing of the RNA after
the removal of introns (Fig. 27.5). The fourth step is movement through the

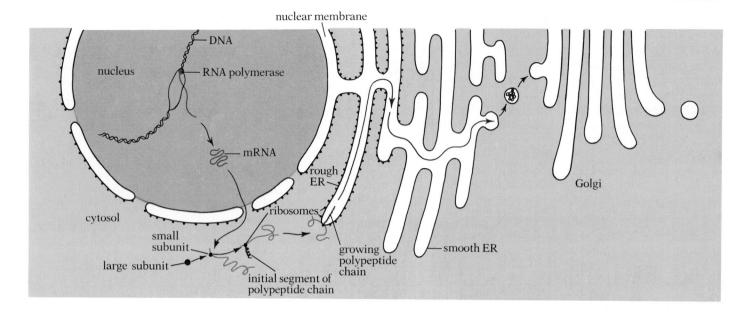

27.10 Synthesis of a hydrolytic enzyme

Though most mRNA is translated freely in the cytosol, when protein products are destined for the ER lumen, the ER membrane, or secretion, translation of mRNA occurs on the rough ER. A gene is transcribed, and the transcript is processed to form mRNA. Next, the mRNA passes through the nuclear pore and binds to a small ribosomal subunit. A large ribosomal subunit then binds and translation begins and continues until a signal sequence on the mRNA or on the growing polypeptide chain binds to the ribosome and interrupts translation. The ribosome complex then binds to the rough ER and translation resumes, with the growing polypeptide chain passing into the lumen of the ER. Finally, the newly synthesized protein, in this case a hydrolytic enzyme, moves to the smooth ER, where it is packaged into a vesicle for transport to the Golgi apparatus. The nuclear membrane provides many possible passageways for mRNA to the cytosol, as can be seen in the micrograph of a small section of nuclear membrane from a frog oocyte.

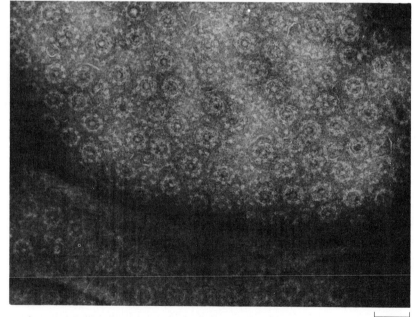

0.2 μm

pores of the nuclear membrane into the cytosol, where unbound ribosomes are available to do translation. Most RNA is translated by ribosomes in the cytosol, but the mRNA for certain proteins is always translated on the rough ER. The mRNA of this type begins the process of translation in the normal way, by binding to a small ribosomal subunit in the cytosol and thereby making possible the binding of the large subunit and the start of translation. But translation ceases almost immediately; probably the mRNA or the partially synthesized protein contains a signal sequence that binds to a receptor on the ribosome, causing the process to stop. The ribosome complex is also altered in some way that allows it to bind to the rough ER. The partially synthesized protein is somehow inserted into a channel, of the ER, and translation then resumes, with the growing chain of amino acids fed into the lumen of the ER. At the end of translation, the ribosomal subunits release the mRNA, separate, and drift free of the ER. As we saw in Chapter 5, the proteins synthesized on the rough ER may become lumen enzymes or ER membrane enzymes, or they may be packaged into secretory or transport vesicles in the smooth ER (Fig. 27.10). Most proteins synthesized on ribosomes in the cytosol, by contrast, remain in the cytosol.

TRANSFER RNA AND ITS ROLE IN TRANSLATION

We have said that in eucaryotes, mRNA is synthesized on DNA in the nucleus, moves to the ribosomes, and functions as a template for protein synthesis. Early researchers, in their attempt to understand translation, tried to ascertain whether the amino acids interact directly with the mRNA, or indirectly by way of some intermediate agent or adapter molecule.

In 1957 Mahlon Hoagland and his associates at Harvard demonstrated that each amino acid becomes attached to some form of RNA *before* it arrives at the ribosome, where it is added to the growing polypeptide. This new kind of RNA, which is neither mRNA nor rRNA, acts, they hypothesized, to transfer amino acids to the ribosomes; they called it transfer RNA. Hoagland and his associates demonstrated conclusively that each tRNA molecule binds a single molecule of amino acid and transports it to a ribosome.

Subsequent investigation by many researchers has yielded considerable information about tRNA structure and function. We now know that at least one form of tRNA is specific for each of the 20 amino acids; the amino acid arginine, for example, combines only with tRNA specialized to transport arginine; leucine combines only with tRNA specialized to transport leucine; and so on. But all forms of tRNA share certain structural characteristics: a length of 73 to 93 nucleotides, a structure consisting of a single chain that is folded into a cloverleaf shape, with internal base pairing (Fig. 27.11), and a CCA termination sequence. When an amino acid binds to its specific tRNA, it always binds at the CCA end.

Specific enzymes (aminoacyl tRNA synthetases) match each amino acid with its appropriate tRNA. Each enzyme is specific for a particular amino acid and its tRNA, and catalyzes a reaction that binds the two together. The tRNA then carries the amino acid to a ribosome that has bound a strand of mRNA. There the tRNA becomes attached to the mRNA by complementary base pairing. We have seen that the codon for each amino acid is a sequence of three bases on the single-stranded mRNA. Each tRNA has, in an exposed position, an unpaired triplet of bases called an **anticodon,** which is complementary to an mRNA codon for its particular amino acid. For example, the

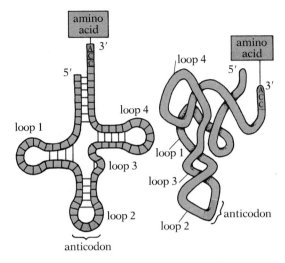

27.11   Structure of a molecule of transfer RNA
The single polynucleotide chain folds back on itself, forming five regions of complementary base pairing, four loops with unpaired bases, and an unpaired terminal portion to which the amino acid can attach. The unpaired triplet that acts as the anticodon is on the second loop. Loop 4 is thought to function in binding the tRNA to the ribosome (probably by binding to rRNA), and loop 1 probably binds an activating enzyme. The molecule is shown flattened at left; its tertiary structure is diagramed at right.

27.12   Translation cycle on the ribosome

After messenger RNA created by transcription binds to a small ribosomal subunit, a large subunit can bind, and translation begins. At this point the tRNA with the anticodon for methionine (which was bound to the small subunit before the two subunits combined) is already paired with the initiation codon on the mRNA at the P-site (A). Another tRNA, bearing the appropriate anticodon, pairs with the codon at the vacant A-site (B). Next the amino acid that was attached to the tRNA at the P-site is detached and bound to the amino acid on the tRNA at the A-site. The P-site tRNA then drifts free (C). Finally, the ribosome moves so that the remaining tRNA occupies the P-site, and the next codon is brought into position at the A-site, ready to accept another complementary anticodon (D). In these drawings part of the small subunit is shown cut away to reveal the mRNA, which lies in the groove between the subunits.

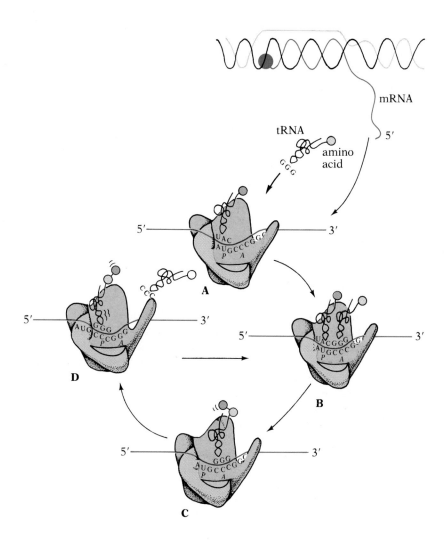

mRNA codon CCG codes for the amino acid proline. One type of tRNA for proline has the anticodon sequence GGC, which is complementary to CCG; similarly, the codon GUA, which codes for valine, has its complement in the tRNA for valine with the anticodon CAU. When a molecule of the proline tRNA with an attached molecule of proline approaches a molecule of mRNA, its exposed GGC triplet can bind to the mRNA only where the mRNA has a CCG triplet; similarly, the exposed CAU triplet of the valine tRNA can bind to the mRNA only at points where a GUA triplet occurs.

EVENTS ON THE RIBOSOME

The actual binding of tRNA molecules and linkage of the amino acids they carry is managed in step-by-step fashion by the large subunit of the ribosome, but only after completion of the preliminaries: first, the tRNA with the anticodon sequence for methionine is bound to the small subunit; in eucaryotes as well as procaryotes this sequence is complementary to the initiation codon on the mRNA, which can now also bind to the small subunit; finally, the large subunit is bound to the small one. As translation then proceeds, the ribosome moves along the mRNA, with the part of the message being translated lying in the groove between the two subunits; two adjacent tRNA binding sites—the *P-site* and the *A-site*—bring the appropriate nucleotide sequences together. When the cycle begins, the anticodon of the first tRNA is in register with the initiation codon of the mRNA at the P-site (Fig. 27.12A). The adjacent A-site then brings a molecule of tRNA with the appropriate anticodon together with the next codon triplet on the mRNA (Fig. 27.12B). Once the new tRNA is bound to the A-site, the enzyme peptidyl transferase moves the amino acid from the P-site tRNA and binds it to the amino acid at the A-site (Fig. 27.12C). The tRNA at the P-site, having relinquished its amino acid, is released, and the cycle is completed as the ribosome moves along the mRNA by one codon, thereby bringing the codon that was at the A-site (and its tRNA, with the growing polypeptide chain) to the P-site (Fig. 27.12D). As a result, the codon to be translated next now occupies the A-site. The translation cycle adds about 15 amino acids per second to the polypeptide chain. It comes to an end when the ribosome encounters one of the termination codons and, with the aid of a protein known as the termination factor, causes the completed protein to be released. The details of how ribosomes work—of how they help "select" the appropriate tRNA from the rich supply of different tRNA molecules, for example—are not known.

The ribosome complex's potential for producing proteins of many sizes and characteristics is enormous. Since the average protein chain is 300–500 amino acids long, a ribosome can produce a chain in about 25–35 seconds. Ribosomes outnumber mRNA by a factor of 10 in most cells, and one mRNA can be translated by many ribosomes simultaneously, so a new protein chain can be generated from each mRNA about every 3 seconds. A typical active eucaryotic cell has on the order of 300,000 mRNA molecules in circulation to direct synthesis of the proteins necessary for self-maintenance; as a result, if sufficient raw materials are available, 100,000 protein chains can be synthesized every second. And some cells—particularly those specialized for secretion or rapid growth—are far more active than this.

A SUMMARY OF INFORMATION FLOW

Let's now review the flow of information in a typical eucaryotic cell. First, when the double-stranded DNA that constitutes a particular gene is activated, RNA polymerase is able to recognize and bind to the promoter sequence. The DNA acts as a template for synthesis of a molecule of single-stranded messenger RNA. Transcription of the DNA begins at a start signal and ends after transcription of a termination sequence. The resulting transcript is capped and tagged with a polyadenine tail to form the primary transcript, and the introns are removed and the remaining segments are spliced—all steps that are unnecessary in procaryotes. The mature mRNA leaves the nucleus and moves into the cytosol, where it becomes associated with ribosomes.

The mRNA, carrying information in the form of three-base codons, acts as the template for synthesis of polypeptide chains. As the ribosomes move along the mRNA, they read the codons, starting at the 5′ end. Amino acids to be incorporated into the polypeptide chains are picked up by molecules of transfer RNA specific for each of the 20 amino acids.

Each molecule of tRNA has an exposed anticodon (a sequence of three unpaired bases) complementary to the mRNA codon (also a sequence of three unpaired bases) that codes for its particular amino acid. After picking up an amino acid from the cytosol with the help of a match-making enzyme, the tRNA moves to a ribosome and attaches to the mRNA at a point where the appropriate codon occurs. This ordering of the tRNAs along the mRNA molecule also orders the amino acids attached to them. As the amino acids are moved into the proper sequence in this way, peptide linkages are formed between them. Their amino acids once unloaded, the tRNAs uncouple from the mRNA and move away to pick up another load. When a ribosome reaches a termination codon, it releases the completed polypeptide chain.

Quite simply, then, the DNA of the gene determines the mRNA, which determines protein structure, which controls chemical reactions and produces the characteristics of the organism.

MUTATION

Types of mutation It is now apparent why mutations—alterations in the sequence of bases in the DNA—can change the information content of genes and thus produce new alleles. As we know from our discussion of DNA repair at the end of Chapter 26, genes are subject to various mutational events. Here we shall examine briefly some of the consequences of these events.

Two types of mutation that are particularly harmful are **additions** and **deletions** of single bases. These usually result in the production of inactive enzymes because at the point of the insertion or deletion the ribosomes begin to translate incorrect triplets, and the original meaning of subsequent codons is lost. Synthesis of the polypeptide chain may even be stopped too early (because the shift in the reading of the codons has created a spurious termination signal) or too late.

A third type of mutation is **base substitution** (also called point mutation), in which one nucleotide is replaced by another. For example, a codon that normally has the base composition CGG may be changed to CAG, which codes for a different amino acid. Base substitution is not as serious as addition or deletion, however, because in most codons a change in the third base does not alter the meaning (Table 27.1). Roughly 30 percent of all base substitutions therefore have no effect. Moreover, even when an amino acid is changed, the substitution usually has relatively little effect if, for example, one nonpolar amino acid replaces another; on the other hand, exchanges that occur between polar and nonpolar amino acids, or between positively charged and negatively charged amino acids, are likely to cause trouble. All in all, just over half of all base substitutions are likely to affect protein function strongly.

Another very common (and until recently, unknown) type of mutation is *transposition*. This phenomenon, which we shall discuss in detail in Chapter 30, results from the insertion of long stretches of DNA from one part of the genome into the middle of another. The majority of easily detectable spontaneous mutations in *Drosophila*, for example, are known to result from transposition.

Mutagenic agents   High-energy radiation, including both ultraviolet light and ionizing radiation such as X rays, cosmic rays, and emissions from radioactive materials, can cause mutations. In addition, a variety of chemicals have been found to be mutagenic. A normal spontaneous mutation rate for a single gene is one mutation in every 10^6–10^8 replications, but the rate can be greatly increased by unusual exposure to mutagenic agents.

Ionizing radiations sometimes induce simple base substitutions, but they also frequently produce large deletions of genetic material, presumably because their high-energy particles collide with the DNA molecules and cause breaks to occur.

As we saw in the last chapter, ultraviolet light most often exerts its mutagenic effect by causing abnormal bonding of thymine bases. When two adjacent thymine units absorb ultraviolet light and are thus energized, they often bond to each other, forming a thymine dimer (Fig. 27.13). An unrepaired dimer can inactivate a strand of DNA; not only can no mRNA be transcribed from it, but—more important for dividing cells—DNA replication cannot take place.

Some mutagenic chemicals produce their effects by directly converting one base into another. For example, nitrous acid (HNO_2) is a very powerful mutagen that deaminates cytosine, changing it into uracil.[3] Other chemical mutagens, of a type called base analogues, are themselves sometimes incorporated into nucleic acids in place of one of the normal bases. An example is 5-bromouracil, an analogue of thymine. When a strand of DNA contains a unit of 5-bromouracil instead of thymine, it is prone to errors of replication, because the 5-bromouracil will sometimes pair with guanine rather than with the requisite adenine. Virtually all these mutations are detected and repaired before they can exert any effect.

There is now a wealth of evidence for a close correlation between the mutagenicity of a chemical and its carcinogenicity—between the potential of a chemical for producing genetic mutations and its cancer-inducing activity. This fact strongly suggests that many cancers are caused, at least in part, by mutations in somatic cells. Thus we see that alterations of the DNA are important not only in germ cells, where they may affect future offspring, but also in somatic cells, whose metabolism or growth they may disrupt, causing disease or degeneration.

The connection between mutagenicity and carcinogenicity is the basis for the widely used Ames test,[4] in which such chemicals as environmental pollutants, reagents used in industrial processes, proposed new drugs, and

27.13   A thymine dimer within a DNA molecule
Two adjacent thymine nucleotides are bonded to each other covalently, so they cannot form hydrogen bonds with the adenine nucleotides of the complementary strand. Such a mutation, often induced by ultraviolet light, inactivates the DNA.

---

[3] Nitrous acid also converts adenine into hypoxanthine and guanine into xanthine.

[4] Named for Bruce N. Ames of the University of California, Berkeley.

food additives are screened for potential carcinogenicity. The compound to be tested is added to a culture of about one billion bacteria; a special mutant strain of *Salmonella* is used, which requires histidine as a nutrient. When the mixture of bacteria and chemical is incubated on a medium deficient in histidine, some cells undergo a mutation that is the reverse of their original mutation, and they thus regain the ability to synthesize histidine and to grow on the deficient medium. After several days a count is made of the number of so-called revertant colonies derived from these cells. Any increase in the mutation rate of the bacteria over the normal spontaneous level is then used to predict the likely cancer-inducing potency of the chemical.

In early screening tests some substances known to be potent carcinogens gave negative results in the Ames test. It was soon realized that many chemicals not themselves mutagens or carcinogens are transformed in the mammalian body, especially in the liver, into derivatives that are mutagenic and carcinogenic. For this reason newer versions of the Ames test add liver homogenate to the bacterial culture, so that the bacteria will be exposed both to the original chemical and to its metabolic derivatives. We shall look more closely at the genetic bases of cancer in a later chapter.

NOW WHAT IS A GENE?

Thomas Hunt Morgan defined a gene as the smallest unit of recombination. This definition, however, assumed that recombination took place only between sequences determining a particular character; evidence now available indicates that recombination can occur between homologous chromosomes at any location, whether within a coding region or not. A more accurate definition is therefore necessary.

In the early 1940s George W. Beadle and Edward L. Tatum at Stanford University created, isolated, and analyzed a series of mutant colonies of the red bread mold, *Neurospora crassa*. They went on to show that each mutant was defective in a single metabolic enzyme, and that the mutation was transmitted by a single gene. Accordingly, they formulated a hypothesis we take almost for granted today: each gene codes for one enzyme—or, since enzymes are proteins, for one protein. Beadle and Tatum won a Nobel Prize for this important work. A little thought, however, will convince you that there are some crucial exceptions to this so-called one gene–one enzyme hypothesis. Hemoglobin, for example, consists of four polypeptide chains (see Fig. 3.25, p. 67); two of the chains are encoded by one gene while the other two are encoded by another. Here, then, synthesis of a single protein involves two genes. Furthermore, some genes do not code for proteins at all, but rather for rRNA, tRNA, and the like. We could say that a gene is a region of the chromosome that codes for an RNA, be it mRNA destined for translation or structural RNA; but in procaryotes a single piece of mRNA sometimes carries the instructions for several proteins.

At present, genes are generally defined as regions that code for single products, whether those products are RNAs to be used directly, enzymes, structural proteins, or polypeptide chains that must combine with other chains to form functional products.

Chapter 28

EXTRACHROMOSOMAL INHERITANCE

In the last two chapters we examined the mechanisms of replication, transcription, and translation of the major chromosome in procaryotes and the nuclear chromosomes of eucaryotes. But there are three other important sources of information: organelle DNA, plasmids, and viral nucleic acids. Plasmids and viruses figure prominently in disease and are utilized in many of the newest techniques of molecular biology, while the study of organelle genes is providing intriguing suggestions about the evolution of eucaryotic genetic mechanisms.

ORGANELLE HEREDITY

As we saw in Chapter 5, mitochondria and chloroplasts share several characteristics with procaryotes, including relatively small ribosomes and (usually) circular chromosomes. The most interesting interpretation of such similarities is that these organelles were once free-living procaryotes —probably purple bacteria in the case of mitochondria, and Cyanobacteria in the case of chloroplasts—which took up symbiotic residence in primitive eucaryotic cells. Most recent discoveries about organelle heredity are consistent with this view.

Organelle DNA   The original observation that led to the discovery that organelles have their own genes came in 1909, when Carl Correns reported that in plants called four-o-clocks (*Mirabilis jalapa*), variegation—the appearance of white patches on green leaves—was transmitted through the cytoplasm of the maternal gamete. Correns correctly guessed that chloro-

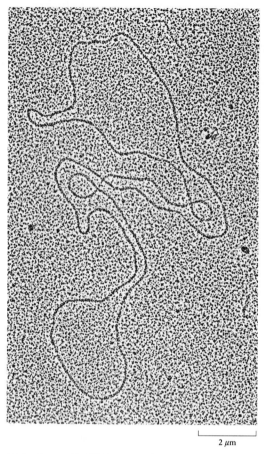

28.1 Organelle DNA
Electron micrograph of mitochondrial DNA from a
liver cell of a chicken.

plasts must contain genes and replicate themselves, and the variegation
occurs when one or more of these genes are defective.

The first evidence for mitochondrial genes did not come until 1938,
when T. M. Sonneborn discovered that some strains of *Paramecium aurelia*
carry a cytoplasmic gene which produces a poison (originally called the
kappa factor) that kills other strains. It subsequently became clear that the
poison is produced by a mitochondrial gene, and that mitochondria, like
chloroplasts, replicate themselves.

We now know that the DNA replication and the division of mitochondria
and chloroplasts occur out of phase with chromosome replication in the
nucleus, though the pace of cell division limits organelle division in some
way. The DNA in these organelles is almost always circular (Fig. 28.1), an
exception being the linear mitochondrial DNA of some fungi and Protozoa.
Each organelle contains several copies of its DNA, and this DNA is more
similar to procaryotic DNA than to the nuclear DNA of eucaryotic cells. It
lacks the histone nucleosome cores characteristic of nuclear DNA, for example, and is attached to the inner membrane of the organelle much as the
procaryotic chromosome is attached to the cell membrane. However, it has
considerably fewer base pairs than bacterial DNA. For instance, the *E. coli*
chromosome encodes about 3,000 products, while a typical animal mitochondrial genome encodes about 40. Since these are insufficient to carry
out organelle synthesis and operation (even in bacteria at least 90 gene
products are required just for replication, transcription, and translation), it
appears that in the course of evolution most of the genes necessary for organelle function have come to reside in the cell's nucleus.

Organelle genetics—recombination Organelle genes can undergo recombination. Organelle recombination is not prominent in animals and
most higher plants because in these organisms the maternal gamete—the
egg—contributes essentially all the cytoplasm, and hence all the organelles. But in other organisms, both parents contribute organelles to the
progeny, and the genes in maternal and paternal organelles are clearly able
to recombine. The recombination apparently occurs when two organelles
fuse, but much remains to be discovered about this phenomenon.

Transcription and translation in organelles On the whole, transcription
and translation in organelles are remarkably similar to the corresponding
processes in procaryotes. For example, since there is no nuclear membrane
to segregate a strand of mRNA from the ribosomes during its synthesis in an
organelle, translation can begin at one end before transcription of the DNA
has been completed at the other, just as it does in a procaryote (see Fig.
27.9, p. 719). The resulting transcript receives neither the 7-methylguano-
sine cap nor the polyadenine tail characteristic of mRNA produced in the
nucleus. Another similarity can be seen in the translation of the initiation
codon, usually AUG: eucaryotic tRNA pairs this codon with an unmodified
methionine, whereas procaryotes begin translation with a tRNA carrying a
formylated methionine; organelles too begin translation with a formylated
methionine. Correspondingly, the rRNAs and ribosomal proteins of organelles differ in size and composition from the rRNAs and ribosomal proteins
used for translation in the cytoplasm (Fig. 28.2), but are much closer in

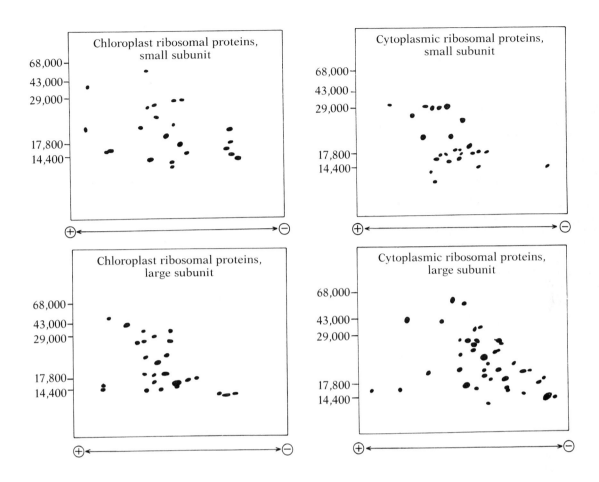

28.2 Proteins from chloroplast and cytoplasmic ribosomes
The proteins from large and small subunits of organelle and cytoplasmic ribosomes have been dissociated and separated by molecular weight (vertical axis) and charge (horizontal axis). It is evident that organelle ribosomes (left) have many fewer proteins than cytoplasmic ribosomes (right); furthermore, few if any of the proteins from the two sources are alike in weight and charge, an indication that they are also not alike in composition.

these respects to those of procaryotes. Indeed, the ribosomes of *E. coli* and chloroplasts have been shown to be essentially interchangeable when used in *in vitro* protein synthesis.

Along with these similarities, there are some differences. One striking feature of the genetic code is that it is the same in procaryotes and eucaryotes; organelles, however, use a slightly different dialect. For instance, the codon UGA, which is a termination signal in eucaryotes and most procaryotes, codes for tryptophan in mitochondria; AUA, which normally codes for isoleucine, designates methionine in mitochondria; and so on. The logic of these differences, if any, is not yet understood. Another complication is that in one respect yeast mitochondrial DNA resembles eucaryotic rather than procaryotic chromosomes: some genes have introns, which must be removed from the transcribed RNA before functional mRNA is produced.

Information flow between the nucleus and organelles As we have already noted, mitochondrial DNA usually encodes about 40 products. The entire base sequence (16,569 nucleotides) of human mitochondrial DNA is now known, as is most of the base sequence of yeast mitochondrial DNA; as a result, the products encoded by these DNAs can now be identified. Both of these kinds of mitochondrial DNA contain instructions for making two of the three rRNAs of the organelle ribosomes, all of their various tRNAs, one of the dozens of ribosomal proteins, one of the nine proteins of the F_1 complex (which makes ATP, utilizing energy stored by the respiratory electron-transport chain), and several other organelle proteins. The other rRNA and ribosomal proteins, the replication, transcription, and translation enzymes, the electron-transport and other F_1 proteins, and so on must be encoded by nuclear genes and transported to the organelles. The nucleus, therefore, contains nearly all the genes for two different kinds of ribosomes. Why don't the organelles themselves contain genes for all the products they require, or—alternatively—why don't the organelles utilize the same kinds of ribosomes, tRNAs, and other enzymes as the rest of the cell? The answers to these questions, when they are discovered, may reveal much about the evolution of eucaryotic life.

PLASMIDS

Plasmids are small circular molecules of DNA, found in both procaryotes and yeast, which came to light when the phenomenon of bacterial recombination was first discovered over twenty years ago (Fig. 28.3). Though yeast plasmids are proving very important in genetic engineering, most studies of plasmids have focused on bacteria.

The sex factor in *E. coli*   Since bacteria are haploid and reproduce asexually, by binary fission, their reproductive behavior was once thought not to include genetic recombination. However, in 1946 Joshua Lederberg and Edward L. Tatum, then at Yale University, demonstrated that bacterial recombination does take place. They isolated two mutant strains of *E. coli*, each lacking the ability to synthesize a particular pair of nutrients; the first mutant was unable to synthesize the amino acid methionine and the vitamin biotin; the second could not make the amino acids threonine and leucine. The mutants could be grown only on a medium that supplied the nutrients they could not synthesize. But when the two mutant strains were mixed on a minimal medium (one lacking all four critical nutrients), some healthy colonies were formed and these could subsist indefinitely on the minimal medium. Apparently the individuals in these colonies could synthesize all four nutrients. It seemed that they had somehow inherited both the first strain's ability to synthesize threonine and leucine and the second strain's ability to synthesize methionine and biotin. In short, they appeared to be the result of recombination of traits from the two original mutant strains. Lederberg and Tatum demonstrated that direct contact between the cells of the two strains was necessary for this recombination to occur.

A few years later several researchers showed that bacterial recombination like that found in *E. coli* is brought about through a process called *conjugation*, which is analogous to sexual mating in higher organisms. Two bacterial cells come to lie very close to each other, and a narrow cytoplasmic bridge, or pilus, extending from one connects to the other. This bridge

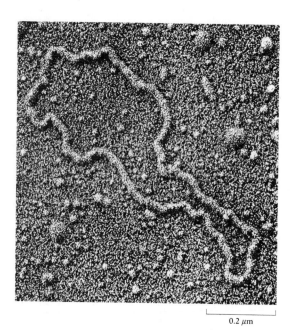

28.3 Electron micrograph of plasmid from *E. coli*
This plasmid carries genetic information for conferring resistance to the antibiotic tetracycline.

0.2 μm

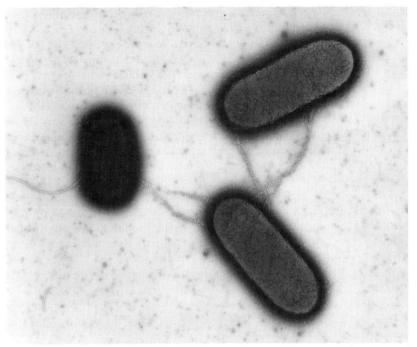

1 μm

28.4 Conjugating bacteria
Long cytoplasmic bridges, or pili, connect the lower cell with two others with which it is conjugating simultaneously.

is visible under the electron microscope (Fig. 28.4). Genetic material can pass through the bridge from one cell to the other.

Conjugation can occur only between cells of different mating types. A type F^- cell cannot conjugate with another F^- cell. Similarly, F^+ cannot conjugate with F^+. But F^- and F^+ can conjugate with each other. F^+ cells differ from F^- cells in containing in their cytoplasm a so-called *sex factor*, which is composed of DNA. Under most circumstances this factor can replicate within a nondividing cell, and a copy can easily be transferred from F^+, which acts as donor and is thus analogous to a male, to F^-, which acts as recipient and is thus analogous to a female. The products of $F^- \times F^+$ crosses are always F^+. If this were the only type of cross possible, all cells should eventually be F^+ and no further conjugation could take place. But conjugation in other types of crosses does not always convert the recipient cell from female into male, as we shall see.

F^+ strains usually include a very few cells that, when they conjugate, do not ordinarily transfer much sex factor to the F^- cells, but do transfer chromosomal DNA. These are called *Hfr* cells (for high-frequency recombination). Pure strains of *Hfr* bacteria have been isolated, and it is these that are generally used in experiments on conjugation. Crosses of the $F^- \times Hfr$ type do not usually convert the F^- cells into F^+ or *Hfr*; the female remains a female.

Bacterial chromosomes, you will recall, differ from those of eucaryotic organisms in being composed almost exclusively of DNA and in being circular. But both electron microscopy and studies of the genetic results of conjugation indicate that when chromosomal material is transferred from

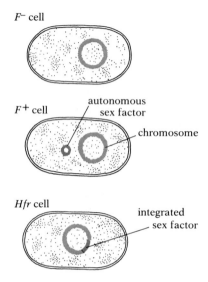

F⁻ cell

F⁺ cell — autonomous sex factor — chromosome

Hfr cell — integrated sex factor

28.5   F⁻, F⁺, and Hfr cells of E. coli compared
F⁻ cells lack the sex factor. *F⁺* cells have the sex factor free in the cytoplasm. *Hfr* cells have the sex factor integrated with the chromosome.

an *Hfr* cell to an *F⁻* cell during conjugation, that material is linear, not circular. These observations have led to the following hypothesis:

The sex factor, which is a type of plasmid, is absent from *F⁻* cells, but present in the cytoplasm of *F⁺* cells. The sex-factor plasmid contains genes that control pilus formation and also insert a component into the cell wall that prevents the attachment of other *F⁺* bacteria. Under most circumstances a copy of the sex factor can easily be transmitted from cell to cell, so mixing *F⁻* and *F⁺* cells usually results in conversion of many of the *F⁻* cells into *F⁺* cells. Hence, the sex factor can be seen as a genetic parasite, reproducing and spreading without necessarily aiding its host. At times, however, it plays a different role. On these occasions the sex factor becomes integrated into the circular main chromosome at a precise location, characteristic for each strain, and the result is an *Hfr* cell (Fig. 28.5). When an *Hfr* cell and an *F⁻* cell begin conjugation, synthesis of a linear chromosome on the template of the circular *Hfr* chromosome is initiated near one end of the sex factor. This newly synthesized linear chromosome moves (by an unknown mechanism) through the conjugation bridge into the *F⁻* cell (Fig. 28.6). Since the linear chromosome always moves during conjugation with most of the sex factor at the rear, and since transfer of the chromosome is seldom complete, an *F⁻* cell crossed with an *Hfr* cell does not ordinarily receive the entire sex factor and is therefore generally not converted into an *F⁺* or an *Hfr* cell.

When two conjugating cells break apart before transfer of the linear chromosome is complete, as they usually do, the *F⁻* cell retains whatever portion of the chromosome has already entered it. Thus the normally haploid cell becomes temporarily partially diploid; it has its own original chromosome plus part of a chromosome donated to it by the *Hfr* cell. Crossing over soon occurs between the *F⁻* cell's own chromosome and the recently acquired chromosomal fragment, so that some genes from the *Hfr* cell become part of the *F⁻* chromosome. The remaining chromosomal fragment, including the tip of the sex factor, is then destroyed, and the *F⁻* cell is restored to the haploid condition.

Because the rate at which the donor chromosome is transferred during conjugation is constant, and because within each strain transfer of the chromosome always begins at the same characteristic point, genes on the bacterial chromosome can be mapped with high precision. Suppose, for example, that we want to locate the gene for the enzyme β-galactosidase (an enzyme involved in lactose digestion in *E. coli*). To find (or create) a strain that lacks this enzyme, we expose bacteria to mutagens, and then look for a strain unable to grow when lactose is the only food source. This mutant strain will serve as recipient, while an *Hfr* strain with a normal gene for β-galactosidase will serve as the donor. When we mix the two strains together, in several groups, conjugation begins. After conjugation has proceeded for a specified period—say, 40 percent of the time required for complete transfer—we interrupt it in one group by shaking the mixture (which breaks the conjugation bridges) and chilling it. We do the same for another group after allowing time for 60 percent of the transfer, for another after allowing time for 90 percent, and so on. Then we see which recipients are able, as a result of recombination with the donor chromosome, to metabolize lactose. If recipients regain the ability to digest lactose after completion of at least 80 percent of a transfer, but not after completion of 60 percent, we can conclude that the β-galactosidase gene must lie somewhere between 60 percent and 80 percent of the way along the linear chromosome. By repeating the process, interrupting conjugation at various times between completion of 60 percent of a transfer and completion of 80 percent, we can zero in on the precise location.

Beyond its utility in mapping, the sex factor is interesting because its behavior differs from that of every other component of the cell so far discussed. At times it is apparently free in the cytoplasm as a very tiny piece of circular DNA. At other times it is incorporated into the chromosome and behaves like other chromosomal genes. François Jacob and E. L. Wollman of the Institut Pasteur in Paris have given the name *episomes* to plasmids that exhibit this dual behavior. All episomes (and most other episomes are variants of the sex factor that simply integrate into the main chromosome at other locations) may exist in the cell in either of two states: the autonomous or detached state, in which their replication is independent of chromosomal replication, and the integrated state, in which they are attached to the chromosome and replicate synchronously with it. Episomes are usually nonessential to the individual cell, as is illustrated by F^- strains of *E. coli*, which lack the sex factor; but when they are present, their effects may be very important. We shall look at similar behavior in viruses and in the curious movable elements in eucaryotic chromosomes called transposons.

Nonepisomal plasmids Most bacterial cells contain, in addition to episomes, plasmids that are never integrated into the main chromosome. In other words, unlike episomes, these plasmids exist only in autonomous form. Some such plasmids carry only one or two genes, while others may be as much as one-fifth the size of the main chromosome and carry many genes.

Especially important are plasmids with genes for resistance to various antibiotics, including streptomycin, tetracycline, and ampicillin (Fig. 28.3). When bacteria with resistance plasmids are exposed to one of these antibiotics, the plasmids immediately begin replicating, and a cell that originally had only two or three may soon have a thousand or more. It is interesting that antibiotic resistance is conferred by changes in the structure of the cell wall analogous to the alterations induced by episomes. On the other hand, plasmids may carry genes that make the bacteria virulent, as in the case of the pneumococci that cause pneumonia. As we shall see, plasmids are an invaluable tool in recombinant DNA technology.

VIRUSES AND VIRAL TRANSDUCTION

In 1892 several scientists demonstrated that the agent responsible for smallpox could pass through porcelain filters that easily trapped bacteria. Similar virulent entities, then called filterable viruses, were found to be responsible for a variety of plant and animal diseases. The reproduction and other details of viral biology, however, remained totally unknown.

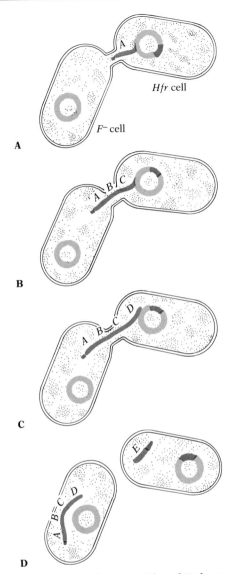

28.6 Conjugation between *Hfr* and *F⁻* bacteria
As a linear chromosome is synthesized on the circular chromosome of the *Hfr* cell, it moves into the *F⁻* cell. Conjugation usually stops before the entire chromosome has moved across; hence much of the sex factor (color) ordinarily remains in the *Hfr* cell. Since the chromosome moves at a fairly steady rate, disrupting conjugation at measured times after its inception permits mapping of the genes on the bacterial chromosome: only genes near the front end can be transferred and recombine if conjugation is interrupted early, while genes toward the rear can recombine if conjugation is permitted to go on longer. Complete transfer takes about 90 minutes.

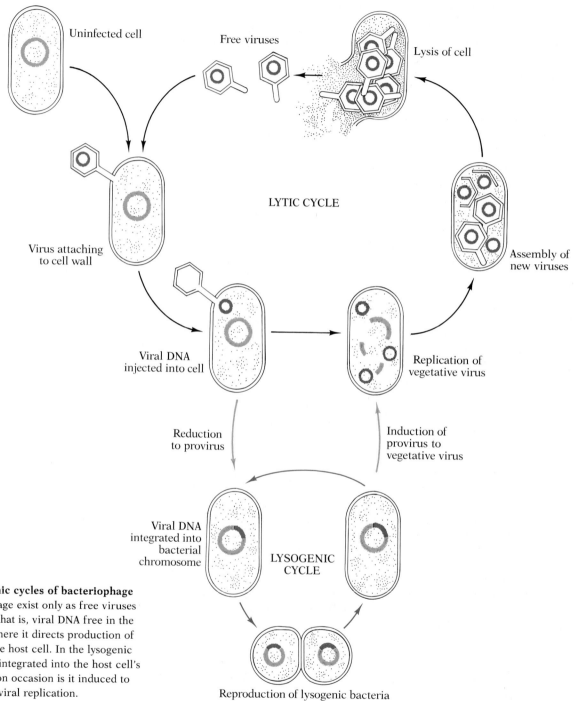

28.7 Lytic and lysogenic cycles of bacteriophage
In the lytic cycle the phage exist only as free viruses or as vegetative virus—that is, viral DNA free in the host cell's cytoplasm, where it directs production of new viral particles by the host cell. In the lysogenic cycle the phage DNA is integrated into the host cell's chromosome, and only on occasion is it induced to break loose and initiate viral replication. Bacteriophage are much smaller relative to their hosts than is indicated here.

REPRODUCTIVE STRATEGIES OF VIRUSES

In the 1940s Max Delbrück at the California Institute of Technology elucidated the interesting reproductive cycle of a common kind of bacteriophage, or phage, a virus that infects and kills bacteria. When he mixed a
number of these phage with a culture of bacteria, the viruses immediately
appeared to vanish, but half an hour or so later a hundred times as many
phage as he had started with suddenly appeared in the culture. Delbrück
hypothesized that phage of this kind enter the host cell, replicate themselves, and then cause the host cell to burst (lyse), releasing a new generation of these infectious particles.

In the years since Delbrück's classic experiment, viruses have become
one of the most valuable tools for studying how genes are organized, and
how they work. In an earlier chapter we described the work of Hershey and
Chase, which elucidated the normal life cycle of most bacteriophage. These
viruses inject their DNA into the bacterial cells they attack, but leave their
protein coats outside. When virulent phage (phage that kill the bacteria
they invade) do this, the viral DNA promptly takes control of the bacterial
cell's metabolic machinery and puts it to work manufacturing new viral
DNA and new viral protein. These two components are then assembled into
new infective phage. About 20–25 minutes after the initial injection of the
viral DNA, the bacterial cell lyses, releasing the new phage, which may then
attack other bacterial cells and start the *lytic cycle* over again (Fig. 28.7).
Since then, a variety of different viral strategies have been recognized: the
genetic material may be single- or double-stranded, circular or linear. For
example, some viruses, bacteriophage among them, contain linear double-
stranded DNA with dozens of genes, and are surrounded by elaborate protein coats (Fig. 28.8A, p. 736). The DNA of others is initially single-stranded,
and must be made double-stranded by the host's own DNA polymerase. Yet
others have no DNA at all, differing from all cellular organisms in having
RNA genes. They may carry single- or double-stranded RNA, which, in conjunction with the appropriate enzymes, can initiate protein synthesis without having to be transcribed first. Some RNA viruses include in their
genome a region that codes for an enzyme, RNA replicase, that catalyzes
production of new copies of the viral RNA (Fig. 28.8B); other RNA viruses,
known as *retroviruses*, carry an enzyme, reverse transcriptase, that catalyzes the formation of a DNA copy of the RNA (known as cDNA because it is
complementary to the RNA from which it is copied) and thus reverses the
usual flow of information from DNA to RNA (Fig. 28.8C). The oddest entities of all are unencapsulated *viroids*, tiny circles of naked RNA that can infect a variety of plant cells. The link between viruses, plasmids, and host
chromosomes is clearest in the so-called temperate viruses, which can infect cells without immediately killing them.

Temperate viruses   In 1953 André Lwoff and his colleagues at the Institut
Pasteur found that if they exposed certain apparently normal strains of bacteria to ultraviolet light or X rays or various chemicals, the bacteria would
lyse within an hour, releasing large numbers of infectious phage. Appar-

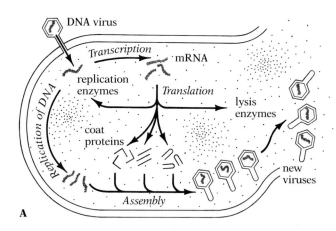

A

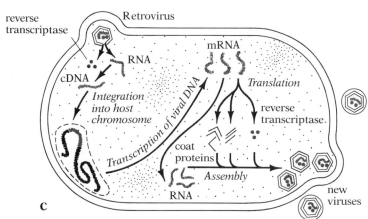

B

C

28.8 Three strategies of viruses

Viruses exploit their hosts in a variety of ways. The most common strategy is for a viral genome consisting of DNA to be both replicated and transcribed, and the transcripts translated, all by means of host enzymes. Transcription and translation generate coat proteins, enzymes that modify host-cell function, and (later) enzymes that lyse the host. Other DNA viruses utilize replication enzymes produced through transcription and translation, as indicated in the drawing (A). For RNA viruses (B) the transcription step is unnecessary, but most encode a special enzyme, RNA replicase, to catalyze replication of the RNA. Some RNA viruses lyse their hosts, as shown, but many others escape from their hosts without lysing them. Retroviruses (C) are an unusual kind of RNA virus. Each retrovirus carries an enzyme, reverse transcriptase, that catalyzes the formation of a cDNA copy of the viral RNA; this copy is then incorporated into a chromosome of the host (often an animal cell). The host's enzymes then take over and accomplish replication (not shown), transcription, and translation of the viral genes. As the drawing indicates, retroviruses may enter the host cell by a kind of endocytosis, losing the protein coat only after entry; similarly, retroviruses generally do not lyse their hosts but instead leave by extrusion, a process similar to budding whereby the virus is enveloped in a segment of the cell membrane as it emerges from the cell.

ently these bacteria had been carrying viruses within their cells, but these viruses did not become active and did not usurp the cells' metabolic machinery until exposed to the inducing action of the ultraviolet light, X rays, or chemicals. Cells that harbor inactive viruses are said to be **lysogenic**.

The discovery by Lwoff that viruses can be present in an inactive state inside their host cells showed that some viruses must be temperate rather than virulent. Virulent phages invariably kill their hosts. Temperate phages may or may not kill their hosts, depending on a variety of conditions. When they do not kill their hosts, their injected DNA usually becomes associated with the bacterial chromosome at a particular location, and the **lysogenic cycle** begins (Fig. 28.7). (The DNA of a few rare viruses survives as a plasmid in the host cell.) While integrated into the bacterial chromosome, the viral DNA behaves as part of that chromosome: it is replicated with the rest of the chromosome; it can be transferred from one cell to another during conjugation; its genes can undergo recombination with bacterial genes; and its genes can even produce phenotypic effects in the host bacterium, such as modifications of colony morphology, changes in the properties of the cell wall, and changes in the production of enzymes. For example, diphtheria bacteria can produce the toxin that causes the disease only if they are carrying a specific type of viral gene. And viral genes confer on the bacterium in which they reside immunity from further infection by the same type of virus. We see, then, that the DNA of temperate viruses has the properties of an episome for the bacterium. It can exist in an autonomous or vegetative state, replicating independently and eventually destroying the cell, or it can exist in the integrated state, as a *provirus*, functioning and replicating as a portion of the bacterial chromosome. It is possible that some plasmids—perhaps even the sex factor of bacteria—may have arisen from temperate viruses that took up residence in the bacteria, and became permanently nonvirulent. In this regard, it is interesting that the plasmids conferring antibiotic resistance are induced to begin proliferating when the cell is subjected to the trauma of damage to the cell wall.

Transduction   Not only do viral genes sometimes become incorporated into bacterial chromosomes, but the reverse may also occur. When temperate viruses are in the vegetative state and have put the bacterial cell to work making more viruses, small fragments of the bacterial chromosome may become enclosed in the new viral coats. If a temperate virus carrying bacterial DNA in this manner infects a new host, it injects both viral and bacterial DNA into this new host. Sometimes the injected bacterial genes undergo recombination with the new host's genes (presumably by a process like crossing over). The virus has thus acted as a vehicle for transferring genes from one bacterial cell to another (Fig. 28.9). This process, called *transduction*, was first described in 1952 by Norton D. Zinder and Joshua Lederberg, then at the University of Wisconsin. Since the likelihood that two bacterial genes will be cotransduced (passed on together to another bacterium by a virus) depends on how close together they are, assessing the frequency of cotransduction has proved a very sensitive way to map bacterial chromosomes.

The movement of genes from one organism to another by viruses is not limited to bacteria. There is good evidence that some of the genes that produce important phenotypic effects in the human body may have been moved into human chromosomes by viruses. These genes may have been transferred from other human beings or even from other species; some of them may be involved in cancer. Viruses may also have been involved in moving organelle genes into the nucleus.

In fact, it may be incorrect to think of viruses as a separately evolved group; instead, it seems likely that most viruses arose as escapees from host chromosomes, spreading because their component genes led, by chance, to their own differential reproduction at the expense of their host. If these possibilities make genetics far more complex and confusing than it seemed a few decades ago, when our knowledge did not go much beyond the Mendelian laws, they also make this subject one of the most exciting and promising of modern biology.

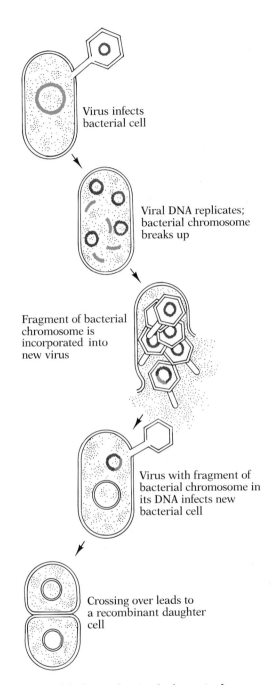

Virus infects bacterial cell

Viral DNA replicates; bacterial chromosome breaks up

Fragment of bacterial chromosome is incorporated into new virus

Virus with fragment of bacterial chromosome in its DNA infects new bacterial cell

Crossing over leads to a recombinant daughter cell

28.9   Model of transduction by bacteriophage

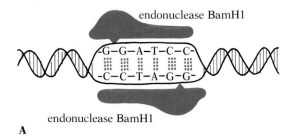

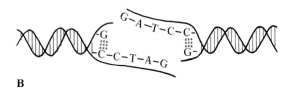

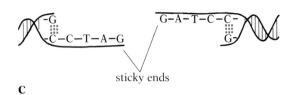

28.10 Endonuclease action
(A) Restriction endonucleases bind to a pair of target sequences—GGATCC in the case of the enzyme BamH1. (B) They break a particular phosphate bond in the backbone of each strand of the DNA. (C) The resulting cut ends with unpaired bases are called sticky ends because under favorable conditions they will pair with cut ends having the complementary sequence of bases. More than 80 endonucleases are known, each with its own specific target sequence. As you may recall, the normal function of endonucleases is to digest infecting viral DNA.

RECOMBINANT DNA

Artificial transformation Nonepisomal plasmids play a major role in the *recombinant DNA* technique for inserting genes from one kind of organism into cells of another kind of organism. This technique has great potential: specific genes can be isolated, introduced into bacteria, and used to produce large quantities of a desirable gene product—insulin, for example; or new genes can be introduced into plants or animals that might benefit from them—genes for nitrogen fixation in plants, let us say, or genes for growth hormones in domestic animals. However, the recombinant DNA technique has become a subject of public debate because of concern over the possibility of accidentally producing new pathogens or developing "genetic monsters." Let us take a look at how the technique works.

Plasmids, like DNA from bacterial chromosomes, can transform bacterial cells. In other words, if bacterial cells of certain species are placed in a medium containing free plasmids, some of the cells will pick up the plasmids, just as some of the nonvirulent pneumococci studied by Griffith picked up DNA that had been released into the medium by dead virulent pneumococci (see p. 695). Thus nonresistant bacteria can be transformed into resistant ones by exposure to a medium in which bacteria with plasmids for antibiotic resistance have been killed. Recombinant DNA technology makes use of this transforming potential of plasmids: purified plasmids are modified by the addition of foreign genetic material; when bacterial cells then pick up the modified plasmids, they acquire the foreign genes.

To follow the procedure in more detail: Bacterial cells containing plasmids are broken up and their DNA is extracted. The DNA is then subjected to differential centrifugation to separate the plasmids from the main bacterial chromosomes. The purified plasmids are next exposed to a restriction endonuclease, an enzyme that cleaves the DNA circle at a certain particular nucleotide sequence (Fig. 28.10). The plasmid DNA is now linear, with "sticky" ends (unpaired bases) where it was cleaved. It is next mixed with fragments of foreign DNA prepared with the same restriction endonuclease and therefore equipped with sticky ends complementary to those of the plasmid DNA. In such a mixture under the appropriate conditions of temperature, pH, and so on, the plasmid DNA and the foreign DNA spontaneously anneal by complementary base pairing, re-forming a circle in the process (Fig. 28.11). The backbones (chains of alternating phosphate and sugar groups) of the DNA circle can then be sealed with the enzyme DNA ligase. The end result is plasmids that contain a graft of foreign genetic material. All that remains to be done is to mix these plasmids with bacterial cells (usually treated with a calcium salt to make them more permeable). The bacterial cells will pick up the modified plasmids, which include both the original plasmid genes and fragments of the foreign genome. In general, plasmids containing genes that confer antibiotic resistance are used in this process; this means that if the experimenter treats the bacterial cells exposed to the recombinant plasmids with antibiotic, only the bacteria that have incorporated hybrid plasmids will survive, and bacteria lacking foreign genes will be eliminated.

One disadvantage of this approach is that the desired combination of plasmid and foreign DNA is not the sole result. After endonuclease treatment, the foreign DNA is left as a mixture of various fragments, of which only one may be of interest. Indeed, the endonuclease may well have found its target sequence within a particular gene, and so cut the gene itself. Hence, rigorous screening is often required to isolate the bacteria with plasmids that have incorporated complete copies of the desired genes. Worse yet, if the foreign DNA is from a eucaryote, as it often is in recombinant DNA research, it will contain introns, which bacteria are unable to remove before translation. This creates a serious problem for commercial applications. To avoid these and other problems, many researchers use yeast, a simple eucaryote. Others use more selective techniques to isolate and incorporate desirable genes into baceria. One of these techniques is gene cloning.

Cloning genes   The crucial step in mass-producing exact copies—clones —of a particular eucaryotic gene is to find cells that specialize in manufacturing that gene's product—pancreatic cells, for instance, if the desired product is insulin. The cytoplasm of such cells will have a high concentration of mRNA molecules coding for their special product, and these mRNAs will, of course, already have undergone intron removal in the nucleus. Various techniques exist to separate out the particular kinds of mRNA on the basis of physical characteristics like weight, so the mRNA found in unusual abundance in the specialist cells can be isolated.

Once the appropriate mRNA has been isolated, the next step is to produce from it the corresponding single-stranded DNA, using reverse transcriptase from a retrovirus. DNA polymerases are then used to replicate the DNA strand by complementary base pairing, a process that supplies the second strand for the double-helical structure of the transcript. The procedures already discussed for inserting foreign DNA into plasmids are then utilized: a restriction endonuclease cuts open the plasmid, creating a pair of sticky ends; the cloned DNA (equivalent to the cDNA produced by normal retroviruses), with a complementary set of sticky ends, is added, and the plasmid DNA and cDNA anneal; a ligase restores the bonds in the DNA backbone; and the plasmid is inserted into a bacterial host. This procedure has two immediate advantages: the expensive process of sorting for bacteria bearing the desired gene is eliminated, since only the appropriate cDNA is used; and introns, which cannot be removed by procaryotes, have already been eliminated in the production of the mRNA. The transformed bacterial cells grow and divide rapidly, creating limitless numbers of bacteria that may synthesize the desired product. Besides insulin, this technique is used to produce large quantities of other hormones—growth hormone is particularly important—that are difficult to synthesize. A naturally produced but poorly understood agent called ***interferon***, which holds great promise in the treatment of various diseases, is also now widely available for research and medical applications as a result of recombinant DNA technology.

Gene cloning makes it possible not only to use host cells as chemical factories to produce substances of medical importance, but also to study the sequencing and activity of genes from eucaryotic cells. Indeed, one of many spinoffs of recombinant DNA technology is a method of mapping genes with enormous precision. A piece of single-stranded DNA is tran-

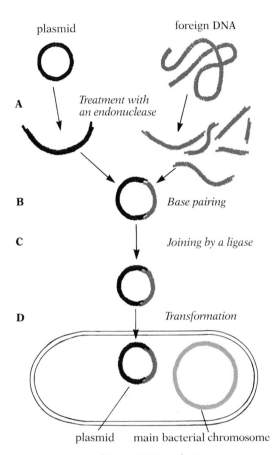

28.11   The recombinant DNA technique
Plasmids, removed from donor cells, are cut by an endonuclease (A). They are then mixed with fragments of DNA (color) from other cells, produced with the same endonuclease. The plasmid DNA and the foreign DNA can join at their sticky ends by complementary base pairing, re-forming a circle in the process (B). The DNA ends are then sealed by treatment with a ligase (C). The modified plasmids, some bearing foreign genes, are added to medium containing live bacterial cells. Some of the cells pick up the plasmids bearing foreign genes and are thus transformed (D). The main bacterial chromosome is much larger than shown here.

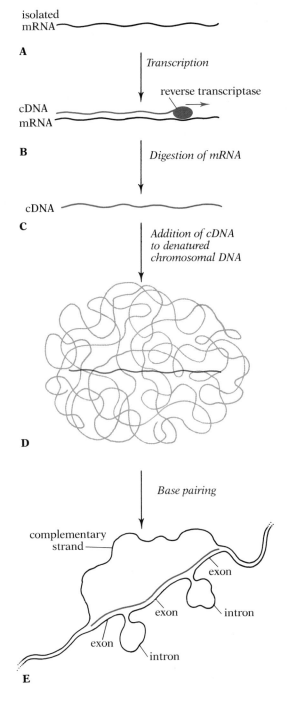

isolated
mRNA

A

Transcription

reverse transcriptase

cDNA
mRNA

B

Digestion of mRNA

cDNA

C

*Addition of cDNA
to denatured
chromosomal DNA*

D

Base pairing

complementary
strand

exon

exon

intron

exon

intron

E

scribed from an mRNA of known function in a medium containing the radioactive isotope ^{32}P. This cDNA is then mixed with a set of chromosomes treated to separate the double helix into single strands. When base pairing is again made possible in the resulting mixture, the cDNA frequently pairs with the corresponding gene on the chromosome (Fig. 28.12). Because the cDNA carries a radioactive label, the exact location on the chromosome of the gene to which it binds is revealed. Other spinoffs include techniques for quickly and precisely identifying nucleotide sequences in DNA, methods of rearranging genes and their components to study gene interaction and gene control, and a technique for creating specific mutations at specific sites in genes. This last procedure makes possible precise investigation of the roles of signal sequences in DNA and mRNA, and also enables the researcher to alter proteins at will to explore the bases of enzyme function.

Looking into the future, some researchers have suggested using cDNA technology to transfer genes for C_4 photosynthesis to some crop plants that normally depend exclusively on C_3 photosynthesis, for the sake of greater productivity under less than optimal growing conditions. Others have proposed transferring genes for the fixation of atmospheric nitrogen to crop plants, thus eliminating the need for application of nitrogenous fertilizers. Both transfers would be formidable undertakings. Genes for disease resistance might also be transferred to susceptible crops. Vaccines against the many viruses for which conventional vaccine techniques fail could be mass-produced; for example, if the virus cannot be cultured in large quantities, a gene coding for a coat protein capable of triggering an immune response might be cloned and the gene product used instead of the entire virus. Gene therapy—the addition of good copies of a gene to organisms with mutant alleles—is also possible, and has even succeeded in preliminary tests in *Drosophila*.

Recognition of the risk of undesirable side effects of recombinant DNA technology has led to strict regulation of how the technology is to be used. Laboratories must be equipped with facilities for sterile handling and for physical containment similar to those long used in medical microbiology laboratories where dangerous pathogens are handled. In addition, there is a consensus that the microorganisms used in recombinant DNA research should be from strains carrying crippling mutations that make them incapable of surviving outside the laboratory, or should be organisms that are not potential pathogens.

28.12 Locating a gene with cDNA
A particular kind of mRNA is isolated (A), and reverse transcriptase is used to make a cDNA transcript incorporating a radioactive label (B). Then RNase is added to digest the mRNA (C). The cDNA is next mixed with denatured chromosomal DNA (D). During base pairing, the cDNA binds to the gene coding for the mRNA, thus showing its location on the chromosome (E). Note that the intron segments loop out. They are not matched on the cDNA because they were absent from the mRNA.

CONTROL OF GENE EXPRESSION

In the process of mitosis every cell in the body of a multicellular organism inherits a complete set of chromosomes; the nucleotide sequences of this DNA carry in full the genetic information that is the cell's evolutionary endowment. But though every cell in the body receives the same set of instructions, individual cells may look entirely different from other cells, and may behave in entirely different ways. In fact, only a particular subset of all the genes a cell contains will ever generate proteins. Furthermore, only a small percentage of this genetic material is active at any one time. We have already seen that many minor, minute-by-minute adjustments of cellular chemistry can be made by shifts in enzyme activity; major adjustments involve altering the expression of a cell's genes as the needs of the cell change with the changing cellular environment. In this chapter we shall examine the logic and mechanisms of gene control, a process we now know involves chemicals that bind directly or indirectly to DNA or mRNA.

CONTROL OF GENE EXPRESSION IN BACTERIA

Early investigators of gene expression in bacteria made several assumptions about the process they sought to understand. First, it was logical to assume that proper control of gene expression would require that only those genes whose products were needed at any given moment be expressed. Second, since most genes code for an enzyme that controls only a single step in a biochemical pathway, the genes coding for several enzymes in the same pathway might be expected to be controlled as a group. And furthermore, since the function of a pathway is to turn a reactant into a product, the availability of the reactant in the cell might be expected to turn on transcription, while the availability of the final product might turn it off.

These assumptions have proven to be correct in many cases of transcription in bacteria. Because bacteria possess only about 3,000 genes (compared to the 30,000 of humans, for instance) and can be grown rapidly in huge numbers, they have been especially useful organisms for study of the control of gene transcription. Much of our current knowledge of this subject comes from research on the intestinal bacterium *Escherichia coli*.

THE JACOB-MONOD MODEL OF GENE INDUCTION

Investigating enzyme synthesis in *E. coli*, the French biochemists François Jacob and Jacques Monod formulated a powerful model of gene regulation in bacterial cells. They worked mainly with the enzyme β-galactosidase, which catalyzes the breakdown of lactose to glucose and galactose, substances both used and produced by other pathways.

Lactose is not continuously available to *E. coli*, and so—as would be expected—the gene for β-galactosidase is normally transcribed at a very low rate; in the absence of lactose there are only about ten β-galactosidase molecules per cell. Jacob and Monod found that the further production of this digestive enzyme is triggered by the presence of a so-called inducer, in this instance allolactose, an isomer of lactose automatically produced in the cell when lactose is present. Normally, then, β-galactosidase is an *inducible enzyme*. But they also found a mutant strain of *E. coli* in which the same enzyme is a *constitutive enzyme*—that is, an enzyme whose production is continuous, apparently uninfluenced by control substances such as inducers.

By means of recombination experiments, Jacob and Monod were eventually able to demonstrate the participation of four genes in the production of β-galactosidase and the two other enzymes involved in lactose breakdown: three so-called *structural genes*, each specifying the amino acid sequence of one of the three enzymes, and a *regulator gene*, which controls the activity of the structural genes. They proposed that the regulator gene, which is located at some distance from the structural genes, normally directs the synthesis of a *repressor* protein that inhibits transcription of the structural genes. The allele of the regulator gene present in the mutant constitutive strain, they concluded, lacks the ability to direct synthesis of an effective repressor; hence it cannot prevent transcription of the structural genes, which are thus left free to direct continuous protein synthesis.

Jacob and Monod also discovered that a special region of DNA contiguous to the structural gene for β-galactosidase determines whether transcription of the structural genes will be initiated; they called this special region the *operator*, and they called the combination of the operator and its three associated structural genes an *operon*. Subsequently it was found that the operator, which does not in itself constitute a gene since it doesn't code for a specific product, is located between the two important sequences of the *promoter*, the region to which RNA polymerase binds (Fig. 29.1). Hence, when the repressor binds to the operator, RNA polymerase cannot physically bind to the promoter, and transcription is blocked (Fig. 29.1A).

If inducer is present, it will bind to the repressor, thus causing a conformational change in the repressor that forces it to dissociate from the operator; in short, the inducer inactivates the repressor. Now free to bind to the promoter, RNA polymerase can initiate transcription of the structural

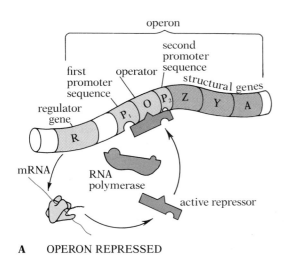

A   OPERON REPRESSED

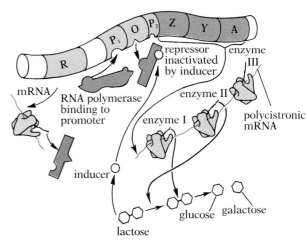

B   OPERON DEREPRESSED

genes and the production of mRNA (Fig. 29.1B). The mRNA carries the instructions of all three structural genes, and is therefore said to be ***polycistronic***. This messenger complexes with ribosomes in the cytoplasm, where its information is translated and the three enzymes necessary for lactose metabolism are synthesized. The number of β-galactosidase enzymes rises to about 5,000 per cell when the operon is not repressed.

According to the Jacob-Monod model, then, the condition of the operator region is the key to whether or not there will be activation of the so-called *lac* operon—the operon responsible for the synthesis of enzymes involved in the breakdown of lactose. If repressor protein is bound to the operator, there will be no transcription. If no repressor is bound to the operator (because the repressor has been inactivated by inducer), transcription can proceed freely.

Notice that the three jointly controlled structural genes of the *lac* operon specify enzymes with closely related functions. It is characteristic for the structural genes of an operon to determine the enzymes of a single biochemical pathway; thus the whole pathway can be regulated as a unit. The adaptive advantage of such coordinated control is obvious.

GENE REPRESSION

In the years since the Jacob-Monod model was first proposed, it has become apparent that not all operons are regulated in the same way as the *lac* operon, which is an inducible operon—that is, one which is inactive until turned on by an inducer substance. Many operons are, instead, continuously active unless turned off by a ***corepressor*** substance. One example is the operon whose five structural genes code for the enzymes necessary to synthesize the amino acid tryptophan. This operon is normally turned on, but when *E. coli* are grown in a medium containing tryptophan, it switches

29.1   The *lac* operon as an example of an inducible operon
(A) The operon consists of a promoter/operator region and three structural genes (Z, Y, and A). For simplicity the structural genes, which are much longer than the promoter/operator region, are shown greatly shortened. Also, the boundaries between the operator and the promoter sequences are drawn to appear quite sharp, though actually the operator sequence overlaps the end of the first promoter sequence and the beginning of the second. The regulator gene codes for mRNA, which is translated on the ribosomes and determines synthesis of repressor protein. When the repressor protein binds to the operator, it blocks the promoter's binding sites for RNA polymerase and thus prevents transcription of the structural genes. (B) Binding of inducer to the repressor inactivates the repressor, and the RNA polymerase can then bind to the promoter regions and initiate transcription of the structural genes. These, transcribed as a unit, determine production of polycistronic mRNA—that is, mRNA coding for more than one gene product. The mRNA then complexes with ribosomes in the cytoplasm and is translated into three enzymes. Enzyme I is β-galactosidase; enzyme II is a permease that helps transport lactose into the cell; and enzyme III is a transacetylase, whose role in lactose utilization is not understood.

HOW CONTROL SUBSTANCES BIND TO DNA

Heroic efforts have revealed many of the details of how a repressor protein may bind to an operator. For the repressor to bind to a particular operator, it must bear active sites that match the DNA substrate exactly. Such matching involves polar amino acids on the repressor and complementary polar groups exposed on the *sides* of the base pairs of the DNA.

The deoxyribose backbones of DNA are attached to the bases slightly off center, and as we saw in Chapter 26, the two chains have opposite polarities, with one running from 3′ to 5′ and the other from 5′ to 3′. The result is that one of the exposed sides, or grooves, of the helix is somewhat wider than the other (Fig. A). The wider opening—the ***major groove***—contains more sequence-specific information, and is therefore thought to be more important in binding the re-

pressor. For example, the major groove of a T–A pair has three polar groups—O, NH, and N, reading from T to A—their respective polarities being −, +, and −. The major groove of the C–G pair has the polar groups NH, O, and N, with polarities of +, −, and −. The ***minor groove*** is slightly less useful: for T–A the polar groups are O and N (− and −). A particular repressor probably has a shape corresponding to the differing physical widths of the operator's major and minor grooves, and polarity patterns complementary to those of the operator's base pairs. Such polarity patterns would enable it to form enough hydrogen bonds to bind to one unique sequence of the DNA (Fig. B). This is probably the mechanism by which repressors and certain other proteins—some hormones, for example—bind to specific operators to control gene transcription.

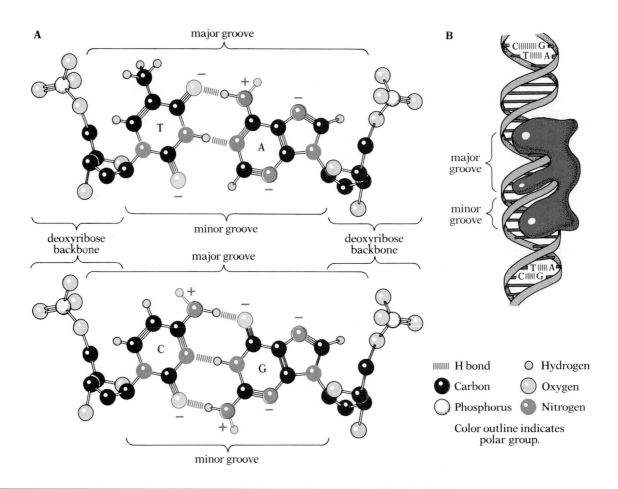

A

major groove

T   A

deoxyribose backbone   minor groove   deoxyribose backbone

major groove

C   G

minor groove

B

C‖‖‖‖G
T‖‖‖A

major groove

minor groove

T‖‖‖A
C‖‖‖‖G

‖‖‖‖ H bond   ○ Hydrogen
● Carbon   ○ Oxygen
○ Phosphorus   ◐ Nitrogen

Color outline indicates polar group.

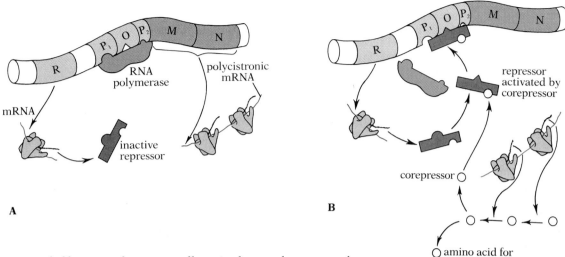

A   **B**

29.2 A repressible operon
The repressor protein encoded by the regulator gene is initially inactive (A). Only if it binds a corepressor molecule (often the end product of a biochemical pathway) can it bind to the operator and block transcription of the structural genes (B). After the operon has been repressed, the concentration of the corepressor falls as it is used in cellular metabolism and no more is produced. When the corepressor becomes scarce, the repressor tends to lose it to metabolic enzymes. As a result, the repressor can no longer bind to the operator, the RNA polymerase binds to the promoter, and transcription resumes. The operon shown here is responsible for the synthesis of an amino acid, which is incorporated into new proteins.

off. Enzymes encoded by genes that are usually active but can be repressed are called *repressible enzymes*. In their case, the repressor protein encoded by the regulator gene is inactive when first produced. Only if a corepressor substance binds to and activates it can it bind to the operator and block RNA polymerase binding (Fig. 29.2). Unlike inducible enzymes, which are synthesized only if their operon is turned on by an inducer, repressible enzymes are automatically synthesized unless their operon is turned off by a corepressor. In tryptophan synthesis, the tryptophan itself activates the repressor protein, enabling it to bind to the operator.

An inducer is often either the first substrate in the biochemical pathway being regulated (that is, the first molecule the synthesized enzyme will bind to) or some substance closely related to that substrate. It is not surprising, therefore, to find that a corepressor is usually the end product of the biochemical pathway being regulated, or a closely related substance. In both substrate induction and end-product corepression, then, gene transcription is regulated by the cellular substances most affected by the transcription—a truly elegant functional arrangement.

POSITIVE CONTROL OF GENE TRANSCRIPTION

Both of the cases we have discussed are examples of *negative control*: a repressor binds to the operator and turns off transcription. Control is effected in one case by deactivating the repressor with an inducer, and in the other case by activating it with a corepressor. Negative-control systems are usually discovered when a mutation increases production of all the proteins encoded by an operon; the increase occurs because the repressor necessary for inactivation of the operon is made less effective by the mutation. Though negative control of one sort or another is the most common way of regulating gene expression in *E. coli*, some systems are regulated by *positive control*. In many of these cases a control protein binds directly to the DNA to activate the operon, and is itself controlled by means of a chemical that binds to it. Systems of this sort are likely to be discovered when produc-

THE LAMBDA SWITCH

The expression of genes is often regulated by systems more complex than any we have described, often involving the interaction of three or more control substances, two or more separate operators, and sometimes other DNA control regions in addition to the regulator genes, operators, and promoters already discussed. The most thoroughly investigated example of the more complex interactions involves the switching of the temperate virus lambda from its lysogenic to its lytic cycle. Though the details that follow describe gene repression in a viral system, they are thought to be very similar to the strategy of gene repression at work in procaryotes.

In Chapter 28 we saw that temperate viruses are able to insert their DNA into a specific section of the host chromosome and remain there without lysing the bacterial cell. Under normal conditions (that is, steady growth of the bacterial colony) this passive kind of infection rarely occurs, but once the virus does enter the host chromosome it can remain there, in the lysogenic cycle, indefinitely. Exposure of an infected bacterium to ultraviolet radiation, however, incites the viral DNA to leave the host chromosome and begin reproduction. In the lambda virus, which parasitizes *E. coli*, two interacting operons on complementary strands of the DNA control this switch from a lysogenic to a lytic cycle.

The lambda system has been studied intensively. Excising the lambda DNA from the *E. coli* chromosome and initiating the destructive lytic cycle is controlled by the so-called *cro* gene and several other genes downstream from the operon. These genes are normally repressed: the *cI* gene on the complementary strand produces a repressor protein that binds to the operator region of the *cro* gene, and thereby blocks transcription (Fig. A). Repression of the *cro* and associated genes is stable until the host bacterium begins to synthesize unusually large quantities of the DNA repair enzyme rec A. Synthesis of this enzyme is brought about by the destructive effect of UV (among other things) on the bacterial DNA. While repairing DNA damage, rec A also digests part of the repressor synthesized by the *cI* gene, freeing the *cro* operator to begin transcription of the normally repressed *cro* and associated viral genes (Fig. B). The *cro* gene codes for a second repressor, which binds to the *cI* operator and turns off transcription of *cI* (Fig. C).

The switch in lambda appears to be irreversible: even the disappearance of the bacterial DNA repair enzyme does not stop production of the

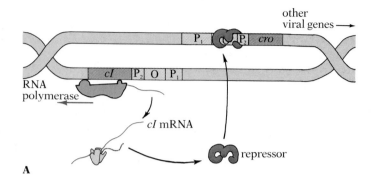

A

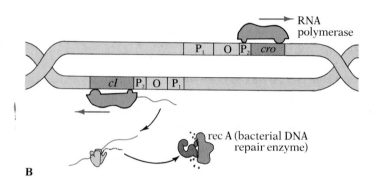

B

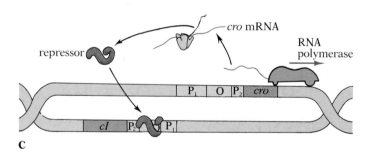

C

second repressor. The result is that the enzymes encoded by the *cro* and associated genes cut the lambda DNA free of its host chromosome and begin directing its reproduction, thus initiating the virulent stage. Apparently a survival strategy encoded in the lambda genome directs the virus to react to the first sign of trouble (UV, in this case) by initiating the synthesis of new phage, which will abandon the damaged bacterium and seek out other hosts.

tion of the proteins encoded by an operon decreases because a mutation has made this control protein less effective in activating transcription.

Let's look at how positive control works. We have seen that RNA polymerase recognizes and binds to the two promoter sequences which flank (and partially overlap) the operator. But few genes have promoters with exactly the same nucleotide sequence. Genes with promoters that differ significantly from the optimal one are rarely transcribed, because the RNA polymerase will bind to them only weakly. In positive control of such genes, a control protein binds at the front edge or even a little "upstream" of the first part of the promoter (Fig. 29.3), and then may help the polymerase bind by actually holding it in place.[1] The presence of the control protein at the upstream end of the operon compensates for the poor promoter sequence and greatly facilitates transcription.

Two versions of positive control are known. In one, the chemical that binds to the control protein causes a conformational change that activates the protein, enabling it to bind to the DNA and facilitate transcription. The *CAP* (catabolic gene activator protein) of *E. coli*, for instance, is normally inactive, but is activated by the by-now familiar intracellular messenger cyclic AMP. Thus when cAMP signals the need for the enzymes of the CAP-controlled operon, the control protein is activated and transcription is switched on (Fig. 29.3).

The other positive-control strategy involves a normally active control protein that is turned off by the binding of a small molecule. Once switched off, the protein cannot help the RNA polymerase bind to the poor promoter sequence, so transcription is minimized.

In summary, control of gene expression in procaryotes can involve (1) interruption of negative control by substrate induction (as in the *lac* operon), (2) initiation of negative control by corepression (as in tryptophan synthesis), (3) induction of positive control (the cAMP-CAP system), and (4) repression of positive control. Some bacteria also control mRNA synthesis by regulating the rate of termination of transcription as well as the rate of initiation.

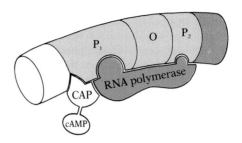

29.3 One model for induction of gene transcription by the cAMP-CAP complex
Binding of cAMP-CAP to the promoter of this operon facilitates binding of RNA polymerase. The cAMP necessary to initiate this process is most abundant in the cell when synthesis of the gene products of this operon becomes important.

CONTROL OF GENE EXPRESSION IN EUCARYOTES

The control of gene expression appears to be more complex in eucaryotes than in procaryotes. For one thing, even the simplest eucaryote can have many more functional genes than a procaryote, and most genes must be turned on or off as needed. And as we have seen, before translation can take place, more than half the eucaryotic primary RNA transcript synthesized from an active gene in the nucleus must be excised by an elaborate en-

---

[1] Another, more complicated possibility is that there may be two overlapping nucleotide sequences, one of which is very effective at binding RNA polymerase but does not allow transcription to begin, and another which, though it binds RNA polymerase less often, does act as a promoter, permitting transcription. When polymerase is bound to the first region, the other will be blocked and transcription will not be possible. The role of the cAMP-CAP in this system would be to bind to a part of the first region that does not overlap the second, preventing it from binding polymerase and thus allowing the promoter to bind the polymerase and thereby initiate transcription.

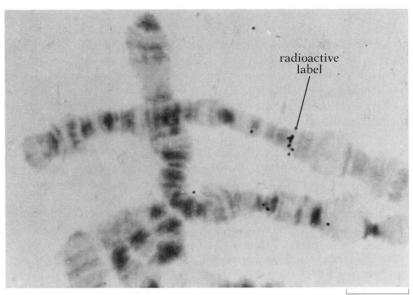

29.4   Locating a gene
In this version of the hybridization technique, *Drosophila* salivary-gland chromosomes have been treated to weaken base pairing, and then a radioactively labeled sample of previously identified DNA has been added and has bound to the complementary region. The dark dots reveal where radioactive decay is taking place, and therefore where this particular gene is located. The giant chromosomes of the salivary gland are suitable for hybridization because they contain multiple copies of all their genes.

radioactive
label

10 μm

zyme-mediated process. A further complication is that the DNA of the eucaryotic chromosome is wrapped on histone-protein cores, wound into tightly packed coils, and organized in loops, forming the nucleoprotein material called **chromatin**; there must be a mechanism by which the nucleosome—DNA wound around a histone-protein core—is unraveled prior to transcription.

THE ORGANIZATION OF EUCARYOTIC CHROMOSOMES

The role of chromosomal proteins   Chromosomal proteins can be categorized either as **histones**, most of which are essential components of nucleosome cores, or as **nonhistone proteins**. There is some indication that histones may be involved in gene expression, but their role would appear to be passive: with the nucleosome cores in place, transcription is not possible, since relaxation of the tightly coiled structure is necessary to allow RNA polymerase access to the DNA.[2] Most recent evidence indicates that the nonhistone proteins may be much more important as selective agents in gene regulation. Some of these acidic proteins are bound directly to the DNA, while others are linked to the nucleosome cores. They exhibit a rich diversity, not only from organism to organism, but also from tissue to tissue within a given organism, and even within a single cell at various times, depending on its developmental stage and its current functional condition. Hence, they may have the specificity necessary for control elements. (Nucleosome cores, on the other hand, appear not to vary with gene identity or activity.) Moreover, the fact that the nonhistone proteins seem to contribute little to chromatin structure suggests that their function is regulatory.

---

[2] By contrast, transcription is known to occur when nucleosome cores are not in place; for example, one nucleosome core is regularly missing in the 5′ side of active insulin genes.

The role of at least some nonhistone proteins, then, may be to bind to specific control regions in the DNA to cause unraveling, or decondensation, of the nucleosomes. They may also play a role in the second step of eucaryotic gene control, in which specific genes or groups of genes on the decondensed loops are activated. As we shall see, activation, rather than inhibition of transcription, appears to be the rule in eucaryotes.

Highly repetitive DNA   The organization of eucaryotic chromosomes has been revealed in large part through a technique known as *DNA hybridization*. The first step in this technique is to heat the chromosomal DNA under certain conditions; this causes the two strands of each helix to separate. In one type of hybridization the next step is to add to the chromosomal DNA a radioactively labeled sample of a particular mRNA (or a cDNA made from it by reverse transcriptase). The mixture is then cooled very slowly. Under these conditions, the mRNA or cDNA will frequently bind to the complementary region of the chromosomal DNA, and because of the radioactive labeling, the gene coding for it can then—in theory—be located on the chromosome. However, this method is effective just with polytene chromosomes,[3] since only when a thousand or more parallel copies of a gene are hybridized is there a significant chance of detecting the radioactivity (Fig. 29.4). Clever variations of this basic method have shown that about 10 percent of most eucaryotic DNA contains base sequences that are found not once but thousands of times in the genome, constituting what is called *highly repetitive DNA*.

There appear to be at least three classes of highly repetitive DNA. The first consists of a vast number of copies of a few kinds of short sequences located near the centromere and in large blocks in chromosome arms (Fig. 29.5A). The function of this DNA, which is never transcribed, is not known; one possibility is that it may be used by enzymes to bring homologous chromosomes into perfect register during meiosis.

Another class of highly repetitive DNA consists of long, tandemly repeated units. These areas contain genes that code for the smallest of the four ribosomal RNAs—5S RNA (Fig. 29.5B). The name is based on the rate at which this RNA sediments (settles) through a sucrose solution; the other three rRNAs—6S, 18S, and 28S—are all larger and heavier by this measure. In one well-studied species of frog, each 5S gene is associated with a nonfunctional region known as a *pseudogene* (so called because its sequence is

---

[3] Polytene chromosomes, typified by those in the salivary glands of *Drosophila*, contain vast numbers of parallel copies of the entire DNA strand of each chromosome.

29.5   Organization of eucaryotic chromosomes
The short highly repetitive DNA (A), long tandem repeats (B), and polytene tandem repeats (D) are confined to different specialized regions of the chromosome. The long interspersed elements (C), moderately repetitive sequences (E), and single-copy DNA (F) are actually intermixed; for clarity, however, they are shown here in isolation on separate loops.

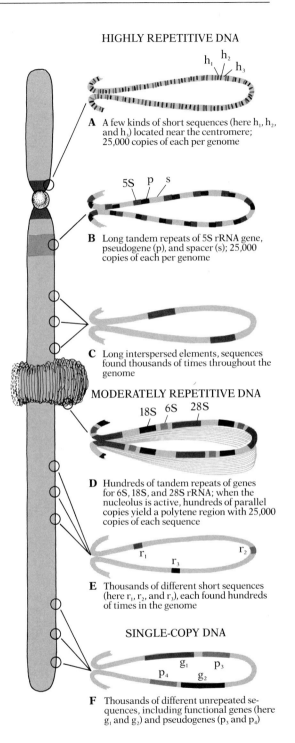

HIGHLY REPETITIVE DNA

A   A few kinds of short sequences (here h_1, h_2, and h_3) located near the centromere; 25,000 copies of each per genome

B   Long tandem repeats of 5S rRNA gene, pseudogene (p), and spacer (s); 25,000 copies of each per genome

C   Long interspersed elements, sequences found thousands of times throughout the genome

MODERATELY REPETITIVE DNA

D   Hundreds of tandem repeats of genes for 6S, 18S, and 28S rRNA; when the nucleolus is active, hundreds of parallel copies yield a polytene region with 25,000 copies of each sequence

E   Thousands of different short sequences (here r_1, r_2, and r_3), each found hundreds of times in the genome

SINGLE-COPY DNA

F   Thousands of different unrepeated sequences, including functional genes (here g_1 and g_2) and pseudogenes (p_3 and p_4)

almost identical to that of a functional gene) and a very long "spacer" region. Neither the pseudogene nor the spacer is ever transcribed. There is a clear reason for the large number of 5S genes: developing eggs need enormous numbers of ribosomes (perhaps 10^{12}, a thousand billion) to handle all the protein synthesis necessary for rapid growth. Repeated transcription of a single copy of this gene would be inadequate to meet the needs of the cell for ribosomal RNA; consequently most eucaryotes have approximately 25,000 copies of this sequence.

A third class of highly repetitive DNA, known as "long interspersed elements," consists of sequences that are several thousand bases long and are found tens of thousands of times in the chromosomes of many eucaryotes (Fig. 29.5C). Their function, if any, is not yet known.

Moderately repetitive DNA   About 20 percent of the typical eucaryotic genome consists of so-called *moderately repetitive DNA*. Each moderately repetitive DNA sequence is found hundreds rather than thousands of times and comes in one of two varieties. The first is a tandem repeat of certain genes; in particular, genes for the other three kinds of rRNA are tandemly repeated in the order 18S rRNA, 6S rRNA, 28S rRNA, untranscribed long spacer, 18S, 6S, 28S, long spacer, 18S, 6S, 28S, long spacer, and so on (Fig. 29.5D). Since every ribosome must have one of each of the four kinds of rRNA, you may wonder how cells manage to get by with only a hundred to a thousand copies of these genes while there are thousands of copies of the 5S gene. The answer is that the portion of the chromosome with the tandem repeat for 18S, 6S, and 28S rRNA can be replicated repeatedly, *independent of the rest of the chromosome,* until there is a polytene *region* (different from an entire polytene chromosome) consisting of 25–250 parallel copies of the repeat region. This process results in a total of up to 25,000 copies of each gene in active cells. The region of the chromosome with the many replicated repeating segments forms a structure we have already discussed— the nucleolus. Moderately repetitive DNA of the other class is more mysterious, and varies widely in size and frequency between species. These sequences, of which there may be as many as 5,000 different types, are only about 300–3,000 bases long—as little as one-tenth the length of the average functional gene. Each type is usually scattered between functional genes throughout the chromosome in 30–500 different locations (Fig. 29.5E). Some of this DNA is clearly derived from mRNA containing, by chance, the promoter sequence for reverse transcriptase; presumably some of the DNA copies "mistakenly" synthesized as an artifact of retroviral infection make their way into the host chromosome. One of the cell's tRNAs appears to be particularly susceptible to this process. Other cases, however, do not fit this model; we shall return to the possible function of this class of moderately repetitive DNA shortly.

Single-copy DNA   Finally, 70 percent of a typical eucaryotic genome consists of *single-copy* sequences (Fig. 29.5F). The majority of this DNA is *never* transcribed; at least in mammals, much of it exists as pseudogenes—nearly identical but untranscribed copies of functional genes. When researchers make calculations for different species, taking into account the introns that are excised after transcription, they typically find that only about 1 percent of eucaryotic DNA ever codes for mRNA that is subsequently translated. This could hardly be more dramatically different from procaryotes, in which well above 90 percent of the genome is translated at one time or another.

CONTROL OF GENE TRANSCRIPTION IN EUCARYOTIC CELLS

In view of the profound differences in organization between procaryotic and eucaryotic chromosomes, researchers have wondered whether the two groups of organisms could possibly share the same control strategies. And indeed, though the various models of transcriptional control in bacteria have provided suggestions and insights for the study of such control in eucaryotic cells, they have not, as was originally hoped, been found directly applicable to eucaryotic cells. The complications of repetitive sequences, pseudogenes, and introns aside, a lack of similarity of control is not really surprising when we remember that gene activation in eucaryotes probably involves two steps: the unwinding of particular sections of the tightly packed nucleosomes, followed by activation of a specific gene or group of genes on the decondensed loops. It may be that different control substances are involved in each step.

Patterns of activity in eucaryotic chromosomes Cytological studies of chromatin suggest that the internal structure of a chromosome is not uniform, and that this variation reflects gene activity. For example, some regions of the giant salivary chromosomes of *Drosophila* stain only faintly when treated with basic dyes, whereas other regions stain intensely. The nonstaining regions are called *euchromatin* and the staining ones *heterochromatin.* Differentiation of chromosomes into euchromatic and heterochromatic regions is now known to be a general characteristic of eucaryotic cells. Chromosomal mapping over the past few years has shown that euchromatic regions contain active genes whereas heterochromatic regions are inactive.

At first, some investigators thought the heterochromatic regions were simply devoid of genes, functioning merely as structural elements of the chromosome. This is probably true of the large region of heterochromatin with highly repetitive base sequences located around the centromere. But most heterochromatic regions do not lack genes; instead, the genes (often long series of genes) are simply inactive. For example, many regions completely heterochromatic in adults were euchromatic at earlier stages in the development of the organism. When an X chromosome of a human female is converted into a Barr body, the entire chromosome becomes heterochromatic.

The existence of extensive regions of heterochromatin indicates a tendency in eucaryotic organisms for functionally related genes (genes concerned with a particular stage of embryonic development, for example) to be grouped loosely together in the same region of the chromosome, where they can be jointly activated or repressed. Despite this general grouping of genes, however, the clumping of a small set of related genes into an operon that is so characteristic of bacteria has not been found in eucaryotes. Indeed, in many instances in which there appears to be simultaneous control of the synthesis of functionally related enzymes, this explanation does not hold; the enzymes are encoded by widely separated genes. The genes that specify two polypeptide chains of a single protein (the α and β chains of hemoglobin, for example) are often on different chromosomes.

Though the "rules" for coordinated activation and repression in such cases are still unknown, the seemingly chaotic organization of eucaryotic genes suggests several possibilities for control mechanisms. Consider, for example, moderately repetitive DNA of the second class that does not appear to have arisen from reverse transcription of RNA: these sequences, some only about 300 bases long, are scattered in as many as 500 locations throughout the chromosome. These regions, which are not translated, could have a role in receiving signals involved in activating genes. By binding different specific activator substances, the 5,000 types of moderately repetitive DNA could act as switches for dozens or hundreds of genes simultaneously, wherever they happened to be located. They could even be involved in sending the control signals; perhaps this is why some moderately repetitive DNA sequences are actually *transcribed*, though they are not *translated*. The function of these sequences is not known, but a similar situation in a procaryote is better understood. In at least one bacterium, one of the RNAs produced by transcription is never translated; instead, it binds to a specific mRNA with the complementary sequence. Since the resulting RNA duplex cannot be translated, the production of this particular RNA serves to regulate the activity of the complementary mRNA. At present there is simply not enough data from eucaryotes to go beyond intriguing speculation.

Lampbrush chromosomes and chromosomal puffs as visible evidence of gene activity Just before the first meiotic division, the oocytes of many vertebrates synthesize very large amounts of mRNA for later use: the mRNA is then ready during the early stages of development, after fertilization,

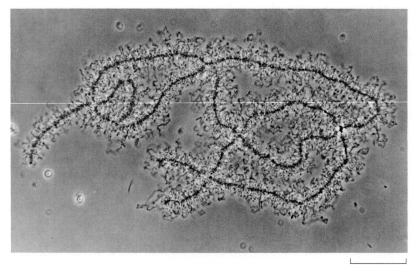

29.6 Lampbrush chromosome from an oocyte of the spotted newt *(Triturus viridescens)*
The many feathery projections from the chromosome are regions bearing genes that are being repeatedly transcribed.

0.5 μm

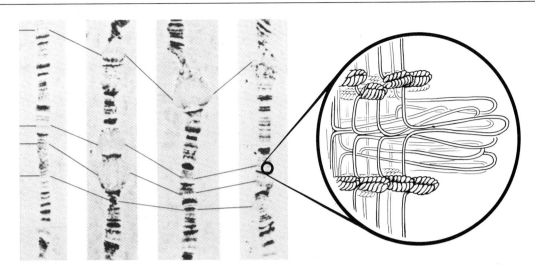

when the chromosomes are so busy with DNA replication in support of rapid cell division that they are largely unavailable for transcription into mRNA. This is but one instance of changing chromosomal activity in a cell at different stages of its development. In vertebrate oocytes this changing activity has certain visible manifestations, which can be observed with a phase-contrast microscope. The parts of the chromosome bearing genes that are being repeatedly transcribed—the euchromatic regions—are looped out laterally from the main chromosomal axis, while the parts bearing repressed genes are tightly compacted; chromosomes in this condition are called *lampbrush chromosomes* (Fig. 29.6).

Some investigators think that each loop contains a single gene, others that it is a series of copies, arranged in sequence, of single genes. The average loop is much longer than the average gene, but we have seen that most eucaryotic DNA does not code for mRNA, so even some long loops may have only one functional gene. Whichever is true, it is certain that the loops, arising in the single-copy DNA, represent those parts of the genome that code for the particular proteins the early embryo will have to produce in quantity. Since some of these proteins will not be needed later in development, some of the chromosomal regions so evident as lampbrush loops in the oocyte will presumably become heterochromatin for most of the life of the organism.

The giant chromosomes of the salivary glands of larval flies (Diptera) are interpreted as lateral arrays of replicated chromosomes (polytene chromosomes) that have remained stuck together. The best estimates indicate that 10 cycles of replication are involved in the creation of a polytene chromosome, yielding 1,024 parallel copies of each gene. The result of this unselective gene amplification is a capacity for rapid synthesis of large amounts of RNA. When certain regions of a giant chromosome are especially active, all the parallel DNA molecules form brushlike loops in those regions (Fig. 29.7). The resulting clusters of lateral loops, called *chromosomal puffs*, are clearly visible under the microscope. The locations of puffs are different on

29.7 Chromosomal puffs

Puffs in polytene chromosomes consist of hundreds of parallel copies of the chromosomal DNA looped out to expose the maximum surface for synthesis of RNA. The interpretive drawing shows only a few of the loops. The series of photographs indicates how the location of puffs in a *Drosophila* chromosome changes dramatically but predictably over time; we can clearly see the varying activation of four different genes or sets of genes on chromosome 3 (each identified by a series of connecting lines) over a period of 22 hours in larval development.

chromosomes in different tissues, and they are different in the same tissue at different stages of development (Fig. 29.7), though at any given time all the cells of any one type in any given tissue show the same pattern of puffing.

If the puffs indicate the location of active genes—as the correlation of puffing pattern with developmental stages suggests—we would expect them to be the primary sites of active synthesis of mRNA. In fact, the unpuffed bands of the chromosome contain mainly DNA and protein, while the puffs contain far more RNA. That this RNA is being synthesized at the puff and has not simply accumulated there after synthesis elsewhere can be shown by injecting radioactive uridine into fly larvae. Uridine is used by the cells as a precursor of uracil, a nitrogenous base that is incorporated into RNA but not into DNA. When this experiment is performed, radioactivity soon appears in the puffs (and in the nucleoli, which synthesize large amounts of ribosomal RNA), but it does not appear in other parts of the cell until much later—an indication that RNA is indeed being synthesized at the puffs. Furthermore, the RNA made in one puff differs chemically from the RNA made in a puff at a different position on the chromosome, as would be expected if each gene codes for a different mRNA.

Puffs thus provide a way of determining visually whether or not changes in the extranuclear environment can alter the pattern of gene activity. As expected, they can. For example, if ecdysone, the hormone that causes molting in insects, is injected into a fly larva, the chromosomes rapidly undergo a shift in their puffing pattern, taking on the pattern characteristically found at the time of molting in normal untreated individuals. If treatment is stopped, the puffs characteristic of molting disappear. If treatment is begun again, they reappear. Or, to give another example, if chromosomes from one type of cell are exposed to the cytoplasm of a different type of cell or of the same type of cell at a different developmental stage, they quickly lose the puffs characteristic of their original cells and develop puffs characteristic of the type and stage of the cells providing the new cytoplasm. Puffing can be prevented entirely by treatment with actinomycin, which is known to be an inhibitor of nucleic acid synthesis.

GENE AMPLIFICATION AS A CONTROL MECHANISM

Ordinarily a single copy of each gene per haploid chromosome set is sufficient to meet the needs of a cell for the substance coded by the gene. The rate of transcription of a single gene is such that in four days it may produce 100,000 messenger RNA molecules, which can lead to the synthesis of as many as 10 billion (10^{10}) protein molecules. But in some cases the demand for a product is so great that a single gene cannot meet it. For example, we've already seen that so much ribosomal RNA is required by a eucaryotic cell for construction of ribosomes that the 5S rRNA gene is repeated tandemly 25,000 times (Fig. 29.5B). The other three rRNA genes are repeated tandemly only 100–1,000 times, but when large numbers of ribosomes are needed, many parallel copies of this region of the chromosome are synthesized (Fig. 29.5D). This region usually loops out from the main axis of one of the chromosomes to form the nucleolus, and rRNA is produced there at a

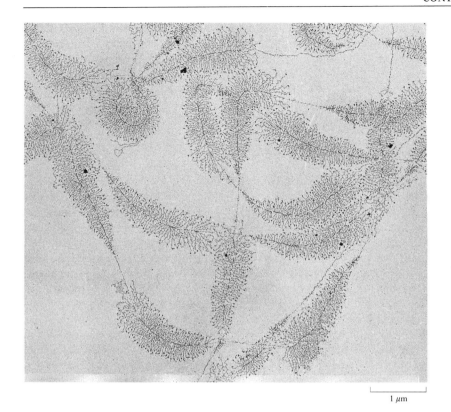

1 μm

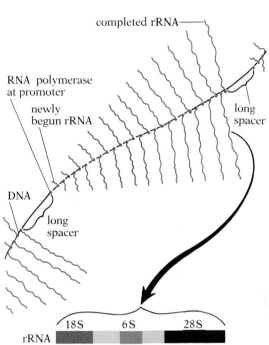

29.8   Transcription of rRNA on DNA in the nucleolus of an oocyte of the spotted newt (Triturus viridescens)

Left: Electron micrograph of portion of the nucleolus. This continuous strand of DNA bears multiple copies of the genes for three RNAs. Strands of rRNA at progressive stages of synthesis feather out from each set of genes. Because the successive gene sets are separated by regions of intergenic DNA spacers where no rRNA is attached, we can see more or less exactly where on the DNA molecule each set of genes begins and ends. Right: Diagram of rRNA synthesis on one gene set. Many molecules consisting of the three kinds of rRNA joined together in a single strand (color) are being synthesized, as RNA polymerase molecules (color circles) specialized for transcribing rRNA genes move along the DNA. Strands of rRNA attached to the DNA near the beginning of the gene set are still short, their synthesis having just begun; strands of rRNA attached at successive points along the gene set are progressively longer.

high rate (Fig. 29.8). The production of these extra copies of the rRNA genes from moderately repetitive DNA is one type of *gene amplification*.

When amplification of the rRNA genes was first discovered, the possibility was immediately suggested that a similar amplification might occur for other sets of genes at various stages of embryonic development. Selective amplification would be one way of influencing the amount of mRNA of particular types that would be produced in the nucleus. But contrary to expectation, additional instances of localized gene amplification have proven relatively rare.

POST-TRANSCRIPTIONAL CONTROL

Our discussion of cellular control mechanisms has so far focused on regulation of the amount of mRNA synthesis, whether by control of DNA transcription or by gene amplification. But there are numerous other points in the flow of information within the cell where control can be exerted. For example, some primary transcripts can be processed in more than one way, yielding different products, depending on the cell's needs. Some eucaryotic mRNAs are synthesized and exported to the cytoplasm, but are not translated until a particular chemical signal is present. The mechanisms underlying alternative processing and translational delay are not understood. In addition, the expression of mRNA may be regulated by the rate at which the messenger is broken down: some sorts of mRNA have long life-spans, while

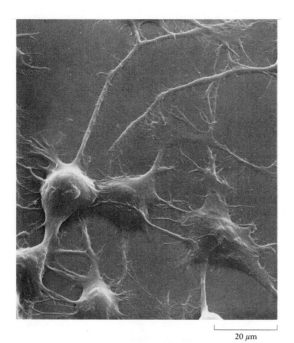

20 μm

29.9 Phase-contrast photomicrograph of cultured neuroblastoma cells from a mouse
These cells, from a malignant tumor of the nervous tissue (neuroblastoma), are growing on a solid surface; they have many elongate processes resembling axons and dendrites.

others are destroyed within minutes or hours. And as we have seen again and again, the activity of the enzymes created by translation is frequently regulated by activators or inhibitors. Even the assembly of many structural proteins (collagen, for example) is regulated by chemical signals. Hence, a gene's expression can in some cases be controlled at every step from before transcription until after translation.

CANCER: A FAILURE OF NORMAL CELLULAR CONTROLS

The most distinctive feature of cancer cells is their unrestrained proliferation, which results in formation of malignant tumors and often in the spread of the cells from the original site of growth to many other parts of the body, a process known as metastasis. Clearly, these are cells in which the normal controls are no longer working properly. Biologists hope to learn more about how normal cellular controls work by studying how they fail—by investigating how the normal controls can become so deranged as to permit cancerous growth. There is the underlying hope, too, of course, that such study will eventually lead to better ways of preventing or treating cancer.

Study of cells in culture   One of the most profitable ways of studying the properties of specific kinds of cells is to remove them from the body of the organism and grow them in tissue culture in the laboratory. Embryonic cells and tumor cells grow best, while fully differentiated cells are poor candidates for culturing. Among nontumor cells, embryonic fibroblasts outgrow all other cells in culture. (Fibroblasts are the cells in the basement membrane that, among other things, produce scar tissue.) The usual procedure is to seed a large number of cells onto (or into) a sterile culture medium to which a variety of nutrients and growth-stimulating factors have been added. The nutrients and other factors are frequently supplied by serum—the liquid portion of blood. Serum contains, for example, the platelet-derived growth factor that signals fibroblasts to initiate the growth of scar tissue; in the body, the growth factor escapes from the circulatory system and binds to receptors on fibroblasts whenever a wound breaks blood vessels. While a cell culture, no matter how elaborately set up and controlled, is far from the normal environment of the cells, some of the cells may nevertheless survive and engage in many of their usual developmental and functional activities, including cell division (Fig. 29.9). Most cultured lineages of animal cells stop dividing when they become crowded or, if crowding is prevented, after a certain number of cellular generations; the number of generations tends to be specific for the species and tissue of origin.

These limitations on the growth of cultured cells exemplify two mechanisms that prevent unlimited cell growth in the organism as well. One is the cell's reaction to the effect of crowding, which we called ***contact inhibition*** in Chapter 4; when receptors in the cell membrane recognize markers in the cell coats (glycocalyces) of adjacent cells of the same type, the receptors signal the nucleus and further cell division is inhibited. This inhibition is not absolute, however: adding high concentrations of growth factors, for example, can often stimulate a further round of cell division even in a crowded culture. The second mechanism, the automatic cessation of cell division after a set number of divisions, is an independent system for limiting cell proliferation; in most tissues it comes into play only if contact inhibition fails to work normally. Cultured cells that continue to divide indefinitely fall into one of two categories: those that have lost the fixed-number-of-divisions control, and will go on growing as long as they are not crowded, and those that have lost both kinds of control. The cells of each category are potentially "immortal" in culture, but the former resemble benign (that is, self-limiting) tumors, while the latter resemble cancers: they will grow indefinitely regardless of crowding to create an ever-larger mass of tissue. The study of cultured cells that have lost both levels of control has been vigorously pursued in recent years.

A recently developed technique called *somatic cell hybridization*, in which two cultured cells, often of different origins, are fused together, has made possible exciting new approaches to the study of cellular control systems. The fusing of cells became relatively easy after the discovery that a virus called Sendai modifies the surfaces of cells in a way that makes them stick to each other and then coalesce. Fusion is now most often promoted instead by treatment with polyethylene glycol or an electric shock.

The nuclei of fused cells usually remain separate, but occasionally when cell division begins only a single spindle forms, and it draws chromosomes from both nuclei. The result is that each daughter cell receives a single nucleus containing two sets of chromosomes, one from each of the two parental lines, which may be different species or different tissues from the same species. In this way hybrid cells containing human chromosomes in combination with chromosomes from a variety of other animal species have been produced.

The cell-fusion technique has been especially valuable in cancer research, because study of fused cancerous and noncancerous cells, and of fused cancerous cells of two different types, helps reveal how the altered control system of one type of cell influences the control system of the other type.

The characteristics of cancer cells   The loss of normal control systems can produce a variety of changes in a cell's anatomy, chemistry, and general behavior. Studies of these changes tell us quite a bit about how cancer cells differ from normal ones, and sometimes suggest potential avenues of treatment. We shall mention a few of the characteristics of cancer cells that these studies have revealed.

One property of cultured cancer cells is that they almost always have an abnormal set of chromosomes. For example, the cells of the human cancer line called HeLa,[4] by far the most widely studied line of cultured human cells, typically have 70–80 chromosomes instead of the normal 46 (Fig. 29.10). Interestingly enough, noncancerous cell lines that pass the "crisis stage" in culture and become potentially immortal also have extra chromosomes; hence it seems likely that possession of extra chromosomes somehow frees cultured cells from some of the normal constraints on pro-

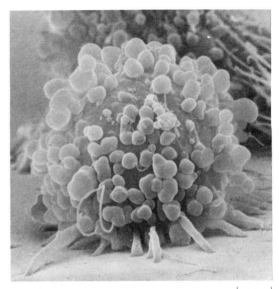

2 μm

29.10 A typical cancer cell
This cell, from the HeLa line, is growing in tissue culture. The cell is rather spherical, and is covered with many small "blisters," called blebs, whose significance is not understood.

---

[4] The HeLa cell line is derived from a carcinoma of the cervix of Henrietta Lacks, who died of her cancer in 1951. This was the first stable, vigorously growing line of cultured human cells used in cancer research.

liferation. In cancerous tissue in organisms, however, chromosome number is usually normal. Instead, there are frequently specific rearrangements of the chromosomes. For example, in 90 percent of patients with Burkitt's lymphoma (a cancer of cells in the immune system) one tip of chromosome 8 has been translocated to the end of chromosome 14. The insertion point in chromosome 14 is next to a gene coding for an immune-system polypeptide (see Fig. 30.10, p. 775), while the translocated region contains a so-called proto-oncogene, which is frequently involved in cancers. The other 10 percent of patients with Burkitt's lymphoma have other translocations from chromosome 8. As we shall see, predictable translocations are also found in the cultured cells of several other sorts of cancer.

Besides differences within the nucleus, cancer cells and normal cells display a host of significant differences pertaining to cell shape and to the nature of the cell surface. Cultured cancer cells, for instance, tend to have a rather spherical shape, one seen in normal cells only during a short period in the G_1 stage, immediately following mitosis. This peculiarity of shape is probably a result of the abnormally small number of functional structure-stabilizing microfibrils (often called stress fibers), which also makes these cells more mobile than normal cells. This characteristic is also related to a striking feature of cultured cancer cells known as anchorage independence: most cells must "cling" to a solid surface in order to grow—a useful constraint, since they will require support later in order to function properly. They cling by means of adhesion plaques connected to the stress fibers. Cancer cells lack this level of control as well, and so can grow on soft surfaces like agar.

Extracellular environmental influences must act first at the surface of the cell, and cancer cells, with their abnormal surfaces, appear to suffer from an inability to react effectually to environmental changes. Cancer cells typically have fewer glycolipids and glycoproteins in their cell coats than normal cells, and those they have are qualitatively different. The anomalous nature of the glycolipids and glycoproteins is probably correlated both with the absence of normal contact inhibition during cell division, and with the apparent inability of cancer cells to recognize other cells of their own tissue type. Whereas normal cells from two different tissues (such as liver and kidney) mixed in culture tend to sort themselves out and reaggregate according to tissue type, cancer cells do not do so. This absence of normal cellular affinities is probably one of the reasons why malignant cells of many cancers can spread in metastasis from their tissue of origin into many other parts of the body.

That the surfaces of cancer cells characteristically bear antigens not found on normal cells constitutes another abnormality of their cell coats. In most human beings the body's normal immunological mechanisms probably respond to the antigens, destroying new cancer cells as fast as they arise. Hence cancerous tumors may require not only a breakdown of normal cellular controls, which causes some cells to become cancerous, but also a failure of the normal immunologic response to such cells.

Many other characteristics distinctive of cancer cells might be cited. For example, cancer cells consume much more glucose than normal cells, metabolize at a high rate, and excrete much lactic acid; and they secrete abnormally high quantities of certain proteolytic enzymes, which may alter both their own surfaces and the surfaces of normal cells near them. Whichever characteristics are singled out, it is important to remember the difficulty of establishing whether they are primary disturbances that create the cancerous condition or merely secondary responses to the primary metabolic changes.

The multistep hypothesis Our current understanding of cancer suggests that the conversion of a normal cell into a cancer cell involves several changes: loss of fixed-number-of-division control, loss or reduction of contact inhibition, loss of anchorage dependence, and, sometimes, tissue-specific cell-surface changes. But in many tissues even these changes result only in benign tumors that grow slowly and then stop, apparently because the cells fail to produce the chemical signal or signals that cause vascularization (the establishment of a capillary system that supplies oxygen and nutrients to growing tissue). Do all these changes take place at once as a result of a single genetic event, or is each change the result of a separate genetic event? Furthermore, if each change resulting in loss of control is independent, what determines whether a specific process such as vascularization will take place? Does this depend on a single event, or are several steps involved?

Health statistics suggested the answer to this question before the actual mechanisms began to be elucidated. As Alfred Knudson of the Institute for Cancer Research in Philadelphia pointed out, if just a single genetic event were required, the probability of an individual's having contracted a particular type of cancer should increase proportionally with age. For example, if the chance of contracting skin cancer were 1 percent per year, then the chance of having contracted it would be 2 percent for a two-year-old, 3 percent for a three-year-old, and so on. In fact, however, the incidence of cancers in the population as a whole does not rise in this linear fashion (Fig. 29.11). Instead, cancer is normally a disease of old age. Knudson concluded that to account for the exponential rise of the incidence of cancer, several independent genetic events are required. The curve would be the product of the individual probabilities of the separate events. If we knew that the different genetic events had exactly equal probabilities of occurring, we could calculate with great precision the number of events required to induce a particular kind of cancer. But since this is an unlikely assumption, and since the curves for different kinds of cancer are not identical, it is not possible to be exact. Nevertheless, it seems almost certain that some cancers (particularly those, like leukemia and immune-system cancers, involving tissues in which tissue-type affinity, vascularization, and a solid surface for growth are not necessary for the normal cells) require only two genetic events, while others require three, four (the most common case), or five. This is in reasonably good agreement with the two to five levels of control thought to be involved—a fixed number of cell divisions, contact inhibition, anchorage dependence, tissue-type affinity, and the need for vascularization.

Additional evidence for a multistep model in at least some cancers comes from Knudson's studies of individuals who inherit a proclivity for certain cancers (retinoblastoma, a cancer of the eye, is a well-documented

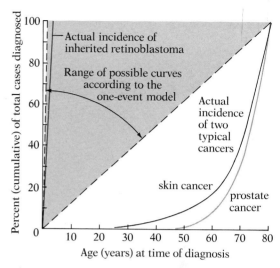

29.11 Incidence of representative cancers
If each type of cancer were caused by a single event, the probability of contracting a particular cancer would increase linearly with age; the range of possible curves is indicated. Actually, the incidence of most cancers increases exponentially, indicating that several contributing events are usually involved. The incidence curve for skin cancer closely approximates a three-event curve, while the curve for prostate cancer approximates a five-event curve. By contrast, the curve for inherited retinoblastoma is linear, approximating a single-event curve.

example). Unlike most people, these individuals usually contract cancer when young, and the incidence curve for their families, whose members share an inborn tendency for some particular form of the disease, shows a linear rise with age, rather than the more common exponential one. The linear curve suggests that in such individuals—those with a predisposition for retinoblastoma, for example—only a single event is necessary to trigger the disease. The inheritance of the genetic susceptibility in this kind of cancer appears to behave as a single-gene characteristic, so contracting it probably requires a total of two transforming events—just one in addition to the one already built into the genome.

These statistical studies alone, however, cannot tell us what constitutes a genetic event leading to cancer (among the possibilities are single-base changes, deletions, insertions, or chromosomal rearrangements); nor do the studies tell us the precise effects of the events. Recent evidence, however, strongly suggests that in at least some cancers each such event is directly linked to the loss of a particular level of control. The best data on this come from studies of oncogenes.

Oncogenes   The discovery of ***oncogenes***—genes that cause cancer—has a complex history. In 1910 Peyton Rous, working at the Rockefeller Institute in New York, found that a viruslike entity, now known as Rous sarcoma virus, could cause cancer in chickens. Rous's conclusions were not widely accepted for decades, and only in 1966 was he awarded a Nobel Prize for his pioneering work. We know now that the causative agent in this particular case is a retrovirus, which, like other retroviruses, produces a cDNA copy of its RNA that is incorporated into the host chromosome (see p. 735). Rous sarcoma virus causes cancer quickly, and at a high rate, because the cDNA brings an oncogene (known as *src*) with it. The expression of this gene in the host chromosome, which does not depend on the site of incorporation, makes the cell cancerous. Ordinary retroviruses—those not containing oncogenes—can also cause cancer, but only at a very low rate, and even then after a long latency. Incorporation of cDNA into the host genome is apparently random, and cancer from ordinary retroviruses occurs only when the incorporation takes place near one of a few particular genes in the host, called ***proto-oncogenes***. Presumably an active control region in cDNA alters the expression of one or more of these genes. Oncogenes, then, are genes that inevitably cause cancer; proto-oncogenes, on the other hand, have the potential to cause cancer, but some change is required to convert them into oncogenes.

Perhaps the most surprising discovery about the oncogene *src* is that it is virtually identical to a normal gene in chickens; indeed, this normal gene (designated *c-src* to indicate that it is cellular in origin) is a proto-oncogene: incorporation of ordinary cDNA near *c-src* can (eventually) cause cancer. Since 1970 nearly two dozen rapidly acting, cancer-inducing retroviruses infecting a variety of birds and mammals have been discovered, each carrying an oncogene almost identical to the normal host proto-oncogene.

The viral cancers are convenient to study, but as yet few forms of cancer in humans have been found to be definitely viral in origin. However, studies of cultured human and nonviral animal tumors have revealed strong similarities to the observed behavior of retroviruses: some of the genetic changes that are responsible for making the cells in these cultures cancerous involve the same oncogenes also found in cancer-inducing retroviruses. Some of the nonviral oncogenes appear to have arisen through a mutation in a proto-oncogene that altered the base sequence through a base change, insertion, or deletion. Others appear to be proto-oncogenes that have been moved (translocated) from their usual locations and so perhaps released from their normal controls, to become oncogenes. Conversely, either mutation or translocation involving a control region next to a proto-oncogene could, in theory, have converted a proto-oncogene into an oncogene. Indeed, when grown in culture, cells of several leukemias, one ovarian cancer, and several other sorts of cancer display predictable chromosomal translocations. These translocations seem to involve moving a structural gene and its control region, which is responsible for its transcription in the tissue at a high rate, to a position next to a proto-oncogene, or moving the proto-oncogene next to a structural gene and its control region.

The various mechanisms that may result in the presence of an oncogene in the genome can be summarized as follows:

I. Altered gene, resulting in altered product
  1. Incorporation of retroviral oncogene
  2. Mutation of normal proto-oncogene to create an oncogene
II. Altered gene expression, in which a gene is transcribed at the wrong time or at too high a rate
  1. Insertion next to a proto-oncogene of retroviral cDNA containing an active control region
  2. Translocation of a proto-oncogene to a position next to an active control region
  3. Translocation of an active control region to a position next to a proto-oncogene
  4. Mutation of a control region next to a proto-oncogene

We have seen that the conversion of a normal cell into a cancer cell requires two or more genetic changes specific to that kind of cell, and at least some of these changes involve oncogenes. We have also seen that some changes that characteristically give rise to cancer cells involve a loss of one or more levels of control. Do oncogenes, then, affect the control of cell division, contact inhibition, anchorage dependence, tissue-type affinities, and vascularization?

The first evidence concerning the actual operation of oncogenes came from work on the *src* oncogene by Raymond Erikson and Marc Collett at the University of Colorado. They found that the enzyme encoded by *src* is one of the protein kinases—enzymes that phosphorylate a particular protein or a particular component of proteins. Phosphorylation is frequently involved in controlling chemical pathways, usually serving to activate an enzyme; the *src* enzyme phosphorylates the amino acid tyrosine. Phosphorylated tyrosine was not previously known to occur. It has since been found in normal cells; however, in cells with *src* oncogenes, tyrosine phosphorylation increases to at least 10 times the normal level. Six other oncogenes are now known to code for tyrosine kinases, while five others are *src*-like, having very similar sequences; what their products actually do is not yet known. All probably evolved from a common precursor kinase gene.

If some oncogenes code for tyrosine kinases, the next step must be to identify the proteins they phosphorylate. Tony Hunter of the Salk Institute and his colleagues have shown that one target is a structural protein called vinculin, which is found at adhesion plaques and is involved in maintaining normal cell shape. Phosphorylation of vinculin and perhaps of other structural proteins may result in the spherical shape that is characteristic of cancer cells but seen in normal cells only in the early G_1 stage of the cell cycle. Phosphorylation of vinculin cannot be a necessary condition for cancer, however, since some Rous sarcoma virus mutants cause cancer without phosphorylating vinculin. Whatever the target of the *src* gene product, it could be that one or more normal control systems work by ending a cell's capacity to alter shape prior to division, thus preventing a return to the spherical shape of early G_1, and the tyrosine kinase oncogenes inactivate one or more of those control systems.

An intriguing clue to the significance of *src*-like oncogenes comes from studies of how cells respond to extracellular chemical signals. One group of these signals consists of growth factors, each of which can stimulate particular kinds of cells to undergo a single additional division. For example, when epidermal growth factor (EGF) binds to the EGF receptors of an epidermal cell (and of certain other cells as well), it can trigger the series of events leading to cell division. The EGF receptor is a transmembrane protein which, when it binds to EGF extracellularly, undergoes an allosteric change intracellularly; the conformational change in the intracellular portion of the receptor allows it to begin catalyzing a specific reaction. The activated EGF receptor is, in fact, then a kinase.

At least one oncogene appears to code for an enzyme that is very similar to the kinase part of the activated EGF receptor, but lacks the extracellular portion that binds to EGF. This oncogene product may act by signaling the cell to divide continuously, whether or not EGF is present. Another oncogene product binds to DNA, and may cause continuous replication of the chromosomes. Yet another activates adenylate cyclase, the enzyme responsible for the second-messenger system of many cells (see pp. 426–27); it presumably signals reception of an extracellular messenger, even though none has arrived.

The modes of action of most oncogenes remain to be discovered. Nevertheless, the pattern that is emerging suggests that understanding cancer in the near future may be a realistic hope: there appear to be only a limited number of ways cancer is triggered, only a limited number of proto-oncogenes in any cell (fewer than a hundred in humans), and an even smaller number of functions for oncogene products. Understanding oncogenes will probably prove the most important step in finding effective treatments for cancer.

Environmental causes of cancer As we have seen, mutations, translocations, and retroviruses can each play a role in triggering cancer. We also know that many, if not all, mutations and translocations are caused by external agents—radiation and mutagenic chemicals. But linking particular chemicals or radiation sources to cancer in humans and evaluating the relative danger of each has proven difficult. Not every mutation occurs in a cancer-causing location, and even a mutation that is in such a location may not trigger all the steps necessary for development of a cancer: we have seen that at least two specific independent genetic events are required; the final step may occur years after the first significant exposure to a mutagenic chemical or radiation.

The absence of strict cause-and-effect relationships between particular mutagenic agents and particular cancers has made the identification and evaluation of causative agents difficult, even in the most obvious cases. For example, though health statistics have shown that cigarette smoking is probably responsible for about 150,000 fatal cancers annually (and for 25 percent of fatal heart disease), critics in the tobacco industry deny the link; while granting that smokers develop lung cancer and heart disease far more often than nonsmokers, the critics point out that many smokers never develop lung cancer, and that a few nonsmokers do. An informed understanding of how oncogenes work and of the probabilistic nature of their development, however, reveals the weaknesses in these arguments: the genetic events necessary for inducing lung cancer must be greatly facilitated by the highly mutagenic "tar" in cigarette smoke (a deadly combination of chemicals that kills two-pack-per-day smokers eight years early on average), but the changes must occur in the correct locations in at least one cell, so even with regular exposure to mutagenic smoke, some smokers will escape unscathed; at the same time, other sources of mutation, such as radiation, are also present, and though less potent, they will sometimes cause the changes necessary to trigger cancer even in nonsmokers.

Another way to evaluate the carcinogenic potential of chemicals and radiation is to expose animals—usually mice—to measured doses of these agents. This procedure is expensive and time-consuming, and in addition we must assume that what causes cancer in mice is equally carcinogenic in humans. Another assay is the Ames test; as described on page 725, this test involves exposing bacteria to potential carcinogens to see if mutations result, the assumption being that what is a mutagen for *E. coli* is a carcinogen for *Homo sapiens*. The Ames test has identified many substances as potential carcinogens. Others, initially having no effect, are converted into mutagens when liver homogenate is added to the bacterial culture, and are presumably also converted thus in the liver of an organism. The coal-tar components of many hair dyes and cosmetics, as well as hexachlorophene soaps, flame-retardant chemicals in children's sleepwear, the seared protein of grilled meat, the smoke from wood fires, and several chemicals in certain vegetables and spices, are just a few of the substances identified by the Ames test as possible carcinogens. When tested on animals each of these has in fact proven carcinogenic. It seems only prudent to assume that all mutagens are carcinogenic, but given the ubiquity of environmental mutagens, avoiding them all is impossible. The only prudent course, then, is to eliminate our exposure to the most potent—cigarette smoke and coal tars—and, weighing the costs and benefits (just as we do each time we decide to take the risk of driving a car), to minimize contact with the rest.

Chapter 30

IMMUNOLOGY AND THE MECHANISMS OF GENETIC VARIABILITY

We all know that once we have had diseases like measles and chicken pox, we cannot contract them in the same form a second time. This immunity develops because the complex set of cells that fights the disease while it is present has acquired the power to recognize and destroy the disease-causing entity far more rapidly in the future. The number of foreign cells and unfamiliar chemicals the immune system can "learn" to recognize—including many that cause disease—is essentially infinite, yet only a few genes are actually involved in encoding the vast number of different proteins, each of which binds specifically to a particular target. If one gene codes for one product, how can this be? As we shall see, the answer lies in a remarkable system for exon selection—a system that may help explain the persistence of those seemingly unnecessary and wasteful complications of the eucaryotic genome that include introns, untranscribed repetitive DNA, and the nonfunctional gene mimics known as pseudogenes.

THE IMMUNE RESPONSE

Essentially all animals have phagocytotic cells that ingest bacteria and dead cells. In vertebrates, this basic line of defense has been elaborated into a highly specific immune system that is probably an adaptation to larger body size and longer life. The immune system directs the manufacture of antibodies that help inactivate or destroy invading antigens—almost always large molecules (usually either proteins or polysaccharides) ordinarily foreign to the organism's own body. The antigens may be free in solution, as are the toxins secreted by some microorganisms, or built into the outer surfaces of viruses or foreign cells like bacteria and pollen. The antigens stimulate certain cells in the immune system to produce highly specific antibodies—chemicals that bind to these antigens exactly as enzymes bind to reactants. The human immune system recognizes as foreign not only nonhuman cells but also nearly all cells from other individuals; this explains why successful organ transplants are so difficult. The vast number of potential antigens means that literally millions or perhaps even billions of different antibodies must exist for the immune system to function properly.

A demonstration of an immunologic reaction   Immunologic reactions have been the subject of much study since the English physician Edward Jenner discovered in 1796 that people develop immunity to smallpox if they are injected with material that induces a very mild form of the disease (actually cowpox). This was the first vaccine (the term is from the Latin *vacca*, "cow"). Further dramatic demonstrations of the immune reaction were made by Louis Pasteur in France during the latter half of the nineteenth century.

The object of one of Pasteur's many investigations was a disease of cattle and sheep called anthrax, which was ravaging the herds of Europe at that time. Persuaded that a certain type of bacterium caused the disease, he exposed bacteria of this type to temperatures that were high enough to weaken but not kill them. When he injected these weakened bacteria into healthy sheep, the sheep became slightly ill, but thereafter they exhibited immunity to further infection by this disease. To convince the skeptics of his day, Pasteur arranged a demonstration attended by his most influential contemporaries. With these as witnesses, he injected weakened bacteria into 25 sheep, leaving 25 others uninjected as controls. Several weeks later, with the witnesses again assembled, he gave all 50 sheep a massive injection of fully active bacteria—more than enough to kill any normal healthy sheep. A few days later, all 25 control sheep were dead, while all 25 of the treated sheep were alive and healthy.

We know now that Pasteur's sheep reacted to the antigens of the weakened anthrax bacteria by producing antibodies against them. The animals' immune systems had thus acquired the ability to recognize and destroy anthrax bacteria. But how does exposure to an antigen stimulate an organism to make antibodies, and how does the immune system "remember"?

The cells of the immune system   The cells that respond to the presence of a foreign antigen are certain white blood cells called **lymphocytes**, which occur not only in the blood but also in the lymph and in lymphoid tissues of the lymph nodes, spleen, liver, tonsils, thymus, and bone marrow, and in more limited lymphoid areas associated with the lungs and the intestinal tract. In most vertebrates, the lymphocytes are derived from lymphoid stem cells located in the thymus and bone marrow. The lymphoid stem cells, in turn, are derived from hemopoietic stem cells (cells that give rise also to the other types of white blood cells and to erythrocytes) located in the red bone marrow of adults, or in the yolk sac of the early embryo or the liver and spleen later in embryonic development (Fig. 30.1). The two main types of lymphocytes are B cells (so called because they mature in the bone marrow) and T cells (which mature in the thymus). The B cells manufacture and secrete antibodies, and are responsible for the ***humoral immune response***, which is particularly effective against pathogenic cells, viruses, and toxins free in the blood and plasma. The T cells are responsible for the ***cell-mediated immune response***, which primarily defends against larger targets, including certain internal parasites, cancer cells, and virus-infected cells.

The humoral immune response Before an organism has encountered a particular antigen, the circulating lymphocytes that can recognize the antigen are small, metabolically quiescent "virgin" cells that can move freely between the blood and the lymphatic tissues by squeezing between the endothelial cells of the walls of blood vessels. Upon stimulation by an antigen, these small lymphocytes grow larger and begin dividing (Fig. 30.2). Let us assume, for the moment, that the lymphocyte we are following begins as a virgin ***B lymphocyte***. Cell division by the stimulated B lymphocyte gives rise

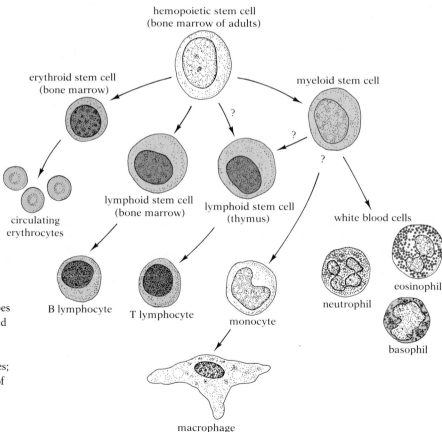

30.1 Hemopoiesis: the origin of blood cells
Hemopoietic stem cells give rise to at least four types of more specialized stem cells: those from which red blood corpuscles (erythrocytes) are derived; those from which B lymphocytes are derived; those that migrate to the thymus and give rise to T lymphocytes; and those that give rise to the various other kinds of white blood cells.

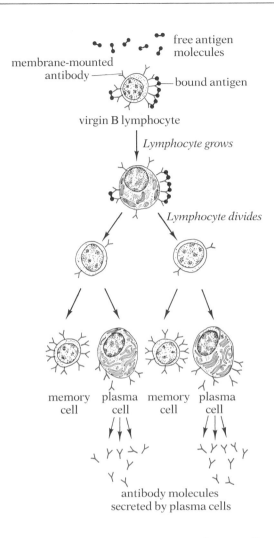

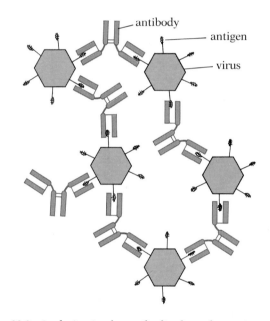

30.2   Stimulation of B lymphocyte by antigen
When antigen molecules bind to membrane-mounted antibodies of a virgin B cell, the lymphocyte first grows larger and then begins a series of cell divisions (only two are shown here). Some of the cells produced by this proliferation are memory cells that resemble the original lymphocyte; others become specialized as plasma cells, which secrete antibodies. The antigen in this example is a toxin.

30.3   Agglutination by antibodies bound to antigens
Each antibody molecule can bind to two antigen molecules; hence the bacteria or viruses bearing the antigens can be held together in large clumps. Here the antigens are surface proteins or carbohydrates on invading viruses. (For clarity, the antibodies and antigens have been enlarged. They are in fact much smaller than the invading viruses.) This agglutination aids in the destruction of antibody-bearing bacteria or viruses by macrophages, killer cells, or complement-system proteins.

over a period of several days to numerous *plasma cells*, and it is primarily these cells that secrete the antibody molecules that bind to the stimulating antigen. A stimulated B lymphocyte also gives rise to other lymphocytes like itself, which serve as *memory cells* and make possible (in a manner to be described later) more rapid response if the same antigen should be encountered again.

Antibodies are Y-shaped molecules with two identical antigen-binding arms. (We shall examine their structure in detail in a later section.) The binding sites on the arms form weak bonds with the antigen molecules in what is basically an enzyme-substrate reaction. Each of the millions of kinds of antibodies has a different set of active sites, so each binds to a different antigen or antigen region. Because each antibody molecule can bind to two antigen molecules, the antibodies tend to agglutinate, or lump together, the antigens and the bacteria or viruses bearing them (Fig. 30.3).

The agglutination can trigger three reactions. First, large phagocytic *macrophages* in the lymph recognize the antigen-bound antibodies, and

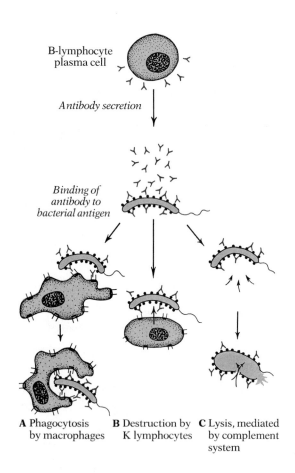

A Phagocytosis by macrophages **B** Destruction by K lymphocytes **C** Lysis, mediated by complement system

30.4 How humoral antibodies facilitate the destruction of bacteria

Once bound to antigens, the antibodies secreted by B lymphocytes can trigger three sorts of reactions. (A) Bound antibodies, and the bacteria to which they are attached, are ingested by phagocytic macrophages; this is also the primary mechanism by which toxins and viruses are eliminated. (B) Bound antibodies are recognized by K lymphocytes, which destroy the marked bacteria by an unknown mechanism. (C) Bound antibodies trigger a cascade reaction by which zymogens of the complement system are activated and catalyze the construction of a membrane channel in the invading bacterium; water moves into the invader through this channel by osmosis, causing the bacterial cell to swell and burst. The action of circulating antibodies on mast cells is not shown in this drawing.

engulf them and their targets (Fig. 30.4A). This reaction is effective against toxins, viruses, and most pathogenic cells. Second, lymphocytes of a different kind—**K lymphocytes**, also called killer cells—recognize the bound antibodies, bind to them, and destroy any bacterial or other foreign cells they have marked (Fig. 30.4B). The mechanism by which K lymphocytes kill is not known. Finally, the bound antibodies can trigger a cascade reaction involving some twenty plasma proteins known as the **complement system**; in this reaction each of these proteins—many of which exist as inactive zymogens—in turn catalyzes the activation of another. Four kinds of complement-system proteins then assemble into an 18-unit membrane channel in the invading cell; the channel allows osmosis of water into the bacterium, which swells and bursts (Fig. 30.4C).

These three reactions, which depend on circulating antibodies secreted by B lymphocytes, are part of the humoral immune response, but they are not the end of the B-lymphocyte story. In addition to binding directly to antigens and thus targeting them for destruction, circulating antibodies can become attached by their tails to mast cells. (Mast cells, as you may recall, secrete local chemical mediators, such as histamine, in response to tissue damage; see p. 431.) The antibodies become mounted on the cell membrane of the mast cell in a way that leaves their antigen-binding region free and exposed to the surrounding medium. When an antigen is bound by a mast-cell–mounted antibody, the mast cell is induced to release histamine and other chemicals. Histamine in turn causes nearby blood vessels to dilate and become "leaky," permitting the plasma, rich in lymphocytes, antibodies, macrophages, and complement-system proteins, to reach the site of the histamine release. The mast-cell/antibody system, then, acts as a sort of cellular burglar alarm to recruit the other elements of the immune system to concentrations of antigens. It is particularly important in the response to roundworms and other parasites.

The cell-mediated immune response Another set of immune reactions is mediated by the **T lymphocyte**, the type of lymphocyte that matures in the thymus. The precise set of events involved in this maturation is not yet understood, but it is known that an array of polypeptide hormones, secreted by the epithelial cells of the thymus and collectively known as thymosin, play an important role. The antibody-like receptor molecules of T lymphocytes are not secreted, but remain instead firmly attached by their tails to the lymphocyte membrane (Fig. 30.5). In this way they resemble the membrane-mounted antibodies of B cells and mast cells. However, as we shall see in more detail shortly, there is a very important difference: while the two arms of a B-cell antibody bind to the same kind of antigen, the two binding sites of a T-cell receptor each bind to a different antigen.

Like their humoral counterparts, T lymphocytes begin as antigen-specific virgin cells. When stimulated by the two particular antigens their receptors bind to, they divide to produce both lymphocytes that are involved in the immediate immune response and memory cells that make future reactions more rapid. Activated T lymphocytes are more complex in their behavior than activated B lymphocytes, which simply secrete antibodies. The simplest variety of activated T lymphocyte is the **cytotoxic T cell**, which acts something like a K lymphocyte in that it destroys an antigen-bearing cell, and then moves on to other targets. However, unlike the K cell, which recognizes any antibodies secreted by B lymphocytes and bound to antigen, a cytotoxic T cell binds directly to antigen-bearing cells with its own antigen-specific surface receptors.

At least two other varieties of T lymphocytes are involved in the immune response—the helper T cell and the suppressor T cell. As we shall see in more detail in a later section, the binding of their receptors to the antigens on the membrane of the corresponding B lymphocyte is essential for the operation of these cells. Once this binding has occurred, the ***helper T cell*** facilitates the immune reaction by secreting local chemical mediators known as ***interleukins*** and ***lymphokines***. These mediators activate nearby macrophages, both T and B lymphocytes, and (indirectly) mast cells; they also stimulate antigen-bound T cells to divide.

The ***suppressor T cell*** has the opposite role: it inhibits the activity of macrophages, lymphocytes, and mast cells. Since suppressor T cells are attracted to the area by antigen-bound helper T cells, the cells of the two types operate simultaneously, together controlling the precise level of each immune response. As we saw in Chapters 16 and 20, this antagonistic strategy, which results in a dynamic balance, has many advantages. There is strong evidence that many allergic reactions—hay fever and asthma, for example —result from a failure of allergen-specific suppressor T cells to limit the immune response. If, as very rarely happens, the release of histamine and other chemicals by mast cells gets out of hand, as in the condition known as anaphylaxis, the swelling caused by the leakage of fluid from blood vessels can even flatten the trachea, leading to death by asphyxiation.

The mechanism of stimulation by antigen   An antigen is almost always a large molecule—usually a protein, polysaccharide, glycoprotein, or glyco-

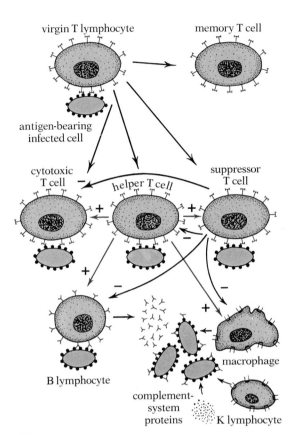

30.5 Immune reactions mediated by T lymphocytes
Each virgin T lymphocyte has membrane-mounted receptors specific to two antigens. (For simplicity, however, this drawing shows each T cell binding to only one class of antigen on target cells, not two.) When the antigens are encountered on an antigen-bearing cell (top left), the T cell divides to produce memory cells (top right) and three varieties of activated T cells (center). Cytotoxic T cells bind to the antigens on the antigen-bearing cells and kill the cells; helper T cells, when bound to the antigens on the target cells (and—most essentially—to the antigens on the corresponding B lymphocytes, as will be shown in Figure 30.8), release local chemical mediators that activate not only other T cells but B lymphocytes, macrophages, and, indirectly, mast cells (not shown). The resulting humoral immune response is shown at the bottom. Helper T cells also activate suppressor T cells; these, when bound to the appropriate antigens, inhibit B lymphocytes, macrophages, the various T cells, and mast cells. The balance between activation and inhibition, which is specific for each combination of antigens, determines the overall level of the immune response. Allergic reactions can occur when antigen-specific suppressor T cells fail to limit operation of the mast-cell/antibody system.

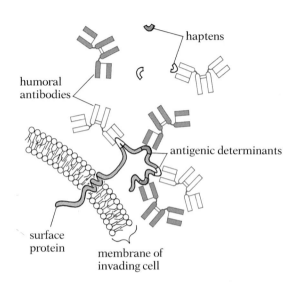

30.6 Antigenic determinants and haptens compared

The surface protein on this invading cell has two different antigenic determinants, each of which occurs twice in each molecule. Specific membrane-mounted antibodies and T-cell receptors can bind to these regions and trigger an immune response, which includes the production of free antibodies (shown). The surface protein and its antigenic determinants may be unique to this kind of invading cell, or they may be shared by cells of other kinds. Isolated antigenic determinants, or haptens, can also be bound by antibodies and stimulate an immune response if the immune system has been previously exposed to a large molecule bearing that particular determinant.

lipid. Not all of the antigen molecule stimulates lymphocytes to begin the immune response. Rather, certain regions (probably, in the case of proteins, about six amino acids long) serve as so-called *antigenic determinants*—sites of interaction with the receptors on lymphocytes. A single large antigen molecule may include several identical antigenic determinants that are bound by only one kind of antibody or T-cell receptor, or, more often, it may have several different kinds of antigenic determinants, which are bound by a corresponding number of different antibodies or receptors (Fig. 30.6). Conversely, different antigen molecules may, by chance, have one or more antigenic determinants in common. The initial reaction to an antigen is triggered only if the antigenic determinant is part of a large molecule, but subsequent reactions can be initiated by the isolated antigenic determinant. Such an isolated determinant is called a *hapten* (Fig. 30.6).

Proper functioning of the immune system is predicated upon the availability of an enormous number of slightly different lymphocytes, each specific for a particular antigenic determinant. Estimates of how many different kinds of lymphocytes exist in any individual run from a few million to as high as 10 billion (10^{10}) or more. Each antigen reacts only with those very few lymphocytes that have antibodies or receptors capable of binding to it, and this binding is necessary to induce proliferation of the appropriate lymphocyte types (Fig. 30.7). In Pasteur's demonstration, the sheep of both groups had lymphocytes specific to anthrax antigen, but only in the 25 that had been previously exposed to anthrax had the lymphocytes proliferated enough to win the race against the invading bacteria.

Each stimulated lymphocyte gives rise to a clone of cells (a group of genetically identical cells descended from a common ancestral cell). Hence the proliferation of the particular lymphocytes that react with a specific antigen is called *clonal selection*. Each of the plasma cells to which a given B lymphocyte gives rise when stimulated may transcribe as many as 20,000 mRNA molecules from its genes for antibody, enabling each of the plasma cells to secrete 2,000 identical antibody molecules per second.

But how does a lymphocyte recognize an antigen in the first place? Recognition takes place through the attachment of the antigen molecule to the lymphocyte surface. Each lymphocyte, whether of the B or the T type, has exposed receptors on its surface. The membrane-mounted antibodies of a B lymphocyte, which serve as that cell's receptors, are essentially identical to the free antibodies its descendant plasma cells will later secrete. The receptor molecules of T cells are similar in structure to B-cell antibodies, except that the two binding sites on each receptor are different. An antigen is bound only to those lymphocytes whose surface antibodies or receptors have binding sites specific for one of its antigenic determinants.

The binding of antigen by lymphocyte receptors allows the bound receptors to cross-link with each other. This cross-linking of receptors (which, by requiring at least two simultaneous antigen-binding events, may serve to reduce false alarms, and may explain why haptens cannot stimulate virgin cells) appears to be the event that induces the lymphocytes to begin dividing (Fig. 30.7). It also has effects in the subsequent generations: in helper T cells, cross-linking causes secretion of interleukins and lymphokine; in the

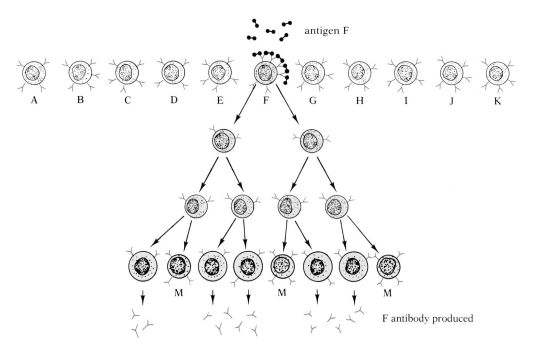

antigen F

A   B   C   D   E   F   G   H   I   J   K

M   M   M

F antibody produced

plasma cells of B lymphocytes, it induces antibody release; in mast cells bound to antibodies, it triggers the release of histamine and other chemicals.

Recognition of self   While the immune system of an organism is able to recognize and destroy almost any foreign antigen or antigen-bearing cell, it must ignore the antigens on its own cells. Tissue-transplantation experiments demonstrate that this self-tolerance is learned early in the embryonic development of the immune system: if a piece of tissue from an organism is transplanted into an unrelated adult animal, the recipient's immune system will almost always "reject" the tissue by mounting an immune reaction against the donor's cell-surface proteins. But tissue transplanted prenatally, while the recipient's immune system is still immature, will be accepted, and transplants from the same donor will be accepted even when the recipient is an adult. Apparently the immature immune system releases millions of virgin B lymphocytes into the blood, and those that bind to an antigen during this stage of development are inactivated, leaving the vast array of virgin cells that are specific for "non-self" antigens capable of proliferating and mounting an immune response. The inactivation of T cells, as we shall see presently, is slightly more complex.

Inactivation is not passive; the continued presence of each "self" antigen is required to maintain it. Unfortunately, even with continuous exposure, the inactivation may not be permanent: should the control over inactivation be loosened, as by infection with certain bacteria or viruses, an au-

30.7 Clonal selection

Antigens are able to react with the membrane-mounted antibodies or T-cell receptors of only a few very specific lymphocytes from among the billions of kinds of lymphocytes in the organism's body. In this example antigen F can be bound only by the B lymphocyte of type F; it does not affect the other lymphocytes (top row). Lymphocyte F, stimulated by the binding of the antigen, proliferates to form a clone of genetically identical cells. Some cells of the clone are memory cells (M); others are cells that actively secrete F antibody. Only B lymphocytes are shown in this drawing; T lymphocytes respond in a similar way, but do not secrete antibody. The cross-linking of the antibodies on lymphocyte F greatly facilitates the proliferation of the F clone.

MONOCLONAL ANTIBODIES

A growing understanding of the workings of the immune system has allowed researchers to develop revolutionary techniques to learn about cellular architecture—techniques that can provide new medical applications while they help reveal the structures of the cell. In one such technique, researchers select and clone a single type of lymphocyte—typically a B lymphocyte—whose antibodies bind to antigen of a particular kind of cell, a particular structure in a cell, or some other particular substance. This monoclonal antibody technique, as it is called, can be used to locate all the tubulin in a cell, or the sodium-potassium pumps, the F_1 complexes of mitochondria, the RNA polymerases, or any of the thousands of enzymes whose distribution and function in a cell are often not yet well understood. The originators of this invaluable research tool, Cesar Milstein of the British Medical Research Council in Cambridge and George J. F. Kohler of the Basel Institute of Immunology, were awarded a Nobel Prize for their work in 1984.

To clone antibodies specific to a particular substance, researchers first inject that substance—whether a chemical or the cells of interest—into mice. Soon there is a substantial increase in the B lymphocytes producing various antibodies specific to the many different antigens on the foreign cell or chemical. Most of these lymphocytes will be producing antibodies to antigens common to many different foreign cells or chemicals, but a few may, by chance, be producing antibodies specific to an antigen unique to the foreign material. If just these lymphocytes could be selected and removed from the mice and then cultured, they would produce a supply of the single antibody desired.

But there are two major problems: the specific antibody-producing cells must be separated from the others, and they must be propagated. In the actual cloning procedure, the second of these problems is solved first. You may recall that normal cells have a set number of cell divisions, after which they age and die, while cancer cells are immortal. If the lymphocytes were cancerous, then, they would multiply indefinitely. Hence the next step is to force cells of the mixed collection of lymphocytes to fuse with cells from a cancerous line of B lymphocytes. Those that fuse become immortal; they are then called hybridomas.

This still leaves the problem of selection. It is

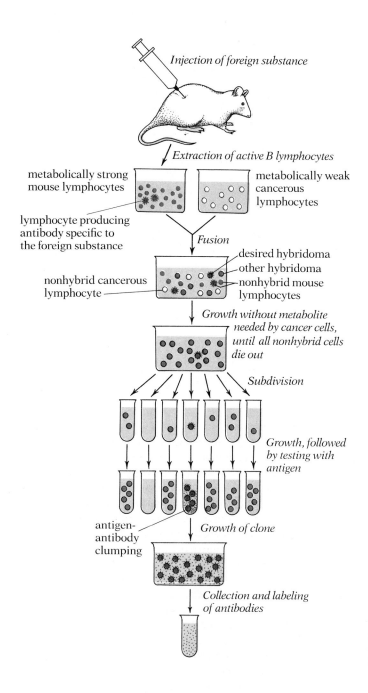

Injection of foreign substance

Extraction of active B lymphocytes

metabolically strong mouse lymphocytes

metabolically weak cancerous lymphocytes

lymphocyte producing antibody specific to the foreign substance

Fusion

desired hybridoma
other hybridoma
nonhybrid mouse lymphocytes

nonhybrid cancerous lymphocyte

Growth without metabolite needed by cancer cells, until all nonhybrid cells die out

Subdivision

Growth, followed by testing with antigen

antigen-antibody clumping

Growth of clone

Collection and labeling of antibodies

necessary to eliminate all but the hydridoma cells, and then to select those producing the desired antibody. The elimination is relatively easy. The nonhybrid lymphocytes from the mouse just die off after completing their set number of cell divisions. Milstein and Kohler solved the more difficult problem of sorting out the nonhybrid cancerous lymphocytes by using a line lacking a crucial synthetic pathway. When these cancer cells are cultured in a medium without the substance they cannot make, they die. Only the hybridomas—immortal by virtue of genes from the cancer cells, and able to synthesize the missing nutrient because of genes from the mouse lymphocytes—will survive.

The next step is to separate the individual hybridoma cells as much as possible, to culture each one separately, and then to test each culture by adding the antigen. If the cells of a particular group are producing the desired antibody, the test will result in antigen-antibody clumping. These cells can then be cultured further, to form a clone of immortal cells producing a continuous supply of a specific antibody.

Researchers use these monoclonal antibodies to tag specific cells, cellular structures, or chemicals by labeling the antibodies with fluorescent dye, the electron-dense substance ferritin, or a radioactive marker. Once a labeled antibody has reached its target, the location of the antigen can be found with a light microscope, an EM, or any of a variety of radioisotope analyses. A labeled antibody can even be used to mark certain kinds of cancer cells selectively for destruction; ovarian cancer cells, for example, can be bound by ferritin-linked antibodies and then destroyed by ferritin-specific T cells. The monoclonal antibody technique, then, has enormous potential both for basic research and for medical technology.

---

toimmune disease results, with the immune system attacking the body instead of protecting it. A well-known example is myasthenia gravis, a neurological disorder involving loss of tolerance of acetylcholine receptors. How self-tolerance is overcome in such ailments is not yet understood.

The antigens involved in reactions like these are generally the membrane-anchored glycoproteins of the glycocalyx (see pp. 109–10). Some are specific to particular tissues, and seem to play a role in cell-type recognition and cell-to-cell adhesion. Others, which are encoded by a group of genes known as the major histocompatibility complex (MHC), appear on the surface of most cells in the body. There are several genes in the complex, and each of these has many alleles. As a result, the MHC products of different individuals generally have only a few antigenic determinants in common. Furthermore, the cells of each individual carry several different antigens produced by the MHC. There are two general classes of MHC products in each individual, MHC I (found on most cells that are not lymphocytes) and MHC II (found only on lymphocytes); the importance of this distinction will become clear presently.

The so-called MHC antigens appear to play a major role in transplant rejection. Some suppressor T cells, as well as most cytotoxic T cells and helper T cells, bind most effectively when there is both a familiar (self) MHC antigen *and* an unfamiliar (foreign) antigen available on a cell sur-

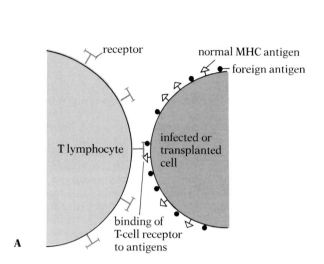

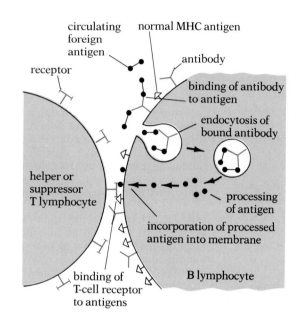

30.8 Binding of T-cell receptors

Most T-cell receptors bind simultaneously to one of an individual's normal MHC antigens and to a foreign antigen. (Cells carry several different MHC antigens, but only one kind is shown here, for simplicity.) (A) In the case of infected cells, the receptor of a cytotoxic T cell binds to one of the cell's normal MHC antigens and to a foreign antigen that an infecting virus causes to be synthesized. In the case of transplanted cells, the foreign antigen is usually another MHC antigen, not shared by the recipient. In either case the T cell then brings about the death of the target cell. (B) This dual-binding strategy is also the basis of T-cell activation and suppression of B cells. In the most plausible model, the B-cell antibody first binds to the antigen, after which the cell encloses the bound antibody by endocytosis, removes the antigen, attaches it to a protein, and inserts the antigen/protein complex into the membrane. When a T-cell receptor binds simultaneously to this processed antigen and to an MHC antigen, the T cell will exert its characteristic helping or suppressing effect.

face. A single T-cell receptor binds to both antigens simultaneously (Fig. 30.8A); one of its two different binding sites is specific for the foreign antigen, while the other is specific for the familiar MHC antigen. Since transplanted tissue will normally have some MHC determinants in common with the host and some that are not, some of the host's T cells will bind to the foreign tissue and trigger rejection. It is now easy to see why inactivation is more involved for T cells than for B cells: apparently, virgin T cells avoid inactivation during the immune system's "learning" phase if they have a receptor with a region that binds to one of the organism's own MHC glycoproteins and a region that does not react to any antigen then present. The selection probably goes on in the thymus during T-cell maturation.

This system seems to allow most cytotoxic T cells, most helper T cells, and some suppressor T cells to recognize when one of an organism's own cells is infected. For example, viruses within a host cell make coat proteins in preparation for self-assembly and lysis. Some of these proteins find their way into the infected cell's membrane. When a cytotoxic T cell binds simultaneously to the viral protein (the antigen) and an MHC antigen, it kills the infected cell before the virus can complete its life cycle. When it encounters a virus by itself, however, it usually ignores the virus because there is no MHC product on its coat. Hence, cytotoxic T cells kill only cells; circulating B-cell antibodies, on the other hand, can bind to viruses and toxins, which are then usually destroyed by phagocytosis.

Mechanism of T-cell helping and suppression We have seen that cytotoxic T cells act directly, by binding to and attacking infected and foreign cells, but helper and suppressor T cells act by modulating the activity of B cells. When a B cell encounters the antigen for which it is specific, its antibodies bind to the antigen, and secretion of circulating antibodies begins. Present evidence suggests that at the same time the bound antibodies on the surface are brought into the B cell by endocytosis; there the antigen is removed, and fragments of it are attached to particular proteins and transported back to the cell membrane (Fig. 30.8B). Once the "processed" antigen is in the membrane, T-cell receptors specific for it can bind simultaneously to this antigen and to the MHC antigen on the membrane, and the T cell (which may also be bound to the infected or foreign cell) can then secrete the local chemical mediators by which it exerts its characteristic activating or suppressing effects.

You may wonder what will happen if a cytotoxic T cell binds to a B cell in the manner just described; fortunately, this cannot happen. Cytotoxic T cells have receptors for MHC I proteins, and so cannot bind to B cells, which have MHC II proteins in their membranes. Most helper and suppressor T cells, on the other hand, have only MHC II receptors, and so can bind to B cells.

The antibody molecule B-lymphocyte antibodies, whether circulating or membrane-bound, are globular proteins, predominantly gamma globulins. Each antibody molecule consists of four polypeptide chains—two identical "heavy" chains and two identical shorter "light" chains; the chains are linked by disulfide bonds (Fig. 30.9).

There are two general classes of light chains; these display no clear functional difference, but are encoded by separate genes (Fig. 30.10). There are

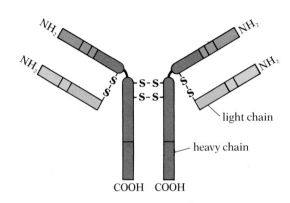

30.9 The arrangement of the four polypeptide chains in a molecule of B-lymphocyte antibody
Variation in the specificity of antibodies is due almost exclusively to differences in amino acid sequences within the regions shown in color, which are at the amino ends of the chains.

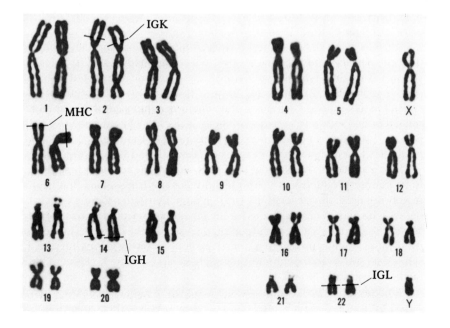

30.10 Location of some immune-system genes in the human genome
The heavy-chain gene (IGH) is near the end of chromosome 14; this is the site of the translocation event that gives rise to most cases of Burkitt's lymphoma (see p. 758). The gene for one version of the light chain (IGK) is located on chromosome 2, while the gene for the other version (IGL) is on chromosome 22. The major histocompatibility complex (MHC), important to the functioning of T lymphocytes, is on chromosome 6.

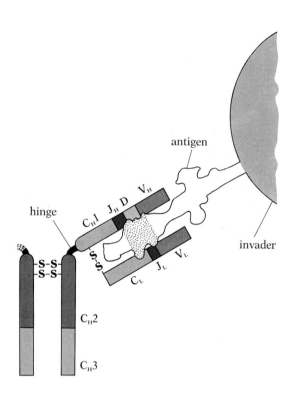

30.11 The antigen-binding site of an antibody molecule

The binding site is a pocket formed by the interaction of a heavy chain and a light chain. The sectors labeled C_H3, C_H2, hinge, C_H1, J_H, D, V_H, C_L, J_L, and V_L on this schematic representation are encoded by separate exons in the genes. The five different classes of heavy chains have different C_H2 and C_H3 regions; class E and class G chains have a fourth C_H region. Only one of the antigenic determinants of the antigen (shaded) is bound by this antibody. Completely different antibodies may bind to separate sites on the same antigen.

five classes of heavy chains—A, D, E, G, and M; these differ in the amino acid sequence at the COOH (tail) ends. The tail regions play no role in antigen specificity, but are involved in determining which reaction of the humoral antibody response will take place. For example, after antigen has been bound and agglutination has occurred, heavy chains with a so-called G tail (by far the most common class of antibodies in higher vertebrates) undergo an allosteric change in conformation that allows them to be recognized by macrophages, which can then perform phagocytosis. Bound antibodies with an M tail, on the other hand, catalyze the activation of the first zymogen in the complement system, which leads to the formation of a membrane channel and lysis of the invader. Antibodies with an E tail become mounted on the membranes of mast cells, signaling them to release histamine when an antigen is encountered.

Regardless of which class of heavy or light chain is incorporated into an antibody, most of each chain has a constant amino acid sequence and structure; the variability crucial to antigen specificity lies mostly at the free amino ends. The binding sites for antigens (two sites on each antibody molecule) are at the ends of the variable portions. Each binding site is a pocket or cleft bounded partly by the heavy chain and partly by the light chain (Fig. 30.11). This region can bind to approximately six amino acids or carbohydrate units of the antigen.

The genetic basis of antibody diversity How is the vertebrate genome able to produce an almost infinite variety of antibodies without employing millions or billions of different genes, one for each potential antigen? The answer to this question involves a remarkable strategy of exon selection, which operates initially during the fetal development of the immune system, and leads to the vast diversity of lymphocytes available to the organism during its lifetime. The process involves selection from a pool of exons in only three genes—one for the heavy chains and one for each of the two classes of light chains.

Let's look at the chromosomes of a developing human lymphocyte before the cell specializes to produce a single type of antibody—that is, before it becomes a virgin B lymphocyte. In the human genome, the part of the DNA coding for the heavy chain is found on chromosome 14; the segment that codes for the constant regions of this chain is composed of 22 exons arranged in five sets, one corresponding to each class of heavy chain—A, D, E, G, and M. Each set contains exons for regions C_H1, hinge, C_H2, C_H3, and, in the sets that code for heavy chains of classes E and G, C_H4 (Fig. 30.11).

This arrangement of exons in the heavy-chain gene seems at first glance to be counterproductive, since so much has to be transcribed that is ultimately not used: thus, even if the virgin lymphocyte will produce only antibodies with class A chains, exons for the other classes would be transcribed as well. Toward the 5′ end of the coding DNA strand, the organization seems even more counterproductive: there are four different exons for the J_H (joiner) region, even though each heavy chain has only one J_H segment. Still farther toward the 5′ end, beyond the exons for the joiner region, are approximately 12 different exons for the D region. Again, each antibody produced by a virgin cell has only one D segment. Still farther toward the 5′

ACQUIRED IMMUNE DEFICIENCY SYNDROME

Acquired immune deficiency syndrome (AIDS) is a devastating disease of the immune system which, according to the best evidence available, first appeared in Africa in the early 1960s. Some researchers believe the viral agent of AIDS is a mutant form of a nonfatal virus that evolved in green monkeys and was transferred to humans through bites. The AIDS virus, HIV-1 (human immunodeficiency virus, type 1) is a retrovirus—an RNA virus that carries its own reverse transcription enzyme. HIV-1 primarily infects T4 lymphocytes (a subclass of helper T cells) where it remains relatively dormant for the duration of its temperate phase. Because of the inactivity of the virus during this period, an infected individual remains asymptomatic, but may pass the virus to others through the exchange of blood and other bodily fluids. Epidemiological studies show that HIV-1 transmission is almost always from the blood or semen of one individual to the circulatory system of another. Hence the disease spreads readily through transfusions of infected blood, reuse of hypodermic needles, and anal intercourse. The virus can also be transmitted through vaginal intercourse, though the rate of this form of transmission appears to be much lower.

As the AIDS virus switches from its temperate to its virulent phase, ending the incubation period, it reproduces rapidly, budding off progeny from the surface of infected lymphocytes, which in turn circulate freely in the blood and infect other T4 cells. The reproduction of HIV-1 reduces the ability of T4 cells to respond to antigens. As a result, the immune system begins to fail and the individual may succumb to one or several illnesses that would ordinarily have had little impact.

The rate of spread of AIDS depends on the average incubation period of the infective agent, the length of the infectious period, the encounter rate (the number of sexual partners), and the chance of communicating the infection with each encounter. Most of this information can now be estimated for the United States. From careful studies most researchers now conclude that the average temperate period is 8 years, the incubation period is quite short, and infected individuals are infectious throughout the temperate *and* virulent phases. All carriers of the virus will probably develop AIDS sooner or later, and

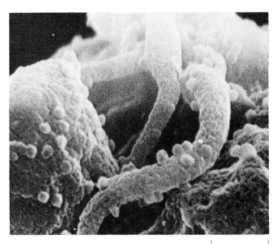

0.5 μm

AIDS viruses budding from a cultured lymphocyte

the average survival from the onset of AIDS symptoms is about a year. Though the death rate has been very nearly 100%, there is hope that new treatments may at least extend the average survival time.

Among American homosexuals and intravenous drug users, the infection rate is as high as 70% in urban areas like San Francisco and New York City where encounter rates are very high. Among heterosexuals, transmission efficiency and encounter rates are lower. At present it is not clear whether there will be an epidemic among heterosexuals. Will the average carrier infect, on the average, more than one other individual (the statistic that defines an epidemic)? In Africa, HIV-1 transmission is primarily heterosexual and risk correlates with the number of sexual partners. The World Health Organization estimates that nearly 60% of all AIDS deaths, which may have totalled nearly 80,000 by the beginning of 1988, have occurred in Africa, where millions more are infected. Clearly, a heterosexual epidemic *is* possible.

Even with a million or more people already infected with HIV-1 in the United States, there is hope that medical advances and behavioral changes among those at risk may yet stop the spread of this terrible disease.

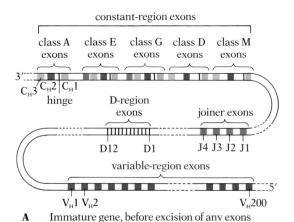

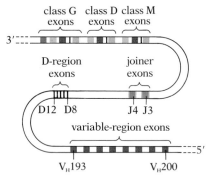

A Immature gene, before excision of any exons

B Mature gene, as it will be transcribed

C mRNA, after processing of the primary transcript

30.12 Organization of immature and fully mature heavy-chain antibody gene, and corresponding mRNA

The immature heavy-chain gene (A) originally contains five sets of constant-region exons (one set for each class of heavy chain), four different joiner exons, approximately 12 different D-region exons, and about 200 different variable-region exons. When the gene is fully mature (B), many of the exons have been removed. The remaining exons and introns are transcribed, but only the exon closest to the 5' end of the primary RNA transcript of each exon group survives processing and appears in the mature mRNA (C). The two steps of exon removal permit a B lymphocyte to produce a unique kind of antibody from millions of possible alternatives.

end lies a string of roughly 200 different V_H exons, only one of which is eventually translated.

How is it, then, that the roughly 240 exons in the original heavy-chain antibody gene give rise to the protein product encoded by only seven or eight exons? The removal of superfluous exons is not yet completely understood, but it seems to be accomplished in two steps. In the first step, which takes place during fetal development, large regions of the gene are cut out of chromosome 14, and the remaining ends are spliced together by specialized enzymes. One such excision, for example, extends a random distance from the D region into the V_H exons (Fig. 30.12). Any number of V_H exons —from 0 to 199—can be removed, but excisions of a relatively small number of exons are more common. The first V_H exon not excised is the one ultimately translated and expressed. During transcription, RNA polymerase copies all the remaining V_H exons on the 5' side, but these superfluous transcribed exons are removed during RNA processing to yield an mRNA with only the first V_H exon that survived excision. As a result, only the V_H exon closest to the D-region exon will be translated (Fig. 30.12). All but one transcribed joiner exon and D-region exon are removed by means of two similar steps. A corresponding process is seen in the gene for the light chain, which has one C_L exon, four alternative joiner exons, and about 300 alternative V_L exons.

By analogy with the V_H, D, and joiner exons, you might guess that some C_H exons would be excised from the chromosome, and all but one of the surviving set of exons removed from the primary transcript. There is debate on this point, but it is clear that something like this must be happening in at least some cells: many fully mature B lymphocytes remain able to switch between C_H classes, or even to produce two kinds of heavy chains simultaneously. Apparently the initial RNA transcript can contain the exons for more than one of the five C_H classes, and all but one set are removed along with introns in the course of RNA processing.

The vast diversity made possible by an active but random assembly of exons like the one just described is impressive. For instance, since the gene for the heavy chain has four alternative joiner exons and 12 alternative D-region exons, there are 48 (4×12) possible joiner/D-region combinations. And since there are roughly 200 V_H alternative exons, a lymphocyte can produce antibodies with one of 9,600 (48×200) forms of the variable region of the heavy chain. Similarly, the light chain, with its four alternative joiner exons and 300 V_L exons, can be produced in 1,200 different forms. Since each antibody consists of one kind of randomly "selected" heavy chain and one kind of light chain, these different forms will give rise to about 12 million ($1,200 \times 9,600$) different antibodies during the development of the immune system. Additional variation is thought to be generated by a poorly understood process of active mutation in the joiner and D-region exons and parts of the V_L and V_H exons. Since the mutations take place in developing B cells, which are somatic cells, they are unique to each cell and are not perpetuated in the germ line.

T-receptor diversity is thought to be generated in much the same way, involving excision from the chromosome and removal from the transcript of all but one of the alternative exons of each exon group. The details of these strategies of exon assembly are still controversial.

MECHANISMS OF GENETIC VARIABILITY

The selective forces that led to the unexpected complexity of the eucaryotic genome and the evolutionary advantages that must serve to maintain it are simply not understood at present. Nevertheless, some tentative hypotheses may help us make some sense of what is a very confusing pattern. Oddly enough, the structure and behavior of certain somatic-cell genes of the eucaryotic genome—the antibody genes—may provide some of the most important clues to how gene organization evolved and continues to evolve in the germ line.

As Darwin discovered, evolution depends on genetic variation and the natural selection of heritable characters. And since the basic stuff of heredity is DNA, there must be a source of variability in DNA that makes evolution possible. So far we have considered the genome as a relatively stable chemical structure that undergoes many small, random mutations, but repairs most of them. With such a genome, the variation necessary for natural selection to operate would come primarily from recombination (as in sexual species) and the rare unrepaired changes in the DNA. It is hard to imagine how these few minor mutations, and the reshuffling of the same old genetic deck, could result in the evolution of major new proteins.

The recent discovery that eucaryotic chromosomes contain introns, exons, seemingly functionless pseudogenes, and repetitive DNA came as a great surprise. Most biologists expected that natural selection would have operated to weed out apparently unnecessary DNA, as it seems to have done in procaryotes. The bacterium *E. coli*, for example, has lost the ability to synthesize the few amino acids always found in abundance in its natural habitat; the genes for the corresponding biosynthetic pathways are missing, and essentially no space in the chromosome of this rapidly multiplying organism is wasted. By contrast, well over 90 percent of eucaryotic DNA lies apparently functionless and dormant. The subsequent discovery that the immune system actively disposes of gene segments was also unexpected, and made it immediately clear that the genome is not necessarily as stable as had been thought. We shall explore how the odd organization of the eucaryotic genome may facilitate evolution, looking at some of the more potent mechanisms of gene evolution—notably duplication and transposition of genes and gene segments.

GENE AND EXON DUPLICATION

The unique structure and organization of antibody genes provides food for thought about the evolution of genes in general. A careful examination of the three C_H exons reveals that they are virtually identical to one another, so the entire constant region of the antibody heavy chain is a series of basically uniform subunits, with a hinge region that provides mechanical flexibility between C_H1 and C_H2 (Fig. 30.11). Moreover, the C_L exon and large portions of the V_L and V_H exons are only slightly different from the C_H exons and from each other. Similarly, the genes for the two versions of the light chain are almost identical, as are the sets of exons for the five classes of heavy chains.

One reasonable hypothesis is that the V and C exons of both the light and

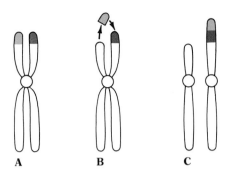

30.13 Duplication of exons through breakage and fusion
Sometimes an end of one chromosome breaks off (A–B) and fuses to the end of another (C). In the example shown here, the fragment fuses to the corresponding end of the twin chromatid. If such an event occurs in the process of meiosis, which leads to gamete formation, one gamete will have two copies of all the exons (and introns) on the fragment.

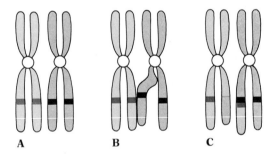

30.14 Exon duplication from unequal crossing over
Sometimes during meiosis two chromatids from homologous chromosomes (A) are misaligned during a crossover event (B). As a result, one chromatid gains a second set of exons (and introns) from the misaligned region of the homologous chromatid (C). This duplication is inherited by one of the resulting gametes.

heavy chains, coding as they do for very similar polypeptides, arose from the duplication and rearrangement of some prehistoric gene sequence. The duplication of exons (and the associated introns) or of entire genes can probably occur in several ways. The simplest is nondisjunction, the complete duplication of chromosomes that results when both members of homologous pairs move to the same pole during meiosis (see p. 688). Another well-established mechanism, also discussed in Chapter 25, involves chromosomal breakage and fusion during meiosis, in which one egg or sperm cell ultimately gets an extra bit of DNA that the other has lost (Fig. 30.13). The exons or genes on the segment that breaks off and fuses to the homologous chromosome are duplicates of those already resident there.

Other mechanisms, as yet less well established, are also thought to give rise to duplication. One is unequal crossing over: on occasion, homologous chromosomes are partially misaligned during crossing over, and one chromatid obtains an extra bit of DNA that duplicates an area already present nearby on that chromatid (Fig. 30.14). This is probably the way the tandem repeats of the rRNA genes arose. Two other mechanisms, as we shall see, are reverse transcription and transposon-mediated duplication.

Reverse transcription You may recall from Chapter 28 that certain RNA viruses—the retroviruses—have an enzyme that catalyzes transcription of their RNA into cDNA, which can be inserted into the host genome. You may have wondered why this reverse transcriptase does not do the same to the host's mRNA. In fact, it sometimes does. On rare occasions, the resulting cDNA—complementary to the host RNA from which it was copied—becomes integrated into the host's chromosomes. If this occurs in a gamete, the resulting duplication can be inherited. The insertions are fairly easy to distinguish from simple duplications: the new DNA sequence has no introns, since those were removed during the processing of the RNA in the nucleus; in addition, the end of the insertion may include some or all of the long poly-A tail that is added to eucaryotic mRNA. A small percentage of the human genome has, in fact, these telltale signs of reverse transcription.

Transposition As we saw in Chapter 28, lysogenic viruses can insert themselves into host chromosomes, and the DNA transcripts of retroviral RNA can also be incorporated into these chromosomes; it now appears that eucaryotic chromosomes can undergo analogous insertion events by themselves, in the absence of viruses, with large sections of chromosomal DNA transposed from one area to another. Sometimes the DNA in question is physically moved, but in other cases one duplication remains at the original location, while another is inserted at a new site.

Transposition of this sort in eucaryotes was discovered more than three decades ago by Barbara McClintock of the Cold Spring Harbor Laboratory. She found that in corn certain genetic elements (which she believed to be genes) will occasionally move, particularly after cells are subjected to trauma, such as exposure to intense UV radiation. These movements produced kernels with unusual colors—colors that could not have resulted from normal recombination. Thirty years ago her results seemed so completely at odds with the prevailing concept of a static gene that the mobile genetic elements she had discovered were considered some sort of anomaly. By 1983, however, when she was awarded the Nobel Prize for her discovery, many such elements—now called ***transposons***—had been discovered, and their possible role in gene evolution was beginning to be recognized.

We now know that transposons can consist of several genes, or only one gene, or just a control element. Transposons probably move within the genome in several different ways, none of which is fully understood. According to one model, a transposon within a chromosome is flanked by identical characteristic sequences, one of which is actually part of the transposon. The transposon moves from its site on the chromosome and finds another site. It may leave no trace except a telltale "scar" consisting of the flanking sequences—one literally left behind, the other a copy of the sequence that is part of the transposon (Fig. 30.15A). Or it may move only after being copied in full by DNA polymerase, so that a complete duplicate of itself remains at its original location (30.15B). In this model, the host chromosome too carries one or more copies of a particular sequence, which serves as a tar-

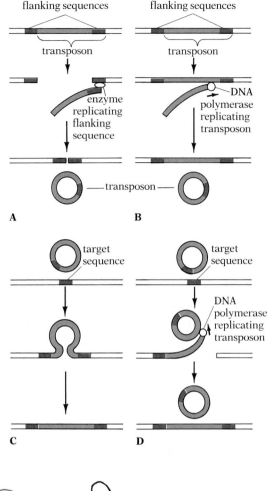

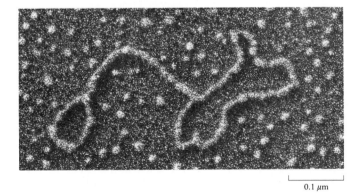

0.1 μm

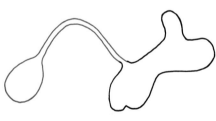

30.15 A model of transposition events While integrated into a chromosome a transposon exists as a stretch of DNA, often flanked by characteristic sequences. At least one of these flanking sequences moved into the chromosome as part of the transposon, while the other may have already been there, and served as a target site. The movement of the transposon may involve either excision from the chromosome (A) or duplication, with one copy of the transposon then left behind, at the original location (B). In both cases, the mobile transposon probably exists as a circle, which can be integrated into any other part of the genome with the target sequence (C). This integration too is frequently accompanied by a duplication, which produces another mobile transposon to continue the cycle (D).

The micrograph shows a transposon integrated into a plasmid. As indicated in the interpretive drawing, the characteristic "stem and loop" structure of the transposon (color) arises from pairing between its two terminal sections within the circular DNA.

SLEEPING SICKNESS: A WAR OF GENE SHUFFLING

Sleeping sickness is one of the most debilitating diseases in the world, and one of the most difficult to treat effectively. Its cause is a protozoan known as a trypanosome, which bloodsucking tsetse flies carry from an infected host to a potential one. The trypanosomes multiply in the blood of the host, and later spread into the nervous system. Symptoms begin with fever and fatigue, and progress to drowsiness during the day and insomnia at night. The victim ultimately becomes too sleepy to eat, then comatose, and finally dies.

Since the trypanosomes are in the bloodstream of the host for weeks before causing death, how is it that the immune system, with its millions of antibodies generated by exon shuffling, fails to react effectively? The answer is that during these weeks the trypanosomes are shuffling their genes.

Like all organisms, trypanosomes have proteins exposed on their cell membranes. But unlike most organisms, an individual trypanosome has only one kind of surface protein. Still, the immune system has one or more antibodies that can bind to the antigen of any particular trypanosome surface protein, so in just a few days a full-scale humoral antibody response is under way, with macrophages, killer cells, and complement-system channels working in concert to destroy the invading organisms.

But by the fifth day after infection, all the surviving trypanosomes have changed their surface proteins. The trypanosome genome has hundreds—perhaps thousands—of surface-protein genes, and a new one is transposed into an active site in the genome roughly every five days. On each trypanosome the new surface protein may be any of many, so the number of trypanosome antigens may increase by the fifth day from the few present at the time of infection to a great many. As a result, the existing highly specific immune reaction becomes irrelevant, and a new one, responsive to a larger set of antigens, must begin essentially from scratch: a new, different set of lymphocytes must now bind to the parasites, to begin again the process in which virgin B

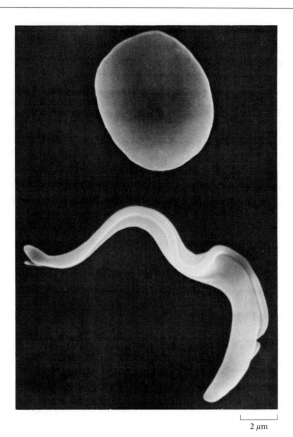

Trypanosome and a red blood corpuscle

lymphocytes are activated and divide, antibodies are produced, and so on.

But again, even if the battle is being won, the surface-protein genes are shuffled. New surface proteins appear, and the number of new antigens multiplies further. Eventually, the body's immune system is simply overwhelmed. Transposition in trypanosomes, then, seems to be a precisely controlled event. Transposons may be more common and more important in the history of disease than most researchers used to think.

get for the transposon, and this sequence pairs with the one copy of the flanking sequence carried on the transposon. A special enzyme, usually encoded by the transposon itself, enables it to recognize and act on the target sequence in the host chromosome. The transposon can then, with the aid of yet another enzyme usually encoded by the transposon, incorporate itself into the host chromosome.

Transposons not only insert and duplicate their genetic material, but can instigate larger changes. For example, once there is a pair of transposons in a single chromosome, the sort of unequal crossing over that was mentioned earlier as a cause of gene duplication becomes more likely, because the transposons may mispair (Fig. 30.16). If the flanking/target sequence is relatively common, then, transposons will tend to proliferate throughout the genome, and genes located between transposons may also be duplicated.

EVOLUTION AND THE GENOME

EVOLUTIONARY CONSEQUENCES OF DUPLICATION

It is fairly easy to see that duplication of genes or exons followed by small-scale evolutionary changes is more likely to lead to functional genes with novel properties than random changes alone would be. Imagine how rarely we could generate a meaningful sentence by randomly arranging letters and spaces, whereas if we *began* with a meaningful sentence (a gene for a functional protein) consisting of words and spaces (exons and introns) and changed a few existing letters, the odds of ending with an intelligible sentence with a new meaning would be fairly high.

Duplication, then, is one possible source of the vast number of multiple copies of similar exons we see in antibody genes. Introns may provide the most suitable points for insertion of new copies of C and V exons, as well as the short but important hinge, joiner, and D-region exons. Why is this so? If exons are inserted into other exons, chances are that the insertion will occur within the codons for particular amino acids, so all subsequent codons may be out of phase and misread during translation. This problem does not arise if an exon is inserted into an intron. As you may recall, during mRNA processing introns are removed with the probable aid of snRNP, which binds to start and end signals that are part of the introns themselves. New exons successfully inserted into introns would probably bring with them enough of their flanking introns to provide the signals for transcript processing.

The consequence of this organization in the antibody gene is that the various exons, though they may have arisen from the same original sequence, can be independently modified through mutation and selection. That is, of course, particularly important in generating the many slightly different alternative exons of the variable region.

EVIDENCE FOR DUPLICATION OF GENES CODING FOR NON-ANTIBODY PROTEINS

Duplicated exons are essential to the versatility of the immune system. But has duplication been even partially responsible elsewhere for the evolution

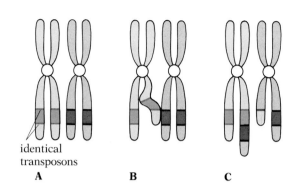

identical transposons

A   **B**   **C**

30.16 Unequal crossing over facilitated by transposons

When two identical transposons exist in the same chromosome (A), there is a chance that they will mispair during crossing over (B). The result is that one chromatid gets not only an extra transposon but also a duplicate copy of all the genetic material between the two transposons (C).

of genes for enzymes and structural proteins? The evidence for the role of duplication is particularly clear in the genes for myoglobin and hemoglobin. As you may remember, myoglobin, the oxygen-storage protein in muscles, consists of a single polypeptide chain, while hemoglobin, which carries oxygen in the blood, has two pairs of chains, for a total for four. The amino acid sequences of the α and β chains in hemoglobin are nearly identical, and they also closely resemble the amino acid sequence of the single chain in myoglobin; the genes for all are thought to have evolved by duplication from a single ancestral gene. This conclusion is reinforced by the discovery that introns are located in the same places in all these genes.

Hemoglobin itself is synthesized in slightly different forms at different times in an organism's life—during embryonic development and during adult life, for instance—and is thus specialized for differing conditions of pH and oxygen concentration. Again, the genes for these alternative forms of hemoglobin, which in human beings lie near one another on chromosome 7, are thought to have originated by duplication, and then to have followed independent evolutionary pathways. This seems all the more likely because the same region contains many pseudogenes with sequences very similar to those of the hemoglobin genes. Perhaps the pseudogenes are duplications whose subsequent evolution never led (or has yet to lead) to improved functional gene products.

The same pattern is seen in many groups of enzymes. For example, the digestive enzymes trypsin, chymotrypsin, and elastase, and the blood-clotting enzyme thrombin, all have different functions, but the genes for them have base sequences and intron locations that are nearly identical. It is highly unlikely that each evolved independently into a near duplicate of the others. Most researchers believe that the genes for these enzymes began as duplications of the gene for some primordial enzyme, and then evolved separately.

There is now considerable evidence, then, that eucaryotic gene evolution depends at least in part on gene duplication followed by changes in base sequence that give rise to functionally different products. But such changes occur slowly, and most random mutations result in genes with products of reduced function—or no function at all—rather than of altered specificity. For every new gene that produces a functional protein, there may be tens or hundreds of incomplete or failed "experiments" involving duplications. In short, unless some process is at work to edit out useless duplications, the eucaryotic chromosome should be full of nonfunctional base sequences with clear similarities to those of functional genes. And indeed, as we have seen, well over 90 percent of the mammalian genome does not code for functional products. It may be that the enormous number of nonfunctional pseudogenes in eucaryotic chromosomes is evidence of past duplications which never evolved into functional genes. Many of the examples of repetitive DNA discussed in Chapter 29 may also fit into this category.

EXON RECOMBINATION

We have seen that a gene can be duplicated and perhaps then evolve so that it codes for a functionally different product. We have also seen that the antibody genes, which appear to have arisen through repeated duplication of gene sequences, contain exons that are selected randomly and combined to generate diverse antibodies. These two observations suggested to Walter Gilbert at Harvard and Colin Blake of Oxford another way in which new genes could evolve: perhaps individual exons from *different* genes could be brought together to produce new combinations.

This highly controversial proposal would make sense only if exons code for "domains" in the resulting protein—distinct subunits that, like building blocks of various shapes, can form new structures when put together in new ways. Careful examination of the genes for dozens of proteins confirms that this is at least sometimes the case. For instance, both introns in the gene for myoglobin occur between regions that code for sections joined at a major turn of this highly folded globular protein. Thus each intron defines a boundary of sorts between compact domains, and each myoglobin exon can be thought of as coding for one of these domains, or subunits. In genes for some other proteins introns fall at the boundaries between regions coding for sections of α helix and regions coding for sections of the β pleated sheet, or the introns flank regions coding for sections of the protein containing the active site. Because the regions encoded by exons form distinct subunits, a recombination of exons would have a chance of generating a working enzyme with novel properties. A new gene could evolve by combining, say, exon 2 of gene *A*, exon 4 of gene *B*, exons 1 and 2 of gene *C*, and exon 2 of gene *D* at a new site (Fig. 30.17). Such a recombination could be effected by a simple movement of the exons from their original genes or pseudogenes. It might involve movement of the exons themselves, or of copies of them. In either case, insertion of exons (with their flanking introns) within an intron region of the new chromosome would improve the chances of generating a functional new gene.

Again, we can reformulate this idea in terms of creating new sentences (functional proteins) either from scratch or by beginning with words (functional building blocks): the odds of obtaining a meaningful sentence are far higher if we recombine words at random than if we recombine letters and spaces at random.

To test this exon-recombination model, we need to know the entire base sequence of a gene (and hence the location of the introns) *and* the three-di-

30.17   Hypothetical recombination of duplicate exons to create a new gene
Duplicates of five exons, from four different genes on a single chromosome (designated chromosome I), are imagined to be inserted near one another in a different chromosome (chromosome II). The intron regions flanking each exon being inserted provide the signals necessary for the new gene's transcript to be processed correctly. The exon duplications on chromosome I could have arisen from unequal crossing over, chromosomal breakage and fusion, transposition, or retroviral insertion. Since many exons code for protein subunits, the newly combined exons are more likely to encode a functional product than they would have been if they began and ended at random points. If their product is at least partially functional, the new gene will survive in the germ line and be subject to improvement through natural selection.

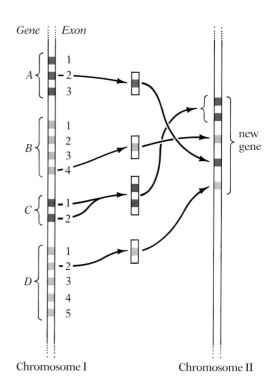

Chromosome I   Chromosome II

mensional conformation of the protein it encodes. The best-known gene fitting these requirements encodes glyceraldehyde 3-phosphate dehydrogenase (GDPDH), the enzyme that catalyzes the reduction of NAD_{ox} during glycolysis by phosphorylating PGAL. One exon of this gene codes for a domain containing an a helix and catalytic site, a functional subunit common to several dehydrogenases; this exon may have been obtained from the gene for another dehydrogenase by exon duplication and movement. The NAD binding site of the enzyme is encoded by a two-exon region that appears twice in the gene; this region closely resembles the region coding for the NAD binding site of other enzymes, and so may also have been obtained by duplication and movement. Only one of the 11 introns of the GDPDH gene does not appear to be dividing functional domains, and so may represent a random insertion.[1] Another example is the gene for the membrane receptors that bind LDL (the complex that transports cholesterol in the blood) and that aggregate to trigger endocytosis of cholesterol (see Fig. 4.25, p. 107). Recent research suggests that the 18 exons of this gene originated as copies of exons from at least three other genes. Of course, before any definitive conclusions can be drawn about whether exon recombination is a major force in evolution, the same sort of pattern must be observed in the genes for several more proteins.

A new gene generated by exon duplication and recombination might survive in the germ line, and, under the influence of mutation and selection, slowly evolve. With the head start that exon recombination could theoretically provide, gene evolution might proceed more efficiently.

We began this chapter by saying that the organization of the immune system might provide hints about how the genome came to be organized as it is. We saw that the exons encoding constant and variable regions of the antibody genes might have arisen through duplication; and we pointed out that many enzymes seem to be encoded by genes with very similar sequences, suggesting that gene duplication may be a major force in evolution. We pointed out that such duplication in the genome would lead to a great many failures, and speculated that this might explain some of the vast number of pseudogenes. We also saw that mechanisms exist which assemble the exons of the immune-system genes in many different combinations, creating enormous antibody diversity; subsequently we saw that the existence of introns and their placement in many cases between sequences coding for separate functional domains in a protein might make possible an analogous reshuffling of exons from different genes to create new genes over evolutionary time. Whether or not these particular highly speculative hypotheses about gene evolution are correct, the old picture of the genome as a static library of unchanging genes is almost certainly incorrect; instead, the genome is probably the active laboratory of evolution by natural selection.

---

[1] Curiously enough, the GDPDH gene in procaryotes—a group of organisms in which introns are very rare—has virtually the same sequence as it does in eucaryotes, minus the introns. This suggests that introns probably existed before procaryotes and eucaryotes diverged in the distant past. Introns may have been a common feature of early genes, and have been largely lost by later procaryotes in response to an unknown selection pressure—for economy, perhaps, at a time when rapid reproduction or a scarcity of nucleotides made an efficient use of nucleic acid a crucial selection factor.

Chapter 31

ORGANIZATION OF DEVELOPMENT IN ANIMALS AND PLANTS

To biologists and nonbiologists alike, probably no aspect of biology is more amazing than the development of a complete new organism from one cell, a development so precisely controlled that the entire intricate organization of cells, tissues, organs, and organ systems characterizing the functioning adult comes into being, each element in just the right place with respect to the others, with rarely a flaw. Moreover, the millionth of a billionth of a gram of DNA in a mouse egg inevitably produces a mouse, while the same minute quantity of DNA in an oak seed produces an oak. The chromosomes of the egg contain all the information necessary to direct the precisely ordered construction of an organism that may ultimately consist of more than 10^{12} cooperating cells.

We have already seen how genes direct the synthesis of proteins, and we have examined many of the ways in which gene expression is controlled. The process of development requires a precisely programed and coordinated series of changes in gene expression, in cell-surface proteins and the recognition and adhesion properties they display, in cell shape, and in cell motility, as cells possessed of all the necessary information interact to create a characteristic species morphology, and give rise to the lines of specialist cells that constitute the various tissues and organs.

Understanding development requires understanding nearly the entire range of biological processes. This chapter will describe the major physiological events of development; the next chapter will examine the molecular and biochemical mechanisms that control and coordinate development.

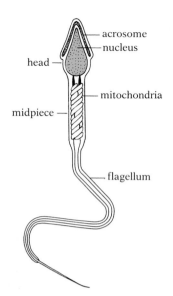

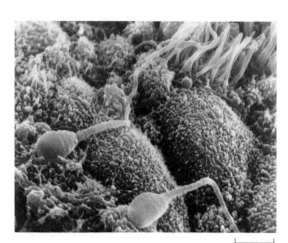

0.02 mm

31.1 Human sperm
Three sections are easily distinguishable: the head,
containing the nucleus; the midpiece, tightly packed
with mitochondria, which produce the ATP that fuels
the sperm's swimming; and the long flagellum. The
chromosomes in the nucleus are complexed with
exceedingly alkaline (arginine-rich) histones and are
thus supercondensed into an almost crystalline,
genetically inactive form. In the apex of the head,
immediately in front of the nucleus, is the acrosome,
a membrane-bounded vesicle derived from the Golgi
apparatus. There is only a tiny amount of cytosol in
the cell. The photograph shows human sperm in
motion; they have not yet reached the egg.

DEVELOPMENT OF A MULTICELLULAR ANIMAL

FERTILIZATION

We saw in Chapter 17 that certain cells are set aside early as egg primordia
in the ovary of a female animal. These cells grow to an unusually large size,
and when they then undergo meiosis, the divisions are unequal and almost
all the cytoplasm is retained in the ripe ovum, the other haploid cells being
the tiny polar bodies that soon deteriorate (see Fig. 23.31, p. 641). The
sperm, on the other hand, is an unusually small cell with very little cytoplasm. The ovum thus furnishes most of the initial cytoplasm for the embryo (hence the much greater importance of the mother than of the father
in the transmission of cytoplasmically inherited traits).

The fusion of the sperm with the egg cell provides a stimulus that causes
the egg to begin development into an embryo. Note that the triggering depends on the fusion of the cell membranes, not the fusion of the sperm
nucleus with the egg nucleus to form a diploid nucleus. Apparently
fertilization—the joining of two haploid sets of chromosomes—is not necessary to induce embryonic development in many animals, even animals
that do not normally reproduce parthenogenetically (from unfertilized
eggs). It is easy to induce unfertilized frog eggs, for example, to begin development in the laboratory by pricking them with a fine needle dipped in
blood. A few such eggs will develop into viable, apparently normal tadpoles.[1] Adult rabbits have been produced from unfertilized eggs by similar
procedures. Unfertilized eggs can even be stimulated to begin developing
by a mild electric shock, or by a change in the salt concentration in the surrounding fluid, or by a physical jolt, but the development aborts after only a
few divisions.

Let us look more closely at the process of fertilization of an egg cell by a
sperm cell. In many mammalian species the egg is initially enclosed in a
thin protective layer of follicle cells. This layer, a barrier to the sperm, is
loosened by hyaluronidase, an enzyme secreted by the sperm; similar enzymes are used by bacteria in penetrating host tissues. But even after the
follicular cell barrier has been loosened, the sperm cell (attracted to the
egg by chemicals released by it) still encounters a membranous barrier.

At this point the acrosome comes into play. This is a membrane-bounded
vesicle located in the apex of the head of the sperm cells of many species
(Fig. 31.1; see also Fig. 17.7, p. 439). Its membrane fuses with the plasma
membrane of the sperm cell and forms a tube that penetrates the jellylike
coat on the egg-cell membrane and then fuses with a microvillus of the egg
cell (Fig. 31.2); the fusion is probably facilitated by enzymes released from
the acrosome. Once this fusion of sperm and egg membranes has occurred,
the electrical potential of the egg-cell membrane changes (thus making fusion with other sperm difficult), and the sperm nucleus moves into the cytoplasm of the egg.

---

[1] Most embryos developed from unfertilized eggs are haploids, and invariably die after reaching, at most, the tadpole stage. Such embryos as survive to reach adulthood are the few that undergo spontaneous chromosomal doubling to become diploid (though the two sets of
chromosomes are identical) and that also lack any recessive lethals. A few embryos even become
$4n$, $6n$, or $8n$; these too may survive.

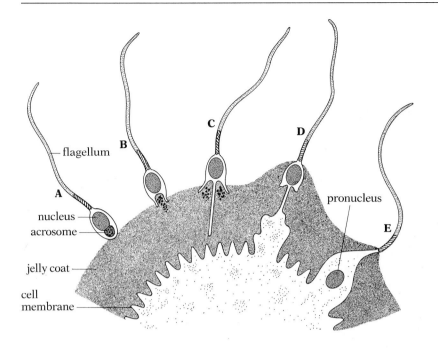

flagellum

A

nucleus
acrosome

jelly coat

cell
membrane

B

C

D

pronucleus

E

31.2 The fertilization process
(A) A sperm cell comes into contact with the jelly
coat surrounding an egg cell. (B) The membrane of
the acrosome fuses with the plasma membrane of the
sperm and ruptures, releasing the acrosomal
contents, which include enzymes that act on the jelly
coat and on the membrane of the egg cell. (C) A
membranous tube forms, which pushes through the
jelly coat. (D) The tube fuses with an enlarged
microvillus of the egg cell; there is now no
membranous barrier between the contents of the
sperm cell and of the egg cell. (E) The sperm
pronucleus moves into the egg cell.

As soon as the sperm and egg membranes have fused and the sperm nucleus, now known as a pronucleus, has moved into the egg, many changes
begin in the egg. First, calcium-containing vesicles in the sperm release
their contents, triggering release of even more calcium from vesicles in the
egg. The resulting dramatic change in the structure of the egg membrane
completely prevents fusion with other sperm, and initiates the earliest steps
in development. (The tricks for inducing development in unfertilized eggs
probably succeed by triggering calcium release.) Another process initiated
immediately after fusion is completion of the oogenic meiosis if that process is not already complete (it is completed prior to sperm fusion in some
species, but not in others, where it is arrested at metaphase of meiosis I or
II, depending on the species). Among the many other changes regularly
seen in the egg after fusion are striking alterations in the permeability of
the plasma membrane (especially a much enhanced permeability to inorganic phosphate) and an increased rate of oxygen consumption.

All the changes we have enumerated are triggered, not by the presence of
the sperm pronucleus in the egg cell, but rather by the interaction of the
sperm with the plasma membrane of the egg. Hence, if a sperm nucleus is
injected into an unfertilized egg with a micropipette, none of the changes
take place and both the egg and the sperm nuclei remain inert; they show
no tendency to move toward each other or to fuse. Some sort of signal from
a receptor in the plasma membrane appears to be essential.

True fertilization—the union of the two gamete nuclei—depends on
some attraction of the sperm pronucleus by the egg pronucleus, the nature
of which is still unknown. If sperm are induced to penetrate immature sea-

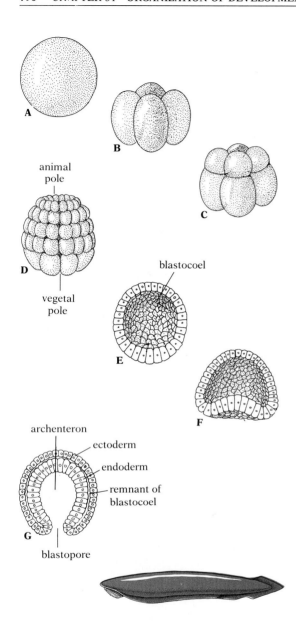

31.3 Early embryology of amphioxus
(A) Zygote. (B–D) Early cleavage stages, culminating in formation of a blastula (D). (E) Longitudinal section through a blastula, showing the blastocoel. (F–G) Longitudinal sections through an early and a late gastrula. Notice that the invagination is at the vegetal pole of the embryo, where the cells are largest. An adult amphioxus is also shown.

urchin eggs in which the chromosomes are still condensed, no attraction of the sperm pronucleus occurs, nor is there any tendency for the supercondensed chromosomes of the sperm pronucleus to undergo decondensation. In short, the mature egg pronucleus must be responsible both for attracting the sperm pronucleus and for inducing the decondensation of its chromosomes. Decondensation of both sperm and egg chromosomes is necessary if the zygote nucleus produced by fusion of the two gamete pronuclei is to carry out DNA replication preparatory to the first mitotic division of embryonic development.

EMBRYONIC DEVELOPMENT

Early cleavage and morphogenetic stages   In normal development the zygote begins a rapid series of mitotic divisions immediately after fertilization has taken place. In many animals these early cleavages are not accompanied by protoplasmic growth. They produce a cluster of cells that is no larger than the single egg cell from which it is derived (Fig. 31.3C). The cytoplasm of the one large cell is simply partitioned into many new cells that are much smaller. But in some animals, notably reptiles and birds, some protoplasmic growth does occur, as nutrients from the yolk (stored food) are consumed.

During this early cleavage stage of development, the nuclei cycle very rapidly between chromosomal replication (the S period of the cell cycle) and mitosis (the M period); the G_1 and G_2 periods are practically absent. Such rapid cycling, in which G_1 and G_2 are skipped, is possible because the ovum already contains huge quantities of the DNA polymerase necessary for catalyzing repeated chromosomal replication as well as most of the mRNA required for synthesis of proteins during early cleavage (at least 50 percent of the proteins synthesized are histones for the new chromosomes; the proteins of growth are produced in only negligible amounts, as would be expected). The brevity of the interval taken up by transcription—because so little new mRNA is needed—allows the rapid cycling between the S and M periods. But note that, since control of this cleavage stage of embryonic development depends largely on mRNA synthesized in the oocyte prior to fertilization, the paternal genes have little input until later in development; the course of the early cleavages is determined almost exclusively by the maternal genes. As you may remember, ova are remarkably large cells. They are so large, in fact, that the ratio of nuclear to cytoplasmic material would be too low for proper control of ordinary cellular activities. The early cleavages of embryonic development, with their minimal cell growth, thus help restore a more normal ratio of nuclear to cytoplasmic material.

As cleavage continues, the newly formed cells (blastomeres) of many species begin to pump sodium ions into the center of the mass of cells; as a result, water diffuses in and the blastomeres come to be arranged in a sphere surrounding a fluid-filled cavity called a ***blastocoel*** (Fig. 31.3E). An embryo at this stage is termed a ***blastula.***

Next begins a series of complex movements important in establishing the definitive shape and pattern of the developing embryo. The establishment of shape and pattern in all organisms is called ***morphogenesis*** (mean-

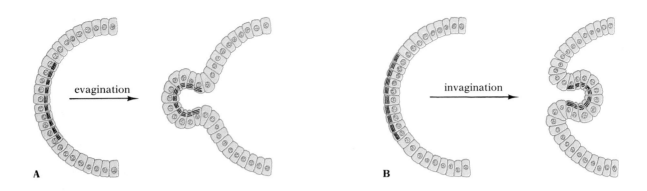

A   evagination   B   invagination

ing "the genesis of form"). Morphogenetic movements of cells in large masses always occur during the early developmental stages of animals.

The mechanism of these movements is still very poorly understood. There are often changes in the shapes of the cells, probably effected by contractile microfilaments or by some microtubular apparatus. The changes in shape may be relatively small (Fig. 31.4), or they may be extensive. Important in some of the movements are changes in the adhesive affinities of the cells for neighboring cells or, in the case of epithelial cells, for the basement membrane on which these cells sit. It may be relatively easy for a group of cells that adhere tightly to each other to slide as a group over the surface of an underlying layer to which their affinity has been at least temporarily reduced.

Since the pattern of cleavages and cell movement is greatly influenced by the amount of yolk in the egg, we shall examine, first, the pattern in an animal whose eggs have little yolk, then that in animals whose eggs have more yolk.

In amphioxus (Fig. 31.3), a tiny marine chordate whose egg has very little yolk, the movements that occur after formation of the blastula (when it is composed of about 500 cells) convert it into a two-layered structure called a *gastrula*. The process of *gastrulation* begins when a broad depression, or invagination, starts to form at a point on the surface of the blastula where the cells are somewhat larger than those on the opposite side (Fig. 31.3F). The differences in cell size are not very great in amphioxus embryos; they are more pronounced in many other animals. The smaller cells make up the *animal hemisphere* of the embryo. The larger cells make up the *vegetal hemisphere*. It is at the pole of the vegetal hemisphere that the invagination of gastrulation occurs in amphioxus. As gastrulation proceeds, the invaginated layer bends farther and farther inward, until eventually it comes to lie against the inside of the outer layer, nearly obliterating the old blastocoel (Fig. 31.3G). In the sea urchin, which has a transparent blastula that permits detailed observation of gastrulation, pseudopodia extend from the inner membranes of the cells where invagination began. The pseudopodia adhere to the inner membranes of other blastula cells; then they shorten, pulling the invaginated layer farther inside, and new pseudopodia form and

31.4 The mechanism of some morphogenetic movements in cells
Contraction of microfilaments (black), asymmetrically positioned in the cells, may change the shapes of the cells and produce evaginations (A), invaginations (B), or other alterations of the arrangement of cells in a developing organ. The contractile microfilaments are of the same type as those involved in amoeboid movement and cytokinesis.

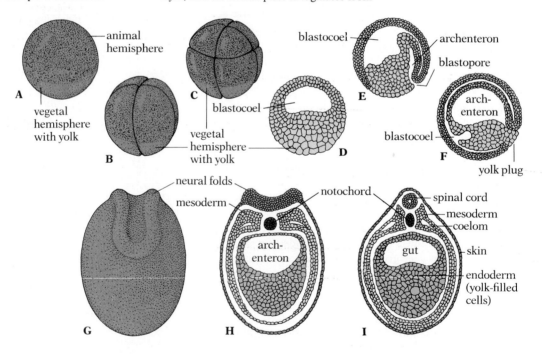

31.5 Neurulation in amphioxus

The cross sections through amphioxus embryos show the progressive formation of mesoderm and the neural tube. (A) When gastrulation has been completed, the dorsal part of the inner layer has already been segregated as presumptive mesoderm— presumptive notochord and mesodermal pouches. Similarly, part of the ectoderm has differentiated as presumptive neural plate. (B–C) The notochord and mesodermal pouches form as evaginations from the inner layer, and the neural plate invaginates from the ectoderm as it begins to form the spinal cord. (D) In this later embryo, both the spinal cord and the mesoderm are taking their definitive form. Notice that there is a cavity (the coelom) in the mesoderm.

31.6 Early embryology of a frog

The large amount of yolk in the frog egg causes its pattern of gastrulation to differ from that in amphioxus. (A) Zygote. (B–C) Early cleavage stages. Note that because the first horizontal cleavage is nearer the animal pole, the cells at the vegetal pole are much larger. (D) Longitudinal section of a blastula. (E–F) Longitudinal sections of two gastrula stages. (G) An early neurula, showing the neural folds and neural groove. (H) Cross section of a neurula after formation of mesoderm. (I) Cross section of a later embryo, showing definitive spinal cord.

continue the process. The embryos of other animals probably gastrulate in a similar way. The resulting gastrula is a two-layered cup, with a new cavity that opens to the outside via the **blastopore**. The new cavity, called the **archenteron**, will become the cavity of the digestive tract, and the blastopore will become the anus.

Gastrulation, as it occurs in amphioxus, first produces an embryo with two primary cell layers, an outer **ectoderm** and an inner layer. The latter subsequently separates into two layers, the **endoderm** and the **mesoderm**; the mesoderm lies dorsally between the ectoderm and the endoderm. In amphioxus the mesoderm originates as pouches flanking a supportive central rod, the **notochord**; these all pinch off from the inner layer (Fig. 31.5). The remaining part, the endoderm, pinches together to form a tube that becomes the digestive tract.

In the amphioxus egg, where the distinction between animal and vegetal hemispheres is only slight, owing to the small amount of yolk in the vegetal hemisphere, the early cleavages are nearly equal. The new cells are thus of nearly the same size, and gastrulation occurs in a direct and uncomplicated manner. But the eggs of many organisms have far more yolk in their vegetal hemisphere, and this deposit of stored food imposes complications and limitations on such processes as cleavage and gastrulation. Generally, the more yolk an egg contains and the more eccentric its cytoplasmic distribution, the more cleavage tends to be restricted to the animal hemisphere and the more gastrulation departs from the pattern in amphioxus.

The frog egg, which contains far more yolk than that of amphioxus but much less than that of a bird, serves as an example of eggs with an intermediate amount of yolk. The first two cleavages, which are perpendicular to each other, cut through both the animal and vegetal poles, producing cells of roughly the same size (Fig. 31.6B). But the next cleavage is equatorial (parallel to the egg's equator) and located decidedly nearer the animal pole (Fig. 31.6C); hence the four cells produced at the animal end of the egg are considerably smaller than the four at the vegetal end. From this stage onward, more cleavages occur in the animal hemisphere of the embryo than in the vegetal hemisphere as the blastula develops. As in amphioxus, there is no increase in total mass during these early cleavage stages (Fig. 31.7).

After the blastula has been formed, the frog embryo begins gastrulation. Simple invagination of the vegetal hemisphere is not mechanically feasible, because of the large mass of inert yolk. Instead, portions of the cell layer of the animal hemisphere move down around the yolk mass and then turn in at the edge of the yolk. This involution begins at what will be the dorsal side of the yolk mass, forming initially a crescent-shaped blastopore at the edge

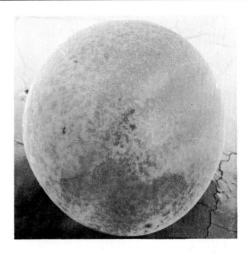

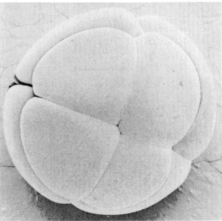

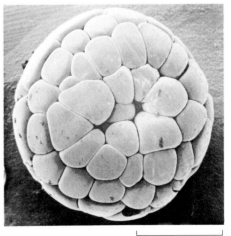

0.5 mm

31.7 Scanning electron micrographs of frog egg and some early cleavage stages
Top: Unfertilized egg. Middle: 8-cell stage. Bottom: 32- to 64-cell stage. All three micrographs are at the same magnification: × 46½. Note that there has been no overall growth in size during these cleavage stages—the 32- to 64-cell embryo is no larger than the egg cell.

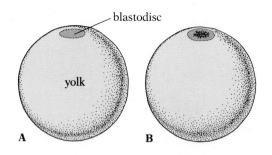

31.8 Egg and early-cleavage embryo of a chick
(A) The zygote. A small cytoplasmic disc—the blastodisc—lies on the surface of a massive yolk. (B) Early cleavage. There is no cleavage of the yolk.

of the yolk (Fig. 31.6E). The infolding slowly spreads to all sides of the yolk, so that the crescent blastopore is converted into a circle. Movement of the other cells around the yolk eventually encloses this material almost completely within the cavity of the archenteron.

Birds' eggs contain so much yolk that the small disc of cytoplasm on the surface is dwarfed by comparison. No cleavage of the massive yolk is possible, and all cell division is restricted to the small cytoplasmic disc, or ***blastodisc*** (Fig. 31.8). (Note that the yolk and the small lighter-colored disc on its surface constitute the true egg cell. The albumin, or "white," of the egg, which is secreted by the oviduct, lies outside the cell.) The gastrulation process is of necessity greatly modified in such eggs. Neither invagination of the vegetal hemisphere, as in amphioxus, nor involution along the edge of the yolk mass, as in a frog, can occur. Instead, outer and inner layers called the ***epiblast*** and the ***hypoblast*** are produced by a splitting of the blastodisc (Fig. 31.9A–C). In the posterior portion of the disc, the cells of the epiblast converge toward the longitudinal midline, giving rise to a clearly visible line or streak on the epiblast (Fig. 31.9D); this streak is, in effect, a very elongate closed blastopore. Individual cells move downward from this region. Some of these cells stay between the epiblast and the hypoblast and give rise to the mesoderm, while others insert themselves into the hypoblast and help form the endoderm.

The fates of cells in different parts of the three primary layers of vertebrates have been determined by staining them with dyes of different colors, or by marking them with carbon or other particles, and then following their movements. As you might expect, the ectoderm eventually gives rise to the outermost layer of the body—the epidermal portion of the skin—and to structures derived from the epidermis, such as hair, nails, the eye lens, the pituitary gland, and the epithelium of the nasal cavity, mouth, and anal canal. As you might also expect, the endoderm gives rise to the innermost layer of the body—the epithelial lining of the digestive tract and of other structures derived from the digestive tract, such as the respiratory passages and the lungs, the liver, the pancreas, the thyroid, and the bladder. The mesoderm gives rise to most of the tissues in between, such as muscle, connective tissue (including blood and bone), and the notochord (the dorsally located supportive rod found in all chordates, at least in the embryonic stages).

One major tissue located topographically between the skin and the gut does not develop from the mesoderm. This is the nervous tissue, which, curiously enough, is derived from the ectoderm. Soon after gastrulation, the ectoderm becomes divided into two components, the epidermis and the neural tube. A sheet of ectodermal cells lying along the midline of the embryo dorsal to the newly formed digestive tract and developing notochord bends inward in a process called ***neurulation***, and forms a long groove extending most of the length of the embryo (Figs. 31.5A–C and 31.6G–H). The dorsal folds that border this groove then move toward each other and fuse, converting the groove into a long tube lying beneath the surface of the back. This neural tube becomes detached from the epidermis dorsal to it, and in time differentiates into the spinal cord and brain (Figs. 31.5D and 31.6I).

We see, then, that the morphogenetic movements of gastrulation and neurulation give shape and form to the embryo, and bring masses of cells into the proper position for their later differentiation into the principal tissues of the adult body. In effect, the movements mold the embryonic mass into the structural configuration on which differentiation will superimpose the finer detail of the finished organism.

Later embryonic development   Much must happen to convert a gastrula into a fully developed young animal ready for birth: the individual tissues and organs must be formed; an efficient circulatory system must quickly come into function (Fig. 31.10); in a vertebrate the four limbs must develop; the elaborate system of nervous control must be established; and so on. The complexity and precision characterizing these developmental changes are staggering to contemplate. To give but one example, approximately 43 muscles, 29 bones, and many hundreds of nervous pathways must form in each human arm and hand. To function properly, all these components must be precisely correlated. Each muscle must have exactly the right origins and insertions; each bone must be jointed to the next bone

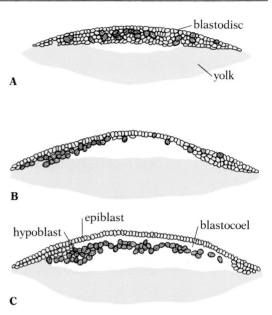

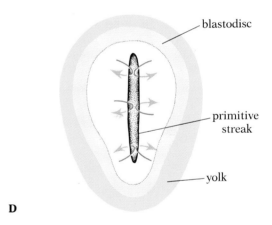

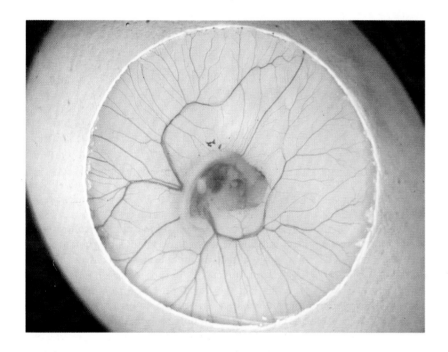

31.10 Chick embryo after four days of incubation
The tiny embryo lies on the surface of the yolk. It has a functional circulatory system, including a beating heart, even at this early stage of its development. Note the long branching blood vessels that run out of the embryo into the yolk; they transport nutrients to the embryo.

31.9 Gastrulation in the chick embryo
(A) Longitudinal section through a blastula. Larger yolk-laden cells are intermixed with smaller cells. (B) The larger cells begin to accumulate on the lower surface of the cell mass. (C) The layer of larger cells separates from the layer of smaller cells to become the hypoblast; the cavity between the two layers is the blastocoel. (D) Surface view of a gastrula. Involution of cells along the midline of the embryo during gastrulation (red arrows) produces a clearly visible primitive streak, which is essentially a very elongate blastopore. Some epiblast cells of the primitive streak move downward to form the mesoderm; others combine with the hypoblast to help form the endoderm.

31.11   Human embryos at successive stages of development

All are seen inside the fluid-filled amnion. (A) The 5-week embryo, ⅖ inch long, shows the beginnings of eyes but no distinctive face; note its mittenlike hands and feet, with no separation between the digits. (B) By contrast, the 7-week embryo, nearly 1 inch long, has a distinct face, and there is separation between its fingers. (C) At 13 weeks the fetus is over 3 inches long and weighs about an ounce, 15 times more than at 7 weeks. (D) By 17 weeks the fetus is over 6 inches long and 7 times heavier than it was a month earlier; all the internal organs have formed.

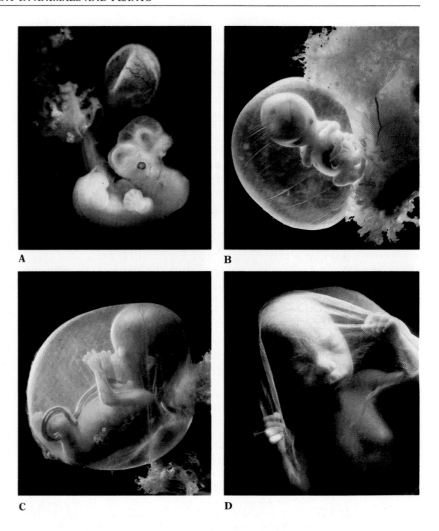

A   B

C   D

beyond it in a certain way; each nerve fiber must have all the proper synaptic connections with the central nervous system and must terminate on the right effector cells. Incredibly sensitive mechanisms of developmental control must operate for such an intricate structure to arise from a mass of initially undifferentiated cells. Yet the developmental processes that produce all these later embryonic changes are the same ones we have seen at work in the early embryo—cell division, cell growth, cell differentiation, and morphogenetic movements. Bursts of mitotic activity in some areas and cessation of cell division in other areas alter the balance between the parts. Special patterns of cell growth produce important changes in size and shape. Through differentiation cells may lose particular capacities, but may become more efficient at performing other functions. Foldings and pouchings establish the primordia of lungs and glands, of eyes and bladder. Even cell death plays an important role in the normal development of the living animal: fingers and toes, for example, become separated by the death of the cells between them (Fig. 31.11B).

DEVELOPMENT OF THE HUMAN EMBRYO

In human beings the egg is fertilized soon after ovulation, while it is still in the upper portion of the oviduct. It begins cleavage and has reached the blastula stage by the time it becomes implanted in the uterine wall 8–10 days after fertilization. During the implantation process the amnion, which will enclose the embryo in a fluid-filled chamber throughout the rest of its development, begins to form.

By the 23rd day, two other embryonic membranes—the chorion and the allantois—have collaborated with maternal tissues of the uterus to give rise to a functional placenta; the chorion contributes the fingerlike villi of the fetal portion of the placenta, and the allantois contributes the blood vessels (see Fig. 17.15, p. 449). By this time, the neural groove is complete; the mesoderm, too, is well developed, and individual segments (somites) can easily be distinguished. The tubular embryonic heart has begun to pulsate weakly.

By the end of the first month of development the embryo, which now exhibits arm and leg buds, is roughly 5 mm (⅕ inch) long. The hands and feet are still mittenlike, with no separations between the individual fingers and toes.

During the second month the embryo grows to a length of approximately 30 mm (a bit more than one inch) and to a weight of about one gram (0.03 ounce). By the end of the month (that is, after eight weeks of development), the embryo, now called a *fetus*, is recognizable as a human. It has a flat face with widely separated eyes, and the digits (fingers and toes) are well separated (Fig. 31.11B). The cerebral cortex has begun to show structural organization, and formation of the sense organs is well advanced. The muscles are sufficiently differentiated for some movement to be possible. Ossification of the skeleton has begun. The liver, which is serving as the chief producer of blood cells, is proportionately much larger than it will be later.

During the first two months of development the embryo is especially sensitive to a wide variety of factors that can cause serious malformations. For example, if during that period the mother should contract rubella, often called German measles, this normally mild viral disease can result in a malformed heart, cataracts of the eyes, or deafness in the fetus. The disastrous effects on the fetus of the tranquilizer thalidomide when taken by the mother early in pregnancy have been widely publicized.

By the end of 13 weeks (Fig. 31.11C) the fetus is about 100 mm long, and the body proportions approach those expected at birth. The sex of the fetus can be determined externally. The cerebral hemispheres have begun to overlap the rest of the brain, and the sense organs are almost complete. The heart, too, has taken on nearly its final form. Spontaneous movements are frequent.

At this point in development the basic pattern of all the physiological systems has been established. The remainder of fetal development is largely a combination of further refinement and elaboration and of growth in overall size (over 90 percent of fetal weight gain takes place in the last four months). In time, the bone marrow takes over from the liver primary responsibility for producing blood cells; the spinal cord and then the brain become myelinated; the eyes become light-sensitive, and later the ears become audio-sensitive; many new brain cells are produced; and numerous new neural circuits are established and become functional up to and past the time of birth.

By the end of 24 weeks the fetus has some chance of survival outside the uterus if it is given respiratory assistance and kept in an incubator. However, it is still so small—about 0.7 kg (1½ pounds)—and so poorly developed that it is subject to many special medical problems, which threaten its life for months. If, on the other hand, birth occurs at full term (on the average, about 280 days after the beginning of the mother's last regular menstrual period), the chances of survival are high.

Several very important developmental changes in the circulatory system occur at the time of birth: the placental circulation is cut off; the ductus arteriosus (the shunt between the pulmonary artery and the aorta) is closed; the lungs are inflated for the first time; blood is forced into the pulmonary system; and production of fetal hemoglobin soon gives way to production of the adult type.

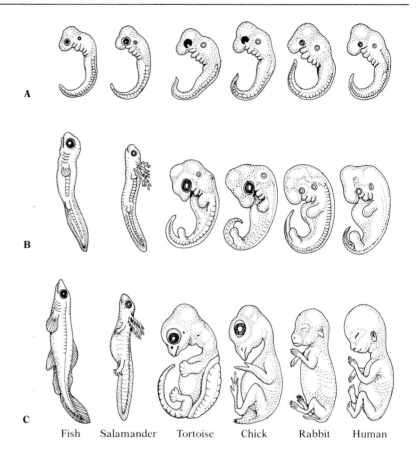

31.12 Vertebrate embryos compared at three stages of development
At stage A, all the embryos—whether fish, amphibian, reptile, bird, or mammal—strongly resemble one another. Later, at stage B, the fish and salamander are noticeably different, but the other embryos are still very similar; note the pharyngeal pouches in the neck region and the prominent tail. By stage C, each embryo has taken on many of the features distinctive of its own species.

Fish   Salamander   Tortoise   Chick   Rabbit   Human

It is beyond the scope of this book to discuss in detail the many events that occur during later embryonic development. Yet these are the events that mold morphologically similar gastrulas into a fish in one instance, a rabbit in another, and a human being in still another, depending on the genetic endowment of the gastrula in question; the developmental events are programed differently for each species. An understanding of how such different programs arise and how they are carried out is one of the important goals of developmental biologists.

One interesting aspect of the differences in the developmental programs of different species should be mentioned here—namely that the early embryos of most vertebrates closely resemble one another. For example, the early human embryo, with its well-developed tail and a series of pouches in the pharyngeal region, looks very much like an early fish embryo (Fig. 31.12); and it looks even more like an early rabbit embryo. Only as development proceeds do the distinctive traits of each kind of vertebrate become apparent.

POSTEMBRYONIC DEVELOPMENT

The extent to which an animal has developed by the time of birth varies greatly among different species. Some young animals are entirely self-suffi-

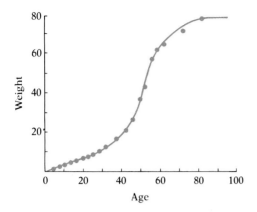

31.13 **A typical S-shaped growth curve**

cient from the time they are born and neither need nor receive parental care. Others—baby chicks and ducks, for example—can run about and feed themselves as soon as they are born, but still benefit from a limited amount of parental care. Yet other animals are born while still at an early stage of development, and are nearly helpless and totally dependent on parental care. Newly hatched robins, for instance, are blind, almost devoid of feathers, and unable to stand.

The extent of development at birth is often (though not always) a reflection of the length of the embryonic period, which is usually correlated in animals that lay eggs with the amount of yolk the eggs contain. Among birds particularly, species that have a short incubation period for the eggs characteristically have altricial—poorly developed—young (see Fig. 43.21, p. 1161), while species that have a longer incubation period characteristically have precocial (well-developed) young; for example, robins incubate their eggs only 13 days, while chickens have a 21-day incubation period. Regardless of their state of development at birth, however, all animals continue to undergo major developmental changes during their postembryonic life.

Growth   Though postembryonic development seldom involves any major morphogenetic movements, there is some cell multiplication and cell differentiation. But the preponderant factor by far in many animals is growth in size. Usually growth begins slowly, becomes more rapid for a time, and then slows down again or stops. This pattern yields the characteristic S-shaped growth curve shown in Figure 31.13. While the general shape of this curve holds for most organisms, its details vary in important ways from species to species. The slope of the curve is different for different species, depending on whether they grow very rapidly for a shorter time or more slowly for a longer time. (Compare the rate of increase in weight of, say, a calf and a child.) The shape of the curve is seldom as smooth as it appears in a generalized growth curve because many factors can affect the rate of growth. For example, in most mammals growth slows down for a while immediately after weaning, and it often varies greatly during puberty; such irregularities are reflected in bumps and dips in the curve (Fig. 31.14). An

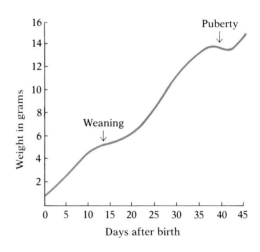

31.14 Growth in weight of a mouse
The rate of growth is slower at weaning and at puberty.

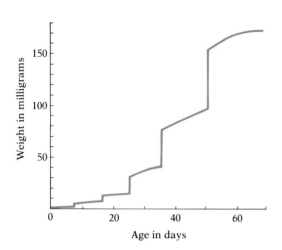

31.15   Growth in weight of an insect
The growth spurts of the water boatman, an aquatic insect, occur at the time of molt, when the old exoskeleton has been shed and the new one has not yet fully hardened.

especially marked departure from the smooth generalized curve is seen in the growth of arthropods. These animals can undergo only limited growth between molts, because the hard exoskeleton that encases their bodies can be stretched only slightly. However, at each molt there is a sharp burst of growth during the short period after the old exoskeleton has been shed and before the new one has hardened. The resulting growth curve shows a step-like pattern (Fig. 31.15).

Growth does not occur at the same rate and at the same time in all parts of the body. It is obvious to anyone that the differences between a baby chick and an adult hen or rooster, or between a newborn baby and an adult human being, are differences not only in overall size but also in body proportions. The head of a young child is far larger in relation to the rest of its body than that of an adult, while a child's legs are much shorter in relation to its trunk than those of an adult. If the child's body were simply to grow as large as an adult's while maintaining the same proportions, the result would be a most unadultlike individual (Fig. 31.16). Normal adult proportions arise because the various parts of the body grow at quite different rates or stop growing at different times (Fig. 31.17).

Two closely related species that differ in size are frequently also quite different in body proportions, not because of any basic difference in the growth patterns of the two species, but simply because a slight increase in overall body size automatically results in disproportionate increases in some parts of the body. In elk, for example, the size of the antlers increases much faster than the overall body size. Consequently, if species A grows slightly larger than species B and natural selection does not intervene, then A will have proportionately much larger antlers.

Larval development and metamorphosis   Growth in size is not always the principal mechanism of postembryonic development. Many aquatic animals, particularly those leading sessile (nonmobile) lives as adults, go through *larval* stages that bear little resemblance to the adult (Fig. 31.18). The series of sometimes drastic developmental changes that convert an

31.16   Changes in body proportions during human fetal and postnatal growth
The head grows proportionately much more slowly than the limbs.

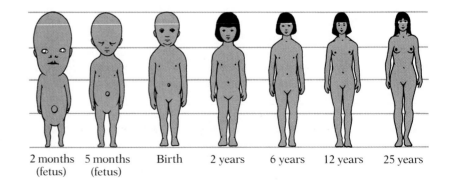

2 months
(fetus)   5 months
(fetus)   Birth   2 years   6 years   12 years   25 years

immature animal into the adult form is called ***metamorphosis***. It often involves extensive cell division and differentiation, and sometimes even morphogenetic movement; growth alone could not accomplish the transformation of a larva into an adult.

In many aquatic animals dispersal of the species depends on the larval stage; the tiny larvae either swim or are passively carried by currents to new locations, where they settle down and undergo metamorphosis into sedentary adults. In other species, such as frogs, where the adult is not sedentary, the adaptive significance of the larval stage (the tadpole in the case of frogs) seems less a matter of dispersal than of exploiting alternative food sources (tadpoles feed primarily on microscopic plant material, while adult frogs are carnivorous and take fairly large prey).

Though a larval stage occurs in the life history of many aquatic animals, probably the most familiar larvae are those of certain groups of terrestrial insects, including flies, beetles, wasps, butterflies, and moths. The young fly, wasp, or beetle is a grub that bears no resemblance to the adult. The young butterfly or moth is a caterpillar. In the course of their larval lives, these insects molt several times and grow much larger, but this growth does not bring them any closer to adult appearance; they simply become larger larvae (Fig. 31.19B–D, p. 802). Finally, after they have completed their larval development, they enter an inactive stage called the ***pupa***, during which they are usually enclosed in a case or cocoon. During the pupal stage most of the old larval tissues are destroyed, and new tissues and organs develop from small groups of cells called ***imaginal discs*** that were present in the larva but never underwent much development. The adult that emerges from the pupa is therefore radically different from the larva; it is the product of an entirely different developmental program—almost a new organism built from the raw materials of the larval body (Fig. 31.19H).

Insects with a pupal stage and the type of development just described are said to undergo ***complete metamorphosis***. The sharp distinction between the larval and adult stages in such insects has meant evolution in two markedly different directions; in general, the larva is more specialized for feeding and growth, while the adult is more specialized for active dispersal and reproduction.

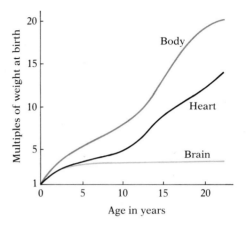

31.17 Graph showing differences in relative growth of body, heart, and brain of a man

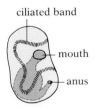   

Ciliated
larva

Bipinnaria
larva

Brachiolaria
larva

Adult

31.18 Three larval stages, and adult, of the sea star *Asterias vulgaris*
The gastrula develops into the ciliated larva, which changes into the bipinnaria larva, which changes into the brachiolaria larva, which metamorphoses into the characteristically shaped sea star, with five arms.

A Monarch butterfly egg

D Full-grown larva going into pupation

B First-stage larva eating its own egg shell

C Later-stage larva eating its own molted skin

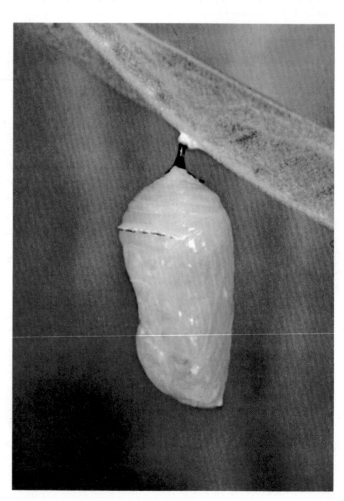

E Early chrysalis

31.19 Developmental stages of a monarch butterfly, an insect with complete metamorphosis
From the egg (A) hatches a small larva (B), which goes through a series of molts, growing larger at each (C). Eventually the full-grown larva attaches to a plant (D) and goes into pupation, forming a case around itself (E); in moths and butterflies the encased pupa is called a chrysalis. During pupation most of the larval tissues are broken down and new adult tissues are formed; in the monarch some of the adult structures can be seen through the case of the late chrysalis (F). The newly formed adult emerges from the pupal case and rests while its wings become inflated (G). It then flies off to feed at flowers—here milkweed (H)—before mating and laying eggs (almost always on milkweed) to start the cycle over again.

F Late-stage chrysalis

G Inflation of wings by newly emerged butterfly

H Adult butterfly at flower

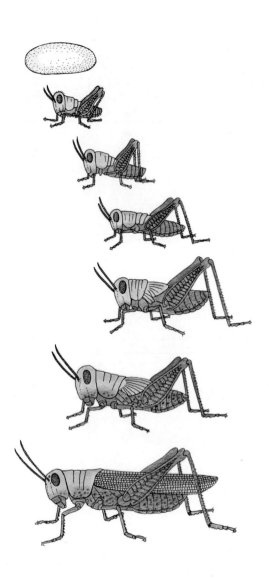

31.20 Gradual metamorphosis of a grasshopper
The insect that emerges from the egg (top) goes through several nymphal stages that bring it gradually closer to the adult form (bottom).

Complete metamorphosis is not characteristic of all insects. Many, such as grasshoppers, cockroaches, bugs, and lice, undergo **gradual metamorphosis** (Fig. 31.20). The young of such insects resemble the adults, except that their body proportions are different (the wings and reproductive organs, especially, are poorly developed). They go through a series of molts during which their form gradually changes and becomes more and more like that of the adult, largely as a result of a differential growth of the various body parts. They have no pupal stage and experience no wholesale destruction of the immature tissues.

Aging and death   Discussions of development often stop with the completely matured adult. But development in its full biological sense does not cease then. The adult organism is not a static entity; it continues to change, and hence to develop, until death brings the developmental process to an end.

The term "aging" is applied to the complex of developmental changes that lead, with the passage of time, to the deterioration of the mature organism (Table 31.1) and ultimately to its death. For many years, little research was devoted to aging, but now it has become a major field of investigation. Modern scientific progress and improved medical techniques have greatly increased our ability to protect ourselves against disease, starvation, and the destructive forces of the physical environment. More and more people are living to an advanced age. And as the life expectancy increases and the proportion of the population in the upper age brackets rises, the changes associated with aging become more obvious and more important to all of us.

Little is known about the factors involved in aging. The process seems to be correlated with specialization of cells for one or a few highly specific functions. Cells that remain relatively unspecialized and continue to divide do not age as rapidly (if at all) as cells that have lost the capacity to divide. Cancer cells, of course, divide continually and are essentially immortal. Bacteria and some other unicellular organisms cannot be said to age, for any cell that is not destroyed eventually divides to produce two young cells; division is thus a process of rejuvenation. Within the body of a multicellular animal, tissues like muscle and nerve, in which cell division has normally ceased, slowly deteriorate, whereas tissues like those of the liver and pancreas, in which active cell division continues, age much more slowly. Furthermore, animals that grow as long as they live seem to show fewer symptoms of aging than, for example, mammals and birds, which cease growing soon after they reach maturity. Within a single species, individuals whose period of growth and development is slowed and extended by a very limited diet are usually older than normal before they begin to show signs of aging.

The same pattern is seen in plants. Many annual plants practically cease growing soon after the flowering period, when their fruit begins to form, and they age and die soon afterward. (In many cases, if the plants are experimentally prevented from flowering, they will live on far beyond their normal time of death.) By contrast, some woody perennials, of which the giant sequoia and the redwood are extreme examples, may live for hundreds or thousands of years, always continuing to grow. In effect, they circumvent the problem of aging by forming vast numbers of new cells each year while equally vast numbers of their older cells die. The specialized cells age and die, but the continued youthful activity of the meristems keeps the living part of the plant young and vigorous. As an organism, a sequoia may be over four thousand years old, but it contains no living cells that are more than a few years old.

Clearly, then, the aging and death of individual cells and the aging and death of the multicellular organism as a whole are two rather different things. Paradoxical as it may seem at first, a plant may retain youthful attributes in part because many of its cells age and die. The death of these cells is functionally necessary for the continued life of the plant; we have seen, for example, that only after the cells of the xylem die can they function in internal transport. Similarly, the death of individual cells, as noted previously, plays an essential role in the development of an animal embryo and in the complete metamorphosis of some insects. And early death of individual red blood corpuscles and epidermal cells is entirely normal even in a young healthy mammal. Aging of the whole organism, therefore, is a matter not simply of the death of its cells, but of the deterioration and death of those cells and tissues that cannot be replaced.

What makes irreplaceable tissues age? We don't know. However, we do know some of the factors that contribute to aging:

1. The replacement of damaged tissue by connective tissue places an increased burden on the remaining cells in that tissue. When cells in a tissue like muscle die as a result of disease or injury, no new muscle cells are formed to replace the ones lost. Wound healing in such cases involves growth of connective (scar) tissue, which serves as a patching material but cannot function like the original muscle cells. As more and more irreplaceable cells die, the increased burden placed on the remaining cells of that type may contribute to their aging.

2. Changing hormonal balance—caused by a drop in the level of sex hormones, for example—may disturb the function of a variety of tissues and perhaps cause them to function less well.

3. As cells become older, they tend to accumulate some metabolic wastes that they apparently cannot expel, and these wastes may contribute to the eventual deterioration of the cells.

Though all these factors are doubtless involved in aging, they are not really explanations but symptoms. The real question is why these changes occur, and to this question scientists cannot as yet give a satisfying answer. Some investigators have suggested that somatic cells slowly cease to function and eventually die as a result of damage by radiation (particularly X rays and cosmic rays). However, all laboratory experiments indicate that radiation damage is greatest to actively dividing cells—the cells that age most slowly. Furthermore, the amount of radiation damage would be proportional to the chronological age of the cells, but we know that aging is a function of physiological age, not of chronological age—a five-year-old rat is physiologically very old indeed, and its tissues show pronounced symptoms of aging, whereas five-year-old tissues in a human being are not yet even mature.

Other investigators put major emphasis on intrinsic rather than extrinsic factors. They believe that the changes characteristic of aging are pro-

TABLE 31.1 *Average decline in a human male from ages 30 to 75*

Characteristic	Percent decline
Weight of brain	44
Number of axons in spinal nerve	37
Velocity of nerve impulse	10
Number of taste buds	64
Blood supply to brain	20
Output of heart at rest	30
Speed of return to normal pH of blood after displacement	83
Number of glomeruli in kidney	44
Glomerular filtration rate	31
Vital capacity of lungs	44
Maximum O_2 uptake during exercise	60

gramed in the genes just like the earlier developmental changes, and that, though extrinsic environmental factors doubtless influence aging, they do so only by speeding up or slowing down processes that would occur anyway. These processes may involve a decline in the production of important enzymes or an altered chemical balance or physical structure, with an ensuing loss of ability to perform certain functions; or they may involve development of autoimmune reactions (allergies against parts of the organism's own body) that result in destruction of essential tissues; or they may involve increased rupture of lysosomes and release of destructive hydrolytic enzymes within the cells. If aging were, indeed, a genetically programed trait, it would have been perpetuated only if it provided a selective advantage; no such universal advantage, applicable to essentially all multicellular organisms, is known.

A particularly interesting proposal is a variation of one already mentioned: that the aging of cells results from somatic-gene mutations leading to the production of defective enzymes. According to this hypothesis, the different rates of aging in different species reflect different inherited capacities for DNA repair. Thus species that age slowly would be genetically well endowed with enzymes that repair DNA damaged by mutagenic agents; hence many of their mutations would be only temporary. Species that age rapidly would have only limited ability to repair mutated DNA.

Whatever the processes of aging, it seems clear that we shall not understand them fully until we know much more about how developmental processes in general are regulated by the interaction of inherited and environmental influences. The basic problems in understanding aging, then, are essentially the same as those of embryonic development or maturation. Aging may be simply another aspect of the general phenomenon of development.

DEVELOPMENT OF AN ANGIOSPERM PLANT

THE SEED AND ITS GERMINATION

The egg cell of an angiosperm plant is retained within the ovary of the maternal plant and is fertilized there by a sperm nucleus from a pollen grain. After fertilization, the zygote undergoes a series of mitotic divisions and develops into a tiny *embryo*. This embryo, together with a food-storage tissue called the *endosperm*, becomes enclosed in a tough protective *seed coat*. The resulting composite structure, made up of embryo, endosperm, and seed coat, is called a *seed*. The embryo in some species, such as peas and beans, absorbs all the endosperm before the seed is released from the parent plant. In other species, such as corn, the embryo does not absorb significant quantities of the endosperm until the seed begins to germinate. The embryonic stages of development usually do not last long, and by the time the ripe seed is released from the parent plant it is quite dry and its embryo has usually become dormant. The seed may last for months or years in this dormant state.

Development of the embryo   Soon after an egg cell has been fertilized, it begins to undergo a series of changes. Its wall, previously very thin,

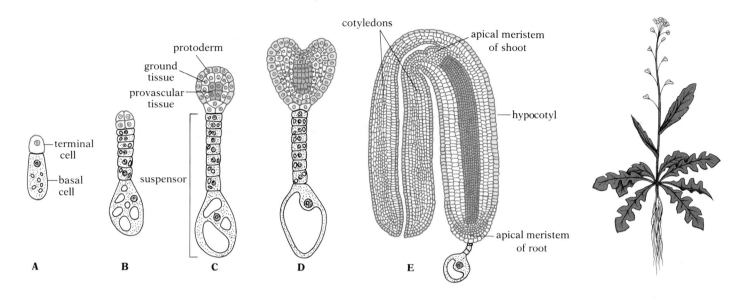

thickens; its endoplasmic reticulum and Golgi become more extensive; and new ribosomes are synthesized.

The first cell division may occur a day or so after fertilization. It invariably gives rise to two cells of unequal size: a smaller nonvacuolated terminal cell and a larger vacuolated basal cell (Fig. 31.21A). The developmental fates of these two cells are very different. The terminal cell gives rise to the embryo proper, while the basal cell, by means of about three transverse-division cycles, forms an elongate **suspensor** structure (Fig. 31.21B), which functions only while the embryo is in the seed—probably in moving nutrients into the embryo. Let us follow the embryonic development of *Capsella*, a much-studied dicot commonly called shepherd's purse.

The terminal cell gives rise to a globular structure, in which three types of tissues begin to differentiate: a surface layer of ***protoderm***, which will form epidermal tissue; an inner core of ***provascular tissue***, which will form cambium and the vascular tissues; and a middle layer of ***ground tissue***, which will form the cortex (Fig. 31.21C). Shortly thereafter, the radial symmetry of the globular-stage embryo begins to disappear as two mounds arise on the portion of the embryo opposite the suspensor (Fig. 31.21D). These mounds, which give the embryo a heart-shaped appearance, will become the ***cotyledons***, or embryo leaves. The part of the embryo proper below the point of attachment of the cotyledons is called the ***hypocotyl***; it will form the first part of the stem of the young plant.

As the cotyledons and hypocotyl of *Capsella* continue to elongate within the very limited space available to them in the seed,[2] the embryo begins to curve back on itself (Fig. 31.21E). Such curvature, though common in angiosperms, does not occur in all species.

[2] More accurately, it is the ovule within which the embryo develops and by which it is constrained.

31.21   Embryonic development of shepherd's purse (*Capsella*)

(A) The first division of the zygote produces a smaller terminal cell (green) and a larger basal cell. (B–C) Divisions of the terminal cell give rise to a globular embryo, whereas divisions of the basal cell give rise to a stalklike suspensor. Cells of the globular embryo soon differentiate to form three tissue types: protoderm on the surface, provascular tissue in the center, and ground tissue between the other two. (D) The formerly globular embryo becomes heart-shaped as two mounds that will develop into cotyledons begin to form. (E) Elongation of the cotyledons and the hypocotyl within the confines of the seed causes the embryo to fold back on itself. A fully grown plant is also shown.

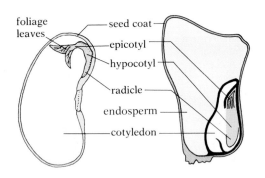

31.22   Diagram of dicot and monocot seeds
The embryos in these seeds consist of epicotyl, hypocotyl, radicle, and cotyledon. Left: A dicot seed (bean) in which one cotyledon has been removed to reveal the remaining parts of the embryo. Right: A monocot seed (corn) has only one cotyledon, but a large endosperm.

As most of the cells of the embryo take on more and more of the characteristics of the tissues to which they will give rise, small clumps of cells at each end of the embryonic axis remain relatively undifferentiated. One clump, located just beyond the point of attachment of the cotyledons, will become the **apical meristem of the shoot**. The other clump, at the pole of the embryonic axis near the suspensor, will become the **apical meristem of the root** (Fig. 31.21E). In some species, cell divisions in these meristems during embryonic development give rise to an **epicotyl**—a region of shoot above the point of attachment of the cotyledons, often bearing the first foliage leaves (the plumule)—and, at the other end of the embryonic axis, to a **radicle**, which will develop into the primary root (Fig. 31.22).

We have now followed the embryo to a point at which its development normally slows down and enters a dormant stage, which will not be broken until the seed germinates. Just what causes this dormancy is not fully understood. Indeed, a number of factors probably interact to bring it on, among them: (1) decline in endosperm, which in some plants is entirely used up as its nutrients are transferred to the cotyledons; (2) lowered concentrations of gibberellin in the tissues surrounding the embryo; (3) changes in the cells of the seed coat, including progressive dehydration, deposition of pigments, and thickening and hardening of the cell walls—changes that tend to reduce light levels inside the seed; and (4) secretion of growth-inhibitor substances by the extraembryonic tissues of the seed.

B

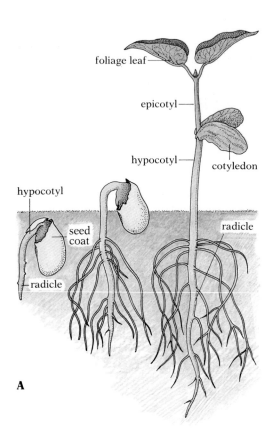

A

31.23   Germination and early development of a bean plant
(A) The hypocotyl and radicle emerge from the seed first (left). As the upper portion of the hypocotyl elongates, it forms an arch that pushes out of the soil into the air; the radicle gives rise to the first root system (middle). The hypocotyl straightens, pulling the cotyledons out of the ground as the epicotyl begins its development (right). (B) The photograph shows a germinating bean plant.

This brief summary points up several major differences between the embryonic development of an angiosperm plant and that characteristic of most animals:

1. Unlike animals, plants have no early developmental stage in which cleavage is unaccompanied by cell growth; the two processes—cell division and growth—typically occur together.[3]

2. Plant cells do not migrate during morphogenesis. The cell walls constrain their shape, and their middle lamellae tend to tie each cell to its neighbors. Also, the plasmodesmata between adjacent cells would be ruptured if the cells were to slide along each other. The form and shape of the plant embryos are established not by movement of cells, as in animals, but by the patterns of cell division and growth.

3. In plants a few cells are early set aside to remain forever embryonic (meristematic).

4. The fully developed plant embryo does not possess, even in rudimentary form, all the organs of the adult plant; organogenesis (the formation of new organs) will continue throughout the life of the plant, as new roots, branches, leaves, and reproductive structures are formed during each growing season.

Seed germination   Germination of a seed begins with the absorption of water, which greatly increases the volume of the seed (sometimes as much as 200 percent). The resulting hydration of the protoplasm increases enzymatic activity, and the metabolic rate of the embryo shows a marked rise. This higher metabolic rate makes possible resumption of active cell division, synthesis of new protoplasm, and increase in cell size by uptake of water. The growing embryo soon bursts out of the seed coat and rapidly assumes a characteristic plant form, with distinguishable shoot and root.

The hypocotyl (with attached radicle) is the first part of the embryo to emerge from the seed. It promptly turns downward no matter what the orientation of the seed may be. By the time the epicotyl begins its rapid development, the radicle, at the lower end of the hypocotyl, has already formed a young root system capable of anchoring the plant to the substrate and absorbing water and minerals. In some dicots (which, as the term "dicot" indicates, have two cotyledons), the upper portion of the hypocotyl elongates and forms an arch, which pushes upward through the soil and emerges into the air (Fig. 31.23). Once the hypocotyl arch is exposed to light, it straightens, and thus pulls the cotyledons and the epicotyl out of the soil. The epicotyl then begins to elongate. In these dicots, of which the garden bean is an example, the shoot of the mature plant is mostly of epicotyl origin, but a short region (usually little more than one centimeter) at the base of the stem is derived from the hypocotyl.

Other dicots, of which the garden pea is an example, show a slightly different pattern of germination (Fig. 31.24). In these plants no hypocotyl arch forms and the cotyledons are never raised above ground. Instead, the epicotyl begins to elongate soon after the young root system has begun to form; it always grows upward and soon emerges from the soil. In such

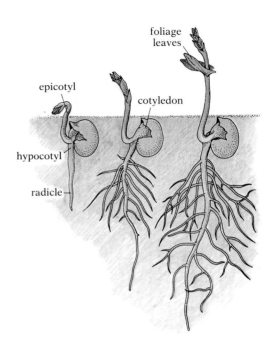

31.24 Germination and early development of a pea plant
The development of peas differs from that of beans in that no hypocotyl arch is formed and the cotyledons remain beneath the soil.

---

[3] However, the zygote is larger than any individual cell in the fully developed embryo will be, because the rate of cell division in the embryo is faster than the rate of cell growth.

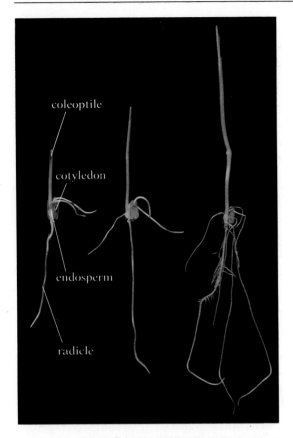

31.25 Germination and early development of a corn plant
As shown in the photograph and drawing, the young shoot is initially enclosed within a coleoptile, a tubular protective sheath.

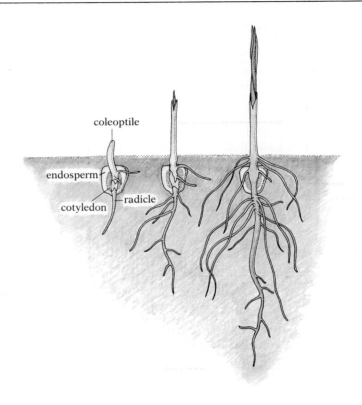

plants the entire shoot is of epicotyl origin. A similar pattern is seen in corn, a monocot, which has only one cotyledon but a large endosperm (Fig. 31.25).

GROWTH AND DIFFERENTIATION OF THE PLANT BODY

The seedling plant grows in length rather slowly at first, then enters a longer period of much more rapid growth, and finally slows down again or even stops as it approaches maturity. If the height (or weight) of an annual plant is plotted against age, the resulting S-shaped growth curve is similar to the one characterizing the growth of animals (Fig. 31.13). The growth curve of a perennial plant differs from that of an annual because perennials continue to grow to some extent throughout their lives, while annuals cease growing after they reach maturity.

Growth of the shoot or root of a young plant involves both cell multiplication and cell elongation. In plants with only primary growth, both processes are ordinarily restricted to a relatively limited region near the apex of the shoot or root. Let us look first at the root.

Growth of the root The extreme tip of a root is covered by a conical root cap consisting of a mass of nondividing parenchyma cells (Fig. 31.26). These cells secrete a gelatinous substance that lubricates the surface of the root cap and facilitates the pushing of the root tip through the soil as the root elongates. As the tip moves through the soil, some of the cells on the

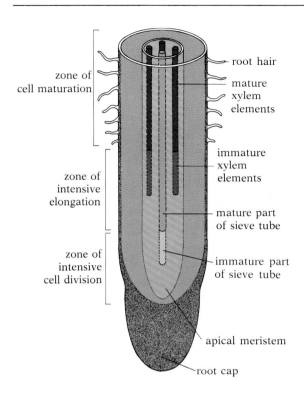

zone of cell maturation

root hair

mature xylem elements

zone of intensive elongation

immature xylem elements

mature part of sieve tube

zone of intensive cell division

immature part of sieve tube

apical meristem

root cap

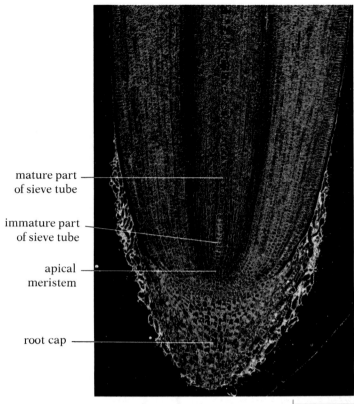

mature part of sieve tube

immature part of sieve tube

apical meristem

root cap

0.1 mm

31.26 Longitudinal sections of root tips
New cells are produced by mitotic divisions in the meristem region just behind the root cap, as shown in this drawing of a tobacco root. There is a zone of especially intensive cell elongation not far behind the meristem. Both phloem and xylem are well differentiated in the region of the root hairs, back of the zone of intensive elongation. Many of the same structures are visible in the photograph of the distal part of the root tip of a corn plant.

surface of the root cap are abraded. They are replaced by new cells added to the cap by the apical meristem. Located just behind the cap, the apical meristem is usually restricted to an area a millimeter long or less and is composed of relatively small, actively dividing cells. Most of the new cells it produces are laid down on the side away from the root cap. These cells are left behind as the meristem lays down additional new cells in front of them and the tip continues to move through the soil. It is these new cells, derived from the apical meristem, that will form the primary tissues of the root.

Cell division and cell enlargement both occur in the meristem, but because the rate of division is high, the rate of enlargement is only sufficient to produce small cells. However, as the new cells become farther removed from the meristem by the deposition of additional intervening cells, mitotic activity in most of them slows down; cell enlargement becomes the dominant process and cell size increases. Most of the enlargement is elongation rather than increase in width. As we saw in Chapter 15, cell elongation is under the control of hormones, particularly auxins. Let us look at the process in more detail.

Since plant cells, unlike animal cells, are enclosed in a boxlike cell wall, cell growth is possible only if the walls can be extended. It is in this process —the enlarging of the walls—that auxin is so important. The walls, you may recall, are composed primarily of polysaccharides, of which cellulose is the one we have mentioned most often. In primary walls (those of growing cells) the cellulose is present in the form of long fibrils, each of which is thought to be a bundle of about 40 cellulose chains aligned parallel to one

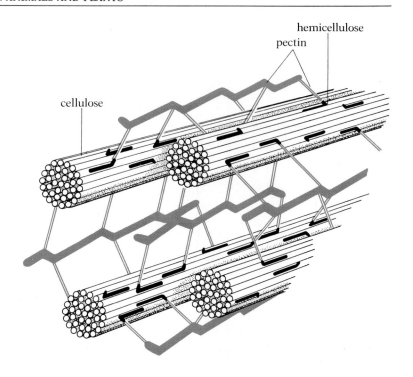

31.27 Molecular arrangement of polysaccharides in a primary cell wall
Each fibril is a bundle of parallel cellulose chains. The fibrils, which are cross-linked by pectin and hemicellulose, form a relatively rigid network.

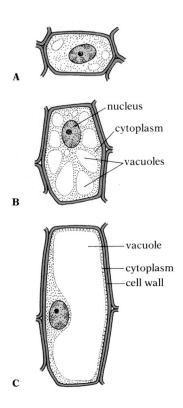

31.28 Elongation growth of a plant cell
As more and more water moves into the cell vacuoles, the wall is stretched, but in only one dimension. Almost no synthesis of new cytoplasm occurs during this kind of growth. The increased volume of the cell is taken up by the expanding vacuoles, which eventually fuse into a single large vacuole (C); in the mature cell the thin band of peripheral cytoplasm may constitute less than 10 percent of the cell's volume.

another and extensively cross-linked by hydrogen bonds (Fig. 31.27). The cellulose fibrils, in turn, are linked to other polysaccharides that make up the matrix of the wall. These matrix polysaccharides, especially those traditionally called pectin and hemicellulose,[4] hold the fibrils in a more or less rigid pattern. For cell growth to occur, two things must happen: some of the cross-linkages must be temporarily broken, so that the wall will become more plastic, and new wall material must be inserted.

A current model holds that auxin plays a fundamental role in inducing both greater wall plasticity and insertion of new wall material. Its effect on plasticity (which is manifested in less than ten minutes), is attributed to auxin-activated membrane receptors that act as proton pumps, transporting hydrogen ions from the cytoplasm into the wall. With an increased concentration of hydrogen ions in the wall, some of the hydrogen bonds between the polysaccharides break, the fibrils loosen, and the walls become more pliable.

Because the cell contents are hypertonic relative to the extracellular fluids, and the sole constraint on further movement of water into the cell has been resistance by the wall, more water begins to move into the cell as soon as the wall becomes more plastic and less resistant to stretching. As more and more water floods into the cell, especially into the vacuole, the hydrostatic pressure inside the cell causes more and more stretching of the wall until the volume of the cell has been increased as much as a hundred-

[4] The so-called pectin fraction of the wall is now thought to be composed largely of two materials, arabinan-galactan and rhamnogalacturonan. The so-called hemicellulose fraction is xyloglucan.

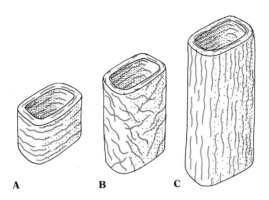

31.29 Change in the orientation of fibrils of the cell wall as a plant cell elongates
(A) Before the start of elongation the cellulose fibrils are arranged horizontally. (B) As elongation begins, and the wall is stretched, the fibrils are displaced and tipped toward a more vertical alignment. (C) In the fully elongated cell the fibrils in the outer, older wall layers are oriented vertically. The recently deposited fibrils in the inner newer layers are horizontal.

fold. But note that this enormous increase in cell size has been achieved with little or no synthesis of new cytoplasm; the cytoplasm has been restricted to a thin layer next to the wall, the greater part of the cellular volume being filled by the expanded cell vacuole (Fig. 31.28). Cell growth in plants thus differs greatly from cell growth in animals, where most increase in size results from formation of new cytoplasm rather than from vacuolation.

The effect of auxin on the second aspect of wall enlargement—insertion of new wall material—is much slower. It is thought to depend on synthesis of new mRNA coding for the enzymes that catalyze addition of further units to the complex polysaccharide structure of the wall.

We have emphasized that growth in length of the root or stem is caused largely by cell elongation (enlargement of cells in other dimensions is, of course, important in the morphogenesis of other parts of the plant). There is presumably some mechanism whereby the effects of auxin can be targeted against the parts of the cell wall along the sides of the cell and not against the parts along the ends, so that the stretching is in only one dimension. The unidimensional stretching is reflected in the way the alignment of the cellulose fibrils changes during cell elongation (Fig. 31.29), but no convincing explanation for the asymmetry has yet emerged. Perhaps the proton pumps are missing from the membrane at the ends of the cells, or perhaps the wall itself differs there in some hitherto undiscovered fashion.

The zone where cell elongation predominates in the root is just behind the meristem. It usually extends only a few millimeters along the root (Fig. 31.26). For example, in a corn seedling the fastest elongation occurs about 4 mm from the root tip, and cells more than 10 mm behind the tip have completed their elongation. The situation is similar in bean seedlings (Fig. 31.30). The elongation in this zone has the effect of pushing the root tip through the soil faster than if it were driven only by the production of new cells in the zone of intensive cell division.

Though cell division has slowed down in the regions behind the apical meristem, it has not ceased entirely. Indeed, the pattern of persistent cell division contributes importantly to the development of the characteristic organization of the root. For example, it is persistent meristematic activity

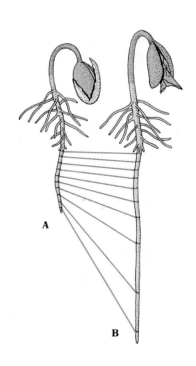

31.30 Growth of a bean root
Note that the parts of the root immediately behind the tip in the earlier stage (A) are those that have undergone the most elongation by the later stage (B).

vascular cell

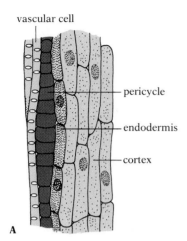

pericycle

endodermis

cortex

A

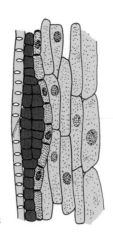

B

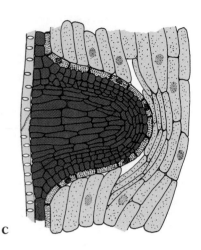

C

31.31 Origin of a new lateral root from the pericycle of an older root
(A) A new lateral root starts to form when a group of cells in the pericycle layer (dark brown) begin to enlarge in a radial direction. (B) These cells then divide tangentially. (C) Continued divisions and elongations in this new direction produce a growing mass that pushes out and through the outer tissue layers of the old root, crushing the cells in its path. Note that a new root forms by a change in the orientation of cell elongation and cell division, not by the morphogenetic movements of cells expected in a comparable developmental event in an animal.

in the pericycle that gives rise to new lateral roots, which, as they grow, push out and through the overlying endodermal, cortical, and epidermal tissues to enter the soil (Figs. 31.31 and 31.32).

Growth of the stem   Growth of the main stem axis is basically similar to growth of the root. New cells are produced by an apical meristem (which is not, however, covered by any structure analogous to the root cap), and these cells then elongate, pushing the apex upward. An obvious difference between growth of the stem and of the root is the lateral production of leaves by the growing tip of the stem. At regular intervals an increase in the rate of cell division under a localized region of the sloping surface of the apical meristem of a stem gives rise to a series of swellings that function as leaf primordia. The point at which each leaf primordium arises from the stem is called a **node**, and the length of stem between two successive nodes is called an **internode** (see Fig. 6.7, p.145). Most increase in length of the stem results from elongation of the cells in the young internodes.

At the tip of the stem is a series of internodes that have not yet undergone much elongation. The tiny leaf primordia that separate these internodes curve up and over the meristem, with the older, larger ones enveloping the younger, smaller ones (Figs. 31.33 and 31.34). The resulting compound structure, consisting of the apical meristem and a series of unelongated internodes enclosed within the leaf primordia, is called a **bud**. In shoots that have periodic (episodic) growth, the bud is protected on its outer surface by overlapping scales, which are modified leaves that grow from the base of the bud.

When a dormant bud "opens" in the spring, the scales curve away from the bud and then fall off, and the internodes that were contained within the bud begin to elongate rapidly. As the nodes become farther and farther apart, mitotic activity (mostly in one plane) in the leaf primordia gives rise to young leaves; the pattern of cell division, characteristic for each species, determines whether the leaves will be entire or lobed, simple or compound.

Before the leaf is fully formed, a small mound of meristematic tissue

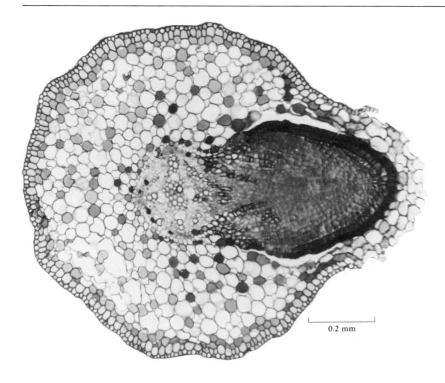

31.32   Origin of a lateral root in *Ranunculus*
The new lateral root has pushed through the cortical and epidermal layers of the old root and is ready to enter the soil.

0.2 mm

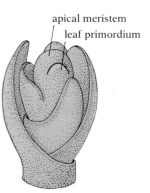

apical meristem
leaf primordium

31.33   A bud

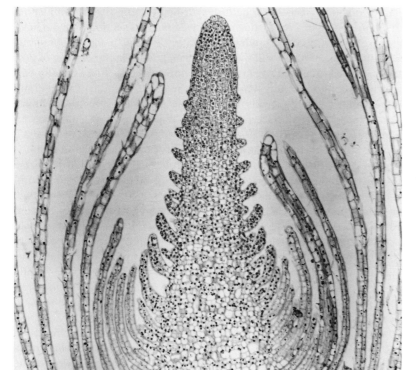

31.34   Photograph of sectioned *Elodea* bud
The stem tip forms a bud as upcurving leaf primordia enclose the apical meristem.

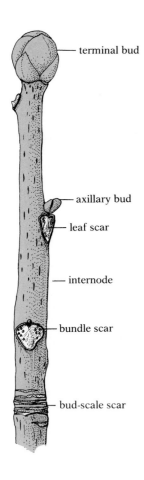

31.35 Portion of a stem
The bud-scale scar shows where the dormant terminal bud of the previous winter was; the length of stem between the bud-scale scar and the terminal bud is one year's growth. Leaf scars show where petioles were attached to the stem. The bundle scars within each leaf scar show where the vascular bundles passed into the petiole. The axillary bud will give rise to a branch stem during the next growing season.

usually arises in the angle between the base of each leaf and the internode above it. Each of these new meristematic regions gives rise to a lateral or *axillary bud* with the same essential features as the terminal buds already described (Fig. 31.35). Elongation of the internodes of the lateral buds produces branch stems. Here, then, is another important way in which growth of the stem differs from growth of the root; in the root there are no surface meristems analogous to the lateral buds, and, as we have seen, the lateral roots arise from cell division in deeper tissue layers.

Differentiation of tissues   Cell division and cell elongation cannot alone produce all the essential features of the fully developed plant, of course. All the new cells produced by the apical meristems are fundamentally alike. Yet some of these cells will become collenchyma, some will become xylem vessels, some will become sieve elements, and so forth. The process whereby a cell changes from its immature form to some one mature form is called *differentiation*.

In the growing root or stem, cells begin to differentiate into the various tissues of the plant while they are still within the meristematic region. As the predominant activities of cell division and cell enlargement run their course, the cells are left to mature into their final form. Hence, though the pattern of histogenesis (formation of tissues) is laid out very close to the tip of the elongating axis, it remains useful to recognize the successive regions along the axis that are related mainly to cell division, cell elongation, and cell maturation (Fig. 31.26).

Three concentric areas can be distinguished even in the region just behind the meristem of a root when it is viewed in cross section (Fig. 31.36A). These are (1) an outer protoderm; (2) a wide area of parenchymatous ground tissue located beneath the protoderm; and (3) an inner core of provascular tissue composed of particularly elongate cells. Just as in the embryo, the protoderm rapidly matures into the epidermis, the ground tissue into the cortex and endodermis, and the provascular core into the primary tissues of the stele: primary xylem, primary phloem, pericycle, and vascular cambium (Fig. 31.36B–C). Differentiation in the growing stem follows a similar pattern, except that there are usually two areas of ground tissue—one between the protoderm and the provascular cylinder, which gives rise to the cortex and endodermis, and a second inside the provascular cylinder, which becomes the pith.

As we saw in Chapter 12, increase in circumference of the root or stem depends on formation of secondary tissues composed of cells derived from lateral meristems, particularly the vascular cambium. As cells of the vascular cambium undergo mitosis, many new cells are produced on the inner face of the cambium and these differentiate into secondary xylem, while other new cells are produced on the outer face of the cambium and differentiate into secondary phloem (Fig. 31.36D–E). As more and more secondary vascular tissue is formed and the circumference increases steadily, the old epidermis and cortex are broken and sloughed off. These are replaced by a secondary protective tissue, the cork, composed of cells derived from a new lateral meristem, the *cork cambium*, which forms from a layer of the old cortex or from the pericycle or even from the older phloem, depending on the species.

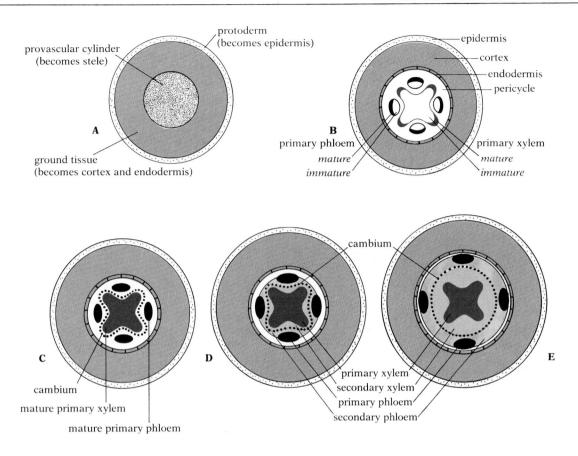

31.36 Differentiation in a young root
(A) Cross section just back of the meristem. Three distinct concentric areas can already be detected. (B) At a slightly later stage of development the protoderm has differentiated into epidermis; the ground tissue has differentiated into cortex and endodermis; and the provascular cylinder has begun to differentiate into primary xylem and primary phloem. (C) Differentiation in the provascular cylinder is complete, and the cambium is about to become active. (D) Divisions in the cambium have given rise to secondary xylem and secondary phloem, which are located between the primary xylem and primary phloem. (E) The areas of secondary tissue continue to thicken as more and more new cells are produced by the cambium.

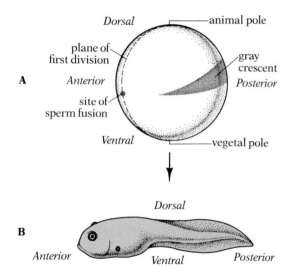

32.1 Polarity of the fertilized frog egg
After a sperm fuses with the egg, a gray crescent develops opposite the point of fusion (A). The crescent, in turn, helps define the anterior/posterior and dorsal/ventral axes of the tadpole (B). The first cleavage divides the embryo's left side from the right. Not all parts of the tadpole arise from corresponding sites on the egg: the anterior half of the egg contributes all of the ectoderm; cells developing from the posterior half fold back into the anterior and give rise to mesoderm.

MECHANISMS OF DEVELOPMENT: DIFFERENTIATION AND PATTERN FORMATION

A single undifferentiated cell, as we have seen, has the genetic information necessary to orchestrate the precise series of events that give rise to a complex multicellular organism. Two basic sets of such events seem to characterize all development: differentiation of cells into ever more specific classes to form various tissues, and spatial patterning of those tissues to form particular morphologies. The various developmental programs must be initiated and coordinated by control mechanisms at the molecular level. Though the molecular mechanisms that underlie differentiation and pattern formation are far from fully understood, great progress is being made; there is reason to believe that they resemble the familiar genetic and biochemical processes we have seen operating in other contexts. We shall examine a few of the more compelling models for differentiation and pattern formation in this chapter.

With few exceptions, all the somatic cells in an organism have an identical set of genes. It is therefore not surprising to find that differentiation and pattern formation depend on the differential activity of various genes, on chemical and electrical signals from other cells, and on information derived from various environmental factors. Signals can serve not only to turn on developmental programs at the right time and in the right place, but also to orient development spatially. Even prior to fertilization, a unique set of genes is active in the egg cell itself. Moreover, environmental cues polarize substances in the cytoplasm of the unfertilized egg in many species; and the resulting polarity subsequently orients development. Let's look at this kind of molecular polarity, as it develops both before and after the fusion of the sperm with the egg.

THE POLARITY OF CELLS AND THE PLANE OF CLEAVAGE

We have already seen that cytoplasm of the unfertilized egg is rarely homogeneous. Most animal eggs contain stored food material, or yolk, which, since it is usually concentrated in one part of the cell, establishes a distinction between animal and vegetal hemispheres, thus polarizing the cell. In the egg cells of many animals—the frog, for example—the animal hemisphere has much more pigment in the cytoplasm than the vegetal hemisphere. Other materials are similarly restricted to certain regions of the cytoplasm—not only in the egg cells of animals but also in those of plants. When an egg cell divides, therefore, the cytoplasmic materials it contains may be distributed unevenly between the two daughter cells, depending on the orientation of the plane of cleavage.

Let us examine an actual example. As soon as the egg cell of a leopard frog fuses with a sperm cell, some of the contents of the egg shift position and a crescent-shaped grayish area appears on the egg opposite the point where the sperm entered (Fig. 32.1). The material in this so-called *gray crescent* will play a very prominent role throughout embryonic development. The first cleavage of the frog zygote normally passes through the gray crescent, so that each daughter cell receives half (Fig. 32.2A). If these two cells are separated, each will develop into a normal tadpole. But if the plane of the first cleavage is experimentally made to pass to the side of the gray crescent, the result of separating the daughter cells will be very different; the cell that contains the gray crescent will develop into a normal tadpole, but the other cell will form only an unorganized mass of cells (Fig. 32.2B). In other words, the way in which the material of the gray crescent is distributed is of utmost importance to the developmental potential of the cells.

As the above experiment confirms, after the first cleavage of the frog zygote, both of the new cells are *totipotent*: they have the full developmental potential of the original zygote (since the polarized cytoplasmic substances are equally distributed in each) and all pathways of differentiation remain open to them. In this respect their development is characteristic of most echinoderms and vertebrates, including humans. Mammalian blastomeres generally remain totipotent until at least the 8-cell stage. In fact, 8-cell-stage embryos of two different strains of mice can be mixed to create a 16-cell *chimera* (after the Chimera, a monster in Greek mythology composed of parts from several different animals) that develops normally. This technique allows researchers to study how cells of one genotype interact with those of another during development. That early human blastomeres should be totipotent was expected, of course, since only animals with this kind of development can give birth to identical twins.[1]

In some other groups of animals, such as annelids and molluscs, the nor-

---

[1] Identical twins develop from the same zygote. Nonidentical (fraternal) twins develop from separate zygotes when two egg cells are released from the ovaries at the same time and are fertilized by different sperm cells. Consequently, identical twins are genetically identical, whereas fraternal twins are no more alike genetically than any two siblings. Fraternal twins are much more common than identical twins.

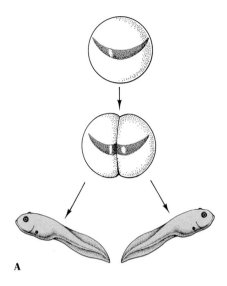

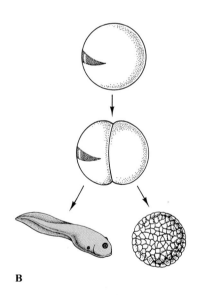

32.2 Importance of the gray crescent in the early development of a frog embryo
(A) If the two cells produced by a normal first cleavage, which passes through the gray crescent, are separated, each develops into a normal tadpole.
(B) If the first cleavage is experimentally oriented so that it does not pass through the gray crescent, and the two daughter cells are separated, the cell with the crescent develops normally, but the other cell develops into an unorganized cellular mass.

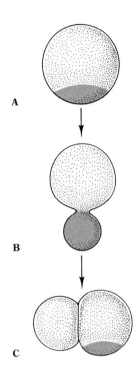

32.3 Cleavage of the fertilized egg of the sea snail *Ilyanassa*
(A) At one pole of the zygote is a region of clear cytoplasm (color). (B) Just before the first cleavage, this polar cytoplasm moves into a large protuberance, the polar lobe. The first cleavage partitions the zygote in such a way that the entire polar lobe goes to one of the daughter cells. (C) The lobe recedes during interphase, but it will form again prior to the next division sequence. Only the cell that receives the polar-lobe material can give rise to a specific set of external structures that are seen in normal larvae.

mal first cleavage partitions critical cytoplasmic constituents asymmetrically; therefore, when the daughter cells are separated, they do not have equivalent developmental potentialities. For example, in some molluscs, such as mussels and sea snails, a protuberance called the polar lobe develops on the fertilized egg cell just before the first cleavage occurs. The plane of cleavage is oriented in such a way that one of the two daughter cells receives the entire polar lobe (Fig. 32.3). If the two daughter cells are separated, the one with the lobe material (which was drawn back into the main body of the cell soon after division was accomplished) will form an embryo possessing two prominent structures—a so-called apical organ and the post-trochal bristles—that are seen on normal larvae; the one with no lobe material forms an aberrant embryo lacking these structures. Something in the polar-lobe material must be essential for formation of the apical organ and the bristles.

Eggs such as those of frogs, which give rise to daughter cells each capable of developing into a complete embryo, have sometimes been called *regulative eggs* because the early embryos they produce can regulate their development to compensate for missing portions of cytoplasmic materials. Eggs such as those of molluscs, whose separated parts produce, in isolation, essentially the same structures they would have formed in the intact embryo, have been called *mosaic eggs*. But this distinction between regulative and mosaic eggs is a little misleading. In almost *all* kinds of eggs, substances that influence the course of cellular differentiation are concentrated in different parts of the cytoplasm and are then apportioned nonuniformly into different cells. In embryos from so-called mosaic eggs, these substances largely specify how the cells containing them will differentiate, while in embryos from so-called regulative eggs interactions between the developing cells are also important in this specification. We have already seen an example of this in the frog; the gray-crescent area contains a determinant without which no embryo can be organized. But in normal development, the part of the embryo at the end away from the gray crescent does indeed become organized; its development depends on interactions with the part of the embryo derived from the crescent.

Sea urchins, which are echinoderms, provide another example. If the animal and vegetal halves of the embryo are separated (along the plane of the third cleavage), the animal half will give rise to an abnormal blastulalike larva with overdeveloped cilia, and the vegetal half will give rise to a different type of abnormal larva with an overdeveloped digestive cavity (Fig. 32.4B). Evidently cytoplasmic determinants promoting different kinds of differentiation are differentially distributed along the animal–vegetal axis of the embryo. By contrast, if the eight-cell embryo is divided along its animal–vegetal axis, each half will develop into a small but normal larva (Fig. 32.4A). In an intact embryo each of these halves would have developed into only half of the normal larva; evidently, as in the frog, interactions between the different regions of the embryo normally act to modify the course of differentiation of neighboring cells, integrating them into a single, properly structured and proportioned larva.

Clearly, then, certain differentially distributed cytoplasmic substances must play a prominent role during early embryonic development. They may function by activating some genes and repressing others in the cells

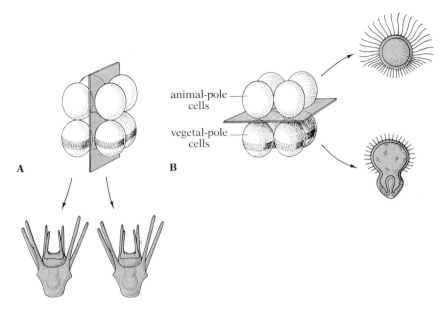

animal-pole cells

vegetal-pole cells

A

B

32.4 Experimental separation of cells after the third cleavage in the embryo of a sea urchin
(A) If the embryo is partitioned meridionally, so that each half receives both animal-pole and vegetal-pole cells, each half develops into a normal larva. (B) If the embryo is partitioned equatorially, so that one half receives only animal-pole cells and the other half only vegetal-pole cells, the animal half develops into an abnormal blastulalike larva with overdeveloped cilia, and the vegetal half develops into an abnormal larva with an overdeveloped digestive cavity.

that come to contain them. Another possibility is that these substances are themselves either maternal messenger RNAs produced by genes expressed during oogenesis or the products resulting from translation of such RNAs. In either case, the nonuniform distribution of these products of previous gene activity along the animal–vegetal axis would lead to the expression of different traits in different parts of the embryo. Whether such substances are arranged in a pole-to-pole gradient (as they apparently are in sea urchins) or in a less symmetrical pattern (as they are in fertilized frog and mollusc eggs, with their gray crescents and polar lobes), cells with different cytoplasmic components are produced early in embryonic development (at the first cleavage in some organisms, at the third cleavage or later in others).

For the normal development of embryos in which localized cytoplasmic substances determine the course of cellular differentiation, these substances must be correctly distributed among the appropriate cells. If they are not, conflicting instructions may get into the same cell. This precise distribution involves an exact correspondence between the location of these determinants in the cytoplasm and the location of the cleavage furrows, which segregate them in different cells. Hence, embryos with mosaic development also display *determinate* cleavage: all embryos in a particular species show identical, and sometimes quite complex, cleavage patterns. Embryos with regulative development have less precise cleavage patterns; they display more or less *indeterminate* cleavage, in which the positions of cleavage furrows are less uniform, especially after the first few cleavages. The nature of the distribution of cytoplasmic determinants in dividing cells, with the correlated orientation of cleavage planes, thus becomes the key to establishing the initial pattern of development.

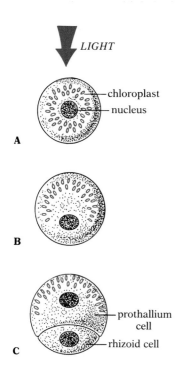

LIGHT

— chloroplast

— nucleus

A

B

— prothallium
cell

— rhizoid cell

C

32.5 Light induction of polarity in the spore of *Equisetum*
(A) The unpolarized spore is exposed to a directed source of light (here shining from above).
(B) Chloroplasts move toward the lighted side, and the nucleus moves toward the opposite side. (C) The first cleavage is unequal and is oriented to produce a large chloroplast-rich prothallium cell and a smaller chloroplast-poor rhizoid cell.

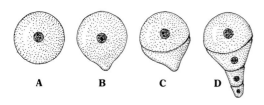

A   B   C   D

32.6 Early development of *Fucus* zygote
(A) The zygote. (B) Formation of a protuberance on one side of the zygote. (C–D) The orientation of the first few cleavages is determined by the location of the protuberance.

But what factors determine the original distribution of substances in the cytoplasm of the egg cell? The detailed mechanisms responsible for patterning the egg cytoplasm are still largely unknown, but some of the stimuli that play a role in guiding the patterning have been studied. We have already mentioned that the point at which a sperm cell fuses with a frog egg determines where on the egg the gray crescent will form. An example in which a physical parameter of the environment plays a crucial role is seen in the spore of the primitive vascular plant *Equisetum*, the horsetail (see Fig. 40.33, p. 1081).

An *Equisetum* spore exhibits little polarity initially (Fig. 32.5A), but when it is exposed to light its chloroplasts (and some other less easily observed cytoplasmic materials) migrate toward the incident light, while the nucleus (and other substances) will move in the opposite direction (Fig. 32.5B). This response to the direction of the incident light establishes an axis of polarity for the spore, one that determines the direction of the first cleavage. The first cleavage produces a large chloroplast-rich cell at the lighted end of the spore axis and a small chloroplast-poor cell at the other end (Fig. 32.5C). The large cell will later give rise to the structure (prothallium) that will form the plant shoot, and the small cell will give rise to the rootlike rhizoids.

The influence of a variety of physical factors on the establishment of embryonic polarity has been demonstrated with particular clarity in the early development of the brown alga *Fucus* (Fig. 40.18, p. 1071), which grows on intertidal rocks along northern coasts. The zygote of *Fucus* is spherical at first and has no visible surface features that would indicate separate cytoplasmic regions. However, shortly after fertilization, the zygote becomes asymmetrical as a small protuberance forms on one side (Fig. 32.6). This protuberance permanently determines the polarization of the embryo that will develop from the zygote. The first cleavage is always oriented in such a way that the side of the zygote bearing the protuberance becomes one daughter cell and the side without the protuberance becomes the other daughter cell. The cell with the protuberance always forms the holdfast, a rootlike structure that anchors the plant to the rocks. The other cell always forms the more erect part of the plant.

D. M. Whitaker of Stanford University studied the factors that determine where on the *Fucus* zygote the protuberance will form. He found that a number of asymmetric environmental conditions play a role, the most important being illumination. The protuberance forms on the side away from the light. When lighting is equal on all sides, but a temperature gradient is maintained across the zygote, the protuberance forms on the warmer side. Similarly, in a pH gradient, the protuberance forms on the side with the lower pH. If illumination, temperature, and pH are all equal on all sides, but other zygotes are present in the culture, the protuberance forms on the side nearest the other zygotes. Finally, if the zygote is spun in a centrifuge, the protuberance forms on the side farthest from the center of the centrifuge. The adaptive significance of some of these reactions is understandable; a holdfast growing away from light and toward the pull of gravity would normally be going in the right direction to encounter the rocks at the bottom of the water.

Before leaving the subject of polarity, we should point out that polarity is by no means restricted to the egg cell and early embryo; it is, in fact, an extremely common aspect of organismic organization. We have encountered a number of earlier examples. The ion-pumping activity of the specialized epithelial cells of osmoregulatory structures depends on the polarity of the cells, with Na^+ and Cl^- ions apparently diffusing passively into the cell on one side and then being actively pumped out on the other (pp. 376–77). In plants, transport of auxin in a stem is strongly polar (Fig. 15.12, p. 387).

That the structures of some developed organisms continue to show polarity in their developmental potential can be seen especially well in the regeneration of plant stems. For example, cuttings of willow stems will form roots at their physiological basal ends and buds at their apical ends, regardless of their orientation in space (Fig. 32.7). They will do so no matter how short the piece of stem may be, apparently as a result of the polarity of the individual cells within the stem.

TISSUE INTERACTIONS IN DEVELOPMENT

The immediate environment of the nucleus of a developing cell is the cytoplasm of that cell, which, as we have seen, can exert a profound influence on differentiation. But developing cells are also subject to influences external to their own cytoplasm. Depending on their location in the embryo, they are likely to be exposed to different combinations of environmental factors, which may help determine their developmental direction.

As countless experiments have shown, developing cells are indeed influenced by neighboring cells. For example, they interact via contact and diffusible chemicals; both their motility and their mitotic activity are influenced when they come into contact with one another. Developing cells are likewise influenced by hormones and by many parameters of their physical environment.

THE ROLE OF ORGANIZERS AND INDUCERS

We have already seen that the gray crescent is important in determining the potentialities of the cells produced during early cleavage in the frog embryo. Let us now return to the gray crescent and examine its role in the course of later development.

The gray crescent is located on the frog egg near the boundary between the animal and vegetal hemispheres. It is along this boundary that involution of cells occurs during gastrulation. In the early gastrula the cells derived from the gray-crescent portion of the egg become the **dorsal lip of the blastopore**. These cells soon move inward and form the **chordamesoderm**, which is at first located in the roof of the newly forming archenteron (see Fig. 31.6E, p. 792), but soon detaches from the archenteron to form the notochord and other mesodermal structures (see Fig. 31.6H). The chordamesoderm also seems to exert a very important influence on the ectodermal tissue lying over it; this tissue folds inward during neurulation, forming the neural tube, which differentiates into the brain and spinal cord, but neurulation does not occur if the chordamesoderm is missing.

In a series of classic experiments performed in 1924, Hans Spemann of the University of Freiburg, Germany, who had previously demonstrated the

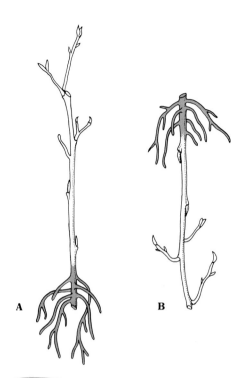

32.7 Polarity of a stem cutting
A cutting of a willow stem will produce new roots at its physiological basal end and new buds and shoots at its physiological upper end, whether it is oriented normally with respect to gravity (A) or upside down (B).

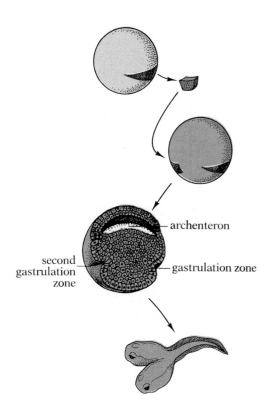

32.8 Spemann's experimental transplantation of the gray crescent
When Spemann transplanted gray-crescent material from a light-colored salamander zygote to a dark-colored one, the zygote with two gray crescents proceeded to form an embryo with two gastrulation zones—one at the dorsal lip of the blastopore, derived from its own gray crescent, and one at the second dorsal lip, derived from the transplanted gray crescent. A double larva of mostly dark-colored tissue was the result.

32.9 Development of optic vesicles and their induction of lenses in a frog
The proximity of the optic vesicle induces the nearby cells of the epidermis to fold inward and form a lens. The infolding presumptive lens tissue, in turn, helps mold the optic vesicle into a two-layered structure called the optic cup. The cup cells then differentiate, the layer adjacent to the lens forming visual receptor cells and nerve cells.

importance of the gray crescent in early cleavage, and his student Hilde Mangold turned their attention to the dorsal lip of the blastopore in the early gastrula stage of a salamander embryo. They transplanted the dorsal lip from its normal position on a light-colored embryo to the belly region of a darker-colored embryo. After the operation, gastrulation occurred in two places on the recipient embryo—at the site of its own blastoporal lip and at the site of the implanted lip. Eventually two nervous systems were formed, and sometimes even two nearly complete embryos developed, joined together ventrally (Fig. 32.8). Most of the tissue in both embryos was dark-colored, an indication that the transplanted blastoporal lip had altered the course of development of cells derived from the host. Similar transplants of tissues from other regions of embryos failed to produce comparable results. The dorsal lip of the blastopore must play a crucial role in determining the form of the early embryo, probably by inducing the formation of the neural groove, which in turn is important in establishing the longitudinal axis of the embryo and in inducing formation of other structures.

Spemann and Mangold called the dorsal lip of the blastopore the *organizer*. They envisioned the entire developmental process as one in which a succession of principal organizer regions, each taking over where the previous one left off, control the differentiation of the major tissues and organs. We now know that induction is not limited to a small number of organizer regions, but is a general phenomenon; inductive tissue interactions are the rule rather than the exception.

Some of the first definitive studies on embryonic induction in an animal were performed in 1905 by Warren H. Lewis of Johns Hopkins University. Lewis worked on the development of the eye lens in frogs. In normal development the eyes form as lateral outpockets from developing brain tissue. When one of these outpockets, or optic vesicles as they are called, comes

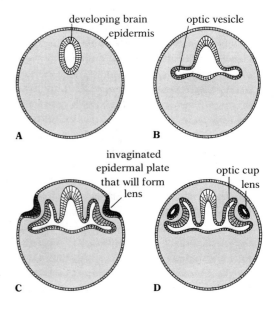

into contact with the epidermis on the side of the head, the contacted epidermal cells promptly undergo a series of changes and form a thick plate of cells that sinks inward, becomes detached from the epidermis, and eventually differentiates into the eye lens (Figs. 32.9 and 32.10). Lewis cut the connection between one of the optic vesicles and the brain before the vesicle came into contact with the epidermis. He then moved the vesicle posteriorly into the trunk region of the embryo. Despite its lack of connection to the brain, the optic vesicle continued to develop, and when it came into contact with the epidermis of the trunk, that epidermis differentiated into a lens. The epidermis on the head that would normally have formed a lens failed to do so. Clearly, the differentiation of epidermal tissue into lens tissue depends on some inductive stimulus from the underlying optic vesicle. Later experiments have shown that if a barrier is inserted between the vesicle and the epidermis no lens develops.

Other experiments have shown that the regulation is not all one-way; the lens, once it begins to form, also influences the further development of the optic vesicle. If epidermis from a species with normally small eyes is transplanted to the sides of the head of a species with large eyes, the eyes that are formed do not have a large optic cup (formed from the optic vesicle) and a small lens, as might be expected. Instead, both the optic cup and the lens are intermediate in size and correctly proportioned to each other. Obviously, each influences the other as they develop together. Feedback is just as important in the control of embryonic development as it is in the nervous or hormonal control of the fully formed organism.

Some cell-to-cell interactions may be an effect of cell contact, as in contact inhibition, but most developmental interactions are probably mediated by substances passed between cells. Such chemical mediation can be demonstrated experimentally. For example, if the presumptive epithelial tissue and the presumptive connective tissue of an embryonic mouse pancreas are separated, the epithelial tissue will not differentiate properly in the absence of the connective tissue. But if the two groups of cells are grown in tissue culture and separated by a filter that will allow macromolecules, but not whole cells or organelles, to pass, then the epithelium will differentiate into normal pancreas tissue. Obviously, the necessary induction has occurred, and it must be attributed to a diffusible chemical that could pass through the filter.

Of the many substances that can act as intercellular inducers in embryonic development, some play a so-called instructive role, others play a more permissive role. The ***instructive inducers*** restrict the developmental potential of the target cell and thus help determine the course of differentiation. The ***permissive inducers*** facilitate the expression of potentialities already determined. For example, a rudimentary organ may form fully committed cells during embryology, but will not complete its development and become functional until acted on by a permissive inducer, which in some instances may be a hormone.

We must stress that the so-called instructive inducers do not give their target cells instructions about the design of the tissues or organs they are to form; they instruct the cells only in the sense that, through repression of some genes and derepression of others, they tell the cells what part of their genetic endowment they are to use. A dramatic example of this principle is

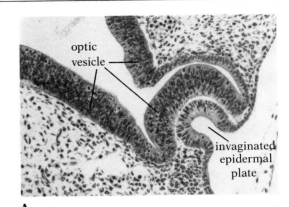

A

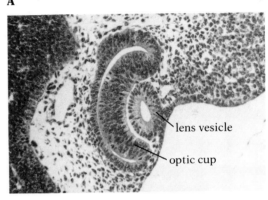

B

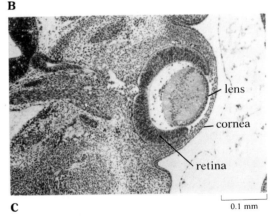

C   0.1 mm

32.10 Development of a mammalian eye
The eye forms when an evagination from developing brain tissue—the optic vesicle—reaches the epidermal layer. The contact with the tip of the optic vesicle causes the epidermis to invaginate (A). The epidermal region differentiates to form the lens vesicle, while the optic vesicle develops into the optic cup (B), which ultimately becomes the retina (C). The events shown here take four days.

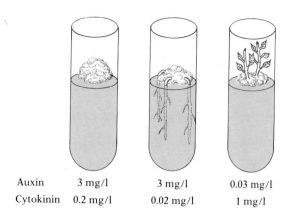

| Auxin | 3 mg/l | 3 mg/l | 0.03 mg/l |
| Cytokinin | 0.2 mg/l | 0.02 mg/l | 1 mg/l |

32.11   The effect of auxin and cytokinin on the development of cultured pith cells from a tobacco root
At certain concentrations of the hormones, only an amorphous callus is formed. At other concentrations roots or shoots develop from the callus.

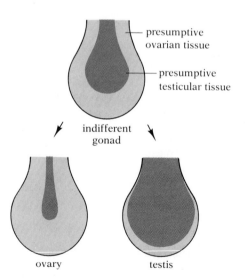

presumptive ovarian tissue

presumptive testicular tissue

indifferent gonad

ovary   testis

32.12   Development of ovary or testis from the indifferent gonad of an amphibian
Depending on the hormonal condition of the embryo, one or the other of the two types of tissue in the indifferent gonad gains developmental ascendancy and the other type is repressed.

the experiment in which Spemann and Oscar E. Schotté transplanted ectoderm from the flank of a frog embryo to the mouth region of a salamander embryo. They found that, though induced by salamander endoderm, the transplanted tissue formed the typical horny jaws of a frog: the constellation of genes for producing a jaw was activated by instructive inducers from the adjacent tissue in the salamander, but the genes, being from frog ectoderm, encoded the morphology of a frog jaw.

HORMONES IN DEVELOPMENT

Plant hormones play a key role in nearly all phases of development. They may fulfill both instructive and permissive functions, as can be shown experimentally when varying concentrations of auxin and cytokinin are added to the basic growth medium of a culture of pith from the root of a tobacco plant. At some concentrations the hormones merely enable cell division to proceed, the result being a mass of undifferentiated cells called a callus (Fig. 32.11, left). Here the hormones are acting permissively; they are in no way restricting the developmental potential of the cells. At other relative concentrations the hormones induce formation of new roots or new stems; because they stimulate the cells to differentiate in a particular way, they are acting instructively.

The inductive action of the hormones varies as a result of differences in other factors influencing the cells. Thus, if pith cells are cultured in a thin layer on a medium containing two parts per million of auxin, they do not divide, but grow to an unusually large size. When the pith tissue is molded into a cylinder, however, and the same concentration of auxin is inserted into one end of the cylinder, the cells undergo many divisions and some of them differentiate into xylem. The difference in response to the same inducing stimulus must be attributed to the different locations—superficial or internal—of the cells.

In vertebrates hormones play a predominantly permissive role in development, but are not, on that account, any less important than instructive inducers. What good would it do an organism if cells were set aside as a presumptive tissue or organ (as a result of determination in response to instructive inducers) and their developmental potential could never be expressed for lack of a necessary permissive hormone? The essential interplay between instructive inducers and permissive hormones can be studied in the gonads of amphibians. Two distinct kinds of cells are early set aside to become gonads in amphibians: peripherally located cortical cells and centrally located medullary cells (Fig. 32.12). The cortical cells have the potential for forming ovarian tissue, and the medullary cells for forming testicular tissue. Only when sex hormone acts on the sexually uncommitted embryonic gonad does one or the other of its potentialities gain ascendancy.

Similarly, the same embryonic primordia give rise to the accessory sexual organs of both sexes in humans (Fig. 32.13). Whether these primordia form male or female structures depends on whether or not male sex hormone is present in the embryo at the critical stages of embryonic development. In the absence of male hormone, the individual develops as a female, no matter what its genetic sex (XX or XY). In birds, where the female is the heterogametic sex, the situation is reversed; the sexual organs will follow a masculine developmental course unless female hormone is present at the critical embryonic stages. Normally, the hormonal condition of the embryo will be the one appropriate to its genetic sex, but in rare instances this is not the case, and an individual will develop organs that, though appropriate to its genetic sex, are poorly formed, or it may develop organs that, though nearly normal, belong to the other sex.

MORPHOGENETIC FIELDS

Not all chemicals influencing the developmental directions of cells are stimulatory. Many, in fact, are inhibitory, or else they are stimulatory to some cells and inhibitory to others. Auxin is an example; in normal development, it stimulates differentiation of xylem but inhibits development of lateral buds.

There is abundant evidence from both plants and animals that as a particular tissue or organ develops it releases substances that inhibit formation of the same kind of tissue or organ in the immediate area. If a leg-forming region from the embryo of a salamander is transplanted (before it has actually taken the form of a leg) to a spot immediately adjacent to a leg-forming region on another embryo, only one leg develops; the leg field has sufficient capacity for self-regulation to ensure that a single normal-sized leg develops, despite the presence of twice the normal number of leg-forming cells. Similar principles seem to operate in determining the position of a new leaf on a growing stem. Each new leaf emerges at a point separated by some critical distance from the nearest older leaves and from the shoot apex. If one of the nearby older leaves is removed, the new leaf develops at a point closer to the position of the removed leaf than it would normally occupy. It seems that each leaf is surrounded by a morphogenetic field that somehow prevents the formation of a new leaf within it, but that, within itself, is capable of regulation. The presence of morphogenetic fields seems to be a general phenomenon.

MECHANISMS OF PATTERN FORMATION

No less important than cell differentiation and tissue formation in the later development of animals is pattern formation. It is crucial, for example, in the development of a vertebrate forelimb, that the bones be arranged in proper order, from humerus at the base to phalanges at the distal end.

Various elegant experiments with *Drosophila* by many different researchers demonstrate that cells have a way of telling where they are: positional information of some sort is clearly available to them. These experiments make use of one of several mutant strains of *Drosophila* in which the developmental fate of particular imaginal discs is altered. As we said in the last chapter, these platelike clumps of cells in insect larvae dif-

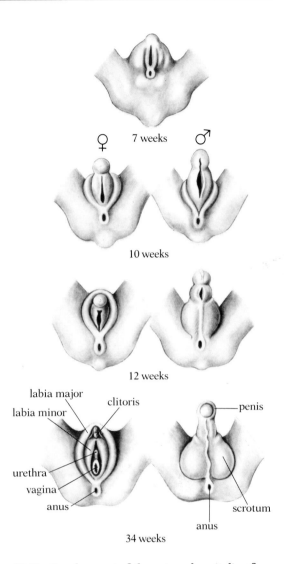

32.13   Development of the external genitalia of human beings

At 7 weeks the genitalia of male and female fetuses are virtually identical. At 10 weeks the penis of the male is slightly larger than the clitoris and labia minor, which form from the same primordium in the female. At 12 weeks these differences are more pronounced, and the male scrotum has formed from the tissue that becomes the labia major in the female. At 34 weeks the distinctive features of the genitalia of the two sexes are fully apparent. It is largely the concentration of male sex hormones that determines which of these developmental pathways will be followed.

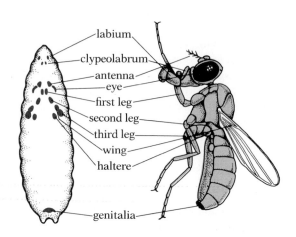

32.14 Imaginal discs of a *Drosophila* larva
Each disc in the larva at left will give rise to part of the adult fly.

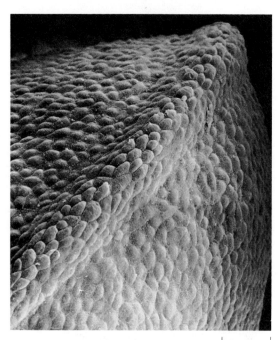

0.02 mm

32.15 Pattern formation in the development of the chick wing
Above: Scanning electron micrograph of the ectodermal ridge of a chick wing bud. Right: Diagram of three stages of development of the wing bud, according to Wolpert's model. (A) Just behind the ectodermal ridge is a progress zone, where new cells are produced. (B–C) The first band of cells derived from the progress zone will become the humerus section of the wing; the second band will become the section containing the radius and ulna; and a third band, derived from the progress zone late in the development of the wing bud, will become the distal part of the wing.

ferentiate during the pupal stage to form the various parts of the adult (Fig. 32.14). A disc's fate is established early in development, so replacing an antennal disc with a leg disc results in a fly with a leg in its head.

The effects of transplantation can be mimicked by so-called homeotic mutations. These single-gene changes can cause an antennal disc to develop into a leg. Since the development of a leg requires the activation of one set of genes, while antennal development requires that a different array be active, homeotic mutations must affect a control gene—a gene that determines which constellation of developmental genes is to be activated. Homeotic mutations apparently alter production of or response to inducer substances, and the resulting changes in the development of the organism confirm what had already been deduced about the role of these substances from transplantation experiments. Less severe homeotic mutations, which only partially transform discs, reveal another aspect of organization. The mutation antennapedia, for instance, usually causes some part of a fly's antenna to develop as leg. The exact location of the part affected is variable, but the pattern of development is constant: if the tip of the antenna is af-

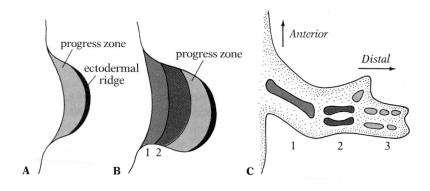

fected, it develops as a foot; if the base is affected, it develops as the proximal part of the leg, and so on. The developing disc, then, must be polarized, and the polarization mechanism must be common to discs for both antennae and legs; in this mutant, both antenna genes and leg genes control development in accordance with the same sort of positional information to create a complex, though misguided, morphological pattern.

Present ideas about pattern formation are largely based on studies of vertebrate limb growth. Lewis Wolpert of Middlesex Hospital Medical School, London, has suggested a highly regarded but still controversial model based on his investigations of wing development in chicks to explain how cells might acquire the positional information they need to ensure proper pattern formation. His model assumes a bicoordinate system: a proximodistal axis (from the body to the end of the extremity) and an anterioposterior (front-to-rear) axis.

The proximodistal coordinate Wolpert proposes is linked with the progress zone for development of the wing, an area associated with an ectodermal ridge running across the tip of the limb bud (Fig. 32.15A). New cells are produced in the progress zone and left behind as the area is pushed farther and farther away from the body—much as the apical meristem of a plant shoot is gradually pushed upward. A cell's proximodistal positional values might be determined by the time it spent in this progress zone (Fig. 32.15B). Cells left behind by the progress zone very early in development of the wing would have a low positional value, which might cause them to develop into the basal part of the wing (the humerus portion); cells left behind a bit later would have an intermediate positional value, appropriate to formation of the middle part of the wing (the radius and ulna portion); and cells left behind late in the development, having spent a long time in the progress zone, would have a high positional value, appropriate to formation of the distal part of the wing. As to the anterioposterior coordinate, Wolpert proposes that a diffusion gradient of a hypothetical control substance (a morphogen) secreted by a small group of cells at the rear margin of the wing bud polarizes the bud. With these two coordinates, the cells could "read" their position with sufficient accuracy to ensure their differentiation into an appropriate structure.

To test the first part of this model, Wolpert grafted the progress zone (the ectodermal ridge and the associated area of actively dividing cells) from an early wing bud onto the end of an older bud whose own progress zone had been removed. The result was development of a wing with two humerus sections and two radius-ulna sections (Fig. 32.16). As Wolpert interprets this result, the cells of the graft had no way of telling they were so far out on the wing that they should form only its distal parts; because they had spent so little time in the progress zone, some of these cells read their position as being near the wing base, and developed into structures appropriate to such a position. The converse experiment of grafting the progress zone from an older bud onto an early bud produces a wing with only the distal parts, the phalanges: the transplanted tissue has no indication that the humerus, radius, and ulna have not yet developed; having spent a long time in the progress zone, it develops structures appropriate to the wing tip.

To test the gradient part of the hypothesis, Wolpert transplanted part of the supposed polarizing region of the bud from the rear (posterior) edge to

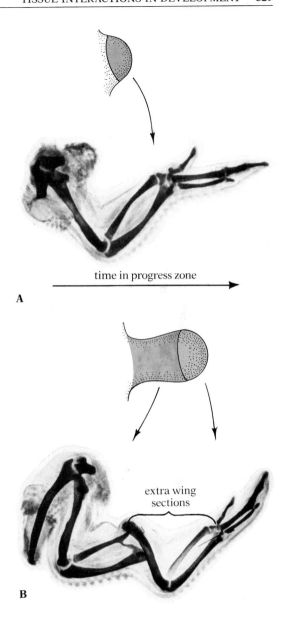

time in progress zone

A

extra wing sections

B

32.16 Results of Wolpert's experiment with a grafted progress zone of a chick wing bud
(A) A normal wing developed from an intact wing bud. The gray area in the bud is the progress zone. (B) The wing that developed when an early wing bud (color) was grafted onto the original bud (gray) after the basal part of the wing had already begun to develop. The wing has extra humerus and radius-ulna sections.

32.17 Results of Wolpert's experiment with a transplanted polarizing region
Tissue from the rear margin of one wing bud (A) is transplanted to the leading edge of another (B). There the morphogen diffusing from the transplant causes cells to differentiate into a second set of phalanges (C).

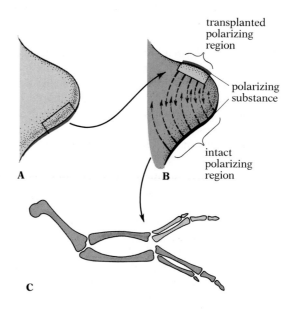

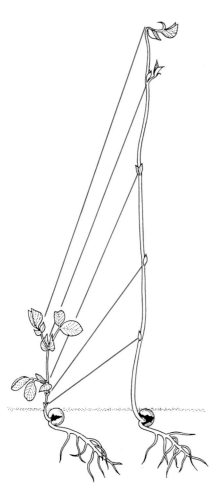

32.18 Development of a pea seedling in the dark
The seedling at left grew normally in the light, but the one at right grew in total darkness. It is abnormally tall and spindly (the color lines link its nodes with corresponding ones of the normal seedling), and it has no chlorophyll.

the front (Fig. 32.17). The results were dramatic: a partial set of mirror-image phalanges developed on the front edge. Their orientation to the transplanted part of the polarizing region was identical to the orientation of the normal set of bones to the polarizing tissue at the rear of the wing, evidently because the morphogen from the transplanted region polarized the front half of the bud, diffusing from front to rear, just as the morphogen from the intact polarizing region on the rear edge polarized the rear half of the bud, diffusing from rear to front. Note that because of the elevated total concentration of morphogen, the level was not low enough anywhere for the normal front digit to develop. Further experiments strongly support the gradient model: if the transplant is large, nearly a full set of reversed digits develops, but if only a small portion of tissue is moved, only a single digit is induced.

EFFECTS OF EXTERNAL FACTORS

Cell-to-cell and tissue-to-tissue interactions are not the only important elements in guiding embryonic development. A number of physical factors from the environment may likewise affect the activity of a developing cell. Among these are temperature, light, humidity, gravity, and pressure. We have previously observed the influence of temperature on the development of pigmentation in Himalayan rabbits and on the vestigial-wing character in *Drosophila* (see p. 661), and we saw in our discussion of plant hormones that light, through its effects on auxin and phytochrome, plays a critical indirect role in such developmental phenomena as increase in shoot length and initiation of flowering. Numerous other examples could be cited. If a seedling plant is grown exclusively in the dark, no mature chloroplasts form; the entire shoot remains colorless, though it usually grows abnormally tall (Fig. 32.18), and the plant eventually dies. Light is an essential factor in inducing the development of chloroplasts and the synthesis of chlorophyll.

A particularly striking example of light-mediated determination of development is seen in ferns. When a spore germinates, it first gives rise to a filamentous structure called a protonema, but normally, as soon as the protonema reaches the light, the pattern of growth changes and a platelike structure called a prothallium develops. The transition from protonema to prothallium depends on short-wavelength (blue) light. If the germinating spore is kept in the dark or in long-wavelength (red) light, only a protonema will develop (Fig. 32.19A), but if the spore is kept from the beginning in blue light (or in white light, which contains sufficient blue-light energy), a prothallium will develop (Fig. 32.19B).

The question arises whether, after the cells have been stimulated by blue light to differentiate into prothallial cells, they are capable of reverting to protonema-type development. To find out, it is only necessary to move the prothallium from the blue light into red light. Very soon a cell at the apex of the prothallium gives rise to a filament that in all respects resembles the protonema normally produced only by a germinating spore (Fig. 32.19C). Here, then, is a case in which an external factor—the wavelength of light—determines which of two markedly different patterns of development will occur.

THE COURSE OF DIFFERENTIATION

THE DEVELOPMENTAL LANDSCAPE MODEL

Let us now look at the picture of the course of development that emerges from the facts and ideas we have discussed so far.

If, as we must assume, all the cells of a single organism usually have the same genetic potential, it follows that other factors determine which potentialities are expressed. The first restrictions on the potential of an embryonic cell are often the result of qualitative (and sometimes also quantitative) differences established during the early cleavages, which may distribute gene products already synthesized into different cells, and may give the nuclei of the different cells different cytoplasmic environments and thus bring about the activation of different genes. As development proceeds, the extracellular environment of the cells becomes less uniform. For example, some cells are located internally and hence are exposed to different factors than cells on the surface of the embryo. Moreover, the various cells differ with regard to their immediate cellular neighbors and hence with regard to the influences—both contactual and chemical—exerted on them by those neighbors. Such differences in the environments of the various cells further intensify the differences in their developmental directions.

As the cells and tissues become more and more differentiated, and perhaps (in animals) undergo morphogenetic movements, they exert an increasing influence on all other cells in their vicinity via chemicals they secrete. Some of the chemicals block pathways the neighboring cells might otherwise have followed; other chemicals tend to induce the neighboring

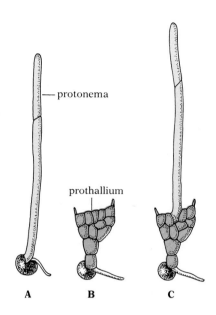

32.19 Effect of different wavelengths of light on the early development of a fern
(A) A germinating fern spore kept in darkness or in red light will give rise to a filamentous protonema. (B) In blue (or white) light a spore will give rise to a platelike prothallium. (C) When a developing prothallium is moved into red light, one of the cells at the summit of the prothallium changes its course of development and produces a protonema.

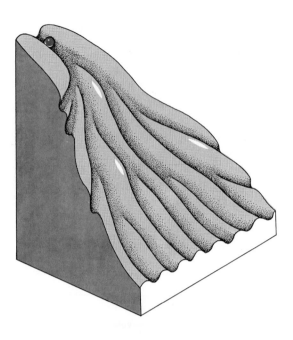

32.20 Developmental landscape model
As the ball rolls down the hill past each branch in the valleys, its potential becomes progressively more restricted.

cells to follow alternative pathways. Still other chemicals may provide, through gradients in their concentration, positional information that helps impose pattern on the development.

As cells and tissues respond developmentally to the host of influences impinging on them, they in their turn alter the environment of the cells and tissues in their vicinity. And so the snowballing effect continues. Each step in the development of one cell alters the influence that cell will have on all other cells. The environment of each cell is constantly changing as development proceeds, and the changes in the environment profoundly affect the activity of the genes.

C. H. Waddington of the University of Edinburgh compares differentiating cells with balls rolling down a slope cut by many valleys (Fig. 32.20). Each ball rolls into one of the valleys. This valley soon branches into two or more separate valleys, and each of these eventually branches in its turn. At each point of branching, the ball enters one of the alternative valleys. As it passes each intersection, the number of alternative pathways still open to it diminishes. Since the ball cannot roll uphill, it cannot normally retrace its course and take a different route. Finally it reaches the bottom of the slope at a point determined by the particular alternative pathways it took at each point of branching.

In a similar way, an early embryonic cell may follow any of a large number of different developmental pathways. Once it has differentiated as ectoderm, however, it cannot ordinarily go back and form a mesodermal or endodermal structure. It has passed the first branching point in the developmental landscape. Now it can form any ectodermal structure, which still leaves it many alternatives. But soon it passes a second important branching point; it either sinks inward as the neural groove forms and differentiates into nervous tissue, or it remains on the surface and differentiates as epidermis. Suppose it follows the former course. Many alternatives are still open to it. It may form part of the brain or part of the spinal cord; it may become part of the somatic or part of the autonomic nervous system; it may differentiate as a multipolar, a bipolar, or a unipolar neuron. But as each branching point is passed, the total number of alternatives still ahead diminishes, until finally the cell has become some one kind of fully differentiated cell. Differentiation is thus a matter of progressive determination, a gradual restricting of development to one of the many initially possible pathways. Note that determination and differentiation, though closely related, are not identical. Determination refers to a cell's fate, regardless of its present morphology; differentiation refers to a cell's specialized chemistry and morphology. Hence, a cell can be determined before it actually differentiates; conversely, some cells differentiate before they are fully determined.

EVIDENCE FROM TISSUE TRANSPLANTS

The gradual course of differentiation, as depicted by Waddington's model, can be demonstrated by transplantation experiments.

If in a very early amphibian gastrula a piece of tissue from the part of the ectoderm that will later participate in the formation of the spinal cord is exchanged with a piece of tissue from the part of the ectoderm that will form epidermis, each transplanted tissue develops into whatever is appropriate for its new location; the tissue that would have formed nerve now forms epidermis, and vice versa. Location is the important factor at this stage.

Entirely different results are obtained if the same kind of exchange is made a short time later. The piece that would have formed nerve in its original location also forms nerve in its new location, and the piece that would have formed epidermis forms epidermis in its new location. Something has happened in the late gastrula that has caused the cells from the two different parts of the ectoderm to set out on different developmental pathways. In the late gastrula, you will recall, chordamesoderm derived from the dorsal lip of the blastopore lies under the part of the ectoderm that will form the spinal cord and brain, but there is no chordamesoderm under the part of the ectoderm that will form epidermis. Induction from the chordamesoderm has started differentiation of some of the ectoderm into nervous tissue, while the rest of the ectoderm, responding to other influences, has begun differentiation into epidermis. Each tissue has lost some of its former potential; hence the presumptive epidermis is no longer competent to respond to the nerve-inducing stimuli of the chordamesoderm.

Now suppose that presumptive spinal cord, instead of being moved to an epidermal location, is moved to the location where the optic vesicles will form. We know from the previous experiment that the tissue has been determined as nerve, but will it form only spinal cord or can it still form other parts of the nervous system? If this experiment is performed very soon after neurulation has begun, the transplanted tissue will form optic vesicle; its location in the developing nervous system is the important factor. But if the experiment is performed after the nervous system has begun to assume its definitive form, the transplanted tissue will form spinal cord, regardless of its new location. It has already passed another critical decision point in its determination, and altering its location does not reverse its development. We see, then, that the tissue is first determined as nervous tissue and later as a particular type of nervous tissue. Determination and differentiation are gradual; they do not occur all at once.

EVIDENCE FROM REAGGREGATION OF CELLS

The gradual course of differentiation can also be demonstrated by the reaggregation of separated cells. In 1907 H. V. Wilson of the University of North Carolina discovered that if he dissociated the cells of the body of a sponge by pressing it through a fine sieve, the dispersed cells migrated over the surface of the culture dish until they encountered one another and clumped together to form multicellular aggregates. These aggregates, if suitably cultured, grew into complete new sponges with the architecture characteristic of the sponge before disruption by the sieve. Later workers showed that if the cells of two sponges of different species (colored differently) were dispersed and then mixed together, they would sort themselves and aggregate by species. These results led other workers to wonder if the capacity of cells for recognizing and aggregating with their own kind was a general phenomenon not limited to sponges. It was soon found that the ability of cells to recognize other cells of similar tissue type is a general phenomenon, but that the recognition of species, demonstrated in sponges, is not so general. For example, retina cells from a chick and retina cells from a mouse will clump together to form a new composite retina.

When the cells of the mesodermal and ectodermal layers of a gastrula are dispersed (by removing calcium ions or raising the pH of the medium) and then mixed with one another, they first reaggregate to form a clump with the two kinds of cells randomly intermixed. Then the cells begin to sort themselves out, rearranging themselves until the ectodermal cells are on the outside and the mesodermal cells are on the inside. Clearly, ectodermal cells can recognize other ectodermal cells and mesodermal cells can recognize other mesodermal cells; and mesoderm and ectoderm can recognize each other and establish the proper inside-outside relationship. Evidently the two types of cells, mesodermal and ectodermal, are already partly differentiated, even at this early stage of development.

If the same experiment is performed with cells from a later embryo, mesodermal cells aggregate not just with any other mesodermal cells, but with mesodermal cells from the same part of the embryo as they. Presumptive cartilage cells clump together and form cartilage; presumptive heart cells clump together and form lumps of beating heart muscle; presumptive kidney cells clump together and constitute characteristic kidney capsules and tubules; and so on. In other words, in this later embryo, the cells have differentiated further and are no longer simply mesodermal as against ectodermal; they are now determined as future cartilage or heart or kidney cells. They have moved farther down the valleys of Waddington's developmental landscape.

IS DIFFERENTIATION REVERSIBLE?

We have seen that as development of a multicellular organism proceeds, the individual cells become more and more committed to one particular course of differentiation. Since this differentiation reflects changes in the activity of the cell's genetic material, the question arises whether any of those changes are reversible. Of course cancerous tissues result from the proliferation of a cell that has lost (or never achieved) its differentiated condition, but such a loss of differentiation requires genetic changes. Even in normal cells, however, gene expression is not static; the cell in its day-to-day, indeed moment-to-moment, activities must be able to alter its pattern of genetic transcription and mRNA translation in response to the constantly fluctuating requirements of its metabolic machinery for enzymes. But what about the more profound changes that occur at critical decision points in the cell's development? Are they invariably permanent?

Reversibility of differentiation in plant cells Some years ago F. C. Steward and his colleagues at Cornell University succeeded in growing whole new carrot plants from single mature cells removed from a carrot root. As Figure 32.11 suggests, the same thing can be done with tobacco. There is, in fact, a wealth of evidence that even fully differentiated plant cells (except, of course, those like sieve elements that have lost their nuclei) can be made to revert to the embryonic state and resume totipotency (unlimited developmental potential). In plants, then, the changes undergone by the nuclei of cells during differentiation appear to be entirely reversible.

Nuclear-transplant experiments with animal cells We saw earlier that the presumptive epidermis and the presumptive neural tissue of an amphibian embryo reach a stage soon after gastrulation when, if their positions are interchanged, one can no longer assume the role of the other. Does this mean that the nuclei have been so altered that they cannot back up and follow a different developmental course? Or might it mean that the whole cells are no longer receptive to the inducing factors encountered in the new locations? One way to find out is to remove the nucleus from a more differentiated cell and insert it into a less differentiated cell, where it will presumably be exposed to cytoplasmic factors more conducive to dedifferentiation.

Some of the earliest nuclear-transplant experiments were performed on frog embryos in 1953 by Robert W. Briggs and Thomas J. King of the Institute for Cancer Research in Philadelphia. Their method was to prick a frog egg with a glass needle to stimulate it to begin developing, remove its nucleus,[2] and then, with a micropipette, insert a nucleus obtained from a partly differentiated cell of a developing embryo. They found that when the transplanted nucleus was from a blastula the egg usually developed normally—an indication that nuclei in blastula-stage cells have not undergone any permanent alteration and retain the capacity to direct all aspects of development. But if the transplanted nucleus was from a late gastrula or later stage, the eggs usually did not develop normally; the embryos produced (except for a small percentage) were severely abnormal, or their development was arrested at some early stage. It seemed that the nuclei of late-gastrula cells had undergone some stable change, which persisted even if the nuclei were subjected to serial transplants through four or five generations. Does this mean that the nuclei of differentiated frog cells, unlike those of plant cells, permanently lose some of their former developmental potential?

This has been a difficult question to answer because, even though most nuclei from late-gastrula cells in Briggs and King's experiments seemed unable to direct development of a normal embryo, a few such nuclei could do so. Moreover, in later experiments with the African clawed toad (*Xenopus*), J. B. Gurdon of Oxford University found that about 2 percent of nuclei taken from intestinal cells of swimming tadpoles could direct completely normal development when injected into enucleated egg cells. The activity of these nuclei from fully differentiated cells should have been far more rigidly determined than that of the gastrula nuclei of Briggs and King; yet they seemed able to revert completely to the egg-stage condition and begin to promote full normal development all over again.[3]

[2] In most recent experiments of this type the nucleus of the egg cell is usually destroyed with UV light.

[3] The possibility of artificial cloning arises for animals in which transplanted nuclei can resume totipotency. Many nuclei, all taken from cells of a single individual, could be transplanted into egg cells. The new individuals produced from these eggs would all be genetically identical copies of the individual from which the nuclei were taken.

32.21 Regeneration of a salamander arm

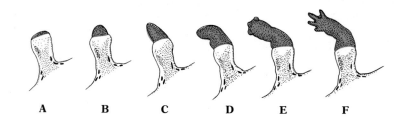

A   B   C   D   E   F

Regeneration of lost body parts Almost all embryonic animals have extensive regenerative capacities. Some animals retain these after they reach maturity, while others lose them. If an adult sea star or hydra is chopped into many pieces, each piece can regenerate all necessary parts to become a whole individual. An earthworm can regenerate a new head or tail. Half a planarian can regenerate the other half. Salamanders and lizards can regenerate new tails. But adult birds and mammals cannot regenerate whole new organs; regeneration in these animals is mostly limited to the healing of wounds, though some internal organs, like the liver, have impressive regenerative capabilities as long as some of the organ is left to initiate the process.

It was shown in the early 1950s that regeneration of structures such as limbs normally depends, at least in vertebrates, on the nerve supply in the regenerative region. For example, if a leg of a salamander is amputated, one of the first things that happen during the healing of the wound is penetration of the wound tissue by nerve fibers growing outward from the spinal cord. Meanwhile a mound of dedifferentiated cells, called a blastema, forms at the end of the limb stump beneath the epidermis. Gradually, as mitotic activity and cellular redifferentiation take place within the blastema, the mound comes to look more and more like the normal limb bud of an embryonic salamander (Fig. 32.21). It slowly elongates, and after several weeks a distinct elbow and digits appear.

This rudimentary limb continues to grow, and its cells continue to differentiate into muscle, tendon, bone, connective tissue, etc., until finally it has become a fully functional new leg. But if the nerves leading to the stump of an amputated leg are removed, no regeneration occurs, and the stump itself shrivels and disappears. Regeneration of the leg will take place only if an abundant supply of nerve fibers is present. Marcus Singer showed that it makes no difference whether the fibers are motor or sensory, or whether they are somatic or autonomic; it is only the total number per unit volume of tissue that is important. Adequate innervation is also necessary for normal maintenance; if the nerve supply to a limb is partly destroyed (as often happens in human spinal injuries), the muscles of that limb usually begin to atrophy.

Discoveries such as these naturally raised the question whether loss of regenerative capacities in the course of development is a result of inadequate nerve supply. To test this possibility, Singer performed a revealing experiment on frogs, which he reported in 1954.

By contrast with adult frogs, young tadpoles have considerable regenerative capabilities. A young tadpole whose leg is amputated quickly grows a new one, but an older tadpole can only partly regenerate a lost leg, and an adult frog cannot regenerate a leg at all. The number of nerve fibers in a tadpole's leg does not increase as rapidly as the volume of other tissues in the leg, and this relative decline in the number of nerve fibers exactly parallels the gradual loss of regenerative capacity. Singer amputated a front leg of a frog. Then he dissected the large sciatic nerve from the corresponding hind leg without cutting its connections to the spinal cord. Next, he pulled this nerve forward under the skin of the thigh, flank, and abdomen to the stump of the amputated front leg and implanted the ends of the nerve in the stump. With this increased nerve supply, the stump proceeded to regenerate a new leg. Singer concluded that many parts of the body probably have the latent power to regenerate and need only sufficient inductive stimulation to do so.

It has generally been held that the regeneration-inducing activity of nerves results from their secretion of some tropic chemical, and that adult frogs have too few limb nerves to provide enough of the chemical unless the supply is augmented, as in Singer's experiment. Many studies, particularly of regeneration in invertebrates, have substantiated that nerve cells in regions of regeneration contain a larger than normal number of secretory granules.

An alternative explanation of regeneration induction is favored by some investigators. For example, since limb regeneration can be induced in adult frogs simply by repeated mechanical trauma to the amputation stump, and since a major effect of such trauma is production of local electric currents, these investigators suggest that the primary initiating stimulus for regeneration may be electric currents of a particular sort resulting from the injury. Indeed, formation of a blastema, with dedifferentiation and increased mitotic activity, can be stimulated by very weak, nonuniform electric currents. Several cases are now known in which injury produces electrical gradients (resulting from the flow of sodium ions from the injury to surrounding skin) that stimulate and orient normal growth. Since the magnitude of the current generated at an amputation site is directly proportional to the amount of innervation at the site, the proponents of the electric-induction hypothesis suggest that the regeneration brought about by Singer's nerve-transplant experiments may have been triggered by increased electrical stimulation rather than increased release of an inducing chemical, though such a chemical may well be involved at a later stage of the process.

Which of these hypotheses is closer to the facts cannot yet be decided; indeed, both chemical and electrical factors may play a role in regeneration. Be that as it may, limb regeneration, with its dedifferentiation and redifferentiation of cells, clearly points to the likelihood that the nuclei of differentiated animal cells have not undergone any irreversible change. Resumption of full embryonic potential seems to be more difficult for them than for the nuclei of differentiated plant cells, but it probably always remains a possibility.

THE ORGANIZATION OF NEURAL DEVELOPMENT

Most aspects of development are illustrated with particular clarity in the development of the nervous system. Nerve cells are "born," migrate to their proper places, send axons to specific target locations, and so come to

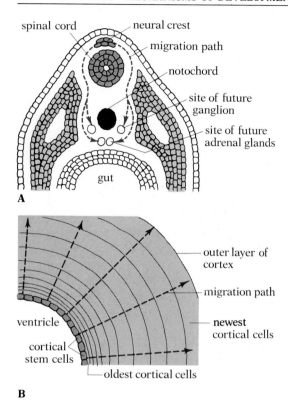

A

B

32.22 Migration of nerve cells

Before nerve cells join together in a network, many must move from their place of origin to their final location. (A) Cells from the neural crest move down and along connective-tissue filaments to positions at the future sites of spinal ganglia and the adrenal glands. (Some neural-crest cells encounter another set of connective-tissue filaments, leading to positions just under the ectoderm; cells following this pathway become pigment cells in the skin.) (B) New cells of the cerebral cortex are generated by a layer of stem cells lining each ventricle and migrate out along glia to form a layer on the outside of the cortex. Cells produced later will move through this layer to positions still farther out.

form a highly integrated functional network that is more complex than any other system in the body.

MIGRATION

The first step in the life of a newly formed presumptive neuron, like that of many other cells in a developing organism, is usually movement from where it was formed to where it is supposed to be. The cells that give rise to the retina, for instance, must migrate in the form of the optic vesicle from the developing brain to where the eyes are to be located, while the cells of the cerebral cortex must move through layers of older cells to get from the core of the brain, where they were engendered, to the outside layer of the cortex, where they belong.

The movement of neuronal cells during development is relatively well oriented; they do not move at random. Consider the cells that give rise to the adrenal glands and to the sympathetic and parasympathetic ganglia. They originate in the neural crest—the region just above the spinal cord—and migrate down to specific spots near the notochord (Fig. 32.22A). They move in the proper general direction from the outset, follow highly predictable pathways, and stop at precise spots. Similarly, cells of the cerebral cortex are generated in a layer of tissue lining cerebral vesicles (called ventricles) and then migrate outward through other, older cortical cells until they reach the outer layer of the cortex (Fig. 32.22B). The three stages the neural-crest and cortical cells undergo—determining the initial direction, following a path, and determining the stopping point—are characteristic of the migration of developing cells. At least three mechanisms appear to be involved, utilizing diffusing chemicals, cell-surface chemicals, and tactile cues.

The gradients of the diffusing chemicals help guide neuronal cells by causing them to move in an amoeboid fashion toward the source of the chemical. This chemotactic behavior, which is evident in monerans and protists (see p. 452) and is very similar to the way in which leukocytes make their way toward a cell that is releasing histamine as a result of an injury, is probably one mechanism of initial orientation—leading neural-crest cells "down," cortical cells "outward," and other cells in an anterior or posterior, or dorsal or ventral, direction. But most path-finding appears to involve the other two mechanisms—local chemical and tactile cues. Neural-crest cells (guided by filaments of connective tissue that lead around the spinal cord and past the notochord) and cortical cells (following a radial array of structural glia cells up through the cortex) apparently recognize their respective pathways by means of the surface chemicals; they partially envelop the guide cells, and then move along them, maintaining intimate tactile contact.

The goal for each type of cell is apparently indicated by another set of chemical markers. When a migrating cell encounters these on the cells it has touched, it stops moving and proceeds to form the cell-to-cell attachments that will anchor it in place. The molecular specificity of nerve-cell surface markers is so precise that each class of neurons (and as few as two cells in an entire animal can constitute a class) can be distinguished by monoclonal antibodies.

FORMATION OF AXONS AND SYNAPSES

Once a neuronal cell has reached its permanent place in the nervous system, it must send axons to specific target cells. Here again, both chemical and tactile information seem to play a role. The leading edge of the developing axon displays an unusual type of structure known as a ***growth cone*** (Fig. 32.23), first described almost a century ago by the great Spanish histologist Santiago Ramon y Cajal. The growth cone continually extends and retracts spikelike pseudopodia called filopodia that probably sample the environment for specific chemicals and for the actual presence of certain guide cells. If the cone encounters the chemical or tactile stimulus of such a guide cell (usually the axon of another nerve), it partially envelops it and grows along it.

The evidence for a chemical "stepping-stone" strategy, first proposed by Roger Sperry of the California Institute of Technology, is especially compelling in the growth of axons. Many neuronal cells, once they have reached their destinations, send axons by circuitous but predictable routes to target cells. Rearrangement of the cellular "terrain" through which the axons must travel can cause equally predictable rerouting. For example, in mice, the axons from the visual area of the thalamus (carrying information from the eyes) normally extend to layer 4 of the visual cortex (Fig. 32.24A). We might suppose that the axons simply grow radially from the thalamus through the lower layers of cortex until they encounter the specially marked cells of layer 4, but this is apparently not what happens. In a strain of mice affected by the so-called reeler mutation, in which the cortical layers are inverted, the axons from the visual area of the thalamus grow up through the inverted cortex, continue past their targets in layer 4 until they encounter some essential chemical marker in layer 6, and then turn and migrate back through the cortex to layer 4 (Fig. 32.24B). This suggests that these axons from the thalamus in mice are programed to find layer 6 of the cortex first, and then to move to layer 4.

In vertebrates, as many as a million axons may project information from the body to one of several large arrays of target cells like those found in the

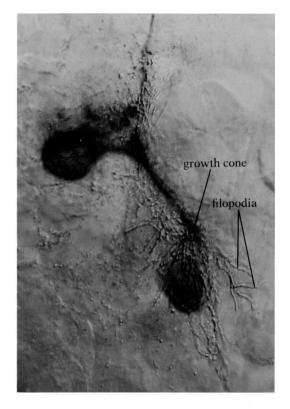

32.23   Growth cone of a developing axon in a grasshopper
The filopodia have made contact with another neuron (lower right).

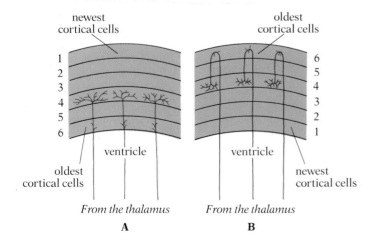

A   **B**

32.24   Connections from the thalamus to the visual cortex
(A) In normal mice the axons from the thalamus extend directly to their targets in layer 4. (B) In reeler mice, whose cortical layers are inverted, the axons travel out to layer 6 and then return to layer 4. This pattern suggests that layer 6 is a necessary intermediate landmark for the development of these axons.

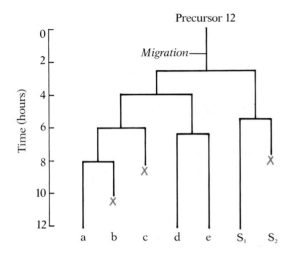

32.25 Cell death during the formation of nematode ganglion 12
This family tree of the cells in the last ganglion of a nematode suggests that the developmental program for the neurons involves a fixed series of divisions regardless of which neurons are needed by the ganglion. For example, the ganglion at the tip of the tail (illustrated here) does not need cells b and S_2, which in other ganglia connect to the next ganglion to the rear. Cell c is also unnecessary except in the middle ganglia of males, where it controls some of the animal's reproductive behavior. Nevertheless, apparently cell a can be produced only if cells b and c are produced as well.

visual, auditory, and tactile areas of the cortex. These axon–target cell connections are spatially organized (see pp. 521–22) and genetically predetermined. It seems unlikely that 10^6 specific molecular labels exist in the visual system to assure that all axons find their correct target cells. Most researchers believe that the development of the visual system depends instead on a developmental strategy similar to the one proposed by Wolpert: after various chemical and tactile cues have guided axons to the appropriate area of the brain, each axon is led close to its final target by a bicoordinate (or multicoordinate) system of chemical gradients.

CELL DEATH AND NEURAL COMPETITION

In most animal nervous systems, many more cells are born than are actually put to use. In vertebrates, for example, identical ganglia containing vast numbers of cells develop next to each of the vertebrae. And yet only the ganglia serving the many muscles and sensory receptors of the arms and legs require so many cells; in the other ganglia, the supernumerary cells die. Apparently it is easier or more efficient in some important way for the developmental program to build all segments alike initially, and then to allow functionless cells to die.

In nematodes, which are the subject of extensive developmental research, a similar pattern can be readily observed. Their simple nervous system consists of a long ventral nerve cord containing twelve ganglia. Each arises during development from one of twelve precursor cells. They mature to control rhythmic swimming movements and mating. The pattern of cell division is the same in all twelve ganglia even though the eventual organization of the ganglia may be very different. Cells that become neurons in only some (or even just one) of the ganglia are nevertheless formed in all twelve, and those not needed are then allowed to atrophy (Fig. 32.25).

An analogous phenomenon occurs in many animals at the synaptic level: many more synapses are formed than survive. When cells wired to a particular target are prevented from responding to external sensory stimuli (either by experimental manipulation or because the cells are misplaced or defective), their synapses seem to be crowded out on the target neuron by synapses from other cells which fire normally. In animals as diverse as crickets and mammals, cells that lose in this competition seem to atrophy and disappear. This process almost certainly serves to fine-tune the connections in large spatial arrays.

PART IV

THE BIOLOGY OF POPULATIONS
AND COMMUNITIES

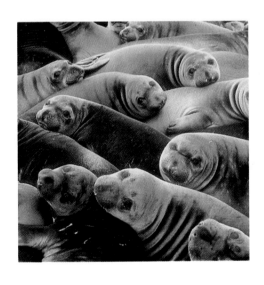

EVOLUTION: ADAPTATION

The subject of evolution was briefly introduced in Chapter 1 as a unifying principle for the study of all the topics that were to follow. That introduction was necessarily far from complete. But the succeeding chapters have provided a background that now enables us to take a closer look at the mechanisms of evolution.

EVOLUTION AS CHANGE IN THE GENETIC MAKEUP OF POPULATIONS

Fundamental to the modern theory of evolution by natural selection are four concepts: that the characteristics of living things differ between individuals of the same species (variability), that many differences are the result of heritable, genetic differences (heritability), that some differences affect how well-adapted an organism is (differential adaptedness), and that some differences in adaptedness are reflected in the number of offspring successfully reared (differential reproduction). You may remember that in Chapter 1 we distinguished evolution—the gradual change of species over time—from natural selection, which was Darwin's brilliant explanation for that change. This distinction is essential. It is also important to keep in mind that evolutionary change is not change in an individual during its lifetime, though such change is a universal and important attribute of life; it is change in the characteristics of populations over the course of many generations. An individual cannot evolve, but a population can. The genetic makeup of an individual is set from the moment of fertilization; most of the changes during its lifetime are simply changes in the expression of the developmental potential inherent in its genes. But in populations both the genetic makeup and the expression of the developmental potential can change. The former—change in the genetic makeup of a population in successive generations—is evolution. And though most evolution is by means of natural selection, we shall see that evolution can result from other processes as well.

GENETIC VARIATION AS THE RAW MATERIAL FOR EVOLUTION

A population is composed of many individuals. With rare exceptions, no two of these are exactly alike. In human beings we are well aware of the uniqueness of the individual, for we are accustomed to recognizing different people at sight, and we know from experience that each has distinctive anatomical and physiological characteristics, as well as distinctive abilities and behavioral traits. We are also fairly well aware of individual variation in such common domesticated animals as dogs, cats, and horses. But we tend to overlook the similar individual variation in less familiar species such as robins, squirrels, earthworms, sea stars, dandelions, and corn plants. Yet even though this variation may be less obvious to our unpracticed eye, it exists in all such species.

The members of a population, then, share some important features, but differ from one another in numerous ways, some rather obvious, some very subtle. It follows that if there is selection against certain variants within a population and selection for other variants within it, the overall makeup of that population may change with time.

Sources of genetic variation We saw in Chapter 30 that new genes arise in a variety of ways, including duplication of preexisting genes followed by mutation, and perhaps even through the movement and recombination of exons from different genes. We also saw in earlier chapters how new alleles of existing genes arise through mutation, giving rise to smaller-scale changes (Fig. 33.1A–B). And quite independent of changes in base sequence, we saw that even simple intergenic recombination provides almost endless genotypic variation in the population—variation resulting from meiosis, crossing over, and recombination at fertilization in biparental organisms (Fig. 33.1C). Together, the processes generating new genes, new alleles, and new combinations of alleles provide the genetic variability on which natural selection can act to produce evolutionary change.

Natural selection, however, can act on genetic variation only when it is expressed in the phenotype. A completely recessive allele never occurring in the homozygous condition would be totally shielded from the action of natural selection; only a variation that affects the way an organism actually functions can be acted on. But in fact, few if any alleles are completely recessive. Most alleles that we ordinarily call recessive probably have some very slight phenotypic effect, either because their gene products have some activity, or because the feedback control mechanisms by which repressors and inducers modulate gene transcription cannot fully compensate for the absence of a second copy of the dominant gene.

Any phenotypic variation within a population may give rise to reproductive differentials between individuals, whether or not the variation reflects corresponding genetic differences. Thus variations produced by exposure to different environmental conditions during development, or produced by

A

B

C

disease or accidents, are subject to natural selection. But though the action of natural selection on all types of variations alters the immediate makeup of a population, only its action on variations reflecting genetic differences has any long-term effect on the population—that is, any effect on subsequent generations. Variation that is exclusively phenotypic is not raw material for evolutionary change.

An understanding of genetics makes it clear that development of athletic prowess by extensive practice, or development of intellectual powers by education, or maintenance of health by correct diet and prompt medical treatment of all ailments cannot alter the genes in the germ cells. The gametes will carry the same genetic information they would have carried in the absence of athletic or mental training or proper health care. In short, selection that acts on variations produced exclusively by practice, education, diet, or medical treatment cannot bring about biological evolution (though it may lead to cultural evolution).

There is also some genetic variation that is not raw material for evolutionary change. This is variation produced by somatic mutations. It would be possible, for example, for an important mutation to occur in an ectodermal cell of an early animal embryo. All the cells descended from the mutant cell would be of the mutant type. The result might be a major change in the animal's nervous system, but the change could not be passed on to the animal's offspring. The ectodermal cells are not the ones that give rise to gametes. Mutations in somatic cells cannot alter the genes in the germ cells, which are the cells that produce the gametes. Hence selection that acts on variations produced by somatic mutations cannot result in evolutionary change in sexually reproducing organisms.

Lacking genetic data, many prominent biologists of the last century and of the early part of this century assumed that exclusively phenotypic variation could serve as evolutionary raw material; that it cannot is far from obvious to many nonbiologists even today. As we saw in Chapter 1, the theory

33.1 Sources of variation

One source of variation is spontaneous mutation. Most wisterias, like the one shown here from the Princeton University campus, have lavender flowers (A). Several decades ago, however, the famous white-flowering Eno wisteria appeared behind the Evolution, Ecology, and Behavior Building on the Princeton campus (B). The white wisteria is almost certainly the result of a spontaneous mutation in a pigment gene of an offspring of one of the many lavender wisterias nearby. Another source of variation is recombination. Offspring of the same parents frequently do not resemble each other or either of their parents. Shown here, with their mother, are the various kittens of a single litter (C).

A

B

33.2   An okapi and a giraffe, two related African herbivores
The okapi (A) and giraffe (B) are thought to have had a common ancestor with a relatively short neck. The long-necked giraffe of today can reach food unavailable to shorter individuals.

of *evolution by natural selection* proposed by Darwin and Wallace had a rival during the nineteenth century in the concept of *evolution by the inheritance of acquired characteristics*—an idea often identified with Lamarck.

The Lamarckian hypothesis was that somatic characteristics acquired by an individual during its lifetime could be transmitted to its offspring. Thus the characteristics of each generation would be determined, in part at least, by all that happened to the members of the preceding generations—by all the modifications that occurred in them, including those caused by experience, use and disuse of body parts, and accidents. Evolutionary changes would be the gradual accumulation of such acquired modifications over many generations. The classic example is the evolution of the long necks of giraffes (Fig. 33.2).

According to the Lamarckian view, ancestral giraffes with short necks tended to stretch their necks as much as they could to reach the tree foliage that served as a major part of their food. This frequent neck stretching caused their offspring to have slightly longer necks. Since these also stretched their necks, the next generation had still longer necks. And so, as a result of neck stretching to reach higher and higher foliage, each generation had slightly longer necks than the preceding generation.

The modern theory of natural selection, on the other hand, proposes that ancestral giraffes probably had short necks, but that the precise length of the neck varied from individual to individual because of their different genotypes. If the supply of food was limited, individuals with longer necks had a better chance of surviving and leaving progeny than those with shorter necks. This means, not that all individuals with shorter necks perished or that all with longer necks survived to reproduce, but simply that a slightly higher proportion of those with longer necks survived and left offspring. As a result, the proportion of individuals with genes for longer necks increased slightly in the succeeding generation. As the proportion of individuals with somewhat longer necks rose, the increased competition for food higher up on the trees resulted in a selective advantage for those with yet longer necks, and so evolution continued.

THE GENE POOL AND FACTORS THAT AFFECT ITS EQUILIBRIUM

The gene pool   To understand evolution as change in the genetic makeup of populations in successive generations, it is necessary to know something about population genetics. Our study of the genetics of individuals was based on the concept of the genotype, which is the genetic constitution of an individual. Our study of the genetics of populations will be based in a similar manner on the concept of the gene pool, which is the genetic constitution of a population. The gene pool is the sum total of all the genes possessed by all the individuals in the population.

As we saw in Chapters 24 and 25, the genome of a diploid individual contains, with a few exceptions,[1] a maximum of only two alleles of any given gene. But there is no such restriction on the gene pool of a population. It

---

[1] Because the multiple copies of the rRNA genes are present in enormous numbers, it is almost inevitable that each individual will carry more than two alleles of these genes.

can contain any number of different allelic forms of a gene. The gene pool is characterized with regard to any given gene by the frequencies of the alleles of that gene in the population. Suppose, to use a simple example, that gene *A* occurs in only two allelic forms, *A* and *a*, in a particular sexually reproducing population. And suppose that allele *A* constitutes 90 percent of the total of both alleles while allele *a* constitutes 10 percent of the total. The frequencies of *A* and *a* in the gene pool of this population are therefore 0.9 and 0.1. If those frequencies were to change with time, the change would be evolution. Evolution, as change in the genetic makeup of populations, is, more precisely then, a change in the allelic frequencies (or genotypic frequencies) within gene pools. Hence it is possible to determine what factors cause evolution by determining what factors can produce a shift in allelic frequencies.

Let us examine more carefully our hypothetical population in which allele *A* has a frequency of 0.9 and allele *a* has a frequency of 0.1. How can we calculate the genotypic frequencies that will be present in this population in the next generation? If we assume that all possible genotypes have an equal chance of surviving, this calculation is not hard to make. If the frequency of *A* in the entire population is 0.9 and the frequency of *a* is 0.1, the alleles carried by sperm and eggs will also appear at these frequencies. Using this information, we can set up a Punnett square much like those we used for crosses between two individuals:

| | | Sperm | |
		0.9*A*	0.1*a*
	0.9 *A*	0.81 *A/A*	0.09 *A/a*
Eggs			
	0.1 *a*	0.09 *a/A*	0.01 *a/a*

Notice that the only difference between this and a Punnett square for a cross between individuals is that here the sperm and eggs are not those produced by a single male and a single female, but those produced by all the males and females in the population, with the frequency of each type of sperm and egg shown on the horizontal and vertical axes respectively. Filling in the square (by combining the indicated alleles and multiplying their frequencies) tells us that the frequency of the homozygous dominant genotype (*A/A*) in the next generation of this population will be 0.81, the frequency of the heterozygous genotype (*A/a* or *a/A*) 0.18, and that of the homozygous recessive genotype (*a/a*) 0.01. Now we want to know whether the frequencies we have found will change in successive generations—in short, whether the population will evolve.

Evolution vs. genetic equilibrium   It is easy to assume that the more frequent allele (in our hypothetical case, *A*) will automatically increase in frequency while the less frequent allele (*a*) will automatically decrease in frequency and eventually be lost from the population. But the assumption that evolution is inevitable is incorrect. The rarity of a particular allele in a population does not doom it to automatic disappearance, as we can show by using the known frequencies for one generation to compute the frequencies for the next generation.

We have said that the genotypic frequencies in the gene pool of the second generation of our hypothetical population will be 0.81, 0.18, and 0.01. We can use these figures to compute the allelic frequencies of the *A* and *a* in this generation. Since the frequency of the *A/A* individuals is 0.81, the frequency of their gametes in the gene pool will be 0.81. All these gametes will contain the *A* allele. Likewise, the frequency of the gametes of the *a/a* individuals will be 0.01, and each gamete will contain an *a* allele. The frequency of the heterozygous (*A/a*) individuals is 0.18 and the frequency of their gametes in the gene pool will be 0.18, but their gametes will be of two types, *A* and *a*, in equal numbers. Hence the frequency of the *A* and *a* alleles in the gametes of the population can be calculated as follows:

Frequency of genotypes	Frequency of *A* gametes	Frequency of *a* gametes
0.81 *A/A*	0.81	0
0.01 *a/a*	0	0.01
0.18 *A/a*	0.09	0.09
	0.9	0.1

We find that the allelic frequencies are 0.9 and 0.1, the same frequencies we started with in the preceding generation.

Since the allelic frequencies are unchanged, the genotypic frequencies in the succeeding generation will again be 0.81, 0.18, and 0.01; and in turn the allelic frequencies will be 0.9 and 0.1. We could perform the same calculation for generation after generation, always with the same results; neither the genotypic nor the allelic frequencies would change. We must conclude, therefore, that evolutionary change is not usually automatic, that it occurs only when something disturbs the genetic equilibrium. This was first recognized in 1908 by G. H. Hardy of Cambridge University and W. Weinberg, a German physician, working independently. According to the **Hardy-Weinberg Law**, *under certain conditions of stability both genetic and allelic frequencies remain constant from generation to generation in sexually reproducing populations.*[2]

Let us examine the "certain conditions" that the Hardy-Weinberg Law says must be met if the gene pool of a population is to be in genetic equilibrium. They are as follows:

 1. The population must be large enough to make it highly unlikely that chance alone could significantly alter allelic frequencies.
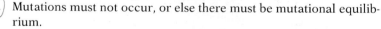 2. Mutations must not occur, or else there must be mutational equilibrium.

---

[2] For this statement to hold true for genotypes, the initial genotypic frequencies must be in equilibrium. If these frequencies are not in equilibrium, they will change in successive generations until the equilibrium is achieved. For example, if genotypes *A/A* and *a/a* were present in a population, but *A/a* were missing (let's suppose, because of human intervention), all three genotypes would appear in the next generation, and thereafter the genotypic frequencies would remain constant.

3. There must be no immigration or emigration that alters allelic frequencies in the population in question.

4. Mating must be totally random with respect to genotype.

5. Reproductive success (that is, the number of offspring and the number of their eventual offspring) must be totally random with respect to genotype.

The Hardy-Weinberg Law takes for granted the existence of both variability and heritability, but shows that these two bases of natural selection cannot alone cause evolution. Despite variability and heritability, if all five conditions of the Hardy-Weinberg Law are met, allelic frequencies will not change, and evolution cannot occur. But, in fact, these conditions are *never* completely met, and so evolution does occur. The value of the Hardy-Weinberg Law is that it provides a base line against which to judge data from actual populations. By defining the criteria for genetic equilibrium, it also indicates when a population is not in equilibrium, and helps to define the possible causative agents of evolution, some of which were by no means obvious to Darwin. The role of the investigator is then to discover the relative contribution of each factor to the evolution of a particular population. We shall look at each factor in turn, and see which ones have been shown likely to be important.

With regard to the first condition, a population would have to be infinitely large for chance to be completely ruled out as a causal factor in the changing of allelic frequencies. In reality, of course, no population is infinitely large, but many natural populations are large enough so that chance alone is not likely to cause any appreciable alteration in the allelic frequencies in their gene pools. A population with more than 10,000 members of breeding age is probably not significantly affected by random change. But allelic frequencies in small isolated populations of, say, fewer than 100 breeding-age members are highly susceptible to random fluctuations, which can easily lead to loss of an allele from the gene pool even when that allele is an adaptively superior one. In the absence of immigration or mutation, such an allele is lost forever. In such populations, in fact, there are relatively few alleles with intermediate frequencies; apparently the tendency is for most alleles either to be soon lost or to become fixed as the only allele present. In other words, small populations tend to have a high degree of homozygosity, while large populations tend to be more variable. Thus chance may cause evolutionary change in small populations, but since this change, called *genetic drift*, is not much influenced by the relative adaptiveness of the different alleles, it is essentially an indeterminate evolution, as likely to take one direction as another (Fig. 33.3). Because genetic drift can cause changes in allelic frequencies independent of natural selection, it is frequently called *neutral selection*.

The second condition for genetic equilibrium—either no mutation or mutational equilibrium—is rarely met in populations. Mutations are always occurring. There is no known way of stopping them. Most genes probably undergo mutation once every million to hundred million replications; the rate of mutation for different genes varies greatly. As for mutational equilibrium, very rarely, if ever, are the mutations of alleles for the same character in exact equilibrium: the number of forward mutations per

33.3 Possible genetic drift in a cichlid fish
Pseudotropheus zebra, one of hundreds of species of cichlid fish living in the rift lakes of Africa, is divided into numerous isolated populations, many of which have evolved their own distinctive morphology. As there is no known selective force that accounts for this diversity, the varied colors are thought by many researchers to be the result of genetic drift in each small population.

THE HARDY-WEINBERG EQUILIBRIUM

On p. 849 we used a Punnett square to calculate the frequencies of the genotypes produced by alleles A and a, whose respective frequencies in a hypothetical population were given as 0.9 and 0.1. The same results can be obtained more rapidly by using an algebraic formula.

Expansion of the binomial expression $(p + q)^2$, where p is the frequency of one allele (in our case, A) and q the frequency of the other allele (a), yields the formula for the Hardy-Weinberg equilibrium:

$$p^2 + 2pq + q^2 = 1$$

Substituting the allelic frequencies 0.9 and 0.1 for p and q respectively, we obtain

$$
\begin{array}{ccccccc}
p^2 & + & 2pq & + & q^2 & = & 1 \\
(0.9)(0.9) & + & 2(0.9)(0.1) & + & (0.1)(0.1) & = & 1 \\
0.81 & + & 0.18 & + & 0.01 & = & 1
\end{array}
$$

The three terms of the Hardy-Weinberg formula indicate the frequencies of the three genotypes:

$$
\begin{array}{lll}
p^2 & = \text{frequency of } A/A & = 0.81 \\
2pq & = \text{frequency of } A/a & = 0.18 \\
q^2 & = \text{frequency of } a/a & = 0.01
\end{array}
$$

These are, of course, the same results we obtained using the Punnett square.

In this example we have assumed that we know the allelic frequencies and want to compute the corresponding genotypic frequencies. But the Hardy-Weinberg formula permits many other sorts of calculations as well. Suppose, for example, that we know a certain disease caused by a recessive allele d occurs in 4 percent of a certain population and that we want to find out what percent are heterozygous carriers of the disease. Since the disease occurs only in homozygous recessive individuals, the frequency of the d/d genotype is 0.04. Letting q^2 stand for the frequency of d/d in the formula, we can write

$$q^2 = 0.04$$

The frequency of allele d, then, is the square root of 0.04:

$$q = \sqrt{0.04} = 0.2$$

If the frequency of allele d is 0.2, the frequency of allele D must be 0.8, because the two frequencies must always add up to 1 (that is, $p + q = 1$). Substituting the frequencies of both alleles in the Hardy-Weinberg formula, we can compute the frequencies of the genotypes:

$$
\begin{array}{ccccccc}
p^2 & + & 2pq & + & q^2 & = & 1 \\
(0.8)(0.8) & + & 2(0.8)(0.2) & + & (0.2)(0.2) & = & 1 \\
0.64 & + & 0.32 & + & 0.04 & = & 1
\end{array}
$$

Since the term $2pq$ stands for the frequency of the heterozygous genotype, which is what we wanted to know originally, our answer is that 0.32 or 32 percent of the population are heterozygous carriers of the allele d that causes the disease we are studying.

Let us now see how this type of reasoning can be applied to calculate changes in allelic frequencies when only phenotypic frequencies can be measured directly. Suppose we find, one spring, in a large population of freely interbreeding plants that 59 percent have yellow blossoms (known to be a dominant phenotype) and 41 percent white blossoms (a recessive phenotype). We return to the same place in the spring of the following year, after a very severe winter, and find 64 percent with yellow blossoms and 36 percent with white blossoms. Clearly, plants with the dominant allele survived better, but we want to know exactly how much the allelic frequencies changed.

Since white blossoms indicate a recessive phenotype, we know that the frequency of the genotype y/y was 0.41 initially. Setting $q^2 = 0.41$, we calculate that $q = 0.64$ (approximately), which means that the frequency of allele y was 0.64. The frequency of the dominant allele Y was therefore 0.36 ($1 - 0.64 = 0.36$). The next spring the frequency of white blossoms had fallen to 36 percent, which gives $q^2 = 0.36$ and $q = 0.60$. The frequency of allele y is therefore 0.60, and the frequency of Y must be 0.40. In summary, the frequency of allele y has changed from 0.64 to 0.60 and the frequency of Y from 0.36 to 0.40 in one year.

The preceding examples involved only two alleles. Similar procedures, even if considerably more complicated mathematically, can be used for situations involving multiple alleles. Thus the Hardy-Weinberg formula for a triallelic situation requires expansion of the trinomial $(p + q + r)^2$, where r is the frequency of the third allele. Similarly, a quadriallelic situation requires expansion of $(p + q + r + s)^2$.

unit time is rarely exactly the same as the number of back mutations.[3] The result of this difference is a **mutation pressure** tending to cause a slow shift in the allelic frequencies in the population. The more stable allele will tend to increase in frequency, and the more mutable allele will tend to decrease in frequency, unless some other factor offsets the mutation pressure. Eventually, of course, the frequency of the more stable allele will become so high that it will undergo the same number of mutations per unit time as the more mutable allele, despite its lower mutation rate, and equilibrium will be achieved. This requires so much time, however, that other events almost always change allelic frequencies before mutational equilibrium is reached. But even though mutation pressure is almost always present, it is seldom a major factor in producing changes in allelic frequencies in a population. Furthermore, each mutation is random; its effect is often to slow the rate of change in a character which is evolving as a result of other factors. Hence, while mutations increase variability and thus provide raw material for evolution, they seldom determine the direction or nature of evolutionary change. Like genetic drift, then, mutations, when they occur, usually result in evolution completely apart from natural selection.

According to the third condition for genetic equilibrium, a gene pool cannot accept immigrants from other populations introducing new alleles or different allele frequencies, and it cannot suffer changes in allelic frequencies by emigration. A high percentage of natural populations, however, probably experience at least a small amount of gene migration, generally called **gene flow**, and this factor, which enhances variation, tends to upset the Hardy-Weinberg equilibrium and lead to evolution. Such evolution—that is, a change in allelic frequencies—need not, at least in the short run, involve natural selection. But there are doubtless populations that experience no gene flow, and in many instances where flow does occur it is probably so slight as to be negligible as a factor causing shifts in allelic frequencies. We can conclude, therefore, that the third condition for genetic equilibrium is sometimes met in nature.

The final two conditions for genetic equilibrium in a population are that mating and reproductive success be totally random. Among the vast number of factors involved in mating and reproduction are choice of a mate, physical efficiency and frequency of the mating process, fertility, total number of zygotes produced at each mating, percentage of zygotes that lead to successful embryonic development and birth, survival of the young until they are of reproductive age, fertility of the young, and even survival of postreproductive adults when their survival affects either the chances of survival or the reproductive efficiency of the young. For mating and reproductive fitness to be totally random, all these factors must be random—that is, they must be independent of genotype, so that natural selection cannot operate. This condition is probably never met in any real population. An organism's genotype almost always influences its choice of a mate, the physical efficiency and frequency of its mating, its fertility when mated with organisms of other genotypes, the total number of zygotes it produces at each mating, the percentage of successful births of embryos,

---

[3] By convention, the mutation from the more common allele to the less common one is called the forward mutation, and the reverse is called the back mutation.

A

B

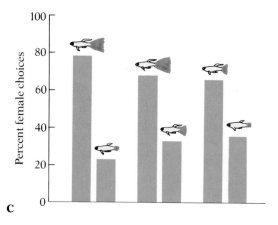

C

33.4 Female-choice sexual selection in guppies
In most species of vertebrates, including guppies, the
female chooses the mate. Male guppies exhibit great
variety in spot size, color, and location, and in tail
size and patterning (A, B). These features are largely
heritable. R. J. Bischoff, J. L. Gould, and D. I.
Rubenstein at Princeton University demonstrated that
females choose males on the basis of their
conspicuousness. Given a choice between males with
tails of two different sizes—large and small, large and
medium (the latter produced by surgical shortening),
or medium and small—females prefer to be near
males with larger tails (C). Subsequent findings
indicate that they also preferentially mate with them.

the survival and fertility of its young, and its postreproductive survival. To
take an obvious case, female guppies do not mate at random, but instead
choose males with large tails (Fig. 33.4). In short, there is probably no
aspect of reproduction that is totally uncorrelated with genotype. Nonrandom reproduction is the universal rule. And nonrandom reproduction, in
the broad sense in which the term has been defined here, is synonymous
with natural selection. Natural selection, then, is always operative in all
populations; there is always *selection pressure* acting to disturb the Hardy-
Weinberg equilibrium and cause evolution, even if selection serves merely
to limit the frequency of deleterious mutations.

In summary, the five conditions necessary to achieve the genetic equilibrium described by the Hardy-Weinberg Law correspond to five possible
causative factors in evolution. The first, genetic drift, is likely to be important in small populations but negligible in large ones, and need not be accompanied by natural selection. The second, mutation, is always at work,
but in the short run is rarely significant. Like genetic drift, mutation is independent of natural selection, at least initially. The third, immigration and
emigration, depends not only on the life history of the population, but also
on the physical environment, which affects the ease of movement between
populations; hence, immigration and emigration too can influence evolution apart from natural selection. The fourth and fifth, selection pressure
from nonrandom mating and from variations in reproductive fitness, are
almost always important in the evolutionary history of a species, though
exactly which phenotypic traits are subject to the ongoing pressures of natural selection must be separately determined for each population.

We shall concentrate on natural selection as by far the most influential
mechanism in bringing about evolution in the natural world, but it is a useful exercise to think about Lamarck's long-necked giraffes again in the context of the Hardy-Weinberg factors. We've already seen that though the
Lamarckian concept of inheritance of acquired characteristics has long
been discredited, the Darwinian explanation of natural selection in favor of
longer necks because they are more adaptive is by no means the only possibility. There are several evolutionary scenarios that might conceivably have
given rise to the present-day giraffe.

854

The most plausible alternative to natural selection is that giraffes have long necks as a result of genetic drift. Suppose that the ancestral population had a wide variety of neck lengths, or that a mutation caused a small subset of the population to have unusually long necks. If we assume that long necks are adaptively neutral—neither advantageous nor disadvantageous —there is no reason according to the Hardy-Weinberg Law for selection pressure to cause neck length in the population as a whole to change. But suppose that the population suddenly declined because of some environmental factor like disease or bad weather, so that only a few individuals survived the crisis. If, by chance, a disproportionate number of the survivors were long-necked, the trait could become established without the intervention of natural selection (Fig. 33.5). This is an example of the founder effect, which we shall examine in the next chapter in the context of speciation.

As this hypothetical example indicates, the phenomenon of evolution does not depend exclusively on any single, particular mechanism, whether natural selection, genetic drift, mutation, or migration. It also underscores the potential error of assuming that all the traits of the living things around us are necessarily the adaptive result of natural selection. Wherever possible throughout this chapter and the next, therefore, we shall consider the alternatives to natural selection. It is important to keep in mind, however, that most of the available evidence indicates that natural selection is by far the most important factor in evolution.

THE ROLE OF NATURAL SELECTION

Changes in individual allelic frequencies caused by natural selection
Let us now return for a moment to our hypothetical population in which the initial frequencies of the alleles A and a are 0.9 and 0.1 and the genotypic frequencies are 0.81, 0.18, and 0.01. According to the Hardy-Weinberg Law, in the absence of any of the five factors that cause evolution, these frequencies will not change with the passage of time. Let's look now at what is by far the most important of these factors in guiding evolution: selection pressure.

Suppose that selection acts against the dominant phenotype in our example, and that this negative selection pressure is strong enough to reduce the frequency of A in the present generation from 0.9 to 0.8 before reproduction occurs. (Of course there will be a corresponding increase in the frequency of a from 0.1 to 0.2, since the two frequencies must total 1.) Now let us set up a Punnett square and calculate the genotypic frequencies that will be present in the zygotes of the second generation:

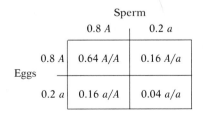

		Sperm	
		0.8 A	0.2 a
	0.8 A	0.64 A/A	0.16 A/a
Eggs			
	0.2 a	0.16 a/A	0.04 a/a

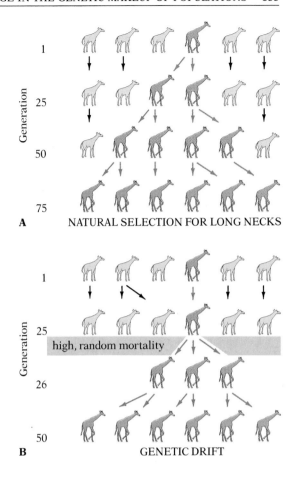

33.5 Natural selection versus genetic drift
In this hypothetical example, a small group of long-necked individuals arises in an otherwise stable population. (For simplicity, each animal in the drawing represents a fraction of the total population.) In evolution by natural selection (A), long-necked individuals become prevalent because they are able to reach more vegetation, and so survive to have proportionately more offspring in succeeding generations. In evolution by genetic drift (B), long-necked individuals become prevalent by chance: the frequency of long-necked individuals does not change until a chance catastrophe—fire, flood, heavy predation, or disease, for example—kills most members of the population. Because only long-necked individuals happen to survive, their offspring multiply to fill the habitat even though their distinctive trait is of no selective advantage.

33.6 An experiment showing that mutation for resistance to penicillin is spontaneous

Bacterial cells were cultured on a normal agar medium; many colonies developed (upper culture dish). Then a block wrapped with velveteen was pressed against the surface of the culture to pick up cells from each of the colonies. The block was next pressed against the surface of a second culture dish, containing sterile medium to which penicillin had been added; care was taken to align the transfer block and the culture dishes according to markers on the block and dishes (black lines). The cells from most of the colonies on the original dish failed to grow on the penicillin medium, but those of a few colonies (two are shown here) did grow. Had the cells of those two colonies spontaneously become penicillin-resistant before being transferred to the penicillin medium, or did exposure to the penicillin induce a mutation for resistance?

Because the transfer block had been aligned with each dish in the same way, according to the markers, it was possible to tell precisely which original colonies had given rise to the two colonies on the penicillin medium. Cells could therefore be taken from those original colonies, which had never been exposed to penicillin, and tested for resistance. They were found to be resistant. Hence the mutation for resistance must have arisen spontaneously; it was not induced by exposure to the drug.

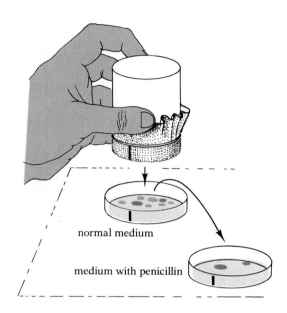

normal medium

medium with penicillin

We find that the genotypic frequencies of the zygotes in the second generation are different from those in the parental generation; instead of 0.81, 0.18, and 0.01, the frequencies are 0.64, 0.32, and 0.04. If selection now acts against the dominant phenotype in this generation, and thereby again reduces the frequency of *A*, the genotypic frequencies in the third generation will be different from those of both preceding generations; the frequency of *A/A* will be lower and that of *a/a* higher. If this same selection pressure were to continue for many generations, the frequency of *A/A* would fall to a very low level and the frequency of *a/a* would rise to a very high one. Thus natural selection would have caused a change from a population in which 99 percent of the individuals showed the dominant phenotype and only 1 percent the recessive phenotype to a population in which very few showed the dominant phenotype and most showed the recessive phenotype. This evolutionary change from the prevalence of one phenotype to the prevalence of another would have occurred without the necessity of any new mutation, simply as a result of natural selection.

Instead of dealing with a hypothetical example, let us consider an actual situation in which selection has produced a radical shift in allelic frequencies. Soon after the discovery of the antibiotic activity of penicillin, *Staphylococcus aureus* (a bacterial species that can cause numerous infections, including boils and abscesses) quickly developed resistance to the drug. Higher and higher doses of penicillin were necessary to kill the bacteria, and the resistant bacteria became a serious problem in hospitals. Clearly, under the influence of the strong selection exerted by the penicillin, the bacterial population has evolved. But many studies have shown that the drug itself does not induce mutations for resistance; it simply selects against susceptible bacteria by killing them (Fig. 33.6). Apparently some genes determining metabolic pathways that confer resistance to penicillin are already present in low frequency in most populations, having arisen earlier as a result of random mutations. Individuals possessing these genes are thus ***preadapted*** to survive the antibiotic treatment, and, since it is they that reproduce and perpetuate the population (the susceptible individuals having been killed), the next generation shows a marked resistance to penicillin. If such genes were not already present in a population exposed to penicillin, no cells would survive and the population would be wiped out.

The primacy of selection pressure does not mean that new mutations cannot improve the resistance; in fact, continued selection with penicillin usually leads to gradually increased resistance, which is almost certainly in part a result of the differential survival of individuals with new mutations that enhance resistance. But it is purely a matter of chance that mutations beneficial in an environment containing penicillin should arise when this drug is administered; the same mutations arise at the same rate in the absence of penicillin, but are not selected for. They nevertheless persist if they have no adverse effect.

Evolution of drug resistance in bacteria is not entirely comparable to evolution in biparental organisms, because intense selection can change gene frequencies much more rapidly in haploid asexual organisms than in biparental ones. The recombination that occurs at every generation in a biparental species often reestablishes genotypes eliminated in the previous generation; this does not happen in asexual organisms. Nevertheless, even very small selection pressures can produce major shifts in gene frequencies in biparental populations over an evolutionarily brief period. J. B. S. Haldane has shown that if the individuals carrying a given dominant allele consistently benefit by as little as 0.001 in their capacity to survive (that is, if 1,000 *A/A* or *A/a* individuals survive to reproduce for every 999 *a/a* individuals that survive to reproduce), then the frequency of the dominant allele could increase from 0.00001 to 0.99 in only 23,400 generations if no other factors intervened. In other words, a positive selection pressure of only 0.001 could cause a very rare allele to become very common in only 23,400 generations. "Only" in conjunction with 23,400 generations may sound incongruous, but remember that many plants and animals have at least one generation a year—*Drosophila*, for example, has more than 30. In very few species is the generation time more than 10 years (humans are among the few exceptions). Hence 23,400 generations often means less than 23,400 years and rarely more than 234,000 years. Both of these are relatively short time spans when measured on the geologic time scale. Recent evidence suggests that many selection pressures in nature are much larger than 0.001; hence major changes in allelic frequencies sometimes probably take less than a century, perhaps even less than 50 years.

Directional selection of polygenic characters   So far, we have discussed idealized situations involving only two clearly distinct phenotypes determined by two alleles of a single gene. But in reality, as we saw in Chapters 24 and 25, the vast majority of characters on which natural selection acts are influenced by many different genes, most of which have multiple alleles in the population; the expression of many characters, moreover, is influenced considerably by environmental conditions. Consequently such characters usually show variation with a range of values, which often tend to have a frequency distribution that, when graphed, approximates the so-called normal, or bell-shaped, curve (Fig. 33.7).

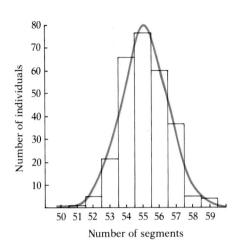

33.7. Frequency distribution of number of body segments in a population of the milliped *Narceus annularis*
The pattern of variation in number of segments (shown by the vertical bars) approximates, but does not exactly fit, the bell-shaped normal curve of probability.

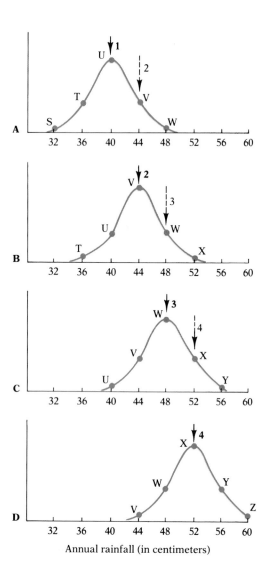

Annual rainfall (in centimeters)

If the environmental conditions should change, creating a consequent shift in the selection pressure, we would expect the curve of phenotypic variation to shift as a result of changing allelic frequencies. To illustrate this, let us assume that the conditions under which a certain plant grows best are genetically determined. The hypothetical case of such a plant is presented in Figure 33.8. The first curve (Fig. 33.8A) shows the annual rainfall at which the various plants in a particular population would grow best. The actual rainfall in the area where this population occurs averages 40 cm, as indicated by arrow 1, though it will vary around this mean from year to year. The population contains a very few plants (S) that would grow best if the annual rainfall were about 32 cm and a very few (W) that would grow best if the rainfall were about 48 cm. Plants that would grow best if the annual rainfall were about 36 cm (T) or 44 cm (V) are fairly common in the population. This phenotypic diversity is maintained because of the variability of the rainfall about the mean, so that in some unusually dry years, for example, the group S plants would have the advantage. In any given year, then, there is selection for an optimum phenotype, but the optimum varies. Since the average rainfall is 40 cm, the U plants are best adapted in the long run, and so are the most common.

Now let us suppose that the average annual rainfall in the area in question slowly increases over a period of years until it is 44 cm (arrow 2). Under these new environmental conditions, the V plants (which grow best when the annual rainfall is about 44 cm) will do better than before; a higher percentage of them can be expected to survive and reproduce, and their frequency should increase. Similarly, the W plants (which grow best when the annual rainfall is about 48 cm) will now grow better than formerly, and they too should increase in frequency. Conversely, the T plants and the U plants will not grow as well as formerly, and they should decrease in frequency. And the S plants, only a few of which managed to survive when the

33.8 Evolutionary change of a hypothetical plant population in response to directional selection by changing rainfall
The various phenotypes (S, T, U, etc.) in a hypothetical plant population reflect different genetically determined growth responses to annual rainfall. The four curves show the frequency distributions of the phenotypes at different times (A, B, C, D), under conditions of average annual rainfall indicated by the solid black arrows. The first curve (A) shows the frequency distribution of phenotypes over a long period of annual rainfall averaging about 40 cm. When the annual rainfall slowly increases to 44 cm (broken arrow 2), the frequency distribution of phenotypes slowly shifts to a new position (B), where it tends to remain as long as the average annual rainfall continues at 44 cm (solid arrow 2). But if rainfall again slowly increases, this time to 48 cm (broken arrow 3), the frequency distribution again shifts (C). Another increase in rainfall to 52 cm (arrow 4) results in still another shift in the population (D). The changing rainfall has exerted directional selection on the plant population, shifting the curve of relative phenotypic frequencies more and more to the right.

annual rainfall was 40 cm, would now be so poorly adapted to the prevalent conditions that none could survive. These changing frequencies, produced by the shift in the selection operating on the population, would give rise to the new curve shown in Figure 33.8B.

If the average annual rainfall continues to increase over a period of years until it reaches 48 cm (arrow 3), the W plants and X plants should increase in frequency, the U plants and V plants should decrease in frequency, and the T plants should disappear; these shifts would give rise to the curve shown in Figure 33.8C. If the average annual rainfall then slowly increases to 52 cm (arrow 4), it should cause further shifts in frequencies, producing the curve shown in Figure 33.8D.

The changing environmental conditions, then, have given rise to what is called *directional selection*, which has caused the population to evolve along a particular functional line. If the population had not been sufficiently variable genetically to have the potential to change when the environment changed, it would have been much reduced, or it might even have become extinct. Before the average rainfall began to increase, there was selection each year in one direction or another, but the adaptive optimum varied, depending on whether the rainfall that year was greater or less than normal. This led to the distribution of plants centered on the 40-cm-per-year average.

The creative role of natural selection   Notice that in our hypothetical plant population the directional selection did not just shift the peak of the curve to the right, as might happen when the variation in the gene pool is small. Instead, it caused the entire curve—the extremes as well as the peak —to shift to the right. The shift was eventually so great, in fact, that a class of plants (X) not even present in the original population became the largest class. But, you might ask, if X plants, Y plants, and Z plants were not present initially, how did they arise in the descendant populations? One possibility is that, purely by chance, new genes or alleles arose that made their possessors grow better in wetter habitats; such novel genes or alleles would have been strongly selected for and would have spread rapidly through the population. But if moisture preference is influenced by many different genes, as is highly likely, new phenotypes such as X, Y, and Z could arise without the necessity of any new genetic variation, simply through the separate increase in frequency of particular alleles already present, which would then be more likely to occur together and produce a new phenotype.

Haldane has calculated how long it would take for a new phenotype to be created in this way. He has shown that if one particular allele of each of 15 independent genes is present in 1 percent of the individuals of a population, then all 15 alleles will occur together in only one of 10^{30} individuals. But there has never been a population of higher organisms containing anywhere near 10^{30} individuals (in fact, individual higher plants have never totaled 10^{30} at any time during the history of life on earth). Hence the chances that all 15 alleles would occur together in even one individual in a real population are exceedingly small—zero for all practical purposes. In other words, the phenotype produced by the combined action of all 15 alleles probably would not exist in the population. But, according to Haldane, if there is moderate natural selection for each of the 15 alleles, it would take

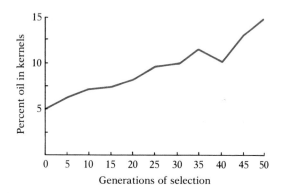

33.9 Results of 50 generations of selection for high oil content in corn kernels

only about 10,000 generations for the frequency of each allele to increase from 1 percent to 99 percent. Once each allele is present in 99 percent of the population, 86 percent of the individuals in it will have all 15 alleles and hence will show the phenotype that was previously nonexistent in the population. Thus selection, even in the absence of new genetic variation, can produce new phenotypes by combining old genes in new ways.

An actual illustration of the sort of change outlined in the preceding hypothetical example is provided by a long-term selection experiment performed on corn by agronomists at the University of Illinois. These agronomists selected for high oil content of the corn kernels; the directional selection was continued for 50 generations. There was a steady increase in oil content throughout most of this period (Fig. 33.9). The kernels of the original stock of corn plants averaged about 5 percent oil; those of the plants in the 50th generation after selection averaged about 15 percent (higher than any individuals in the first generation), and there was no indication that a maximum had been reached.

That this steady change over the course of 50 generations must have resulted primarily from the formation of new genetic combinations through selection rather than from the occurrence of a series of new mutations can be seen from a few simple calculations. The agronomists raised between 200 and 300 corn plants in each generation. In 50 generations, then, they raised between 10,000 and 15,000. But the usual rate of mutation per gene in corn is never greater than one in 50,000 plants, and it is usually lower. Hence it is unlikely that even one mutation contributing to an increase in oil content occurred in any particular gene affecting this phenotype during the experiment. And if several genes are involved in oil production, it is unlikely that a series of mutations of these genes could have been a major factor. The gradual increase in oil content during the 50 generations of directional selection must have resulted largely or entirely from the formation of new genetic combinations.

The new genetic combinations produced by selection sometimes have the result of changing formerly recessive alleles into dominant ones. As we saw in Chapter 24, many alleles are not automatically either dominant or recessive; the genetic background against which they must function (that is, the other genes present in the same individual) may determine their activity. When that background changes through shifting allelic frequencies, the enzyme-making activity of individual alleles may also change, because many genes influence the activity of other genes to some extent. We saw earlier that most new mutant alleles are recessive, and that they may be carried in low frequency in the population indefinitely without being expressed phenotypically. When, generations later, selection alters the genetic background in such a manner that the allele becomes dominant (that is, more active), the phenotype it produces (which is new to the population, even though the allele is not) provides new variation as raw material upon which selection may act. Thus the evolution of dominance as a result of selection is in a very real sense a creative process.

To summarize: In biparental populations selection—whether natural or artificial—determines the direction of change largely by altering the frequencies of alleles (and genes) that arose through duplication, transposition, and random mutation many generations before, thus establishing new

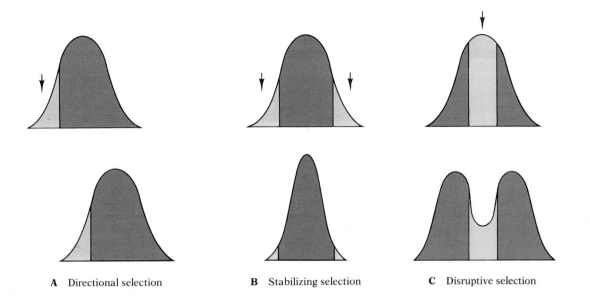

A Directional selection   **B** Stabilizing selection   **C** Disruptive selection

genetic combinations and gene activities that produce new phenotypes. The processes that create new alleles and genes are not usually major *directing* forces in evolution; the principal evolutionary role of genetic changes consists in replenishing the store of variability in the gene pool and thereby providing the potential upon which future selection can act.

Disruptive selection   It can sometimes happen that a polygenic character of a population is subject to two (or more) directional selection pressures favoring the two extremes of the distribution. Suppose, for example, that a certain population of birds shows much variation in bill length. Suppose, further, that as conditions change there are increasingly good feeding opportunities for the birds with the shortest bills and also for those with the longest bills, but decreasing opportunities for birds with bills of intermediate length. This might happen if the population of plants producing fruits suitable for intermediate-sized bills began to decline, or if a competing species more efficient at harvesting such fruits were to immigrate and become established. The effect of such selection, at least in the short run, would be to divide the population into two distinct types, one with short bills and one with long bills. The combined action of the opposing directional pressures would thus disrupt the smooth curve of phenotypes in a population (Fig. 33.10C); hence this sort of selection is called ***disruptive selection***.

Stabilizing selection   Most often, when a polygenic character in a population is subject to two or more opposing directional selection pressures operating simultaneously, the pressures select against the two extremes of the distribution—plants that are too tall to resist high winds, for example, and those of the same species that are so short that they are shaded by other plants, and so lose essential sunlight. Similarly, unusually severe winter

33.10 Directional, stabilizing, and disruptive selection compared
Each graph indicates the relative abundance of individuals of various heights in a population. In each instance the original condition of the population is shown above, the later condition, after the specified selection, below. (A) Directional selection acts (arrow) against individuals exhibiting one extreme of a character (here the shortest individuals —represented by the gray area under the upper curve). The eventual result (bottom curve) is that the distribution of heights in the population has shifted to the right, indicating that the population has evolved in the direction of greater height.
(B) Stabilizing selection acts against both extremes, culling individuals that deviate too far from the mean condition and thus decreasing diversity and preventing evolution away from the standard condition. (C) By contrast, disruptive selection acts against individuals in the mid-part of a distribution, thereby favoring both extremes; in our example both the shortest and the tallest individuals would be favored, but individuals of medium height would be selected against. The result is a tendency for the population to split into two contrasting subpopulations.

33.11 Male and female guppies
The showy coloration of the male (top) contrasts with the drab gray of the female.

storms frequently kill a disproportionate number of the largest and smallest birds in a population. When selection operates against individuals at the two ends of the distribution for a polygenic trait, the process is called ***stabilizing selection*** (Fig. 33.10B).

Stabilizing selection goes on all the time, operating on a much larger scale than that of the single characters we have been discussing, and as a result plays an extremely important conservative role. Each species, in the course of its evolution, comes to have a constellation of genes that interact in very precise ways in governing the developmental, physiological, and biochemical processes on which the continued existence of the species depends. Anything that disrupts the harmonious interaction of its genes is usually deleterious to the species. But in a sexually reproducing population, favorable groupings of alleles tend to be dispersed and new groupings formed by the recombination that occurs when each generation reproduces. Most of these new groupings will be less adaptive than the original grouping (though a few may be more adaptive). And the vast majority of new genetic variations tend to disrupt rather than enhance the established harmonious relationships among the genes. If unchecked, forces like recombination and random mutation would therefore tend to destroy the favorable genetic groupings on which the success of members of the population rests. Selection, by constantly acting to eliminate all but the most favorable genetic combinations, counteracts the disrupting, disintegrating tendency of recombination, mutation, and the like, and is thus the chief factor maintaining stability where otherwise there would be chaos.

Effective selection pressure as the algebraic sum of numerous separate selection pressures Probably many characteristics benefit the organisms that possess them in some ways and harm them in others. The evolutionary fate of such characteristics depends on whether or not the various positive selection pressures produced by their advantageous effects outweigh the negative selection pressures produced by their harmful effects. If the algebraic sum (an addition taking into account plus and minus signs) of all the separate selection pressures is positive, the trait will increase in frequency, but if the algebraic sum is negative, the trait will decrease in frequency.

As an example of the determination of a complex character having both beneficial and deleterious effects, let us consider the selection pressures on showiness in male guppies, a fish native to the freshwater streams of South America (Fig. 33.11). As we have already seen, males with larger tails are more likely to mate, evidently because females prefer such males. John Endler of the University of Utah has found that females also prefer males with large, bright spots. Competition for females should result in very showy males, since males with alleles for large tails and bright spots will leave more offspring; and, indeed, males in large aquarium populations become increasingly showy with succeeding generations. As we saw in Chapter 22, this process, which is acknowledged by most biologists, is known as female-choice sexual selection.

No such selection pressure has operated on the females. These gray, nondescript fish closely resemble one another and are much less easily seen by predators than the males, whose bright markings and showy tails make them more subject to predation. This liability causes strong selection against such features; indeed, Endler has shown that in laboratory situations, predators capture the showiest males first. As we might expect, males found in the wild are much less conspicuous than their counterparts in predator-free aquaria. The showiness of males, then, exemplifies stabilizing selection—a balance between selection for greater showiness (by the females, who confer reproductive success on the most "attractive" males) and selection for inconspicuousness (by the predators, who can terminate any prospect for further mating if a male is too visible).

Just as polygenic traits often have both advantageous and disadvantageous effects, the alleles of a single gene also usually have multiple effects (pleiotropy), and it is most unlikely that all of them will be advantageous. Whether an allele increases or decreases in frequency is determined, as in the case of polygenic traits, by whether the sum of the various selection pressures favoring it is greater or smaller than the sum of the selection pressures acting against it.

Many instances are known in which the effects of a given allele are more advantageous in the heterozygous than in the homozygous condition. In some parts of Africa, for example, the allele for sickle-cell anemia occurs in humans much more often than we might expect in view of its highly deleterious effect when homozygous (Fig. 33.12). This is because the allele, when heterozygous, confers on the possessor a partial resistance to malaria. The equilibrium frequency of the sickle-cell allele is thus determined by at least four separate selection pressures: (1) the strong selection against the recessive homozygotes, who suffer the full debilitating effects of sickle-cell anemia; (2) the weaker selection against the heterozygotes as a result of their mild anemia; (3) the selection against the dominant homozygotes, who are more susceptible to malaria; and (4) the fairly strong selection favoring the heterozygotes as a result of their resistance to malaria.

Balanced polymorphism and the maintenance of genetic variability

Polymorphism is the occurrence in a population of two or more distinct forms, or morphs, of a genetically determined character. These phenotypes do not grade into one another like those represented by a bell-shaped curve for height, where "short" and "tall" designate not distinct morphs, but rather the extremes of a continuum. Instead, individuals fall into separate categories known as *discontinuous phenotypes*, and intermediates are rare or absent. For example, Mendel's peas were polymorphic with regard to the color of their flowers: some plants had red flowers, others white, but none had pink. Human populations are polymorphic with regard to blood groups: the same population usually includes type A, type B, type AB, and type O individuals. Polymorphism is common in wild populations too: several species of snails, for instance, occur in banded and unbanded forms; the red fox has both red and silver-colored morphs; the fish *Xiphophorus maculatus* shows a variety of patterns of spotting, of which three are illustrated in Figure 33.13. In some instances the genetic basis of the polymorphism is known, as in Mendel's peas and human blood groups. But in many other instances, especially when the traits are polygenic, it is not known.

When the relative frequencies of the different morphs in a population are stably balanced over time, we speak of *balanced polymorphism*. Sometimes the balance is maintained because the polymorphism itself is advanta-

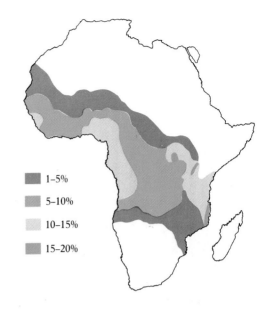

▪	1–5%
▪	5–10%
▪	10–15%
▪	15–20%

33.12   The distribution of sickle-cell anemia in Africa

The various colors indicate the percentage of the population in each area that suffers from the disease.

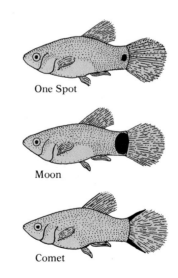

One Spot

Moon

Comet

33.13   Polymorphism in the fish *Xiphophorus maculatus*

Notice the differences in the black spots at the base of the tail.

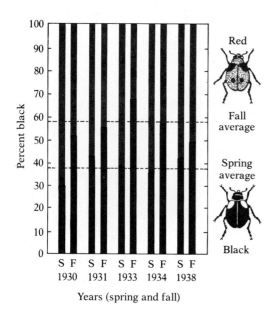

33.14 Polymorphism in the ladybird beetle *Adalia bipunctata*

The frequency of black morphs (black bars) is much higher in the fall than in the spring; conversely, the red morphs (red bars) are much more frequent in the spring than in the fall. Apparently the black morphs are better adapted for life in the summer, and the red morphs are better adapted for life in the winter.

geous. Thus, if a polymorphic species lives in an environment subdivided into many local areas where different conditions prevail, one of the morphs may do better in one area, others may do better in other areas. If the various morphs are all present among an individual's descendants, these can exploit more completely the subdivisions of the variable environment. Sometimes one morph is adaptively superior at one time of year, and another is superior at another time of year (Fig. 33.14); hence an individual's descendants may have a better chance of survival if they are polymorphic than if all belong to a single form. Polymorphism of this type, accordingly, may be the result of a kind of disruptive selection; each of several contrasting forms is being selected for.

But there are many instances of balanced polymorphism in which some of the contrasting morphs, instead of being themselves selected for, are the unavoidable by-products of selection for a heterozygous phenotype. As we saw with sickle-cell anemia in Africa, the heterozygotes are sometimes better adapted than the homozygotes—a condition known as ***heterozygote superiority*** (or sometimes heterosis or overdominance).

Heterozygote superiority favors balanced polymorphism, because it makes for retention in the population of both alleles of a given gene at frequencies higher than would be predicted from the selection acting on the homozygous phenotypes. Thus, if A/a individuals are adaptively superior to both A/A and a/a individuals, both allele A and allele a will be retained in fairly high frequency in the population; neither allele will be eliminated, as might tend to happen if one of the homozygotes were far superior. Therefore all three possible genotypes—A/A, A/a, and a/a—will occur frequently in each generation, and if each of these produces a noticeably different phenotype the population will be polymorphic. The relative frequencies of A and a, and of the three morphs they produce, will be determined, as in sickle-cell anemia, by the balance between the several selection pressures acting on the system.

ADAPTATION

Every organism is, in a sense, a complex bundle of immense numbers of adaptations. We have already examined a host of adaptations in earlier chapters—adaptations concerned with nutrient procurement, gas exchange, internal transport, regulation of body fluids, hormonal and nervous control, effector activity, reproduction, development, and behavior. We should pause here to say more explicitly what is meant by adaptation.

In biology, an adaptation is any genetically controlled characteristic that increases an organism's fitness. ***Fitness***, as the term is used in evolutionary biology, is an individual's (or allele's or genotype's) probable genetic contribution to succeeding generations. An adaptation, then, is a characteristic that enhances an organism's chances of perpetuating its genes, usually by leaving descendants. Notice that we did not say adaptations increase the organism's chances of surviving, as is sometimes erroneously stated. While an adaptation, if it is to enhance the production of descendants, will ordinarily also enhance prereproductive survival, it will not necessarily enhance postreproductive survival. In many species it is, in fact, adaptive for the adults to die soon after they have completed reproduction, as we shall see presently in the case of viceroy butterflies.

Adaptations may be structural, physiological, or behavioral. They may be genetically simple or complex. They may involve individual cells or subcellular components, or whole organs or organ systems. They may be highly specific, of benefit only under very limited circumstances, or they may be general, of benefit under many and varied circumstances.

A population may become adapted to changed environmental conditions with extreme rapidity. A good example is provided by a study published in 1937 by W. B. Kemp of the Maryland Agricultural Experiment Station. The owner of a pasture in southern Maryland had seeded the pasture with a mixture of grasses and legumes. Then he divided the pasture into two parts, allowing one to be heavily grazed by cattle, while protecting the other from the livestock and leaving it to produce hay. Three years after this division, Kemp obtained specimens of blue grass, orchard grass, and white clover from each part of the pasture and planted them in an experimental garden where all the plants were exposed to the same environmental conditions. He found that the specimens of all three species from the heavily grazed half of the pasture exhibited dwarf, rambling growth, while specimens of the same three species from the ungrazed half exhibited vigorous, upright growth. In only three years the two populations of each species, known to have been identical initially because one batch of seed was used for the entire pasture, had become markedly different in their genetically determined growth pattern. Apparently the grazing cattle in the one half of the pasture had devoured most of the upright plants, and only plants low enough to be missed had survived and set seed. There had been, in short, intense selection against upright growth in this half of the pasture and correspondingly intense selection for the adaptively superior dwarf, rambling growth. In the half of the pasture where there was no grazing, by contrast, upright growth was adaptively superior, and dwarf plants were unable to compete effectively.

Experimental tests of adaptiveness are usually not as easy to design as straightforward laboratory evaluations of cause and effect. There may be several alternative explanations for why a trait may be adaptive. It is important to remember, furthermore, that not all characters present in an organism need be adaptive in the first place. The devising of simple experiments that will generate compelling evidence requires ingenuity and persistence. The ethologist Niko Tinbergen was one of the first scientists to insist on putting his evolutionary theories to the test. When Tinbergen wondered, for instance, why ground-nesting gulls meticulously remove broken eggs from their nests (Fig. 33.15A), he formulated a variety of possible explanations: the damaged eggs might be a source of disease, which would infect the newly hatched young; the jagged edges of the broken shells, too, might endanger the chicks; the unrelieved white of the exposed interiors of the broken shells might nullify the camouflage provided by the olive-drab exteriors, and so attract predators.

Tinbergen solved this problem (and many similar ones) in two steps: first, species comparisons; then, experimental tests. The species-comparison step allowed him to isolate the most likely hypothesis for testing. In the case of the damaged eggs, he observed that kittiwake gulls, which live on

A

B

33.15 Eggshell removal by gulls
Ground-nesting gulls remove broken eggshells from their nest and carry them at least a meter away (A), while cliff-nesting species like the kittiwake do not remove eggshells (B). The nesting habit of the kittiwakes protects them from predation, while ground-nesting gulls must instead keep their nests inconspicuous.

A

B

C

D

33.16 A variety of plant pollinators
(A) A bat (*Leptonycteris*) at a flower, its face liberally
dusted with pollen. (B) A honey possum feeding on
nectar from flowers. (C) A bird, called the honeyeater,
probing its bill into a flower as pollen adheres to its
face. (D) A butterfly inserting its long proboscis into a
flower to obtain nectar. The proboscis is likely to be
dusted with pollen, which the butterfly may carry to
the next flower it visits.

cliffs and are therefore exposed to virtually no predation, do not remove
broken eggshells (Fig. 33.15B). Since disease and cuts, if they posed significant threats, ought to be as dangerous for the kittiwakes as for ground-nesting gulls, Tinbergen decided to test the predation hypothesis first. He did
this by setting out an array of nests containing normal eggs, with broken
eggs placed at varying distances from them. The results were rewardingly
clear-cut: broken eggshells nearby called the attention of predators to an
otherwise inconspicuous nest (Table 33.1). By such experiments Tinbergen
provided new insight into the evolution of adaptive behavior. In the absence of direct experimental tests or compelling species comparisons,
however, "explanations" of adaptiveness remain highly speculative, and
should simply serve as working hypotheses to stimulate research.

Let us now look at some other particularly striking examples of adaptations, which help clarify the processes of evolution.

ADAPTATIONS FOR POLLINATION IN FLOWERING PLANTS

Flowering plants depend on external agents to carry pollen from the male
parts in the flowers of one plant to the female parts in the flowers of an-

TABLE 33.1 *Survival Value of Eggshell Removal*[a]

Distance from egg to eggshell (cm)	Percent eggs taken by predators
5	65
15	42
100	32
200	21
no eggshell	22

[a] N. Tinbergen et al., Egg Shell Removal by the Black-headed Gull, *Behaviour*, vol. 19, 1963.

other plant (Fig. 33.16). The flowers of each species are adapted in shape, structure, color, and odor to the particular pollinating agents on which they depend, and they provide an especially clear illustration of the evolution of adaptedness. Evolving together, the plants and their pollinators became more finely tuned to each other's peculiarities—a process often termed *coevolution.* We have already encountered examples of coevolution, like the ruminant mammals and their symbiotic microorganisms.

There are indeed striking correspondences between the pollinators and the species they pollinate. Bees are attracted innately to bright colors, ultraviolet bull's-eye patterns, and sweet, aromatic, or minty odors; they are active only during the day, and they usually alight on a petal before moving into the part of the flower containing the nectar and pollen. Bee flowers have showy, brightly colored petals that are usually blue or yellow but seldom red (bees can see blue or yellow light well, but they cannot see red at all); indeed, most bee-pollinated flowers have a UV bull's-eye, a sweet, aromatic, or minty fragrance, daytime opening or nectar production, and a special protruding lip or other suitable landing platform. But these observations tell us only that correspondences exist between certain flowers and the preference of bees. They do not indicate how these correspondences may have come about: whether the preferences of the pollinators provided the exclusive selective force on flower morphology and odor, or whether, instead, early flowers provided the selection pressures leading to the innate preferences of modern bees, or whether both factors have been at work. A look at other species of pollinators provides some clues.

Hummingbirds, for example, can see red well but blue only poorly; they have a weak sense of smell; and they ordinarily do not land on flowers, but hover in front of them while sucking the nectar. Flowers pollinated primarily by hummingbirds are usually red or yellow, are nearly odorless, and lack any protruding landing platform. Since flowers of the same genus can have very different morphologies to suit different pollinators (Fig. 33.17), it is probably the flowers that have done most of the adapting. However, pollinators have probably been adapting to flowers too, though to a lesser extent; different species of bees, for example, can have very different tongue lengths, suitable for different flower morphologies.

This pattern of coadaptation between pollinators and flowers extends to other nectar-feeding species as well. For example, in contrast to both bees and hummingbirds, moths and bats are generally most active at dusk and during the night, and the flowers they pollinate are mostly white and are

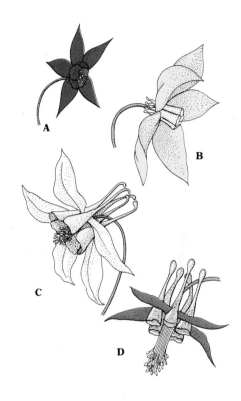

33.17 Characters of columbine flowers correlated with their pollinators
(A) *Aquilegia ecalcarata,* pollinated by bees.
(B) *A. nivalis,* pollinated by long-tongued bees.
(C) *A. vulgaris,* pollinated by long-tongued bumble bees.
(D) *A. formosa,* pollinated by hummingbirds. The length and curvature of the nectar tubes of the flowers are correlated with the length and curvature of the bees' tongues and the hummingbirds' bills.

A   **B**

33.18 Pollinator tracking in scarlet gilia (*Ipomopsis aggregata*)
During the early part of the summer, when these plants are pollinated by both hummingbirds and hawk moths, they produce about twice as many red flowers (preferred by hummingbirds) as white flowers (A). Later in the summer, as the hummingbird populations leave the area, scarlet gilia plants shift over to producing pink and—especially—white flowers, which are more attractive to the sole remaining pollinator, a local species of hawk moth (B).

33.19 An orchid flower that resembles a fly
The flowers of this species (*Ophrys insectifera*) look enough like female flies to attract some male flies to land on them. The males thus become dusted with pollen, which they may carry to other flowers.

open only during the late afternoon and night. These flowers often have a heavy fragrance that helps guide the moths and bats to them.

Moths play a role in a particularly interesting adaptation of plants to their pollinators. The flowers produced by scarlet gilia plants near Flagstaff, Arizona, range from red through pink to white. The dark-red flowers are most effective in attracting hummingbirds, but these pollinators emigrate a month after the season begins; the white flowers are most effective in attracting hawk moths, the pollinators available throughout the blooming season. The plants compensate for this shift in relative pollinator abundance by doubling the production of white flowers late in the season, while at the same time ceasing to produce any red blossoms (Fig. 33.18).

Unlike bees and moths, the short-tongued flies (which feed primarily on carrion, dung, humus, sap, and blood) are attracted by rank rather than sweet odors, and they rely very little on vision in locating food. The flowers of plants that depend on these flies for pollination are usually dull-colored and very ill-smelling. These flowers are sometimes shaped in such a way that they temporarily entrap the flies that enter them; thus they ensure that the flies become covered with pollen before they escape and fly to another flower. Trapping mechanisms are also common in flowers pollinated by beetles.

A particularly dramatic example of adaptation for pollination is seen in some species of orchids, where the flowers resemble in shape, odor, and color the females of certain species of wasps, bees, or flies (Fig. 33.19). The male insect is stimulated to attempt to copulate with the flower and becomes covered with pollen in the process. When he later attempts to copulate with another flower, some of the pollen from the first flower is deposited on the second. So complete is the deception that sperm have actually been found inside the orchid flowers after a visit by the male insect.

Flowers pollinated by wind rather than animals characteristically lack bright colors, special odors, and nectar. In fact, most of them have no petals, and their sexual parts are freely exposed to the air currents. The pollen grains produced by these flowers are particularly small and light, and it is not unusual for them to be blown hundreds of miles.

We see from this species comparison, then, that the characteristics of flowers are not simply pleasing curiosities of nature that serve no practical function. They are important adaptations that evolved in response to fundamental selection pressures.

DEFENSIVE ADAPTATIONS OF ANIMALS

Defensive secretions of arthropods Many arthropods possess glands whose secretions act as repellents against predators. In some species the secretions are merely released as a liquid ooze when the animal is disturbed, while in others the secretion is forcibly expelled as a spray that may be aimed very precisely toward the source of the disturbance. The secretions are usually odorous and irritating, particularly if they hit a sensitive part of the predator, such as the mouth, nose, or eyes. For example, the bombardier beetle *Brachinus* sprays a noxious compound from the tip of its abdomen, which it can aim precisely enough to hit a single ant attacking one of its legs (Fig. 33.20). Ants hit by the spray are instantly repelled. The spray is also effective against some vertebrate predators, but others (notably skunks and one species of mouse) have evolved specialized innate behavior patterns that cause the spray to be discharged harmlessly, and they can then eat the beetles.

In some cases defensive chemicals are not actually synthesized by the arthropod that uses them, but are obtained from the plants on which the animals feed. The plants' production of these chemicals has usually evolved

33.20 A bombardier beetle (*Brachinus*) spraying its defensive secretion
Note how accurately the beetle can aim its spray at the offending object, in this case forceps grasping a rear leg (left) or a front leg (right).

33.21   Defensive behavior of the sawfly *Pseudoperga guerini*

Left: The mother sawfly is protecting her very young larvae, which have not yet developed their own chemical defenses. She would not fly away, no matter how much the photographer disturbed her. Right: After her death, the larvae can protect themselves by regurgitating a repellent substance (yellow droplets), which they have obtained from the *Eucalyptus* leaves on which they feed. In the circle they form, only the individuals on the periphery, all of which face outward, release the repellent when disturbed.

as a defense against vertebrate and invertebrate herbivores. In turn, certain animal species have evolved the ability to tolerate these toxic compounds, thereby opening the way to a competition-free food supply and a potentially powerful defense system. A particularly striking example is shown in Figure 33.21. The larvae of the sawfly *Pseudoperga guerini* feed on *Eucalyptus* leaves, from which they obtain repellent chemicals that they use to ward off predators.

It must be emphasized that arthropods with defensive secretions are sometimes injured or even killed before a predator is repelled. Such events, however, do not indicate that the defensive secretion is an adaptive failure. A defensive mechanism need not be 100 percent effective to give the individuals that possess it a significantly greater chance to survive to reproduce. In this case the insects with defensive secretions have a far better chance of being dropped by predators before being seriously injured than if they had no secretion. Moreover, some predators may learn from one or two unpleasant experiences to avoid the prey in question on sight alone. This "education" of the predator enhances the fitness (probable genetic contribution) of the individual who dies in the process only if the beneficiaries—the members of the species less likely to be killed in the future—are relatives. We discussed this indirect kind of selection (kin selection) in Chapter 22.

Cryptic appearance   Many animals blend into their surroundings so well as to be nearly undetectable. Frequently their color matches the background almost perfectly (Fig. 33.22). In some cases the animals even have the ability to alter the condition of their own pigment cells and change their appearance to harmonize with their background (Figs. 33.23 and 33.24).

33.22 Cryptic coloration of aquatic animals
Left: The sargassum crab, which lives in dense growths of the brown alga *Sargassum* off Bermuda, is the same color as the alga, and its rounded body resembles the floats of the alga. Right: Patterning on the body of the hawkfish mimics the coral of its habitat in the Red Sea.

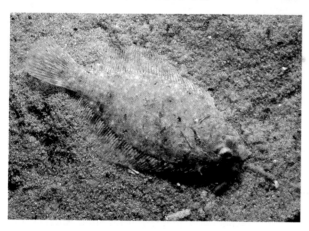

33.23 Flounders on two different backgrounds
The fish can change color to match the background, whether it is light-colored (top) or dark (bottom).

33.24 Color change by the frog *Hyla versicolor*
The brownish individual has been on the tree trunk for some time and matches it well. The green one has just been moved there from a pond, where it was among green duckweed; it has not had time to change color.

33.25   Leaflike animals
Left: The mantis (at top in picture) looks strikingly like the green leaves below it. Even the relationship of its thorax and abdomen reflects the way the leaves often occur in pairs. Right: A frog that resembles a leaf.

33.26.   A moth that looks like a broken twig
The perfection of the cryptic appearance of this moth (*Phalera bucephala*) is a triumph of evolutionary adaptation.

33.27   A nudibranch mollusc that strikingly resembles a shell
The nudibranch, which does not itself have a shell, is at lower right; the shell is at upper left.

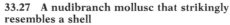

33.28   A frog from Ecuador that resembles bird droppings

Often, rather than match the color of the general background, the animals may resemble inanimate objects commonly found in their habitat, such as leaves (Fig. 33.25), twigs or sticks (Fig. 33.26), shells (Fig. 33.27), or even bird droppings (Fig. 33.28).

Careful studies have confirmed that cryptic appearance is an adaptive characteristic that helps animals escape predation. One such study was conducted by F. B. Sumner of the Scripps Institution of Oceanography in California. Sumner investigated predation by Galápagos penguins on mosquito fish (*Gambusia partuelis*), which can contract or expand their pigment cells to become lighter or darker, depending on their background. He established that the penguins caught 70 percent of the fish that contrasted with their background but only 34 percent of the fish that resembled their background. Sumner also exposed mosquito fish to large sunfish and found that the sunfish captured 53 percent of the contrastingly colored prey, but only 25 percent of the cryptically colored prey. In a similar experiment F. B. Isely of Trinity University in San Antonio, Texas, studied predation by chickens, turkeys, and native birds on grasshoppers of various colors on differently colored backgrounds. He found that 88 percent of the nonprotected grasshoppers were eaten whereas only 40 percent of the cryptically colored ones were eaten.

Predators as well as prey may exhibit cryptic coloration. We are all familiar with the camouflaging stripes or spots of carnivores like tigers and leopards. Patterns and colors that match a particular background are found also in other predatory species, large and small (Fig. 33.29).

One of the most extensively studied cases of cryptic coloration is the so-called industrial melanism of moths. Since the mid-1880s many species of moths have become decidedly darker in industrial regions. This is actually a case of polymorphism in which the less frequent of two forms has become the more frequent, and vice versa; the originally predominant light form in these species of moths has given way in industrial areas to the dark (melanic) form. In the Manchester area of England, the first black specimens of the species *Biston betularia* were caught in 1848; by 1895 melanics constituted about 98 percent of the total population in the area. For such a remarkable shift in frequency to have occurred in so short a time, the melanic form must have had at least a 30 percent advantage over the light form.

In 1937 E. B. Ford of Oxford University proposed the following explanation for this striking evolutionary change. The various species of moths exhibiting the rapid shift to melanism, though unrelated to one another, all habitually rest during the day in an exposed position on tree trunks or rocks, being protected from predation only by their close resemblance to their background. In former years the tree trunks and rocks were rather light-colored and often covered with light-colored lichens. Against this background the light forms of the moths were astonishingly difficult to see, whereas the melanic forms were quite conspicuous (Fig. 33.30). Under these conditions it seems likely that predators captured melanics far more easily than the cryptically colored light moths. The light forms would thus have been strongly favored, and they would have occurred in much higher frequency than melanics. But with the advent of extensive industrialization, tree trunks and rocks were blackened by soot, and the lichens, which are particularly sensitive to such pollution, disappeared. In this altered environment the melanic moths resembled the background more closely than

33.29 Cryptic coloration in a crab spider
Matching the color of the flower on which it waits, a crab spider remains unseen long enough to trap an unwary honey bee.

33.30 Cryptic coloration of peppered moths
Top: Light and dark morphs of *Biston betularia* at rest on a tree trunk in unpolluted countryside. Bottom: Light and dark morphs on a soot-covered tree trunk. Here the light form is easier to see.

33.31 Aposematic coloration
The bright color of the poison-arrow frog (from which South American Indians obtain poison for their arrow tips) makes it easily recognizable by predators, which carefully avoid it.

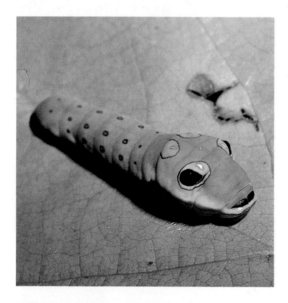

33.32 An example of Batesian mimicry
The prominent imitation eyes and mouth of the spicebush swallowtail larva give it the appearance of a predator, to be avoided.

the light moths. Thus selection would have been reversed and would now have favored the melanics, which would consequently have increased in frequency.

Though plausible, Ford's hypothesis was not tested until the mid-1950s, when H. B. D. Kettlewell of Oxford released approximately equal numbers of the light and melanic forms of *Biston betularia* onto trees in a rural area in the country of Dorset, England, where the tree trunks were light-colored, lichens were abundant, leaf washings revealed very little pollution, and the wild population of the moth was about 94.6 percent light-colored. A direct watch on the resting moths was maintained, with the help of binoculars, from blinds where the observer could not be seen by potential bird predators. It was found that several species of insectivorous birds do prey by sight on the moths and that, of 190 moths observed to be captured by birds, 164 were melanics and only 26 were light forms. Furthermore, of approximately 500 marked individuals of each color form released in another experiment, roughly twice as many light moths as melanic moths were recaptured in traps set up in the Dorset woods, an indication that more of the light moths had survived. It was plain that predation was an important selective factor among the moths, and that melanics were more subject to predation under the conditions prevailing in the Dorset woods than light-colored moths.

Taken alone, however, these experiments did not prove conclusively that the factor favoring the light moths over the melanics was their resemblance to their background. The results could be explained by assuming, for example, that the birds preferred the melanics because of some difference in flavor. Therefore Kettlewell duplicated the experiments under the reverse environmental conditions—in woods near Birmingham, England, where the tree trunks were blackened with soot, lichens were absent, leaf washings revealed heavy pollution, and the wild population of the moth was about 85 percent melanic. The results of these experiments were the reverse of those in the Dorset experiments. Now birds were observed to capture nearly three times as many light moths as melanics, and roughly twice as many melanics were recaptured in traps. Here also predation was an important selective factor, but here the melanics had the adaptive advantage. These experiments by Kettlewell prove that birds do hunt by sight, and that those moths that most closely resemble the background on which they rest have much the best chance of escaping predation.

Warning coloration Whereas some animals have evolved cryptic coloration, others have evolved colors and patterns that contrast boldly with their background and thus render them clearly visible to potential predators (Fig. 33.31). Many of these animals are in some way disagreeable to predators; they may taste bad, or smell bad, or sting, or secrete poisonous substances. In other words, they are animals that a predator will usually reject after one or two unpleasant encounters with them. Such animals benefit by being gaudily colored and conspicuous because predators that have experienced their unpleasant features learn to recognize and avoid them more easily in the future. Their flashy appearance is protective because it warns potential predators that they should stay away; such a warning appearance is said to be ***aposematic***.

The warning is sometimes so effective that, after unpleasant experiences with one or two warningly colored insects, some vertebrate predators simply avoid all flashily colored insects, whether or not they resemble the ones they encountered earlier. G. D. H. Carpenter demonstrated this by offering over 200 different species of insects to an insectivorous monkey. The monkey accepted 83 percent of the cryptically colored insects but only 16 percent of those with the warning coloration, even though many of the insects belonged to species the monkey had probably not previously encountered. It is possible that such avoidance of aposematic insects does not depend entirely on learning. Predators that have a genetic predisposition to avoid brightly colored prey would probably have an adaptive advantage over predators that waste time and energy pursuing inedible prey; hence there may be a tendency for predators to evolve an avoidance response to prey displaying warning coloration.

Mimicry   Species not naturally protected by some unpleasant character of their own may closely resemble (mimic) in appearance and behavior some dangerous or unpalatable aposematic species (Fig. 33.32.). Such a resemblance can be adaptive: the mimics may suffer little predation because predators cannot distinguish them from their models, which the predators have learned are unpleasant. This phenomenon is called ***Batesian mimicry***.

Convincing evidence for the effectiveness of this type of mimicry in protecting the mimic species comes from the elegant experiments of Jane van Z. Brower, then at Oxford. In one experiment she fed specimens of the viceroy butterfly (*Limenitis archippus*) to a group of caged jays, which accepted the viceroys readily. Then she offered the jays specimens of the distasteful monarch butterfly (*Danaus plexippus*), which the viceroy mimics (Fig. 33.33). After a few trials, the jays refused to eat the monarchs. When they were again offered viceroys, the jays also refused to eat these, even though they had earlier eaten them readily. A few unpleasant encounters with the monarchs had caused the jays to reject their mimics, the viceroys. (The name viceroy is apt; it comes from the Middle French *vice-roi*, meaning "someone who takes the place of the monarch.")

In another series of experiments Brower produced an artificial model-mimic system with starlings as the predators and mealworms, which starlings ordinarily eat voraciously, as prey. The first step was to paint a green band on some of the mealworms and an orange band on the others. The paint itself was not distasteful to the birds. Some of the green-banded worms were then dipped into a solution of a chemical very distasteful to starlings; these worms were used as the "models." The rest of the green-banded worms were dipped only into distilled water and therefore remained palatable; these worms were the "mimics." None of the orange-banded worms were made distasteful. Green was chosen as the "warning color" for the distasteful worms because in nature this color, usually associated with cryptically colored rather than with aposematic species, was not likely to have been associated by the birds with an unpleasant experience.

Each starling was given 10 trials per day for about 16 days, each trial consisting of two mealworms—one orange-banded and one green-banded. Of the green-banded worms, different groups of birds received a different

33.33 Batesian mimicry in butterflies
The monarch butterfly (top) is a distasteful species. The viceroy (bottom) mimics the monarch; species in the group to which the viceroy belongs ordinarily have a quite different appearance.

proportion of "mimics" (palatable) to "models" (unpalatable because of having been dipped into the distasteful chemical); the proportion of mimics for the different groups was, respectively, 10, 30, 60, and 90 percent for the 16 days of the experiment.

After a few unpleasant encounters with the models, the birds learned to recognize and avoid green-banded worms as distasteful, with the result that the mimics among them also escaped predation, particularly when the percentage of mimics presented to the starlings was 60 percent or less. Even when the percentage of mimics was as high as 90 percent and the percentage of models only 10 percent, 17 percent of the mimics escaped predation. With one exception, all the starlings readily ate the orange-banded worms throughout the experiments—a demonstration that the birds had not simply learned to avoid all mealworms and that their avoidance of the palatable green-banded ones (mimics) was a result of the resemblance of these worms to the unpalatable green-banded ones (the models).

Brower's original experimental proof of the adaptiveness of mimicry, published in 1960, stimulated a vigorous new quantitative approach to this intriguing subject. And her experiments made it necessary to revise some long-held ideas about mimicry. For example, it was generally believed that for Batesian mimicry to be effective the model species must be much more common than the mimic species, so that in any given trial a predator would be far more likely to get an unpalatable than a palatable mouthful. Otherwise, it was thought, the predator would not readily learn to associate the appearance of the prey with an unpleasant rather than a pleasant experience. Brower's experiments showed, however, that, if the model is distasteful enough, mimicry is still very effective when the ratio of mimics to models is as high as 60 to 40 and that mimics benefit to some degree even when the ratio is 90 to 10. Even so, the viceroy, like other palatable mimics, has evolved behavioral patterns that further stack the cards in its favor. Its flight, for example, is less flamboyant—and therefore less noticeable—than that of its toxic model. In addition, the viceroy dies soon after it reproduces, thus giving its offspring's potential predators less chance to discover the ruse. The monarch, on the other hand, lives on, increasing the chances that it will teach its offspring's potential predators to avoid attacking monarchs.

So far we have discussed only Batesian mimicry, which is based on deception—mimicry of a distasteful or dangerous species by individuals of a species that is neither. A second kind of mimicry, called **Müllerian mimicry**, involves the evolution of a similar appearance by two or more distasteful or dangerous species. In this type of mimicry, individuals of each species act as both model and mimic. The members of each species have some defensive mechanism, but if each species has its own characteristic appearance, the predators would have to learn to avoid each of them separately; the learning process would thus be more demanding, and would involve the death of some individuals of each prey species. If, however, several protected species evolve more and more toward one appearance type, they come to constitute a single prey group from the standpoint of the predators, which accordingly learn avoidance more easily. This may explain the similar markings of many unrelated species of wasps and bees, or of the group of poisonous reptiles known as coral snakes. If avoidance involves any genetic predisposition, then resemblances among the prey would facilitate more rapid selection for improved prey-recognition mechanisms in the predators. In fact, there is evidence that some predators have evolved the ability to recognize coral snakes innately; perhaps only Müllerian mimicry can provide a strong enough selection pressure to cause the evolution of such specialized recognition—a recognition that benefits individuals of both predator and prey species.

Ambiguity of body orientation Cryptic appearance and mimicry function in deceiving potential predators; so do a variety of appearances that make it difficult for the predator to tell which is the front end of the prey animal. For example, the real eye is often obscured by the pattern of coloration on the animal's head, but a prominent fake eyespot is located near the animal's tail (Fig. 33.34). Or the tail looks more like a head than the head itself (Fig. 33.35). The result may be that the predator often aims its attack incorrectly, expecting the prey to move in one direction while it is in fact escaping in the opposite direction.

SYMBIOTIC ADAPTATIONS

The term "symbiosis" is used in a variety of ways in the biological literature. Some authors apply it only to cases in which two species live together to their mutual benefit. We use the word in a broader sense.

Etymologically, symbiosis simply means "living together," without any implied value judgments. This is the meaning it was given when it was first introduced into biology, and this is the meaning it will have in this book. We shall, however, recognize three categories of symbiosis. The first is *commensalism*, a relationship in which one species benefits while the other receives little or no benefit or harm. The second is *mutualism*, in which both species benefit. The third is *parasitism*, in which one species benefits and individuals of the other species are harmed. We can summarize the distinctions as follows (a plus sign means benefit, a minus sign harm, and a zero no significant benefit or harm):

Relationship	Species A	Species B
Commensalism	+	0
Mutualism	+	+
Parasitism	+	−

Commensalism The advantage derived by the commensal species from its association with the host often involves shelter, support, transport, or food, or several of these. For example, in tropical forests numerous small plants, called epiphytes, usually grow on the branches of the larger trees or in forks of their trunks (see Fig. 36.25, p. 974). These commensals, among which species of orchids and bromeliads are prominent, are not parasites. They use the host trees only as a base of attachment and do not obtain nourishment from them. They apparently do no harm to the host except very rarely when so many of them are on one tree that they stunt its growth or cause limbs to break. A similar type of commensalism is the use of trees as nesting places by birds.

33.34 A fish with a fake eyespot near the tail
This butterfly fish has a prominent fake eyespot; its real eye is smaller and much less conspicuous within the stripe that runs through it.

33.35 Ambiguity of body orientation in Malaysian moth caterpillars
Contrary to appearances, all but the two caterpillars at the upper right are facing to the left. Their deceptive markings not only mislead predators but also scare them off: caterpillars that have been disturbed all raise up their false heads together, and intimidate potential predators.

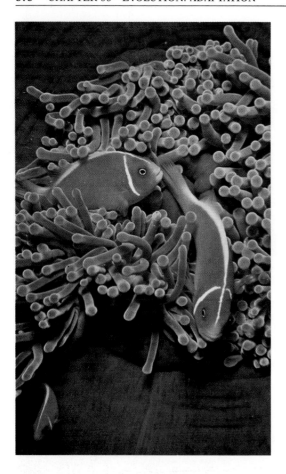

33.36 Three anemone fish living among the tentacles of a sea anemone from the Palau Islands

Sometimes it is difficult to tell what benefit is involved in a commensal relationship. For example, certain species of barnacles occur nowhere except attached to the backs of whales, and other species of barnacles occur nowhere except attached to the barnacles that are attached to whales. Just what advantages either of these groups of barnacles enjoys is not clear. They do, of course, get a relatively unoccupied base for attachment, and they get transport that increases the dispersal of their offspring. But it is hard to see how these benefits alone would have sufficed for the evolution of such specificity.

In some cases of commensalism, however, the benefit is dramatically obvious. For example, certain species of fish live in association with sea anemones, deriving protection and shelter from them and sometimes stealing some of their food (Fig. 33.36). These fish swim freely among the tentacles of the anemones, even though those tentacles quickly paralyze other fish that touch them. The anemones regularly feed on fish; yet the particular species that live as commensals with them sometimes actually enter the gastrovascular cavity of their host, emerging later with no apparent ill effects. The physiological and behavioral adaptations that make such a commensal relationship possible must be quite extensive. Another striking example is a small tropical fish that lives in the respiratory tree of a particular species of sea cucumber. The fish emerges at night to feed and then returns to its curious abode by first poking its host's rectal opening with its snout and then quickly turning so that it is drawn tail first through the rectal chamber into the respiratory tree. Still another example is a tiny crab that lives in the mantle cavity of oysters. The crab enters the cavity as a larva and eventually grows too big to escape through the narrow opening between the two valves of the oyster's shell. It is thus a prisoner of its host, but a well-sheltered prisoner. It steals a few particles of food from the oyster, but apparently does it no significant harm.

Mutualism Symbiotic relationships beneficial to both species are common. Figure 33.37 illustrates two instances of the widespread phenomenon called cleaning symbiosis, which is patently mutualistic. We mentioned several other examples of mutualism in earlier chapters: the relationship between a termite or a cow and the cellulose-digesting microorganisms in its digestive tract, or between a human being and the bacteria in his intestine that synthesize vitamin B_{12}. The plants we call lichens are actually formed of an alga or a cyanobacterium and a fungus united in such close mutualistic symbiosis that they give the appearance of being one plant (see Fig. 41.8, p. 1100). Apparently the fungus benefits from the photosynthetic activity of its "guest," and the alga or bacterium benefits from the water-retaining properties of the fungal walls.

As is apparent from this discussion of commensalism and mutualism (particularly the former), the division of symbiosis into three subcategories is in many ways arbitrary. Commensalism, mutualism, and parasitism are all parts of a continuous spectrum of possible interactions. Fortunately, it really isn't very important which category we apply to most cases. The categories are only devices to help us organize what we know about nature and to form testable hypotheses. What is important is to keep in mind how commensalism, mutualism, and parasitism grade into each other, and to recognize that each case of symbiosis is different from all others and must be studied and analyzed on its own.

Parasitism   Just as there are no sharp boundaries between parasitism and commensalism, or even between parasitism and mutualism, there is no strict line between parasitism and predation. Mosquitoes and lice both suck the blood of mammals, yet we usually call only the latter parasites. Foxes and tapeworms may both attack rabbits, but foxes are called predators and tapeworms are called parasites. The usual distinction is that a predator eats its prey quickly and then goes on its way, while a parasite passes much of its life on or in the body of a living host, from which it derives food in a manner harmful to the host. Parasites generally do not kill their hosts; those that eventually do so are called *parasitoids* (Fig. 33.38). Obviously the distinction between predator and parasite is not always clear. How long must one organism live on the body of another to be classed as a parasite? But though there will always be intermediate cases, it is profitable to distinguish between predation and parasitism, because each of these is a mode of existence followed by many kinds of organisms and each involves its own characteristic sorts of adaptations.

Parasites are customarily divided into two types: external parasites and internal parasites. The former live on the outer surface of their host, usually either feeding on the hair, feathers, scales, or skin of the host or sucking its blood. Internal parasites may live in the various tubes and ducts of the host's body, particularly the digestive tract, respiratory passages, or urinary ducts; or they may bore into and live embedded in tissues such as muscle or liver; or, in the case of viruses and some bacteria and protozoans, they may actually live inside the individual cells of their host.

Internal parasitism is usually marked by much more extreme specializations than external parasitism. The habitats available inside the body of another living organism are completely unlike those outside, and the unusual problems they pose have resulted in evolutionary adaptations entirely different from those seen in free-living forms. For example, internal parasites

33.37   Cleaning symbiosis
Top: A grouper being cleaned by cleaner fish. Bottom: Yellow-billed oxpeckers search for parasitic insects on a sable bull. In both cases the symbiosis is mutualistic: the cleaner obtains food, and the host gets rid of parasites that could endanger its health.

33.38   A caterpillar (tomato hornworm) with numerous pupae of a parasitoid wasp attached to its body
The wasp laid her eggs inside the body of the host caterpillar, and the larvae fed on it until they emerged and pupated.

33.39 A parasitic ant
Queens of the "ultimate ant," *Teleutomyrmex*, ride on the host queen and are fed by a host worker (lower right); they dispense their eggs along with those of the host queen. Their offspring, all reproductives, mate in the colony.

have often lost organs or whole organ systems that would be essential in a free-living species. Tapeworms, for instance, have no digestive system. They live in their host's intestine, where they are bathed by the products of the host's digestion, which they can absorb directly across their body wall without having to carry out any digestion themselves.

Because of their frequent evolutionary loss of structures, certain parasites are often said to be degenerate. "Degenerate," of course, implies no value judgment, but simply refers to the lack, common in internal parasites, of many structures present in their free-living ancestors. The ant species *Teleutomyrmex* (Fig. 33.39) is a degenerate parasite living in the nests of another species of ant. Though not an internal parasite, it is degenerate to the extent that it cannot survive outside its hosts' nest: it has lost many of its glands, its sting, its pigmentation, its ability to digest anything but liquid fed to it by its hosts. Even its brain is degenerate. But from an evolutionary point of view, loss of structures useless in a new environment is an instance of specialized adaptation.

Specialization, then, does not necessarily mean increased structural complexity; it only means the evolution of characteristics particularly suited to some special situation or way of life. In internal parasites—or cave animals, which frequently lack eyes—the development and maintenance of structures that no longer serve a useful function would require energy that the organism might use to more advantage in some other way. And some useless structures, such as eyes in both internal parasites and cave animals, might well be a handicap in these special environments, because they would be a likely point of infection. It is readily understandable, therefore, that natural selection might favor those individuals in which such useless organs are either relatively small or lacking entirely. Alternatively, the unused structures could have been adaptively neutral and simply lost through genetic drift. In either case, the concept of evolutionary loss of unused structures has nothing to do with the erroneous Lamarckian notion that use and disuse can directly influence the size of a structure in an organism's progeny. Only the superior qualifications of a genotype lacking the structure, or the spontaneous loss of the genotype through genetic drift, could eliminate a structure from a population.

Structural degeneracy is far from being the only sort of special adaptation commonly seen in internal parasites. They often have body walls highly resistant to the destructive enzymes and antibodies of the host. Tapeworms, for example, are constantly bathed by the potent digestive juices of their host; yet their enzyme-resistant cuticle protects them from being digested. And tapeworms have very specialized heads, with hooks and suckers, that enable them to anchor themselves and avoid being expelled by the often vigorous peristaltic contractions that move the other contents of the host's intestine (see Fig. 42.26, p. 1115).

Perhaps the most striking of all adaptations of internal parasites concern their life histories and reproduction. Individual hosts don't live forever. If the parasitic species is to be perpetuated, therefore, a mechanism is needed for changing hosts. At some point in their life cycle, then, all internal parasites move from one host individual to another. But this is seldom simple. Rarely can a parasite move directly from one host to another of the same species. For example, consider the life cycle of a beef tapeworm.

The eggs of a beef tapeworm living in a human intestine are shed in the host's feces. A cow eats plants contaminated with human feces, and the tapeworm eggs hatch in the cow's intestine. The young larvae bore through the wall of the cow's intestine, enter a blood vessel, and are carried by the blood to a muscle, where they encyst (become surrounded by a bladderlike case and lie inactive). If a human being then eats the raw or insufficiently cooked beef, the tapeworm larvae become activated in the person's intestine; the head of a larva attaches to the intestinal wall, and a mature worm develops. The beef tapeworm thus passes through two hosts during its life cycle: an intermediate host (cow), in which it undergoes some of its early development, and a final host (human) in which it matures.

As life cycles of internal parasites go, the one just described is rather simple. It is not unusual for a life cycle to include two or three intermediate hosts and/or a free-living larval stage. But such a complex development makes the chances that any one larval parasite will encounter the right hosts in the right sequence exceedingly poor. The vast majority die without completing their life cycle. It is therefore understandable that internal parasites should characteristically produce huge numbers of eggs. Though they may be degenerate in other ways, they usually have extremely well-developed reproductive structures. In fact, some internal parasites seem to be little more than a sac of highly efficient reproductive organs.

As the parasites evolve, the host species is also evolving, and there is strong selection pressure for the evolution of more effective defenses against the ravages of parasites. There is thus a constant interplay between host and parasite. Those individuals of the host species with superior defenses will be better able to survive and reproduce. Correspondingly, those individuals of the parasite species with the best ways of counteracting these defenses will be most likely to prosper. In turn, their counteractions will lead to pressure on the host to evolve still better defenses, against which the parasite may then evolve new means of surviving, and so forth. Though this sort of coevolution will continue as long as the host-parasite relationship exists, a dynamic balance is usually reached eventually, the host surviving without being seriously damaged and the parasite prospering moderately well too.

Probably most long-established host-parasite relationships are balanced ones. Relationships that result in serious disease in the host are usually relatively new, or they are relationships in which a new and more virulent form of the parasite has recently arisen, or in which the host showing the serious disease symptoms is not the main host of the parasite. American Indians, for example, suffered severely when exposed to pathogens first brought to North America by European colonists, even though some of those same pathogens caused only mild symptoms in the Europeans, who had been exposed to the pathogens for many centuries and in whom the host-parasite relationship had nearly reached a balance. Many examples are known in which humans are only occasional hosts for a particular parasite and suffer severe disease symptoms, though the wild animal that is the major host shows few ill effects from its relationship with the same parasite.

Chapter 34

EVOLUTION: SPECIATION AND PHYLOGENY

SPECIES AND SPECIATION

We have so far discussed only one major aspect of evolution, the change of a given population through time. Now we must turn to another of its major aspects, the processes by which a single population may split, giving rise to two or more different descendant populations. But before we can discuss this topic meaningfully, we must examine more carefully the populations that we have so far casually taken for granted. With reference to sexually reproducing organisms, we can define a ***population*** as a group of individuals that interbreed and so share a common gene pool.

UNITS OF POPULATION

Demes A *deme* is a small local population, such as all the deer mice or all the red oaks in a certain woodland, or all the perch or all the waterstriders in a given pond. Though no two individuals in a deme are exactly alike, the members of a deme do usually resemble one another more closely than they resemble the members of other demes. There are at least two reasons for this: (1) the individuals in a deme are more closely related genetically, because pairings occur more frequently between members of the same deme than between members of different demes; and (2) the individuals in a deme are exposed to more similar environmental influences and hence to more nearly the same selection pressures.

We can see that demes are not clear-cut permanent units of population. Though the deer mice in one farm's woodlot are more likely to mate among themselves than with deer mice in the next woodlot down the road, there will almost certainly be occasional matings between mice from different woodlots. Similarly, though the female parts of a particular red oak tree are more likely to receive pollen from another red oak tree in the same wood-lot, they will sometimes receive pollen from a tree in another nearby wood-lot. And the woodlots themselves are not permanent ecological features. They have only a transient existence as separate and distinct ecological units; neighboring woodlots may fuse after a few years, or a single woodlot may become divided into two or more separate smaller ones. Such changes in ecological features will produce corresponding changes in the demes of deer mice and red oak trees. Demes, then, are usually temporary units of population that intergrade with other similar units.

Species   Notice that intergradation is between "similar" demes. We ex-pect some interbreeding between deer mice from adjacent demes, but we do not expect interbreeding between deer mice and house mice or between deer mice and black rats or between deer mice and gray squirrels. Nor do we expect to find crosses between red oaks and sugar maples or even be-tween red oaks and pin oaks, even if they occur together in the same wood-lot. In short, we recognize the existence of units of population larger than demes and both more distinct from each other and longer-lasting than demes. One such unit of population is that containing all the demes of deer mice. Another is that containing all the demes of red oaks. These larger units are known as species.

For centuries it has been recognized that plants and animals seem to fall naturally into many separate and distinct "kinds," or species. This does not mean that all the individuals of any one species are precisely alike—far from it. Any two individuals are probably distinguishable from each other in a variety of ways. But it does mean that all the members of a single spe-cies share certain biologically important attributes and that, as a group, they are genetically separated from other such groups. That such groups exist in nature has been recognized even by tribal peoples. Ernst Mayr of Harvard University cites a tribe in New Guinea that had 136 different names for what biologists later showed to be 137 species of local birds; the natives had confused only two species.

But though the existence of discrete clusters of living things that can be called species has long been recognized, the concept of what a species is has changed many times in the course of history. One idea widely held by nonbiologists, and once popular among biologists as well, is that each spe-cies is a static, immutable entity typified by some ideal form, of which all the real individuals belonging to that species are rough approximations; in-dividual variation is supposed to result from the imperfection with which the individuals reflect the ideal characteristics. This static, typological con-cept contradicts all that we have learned about evolution. The modern concept of species rejects the notion that there is some immutable ideal type for every species. A ***species***, in the modern view, is a genetically dis-tinctive group of natural populations (demes) that share a common gene pool and that are reproductively isolated from all other such groups. Or, to put it another way, a species is the largest unit of population within which effective gene flow (exchange of genetic material) occurs or can occur. The

A

B

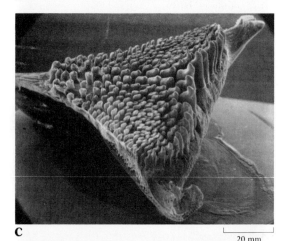

C _____ 20 mm

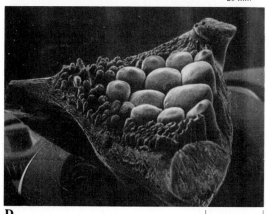

D _____ 20 mm

key word here is "effective"; we shall see later why two species whose members mate but produce infertile hybrids are not classified as a single species.

Notice that the modern concept of species says nothing about how different from each other two populations must be to qualify as separate species. Admittedly, most species can be separated on the basis of fairly obvious anatomical, physiological, or behavioral characters, and biologists often rely on these in determining species. But the final criterion for living species is always reproduction—whether or not there is actual or potential gene flow.[1] If there is complete intrinsic reproductive isolation between two outwardly almost identical populations—that is, if there can be no gene flow between them—then those populations belong to different species despite their great similarity. On the other hand, if two populations show striking differences, but there is effective gene flow between them, those populations belong to the same species (Fig. 34.1). Anatomical, physiological, or behavioral characters simply serve as clues toward the identification of reproductively isolated populations; they do not in themselves determine whether a population constitutes a species.

Intraspecific variation   We have already discussed the sorts of variation that may occur between individuals of a single deme as a result of mutation and recombination, particularly the latter. And we have seen that this variation is very important biologically, whether it involves almost imperceptible and intergrading differences or striking polymorphic discontinuities. But there is another sort of intraspecific variation that we have not yet discussed. This is variation between the demes of a single species, variation that is often correlated with geographical distribution.

There is usually so much gene flow between adjacent demes of the same species that differences between them are slight. Thus the frequencies of alleles A and a may be 0.90 and 0.10 in one deme and 0.89 and 0.11 in the adjacent deme. But the farther apart geographically two demes are, the

[1] Because paleontologists deal almost exclusively with fossils, they must rely to a great extent on anatomical criteria in distinguishing between species.

34.1   Variations in morphology between populations within a species

In some species, populations within a single area show easily observable morphological variations, as well as less obvious ones. A male of the cichlid *Cichlasoma minckleyi*, found in the Cuatro Ciénegas basin in Mexico, exhibits the so-called deep-bodied form, or morph (A), while another male of the same species from the same basin exhibits the slender-bodied morph (B). An independent variation in the lower pharyngeal jaw of males of this species also exists. Some males, regardless of body form, display the papilliform morph (C), while others display the molariform morph (D). Despite these major differences, individuals of the various morphs readily interbreed and produce fertile offspring; hence they are all considered members of the same species.

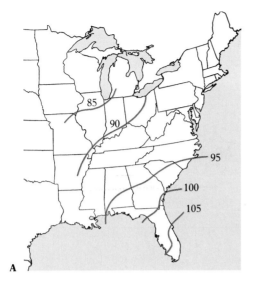

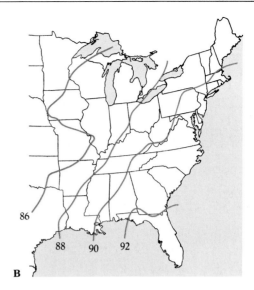

A

B

34.2 Clinal variation
(A) Map showing by means of isophene lines (lines connecting equal values) the geographic variation in the mean number of subcaudal scales of the snake *Coluber constrictor* (the racer). (B) Isophene map showing geographic variation in the apical taper of leaves of the milkweed *Asclepias tuberosa*. The numbers represent degree of apical taper.

smaller the chance of direct gene flow between them, and hence the greater the likelihood that the differences between them will be more marked. If, for example, we collect samples of 500 deer mice each from Plymouth County, Massachusetts, Crawford County, Pennsylvania, and Roanoke County, Virginia, we will find numerous differences that enable us to distinguish between the three populations quite readily—much more readily than we could between populations from three adjacent counties in Massachusetts or from three adjacent counties in Pennsylvania. Some of this geographic variation may reflect chance events such as genetic drift or the occurrence in one deme of a mutation that would be favorable in all demes but has not yet spread to them. But much of the geographic variation probably reflects differences in the selection pressures operating on the populations as a result of the differences between the environmental conditions in their respective ranges. (A species' range is the geographical area in which members of the species as a whole live or travel in the course of their normal activities, an area that may encompass an entire continent; individual animals generally occupy only a restricted part of this area—a home range. A species' habitat, by contrast, is where its members live—ponds, grasslands, forests in general, or only oak trees, for example.) In other words, much geographic variation is adaptive. Each local population or deme tends to evolve adaptations to the specific environmental conditions in its own small portion of the species range. Such geographic variation is found in the vast majority of animal and plant species.

Environmental conditions often vary geographically in a more or less regular manner. There are changes in temperature with latitude or with altitude on mountain slopes, or changes in rainfall with longitude, as in many parts of the western United States, or changes in topography with latitude or longitude. Such environmental gradients are usually accompanied by genetic gradients—gradients of allelic frequency—in the species of animals and plants that inhabit the areas involved. Most species show north–south gradients in many characters; east–west gradients (Fig. 34.2B) and

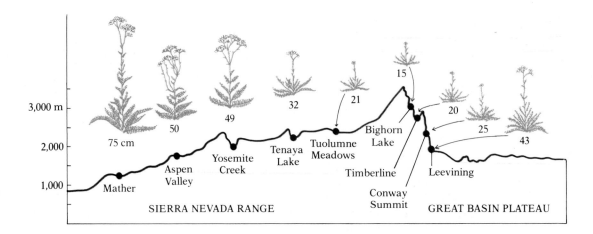

34.3 Altitudinal cline in height of the herb *Achillea lanulosa*
The higher the altitude, the shorter the plants. This variation was shown to be genetic (as opposed to merely environmental) by collecting seeds from the locations indicated and planting them in a test garden at Stanford where all were exposed to the same environmental conditions. The differences in height were still evident in the plants grown from these seeds.

altitudinal gradients (Fig. 34.3) in various characters are not uncommon.

When a character of a species shows a gradual variation correlated with geography, we speak of that variation as forming a *cline*. For example, many mammals and birds exhibit north–south clines in average body size, being larger in the colder climates farther north and smaller in the warmer climates farther south. Similarly, many mammalian species show north–south clines in the size of such extremities as the tails and ears: these exposed parts are smaller in the demes farther north.[2] A single widespread species often has many characters that vary clinally, but the several clines frequently do not coincide in direction, location, or intensity; one character may show clinal variation from north to south, another from east to west, and still another from northwest to southeast.

Sometimes geographically correlated genetic variation is not as gradual as in the clines discussed above. There may be a rather abrupt shift in some character in a particular part of the species range. When such an abrupt shift in a genetically determined character occurs in a geographically variable species, some biologists designate the populations involved as *subspecies* or *races*. These terms are also sometimes applied to more isolated populations—such as those on different islands or in separate mountain ranges or, as in fish, in separate rivers—when the populations are recognizably different genetically but are potentially capable of interbreeding freely. Subspecies or races (the two terms, as used here, are equivalent) may be defined, then, as groups of natural populations within a species that differ genetically and that are partly isolated from each other because they have different ranges (Fig. 34.4).

Note that two subspecies of the same species cannot, by definition, long occur together geographically, because it is only the limitation on inter-

---

[2] Increase in average body size with increasing cold is so common in homeothermic animals that this tendency has been called Bergmann's rule; the tendency toward decrease in the size of the extremities with increasing cold has been called Allen's rule. The adaptive significance of these clines seems obvious when one considers the role of surface-to-volume ratios in heat exchange.

34.4 Two subspecies of Canada goose
These two subspecies of Canada goose have different breeding grounds and different ranges. *Branta canadensis maxima* (left) breeds in the central and south-central United States. *Branta canadensis moffitti* (right) breeds largely in central and western Canada.

breeding imposed by distance that keeps them genetically distinctive. If they occurred together, they would interbreed and any distinction between them would quickly disappear. Many biologists have argued against the formal recognition of subspecies or races. One reason is that the distinctions between them are often made arbitrarily on the basis of only one morphologically obvious character while other less obvious characters may form entirely different patterns of variation that are ignored (Fig. 34.5). Another reason is that most groups so recognized probably have only a transitory existence as separate populations and do not, as was once thought, proceed inevitably to become fully separate species. Nevertheless, since assigning names to distinguish separate populations is frequently a great convenience, the concept of race or subspecies will probably continue to be used.

As a result of intraspecific geographic variation (whether irregular, clinal, or racial), two populations belonging to the same species but occurring in two widely separated localities often show no more resemblance to each other than to populations belonging to other species. Such intraspecific dissimilarity serves to emphasize the point made earlier, that it is not the degree of morphological resemblance that determines whether or not two populations belong to the same species; it is whether they are reproductively isolated from each other. There are even instances where two widely separated populations are regarded as belonging to the same species even though the respective individuals, when brought together, are incapable of producing viable offspring. They are considered members of the same species because the populations are connected by an unbroken chain of intermediate populations that permit gene flow between them.

SPECIATION

In considering how species originate—the phenomenon of speciation—we shall concentrate particularly on divergent speciation, the process by

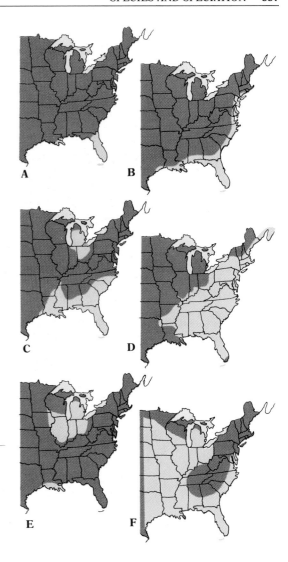

34.5 Discordant geographic variation in six characters of the snake *Coluber constrictor* in the eastern United States
Since no two of the characters vary together, selection of any one of them as a criterion for recognition of subspecies or races is largely arbitrary. (A) Areas where red eyes are found in juveniles. (B) Areas where red ventral spots are found on juveniles. (C) Areas where the loreal scale is in contact with the first supralabial scale in at least 10 percent of the specimens. (D) Areas where black adults are found. (E) Areas where dark postocular stripes are found. (F) Areas where full-grown adults have white chins.

which one ancestral species gives rise to two or more descendant species, which grow increasingly unlike (diverge) as they evolve.

The role of geographic isolation   Since species are defined in terms of reproductive isolation, the fundamental question of divergent speciation must be: How do two sets of populations that initially share a common gene pool come to have completely separate gene pools? That is, how does the possibility of effective gene flow between the two sets of populations disappear? How do barriers to the exchange of genes arise?

Most biologists agree that in the majority of cases the initiating factor in speciation is geographic separation. As long as all the populations of a species are in direct or indirect contact, gene flow will normally continue throughout the system and splitting will not occur, though various populations within the system may diverge in numerous characters and thus give rise to much intraspecific variation of the sorts discussed above. But if the initially continuous system of populations is divided by some geographic feature that constitutes a barrier to the dispersal of the species, then the separated population systems will no longer be able to exchange genes and their further evolution will therefore be independent. Given sufficient time, the two separate population systems may become quite unlike each other as each evolves in its own way. At first, the only reproductive isolation between them will be geographic—isolation by physical separation—and they will still be potentially capable of interbreeding; according to the modern concept of species, they will still belong to the same species. Eventually, however, they may become genetically so different that there would be no effective gene flow between them even if they should again come into contact. When this point in their divergence has been reached, the two population systems constitute two separate species.

There are at least three factors that can make geographically separated population systems diverge:

1. Chances are that the two systems will have somewhat different initial gene frequencies. Because most species exhibit geographic variation, it is highly unlikely that a geographic barrier would divide a variable species into portions exactly alike genetically; the barrier would be much more likely to separate populations already genetically different, such as the terminal portions of a cline. Separation can occur in other ways than through the splitting of a once continuous distribution by a new geographic barrier. For example, small numbers of individuals frequently manage to cross an already existing barrier and found a new geographically isolated colony. Regardless of what causes the population to become divided, if one group is relatively small, its members will, of course, carry with them in their own genotypes a relatively small percentage of the total genetic variation present in the gene pool of the parental population, and the new colony will therefore have allelic frequencies very different from those of the parental population; this is the special form of genetic drift, mentioned in the last chapter, called the *founder effect*. Obviously, if from the moment of their separation two populations have different genetic potentials, their future evolution may follow different paths. The present consensus is that most cases of geographic isolation probably involve small populations that exhibit the founder effect and genetic drift.

2. Separated population systems will probably experience different mutations. Mutations are random (though some are more probable than others), and the chances are good that some mutations will occur in one of the populations and not in the other, and vice versa. Since there is no gene flow between the populations, a new mutant gene arising in one of them cannot spread to the other.

3. Isolated populations will almost certainly be exposed to different environmental selection pressures, since they occupy different ranges. The chances that two separate ranges will be identical in every significant environmental factor are essentially nil.

The barriers that can cause the initial spatial separation leading to speciation are of many different types. A barrier is any physical or ecological feature that prevents the movement across it of the species in question. What is a barrier for one species may not be a barrier for another. Thus a prairie is a barrier for forest species but not, obviously, for prairie species. A mountain range is a barrier to species that can live only in lowlands, a desert is a barrier to species that require a moist environment, and a valley is a barrier to montane species. On a grander scale, oceans and glaciers have played a role in the speciation of many plants and animals. Let us look at a few actual examples of geographic isolation leading to speciation.

One of the most frequently cited examples is that of the Kaibab squirrel, which occurs on the north side of the Grand Canyon, and of the Abert squirrel, which occurs on the south side. The two are clearly very closely related and doubtless evolved from the same ancestor, but they almost never interbreed at present, because they do not cross the Grand Canyon. Biologists are not agreed whether these two morphologically distinct groups of squirrels have reached the level of full species or whether they should be considered well-marked geographic variants of a single species, but the fact remains that the Grand Canyon has acted as a barrier separating the two sets of populations, and that those populations have, as a result, evolved divergently until they have at least approached the level of fully distinct species (Fig. 34.6). The Grand Canyon also separates the range of the gray-tailed antelope squirrel from that of the closely related white-tailed antelope squirrel, and it separates the range of the rock pocket mouse from that of the long-tailed pocket mouse.

On islands of the Pacific, in many instances, two closely related species of snails, clearly descended from the same ancestral population, live in valley woodlands separated by treeless ridges that the snails apparently cannot cross. Blind cave beetles (genus *Pseudanophthalmus*) living in different caves in the eastern United States have often diverged to the level of full species.

Intrinsic reproductive isolation According to the model of divergent speciation just outlined, the initial factor preventing gene flow between two closely related population systems is ordinarily an extrinsic one—geography. Then, the model says, as the two populations diverge, they accumulate differences that will lead, given enough time, to the development of intrinsic isolating mechanisms—biological characteristics involving morphology, physiology, chromosomal compatibility, or behavior that prevent the two populations from occurring together or from interbreeding effectively

34.6 Squirrels of the Grand Canyon
Two populations of squirrels that live in different ranges of the Grand Canyon in Arizona are morphologically distinct. The Kaibab squirrel (top), which lives on the Kaibab Plateau on the northern rim of the canyon, is darker than the Abert squirrel (bottom), of the related population that inhabits a range on the southern rim.

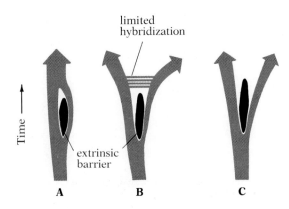

34.7 Model of geographic speciation

An ancestral population is split by an extrinsic (geographic) barrier for a time, and diverges with regard to various traits (only one of which is considered in these hypothetical examples). (A) If the barrier breaks down before the two subpopulations have been isolated long enough to have evolved intrinsic reproductive isolating mechanisms, the populations will fuse back together. (B) Two populations are isolated by an extrinsic barrier long enough to have evolved incomplete intrinsic reproductive isolating mechanisms. When the extrinsic barrier breaks down, some hybridization occurs. But the hybrids are not as well adapted as the parental forms; hence there is a strong selection pressure favoring forms of intrinsic isolation that prevent mating, and the two populations diverge more rapidly until mating between them is no longer possible. This rapid divergence is called character displacement. (C) Two populations are isolated by a geographic barrier so long that by the time the barrier breaks down they are too different to interbreed. In most cases, one population is much smaller than the other, as indicated by the width of the "branches" where they separate. The smaller population—often quite small—usually diverges more from the common ancestor than does the larger population. This greater divergence is the result of the founder effect, a greater tendency for genetic drift, and a smaller and perhaps more specialized habitat.

when (or if) they again occur together. In other words, speciation is initiated when through external barriers the two population systems become entirely **allopatric** (come to have different ranges), but is not completed until the populations have evolved intrinsic mechanisms that will keep them allopatric or that will keep their gene pools separate even when they are **sympatric** (have the same range) (Fig. 34.7). Let us now examine the various kinds of intrinsic isolating mechanisms that may arise.

1. *Ecogeographic isolation* Two population systems, initially separated by some extrinsic barrier, may in time become so specialized for different environmental conditions that even if the original extrinsic barrier is removed they may never become sympatric, because neither can survive under the conditions where the other occurs. In other words, they may evolve genetic differences that will maintain their geographic separation. An example is seen in two well-known tree species of the genus *Platanus*: *P. occidentalis* (the sycamore or buttonwood tree), which occurs in the eastern United States, and *P. orientalis* (the Oriental plane tree), which occurs in the eastern part of the Mediterranean region. They can be artificially crossed and the hybrids are vigorous and fertile. But each species is adapted to the climate in its own native range, and the climates in the two ranges are so different that neither species will long survive under natural conditions in the range of the other. Thus there are genetic differences that under natural conditions would prevent gene flow between the two species. Their separation is not merely geographic; it is both geographic and genetic.

2. *Habitat isolation* When two sympatric populations occupy different habitats within their common range, the individuals of each population will be more likely to encounter and mate with members of their own population than with members of the other population. Their genetically determined preference for different habitats thus helps keep the two gene pools separate. There are numerous examples of such habitat isolation. *Bufo woodhousei* and *B. americanus* are two closely related toads that can cross and produce viable offspring. But in those areas where the ranges of the two toads overlap, *B. woodhousei* normally breeds in the quieter water of streams, while *B. americanus* breeds in shallow rainpools. The dragonfly *Progomphus obscurus* lives in northern Florida, while its close relative *P. alachuensis* lives in southern Florida. The ranges of the two species overlap in north-central Florida, but there the two species occupy different habitats, *P. obscurus* being restricted to rivers and streams and *P. alachuensis* to lakes. In California the ranges of *Ceanothus thyrsiflorus* and *C. dentatus*, two species of wild lilacs, overlap broadly, but *C. thyrsiflorus* grows on moist hillsides with good soil, while *C. dentatus* grows on drier, more exposed sites with poor or shallow soil.

3. *Seasonal isolation* If two closely related species are sympatric, but breed during different seasons of the year, interbreeding between them will be effectively prevented. For example, *Pinus radiata* and *P. muricata*, two species of pine, are sympatric in some parts of California. They are capable of crossing, but rarely do so under natural conditions because *P. radiata* sheds its pollen early in February while *P. muricata* does not shed its pollen until April. *Reticulitermes hageni* and *R. virginicus*, two closely related spe-

cies of termites, are sympatric in southern Florida, but the mating flights of the former occur from March through May while those of the latter occur in the fall and winter months. Five species of frogs belonging to the genus *Rana* are sympatric in much of eastern North America, but the period of most active mating is different for each species (Fig. 34.8).

4. ***Behavioral isolation*** In Chapter 22 we discussed the immense importance of behavior in courtship and mating, particularly with respect to species recognition. We noted the complex interplay of behavioral patterns during the courtship of queen butterflies (see Fig. 22.3, p. 584). Each species of butterfly has its own courtship pattern, and where two species are sympatric, crosses rarely occur, because a courtship between members of different species involves so many wrong responses that it is unlikely to proceed all the way to copulation.

A particularly interesting example of the functioning of visual displays in species recognition has been reported by Jocelyn Crane of the New York Zoological Society. She found twelve species of fiddler crabs of the genus *Uca* actively courting on the same small beach (only about 56 m²) in Panama. Each species had its own characteristic display, which included waving the large claw (cheliped), elevating the body, and moving around the burrow (Fig. 34.9). Crane found that the displays were so distinctive that she could recognize each species from a considerable distance merely by the form of its display; presumably the female crabs do likewise.

Auditory stimuli are important in species recognition among many animals, particularly birds and insects, and help prevent mating between related species. In several instances specialists have noticed that two or three very different songs were sung by what had been considered members of a single species of cricket. On investigation, they found that each song was, in fact, sung by a different species, but the species were so similar morphologically that no one had previously distinguished among them. Despite the morphological similarity of these closely related crickets, they do not hybridize in nature even when they are sympatric, because females do not respond to the stridulation of a male of a different species. ***Sibling species*** —species so closely related that humans can hardly distinguish them, at least until some diagnostic character, such as the song of the crickets, is found—may be a fairly common phenomenon.

5. ***Mechanical isolation*** If structural differences between two closely related species make it physically impossible for matings between males of one species and females of the other to occur, the two populations will obviously not exchange genes. If, for example, one species of animal is much larger than the other, matings between them may be very difficult, if not impossible. Or if the genital organs of the males of one species and the females of the other do not fit, mating will be prevented. The observation that the copulatory organs vary greatly from species to species in many animal groups led long ago to the so-called ''lock-and-key'' hypothesis, which holds that the male and female genitalia are so precisely fitted to each other that even slight changes in the structure of either would make copulation impossible. This supposition has proved incorrect for some groups. For example, even though the male copulatory organs (gonopods) of millipeds are usually so different in different species that they have served as the basis for

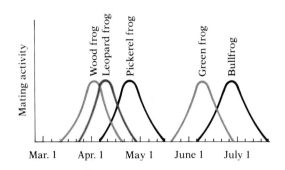

34.8 Mating seasons at Ithaca, New York, for five species of frogs of the genus *Rana*
The period of most active mating is different for each species. Where the mating seasons for two or more species overlap, different breeding sites are used.

34.9 A male fiddler crab (*Uca*)
The animal waves its large cheliped in the air as a courtship display. Details of the display differ among the various species of fiddler crabs.

  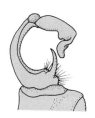

34.10   Male gonopods of six species of millipeds of the genus *Brachoria*
The shape of the distal portion of the male gonopods is distinctive for each species; yet the genitalia of the females of these species are nearly indistinguishable from one another. Since the animals are eyeless, the male gonopods can play no role in visual species recognition; no functional or evolutionary explanation for the differences in these structures has been found thus far. They may be the result of genetic drift.

much of the classification of these animals (Fig. 34.10), the female genitalia of related species are often almost indistinguishable—an indication that the major isolating mechanism between the species must be something other than a lock-and-key lack of fit. Nonetheless, a lock-and-key mechanism probably does operate in other animal groups, among them several subfamilies of snails in which interspecific matings are physically very difficult.

Mechanical isolation is probably more important in plants than in animals, particularly in plants that depend on insect pollinators. Consider milkweeds, for example. In these plants the pollen is contained in small sacs that stick to the legs of insects. The female part of each flower (stigma) has slits in it into which the sac must be inserted if pollination is to take place. This insertion is not easy, and few of the sacs carried from one flower to another on the legs of an insect are ever actually introduced into a stigma. Insertion of a pollen sac from one species of milkweed into the stigma of a different species is essentially impossible, because both the sacs and the slits in the stigma vary in shape from species to species. Hence even though several closely related species of milkweeds are sympatric in many parts of the world, hybridization between them is almost nonexistent. Consider also *Salvia apiana* and *S. mellifera,* two closely related species of sages with overlapping ranges in California. They are reproductively isolated by differences in habitat and flowering season and by the behavior of their pollinators. In addition, mechanical features play a role. Whereas *S. mellifera* is pollinated by relatively small bees, the flowers of *S. apiana* can be entered only by very large bees whose weight is sufficient to cause the landing platform (lower lip of corolla) to unfold and permit free entrance into the flower (Fig. 34.11).

6. ***Gametic isolation***   If individuals of two animal species mate or if the pollen from one plant species gets onto the stigma of another, fertilization still may not take place. Some 68 different interspecific combinations are known in tobacco in which no cross-fertilization will occur even when the pollen is placed on the stigma, because the sperm nucleus from the pollen is unable to reach the egg cell in the ovule. In *Drosophila,* if cross-insemination occurs between *D. virilis* and *D. americana,* the sperm are rapidly immobilized by the unsuitable environment in the reproductive tract of the female and they never reach the egg cells. In other species of *Drosophila,* interspecific matings cause an antigenic reaction in the genital tract of the female; the walls of the vagina swell enormously and kill the sperm before they reach the eggs.

(The mechanisms we have discussed so far exact little cost from the individual organisms: the machinery of isolation is either imposed from without, or generated internally from existing behavior, physiology, and morphology. As a result, no great loss of fitness is involved in maintaining isolation. The remaining four mechanisms, though effective, do exact a significant cost from the individual, and this cost acts as a selection pressure favoring evolution of the more efficient mechanisms of reproductive isolation we have already mentioned.)

7. *Developmental isolation*   Even when cross-fertilization occurs, the development of the embryo is often irregular and may cease before birth. The eggs of fish can often be fertilized by sperm from a great variety of other species, but development usually stops in the early stages. Crosses between sheep and goats produce embryos that die long before birth.

8. *Hybrid inviability*   Hybrids are often weak and malformed and frequently die before they reproduce; hence there is no gene flow through them from the gene pool of the one parental species to the gene pool of the other parental species. An example of hybrid inviability is seen in certain tobacco hybrids, which develop tumors in their vegetative parts and die before they flower.

9. *Hybrid sterility*   Some interspecific crosses produce vigorous but sterile hybrids. The best-known example is, of course, the cross between a female horse and a male donkey, which produces the mule. Mules have many characteristics superior to those of both parental species, but they are sterile. No matter how many mules are produced, the gene pools of horses and donkeys remain distinct, because there is no gene flow between them. The same is true of horses and zebras, which can hybridize to produce sterile zebroids.

10. *Selective hybrid elimination*   The members of two closely related populations may be able to cross and produce fertile offspring. If those offspring and their progeny are as vigorous and well adapted as the parental forms, then the two original populations will not remain distinct for long if they are sympatric, and it will no longer be possible to regard them as full species. But if the fertile offspring and their progeny are less well adapted than the parental forms, then they will soon be eliminated. There will be some gene flow between the two parental gene pools through the hybrids, but not much. The parental populations are consequently regarded as separate species.

These last four mechanisms do effectively isolate species from one another, but their cost is significant: valuable metabolic resources are invested in doomed embryos or in frail or possibly sterile young, and the seasonal nature of many reproductive cycles may preclude a second chance for the individuals involved to mate and rear young. It is important to remember that success in finding a mate or in copulating does not necessarily lead to reproductive success—the production of fit offspring. Individuals that tend to mate with members of the wrong species will leave fewer descendants than those that mate with members of their own species. Wrong matings waste gametes, whether fertilization takes place or not, and whether the hybrids are viable or not. Selection, therefore, will strongly favor individuals whose behavior, morphology, or physiology reduces the chance of a mismatch in the first place, and if the parental populations are sympatric, there will be strong selection for the evolution of more effective

34.11   Pollination of *Salvia apiana* by a bumble bee
The flower is inaccessible to lighter pollinators like honey bees, whose weight is insufficient to lower the landing platform and permit free entrance. However, a bumble bee has little trouble gaining entrance.

TABLE 34.1   *Intrinsic isolating mechanisms*

Mechanisms that prevent mating	1. Ecogeographic isolation 2. Habitat isolation 3. Seasonal isolation 4. Behavioral isolation 5. Mechanical isolation	Mechanisms operative in the parents, preventing fertilization
Mechanisms that prevent production of hybrid young after mating	6. Gametic isolation 7. Developmental isolation	
Mechanisms that prevent perpetuation of hybrids	8. Hybrid inviability 9. Hybrid sterility 10. Selective hybrid elimination	Mechanisms operative in the hybrids, preventing their success

intrinsic isolating mechanisms. Gene combinations that lead to correct mate selection will increase in frequency, and combinations that lead to incorrect selection will decrease, until eventually all hybridization ceases. The tendency of closely related sympatric species to diverge rapidly in characteristics that reduce the chances of hybridization and/or minimize competition between them is called *character displacement* (see Fig. 34.7). Table 34.1 summarizes the different intrinsic isolating mechanisms.

Situations in which only one of the ten isolating mechanisms is operative are extremely rare. Ordinarily several contribute to keeping two species apart. For example, closely related sympatric plant species often exhibit habitat and seasonal isolation in addition to some form of hybrid incapacity. And as we have seen, sympatric species, whether plant or animal, tend rapidly to evolve one or more of the forms of isolation that prevent mating (habitat, seasonal, behavioral, mechanical) rather than depend only on those forms that prevent the birth or perpetuation of hybrids.

Speciation by polyploidy   The model of speciation discussed above involves the divergence of geographically separated populations. There are other ways in which new species may arise. Speciation that does not involve geographic isolation is called *sympatric speciation*. One important example is speciation by polyploidy—an almost instantaneous process that makes it entirely possible for a parent to belong to one species and its offspring to another. (Other forms of sympatric speciation will be discussed later.) Speciation by polyploidy has apparently been common in plants but rare in animals.

One type of polyploid speciation, called *autopolyploidy*, involves a sudden increase in the number of chromosomes, usually as a result of the nondisjunction of chromosomes during meiosis. An example of this type of polyploidy was discovered by Hugo De Vries, one of the early geneticists, while he was making extensive studies of the evening primrose, *Oenothera lamarckiana*. This diploid species has 14 chromosomes. During De Vries' studies, a new form suddenly arose. This new form, to which he gave the species name *Oenothera gigas*, had 28 chromosomes. This tetraploid was reproductively isolated from the parental species because hybrids between *O. lamarckiana* and *O. gigas* were triploid (they received one of each type of chromosome from their *O. lamarckiana* parent and two of each type from their *O. gigas* parent), and triploid individuals, because of the highly irregular distribution of their chromosomes at meiosis, are sterile. It is characteristic of autopolyploidy that the polyploids are fertile and can breed with each other, but cannot cross with the diploid species from which they arose. Hence polyploid populations fulfill all the requirements of the modern definition of species—they are genetically distinctive, and they are reproductively isolated—though botanists do not always choose to give each polyploid population a formal species name.

The reproductive isolation of the polyploid daughter species from the ancestral stock sometimes permits adaptive divergence that would not otherwise be possible. For example, new polyploid species of certain plants have become adapted to mineral soils like mine tailings and serpentine outcrops. If the plants were not reproductively isolated, gene flow from the large surrounding population of normal diploids would probably prevent any local adaptation to these special soils. Polyploidy, then, substitutes for geographic isolation, and is one way in which speciation can occur sympatrically.

A second type of polyploid speciation, called **_allopolyploidy_**, involves a multiplication (usually a doubling) of the number of chromosomes in a hybrid between two species. The hybrid has one set of chromosomes from each of the two species. Unless these two species are so closely related that they have homologous chromosomes capable of pairing (synapsing) in meiosis, the hybrid will almost certainly be sterile because of an inability to produce gametes. But if the hybrid undergoes chromosome doubling before meiosis, it will have a complete diploid set of chromosomes from each species, and viable gametes are far more likely to be produced. This type of polyploidy has probably been far more important in speciation than autopolyploidy. Allopolyploid individuals will be able to breed freely among themselves, but they will be unable to cross with either of the parental species. Consequently the allopolyploid population must be regarded as a distinct species.

Allopolyploid plants are only rarely more vigorous than the parental diploid plants, probably because each parental species is a delicately balanced mixture of genes, while the allopolyploid mixes together two sets of gene products, metabolic pathways, and control systems. But occasionally allopolyploids, because of the combination of genes from two different species, can grow in habitats which neither parental species can colonize. Allopolyploid speciation, therefore, has probably played an important role in the perpetuation of some plant groups during periods of widespread environmental change.

Allopolyploidy has also proven of great importance in the production of valuable new crop plants. As soon as plant breeders realized that many of

34.12 Sympatric speciation in tree hoppers
Two sympatric populations of the tree hopper *Enchenopa binotata* have evolved adaptations to different host plants. The tree hopper on the left lives on bittersweet, while the one on the right lives on butternut. Host specificity may take the place of physical separation (allopatry) in preventing these two populations from interbreeding.

our most useful plants, such as oats, wheat, cotton, tobacco, potato, banana, coffee, and sugarcane are polyploids, they began trying to stimulate polyploidy, and obtained many new varieties. It was found that the chemical colchicine would readily induce polyploidy. One of the first artificially produced allopolyploids came from a cross made in 1924 between the radish and the cabbage. Unfortunately it had the root of the cabbage and the shoot of the radish. Other crosses have yielded more desirable results (see Fig. 25.16, p. 689).

Chromosome doubling creates an obvious problem: if tetraploids cannot breed with diploids, they are restricted to breeding with each other. Hence, the rare diploid pollen grain must find an equally rare diploid egg cell to produce a fertile tetraploid. And, most likely, the resulting plant must then fertilize itself, since other tetraploids will be exceedingly rare. The same is true of allopolyploids: the sole hybrid is extremely unlikely to find a genetically compatible plant, and so will probably have to fertilize itself. The result in both cases will be severe inbreeding of the new species, a strong founder effect, and rapid genetic drift. The new population will persist only if it has a strong competitive advantage in the habitat.

As we have seen, polyploid speciation generates new species almost instantaneously. Some researchers believe that the same principle operating on a smaller scale may also be a powerful but unrecognized mechanism of speciation. Chromosomal rearrangements, whether the result of a few major breakage-and-fusion events or of smaller but more numerous transpositions, duplications, and deletions, might give rise to individuals or small populations genetically incompatible with the rest of their species. If the variant organisms are at a competitive advantage, selection would favor further intrinsic isolation, and sympatric speciation might result.

Nonchromosomal sympatric speciation   Though most speciation not involving gross chromosomal changes is certainly allopatric (requiring a period of geographic isolation), there is increasing evidence that sympatric speciation can occur without polyploidy or other major chromosomal rearrangements. Reproductive isolation is just as essential to these types of sympatric speciation, but is effected by other means. For example, relatively small changes in habitat preference may produce habitat isolation with effects sufficient to compensate for the absence of geographic separation in the development of isolating mechanisms. Thus a species of clover adapted to soil containing mine tailings is now reproductively isolated from the widespread and closely related species from which it evolved, not by polyploidy but by virtue of having a different flowering season. Another alternative to geographic isolation, strongly implicated in arthropods like tree hoppers, is related to host specificity. These insects often mate on the plants on which they feed, so races adapted for a particular species of host plant will tend to inbreed. According to this controversial scenario, selection would favor those individuals that breed strictly on the host species; consequently, intrinsic isolating mechanisms might develop which would serve to isolate tree hoppers with adaptations to different host plants. In effect, sympatric speciation in tree hoppers would become possible because host specificity could create reproductive isolation even in the absence of true geographic isolation (Fig. 34.12).

Yet another mechanism that may contribute to sympatric speciation is the behavioral phenomenon known as sexual imprinting (see p. 571). You may recall that the members of one sex of many species automatically memorize while young a particular feature or set of features displayed by their parents or siblings. The features may be visual, auditory, or olfactory. The memorization enables these individuals later to identify suitable mates with great precision. The lasting effect of sexual imprinting in birds was shown in experiments by Klaus Immelmann involving zebra finches and Bengalese finches. Birds of these two species are physiologically capable of interbreeding, but imprint so forcefully on their own species that they do not. In Immelmann's experiments, male zebra finches raised by Bengalese parents invariably courted female Bengalese finches when given a choice. Now, if the parent on which the young imprint displays a mutation in the feature being committed to memory (a brightly colored eye ring, for example, or novel elements in the courtship song), and if the young are later able to locate mates displaying the same mutation, the mutant individuals are likely to pair with each other. The result could be the founding of a population that does not interbreed with the rest of the species. In birds, then, instant reproductive isolation based on imprinting may play an important role in shaping populations.

Adaptive radiation   One of the most striking aspects of life is its extreme diversity. A bewildering array of species now occupies this globe. And the fossil record shows that of the species that have existed at one time or another those now living represent only a tiny fraction (probably less than one-tenth of 1 percent, all the other species being now extinct). Clearly, then, divergent evolution—the evolutionary splitting of species into many separate descendant species—has been exceedingly frequent. Is it possible to account for such a degree of evolution radiation by the models outlined above? In particular, how could opportunities for geographic isolation—thought to be by far the most common precursor to speciation—have been sufficient to lead to all the speciation not caused by sympatric speciation? After all, it is not unusual for a complex of four, five, or more closely related species to occur within a rather limited area. For instance, 28 species of the milliped genus *Brachoria,* many of them sympatric, are confined to a small portion of the deciduous forests of the eastern United States (Fig. 34.13).

34.13 Distribution of 28 species of millipeds of the genus *Brachoria*
The color dots indicate all known localities for these millipeds, many of which are sympatric. All the speciation in this genus must have occurred within this very limited area in the eastern United States.

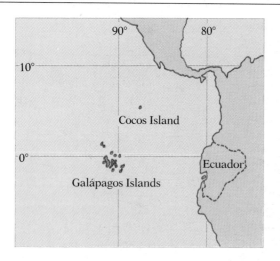

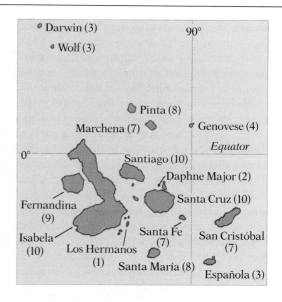

34.14 The Galápagos Islands
Left: The islands are located about 950 km off the coast of Ecuador. Cocos Island is about 700 km northeast of the Galápagos. Right: The islands shown in greater detail. The number in parentheses after each name indicates the number of species of Darwin's finches that occur on the island.

How could so much speciation have occurred in so small an area if geographic isolation was a necessary factor? In an attempt to answer such questions, let us turn to a particularly instructive and historically important example—the finches on the Galápagos Islands, which (along with tortoises and mockingbirds on those islands) played a major role in leading Charles Darwin to formulate his theory of evolution by natural selection.

The Galápagos Islands lie astride the equator in the Pacific Ocean roughly 950 km west of the coast of Ecuador, the country to which they now belong (Fig. 34.14). These islands have never been connected to each other. They apparently arose from the ocean floor as volcanoes approximately five million years ago. At first, of course, they were completely devoid of life, and were therefore open to exploitation by whatever species might chance to reach them from the mainland. Relatively few species did manage to reach the islands and become established. The only land vertebrates present on the islands before human beings got there were at least seven species of reptiles (one or more snakes, a huge tortoise, and at least five lizards, including two very large iguanas), seven species of mammals (five rats and two bats), and a limited number of birds (including two species of owls, one hawk, one dove, one cuckoo, one warbler, two flycatchers, one martin, four mockingbirds, and the famous Darwin finches).

The 14 species of Darwin's finches constitute a separate subfamily found nowhere else in the world. Thirteen of them are believed to have evolved on the Galápagos Islands, and one on Cocos Island, from some unknown finch ancestor that colonized the islands from the South American or Central American mainland. It is readily understandable that the descendants of the geographically isolated colonizers should have undergone so much evolutionary change as to become, in time, very unlike their mainland ancestors. More perplexing at first glance is the manner in which the descendants of the original immigrants split into the separate populations that gave rise to today's 14 species.

The point to remember is that we are dealing not with a single island but with a cluster of more than 25 separate islands. The finches will not readily fly across wide stretches of water, and they show a strong tendency to remain near their home area. Hence a population on any one of the islands is effectively isolated from the populations on the other islands. We suppose that the initial colony was established on one of the islands where the colonizers, perhaps blown by high winds, chanced to land. Later, stragglers from this colony wandered or were blown to other islands and founded new colonies. The allelic frequencies in the new colonies differed from those in the original colonies from the moment they started, because of the founder effect. In time, the colonies on the different islands diverged even more, for the reasons already outlined in the model of geographic speciation (different mutations, different selection pressures, and, in such small populations as some of these must have been, genetic drift). What we might expect, therefore, is a different species, or at least a different race, on each of the islands. But this is not what has actually been found; most of the islands have more than one species of finch, and the larger islands have 10 (Fig. 34.14, right). What is the explanation?

Let us suppose that form A evolved originally on the island of Santa Cruz and that the closely related form B evolved on Santa María. If, later, form A had spread to Santa María before the two forms had been isolated long enough to evolve any but minor differences, the two forms might have interbred freely and merged with each other. But if A and B had been separated long enough to have evolved major differences before A invaded Santa María, then A and B might have been intrinsically isolated from each other, having developed into separate species, and they might have been able to coexist on the same island without interbreeding (Fig. 34.15). If they formed occasional hybrids, those hybrids might well have been less viable than the parental forms. Accordingly, natural selection would have favored individuals that mated only with their own kind, and this selection pressure would have led rapidly to more effective intrinsic isolating mechanisms, which would prevent the gamete wastage involved in cross-matings. It has been shown, in fact, that Darwin's finches readily recognize members of their own species and show little interest in members of a different species.

We have now arrived at a point in our hypothetical example where Santa Cruz is occupied by species A and Santa María by both A and B. It would be highly unlikely that A and B could coexist indefinitely if they used the same food supply; the ensuing competition would be very severe, and the less well adapted species would tend to be eliminated by the other unless it evolved differences that minimized the competition. In other words, wher-

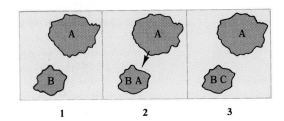

1   2   3

34.15 Model of speciation on the Galápagos Islands An ancestral form colonized the larger of these two hypothetical islands. Later, part of the population dispersed to the smaller island. (1) Eventually the two populations, being isolated from each other, evolved into separate species A and B. (2) Some individuals of A dispersed to B's island. The two species coexisted, but intense competition between them led to rapid divergent evolution. (3) This rapid evolution of the population of A on B's island caused it to become more and more different from the original species A, until eventually it was sufficiently distinct to be considered a full species, C, in its own right. At the same time, the selection pressure imposed by the small invading population caused the large population of species B to evolve to a small degree as well.

34.16 Darwin's finches

Darwin's finches fall into four genera: birds 1, 3, 4, 5, 6, and 10 are the tree finches (*Camarhynchus*); birds 7, 8, 11, 12, 13, and 14 are the ground finches (*Geospiza*); bird 2 is the unfinchlike warbler finch; bird 9 is the one species inhabiting Cocos Island.

1. Vegetarian tree finch (*C. crassirostris*)
2. Warbler finch (*Certhidea olivacea*)
3. Large insectivorous tree finch (*C. psittacula*)
4. Medium insectivorous tree finch (*C. pauper*)
5. Mangrove finch (*C. heliobates*)
6. Small insectivorous tree finch (*C. parvulus*)
7. Large cactus ground finch (*G. conirostris*)
8. Cactus ground finch (*G. scandens*)
9. Cocos finch (*Pinaroloxias inornata*)
10. Woodpecker finch (*C. pallidus*)
11. Large ground finch (*G. magnirostris*)
12. Sharp-beaked ground finch (*G. difficilis*)
13. Medium ground finch (*G. fortis*)
14. Small ground finch (*G. fuliginosa*)

ever two or more very closely related species occur together, competition will lead either to the extinction of one, or to character displacement—in this instance the evolution of different feeding specializations.

Character displacement is indeed what we find in Darwin's finches (Fig. 34.16). The 14 species form four groups (genera). One group contains six species that live primarily on the ground; of these, some feed primarily on seeds and others mostly on cactus flowers. Of the species that feed on seeds, some feed on large seeds, some on medium-sized seeds, and some on small seeds. These feeding preferences result from the morphological specialization of the beaks: small beaks are most efficient at handling small seeds, while larger beaks can crack large seeds. (The specialization is not symmetrical: large beaks can handle small seeds, though not very efficiently, but small beaks cannot crack large seeds at all.) From a series of careful beak measurements, David Lack of Oxford University was able to find clear evidence of character displacement. For example, when the small and medium ground finches coexist on the larger islands, their beak sizes are widely separated, with depths averaging about 8.4 and 13.2 mm respectively. But on small islands where only one of the two species exists, the birds of that species tend to have beaks of intermediate size, on the order of 9.7 mm (Fig. 34.17).

The second group of finches contains six species that live primarily in trees. Of these, one is vegetarian and the others eat insects, but the insect eaters differ from one another in the size of their prey and in where they catch them (Fig. 34.18). A third group contains only one species, which has become very unfinchlike and strongly resembles the warblers of the main-

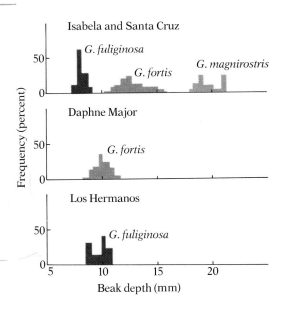

34.17 Beak sizes of ground finches
On large islands like Isabela and Santa Cruz, three species of ground finch coexist. Though the beak sizes of individuals within each species show the sort of variation essential for evolution by natural selection, they fall into three separate ranges, specialized respectively for efficient feeding on small, medium-sized, and large seeds. However, on small islands where only one species exists (either the small or the medium ground finch) the beak sizes fall into an intermediate range regardless of the size specialization found in the same species on the larger islands. In the face of competition for food on the larger islands, character displacement has taken place.

34.18 The tool-using finch
One of the insectivorous tree finches of the Galápagos has evolved most unusual feeding habits. Sometimes called the woodpecker finch, it chisels into wood after insects, but it lacks the long tongue that a woodpecker uses to probe insects out of a crack. Instead, it pokes into the crack with a cactus spine or twig that it holds in its beak. The mangrove finch does the same thing in a different habitat. These are two of the few known cases of tool use by a bird.

34.19 Beak differences in Darwin's finches on the central islands

The differences may appear slight at first glance, but they have important functional implications for the birds' diets and for species recognition in mating. Top row (diagonally downward from left to right): *Geospiza magnirostris, G. fortis, G. fuliginosa*. Second row: *G. difficilis, G. scandens, Camarhynchus crassirostris*. Third row: *C. psittacula, C. parvulus, C. pallidus*. Fourth row: *C. heliobates, Certhidea olivacea, Pinaroloxias inornata*.

land. The fourth group contains only one species, restricted to Cocos Island, which is about 700 km northeast of the Galápagos Islands and about 500 km from Panama. Correlated with the differences in diet among the species are major differences in the size and shape of their beaks (Fig. 34.19). (Noteworthy variations in beak size and shape among birds thought to share a common ancestry can be found in other island habitats as well, as shown, for example, in Figure 34.20. These characteristics of the beak are an important means by which the birds recognize other members of their own species; song is another.)

Now, if selection on Santa María favored character displacement between species A and B, the population of species A on Santa María would become less and less like the population of species A on Santa Cruz (Fig. 34.15). Eventually these differences might become so great that the two populations would be intrinsically isolated from each other and would thus be separate species. We might now designate as species C the Santa María population derived from species A. The geographic separation of the two islands would thus have led to the evolution of three species (A, B, and C) from a single original species. The process of island hopping followed by divergence could continue indefinitely and produce many additional species. It was doubtless such a process, involving initial divergence on separate islands followed by intensification of differences when sympatry developed as a result of subsequent invasions, that led to the formation of the 14 species of Darwin's finches.

In Chapter 21 we distinguished between generalist species, which feed on a wide range of foods, and specialist species, which require foods within a very narrow range. We said that most species fall somewhere between these extremes on the continuum of lifestyles. In Darwin's finches, we can see how competition leading to character displacement can contribute to the evolution of specialist morphology, physiology, and behavior. On the large islands where two or more species of ground finch coexist, the beak size of each species reflects a relative specialization for small, medium-sized, and large seeds, whereas on the smaller islands with only one species, the beak size is intermediate, allowing the individuals to be comparative generalists, taking both small and medium-sized seeds. (Large seeds are relatively rare.) Thus birds of the same species (*G. fuliginosa*, for instance), can be relative specialists on one island and relative generalists on another, depending on competition from other species (Fig. 34.17). The evolution of such intraspecific variations by means of natural selection is a crucial first step in allopatric speciation.

Now let us apply the principles learned from Darwin's finches to the case of the 28 species of *Brachoria* millipeds confined to a small area in the eastern United States (see Fig. 34.13). These animals live in the humus layer on the floor of deciduous forests. They are rather sluggish and seldom move very far. It would have been easy for populations to become isolated in local forested areas separated by less hospitable regions. Such allopatric populations could have become sufficiently different so that, when conditions changed and they became sympatric, they would behave as full species. Clearly, the sorts of processes seen on islands can account also for radiation in a continental area. And on a somewhat larger geographic scale, the same processes can account for the observed adaptive radiation in insects, fish, reptiles, birds, mammals, and many plant groups. In short, adaptive radiation on islands, like that of Darwin's finches, is dramatic and lends itself particularly well to analysis, but it does not differ in principle from adaptive radiation under other circumstances. It helps show that the model of speciation we have outlined can account for the great amount of divergence necessary to produce the immense diversity among living things.

It should be clear, as Darwin himself pointed out, that the rate of evolutionary divergence is not always constant. Almost surely, when the first colonizing finches reached the Galápagos Islands, they encountered environmental conditions quite unlike those they had left behind in Central or South America. Hence the selection pressures to which they were subjected were probably quite different from those in their former home; differences in the resources available, for example, may have led to selection for different morphology, physiology, and behavior. On the other hand, such differences may not have been immediately important: if initially there was little or no competition from specialist species, the result would have been a temporary relaxation of selection pressures. Only when the new habitat became saturated with finches would intraspecific competition for the available resources have become important. And so, sooner or later, selection pressures must have led to very rapid divergence from the ancestral population. Later, as the finches became increasingly well adapted to conditions on the Galápagos, the rate of evolutionary change probably slowed down.

In general, when conditions change radically and organisms have new evolutionary opportunities for which they are at least modestly preadapted, they may undergo an evolutionary burst—a period of rapid adaptational change—which may then be followed by a more stable period during which any further evolutionary changes are merely fine-tuning of their already well-adapted characteristics. Such bursts of rapid evolutionary divergence probably characterized the tremendous radiation of amphibians when they moved onto land for the first time, and the explosive radiation of mammals when the demise of the dinosaurs left many adaptive slots unoccupied.

How important is competition?   In the last chapter we contrasted the roles played by genetic drift (chance) and natural selection in the evolution of populations, emphasizing that evolution is a vital, ongoing phenomenon, while drift (potentially important only in small populations) and selection are two contributing—and clearly demonstrated—mechanisms by which it occurs. In our discussion of speciation we have touched on a similar theme: barriers that contribute to reproductive isolation can arise by chance (primarily genetic drift), by natural selection, or by a combination of the two. Throughout our discussion we have emphasized the role of competition in the formation of the species: even reproductively isolated populations may not be able to coexist indefinitely if they compete for precisely the same food, since just a slight but systematic superiority of one will tend to lead to the extinction of the other.

G. F. Gause of the University of Moscow, who first observed this phenomenon in the laboratory in the 1930s, formulated the competitive exclusion principle, an early version of what we referred to in Chapter 22 as the Niche

34.20 Beak differences in Hawaiian honeycreepers Differences in beak size and shape are apparent in related species of Hawaiian honeycreepers, which, like Darwin's finches, are thought to have evolved from a common finchlike ancestor.

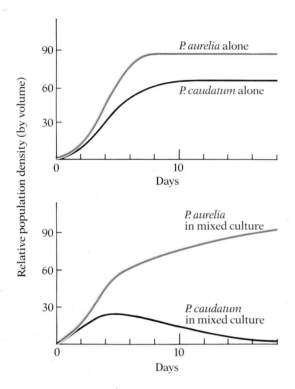

34.21 Competition and extinction
When a few individuals of either species of *Paramecium* are introduced into a tank alone (top), they multiply until they reach a limiting density. But when individuals of both species are introduced together (bottom), they multiply independently for the first three days, then begin to compete for resources. *P. aurelia* is in some way more efficient under these conditions, and drives *P. caudatum* to extinction in only three weeks.

Rule: no two species occupying the same niche can long coexist (Fig. 34.21). Only the character displacement that natural selection produces, resulting in changes in food preference, habitat choice, and the like, will allow closely related species to coexist. This Darwinian interpretation of how separate species form, with its emphasis on reproductive isolation and character displacement in the face of competition, probably accounts for much of what we can observe of speciation. But during the last decade the role of competition in speciation has come under increasingly close scrutiny. Today, competition is no longer thought to be as widespread and important as was formerly believed, while chance is now seen as clearly a greater force than Darwin and most scientists since him have thought.

There are at least two ways in which chance can take precedence over competition in causing species to diverge. The first we have already discussed: in small populations, genetic drift can be very powerful, driving alleles to extinction before the more gradual process of natural selection can produce stability. The other process, however, is more fundamental to our understanding of evolution. Most Darwinian analyses tacitly assume that the selection pressures organisms face change relatively slowly, and that large continuous populations therefore have ample time, generation by generation, to evolve specialized adaptations to their environments. But do conditions really remain sufficiently constant for a species to achieve equilibrium—a stable set of allelic frequencies—in the face of all the selection pressures affecting it?

Apparently, many populations are not always at equilibrium. In one particularly thorough study Peter Grant of Princeton University and his colleagues tagged 1,500 medium ground finches on one of the smaller Galápagos islands, Daphne Major. In 1977 only 20 percent of the normal amount of rain fell, and the plants on the island produced many fewer seeds than usual. Relatively dry or wet seasons, a surprisingly early or late hard frost, or especially warm or cold years are familiar conditions which organisms can generally take in stride, but the effects on plants and animals of extreme once-a-century events can be drastic.

On Daphne Major, the consequences of the reduction in food were dramatic for the medium ground finch: with only 35 percent of the usual amount of food available, 387 of the 388 nestlings of 1976 died, as did the majority of adults. Furthermore, the survivors were not a representative cross section of the former population: roughly 180 of the 500 adult males survived, but only about 30 of the 500 females, and the birds that did survive the year of starvation were significantly larger than those that succumbed. Exceptionally intense natural selection in favor of large size had occurred.

Thus one of several rare, unpredictable, but inevitable crises finches face sharply altered the selection pressures for a year, and reduced the population to a level at which genetic drift suddenly became a real possibility. What had seemed to be an adaptive equilibrium was upset by a chance disturbance, and a population with altered characteristics emerged. The finch population had been forced through a period of environmental crisis. When such a crisis is so severe as to cause major changes in allelic frequencies in a population, it is called an ***evolutionary bottleneck***.

The assumption that interspecific and intraspecific competition, leading gradually to character displacement and to new species, is the major mode of speciation is currently being challenged by the view that chance crises may occur often enough to upset whatever stability exists in the allelic frequencies within a species. Crises can be caused by any environmental factor—pestilence, or extreme weather, for example—that severely affects an isolated population, or that itself serves to isolate one part of a population from the remaining body. During such a crisis, as an isolated population passes through the evolutionary bottleneck produced by extraordinary environmental conditions, one character or another may gain ascendancy. Having been selected by a founder effect, such a character is still at risk if the surviving population is small, because it may decline in frequency or even become extinct through genetic drift. Alternatively, if a character that was not adaptive prior to the crisis proves adaptive thereafter, it may increase in frequency in the population. Though most researchers still believe that competition is the predominant force leading to speciation over time, many studies are beginning to suggest that the catastrophic effects of rare crises may have been important in the evolution of certain populations.

Punctuated equilibrium   One consequence of the increasing evidence that natural catastrophes can affect the allelic frequencies of populations has been a growing interest in ***punctuated equilibrium***. This much-debated hypothesis concerning the mode and tempo of speciation was originally formulated by Niles Eldredge of the American Museum of Natural History and Stephen Jay Gould of Harvard. Based on careful study of certain fossil records, their hypothesis suggests that most allopatric speciation events are geologically "instantaneous," the result of crises that punctuate long periods of equilibrium, or ***stasis***, in which the morphology of the species remains relatively constant. Gould and Eldredge maintain that the fossil record does not support ***gradualism***—the view, which they believe was Darwin's, that speciation occurs as a gradual accumulation of morphological and physiological changes (Fig. 34.22). "Instantaneous" speciation, however, is by no means as rapid as it sounds: they assert that it takes thou-

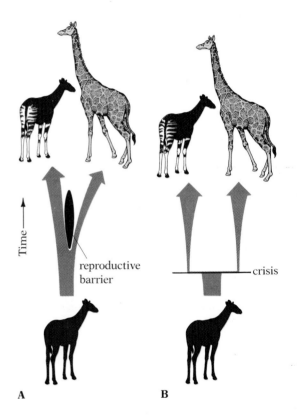

34.22   Gradualism versus punctuated equilibrium
These drawings represent two hypotheses about how different species—here, the okapi and the giraffe—might have evolved from a common ancestor, the pre-okapi. Forms intermediate between the pre-okapi and the okapi and giraffe are not shown. (A) In the conventional interpretation of natural selection, a population is imagined to evolve gradually (as suggested by the gentle leftward movement of the lower part of the arrow, indicating a slow change in one trait) until a reproductive barrier (usually geographic isolation) separates two parts of the population. Generally one of the subgroups is very small, as indicated by the narrowness of the right-hand branch; it may diverge slowly from the larger subgroup through selection or genetic drift or both. (B) According to the punctuated-equilibrium model, there is little or no gradual change of a trait over time. Instead, a crisis or other event selects or gives rise by chance to one or more populations with traits very different from those of the ancestral population.

sands of years, requiring perhaps one-tenth of the lifetime of a species, or up to 100,000 years, to complete. Perhaps most important, because the population that survives and evolves after the crisis would seem to arise suddenly in the fossil record, Eldredge and Gould's theory provides an explanation of the gaps that exist in the record. According to the supporters of punctuated equilibrium, more gradual processes are of relatively minor importance in speciation, and serve mostly therefore to fine-tune a species to its environment.

The debate between punctuated equilibrium and the more conventional gradualistic perspective is fueled to a great extent by the many differences in how the fossil record is interpreted. One difference arises because there is variation in rates of morphological change. Current evidence suggests that in some species certain morphological features may change rapidly in response to the selection pressures upon them, while others may remain relatively constant. A researcher who examines or is only able to discern the characters that remain constant may overlook this so-called *mosaic evolution* and assume that the species exhibits stasis—no change at all. On the other hand, supporters of gradualism often asssume that slow morphological or physiological change is occurring even when it is not obvious from the record; of course, neither physiological nor morphological changes in most soft body parts are preserved in the fossil record.

Another difference in interpretation concerns the great discontinuities in the record. The conventional view of these gaps is that they are, in most cases, just anomalies, and that the missing pieces in the picture of gradual evolutionary change might, if found, provide valuable information about transitions between the species preceding and following the gaps. The supporter of punctuated equilibrium, by contrast, regards the gaps as the norm, and as proof of the "instantaneous" speciation events the hypothesis postulates. But gaps in the record may frequently result from the kinds of environmental crises we have already discussed. Consider what happens when a lake, whose sediments have provided an excellent record of its organic inhabitants for hundreds of thousands of years, suddenly dries up as a result of a major climatic change. The formation of the fossil record in this location is terminated just at the time when the inhabitants of the lake are faced with extraordinary selection pressures and potential extinction. Also, since the gaps frequently cover a much longer time span than the fossil records being studied, doubters question whether any useful speculation is possible about evolutionary occurrences during these periods. In view of these sorts of paleontological puzzles, and the difficulty of designing tests to verify evolutionary hypotheses, a definitive resolution in favor of either the gradual-change or the punctuated-equilibrium model seems unlikely in the near future.

Nevertheless, studies of particularly complete fossil records are beginning to provide intriguing clues about the tempo of speciation. P. G. Williamson of Harvard, for instance, analyzed in detail an unusually complete four-million-year sequence of mollusc fossils in East Africa. He found that in this case speciation took place over geologically brief periods of 5,000 to 50,000 years, and that these periods interrupted long periods of little change in external morphology. To be sure, even 5,000 years hardly seems instantaneous, but the point is that changes in shell morphology were remarkably slow except in these clearly defined periods, during which change may very well have been caused by environmental crises leading to evolutionary bottlenecks. Of course, this record—like the fossil record in general—is silent on changes in physiology and internal structure, as well as on the matter of reproductive isolation, which is crucial to the classification of living species.

The evolution of some organisms is evidently less uniform and gradual than many present-day biologists have supposed. Indeed, Darwin himself pointed out that both local and global environmental phenomena must redefine and magnify selection pressures enormously and hence alter the direction and rate of evolution. The usual tempo of speciation may turn out to lie somewhere between the gradual-change and the punctuated-equilibrium models.

THE SPECIES PROBLEM

We have been discussing speciation without worrying about any potential problems in identifying species as opposed to races, populations in a cline, and so on. But it would be wrong to leave the impression that the modern definition of species can be applied without difficulty in all cases, or, indeed, that it is valid in all cases. The details of the definition itself are controversial and can provoke heated argument among biologists. Though most of them accept the major ideas on which the definition rests, and though most of them, if they were to study the same set of natural populations, would probably agree in the great majority of instances about which populations represent full species and which do not, there would be a small percentage of populations on which they could not agree. These are the cases where the modern definition of species is hard to apply or invalid. Let us examine a few such cases.

Asexual organisms   Since the modern definition of species assumes interbreeding, it obviously does not apply to asexual organisms. Though most so-called asexual organisms actually do have provision for occasional sexual recombination, a few lack sexual mechanisms altogether. Dandelions, for instance, despite their flowers, are totally asexual (Fig. 34.23). Can such truly asexual organisms be said to form species in any sense? And would such species be comparable to sexual ones?

Asexual organisms do seem to form recognizable groups or kinds, even though the members of a group cannot exchange genes. Gaps, or discontinuities in the variation, occur between the various kinds just as they do between sexual species. One possible explanation for the groupings is that each group of asexual organisms evolved from a sexual species. In this way, the flowers of dandelions, which now have infertile pollen and diploid eggs that go on to form seeds without fertilization, might have originated in sexually reproducing ancestors. Since not all the variations that occur over the course of time are likely to be equally well adapted, asexual organisms like dandelions would continue to form recognizable groups: only those individuals whose genotypes produced well-adapted phenotypes would survive in significant numbers; hence there would be a limited number of super-

34.23   Flower and seeds of the dandelion, an asexual organism
Despite appearances to the contrary, the dandelion reproduces asexually at all times. The flower of the dandelion, which is thought to have been inherited from a sexually reproducing precursor, now has infertile pollen and diploid eggs that go on to form seeds without fertilization.

iorly adapted "types," and all individuals falling within the bounds of one such type would constitute a natural group that could be called a species, while all individuals falling within the bounds of another adaptive type would constitute a second species. The asexual species thus determined would resemble sexual species in that the latter, too, represent adaptive peaks. The two kinds of species, sexual and asexual, therefore play comparable ecological roles, and both would be subject to natural selection. In the long run, however, the inability of asexual organisms to exploit the potential of genetic recombination to meet the challenges of changing conditions makes them less able to survive the vicissitudes of environmental change.

Fossil species   The modern definition of species can be applied formally only to organisms that coexist, since the criterion of interbreeding cannot be used when comparing an organism with its likely ancestors of a million years earlier. Therefore, paleontologists can compare organisms from different periods in the earth's history only by using morphological criteria and geographic distribution, classifying two forms as separate species when they differ to about the same degree as do related organisms from the present day that are known to constitute reproductively isolated species. For practical purposes, paleontologists usually regard gaps in the fossil record as breaks between species, even though they are fully aware that no gaps actually occurred in the lineages of the organisms.

Populations at an intermediate stage of divergence   Our model of allopatric speciation assumes that geographically isolated populations will slowly diverge by essentially imperceptible stages until they have reached the level of full species. The intrinsic reproductive isolation that makes them full species itself evolves gradually. Hence there is no precise point at which the diverging populations suddenly reach the level of full species. There will be a period in the history of any two diverging lineages when the populations are in a hazy intermediate state between obviously belonging to the same species and obviously belonging to two separate species. But our definition of species makes no provision for such intermediate stages. Consequently intermediate stages, when they are encountered, must always pose a problem to any biologist intent on rigid categorization of what in nature is a fluid system. But the existence of intermediate stages does not invalidate the concept of speciation, because that very concept, in its modern form, predicts them.

Allopatric species   One of the most obvious and frequently encountered problems in applying the modern definition of species arises when two populations are closely related and completely allopatric. Since they are allopatric, they are obviously not exchanging genes. But the definition of species is not based on extrinsic isolation; it is based on intrinsic isolation. There must be neither actual nor potential effective gene flow if the two populations are to be regarded as separate species. How can potential gene flow be determined? One way that immediately comes to mind is to release a large sample of individuals from one population in the range of the other and then see whether free interbreeding takes place and, if it does, whether the hybrids are as viable as the parents. But there are obvious reasons why wholesale introduction of foreign plants and animals is seldom desirable; in fact, in many cases it is illegal. An alternative would be to bring individuals from the two allopatric populations together in the laboratory and see if they will interbreed. Sometimes this procedure is useful. If the individuals will breed freely with other members of their own population but not with members of the other population, then we may reasonably conclude that the two populations are intrinsically isolated and should be considered separate species.

But what if interbreeding occurs freely between members of different populations in the laboratory? Are the two populations then to be regarded as belonging to the same species? No, the interbreeding simply demonstrates that certain types of intrinsic isolation do not exist between the populations. It says nothing about other types of intrinsic isolation. For example, under natural conditions ecogeographic or habitat isolation may exist, but these might very well be inoperative under laboratory conditions. Or behavioral isolation may be operative in nature but not in the laboratory. Many species of animals that will have nothing to do with each other in the wild, because of important differences in their behavior patterns, will mate in the laboratory, where their normal behavior patterns break down. Lions and tigers, for example, are distinct allopatric species in the wild, and never mate in their small region of range overlap; in the unnatural environment of a zoo, however, they will mate, and produce living offspring. Clearly, when members of two different allopatric populations mate in the laboratory and produce viable offspring, the question of whether they belong to the same or to different species remains unanswered. The same ambiguity exists for the many organisms that will not breed at all under laboratory conditions: after all, males and females of the *same* species often refuse to mate outside of their natural habitats.

In many cases, then, there is no good test for determining whether two allopatric populations belong to the same or to different species. The usual practice in such cases is to determine the extent to which the two populations differ and then to compare this degree of difference with that seen in related sympatric species. If the differences between the allopatric populations are of the same order of magnitude as (or greater than) those that distinguish sympatric species, the allopatric populations are considered fully separate species; if the differences are less than those that usually distinguish sympatric species, the two allopatric populations are likely to be regarded as belonging to the same species.

But degree of difference leaves much to be desired as an index to the probable presence or absence of intrinsic reproductive isolating mechanisms between allopatric populations. The assumption that allopatric populations are intrinsically isolated only if they differ from each other at least as much as related sympatric species do is often not valid. Owing to the frequency of character displacement in sympatric species, the differences between them are apt to be greater than between allopatric species. Therefore, taking the differences between sympatric species as a yardstick against which to measure differences between allopatric populations probably often results in classifying allopatric populations as members of the same species when in reality they are intrinsically isolated and should be regarded as separate species.

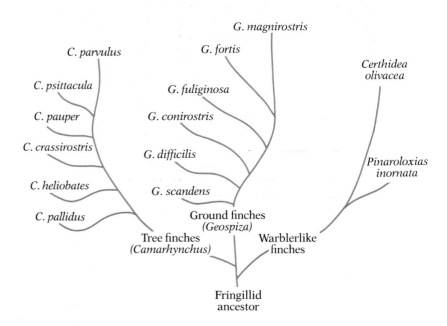

34.24   A phylogenetic tree for Darwin's finches
This tree is based on both the degree and the nature of the morphological differences between species. The distances between the branch points—the places where individual branches originate—reflect degrees of similarity.

THE CONCEPT OF PHYLOGENY

Evolution implies that many unlike species have a common ancestor and that all forms of life probably stem from the same remote beginnings. Hence one of the tasks it sets for biologists is to discover the relationships among the species alive today and to trace the ancestors from which they descended (Fig. 34.24).

DETERMINING PHYLOGENETIC RELATIONSHIPS

When systematists,[3] also known as taxonomists, set out to reconstruct the evolutionary history—the *phylogeny*—of a group of species that they think are related, they usually have before them only the species living today. They cannot observe their phylogenetic history. To reconstruct it as closely as possible, they must make inferences based on observational and experimental data that appear relevant. There are at least four major approaches to systematics—classical evolutionary taxonomy, phenetics, cladistics, and molecular taxonomy. Each uses different techniques.

Classical evolutionary taxonomy   Classical systematics, the most popular approach, depends more than any other on experience and subjective judgment. The usual procedure in reconstructing phylogenies by the classical method is to examine as many independent characters of the species in question as possible and to determine in which characters they differ and in which they are alike. The assumption is that the differences and resemblances will reflect, at least in part, their true phylogenetic relationships. Ordinarily, as many different types of characters as possible are used in the

---

[3] Systematics, or taxonomy, in the words of G. G. Simpson, is "the scientific study of the kinds and diversity of organisms and of any and all relationships among them."

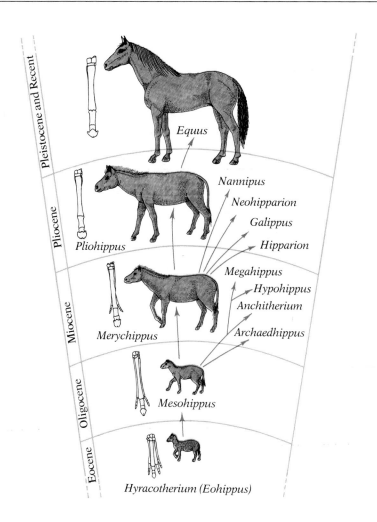

Pleistocene and Recent

Pliocene

Miocene

Oligocene

Eocene

Equus

Pliohippus

Nannipus

Neohipparion

Galippus

Hipparion

Megahippus

Hypohippus

Anchitherium

Archaedhippus

Merychippus

Mesohippus

Hyracotherium (Eohippus)

34.25 Presumed evolution of horses
The fairly complete fossil record of horses has enabled paleontologists to work out a reasonable picture of the evolutionary history of the group. The emphasis here is on the direct ancestors of the modern horse; many branches left no modern descendants. *Hyracotherium* lived in the Eocene epoch about 55 million years ago. It was a small animal, only about the size of a fox terrier. It had four toes on each front foot (shown here) and three on each rear foot. It was a browser, feeding on trees and bushes. *Mesohippus*, which lived during the Oligocene epoch about 35 million years ago, was a bit larger, and its front feet, like the rear, had only three toes. *Merychippus*, a grazer, lived during the Miocene about 25 million years ago. It had three toes on each foot, the middle one much larger than the others, which were short and thin and did not reach the ground. *Pliohippus*, of the Pliocene, often had only one toe on each foot, though in some individuals tiny remnants of other toes persisted. *Equus*, the modern horse, is much larger than the ancestors shown here. It has only one toe on each foot.

hope that misleading data from any single character will be detected by a lack of agreement with the data from other characters.

The most easily studied and widely used characters pertain to morphology—including external morphology, internal anatomy and histology, and the morphology of the chromosomes in cell nuclei. It is particularly helpful, of course, when morphological characters of living species can be compared with those of fossil forms. The fossil record is the most direct source of evidence about the stages through which past forms of life passed, but unfortunately that record is usually very incomplete and is subject to the same sorts of errors of interpretation as the characters of living species. For many groups of organisms there is no suitable fossil record for working out the relationships between species; at best, the fossils may suggest the broad outlines of the evolution of major groups. Even in those groups where fossils are abundant, only the hard parts of the organisms' bodies have usually been preserved. Nevertheless, in some groups, notably the horses, the fossil record has provided much phylogenetic information that could have been obtained from no other source (Fig. 34.25).

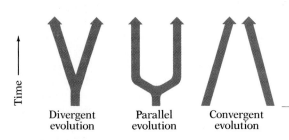

Divergent evolution   Parallel evolution   Convergent evolution

34.26   Patterns of evolution

In divergent evolution one stock splits into two, which become less and less like each other as time passes. In parallel evolution two related species evolve in much the same way for a long period of time, probably in response to similar environmental selection pressures. Convergent evolution occurs when two groups that are not closely related come to resemble each other more and more as time passes; this is usually the result of their occupation of similar habitats, and so the presence of similar selection pressures.

Another frequently used source of information is embryology. Morphological characters are often easier to interpret if the manner in which they develop is known. For example, if a particular structure in organism A and a structure of quite different appearance in organism B both develop from the same embryonic primordium, the resemblances and differences between those structures in A and B take on a phylogenetic significance that they would not have had if they had developed from entirely different embryonic primordia. Embryological evidence often allows biologists to trace the probable evolutionary changes that have occurred in important structures and helps them reconstruct the probable chain of evolutionary events that led to the modern forms of life. For example, the brief appearance of pharyngeal gill pouches during the early embryology of mammals, including humans, is thought to indicate that the distant ancestors of land vertebrates were aquatic.

Life histories have also played an important role in classical phylogenetic studies. The stages through which plants pass during their life cycles are particularly important sources of information, as we shall see when we examine the algae and the vascular plants, for example.

The problem of convergence   The classical approach, then, is based on evaluating similarities in a range of characters. But similarity, by itself, does not necessarily indicate common evolutionary descent. A particular similarity might simply reflect similar adaptation to the same environmental situation. The latter phenomenon is common in nature and is a serious source of confusion in phylogenetic studies.

When organisms that are not closely related become more similar in one or more characters because of independent adaptation to similar environmental situations, they are said to have undergone convergent evolution and the phenomenon is called *convergence* (Fig. 34.26). Whales, which are mammals descended from terrestrial ancestors, have evolved flippers from the legs of their ancestors; those flippers superficially resemble the fins of fish, but the resemblances result from convergence and they do not indicate a close relationship between whales and fish. Both arthropods and terrestrial vertebrates have evolved jointed legs and hinged jaws, but these similarities do not indicate that arthropods and vertebrates have evolved from a common ancestor that also had jointed legs and hinged jaws; there is good reason to think that these two groups of animals evolved their legs and jaws independently and that their legless ancestors were not closely related. The "moles" of Australia are not truly moles but marsupials (mammals whose young are born at an early stage of embryonic development and complete their development in a pouch in their mother's abdomen); they occupy the same habitat in Australia as do the true moles in other parts of the world and have, as a result, convergently evolved many startling similarities to the true moles. The marsupial mole is but one of a vast array of Australian marsupials that are strikingly convergent with placental mammals of other continents (Fig. 34.27).

The preceding discussion makes it evident that when classical systematists find similarities between two species, they must try to determine whether the similarities are probably *homologous* (inherited from a common ancestor) or merely *analogous* (similar in function and often in su-

A

B

C

D

34.27 Australian marsupials that are convergent with placental mammals of other continents
(A) A marsupial mouse. (B) A marsupial glider, convergent with placental flying squirrels. (C) The tiger cat (*Dasyurus*), a marsupial carnivore. (D) A cuscus, a marsupial monkey of New Guinea with a prehensile tail.

perficial structure, but of different evolutionary origins). Thus the wings of robins and those of bluebirds are considered homologous, since the evidence indicates that both were derived from the wings of a common avian ancestor. But the wings of robins and those of butterflies are only analogous, because, though they are functionally similar structures, they were not inherited from a common ancestor; they evolved independently and from different ancestral structures.

It is always important to indicate in what sense two structures are considered homologous or analogous. For example, the wings of birds and of bats are not homologous as wings, for they evolved independently, but they contain homologous bones, both types of wings having evolved from the forelimbs of ancient vertebrates that were ancestors to both birds and mammals. In short, the wings of birds and bats are analogous as wings and homologous as forelimbs. Similarly, the flippers of whales and seals evolved independently of each other, but both evolved from the front legs of land-mammal ancestors. Hence the flippers are homologous in the sense that both are forelimbs, with the same basic bone structure as that of other vertebrate forelimbs, but the modifications that make them flippers are analogous, not homologous.

Phenetics   As we have seen, classical taxonomists must utilize personal judgment in deciding which characters should be considered and how they should be weighted. To be sure, intuition does play an important role in the scientific process, as was pointed out in Chapter 1, but the unusual degree of subjectivity evident in classical taxonomy has motivated many systematists to attempt to develop more objective methods. One, less popular now than a decade ago, is phenetics. This approach to taxonomy uses as many characters as possible, weights all characters equally, and ignores the issue of analogy versus homology. The expectation is that if enough characters are compared, the subjective judgments necessary for making relative weightings and identifying cases of analogy will be unnecessary; any errors will be canceled out, or will be simply swamped by the mass of other data. In some cases this approach is fruitful. However, phenetics (often called numerical taxonomy) encounters serious problems when, as frequently happens, its two assumptions—that all characters are equally useful in determining phylogeny, and that little convergent evolution has taken place —are invalid. For example, phenetics, strictly applied, would classify the true mole and the marsupial mole as close relatives, a conclusion pheneticists themselves recognize cannot be correct.

Cladistics   Cladists, too, seek objectivity by ignoring the issue of analogy versus homology; they focus instead on what they call ***shared derived characters***—traits that are common to the several species in question and that are of relatively recent rather than ancient origin. Hence, in classifying two species of bats, the traits shared by mammals in general would not be considered; phenetics, on the other hand, would include any measurable trait. The cladistic approach is helpful in providing a rule for ignoring the absence (secondary loss) of traits—the hind limbs of aquatic mammals, for example—that causes many species to stand out as obvious exceptions to the general taxonomic patterns of their group. Cladists weight equally each of the traits they consider, but by ignoring features presumed to be shared at the point at which the speciation event in question occurred, they can be more selective than pheneticists: they have a criterion, or rule, for choosing the traits to be included. In the view of critics, however, deciding what traits are shared derived characters still calls for subjective judgment. And cladistic analysis, like phenetics, assumes that analogous characters will be outnumbered by homologous ones, and so false similarities will not distort the analysis. The result is controversial: cladistic analysis places the crocodiles with the birds, and the lungfish with the mammals, for example. It seems unlikely that apparent objectivity can prevail against the strong subjective judgment that crocodiles are reptiles and lungfish are fish.

Molecular taxonomy   The most promising new approach at present is molecular taxonomy. This approach avoids the analogy/homology issue with respect to particular traits by focusing instead on the molecular level. Since neutral base changes (those that do not significantly alter the activity of the gene product) are thought to accumulate at a relatively constant frequency independent of selection, the number of neutral differences between equivalent gene sequences in two species is a rough measure of the time since they had a common ancestor. In the same way, the amino acid sequences of proteins that are found in many different organisms can be compared, either directly or by sophisticated immunological means. Molecular taxonomy has already helped systematists make some very difficult decisions—placing the flamingoes with the storks rather than with the geese, for example, and settling a heated debate over the ancestry of the Tasmanian wolf: classical taxonomists correctly put this species with the Australian line of marsupial carnivores, while cladists mistakenly classified it with a group of South American marsupials. Molecular taxonomy may replace cladistics as the most popular alternative to classical taxonomy in the classification of living species, and even of many extinct forms where some of the DNA or, more often, protein (usually collagen) is adequately preserved. Whether it will displace classical taxonomy itself, however, is more difficult to say. Much may depend on whether this relatively new approach requires any major, subjectively unacceptable changes to the current phylogenetic picture.

PHYLOGENY AND CLASSIFICATION

Over a million species of animals and over 325,000 species of plants are known. To deal with this vast array of organic diversity, we obviously need some sort of system by which species can be classified in a logical and meaningful manner. Many different kinds of classification are possible. We could, for example, classify flowering plants according to their color: all white-flowered species in one group, all red-flowered species in a second group, all yellow-flowered species in a third group, and so on. This sort of system has its usefulness, and indeed provided a general basis for classification before the goal of recognizing evolutionary affinities had become clear. But the information such categories convey—about flower color, for example—is of an incidental kind that fails to set apart fundamentally different organisms. The classification system used in biology today is strictly

*Sequence of amino acids in cytochrome c for **28** organisms*[a]

Position		1 5 10 15 20 25 30 35 40 45 50
Mammals	Human, chimpanzee	GDVEK GKKIF IMKCS QCHTV EKGGK HKTGP NLHGL FGRKT GQAPG YSYTA
	Rhesus monkey	GDVEK GKKIF IMKCS QCHTV EKGGK HKTGP NLHGL FGRKT GQAPG YSYTA
	Horse	GDVEK GKKIF VQKCA QCHTV EKGGK HKTGP NLHGL FGRKT GQAPG FTYTD
	Donkey	GDVEK GKKIF VQKCA QCHTV EKGGK HKTGP NLHGL FGRKT GQAPG FSYTD
	Cow, pig, sheep	GDVEK GKKIF VQKCA QCHTV EKGGK HKTGP NLHGL FGRKT GQAPG FSYTD
	Dog	GDVEK GKKIF VQKCA QCHTV EKGGK HKTGP NLHGL FGRKT GQAPG FSYTD
	Rabbit	GDVEK GKKIF VQKCA QCHTV EKGGK HKTGP NLHGL FGRKT GQAVG FSYTD
	California gray whale	GDVEK GKKIF VQKCA QCHTV EKGGK HKTGP NLHGL FGRKT GQAVG FSYTD
	Great gray kangaroo	GDVEK GKKIF VQKCA QCHTV EKGGK HKTGP NLNGI FGRKT GQAPG FTYTD
Other vertebrates	Chicken, turkey	GDIEK GKKIF VQKCS QCHTV EKGGK HKTGP NLHGL FGRKT GQAEG FSYTD
	Pigeon	GDIEK GKKIF VQKCS QCHTV EKGGK HKTGP NLHGL FGRKT GQAEG FSYTD
	Pekin duck	GDVEK GKKIF VQKCS QCHTV EKGGK HKTGP NLHGL FGRKT GQAEG FSYTD
	Snapping turtle	GDVEK GKKIF VQKCA QCHTV EKGGK HKTGP NLNGL IGRKT GQAEG FSYTE
	Rattlesnake	GDVEK GKKIF TMKCS QCHTV EKGGK HKTGP NLHGL FGRKT GQAVG YSYTA
	Bullfrog	GDVEK GKKIF VQKCA QCHTC EKGGK HKVGP NLYGL IGRKT GQAAG FSYTD
	Tuna	GDVAK GKKTF VQKCA QCHTV ENGGK HKVGP NLWGL FGRKT GQAEG YSYTD
	Dogfish shark	GDVEK GKKVF VQKCA QCHTV ENGGK HKTGP NLSGL FGRKT GQAQG FSYTD
Insects[b]	Tobacco hornworm moth	GNADN GKKIF VQRCA QCHTV EAGGK HKVGP NLHGF FGRKT GQAPG FSYSN
	Fruit fly (*Drosophila*)	GDVEK GKKLF VQRCA QCHTV EAGGK HKVGP NLHGL IGRKT GQAAG FAYTN
Fungi[b]	Baker's yeast	GSAKK GATLF KTRCE LCHTV EKGGP HKVGP NLHGI FGRHS GQAQG YSYTD
	Red bread mold	GDSKK GANLF KTRCA ECHGE GGNLT QKIGP ALHGL FGRKT GSVDG YAYTD
Plants[b]	Wheat	GNPDA GAKIF KTKCA QCHTV DAGA GHKQGP NLHGL FGRQS GTTAG YSYSA
	Sunflower	GDPTT GAKIF KTKCA QCHTV EKGA GHKQGP NLNGL FGRQS GTTAG YSYSA
	Castor bean	GDVKA GEKIF KTKCA QCHTV EKGA GHKQGP NLNGL FGRQS GTTAG YSYSA
Number of different amino acids		1 3 5 5 4 1 3 3 4 1 4 3 2 1 3 3 1 1 2 3 3 4 2 3 4 2 1 4 1 1 2 1 5 1 3 2 1 1 3 2 1 3 3 6 1 2 3 1 2 4

[a] Adapted from M. O. Dayhoff, ed., *Atlas of Protein Sequence and Structure* (Washington, D.C.: National Biomedical Research Foundation, 1972), vol. 5; and R. E. Dickerson, The structure and history of an ancient protein, *Sci. Am.*, April 1972, copyright © 1972 by Scientific American, Inc.; all rights reserved.

[b] In cytochrome c from insects, fungi, and plants, a few (4–8) amino acids are usually ahead of what is here labeled Position 1; these are omitted from the table.

NUCLEIC ACIDS AND PROTEINS AS TAXONOMIC CHARACTERS

A mutation that changes a single base in a gene may affect the gene product to varying degrees. At one extreme, it may not affect the gene product at all; if the change is in the third base of a codon, the new codon will often specify the same amino acid as the old one (see Table 27.1, p. 717). In most cases, however, the change may result in a codon for either a similar amino acid (one hydrophobic amino acid instead of another, for example) or a very different one. Very rarely, the change may create a termination codon, which will cause translation to end prematurely. Base changes that survive are of two major types: neutral mutations that do not significantly alter the activity of the gene product, and mutations that are fixed by selection because they improve the gene product, making it better suited to the needs of a particular group of organisms. In practice, distinguishing between neutral and selectively advantageous changes is not easy.

Mutations that do not alter the meaning of the codon are probably nearly neutral, and probably accumulate randomly with time after two groups of organisms have diverged from a common ancestor. By comparing the extent of single-base changes of this type in the sequences of genes common to a wide range of species—genes for the enzymes of glycolysis, the citric-acid cycle, or the electron-transport chain, for example—we might get some measure of the time elapsed since divergence. As a tool for dating evolutionary events, therefore, neutral single-base changes have great potential. At the moment, however, very few genes have been fully sequenced in a significant number of unrelated species.

Another approach is to compare the amino acid sequences of the gene products. Consider, for example, cytochrome c, an essential component of the respiratory chain in mitochondria. The complete amino acid sequence of this enzyme has been worked out for a variety of organisms, both plant and animal. The table shows the sequence for some of the species so far examined, with the various functional groupings of amino acids (as determined by their R groups) indicated by a color code.

Perhaps the most obvious feature of this table is that cytochrome c is remarkably similar in all the species, even though some of them have probably not had a common ancestor for more than a

	55	60	65	70	75	80	85	90	95	100	104

```
A N K N K G I I W G E D T L M E Y L E N P K K Y I P G T K M I F V G I K K K E E R A D L I A Y L K K A T N E
A N K N K G I T W G E D T L M E Y L E N P K K Y I P G T K M I F V G I K K K E E R A D L I A Y L K K A T N E
A N K N K G I T W K E E T L M E Y L E N P K K Y I P G T K M I F A G I K K K T E R E D L I A Y L K K A T N E
A N K N K G I T W K E E T L M E Y L E N P K K Y I P G T K M I F A G I K K K T E R E D L I A Y L K K A T N E
A N K N K G I T W G E E T L M E Y L E N P K K Y I P G T K M I F A G I K K K G E R E D L I A Y L K K A T N E
A N K N K G I T W G E E T L M E Y L E N P K K Y I P G T K M I F A G I K K T G E R A D L I A Y L K K A T K E
A N K N K G I T W G E E T L M E Y L E N P K K Y I P G T K M I F A G I K K K D E R A D L I A Y L K K A T N E
A N K N K G I T W G E E T L M E Y L E N P K K Y I P G T K M I F A G I K K K G E R A D L I A Y L K K A T N E
A N K N K G I I W G E D T L M E Y L E N P K K Y I P G T K M I F A G I K K K G E R A D L I A Y L K K A T N E

A N K N K G I T W G E D T L M E Y L E N P K K Y I P G T K M I F A G I K K K S E R V D L I A Y L K D A T S K
A N K N K G I T W G E D T L M E Y L E N P K K Y I P G T K M I F A G I K K K A E R A D L I A Y L K Q A T A K
A N K N K G I T W G E D T L M E Y L E N P K K Y I P G T K M I F A G I K K K S E R A D L I A Y L K D A T A K
A N K N K G I T W G E E T L M E Y L E N P K K Y I P G T K M I F A G I K K K A E R A D L I A Y L K D A T S K
A N K N K G I I W D D T L M E Y L E N P K K Y I P G T K M I F T G L S K K K E R T N L I A Y L K E K T A A
A N K N K G I T W G E D T L M E Y L E N P K K Y I P G T K M I F A G I K K K G E R Q D L I A Y L K S A C S K
A N K S K G I V W N N D T L M E Y L E N P K K Y I P G T K M I F A G I K K K G E R Q D L V A Y L K S A T S -
A N K S K G I T W Q Q E T L R I Y L E N P K K Y I P G T K M I F A G L K K K S E R Q D L I A Y L K K T A A S

A N K A K G I T W Q D D T L F E Y L E N P K K Y I P G T K M V F A G L K K A N E R A D L I A Y L K Q A T K -
A N K A K G I T W Q D D T L F E Y L E N P K K Y I P G T K M I F A G L K K P N E R G D L I A Y L K S A T K -

A N I K K N V L W D E N N M S E Y L T N P K K Y I P G T K M A F G G L K K E K D R N D L I T Y L K K A C E -
A N K Q K G I T W D E N T L F E Y L E N P K K Y I P G T K M A F G G L K K D K D R N D I I T F M K E A T A -

A N K N K A V E W E E N T L Y D Y L L N P K K Y I P G T K M V F P G L K K P Q D R A D L I A Y L K K A T S S
A N K N M A V I W E E N T L Y D Y L L N P K K Y I P G T K M V F P G L K K P Q E R A D L I A Y L K T S T A -
A N K N M A V Q W G E N T L Y D Y L L N P K K Y I P G T K M V F P G L K K P Q D R A D L I A Y L K E A T A -
```

1 1 2 5 2 3 2 6 1 6 4 3 2 2 5 3 1 1 3 1 1 1 1 1 1 1 1 1 1 1 1 1 3 1 5 1 2 2 1 6 9 2 1 7 2 2 2 2 2 2 1 6 4 3 5 4

Symbol	Amino acid
	NONPOLAR
G	glycine
A	Alanine
V	Valine
L	Leucine
I	Isoleucine
M	Methionine
F	Phenylalanine
W	Tryptophan
P	Proline
	POLAR
S	Serine
T	Threonine
C	Cysteine
Y	Tyrosine
N	Asparagine
Q	Glutamine
	ACIDIC
D	Aspartic acid
E	Glutamic acid
	BASIC
K	Lysine
R	Arginine
H	Histidine

G is printed in color, because glycine, despite its technically nonpolar R group, behaves like a polar amino acid.

billion years. For example, all the cytochromes have the same amino acid sequence from positions 70 through 80. In fact, cytochrome c is an evolutionarily conservative protein; its amino acid sequence has changed at a considerably slower rate (about 20 million years for a 1 percent change) than, for example, the amino acid sequence of hemoglobin (5.8 million years) or that of fibrin (only one million years). The minimal change in cytochrome c suggests that only minor alterations can be tolerated if the enzyme is to continue functioning properly. Notice that even at points along the chain where there are differences, the amino acids are often functionally similar ones (with one polar amino acid substituted for another, one nonpolar amino acid for another, and so on). Some of the alterations may be neutral, having essentially no effect on the activity of the gene product, while others may represent minor but adaptive species-specific modification of the protein. But still others almost certainly indicate major changes in the gene product. Among these are substitution of a polar for a nonpolar amino acid and alterations involving proline (which induces turns in polypeptide chains) or cysteine (which forms strong covalent bonds with other cysteines).

If we compare various species in the table with one another, we find that the number of differences in amino acids usually agrees reasonably well with the presumed evolutionary distances among the species. Thus the mammals differ less among themselves than any of them differ from the fish. Human beings and chimpanzees do not differ at all; both differ by one amino acid from the rhesus monkey, by an average of 10.4 amino acids from the other mammals, by an average of 14.5 from the reptiles, by 18 from the amphibian, and by an average of 22.5 from the fish. This is an accurate reflection of the generally accepted evolutionary sequence of fish $\longrightarrow$ amphibian $\longrightarrow$ reptile $\longrightarrow$ mammal.

Because cytochrome c is evolutionarily so conservative, its value as a taxonomic character is limited to studies of the relationships among evolutionarily distant organisms. It cannot be used in comparing families or genera. But more rapidly changing proteins, such as fibrin, may prove useful in cases of closely related species. In any comparison, reliable conclusions depend on using several different proteins; because of unusual selection pressures or an unusually large or small number of mutations in the gene, the rate of change in one particular protein may be too great, or too small, to be representative.

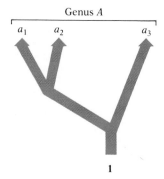

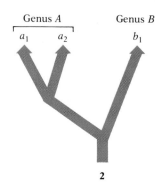

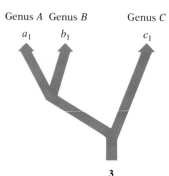

34.28 Alternative generic grouping for three related species
Biologists try to group species in a way that will indicate their phylogenetic relationships. Thus a genus is a group of related species. But how closely related? There is no absolute answer to this question. Some biologists (the "lumpers") like large genera containing many subdivisions (subgenera or "species groups"); others (the "splitters") prefer small compact genera, containing only species that are very closely related. In the drawing, three alternative ways of grouping three related species are shown. The first recognizes only one genus, the second recognizes two, and the third recognizes three.

an attempt to encode the evolutionary history of the organisms; it is thus often a means of conveying information about many of their characteristics to those familiar with the various taxonomic groups (taxa) to which the organisms are assigned.

The classification hierarchy Suppose you had to classify all the people on earth on the basis of where they live. You would probably begin by dividing the world population into groups based on country. This subdivision separates inhabitants of the United States from the inhabitants of France or Argentina, but it still leaves very large groups that must be further subdivided. Next, you would probably subdivide the population of the United States by states, then by counties, then by city or village or township, then by street, and finally by house number. You could do the same thing for Mexico, England, Australia, and all the other countries (using whatever political subdivisions in those countries correspond to states, counties, etc. in the United States). This procedure would enable you to place every individual in an orderly system of hierarchically arranged categories, as follows:

Country
State
County
City
Street
Number

Note that each level in this hierarchy is contained within and is partly determined by all levels above it. Thus, once the country has been determined as the United States, a Mexican state or a Canadian province is excluded. Similarly, once the state has been determined as Pennsylvania, a county in New York or in California is excluded.

The same principles apply to the classification of living things on the basis of phylogenetic relationships. Again a hierarchy of categories is used, as follows:

Kingdom
Phylum or Division
Class
Order
Family
Genus
Species

Each category (taxon) in this hierarchy is a collective unit containing one or more groups from the next-lower level in the hierarchy. Thus a genus is a group of closely related species (Fig. 34.28); a family is a group of related genera; an order is a group of related families; a class is a group of related orders; and so on. The species in any one genus are believed to be more closely related to each other than to species in any other genus; the genera in any one family are believed to be more closely related to each other than to genera in any other family; the families in any one order are believed to be more closely related to each other than to families in any other order; and so on.

TABLE 34.2 *Classification of six species*

Category	Haircap moss	Red oak	House fly	Herring gull	Wolf	Human
Kingdom	Plantae	Plantae	Animalia	Animalia	Animalia	Animalia
Phylum or Division	Bryophyta	Tracheophyta	Arthropoda	Chordata	Chordata	Chordata
Class	Musci	Angiospermae	Insecta	Aves	Mammalia	Mammalia
Order	Bryales	Fagales	Diptera	Charadriiformes	Carnivora	Primata
Family	Polytrichaceae	Fagaceae	Muscidae	Laridae	Canidae	Hominidae
Genus[a]	*Polytrichum*	*Quercus*	*Musca*	*Larus*	*Canis*	*Homo*
Species[a]	*commune*	*rubra*	*domestica*	*argentatus*	*lupus*	*sapiens*

[a] The name of a particular species consists of the genus name and the species designation, both customarily in italics, as shown in this table.

Table 34.2 gives the classification of six species. Notice that the table shows us immediately that the six species are not closely related, but that a human being and a wolf are more closely related to each other, both being Mammalia, than either is to a bird such as the herring gull. And it shows us that the mammals and the bird are more closely related to each other than to the house fly, which is in a different phylum, or to the moss or the red oak, which are in a different kingdom. These relationships are correlated with similarities and differences in the morphology, physiology, and ecology of the six species.

The six species of Table 34.2 are all well known, and the relationships between them are probably intuitively clear to you. But many species are not so well known, and the relationships between them not so clear. Much research may be necessary before they can be fitted into the classification system with any degree of certainty, and their assignment to genus or family (or even order) may have to be changed as more is learned about them.

Hierarchical classification systems similar to the one in current use have been employed by naturalists for many centuries. The current system dates from the work of the great Swedish naturalist Carolus Linnaeus (1707–1778), who wrote extensively on the classification of both plants and animals. His system used kingdoms, classes, orders, genera, and species; the phylum and family categories were added later. Now, the rationale on which Linnaeus based his system was necessarily very different from the phylogenetic one employed today. He worked a century before Darwin, and he had no conception of evolution, doubtless conceiving of each species as an immutable entity, the product of a divine creation. He was simply grouping organisms according to similarities, primarily morphological. That his results were so similar to those obtained today is a reflection of the fact that morphological characters, being products of evolution, tell us much about evolutionary relationships. The Linnaean system, however, produces results quite different from those of the modern phylogenetic system whenever it has to deal with convergence or with gross morphological similarities that may be misleading as to actual phylogenetic relationships.

An outline of a modern classification of living things is given on pp. A1–A6.

Nomenclature The modern system of naming species also dates from Linnaeus. Before him, there had been little uniformity in the designation of species. Some species had a one-word name, others had two-word names, and still others had names consisting of long descriptive phrases. For example, the pre-Linnaean name for the common carnation was *dianthus floribus solitariis, squamis calycinis subovatis brevissimis, corollis crenatis,* and the name for the honey bee was *Apis pubescens, thorace subgriseo, abdomine fusco, pedibus posticis glabris utrinque margine ciliatis.* Linnaeus simplified things by giving each species a name consisting of two words: first, the name of the genus to which the species belongs and, second, a designation for that particular species. Thus the above-mentioned species of carnation became *Dianthus caryophyllus,* and the honey bee became *Apis mellifera.* Other species in the genus *Dianthus* have the same first word in their names, but each has its own specific designation (*Dianthus prolifer, Dianthus barbatus, Dianthus deltoides*). No two species can have the same name.[4] Notice that the names are always Latin (or Latinized) and that the genus name is capitalized while the specific name is not.[5] Both names are customarily printed in italics (underlined if handwritten or typed). The correct name for any species, according to the present rules, is usually the oldest validly proposed name.[6]

The same Latin scientific names are used throughout the world. This uniformity of usage ensures that each scientist will know exactly which species another scientist is discussing. There would be no such assurance if common names were used, for not only does a given species have a different common name in each language, but it often has two or three names in a single language. For example, the plant *Bidens frondosa* is known by all the following English names: beggar-ticks, sticktight, bur marigold, devil's bootjack, pitchfork weed, and rayless marigold. To confuse matters further, a single common name is frequently applied to several species. For example, "gopher" is the name of a turtle in Florida and of a rodent in Kansas, and "raspberry" is the common name for more than a hundred species of plants.

---

[4] More precisely, no two species of plants can have the same name, and no two species of animals can have the same name. Since the International Rules of Botanical Nomenclature and the International Rules of Zoological Nomenclature are completely separate, it is possible for a plant and an animal to have the same name. There is also a separate International Bacteriological Code of Nomenclature.

[5] This rule always holds for zoological names, but specific botanical names are sometimes capitalized when they are derived from the name of a person or from other proper nouns.

[6] For purposes of priority, botanical naming dates from the publication of Linnaeus' *Species Plantarum* in 1753, and zoological naming dates from the publication of the 10th edition of his *Systema Naturae* in 1758.

Chapter 35

ECOLOGY

Ecology is usually defined as the study of interactions between organisms and their environment. "Environment," given a very broad meaning here, embraces all those things extrinsic to the organism that in any way impinge on it. It includes not only light, temperature, rainfall, humidity, and topography, but also parasites, predators, mates, and competitors. Anything affecting a particular organism and not an integral part of it is part of that organism's environment.

Life is characterized by many different levels of organization. Much of the first half of this book dealt with life on the molecular, cellular, tissue, organ, organ-system, or individual level. Ecology is concerned with phenomena on many of these levels—with how, for example, enzymes and organ systems affect the way a species is distributed over the range of habitats potentially available to it. The interaction between an organism's biochemistry or general physiology and its environment is called *physiological ecology*; in our exploration of ecology, we shall often stress the relationship between an organism's physiological systems and its interactions with its environment. Ecology is also concerned with three higher levels of organization: *populations*, which are groups of individuals belonging to the same species; *communities*, which are units composed of all the populations living in a given area; and *ecosystems*, which are communities and their physical environments considered together. Each of these designations may be applied to a small local entity or to a large widespread one. Thus the sycamore trees in an isolated patch of forest may be regarded as a population, and so may all the sycamore trees in the eastern United States. Similarly, a small pond and its inhabitants, or the forest in which the pond is located, may be treated as an ecosystem.

The various ecosystems are linked to one another by biological, chemi-

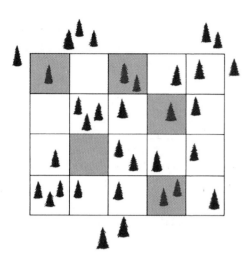

35.1   The quadrat sampling method
A grid is set up in the study area, and counts are made in a sampling of the squares thus delimited. The results permit an estimate of the population of the entire study area. The five squares sampled here (color) contain a total of six trees; hence it is estimated that the 20 squares of the entire grid will contain approximately 24 trees. In this example the selection of squares was random, but in some situations, as when obvious gradients run through the study area, other sampling designs may be more suitable.

cal, and physical processes. Inputs and outputs of energy, gases, inorganic chemicals, and organic compounds can cross ecosystem boundaries through meteorological factors such as wind and precipitation, geological ones such as running water and gravity, and biological ones such as the movement of animals and the dispersal of pollen and seeds. Thus the entire earth is itself a true ecosystem, in that no part is fully isolated from the rest. The global ecosystem is ordinarily called the *biosphere*.

The biosphere contains all living organisms and their environments. It forms a relatively thin shell around the earth, extending only a few kilometers above and below sea level. Except for energy, it is self-sufficient; all other requirements for life, such as water, oxygen, and nutrients, are supplied by utilization and recycling of materials already contained within the system.

POPULATIONS AS UNITS OF STRUCTURE AND FUNCTION

We shall begin our study of ecology by focusing on the same level of organization that claimed our attention in the last chapter, namely populations—groups of individuals belonging to the same species or to the same local subdivision of a species. We shall be especially interested in the dynamics of populations and in the environmental factors that help regulate them.

POPULATION SIZE AND DISTRIBUTION

Ecologists sometimes need to know the number of individuals in a population. When they consider endangered species, for example, the size of the surviving population is of crucial importance in the design of proper management procedures. Thus conservation authorities must have an accurate count (or at least a good estimate) of just how many whooping cranes or blue whales still exist on the earth.

But more often ecologists are concerned, not with the total number of living individuals of a species, but with the density of the population in a given region—the number of individuals per unit area or volume (50 pine trees per acre, for instance, or 5,000 diatoms per liter of water). In some situations, especially when the size of the individuals in a population is extremely variable, ecologists find that biomass—the total weight of all the individuals per unit area or volume, or its energy equivalent in calories, is a more useful index to the population's importance in the ecosystem.

Estimating population density   However population density is expressed, measuring it is not always simple, and different methods may yield widely varying results. Making total counts in an area under study often works well when the organisms being counted are large or conspicuous, and not excessively abundant. An alternative procedure is to count only a limited number of small sampling blocks (quadrats) and then estimate the density for the whole study area from these (Fig. 35.1). To avoid large errors in estimating population density, the sampling blocks must be as representative of the entire study area as possible.

Another method sometimes used in estimating population densities of animals is the mark-and-recapture technique. Here a limited number of individuals (let's say 20) are captured at random, marked with a tag or dye or some other device, and then released back into the same population. At some later time a second group of animals is captured at random from the population, and the percentage of marked individuals is determined; if 10 percent of the animals in this second group are marked, the investigator may conclude that the original 20 marked individuals represented 10 percent of the population in the study area, and hence that the total population is about 200.

Spacing Within any given area the individuals of a population can be distributed uniformly, randomly, or in clumps (Fig. 35.2). Uniform distributions are not common; they occur only where environmental conditions are fairly uniform throughout the area and where, in addition, there is intense competition or antagonism between individuals. For example, creosote bushes are often spaced almost uniformly over a desert area because the roots of each bush give off toxic substances that prevent germination of seedlings in a circular zone around the base of the bush. Other aromatic shrubs give off volatile growth inhibitors that prevent root growth of seedlings in the vicinity, and this emanation sometimes produces quite regular spacing of the shrubs. Animals of many species frequently create a nearly uniform distribution of individuals by establishing and defending exclusive territories (Fig. 35.3).

Random distributions are also relatively rare. They occur only where the environmental conditions are uniform, where there is no intense competition or antagonism between individuals, and where, moreover, there is no tendency for the individuals to aggregate. Since the chances that all three of these conditions will be met simultaneously are obviously poor, ecologists must usually work with nonrandom distributions, which complicates the sampling procedures and statistical tests they may use. Hypothetical random distributions represent a useful reference point, however, from which deviations toward uniform or clumped distributions can be measured.

Clumping, or aggregation, is by far the most common distribution pattern for both plants and animals in nature, for several reasons:

1. Environmental conditions are seldom uniform throughout even a relatively small area. Variations in soil, in topography, in the distributions of other species, and in such microclimatic factors as moisture, temperature, and light may produce important habitat differences within the area. Organisms will obviously tend to occur in those spots where such conditions are most favorable.

2. Reproductive patterns often favor clumping. This is particularly true in plants that reproduce vegetatively (asexually) and in animals whose young remain with the parent.

3. Animals often exhibit behavior patterns that lead to active congregation in loose groups or in more organized colonies, schools, flocks, or herds. And even sexual attraction produces a departure from the theoretical conditions necessary for a completely random distribution.

A clumped distribution may increase competition for nutrients, food, space, or light, but this deleterious effect is often offset by some beneficial ones. For example, trees growing together in a hedgerow on the Great Plains may compete more intensely for nutrients and light than if they were

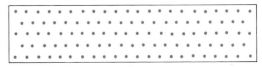
Uniform

Random

Clumped, but groups random

35.2 Uniform, random, and clumped distributions

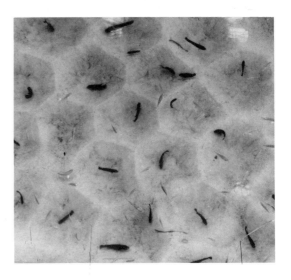

35.3 Territories of male cichlids
These male cichlids, known as black tilapias, establish clearly defined breeding territories in the bottom of their stream. In this photograph, they have established breeding territories in an artificial environment. In both cases, the result is an unusually uniform distribution of individuals.

widely separated, but, because they shelter one another, they may be better able to withstand strong winds. And the clump, which has less surface area in proportion to mass than an isolated, exposed tree, may create its own, more favorable microenvironment. Within a clump of trees, for example, shading and protection from the wind help insulate against temperature extremes and reduce evaporation from the soil; fallen leaves are more likely to be trapped and help mulch and enrich the soil; and so on. Also, the existing trees may provide essential shelter that enables saplings on the periphery to grow, thus steadily enlarging the clump. Aggregations of animals, too, often reduce the rate of temperature change in their midst—an effect that is particularly important in cold weather. Honey bees, for example, overwinter in self-warmed groups of many thousands of individuals, and so are able to begin efficient foraging in the early spring before other pollinators are active. A group of animals may also have an advantage over isolated individuals in locating food and in withstanding attacks by predators. Light-sensitive animals often survive exposure to illumination longer if they are in a group. Thus we see that the optimum density for population growth and survival is often an intermediate one; undercrowding may be as deleterious as overcrowding.

POPULATION GROWTH

The exponential growth curve   One way to understand the dynamics of real populations is to find out what to expect of a population under ideal conditions, and then to try to determine how actual conditions modify this expected pattern. Let us assume we can study a population that has a stable age distribution, faces no predation, parasitism, or competition, experiences no immigration or emigration, and exists in an environment with unlimited resources. For any species the population growth rate under such ideal conditions is enormous. Consider what would happen if a pair of organisms produces a full complement of offspring and all those offspring survived to produce a full complement of offspring in their turn, and so on for a number of generations. For example, one pair of house flies starting to breed in April could have 191,010,000,000,000,000,000 descendants by August if all their eggs hatched and if all the resulting young survived to reproduce. Similarly, a biologist[1] has calculated that if a population of 100 California sea stars (starfish) were allowed to reproduce unimpeded and all offspring survived, in only 15 generations the number of starfish would exceed the number of electrons in the visible universe!

But what about species in which the females produce relatively few young? Darwin made the following estimates on the reproductive potential of elephants:

> The elephant is reckoned the slowest breeder of all known animals, and I have taken some pains to estimate its probable minimum rate of natural increase; it will be safest to assume that it begins breeding when 30 years old, and goes on breeding till 90 years old, bringing forth six young in the interval, and surviving till 100 years old; if this be so, after a period of from 740 to 750 years there would be nearly 19 million elephants alive descended from the first pair.[2]

[1] E. O. Dodson, *Evolution: Process and Product*, Reinhold Publishing Corp., 1960, p. 4.

An extension of Darwin's calculations shows that in 100,000 years one pair of elephants would have so many living descendants that they would fill the visible universe.

We have here mentioned only house flies, sea stars, and elephants, but the same considerations apply to all organisms, whether plant or animal, unicellular or multicellular. All have the potential for explosive growth, so that in the absence of environmental limitations, their growth curve would be exponential (Fig. 35.4).

To examine exponential growth more carefully, let us designate I the rate of increase in the number of individuals in the population, b the average birth rate per individual in the population, d the average death rate per individual, and N the number of individuals in the population at any given moment.[3] We can then write the equation for a population growth curve like that of Figure 35.4 as follows:

$$I = (b - d)N$$

From this formulation it is immediately apparent that a population will grow only if the average birth rate exceeds the average death rate, so that the term $(b - d)$ is greater than zero. Conversely, the population will decline if the average birth rate is less than the average death rate, so that the term $(b - d)$ is less than zero. If the birth and death rates are equal, so that $b = d$, then $(b - d) = 0$, and therefore $I = 0$; such a population would be neither growing nor decreasing. In summary, then, it is the value of $(b - d)$, the difference between the birth and death rates, that determines whether a population will grow, be stable, or decline. This difference, the net rate of population change per individual at a given moment, is designated r; hence $(b - d) = r$, and we can rewrite the equation for exponential population change as follows:

$$I = rN$$

Figure 35.4 illustrates a case in which r is greater than zero.

In the hypothetical house-fly or elephant populations living under unlimited environmental conditions and undergoing a population explosion, r is at its maximum for the species. Under these conditions of minimum death rate and maximum birth rate, r is designated r_{max}; it represents the **intrinsic rate of increase** of the population. Clearly r_{max} varies between species—it is much larger for house flies than for elephants, for example.

In the exponential model, the rate of population growth is a function not just of r but also of N, the population size, which translates in this context into the number of individuals able to reproduce. Since N becomes larger with each successive generation, the rate of increase, I, also becomes larger with each generation. It is because of this accelerating rate of increase that the slope of the curve in Figure 35.4 becomes steeper and steeper.

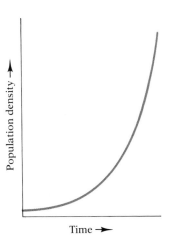

35.4   The exponential growth curve
The rate of increase steadily accelerates until, in theory, the population density increases at an infinitely high rate. Clearly, no real population can for long continue increasing exponentially.

---

[2] Darwin attributed a somewhat greater life-span to elephants than they probably have; most do not live past 50, and the highest authenticated age is 77. Since they begin breeding well before age 30, however, Darwin's calculations remain essentially correct.

[3] Put another way, I is the change (Δ) in N per unit time—$\Delta N/\Delta t$ or, in the notation of calculus, $\delta N/\delta t$.

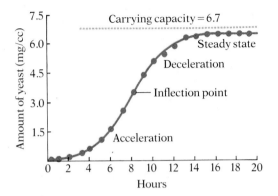

35.5 A representative logistic growth curve
Shown here is the growth curve of a laboratory population of yeast cells. Often called an S-shaped growth curve, the logistic curve exhibits an accelerating rate of growth at low densities, but eventually reaches an inflection point, where the rate change shifts from acceleration to deceleration. The deceleration continues as the population density approaches the carrying capacity of the environment. When the carrying capacity is reached, there is no further increase in density, and the population continues in steady state. This curve closely approximates the hypothetical logistic curve.

The logistic growth curve No real population is expanding at infinite speed, and we are not buried in house flies or elephants. The reason must be that the exponential growth of real populations is checked. In many instances the result is a growth curve like that in Figure 35.5, which shows initial rapid expansion of the population when it is at low densities, then decelerating growth at higher densities, and an eventual leveling off as the density approaches what we called in Chapter 22 the *carrying capacity* (K) of the environment. The carrying capacity is the maximum density of population that the environment can support over a sustained period without lasting damage to the environment itself.

A population growth curve such as that in Figure 35.5, an S-shaped logistic curve, reflects a changing relationship between births and deaths. During the acceleration stage births greatly exceed deaths; during the deceleration stage the value of r is steadily falling (because the birth rate is declining, or the death rate is rising, or both), though it still is greater than zero; and when the curve levels off, births and deaths are in balance and $r = 0$. But for such changes in the relationship between births and deaths to occur—for the birth rate to decline (or the death rate to increase) as the population density rises—something must be limiting population growth in such fashion that the effectiveness of the limitation is proportional to the population density; that is, the limitation must become more severe as the density approaches the carrying capacity of the environment. We can express this *density-dependent limitation* as

$$\frac{K - N}{K}$$

This ratio represents the fraction of the total carrying capacity that remains to be filled. Inserting this limiting term into the equation for maximum exponental growth, we obtain

$$I = r_{max}\left(\frac{K - N}{K}\right)N$$

in which r_{max} and K are constants for a given species in a given environment, and N, the population size, changes with the passage of time. This insertion has converted the equation into one for logistic growth, in which the growth is no longer a function solely of the intrinsic rate of increase, r_{max}, and the population size, N, but also the ratio between ($K - N$) and the carrying capacity, K.

Let us now see how insertion of the limiting term ($K - N$)/K results in the S-shaped logistic growth curve of Figure 35.5. At low population densities, where N is much smaller than the carrying capacity, the value of the limiting term is essentially K divided by K, or approximately one, which means that growth is primarily a function of $r_{max} \cdot N$ and hence increases almost exponentially. But as growth continues, and N becomes larger and larger, the value of the limiting term steadily declines from one and acts as a brake on the rate of further growth. Finally, when the carrying capacity is reached and N equals K, the value of the limiting term becomes zero and no further growth is possible:

$$I = r_{max}\left(\frac{K - N}{K}\right)N = r_{max} \cdot 0 \cdot N = 0$$

A population that has reached this steady-state level, in which births and deaths are in balance, is said to have *zero population growth*.

The growth curves for some real populations approach the idealized logistic curve almost exactly, as we saw in Figure 35.5, but more often they are only rough approximations and exhibit considerable fluctuation around the carrying capacity, sometimes overshooting it temporarily and sometimes falling well below it (Fig. 35.6). Note that whenever the population density fluctuates above the carrying capacity, so that N is greater than K, the limiting term $(K - N)/K$ takes on a negative value, with the result that I also becomes negative. This means that the population density will tend to decrease instead of increasing, until it returns to the carrying capacity or below. In short, the limiting term provides feedback control, usually holding the population density near the steady-state level.

Maximum sustained yield   Figure 35.5 shows us that the rate of increase of a population is greatest at the inflection point (the point of transition from an accelerating to a decelerating rate), rather than when the population density has reached its higher steady-state level. It follows that the density at which the inflection point (sometimes called the point of *maximum sustained yield*) occurs is of very great importance in the informed management of such organisms as game animals or commercially valuable fishes. If harvesting of the organisms reduces their population density only to the point of maximum sustained yield, one would expect no lasting damage to the resource population. If, however, the resource population is over-exploited, so that its density is reduced too far below the point of maximum sustained yield, then the recovery of the population may be endangered. In short, cropping resource populations to their point of maximum sustained yield would seem to be an optimum strategy in terms of both human benefit and perpetuation of the resource.

Unfortunately things are rarely that simple. Conclusions based only on the logistic growth curve fail to take into account such important variables as age and size distributions in the population. For example, when a laboratory population of flour beetles was cropped to a level one-tenth as dense as K, far below the point of expected maximum sustained yield at $K/2$, the sustained yield actually increased because, though the number of young beetles had been reduced, the number of reproductives had not been significantly altered. Moreover, in deciding on optimal cropping rates for resource populations, economic yield must also be considered. Thus, though there may be a maximum yield in terms of numbers of animals caught if a fish population is cropped to the $K/2$ level, economic yield may be greater if cropping is less severe, so that the fish will grow to larger size; fewer larger fish may bring more money to the fisherman than more smaller ones. In short, the point of optimal yield that managers of fish and game must strive to find is not necessarily the same as the point of maximum sustained yield.

The exponential growth curve with sudden crash   Not all growth curves of real populations assume the logistic form of Figures 35.5 and 35.6. The population densities of many small short-lived organisms, or of organisms that live in disturbed or transient habitats, never reach the carrying capacity of the environment before they crash, often abruptly (Fig. 35.7). For ex-

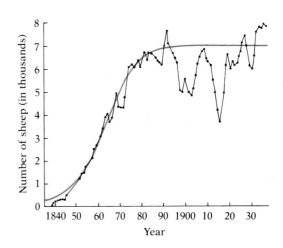

35.6   Growth curve of the sheep population of South Australia
The smooth curve (color) is the hypothetical logistic curve about which the real curve seems to fluctuate.

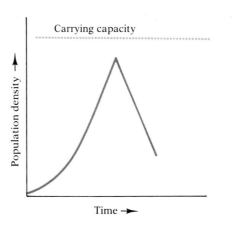

35.7   A growth curve in which an exponentially growing population suddenly declines before reaching the carrying capacity of its environment
This type of growth curve, sometimes called a boom-and-bust curve, is most characteristic of populations of small organisms limited by density-independent factors.

35.8 Pea aphids on alfalfa
Aphid populations are normally limited by density-independent factors like weather. Because the aphids produce winged offspring that disperse as the density of insects on a plant rises, overcrowding does not usually occur; a few winged morphs are visible here.

ample, a population of pea aphids in a field of alfalfa in spring may grow at an exponential rate if the weather is cool and moist, but if the weather then becomes hot and dry most of the aphids will die, and hence population density will fall precipitously to a very low level. This crash will occur when the weather changes even if the density of aphids is far below the carrying capacity of the alfalfa field. Note, then, that the weather is here exerting a ***density-independent limitation*** on the aphid population; its operation does not depend significantly on the density of the aphids (Fig. 35.8). In addition to the weather, other environmental factors that may exert density-independent limitation on populations of organisms are sudden floods, fire, or physical disruption of the habitat.

Mortality and survivorship Our discussion of growth curves so far has been based on several simplistic assumptions. We have supposed that the potential (maximum) life-span, the average life expectancy, the average age of reproduction, and the age distribution of a population can be ignored. When we think in terms of bacteria or *Paramecium* or yeast, in which each cell divides to produce two independent organisms, these assumptions may be reasonable. But like the flour beetles in the experiment on maximum sustained yield, most larger organisms, whether plants, animals, or fungi, do not reproduce at a steady rate. Instead, they live a significant part of their lives before beginning to reproduce, and after they have reached that point their reproductive potential frequently varies with age—peaking and then declining in some species, such as ours, but increasing steadily in many others, as different as trees and vertebrates. If the potential life-span is long and the age of first reproduction low, more generations will be living concurrently than if the potential life-span is short and the age of first reproduction high. If the potential life-span is long but the average life expectancy is low, the age distribution in the population will be very different from what it is if the potential life-span is long and the average life expectancy also long.

In attempting to understand the dynamics of a population, it is often useful to determine the mortality rates for the various age groups in the popu-

35.9 Five types of survivorship curves
For an initial population of 1,000 individuals, the curves show the number of survivors at different ages from birth to the maximum possible age for that species. Curves I, II, and III represent three basic types of survivorship. A type I curve is typical of certain organisms for which density-independent events like a heavy frost are the major source of mortality. It can be simulated in the laboratory by raising fruit flies in a closed container without food. The two human curves represent survivorship in developed (A) and developing (B) nations. The initial downward slope of the human B curve has been exaggerated for clarity.

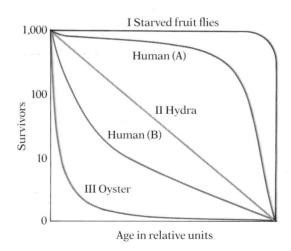

lation. Such data show what stages in the life cycle are most susceptible to environmental control, and make it possible to compute the percentage of individuals that will still be alive at the end of each age interval. The results may be graphed as a survivorship curve (Fig. 35.9).

The curves in Figure 35.9 illustrate several survivorship patterns. Curve I approaches the pattern that would be expected if all the individuals in the population lived as long as possible. There would be full survival through all the early age intervals (as shown by the horizontal portions of the curve), and then all the individuals would die more or less at once and the curve would fall suddenly and precipitously. Curve III approaches the other extreme, where the mortality is exceedingly high among the very young, but any individual surviving the earliest life stages has a good chance of surviving for a long time thereafter. Between these two extremes is the condition represented by curve II, where the mortality rate of all ages is constant.

The survivorship curves for most wild-animal populations are probably intermediate between types II and III, and the curves for most plant populations are probably near the extreme of III. In other words, high mortality among the young is the general rule in nature.

Changes in environmental conditions may radically alter the shape of the survivorship curve for any given population, and the altered mortality rates, in turn, may have profound effects on the dynamics of the population and on its future size. Let's consider the age distribution of three human populations. In Sweden (Fig. 35.10A), the age distribution is nearly flat, with the percentage of the population in each age class approximately

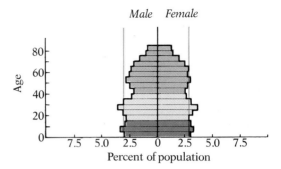

A   Sweden

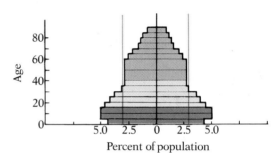

B   United States

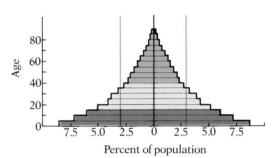

C   India

35.10   Age distribution in three human populations
(A) In Sweden, the size of each age class from prereproductive through early postreproductive ages is roughly equal. Because the birth rate roughly equals the death rate, and has for many years, the population is not growing; reproductive individuals are simply replacing themselves. Were this not so, the percentage of the population in each of the prereproductive age classes (0–15 years) would be greater than the percentage in each of the reproductive age classes (15–40 years). The vertical lines at 3 percent suggest what a completely stable distribution of age classes would be. Note that since older individuals of our species do not reproduce, any extension of the life-span beyond about age 50 has no direct impact on birth rate as long as the population is below the carrying capacity. (B) In the United States, the size of the youngest group (0–5 years) is roughly in equilibrium with that of the younger reproductive classes. However, since there is a bulge well beyond the 3 percent line in some of the age classes that have not finished reproducing (5–30 years), the population will continue to rise until 1990 or 2000 in spite of the low birth rate. (C) In India, the population is heavily weighted toward the younger age classes as a result of the short life-span of many individuals and of a high birth rate. Even if the birth rate were to fall to two children per family, the population would continue to soar for at least three decades.

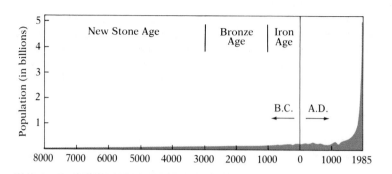

35.11   Growth of the human population of the world
The growth was slow for many thousands of years, but has become very rapid in the past century.

equal except for the oldest classes. The survivorship curve of the Swedish population (human A in Figure 35.9) is close to type I. The birth rate equals the death rate, and the population size is stable; since each couple, on the average, has two children, individuals are only replacing themselves.

The population in the United States is slightly different (Fig. 35.10B). Though the birth rate is now down to the replacement rate, this equilibrium has been achieved only recently. As a result, the population includes a disproportionate percentage of prereproductive individuals, born when the average family size was greater than two. The population of the United States, then, will continue to grow until the reproductive and prereproductive age classes near equilibrium. This should happen about 1990, when the population reaches about 300 million.

In India, by contrast (Fig. 35.10C), the age distribution is heavily weighted toward the bottom, with a relatively large percentage of the population in the younger age classes. This could mean that the survivorship curve of the Indian population is type III, with high infant mortality, or that the average family size is large and the population is growing rapidly. Actually, both factors are responsible; in India the Indian survivorship curve falls between type III and type II (human B in Figure 35.9). It should be clear, as Stanford ecologist Paul Ehrlich pointed out more than two decades ago, that in the absence of a dramatic reduction in birth rate, advances in reducing infant mortality in developing countries (which have resulted from improvements in sanitation, nutrition, and medical care) will inevitably result in explosive population growth, accompanied by widespread famine and disease.

Revolutions in medical and in agricultural technology have simultaneously raised both infant survival rates and the carrying capacity of the earth. As Figure 35.11 shows, for many thousands of years the human population of the world increased very slowly, even though the birth rate was probably high. Specialists estimate that there were approximately 5 million people on the entire earth ten thousand years ago, and that the number had risen to only about 250 million by A.D. 1 and to 500 million by the year 1650. Until three hundred years ago, then, the human population doubled approximately every 1,600 years. Now, by contrast, the population of the world is doubling every 35 years; the present total is over 4 billion.

How much longer the human species can continue with such a rate of increase, with such an imbalance between births and deaths, is one of the most pressing questions of our time. Some have argued, in fact, that it is *the* most important question; all other aspects of the so-called ecological crisis —hunger, poverty, crowding, pollution, accumulation of wastes, destruction of the environment on which all life depends—are the inexorable consequences of a continuing rise in the number of human beings. We shall examine the biological bases of some of these problems in the next chapter.

POPULATION REGULATION

As we have seen, populations cannot grow indefinitely; one or more factors always work to limit population size. Most populations can be usefully designated according to which type of growth curve best characterizes them. At one extreme are populations that can grow almost exponentially. Because their growth rate closely approximates the intrinsic rate of increase, r_{max}, organisms with this kind of population growth are frequently called **r-selected species**. Their population normally does not reach the carrying capacity of the habitat, which is usually in a transitional, disturbed, or otherwise unpredictable environment. The rapid multiplication of mosquitoes in springtime typifies the behavior of an *r*-selected species; fortunately for us, the size of the mosquito population is generally held below saturation by environmental factors like drought and cold.

Indeed, precarious environmental circumstances like these dictate many of the behavioral and physiological adaptations in *r*-selected species. For example, *r*-selected individuals usually produce large numbers of small, quickly maturing offspring, since in the absence of severe competition, the fitness of individuals in an *r*-selected species is heavily dependent upon producing as many young as possible before some environmental disruption brings on a sudden crash. The survivorship curve of such a species usually resembles type III (Fig. 35.9).

At the other extreme of the spectrum are populations that are usually at or near the carrying capacity (*K*) of their habitats, and so grow slowly or not at all. These relatively stable populations are characteristic of **K-selected species**, of which most large mammals, such as elephants, are examples. *K*-selected species are typically found in fairly stable or predictable environments. Because their fitness depends less on rapid reproduction than on their ability to compete effectively for limiting resources, individuals of such species usually produce only a few large, longer-lived, slowly maturing young. Consequently, the survivorship curve for *K*-selected species usually falls between type I and type II.

The two extremes of the *r*–*K* continuum illustrate dramatically the ways in which the various sorts of population controls operate. As we have seen, the density of the population in *r*-selected species is generally too low for its growth to be limited by the availability of essential resources like food, water, or nesting sites. And though *r*-selected organisms may be subject to severe predation, they reproduce so quickly and profusely that this factor is rarely limiting. More important are the many density-*independent* factors such as weather, fire, or flooding that can cause the population to crash periodically (Fig. 35.7). The population of *K*-selected species, on the other hand, is usually kept to a specific density (the carrying capacity) by such factors as the limited availability of one or more resources. Hence these

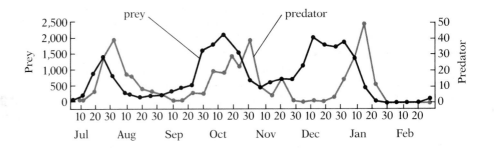

35.12 Linked fluctuations in predator and prey populations
The size of the population of a predatory mite tends to reflect (after a delay) fluctuations in the population of its prey (another mite). Predation is but one density-dependent limiting factor operating on the prey. The predator population, however, is limited mainly by the availability of prey.

species are subject to both density-independent and density-dependent factors.

Let's look now at the various environmental factors that can limit populations in a density-dependent fashion.

Predation and parasitism as density-dependent limiting factors Both predation and parasitism usually influence the prey (or host) species in a density-dependent manner. As the population of the host or prey increases, a higher percentage of the population is usually victimized, because the individuals—having perhaps been forced into less favorable situations or being weaker on account of the greater drain on available resources—become easier to find and attack. In the same way, the probability that a parasite will find a suitable host is also usually density-dependent.

When the relative densities of prey shift, predators that take a variety of prey species tend to change their search images and alter their hunting patterns so as to concentrate on the most common species. If the density of a prey species increases, the density of the predators feeding on it often increases also. This increase in predators, together with their increased concentration on the particular prey species, may be one factor that causes the density of the prey to fall again. But as the density of the prey falls, there is usually a corresponding, but slightly later, fall in the density of the predator. The result may be a series of density fluctuations like those shown in Figure 35.12. Such linked fluctuations of predator and prey suggest that the major limiting factor for the predator is the availability of its food, and therefore that predation is probably one significant limiting factor for the prey.

In stable predator-prey systems, the density-limiting function of the predation is often decidedly beneficial to the prey population as a whole, even though it is destructive to individuals. By keeping the size of the population in check, predators (like the wolves on Isle Royale, p. 591) prevent overgrazing and periodic famine among the prey.

Ironically, predator-prey stability can be destroyed inadvertently by well-intended control measures. For example, the application of certain insecticides to strawberries in an attempt to destroy cyclamen mites that were damaging the berries killed both the cyclamen mites and the carnivorous mites that preyed on them. But the cyclamen mites quickly reinvaded the strawberry fields, while the predatory mites did not. The result was that the cyclamen mites, now free of their natural predators, rapidly increased in density and did more damage to the strawberries than if the insecticides had never been applied. This pattern of greater pesticide susceptibility in predatory (beneficial) arthropods than in prey (pest) arthropods is quite general, and accounts in part for current widespread doubts about the long-term value of heavy pesticide use in agriculture.

As the example of Isle Royale suggests, predator-prey relationships in a stable ecosystem, like long-established host-parasite relationships, tend to evolve toward a dynamic balance in which the predation is an important regulatory influence in the life of the prey species, but not a real threat to its survival. Factors affecting the predator-prey balance include the relative numbers and sizes of the two species, the vulnerability of the prey to the predator, the extent to which the predator can (or does) use other food sources, and the amount of energy obtained by the predator from consuming one prey individual relative to the energy expended in its capture.

Intraspecific competition as a density-dependent limiting factor
Competition is one of the chief density-dependent limiting factors. The continued healthy existence of most *K*-selected organisms depends on utilization of some environmental resources that are in limited supply, such as food, water, space, or light. Each member of a population shares the same basic requirements; each needs the same sort of food, shelter, and mates. Unless some other force, such as predation, holds a population below the carrying capacity of its environment, individuals in *K*-selected populations must inevitably compete for resources.

Duikers, a group of small antelope species found mostly in Africa, provide a clear-cut example of this sort of competition. Pairs of duikers compete for territory, dividing up the particular patch of forest that provides suitable concealment and a reliable food supply (Fig. 35.13). Each of the vigorously defended territories is barely large enough to supply the young leaves and berries that the resident pair consume through most of the year. Even during the breeding season there is not always more than the minimum amount of food required by the parents, and therefore the surplus necessary for the successful rearing of offspring is sometimes not available. Any offspring that are successfully reared must be chased out of the territory lest the parents starve. Thus the fact that rearing young requires a surplus of essential resources limits population growth when the size of a population approaches the carrying capacity of the environment. As the population density increases, the competition among members of the population for these limited resources becomes more intense: limited supplies effectively limit population growth.

Similarly, many plants compete for light, water, and nutrients, and a substantial surplus must be available if they are to produce seeds, which are metabolically costly. To take a familiar example, if flowers or vegetables are planted too close together in a garden, the densely packed plants will be weak and spindly, and will produce few if any blossoms. Only if they are thinned, either artificially or by the natural death of the weakest individuals, will the surviving plants thrive. Competition for resources also limits the surviving plants' ability to produce seeds or runners. The same sort of competition for space, light, water, and nutrients operates in a forest and keeps down the density of the trees.

35.13 A mated pair of duikers

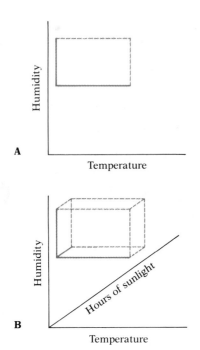

35.14 Graphical representations of the ecological niche of a species
(A) The ranges of toleration by a species for two environmental variables are plotted on a coordinate system, and a rectangular area is determined. (B) If a third variable is plotted at right angles to the other two, a volume is determined. Though the procedure is difficult to carry further graphically, it can be continued mathematically to define a species' niche in multidimensions—or as a so-called hypervolume. The more the niche hypervolumes of two species overlap, the greater the potential for competition between the species.

Interspecific competition as a density-dependent limiting factor We saw in the last two chapters how potent a force competition between species can be in evolution. We saw that when two species compete for precisely the same limited resources—that is, when their niches are identical—the superiority of one species will inevitably lead to the extinction of the other. This is the competitive exclusion principle developed by Gause, which was discussed in the last chapter. In any particular instance, the extent of this competition depends largely on the density of the competing species relative to the resources in question.

An organism's **niche** is determined by its genetic endowment, and includes not merely what it eats, but how and where it finds and captures its food; what extremes of heat and cold, dry and wet, sun and shade, and other climatic factors it can withstand, and what values of these factors are optimal for it; what its parasites and predators are; where, how, and when it reproduces; and so forth. In addition, the time of year and the time of day when it is most active are important: the niche of night-hunting owls preying on small rodents, for instance, is different from that of small day-hunting hawks, not to mention small ground-hunting mammalian carnivores, rodent parasites, and rodent diseases; hence under normal conditions these species can coexist, even though all utilize the same food resource. Similarly, a species of plant that flowers before another species can "exploit" the same pollinator without in fact directly competing for it. Since every aspect of an organism's existence helps define that organism's niche, we cannot completely describe or measure it; hence the concept of niche is an abstraction. It is defined by a species' lifestyle rather than existing as a preformed mold into which a species must fit itself as best it can.

One way that we can get a concrete grasp of a species' niche is by determining the acceptable limits of variables that affect the species. Thus, for a particular plant species, we can begin with an environmental variable such as temperature, and determine the high and low extremes the species in question can tolerate. We can plot these on a coordinate grid, and connect them by a line. We might then do the same for a second environmental variable, such as humidity, and draw its line at right angles to the first one. These two lines together determine a rectangular surface (Fig. 35.14A).

Now we can determine the values for a third environmental variable, such as hours of sunlight, and connect them by a line oriented at right angles to the other two. The three lines together determine a volume within which every point corresponds to some combination of values for the three variables that would permit the species to survive (Fig. 35.14B).[4]

If we were to continue this procedure, adding more and more variables each determining a different dimension, we would obtain a multidimensional hypervolume. If we included a dimension for every variable relevant

---

[4] There is an important consideration that complicates this procedure: the environmental variables are seldom completely independent (for example, the temperature extremes a species can tolerate may be different at different humidities), and actual ecological niche measurements must take these variables into account simultaneously. In the hypothetical model, the "corners" are points at which two or more factors are at their individual limits; but in actuality, when one condition is not optimal, the limits of tolerance for some other condition may be reduced, and consequently the resulting three-variable volume will be more ellipsoidal than rectangular.

to the species in question, the resulting multidimensional hypervolume would represent the niche of the species.

The more similar two niches are—the more the hypervolumes for two species overlap—the more likely it is that both species will use in common and in the same way at least one limited resource (food, shelter, nesting sites, or the like) for which they will compete. According to the principle of limiting similarity—a modification of the competitive exclusion principle—there is a limit on the amount of niche overlap compatible with coexistence. Competition for the one most limited resource therefore usually leads to one (or two) of four possible outcomes:

1. One species may simply become extinct. Usually this will be the result of the superiority of one of the rival species under ordinary circumstances. This superiority need not be in the ability to locate and harvest food; indeed, the species inferior in this regard may well be the one to survive if it is less susceptible to predation, disease, or extinction from the stresses of a fluctuating environment. Alternatively, as we said in the last chapter, chance may play a role. In some crises even major differences in adaptiveness will be irrelevant, and extinction of the superior species may result.

2. One species may be superior in some regions, and the other may be superior in other regions with different environmental conditions, with the result that one is eliminated in some places and the other is eliminated in other places; as a result, sympatry disappears, but both species survive in allopatric ranges.

3. One species may be superior under normal conditions, but at a strong disadvantage during periodic crises. The crises, then, will reduce the population size of the superior species, and the competition will begin anew, as long as the crises are frequent enough to prevent the extinction of the normally inferior species from happening first.

4. Given time and a slight difference in niche, selection may act to produce character displacement, which reduces competition. For example, we saw how interspecific competition for the same seeds in Darwin's finches on the larger islands led to character displacement in beak size, with the result that the several species tend to have small, medium, or large beaks—specialized for seeds of different sizes—while on smaller islands the one resident species has an intermediate size beak. Through character displacement, then, the two initially competing species may evolve greater differences in their niches.

A host of factors too complex to analyze here determines whether extinction, range restriction, character displacement, or a combination of the last two, will be the outcome in any given instance of intense interspecific competition.

Let us examine a few well-studied examples of competition and its consequences. We saw in the last chapter that two species of *Paramecium* could not coexist in the same container, though either could survive under the same conditions on its own. In a related experiment, Thomas Park and his colleagues at the University of Chicago worked with flour beetles of the genus *Tribolium*. When *T. confusum* and *T. castaneum* were kept together in the same container of flour, one or the other species always became extinct. The conditions of temperature and humidity under which the competition took place greatly influenced which species would win. *T. casteneum*

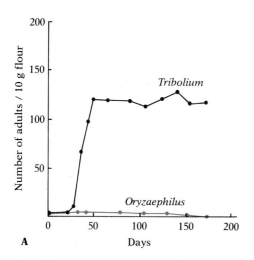

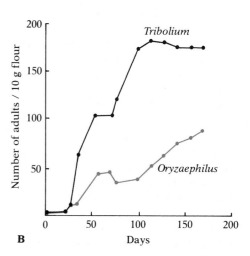

35.15 Role of habitat complexity in the survival of competing species of flour beetle
In a uniform habitat of pure flour, *Tribolium* always overcame *Oryzaephilus*, which became extinct (A). If pieces of fine glass tubing were added to the culture to make it less uniform, both species survived and increased in numbers (B).

usually won under hot-wet conditions, while *T. confusum* usually won under cool-dry conditions. What would happen, then, if the environment were made less uniform, if the experimenter added vegetation, for instance, or rocks, to stimulate natural variations in the habitat? Perhaps in a more heterogeneous environment, both species would survive. A. C. Crombie of Cambridge University put this expectation to the test by pitting two other species of flour beetle against each other. In a simple environment of pure flour, the *Tribolium* species always drove the *Oryzaephilus* beetle to extinction (Fig. 35.15A), but when Crombie made the environment more heterogeneous by adding pieces of fine glass tubing, both species survived together (Fig. 35.15B). Apparently each species enjoyed a competitive superiority in different parts of the habitat, even though they depended on the same food supply. Since the environment in any habitat is rarely perfectly uniform, competitive differences in adaptation to different **microhabitats** frequently allow several species with similar food requirements to coexist.

J. H. Connell's study of barnacles on the Scottish coast provides another good example of competition between two species whose niches overlap but are sufficiently distinct for both species to survive by occupying slightly different habitats. A species of the genus *Chthamalus* occupies the upper part of the intertidal zone, and a species of *Balanus* occupies the lower part of the intertidal zone (Fig. 35.16). The boundary between the two distributions is roughly the level of the mean high neap tide.[5] Casual examination of the distribution of these two species of barnacles on the intertidal rocks would not reveal whether they were kept separate by competition or by different responses to physical factors such as the percentage of time out of water.

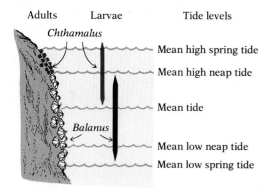

35.16 Effect of competition between two species of barnacles
Though there is a broad area in which the larvae of both species can settle successfully, competition eliminates most of the overlap by the time the adult stage is reached, *Chthamalus* being largely restricted to the zone above the level of the mean high neap tide.

To find the answer, Connell artificially kept one study area clear of *Balanus* and another clear of *Chthamalus*. He found that the larvae of *Chthamalus* would settle and grow in the upper portion of the zone normally occupied by *Balanus* as long as *Balanus* was not there. In the reciprocal ex-

[5] Neap tides are the lowest tides of a lunar month.

periment, *Balanus* larvae would settle in the *Chthamalus* zone but could not survive there even if no *Chthamalus* were present. Apparently each species could occupy a portion of the intertidal zone (roughly between the levels of the mean high spring tide and the mean high neap tide for *Chthamalus* and below the mean tide for *Balanus*) from which the other was barred by physical factors. But in the intermediate zone where physical factors permitted each species to survive (roughly between the levels of mean high neap tide and mean tide), it was competition that kept them separate, *Balanus* usually eliminating *Chthamalus*.

It is not always easy to detect the differences between the niches of two or more closely related sympatric species. At first glance, the species may appear to be occupying the same niche in a stable way and thus to discredit the exclusion principle. But closer study usually reveals differences of fundamental importance. When Robert MacArthur of Princeton University studied a community where several closely related species of warblers (small insect-eating birds) occurred together, he found that their feeding habits were significantly different. Myrtle warblers fed predominantly among the lower branches of spruce trees; bay-breasted warblers fed in the middle portions of the trees; and Cape May warblers fed toward the top of the same trees and on the outer tips of the branches (Fig. 35.17). We noted equivalent differences in feeding habits between sympatric Galápagos finches.

Cormorants and shags are closely related sympatric species of birds whose habits and ecological requirements appear very similar. But David Lack, who studied them, found out that, though both nest on cliffs and feed on fish, cormorants nest on broad ledges and feed chiefly in shallow estuaries and harbors, whereas shags nest on narrow ledges and feed mainly at sea. Their niches, then, are very different, and there is little competition between them.

Emigration as a density-dependent limiting factor In some animals crowding induces physiological and behavioral changes that result in increased emigration from the crowded region. Such changes can be observed in many species of aphids. During seasons of the year when conditions are favorable, they are represented largely by wingless females reproducing parthenogenetically. But when conditions deteriorate and competition becomes intense, winged females develop, and these may gain

Cape May warbler

Bay-breasted warbler

Myrtle warbler

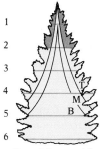
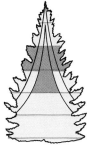
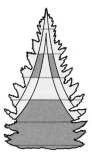

Cape May warbler   Bay-breasted warbler   Myrtle warbler

35.17 Differences in the feeding niches of three species of warblers living in the same community
Left: The dark color areas are the parts of the tree where the birds spend half their total feeding time. T: terminal parts of branches; M: middle; B: base. The height zones (1–6) are measured in 3-m units. Above: The Cape May warbler feeds on the tips of the highest branches of the spruce tree; the bay-breasted warbler feeds in the middle part of the tree; and the myrtle warbler feeds predominantly (though not exclusively) among the lower branches.

a large competitive advantage by moving out of the area in which they were born. But given the high risk involved in searching for a new habitat—a suitable plant host—conditions must be intolerable before emigration makes sense. Similarly, in the fungus fly *Oligarces paradoxus*, individuals reproduce parthenogenetically without ever fully maturing to the winged adult stage as long as there is an abundant supply of fresh food. But when the food supply deteriorates, in either quantity or quality, and competition becomes intense, the insects go through a complete developmental sequence leading to winged adults, which fly in search of other, more hospitable areas. Many rodents (including the legendary lemming) are also programed to emigrate in search of more favorable habitats when the population density, or, more often, some important secondary effect of crowding such as changes in hormone levels, exceeds a certain threshold.

One of the best-known examples of physiological and behavioral changes induced by crowding is that of the solitary and migratory phases seen in several species of locusts, particularly *Locusta migratoria* in Eurasia. Individuals of the migratory phase have longer wings, a higher fat content, a lower water content, and a darker color than solitary-phase individuals, and they are much more gregarious and more readily stimulated to marching and flying by the presence of other individuals. The solitary phase is characteristic of low-density populations, and the migratory phase of high-density ones. As the density of a given population rises, the proportion of individuals developing into the migratory phase also rises; the sight and smell of other locusts seem to play an important role in triggering this line of development. When the proportion of migratory-phase individuals has risen sufficiently high, enormous swarms emigrate from the crowded area, consuming nearly all the vegetation in their path and often completely devastating agricultural crops (Fig. 35.18).

35.18 Locusts in the migratory phase, in Ethiopia

Mutualism as a density-dependent limiting factor We have seen how prey density can limit predator density, and vice versa. So too the relative numbers of two competing species, or of parasites and their hosts, are clearly linked. But there are also density-dependent interspecific interactions in which the two species, instead of playing antagonistic roles, act in a mutually beneficial fashion. One of the more remarkable examples of such mutualism is the finely tuned cooperation between certain ants and the acacia trees in which they live, a phenomenon first discovered about 1870 by Thomas Belt. At first glance the ants of this species seem to be exploiting their hosts. Not only do they harvest the tree's nectar and special "Beltian bodies" on the leaves, but they burrow into its thorns and stems to make their nests. A closer look, however, reveals that this acacia has evolved special adaptations that accommodate the ants. Compared to the thorns of the *Acacia* species not occupied by ants, its thorns are large and hollow (Fig. 35.19A), while the Beltian bodies that the ants eat are quite different from the tree's ordinary leaves, and extraordinarily nutritious (Fig. 35.19B).

Comparisons between normal "occupied" acacias and acacias from which the ants have been removed reveal that the ants are almost essential for the trees' survival. The ants keep acacias relatively free from plant-eating insects, and swarm out at the slightest disturbance to attack and repel browsing mammals as well. Further, the ants prevent damaging vines from growing on the tree, and literally defoliate nearby trees that shade their host. And the acacia is equally important to the species of ant it feeds and shelters. As in predator-prey systems and interspecific competition, the density of each affects the density of the other; but in this instance greater density, by increasing the probability that a sapling will be found and colonized by a fertile queen, has a positive effect on both. Other examples of mutualism, such as the interdependence of many flowering plants and honey bees, abound in nature.

The physiological bases of density-dependent limiting factors   Most of the density-dependent factors we have examined have clear physiological bases: emigration can be triggered by internal hormonal signals, for example, while the outcome of interspecific competition often turns upon slight differences in physiological adaptations to the environment. The severe intraspecific competition for resources that results from overcrowding may leave most individuals in a population with less food or water than they need, or weakened by the physical demands of competition. Already easier marks for predators and disease-causing agents because they are crowded together, these individuals become even more vulnerable as a result of such physiological stresses. Indeed, numerous experiments have shown that when population density is increased, the members of the population may exhibit marked depression of both inflammatory responses and antibody formation, with a resultant loss of resistance to infection or parasitism. There may also be increased susceptibility to the harmful effects of various toxic substances.

The effects of increasing population density have been observed in numerous laboratory experiments with mice. There is hypertrophy of the adrenal cortex and degeneration of the thymus. Somatic growth is suppressed, sexual maturation delayed (or even totally inhibited at very high densities), and reproduction by mature mice diminished. The effects on reproduction include delayed spermatogenesis in males and, in females, prolonged estrous cycles, reduced rate of uterine implantation, and inadequate lactation. There is also evidence of increased intra-uterine mortality of the embryos. There would seem, then, to be an endocrine feedback mechanism that alters the individual reproductive rate. Presumably, as the density rises and aggressive behavior increases, endocrine disturbance rises and the reproductive rate falls; conversely, as the density and aggressive behavior decrease, the reproductive rate rises. The existence of such a complex system suggests that it must in some way benefit the individuals involved, perhaps by saving females from the physical stresses of pregnancy, birth, and lactation when the chances for their offsprings' survival are low.

The idea that endocrine changes act as major density-dependent limiting factors in nature has been severely criticized on the ground that it is largely based on laboratory work with densities much higher than those that would actually occur in nature. To meet this criticism, attempts have been made to duplicate the intermittent crowding more likely to occur in nature. These have shown that mice exposed to a few short periods of crowding every day in the laboratory actually show greater hypertrophy of the adrenal cortex than mice exposed to continuous crowding. Nevertheless, endocrine

A

B

35.19 The mutualistic association of ants and acacia trees
The acacia (*Acacia corigera*) provides overgrown hollow thorns that the ants (*Pseudomyrmex nigrocincta*) inhabit (A), and specialized yellow, protein-rich leaves called Beltian bodies, which the ants eat (B).

changes have seldom been found in wild animals undergoing population stress, and the importance of such changes in population regulation remains unclear.

THE BIOTIC COMMUNITY

In our preceding discussion of populations, we considered species either alone or in pairs—as predator and prey, mutualists, or species with overlapping niches. Moreover, we viewed the environment as a more or less static backdrop against which various interspecific and intraspecific interactions take place. For many purposes, this narrow picture is reasonably accurate. In the broader view, however, we must recognize that a species usually interacts with several others and is indirectly affected by many more. Furthermore, the environment is anything but static: habitats are constantly changing, sometimes literally overnight, and as a consequence vastly different selection pressures are introduced. In this section we shall consider how these complex factors affect the biology of populations. We shall begin with the interactions among the members of a biotic community—the many plant, animal, and other species that occupy the same general area, "sharing" some resources and competing for others. Biotic communities can be considered units of life, with their own characteristic structure and functional relationships. Then we shall consider how these communities respond to environmental change, often in predictable and economically important ways.

DOMINANCE, DIVERSITY, AND COMMUNITY STABILITY

As we have seen, the species that make up a community influence one another in countless ways, for both good and ill. Thus a predator may, if overabundant, represent a threat to the continuation of a prey population, but under more balanced conditions the activity of the same predator may, by regulating the density of the prey population, help keep it vigorous and healthy. There are also numerous interactions of the populations in the community with the physical environment. Thus plants extract mineral nutrients from the soil, but, when they die, their substance may contribute to the organic content of the soil.

Species interactions that are far from obvious may yet be of crucial importance to a community. Consider an experiment with an intertidal community of 15 species of marine invertebrates conducted by R. T. Paine of the University of Washington at Mukkaw Bay. The top predator in this community was a sea star, *Pisaster ochraceus*. When Paine excluded the sea star from one area but allowed it to remain in another, the result was a radical change in the species diversity of the first area. Of the original 15 species, only eight remained. One of two competing species of barnacles was eliminated, because the sites for its attachment were not cleared of other organisms by the sea star, and the other barnacle was a better competitor for what little space there was. A sponge and its predator also disappeared; apparently the sea star had some sort of indirect influence on them, probably by clearing space for the sponge. Dramatic as these effects of the removal of

Pisaster were, they would have been virtually impossible to predict before-hand.

Efforts to eliminate undesirable species from a community often reveal hidden linkages to other organisms, and many dramatically demonstrate the complex interactions on which community stability rests. An unintended cautionary example was provided by the World Health Organization (WHO) in a campaign to eradicate malaria-carrying mosquitoes in the Borneo states of Malaysia, where as many as 90 percent of the population of some areas suffered from the disease. Mosquito control was achieved by spraying the insides of the village huts with DDT and dieldrin, two powerful contact insecticides, and malaria was indeed eradicated. But soon the villagers began to notice that the thatch roofs of their huts were rotting and beginning to collapse. Investigation showed that the deterioration, which occurred only in huts sprayed with DDT, was inflicted by the larvae of a moth that normally lives in small numbers in the thatch roofs. Whereas the thatch-eating moth larvae avoided food sprayed with DDT, the moth's natural enemy, a parasitic wasp, was very sensitive to it. The net result was a substantial increase in the population of the larvae eating the thatch.

The collapse of thousands of roofs would have been tragedy enough, but there was yet another side effect potentially more serious. Cockroaches and a small house lizard, the gecko, are two normal inhabitants of the village huts. DDT-contaminated cockroaches were eaten by the geckos, which were in turn eaten by house cats (as were some cockroaches). The cats, poisoned by the accumulation of the insecticide, died. What ensued was a population explosion of rats, which are potential carriers of such diseases as leptospirosis, typhus, and plague. In an attempt to restore the cat population, WHO and the Royal Air Force undertook a remarkable venture, "Operation Cat Drop," in which they parachuted cats into the villages. With the cat population restored, the rats and the consequent threat of serious disease subsided.

Fortunately, not every pest-control effort entails such complications, though unanticipated results are very common. Insecticides rapidly broken down in the environment and more selective in their toxicity have been developed, and these are less disruptive to biological communities. Much to be preferred, however, is biological control through the use of the natural predators, parasites, or pathogens of the pest.

Recognizing the crucial role played in a community by one or two species, such as the sea star at Mukkaw Bay or the parasitic wasp on Borneo, ecologists would like to be able to assign some sort of importance value to each species in a community. Such a procedure might help them predict what environmental impact a proposed human disturbance would have, for example. But what is the meaning of "importance" in this context? Many answers are possible, and each, unfortunately, leads to a different conclusion.

If the emphasis is placed on density, the highest importance value will be assigned to the densest species. But in a forest the densest plant species may be a small shrub growing below the large trees; yet the large trees, not the shrubs, are clearly the dominant species, in the sense of determining the nature of the whole community. An alternative would be to emphasize productivity, which, as we shall see in the next chapter, is a much better meas-

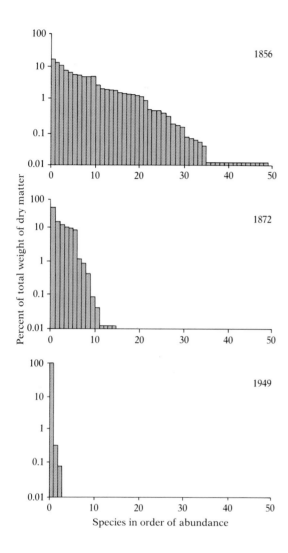

35.20 An experiment showing the effect of human intervention on species diversity in an ecosystem

Nitrogen fertilizer has been applied yearly since 1856 to an experimental grass plot at the Rothamsted Experimental Station in England. When the experiment began, 49 species were growing in the plot (top). By 1872 the number had fallen to 15 (middle). Only 3 species remained in 1949 (bottom). This sequence resembles an early succession run backward. The community probably shrank because one ecological factor was emphasized at the expense of all the others.

ure of the species' position in the flow of energy and materials through the community. But this approach fails to give adequate recognition to the regulatory role played by predators such as the sea star or parasitic wasp mentioned earlier. Though relatively few in number, these organisms are clearly key members of their communities, and their removal resulted in sweeping community changes.

In each of our two examples the removal of a single species caused major community upheavals. In the past, many ecologists held that in more complex communities, where there are more species and therefore more alternative interactions, there will be greater stability—a superior ability to withstand perturbations such as the removal of a species. In such a community, for example, the predators often have many prey species available and, because they can switch from one to another, they were thought to be less sensitive to variations in the abundance of any one. Reciprocally, the prey may be subject to several different predators, its density regulation thus being less dependent on any single predator.

However, theoretical studies pioneered by Robert M. May of Princeton University show just the opposite trend: as species are added to a hypothetical community, stable configurations become progressively rarer. Moreover, as we have come to know more about tropical communities, the most complex and species-rich on earth, it has become increasingly apparent that these communities are no more resilient than temperate-zone communities, and perhaps less so.

Most ecologists now reject the notion that greater species diversity and complexity of interactions necessarily make for increased community stability. The response of a simple community to a disturbance may be more violent and immediate, but such a community may recover more quickly and settle more rapidly into a new functional mode. The complex community, by contrast, may sometimes respond less dramatically to the disturbance, but, because of the multiplicity of interactions within it, the effects may continue to ripple through the system for a long time, causing numerous small, but nonetheless important, dislocations and distortions. Which community, then, is the more stable—the simpler one, which responds violently but recovers quickly, or the more complex one, which responds less violently but continues to show effects over a longer period? The answer, of course, lies in how one defines "stability." The point to remember is that species diversity and community complexity do indeed influence the way a community will respond to disturbance, though not in any direct and simple way.

Another kind of complexity likewise needs mention here—complexity of the physical environment. In simple laboratory cultures of several competing species or of predator-prey systems, extinction of at least one of the species virtually always ensues, as we saw earlier in the experiments with *Paramecium* and with flour beetles. But if the physical environment in the culture is made more diverse, the wild fluctuations and oscillations typical of simple laboratory systems can often be reduced, and as was demonstrated with the flour beetles in Figure 35.15B, extinction can be postponed indefinitely. Thus structural diversity of the environment seems to favor community stability.

Human intervention in biological communities nearly always has the effect of simplifying them, in terms both of reduced species diversity (Fig.

35.20) and of diminished structural complexity. The result is to make the communities far more prone to extreme fluctuations in response to changing conditions. Prime examples are the highly artificial communities created by modern agriculture, which has tended more and more to emphasize monoculture—the planting of a single crop species in enormous fields from which all other plant species are systematically excluded. These communities are very unstable and can be maintained only at the price of constant vigilance and the investment of much energy in curbing insect and mite infestations and outbreaks of disease, in maintaining soil fertility, and in cleaning out or killing invading weeds. As experiments have demonstrated, multiple-species crops have greater resistance to pests and disease, but the difficulty and expense of adapting conventional farm machinery to the planting of such crops has made it impractical for large-scale farming so far.

Pollution, though less extreme in its simplification of communities than monoculture farming, also reduces species diversity and thereby shortens food chains and increases community instability. Top predators are frequently the first species to disappear. With the predators gone, one or more prey species may multiply unchecked and further degrade the habitat, often making it unfit for other species and sometimes actually rendering it inadequate for the support of even the exploding species itself.

ECOLOGICAL SUCCESSION

The nature and causes of succession   Succession is a more or less orderly process of community change. It involves replacement, in the course of time, of the dominant species within a given area by other species. If a farmer's field is allowed to lie fallow, a crop of annual weeds will grow in it during the first year. Many perennial herbs appear in the second year and become even more common in the third year. Soon, however, these are superseded as the dominant vegetation by woody shrubs, and they may in turn be replaced eventually by trees. Or a lake dries up, and its sandy bottom becomes covered with grass, which later gives way to trees.

But why the change? Why does succession occur? The cause cannot be climate, because succession will occur even if the climate remains the same. Shifts in climate can initiate changes that leave some resources underutilized and so pave the way for succession, and climate may be a major factor in determining what sorts of species will follow one another, but the phenomenon of succession itself must result from other causes. The traditional view has been that the most important cause is the modification of the physical environment produced by the community itself. In this view, successional communities tend to alter the area in which they occur in such a way as to make it less favorable for themselves and more favorable for other communities. In effect, each community in the succession sows the seeds of its own destruction.

Consider the alterations initiated by pioneer communities on land. Usually these communities will produce a layer of litter on the surface of the soil. The accumulation of litter affects the runoff of rainwater, the soil temperature, and the formation of humus (a kind of decomposed organic material). The humus, in turn, contributes to soil development and thus alters the water relations, the pH and aeration of the soil, the availability of nutrients, and the sorts of soil organisms that will occur. But the organisms characteristic of the pioneer communities that produced these changes may not prosper under the new conditions, and they may be replaced by invading competitors that do better in an area with the new type of soil.

Though much of the modification of the habitat that produces succession results from the action of living organisms, other forces may also contribute. Among these are physiographic changes. For example, greater depth in the channel of a stream may result in better drainage of a swamp; or an overflowing stream may deposit rich silt on nearby bottomland. Such changes will soon be followed by changes in vegetation.

There has been growing recognition that modification of the habitat—whether by the action of organisms or by physiographic changes—is not the entire explanation for ecological succession. To some degree, changes in vegetation merely demonstrate that some species are more easily dispersed and grow more rapidly than others. Hence it is only to be expected that annual herbs will be far more important members of a pioneer community on recently abandoned farmland than the seedlings of slow-growing trees, even though the trees may eventually become the dominant plants.

Successions in different places and at different times are not identical; the species involved are often completely different, and the climatic and substrate conditions vary widely. Moreover, the sequence of changes in *primary succession*, in which communities are established in newly formed habitats like sand dunes, lava flows, or bare rock, is often longer and slower than in *secondary succession*, in which communities are reestablished in areas where they were destroyed, as in fields where the original forests were cleared for farming.

Some examples of succession   One of the first examples of primary succession studied in detail (by H. C. Cowles of the University of Chicago) was of sand-dune vegetation at the southern end of Lake Michigan. The lake once extended much farther south than it does today. As the lakeshore gradually receded northward, it left exposed a series of successively younger beaches and sand dunes. Hence someone who starts at the water's edge and walks south for several miles will pass through a series of communities (Fig. 35.21) that represent various successional stages beginning with bare beach and culminating in an old well-established forest dominated by beech and sugar-maple trees. Let us examine these stages in more detail.

1. The lower beach near the water's edge has no land life because it is frequently exposed to the destructive action of waves.

2. The middle beach is ordinarily dry in summer, but is occasionally washed by the waves produced by severe winter storms. Conditions of life are very severe and exclude biennial and perennial plants, but a few succulent annuals similar to those that inhabit deserts grow there in summer.

3. Conditions on the upper beach are much less severe than on the lower and middle beaches and the flora is richer, but vegetation is still very sparse. Decay of driftwood has begun adding some organic matter to the substratum.

4. Behind the upper beach is the foredune community, a pioneer com-

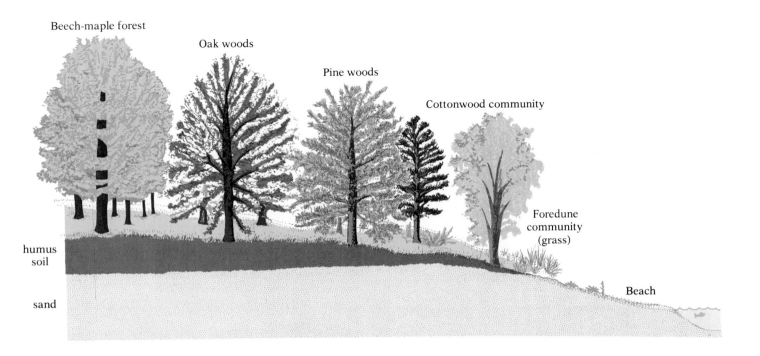

Beech-maple forest

Oak woods

Pine woods

Cottonwood community

Foredune community (grass)

humus soil

sand

Beach

35.21 Successional stages at the southern end of Lake Michigan
In a sequence of this sort the distance from the water to the beech-maple forest can be several miles.

munity dominated by sand-binding beach grasses. Tiger beetles, grasshoppers, and burrowing spiders are the characteristic animals of this community.

5. The first tree-bearing stage in this succession is a cottonwood community, a loosely organized pioneer community on sands that have usually been partly bound by the roots of grasses, but are still subject to considerable shifting by the wind.

6. The cottonwood community grades into a community dominated by jack pine, juniper, and bearberry, where the soil has a considerable humus component.

7. The pine community grades, in turn, into a forest dominated by oaks.

8. The oak community grades into a forest dominated by sugar-maple and beech trees growing in a deep humus-rich moist soil.

During the walk from the lakeshore to the beech-maple forest, an observant person would notice many changes in the physical environment—a progressive decrease in total light intensity at ground level, a decrease in wind velocity and in the rate of evaporation, an increase in soil moisture and in relative humidity, and an increase in the amount of humus in the soil and the amount of leaf mold on its surface. Presumably the series of communities and environmental changes seen as one moves from the lake to the forest duplicates approximately the series of successional stages through which the area now covered by the beech-maple forest must have passed since the time when it was a wave-washed beach.

A A newly formed pond near the beach has sandy borders bare of vegetation.

B After two years such a pond is ringed by low vegetation, including cottonwood saplings.

C. A 50-year-old pond is bordered by mature cottonwood trees. So much sediment is produced by organisms growing in the pond that only a small area of water, choked with weeds, remains.

D After 150–250 years the area that was once a pond has become a meadow.

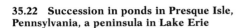

35.22 Succession in ponds in Presque Isle, Pennsylvania, a peninsula in Lake Erie

Another much-studied type of succession is that in ponds (Fig. 35.22). Sediments washed from the surrounding land begin to fill the pond, and the dead bodies of planktonic organisms add organic material. Soon pioneer submerged vascular plants appear in the shallower water near the margins of the pond. Their roots hold the silt, and the pond bottom is built up faster where they are growing. In addition, as these plants die, their bodies accumulate faster than decomposers can break them down. Soon the water is shallow enough for broad-leaved floating pondweeds to displace the submerged species, which now become established in a zone farther out in the pond, where conditions are more favorable for them. But as the bottom continues to build up, the floating pondweeds are in their turn displaced by emergent species (plants that have their roots in the mud of the bottom, but their shoots extending into the air above the water), such as cattails, bulrushes, and reeds. These plants grow very close together and hold the sediment tightly, and their great bulk results in rapid accumulation of organic material. Soon conditions are dry enough for a few terrestrial plants to gain a foothold. Now an area that was formerly part of the pond is newly formed dry land. This entire sequence can sometimes be seen as a nearly continuous series of zones girdling a pond or lake. With the passage of the years, the pond becomes smaller and smaller as the zones move nearer and nearer its center. Eventually nothing of the pond remains.

Successions need not begin with land reclaimed from lakes or ponds as in the preceding examples. Consider the bare rock surfaces that the volcanically produced Galápagos must once have presented, or that were left by the volcanic explosion of Krakatoa more recently. In such areas the first pioneer organisms of the primary succession may be lichens (Fig. 35.23) and nitrogen-fixing Cyanobacteria, which grow during the brief periods when the rock surface is wet and lie dormant during periods when the surface is dry. The lichens release acids and other substances that corrode the rock. Dust particles and bits of dead bacteria may collect in the tiny crevices thus formed, and pioneer mosses may gain anchorage there. The mosses grow in tufts or clumps that trap more dust and debris and gradually form a thickening mat. A few fern spores or seeds of grasses and annual herbs may land in the mat of soil and moss and germinate. These may be followed by perennial herbs. As more and more such plants survive and grow, they catch and hold still more mineral and organic material, and the new soil layer thus becomes thicker. Later, shrubs and even trees may start to grow in the soil that now covers what once was a bare rock surface.

Secondary succession on abandoned croplands, unused railway rights-of-way (Fig. 35.24), plowed grasslands, or cutover forests often proceeds relatively quickly in its initial stages, because the effects of the previous communities have not been wholly erased and the physical conditions are not as bleak as on a beach or a bare rock surface. Let us take an abandoned cornfield in Georgia as an example. The very first year it will be covered with annual weeds, such as ragweed, horseweed, and crabgrass. In the second year ragweed, goldenrod, and asters will probably be common, and there will be much grass. The grass will usually be dominant for several years, and then more and more shrubs and tree seedlings will appear. The first tree seedlings to grow well in the unshaded field will be pines, and eventually a pine forest will replace the grass and shrubs. But pine seedlings do not grow well in the shade of older pines. Seedlings of oaks, hickories, and other deciduous trees are more shade-tolerant, and these trees will gradually develop in the lower strata of the forest beneath the old pines, eventually replacing them. The deciduous forest thus formed is more stable and will ordinarily maintain itself for a very long time.

Forests often become stratified into more or less distinct layers, each of which has its own populations and interactions. The forest floor, or herb layer, the shrub level, the short-tree level, and the canopy level are com-

35.23 An early successional stage on a bare rock surface
Lichens, shown here bearing fruiting bodies, are often the first multicellular organisms to gain a foothold on rock. Chemicals produced by the lichens corrode the rock surface and help prepare the way for later successional stages.

35.24 Secondary succession on an abandoned railway right-of-way
The ties are rotting, and vegetation is taking over in the formerly cleared area.

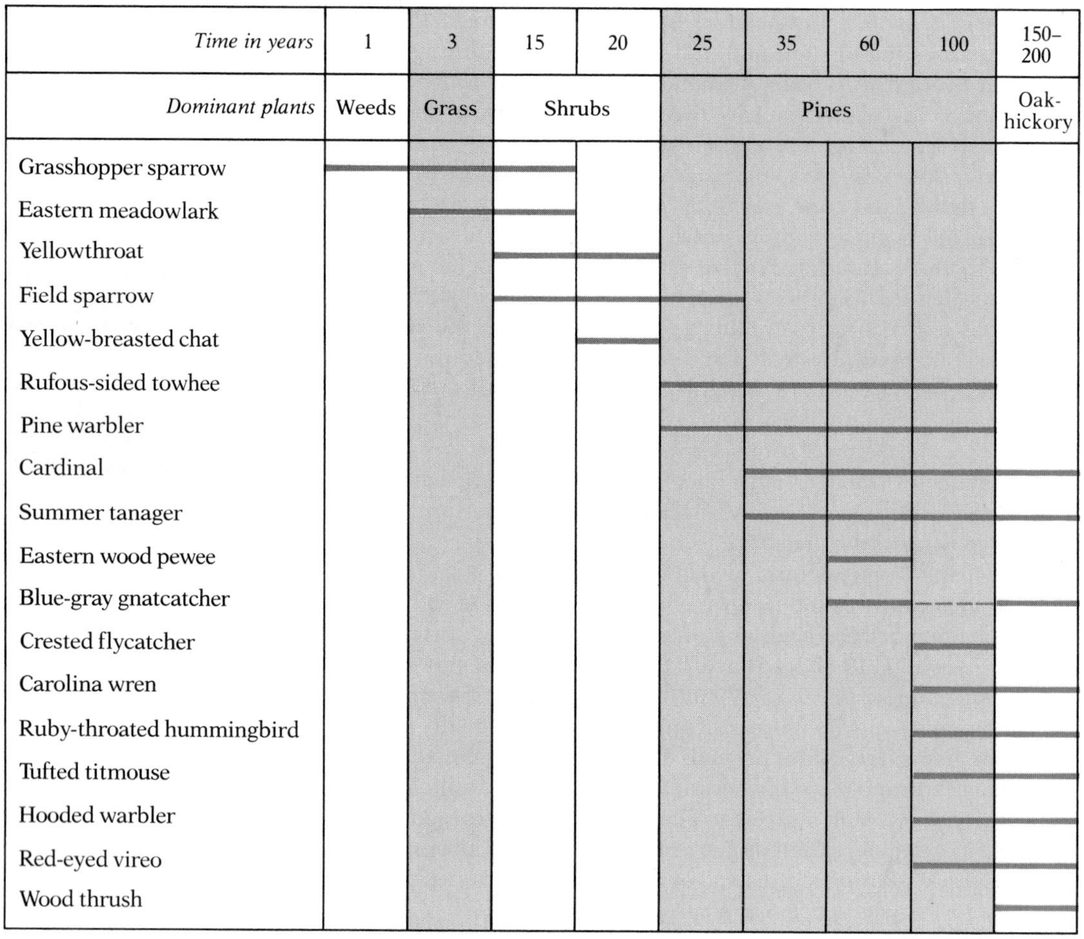

Time in years	1	3	15	20	25	35	60	100	150–200
Dominant plants	Weeds	Grass	Shrubs		Pines				Oak-hickory
Grasshopper sparrow									
Eastern meadowlark									
Yellowthroat									
Field sparrow									
Yellow-breasted chat									
Rufous-sided towhee									
Pine warbler									
Cardinal									
Summer tanager									
Eastern wood pewee									
Blue-gray gnatcatcher									
Crested flycatcher									
Carolina wren									
Ruby-throated hummingbird									
Tufted titmouse									
Hooded warbler									
Red-eyed vireo									
Wood thrush									

35.25 Bird succession on abandoned upland farmland in Georgia
The bars indicate when each of the bird species was present in a density of at least one pair per 10 acres. In the early (weed and grass) stages grasshopper sparrows and eastern meadowlarks were the dominant bird species. During the shrub stage yellow-throats and field sparrows became dominant. Pine warblers and rufous-sided towhees dominated the young pine forests, and red-eyed vireos, wood thrushes, and cardinals dominated the oak-hickory forests.

mon strata of deciduous and tropical forests. The canopy species capture most of the sunlight, but much of the energy they assimilate must be used to build and maintain woody supporting tissues. The herb layer, on the other hand, receives as little as 1 percent of the available sunlight, but the plants have no wood, and can use all the energy from photosynthesis for maintenance and reproduction.

As Figure 35.25 indicates, changes corresponding to the succession of dominant plants also take place in the animal populations of the abandoned cropland community.

A summary of common trends in succession   Despite numerous differences between various instances of succession, and especially between primary and secondary succession, some generalizations tend to hold true in most cases where both autotrophs and heterotrophs are involved:

1. The species composition changes continuously during the succession (Fig. 35.26), but the change is usually more rapid in the earlier stages than in the later ones.

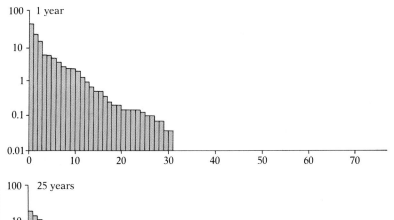

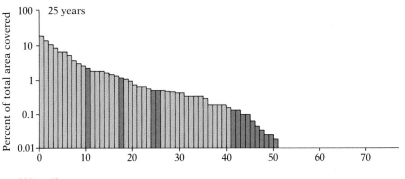

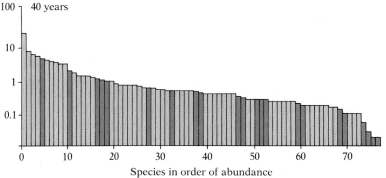

Species in order of abundance

35.26 The changing pattern of species diversity and abundance in the early successional plant communities on an abandoned agricultural field in Illinois
After one year there were 31 species of plants, all herbs (green bars), growing in the field. After 25 years there were 51 species, including some shrubs (gold bars) and trees (purple bars). After 40 years there were 77 species, and shrubs and trees had greatly increased in relative abundance.

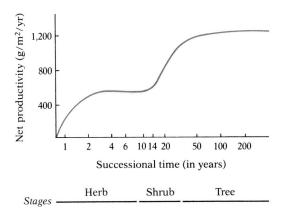

Successional time (in years)

Stages ──── Herb ──── Shrub ──── Tree ────

35.27 Change in net primary productivity during plant succession on an area cleared of an oak-pine forest in Brookhaven, New York
The first rise represents the invasion of the area by herbs. The later rise (after about 14 years) reflects the entrance of larger woody plants into the community.

2. The total number of species represented increases initially, then sometimes declines slightly, and finally becomes more or less stabilized in the older stages. This trend applies particularly to the heterotrophs, whose variety is usually much greater in the later stages of the succession.

3. Net primary productivity (the amount of energy converted into products of photosynthesis by autotrophs, and available to heterotrophs) increases until it reaches a stable high level (Fig. 35.27).

4. The store of inorganic nutrients held in the organisms and soil of the ecosystem increases, and an increasing proportion of this store is held in the tissues of plants.

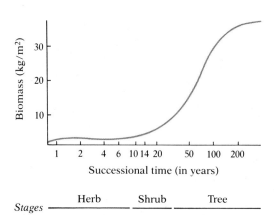

35.28 Change in biomass during succession in the Brookhaven study area
The total biomass remained low during the early years, when herbs were the dominant plants in the community, but increased later when shrubs and trees became more prominent.

5. Both the total biomass in the ecosystem (Fig. 35.28) and the amount of nonliving organic matter increase during the succession until a more stable stage is reached.

6. The height and massiveness of the plants in the community increase and lead to greater differentiation of vertical strata.

7. With more extensive aboveground plant cover, the microclimate within the community becomes increasingly determined by the community itself.

8. The food webs (to be discussed in the next chapter) become more complex, and the relations between species in them become better defined or more specialized. As a result, the efficiency of resource utilization at the various levels usually rises.

In summary, if these generalizations are correct, the trend of most successions is toward a more complex and longer-lasting ecosystem, in which less energy is wasted and hence a greater biomass can be supported without further increase in the supply of energy.

THE CONCEPT OF CLIMAX

If some disruptive factor does not interfere, succession—as the previous discussion will have suggested—eventually reaches a stage far more stable than those that preceded; the important species populations attain a steady state, balancing births and deaths, and both energy flow and biomass likewise attain equilibrium, with gross primary productivity equaled by total respiration. The community of this stage is called the *climax community*. It has much less tendency than earlier successional communities to alter its environment in a manner injurious to itself. In fact, its more complex organization, larger organic structure, and more balanced metabolism enable it to buffer its own physical environment to such an extent that it can be self-perpetuating. Consequently it may persist for centuries, not being replaced by another stage so long as climate, geography, and other major environmental factors remain essentially the same (Fig. 35.29).

It must be emphasized, however, that a climax community is not static; it does slowly change, and will change rapidly if there are major shifts in the environment, either physical or biotic. For example, sixty years ago chestnut trees were among the dominant plants in the climax forests of much of eastern North America, but they have been almost completely eliminated by a fungal blight, and the present-day climax forests of the region are dominated by other species. Thus there can be no absolute distinction between climax and the other stages of succession; the difference between them is relative.

A view that held sway among some American ecologists during much of this century was that all succession in a given large climatic region will converge to the same climax type—that there is only one type of climax community for the region and that any sites dominated by other communities have not yet reached climax, no matter how stable and long-lasting they may seem. Thus a beech-maple forest was considered the climax for much of the northeastern United States, and a white-spruce–balsam-fir forest was considered the climax for much of Canada. The proponents of this so-called monoclimax hypothesis regarded plant communities as distinct entities comparable to species—describable in analogous ways and amenable to categorization and classification.

Most modern ecologists, however, reject this view. Pointing to evidence that each species is distributed according to its own particular biological potential, they contend that the aggregation of species characterizing any given community is the fortuitous product of local environmental conditions and of whatever plant and animal species happen to be available in the area. Since environmental conditions such as temperature, humidity, soil characteristics, topographic features, wind patterns, and so on, vary continuously in both space and time, vegetation likewise varies continuously in both space and time. Boundaries between communities are seldom distinct, because the distributions of the various species composing communities are not well correlated with one another. If a traveler along the Mississippi River basin carefully noted which tree species occur where, he would see that the different species drop out or appear at different places, with little apparent correlation with one another. There would be no place where he would notice any abrupt change in the community; yet the small changes from mile to mile would be cumulative, so that after traveling many miles he would find himself in a community with a species composition almost completely different from that of the community in which his journey began.

This view of communities as parts of a gradually changing continuum, whose characteristics at any specific place are uniquely determined by a combination of local physical conditions, local biotic factors, local species distributions, and a considerable element of chance, leads to the conclusion that similarities between climax communities in different places are only approximate. Hence there is no absolute climax for any region. Climax, according to this view, has meaning only in relation to the individual site and its environmental conditions. The climax for a given spot should be determined, not by referring to some theoretical regional climax, but by actually observing what populations replace others and then maintain themselves in a stable condition. Rather than try to fit all the climax communities of a large region into a single regional climax, it is more productive in most cases to emphasize study of the gradients between the individual local climaxes, in an attempt to learn how changes in the component parts of communities are correlated with changing environmental conditions.

35.29 Sequoia forest
The long-term stability of a climax forest is illustrated by this sequoia forest in south-central California. The exact age of the forest is not known, but some of the trees still standing are more than two thousand years old.

Chapter 36

ECOSYSTEMS AND BIOGEOGRAPHY

The discussion of ecology in the last chapter was concerned primarily with interactions between individuals within a population, between species within a community, and between the community and its environment. But how do communities come to be located where they are, how do they spread, and how do energy and nutrients essential to the continued survival of the community cycle between its members and the environment? Further, how do characteristics of the physical environment and the cycling of nutrients influence the organization of communities and the dispersal of organisms around the globe? This chapter will explore some of the ways in which the movement of energy and materials binds together the community and the physical environment as a functioning system; it is concerned with the energy and nutrient "economics" of ecosystems and the biogeography of communities.

THE ECONOMY OF ECOSYSTEMS

THE FLOW OF ENERGY

Life depends, ultimately, on energy from the sun. Making use of the chemical pathways described in Chapters 7 and 8, all forms of life, with the exception of the chemosynthetic organisms, obtain their high-energy organic nutrients either directly or indirectly from photosynthesis. The total amount of energy converted into the products of photosynthesis, which accounts for less than one-tenth of 1 percent of the solar energy reaching the surface of the earth, is called ***gross primary productivity***. Plants use from 15 to 20 percent of their gross productivity in their own respiration. What is left over is known as ***net primary productivity***. The total net primary productivity of the biosphere, estimated at about 6×10^{20} gram calories of energy per year,[1] constitutes the energy base for heterotrophic life on earth. Heterotrophic organisms—most bacteria and protists, and all fungi and animals[2]—obtain the energy they need by feeding on autotrophic organisms, on other heterotrophs that fed on autotrophs, or on the ***detritus*** (waste products or bits of dead tissue) of other organisms.

The sequence of organisms through which energy may move in a community is customarily called a ***food chain***. In most real communities there are so many possible food chains, so complexly intertwined, that together they form a community ***food web*** (Fig. 36.1). No matter how long a food

[1]The energy content of organic materials is determined by burning them in pure oxygen and measuring the heat liberated. Organic matter has a relatively uniform energy content of about 4.25 kcal per dry gram of plant tissue and 5.0 kcal per dry gram of animal tissue. Some ecologists, therefore, give primary productivity values in dry-weight units. In these terms, the total net primary productivity for the biosphere is about 164 billion dry tons of organic matter per year.

[2]Certain fungi (in lichens) and animals (mostly certain corals) obtain at least some of their energy from photosynthetic symbionts.

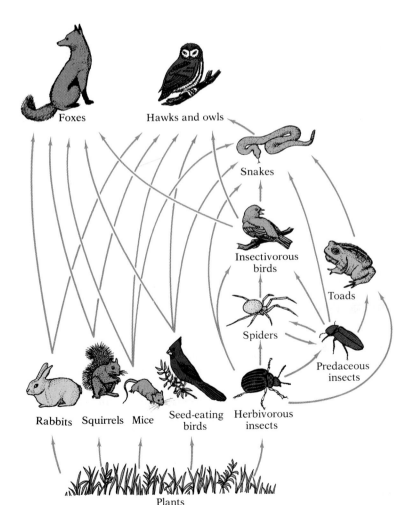

36.1 Diagram of a hypothetical food web
No real food web would be as simple as this one.

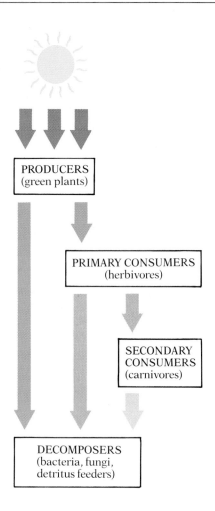

36.2 Flow of energy through the principal trophic levels in an ecosystem
The green plants are the producers, which are eaten by the herbivores, the primary consumers. The primary consumers may in turn be eaten by parasites or carnivores, the secondary consumers. Producers or consumers may die and become food for decomposer organisms.

chain or how complex a food web may be, however, certain basic characteristics are always present. Every food chain or web begins with the autotrophic organisms (green plants in the vast majority of cases) that are the **producers** for the community. And every food chain or web ends with **decomposers**, the organisms of decay, which are usually (though not exclusively) bacteria and fungi; animals like lobsters, clams, and some catfish also feed at least partially on detritus. Decomposers, like all heterotrophs, release simple substances—CO_2 for example—reusable by the producers. The links between the producers and the decomposers are variable. The producers may die and be acted upon directly by the decomposers, in which event there are no intermediate links. Or the producers may be eaten by **primary consumers**, the herbivores. These, in turn, may be either acted upon directly by decomposers or fed upon by **secondary consumers** such as carnivores or parasites or scavengers (Fig. 36.2).

Ecologists speak of the successive levels of nourishment in the food chains of a community as **trophic levels**. Thus all the producers together constitute the first trophic level, the primary consumers (herbivores) the second trophic level, the herbivore-eating carnivores the third trophic level, and so on. The species that comprise each trophic level differ from one community to another. Moreover, the trophic levels themselves are not hard-and-fast categories, since many species that eat a varied diet—especially omnivores—may function at two or more trophic levels within a single food web. For example, a chickadee, which eats seeds, herbivorous insects, and carnivorous insects, functions at the second, third, and fourth trophic levels. Despite these complications, the concept of trophic levels remains a useful one in community analysis.

At each successive trophic level there is loss of energy from the system. The loss results in part from the consumer population's inability to harvest more than a fraction of the available biomass; in part from a failure of assimilation (with the exception of the ruminants and termites, most herbivores cannot metabolize the cellulose walls of plant cells, for example); and in part from respiration and the consequent dissipation of energy as heat, in accordance with the Second Law of Thermodynamics (every energy transfer involves loss of some usable energy, usually as heat; see p. 71). As a result, only about 10 percent of the energy at one trophic level can usually be passed on to the next level. In other words, the productivity (energy bound into new organic matter per unit area per unit time) at each trophic level is only about 10 percent of that at the preceding level. There is less productivity from the herbivores of a community than from the plants of that community, and there is less productivity from the carnivores than from the herbivores, and so on. Hence the distribution of productivity within a community can be represented by a pyramid, with the first trophic level (producers) at the base and the last consumer trophic level at the apex (Fig. 36.3). Because of the rapid fall in productivity from one trophic level to the next, there are seldom more than four or five levels in a food chain; the fifth level rarely has more than about 0.0001 the productivity of the first, and the productivity possible for any subsequent levels normally becomes too low to be effective.

The **pyramid of productivity** (also called the pyramid of energy flow) just described is a characteristic of all ecosystems. Several other attributes of ecosystems may fit a pyramidal model because they are related to the flow of energy through the system, but they are only secondary consequences of the distribution of productivity, and so may well deviate from the pyramidal model. One example is the ***pyramid of biomass*** (Fig. 36.4A). In general, the decrease of energy at each successive trophic level means that less biomass can be supported at each level. Hence the total mass of carnivores in a given community is almost always less than the total mass of herbivores. But the size, growth rate, and longevity of the species at the various trophic levels of a community are important in determining whether or not the pyramidal model will hold for the biomass of that community. For example, in some aquatic communities where the producers are small algae with high metabolic and reproductive rates, there may be a greater biomass of consumers than of producers at any given moment, but the total mass of all the algae that live during the course of a year will be greater than the total mass of consumers that live during that year.

The interrelationships between the organisms at different trophic levels exert some influence on the size of the organisms. Thus carnivores are often larger and stronger than their herbivorous prey. And secondary carnivores are often larger than the primary carnivores on which they feed. Now, since total biomass tends to decline at successive trophic levels, it follows that, if the size of individuals increases at successive levels, the number of individuals must decline at each level (except at the decomposer level). Consequently some communities show a ***pyramid of numbers***, there being fewer individual herbivores than plants, and fewer individual carnivores than herbivores (Fig. 36.4B). Indeed, it is entirely understandable why ***top predators*** (predators at the top of their food chains), such as lions or wolves or killer whales, are not themselves preyed on: there are too few of them, they are too widely scattered, and they contain too little energy to make the effort worthwhile.

Many communities, however, have no pyramid of numbers. For example, there may be many more individual insect consumers than plants, even though their biomass may be less, because plant-eating insects are often far smaller than their food plants; a common example would be a single large tree in spring on which thousands of leaf-eating caterpillers and boring insects may be feeding. Food chains involving parasites also tend to have inverted population–size relationships, because the parasites are smaller and usually more numerous than the hosts.

Given the inefficiency of energy transfer from one trophic level to the next, it might seem that the earth could support more humans if we all stopped being omnivorous, and lived on a wholly vegetable diet instead of the combined animal and vegetable diet for which we inherited many specializations from our distant ancestors. This popular view has several flaws, however. One is that large areas of the world—much of Argentina, Australia, Africa, and the western and southwestern United States, for example—can support only low-quality pasturage plants that are unsuitable for human consumption, but can sustain large herbivores. Another flaw involves human nutritional needs: vegetarian diets usually require some supplemental animal protein, most commonly in the form of dairy products. To be sure, most individuals in Western nations eat more animal protein than they need for survival, but even if they did not, a substantial portion of the earth's cultivable land might always be needed to supply fodder for milk cows, which require high-quality grass to maintain lactation.

Another persistent problem is the Malthusian dilemma: without a drastic reduction in the human birth rate in the developing nations, any increase in

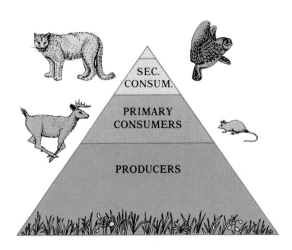

36.3 The pyramid of productivity
There is much more productivity at the producer level in an ecosystem than at the consumer levels, and there is more at the primary consumer level than at the secondary consumer level.

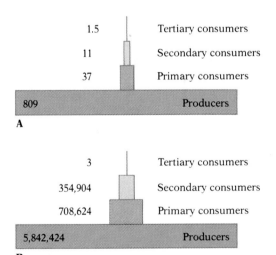

36.4 Examples of the pyramids of biomass and numbers
(A) Pyramid of biomass in the aquatic ecosystem of Silver Springs, Florida. Figures represent grams of dry biomass per square meter. (B) Pyramid of numbers in a bluegrass field.

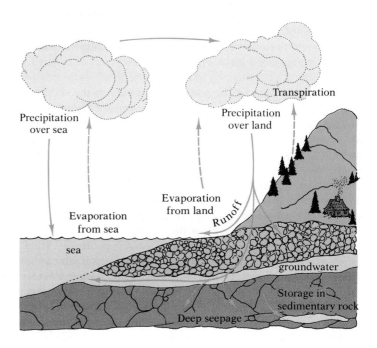

36.5 The water cycle
This diagram shows most of the major pathways of water movement through the ecosystem—but not the more recent ones created by human beings.

the world food supply may only serve to increase the number of people who must ultimately go hungry.

CYCLES OF MATERIALS

We have seen that energy is steadily drained from the ecosystem as it is passed along the links of a food chain. The system cannot continue functioning without a constant input of energy from the outside. In other words, there is no such thing as an energy cycle. But this is not the case with most materials. The same materials can and must be used over and over again, and hence can be passed round and round through the ecosystem indefinitely. We can, therefore, speak of cycles of materials. Let us examine several such cycles.

The water cycle   When rainwater falls on the land, some of it quickly evaporates back into the atmosphere. Of the water that does not immediately evaporate, some is absorbed by plants or is drunk by animals, some runs off the surface of the land into streams and lakes, and some percolates down through the soil, to accumulate as groundwater (Fig. 36.5). The water in the streams and lakes, and the subsurface groundwater, eventually find their way to the ocean. There is constant evaporation from streams, lakes, and oceans, and also from the bodies of plants and animals. The energy for most of this evaporation comes either directly or indirectly from solar radiation.

The endless cycling of water to earth as rain, back to the atmosphere through evaporation, and back again to earth as rain, maintains the various freshwater environments and supplies the vast quantities of water necessary for life on land. The water cycle is likewise a major factor in modifying temperatures, and it transports many chemical nutrients through ecosystems. The tremendous importance of rainfall for terrestrial and freshwater life requires no elaboration here; one need only visualize a desert on the one hand and a lush tropical forest on the other.

The carbon cycle   The carbon dioxide contained in the atmosphere or dissolved in water constitutes the reservoir of inorganic carbon from which almost all organic carbon is derived. Most of the rest is derived from sedimentary deposits of limestone ($CaCO_3$), which form from the shells of marine organisms like mussels and clams; the erosive action of water on limestone leaches CO_2 into the water. It is photosynthesis, largely by green plants, that extracts the carbon from this inorganic reservoir and incorporates it into the complex organic molecules characteristic of living substance (Fig. 36.6). Some of these organic molecules are soon broken down again, and their carbon is released as CO_2 by the plants in the course of their own respiration. But much of it remains in the plant bodies until they die or are eaten by animals. The carbon obtained from plants by animals may be released as CO_2 during respiration, or it may be eliminated in more complex compounds in the body wastes, or it may remain in the animals until they die. Usually the wastes from animals and the dead bodies of both plants and animals are broken down (respired) by the decomposers, and the carbon is released as CO_2.

Notice that whether the carbon follows a short pathway involving only one or two trophic levels or a longer pathway involving three, four, or more, most of it eventually returns as CO_2 to the air or water whence it started. This is, then, a true cycle (or rather a complex of interlocking cycles); carbon is constantly moving from the inorganic reservoir to the living system and back again.

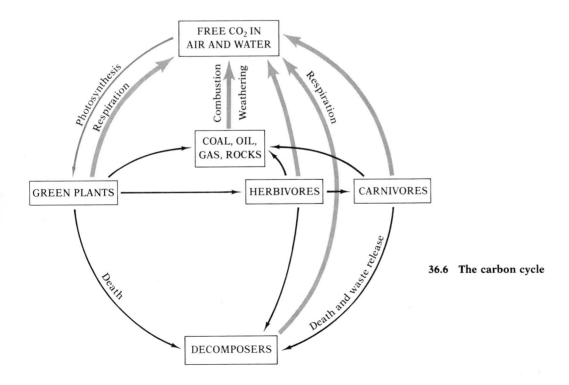

36.6   The carbon cycle

The pathways just outlined are all pathways through which carbon moves rather rapidly. Complete passage through the system may take only minutes or hours, or at most a few years. There are alternative pathways, however, that take much longer. The dead bodies of organisms occasionally fail to be decomposed promptly and are converted instead into coal, oil, gas, or rock (particularly limestone). Carbon in these forms may be removed from circulation for very long periods, perhaps permanently; but some of it may eventually return to the inorganic reservoir if the coal, oil, and gas (the fossil fuels) are burned or if the rocks are sufficiently weathered. Humans have of course greatly accelerated the return of such carbon to the active cycle.

Of the CO_2 released by the burning of fossil fuels, about half remains in the atmosphere and the rest is absorbed by the ocean waters. The CO_2 reservoir in the atmosphere is also being increased through the oxidation of organic materials once incorporated into the plants that grew in areas subsequently cleared for roads or buildings or being used for agriculture. (Agricultural crops usually fix less CO_2 than the natural vegetation they displace, because they are highly productive only for a relatively short time. Meanwhile the reserve of organic litter built up in the soil by the native vegetation slowly decomposes, releasing its carbon into the atmosphere as CO_2.) These human activities have increased atmospheric CO_2 levels by 15 percent in the last hundred years; it is entirely possible that they will double the CO_2 concentration in the next hundred years.

CO_2 plays an integral role in the regulation of temperatures on the surface of the earth. Heat radiated from the earth is absorbed by CO_2 in the atmosphere and, radiated back to the surface, tends to warm the earth in a so-called greenhouse effect. Hence a rise in the CO_2 levels in the atmosphere should cause the temperature of the earth to increase, were it not that cloud patterns and the amount of water vapor and atmospheric particulate matter—which are also changing as a result of human activities—likewise influence the temperature balance of the earth. Some current estimates indicate that the average temperature of the earth will indeed rise, by one or two degrees Celsius over the next few decades. This minute increase in global temperature would be enough to melt a significant amount of polar and other ice, thereby raising the sea level by perhaps as much as a few meters. The consequences for coastal areas, particularly major port cities like New York, could be serious.

The nitrogen cycle Another critical element in community metabolism is nitrogen, a constituent of amino acids and nucleic acids. The reservoir of inorganic nitrogen is the gaseous N_2, which constitutes roughly 78 percent of the atmosphere. But N_2 has very little biological activity. It enters the bodies of all organisms, but comes back, in most cases, without having played any significant role in their life processes. Some microorganisms, however—a few bacteria (particularly the Cyanobacteria)—can use N_2 in the synthesis of substances usable by other organisms. This process is known as *nitrogen fixation*. Though some nitrogen fixation may also occur as a result of electrical discharges, such as lightning, and though more and more nitrogen is now supplied in commercial fertilizers manufactured by industrial processes, it is biological nitrogen fixation by microorganisms that provides most of the usable nitrogen for the earth's ecosystems (Fig. 36.7).

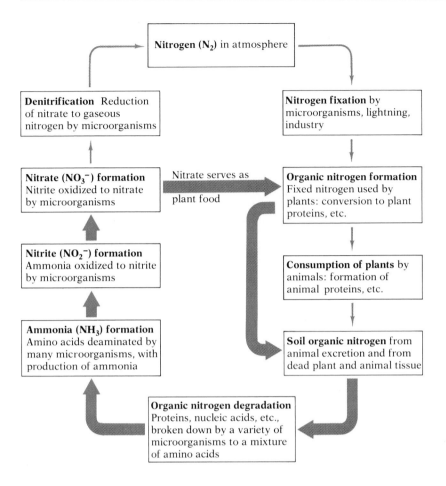

Nitrogen (N₂) in atmosphere

Denitrification Reduction of nitrate to gaseous nitrogen by microorganisms

Nitrogen fixation by microorganisms, lightning, industry

Nitrate (NO₃⁻) formation Nitrite oxidized to nitrate by microorganisms

Nitrate serves as plant food

Organic nitrogen formation Fixed nitrogen used by plants: conversion to plant proteins, etc.

Nitrite (NO₂⁻) formation Ammonia oxidized to nitrite by microorganisms

Consumption of plants by animals: formation of animal proteins, etc.

Ammonia (NH₃) formation Amino acids deaminated by many microorganisms, with production of ammonia

Soil organic nitrogen from animal excretion and from dead plant and animal tissue

Organic nitrogen degradation Proteins, nucleic acids, etc., broken down by a variety of microorganisms to a mixture of amino acids

36.7   The nitrogen cycle

The many steps at which microorganisms play a major role show to what extent higher forms of life depend on bacteria (particularly Cyanobacteria) and fungi for their own continued existence. Most of the nitrogen used by organisms is cycled within the soil and ocean, and so never enters the atmosphere.

36.8   Photograph of roots of a peanut plant, showing nodules

As we saw in Chapter 9, some of the nitrogen-fixing bacteria (genus *Rhizobium*) live in a close mutualistic relationship with the roots of higher plants, where they occur in prominent *nodules* (Fig. 36.8). The legumes (plants belonging to the pea family—bean, clover, alfalfa, lupine, and the like) are particularly well known for their numerous root nodules, but plants of some other families have them also. Other nitrogen-fixing microorganisms live free in soil or water. All of these nitrogen-fixing microorganisms can reduce N₂ to ammonia (NH₃), which is often in the form of ammonium ions (NH₄⁺). They then either use the ammonium in the synthesis of organic nitrogen-containing compounds or excrete it into the soil or water in which they live.

The bacteria in root nodules promptly release much of the fixed nitrogen they produce into the host plant's cytoplasm, primarily in the form of amino acids. As much as 90 percent of the fixed nitrogen can be liberated into the host's cytoplasm in this way, there being little or no storage of fixed nitrogen within the bacteria or the nodules. Consequently legumes can grow well in soils very poor in available nitrogen. The bacteria in the nodules not only supply the plants with all the fixed nitrogen they need, but actually produce a surplus, some of which is excreted from the roots of the legumes into the soil.

Cyanobacteria are the most important nitrogen-fixing microorganisms

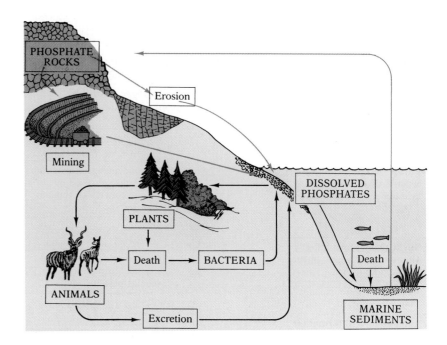

36.9 A simplified version of the phosphorus cycle
Phosphate from rock dissolves very slowly (unless the process is speeded up by human intervention). The dissolved phosphate can be used by plants, which may pass it to animals. Some of the phosphate is excreted by animals and goes immediately into the dissolved pool. When plants or animals die, phosphate is released by bacteria from organic compounds like nucleic acids that are present in the bodies.

Each year huge quantities of dissolved phosphate are carried into the sea in runoff water. Though the formation of new rocks from marine sediments, where the phosphorus eventually comes to rest, is a very slow process, it is unlikely that we shall soon run out of phosphate rock, because the known reserves are large.

that live free in the soil or water, though some other bacteria (*Azotobacter* and *Clostridium*, for example) also play a role. These organisms, which may fix between 20 and 50 pounds of nitrogen per acre annually, release ammonia into the surrounding medium, and when they die the fixed nitrogen in their cells is broken down to ammonia by decomposer organisms. The decomposers act in the same way upon the organic nitrogen compounds in the bodies of green plants or animals or other microorganisms when they die, and upon the nitrogen compounds in the urine and feces of animals. Some of this free ammonia is picked up as ammonium ions by the roots of higher plants, particularly certain grasses and forest trees, and incorporated into more complex compounds. But most flowering plants use nitrate in preference to ammonia. Nitrate seems to be the main source of nitrogen for higher plants. It is produced from ammonia in the soil by nitrifying bacteria.

The process of *nitrification* is usually accomplished by two different groups of bacteria, working in sequence. The first group converts ammonium ions into nitrite (NO_2^-), and the second group converts this nitrite into nitrate (NO_3^-) and releases it into the soil, where it can be picked up by the roots of plants.[3] Most of the nitrate in the roots is quickly incorporated into organic nitrogen compounds and then either stored, primarily in cell vacuoles, or transported to other parts of the plant body through the vascular tissue.

The nitrogen compounds in the plant body may eventually again be broken down to ammonia by decomposers when the plant dies or when an ani-

---

[3]Among the most important genera of nitrite bacteria are *Nitrosomonas*, *Nitrosocystis*, and *Nitrosospira*. Among the most important genera of nitrate bacteria are *Nitrobacter*, *Nitrocystis*, and *Bactoderma*.

mal that ate the plant dies or excretes it. Notice that nitrogen can cycle repeatedly from plants to decomposers to nitrifying bacteria to plants without having to return to the gaseous N_2 state in the atmosphere. In this respect, the nitrogen cycle differs from the carbon cycle, where every turn of the cycle includes a return of CO_2 to the air or water.

Though nitrogen need not return to the atmosphere at every turn of the cycle, there is a steady drain of some of it away from the soil or water and back to the atmosphere. This is because some bacteria carry out a process of *denitrification*, converting ammonia or nitrite or nitrate into N_2 and releasing it. In short, the denitrifying bacteria remove nitrogen from the soil–organism part of the nitrogen cycle and return it to the atmosphere, while the nitrogen-fixing microorganisms do the reverse: they take nitrogen from the atmosphere and add it to the soil–organism part of the cycle.

The phosphorus cycle Another mineral essential to life is phosphorus. Like nitrogen, it is one of the chief ingredients in commercial fertilizers. Unlike carbon and nitrogen, for which the major reservoir is the atmosphere, phosphorus has its reservoir in rocks (Fig. 36.9).

Under natural conditions much less phosphorus is available to organisms than nitrogen; in natural waters, for example, the ratio of phosphorus to nitrogen is about 1:23. However, the mining of roughly three million tons each year has greatly accelerated the movement of phosphorus from the rocks to the water–organism part of the cycle. This mineral, the normal limiting resource for algae in many freshwater lakes, is now being poured into the aquatic environment in enormous quantities in sewage and in run-off from inorganic fertilizers used in farming. One consequence is extensive algal blooms that cover the surface of the water with scum and foul the shores with stinking masses of rotting organic matter (Fig. 36.10).

The increased photosynthetic productivity associated with the algal blooms might be expected to make more food available for higher links in food webs and thus be of benefit to the biotic community. But excessive growth of algae actually causes destruction of many of the higher links in the food webs. At the end of the growing season, many of the algae die and sink to the bottom, where they stimulate massive growth of bacteria the following year. The bacterial decomposers are so active that they consume most of the oxygen of the deeper, colder layers of lakes, with the result that cold-water fish such as trout, cisco, whitefish, pike, and sturgeon are asphyxiated, and are replaced by less valuable species such as carp and catfish. Deoxygenation of the water also causes chemical changes in the bottom mud that produce increased quantities of odorous, sometimes toxic, gases. These changes further accelerate *eutrophication*, or nutrient enrichment and associated "aging," of the lake.[4]

Extensive use of tertiary treatment of sewage, combined with the recent reduction in the phosphate content of detergents, would slow the undesirable changes in the appearance and biotic composition of lakes, but would not ultimately prevent them, in part because at least 30 percent of the pol-

36.10 A field experiment demonstrating the importance of phosphorus in the eutrophication of a lake

The two basins of a lake were separated by a plastic curtain. The far basin was fertilized with phosphorus, carbon, and nitrogen. The near basin, used as a control, received only carbon and nitrogen. Within two months the far basin was covered by a heavy bloom of algae, whereas the control basin showed no change in organic production.

---

[4]The term "eutrophication" was originally applied to the accumulation of nutrients and increase in organic matter that are thought to be a natural part of the aging of lakes. Recently it has been applied not so much to the natural process as to the greatly accelerated one resulting from human interference.

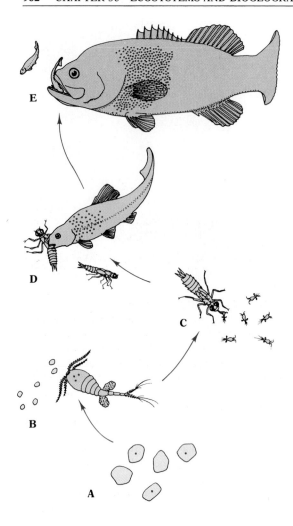

36.11 An example of biological magnification
(A) Some individuals of a single-celled plant species at the bottom of a food chain have picked up a small amount of a stable nonexcretable chemical (red). (B) *Cyclops*, a small crustacean, incorporates the chemical from the plants it eats into its own tissues. Like the other organisms in the chain, it lacks the biochemical pathways necessary to metabolize or excrete the novel substance. (C) A dragonfly nymph stores all the chemical acquired from the numerous *Cyclops* it eats. (D) Further magnification occurs when a minnow eats many of the dragonfly nymphs that have stored the chemical. (E) When a bass, the top predator in this food chain, eats many such minnows, the result is a very high concentration in its tissues of a chemical that was much less concentrated in the organisms lower in the chain.

luting phosphorus comes from agricultural sources; this mineral is essential to the production of the large crops needed to feed a burgeoning population. Furthermore, the tertiary sewage treatments devised so far, though quite effective in removing phosphates, are much less so in removing nitrates, which are probably the major natural limiting nutrient in undisturbed estuaries. Hence the change from phosphate detergents to nitrogen detergents, which has become popular in the name of "better ecology," may simply shift the pollution problem from lakes to estuaries. Unless new detergents free of nutrients that are difficult to remove can be developed, or more effective sewage treatment devised, a return to soaps which cleanse but contain neither phosphate nor significant nitrogen may become imperative.

Commercial chemicals Modern industry and agriculture have been releasing vast quantities of new or previously rare chemicals into the environment. The U.S. Food and Drug Administration has estimated their number in the environment at about half a million, with hundreds of new ones created each year. The pathways through ecosystems are known for only a few. Most will probably be incorporated into the natural biogeochemical cycles and degraded to harmless simpler substances. But many are so different from any naturally occurring substances that we have no idea as yet what their eventual fate or their effects on the biosphere may be. Many of the by-products of industrial processes have so far not even been fully characterized chemically. Will some of these prove harmful to life? The answer is surely yes. But which will be harmful, and how harmful, and to what organisms? It is with these questions that much ecological research in the future must deal.

The matter is further complicated because a substance that is not harmful in the form in which it is released may be changed by microorganisms, or by natural physical processes, into some other substance with vastly different properties. A case in point is the mercury pollution of bodies of water near some plastics factories. The mercury was originally released in an insoluble and nontoxic form that was thought to be stable. When it settled in the bottom mud, however, microorganisms converted it into methyl mercury, a water-soluble compound that accumulates in organisms. The mercury poisoning that resulted was most severe in human beings and other top predators.

The harmful effect on top predators of the mercury from plastics factories (and of the mercury, principally from fungicides, that has recently caused the fish in many streams and lakes to be declared unsafe for human consumption) is an example of a common phenomenon called *biological magnification*. If a persistent chemical,[5] when ingested, is retained in the body rather than excreted, that chemical will tend to become more and more concentrated as it is passed up the food chain (Fig. 36.11).

DDT (a persistent insecticide that has become so pervasive in the biosphere that it can now be found in the fatty tissues of nearly every living organism) has had more severe effects on predatory birds such as the bald eagle, the peregrine falcon, and the osprey than on seed-eating birds be-

[5]A persistent chemical is one that is comparatively stable under natural biological and environmental conditions.

cause of biological magnification. Some investigators have reported that the reproductive rate of these birds has been calamitously reduced because DDT—and its metabolites, DDD and DDE—interfere with deposition of calcium in the eggshells, with the result that the thin-shelled eggs are easily broken and few birds hatch (Fig. 36.12).

This brief summary of some of the biogeochemical cycles will have suggested the complexity of the movement of materials through an ecosystem, the interdependence of the different species within it, and, in particular, the fundamental and essential role played by microorganisms. Because the microorganisms are not observed as easily as the larger plants and animals, we tend to forget about them, or to think only of the harmful ones, especially those that cause diseases; and thus we overlook the others, many of which are indispensable to our continued existence.

THE ROLE OF SOILS

Soil is an essential link in the water cycle, on which terrestrial plants depend. Plants require water, of course, but they also depend on the cycling of water for nitrogen and phosphorus—and for calcium, sulfur, potassium, and other ions whose cycles we have not discussed—because these come to them dissolved in soil water. It follows that the properties of soils, including particle size, amount of organic material, and pH, play a very important role; these factors help determine the availability of water and minerals to the plants, and the rapidity with which these materials move through the soil.

Most soils are a complex of mineral particles, organic material, water, soluble chemical compounds, and air. In this complex the dominant components by far are the mineral particles, which are composed largely of compounds of silicon and aluminum. They vary in size from tiny clay particles (less than 0.002 mm in diameter), through silt particles (0.002–0.05 mm), to sand grains (0.05–2.0 mm). The proportions of clay, silt, and sand particles in any given soil determine many of its other characteristics. For

36.12 Effect of DDT on osprey
In 1950 there were over 200 mating pairs of osprey nesting at the mouth of the Connecticut River, where these pictures were taken. By 1970 only six mating pairs were observed, the decline in the local population being attributed to the detrimental effects of DDT and related hydrocarbons on the calcification of eggshells produced by these birds. The hydrocarbons had been introduced into the ecosystem in insect-retarding plant sprays that were subsequently ingested by the osprey's prey—fish exposed to the polluted runoff of local streams and rivers. The resulting weakening of the shells caused a high percentage of the eggs to break during incubation; approximately 10 eggs, or two to three nestings, were needed to produce one single offspring. A nest with two eggs and a broken shell is seen at left; at right is a female osprey with some of the few young successfully hatched during this period. Since 1970 the local osprey population has grown considerably, largely because of a ban on the use of the chemicals.

36.13 Humus
Organic decomposition products containing high
quantities of cellulose and lignin create humus. This
underlying dense layer of humus, formed from
decaying woody plants on a forest floor in West
Virginia, improves the drainage and aeration of the
soil and provides inorganic substances for the living
plants the soil supports.

example, very sandy soils, which contain less than 20 percent of silt and
clay particles, have many air-filled spaces, but they are so porous and their
particles have so little affinity for water that it rapidly drains through them
and they are unsuitable for growth of many kinds of plants. As the percentage of clay particles increases, the water retention of the soil also increases
until, in soils with an excessive amount of clay, the drainage is so poor and
the water is held so tightly to the particles that the air spaces become filled
with water; few plants are adapted to grow in such water-logged soil.
Though different species of plants are adapted to different soil types, most
do best in soils of the type known as *loam*, which contain fairly high percentages of particles of each size (for example, 24 percent clay, 29 percent
silt, 30 percent fine sand, and 17 percent coarse sand). In such soil, drainage is good but not excessive and there is good aeration; the soil particles
are surrounded by (or contain) a shell of water, but there are numerous air-
filled spaces between them.

Loams usually also contain considerable amounts of organic material
(roughly 3–10 percent), mostly of plant origin. As this material decomposes, inorganic substances required for good plant growth are released
into the soil. The organic material thus contributes to soil fertility. But it
also plays another very important role. Since it usually has a rather porous
spongy texture, it helps loosen soils with high clay content and increase the
proportion of pore spaces, thus promoting drainage and aeration. This is
particularly true when the organic material is in the form of *humus*, which
consists mostly of decomposition products from cellulose and lignin (Fig.
36.13). It is interesting that humus has the opposite effect on sandy soils,
where it tends to reduce pore size by binding the sand grains together,
thereby increasing the amount of water held in the soil. Thus we see that
organisms not only are influenced by their physical environment but, in
turn, modify that environment.

The proportion of clay particles affects not only the physical structure of
the soil and its aeration and water-holding capacity, but also the amounts
and the availability to plants of certain minerals—in part because of the influence of the clay particles on water movement. If, for example, water
percolates downward through the soil very rapidly and in large quantities,
it will tend to leach many important ions from the soil, carrying them deep
into the underlying rock layers, where roots cannot reach them. Nitrate
ions are especially susceptible to leaching, and sulfate, calcium, and potassium ions may also be rapidly removed from the soil.

Excessive removal of calcium is particularly serious because it tends to
make the soil more acid. Though many plants grow best in slightly acid
soils, most do not do well in strongly acid ones (however, some species,
such as rhododendrons and cranberries, are well adapted to very acid conditions). The acidity of the soil influences the solubility, and hence the availability, of iron, manganese, phosphate, and some other ions (see Fig. 9.9, p.
233), as well as the activity of soil organisms, many of which are inhibited
by high acidity.

We have repeatedly spoken of the "availability" of ions to plants. Chemical analyses that give the total amount of the various ions present in soils
can be misleading, since certain proportions of these ions are not free, and
hence not available to plants. A complex equilibrium generally exists between ions free in the soil water and ions adsorbed on the surface of colloi-

dal clay and organic particles. Many factors, of which acidity is a prime example, can shift this equilibrium, either increasing the proportion of ions bound to the particles, and thus reducing availability, or increasing the proportion of free ions available in the soil solution.

Even air conditions may influence the ionic makeup of soil. An extensive study of an experimental forest at Hubbard Brook, New Hampshire, showed that the rain often contained appreciable quantities of sulfuric acid, probably because of industrial release of sulfur dioxide into the air. Hydrogen ions from the acid tended to displace nutrient cations, such as Ca^{++}, from negatively charged sites on the soil particles, with the result that these nutrients were leached more rapidly from the soil into streams and lakes. There is increasing evidence that such loss of soil fertility as a consequence of this so-called *acid rain* is distressingly widespread and represents a hitherto unrecognized cost of air pollution that now threatens forests in much of eastern North America and northern Europe (Fig. 36.14). The leaching of nutrients into lakes may also accelerate eutrophication, and so lead to major changes in the character and abundance of species in many lakes. Control of sulfur emissions is an extraordinarily difficult economic problem: to begin with, availability of alternatives to the burning of high-sulfur fuels—alternatives that include nuclear-powered generating facilities, natural gas, and low-sulfur oil—is uncertain; in addition, the production of sulfur dioxide wastes is often far removed from the areas affected by acid rain—literally in a different country in some cases. Then, too, there is considerable controversy about the possible role of ozone (O_3), a by-product even of low-sulfur combustion: ozone can drastically reduce the efficiency of photosynthesis, and may be far more important than acid rain in damaging forests.

The various characteristics of soils have a bearing not only on how many and what kinds of plants are likely to grow in any given region, but also on the occurrence of soil animals. Earthworms, nematode worms, millipeds, and mites, for example, are all sensitive to the structure, drainage, acidity, and chemical composition of soils. And animals that do not live in the soil are, of course, indirectly influenced in their distributions by soil types, because of their dependence on plants as a source of high-energy nutrients.

Conversely, the soils themselves can be fully understood only by considering the effects on them of the plants and animals living in and on them. Plant roots break up the soil in which they grow, and they remove substances from the soil and add other substances to it. The plant shoots shield the soil beneath them, thereby altering the patterns of rainfall, humidity, light, and wind to which the soil is subject. And when the plants die, their substance adds organic material to the soil, changing both its physical and chemical makeup. Microorganisms in the soil alter its composition profoundly. Soil animals, such as earthworms and millipeds, also have a marked effect; they constantly work the soil, breaking down its organic components and moving materials between soil layers.

Some of the effects of vegetation on the soil were dramatically demonstrated in the Hubbard Brook study. The investigators, after first obtaining accurate measurements of the nutrient input and output of a particular watershed over a period of several years, cleared the watershed of all its vegetation, and again monitored input and output. They found that the volume of runoff water promptly rose to levels many times greater than before

36.14 Forest damage attributed to acid rain
These Norway spruce in the Black Forest of West Germany have discolored needles, a characteristic sign of a condition (popularly known in Germany as *Waldsterben*, or "forest death") widely attributed to the effect of acid rain on forest soil. Whether, instead, elevated levels of ozone, which reduce the photosynthetic efficiency of leaves, may account for some or even most of the damage is the subject of much debate.

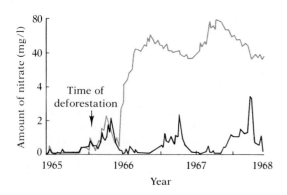

36.15   Change in the runoff of nitrate as a result of deforestation in the Hubbard Brook experimental forest

The color curve indicates the output of nitrate in stream water in the deforested area, and the black curve the output in an undisturbed area. Shortly after deforestation the output in the experimental area rose to about 40 times its previous level, whereas the output in the undisturbed control area remained at about the same level. (Note that the vertical axis is not linear.)

(during one period it was actually 418 percent greater)—a predictable result of disrupting the water cycle by removing all transpiring surfaces. Less predictable was an extraordinary loss of soil fertility. The runoff output of nitrate rose steeply (as much as 45 times higher than in undisturbed watersheds) (Fig. 36.15), and there was a drastic increase in net losses of such nutrients as potassium (21 times greater) and calcium (10 times greater). Apparently removal of the vegetation had so altered the chemistry of the soil that nutrients were bound less tightly to soil particles and hence were rapidly leached away.

It has become evident, then, that wholesale human destruction of vegetation results not only in increased erosion by wind and water, a long-familiar consequence, but also in severe loss of fertility in the soil that remains. The stability of the physical part of an ecosystem clearly depends on the production and decomposition of organic matter, and on an orderly flow of nutrients between the living and the nonliving components of the system.

Cutting down forests is not the only way to ruin the soil; overgrazing and other poor farming practices have caused permanent damage in areas where forests never stood. The valleys of the Tigris and the Euphrates once supported the Sumerian civilization; later they were the granary of the great Babylonian Empire. But poor farming practices led to such extensive erosion and to such a buildup of salt in the soil that the amount of cultivable land today is under 20 percent; the ancient irrigation works are filled with silt, and so much soil has been washed into the Persian Gulf that the ancient seaport of Ur is now 240 km from the coast, its buildings buried under 10 m of silt (Fig. 36.16). Similar conditions prevail in Syria, where human activity has reduced more than a million acres to desert. Closer to home, overgrazing and the plowing-under of native grasses, combined with a decade of drought, led to the dust-bowl conditions of the 1930s in parts of the United States, when clouds of topsoil were blown hundreds of kilometers and vast areas were left barren and useless. Such examples from the past show us that the ecological crisis is not entirely the result of modern technology, but it is disconcerting that, despite increased knowledge of soil dynamics, wholesale destruction of the earth's soils and the vast numbers of species they support should continue. In the tropics today, for example, trees are being cut down at a high rate; sometimes they are then burned and plowed under in an effort to enrich the soil for agriculture. But the land is planted for a few years at most, and finally abandoned as the nutrients in the soil are leached out. Because the leaching leaves the soil very poor, many of the former forests, which occupied one-sixteenth of the earth's land, will probably never be regenerated. This vast feature of our planet, with its hundreds of thousands of unique species and its perhaps crucial role in global climate, seems doomed to disappear within our lifetimes unless vigorous and immediate efforts are undertaken to save it.

Irrigation has been viewed as a way of greatly increasing the productivity of dry areas. However, it is often a short-term remedy, likely to be extremely destructive in the long run. In many cases it leads simply to accelerated erosion; in the United States, for example, an estimated 2,000 irrigation dams are now useless impoundments of silt, sand, and gravel. In other cases irrigation leads to rapid deposition of salt in the soil—salinization—until eventually there is so much salt that plants cannot grow. The sa-

36.16 Partially uncovered ruins of the ancient port city of Ur
Irrigation of the lands surrounding the Tigris and Euphrates rivers over the centuries has brought about a great extension of the fluvial plain, and the burial in silt of the ancient port city of Ur.

linization may occur because adding water to land overlying salty groundwater causes the water table to rise, the salt thus being carried into the topsoil, or because salts, originally present in low concentrations in the irrigation water, accumulate in the soil as the water evaporates (Fig. 36.17). The Indus Valley of Pakistan, the largest irrigated region in the world—over 23 million acres watered by canals—fell victim to salinization. The irrigation system seemed very promising: the soil was good, and the addition of adequate water was expected to make it produce abundant crops. But as one observer has said, "The result was tragically spectacular. In flying over large tracts of this area one would imagine that it was an Arctic landscape because the white crust of salt glistens like snow." There are indications that the irrigation system made possible by the Aswan High Dam in Egypt may similarly produce salinization of the soil; one ecologist has commented, "The Aswan High Dam is designed to bring another million acres of land under irrigation, and it may well prove to be the ultimate disaster for Egypt."

One promising development is the use of "drip irrigation," a method of reducing salinization by dripping water onto the soil from overhead pipes at a rate calculated to ensure that virtually all of it penetrates into the soil instead of being lost by evaporation. Of particular value in drip irrigation is the use, where possible, of water from treated sewage; mineral nutrients are thus recycled to the land, where they are needed, rather than released into streams and lakes, where they are ecologically damaging.

This brief discussion of soils and what can happen to them illustrates how complex, and at the same time how fragile, the earth's ecosystems are. In the past, people have usually simply trusted to luck in their attempts to manipulate them. Someone has said that human beings refuse to try to understand any system they did not design themselves. Surely the time for

36.17 Signs of salinization in Iran
A river on the island of Hormuz in the Persian Gulf shows obvious signs of salinization. Irrigation waters raised the water table of salty groundwater and brought in additional salt, much of which was eventually deposited in the topsoil. The salinization is made obvious here by evaporation, which has left an encrustation of salt on the banks.

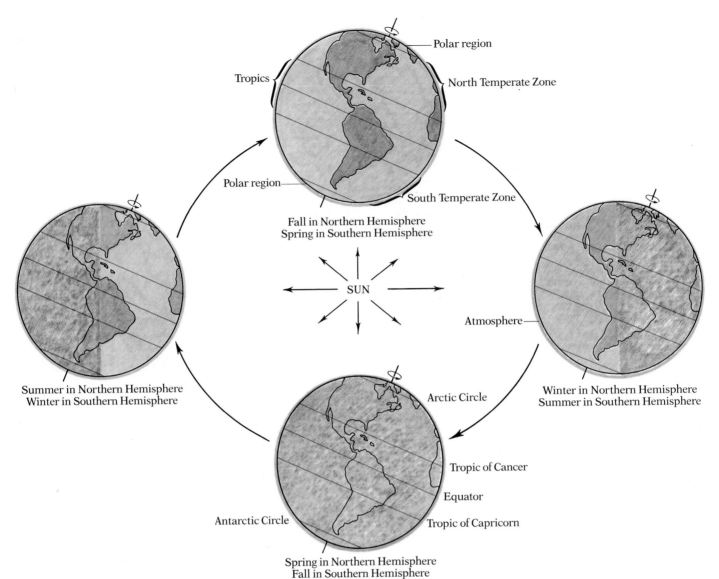

Polar region

Tropics

North Temperate Zone

Polar region

South Temperate Zone

Fall in Northern Hemisphere
Spring in Southern Hemisphere

SUN

Atmosphere

Summer in Northern Hemisphere
Winter in Southern Hemisphere

Arctic Circle

Winter in Northern Hemisphere
Summer in Southern Hemisphere

Tropic of Cancer

Equator

Antarctic Circle

Tropic of Capricorn

Spring in Northern Hemisphere
Fall in Southern Hemisphere

36.18 The source of the seasons
Because the earth's axis of rotation is inclined 23.5°
with respect to its orbital plane, the Northern
Hemisphere is tipped toward the sun during the
summer (left) but away during the winter (right). The
part of the earth tipped toward the sun is illuminated
more vertically, so the energy received per unit area
is higher in summer than in winter. The number of
hours of sunlight is also greater in summer. The
tropics, by contrast, receive strong, relatively vertical
illumination all year, while the polar regions, on
average, receive their sunlight more obliquely, and so
spread out over more surface area. The absorption of
sunlight by the atmosphere magnifies these seasonal
and latitudinal differences in illumination: sunlight
falling on the polar regions must travel through more
air than sunlight in the tropics; hence more of its
energy is dissipated before reaching the surface.

indulging such attitudes is past. Gaining knowledge of how ecosystems
function, and using such knowledge, have virtually become ethical impera-
tives if human beings are to continue to inhabit the earth.

THE BIOSPHERE

The interactions of the various communities on earth among themselves
and with the environment are subtle and pervasive. Indeed, as we have
said, the earth is itself an ecosystem, usually called the biosphere—a complex array of communities and their environments, bound together by interrelated biological, chemical, and physical forces. In this section we shall
consider how physical forces create regional variations in climate and thus
profoundly influence the dispersal of species and the distribution of communities on earth.

CLIMATE AND THE SUN

The sun is the ultimate energy source for life, and the distribution of its energy in large part determines the distribution of living things. The sun bathes the earth in warming radiation, but because the earth's surface curves away from the path of incident light, areas at different latitudes receive different amounts of sunlight, and consequently have different ranges of temperature. The tropics, for instance, receive almost five times as much energy per unit area as the mid-polar latitudes. Moreover, since the earth's axis of rotation is tilted with respect to the sun it orbits, mid-temperate latitudes receive more than twice as much solar energy at the beginning of summer as at the beginning of winter (Fig. 36.18).

In addition to a latitudinal gradient of temperature, the uneven distribution of sunlight has another set of consequences. The warm air of the tropics tends to rise, drawing behind it along the surface cooler air from the temperate zones. This dramatically affects the distribution of rain. The capacity of air to hold moisture decreases as the temperature falls, and the temperature of the atmosphere falls with increasing altitude. As a result, the moisture in the rising tropical air condenses, bathing the tropics in rain. As this air cools and loses its moisture, it moves up and away from the Equator, finally descending near the Tropic of Cancer or the Tropic of Capricorn (Fig. 36.19). These areas of the world, which include Australia, Saudi Arabia, the veldt of South Africa, and the Sahara of northern Africa, tend as a result to be extremely dry.

The presence of mountains, by the way, may give rise to similarly radical variations in climate. Just as the moisture in tropical air condenses when it is carried up to cooler altitudes, so the moisture in winds blowing up and

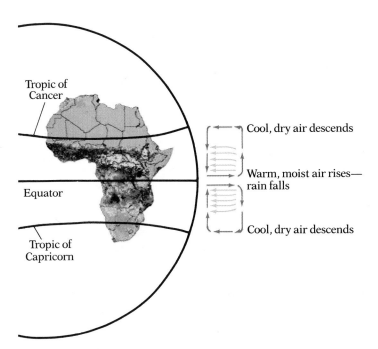

36.19 Latitudinal patterns of rainfall
Warm, moist air rises in the tropics, cools at higher altitudes, and—still over the tropics—releases its moisture as rain. As a result, when this air descends (roughly at the Tropic of Cancer and the Tropic of Capricorn), it is unusually dry, and contributes to the formation of deserts. This circulation of air is shown in cross section at right. Its effect on the continent of Africa is evident in the ultraviolet NASA photograph. Vegetation is heaviest in the white areas, still abundant in the red areas, and somewhat less so in the green ones. The brown areas are very dry and often barren.

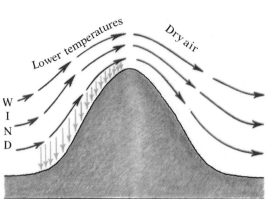

36.20   Effects of mountains on local climate

When moist air is forced up to cooler altitudes by mountains, the moisture frequently condenses as rain, and the result may be lush vegetation on the windward side, as found, for example, along the western border of the Sierra Nevada in California (left). The dry air descending on the other side of the mountain creates a more arid environment on the leeward side; the Mojave Desert, on the eastern border of the Sierra Nevada (right), is typical of the deserts that may occur downwind of a mountain range.

across a mountain tends to condense at higher altitudes on the windward side. As a result, the side of the mountain facing the prevailing winds is usually much more lush than the side swept by the dry descending air (Fig. 36.20).

BIOMES

The many interactions of temperature, wind, humidity, latitude, altitude, and topography result in large climatic regions, usually called **biomes**. Each type of biome is characterized by different sorts of plants and animals. Let's briefly survey some of the world's major biomes (Fig. 36.21).

Tundra   In the far-northern parts of North America, Europe, and Asia is the tundra. It is the most continuous of the earth's biomes, forming a circumpolar band interrupted only narrowly by the North Atlantic and the Bering Sea. It corresponds roughly to the region where the subsoil is permanently frozen. The land has the appearance of a gently rolling plain, with many lakes, ponds, and bogs in the depressions (Fig. 36.22, p. 972).

Even though the Russian word *tundra* means "north of the timberline," there are, in fact, a few trees on the tundra, but they are small, widely scattered, and clearly not the dominant vegetation except locally. Much of the ground is covered by mosses (particularly sphagnum), lichens (particularly so-called reindeer moss), and grasses. There are numerous small perennial herbs, which are able to withstand frequent freezing and which grow rapidly during the brief cool summers, often carpeting the tundra with brightly colored flowers.

Reindeer, caribou, arctic wolves, arctic foxes, arctic hares, and lemmings are among the principal mammals; polar bears are common on parts of the tundra near the coast. Vast numbers of birds, particularly shorebirds (sandpipers, plovers, etc.) and waterfowl (ducks, geese, etc.), nest on the tundra in summer, but they are not permanent residents and migrate south for the winter. Insects, particularly flies (including mosquitoes), are incredibly abundant. In short, far from being the barren lifeless land that many

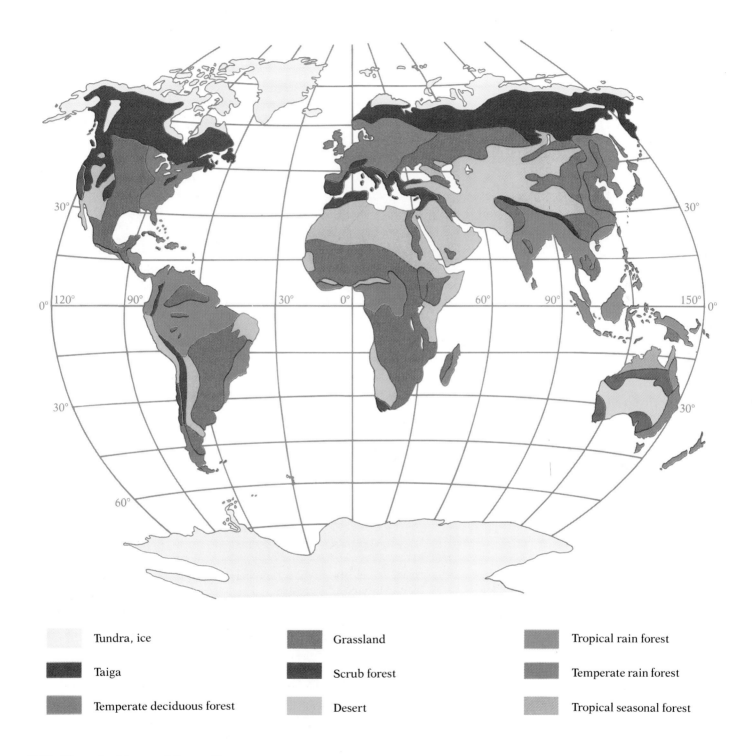

	Tundra, ice		Grassland		Tropical rain forest
	Taiga		Scrub forest		Temperate rain forest
	Temperate deciduous forest		Desert		Tropical seasonal forest

36.21   The major biomes of the world

36.22   The tundra
Right: A tundra in Alaska, seen in spring.
Above: A caribou bull, distinctive animal of the
tundra.

36.23   The taiga
Right: Coniferous forests, such as this one in
Norway, cover extensive areas in the northern
part of North America, Europe, and Asia.
Above: Wolves, one of the important mammals
of the taiga.

36.24   A temperate deciduous forest along the Housatonic River, Connecticut, in early autumn

people imagine, the tundra teems with life. It is true, however, that though the number of individual organisms on the tundra is often very large, the number of species is quite limited.

Taiga   South of the tundra, in both North America and Eurasia, is a wide zone dominated by coniferous (evergreen) forests. This is the taiga (Fig. 36.23), also called the Boreal forest. Like the tundra, it is dotted by lakes, ponds, and bogs. And like the tundra, it has very cold winters. But it has longer and somewhat warmer summers, during which the subsoil thaws and vegetation grows abundantly.

The number of species living in the taiga is larger than on the tundra, but considerably smaller than in biomes farther south. Though conifers (including spruce, fir, and tamarack) are the most characteristic of the larger plants in the taiga, some deciduous trees, like paper birch, are also common. Moose, black bears, wolves, lynx, wolverines, martens, porcupines, and many smaller rodents are important mammals in the taiga communities. Birds are abundant in summer.

Deciduous forests   The biomes south of the taiga do not form such definite circumglobal belts as the tundra and the taiga. There is more variation in the amount of rainfall at this latitude, and consequently more longitudinal variation in the types of communities that predominate.

In those parts of the temperate zone where rainfall is abundant and the summers are relatively long and warm, as in most of the eastern United States, most of central Europe, and part of eastern Asia, the major communities are frequently dominated by broad-leaved trees. Such areas, in which the foliage changes color in autumn and drops, constitute the deciduous-forest biomes (Fig. 36.24). They characteristically include many more plant species than the taiga to the north, and show more vertical stratification. Among the common mammals in this biome are squirrels, deer, foxes, and bears.

36.25   Tropical rain forest
Left: A forest in Queensland, Australia. Right:
Epiphytic bromeliads (distinctive plants of the tropics)
growing on a branch of a tree in Costa Rica.

**36.26   A grassland in Kenya, with zebras,
characteristic animals of this biome**

Tropical rain forests   Tropical areas with abundant rainfall are (or, more correctly, were) usually covered by rain forests, which include some of the most complex communities on earth. The diversity of species is enormous; a temperate forest is composed of two or three, or occasionally as many as ten, dominant tree species, but a tropical rain forest may be composed of four hundred or more. It may actually be difficult to find any two trees of the same species within an area of many acres. In a 13-km² rain-forest pre-serve in Costa Rica, biologists have so far recorded 450 species of trees, more than 1,000 other plant species, 400 species of birds, 58 species of bats, and 130 species of amphibians and reptiles.

The dominant trees in rain forests are usually very tall, and their inter-lacing tops form a dense canopy that intercepts much of the sunlight, leav-ing the forest floor only dimly lit even at midday (Fig. 36.25). The canopy likewise breaks the direct fall of rain, but water drips from it to the forest floor much of the time, even when no rain is actually falling. It also shields the lower levels from wind and hence greatly reduces the rate of evapora-tion. The lower levels of the forest are consequently very humid. Tempera-tures near the forest floor are nearly constant. The pronounced differences in the microenvironmental conditions at different levels within such a for-est result in a striking degree of vertical stratification; many species of ani-mals and epiphytic plants (plants growing on the large trees) occur only in the canopy, others only in the middle strata, and still others only on the for-est floor. Some vertical stratification is found in any community, particu-larly any forest community, but nowhere is it so extensively developed as in a tropical rain forest.

Grasslands   Huge areas in both the temperate and the tropical regions of the world are covered by grassland biomes (Fig. 36.26). These are typically areas where either relatively low total annual rainfall (25–30 cm) or uneven seasonal occurrence of rainfall makes conditions inhospitable for forests but suitable for often luxuriant growth of grasses. The grasslands of tem-perate regions characteristically undergo an annual warm-cold cycle, whereas the grasslands of the tropics (often called savannas) undergo a wet-dry cycle instead.

Temperate and tropical grasslands are remarkably similar in appearance, though the particular species inhabiting them may be very different. Both usually contain vast numbers of large and conspicuous herbivores, often including ungulates, like bison and pronghorn antelope in the United States, or wildebeest and gazelle in Africa. Burrowing rodents or rodent-like animals, like prairie dogs in the western United States, are often common.

Deserts   In places where rainfall is often less than 25 cm (10 inches) per year, not even grasses can survive as the dominant vegetation, and desert biomes occur. Deserts are subject to the most extreme temperature fluctuations of any biome type; during the day they are exposed to intense sunlight, and the temperature of both air and soil may rise very high (to 40°C or higher for air temperature and to 70°C or higher for surface temperature), but in the absence of the moderating influence of abundant vegetation, heat is rapidly lost at night, and a short while after sunset, searing heat has usually given way to bitter cold.

Some deserts, such as parts of the Sahara, are nearly barren of vegetation, but more commonly there are scattered drought-resistant shrubs (sagebrush, creosote bush, and mesquite, for example) and succulent plants like cactus that can store much water in their tissues (Fig. 36.27). In addition, there are often many small rapid-growing annual herbs with seeds that will germinate only when there is a hard rain; once they germinate, the young plants shoot up, flower, set seed, and die, all within a few days.

B

A

C

36.27 Deserts
(A) Death Valley, California; vegetation is extremely sparse. (B) Cacti and other thorny, drought-resistant plants, abundant in many deserts, growing in Picacho Park, Arizona. (C) This desert toad of western Australia secretes fluid through its skin, which dries to form a moisture-proof film. It can then survive lengthy periods of drought buried in hardened mud.

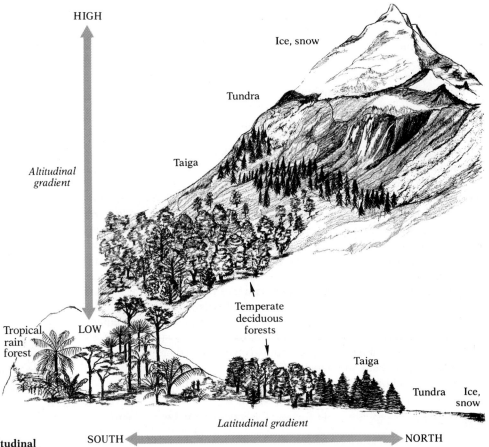

HIGH

Ice, snow

Tundra

Taiga

Altitudinal gradient

Temperate deciduous forests

Tropical rain forest

LOW

Taiga

Tundra   Ice, snow

Latitudinal gradient

SOUTH ⟷ NORTH

36.28 The correspondence between latitudinal and altitudinal life zones in North America

Most desert animals are active primarily at night or during the brief periods in early morning and late afternoon when the heat is not so intense. During the day they remain in cool underground burrows or in cavities in plants or, in the case of some spiders and insects, in the shade of the plants. Among the animals often found in deserts are rodents like the kangaroo rat, snakes, lizards, a few birds, arachnids, and insects. Most show numerous remarkable physiological and behavioral adaptations for life in their hostile environment (Fig. 36.27C).

Altitudinal biomes We have seen a series of different biomes, largely the result of changing temperature, as we move from north to south in the Northern Hemisphere. But higher altitudes, like higher latitudes, tend to be colder than lower ones, and the result—if rainfall and humidity are equivalent at the corresponding latitudes and altitudes—is that the changing pattern of vegetation seen on a mountainside as elevation increases resembles the pattern observed as one goes toward higher latitudes (Fig. 36.28). Thus arms, or isolated pockets, of the taiga extend far south in the United States on the slopes of the Appalachian Mountains in the east and of the Rockies and Coast Ranges in the west (Fig. 36.29). There are even tundralike spots on the highest peaks.

Aquatic ecosystems   So far, we have concentrated on terrestrial ecosystems, but we should not forget that many of the earth's biotic communities are found in aquatic environments, and that these too vary in type with varying physical conditions. Thus the communities in lakes differ from those in the flowing waters of rivers and streams, and even those in a single stream differ from one another, depending on whether they are in rapids, where fast-flowing water is made turbulent by rocks and sudden falls, or in water flowing slowly and calmly over a smooth bottom.

Let us consider the largest bodies of water—the oceans. There are several ways of classifying the ecosystems found in them (Fig. 36.30). It is often useful, for example, to distinguish between a ***benthic division*** comprising the ocean bottom together with all bottom-dwelling organisms and a ***pelagic division*** consisting of the water above the bottom and all the organisms in it—both those that are free-swimming (the nekton) and those that float (the plankton) and are carried passively with the water currents. If the criterion is the distribution of light in the water, it is possible to distinguish between an upper lighted ***photic zone*** and a deeper lightless ***aphotic zone*** (usually below about 200 m). Still another possible distinction is between the ***neritic province*** above the continental shelves and the ***oceanic province*** of the main ocean basin.

The various subdivisions of the oceans are not fully analogous to the terrestrial biomes, because they are more biologically interdependent. For example, except for isolated colonies of chemosynthetic autotrophs found

36.29   An alpine meadow in the Sierra Nevada of California, with paintbrush in bloom
Trees can be seen in the lower, more protected areas. Most of the meadow is above the timberline. The upper limit of tree growth in mountainous regions (usually between 3,000 and 3,500 m) is determined largely by temperature, but is also influenced by soil conditions and rainfall. More barren areas can be seen on the higher peaks in the distance.

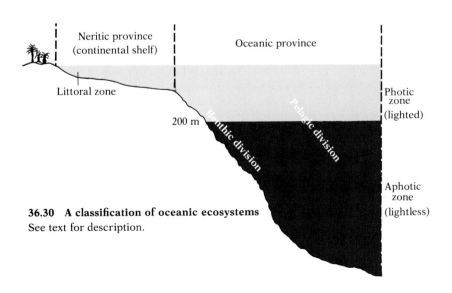

36.30   A classification of oceanic ecosystems
See text for description.

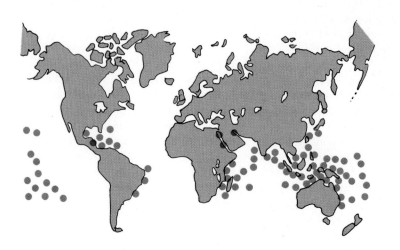

36.31 The distribution of present-day coral reefs
The intensity of solar radiation, which is greatest in the equatorial latitudes, is the major factor controlling the distribution of present-day coral communities in the earth's oceans.

near underwater volcanic vents, benthic communities in the aphotic zone are made up of heterotrophs (including animals, fungi, and heterotrophic protists and bacteria), and they must therefore depend on the photosynthetic organisms of the photic portion of the pelagic division above them for the primary productivity that makes their existence possible; they receive their nutrients as a rain of dead organisms (detritus) from the water above. When the flow of energy and materials is considered, then, the communities in the dark aphotic zone cannot be understood apart from those in the photic zone, and, more broadly, the benthic division cannot be understood apart from the pelagic division.

The most complex oceanic communities occur in the shallow waters of the neritic province, especially that portion called the *littoral zone*, which extends from the beach to a point where the water is deep enough so that it is no longer completely stirred by the action of waves or tides (Fig. 36.30). Here high primary productivity by both free-floating and bottom-anchored algae makes possible much niche diversification among herbivores. Because the littoral zone is subject to far more variation in temperature, water turbulence, salinity, and lighting than any other portion of the ocean, the littoral communities in different locations often vary greatly.

The coral reef Since the intense solar radiation falling on the equatorial latitudes reaches oceanic as well as terrestrial ecosystems, it comes as no surprise that the littoral and associated benthic regions of the tropical zone have for the past several million years provided environments for some of the most stable and diverse communities on earth (Fig. 36.31). The cornerstones, quite literally, of these communities are diverse species of stony corals—colonial cnidarians that feed in a multitude of ways on zooplankton, detritus, organic solutions taken in through osmosis, and the sugars and amino acids produced by symbiotic photosynthetic algae called zooxanthellae (Fig. 36.32). Stony corals (as well as hard corals of other orders, and various species of algae) lay down calcareous skeletons, which constitute the basic structure of a reef and provide a multitude of microhabitats for the organisms it supports. Like the rain forests to which they are

A

B

often compared, coral reefs are now inhabited by animals of virtually every phylum as well as the representatives of most of the plant divisions. And like the rain forests, they are threatened both by local human activities and by human interference with the biosphere; if the burning of fossil fuels results in a general heating of the earth's surface and the melting of the polar ice caps, as some have predicted, the consequent changes in water temperature and water level may irreversibly damage these fragile ecosystems.

EVOLUTION OF BIOGEOGRAPHIC REGIONS

Today's climate in the various biomes is one important element in the distribution of living things. Another important element is history: biomes are by no means static. The earth itself and the organisms on it are constantly changing, and the present distribution is in large part the result of past conditions—conditions often quite different from those now prevailing. For example, a knowledge of present conditions alone would be insufficient to explain why certain animals occur in South America, Africa, and southern Asia, but nowhere else. Only by combining knowledge of present conditions with evidence from the fossil record and with other geographical evidence of the past can ecologists hope to gain insight into the present geography of life.

Continental drift   Not many years ago both geologists and biologists assumed that the distribution of the earth's major land masses had been more or less the same throughout the history of life. That assumption has since been summarily dismissed by geologists, who are now convinced that some

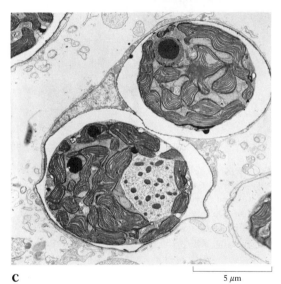

C   5 μm

36.32 Corals
(A) Hard corals growing in abundance in tropical waters provide diverse habitats for marine creatures. Among the corals found in waters surrounding the Philippines are fire coral (left foreground) and brain coral (right foreground). (B) The corals in this photograph are colonial assemblies of individual polyps. (C) The formation of the coral skeleton is enhanced by the photosynthetic activities of single-celled endosymbiotic algae, or zooxanthellae, two of which can be seen in this electron micrograph of the tissues of a hard coral.

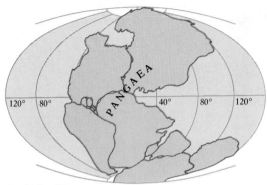

A   225 million years ago

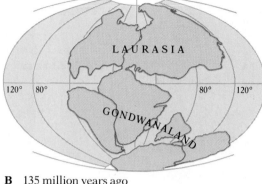

B   135 million years ago

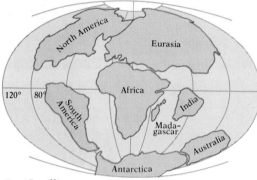

C   65 million years ago

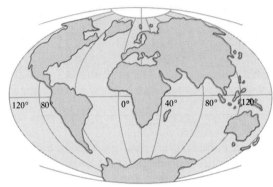

D   Present

36.33   The origin of the modern distribution of continents through continental drift

(A) Early in the Mesozoic era, about 225 million years ago, all the earth's major land masses were united in a single massive supercontinent called Pangaea. (B) About 135 million years ago, at the start of the Cretaceous period, Pangaea had broken into a northern supercontinent, Laurasia, and a southern supercontinent, Gondwanaland; Gondwanaland itself had also begun to break up. (C) By 65 million years ago the breakup of Gondwanaland was complete, and the future South America, Africa, Madagascar, India, Antarctica, and Australia were drifting apart. (D) The present continental arrangement.

225 million years ago, early in the Mesozoic era (see Table 37.1, p. 1012), there was a single massive supercontinent, called ***Pangaea*** (Fig. 36.33A).

Pangaea is thought to have broken up in the course of the Mesozoic as the present-day continents (and some major islands like Madagascar and Greenland) drifted slowly apart. The first major break was an east-west one, separating a northern supercontinent called ***Laurasia*** (composed of the land masses that would later become North America, Greenland, and Eurasia minus India) from a southern supercontinent called ***Gondwanaland*** (composed of the future South America, Africa, Madagascar, India, Antarctica, and Australia). Soon thereafter Gondwanaland began to break up, as India drifted off to the north and an African–South American mass separated from an Antarctic-Australian mass. By the start of the Cretaceous period, roughly 135 million years ago, the distribution of the continents was probably that shown in Figure 36.33B.

Some 70 million years later, about 65 million years ago, South America had split from Africa and was drifting westward; India had moved farther northward, but had not yet collided with the rest of Asia; and Australia had split from Antarctica and begun drifting northeastward (Fig. 36.33C). The Laurasian supercontinent was still intact, however; no split had occurred yet between North America and Eurasia. This split was one of the last to take place as the distribution of continents we know today slowly emerged (Fig. 36.33D).

The conclusion that the earth's major land masses (or plates, as geologists now call them) can drift from one place to another (at roughly 2.5 cm per year) is supported by evidence of many different kinds. Students of the geological subdiscipline of plate tectonics have found that along some rifts in the earth's crust molten rock is slowly welling to the surface, forcing the plates on the two sides of the rift apart. At other places the crust is slowly folding down and sinking into the earth's interior. It is this process of upwelling of new crust in some places and sinking of old crust in others that provides the force for continental drift. Evidence that such drift must have occurred comes from studies of the geological features of regions now separated but thought to have once been in contact; striking similarities are frequently found in such regions. For example, the rocks along part of the east coast of Brazil match almost exactly those in Ghana, on the west coast of Africa. The complementarity of the shapes of the east coast of South America and the west coast of Africa had, of course, suggested long ago that the two continents might once have been joined, but until plate tectonics supplied a plausible mechanism for continental drift the possibility had been discounted.

If continents move in the course of millions of years, so that their distances from the earth's poles and from the Equator change, then their climates must have undergone major shifts. India, for example, has moved from a position next to Antarctica all the way across the equator to its present location in the tropics of the Northern Hemisphere. Australia, too, has moved steadily northward. Since shifts in climate would result in altered selection pressures, evolutionary forces must have changed the organisms in these regions gradually but dramatically.

Climatic changes brought about by causes other than continental drift have also occurred during the history of life on earth. Antarctica, for example, though probably always near the South Pole, has not always been the bleak ice-covered land it is today, for fossils of amphibians and reptiles have been found there. During at least part of the Mesozoic, Antarctica must have been reasonably warm, and it was probably warm again about 50 million years ago, when tropical and subtropical climates were far more widespread on the earth than they are today (Fig. 36.34). By contrast, the earth

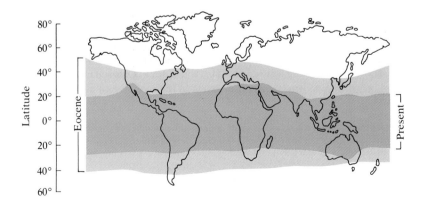

36.34 Distribution of tropical and subtropical forests during the Eocene (about 50 million years ago) and now, shown on a modern map
Warm conditions extended much farther north and south during the Eocene. The possible causes of global warming and cooling cycles are hotly debated.

36.35   Distribution of glaciation during the most recent Pleistocene ice age, 20,000–15,000 years ago

36.36   Kangaroos
Though their appearance and behavior do not immediately suggest it, kangaroos are Australia's ecological equivalents of the ungulate grazers of other continents.

was much colder only a few thousand years ago, during the periods of extensive glaciation (Fig. 36.35). Thus both the past configurations of the land masses and their past climates are important in explaining the distributions of organisms on the earth.

The island continents   The biota (flora and fauna) of the **Australian region** (Australia, New Zealand, and adjacent islands) is by all odds the most unusual found on any of the earth's major land masses. Many groups common in Australia occur nowhere else. Conversely, many taxa widespread in the rest of the world are absent from Australia. As we have seen, Australia has had no land connection to Eurasia or Africa since Pangaea began to break up more than 135 million years ago. The ancestors of some of the more ancient groups of organisms now living in the Australian region were probably there before the breakup of Pangaea, but the ancestors of more recent groups must have crossed a water barrier to get there.

That water barrier, however, was dotted by islands (now Indonesia), and organisms could have spread from one island to the next over a period of millions of years; it is not necessary to assume that a great expanse of ocean was crossed at any one time. Furthermore, most of the western islands of Indonesia were probably interconnected as an extension of the Asian land mass several times in the not-too-distant past; hence the distance from Asia to Australia has not always been as great as it is today.

Perhaps the best-known aspect of Australia's curious biota is its mammalian fauna, which is completely unlike that of any other continent. Except for wild dogs (dingoes), which were probably human imports of prehistoric times, the only placental mammals present in the Australian region before European explorers landed there were a number of species of rodents belonging to a single family and a variety of bats. Bats can, of course, fly across water barriers and would be expected to reach most oceanic islands. The rodents of Australia are apparently relatively recent arrivals, having come from Asia by island-hopping through Indonesia. Most of the ecological niches that on other continents would be filled by placental mammals are in Australia filled by marsupials.[6]

Apparently the marsupials reached Australia very early and, encountering no competition from placental mammals, underwent extensive evolutionary radiation. Since they were filling niches similar to those filled in the rest of the world by placentals and were therefore subject to similar selection pressures, they evolved striking convergent similarities to the placentals. Certain of the marsupials resemble placental shrews, others placental jumping mice, weasels, wolverines, wolves, anteaters, moles, rats, flying squirrels, groundhogs, bears, and so on (see Fig. 34.27, p. 913). The uninitiated visitor to an Australian zoo finds it hard to believe when first looking at an assemblage of these animals that he is not seeing close relatives of the mammals familiar to him from other parts of the world. Not all marsupials

[6]Whereas the young of placental mammals undergo their entire embryonic development in the mother's uterus, the young of marsupial mammals develop only a short while in the uterus; after birth, they move to a pouch on the mother's abdomen, attach to a nipple, and there complete their development.

look like their ecologically equivalent placentals, however; the kangaroos are markedly different (Fig. 36.36), though some of them play an ecological role very similar to that of horses and other large placental grazers. The Australian marsupials are a fascinating biological development, but most of them are probably doomed to extinction, at least in the wild. The changes brought by civilization have not been favorable for many of them. And most marsupials seen unable to compete successfully with the placental mammals subsequently introduced by humans.

Another island continent, South America, which is known to biologists as the *Neotropical region* (meaning "new tropics"), has had a history similar to that of Australia. It, too, has been unconnected to the other major land masses of the world through much of its history. And it, too, had an early mammalian fauna that included a great variety of marsupials, as the fossil record shows. The prevalence of marsupials in both Australia and South America probably means that these organisms were present throughout the region comprising Australia, Antarctica, and South America at a time when there were only small water breaks (if any) in this land mass (Fig. 36.33C) and when the climate was warmer than it is now.

Later, a variety of placentals reached South America, probably during a short period of connection to North America via a Central American land bridge during the early part of the Age of Mammals (about 60 million years ago). After this land bridge disappeared, both the marsupials and the placentals of South America evolved, in isolation, many characteristics convergent to those evolved by placentals on the so-called World Continent (Fig. 36.37). There were times during this period of isolation lasting some 60 million years when the water barrier between South America and what is now northern Central America was not so wide. A few additional placental mammals chanced to get across into South America at such times; among these were the ancestors of the modern New World monkeys and a number of rodents.

Though, by five million years ago, there were 23 families of mammals in South America, not one of these was represented in North America, where a different group of mammalian families lived. But then a land connection to North America (the Isthmus of Panama) was reestablished, and many additional immigrants arrived in South America. Some species also moved in the opposite direction, from South America to North America; the opossum, the porcupine, and the armadillo are examples. But the movement was predominantly southward, apparently because the groups that had originated in the north were competitively superior and could displace the older South American fauna, much of which suffered extinction.

Central America has never been more than a narrow bridge. Because its climate and rugged terrain have not been hospitable to many northern species, several groups of organisms have never been able to move between the North American and South American continents. The Central American land bridge has been a selective filter, through which some species but not others could pass. For this reason, South America, like Australia, has been an island continent through much of its history, but its nearness to North America (Mexico) and its recent direct connection to North America via the Central American land bridge have given it a more diverse biota with more similarities to that of the World Continent.

South American rodents

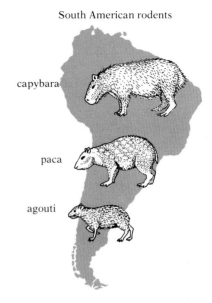

capybara

paca

agouti

African ungulates

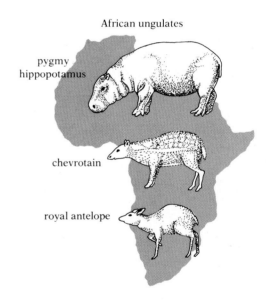

pygmy hippopotamus

chevrotain

royal antelope

36.37 Convergence of South American rodents and African ungulates
There are remarkable similarities between the two groups of animals shown here, even though rodents and ungulates comprise separate mammalian orders that are not closely related.

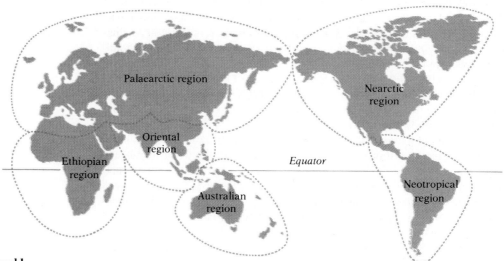

36.38   Biogeographic regions of the world

The World Continent   Europe, Asia, Africa, and North America have formed a relatively continuous land mass, known as the World Continent, throughout most of geological time. Consequently their biotas are more alike in many aspects than they are like those of the two island continents. Nevertheless, biologists customarily divide the World Continent into four biogeographical regions: the ***Nearctic*** ("new northern"), which is most of North America; the ***Palaearctic*** ("old northern"), which is Europe, northernmost Africa, and northern Asia; the ***Oriental,*** which is southern Asia; and the ***Ethiopian,*** which is Africa south of the Sahara (Fig. 36.38).

The geological feature that looks as if it would be the main present-day barrier between the Palaearctic and Ethiopian regions—the Mediterranean Sea—is not really the important barrier. Many species can move between Europe and North Africa by circling around the eastern end of the Mediterranean. The real barrier to species dispersal is the Sahara. Africa north of the Sahara is part of the Palaearctic region.

The Oriental region—tropical Asia—is separated from Palaearctic northern Asia in most places by east–west mountain ranges, of which the Himalayas between India and China are part. These east–west mountains constitute important breaks between climatic regions, and they therefore act as both topographic and climatic barriers between cold-adapted and warm-adapted species. By contrast, north–south mountains such as those in North America tend to facilitate mixing of cold-adapted and warm-adapted species.

That the Palaearctic, Oriental, and Ethiopian regions constitute an essentially continuous land mass is obvious, but you may wonder why the Nearctic is considered part of the same land mass. The answer is that North America and Eurasia have been connected through much of their geological history (Fig. 36.33B, C). Even after they broke apart as the northern part of the Atlantic Ocean formed, a new connection, the Siberian land bridge between what are now Alaska and Siberia, provided a link; part of this bridge is beneath water at the present time. Because the climate of Alaska and Siberia is so forbidding, you might not expect many organisms to use a bridge in that region. But again present conditions are misleading. Fossils of many temperate and even subtropical species of plants and animals are abundant in Alaska. Indeed, all the evidence points to much movement, on many occasions, between Asia and North America via the Siberian land bridge. In fact, the Nearctic and Palaearctic regions remain biologically so similar that many biologists regard them as a single region, which they call the Holarctic.

The history of the major land masses and of their changing climates helps explain present distribution patterns. Consider the disjunct distribution referred to earlier, in which a species occurs in South America, Africa, and southern Asia. If the species belongs to a group of animals of very ancient origin—the cockroaches, for example—this distribution may indicate that the species occurred throughout the old Gondwanaland supercontinent and continued to survive in South America, Africa, and India after these drifted apart. But if the species belongs to a more recently evolved group, like the majority of modern mammalian and bird families, which arose after Gondwanaland had broken up, it may be assumed that the species moved between the New World and the Old World via either the North Atlantic or the Siberian land bridge and between North and South America via Central America, and that it then became extinct in the north, either because of climatic changes or because of intense competition. The fossil record shows that this pattern of dispersal between southern regions by way of the northern continents has occurred again and again. For example, members of the camel family occur today in South America (llamas, alpacas, vicuñas, etc.), northern Africa, and central Asia, but fossils indicate that the family originated in North America, spread to South America via Central America and to the Old World via Siberia, and later became extinct in North America; hence the disjunct distribution we see today.

THE DISPERSAL OF SPECIES

It will seem rather obvious by now that the many species of plants and animals inhabiting the earth do not often remain in the same location indefinitely; indeed, given the drastic changes that occur in climate, most terrestrial species would face inevitable extinction without the ability to disperse to new areas. Before a species can successfully spread into a new area, it must meet at least three major conditions: (1) It must possess the *physiological potential* to survive and reproduce in the new area. (2) It must have the *ecological opportunity* to become established in the new area. And (3) it must have *physical access* to the new area. Let us examine each of these conditions in more detail.

Physiological potential for survival in a new area   Any new area into which colonizing members of a species move will be different at least to some degree from the area the colonizers left. Hence the colonizing population will immediately be subject to selection pressure for evolution of better adaptations for the new environment. But since such evolutionary improvement comes after colonization, the colonization itself is possible only if the colonizers are already at least minimally preadapted for survival under the new environmental conditions. "Preadapted," of course, does not imply any foresight or any intentional preparation for the move into strange territory; it simply means that characteristics evolved in the previous habitat are at least minimally suited to the new habitat. For example, the colonizers must be able to use some source of food in the new area, and they must be able to withstand the rigors of the new climate.

Let us see how climate limits the distributions of organisms. We have already touched on broad regional differences in temperature, humidity, rainfall, and other meteorological factors; the average pattern of such factors over an extended period constitutes the climate of a region.

Though the climate of a desert is dry, at any given time rain may be falling. Thus we distinguish between climate, defined above, and weather, which is the immediate pattern of meteorological factors. Weather has obvious day-to-day effects on living organisms: a late frost after a warm spell can destroy the potential productivity of a plant seedling; the reproductive success of many animals will be very different in unusually dry and unusually wet years. In other words, though the climate of an area is important in determining distributions of species, the weather is what is important to the organism at any particular moment.

Indeed, the extremes of the weather—the highest or lowest temperatures of the year, for example, or the longest period without rainfall—are often most important in limiting the distribution of organisms. A plant that cannot tolerate temperatures below 0°C will be unable to survive in a region renowned for its "warm" climate (with an average annual temperature of, say, 25°C) if the temperature falls below 0°C even one day a year. All other conditions in the region may be optimal for the plant, but the one condition it cannot tolerate prevents it from growing there.

Since the environmental conditions affecting an organism and limiting its distribution are those impinging directly on it, ecologists must analyze the microenvironmental conditions—the microhabitat—to obtain a full understanding of the physical factors with which the organisms they study must cope. Variations in the weather can be very local. For example, within a single small area, temperature differences of several degrees may occur between sheltered and exposed places, or between points near the ground and points at various elevations above the ground. Since the same sorts of highly local variations may be found in humidity, wind velocity, amount of sunlight, soil type, etc., we should not expect to find exactly the same kinds of organisms living at all points within even a very small area. Plants and animals don't live under generalized regional conditions; they live within some range of microenvironmental conditions that may vary radically over a given region. Hence when ecologists say that plant species A ranges from South Carolina in the south to central New York in the north, and from the Atlantic coast to central Ohio in the west, they mean that the species occurs in the microhabitats that support it within the overall range, and does not occur in any microhabitats outside that range.

Our focus has been on climate and weather, but the same sorts of considerations apply to other aspects of the environment. The climate of a region might be ideal for a particular species of plant, and the soil rich in nitrogen and potassium, but if there is less phosphorus in a part of the region than the plant requires, it cannot grow there. The importance of single environmental factors was recognized as long ago as 1840 by Justus von Liebig of the University of Giessen, Germany. Liebig's so-called **Law of the Minimum** states that growth of a plant will be limited by whichever requisite factor is most deficient in the local environment. Later, V. E. Shelford of the University of Illinois expanded Liebig's Law, applying it also to animals and taking into account that too much may be as bad as too little. Shelford's **Law of Tolerance** states that the distribution of a species will be limited by that environmental factor for which the organism has the narrowest range of adaptability (Fig. 36.39).

Though the principle behind both Liebig's and Shelford's laws is important in ecology, the assumption that a single factor is always limiting should be viewed with caution. In real-life situations (as contrasted with laboratory models), the various environmental factors interact in so many complex ways that it is often impossible to describe any one factor as *the* limiting one. For example, as we saw in the last chapter (p. 934), when one condition is not optimal—though tolerable—for a species, the limits of tolerance for some other factor may be reduced (the result being that an ellipsoid rather than a rectangular volume will most accurately represent a species' ecological niche). Moreover, unless the Law of Tolerance is extended to such biotic limiting factors as predation and competition, in addition to the more physical ones included in its original formulation, it is of only restricted applicability. As an illustration, consider the case of Klamath weed *(Hypericum perforatum).*

This plant, which is poisonous to cattle, was brought from Europe to the western United States, where it quickly spread to millions of acres of valuable range land. To control the Klamath weed, a species of flea beetle that feeds on it was introduced. The beetle soon eliminated the weed from the open range lands, though some small populations persist in shaded places in forests, where the beetle does not feed on it as much. It is clearly the beetle, not lack of tolerance to some physical factor of the environment, that today limits the range of the Klamath weed to forests; yet an observer who did not know the history of the weed–beetle interaction would hardly be able to determine what keeps the weed from spreading into the range lands and might erroneously suspect that the seed requires a shaded habitat.

Ecological opportunity to become established in a new area This requirement often means that the colonizing species must encounter little competition or danger from natural enemies at first. There must be underutilized resources it can exploit. The reason is simple: even if the colonizer has the physiological potential for surviving in the new habitat, chances are that it will be less well adapted to the new conditions than species that have been in the area for a longer time. If the niche of one of the established species is very similar to potential niches of the colonizer, the established species will probably have the competitive advantage and be able to prevent the colonizer from taking hold. This is not always the case, however; if conditions in the new range are very similar to those in the old, a colonizing species may sometimes be competitively superior to an established species and be able to supplant it.

Ecological opportunity is also affected by the size of the new area. The

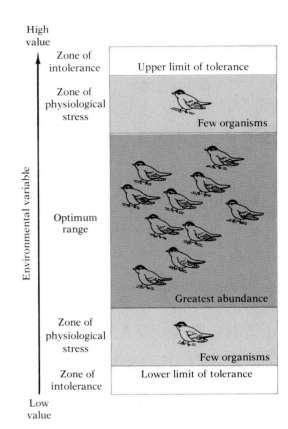

36.39 A diagrammatic illustration of the Law of Tolerance

The species in question is most abundant in areas where the environmental variable is within the optimum range for that species. The species is rare in areas where it experiences physiological stress because the environmental variable has either too high or too low a value. And the species does not occur at all in areas beyond its upper and lower limits of tolerance.

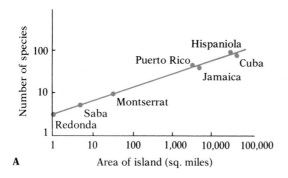

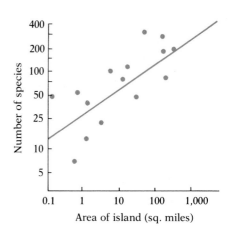

36.40   Two examples of area–species curves

(A) Number of species of amphibians and reptiles on islands in the West Indies. (B) Number of plant species on the Galápagos Islands. In each case the logarithms of the actual counts of species increase roughly as a linear function of the logarithms of the area of the islands.

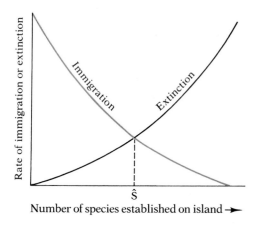

36.41   The equilibrium model of species diversity on an island

As the number of species established on the island increases, the rate of immigration of species not already established falls and the rate of extinction of established species rises. When the point is reached at which immigration and extinction are equal (the point at which the two curves cross), the island has the equilibrium number of species (Ŝ); it cannot support more species unless the immigration rate increases or the extinction rate decreases.

importance of island size has been especially well illustrated in the studies of island biogeography pioneered by Robert MacArthur of Princeton and E. O. Wilson of Harvard. We might reasonably expect that a small island would support fewer species of any given group of organisms. This has, in fact, been found to be true; the number of species on islands has repeatedly been shown to be related to the area of the islands (Fig. 36.40). Though the slopes of area–species curves differ, depending on ecological conditions and on the taxonomic group of organisms studied, a rough generalization is that for each tenfold increase in area the number of species approximately doubles; thus an island with an area of 100 km² will, other things being equal, support roughly twice as many species as an island with an area of 10 km². In part this is because larger islands provide more diverse habitats and therefore a greater assortment of resources and opportunities for species interactions. However, species number on small islands is also kept down by the small size of the various populations. Chance, perhaps in the form of unusual weather, is more likely to wipe out every individual in a small population than in a larger one, and larger islands tend to support larger populations of each species.

Not only size but also distance from the source of colonists profoundly affects the number of species found on an island. As we shall see in the next section, species differ in their ability to cross climatic and geographical barriers; for many years, biogeographers attributed the higher species diversity of near-shore islands compared with mid-ocean islands of similar size and topographic complexity to the relative accessibility of the near-shore islands. But there was a problem with this explanation: we would expect that given sufficient time, the diversity of organisms that have good dispersal ability, such as birds, eventually would reach about the same level on similar islands, regardless of distance from the source of colonists. Yet even among birds and other mobile organisms, the "distance effect" is strong.

MacArthur and Wilson suggested a solution to this seeming paradox. They proposed that, just as a population with access to only limited re-

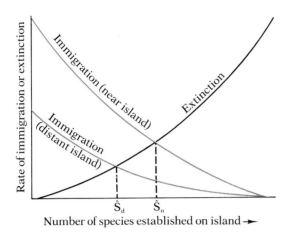

Number of species established on island →

36.42 The effect of the distance of an island from potential sources of colonizers on the equilibrium number of species on the island
The rate of immigration of species not already established is lower for a distant island than for a near island. Hence species equilibrium is reached at a lower number of species ($\hat{S}_d$) for the distant island than for the near island ($\hat{S}_n$).

sources must eventually reach an equilibrium between births and deaths, so the number of species on an island eventually reaches an equilibrium between the establishment of new species (immigration) and the extinction of species already resident on the island. The size, climate, and topography of the island determine a probability of extinction that is independent of island isolation, but the rate of immigration of new species varies with the distance from the source of colonists. The inevitable result is that the equilibrium number of species on otherwise similar islands declines with increasing isolation, because the replacement of extinct species through immigration takes longer. It is important to remember that when an island has reached equilibrium in terms of species diversity, there is still a slow but steady *turnover* of resident species.

We can construct a graphic model of this theory. As Figure 36.41 shows, the rate of immigration of species not already established on an island falls as the number of established species on the island rises: because more and more of the immigrants from source islands or continents represent species that have previously become established on the island, the percentage of immigrant individuals from species new to the island decreases. On the other hand, the rate of extinction, measured as the number of species becoming extinct per unit time, rises as the number of species on the island increases. In part this is because the total number of species is greater, but another reason (the one that makes the extinction curve in Figure 36.41 bend upward) is that as the number of species on the island grows, so does interspecific competition—a reflection of the increased overlap in resource use. The point at which the immigration and extinction curves cross represents a state of equilibrium for species diversity; both immigration and extinction continue, but the total number of resident species does not change significantly.

Figure 36.42 shows how the distance effect comes about. Two islands are compared: one is near the source of colonists, and so has a higher rate of immigration; the other is more distant, and consequently has a lower rate. Notice that the two immigration curves meet the horizontal axis at the same point: if there were no extinction on the islands, the maximum number of established species would eventually be the same for each,

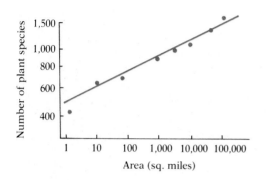

36.43 An area–species curve for flowering plants in England

Areas of different sizes were sampled, from a plot of only one square mile in Surrey to the whole of England.

though that number would be reached more slowly on the distant island because the probability of a potential immigrant's reaching that island is so much lower per unit time. Since the two islands are assumed to be similar in size, topography, and climate, a single extinction curve applies to both. However, the intersection of the extinction curve with the near-island immigration curve is to the right of its intersection with the distant-island curve. In other words, the equilibrium number of species on the near island, $\hat{S}_n$, is greater than the equilibrium number for the distant island, $\hat{S}_d$.

Ecologists have zealously studied island biogeography because it provides a clear model of the factors governing species dispersal in general. Freshwater lakes or springs can be considered biogeographic islands; so can forests separated from one another by grasslands, or deserts separated by wetter areas, or mountaintops, or even some parks. These habitat "islands," like real islands, obey the area–species rule: the greater the area, the greater the species diversity (Fig. 36.43).

Physical access to a new area   It is useless for a species to have the physiological potential and ecological opportunity for surviving in a new range if it has no way of getting there. Doubtless many common North American mammals could survive and prosper in Australia, but unless they have some way of reaching that continent this potential range extension will be unrealized.

There are many ways organisms may disperse or be dispersed from one place to another. Most obvious for many animals is active locomotion—walking, crawling, swimming, or flying. Even many sedentary marine animals have a free-swimming larval stage. Active locomotion may carry the members of a single generation only a short distance from their point of origin, but over many generations the cumulative effect may spread the species over hundreds or thousands of miles. This progressive dispersal in the course of a long series of generations may at first glance seem too slow to be of much significance, but it is not. Probably all organisms can disperse fast enough so that, if favorable habitats were continuous, they could spread over the entire earth in a relatively short time, as measured on the geological and evolutionary time scale.

As large and heavy animals, we tend to think of active locomotion as the principal means of dispersal. But for plants and for many very small animals, passive transport is the chief means of dispersal. For example, the seeds or spores of many plants may be blown very long distances, and insects, spiders, small molluscs, and other invertebrates have been known to be blown hundreds of miles by the strong winds of storms. Pilots sometimes encounter large numbers of insects being swept along by fast-moving air currents at high altitudes. Aquatic organisms may similarly be swept along by strong water currents. Even some fairly large terrestrial plants and animals may be carried across many miles of water on floating logs or rafts of matted vegetation. Many such logs and rafts are swept out to sea by large rivers like the Amazon and the Congo, particularly during floods. A raft about 100 m², composed of soil and decaying organic matter laced together by roots, was sighted in the Atlantic Ocean off the coast of North America in 1892. Many shrubs and several trees 10 m tall were growing on it. This raft, which looked like a floating island, is known to have drifted at least 1,600 km.

A

B

36.44 Krakatoa

All life on the island of Krakatoa was destroyed by a volcanic eruption in 1883. The effect of such an eruption can be seen in the photograph of an ash-covered beach on Krakatoa, largely barren of multicellular organisms, with the exception of an occasional sprouting coconut (A). Now, however, large areas of the island have been recolonized (B).

Some plants and small animals are dispersed by birds and mammals. For example, the seeds of many plants pass through the digestive tracts of higher animals without being harmed, and may germinate and grow if the animals' feces are deposited in a favorable place; indeed, some seeds simply will not germinate without first passing through a vertebrate gut. (Edible fruits may be adaptive, in fact, because the vertebrates that eat them disperse their seeds and, when defecating, even supply fertilizer for them; the laxative effect of fruit may be adaptive in that it gets the seeds out of the gut before they are damaged by digestive enzymes.) Birds sometimes transport seeds long distances in this way. There is also some evidence that plant seeds and the eggs and larvae of some small aquatic animals may be transported on the feathers or feet of swimming or wading birds. Of course humans are now important agents of dispersal as well.

The recent history of Krakatoa, a small island between Java and Sumatra in the southwest Pacific, constitutes a natural experiment in colonization of new territory by both plants and animals, and well illustrates the types of dispersal enumerated here. On August 27, 1883, a violent volcanic explosion destroyed much of the island and left the rest completely covered by a layer of hot ashes and pumice 6 to 60 m deep. There is no evidence that any life remained on the island. The island nearest to Krakatoa was 19 km away; most of the life on it was destroyed by toxic gases and a thick layer of ashes produced by the explosion. The next-closest island was about 40 km away. In short, Krakatoa suddenly became an island devoid of life, separated by 40 km of open ocean from the nearest significant source of potential colonizers (Fig. 36.44).

Nine months after the eruption a biologist visited the island, and though he searched diligently, the sole living thing he could find was a lone spider busily spinning a web; the spider had almost certainly been blown to the island. Nevertheless, there were probably dense colonies of photosynthetic Cyanobacteria thriving in crevices near the shoreline. Only three years later, numerous plants were found growing along the beaches, and several species of ferns and grasses were growing farther inland. The beach plants were of the kind found on the beaches of almost all tropical Pacific islands —plants whose seeds are highly resistant to seawater and are regularly carried long distances by ocean currents. The ferns found growing so soon on Krakatoa reproduce by means of very light spores that can easily be carried by even gentle air currents.

By 1896, thirteen years after the explosion, the island was fairly well covered with vegetation, but plants still were more abundant near the shores than in the interior. Most of the plants were ones distributed by sea currents and wind, but about 9 percent must have arrived by other intermediaries, probably birds. By 1906 the island was densely covered with plants, and there were 263 species of animals living there; most were insects, but there were four species of land snails, two species of reptiles, and 16 species of birds. Many of the insects either flew or were blown to the island, but some of them (and perhaps the reptiles also) probably arrived on floating logs or rafts. The number of bird species on Krakatoa reached saturation by 1921, and has since undergone the slow turnover predicted by MacArthur and Wilson, with extinction of established species balanced by the establishment of immigrant species. The number of plant species continued to rise at least until the late 1930s.

Organisms that are spread by wind and ocean currents, and actively flying animals such as birds, bats, and insects, continue to dominate Krakatoa today, but the percentage of organisms that must have arrived by other means has risen. If a person with no knowledge of the island's history were to visit Krakatoa, he would hardly guess that approximately a hundred years ago it was a lifeless mass of ash and steaming lava.

Had Krakatoa been farther away from areas that could act as sources of colonizers, the species diversity would have increased more slowly. In addition, because of the distance effect, equilibrium would have been reached at a lower species number; as we have seen, the more distant an island is from a major source of new colonists, the lower its species diversity will be at equilibrium, because distance alone reduces the immigration rate.

Whatever the distance to be crossed, some species have readier physical access than others to a region that they might potentially colonize; thus it was easier for small animals that could be blown by wind or carried on rafts to get to Krakatoa than it was for large mammals to get there. It is obvious that the geological or ecological zones that intervene between any two regions will be much more effective as barriers to dispersal for some species than for others. A wide expanse of ocean may almost completely prevent movement of horses or elephants, but coconut palms may cross it in fair numbers, because their large water-resistant seeds can float in seawater for many weeks without harm. A grassland separating two forested areas may be an almost insuperable barrier for some forest animals and prevent them from moving from one forest to the other. Other forest animals may have difficulty crossing the grassland, but manage to do so occasionally, and still others may cross the grassland freely. In short, what is a barrier to dispersal for one species may be a possible but difficult route for another and an easily negotiated path for a third. A knowledge of the sorts of routes and barriers that are effective for different species is a prerequisite to understanding the distribution patterns of organisms on the earth's surface.

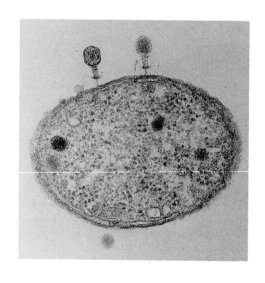

PART V

THE GENESIS AND
DIVERSITY OF ORGANISMS

Chapter 37

THE ORIGIN AND EARLY EVOLUTION OF LIFE

Few problems have exercised the human imagination like the problem of the origin of life. Religion, mythology, and philosophy have proposed a great variety of answers to it, but different though these answers have been, most of them share the assumption that the phenomenon must be attributed to an agency outside nature, a creator. In the same way, the diversity of species was conceived as resulting from separate, deliberate acts of creation. Not until the latter part of the nineteenth century was the theory of evolution able to account for the origin of species without invoking a supernatural agency. Can twentieth-century science do the same for the origin of life itself?

THE ORIGIN OF LIFE

We discussed the principle of biogenesis—that life can arise only from life —in Chapter 4. And we cited Pasteur's classic experiment in support of this principle. Pasteur's work effectively put to rest, as far as most biologists were concerned, the long-held idea of spontaneous generation. No longer could scientists seriously entertain the notion that the maggots in decaying meat arise *de novo* from the meat, or that earthworms arise from the soil during heavy rains, or that mice arise from sweaty shirts placed in a dark corner and sprinkled with wheat, as Jean-Baptiste van Helmont had suggested, or even that microorganisms appear spontaneously in spoiling broth. It may seem strange, then, that in the last quarter of the twentieth century spontaneous generation should be a topic of major interest in biology. But there is a radical difference between the modern ideas of sponta-

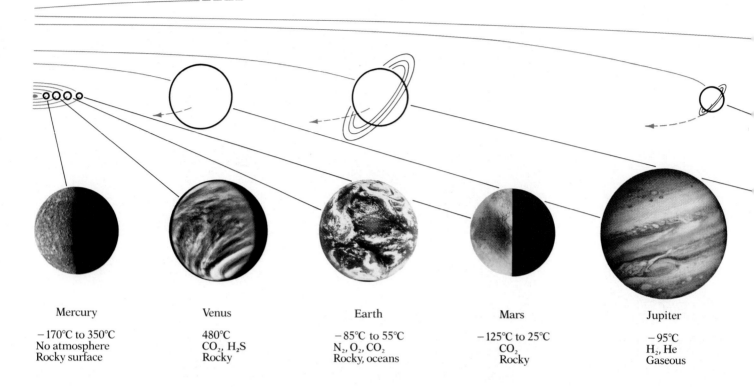

Mercury	Venus	Earth	Mars	Jupiter
−170°C to 350°C	480°C	−85°C to 55°C	−125°C to 25°C	−95°C
No atmosphere	CO_2, H_2S	N_2, O_2, CO_2	CO_2	H_2, He
Rocky surface	Rocky	Rocky, oceans	Rocky	Gaseous

37.1 The present-day solar system
The planetary diameters in the upper part of the
figure are exaggerated by 1,500 times relative to the
orbital diameters, in order for the planets to be
visible in this drawing. The sun comprises 99.9
percent of the solar system's mass.

neous generation and those of Pasteur's day. The modern theorists do not
suggest that life can arise spontaneously under present conditions on earth;
indeed, most of them are convinced that this cannot happen. What they do
suggest is that life could and did arise spontaneously from nonliving matter
under the conditions prevailing on the early earth, and that it is from such
beginnings that all present earthly life has descended.

It is one of the purposes of this chapter to outline a theory of the origin of
life now widely held by scientists. The basis for this theory was first enunciated clearly and forcefully by the Russian biochemist A. I. Oparin in 1936.[1]
Although the broad outlines of the theory have wide support, many of the
details are disputed; there is no direct evidence concerning the origin of
life.

Formation of the earth and its atmosphere   We are not certain how the
solar system formed; we have only hypotheses. But as astronomers probe
ever deeper into the secrets of the universe and gather more evidence, these
hypotheses become increasingly convincing. The one most widely held
today is that the universe is about 20 billion years old, and that the sun and

[1] Oparin had published a brief explanation of his theory earlier (1924), but this book was never
translated from the Russian. His ideas did not have a major impact on scientific thought until *The
Origin of Life on Earth* appeared in 1936 (first English edition, 1938).

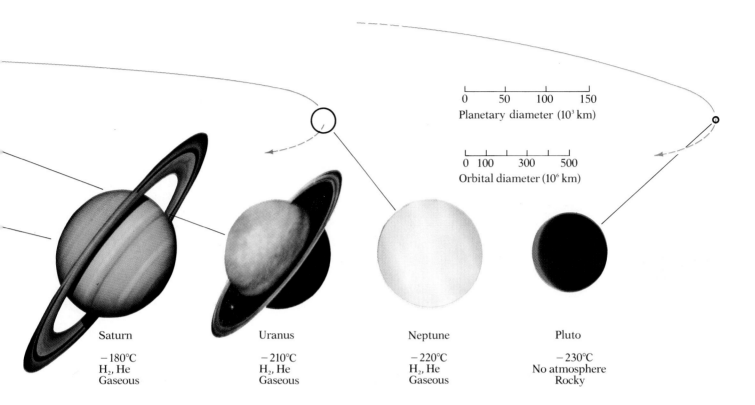

Planetary diameter (10^3 km)

Orbital diameter (10^6 km)

Saturn

$-180°C$
H_2, He
Gaseous

Uranus

$-210°C$
H_2, He
Gaseous

Neptune

$-220°C$
H_2, He
Gaseous

Pluto

$-230°C$
No atmosphere
Rocky

its planets formed between four and a half and five billion years ago from a cloud of cosmic dust and gas. Most of this material condensed into a single compact mass, producing enormous heat and pressure, which initiated thermonuclear reactions and converted the condensed mass into the sun. Within the remainder of the dust and gas cloud, which now formed a disc held in the gravitational field of the newborn sun, lesser clouds of condensation began to form. These became the planets, of which the earth is one (Fig. 37.1).

As the earth condensed, a stratification of its components took place, the heavier materials, such as iron and nickel, moving toward the center and the lighter substances becoming more concentrated nearer the surface. Among these lighter materials must have been hydrogen, helium, and the noble gases, which formed the first atmosphere. But unlike larger planets such as Jupiter and Saturn, the earth was too small, and its gravitational field too weak, to retain this first atmosphere; eventually all the gases escaped into space, leaving a bare rocky globe with neither oceans nor atmosphere.

As time passed, however, gravitational compression of the earth, together with radioactive decay, generated an enormous amount of heat, and the interior of the earth became molten. The effect was further stratification into a core of iron and nickel, a mantle some 4,700 km thick composed of dense silicates of iron and magnesium, and an outer crust 8–65 km thick

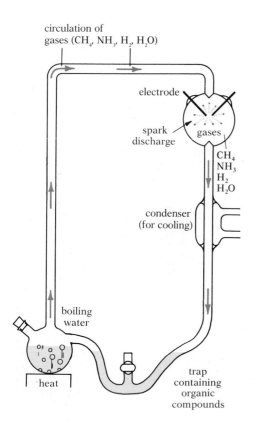

circulation of
gases (CH₄, NH₃, H₂, H₂O)

electrode

spark
discharge   gases

CH₄
NH₃
H₂
H₂O

condenser
(for cooling)

boiling
water

heat

trap
containing
organic
compounds

37.2 A simplified drawing of Stanley L. Miller's apparatus for synthesizing organic molecules under abiotic conditions

The closed system contained CH_4, NH_3, H_2, and H_2O, the gases thought to have been abundant if the early atmosphere was reducing. Water in the lower flask was boiled to circulate gases past the electrical sparks in the upper flask. As the products of the reaction passed through the condenser, they were cooled and condensed into liquid, which accumulated in the trap, forming compounds like urea, hydrogen cyanide, acetic acid, and lactic acid. Similar results have been obtained with volcanic gases.

composed mainly of lighter silicates. But the intense heat in the interior of the earth also tended to drive out various gases, which escaped primarily by volcanic action. These gases formed a second atmosphere for the earth.

We must know something about the probable early composition of this secondary atmosphere to understand the conditions under which life arose. The present atmosphere contains about 78 percent molecular nitrogen (N_2), 21 percent molecular oxygen (O_2), and 0.033 percent carbon dioxide (CO_2), as well as traces of rarer gases such as helium and neon. But available evidence indicates that when the atmosphere first formed it contained virtually no free oxygen and was therefore not an oxidizing atmosphere, as the present one is. Two principal models for the composition of the early atmosphere have been proposed in recent years; both are compatible with the modern hypothesis of the origin of life.

According to Oparin's model, the early atmosphere was a reducing one, containing much hydrogen (H_2). Consequently some of the atmospheric nitrogen was probably in the form of ammonia (NH_3); oxygen would have been in the form of water vapor (H_2O); and carbon would have been primarily in the form of methane (CH_4). Ultraviolet light from the sun would have caused most of the atmospheric ammonia to split into hydrogen and nitrogen gas. Note that methane is a simple hydrocarbon—an organic compound; hence, if this model is correct, the atmosphere of the early earth contained organic molecules long before there were any organisms.

The second model, which enjoys wider acceptance, assumes that the earth's early atmosphere was made up primarily of the gases known to occur in the present-day outgassing of volcanoes. These gases are H_2O, CO, CO_2, N_2, H_2S, and H_2; in such a mixture, hydrogen cyanide (HCN) is easily formed and would also have been present in the atmosphere. Note that this model, like the first, envisions an atmosphere in which there is little or no free oxygen, but with less free hydrogen such an atmosphere would not be reducing.

Initially, most of the earth's water was probably present as vapor in the atmosphere, a condition leading to torrential rains. These would have filled the low places on the crust with water and given rise to the first oceans. As rivers rushed down the slopes, they must have dissolved away and carried with them salts and minerals (iron and uranium being of particular importance), which slowly accumulated in the seas. Atmospheric gases probably also dissolved in the waters of the newly formed oceans. Whatever free oxygen might have existed in the atmosphere would have been rapidly removed by dissolved ions in the oceans; the extensive deposits of uraninite (UO_2) dating from that period must have been formed in this way.

Formation of small organic molecules If the earth had an oxygen-poor atmosphere, as many astronomers and geochemists believe, and the primitive seas contained a mixture of salts, CO_2, H_2S, HCN, and N_2, how were complex organic molecules formed? The mixture thought to have been present in the primitive seas is thermodynamically stable; there is no tendency for these substances to react with each other to form other compounds. Yet for life to have arisen it would seem at the very least that the critical building-block materials, particularly amino acids and the purine and pyrimidine bases, would have been necessary. How might these compounds have been formed on the abiotic primitive earth?

If more complex organic compounds were produced by reactions in a stable mixture, clearly some external source of energy must have been acting on the mixture. During the early history of the earth there was no shortage of such energy. One source must have been solar radiation, including visible light, ultraviolet light, and X rays; of these, ultraviolet light would probaby have been the most important. A second important source was probably energy from electrical discharges such as lightning. A third was heat from the earth's core and from the sun. There are other possible sources of energy, such as cosmic rays, radioactive disintegration of elements in the earth's interior, and volcanic explosions, but any role played by these in organic synthesis is likely to have been minor.

How can we know if ultraviolet radiation, electrical discharges, heat, or a combination of these are capable of causing reactions that produce complex organic compounds? An answer was provided in 1953 by Stanley L. Miller, who was then a graduate student working under Harold C. Urey at the University of Chicago. Miller set up an airtight apparatus in which a mixture of ammonia, methane, water, and hydrogen was circulated past electrical discharges from tungsten electrodes. He kept the gases circulating continuously in this way for one week, and then analyzed the contents of his apparatus (Fig. 37.2). He found that an amazing number and variety of organic compounds had been synthesized. Among these were some of the biologically most important amino acids and also such subtances as urea, hydrogen cyanide, acetic acid, and lactic acid. To rule out the possibility that microorganisms had contaminated his gas mixture and synthesized the compounds, Miller circulated the gases in the same way, but without any electrical discharges; no significant yield of complex organic compounds resulted. In still another experiment, he prepared the apparatus with the gas mixture inside and then sterilized it at 130°C for 18 hours before starting the sparking. The yields of complex compounds were the same as in his first experiments; a great variety of organic compounds were formed. Clearly, the synthesis was not brought about by microorganisms, but was abiotic—a synthesis in the absence of any living organisms, a synthesis under conditions possibly similar to those on the primordial earth. This experiment by Miller, which gave the first conclusive evidence that some of the steps hypothesized by Oparin could really occur, marked a turning point in the scientific approach to the problem of how life began.

In the years since 1953 many investigators have used mixtures of gases characteristic of volcanic emissions, plus hydrogen cyanide—on the more widely accepted assumption that the early atmosphere contained these or similar gases—and have also achieved positive results (Fig. 37.3). Those who have turned to other energy sources, such as ultraviolet light or heat or both, have also obtained large yields; these results are important, because on the early earth ultraviolet light was probably more available as a source of energy than lightning discharges. Significantly, the amino acids most easily synthesized in all these experiments performed under abiotic conditions are the very ones that are most abundant in proteins today, and the most

37.3 Volcanoes and lightning
One of the effective combinations for producing complex organic molecules abiotically is volcanic gas and electrical discharges. Here nature performs a similar experiment during the eruption of Surtsey, off the coast of Iceland.

$$NH_3 \; + \; CH_4 \; + \; \text{energy} \; \longrightarrow \; HCN \; + \; 3\,H_2$$

Ammonia   Methane   Hydrogen   Hydrogen
                                                            cyanide   gas

A

37.4 Abiotic production of a nitrogenous base
Mixing methane and ammonia in the presence of
external energy of the sort available during the early
history of the earth results in the production of,
among other things, hydrogen cyanide (A). Even in
the absence of a catalyst, this product can react with
itself at a low rate to produce adenine (B), an
important component of ATP, NAD, RNA, and a
variety of other compounds.

$HC \equiv N$
$HC \equiv N$
$HC \equiv N$
$HC \equiv N$
$HC \equiv N$

Hydrogen
cyanide

Adenine

B

important nitrogenous base is also the one most readily produced abiotically (Fig. 37.4).

The wide variety of conditions under which abiotic synthesis of the organic compounds essential to life has now been demonstrated makes it safe
to conclude that, even if conditions on the primitive earth were only
roughly similar to the ones postulated in either of the models outlined earlier, these compounds actually appeared and became dissolved in the
waters of the seas.

But though organic compounds were synthesized abiotically on the primordial earth, might they have been destroyed too fast to accumulate in
quantities sufficient for the later origin of living things? After all, most of
these organic compounds are known to be highly perishable. But why are
they perishable? One reason is that they tend to react slowly with molecular
oxygen and become oxidized. Another is that they are broken down by organisms of decay, primarily microorganisms. But the prebiotic atmosphere
contained virtually no free oxygen, and there were no organisms of any
kind. Therefore neither oxidation nor decay would have destroyed the organic molecules, and they could have accumulated in the seas over
hundreds of millions of years. No such accumulation would be possible
today.

Formation of polymers   Let us suppose, then, that a variety of hydrocarbons, fatty acids, amino acids, purine and pyrimidine bases, simple sugars,
and other relatively small organic compounds slowly accumulated in the
ancient seas. That is still not a sufficient basis for the beginning of life. Macromolecules are needed, particularly polypeptides and nucleic acids. How
could these polymers have formed from the building-block substances
present in the "soup" of the ancient oceans? This question is not easy to answer, and several hypotheses are currently supported by different investigators.

Some think that the concentration of organic material in the seas was high enough for chance bondings between simpler molecules to give rise, over a period of hundreds of millions of years, to considerable quantities of macromolecules. They point out that, even though each such polymerization reaction is rather unlikely in the absence of protein enzymes, nevertheless on the time scale here involved enough rare and unlikely events may occur to produce, collectively, a major change. As George Wald of Harvard University has said, "Given so much time, the 'impossible' becomes possible, the possible probable, and the probable virtually certain."

Other investigators, however, have been unwilling to agree that organic material in the early oceans was sufficiently abundant for the occurrence of chance polymerizations. They have suggested concentration mechanisms that would have speeded up chemical reactions. One such mechanism might have been adsorption of the building-block compounds on surfaces such as those of clay minerals. Another might have involved the accumulation of small amounts of dilute solution of building-block compounds in puddles on the beaches of lagoons and ponds. The heat of the sun would have evaporated most of the water, thus concentrating the organic chemicals, and providing energy for polymerization reactions. The resulting polymers might then have been washed back into the pond. Such a process could slowly have built up a supply of macromolecules in the pond. The hypothesis seems a reasonable one, for Sidney W. Fox of the University of Miami has shown that, if a nearly dry mixture of amino acids is heated, polypeptide molecules are rapidly synthesized (particularly if phosphates are present). Alternatively, after condensation by evaporation, the energy for polymerization reactions in the puddles might have come from ultraviolet radiation rather than heat.

Though various concentrating mechanisms may well have played a role in prebiotic polymerization reactions, too much may be made of the difficulty of incorporating building-block compounds into polymers, at least in the case of amino acids. In experiments of the Miller type, various researchers have observed that spherules yielding amino acids upon hydrolysis are formed after only 48 hours, whereas free amino acids do not appear in appreciable quantities until much later. Similarly, in some of their experiments on thermal synthesis, Fox and his associates obtained polymers first, and amino acids only later, *after hydrolysis* of the polymers. In short, abiotic synthesis may differ fundamentally from the more familiar biochemical syntheses in that polymers may be formed first, rather than the expected monomers.

Formation of molecular aggregates and primitive cells   We have now reached a point in our model for the origin of life in which the "soup" of the ancient seas, or at least that in some estuaries and lagoons, contains a mixture of salts and organic molecules, including polymers such as polypeptides and perhaps nucleic acids. We must next ask how the orderliness that characterizes living things emerged from this mixture.

Oparin pointed out that, under appropriate conditions of temperature, ionic composition, and pH, colloids of macromolecules tend to give rise to complex units called ***coacervate droplets***. Each such droplet is a cluster of macromolecules stabilized by hydrophobic interactions and surrounded by a shell of water (see Fig. 2.18, p. 41). There is thus a definite demarcation or

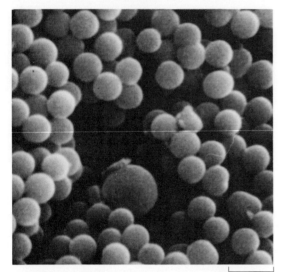

2 μm

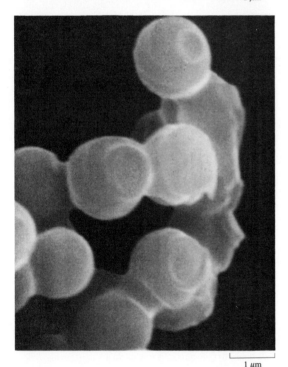

1 μm

37.5 Scanning electron micrographs of proteinoid microspheres

Top: The spheres are remarkably uniform in size, though a few are as much as four times larger than the others. Bottom: At higher magnification, connecting bridges between the spheres can be seen. The scars on the surface of some of the spheres indicate locations where bridges have been broken.

interface between the coacervate droplet and the liquid in which it floats. In a sense, the shell of oriented water molecules forms a membrane around the droplet.

Now, coacervate droplets have a marked tendency to adsorb and incorporate various wholly or partially hydrophilic substances from the surrounding solution; sometimes this selective tendency is so pronounced that the droplets may almost completely remove some materials from the medium. In this way the droplets grow at the expense of the surrounding liquid. And, again as a result of hydrophilic, hydrophobic, and ionic interactions, coacervate droplets have a strong tendency toward formation of definite internal structure—that is, the molecules within the droplet tend to become arranged in an orderly manner instead of being randomly scattered. As more and more different materials are incorporated into the droplet, a membrane consisting of surface-active substances may form just inside the shell of oriented water molecules, the permeability of the boundary of the droplet thus becoming even more selective than before. Thus, though coacervate droplets are not alive in the usual sense of the word, they do exhibit many properties ordinarily associated with living organisms, protecting their internal chemistry from the surrounding environment. In fact, the droplets look so much like organisms when viewed under a light microscope or even in an electron micrograph that experienced biologists have on occasion mistaken them for bacteria and attempted to assign them to species!

Fox, like Oparin, has envisioned the prebiological systems (prebionts) that led to development of the first cells as microscopic multimolecular droplets, but he suggests ***proteinoid microspheres*** rather than coacervate droplets. Fox's microspheres are droplets that form spontaneously when hot aqueous solutions of polypeptides are cooled. The microspheres may exhibit many properties characteristic of cells, including swelling in a hypotonic medium and shrinking in a hypertonic one, formation of a double-layered outer boundary, an internal movement reminiscent of cytoplasmic streaming, growth in size and increase in complexity, budding in a manner superficially similar to that seen in yeasts, and a tendency to aggregate in clusters of various types resembling those seen in many bacteria (Fig. 37.5). In fact, Fox has recently discovered that when these microspheres are illuminated with sunlight, they develop an electrical potential across their membranes not unlike the potential seen in essentially all living cells.

Since either type of droplet—complex coacervate or proteinoid microsphere—is structurally organized and sharply separated from the external medium, the chemical reactions that take place within the droplet depend not only on the conditions of the medium, but also on the physicochemical organization of the droplet itself. Because various substances may be more concentrated in the droplet, the probability of their taking part in chemical reactions is increased; and because of the organization within the droplet, each reaction that takes place will influence other reactions in ways that are most unlikely when the substances are free in the external medium. Furthermore, catalytic activity of both inorganic substances like metallic compounds and organic ones like proteins is enhanced by the regular spatial arrangement of molecules within the droplet. In short, the special conditions within the droplet will exert selective and regulative influence over the chemical reactions taking place there.

Vast numbers of different prebiological systems of this kind may have arisen in the seas of the early earth. Though most would probably have been too unstable to last long, some proteinoid microspheres have been stable in the laboratory for more than six years, long enough for countless chemical reactions to occur, for many "generations" of growth and budding to take place, and for natural selection to operate. Some of the early droplets must have contained particularly favorable combinations of materials, especially complexes with catalytic activity, and may thus have developed unusually harmonious interactions between the reactions occurring within them. As such droplets increased in size, they would have been more susceptible to physical fragmentation, which would have produced new smaller droplets with composition and properties essentially similar to those of the original droplet. These, in turn, would have grown and fragmented again. This primitive reproduction would not initially have been under the control of nucleic acids, even though these compounds could have been synthesized under abiotic conditions and may well have been incorporated into some of the prebionts. The nucleic acids might, of course, have been able to reproduce themselves exactly if a sufficient pool of nucleotides was present and if there were appropriate catalysts (even weak ones). Recent evidence suggests that some present-day RNAs, like the presumed constituents of the first primitive genes, can catalyze specific reactions. Their specificity, like that of enzymes, is probably a consequence of their ability to fold spontaneously. Therefore, RNA may have served originally as both genes and catalysts, only later giving up those roles to DNA and enzymes.

Clearly, if the new droplets formed upon fragmentation of the most stable prebionts could have incorporated more exact replicates of the favorable features of the parent droplet, their chances of survival would have been greater than without a genetic control system. In other words, chances of survival would have been increased if the catalytically active sequences of RNA could have come to code for RNA molecules that catalyzed replication and transcription of themselves. The next step—and the most difficult one for theorists—is the evolution of translation. Though the selective advantage to living systems of using proteins rather than RNA as enzymes and structural elements is clear, the gap between the processes of transcription and translation appears enormous. Just how a correlation between nucleotide sequences in nucleic acids and amino acid sequences in proteins could have arisen remains a mystery. There are, however, some suggestive leads to a possible solution. The first is that, so far as is known, the genetic code is almost universal; with minor exceptions, the codons have the same amino acid translations in all organisms, from microorganisms to human beings. That so little variation should have arisen in the whole course of cellular existence argues against mere happenstance dictating which codon would go with which amino acid. Moreover, a careful look at the dictionary of the genetic code (p. 717) shows that all codons in which U is the second letter code for hydrophobic amino acids, and that all five electrically charged amino acids (see Fig. 3.16, p. 58) have codons in which the middle letter is a purine (A or G). In short, some chemical system seems to underlie the pairing of codons and amino acids. Though the chemical explanation for that system is unknown, the presumption is that, as it gradually came into being, it made possible both more accurate duplication

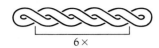

6×

37.6   Viroids

Potato spindle-tuber viroids appear at this magnification to be short rods. But they are in fact circular, single-stranded bits of RNA which, because they have several lengths of complementary base sequences, are able to assume the extended, double-stranded hairpin form seen in the drawing; they actually have about six times as many turns as are shown. Their double-stranded form may protect them from digestion by certain host-cell enzymes. A segment of DNA from the bacterial virus T7 runs through the bottom left of the photograph and provides a sense of scale.

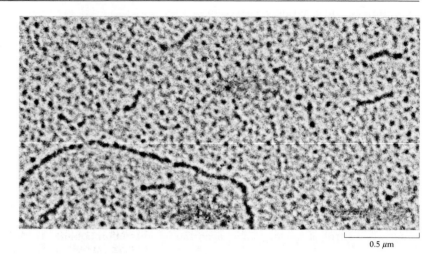

0.5 μm

during the reproductive process and more precise control over the chemical reactions taking place within the droplets.

The next step must have been that a small percentage of prebionts with particularly favorable characteristics slowly developed into the first primitive cells. Notice that there was almost certainly no abrupt transition from "nonliving" prebionts to "living" cells. The attributes we normally associate with life were acquired gradually. At this stage, the boundary between living and nonliving is an arbitrary one.

Not all biologists accept the sequence outlined above. Some think it more likely that the first "living" things were self-replicating macromolecules such as nucleic acids—"naked genes," if you will. The first cells would then have arisen as these macromolecleues slowly accumulated a shell of other substances (a primitive cytoplasm) around themselves. In other words, while the first model for the genesis of cells suggests that prebionts capable of a very primitive and imprecise form of reproduction arose first and then slowly developed a genetic control system, the suggestion here is that the control system arose first and that cytoplasm and a membrane then developed around it.

The closest things to naked genes found in nature today are the viroids, which are short circles of unprotected RNA (Fig. 37.6); viruses, which consist of DNA or RNA enclosed in a thin shell of protein, are a close second. Viruses were once thought by some researchers to be modern representatives of one of the earliest steps in the evolution of cells from naked genes. This possibility has been disputed on the ground that all modern viruses, as well as the more recently discovered viroids, are obligatorily parasitic— that they can reproduce only inside living cells—and consequently that they could not have existed before there were any cells. According to this widely accepted view, viruses and viroids are of much more recent origin, having probably evolved as genetic escapees from their present hosts or (in some cases) other organisms. The chief obstacle to the hypothesis that viruses or viroids evolved first is the absence of enzymes to facilitate replication; even if the "soup" contained all other necessary "nourishment" for these organisms, reproduction would have been impossible in the absence of replication enzymes.

Evolution of more complex biochemical pathways According to the view now most widely accepted, the earliest organisms arose well over three billion years ago, and were heterotrophs. They used as nutrients the carbohydrates, amino acids, and other organic compounds free in the environment in which they lived. In other words, they depended on previous abiotic synthesis of organic compounds. But as the organisms became more abundant and more efficient at removing preformed nutrients from the medium, they must have begun to deplete the supply of nutrients. The rate of spontaneous, extracellular formation of organic matter from inorganic raw materials was most likely never very high, and it must have taken many millions of years for a moderate supply of nutrients to accumulate. Now, that supply was probably being used up at an ever-increasing rate and, with the drain on available resources becoming more and more severe, the competition between organisms must have increased too. Forms inefficient at obtaining nutrients doubtless perished; those more efficient survived in greater numbers. Natural selection would have favored any new mutation that enhanced the ability of its possessor to obtain, process, or synthesize food.

At first, the primitive organisms probably carried out relatively few complex biochemical transformations. They could obtain most of the materials they needed ready-made. But it would have been these very materials—materials that could be used directly with little alteration—that would have dwindled most rapidly. Hence there would have been strong selection for any organisms that could use alternative nutrients. Suppose, for example, that a compound A, which was necessary for the life of cells, was initially available in the medium, but that its supply was being rapidly exhausted. If some cells possessed a mutant gene that coded for an enzyme a that catalyzed synthesis of A from another compound, B, in greater supply in the medium, then those cells would have had a marked adaptive advantage over cells that lacked the mutant gene. They could survive even when A was no longer available in the medium by carrying out the reaction

$$B \xrightarrow{a} A$$

But then there would have been increasing demand for free B, and the rate of its utilization would soon have exceeded the rate of its abiotic synthesis. Thus the supply of B would have dwindled, and there would have been strong selection for any cells possessing a second mutant gene that coded for enzyme b, catalyzing synthesis of B from C. These cells would not be dependent on a free supply of either A or B, because they could make both A and B for themselves as long as they could obtain enough C:

$$C \xrightarrow{b} B \xrightarrow{a} A$$

This general process of evolution of synthetic ability, first proposed by N. H. Horowitz of the California Institute of Technology, might have continued until eventually most cells made all the complex compounds they required by carrying out a long chain of chemical reactions:

$$G \xrightarrow{f} F \xrightarrow{e} E \xrightarrow{d} D \xrightarrow{c} C \xrightarrow{b} B \xrightarrow{a} A$$

In this way, the primitive cells would slowly have evolved more elaborate biochemical capabilities.

It seems most unlikely that much synthetic ability could have arisen unless the cells possessed some mechanism for handling the chemical energy released from such catabolic reactions as splitting molecular hydrogen or hydrolyzing organic compounds. The fact that all living things, from bacteria to human beings, employ ATP as their principal energy currency strongly suggests that use of ATP was an early evolutionary development. But what is the evidence that ATP was available to early organisms? The two organic precursors of ATP are adenine and the five-carbon sugar ribose (Fig. 37.7). Both these compounds also occur in nucleic acids, and experiments have demonstrated that both can be synthesized abiotically under presumed prebiological conditions. Indeed, of the five nitrogenous bases that occur in DNA and RNA, adenine is the one that forms most easily under simulated prebiological conditions. It is probably no accident, therefore, that adenine is the base found in a host of biologically critical compounds—not only ATP (and ADP, AMP, and cAMP), but also the electron carriers NAD, NADP, and FAD (Fig. 37.7). ATP would probably have been available for coupling exergonic and endergonic reactions as soon as the primitive cells could perform the reactions.

The earliest form of metabolism using ATP, which probably arose about 3.5 billion years ago, was almost surely fermentation, the anaerobic process universal in living organisms today; of the three main energy-yielding systems—fermentation, respiration, and photosynthesis—only fermentation is found in all bacteria and all eucaryotes.

Evolution of autotrophy Even though the primitive heterotrophs probably evolved more and more elaborate biochemical pathways that enabled them to use a greater variety of the organic compounds free in their environment, and even though some of them probably evolved other, purely heterotrophic methods of feeding, such as saprophytism, parasitism, and predation, nevertheless life would eventually have ceased if all nutrition had remained heterotrophic. The reason is not only that nutrients must have been used up much faster than they were being synthesized, but also that the organisms themselves must have been altering the environment in ways that decreased the rate of abiotic synthesis of organic compounds. For example, their metabolism, which would have been fermentative in the absence of molecular oxygen, would have released carbon dioxide into the atmosphere. Abiotic synthesis of complex organic compounds from CO_2 is much less likely than from methane or hydrogen cyanide.

37.7 Adenine and its derivatives
The nitrogenous base adenine and the sugar ribose combine to form the structural basis for AMP, ATP, acetyl-CoA, NAD, FAD, NADP, and other important biological intermediates.

Life did not become extinct as the supply of free organic compounds dwindled, because some of the primitive heterotrophs evolved autotrophic pathways. The first such pathways were almost certainly chemosynthetic, utilizing for the most part the energy in the covalent bonds of molecular hydrogen (see p. 168). This energy was used to synthesize many of the organic compounds no longer available from the "soup," as well as many novel compounds. Chemosynthetic autotrophs are still found today, particularly in bogs and at volcanic vents on the ocean floor.

The next autotrophic strategy to evolve—cyclic photophosphorylation —was far more important to the history of life on earth. In cyclic photophosphorylation, light of visible wavelengths is used as an energy source in the synthesis of ATP by cells. The present-day anaerobic photosynthetic bacteria may be the direct descendants of those first nonchemosynthetic autotrophs. Later, the much more complex pathways of noncyclic photophosphorylation and CO_2 fixation, in which energy from sunlight is used in synthesis of carbohydrate from CO_2 and water (or some other electron source) evolved, probably appearing first in organisms now interpreted as primitive Cyanobacteria. From this time onward, the continuation of life on earth depended primarily on the activity of the photosynthetic autotrophs.

The evolution of photosynthesis based on water as the electron donor probably administered the coup de grâce to significant abiotic synthesis of complex organic compounds. An important by-product of such photosynthesis is molecular oxygen, which is highly electronegative. The O_2 released by photosynthesis must have first entered the water cycle, and reacted with many of the minerals, including iron, dissolved in the oceans; this led to the precipitation of iron as Fe_3O_4 and the consequent deposition of vast sediments known as "banded iron." With dissolved iron removed from the water, free oxygen would have been able to accumulate there and then in the atmosphere. Once the layer of ozone (O_3) now present high in the atmosphere had been formed by some of the O_2, this layer effectively screened out most of the ultraviolet radiation from the sun and allowed very little high-energy radiation to reach the earth's surface. In other words, living organisms, once they arose, changed their environment in a way that destroyed the conditions that had made possible the origin of life; they caused what has sometimes been called an oxygen revolution.

Once molecular oxygen became a major component of the atmosphere, both heterotrophic and autotrophic organisms could use the biochemical pathways of aerobic respiration, by which far more energy can be extracted from nutrient molecules than by fermentation alone.[2] The progressive increase in atmospheric O_2 and the formation of the ozone shield were probably the major factors in permitting organisms to leave the UV-absorbing oceans and to move onto the land.

---

[2] As we saw in Chapters 7 and 8, the electron-transport chains of aerobic respiration, cyclic photophosphorylation, and noncyclic photophosphorylation are closely related. Which evolved first is not known, but since even small abiotically produced quantities of oxygen (or alternative electron acceptors like nitrates, sulfates, and carbonates) would have given heterotrophs with an electron-transport chain enormous advantages over obligate fermenters, most researchers believe that at least some elements of the respiratory electron-transport chain arose prior to the photosynthetic pathways.

37.8 Prospects for life on neighboring planets in the solar system

Only the earth and Mars support the planetary temperatures we have come to associate with life, temperatures between 0°C and 100°C. Venus, because of its thick, cloudy atmosphere and the resulting greenhouse effect, remains at a nearly constant 480°C, with an atmospheric pressure at ground level nearly 100 times that of the earth. Its surface, as revealed in a photo from a short-lived Soviet spacecraft, is rocky and barren (A). Mars, on the other hand, has a very thin, cool CO_2 atmosphere, but lacks water. Pictures from spacecraft in orbit around Mars reveal what appear to be ancient streams and riverbeds (B), suggesting that water may once have been plentiful, and may now be stored as ice in the soil. Pictures and tests on the planet's surface taken by the Viking Lander II reveal no signs of present life (C).

The possibility of life on other planets If life could arise spontaneously from nonliving matter on the primordial earth, might it also have arisen elsewhere in the universe? Few scientists concerned with the origin of organisms think that life is unique to the planet earth; most are convinced that life has probably arisen many times in many places. They point out that no unduplicable event was necessary to the origin of life on earth. On the contrary, all the events now hypothesized and all the known characteristics of life seem to fall well within the general laws of the universe. Indeed, some have argued recently that biochemical evolution and life, when conditions permit, are inevitable parts of the overall evolution of matter in the universe. Given the immense size of the universe, they argue, it would actually be unreasonable to think that life is restricted to one small planet in one minor solar system. Within our own solar system, however, the prospects for life on planets other than earth are small (Fig. 37.8).

One interesting series of calculations on the probability of life outside our solar system was made some years ago by Harlow Shapley of Harvard University. Shapley said that at least 10^{20} stars are visible to us with present-day telescopes (to say nothing about the vast numbers beyond the reach of our telescopes). Many of these are binary (double) stars in which two suns orbit one another. Because of the irregular gravitational forces and extreme temperature cycles this arrangement generates, life is unlikely in these systems. Even among the single stars, many are too short-lived, too bright, or too dim; and many of the potentially suitable stars must lack planets. Shapley thought it reasonable to assume that at least one star in every thousand has a planetary system, which gives a total of 10^{17} stars with planets. Now, if life wherever it occurs is at least roughly similar to earthly life in its basic chemistry, then only planets with moderate temperatures could support life. The planetary systems of many stars may not include a planet orbiting at the right distance. Shapley suggested that, by a modest estimate, at least one in a thousand of the stars with planets has an appropriate planet; this gives a total of 10^{14} stars with at least one planet of the right temperature. But not only must a planet have moderate temperatures if it is to support life, it must also be within a certain size range to hold a suitable atmosphere. If one out of every thousand of the planets of the right temperature is also the right size, this gives a total of 10^{11}. Even if a planet has an appropriate temperature and atmosphere, though, life still might not have arisen, for any number of reasons. Again Shapley used an estimate of one in a thousand, which gives 10^8 (one hundred million) planets on which life may well have arisen. Today biologists working in this field consider Shapley's estimate too conservative; they suggest that the figure should be 10^{16} or more.

No matter how many times life might have evolved on other planets in other galaxies, however, human beings probably could not have evolved on any planet but earth. Totally separate evolutionary developments can never exactly duplicate each other. Indeed, if human beings should become extinct on earth, and intelligent life arise a second time, that life would almost certainly not be human. Loren Eiseley of the University of Pennsylvania has put the matter eloquently: "There may be wisdom; there may be power; somewhere across space great instruments, handled by strong, manipulative organs, may stare vainly at our floating cloud wrack, their owners yearning as we yearn. Nevertheless, in the nature of life and in the principles of evolution we have had our answer. Of men, elsewhere and beyond, there will be none forever."

PRECAMBRIAN EVOLUTION

THE FOSSIL RECORD

The oldest known fossils (as of this writing) are from deposits in Western Australia, and are dated at 3.5 billion years (Fig. 37.9). They appear to be bacteria—procaryotic cells, as would be expected of the oldest fossils, since these cells are assumed to be more primitive than eucaryotic ones. Most authorities think it probable that the first cellular organisms, living at a time when all nutrition was chemosynthetic or heterotrophic, were of the bacterial type, and that it was the evolution of Cyanobacteria from these roughly 2.3 billion years ago that initiated the oxygen revolution. These "blue-green algae," like true plants but unlike other photosynthetic bacteria, use water as the electron source in noncyclic photophosphorylation and hence release molecular oxygen as a by-product.

More and more fossils of procaryotic organisms from the first two and a half billion years or so of life are being found, and some fossils of eucaryotic cells are thought to date back at least 1.5 billion years. But the oldest geologic period from which fossils of higher forms of life are fairly abundant is

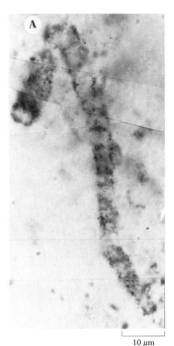

10 μm

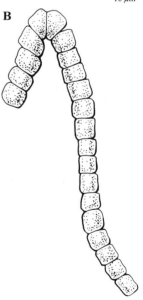

37.9 Photograph (A) and drawing (B) of a filamentous procaryotic microfossil from Western Australia
This bacteriumlike organism lived about 3.5 billion years ago.

TABLE 37.1 *The geologic time scale*

Era	Period	Epoch	Millions of years (approx.) from start of period to present	Plant life	Animal life
CENOZOIC	Quaternary	Recent	0.01	Increase in number of herbs	Rise of civilizations
		Pleistocene	2.5		First *Homo*
	Tertiary	Pliocene	7	Dominance of land by angiosperms	First humans
		Miocene	26		
		Oligocene	38		Dominance of land by mammals, birds, and insects
		Eocene	54		
		Paleocene	65		
			—BUILDING OF ANCESTRAL ROCKY MOUNTAINS—		
MESOZOIC	Cretaceous		136	Angiosperms arise and expand as gymnosperms decline	Last of the dinosaurs; second great radiation of insects
	Jurassic		190	Gymnosperms (esp. cycads and conifers) still dominant, last of the seed ferns	Dinosaurs abundant; first birds
	Triassic		225	Dominance of land by gymnosperms; further decline of lycopsids and sphenopsids	First mammals First dinosaurs
			—BUILDING OF ANCESTRAL APPALACHIAN MOUNTAINS—		
PALEOZOIC	Permian		280	Precipitous decline of lycopsids, sphenopsids, and seed ferns	Great expansion of reptiles; decline of amphibians; last of the trilobites
	Carboniferous[a]		345	Great coal forests, dominated at first by lycopsids and sphenopsids, and later also by ferns and gymnosperms	Age of Amphibians; first reptiles; first great radiation of insects
	Devonian		395	Expansion of primitive tracheophytes; origin of first seed plants toward end of period; first liverworts	Age of Fishes; first amphibians and insects
	Silurian		430	Invasion of land by the first tracheophytes toward end of period	Invasion of land by a few arthropods
	Ordovician		500	Marine algae abundant	First vertebrates (Agnatha)
	Cambrian		570	Primitive marine algae (esp. Cyanobacteria and probably Chlorophyta)	Marine invertebrates abundant (including representatives of most phyla)
			—INTERVAL OF GREAT EROSION—		
PRECAMBRIAN					Primitive marine life

[a]In North America, the Lower Carboniferous is often called the Mississippian period, and the Upper Carboniferous is called the Pennsylvanian period.

37.10 Fossil of a Precambrian animal (*Dickinsonia costata*) from South Australia
This animal, which appears to have been a segmented worm of some sort, lived some 600 million years ago. Reproduced natural size.

the **Cambrian**, which began nearly 600 million years ago (Table 37.1). Many of the Cambrian fossils are of relatively complex organisms—most of the animal phyla extant today are represented. The few Precambrian fossils of eucaryotic organisms are mostly of simple algae and a few invertebrates, whose relationships are poorly understood. We therefore have very little evidence concerning a most fascinating evolutionary development—the early radiation of life and the origin of the major animal phyla and plant divisions.

It is not clear why there are so few Precambrian fossils. Among the many explanations that have been offered, one is that most Precambrian organisms were soft-bodied and hence did not readily form fossils; it is the hard parts of plants and animals that are most often fossilized, because these are the most likely to resist decay and become buried. This explanation, though doubtless partly correct, is insufficient. It seems most unlikely that the many hard-bodied animals present at the start of the Cambrian period arose suddenly at that time; they almost certainly descended from Precambrian ancestors that gradually developed shells or exoskeletons or other hard parts. And, indeed, a few recently discovered beds of Precambrian fossils, notably in Australia, have confirmed the existence of hard-bodied invertebrate animals before the Cambrian (Fig. 37.10). It is possible that the ranges of these organisms were restricted to areas with environmental conditions that made fossilization unlikely. Furthermore, we must remember that fossils may be destroyed by normal geological processes; the greater the lapse of time, the more likely the destruction. A high percentage of any fossils formed as long ago as the Precambrian is likely to have been destroyed long before there were human beings to study them.

THE ORIGIN OF EUCARYOTIC CELLS

The scarcity of fossils from the Precambrian period leaves us with virtually no direct evidence concerning the evolution of the first eucaryotic cells

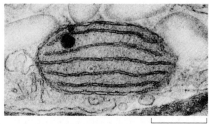

A

0.5 μm

37.11   Chloroplast of a red alga (A) compared with a whole cyanobacterium (B)
The photosynthetic lamellae in the chloroplasts of red and brown algae are not arranged in the stacks, called grana, that characterize the chloroplasts of most tracheophytes; they are more like the stroma lamellae of tracheophytes. In this respect a red algal chloroplast (A) resembles an entire cyanobacterium; in such a cell (B), the lamellae, which show no granum arrangement either, are free in the cytoplasm, for there is no plastid membrane.

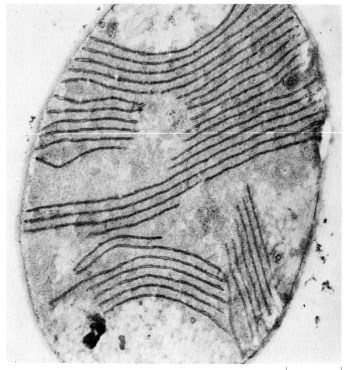

B

0.5 μm

from procaryotic progenitors. The oldest fossil eucaryotes are already relatively complex, and hence shed little light on the problem of their derivation.

Anatomical evidence of similarities between the subcellular organelles of eucaryotes and certain procaryotes has gradually accumulated over the years. As long ago as the nineteenth century, when chloroplasts were first studied under the microscope, investigators noticed that they resembled certain free-living Cyanobacteria, and suggested that they might have many features in common with procaryotic organisms (Fig. 37.11). A bacterial origin for mitochondria was first proposed in the 1920s.

As we have seen in previous discussions of cell biology and genetics, there has been growing molecular evidence in the past few decades to support the view that present-day eucaryotes may have arisen from precursors, now extinct, called **urcaryotes**, which had developed a symbiotic relationship with certain procaryotes. Let's review and summarize here the evidence for the endosymbiotic hypothesis, as put forward by Lynn Margulis of Boston University.

Chloroplasts and mitochondria are self-replicating bodies (Figs. 37.12 and 37.13), which contain genetic material and carry out protein synthesis on ribosomes of their own making (see p. 136). They are about the size of procaryotes (1–10 μm), lack nuclear membranes, and have a single, usually

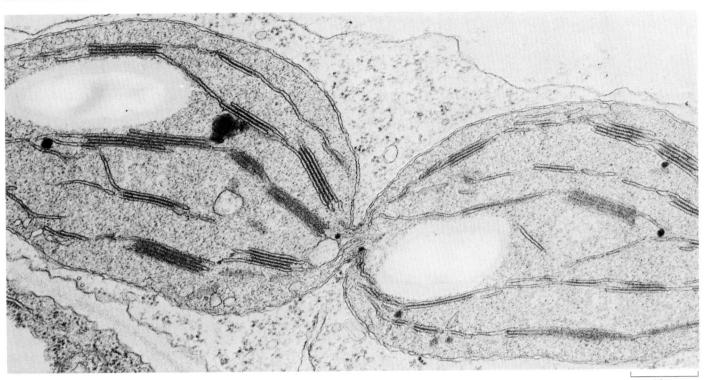

0.5 μm

37.12 Electron micrograph of a dividing chloroplast in a tobacco leaf
The large white areas in each of the daughter chloroplasts are starch granules.

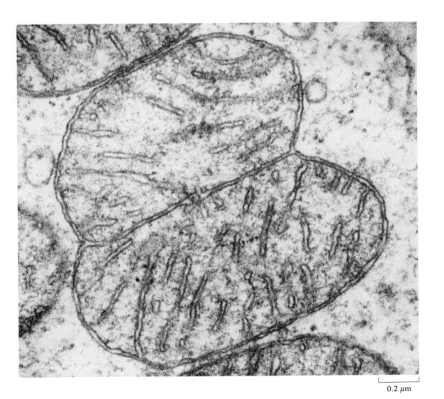

0.2 μm

37.13 Electron micrograph of a dividing mitochondrion
The partition between the two daughter mitochondria is nearly complete.

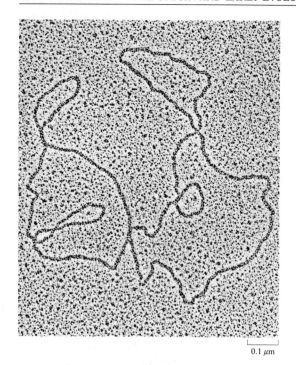

37.14 Electron micrograph of the circular chromosome of a mitochondrion from an oocyte of _Xenopus_

circular chromosome which is not packaged on nucleosomes (Fig. 37.14). They are also like procaryotes in that their ribosomes are small (even interchangeable with procaryotic ribosomes in some cases), and they use the same control-sequence codes and initial amino acid as most procaryotes. Moreover, gene expression in both procaryotes and organelles is controlled primarily by negative control strategies and the genetic material is almost entirely free of introns; eucaryotes usually employ positive control and generally have several introns per gene. The RNA and DNA polymerases of procaryotes and organelles are similar, but qualitatively different from those of eucaryotes. Of course, procaryotes and organelles both accomplish protein synthesis, metabolism, and internal transport without the ER, Golgi, lysosomes, microtubules, and subcellular organelles typical of eucaryotes.

The capacity of unicellular organisms to live as endosymbionts inside the cells of other organisms—as the ancestors of chloroplasts and mitochondria are presumed to have done—can be amply demonstrated by present-day examples. A variety of protists contain endosymbiotic single-celled algae, and the gastrodermal cells of _Chlorohydra_ and of several species of corals and sea anemones also contain algal cells (see Fig. 36.32, p. 979). Some heterotrophic protists that inhabit the guts of termites harbor bacteria that digest the cellulose the termites eat; neither the termites nor the protists can live without them (Fig. 37.15). Even some higher animals, especially several species of molluscs, regularly have intracellular algal symbionts. And intracellular symbiotic bacteria occur widely in both the plant and animal kingdoms.

Though both anaerobic photosynthetic bacteria and Cyanobacteria probably arose long before the first eucaryotic cells, the Cyanobacteria seem the more likely progenitors of chloroplasts,[3] since they have the same basic type of chlorophyll _a_ and use the same noncyclic photophosphorylation pathways. Mitochondria may have been derived from aerobic bacteria, though the reverse sequence is also possible. In either case, the inner mitochondrial membrane may be viewed as the homologue of the bacterial plasma membrane, since the two are biochemically similar, particularly in

[3] But see discussion of the Prochlorophyta on pp. 1039–40.

37.15 Endosymbiosis in the termite gut
The single-celled intestinal protist of a termite shown here harbors within it a bacterium that is capable of digesting the cellulose in the termite's food. Neither the protist nor the termite could survive without the presence of the bacterium.

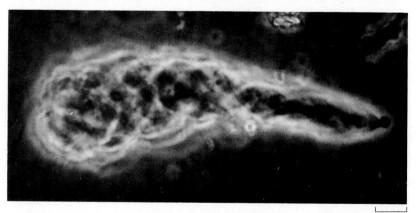

their capacity for electron transport. The outer mitochondrial membrane resembles the endoplasmic reticulum, and hence was probably derived from the host cell during the period of common evolution. If chloroplasts and mitochondria did indeed arise as symbionts, the symbiosis has clearly been a mutualistic one.

Though the organelles discussed here contain DNA, and divide, grow, and differentiate partly on their own, they are not fully autonomous entities. Many of their proteins are specified by nuclear genes. Presumably the symbionts gave up much of their genetic control to the host during the hundreds of millions of years since they last lived as free cells. This surrender of control to the nuclear genes is probably the reason why the mini-chromosomes of chloroplasts and mitochondria are invariably less than one-tenth the size of typical procaryotic chromosomes.

In contrast to the wealth of data and speculation on chloroplasts and mitochondria, there is little evidence or even conjecture bearing on the origin of the nuclear membrane and the associated endoplasmic reticulum. Since there is membrane flow in fully developed eucaryotic cells from the nuclear membrane to the ER to the Golgi apparatus (see Fig. 5.9, p. 119), the nuclear membrane may have been the first part of the intracellular membrane system to evolve, with the other parts arising from it later (Fig. 37.16), but how the nuclear membrane itself arose is still a mystery. Most researchers believe that the nuclear membrane evolved in urcaryotes, the organism believed to have given rise to the eucaryotic line.

Despite its current prestige, the endosymbiont model of the origin of eucaryotic cells is far from proven. Indeed, a variety of other models have been proposed in recent years; two, by Bogorad and by Uzzel and Spolsky, are described in articles listed in the readings for this chapter.

THE KINGDOMS OF LIFE

Because the fossil record is nearly blank for the long span of time when the basic pattern of organismic diversity was coming into being, ideas about the evolutionary relationships between the major phyla and divisions of organisms are rather vague. There is little evidence concerning the relationships between the major groups of algae. It is not known whether fungi evolved from photosynthetic green algae, or directly from heterotrophic organisms such as bacteria, or from some other stock. It is uncertain how protists are related to multicellular plants or to multicellular animals.

Ignorance of the relationships between major groups of organisms apparently never hindered the age-old attempt to assign all living things to one or another of a few large categories called kingdoms. One of the oldest and most widely used classifications recognizes only two kingdoms—one for plants and the other for animals. This dichotomy works well as long as the organisms to be classified are the generally familiar ones. Dandelions, grasses, daffodils, roses, and oak trees can easily be recognized as plants. Similarly, cats, horses, chickens, earthworms, and house flies can easily be recognized as animals. Things become a bit more difficult when the organisms in question are bread molds or sponges. These don't fit quite so neatly within the common intuitive concept of "plant" and "animal." Nevertheless, most biologists managed to convince themselves that bread molds,

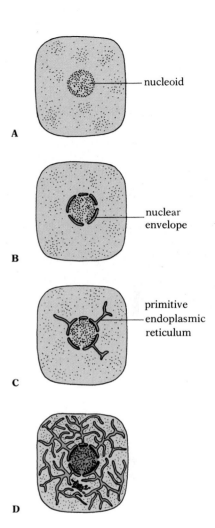

nucleoid

nuclear envelope

primitive endoplasmic reticulum

37.16 Model for the evolution of the nuclear envelope and the cytoplasmic membrane system
(A) The ancestral cell is presumed to have been procaryotic—having a nucleoid with no boundary membrane. (B) Membranous vesicles may have formed as new structures surrounding the nuclear area, giving rise to the double nuclear envelope. (C) Evaginations from the outer membrane of the nuclear envelope may have given rise to the endoplasmic reticulum, which itself became elaborated (D) into an extensive cytoplasmic membrane system.

when their characteristics were carefully examined, seemed definitely more "plantlike" than "animal-like," despite their lack of chlorophyll, and that sponges are "animals," despite their sedentary way of life.

With the advent of electron microscopy and more detailed study of cellular structure, however, it soon became clear that the bacteria differ from all other forms of life in a very fundamental way: they are procaryotic, whereas all other organisms are eucaryotic. Though bacteria had traditionally been assigned to the plant kingdom, this procedure was plainly no longer acceptable. The practice of setting procaryotes apart as a separate kingdom, the **Monera**, rapidly gained ground; it is a practice we shall follow.

More recently, there has been a growing realization that the fungi—traditionally regarded as plants—differ from true plants in a variety of basic characteristics. First, they are not photosynthetic; indeed, as heterotrophs, they are more animal-like than plantlike in their nutrition. Second, their cell walls are chemically different from those of most plants in that the primary component of the walls is usually not cellulose. Third, they are not multicellular in the sense that both plants and animals are; the partitions between adjacent fungal cells, if present at all, tend to be incomplete, the cytoplasm thus being continuous. In short, the fungi seem to represent a separate evolutionary development, coordinate with that of plants and that of animals. We shall, therefore, follow the growing practice of recognizing a kingdom **Fungi**.

Putting the fungi in a kingdom of their own makes it possible to understand the tripartite evolution of higher organisms in terms of exploitation of three different modes of nutrition—photosynthetic autotrophism by plants, absorptive heterotrophism by fungi, and ingestive heterotrophism by animals.

Recognizing the kingdoms Monera and Fungi resolves two serious anomalies in the older classifications. But another major problem remains: the classification of eucaryotic unicellular organisms. The ones zoologists have traditionally called Protozoa have been particularly troublesome to those who insist on a neat separation between plants and animals. This is true especially of the group of protozoans known as flagellates. These creatures have long flagella that enable them to swim actively in a manner intuitively felt to be "animal-like." Yet some of them possess chlorophyll and carry out photosynthesis, a characteristic ordinarily considered decidedly "plantlike." How can organisms such as these be classified? One possibility would be to rule arbitrarily that all those with chlorophyll are to be considered plants, and all lacking chlorophyll animals. But this seemingly simple procedure has serious drawbacks. Some green flagellates are clearly very closely related to colorless species; yet the proposed rule would put the green ones in a different kingdom from their colorless close relatives. Furthermore, there are some species of flagellates in which both green and colorless races sometimes occur; the suggested rule is unable to deal effectively with such species, and violates the whole purpose of modern taxonomy—the codification of evolutionary relationships. By the same token, we cannot simply suppose today that since most protists lack chlorophyll and are rather animal-like, they are all—even the green ones—animals. Many unicellular green flagellates are obviously very closely related

TABLE 37.2 *Various kingdom classifications*

System 1	System 2	System 3	System 4	System 5	System 6	System 7	System 8
PLANTAE	MONERA	PROTISTA	PROTISTA[a]	MONERA	MONERA	MONERA	ARCHAEBACTERIA
Bacteria	Bacteria	Bacteria	Bacteria	Bacteria	Bacteria	Bacteria	
Chrysophytes		Protozoa	Protozoa				EUBACTERIA
Green algae	PLANTAE	Slime molds	Chrysophytes	PROTISTA[a]	PLANTAE	PROTISTA	
Brown algae			Green algae				PROTISTA
Red algae	Chrysophytes	PLANTAE	Brown algae	Protozoa	Chrysophytes	Protozoa	
Slime molds	Green algae		Red algae	Chrysophytes	Green algae	Chrysophytes	Protozoa
True fungi	Brown algae	Chrysophytes	Slime molds	Green algae	Brown algae	Slime molds	Chrysophytes
Bryophytes	Red algae	Green algae	True fungi	Brown algae	Red algae		
Tracheophytes	Slime molds	Brown algae		Red algae	Bryophytes	PLANTAE	SLIME MOLDS
	True fungi	Red algae	PLANTAE	Slime molds	Tracheophytes		
ANIMALIA	Bryophytes	True fungi		True fungi		Green algae	PLANTAE
	Tracheophytes	Bryophytes	Bryophytes		FUNGI	Brown algae	
Protozoa		Tracheophytes	Tracheophytes	PLANTAE		Red algae	Green algae
Multicellular	ANIMALIA				Slime molds	Bryophytes	Brown algae
animals		ANIMALIA	ANIMALIA	Bryophytes	True fungi	Tracheophytes	Red algae
	Protozoa			Tracheophytes			Bryophytes
	Multicellular	Multicellular	Multicellular		ANIMALIA	FUNGI	Tracheophytes
	animals	animals	animals	ANIMALIA			
					Protozoa	True fungi	FUNGI
				Multicellular	Multicellular		
				animals	animals	ANIMALIA	True fungi
						Multicellular	ANIMALIA
						animals	
							Multicellular
							animals

[a] When multicellular groups are included, the Protista are sometimes rechristened Protoctista.

to species of simple multicellular green algae. Evolutionary patterns within the green algae are best interpreted, in fact, if one includes unicellular green flagellates; transferring them from the algae to the animal kingdom is unsatisfactory, because it clearly separates closely related organisms.

Whatever the criteria chosen, it is impossible to make a clean separation between plants and animals at the unicellular level. The reason is obvious. Unicellular organisms (and some multicellular ones) are at an evolutionary stage where it is essentially meaningless to talk about "plants" and "animals"—artificial human categories not dictated by the rules of nature. At the lowest evolutionary levels, about the only distinction between plants and animals that stands up is this: plants are living things studied by people who say they are studying plants (botanists), and animals are living things studied by people who say they are studying animals (zoologists). Facetious as this distinction sounds, it is basically accurate. The unicellular green flagellates may reasonably be classified as both plants and animals because they are studied by both botanists and zoologists, the first calling them algae and the second calling them Protozoa.

An alternative that has won much support is to assign the eucaryotic unicellular (and very primitive multicellular) organisms to a separate kingdom, designated **Protista**. As Table 37.2 shows, the limits of this kingdom vary greatly among the many different kingdom classifications that may be encountered in textbooks today. In some classifications the Protista take in only unicellular organisms (with or without the Monera); in others they include all multicellular algae (and at times fungi) as well. In other words, the category Protista is sometimes restricted to organisms that are unicellular during much or all of their lives, and sometimes it is broadened—the name then used may be Protoctista—to include all plantlike organisms whose bodies show relatively little distinction between tissues.

Both usages of Protista have the advantage of allowing a clear separation between the plant and the animal kingdom, all the troublesome forms being lumped together under Protista. But the first usage, which restricts Protista to unicellular organisms, separates green flagellates from their close multicellular relatives in the algae. And the second usage, which includes all algae in the Protista, leaves only bryophytes and vascular plants in the plant kingdom—an arrangement accepted by few botanists; moreover, it combines groups that are probably not at all closely related—notably the ciliate Protozoa and the brown algae—thus making the Protista a phylogenetically meaningless assemblage. In brief, both usages of Protista get around some of the difficulties inherent in the two-kingdom system, but both create other problems.

Nonetheless, there is a growing tendency to recognize a kingdom Protista, generally limited to groups entirely or primarily unicellular. The Protista so conceived figure in the five-kingdom system proposed by R. H. Whittaker of Cornell University (System 7 of Table 37.2), which likewise includes, in addition to Plantae and Animalia, the kingdoms Monera and Fungi.

This system can be interpreted in two ways: the protists can be regarded as actually consisting of three separate groups, which represent the *precursors* of plants, animals, and fungi (Fig. 37.17A); or the protists can be placed in a separate line that diverged *before* the other three eucaryotic kingdoms began to emerge as distinct groups (Fig. 37.17B). But in either interpretation the system presents several difficulties that have yet to be resolved satisfactorily. For one, the Archaebacteria—a group of diverse but related bacteria which includes methogens, halophiles, and thermoacidophiles, discussed in the next chapter—appear to be only distantly related to other bacteria. Then, too, the slime molds do not fit well into any of the systems discussed so far. Another scheme, a seven-kingdom system proposed by Carl R. Woese of the University of Illinois, seems to address most problems of this sort. Woese, who pioneered the biochemical studies that have demonstrated that Archaebacteria are fundamentally different from other bacteria, has used recent biochemical data like the variations of amino acid sequences in cytochrome c to reconstruct likely evolutionary affinities. Though this evidence is not yet complete, he suggests that the Archaebacteria and true bacteria (Eubacteria) deserve separate kingdoms (Fig. 37.17C), that modern protists are a separate evolutionary line, and that slime molds represent a separate kingdom. (This scheme is System 8 in Table 37.2.) If biochemical, genetic, and microanatomical evidence continues to fall into the present pattern, some version of the seven-kingdom system is likely to supersede all others as the best representation of evolutionary relationships.

Another popular modern scheme, proposed in 1974 by Gordon R. Leedale of the University of Leeds, recognizes no separate Protista kingdom, but rather a protist level within Plantae, Animalia, and Fungi (System 6 in

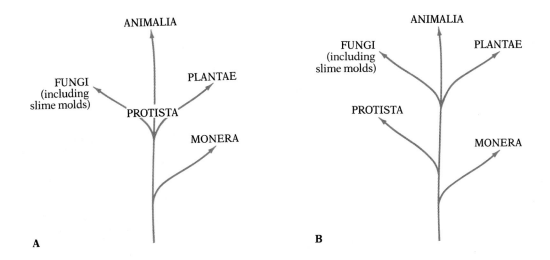

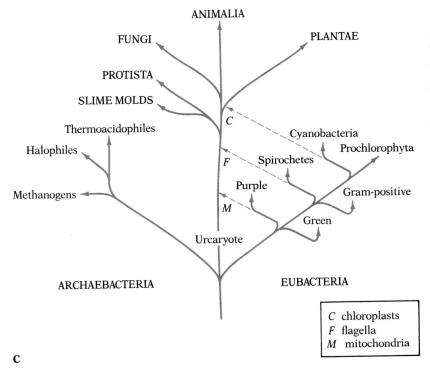

A

B

C

C chloroplasts
F flagella
M mitochondria

37.17 Three modern classifications
Two versions of the five-kingdom system are shown, as well as a seven-kingdom system. (The placement of monerans and protists at lower levels than animals, plants, and fungi in these diagrams does not mean that they are less advanced or less specialized.) In A, the Protista are recognized as a separate kingdom consisting of three groups derived from a common ancestor and leading to Plantae, Animalia, and Fungi. In B, the Protista are seen as a group that diverged from the rest of the eucaryotes *before* the divergence of Plantae, Animalia, and Fungi. Neither version is wholly satisfactory. The seven-kingdom system, shown in C, which is based on preliminary biochemical evidence, divides the procaryotes into two separate kingdoms (Archaebacteria and Eubacteria), assumes a separate, early divergence for modern protists, separates the slime molds from the fungi, and assigns origins, indicated by dotted lines, for eucaryotic mitochondria, flagella, and chloroplasts.

Table 37.2). Biochemical evidence, however, supports the view that many modern protists are part of a separate line, as reflected in Figures 37.17B and 37.17C. Currently, most biologists accept the five-kingdom system as the most accurate working scheme, and this is the system we shall follow in discussing the diversity of life.

Chapter 38

VIRUSES AND MONERA

Viruses are not ordinarily given a place in formal classifications of living organisms. Yet biologists and laymen alike tend to think of them as living, using such expressions as "live-virus vaccine" and "killed-virus vaccine"—curious phrases to apply to nonliving things. The ambivalence is excusable. On the one hand, viruses lack all metabolic machinery and cannot reproduce in the absence of a host. On the other, they possess nucleic acid genes that encode sufficient information for the production of new viruses with the same characteristics as those that supply the templates; and reproduction with gene-controlled heredity is the most basic attribute of life.

Whether viruses are classed as living, as nonliving, or as something in between, investigation of viruses is intimately bound to the study of procaryotic organisms, especially bacteria. Not only do viruses and procaryotes have similar genetic material, but their life cycles are often intertwined. We shall therefore consider both these groups in the present chapter.

VIRUSES AND VIROIDS

The discovery of viruses   It was clear by the latter part of the nineteenth century that many diseases are caused by microorganisms. Pioneer bacteriologists such as Louis Pasteur and Robert Koch had isolated the pathogens for a number of diseases that afflict humans and domestic animals. But for some diseases, notably smallpox, biologists, try as they might, could find no causal microorganism. In 1796 Edward Jenner had discovered that smallpox could be induced in a healthy person by something in the pus from a smallpox victim, and demonstrated that a person vaccinated with material from cowpox lesions developed an immunity to smallpox. Yet no bacterial agent could be found.

A crucial experiment was performed in 1892 by a Russian biologist, Dmitri Ivanovsky, who was studying a disease of tobacco plants called tobacco mosaic. The leaves of plants with this disease become mottled and wrinkled. If juice is extracted from an infected plant and rubbed on the leaves of a healthy one, the latter soon develops tobacco mosaic disease. If, however, the juice is heated nearly to boiling before it is rubbed on the healthy leaves, no disease develops. Concluding that the disease must be caused by bacteria in the plant juice, Ivanovsky passed juice from an infected tobacco plant through a very fine porcelain filter in order to remove the bacteria; he then rubbed the filtered juice on the leaves of healthy plants. Contrary to his expectation, the plants developed mosaic disease. What could the explanation be?

Ivanovsky suggested two possibilities. Either bacteria in the infected plants secrete toxins, and it is these rather than the bacterial cells themselves that are present in infectious juice, or the bacteria that cause this disease are much smaller than other known bacteria and can pass unharmed through a fine porcelain filter. When it was later demonstrated that the infectious material in filtered juice could reproduce in a new host, Ivanovsky abandoned his first explanation in favor of the second—that some type of extremely small bacterium was the causal agent of the disease. During the next several decades many other diseases of both plants and animals were found to be caused by infectious agents so small that they could pass through porcelain filters and could not be seen with even the best light microscopes. These microbial agents of disease came to be called filterable viruses, or simply viruses (from the Latin word for "poison"). They were still assumed to be very small bacteria.

There were, however, a few hints that viruses might be something quite different from bacteria. First, all attempts to culture them on media customarily used for bacteria failed. Second, the virus material, unlike bacteria, could be precipitated from an alcoholic suspension without losing its infectious power. Finally, in 1935 W. M. Stanley of the Rockefeller Institute isolated and crystallized tobacco mosaic virus, conclusively demonstrating that viruses and bacteria are two very different things. If the crystals were injected into tobacco plants, they again became active, multiplied, and caused disease symptoms in the plants. That viruses could be crystallized showed that they were not cells but must be much simpler chemical entities.

The structure of viruses We saw in Chapter 26 that a free virus particle consists of a protein coat and a nucleic acid core. The protein coat, or *capsid*, may be complicated, as in the bacteriophage T4 (Fig. 38.1A), with its tail and long leglike fibers, or it may be a simple polyhedron or rod (actually a helix of protein molecules) (Fig. 38.1B–C). Many animal viruses, some plant viruses, and a very few bacteriophages have a membranous *envelope* surrounding the capsid; sometimes this envelope is derived from the plasma membrane of the host cell in which the virus was produced, and sometimes it is synthesized in the host cell's cytoplasm, but in either case it usually contains some virus-specific proteins. In addition to the proteins of the capsid and envelope, some viruses—the retroviruses in particular—possess a limited number of enzymes (though never any multienzyme systems).

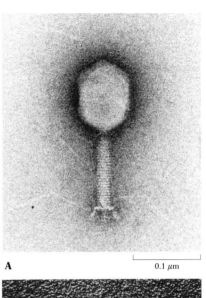

A   0.1 μm

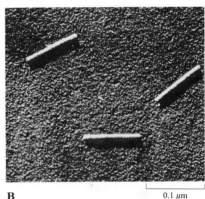

B   0.1 μm

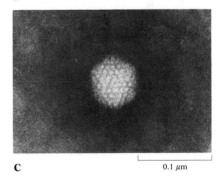

C   0.1 μm

38.1 Electron micrographs of three types of viruses
(A) T4, a complex virus with a "head" containing its DNA, a "tail," and six leglike fibers which aid in attaching it to a host cell (normally *E. coli*).
(B) Tobacco mosaic virus, a rod-shaped virus.
(C) Adenovirus, a large polyhedral virus that multiplies in the upper respiratory tract to produce coldlike symptoms. The adenovirus is actually an icosahedron, having 20 equal sides.

The nucleic acid is usually a single molecule[1] consisting of as few as 3,500 nucleotides[2] or as many as 600,000, depending on the kind of virus. If 1,000 nucleotides is taken as a reasonable estimate of the average length of a gene, then the total number of genes in a virus particle ranges from fewer than five to several hundred.

Viruses show a great diversity in the form of their nucleic acid. Thus many have double-stranded DNA, but some have single-stranded DNA. In either case the DNA may be linear or circular. Many other viruses differ fundamentally from all cellular organisms in having RNA genes. In most instances the RNA is single-stranded, but there are some RNA viruses in which it is double-stranded.

The reproduction of viruses   Because viruses lack multienzyme systems and cannot generate ATP, and because they lack raw materials for synthesis, they cannot reproduce themselves in the same sense as most living organisms. It is the host cell, not the virus, that manufactures new virus particles when the old virus provides the instructions. Hence viruses cannot be cultured on artificial media; they require living host cells. Pharmaceutical companies and research laboratories often grow them in bacteria, in fertilized chicken eggs, or in cells in tissue cultures.

Before examining the diverse reproductive adaptations of viruses, let us briefly review our earlier discussion of the reproduction of bacteriophage viruses (pp. 696, 735). You may recall that a bacterial virus becomes attached by the tip of its tail to the wall of a bacterial cell and that its nucleic acid is injected into the host while the protein coat remains outside (Fig. 38.2). The energy for this injection comes from hydrolysis of about 140 ATP molecules bound (together with 140 Ca^{++} ions) in the tail of the phage. Once inside the bacterial cell, the phage DNA provides genetic information for synthesis of new viral DNA and protein. Among the viral proteins synthesized are not only structural proteins for new viral capsids, but also enzymes that aid in the synthesis and processing of viral components.

Eventually, after the new viral nucleic acid and proteins have been manufactured, they are assembled into new bacteriophage and released by lysis of the bacterial cell through the agency of a phage-induced enzyme that attacks the wall of the bacterial cell. This series of events is known as the lytic cycle.

In Chapter 28 we saw that at times viral DNA does not immediately take control of the host cell's metabolic machinery and put it to work making new virus particles, but is instead integrated into the bacterial chromosome and reproduced with the chromosome for an indefinite number of generations; it is then known as provirus. This so-called lysogenic cycle (see Fig. 28.7, p. 734) occurs when a viral gene induces production of a repressor substance that blocks expression of the viral genes governing reproduction. As long as repression continues, the viral nucleic acid may remain integrated in the host's chromosome. The repressor substance also confers on the host cell immunity to virulent infection by other viruses of

[1] A few viruses like influenza virus have their nucleic acid in several pieces (six to eight for influenza virus).

[2] Some so-called defective viruses have even fewer nucleotides; these viruses can reproduce only when the host cell is already infected with another "helper" virus.

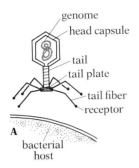

genome
head capsule
tail
tail plate
tail fiber
receptor

A

bacterial
host

B

C

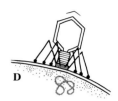

D

38.2 Lytic cycle of a typical bacteriophage

The bacteriophage T4 has receptors at the tips of its "legs" (A). When the receptors encounter the host cell wall, they bind to the bacterium (B) and bring the tail plate into contact with the host (C). This triggers contraction of the tail, and the viral genome is injected into the bacterium (D). The phage DNA is then both replicated and transcribed by host enzymes. The viral mRNA is translated to make coat proteins, regulatory and assembly enzymes, and, at the end of the cycle, lysis enzymes such as lysozyme (E). The head capsule is assembled around the replicated DNA, and the tail is added, followed by the tail plate and fibers (F). Finally, the host is lysed and mature bacteriophage are released (G).

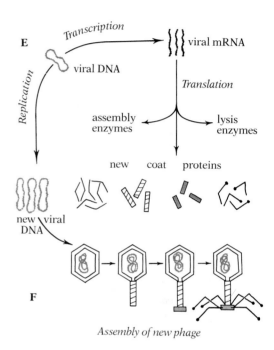

E *Transcription* viral mRNA

Replication viral DNA

Translation

assembly enzymes

lysis enzymes

new coat proteins

new viral DNA

F

Assembly of new phage

the same or similar types. Indeed, a standard test for finding out whether cells are carrying provirus of a given type is to attempt to infect them with the virus in question.[3]

As this discussion makes clear, viruses can exist in three different states: as free infectious particles; as vegetative viruses directing synthesis of new viral components in a host cell; and as proviruses, with their viral nucleic acid integrated with host DNA and replicated in synchrony with it.

Not all viruses inject their nucleic acid into their host cells in the manner of bacteriophages. Often, in fact, the entire virus enters the cell. Animal viruses of this type first become attached to the cell membrane at special glycoprotein adsorption sites for which they have a high affinity; much of the host specificity of such viruses comes from their differing affinities for the various adsorption sites. Most plant viruses depend on insect vectors to inject them through the thick cell walls into the host-cell cytoplasm. Once a virus has entered the cytoplasm of the host cell, its nucleic acid is promptly freed.

If the nucleic acid of a plant or animal virus is DNA, it enters the nucleus of the host cell and serves as a template for the synthesis of both mRNA and new viral DNA.[4] But if the viral nucleic acid is RNA, the virus either brings with it into the host cell special enzymes, RNA replicases, capable of using

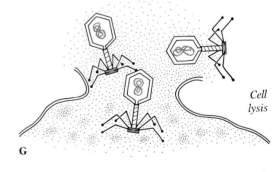

G

Cell lysis

---

[3] A more detailed account of viral repression is presented in the box on the lambda switch (p. 746).

[4] The pox viruses, which have double-stranded DNA as their genetic material, are an exception. Their DNA does not enter the nucleus; instead, it functions entirely in the cytoplasm, and the viruses must bring their own DNA and RNA polymerases into the cells.

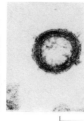

0.1 μm

38.3   Budding in a retrovirus
Particles of Visna virus are shown, budding off cells of infected sheep.

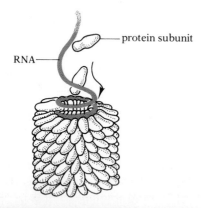

RNA

protein subunit

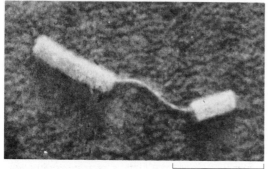

0.01 μm

38.4   Tobacco mosaic virus
Top: Diagram of part of a tobacco mosaic virus, showing its structure. At the center of the cylindrical virus is a long RNA molecule. Protein subunits are packed in a helical array around the coiled RNA; they are added in sequence as a virus is assembled. Bottom: Scanning electron micrograph of the rod-shaped tobacco mosaic virus. Part of the protein coat has been stripped away from the center portion of the virus, exposing the nucleic acid core.

an RNA molecule as a template for synthesis of new RNA; or else part of the viral RNA first functions as mRNA on the host cell's ribosomes, providing instructions for synthesis of RNA replicase by the cell (see Fig. 28.8, p. 736). In either case, then, the secret of replication by RNA viruses in the cytoplasm of their host cells is RNA replicase, which enables sequences of RNA nucleotides to play the role of genes.

Some RNA viruses have another mechanism of replication. A group of RNA viruses called retroviruses do not carry out RNA ⟶ RNA transcription. Instead, their RNA is transcribed into DNA in a reaction catalyzed by the enzyme *reverse transcriptase*, which the viruses bring with them into the host cells. The newly formed DNA then becomes integrated as provirus into the host's chromosomes, where it may remain for an extended period, even being passed to the host's descendants, generation after generation (Fig. 28.8C). Sometimes the provirus remains replicatively silent, producing no new virus, but in other instances the proviral DNA is transcribed to produce both new viral RNA and mRNA coding for viral proteins; when new viruses are assembled, they can be released from the cell by a budding process, which, unlike lysis of host cells by bacteriophage, does not kill the cell (Fig. 38.3). As far as is known, RNA ⟶ DNA transcription is unique to the life cycle of the retroviruses; cells not infected by retrovirus do not contain reverse transcriptase.

Let us now turn to the assembly of new viruses in infected host cells. The traditional view has been that, once new viral components have been synthesized by the host cell, a new virus is produced by self-assembly of these components. This process is exemplified by tobacco mosaic virus (TMV), whose 2,130 identical protein subunits have traditionally been envisioned as spontaneously forming a helix, with a long RNA molecule inside (Fig. 38.4).[5] Even the most complex bacteriophages, with capsids composed of more than 50 kinds of proteins, have been thought to form by self-assembly. But a growing body of evidence now suggests that the capsids of only a few of the simplest viruses (including TMV, but not the DNA phages) are formed entirely through self-assembly. Though some self-assembly seems to be involved, in most cases there is also some enzymatic processing of the proteins, so that new binding sites are sequentially exposed by cleavage of large polypeptides into smaller ones after they have been brought into position in the growing capsid structure.

The new viruses, once assembled, are not always released by lysis of the host cell. Like the retroviruses already mentioned, many viruses are released by extrusion, a process something like budding whereby the virus becomes enveloped in a small piece of cell membrane (Fig. 38.5).

Viroids   Until 1970 viruses were thought to be the simplest self-reproducing entities in nature. In 1971, however, T. O. Diener of the U.S. Department of Agriculture demonstrated that the infectious agent responsible for potato spindle-tuber disease was a small circular RNA altogether lacking a protein coat (see Fig. 37.6, p. 1006). Since then more than a dozen diseases

---

[5] Recent evidence suggests that TMV capsids may actually be built initially as stacks of two-layered discs, which later undergo a conformational change that generates the characteristic helix of the completed capsid.

of higher plants have been shown to be caused by these naked bits of RNA, which Diener named *viroids.* Because of extensive stretches of complementary bases, viroid RNA can pair internally; this self-pairing results in a stable, double-stranded conformation (Fig. 38.6), just as self-pairing helps stabilize the structure of tRNA, rRNA, and the tail of mRNA (see Figs. 27.4, 27.7, and 27.11, pp. 712–21).

The viroid genome is extremely short—even the largest could code only for a small protein of about 100 amino acids—and there is no evidence that any enzyme or other product is generated by viroids. Apparently viroids contain only enough information to allow an RNA polymerase to copy them. But how might viroids be replicated? Apparently, since RNA polymerase is found only in the nuclei of eucaryotes, some mechanism, perhaps involving a signal sequence on the viroid RNA, must effect transport into and out of the nucleus. And then, since RNA polymerase normally recognizes a promoter on the DNA of the nucleus, a viroid sequence must somehow mimic the host's DNA promoter. The extensive double-stranded structure of viroids may protect them from enzymatic degradation and provide the physical structure—the major and minor grooves—that, as we saw in Chapter 29, is necessary in order for regulatory proteins to bind to operator regions, and for polymerases to bind to promoters in the process of transcription.

In short, viroids appear to possess all the necessary information to be reproduced by the complex eucaryotic genetic system, but not enough to

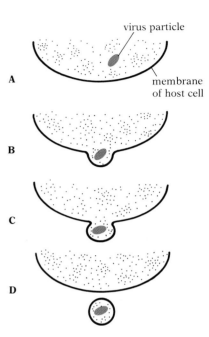

38.5 Extrusion of a virus from the host cell
In some cases a newly assembled virus particle becomes closely associated with the membrane of the host cell (A). The membrane then forms an evagination into which the particle moves (B, C). Finally the evaginated membrane is pinched off as an envelope around the free virus particle (D).

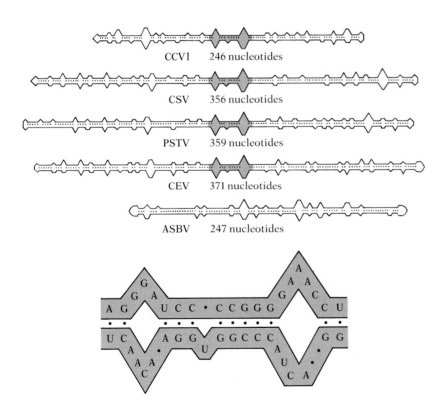

38.6 Structure of viroids
All five of these viroids have a similar self-complementary structure that gives rise to their double-stranded conformation. In addition, a specific sequence in the central regions (shown here in color and enlarged) is common to four of the five, suggesting that it plays an important signaling role. The dots indicate bonds between complementary bases. CCV1 infects coconuts; CSV infects chrysanthemums; PSTV, potatoes; CEV, citrus trees; and ASBV, avocados. This diagram is courtesy of T. O. Diener, *Am. Sci.,* vol. 71, 1983.

produce a protein "message"—rather like an envelope correctly addressed and stamped, but empty. Of the viroids discovered so far, all are known only because, as a side effect of their presence, they cause visible disease symptoms in commercially valuable plants—tumors on potatoes, for example. Viroids may be widespread in nature, but since they do not lyse the cells that act as hosts, their presence may go undetected. Very little is known about how they spread.

Prions Though Ivanovsky abandoned the bacterial-toxin hypothesis of viral action, diseases have since come to light for which something similar —a self-replicating protein—may be near the truth. For instance, the infectious agent responsible for scrapie, a degenerative disease of the central nervous system of sheep and goats, is almost certainly a protein. How such an isolated protein agent—now called a ***prion***—could cause the production of more infectious proteins is not known. Two rare neurological diseases of humans fall into the same pattern. A system in which an infectious protein elicits the production of other infectious proteins might work if the protein modified a preexisting product in the host cell to create new versions of itself, in something like the way the digestive enzyme chymotrypsin catalyzes one step in its own activation. Furthermore, if the normal negative-feedback mechanism that regulates the production of enzymes were not sensitive to the second one in this pathway—the modified infectious version—additional normal proteins would be synthesized to "replace" those converted into the infectious form; these would then be converted in their turn, so that this sort of self-catalyzation could go on indefinitely. Other possible mechanisms have been suggested, but the way prions work remains an intriguing mystery.

The origin of viruses Viruses are not cells; even the most complex viruses, which carry with them an array of enzymes, lack the metabolic machinery for the basic cellular function of energy generation, and they never have the ribosomes required for protein synthesis. Their precise evolutionary origin has been the subject of considerable discussion. Of the various hypotheses of viral origin that have been proposed, the three outlined below have received serious consideration.

The first hypothesis suggests that viruses are organisms that have reached the extreme of evolutionary specialization for parasitism. Its proponents point out that loss of structures is commonly seen in internal parasites, and that intracellular parasites might conceivably lose everything but their nucleus. A virus particle resembles a cellular nucleus; hence they conclude that viruses arose from cellular ancestors by a process of gradual loss of all other cellular components.

The second hypothesis suggests that the ancestors of modern viruses were free-living noncellular predecessors of cellular organisms. When the organic nutrients of the primordial seas disappeared, those of their descendants that remained noncellular survived by becoming parasites on the cellular organisms that had arisen by that time. According to this view, modern viruses are representatives of an early "nearly living" stage in the origin of life.

The third, and most likely, hypothesis suggests that viruses are neither primitive nor specialized organisms, but fragments of genetic material derived from cellular organisms. They might originally have existed as DNA-containing cellular organelles, but that seems unlikely because no organelle bounded by anything resembling a capsid (which is unlike a membrane) is known. Alternatively, they might have begun as bare nucleic acid similar to viroids or episomes, but later evolved the capacity of causing their host cells to synthesize a protein shell in which replicates of the nucleic acid could be enclosed when moving from cell to cell. Some viruses may have arisen as fragments of bacterial DNA, others as fragments of plant nucleic acid, and still others as fragments of the genetic material of higher animals. The host specificity of viruses would then normally be a reflection of their origin; a given type of virus might be able to parasitize only species fairly closely related to the one from which the virus was originally derived. Given the likelihood of this third hypothesis, any attempt at a taxonomic classification of viruses based on likely evolutionary relationships is pointless, since most viruses would have evolved independently of each other as "offspring" of their hosts.

Viral disease Among the many human diseases caused by viruses are chicken pox, mumps, measles, smallpox, yellow fever, rabies, influenza, viral pneumonia, the common cold, poliomyelitis (infantile paralysis), fever blisters, several types of encephalitis, infectious hepatitis, and AIDS (acquired immune deficiency syndrome). We have already discussed the association of viruses with some kinds of cancers (pp. 760–61).

Unfortunately most viral infections do not respond to treatment with the sulfa drugs or antibiotics that have been so effective against bacterial diseases. But some of the most pernicious virus diseases, notably smallpox and polio, are preventable by means of vaccines (see p. 1042); indeed, smallpox has probably been eradicated as a result of intensive worldwide vaccination programs.

Immune responses, though important in preventing reinfection, ordinarily appear too late to account for recovery from viral diseases. The evidence suggests that at least one important factor in recovery is a protein called **interferon**, which is produced by host cells in response to invading viruses. The interferon cannot save those cells, but when it is released into the medium and encounters uninfected cells, it interacts with a receptor site on their membranes and may confer on them a resistance to viral infection, probably by inducing production of an enzyme that blocks translation of the mRNA transcribed from the viral nucleic acid. In other words, the interferon acts as a messenger from infected cells to uninfected cells, telling them to mobilize their defenses against viral infection.

Shortly after the discovery of interferon in 1957, many researchers thought it might provide an effective chemotherapy by speeding up the response of the immune system to viral diseases, but this hope has not yet been realized; to date there have been almost equal numbers of successes, failures, and undesirable side effects from its use. Much remains to be learned about the natural role and molecular biology of interferon.

MONERA

The kingdom Monera includes two sections: Archaebacteria, or "ancient" bacteria, and Eubacteria, the much more common "true" bacteria (Fig.

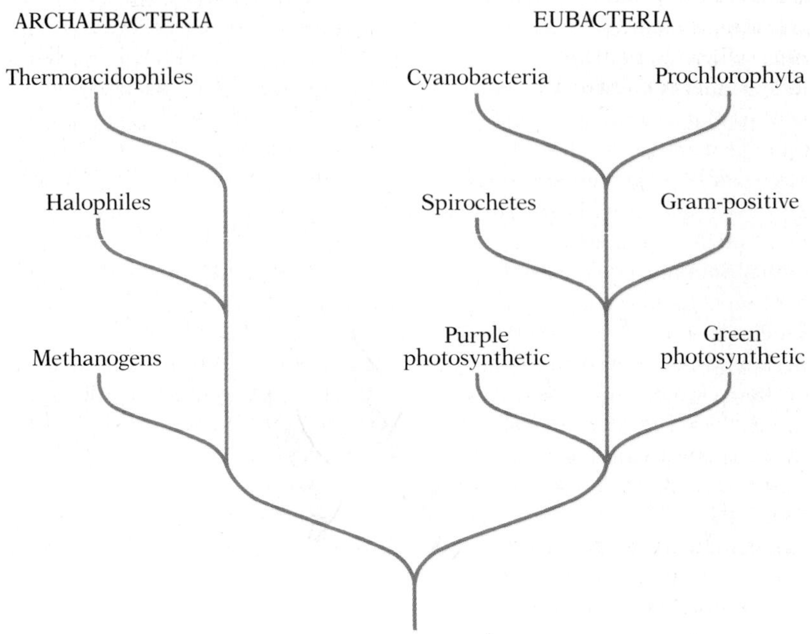

ARCHAEBACTERIA   EUBACTERIA

Thermoacidophiles   Cyanobacteria   Prochlorophyta

Halophiles   Spirochetes   Gram-positive

Methanogens   Purple   Green
                               photosynthetic   photosynthetic

38.7 Probable phylogenetic relationships of modern bacteria

38.7). The cells of both groups are procaryotic—that is, they lack a nuclear membrane, mitochondria, endoplasmic reticulum, Golgi apparatus, and lysosomes. Monerans divide by fission rather than by mitosis or meiosis; the terms "Monera" and "bacteria" are interchangeable. Some of the distinctive characteristics of the various divisions of the two sections of Monera will become clear as we discuss them. Because the Eubacteria are by far the more numerous and well studied, we shall pay particular attention to their characteristics.

ARCHAEBACTERIA

Archaebacteria differ from Eubacteria in a number of ways: their cell walls have a markedly different chemical composition; the lipids in their cell membranes are branched rather than straight as in Eubacteria; their translation system differs in several small but consistent ways.

The Archaebacteria typically occupy very challenging habitats; conditions in some of them are thought to be similar to the conditions under which life evolved on the early earth. The methanogens, the group with the least extreme niche, are anaerobic chemosynthesizers, living in bogs and other habitats rich in decaying vegetation, and produce methane ("marsh gas") as a by-product of their metabolism. The second of the three groups is halophilic ("salt-loving"), and requires highly concentrated salt water like that found in the (otherwise) Dead Sea; because of the high free energy of their surroundings, these photosynthetic bacteria can maintain an enormous electrochemical gradient across their membrane and use it to help make ATP. The third group is thermoacidophilic ("heat- and acid-loving"),

38.8 Scanning electron micrograph of thermoacidophilic Archaebacteria
This dense mat is formed by chemosynthetic bacteria growing on a mussel shell near an underwater volcanic vent in the Pacific Ocean.

10 μm

with one genus that lives in hot sulfur springs at 80°C and pH 2, and another that thrives near deep-sea volcanic vents (Fig. 38.8).

EUBACTERIA

The section Eubacteria includes at least six divisions: the purple bacteria; the green bacteria; two divisions of heterotrophic bacteria, the so-called Gram-positive bacteria and the spirochetes; and two divisions of aerobic photosynthetic bacteria, the Cyanobacteria and the Prochlorophyta (Fig. 38.7). There are almost certainly other divisions.

The first two divisions, the purple and green bacteria, probably evolved as chlorophyll-based photosynthetic anaerobes, which, however, lacked chlorophyll *a*. Some members of these divisions evidently lost their photosynthetic pigments and gave rise to aerobic heterotrophs, of which *E. coli*, a bacterium closely related to the purple photosynthetic species, is a notable modern example. Proponents of the endosymbiotic hypothesis believe that one species of heterotrophic purple bacterium became symbiotically associated with a primitive cell—the urcaryote—and was thus the source of the mitochondria of modern eucaryotes. Members of the third division, the Gram-positive bacteria, are grouped together because a chemical in their cell walls reacts with Gram's solution, identifying them as a chemically distinct group. The bacteria of the other division of heterotrophs, the spirochetes, are Gram-negative. All members of the fifth and sixth divisions, the Cyanobacteria and the Prochlorophyta, have chlorophyll *a*.

BACTERIAL ANATOMY

Most bacteria, whether Archaebacteria or Eubacteria, are very tiny, far smaller than the individual cells in the body of a multicellular plant or ani-

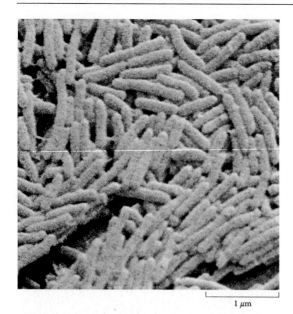

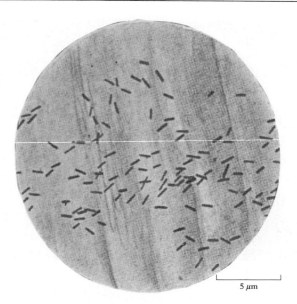

38.9 *Escherichia coli,* **a rod-shaped eubacterium**
Left: Electron micrograph of a colony. Right: Stained cells as they appear under a high-power light microscope.

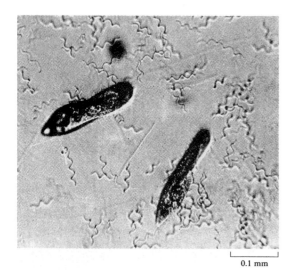

38.10 *Spirillum volutans,* **a helically coiled eubacterium**
The two paramecians (which are Protozoa) are included for size comparison.

mal. In fact, some Eubacteria, like the Rickettsiae and the Chlamydiae, are as small as some of the largest viruses. However, even the smallest Monera are fundamentally different from viruses in that they are cellular. They always contain both RNA and DNA, whereas viruses contain one or the other but never both; they have ribosomes, whereas viruses do not; they always possess integrated multienzyme systems, whereas viruses carry, at most, only an assortment of individual enzymes; they can generate ATP and use it in the synthesis of many other organic compounds, whereas viruses cannot; they provide both the raw material and all the metabolic machinery for their own reproduction, whereas viruses do not.

Cell shape   The cells of most Monera have one of three fundamental shapes: spherical or ovoid, cylindrical or rod-shaped (Fig. 38.9), or helically coiled (Fig. 38.10). Spherical Eubacteria are called *cocci* (singular, coccus); rod-shaped ones are called *bacilli*; and helically coiled ones are called *spirilla.*

When cell division takes place, the daughter cells of some species remain attached and form characteristic aggregates. Thus cells of the eubacterium that causes pneumonia are often found in pairs (diplococci). The cells of some spherical species form long chainlike aggregates (streptococci) (Fig. 38.11), while others form grapelike clusters (staphylococci) (Fig. 38.12). Each of the cells in a diplococcal, streptococcal, or staphylococcal aggregate is an independent organism, but some Monera, notably some photosynthetic Cyanobacteria, form either coenocytic or multicellular filaments. A basic difference between a chain of independent cells and a multicellular filament is that adjacent cells of a filament have a common cell wall. Many Actinomyces, among other filamentous forms, strikingly resemble molds (which are fungi), but this is a case of evolutionary convergence and not an indication of close relationship: the cells of Actinomyces, like those of other Monera, are procaryotic whereas those of fungi are eucaryotic (Fig. 38.13).

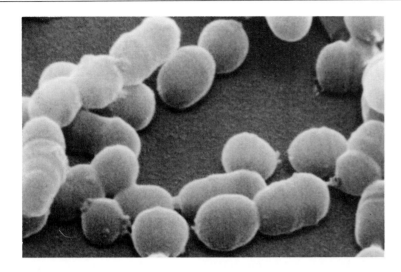

38.11 *Streptococcus salivarius*, a bacterium common in the human mouth
All streptococci are spherical (coccal) bacteria; they are normally grouped in chainlike clusters.

20 μm

38.12 *Staphylococcus aureus*, a bacterium common in the human nasal cavity
The round cells occur in irregularly shaped clusters.

Cell walls Like plant cells, most bacterial cells are enclosed in a cell wall, which protects the cell both from physical damage and from osmotic disruption. But the walls of bacteria differ from those of plants in important aspects of their composition. The walls of eucaryotic cells derive their tensile strength largely from cellulose and related compounds (or from chitin in fungi), whereas those of Eubacteria derive theirs from murein, a huge polymer composed of polysaccharide chains covalently cross-linked by short chains of amino acids; muramic acid, one of the principal polysaccharide constituents of murein, never occurs in the walls of eucaryotic cells. The walls of Archaebacteria are unlike those of either Eubacteria or eucaryotes. These important differences between the chemical composition of bacterial and eucaryotic cells are the basis for the selective activity of some drugs, such as penicillin. Nontoxic to plants and animals (or to resting bacterial cells), penicillin is toxic to many growing Eubacteria, because it inhibits formation of murein, and thus interferes with most bacterial multiplication.

Most Monera—halophiles being an obvious exception—are hypertonic relative to the fluid medium in which they live; hence they would swell and burst if they did not have well-developed walls. However, many species of Eubacteria can be cultured in the laboratory on a medium with an elevated osmotic pressure. When treated with penicillin under such conditions, these bacteria survive, growing as so-called L forms, which have very poorly developed walls. Their shape is usually very different from normal, as would be expected.

The L forms are human artifacts, but there are some naturally occurring Eubacteria that lack walls and can live only as parasites of plants and animals—that is, in environments where their osmotic relations are such that the absence of a mechanically strong cell wall is not lethal. These are the mycoplasmas,[6] which are the smallest known cellular organisms. Not only

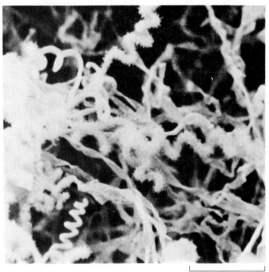

10 μm

38.13 Scanning electron micrograph of a colony of *Streptomyces*, one of the Actinomyces
The Actinomyces have a much more complicated morphology than most other Monera.

---

[6] The mycoplasmas were formerly known as pleuropneumonialike organisms (PPLO).

38.14 Electron micrograph of a sporulating bacillus
The spore is the dark oval at the right side of the cell.
The developing spore coat is clearly visible. The
white areas at the left side of the cell are not
vacuoles, but areas filled in with fatty material.

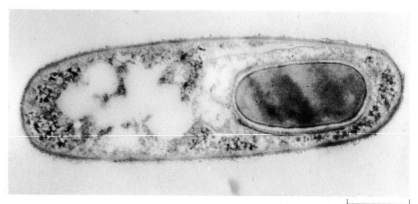

0.5 μm

are the cells at the lower limit of resolution of the light microscope, but they
also have less than half as much DNA as most other Monera. The informa-
tional content of the DNA is probably near the lower limit of what is necessary to code for the essential metabolic machinery of a cell. Some of the
mycoplasmas cause severe human diseases.

Differences in the relative amounts of certain components in the walls of
different types of bacteria make the cells show characteristic reactions to a
variety of stains. Since there are few visible morphological characters that
can be used in identifying Monera, diagnostic staining is an important laboratory tool.

Many eubacterial cells secrete polysaccharide mucoid materials that accumulate on the outer surface of the cell wall and form a *capsule*. The capsule makes the cell more resistant to the defenses of host organisms; hence
encapsulated strains of a given bacterial species are more likely to cause
disease than unencapsulated strains.

Some Eubacteria (mostly rod-shaped ones) can form special resting cells
called *endospores*, which enable them to withstand conditions that would
quickly kill the normal active cell. Each small endospore develops inside a
vegetative cell and contains DNA plus a limited amount of other essential
materials from that cell (Fig. 38.14). It is enclosed in an almost indestructi-
ble spore coat. Once the endospore has fully developed, the remainder of
the vegetative cell in which it formed may disintegrate. Because of their
very low water content and refractile coats, spores of many species can
survive an hour or more of boiling or an hour in a hot oven. They can be
frozen for decades or perhaps for centuries without harm. They can survive
long periods of drying. And they can even withstand treatment with strong
disinfectant solutions. When conditions again become favorable, the
spores may germinate, giving rise to normal vegetative cells that resume
growing and dividing. Fortunately, few disease-causing bacteria can form
endospores.

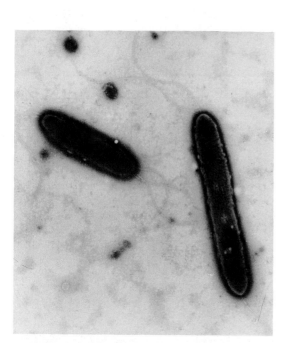

**38.15 Electron micrograph of Eubacteria,
showing their long helical flagella**
Though bacterial flagella are organelles of
locomotion, most differ from the flagella of
eucaryotic cells in lacking internal microtubules and,
therefore, in their mechanism of movement.

BACTERIAL CELL MOVEMENT

Many Monera are motile, and so can move about actively. In most cases the
motion is produced by the rotation of flagella (Fig. 38.15). Virtually all bac-

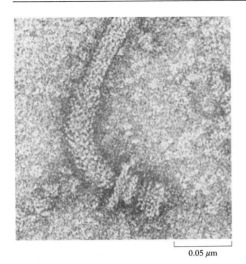

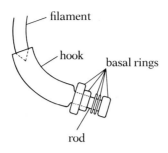

filament

hook

basal rings

rod

0.05 μm

38.16 The basal portion of a typical eubacterial flagellum

The electron micrograph shows the base of a flagellum from *Caulobacter crescentus*. The interpretive drawing elucidates the complex system of rings, rod, hook, and filament. Unlike the flagella of eucaryotes, which beat back and forth, nearly all eubacterial flagella rotate. Their complicated structure provides the necessary anchor, joints, and couplings for such rotational motion.

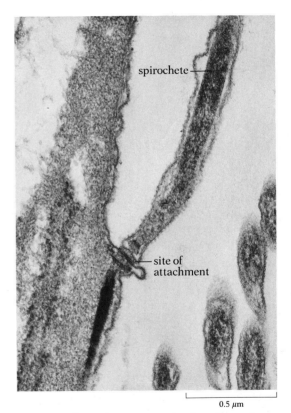

spirochete

site of attachment

0.5 μm

38.17 Symbiosis between a spirochete bacterium and a protistan

Flagella of spirochetes appear to have microtubules and to move in a manner similar to eucaryotic flagella and cilia—of which they may have been the evolutionary precursor. Here, a spirochete specialized for attachment to eucaryotic cells is bound to a nonmotile protistan which inhabits the gut of a termite; as a result of this attachment, the protistan is able to move.

terial flagella are structurally very different from the flagella of eucaryotic cells. They are not enclosed within the cytoplasmic membrane; they do not contain the nine peripheral and two central microtubules found in all flagella of eucaryotic cells; and they do not contain the protein tubulin. They arise from basal granules, but these granules are much smaller than the basal bodies of eucaryotic cells and are probably not homologous with them. A bacterial flagellum has approximately the same diameter as one of the tubules from a eucaryotic flagellum. Its motion is not the whiplike beating characteristic of eucaryotic flagella, but rather a rotary motion made possible by an amazingly complex attachment, which is unlike anything seen elsewhere in the living world (Fig. 38.16). The energy for rotation is supplied by hydrogen ions moving down an electrochemical gradient through special channels at the base of each flagellum; this is analogous to the gradient in both chloroplasts and mitochondria in which the movement of hydrogen ions provides energy for the synthesis of ATP from ADP and inorganic phosphate. Whether the similarity runs deeper is not yet known.

An exception to the characteristic bacterial flagella is found in certain species of spirochetes; in these bacteria, flagella appear to contain microtubules and beat rather than rotate. In fact, there is reason to believe that a spirochete or spirochete ancestor gave rise to the eucaryotic flagellum (Fig. 38.17).

Some Eubacteria (the myxobacteria) that lack flagella exhibit a peculiar gliding movement that does not involve any visible locomotor organelles; the mechanism of this movement has not yet been discovered.

BACTERIAL REPRODUCTION

As we have already seen, Monera lack membrane-bounded nuclei, but electron microscopy reveals the presence of a nuclear area, called a nucleoid, in which genes are arranged in sequence along a single circular chromosome composed of DNA (see p. 617). The single chromosome is

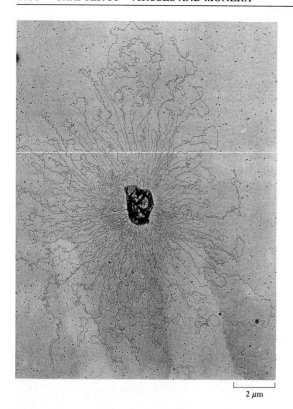

2 µm

38.18   Electron micrograph of the DNA of a bacterial cell (*E. coli*) spread onto a surface
All the many coils of DNA seen here are parts of a single chromosome. It is hard to imagine how so very much DNA could have been packed into so small a cell.

surprisingly long, considering that it is packed into so tiny a cell (Fig. 38.18).

Most bacteria reproduce by **binary fission**, a type of cell division in which two equal daughter cells with characteristics essentially like those of the parent cell are produced without mitosis (see Fig. 23.5, p. 617). When a bacterial cell undergoes fission, each daughter cell receives a complete chromosome with a full set of genes. The details of nuclear division in procaryotic cells have been discovered only recently (see Fig. 23.4, p. 617). The DNA, which is attached to the cell membrane, is replicated, and the two chromosomes move apart into separate nuclear areas long before division of the cytoplasm occurs. When the plasma membrane grows inward, it simply partitions the binucleate parent cell into two daughter cells, each of which already has its own nucleoid. In many Monera, a much-convoluted inward extension of the plasma membrane, called a mesosome, is seen at or near the site of division. Its functional role in the division process is not yet well understood.

Monera commonly have an enormous reproductive potential. Many species may divide as often as once every 20 minutes under favorable conditions. If all the descendants of a cell of this type survived and divided every 20 minutes, the single initial cell would have about 500,000 descendants at the end of 6 hours, and by the end of 24 hours the total weight of its descendants would be nearly 2,000,000 kg. Though increases of this magnitude do not actually occur, the real increases are frequently huge, which helps explain the rapidity with which food sometimes spoils or a disease develops.

Though the reproductive process itself is asexual, genetic recombination does occur occasionally, at least in some bacteria. As we saw in Chapter 28, three mechanisms of recombination are known: conjugation, in which part of a chromosome is transferred from a donor cell to a recipient; transformation, in which a living cell picks up fragments of DNA released into the medium from dead cells; and transduction, in which fragments of DNA are carried from one cell to another by viruses. When a normally haploid bacterial cell receives extra DNA by one of these procedures, it becomes partly diploid (usually only partly, because it is very rare for a cell to receive an entire extra chromosome). The diploidy is only temporary, however; haploidy is soon reestablished by elimination of all genes that do not become incorporated into the new recombinant chromosome.

BACTERIAL NUTRITION

Most Monera are heterotrophic, being either saprophytes or parasites. Like animals, the majority of these bacteria are aerobic—that is, they cannot live without molecular oxygen, which they use in the respiratory breakdown of carbohydrates and other food materials to carbon dioxide and water. Aerobic respiration is carried out with the help of an electron-transport system, built (since bacteria lack mitochondria) into the inner surface of the cell membrane or its invaginations. As we have said, aerobic purple bacteria, after secondary loss of cyclic photophosphorylation, may have been the source of eucaryotic mitochondria.

To some Monera that obtain all their energy by fermentation, however, oxygen is lethal. Such bacteria are called **obligate anaerobes**. *Clostridium botulinum*, the causal agent of the most dangerous kind of food poisoning,

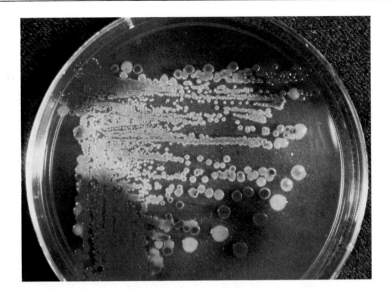

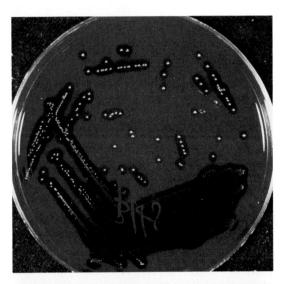

38.19 Colonies of two kinds of Eubacteria growing on the same culture medium, with different effects on the medium
Colonies of *E. coli* on Endo agar carry out fermentation, making the medium red, whereas colonies of *Salmonella* do not carry out fermentation and the medium is colorless.

botulism, is an example of an obligate anaerobe; it grows well in tightly closed food containers that were not properly sterilized before being filled.

Other Monera, called *facultative anaerobes*, can live either in the presence or in the absence of molecular oxygen. Some of the facultative anaerobes are simply indifferent to O_2, obtaining all their energy from fermentation whether O_2 is present or not; others obtain their energy by fermentation when O_2 is not available, but carry out respiration (via the Krebs cycle and the electron-transport chain) when O_2 is present, and may grow faster under these conditions. Respiration is, of course, a much more efficient energy-yielding process than fermentation. Certain types of facultative anaerobes do not rely on fermentation when O_2 is absent; they continue to carry out complete respiration by using substitute inorganic substances (such as nitrates, sulfates, or carbonates) as ultimate electron acceptors in place of O_2.

Besides lactic and alcoholic fermentations—the most common fermentative processes in living organisms—at least ten other types of fermentation occur in different groups of Monera. The products include acetic acid, butylene glycol, butyric acid, and propionic acid. All fermentations are alike, however, in that they are energy-yielding biological oxidation-reduction sequences in which organic molecules serve as the final electron acceptors. ATP is generated in fermentations by phosphorylation.

Monera differ considerably in the sorts of molecules they can use as energy sources and in the specific amino acids and vitamins they require. These differences provide valuable diagnostic characters for workers attempting to identify unknown bacteria. Samples of the organisms to be identified are placed on a variety of nutrient media and cultured at standard temperatures. By determining on which of the media the organisms will grow and on which they will not, and, when they grow, by comparing the color, texture, and other characteristics of the colony they produce with data established for known species, it is often possible to assign the unknown organisms to the proper group or even to the proper species (Figs. 38.19 and 38.20).

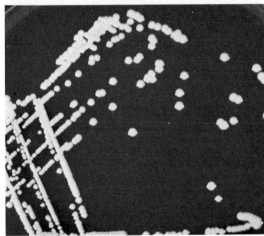

38.20 Different appearances of colonies of two kinds of Eubacteria growing on blood agar
Top: Colonies of *Streptococcus* are surrounded by black zones of hemolysis (breakdown of red blood corpuscles). Bottom: Colonies of *Sarcina lutea* have no surrounding zones of hemolysis.

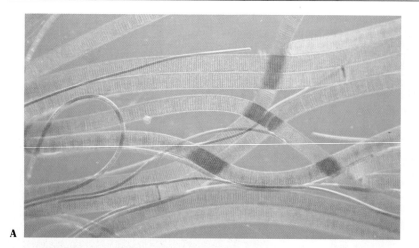

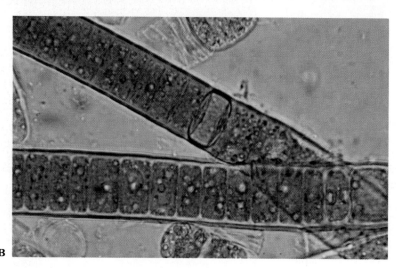

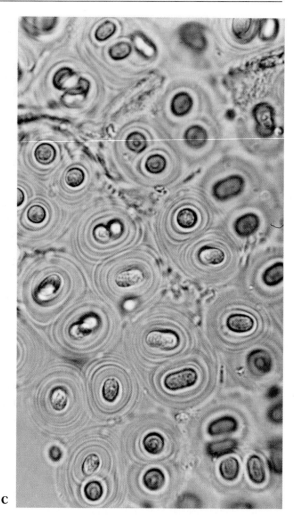

38.21 Some representative Cyanobacteria
(A) *Oscillatoria,* a filamentous bacterium without a sheath. Members of this genus exhibit an odd oscillatory movement. (B) *Scytonema,* another filamentous bacterium, has oval, nearly rectangular cells. The lighter colored cell in the diagonal filament is a heterocyst, a cell specialized for carrying out nitrogen fixation. (C) *Gloeothece,* a unicellular bacterium in which groups of nearly spherical cells are enclosed in layers of gelatinous material.

BACTERIAL PHOTOSYNTHESIS

Most Monera, as we said, are heterotrophs, but some are either chemosynthetic or photosynthetic autotrophs. The chemosynthetic bacteria oxidize inorganic compounds, such as ammonia, nitrite, sulfur, hydrogen gas, or ferrous iron, and trap the released energy. The mechanisms of some of these oxidations were discussed on pp. 168–71. The Monera that oxidize ammonia or nitrite are the nitrifying bacteria, important in the nitrogen cycle.

There are several kinds of photosynthetic bacteria, which differ in the type of bacteriochlorophyll they possess. The green bacteria and the purple bacteria—both Eubacteria—are similar in that neither group has chlorophyll *a,* the chief light-trapping pigment in higher plants. And unlike higher plants, neither group ever uses water as the ultimate electron donor in photosynthesis; hence neither produces molecular oxygen. Indeed, photosynthesis for the green and the purple bacteria is a strictly anaerobic process; it cannot take place in the presence of O_2. Depending on the species of bacteria, the electron (and hydrogen) donor for reduction of NADP is molecular hydrogen, reduced sulfur compounds (such as H_2S), or organic compounds, but the oxidation of these substances is not a light-driven process. In fact, these photosynthetic bacteria possess only Photosystem I (and carry out only cyclic photophosphorylation); they have no Photosystem II.

The pigments and enzymes of the light-trapping process in anaerobic photosynthetic bacteria are located in chromatophores, organelles composed of vesicles or paired membranous lamellae, which differ in structural details between the purple and the green bacteria. The chromatophores are not contained within any membrane-bounded structure that could be interpreted as a chloroplast. Hence the enzymes involved in the "dark" reactions of photosynthesis—carbon fixation via the Calvin cycle—which are in the stroma portion of plant chloroplasts, are assumed to be in the general cytoplasm of the bacterial cell.

The halophilic Archaebacteria, which live in salt lakes and brines, carry out a unique type of photosynthesis that does not use chlorophyll. When living under anaerobic conditions, they synthesize a carotenoid pigment very similar to the rhodopsin of the vertebrate retina. This pigment, located in the plasma membrane, to which it imparts a purplish color, is bleached by light in a process that releases protons from the cell; the proton gradient thus established across the plasma membrane provides energy for ATP synthesis by the chemiosmotic process (pp. 185–87).

The major group of aerobic photosynthetic bacteria are the Cyanobacteria (Fig. 38.21). All Cyanobacteria possess photosynthetic pigments located in flattened membranous vesicles called thylakoids (Fig. 38.22). These structures are similar to the chromatophores of the photosynthetic green and purple bacteria and, like the chromatophores, are not contained within chloroplasts.

The chlorophyll of the Cyanobacteria is chlorophyll *a*, the pigment also found in higher plants, rather than the chlorophyll of the anaerobic photosynthetic bacteria; and like the higher plants, Cyanobacteria generate molecular oxygen as a by-product of their photosynthesis. As we saw in Chapter 37, it was probably the Cyanobacteria that contributed most to the creation of an oxidizing atmosphere some 2.3 billion years ago. According to the endosymbiotic hypothesis, the Cyanobacteria may have been the source of eucaryotic chloroplasts.

In addition to chlorophyll and various carotenoids, these organisms contain **phycocyanin** (a blue pigment) or sometimes **phycoerythrin** (a red pigment). It is the presence of phycocyanin with the chlorophyll that gives these bacteria their characteristic blue-green color. However, not all "blue-green algae" are blue-green; black, brown, yellow, red, grass green, and other colors also occur. The periodic redness of the Red Sea is caused by a species that contains a particularly large amount of phycoerythrin.

The discovery of a new division (phylum) of photosynthetic monerans was announced in 1976 by R. A. Lewin of the Scripps Institution of Oceanography. These bright green procaryotic organisms were first thought to be Cyanobacteria, but analysis of their pigment system revealed them to be quite different. They have both chlorophyll *a* and *b*, whereas Cyanobacteria have only chlorophyll *a*. And their accessory pigments are limited to the

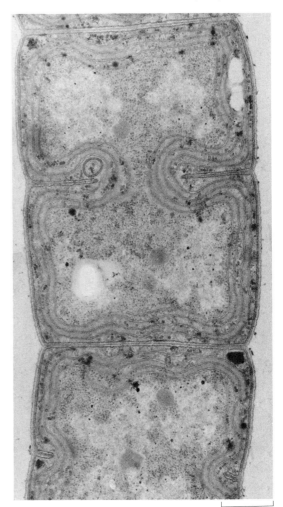

0.5 μm

38.22 A cyanobacterium, *Plectonema boryanum*, undergoing cell division
The photosynthetic membranes (thylakoids) lie near the periphery of the cell. Walls have begun to grow inward at a point midway along the cell's length.

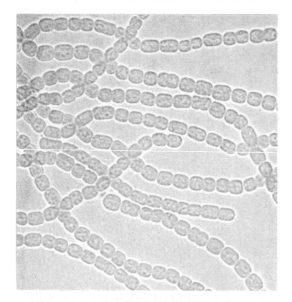

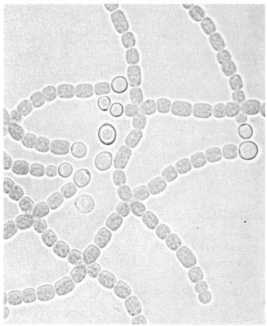

38.23   The effect of environmental nitrogen on heterocyst formation by a cyanobacterium, *Anabaena*

Top: When grown in a culture containing sodium nitrate, the filaments form no heterocysts. Bottom: On a medium containing N_2 as the only nitrogen source, the filaments form numerous heterocysts.

standard carotenoids found in higher plants, with no trace of the red or blue pigments so characteristic of the Cyanobacteria. In short, these organisms, designated Prochlorophyta by Lewin, have a pigment system like that of the green algae and the higher land plants, and are thought by some to be a more likely ancestor of eucaryotic chloroplasts than Cyanobacteria.

NITROGEN FIXATION IN BACTERIA

One particularly important property of many Cyanobacteria is their ability to fix atmospheric nitrogen (N_2). This process depends on an enzyme, nitrogenase, that functions only under anaerobic conditions. Indeed, O_2 does more than inhibit the activity of nitrogenase: it poisons the enzyme irreversibly. But Cyanobacteria are oxygen-generating photosynthesizers; how can they fix nitrogen as well?

The species of Cyanobacteria most active in N_2 fixation are, with few exceptions, filamentous forms that produce a few highly specialized cells called *heterocysts*, where the N_2 fixation takes place (Fig. 38.21B). A heterocyst is a cell with exceptionally thick walls whose nucleus has disappeared. Because O_2, it seems, is largely excluded from the interior of the heterocyst, nitrogenase can function there. An interesting adaptation for maintaining anaerobic conditions inside the heterocyst is its loss of Photosystem II, the oxygen-generating component of the photosynthetic apparatus. Equipped only with Photosystem I, the heterocyst can generate ATP in the light, but cannot manufacture carbohydrate and must therefore depend on neighboring cells to supply it with this essential material, presumably via plasmodesmata. It is presumably also via plasmodesmata that the heterocyst exports fixed nitrogen to the other cells of the filament. Heterocysts are rarely present when the bacteria are growing in an environment where a supply of already fixed nitrogen like sodium nitrate is ample, but they form quickly if the fixed nitrogen becomes limiting (Fig. 38.23).

Nitrogen fixation by Cyanobacteria proceeds best when the heterocysts are in an atmosphere with less than 10 percent free O_2. At higher concentrations, some O_2 can leak into the heterocysts and begin to inhibit the nitrogenase. It is interesting that photosynthesis, too, is inhibited in Cyanobacteria when the O_2 concentration is above 10 percent. It has been suggested that the metabolism of Cyanobacteria functions best with low O_2 concentrations because these organisms are relicts of nearly two billion years of evolution under conditions of an oxygen-poor earthly atmosphere in pre-Silurian times. What lends credence to this idea is that even many species that do not form heterocysts and do not normally produce nitrogenase apparently have a gene that codes for nitrogenase; these species can synthesize the enzyme and fix N_2 when conditions are strictly anaerobic.

BACTERIA AS AGENTS OF DISEASE

Perhaps the bacteria best known to most people are the ones that cause diseases in human beings, domesticated animals, and cultivated plants. Most so-called germs are either Eubacteria or viruses (though a few are fungi, protozoans, or parasitic worms). The idea that bacteria can cause disease —often called the Germ Theory of Disease—was first developed by Louis

Pasteur in the late nineteenth century. Initially scorned, the idea soon gained the support of such prominent scientists and physicians of the day as Joseph Lister, Robert Koch, Thomas Burrell, and Ferdinand Cohn. Lister, an English surgeon, was one of the first to realize the implications of Pasteur's discoveries for surgical procedures. He initiated use of antiseptic techniques in the operating room, using carbolic acid solution as a disinfectant. Koch, a German physician, showed that a bacillus was the cause of anthrax in horses, cows, sheep, and human beings, and he later demonstrated that another bacillus caused tuberculosis in humans. Burrell, an American botanist at the University of Illinois, showed that a plant disease —fire blight of pears—was caused by bacteria. Cohn, a German botanist who published a classic text on bacteria, is often considered the father of modern bacteriology.

In the course of his investigations of anthrax and tuberculosis, Robert Koch formulated the rules of procedure for proving that a particular microorganism is the cause of a particular disease. A slightly modified version of these rules, traditionally called **Koch's postulates**, is still used today:

1. It must be shown that the microorganism in question is always present in diseased hosts.
2. The microorganism must be isolated from the diseased host and grown in pure culture—that is, in a culture containing only that one species of microorganism.
3. Microorganisms obtained from the pure culture, when injected into a healthy susceptible host, must produce the disease in that host.
4. Microorganisms must be isolated from the experimentally infected host, grown in pure culture, and compared with the microorganisms in the original culture.

The procedures outlined by Koch have been followed by hundreds of bacteriologists, and Eubacteria have been shown to cause a long list of human diseases, including bubonic plague, cholera, diphtheria, syphilis, gonorrhea (Fig. 38.24), leprosy, scarlet fever, tetanus, tuberculosis, typhoid fever, whooping cough, bacterial pneumonia, bacterial dysentery, meningitis, strep throat, boils, and abscesses. To this list we can now add such diseases as typhus fever and Rocky Mountain spotted fever, which are caused by rickettsias and are spread to humans by the bites of arthropods like ticks, mites, and fleas, and also such diseases as ornithosis, lymphogranuloma, and trachoma, which are caused by the chlamydias. Both the rickettsias and the chlamydias are obligate intracellular parasites once thought to be intermediate between viruses and Monera, but now known to be simply very small, specialized Eubacteria. Equally long lists could be compiled of bacterial diseases of other animals or of plants.

Microorganisms cause disease symptoms in a variety of ways. In some cases their immense numbers simply place such a tremendous material burden on the host's tissues that they interfere with normal function. In other cases the microorganisms actually destroy cells and tissues. In still other cases bacteria produce poisons, called **toxins**. These may be exotoxins, which are poisons released from the living bacterial cell into the host's tissues, as in diphtheria or tetanus, or they may be endotoxins, which

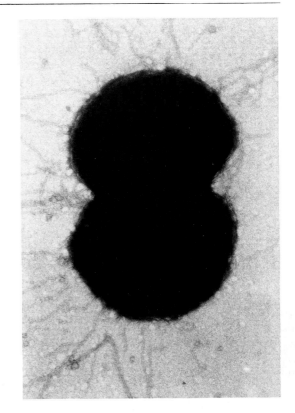

38.24   The gonorrhea bacterium
A diplococcal pair of the gonorrhea bacterium, *Neisseria gonorrhoeae,* showing many pili, which appear to play a role in the attachment of the bacterium to human mucosal cells. This bacterium infects at least one million people in the United States annually. Unchecked infections can cause blindness in newborns and sterility in adults.

are poisons retained in the cells of the bacteria that produce them and are only released into the host when the bacteria die and disintegrate. There are even cases, as we mentioned in Chapter 30, in which the disease symptoms are not caused directly by microorganisms, but result from an excessive immune response by the host's body.

We discussed in Chapter 30 the ways in which the human body resists attacks by pathogenic microorganisms. The first line of defense, once the pathogens have gotten into the body past the protective epidermal tissues, is phagocytic action by certain kinds of white blood cells; the pathogens are engulfed and destroyed. The second line of defense is production of antibodies that react with the antigens of the pathogens. Production of antibodies the first time a host individual is exposed to a particular antigen is a rather slow process, which may take days or even weeks. However, in most cases the immunity thus built up is relatively long-lasting. For example, a person who has once had whooping cough or chicken pox usually remains immune for life to further infection by the pathogens of those diseases. Immunity to some diseases, among them most respiratory infections, does not last for life, but may have a duration of several months.

Modern medicine often takes advantage of the body's antigen-antibody reaction to induce prophylactic immunity—immunity that prevents a first case of disease. The patient is inoculated with either a vaccine or an antiserum. A *vaccine* is material containing antigen from the pathogen. Sometimes the antigen consists of dead microorganisms; sometimes it consists of attenuated microorganisms—that is, microorganisms that are alive, but have been treated in a manner that weakens them sufficiently to prevent their causing disease; and sometimes it consists either of a small amount of active bacterial toxin (enough to induce formation of antibodies, but not enough to produce disease) or of inactivated toxin, called toxoid, as in tetanus toxoid. Vaccines, whatever the kind, induce active immunity in the patient, stimulating the patient to produce his own antibodies. They are therefore rather slow-acting, but their effects are long-lasting. By contrast, inoculation with *antiserum* produces almost immediate immunity, but the immunity lasts only a short time, because an antiserum contains presynthesized antibodies instead of antigens, and hence produces passive rather than active immunity. An antiserum is made by injecting antigen into some other animal, usually a horse, waiting until the animal has produced antibodies specific for that antigen, and then removing blood serum containing the antibodies from the animal.

The immunity that a newborn baby acquires from its mother is an example of a natural passive immunity. The milk produced in the first few days after birth consists of *colostrum*, a thin yellowish fluid that is rich in protein and contains numerous antibodies. These maternal antibodies provide the baby with an immediate short-term protection against certain microorganisms; babies fed exclusively on bottle formulas lack this protection.

BENEFICIAL BACTERIA

Contrary to the popular impression, beneficial Monera outnumber harmful ones. In Chapter 36 we mentioned the nitrogen-fixing bacteria and the role of Monera as organisms of decay—a process that not only prevents the accumulation of dead bodies and metabolic wastes but also converts materials such as the nitrogen of proteins into a form usable by other living things. We have also mentioned the Eubacteria in the intestine that synthesize vitamins absorbed by the body and that aid in the digestion of certain materials. Anyone who has been given doses of antibiotics massive enough to exterminate his intestinal flora can testify to the ensuing disturbances in normal intestinal activity.

Monera are also of great importance in many industrial processes. Manufacturers often find it easier and cheaper to use cultured microorganisms in certain difficult syntheses than to try to perform the syntheses themselves. Among the many substances manufactured commercially by means of bacteria are acetic acid (vinegar), acetone, butanol, lactic acid, and several vitamins. Bacteria are also used in the retting of flax and hemp, a process that decomposes the pectin material holding the cellulose fibers together; the fibers, once freed, may be used in making linen, other textiles, and rope. Commercial preparation of skins for making leather goods often involves use of bacteria, as does the curing of tobacco.

Many branches of the food industry depend on bacteria. You are probably aware of the central role of bacteria in the making of dairy products, such as butter and the various kinds of cheeses; the characteristic flavor of Swiss cheese, for example, is due in large part to propionic acid produced by bacteria.

Many farmers depend on bacterial action in the making of silage for use as cattle feed. Also of considerable interest to farmers is the possibility that bacteria pathogenic for destructive insects may eventually replace insecticides.

Particularly interesting is the use of bacteria in the production of antibiotics that can help control other bacteria. Most of the antibiotic drugs in use today (but not penicillin) are produced by various species of bacteria of the Actinomycetes group or, if synthesized artificially, were discovered in these organisms. Among these drugs are streptomycin, Aureomycin, Terramycin, and neomycin.

Most recently, bacteria—particularly *E. coli*—have become the workhorses of genetic engineering, turning out large quantities of enzymes or other products like insulin encoded by genes removed from other organisms and inserted into a bacterial host.

Chapter 39

THE PROTISTAN KINGDOM

The delimitation of a kingdom Protista has had a checkered history, as we found in Chapter 37 (see Table 37.2, p. 1019). At times the kingdom has included both procaryotic and eucaryotic organisms; at other times it has been restricted to eucaryotic members. At times it has taken in both primarily unicellular groups, such as Protozoa and Euglenophyta, and groups primarily multicellular but lacking extensive tissue differentiation (Phaeophyta and Rhodophyta, for example); at other times it has comprised only the unicellular ones. It has sometimes included the fungi and sometimes not. In short, the need has often been felt to set aside a kingdom for organisms at an evolutionary level offering little basis or justification for distinguishing between plant and animal; but viewpoints have varied as to where the dividing lines should be drawn.

Recent opinion has increasingly favored a definition of Protista as taking in primarily unicellular or colonial eucaryotic groups (or very simple multicellular or multinucleate ones). Though some of these groups tend to be plantlike, others funguslike, and still others animal-like, they share many characteristics. Thus all the plantlike protistan groups, though primarily photosynthetic, contain nonphotosynthetic members nutritionally similar to the animal-like protozoans. The slime molds, though funguslike at some stages of their life cycles, are remarkably like amoeboid protozoans at other stages. And some of the plantlike or funguslike protistans are very motile, while some of the animal-like ones are sedentary.

As we saw in Chapter 37, there are conflicting interpretations of protistan evolution. One interpretation places the protistans at a level of organization representing the three separate precursors of multicellular plants, animals, and fungi (Fig. 39.1A). In another five-kingdom classification, the protistans are seen as a separate evolutionary line that arose prior to the appearance of multicellular organisms (Fig. 39.1B). Another interpretation recognizes seven kingdoms, with modern protistans on a separate evolutionary line that diverged before the common eucaryotic ancestors segregated into plants, animals, and fungi; advocates of this system often recognize slime molds as a separate kingdom, and divide the Monera into Archaebacteria and Eubacteria (Fig. 39.1C).

The separate-evolution interpretations are more consistent with comparisons of amino acid and nucleic acid sequences, whereas the precursor interpretation reflects the morphological similarities between animals and the so-called animal-like protistans, and between plants and the so-called plantlike protistans. The basis of this conflict of interpretations is exemplified by "plantlike" protistans that appear on the one hand to be more closely related chemically to "animal-like" protistans than to plants, but on the other to be more closely related morphologically to plants than to other protistans. Only further research will resolve these questions of evolutionary descent; it may be that some organisms currently termed protistans will turn out to belong to a separately evolved kingdom, while others will be found to be part of the evolutionary lines that led to the other kingdoms. In organizing this chapter we have adopted the precursor perspective: we recognize a separate protistan kingdom that encompasses three groups leading to plants, animals, and fungi (Fig. 39.1A).

ANIMAL-LIKE PROTISTA

Protozoans are usually said to be unicellular. However, as Libbie Henrietta Hyman of the American Museum of Natural History pointed out, "Each protozoan is to be regarded not as equivalent to a cell of a more complex animal but as a complete organism with the same properties and characteristics as cellular animals." Though they "necessarily lack tissues and organs, since these are defined as aggregations of differentiated cells," many do exhibit "a remarkable degree of functional differentiation." Instead of organs, they have functionally equivalent subcellular structures called *organelles*. In recognition of the complexity of Protozoa, which often far exceeds that of other individual cells, Hyman and many other biologists prefer to call them *acellular* organisms, thereby emphasizing that these single-celled organisms function like a collection of different specialist cells.

Protozoans occur in a great variety of habitats, including the sea, fresh water, soil, and the bodies of other organisms—in fact, wherever there is

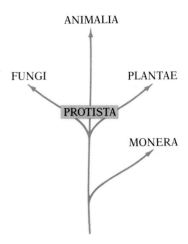

A

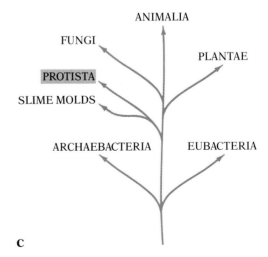

B

C

39.1 Contrasting views of protistan evolution
In a system emphasizing morphological similarities (A), protistans are seen as precursors of the multicellular plants, animals, and fungi. In systems relying more on biochemical evidence (B–C), the modern protistans are seen instead as belonging to a separate evolutionary line; slime molds, Archaebacteria, and Eubacteria may each be regarded as a separate kingdom.

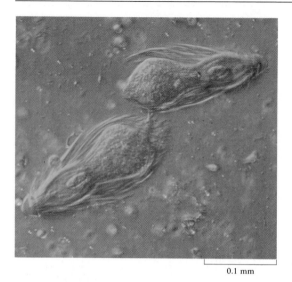

39.2 *Trichonympha*, a flagellate that inhabits the gut of termites
The flagellate helps digest the cellulose in the termite's diet.

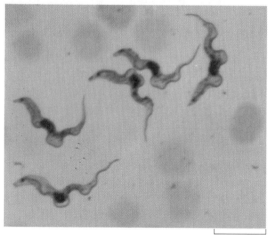

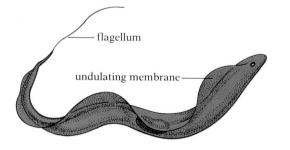

moisture. Most are solitary, but some are colonial. Many are free-living, but others are commensalistic, mutualistic, or parasitic. They are typically heterotrophic. Though some flagellated organisms that possess chlorophyll and are photosynthetic are placed in the Protozoa by many authors, we shall assign these plantlike forms to various protistan algal groups, to be discussed later in this chapter, or, in some cases, to the green algae, discussed in Chapter 40.

The heterotrophic protozoans usually digest food particles in food vacuoles (see pp. 250–51 and Figs. 10.7 and 10.8). There are no special organelles for gas exchange, the general cell membrane serving as the exchange surface. Many species, particularly those living in hypotonic media such as fresh water, possess contractile vacuoles, which function primarily in osmoregulation (see pp. 362–64 and Figs. 14.9–14.11). Small amounts of nitrogenous waste may also be expelled by the contractile vacuoles, but most of it is released as ammonia by diffusion across the general cell surface. Locomotion is by formation of pseudopodia or by means of beating cilia or flagella (see pp. 550–52). Where a single individual has many cilia, their action is coordinated by a system of fibrils connecting their basal bodies; these fibrils or the cell membrane have conductile properties like those of nerves in multicellular animals. Reproduction is sometimes asexual and sometimes sexual. Most freshwater and parasitic protozoans can encyst when conditions are unfavorable; they secrete a thick resistant case around themselves and become dormant.

The Protozoa have been divided into five groups: Mastigophora, Sarcodina, Sporozoa, Cnidospora (which we shall not discuss), and Ciliata. Some of these are heterogeneous assemblages of structurally similar, but probably not closely related, organisms. The relationships of the five groups to one another are unclear. Biologists who treat the protozoans as Protista usually assign them full phylum rank. We shall follow this system.

MASTIGOPHORA

The Mastigophora (also called Zooflagellata) are protozoans that possess flagella as their principal locomotor organelles. They appear to be the most primitive of all the Protozoa, and it seems likely that some (and possibly all) of the other protozoan groups arose from them. Many zoologists (though certainly not all) believe that the flagellate protozoans were also the ancestors of the multicellular animals; similarly, most botanists regard the flagellated photosynthetic protistans as the ancestors of the multicellular plants. There is good reason to think, then, that flagellated unicells played a key role in the evolution of life on earth. This is not to suggest that any flagellate species still living today was the ancestor of these other organisms, but simply that the ancestral flagellates from which modern flagellate protozoans

39.3 *Trypanosoma gambiense*, the cause of African sleeping sickness
Top: Photograph of the protozoans among red blood corpuscles. Bottom: Drawing showing the general structure of the organism.

flagellum

undulating membrane

and modern flagellate protistan algae are descended may have given rise to the other protozoans and to multicellular plants and animals.

A few of the Mastigophora are free-living aquatic organisms, but most live as symbionts in the bodies of higher plants or animals. Several species, for instance, are found in the gut of termites, where they participate mutualistically in the digestion of the cellulose consumed by the termite (Fig. 39.2).

Trypanosoma is a genus of parasitic zooflagellates that cause several severe diseases in human beings and domestic animals. *Trypanosoma gambiense* (Fig. 39.3), for example, is the causative agent in African sleeping sickness.[1] The trypanosomes live in the blood of their host, where they multiply and release a poisonous by-product of their metabolism. In humans or domestic animals (but not in the native wild mammals of Africa), they eventually invade the nervous system, causing lethargy and finally death. The trypanosomes are spread from host to host by blood-sucking tsetse flies (genus *Glossina*). When a tsetse fly sucks blood from an infected animal, some of the trypanosomes are sucked into its intestine, where they multiply and undergo several developmental changes. They then migrate to the fly's salivary glands, where they undergo additional developmental changes and continue to multiply. If the fly now bites an uninfected vertebrate, some of the trypanosomes are injected from the salivary glands into the vertebrate host. Sleeping sickness, which makes large parts of Africa nearly uninhabitable for humans, is difficult to control, because the many wild animals serve as a constant reservoir of trypanosomes. It may eventually be eradicated by extermination of the tsetse flies, but that is a huge undertaking. Treatment for sleeping sickness is complicated, as we have seen, by the ability of the trypanosome to shuffle its genes periodically, thereby altering the surface proteins to which the immune system is specific (see p. 782).

SARCODINA

The Sarcodina are the amoeboid Protozoa. They are thought to be more closely related to the zooflagellates than to the other protozoan groups, because some zooflagellates undergo amoeboid phases, and, conversely, some Sarcodina have flagellated stages.

The most familiar sarcodines are the naked freshwater species of the genera *Amoeba* and *Pelomyxa* (see Fig. 6.24, p. 157), which have asymmetrical bodies that constantly change shape as new pseudopodia are formed and old ones retracted. These pseudopodia, which are large and have rounded or blunt ends, function both in locomotion (see p. 550 and Fig. 20.29) and in feeding by phagocytosis (see p. 105 and Fig. 4.22; also p. 249). The food consists of small algae, other protozoans, and even some small multicellular animals such as rotifers and nematode worms.

Also included in the Sarcodina are several groups of protozoans that secrete hard calcareous or siliceous shells around themselves. These shells, often quite elaborate and complex, can be used in species identification. The pseudopodia of the shelled Sarcodina are usually thin and pointed (Fig. 39.4); in some forms they have no locomotory function, being exclusively

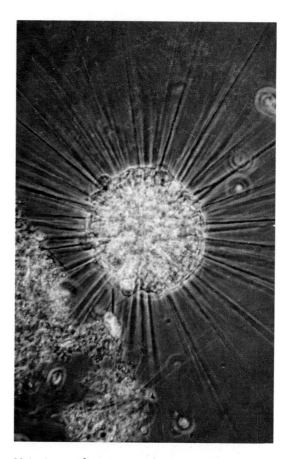

39.4   A sarcodine (*Actinosphaerium*) with long pointed pseudopodia

---

[1] African sleeping sickness should not be confused with ordinary sleeping sickness, or encephalitis, which is caused by a virus.

1 mm

39.5 A group of calcareous foraminiferan shells

0.1 mm

39.6 A group of siliceous radiolarian shells

feeding devices. This type of pseudopodium does not flow around and engulf the prey, but instead functions as a trap. When prey organisms touch it, they become stuck in a mucoid adhesive secretion that coats the pseudopodium surface. The secretion apparently contains proteolytic enzymes that initiate digestion of the prey, which is eventually enclosed in a food vacuole and drawn toward the interior of the cell.

Two groups of shelled sarcodines, the Foraminifera and the Radiolaria, have played major roles in the geologic history of the earth. Both groups are extremely abundant in the oceans, and when the individuals die their shells become important components of the bottom mud. The shells of foraminiferans (Fig. 39.5), which are calcareous, are especially prevalent in the mud at depths of 2,500 to 4,500 m; at greater depths they tend to dissolve, because of the increased carbon dioxide content of the water. The bottom ooze in deeper parts of the ocean is composed chiefly of the siliceous shells of radiolarians (Fig. 39.6). Both these groups have been abundant for a very long time; fossils of foraminiferans go back to the Cambrian period, and fossils of radiolarians occur in Precambrian rocks. Much of the limestone and chalk now present on the earth was formed from deposits of foraminiferan shells, and radiolarian shells have contributed to the formation of siliceous rocks such as chert.

SPOROZOA

The Sporozoa, as their name implies, usually have a sporelike infective cyst stage in their life cycles. They lack special locomotor organelles (except in the male gametes). All are internal parasites, and they usually have complex life cycles. They cause a variety of serious diseases including malaria in humans, and equally serious conditions in both domestic and wild animals.

As you probably know, malaria, which is caused by species of the genus *Plasmodium*, is transmitted from host to host by female *Anopheles* mosquitoes. When an anopheline mosquito bites a person and starts to suck blood, it releases saliva containing a chemical that prevents coagulation of the blood and often, too, *Plasmodium* cells of a stage called sporozoites (Fig. 39.7:5). The sporozoites enter the victim's bloodstream and are carried to the liver, where they enter liver cells and grow for 5–15 days. Then each sporozoite divides, producing a large number of new cells called merozoites. These are released from the liver cells and penetrate red blood corpuscles, where they reproduce asexually, producing additional merozoites. At regular intervals (48 hours in some types of malaria; in one type, 72 hours), all infected red corpuscles burst, releasing the merozoites, which enter new blood corpuscles and repeat the asexual reproductive process. Thus, in the most common form of malaria, hordes of merozoites are released into the bloodstream from ruptured red corpuscles every 48 hours. The host experiences attacks of chills and fever each time such a release of merozoites occurs; these symptoms apparently stem from toxins discharged into the blood by the rupturing red corpuscles.

Eventually some of the merozoites develop into special sexual cells capable of becoming either male or female gametes; they will not mature as gametes, however, as long as they remain in the blood of the human host. But if an anopheline mosquito sucks blood containing these cells, they

SEXUAL REPRODUCTION PHASE IN MOSQUITO

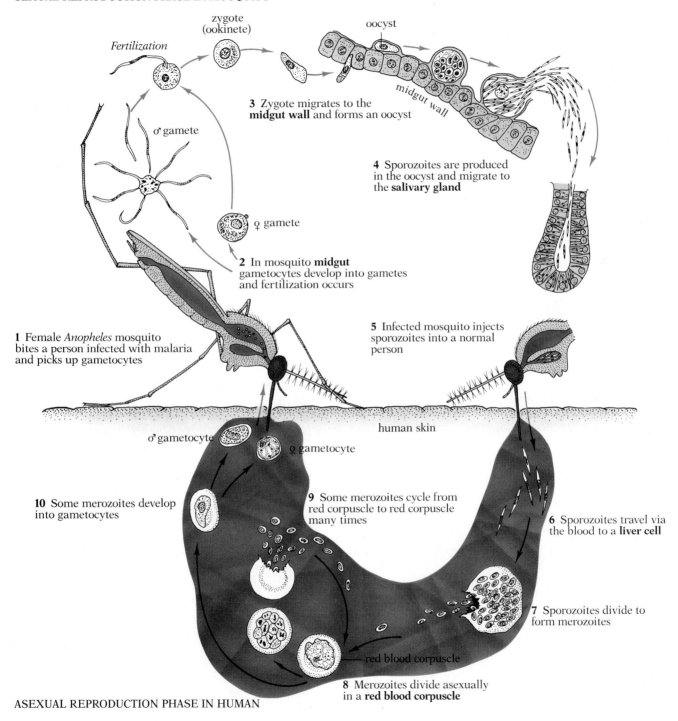

zygote (ookinete)

oocyst

Fertilization

♂ gamete

3 Zygote migrates to the **midgut wall** and forms an oocyst

midgut wall

4 Sporozoites are produced in the oocyst and migrate to the **salivary gland**

♀ gamete

2 In mosquito **midgut** gametocytes develop into gametes and fertilization occurs

1 Female *Anopheles* mosquito bites a person infected with malaria and picks up gametocytes

5 Infected mosquito injects sporozoites into a normal person

human skin

♂ gametocyte

♀ gametocyte

10 Some merozoites develop into gametocytes

9 Some merozoites cycle from red corpuscle to red corpuscle many times

6 Sporozoites travel via the blood to a **liver cell**

7 Sporozoites divide to form merozoites

red blood corpuscle

8 Merozoites divide asexually in a **red blood corpuscle**

ASEXUAL REPRODUCTION PHASE IN HUMAN

39.7 The life cycle of *Plasmodium*, the malarial parasite

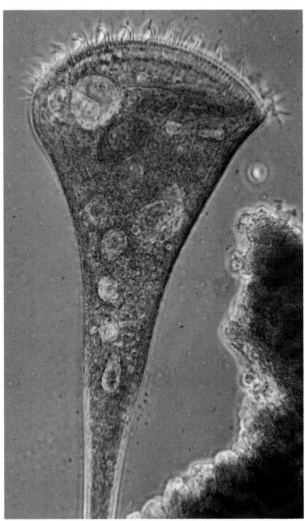

A

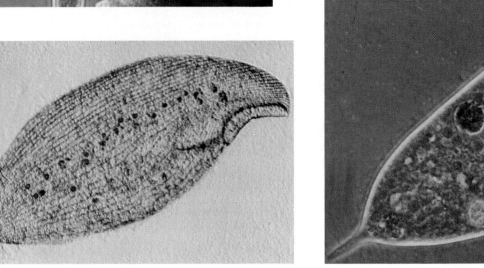

B

Colpoda cucullus

Vorticella

39.8 Representations of two ciliate protozoans
The drawings illustrate the remarkable complexity of these unicellular organisms, to which the term "acellular" applies especially well.

39.9 Photographs of three representative ciliates
(A) *Stentor.* (B) *Loxoda.* (C) *Dileptus.*

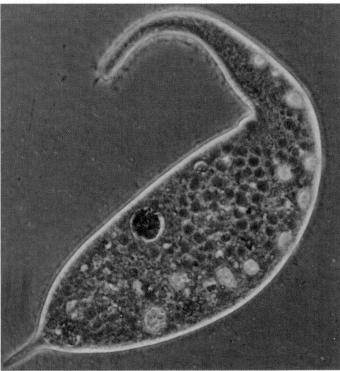

C

39.10 A living ciliate viewed with Nomarski optics
This technique permits study of internal detail without damage to the organism. The large nucleus is visible near the center of the cell. Below the nucleus, and to the left, are contractile vacuoles of the sort filled by small fusion vesicles, some of which can be seen around the periphery of the largest vacuole. The cell contains numerous food vacuoles of various shapes containing several kinds of algae, including diatoms and filamentous forms. An array of undischarged trichocysts can be seen just inside the plasma membrane at right.

complete their development in the midgut of the mosquito. The male gametes then fertilize the female gametes, and the zygote thus produced, which is amoeboid, works its way into the wall of the gut and encysts. Within the cyst, a series of divisions ultimately produces new sporozoites, which are released when the cyst ruptures. These sporozoites migrate to the salivary glands of the mosquito, from which they are discharged into a new vertebrate host when the mosquito feeds. Efforts at malarial control have largely been attempts to eradicate *Anopheles* mosquitoes, either directly by use of insecticides or indirectly by destroying their breeding places.

CILIATA

The phylum Ciliata is the largest of the five protozoan phyla and also the most homogeneous. It differs markedly from the other phyla, and its relationship to them is unclear. A possible derivation from the Mastigophora is indicated by a small group of internal symbionts of frogs (the class Opalinata, now usually placed in the Mastigophora) that shows a combination of some ciliate and some flagellate traits.

As their name implies, ciliates possess numerous cilia as locomotory organelles (Figs. 39.8 and 39.9). In most species the cilia are present throughout life, but in a few (Suctoria) they are absent in the adult stages. The ciliates exhibit the greatest elaboration of subcellular organelles of any Protozoa (Fig. 39.10; see also Fig. 6.23, p. 157, and Fig. 14.11, p. 364), and it is to them that the term "acellular" applies best. As we saw with *Paramecium* (see Fig. 10.7, p. 250), they may have a special oral groove and cytopharynx into which food particles are drawn in currents produced by beating cilia, and they often have an anal "pore" through which indigestible wastes are expelled from food vacuoles. Conductile fibrils connect the

39.11 Two *Stylonychia* in conjugation

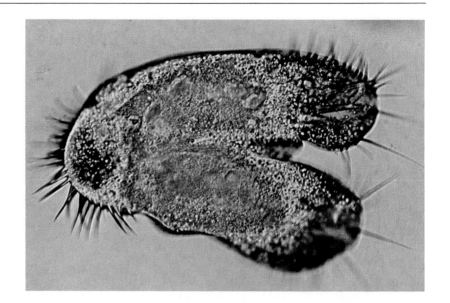

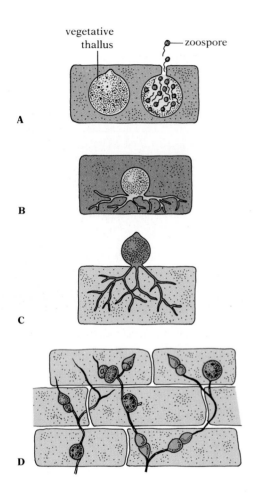

39.12 A variety of thallus forms of chytrids
(A) Some species have a simple saclike thallus (body) located entirely within the host cell. When flagellated zoospores are produced, a tube grows out to the membrane of the host cell and then ruptures, the zoospores being released to the exterior. (B) Other species that live entirely inside their host cells have rootlike rhizoids for nutrient absorption. (C) Still other species, in which the main part of the thallus is outside the host cell, send rhizoids into the host. (D) Some species exhibit a simple form of multicellularity, and their threadlike bodies exploit several host cells simultaneously.

bases of the cilia, and there may be a system of contractile fibers (sometimes striated) analogous to the muscular system of multicellular animals. Stiffened plates occasionally found in the pellicle of the cell together constitute a "skeleton." In some species there is a long stalk by which the individual may attach itself to the substrate (Figs. 39.8 and 39.9A). A few species have tentacles for capture of prey. Some can discharge toxic threadlike darts called ***trichocysts*** that resemble the nematocysts of coelenterates (Fig. 39.10); these may function in defense against predators, in capturing prey, or in anchoring the organism to the substrate during feeding.

Ciliates differ from all other protozoans in having two types of nuclei: a large macronucleus and one or more small micronuclei (see Figs. 6.23 and 10.7). The macronucleus, which is polyploid (about 860-ploid in *Paramecium aurelia*), controls the normal metabolism of the cell. This is a special sort of gene amplification. The diploid micronuclei are concerned only with reproduction and with giving rise to the macronucleus. During asexual reproduction, which is usually by transverse binary fission, the micronuclei divide mitotically.[2] The macronucleus, however, divides amitotically—that is, it forms no spindle and appears to divide by simple constriction. Since division in the macronucleus is preceded by DNA replication and each daughter nucleus receives a full set of genetic material, the division process must be more precise than it appears.

Many ciliates reproduce occasionally by a sexual process called ***conjugation***. Two individuals of appropriate mating types come together and adhere in the oral region; there is some fusion in the area of contact (Fig. 39.11). Next, the micronuclei divide by meiosis. Then all but two of the resulting haploid micronuclei in each cell disintegrate; the macronucleus also usually disintegrates. One of the two nuclei in each cell remains stationary and functions as the female nucleus. The second moves into the

[2] The nuclear membrane of the micronuclei of ciliates does not disappear during mitosis, and the spindle forms inside the membrane-bounded nucleus.

other cell and fuses with that cell's female nucleus in the process of fertilization. Thus each cell acts as both male and female, donating a nucleus and receiving one in return, and when the two cells part each has a new recombinant diploid nucleus. This nucleus then undergoes one or more divisions, and some of the new nuclei thus produced develop into macronuclei. Following a variable number of cytoplasmic cleavages, the normal number of micro- and macronuclei per cell is restored.

FUNGUSLIKE PROTISTA

Among the several major groups of fungal-type organisms that are at the protistan level of complexity, we can, for convenience, recognize two categories: (1) those divisions[3] considered true fungi by more traditional classifications, but containing only unicellular or very simple multicellular organisms, and (2) those divisions traditionally set apart from the true fungi and called slime molds.

PROTOMYCOTA (THE TRUE FUNGUSLIKE PROTISTS)

The two groups included here, the chytrids and hyphochytrids, are small organisms of a more or less typical fungal type, but clearly at the protistan level of complexity. Many are aquatic, living as parasites or saprophytes in or on algae or other plants, but some are found in soil and a few are internal parasites of such animals as mosquito larvae, nematode worms, or liver flukes.

The haploid bodies of these organisms may be simple sacs living entirely inside a cell of the host (Fig. 39.12A), or sacs with rootlike nutrient-absorbing protuberances called rhizoids (Fig. 39.12B, C), or sometimes filamentous forms with saclike reproductive structures (Fig. 39.12D). Reproduction begins when the sacs become multinucleate upon repeated mitotic division of the nucleus without accompanying cytokinesis. Eventually the cytoplasm is partitioned in such a way that each nucleus receives some. The newly formed cells, each of which develops a single whiplash flagellum (located posteriorly in the chytrids and anteriorly in the hyphochytrids), are then released into the surrounding medium. Under some circumstances these flagellated cells function as gametes in sexual reproduction. But more often they function as asexual reproductive cells called *zoospores*, settling down in a suitable location and developing into a new sac or filament, depending on the species.

GYMNOMYCOTA (THE SLIME MOLDS)

The slime molds are curious organisms decidedly animal-like at some stages in their life cycle and plantlike at others. Often placed among the Protozoa in the animal kingdom by zoologists, they have been placed among the fungi in the plant kingdom in traditional botanical classifications and are even given their own kingdom in some systems (Fig. 39.13). When only four or five kingdoms are recognized, they seem to fit best in the Protista.

39.13   Plasmodium of a yellow slime mold (*Hemitrichia stipitata*)

---

[3] "Division" and "phylum" are equivalent terms. Whereas "phylum" is applied to animals and animal-like organisms, "division" is applied to plants and plantlike organisms.

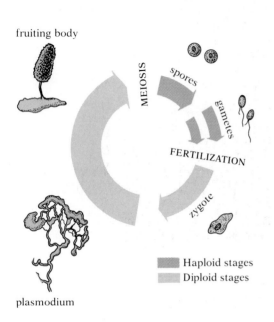

fruiting body

MEIOSIS

spores

gametes

FERTILIZATION

zygote

Haploid stages
Diploid stages

plasmodium

39.14   Life cycle of a true slime mold
See text for description.

The slime molds are generally found growing on damp soil, rotting logs, leaf mold, or other decaying organic matter in moist woods, where they look like glistening viscous masses of slime. They are sometimes white, but are often colored red or yellow (Fig. 39.13).

In the so-called *true slime molds* (division Myxomycota), the vegetative phase of the life cycle is a large diploid multinucleate (coenocytic) amoeboid mass called a *plasmodium*, which moves about slowly and feeds on particles of organic material by phagocytosis. The behavior of the naked plasmodium is thus animal-like. Under certain conditions, however, the plasmodium becomes stationary and develops *fruiting bodies*, which may be either simple rounded masses or elaborate stalked organs; at this stage the appearance and behavior of the organism are plantlike. Meiosis occurs within the *sporangia* (spore-producing structures) of the fruiting bodies, and the haploid cells thus formed are released as spores, whose walls contain cellulose. When the spores germinate, they produce naked flagellated gametes. These fuse in pairs to form zygotes, which soon lose their flagella and become amoeboid. As this amoeboid form flows along the substrate, engulfing bacteria and other organic particles and digesting them in vacuoles, its diploid nucleus undergoes repeated mitotic divisions without accompanying cytokinesis. In this way the zygote develops into a multinucleate plasmodium, which may grow to a length of 25 cm (though 5–8 cm is a more common size). Some growth may also occur by fusion of the cytoplasms of two or more zygotes or young plasmodia. In summary, then, the life cycle proceeds from diploid amoeboid plasmodium, to stationary spore-producing plasmodium, to haploid spores, to flagellated gametes, to zygote, and back to amoeboid plasmodium (Fig. 39.14).

The life cycle of the *cellular slime molds* (division Acrasiomycota) is quite different from that of the true slime molds. Current evidence suggests that the two groups are probably not closely related and should be placed in separate divisions.[4] Their relationships to other organisms are unclear.

In the cellular slime molds, the spores do not develop into flagellated gametes, but instead give rise to free-living soil-inhabiting amoeboid cells, each with a single haploid nucleus (Fig. 39.15). The amoebae feed on bacteria and other organic matter. During this feeding stage the amoebae divide repeatedly (both mitosis and cytokinesis occur), producing independent uninucleate daughter cells rather than a multinucleate plasmodium. As the local food supply diminishes, the behavior of the amoebae suddenly changes. Two quite different cycles—one sexual and the other asexual— are possible, depending on environmental conditions. Most often the amoebae enter the asexual cycle: They cease feeding and begin to aggregate at central collecting points where they clump together to form a sluglike *pseudoplasmodium*. The individual haploid cells retain their separate identities within the slug; they do not fuse. The pseudoplasmodium of the cellular slime molds is thus very different from the diploid multinucleate

---

[4] Two other small groups, the Plasmodiophoromycota and the Labyrinthulomycota, are also usually put in the Gymnomycota, though they are probably not closely related to either the Myxomycota or the Acrasiomycota. Thus Gymnomycota is more a grouping of convenience than a true phylogenetic category. Authors who wish to emphasize the animal characteristics of the slime molds often use the name ''Mycetozoa.''

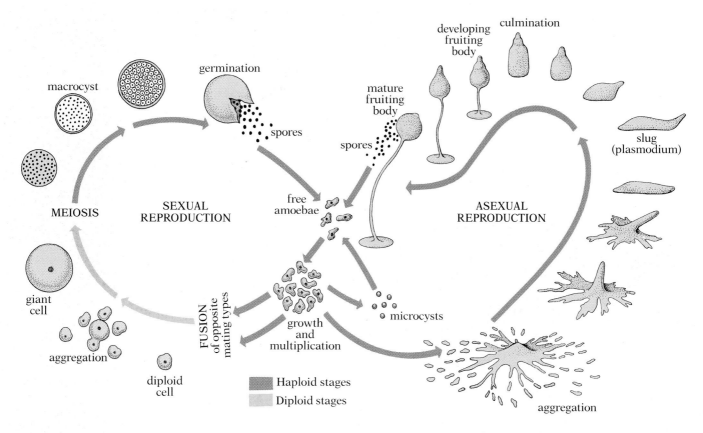

39.15   Life history of a cellular slime mold
See text for discussion.

plasmodium of the true slime molds. The slug may move around as a unit for a while, but eventually it becomes sedentary and in the culmination of the asexual phase forms a stalked fruiting body in which new spores are produced. Notice that this part of the life cycle does not include any sexual events, and that all cells are apparently haploid.

Under very moist conditions, when the soil is so wet that the fruiting bodies would probably be under water, making spore dispersal difficult, the amoebae enter a different cycle, in which a sessile aggregation of free-living amoebae can form. If by chance the first two amoebae to begin this aggregation are of different mating types, a series of events that lead to what seems to be sexual recombination takes place. The two nuclei combine, and as new cells aggregate they are ingested by phagocytosis. Whether their nuclei fuse with the now-diploid nucleus formed by the original pair of amoebae is not known, but the resulting giant cell clearly grows and grows. Meiosis then restores the haploid state, and continued replication and division lead to a spore-filled macrocyst. When conditions become favorable, the cyst releases its spores in what may be termed germination.

Cellular slime molds can easily be grown in the laboratory, and they have been used extensively in studies of the factors influencing development. During the amoeboid stage, for instance, individual amoebae are not conspicuously polar, each having an irregular shape with no distinct anterior

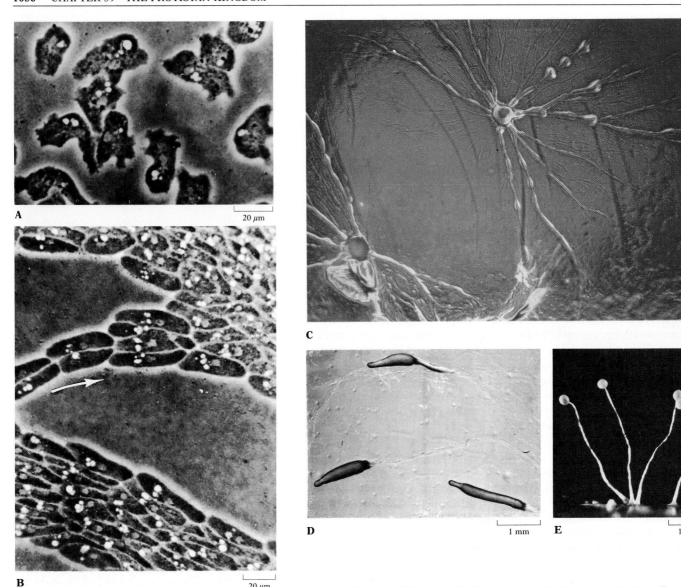

A   20 μm

B   20 μm

C

D   1 mm   E   1 mm

39.16 *Dictyostelium*, a cellular slime mold
(A) In the free-living stage each cell is an irregularly
shaped amoeba. (B) When aggregation begins, the
cells take on an elongate shape, become polar, with
distinct anterior and posterior ends, and begin to
move in an oriented direction (indicated here by the
arrow). (C) Two aggregations (pseudoplasmodia) are
in process of formation on a glass surface, with long
streams of amoebae funneling into the centers of
aggregation. (D) Three pseudoplasmodia moving on a
glass surface. (E) Fruiting bodies, the culmination of
the asexual phase.

or posterior end (Fig. 39.16A). However, as the food supply dwindles, they
become polar, and some begin to secrete cyclic AMP into the medium.
Other polar amoebae, detecting the cAMP at their anterior ends, turn
toward the signal and at the same time secrete cAMP, thus stimulating still
other amoebae to orient in the same direction. The amoebae often secrete
the cAMP in pulses rather than in a steady stream, and each time an oriented amoeba detects a pulse it moves forward for about 100 seconds. In
this fashion more and more amoebae are recruited and move toward the
aggregation center, where, depending on conditions, either a pseudoplasmodium or a giant cell is formed. It is especially intriguing to biologists that
the substance responsible for the remarkable behavioral coordination in
aggregating slime-mold amoebae should be cAMP, which, as we have seen,
is involved in many sorts of intracellular control in animals.

PLANTLIKE PROTISTA

Several groups of primarily unicellular, often flagellated, organisms have traditionally been regarded by botanists as algal members of the plant kingdom, because many of their members possess chlorophyll and many have cell walls. They exemplify, perhaps better than any other living organisms, the difficulty of distinguishing between plants and animals at the unicellular level. Their assignment to the kingdom Protista makes such distinction unnecessary, though it does not, however, clarify their evolutionary position.

EUGLENOPHYTA (THE EUGLENOIDS)

The euglenoids are unicellular organisms that show a combination of plantlike and animal-like characteristics. They are plantlike in that many species have chlorophyll and are photosynthetic; they are animal-like in lacking a cell wall and being highly motile, and the species that lack chlorophyll are heterotrophic, like animals. Though their pigmentation (chlorophylls *a* and *b* and carotenoids) is like that of the green algae and land plants, they seem to have no close relatives among the algae. Botanical classifications usually put them in a division by themselves. There are about 25 genera of euglenoids, containing approximately 450 species. Most live in fresh water, but a few are found in soil, on damp surfaces, or even in the digestive tracts of certain animals.

A representative genus is *Euglena*. A typical cell of *Euglena* is an elongate ovoid, with a long flagellum emerging from an anterior invagination (Fig. 39.17).[5] Since the cell lacks a wall, it is fairly flexible, and its shape may change somewhat as it swims about; however, the pellicle, a proteinaceous layer just beneath the plasma membrane, prevents excessive alterations of its shape. The large nucleus contains a prominent nucleolus that is unusual in not disappearing during mitosis. The nuclear membrane too remains intact during mitosis, instead of disappearing as it does in most cells. An orange granule, called the **stigma** (or eyespot), is located near the anterior end of the cell and functions in light detection and phototactic responses. (There is no stigma in nonphotosynthetic euglenoid species.) Most green euglenoids have a special organelle, the **pyrenoid**, that functions in the production of paramylum, a polymer of glucose that these organisms use as a storage product instead of starch or glycogen; paramylum is unique with the euglenoids. A large contractile vacuole lies near the anterior end of the cell.

[5] *Euglena* possesses a second shorter flagellum that does not emerge from the anterior invagination. In some other euglenoids both flagella are emergent.

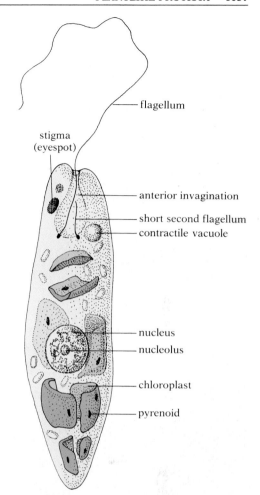

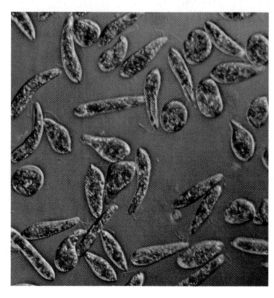

39.17 *Euglena*
Top: For a full description of the structures shown in this drawing, see text. Bottom: Photograph of a group of *Euglena*.

39.18 A unicellular chrysophyte with characteristically unequal flagella

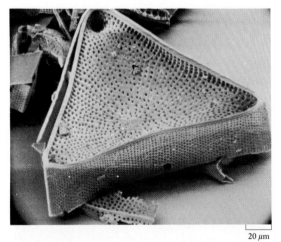

20 μm

39.19 Scanning electron micrograph of a diatom (*Trinacria regina*)
The organism has a distinctly boxlike structure.

The euglenoid species that lack chlorophyll are obligate heterotrophs. The species that have chlorophyll are facultative heterotrophs and can survive in the dark if they have a source of organic nutrients; indeed, even in the light they are not entirely autotrophic, since they require one or more vitamins in their diet. It is easy to destroy the chloroplasts of *Euglena* by treatment with streptomycin or heat (or by keeping the cells in the dark for a long time). A colorless strain of *Euglena* can be produced in this way, since chloroplasts, as self-perpetuating organelles, cannot be regained once they are lost. It is thus possible to make a lineage of organisms cross the traditional boundary (presence or absence of chlorophyll) between the plant and animal kingdoms.

Reproduction in euglenoids is by longitudinal mitotic cell division. Sexual reproduction has never been conclusively demonstrated in them.

CHRYSOPHYTA (THE YELLOW-GREEN AND GOLDEN-BROWN ALGAE AND THE DIATOMS)

As the names of many algal divisions indicate, the earliest classifications of algae were based on color, which in turn depends on the sorts of pigments the cells contain. Fortunately, later study of other important characters—particularly the details of flagellation, the type of energy-storage materials produced, and the chemistry of the cell wall—showed that algae of like pigmentation usually share such characters as well and that the old color classification was still acceptable. Thus most of the species in the six classes of algae placed together in the division Chrysophyta are some shade of yellow or brown (caused in part by a predominance of carotenoids), and they also resemble each other in possessing chlorophylls *a* and *c*, but no *b*, and in using a polysaccharide called chrysolaminaran[6] instead of starch as their storage material. The walls of many species are impregnated with silica or calcium. Two anteriorly attached flagella of unequal length are common (Fig. 39.18), but some species have no flagella, some have one, and some have two that are equal.

The majority of the Chrysophyta are unicellular or colonial, though a few have small simple multicellular bodies. Reproduction is usually asexual but occasionally sexual. Most of the yellow-green and golden-brown algae live in fresh water, but a few are marine. Diatoms are abundant in both freshwater and saltwater habitats.

The diatoms, which some authorities assign to a division of their own, separate from the other Chrysophyta, are of special interest for several reasons. They are unusual in that the vegetative cells are ordinarily diploid—not haploid, as might be expected in such simple and seemingly primitive plantlike organisms. Unlike most other Chrysophyta, diatoms lack flagella; some species, however, produce flagellated sperm cells. Silica-impregnated glasslike walls, composed of two pieces that fit together like a box with its lid (Fig. 39.19), often give the cells a jewel-like appearance; the different species exhibit a great variety of shapes and ornamentations (Fig. 39.20). The classification of the diatoms is based almost entirely on the characters of the walls, or shells, as they are commonly called.

---

[6] Formerly called leucosin.

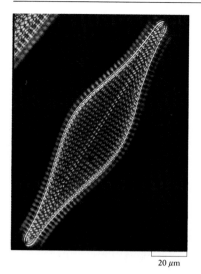

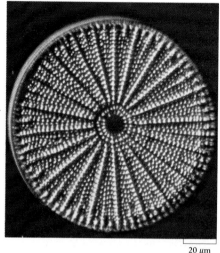

20 μm   20 μm   20 μm

39.20 Some representative diatoms
Left: The disclike diatom *Raphoneis*. Middle:
Triceratium pentacrinus, a star-shaped diatom. Right:
Arachnoidiscus ehrenbergi, a wheel-like organism.

When the cells of diatoms die, their shells sink to the bottom, where they may accumulate in large numbers, forming deposits of a material called diatomaceous earth. This material is used as an ingredient in many commercial preparations, including detergents, polishes, paint removers, decolorizing and deodorizing oils, and fertilizers. It is also extensively used as a filtering agent and as a component in insulating and soundproofing products.

The diatoms play an extremely important role in aquatic food webs, in both freshwater and marine habitats. They are the most abundant component of marine plankton, for example; it is not unusual for a gallon of seawater to contain as many as one or two million diatoms. *Plankton* consists, by definition, of small organisms floating or drifting near the surface. Planktonic organisms are generally divided into two groups—phytoplankton (plant plankton) and zooplankton (animal plankton). The organisms constituting phytoplankton are the principal photosynthetic producers in marine communities.

PYRROPHYTA (THE DINOFLAGELLATES)

The dinoflagellates are small, usually unicellular, organisms (Fig. 39.21). A cell wall may or may not be present; if present, it is composed largely of cellulose. Photosynthetic species possess chlorophylls *a* and *c*; they usually have a yellowish-green to brown color as a result of an abundance of carotenoids and xanthophylls, several of which are unique in these organisms. There are many colorless species of dinoflagellates as well; some feed on particulate organic matter, and others live as parasites in a variety of marine invertebrates. The energy-storage material is either starch or oil.

The nuclei of dinoflagellates are unlike those of any other organisms. The chromosomes lack centromeres and remain permanently in the condensed configuration, even during interphase. The nucleolus and nuclear membrane persist during cell division, and no spindle is formed. Long cytoplas-

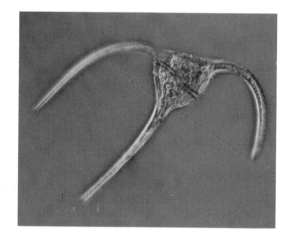

39.21 Photograph of *Ceratium*, a dinoflagellate

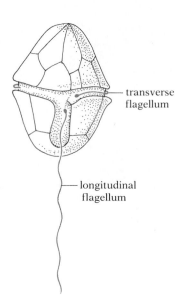

transverse
flagellum

longitudinal
flagellum

**39.22 Drawing of a freshwater dinoflagellate
(*Glenodinium cinctum*)**
The two flagella lie in distinctive grooves on the
surface of the cell.

mic channels, probably containing microtubules, intrude into the nucleus during division, and these may replace the spindle in the movement of the chromosomes.

Most species possess two very unequal flagella, which are attached laterally. One of these runs along a groove to the posterior end of the cell and extends beyond the cell like a tail; the other lies in a groove that encircles the midportion of the cell like a belt (Fig. 39.22).

Many dinoflagellates produce trichocysts similar to those of ciliate protozoans such as *Paramecium*. A few even produce nematocysts, which are so much like those of coelenterates that it was once thought they were acquired by the dinoflagellates from coelenterates.

Some species of dinoflagellates can produce light and are responsible for much of the luminescence often seen in ocean water at night.

Dinoflagellates are second only to the diatoms as primary producers of organic matter in the marine environment; they play a lesser role in fresh water. Not only are they important as food, but they also function as endosymbionts for an amazing variety of marine invertebrate animals that in some cases appear to be unable to survive without them. Some types of corals take as much as 60 percent of the carbon their dinoflagellate symbionts (zooxanthellae) fix in photosynthesis; it is estimated that such corals would deposit calcium in their skeletons only one-tenth as fast if they lacked the symbionts (see Fig. 36.32, p. 979).

A number of species of dinoflagellates are poisonous. Some of these contain red pigments, and when they occur in great abundance they produce the so-called red tides that sometimes kill many millions of fish. Red tides are fairly common in the Gulf of Mexico off the coast of Florida.

Chapter 40

THE PLANT KINGDOM

The various divisions of the plant kingdom have traditionally been separated into two groups: the **Thallophyta** and the **Embryophyta**. ("Phyte" is from the Greek *phyton*, "plant.") Though not usually recognized as taxonomic categories in modern classifications, the groupings are useful in that they include plants at similar levels of structural complexity. The Thallophyta are the more primitive plants, the Embryophyta the more advanced. We can summarize the chief distinctions between them as follows:

1. Thallophytes usually show little if any tissue differentiation: hence there is no anatomical basis for distinguishing roots, stems, or leaves, and the entire plant is known as a **thallus**. There is far more differentiation in embryophytes; the higher embryophytes have distinct roots, stems, and leaves.

2. The reproductive structures of thallophytes are often unicellular and, whether unicellular or multicellular, lack a protective wall or jacket of sterile (i.e., nonreproductive) cells.[1] The reproductive structures of embryophytes, by contrast, are multicellular and have a jacket of sterile cells.

3. The zygotes of thallophyte plants do not develop into embryos within the female reproductive organs where they are produced,[2] whereas the early stages of embryonic development in embryophytes take place while the embryo is still contained within the female reproductive organ (hence the name "embryophyte").

In older classification systems the Thallophyta included both the various groups called algae (the photosynthetic thallophytes) and the groups called fungi (the nonphotosynthetic thallophytes). In systems like the one in this

---

[1] There are some seeming exceptions, in which the reproductive structures appear superficially to have jacket cells.

[2] Except in some red algae.

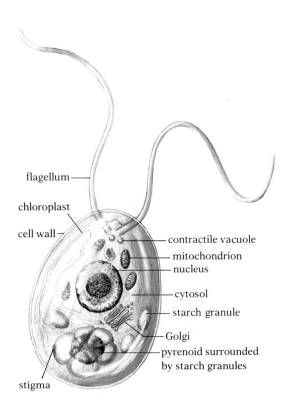

flagellum

chloroplast

cell wall

contractile vacuole

mitochondrion

nucleus

cytosol

starch granule

Golgi

pyrenoid surrounded by starch granules

stigma

40.1 Mature cell of *Chlamydomonas*

text, in which the fungi have kingdom status and the various unicellular algae are assigned to either the kingdom Protista or the kingdom Monera, only algal groups with many multicellular members remain in the thallophyte portion of the plant kingdom. A formal outline of the classification of the plant kingdom is found in the Appendix. The following list relates the traditional groupings to the formal classification:

Algal Thallophyta
Division Chlorophyta
Division Phaeophyta
Division Rhodophyta

Embryophyta
Division Bryophyta
Division Tracheophyta

CHLOROPHYTA (THE GREEN ALGAE)

The green algae are of particular interest, because they are generally regarded as the group from which the land plants arose. They are thus probably the only algal division that has not been a phylogenetic dead end. Like land plants, the green algae possess chlorophylls *a* and *b* and carotenoids; unlike many other algae, they have no unusual chlorophylls. The majority of green algae live in fresh water, but some live in moist places on land, and there are many marine species.

Many divergent evolutionary tendencies, all probably beginning with walled and flagellated unicellular organisms, can be traced in the Chlorophyta: (1) the evolution of motile colonies; (2) a change to nonmotile unicells and colonies; (3) the evolution of extensive tubelike bodies with numerous nuclei but without cellular partitions (coenocytic organisms); (4) the evolution of multicellular filaments and even three-dimensional leaflike thalluses.

These evolutionary trends can be studied especially well in the green algae, because many unicellular and rudimentarily multicellular members of the group are still extant. But for that very reason the green algae are difficult to place in a classification system that separates Protista from Plantae. It makes no evolutionary sense to assign the unicellular green algae to one kingdom and their multicellular relatives to another. Hence we shall treat the whole group as true plants whose unicellular representatives have remained at the protistan level.

***Chlamydomonas* as a representative unicellular green alga** *Chlamydomonas* is a genus of unicellular green algae that probably resemble the ancestral organisms from which the rest of the plant kingdom arose. Its many species are common in ditches, pools, and other bodies of fresh water and in soils. The individual organism is an oval haploid cell with a wall composed largely of glycoprotein; it differs from the walls of many other green algae in lacking cellulose.[3] There are two anterior flagellae of equal length

---

[3] In the absence of cellulose from its cell walls, *Chlamydomonas* probably differs from the organisms ancestral to the rest of the plant kingdom.

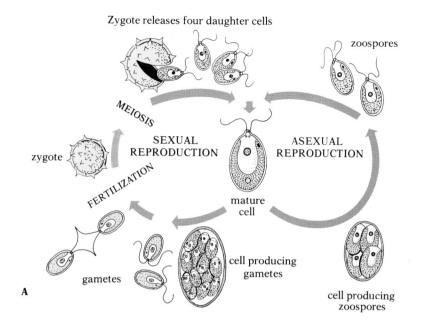

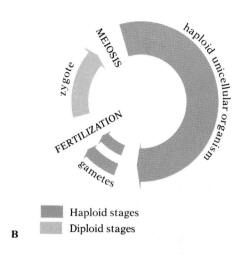

40.2 Life history of *Chlamydomonas*
(A) Diagram showing all stages of both the sexual and asexual cycles. (B) Schematic representation of life cycle for comparison with those of other organisms. Note that the zygote is the only diploid stage. This type of life cycle was probably characteristic of the first sexually reproducing unicellular organisms, and it may thus be the type from which all other types arose.

and a single large cup-shaped chloroplast that fills from one-half to two-thirds of the basal portion of the cell (Fig. 40.1). Inside the chloroplast are numerous chlorophyll-bearing lamellae, often arranged in stacks rather like the grana of higher plants. A conspicuous pyrenoid in the basal portion of the chloroplast functions as the site of starch synthesis. A stigma, or eyespot, which helps mediate the organism's positive phototaxis (but is not essential for that behavior), is also located inside the large chloroplast. The cell has no large central vacuole such as is seen in mature cells of higher plants. Two small contractile vacuoles lying near the base of the flagella discharge alternately and rhythmically.[4]

Asexual reproduction is common in *Chlamydomonas* (Fig. 40.2). A vegetative cell resorbs its flagella; then mitotic division of the nucleus and cytokinesis take place. This process gives rise to two daughter cells, both of which lie within the wall of the original cell. In some species the two daughter cells are promptly released by breakdown of the wall; in other species the daughter cells themselves divide while still inside the wall of the parent cell, and a total of 4 (as in the figure), 8, 16, or more daughter cells is produced, depending on the species and conditions of growth. The daughter cells each develop a wall and flagella just before they are released as free ***zoospores***, which are motile asexual reproductive cells—motile reproductive cells not specialized as gametes. In *Chlamydomonas* the zoospores are smaller than mature vegetative cells, but otherwise indistinguishable from them; in many species of algae, however, there are noticeable morphological differences between the zoospores and the mature cells. The free zoospores soon grow to full size, completing the asexual reproductive cycle.

[4] The descriptions of *Chlamydomonas* and other algae given in this chapter are based largely on material in J. M. Kingsbury, *Biology of the Algae*, published privately, Ithaca, N.Y., 1963, and on discussions with Kingsbury.

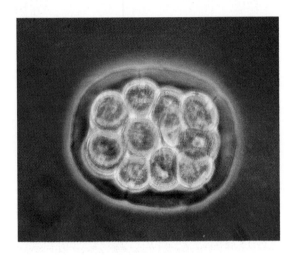

40.3 *Pandorina* **colony**
The cells are embedded in a gelatinous matrix.

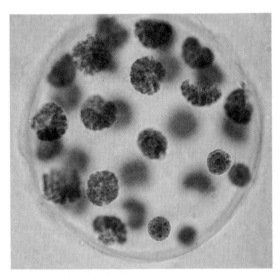

40.4 Photograph of a *Pleodorina* colony

Under certain conditions, especially when the concentration of nitrogen in the medium is low, *Chlamydomonas* may reproduce sexually. A mature haploid vegetative cell divides mitotically to produce several gamete cells, which develop walls and flagella and are released from the parent cell. Gametes (usually of two different mating types) are attracted to each other and form large clusters. Eventually the clustered cells move apart in pairs. The members of a pair are positioned end to end, with their flagella, which bear species-specific and mating-type–specific attractant sites at their tips, in close contact. The cells then shed their walls, and their cytoplasms slowly fuse. Finally, their nuclei unite in the process of fertilization, which produces a single diploid cell, the zygote. The zygote sheds its flagella, sinks to the bottom, and develops a thick protective wall. It can withstand unfavorable environmental conditions, such as the drying up of the pond or the cold of winter. When conditions are again favorable, it germinates, dividing by meiosis to produce 4 (or 8) new flagellated haploid cells, which are released into the surrounding water. The new cells quickly mature, thus completing the sexual reproductive cycle.

Because sexual reproduction in most species of *Chlamydomonas* is at a very simple level, it can give us insight into the way sexuality probably arose. There are no separate male and female individuals. Furthermore, though the gametes usually differ in their mating-type–specific attractant sites, they are usually morphologically indistinguishable. They cannot be separated into male gametes (sperm) and female gametes (eggs). Such a condition, in which all gametes are alike, is called *isogamy*; it is probably the primitive (ancestral) condition in plants. The isogametes of *Chlamydomonas* are indistinguishable from vegetative cells; they may be viewed simply as small vegetative cells that tend to fuse and act as gametes under certain conditions.[5] This, too, is probably the primitive condition; the specialization of gametes as morphologically distinctive cells—a characteristic of most higher plants and animals—is surely a later evolutionary development.

Notice that the haploid stages of the life cycle of *Chlamydomonas* are the dominant ones; the only diploid stage is the zygote. Dominance of the haploid stages is characteristic of most very primitive plants, and it seems clear that this was the ancestral condition.

The volvocine series As an example of one of the evolutionary tendencies that can be traced in the Chlorophyta, let us examine the so-called volvocine or motile-colony series. This is a series of genera showing a gradual progression from the unicellular condition of *Chlamydomonas* to an elaborate colonial organization.

Gonium may be taken as an example of the simplest colonial stage. Each colony of *Gonium* is made up of 4, 8, 16, or 32 cells (depending on the species), each of which is morphologically similar to *Chlamydomonas*. The cells are embedded in a mucilaginous matrix and are arranged in a flat or slightly curved plate. In some species delicate cytoplasmic strands run between the cells; these may provide a route for direct interaction and coordination between the cells of the colony. That some sort of coordination does

---

[5] A few species of *Chlamydomonas* are anisogamous (one gamete is larger than the other), and a few are even oogamous (one of the gametes is a nonmotile egg cell).

indeed exist is shown by the organized fashion in which the flagella of all the cells beat together and thus enable the colony to swim as a unit.

When asexual reproduction occurs in *Gonium*, all the cells in a colony divide simultaneously. When each cell of the parent colony has divided enough times to contain within its wall the same number of daughter cells as in the parent colony, its wall disintegrates and the daughter cells are released. However, the daughter cells from a single parent cell do not swim off independently as separate zoospores as in *Chlamydomonas*. Instead, they remain together in a common matrix and mature into a new *Gonium* colony. Thus each cell of the parent colony gives rise to a complete new colony.

Sexual reproduction in *Gonium* is similar to that in *Chlamydomonas*. Individual free-swimming cells are released from a colony and function as gametes, fusing in pairs to form zygotes. The gametes are isogamous. As in *Chlamydomonas*, the zygote is the only diploid stage in the life cycle.

Pandorina is a genus of colonial forms slightly more complex than *Gonium*. Each colony is a hollow sphere in which 8, 16, or 32 cells are arranged in a single layer about the periphery, their flagella oriented to the outside of the sphere (Fig. 40.3). Three main advances over *Gonium* are noticeable: (1) The colony shows some regional differentiation; it has definite anterior and posterior halves (detectable both by the orientation of the colony when it is swimming and by the larger size of the stigma in the anterior cells than in the posterior ones). (2) The vegetative cells of the colony are so dependent on one another that they cannot live apart from the colony, and the colony itself cannot survive if disrupted or broken. (3) Sexual reproduction is **heterogamous**—that is, it involves two kinds of gametes. In *Pandorina* the male gametes are smaller than the female gametes, but both types of gametes have flagella and are free-swimming. This type of heterogamy, where the only morphological difference between the gametes is one of size, is called **anisogamy**.

Eudorina is a still more advanced genus. The spherical colonies contain 16 or 32 cells. The differences between the anterior and posterior portions of the colony are greater than in *Pandorina*, and the heterogamy is more pronounced in that the large female gametes are not released, but remain embedded in the matrix of the colony and are fertilized there by the much smaller free-swimming male gametes.

A still more advanced genus is *Pleodorina* (Fig. 40.4), whose spherical colonies are composed of 32 to 128 cells. These large colonies exhibit considerable division of labor. The anterior cells are vegetative, almost never participating in reproduction. The posterior cells, which function in both asexual and sexual reproduction, are much larger. Sexual reproduction is heterogamous and, as in *Eudorina*, the female gametes are retained within the parental colony; in fact, in some species the large female gametes lose their flagella and thus become true nonmotile egg cells. This type of advanced heterogamy, where only the male gamete is motile and the female gamete is a nonflagellated nonmotile egg cell, is called **oogamy**.

The culmination of the evolutionary series traced here is represented by the genus *Volvox* (Fig. 40.5). Its spherical colonies are very large, consisting of about 500 to 50,000 cells. Delicate cytoplasmic strands between cells make possible some intercellular communication (Fig. 40.6). Most of the

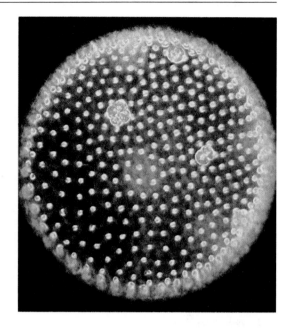

40.5 A *Volvox* colony
The colony is a sphere with a single layer of cells embedded in gelatinous material; the interior of the sphere is filled with a watery mucilage. Some small daughter colonies can be seen developing in this colony.

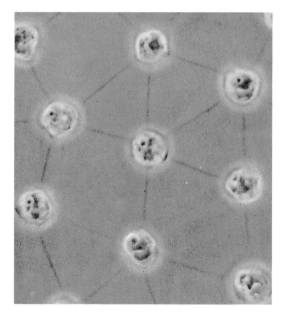

40.6 Cells of *Volvox* interconnected by protoplasmic strands

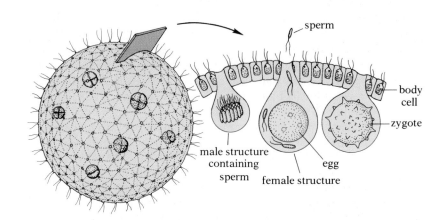

40.7   Reproduction in *Volvox*
Left: The colony is very large, containing 500 to 50,000 vegetative cells. Six daughter colonies at various stages of development can be seen still embedded in the matrix of the parent colony. Right: Section through the surface of a colony showing male and female reproductive structures. Sperm released by the male structure enter the female structure and fertilize the egg. After a period of inactivity, the zygote divides meiotically, and the haploid cells thus formed then divide mitotically, producing a new daughter colony, which is eventually released. The colony shown here is producing both male and female gametes; in some species or strains the sexes are separate, and a given colony produces only one kind of gamete.

cells are exclusively vegetative. A few cells (between 2 and 50), scattered in the posterior half of the colony, are much larger than the others and are specialized for reproduction. Each of the female reproductive cells can give rise to an entire new daughter colony (Figs. 40.5 and 40.7). Sexual reproduction is always oogamous.

We can summarize the major lines of evolutionary changes manifest in this series as follows: (1) a change from unicellular to colonial life, and a tendency for the number of cells in the colonies to increase; (2) increasing coordination of activity among the cells; (3) increasing interdependence among the vegetative cells, so that they cannot live apart from the colonies and the colonies cannot survive if disrupted; (4) increasing division of labor, particularly between vegetative and reproductive cells; (5) a gradual change from isogamy to anisogamy to oogamy. In both the oogamous algae and the oogamous higher plants the female gametes are characteristically retained within the parental organism or colony, and the meeting of gametes becomes a less random process. Hence fewer female gametes need be produced, and more energy can be devoted to providing a large store of nutrients in those few.

In tracing the series *Chlamydomonas–Gonium–Pandorina–Eudorina–Pleodorina–Volvox,* we do not mean to imply that each genus evolved from the preceding one; the available evidence will not allow us to decide whether it did or not. But it does seem likely that each of these genera evolved from an ancestor that resembled in many important ways the modern genus placed just before it in this series, and therefore that the actual evolutionary progression from some unicellular ancestor to *Volvox* involved a series of stages similar to those represented by the modern genera discussed here. Study of this series suggests how complex colonial forms may have evolved, and indicates one possible way in which multicellularity may have arisen in plants. After all, it is largely an arbitrary decision whether one calls *Volvox* colonial or multicellular.

Although multicellular animals certainly did not evolve from *Volvox* or any of the other genera discussed here, a similar evolutionary series, beginning with a nonwalled unicellular organism, may have been the beginning of multicellularity in the animal kingdom.

Some multicellular green algae Many green algae have a multicellular stage in their life cycle. In most cases this stage is a filamentous thallus, which may be either nonbranching or branching, depending on the species (Fig. 40.8). Let us take *Ulothrix* as a first example.

The species of *Ulothrix* are unbranched filamentous forms, most of which live in fresh water although a few are marine (Fig. 40.9). The filament of each plant is a very small threadlike structure attached to the substratum by a specialized cell called a ***holdfast***. Except for the holdfast cell, all the cells of the filament are identical and are arranged end to end in a single series. The filament increases in length as its cells divide horizontally and as the new cells thus added to the chain grow to mature size. Adjacent cells have common end walls—a basic step in the evolution of multicellularity in algae. Each cell contains a single nucleus and a single large chloroplast.

Ulothrix may reproduce by fragmentation (each fragment growing into a complete plant), by asexually produced zoospores, or by sexual processes. In asexual reproduction any cell of the filament except the holdfast may act as a ***sporangium***, producing zoospores, each of which has four flagella. After the zoospores are released, they swim about for a short while, and then settle down and give rise to a new filament.

Sexual reproduction is isogamous. The zygote (stage 4 in Fig. 40.9), formed by the union of two of the biflagellate gametes, develops a thick wall

40.8 *Stigeoclonium*, a branching filamentous green alga

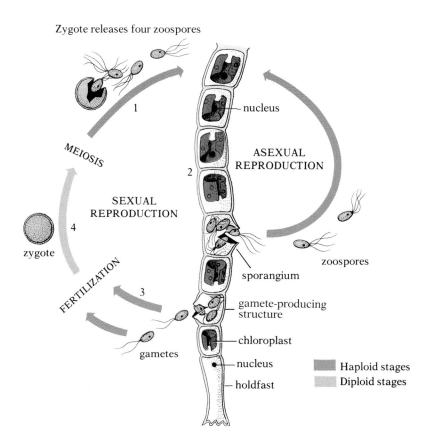

40.9 Life history of *Ulothrix*
The haploid plant may reproduce either asexually or sexually (though a single filament would never reproduce in both ways at once as shown here). Asexual reproduction is the more common; certain cells of the filament develop into sporangia (spore-producing structures) and produce zoospores, which settle down and develop into new filaments. Under certain environmental conditions the filament may cease reproducing asexually and begin reproducing sexually; a cell becomes specialized as a gametangium (gamete-producing structure) and produces isogametes. Two such gametes may fuse in fertilization, producing a zygote, which divides meiotically and releases zoospores.

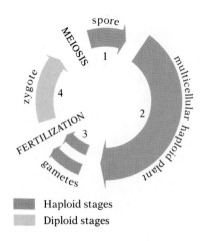

Haploid stages
Diploid stages

40.10 Life cycle characteristic of most multicellular green algae
Note that multicellularity is present only in the haploid phase.

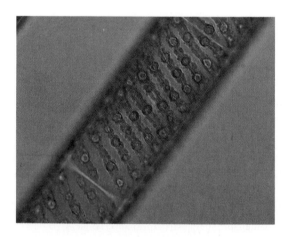

40.11 Photograph of living *Spirogyra*

and functions as a resting stage capable of withstanding unfavorable environmental conditions. At germination the zygote divides by meiosis, producing haploid zoospores (stage 1), each of which will grow into a new filament (stage 2). The main difference, then, between this life cycle and that of *Chlamydomonas* is the addition of the haploid multicellular stage (stage 2). As in *Chlamydomonas*, the only diploid stage is the zygote. This type of life cycle is diagramed in a more generalized form in Fig. 40.10.

Spirogyra (Fig. 40.11), a genus of rather odd filamentous green algae occurring in a great variety of freshwater habitats, is perhaps the most widely distributed member of the Chlorophyta. It has a sexual life cycle similar to that of *Ulothrix* except that the gamete cells are not flagellated and are not released from the plant that produces them. Instead, two filaments come to lie side by side; protuberances develop on the sides of the cells where they are in contact; the walls between the protuberances of each pair of cells disintegrate; then one cell becomes amoeboid, moves through the conjugation tube, and fuses with the other cell, forming a zygote (Fig. 40.12). Relatively few green algae reproduce by this conjugation process.

Ulva, or sea lettuce, is an example of a green alga with an expanded leaflike thallus two cells thick (Fig. 40.13). Its sexual life cycle is more complex than those of the other green algae discussed here in that it includes both multicellular haploid and multicellular diploid stages (stages 2 and 5 in Fig. 40.14). The entire cycle can be summarized as follows: Haploid zoospores (stage 1) divide mitotically to produce the haploid multicellular thalluses of stage 2. These may reproduce either asexually by means of zoospores or sexually by means of gametes (stage 3). Fusion of pairs of gametes (fertilization) produces diploid zygotes (stage 4). Upon germination, the zygotes divide mitotically (not meiotically as in the green algae previously discussed), producing diploid multicellular thalluses (stage 5). Eventually certain reproductive cells (sporangia) of these diploid plants divide by meiosis, producing haploid zoospores, which begin a new cycle. A life cycle of this type is said to exhibit **alternation of generations** in that a haploid multicellular phase alternates with a diploid multicellular phase. The haploid multicellular stage is customarily called a **gametophyte** (meaning that it is a plant that can produce gametes), and the diploid multicellular stage is called a **sporophyte** (meaning that it is a plant that can reproduce only by spores).

We have seen that multicellularity in plants arose first in the gametophyte, and that many green algae have no sporophyte stage. *Ulva* shows a more advanced life cycle in that both gametophyte and sporophyte stages are present. Furthermore, the two stages are equally prominent in *Ulva*, being nearly equal in duration and almost identical in appearance; in other words, the haploid portion of the life cycle is no longer dominant over the diploid.

PHAEOPHYTA (THE BROWN ALGAE)

The brown algae are almost exclusively marine, the few freshwater species being quite rare. Many of the plants called seaweeds are members of this division. They are most common along rocky coasts of the cooler parts of the oceans, where they normally grow attached to the bottom in the littoral (intertidal) and upper sublittoral zones. They may be seen in great abun-

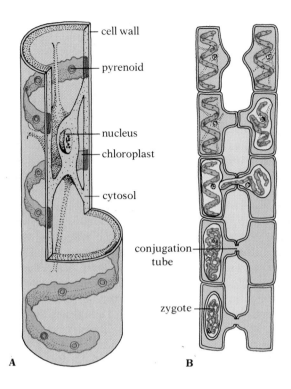

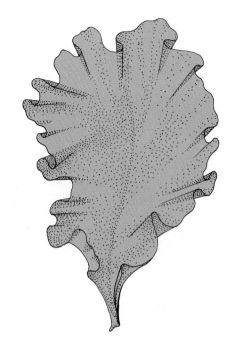

40.13 *Ulva*, a marine green alga with a three-dimensional leaflike thallus

40.12   Diagrammatic representations of *Spirogyra*
(A) A single vegetative cell removed from the filament and partially sectioned. Note the unusual spiral chloroplast that runs the length of the cell (some species have more than one chloroplast); numerous pyrenoids are associated with the chloroplast. The cell has a large central vacuole, in which the nucleus is suspended by cytoplasmic threads; these threads connect to the peripheral cytosol, which forms a layer just inside the cell wall. (B) Conjugating filaments. The two filaments lie side by side, and a conjugation tube develops between each pair of cells. One cell acts as the sperm, moving through the tube to fuse with the other cell (see middle pair of cells). The zygote thus formed is the only diploid stage in the life cycle.

40.14   Life cycle of *Ulva*
The gametophyte (multicellular haploid) and sporophyte (multicellular diploid) stages are equally prominent. *Ulva* and its close relatives are unusual among the green algae in having a life cycle of this sort; most of the Chlorophyta have no alternation of generations, the sporophyte stage being absent.

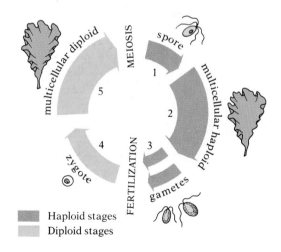

40.15 Brown algae (mostly *Fucus*) growing on rocks exposed at low tide

40.16 A giant kelp, *Macrocystis pyrifera*
At the base of each of the flattened blades is a gas-filled bladder that functions as a float.

dance covering the rocks exposed at low tide along the New England coast (Fig. 40.15). A few species occur in warmer seas, and some of these differ from the majority of brown algae in being able to live and grow when detached from the substratum; for example, some species of *Sargassum* form dense floating mats that cover much of the surface of the so-called Sargasso Sea, which occupies some six and a half million square kilometers of ocean between the West Indies and North Africa.

All brown algae are multicellular, and most are macroscopic, some growing as long as 45 m or more. The thallus (plant body) may be a filament, or it may be a large and rather complex three-dimensional structure (Figs. 40.16 and 6.25, p. 158). The latter type of thallus has apparently arisen several times independently; in some species it develops from interwoven and tightly compacted filaments, and in others it results from cell divisions in more than one plane. The individual cells have cell walls composed of cellulose and gummy material called alginic acid. The walls sometimes have pits through which plasmodesmata pass. The cells usually contain a large vacuole, one or several plastids, and sometimes a pyrenoid. Unlike the cells of most higher land plants, they usually have centrioles.

Like all photosynthetic plants, the Phaeophyta possess chlorophyll *a*. However, they have chlorophyll *c* instead of the chlorophyll *b* found in green algae and in land plants. Large amounts of a xanthophyll carotenoid called **fucoxanthin**, which are also present, give the characteristic brownish color to these algae. A polysaccharide called laminaran is the principal storage product.

Asexual reproduction is most often by flagellated zoospores. Sexual reproduction often involves specialized multicellular sex organs called **gametangia**. In isogamous species all the gametangia are alike. But in heterogamous species they are of two kinds. When the heterogamy is of the

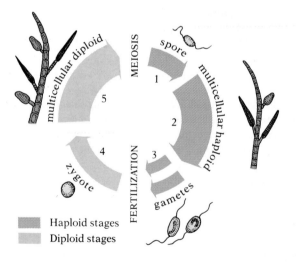

40.17 Life cycle of *Ectocarpus*
The gametophyte (multicellular haploid) and sporophyte (multicellular diploid) stages are equally prominent.

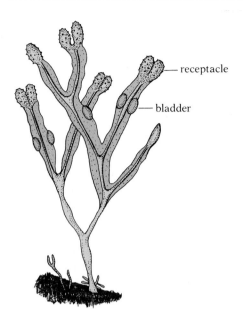

40.18 *Fucus* **(often called rockweed), a brown alga common along northern coasts**
Left: Each thallus is flattened and characterized by repeated dichotomous branching. Each younger axis consists of a midrib and thin paired wings. In some species (including the one shown here), there are bladders (floats) at intervals along the wings. The tips of fertile thalluses develop swollen reproductive structures called receptacles, whose surface is pocked by numerous tiny openings that lead into cavities (conceptacles) where the sex organs are located. In some species each individual has both male and female organs; in others the sexes are separate. Right: The receptacles can be seen especially well here.

oogamous type, the gametangia that produce sperm are called *antheridia* and those that produce eggs *oogonia*.

In brown algae the sex organs are ordinarily not enclosed by a protective layer of sterile jacket cells. The life cycle is usually characterized by an alternation of gametophyte (haploid) and sporophyte (diploid) multicellular generations. In many forms, such as *Ectocarpus*, the gametophyte and sporophyte (stages 2 and 5 in Fig. 40.17) are essentially similar in structure, and neither can be said to be dominant. In other forms, such as *Laminaria*, the haploid gametophyte is reduced and the diploid sporophyte is much larger and more prominent. In a few, such as *Fucus* (Fig. 40.18), reduction of the haploid stages has progressed so far that there is no longer any multicellular haploid gametophyte and the only haploid cells in the life cycle are the gametes (Fig. 40.19); such a life cycle, which is very rare in plants, is the sort seen in animals.

Let us look more closely at a few representative genera of brown algae. *Ectocarpus* has a branching filamentous thallus. The diploid sporophyte plants (stage 5 in Fig. 40.17) sometimes bear small unicellular sporangia, in which haploid zoospores (stage 1) are produced by meiosis. After swimming about for a while, the zoospores settle down and develop into haploid multicellular gametophyte plants (stage 2). These plants may bear multicellular gametangia, in which morphologically isogamous gametes (stage

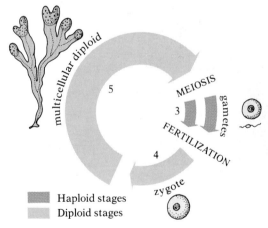

Haploid stages
Diploid stages

40.19 **Life cycle of *Fucus***
Fucus and its close relatives have a very unusual life cycle. They are the only multicellular plants in which the multicellular haploid stage (the gametophyte) is completely absent and meiosis produces gametes directly (that is, both stage 1 and stage 2 are absent). In this respect their life cycle is like that of animals.

40.20 Two examples of red algae
Top: *Kallimenia reniformis*, a species with a flattened
bladelike thallus. Bottom: The seaweed *Scinaia
furcellata.*

3) are produced. Two gametes (from different plants) may fuse in fertilization to form a zygote (stage 4), which is motile at first, but soon settles down and germinates, giving rise to a new diploid multicellular sporophyte plant (stage 5), thus completing the cycle.

Laminaria belongs to the group of brown algae commonly called kelps. The sporophyte thallus is large (about 2 m long in *Laminaria*; 45 m long or more in some kelps) and consists of a rootlike **holdfast**, a stemlike **stipe**, and an expanded leaflike **blade** (see Fig. 11.1, p. 273). Although thallophyte plants usually lack tissue differentiation, the stipe of some kelps has an outer surface tissue (epidermis), a middle tissue (cortex) containing many plastids, and a central core tissue (medulla); it may even have a meristematic layer similar to the cambium of higher vascular plants and, in a few genera, a phloemlike conductive tissue in the medulla. In short, these brown algae are complex plants that have convergently evolved many similarities to the vascular plants. However, none of them has a protective layer of sterile jacket cells around their reproductive organs; none develops multicellular embryos inside the oogonia; none has a cuticle; and none has xylem.

RHODOPHYTA (THE RED ALGAE)

The red algae are mostly marine seaweeds (Figs. 40.20 and 6.26, p. 158), but a few live in fresh water or on land. They often occur at greater depths than the brown algae. Most are multicellular and are attached to the substratum, but a few species are unicellular. No red algae attain the very large sizes often seen in brown algae.

The cell walls contain cellulose and also large quantities of mucilaginous material. The reserve product is not starch, but a polysaccharide similar to it called floridean starch. Red algae are an important source of commercial colloids—among others, agar used in culturing bacteria; suspending agents used in chocolate milk and puddings; stabilizers used in ice creams, some cheeses, and salad dressings; and moisture retainers used in icings, creams, and marshmallows.

In addition to chlorophyll *a*, which is found in all photosynthetic organisms except some photosynthetic bacteria, the Rhodophyta often possess chlorophyll *d*, which is not found in any other group of plants. They also contain phycocyanins and phycoerythrins. ("Phyco" is from the Greek *phykos*, "seaweed.") It is the phycoerythrins that give many of these algae their characteristic reddish color (Fig. 40.20). It should be emphasized, however, that "red algae" are not always red; many are black, violet, brownish, yellow, or even green.

The accessory pigments of the Rhodophyta play an important role in absorbing light for photosynthesis. The wavelengths preferentially absorbed by chlorophyll *a* for use in photosynthesis are among those at the ends of the visible spectrum (recall the discussion of the absorption spectrum of chlorophyll, p. 200). But those wavelengths do not penetrate to the depths at which the red algae grow, partly because of the imperfect transparency of the water, partly because they are selectively absorbed by pigmented phytoplankton. The wavelengths that do penetrate deep enough are mostly those of the central portion of the spectrum, which are not readily absorbed by chlorophyll *a*. But these wavelengths can be absorbed by the accessory pigments of the Rhodophyta, which then pass the energy to chlorophyll *a*. Thus the accessory pigments make it possible for red algae to live at depths where other algae, lacking these pigments, cannot survive.

The life cycles of red algae are usually very complex, and only a few have been worked out in detail. There is commonly some sort of alternation of generations. Flagellated cells never occur; even the sperm cells lack flagella and must be carried to the egg cells by water currents.

THE MOVEMENT ONTO LAND

Let us now turn to the Embryophyta, which have evolved numerous adaptations for life on land. We have seen that life probably arose in water, and that many plants, notably the algae, are still largely restricted to the aquatic environment; the few algae that live on land are not truly terrestrial, occurring, as they do, only in very moist places and actually living in a film of moisture. The evolutionary move from an aquatic mode of existence to a terrestrial one was not simple, for the terrestrial environment is in many ways hostile to life. Among the many problems faced by land plants are the following:

1. Obtaining enough water when fluid no longer bathes the entire surface of the plant body.
2. Transporting water and dissolved substances from restricted areas of intake to other parts of the plant body, and transporting the products of photosynthesis to those parts of the plant that no longer carry out this process for themselves.
3. Preventing excessive loss of water by evaporation.
4. Maintaining a sufficiently extensive moist surface for gas exchange when the surrounding medium is air instead of liquid.
5. Supporting a large plant body against the pull of gravity when the buoyancy of an aqueous medium is no longer available.
6. Carrying out reproduction when there is little water through which flagellated sperm may swim and when the zygote and early embryo are in severe danger of desiccation.
7. Withstanding the extreme fluctuations in temperature, humidity, wind, light, and other environmental parameters to which terrestrial organisms are often subjected.

Much of the evolution of the embryophyte plants can best be understood in terms of adaptations that help solve these problems.

As we indicated earlier in this chapter, embryophyte plants characteristically have multicellular sex organs with an outer layer of sterile (nonreproductive) jacket cells that help protect the enclosed gametes from desiccation; male and female sex organs of this type are known as ***antheridia*** and ***archegonia***.[6] The sporangia in embryophytes are also multicellu-

---

[6] Like "oogonium," "archegonium" denotes an organ producing female gametes, but most botanists restrict "archegonium" to the jacketed female reproductive organs of embryophytes, "oogonium" to the unjacketed ones of thallophytes.

TABLE 40.1 *A comparison of the major plant divisions*

Characteristics	Chlorophyta	Phaeophyta	Rhodophyta	Bryophyta	Tracheophyta
Sperm usually flagellated	+	+	−	+	+ or −
Chlorophyll *b*	+	− (*c* instead)	− (*d* instead)	+	+
Principal reserve material usually starch	+	−	−	+	+
Sporophyte equal or dominant to gametophyte in most species	*	+	*	−	+
Sex organs usually multicellular with jacket cells	−	−	−	+	+
Embryo development within archegonium	−	−	−	+	+
Cuticle usually present	−	−	−	+	+
Both xylem and phloem present in most species	−	−	−	−	+

* Among Chlorophyta and Rhodophyta species for which the full life cycle is known, some have a dominant gametophyte and some have a dominant sporophyte. Since many species have not yet been studied, it is not possible to say which condition is the more usual.

lar, and they too have a layer of jacket cells. All embryophytes are oogamous, and the egg cells are fertilized while they are still contained within the archegonia. Each zygote develops into a multicellular diploid embryo while still inside the archegonium. The embryo obtains some of its water and nutrients from the parent plant and is thus a parasite on it. This type of embryonic development, which is clearly an adaptation permitting the stages of development most susceptible to desiccation to take place in a favorably moist microenvironment, is strongly reminiscent of the internal gestation of mammals.

The surfaces of the aerial parts of the plant bodies of embryophytes are usually covered by a waxy cuticle, which waterproofs the epidermis and helps prevent excessive water loss.

The principal pigments in embryophytes are chlorophylls *a* and *b* plus carotenoids, and the reserve material is starch (Table 40.1). In other words, these plants are biochemically similar to the Chlorophyta, from which they almost certainly arose.

BRYOPHYTA (THE LIVERWORTS, HORNWORTS, AND MOSSES)

The bryophytes are relatively small plants that grow in moist places on land—on damp rocks and logs, on the forest floor, in swamps or marshes, or beside streams and pools. Some species can survive periods of drought,

40.21 A young moss plant
The spore (yellow) gives rise to a filamentous plant (called a protonema) that strikingly resembles a green alga. The protonema develops into the mature moss plant.

but only by becoming dormant and ceasing to grow. In short, the bryophytes live on land, but they have never become fully emancipated from their ancestral aquatic environment, and they have therefore never become a dominant group of plants. Their great dependence on a moist environment is linked to two characteristics: they retain flagellated sperm cells, which must swim to the egg cells in the archegonia; and most lack vascular tissues—and hence the means for efficient long-distance internal transport of fluids. A few bryophytes, however, possess cells that resemble the sieve cells of phloem, and some bryophytes possess elongated cells that sometimes transport water and are thus functional analogues of tracheids. The absence of true xylem cells with secondarily thickened walls, which function as major supportive elements in vascular plants, has probably also limited the size the bryophytes can attain.

The bryophytes are thought by some botanists to have arisen from filamentous green algae. Indeed, a very young moss plant, called a protonema (Fig. 40.21), often closely resembles a green algal filament. As the plant grows, it forms some branches (rhizoids) that enter the ground and function like roots, anchoring the plant and absorbing water and nutrients; different branches form upright shoots with stemlike and leaflike parts.

Other botanists hypothesize that the bryophytes arose from true vascular plants by "downgrade" evolution, that is, by secondary evolutionary loss of structures. They point to the rudimentary vascular tissue of some bryophytes, which they regard as vestiges of tissues well developed in ancestral forms rather than as newly evolved structures, and they also point to the nonfunctional guard cells seen in some bryophytes, which may be indicative of the ancestral condition.

We noted earlier among the larger and more complex algae, most of which exhibit alternation of generations, an apparent evolutionary tendency, in many instances, toward reduction of the gametophyte (multicellular haploid) stage and increasing emphasis of the sporophyte (multicellular diploid) stage. In brown algae, for example, the sporophyte is at least as prominent as the phylogenetically older gametophyte, and it is often much more prominent. As we shall see later, the same evolutionary tendency is found in the vascular plants. But this tendency is not apparent in the bryophytes, where the haploid gametophyte (stage 2 in Fig. 40.22) is clearly the dominant stage in the life cycle. The "leafy" green moss plant or liverwort is the gametophyte. These plants bear antheridia and archegonia

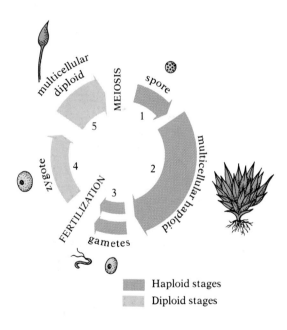

40.22 Life cycle of a bryophyte
Both gametophyte (stage 2) and sporophyte (stage 5) are present. The former is dominant.

in which gametes (stage 3) are produced by mitosis. The flagellated male gametes (sperm) are released from the antheridia and swim through a film of moisture, such as rain or heavy dew. Responding to chemical attractants, they swim to archegonia, where they fertilize the egg cells, producing zygotes (stage 4). Each zygote then divides mitotically, producing a diploid sporophyte (stage 5).

In a moss this sporophyte is a relatively simple structure consisting of three parts: a foot embedded in the "leafy" green gametophyte, a stalk, and a distal capsule, or sporangium (Figs. 40.23 and 40.24). The sporophyte has chloroplasts and carries out some photosynthesis, but it also obtains nutrients parasitically from the gametophyte to which it is attached. Meiosis occurs within the mature capsule of the sporophyte, producing haploid spores (stage 1), which are released. These spores, encased in highly specialized walls that are extremely resistant to degradation, may remain inactive for a long time (sometimes many years) if conditions are unfavorable. When they do germinate, they develop into protonemata and eventually into mature gametophyte plants (stage 2), thus completing the life cycle.

The gametophyte plants of some liverworts resemble mosses except that the "leaves" are scaly in appearance, and the "stem" is prostrate. Other liverworts grow as flat green structures lying on the substratum. In some species the antheridia and archegonia are borne in receptacles located at the top of stalks that arise from the flat part of the plant body (Fig. 40.25); in other species there is no receptacle or stalk (sometimes only the stalk is missing), and the reproductive structures are embedded in the upper por-

40.23 Gametophyte and sporophyte stages of a moss

The haploid gametophyte (green) is the lower "leafy" plant; the "leaves," except at their midrib, are only one cell thick. The diploid sporophyte plant (reddish gold), which consists of a foot (not visible here), a stalk, and a capsule, is attached to the gametophyte and is to some degree parasitic on it. (The capsule of the sporophyte is here covered by a cap—the calyptra—derived from the archegonium of the gametophyte; in time it will fall away, leaving the capsule fully exposed.)

40.24 Photograph of moss plants showing both the leafy green gametophytes and the gold-stalked sporophytes

40.27 Gemmae cups
Two cups can be seen here on the gametophyte of *Marchantia*. The gemmae cups function in asexual reproduction.

tion of the prostrate "leaf." The life cycle is much like that of mosses, except that the sporophyte is even simpler (Fig. 40.26). Asexual reproduction sometimes occurs by production of special cells called gemmae, usually borne in cuplike structures located on the surface of the flat gametophyte (Fig. 40.27). When detached from the parent plant, the gemmae can grow into new gametophytes.

TRACHEOPHYTA (THE VASCULAR PLANTS)

Though most bryophytes live on land, in a sense they are not fully terrestrial. The tracheophytes, by contrast, have evolved a host of adaptations to the terrestrial environment that have enabled them to invade all but the

40.25 Liverworts (*Marchantia*) with stalked receptacles bearing archegonia

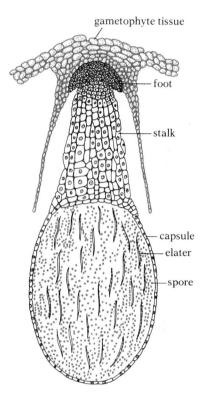

40.26 Sporophyte of *Marchantia*
The sporophyte of this liverwort is a small structure consisting of a foot, a short stalk, and a capsule. The foot remains embedded in the gametophyte plant, in the tissue of the undersurface of the umbrella-shaped receptacle (see Fig. 40.25). The mature capsule contains spores and elaters, which are elongate cells with spirally thickened walls. Eventually the wall of the capsule dries and bursts, releasing the spores. Ejection of the spores is aided by the elaters, which twist and jerk as they dry, thus throwing the spores from the capsule.

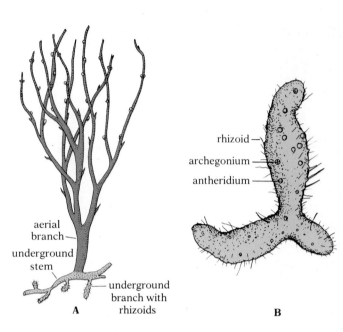

40.28 *Psilotum*
(A) Sporophyte. (B) Gametophyte: This stage is entirely subterranean. The two plants are not drawn to the same scale; the sporophyte is actually much larger than the gametophyte. (C) Photograph of a living *Psilotum*.

most inhospitable land habitats. In the process, they have diverged sufficiently from one another for botanists to classify them in five subdivisions:

Division Tracheophyta
    Subdivision Psilopsida (psilopsids)
    Subdivision Lycopsida (club mosses)
    Subdivision Sphenopsida (horsetails)
    Subdivision Pteropsida (ferns)
    Subdivision Spermopsida (seed plants)

All members of this division (with a few minor exceptions) possess four important attributes lacking in even the most advanced and complex algae: a protective layer of sterile jacket cells around the reproductive organs; multicellular embryos retained within the archegonia; cuticles on the aerial parts; and xylem (see Table 40.1). All four are obviously fundamental adaptations for a terrestrial existence. Many other such adaptations, absent in the earliest tracheophytes, appear in more advanced members of the division; a history of the evolution of these adaptations is a history of the increasingly extensive exploitation of the terrestrial environment by vascular plants. Let us briefly trace the history of adaptation to life on land.

PSILOPSIDA

The oldest undisputed fossil representatives of the vascular plants can be placed late in the Silurian period, which means that they lived more than 395 million years ago (see Table 37.1, p. 1012). They are classified in the subdivision Psilopsida, most of whose members lived during the Devonian period and then became extinct. Two living genera, *Psilotum* and *Tmesipteris*, have traditionally been regarded as members of this ancient group, but recent evidence from embryology and from the morphology of the gametophyte—as D. W. Bierhorst of the University of Massachusetts has pointed out—suggests that they may actually be very primitive ferns. If this is so, then Psilopsida contains only extinct species. Whether *Psilotum* and *Tmesipteris* should be retained in the Psilopsida despite the differences between them and the ancient members of that class, from which they are separated by about 400 million years with no intervening fossils, or whether they should be transferred to the ferns, despite their lack of true roots and the presence of other primitive features, is still an open question.

The psilopsid sporophytes are simple dichotomously branching plants that lack leaves[7] and have no true roots, although they do have underground stems that bear unicellular rhizoids similar to root hairs (Fig. 40.28A). The aerial stems are green and carry out photosynthesis. There is no cambium, and hence no secondary growth. Sporangia develop at the tips of some of the aerial branches. Within the sporangia meiosis produces haploid spores.

In *Psilotum* and *Tmesipteris* the spores give rise to minute subterranean gametophytes (Fig. 40.28B). Each gametophyte bears both archegonia and antheridia and thus produces both eggs and sperm. When the gametes unite in fertilization, they form diploid zygotes that develop into the sporophyte plants described above, thus completing the life cycle. Note that although the diploid sporophyte (stage 5) is more prominent in the modern genera and hence may be said to be dominant, the haploid gametophyte (stage 2) is still relatively large. It seems possible that the earliest Psilopsida had an alternation of coequal generations.

Botanists, noting the resemblance between the Psilopsida and certain branching filamentous green algae, have assumed it was from such algae that these primitive vascular plants arose. Since it is the psilopsid sporophyte, not the gametophyte, that most resembles the green algae, it must be concluded that, if the vascular plants did indeed evolve from green algae, those algae probably had a prominent sporophyte stage, at least coequal with the gametophyte.

40.29   *Lycopodium* **plants with strobili**

LYCOPSIDA (THE CLUB MOSSES)

The first representatives of the subdivision Lycopsida appeared in the middle of the Devonian period, almost 10 million years after the first Psilopsida. During the late Devonian and the Carboniferous periods these were among the dominant plants on land. Some of them were very large trees that formed the earth's first forests. Toward the end of the Paleozoic era, however, the group was displaced by more advanced types of vascular plants, and only five genera are alive today. One of these, *Lycopodium* (often called running pine or ground pine), is common in many parts of the United States and is frequently used in Christmas decorations (Fig. 40.29).[8]

---

[7] Some Psilopsida have scalelike structures that superficially resemble leaves.

[8] The other living genera of lycopsids are *Phylloglossum, Selaginella, Isoetes,* and *Stylites.*

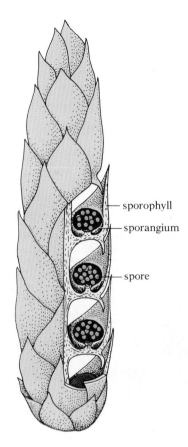

40.30   A strobilus of a lycopsid, partially sectioned to show the arrangement of sporangia on the sporophylls

sporophyll

sporangium

spore

40.31   Carboniferous swamp forest
Note the tree sphenopsids with their jointed stems and whorls of leaves. The trunk at right is a lycopsid.

Unlike the psilopsids, lycopsids have true roots. It is generally supposed that these arose from branches of the ancestral algae that penetrated the soil and branched underground. Lycopsids also have true leaves, which are thought to have arisen as simple scalelike outgrowths (emergences) from the outer tissues of the stem. Certain leaves that have become specialized for reproduction bear sporangia on their surfaces. Such reproductive (fertile) leaves are called *sporophylls*. In many lycopsids the sporophylls are congregated on a short length of stem and form a conelike structure (strobilus) (Fig. 40.30). The cone is rather club-shaped; hence the name "club mosses" for the lycopsids (note, however, that lycopsids are not related to the true mosses, which are bryophytes).

The spores produced by *Lycopodium* are all alike, and each can give rise to a gametophyte that will bear both archegonia and antheridia. However, some lycopsids (such as *Selaginella*) have two types of sporangia, which produce different kinds of spores. One type of sporangium produces very large spores called *megaspores*, which develop into female gametophytes bearing archegonia; the other type produces small spores called *microspores*, which develop into male gametophytes bearing antheridia. Plants like *Lycopodium* that produce only one kind of spore, and hence have only one kind of gametophyte that bears both male and female organs, are said to be *homosporous*. Plants like *Selaginella* produce both megaspores (female) and microspores (male); such plants, in which the sexes are separate in the gametophyte generation, are said to be *heterosporous*.

40.32 A fossil of a sphenopsid
A whorl of leaves is located at each joint of the stem.

SPHENOPSIDA (THE HORSETAILS)

The sphenopsids first appear in the fossil record late in the Devonian period. They became a major component of the land flora during the Carboniferous period and then declined. Members of the one living genus, *Equisetum*, are commonly called horsetails or scouring rushes. Though most of these are small (less than 1 m), some of the ancient sphenopsids were large trees (Fig. 40.31). Much of the coal we use today was formed from the dead bodies of these plants.

Like the lycopsids, sphenopsids possess true roots, stems, and leaves. The stems are hollow and are jointed. Whorls of leaves occur at each joint (Fig. 40.32). Many of the extinct sphenopsids had cambium, and hence secondary growth, but the modern species do not. Spores are borne in terminal cones (strobili) (Fig. 40.33). In *Equisetum* all spores are alike (that is, the plants are homosporous) and, since the sexes are not separate, they give rise to small gametophytes that bear both archegonia and antheridia.

PTEROPSIDA (THE FERNS)

In the opinion of many botanists, the ferns evolved from the Psilopsida. They first appeared in the Devonian period and greatly increased in importance during the Carboniferous. Their decline late in the Paleozoic era was much less severe than that of the psilopsids, lycopsids, and sphenopsids, and, as you doubtless know, there are many modern species.

The ferns are fairly advanced plants with a very well developed vascular system and with true roots, stems, and leaves. The leaves are thought to have arisen in another way than those of the lycopsids. Instead of emergences, they are probably flattened and webbed branch stems—that is, a group of small terminal branches probably became arranged in the same plane (planated), and the interstices filled with tissue.[9] Such leaves are larger, and provide a much greater surface area for photosynthesis, than the emergence leaves of the lycopsids and sphenopsids.

The leaves of ferns are sometimes simple, but more often they are com-

40.33 *Equisetum*
Three stalks bearing mature reproductive cones can be seen.

[9] Leaves arising as emergences are called microphylls. Those arising as planated and webbed branch systems are called megaphylls.

THE EVOLUTION OF
SPOROPHYTE DOMINANCE

An obvious question, in view of the tendency toward increasing dominance of the sporophyte generation in so many major groups of plants is why this shift of emphasis in the life cycle should have occurred. Why should diploidy be adaptively superior to haploidy?

The question can be answered only tentatively. In haploid organisms there can be no such thing as dominance and recessiveness; the one copy of each gene must be phenotypically expressed. This means that selection is very rigorous; deleterious new mutations cannot easily be covered up, and the effects of beneficial new mutations are immediately felt. In diploid organisms, by contrast, recessive genes harmful or nearly neutral in their effects may be carried in populations for a long time and may, if conditions change, become more beneficial in the future; such recessive genes thus represent a pool of genetic potential that may allow greater flexibility of adaptational response. Another source of evolutionary potential is the heterozygosity that diploidy makes possible; as we have seen earlier, heterozygous phenotypes are often adaptively superior to either of the corresponding homozygous phenotypes.

There is reason to believe, then, that the superiority of diploidy is the greater evolutionary flexibility it confers in the face of predation, intense competition, unstable environmental conditions, and other situations that make for rigorous selection.

40.34 Marsh ferns in autumn
The large leafy fern plant is the diploid sporophyte phase.

pound, being divided into numerous leaflets that may give the plant a lacy appearance.[10] In a few ferns like the large tree ferns of the tropics, the stem is upright, forming a trunk. But in most modern ferns, especially those of temperate regions, the stems are prostrate on or in the soil, and the large leaves are the only parts normally seen.

The large leafy fern plant is the diploid sporophyte phase (Fig. 40.34). Spores are produced in sporangia located in clusters on the underside of some leaves (sporophylls) (Fig. 40.35). In some species the sporophylls are relatively little modified and look like the nonreproductive leaves. In other species the sporophylls look quite different from vegetative leaves; sometimes they are so highly modified that they do not look like leaves at all, forming spikelike structures instead (Fig. 40.36).

Most modern ferns are homosporous—that is, all their spores are alike. After germination, the spores develop into gametophytes that bear both archegonia and antheridia (Fig. 40.37). These gametophytes are tiny (less than 1 cm wide), thin, and often more or less heart-shaped. Although most people are familiar with the sporophytes of ferns, few have ever seen a gametophyte, and even fewer would guess that it had anything to do with a fern. Small and obscure as it is, however, the fern gametophyte is an independent photosynthetic organism. Here, then, is a life cycle in which all five principal stages are present, but in which the multicellular haploid stage has been much reduced and the multicellular diploid stage emphasized (Fig. 40.38).

In some respects, the ferns (and also the three primitive groups of vascular plants discussed above) are no better adapted for life on land than the bryophytes. Their vascularized sporophytes can live in drier places and

[10] When a fern leaf, or frond, is divided into leaflets, the leaflets are called pinnae. The pinnae may themselves be subdivided into pinnules.

40.35 Undersurface of fertile leaf (sporophyll) of polypody fern
Each of the round dots is a sorus, which is a cluster of many tiny sporangia.

40.36 Leaves of sensitive fern
The sterile leaves have expanded blades, but the fertile leaves (sporophylls) are spikes bearing grapelike clusters of reproductive organs.

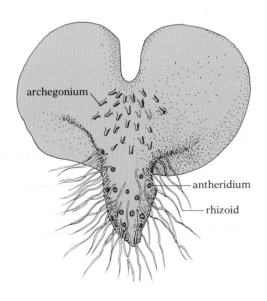

archegonium

antheridium

rhizoid

40.37 Fern gametophyte
This is a much-magnified view of the undersurface of the tiny heart-shaped organism.

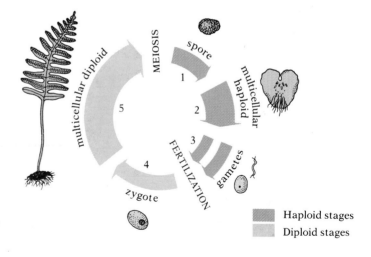

MEIOSIS

spore

multicellular diploid

multicellular haploid

1

2

5

3

4

FERTILIZATION

gametes

zygote

Haploid stages

Diploid stages

40.38 Life cycle of ferns
Both gametophyte (stage 2) and sporophyte (stage 5) are present; the latter is much the more prominent. Compare this life cycle with that of bryophytes, shown in Figure 40.22.

TABLE 40.2   *A comparison of the subdivisions of Tracheophyta*

Characteristics	Psilopsida	Lycopsida	Sphenopsida	Pteropsida	Spermopsida	
					Gymnosperms	Angiosperms
Vascular tissue	+	+	+	+	+	+
True roots and leaves	−	+	+	+	+	+
Megaphyllous leaves	−	−	−	+	+	+
Gametophyte retained in sporophyte tissue	−	−	−	−	+	+
Sperm cells without flagella	−	−	−	−	+ (− in primitive groups)	+
Production of seeds	−	−	−	−	+	+
Flowers and fruit	−	−	−	−	−	+

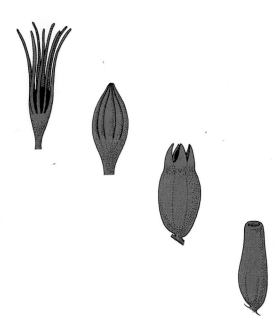

40.39   A model for the possible evolution of the seed
Shown here are the seeds of four species of extinct seed ferns (pteridosperms); note the progressive development of the integument. This sequence is thought to be similar to the one by which seeds of other plants evolved.

grow bigger, but for a number of reasons—because their nonvascularized free-living gametophytes can survive only in moist places, because their sperm are flagellated and must have a film of moisture through which to swim to the egg cells in the archegonia, and because the young sporophyte develops directly from the zygote without passing through any protected seedlike stage—these plants are most successful in habitats where there is at least a moderate amount of moisture.

SPERMOPSIDA (THE SEED PLANTS)

The seed plants have been by far the most successful in fully exploiting the terrestrial environment. They first appeared in the late Devonian; in the Carboniferous they soon replaced the lycopsids and sphenopsids as the dominant land plants, a position they still hold today. In these plants the gametophytes are even more reduced than in the ferns—they are not photosynthetic or free-living—and the sperm of most modern species are not independent free-swimming flagellated cells. In addition, the young embryo, together with a rich supply of nutrients, is enclosed within a desiccation-resistant seed coat and can remain dormant for extended periods if environmental conditions are unfavorable (Fig. 40.39). In short, the aspects of the reproductive process that are most vulnerable in more primitive vascular plants have been eliminated in the seed plants (Table 40.2).

The seed plants have traditionally been divided into two classes, the Gymnospermae and the Angiospermae. In recent years, however, it has become increasingly clear that the relationships between the five groups bracketed together as the gymnosperms are not particularly close and that these groups differ from one another at least as much as they differ from the angiosperms. Consequently many modern classifications recognize each of the gymnospermous groups as a separate class. We have adopted this procedure in the technical classification given in the Appendix and outlined below, but shall discuss the gymnospermous groups together.

Subdivision Spermopsida
    Class Pteridospermae ⎫
    Class Cycadae   ⎪
    Class Ginkgoae   ⎬ Gymnosperms
    Class Coniferae   ⎪
    Class Gneteae   ⎪
    Class Angiospermae ⎭

The gymnosperms   The first gymnosperms appear in the fossil record in the late Devonian, some 350 million years ago. Many of those first seed plants had bodies that closely resembled the ferns, and indeed for many years their fossils were thought to be fossils of ferns. Slowly, however, evidence accumulated that some of the "ferns" that were such important components of the coal-age forests produced seeds, not spores. Today these fossil plants—usually called the seed ferns—are grouped together as the class Pteridospermae of the subdivision Spermopsida. No members of this class survive today.

Another ancient group, the cycads and their relatives (class Cycadae), may have arisen from the seed ferns. These plants first appeared in the Permian period and became very abundant during the Mesozoic era. They had large palmlike leaves; the palmlike plants so often shown in pictures of the dinosaur age are usually cycads, not true palms. The cycads declined after the rise of the angiosperms in the Cretaceous period, but nine genera containing over a hundred species are in existence today (Fig. 40.40). They are generally called sago palms and are fairly common in some tropical regions. One genus (*Zamia*) occurs in Florida.

The Ginkgoae are still another once widespread group now nearly extinct. There is only one living species, the ginkgo or maidenhair tree, often planted as a lawn tree, but almost unknown in the wild.

By far the best-known group of gymnosperms is the conifers (class Coniferae), which include such common species as pines, spruces, firs, cedars, hemlocks, yews, and larches. The leaves of most of these plants are small evergreen needles or scales (Fig. 40.41), with an internal arrangement of

40.40   A living cycad
Though often called "sago palms," these plants are not really palms at all, but members of an ancient gymnosperm group.

40.41   Branch of a larch tree in spring
The leaves are needlelike. At this time of year the new female cones have a pinkish coloration.

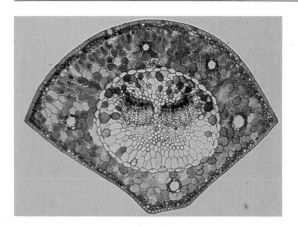

40.42   Cross section of a pine needle
Large resin ducts can be seen outside the prominent
endodermis that bounds the stele.

**40.43   Photographs of sections of male and female
pine cones**
(A) Female cone. Ovules can be seen on the surface
of the sporophylls near their base. (B) Male cone.
Each sporophyll (cone scale) bears a large
sporangium that becomes a pollen sac. (C) Male and
female cones can occur together. A female cone, or
strobilus, is seen at the bottom in this photo, with
male cones just above it.

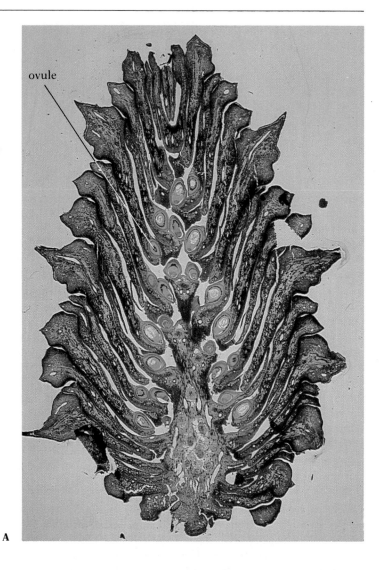

ovule

A

tissues (Fig. 40.42) that differs somewhat from that in angiosperms, which
we examined in earlier chapters. This group first arose in the Carboniferous
period and was very common during the Mesozoic era. It remains an important part of the earth's flora.

Let us follow in some detail the life cycle of a pine tree as an example of
the seed method of reproduction. The large pine tree is the diploid sporophyte stage (stage 5). This tree produces reproductive structures called
cones, of which there are two kinds: large female cones, in whose sporangia
meiosis gives rise to haploid megaspores (stage 2), and small male cones, in
whose sporangia meiosis gives rise to haploid microspores (Fig. 40.43).
(Production of distinctive male and female spores—heterospory—is characteristic of all seed plants, both gymnosperms and angiosperms.) In both
kinds of cones the sporangia are produced by highly modified leaves (sporophylls).

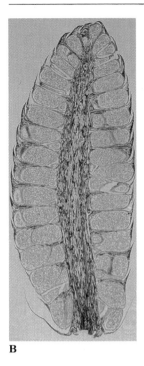

B

C

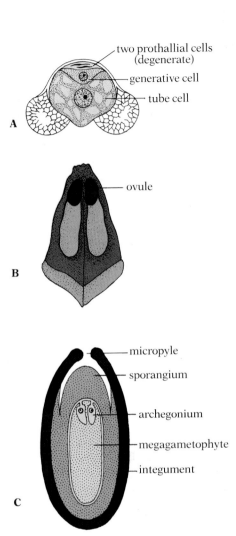

A

two prothallial cells (degenerate)

generative cell

tube cell

B

ovule

C

micropyle

sporangium

archegonium

megagametophyte

integument

40.44 Pollen grains and ovules of pine

(A) Pollen grain, composed of four cells, two of which are degenerate. (B) Scale from female cone. The two ovules, each containing a sporangium, lie on the surface of the scale near its point of attachment to the cone axis. (C) Section of an ovule.

Each scale of a female cone bears two sporangia on its upper (adaxial) surface (Fig. 40.44B). Each sporangium is encased in an integument with a small opening, the *micropyle*, at one end (Fig. 40.44C). Meiosis takes place inside the sporangium, producing four haploid megaspores, three of which soon disintegrate. Next, the single remaining megaspore gives rise, by repeated mitotic divisions, to a multicellular mass, which is the female gametophyte (megagametophyte). When mature, the female gametophyte produces two to five tiny archegonia at its micropylar end. Egg cells develop in the archegonia. Note that the megaspore is never released from the sporangium, and that the female gametophyte derived from it remains embedded in the sporangium, which is still attached to the cone scale. The composite structure consisting of integument, sporangium, and female gametophyte is called an *ovule*.

Each of the many microspores produced by meiosis in a sporangium of a male cone becomes a *pollen grain*. It develops a thick coat, which is highly resistant to loss of water, and winglike structures on each side, which doubtless aid its dispersal by wind. Within the pollen grain the haploid nucleus divides mitotically several times, and walls develop around each nucleus. In this manner the pollen grain becomes four-celled (Fig. 40.44A). Two of the cells soon degenerate;[11] the two cells that remain are called the generative cell and the tube cell. The mature pollen grain is released from

[11] The two cells that degenerate are called prothallial cells.

the cone when the sporangium bursts. A single male cone may release millions of tiny pollen grains, which may be carried many miles (sometimes as many as a hundred) by the wind. Note that the pollen grains are multicellular haploid structures (if four cells may be said to be "multi") and that they constitute the male gametophyte (microgametophyte) (stage 2 in Fig. 40.45).

Most of the millions of pollen grains released by a pine tree fail to reach a female cone. But of the few that sift down between the scales of a female cone, some land in a sticky secretion near the open micropylar end of an ovule. As this secretion dries, it is drawn through the micropyle, carrying the pollen grains with it. The arms of the integument around the micropyle then swell and close the opening. When a pollen grain comes in contact with the end of the sporangium just inside the micropyle, it develops a tubular outgrowth, the **pollen tube**. The nucleus of the tube cell enters the tube, followed by the generative cell. The generative cell then divides, giving rise to a sterile cell and a spermatogenous cell. The latter of these daughter cells divides again, producing two sperm cells. Thus a germinated pollen grain contains four active nuclei plus the two nuclei of the degenerate cells; this six-nucleate condition is as far as the male gametophyte of pine ever develops toward multicellularity.

The pollen tube grows down through the tissue of the sporangium and penetrates into one of the archegonia of the female gametophyte.[12] There it discharges its sperm cells, one of which fertilizes the egg cell. The resulting zygote (stage 4) then divides mitotically to produce a tiny embryo sporophyte consisting of a hypocotyl and an epicotyl. The embryo is still contained in the female gametophyte, which is itself contained in the

[12] Since an ovule contains several archegonia, several embryos may begin development, but usually only one completes it.

40.45 Life cycle of pine
The familiar tree is the sporophyte (multicellular diploid) stage. The gametophytes (multicellular haploid) are very tiny and cannot lead an independent existence. In all the haploid stages the sexes are separate.

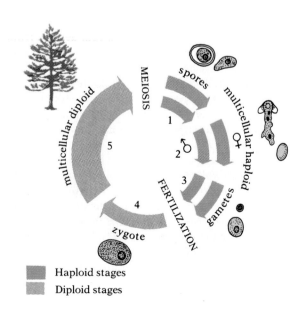

Haploid stages
Diploid stages

EVOLUTION OF PINE CONES

A male pine cone is a spiral cluster of cone scales, which are highly modified reproductive leaves (sporophylls) on a short section of stem (Fig. 40.43B). The same description was generally thought to apply to the larger female cone, too, until the fossil evidence recently forced a revision of that assumption.

Apparently the earliest female cones were compound structures composed of a section of stem bearing a series of modified nonreproductive leaves called bracts (Figure A). In the axil where each bract joined the stem was a very short budlike lateral branch; a few of its spirally arranged scalelike leaves bore sporangia—that is, they were sporophylls. In the course of evolution, all the dwarf lateral branches moved close together, and each was reduced to one to three sporophylls, which fused with several tiny sterile leaves to form a single compound structure called an ovuliferous scale. Each ovuliferous scale, in turn, partly fused with the bract in whose axil it developed (Figure B). Thus each "scale" of a modern female pine cone consists of a bract (which is a modified leaf) and an ovuliferous scale (derived from a dwarf branch consisting of fused sporophylls and sterile leaves) (Fig. 40.43A).

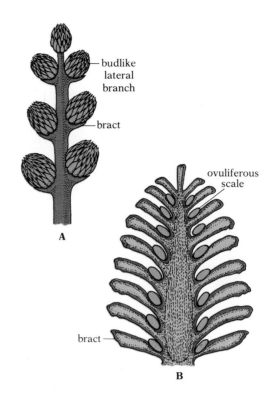

sporangium. Finally, the entire ovule is shed from the cone as a **seed**, which consists of three main components: a seed coat derived from the old integument, stored food material derived from the tissue of the female gametophyte, and an embryo.[13]

According to the fossil evidence, the gymnosperms did not evolve from the ferns; the two groups appear to have arisen independently from an ancestral group, probably the long-extinct progymnosperms. It is nonetheless instructive to summarize the advances in the life cycle of pine as compared with that of a typical fern:

1. The sporophylls are more highly modified and less leaflike.
2. There are two types of sporangia, which produce two types of spores—microspores (male) and megaspores (female).
3. Two kinds of gametophytes are derived from the two types of spores —that is, the sexes are separate in the gametophyte stage.

[13] As a rule, the sporangium (or nucellus, as botanists generally call the sporangium in the ovule) eventually disintegrates and is not present in the seed, but in a few species it may be preserved, usually as the inner layer of the seed coat.

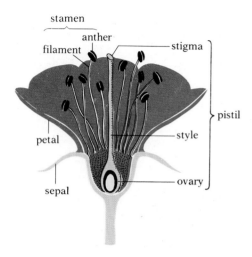

40.46 The parts of a flower

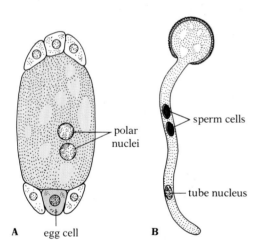

40.47 Gametophytes of an angiosperm
(A) Female gametophyte (embryo sac), which is composed of seven cells. One cell is much larger than the others and contains the two polar nuclei. (B) Male gametophyte (pollen grain and tube).

4. The gametophytes are much further reduced than those of ferns; they do not possess chlorophyll and are not free-living. A male gametophyte consists only of the six-nucleate pollen grain and tube. A female gametophyte is only a mass of haploid tissue that remains in the sporangium and is parasitic on it.
5. There are usually no flagellated sperm cells.[14]
6. The young embryo is contained within a seed.

The angiosperms The first undisputed representatives of the angiosperms are from the early Cretaceous. The group underwent great expansion later in the Cretaceous, and these plants became the dominant land flora of the Cenozoic era, as they are today.

We have seen that the reproductive structures of gymnosperms are cones (or similar structures) and that the ovules, which later become the seeds, are borne naked on the surface of the cone scales. The reproductive structures of angiosperms, by contrast, are flowers, and the ovules are enclosed within modified leaves called carpels.

A flower is generally interpreted as a short length of stem with modified leaves attached to it. The modified leaves of a typical flower (Fig. 40.46) occur in four sets attached to the enlarged end (receptacle) of the flower stalk: (1) The *sepals* enclose and protect all the other floral parts during the bud stage. They are usually small, green, and leaflike, but in some species they are large and brightly colored. All the sepals together form the *calyx*. (2) Internal to the sepals are the *petals*, which together form the *corolla*. The calyx and corolla together constitute the *perianth*. In flowers pollinated by insects, birds, or other animals, the petals are usually quite showy, but in those pollinated by wind they are often small or even absent. (3) Just inside the circle of the corolla are the *stamens*, which are the male reproductive organs—that is, they are the sporophylls that produce the microspores.[15] Each stamen consists of a stalk, called a *filament*, and a terminal ovoid pollen-producing structure called an *anther*. (4) In the center of the flower is the female reproductive organ, the *pistil* (some species have more than one pistil per flower). Each pistil consists of an *ovary* at its base, a slender stalk (more than one in some species) called a *style*, which rises from the ovary, and an enlarged apex called a *stigma*. The pistil is derived from one or more sporophylls, which in flowers are called *carpels*.[16] All four kinds of floral organs—sepals, petals, stamens, and pistils—are

[14] The pollen tubes of primitive gymnosperms, such as ginkgoes and cycads, produce flagellated sperm cells, but these have only a very short distance to swim within the cytoplasm of the pollen tube to reach the egg cells, the pollen grains having already been carried to the ovules by the wind.

[15] It is not entirely certain that the stamens and carpels evolved directly from individual leaves. Some investigators consider them compound structures derived from the fusion of a bract leaf and a short reproductive shoot. Whether they are right, or whether, in accordance with the older, more widely held view, the stamens and carpels are highly modified single leaves, it is probably safe to regard the actual spore-producing parts as sporophylls.

[16] A simple pistil is composed of only one sporophyll, or carpel. A compound pistil is composed of several fused carpels.

present in so-called complete flowers, but some flowers, which are said to be incomplete, lack one or more of them.[17]

Within the ovary are one or more (at least one for each carpel) sporangia, called ***ovules***, which are attached by short stalks to the wall of the ovary. Meiosis occurs once in each ovule, with formation of four haploid megaspores, three of which usually soon disintegrate. The remaining megaspore then divides mitotically several times, producing, in most species, a structure composed of seven cells, one of which is much larger than the others and contains two nuclei, called polar nuclei (Fig. 40.47). This haploid seven-celled eight-nucleate structure is the much-reduced female gametophyte (often called an embryo sac).[18] One of the cells located near the micropylar end will act as the egg cell.

Each anther has four sporangia in each of which many cells undergo meiosis, producing numerous haploid microspores. The wall of each microspore thickens, and the nucleus divides mitotically, producing a generative nucleus and a tube nucleus. The resulting thick-walled two-nucleate structure is a pollen grain—a male gametophyte—which is released from the anther when the mature sporangium (Fig. 40.48) splits open.

A pollen grain germinates when it falls (or is deposited) on the stigma of a pistil, which is usually rough and sticky. A pollen tube begins to grow, and the two nuclei of the pollen grain move into it. The generative nucleus (which is surrounded by a plasma membrane and is thus technically a cell, though with virtually no cytoplasm) then divides, giving rise to two sperm cells (Fig. 40.47B).[19] The pollen tube grows down through the tissues of the stigma and style and enters the ovary (Fig. 40.49). When the tip of the pollen tube reaches an ovule, it enters the micropyle and then discharges the two sperm cells into the female gametophyte (embryo sac). One of the sperm fertilizes the egg cell, and the zygote thus formed develops into an embryo sporophyte. By the time fertilization occurs, the two polar nuclei of the fe-

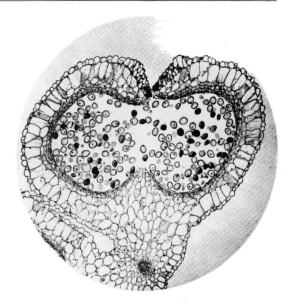

40.48 Pollen chamber of a lily anther
When the pollen chamber (derived from a sporangium) opens, the numerous pollen grains within it will be released.

---

[17] In some species, such as corn, willow, oak, and walnut, the stamens and pistils are in separate flowers. Incomplete flowers of this type, in which only one of the two kinds of reproductive structures is present, are called imperfect flowers. Flowers with both stamens and pistils (whether complete or incomplete) are called perfect flowers.

[18] The embryo sac of some species has more than eight nuclei, and that of a limited number of other species has fewer than eight. Furthermore, in some species no cytokinesis occurs, and all the nuclei lie in the same mass of cytoplasm. However, the most common sort of embryo sac is the seven-celled eight-nucleate type described here.

[19] The pollen grains in most species are released in the two-nucleate condition and the division of the generative nucleus does not take place until germination; but in some species this division occurs earlier and the pollen grains are released from the anthers in the three-nucleate condition.

40.49 Fertilization of an angiosperm
Pollen grains land on the stigma and give rise to pollen tubes that grow downward through the style. One of the pollen tubes shown here has reached the ovule in the ovary and discharged its sperm cells into it. One sperm will fertilize the egg cell, and the other will unite with the diploid fusion nucleus (derived from the two polar nuclei) to form a triploid nucleus, which will give rise to endosperm.

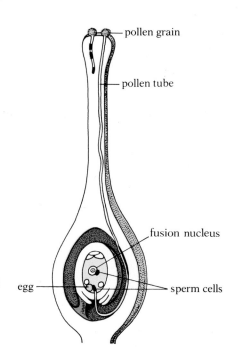

— pollen grain

— pollen tube

fusion nucleus

egg —

— sperm cells

40.50   Thistle of the milkweed
The milkweed thistle is a fruit and bears many seeds, each of which is attached to a set of fibers that act as a wind-borne sail.

male gametophyte have combined to form a diploid *fusion nucleus*, with which the second sperm unites to form a triploid nucleus. This nucleus undergoes a series of divisions, and a triploid tissue, called *endosperm*, is formed. The endosperm functions in the seed as a source of stored food for the embryo.

After fertilization, the ovule matures into a seed, which, as in pine, consists of seed coat, stored food, and embryo. However, the angiosperm seed differs from that of pine in being enveloped by the ovary. It is the ovary that develops into the *fruit*, usually enlarging greatly in the process. Sometimes other structures associated with the ovary, such as the receptacle, are also incorporated into the fruit. The ripe fruit may burst, expelling the seeds, as in peas (where the pod is the fruit). Or the ripe fruit with the seeds still inside may fall from the plant, as in tomatoes, squash, apples, peaches, and acorns. The fruit not only helps protect the seeds from desiccation during their early development, before they have fully ripened, but often also facilitates their dispersal by various means—the wind, say (Fig. 40.50), or an animal which, attracted by the fruit, carries it to other locations or eats both fruit and seeds and later releases the unharmed seeds in its feces.

The main features in which the angiosperm life cycle (Fig. 40.51) differs from that of gymnosperms can be summarized as follows:

1. The reproductive structures are flowers instead of cones. The sporophylls (stamens and pistils) of flowers are even less leaflike than those of cones.
2. The ovules are embedded in the tissues of the female sporophylls instead of lying bare on their surface.
3. The gametophytes are even more reduced than those of gymnosperms. The male gametophyte (pollen grain and tube) has only three nuclei. The female gametophyte usually has only eight nuclei.
4. In pollination the pollen grains are not deposited close to the opening of the ovule, as in gymnosperms, but are deposited on the stigma instead. The pollen tube thus has much farther to grow in angiosperms.
5. Angiosperms have "double fertilization," one sperm fertilizing the egg cell and the other uniting with the fusion nucleus to give rise to a triploid endosperm.[20] Gymnosperms, by contrast, have single fertilization. One sperm fertilizes the egg, and the other soon deteriorates. The stored food in the seed of gymnosperms is the haploid tissue of the female gametophyte and is thus developmentally quite different from the triploid endosperm of the angiosperms.
6. The seeds of angiosperms are enclosed in fruits that develop from the ovaries and associated structures; gymnosperms have no fruit.

Angiosperms also differ from gymnosperms in other than reproductive characteristics—in the structure of the xylem, for example, which in angiosperms often contains vessels, but seldom does in gymnosperms.

The class Angiospermae is customarily divided into two subclasses, the

[20] In some angiosperms—lilies, for example—the endosperm is pentaploid ($5n$).

Dicotyledoneae and the Monocotyledoneae. The dicots include oaks, maples, elms, willows, roses, beans, clover, tomatoes, asters, and dandelions. The monocots include grasses, corn, wheat, rye, onions, daffodils, irises, lilies, and palms. Most spring-flowering bulbs are monocots. There are certain basic differences between the two groups:

1. As the names imply, the embryos of dicots have two cotyledons, whereas those of monocots have only one.
2. Dicots often have vascular cambium and secondary growth; monocots usually do not.
3. The vascular bundles in the stems of young dicots are arranged in a circle or fused to form a tubular vascular cylinder; monocots have more scattered vascular bundles.
4. The leaves of dicots usually have net venation; those of monocots usually have parallel venation.
5. Dicot leaves generally have petioles; monocots generally do not. Many monocots can be recognized by the way the leaf base clasps the stem (as in corn).
6. The flower parts of dicots usually occur in fours or fives or multiples of these (for example, four sepals, four petals, four stamens); those of monocots usually occur in threes or multiples of three (Fig. 40.52).

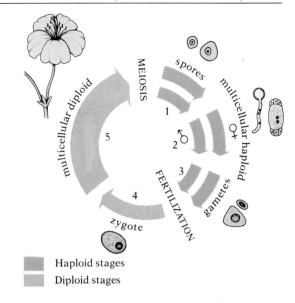

Haploid stages
Diploid stages

40.51   Life cycle of angiosperms
The mature diploid plant produces gametes—pollen, which is dispersed, and eggs, which are retained in the ovary. Once fertilization has occurred, a seed is produced, from which the new plant develops.

40.52   Dicot and monocot flowers compared
Above: In a nasturtium flower, a dicot, the main parts occur in fives. Right: In a trillium flower, a monocot, the parts occur in threes.

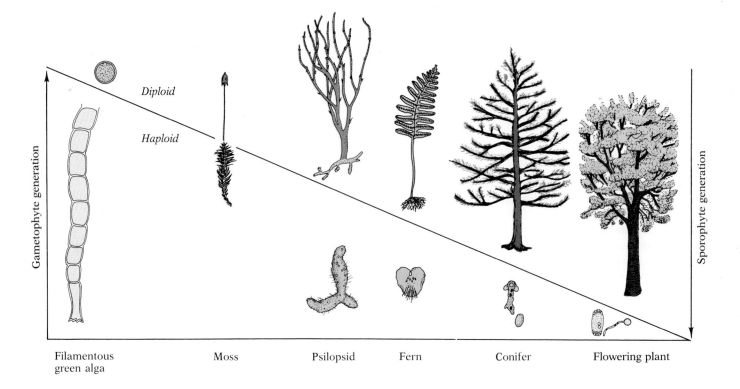

Gametophyte generation

Diploid

Haploid

Sporophyte generation

| Filamentous green alga | Moss | Psilopsid | Fern | Conifer | Flowering plant |

40.53 Transition from gametophyte to sporophyte dominance as shown by representative plant groups
The gradual decrease in the size and importance of the gametophyte and the corresponding increase in the size and importance of the sporophyte is shown. In filamentous green algae and in mosses, the gametophyte state is dominant and the sporophyte state is much reduced. In the other plant groups shown, the sporophyte is dominant and the gametophyte becomes progressively smaller. Psilopsids have a large underground gametophyte, ferns have a small but independent (photosynthetic) gametophyte, and conifers and flowering plants have tiny female gametophytes and male pollen grains, both dependent on the sporophytes. The organisms are not drawn to scale.

SUMMARY OF EVOLUTIONARY TRENDS

Let us summarize some evolutionary trends and advances among the plant groups. One major trend, shown diagrammatically in Figure 40.53, is the progressive decrease in the size and importance of the gametophyte. Along with the reduction in size has come a gradual loss of independence. At the same time, the sporophyte has become increasingly larger and more independent.

Another evolutionary trend in higher plants has been the change from the homosporous to the heterosporous condition. The early vascular plants were largely homosporous, producing only one type of spore, which developed into gametophytes containing both archegonia and antheridia. The more advanced vascular plants, however, are heterosporous, producing two types of spores—microspores and megaspores. The microspores give rise to male gametophytes, and the megaspores to female gametophytes. Here the sexes are separate in the gametophyte generation.

Many of the evolutionary advances can be viewed as adaptations for life on land. The evolution of vascular tissue in the primitive tracheophytes enabled plants to grow larger and survive in drier environments. However, these plants were not completely freed from their dependence on an aquatic environment; water was still required for reproduction. The angiosperms and gymnosperms solved this problem by the evolution of pollen and seeds. The development of the pollen grain eliminated the need for external water for fertilization. And the seed, with its protective coat and stored food, protected the embryo and enabled it to withstand unfavorable conditions. The development of the seed method of reproduction is probably responsible for the dominance of the seed plants today.

Chapter 41

THE FUNGAL KINGDOM

The several divisions of organisms that we here place in the kingdom Fungi were in the past often grouped together in a single division, the Eumycophyta, or true fungi.

The fungi are eucaryotic organisms, and they are primarily multicellular or multinucleate. Their multicellularity, however, differs from that of plants and animals in that partitions between nucleated compartments, or "cells," are generally either absent or only partial; thus the cytoplasm is continuous. Moreover, each compartment often has more than one nucleus. For many fungi, therefore, "multinucleate" is perhaps a more accurate description than "multicellular." The "cells" are usually organized into branched filaments called **hyphae**, which form a mass called a **mycelium**. In most fungi the basic component of the cell wall is not cellulose (though a small amount may be present in a few species), but chitin, a derivative polysaccharide containing nitrogen—also found in the exoskeletons of arthropods.

The fungi are parasitic or saprophytic, though a few are at times predatory (see Fig. 10.5, p. 248). Most saprophytic fungi secrete digestive enzymes onto their food material and absorb the products of the extracellular digestion by means of rootlike rhizoids or haustoria (see p. 247). Parasitic fungi may carry out extracellular digestion, or they may directly absorb materials produced by the body of their host.

Some fungi are parasitic on or in animals, including humans; many skin diseases, including "ringworm" and athlete's foot, are caused by fungi, and there are several serious fungal diseases of the lungs. Other fungi are parasitic on plants, and some of these cause annual losses of hundreds of millions of dollars when they attack agricultural crops. Still others cause spoilage of bread, fruit, vegetables, and other foodstuffs, and deterioration of leather goods, fabrics, paper, lumber, and other valuable products.

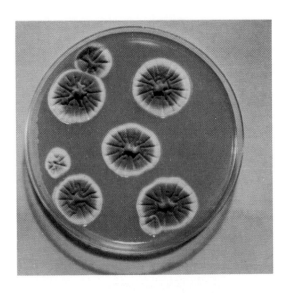

41.1 Seven colonies of *Penicillium chrysogenum* growing on culture medium in a Petri dish
Almost the entire world's supply of commercial penicillin is produced by this fungus in pharmaceutical laboratories.

The numerous pathogenic or destructive fungi should not cause us to forget the many others that are beneficial. For instance, most plant roots have symbiotic fungi living in their cortex, sending hyphae into the soil; the hyphae can significantly enhance a plant's ability to extract water and nutrients, particularly phosphorus, from the soil. These associations, which we discussed earlier (see p. 234), are called mycorrhizae. Yeasts are used extensively in the manufacture of alcoholic products and to make bread dough rise. The antibiotic penicillin is obtained from a fungus (Fig. 41.1). Fungi are important in the manufacture of many cheeses; and certain mushrooms are regularly used as food. Fungi, together with bacteria, decompose vast quantities of dead organic matter that would otherwise rapidly accumulate and make the earth uninhabitable.

Reproduction in the fungi may be either asexual or sexual, but in both cases the haploid stages are usually dominant (the Oomycota are an exception). The characteristics of sexual reproduction are especially important in distinguishing the four divisions of true fungi currently recognized.[1]

OOMYCOTA (THE WATER MOLDS)

As their common name "water molds" suggests, most of the Oomycota are aquatic. Some, however, live in the soil, and some are parasites of higher plants. Among the latter is the species *Phytophthora infestans*, the cause of late blight of potatoes, a historic scourge responsible for such devastations as the Irish potato famine of 1845–47, which did much to change the course of Irish history. Other members of this group of fungi—notably the agents of downy mildew of sugar beets and grapes—can also cause severe economic damage.

There are a number of striking differences between the Oomycota and the other three groups of true fungi, which have led some authorities to postulate that they may represent an entirely separate evolutionary development: (1) They are the only true fungi that typically have flagellated zoospores. (2) Their sexual reproduction is oogamous. (3) Unlike other fungi, they are diploid during most of their life cycle. (4) They are unusual among the fungi in lacking chitin in their cell walls; though some cellulose is present, the principal component of their walls is another polysaccharide with linkages between adjacent sugar units different from those in cellulose. In most instances the hyphae lack cellular partitions (septa); the multinucleate cytoplasm is therefore a continuous (coenocytic) system.

ZYGOMYCOTA (THE CONJUGATION FUNGI)

The hyphae of the Zygomycota characteristically lack cross walls, although they contain many haploid nuclei—that is, they are coenocytic. Cross walls appear only during the formation of reproductive structures. Neither the spores nor the gametes are motile. Sexual reproduction is accomplished by

---

[1] Other bases of classification are generally unsatisfactory. Hence it is often difficult to classify species for which sexual stages have not been found. Such species are customarily assigned to a category called Fungi Imperfecti; the "imperfection," of course, is in our understanding rather than in the fungus.

the conjugative fusion of morphologically indistinguishable cells from hyphae of two different mycelia (Fig. 41.2B).

A typical example of a member of this division is the common black bread mold, *Rhizopus*. The hyphae of this mold form a whitish or grayish mycelium on the bread. The mycelium includes three types of hyphae: hyphae (called stolons) that form a network on the surface of the bread; rootlike hyphae (called rhizoids) that penetrate into the bread and function both in anchoring the plant and in absorbing nutrients; and hyphae (called sporangiophores) that grow upright from the surface and bear globular sporangia on their ends (Fig. 41.2A; see also Fig. 6.30, p. 161). Thousands of asexual spores are produced in each sporangium. The spores, which have no flagella, are very tiny and light, and when liberated at maturity (by disintegration of the wall of the sporangium) they may be carried long distances by wind, rain, or animals. If a spore lands in a suitable location, where conditions are warm and moist, it germinates and soon gives rise to a new mass of hyphae, thus completing the asexual cycle.

Sexual reproduction in *Rhizopus* resembles that of the green alga *Spirogyra* (see Fig. 40.12, p. 1069). Short branches from two hyphae (which must be of different mating types or sexes) contact each other at their tips (Fig. 41.2B). Cross walls soon form just back of the tips of these hyphal branches. The gamete cells thus delimited then fuse to form a zygote, which develops a thick protective wall and enters a period of dormancy usually lasting from one to three months. At germination the nucleus of the zygote undergoes meiosis, and a short hypha grows from the zygote. This haploid hypha promptly produces a sporangium, which releases asexual spores that grow into new mycelia. Note that the only diploid stage in the entire sexual cycle is the zygote (Fig. 41.3).

Members of the Zygomycota are widespread as saprophytes in soil and dung; they are thus mostly terrestrial, only a few species living in aquatic habitats. Some, however, are parasites on plants, animals, or other fungi.

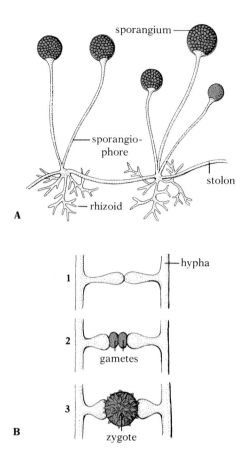

41.2 *Rhizopus*
(A) Hyphae with sporangia. (B) Sexual reproduction by conjugation: (1) Short branches from two different hyphae meet. (2) The tips of the branch hyphae are cut off as gametes. (3) The gametes fuse in fertilization to form a zygote with a thick spiny wall.

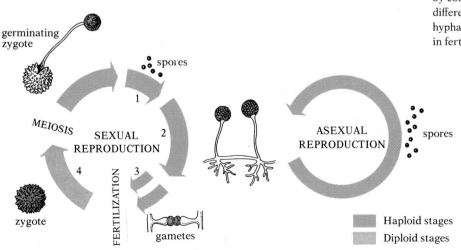

41.3 Life cycle of *Rhizopus*
See text for description.

41.4 The cup fungus *Peziza* growing on a log in a rain forest in Central America

ASCOMYCOTA (THE SAC FUNGI)

The members of this large division are very diverse, varying all the way from unicellular yeasts through powdery mildews and cottony molds to complex cup fungi. These last form a cup-shaped structure composed of many hyphae tightly packed together (Fig. 41.4). The vegetative hyphae of Ascomycota, unlike those of Zygomycota, are septate—that is, they possess cross walls—but the septa are usually incomplete, having large holes in their centers. The cytoplasm of adjacent cells is thus continuous.

Though their vegetative structures differ, all Ascomycota resemble each other in forming a reproductive structure called an *ascus* during their sexual cycle. An ascus is a sac within which haploid spores (usually eight, but sometimes four) are produced; all the spores in an ascus are derived from a single parent cell. The events leading to the formation of a mature ascus are shown in Figure 41.5.

Most Ascomycota also reproduce asexually by means of special spores called *conidia* (Fig. 41.6). Conidia are produced in chains at the end of conidiophore hyphae (but not inside sporangia). Each conidium can grow into a new fungal plant.

It may seem strange that yeasts are considered members of the Ascomycota. They are unicellular, and they reproduce asexually by *budding* (Fig. 41.7), not by conidia formation. However, under certain conditions a single yeast cell may function as an ascus, producing four spores. The spores are more resistant to unfavorable environmental conditions than vegetative

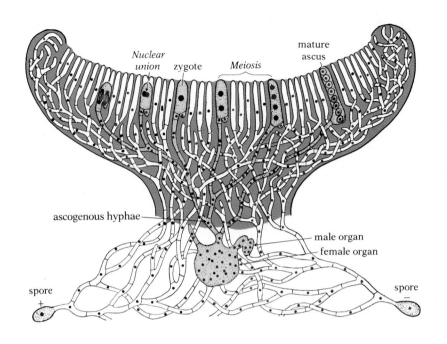

41.5 Diagram of a magnified section through a cup fungus

Hyphae of two haploid mycelia, one derived from a plus (female) spore and the other from a minus (male) spore, participate in forming the cup structure and in producing the spores. The plus mycelium bears a female organ (ascogonium), and the minus mycelium bears a male organ (antheridium). A tube grows from the female organ to the male organ, and then minus nuclei (color dots), acting as male nuclei, move into the female organ and become associated with plus nuclei (black dots). Next, hyphae grow from the female organ; each cell in these hyphae contains two associated nuclei, one minus and one plus. Such a cell, containing two nuclei, is said to be dikaryotic; when, as here, the nuclei are genetically different (heterokaryotic), the cells are effectively diploid even though the individual nuclei are haploid. The terminal cells of the dikaryotic hyphae eventually become elongate, and their nuclei unite, forming a zygote nucleus (dark color). The zygote nucleus promptly divides meiotically, and each of the four haploid nuclei thus formed then divides mitotically, producing a total of eight small spore cells, which are still contained within the wall of the old zygote cell, now called an ascus. When the mature ascus ruptures, the spores are released.

41.6 Asexual reproductive structures of *Aspergillus,* a member of the Ascomycota

(A) Two conidiophores arise from a mycelium. (B) An enlarged section through a conidiophore shows that the structure bears numerous spores, called conidia, arranged in chains.

41.7 Budding in brewer's yeast

A small new cell is pinched off the larger parent cell.

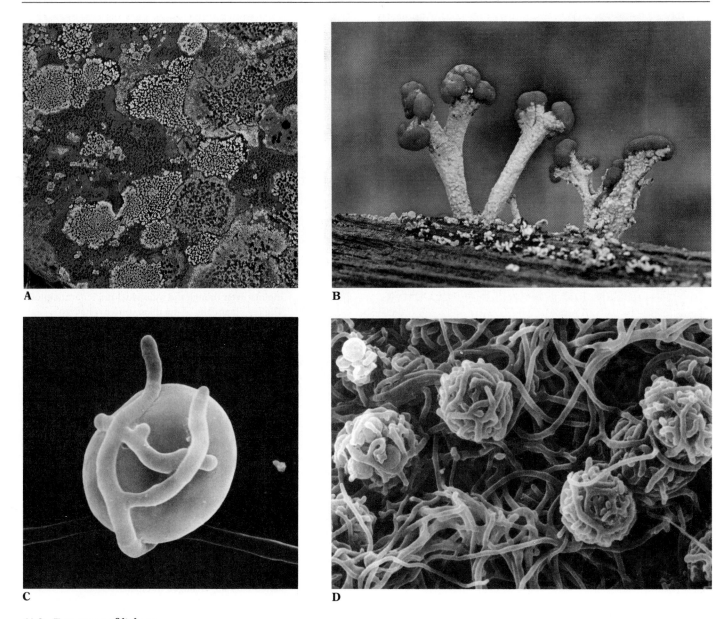

41.8 Two types of lichens

(A) Both *Lecidea atrata* (rust-red) and *Lecidea lithophila* (gray-black) grow as a crust, shown here on a boulder in Cairngorms, Scotland. (B) British soldiers, with their bright red tops, have a more upright growth. (C) A resynthesis of the algal–fungal association can be generated through the culturing of British soldiers. Here a fungal hypha encircles an algal cell. (D) In a later stage of development, many algal cells are held together by a network of hyphae.

41.9 Two representative Basidiomycota
Left: Fly agaric mushroom (*Amanita muscaria*), a deadly poisonous species. The spore-carrying gills on the underside of the cap, which greatly enlarge the spore-bearing surface, are clearly visible in the specimen that has fallen on its side. Right: One of the many kinds of edible mushrooms, *Lepiota procera*.

cells, and they may enable yeasts to survive temperature extremes or periods of prolonged drying.

In previous chapters we mentioned the ***lichens***, plants composed of a fungus and an alga or photosynthetic bacterium growing together in a complex symbiotic relationship (Fig. 41.8). The fungal components of most lichens are Ascomycota, though in a few tropical forms the fungus is a member of the Basidiomycota, discussed below. The photosynthetic components are usually green algae (Chlorophyta), though they may be Cyanobacteria.

BASIDIOMYCOTA (THE CLUB FUNGI)

Many of the largest and most conspicuous fungi—puffballs, mushrooms, toadstools, and bracket fungi—are Basidiomycota (Fig. 41.9). Though the above-ground portion of these plants looks like a solid mass of tissue, and in some is differentiated into a stalk and a prominent cap, it is nevertheless composed of hyphae, as are all fungi. The above-ground portion, or fruiting body, of many mushrooms is only a small part of the total plant; there is an extensive mass of hyphae in the soil. The hyphae are septate.

The members of this class are distinguished by their club-shaped repro-

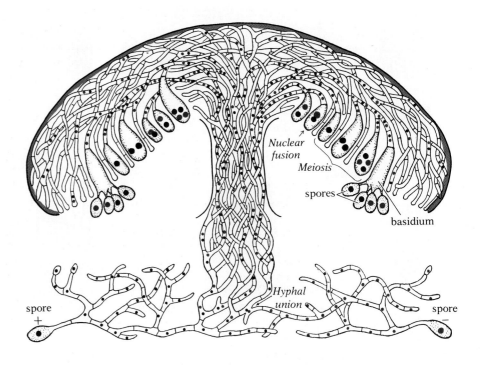

41.10 Diagram of section through a mushroom
Hyphae from two uninucleate mycelia—one of the plus, the other of the minus, mating type—unite and give rise to dikaryotic hyphae, which then develop into the above-ground part of the mushroom. The entire stalk and cap are composed of these hyphae tightly packed together. Spores are produced by basidia on the lower surface of the cap.

ductive structures called **basidia** (Fig. 41.10). The cells of the hyphae that produce basidia are binucleate (dikaryotic), containing one haploid male nucleus and one haploid female nucleus (usually designated as minus and plus). Certain terminal cells of these hyphae, located in rows, or "gills," on the undersurface of the cap of those Basidiomycota called gill fungi (Fig. 41.10), become zygotes when their two nuclei fuse in fertilization. The zygote then becomes a basidium. Its diploid nucleus divides by meiosis, producing four new haploid nuclei. Four small protuberances develop on the end of the basidium, and the haploid nuclei migrate into these. The tip of each protuberance then becomes walled off as a spore, which is usually ejected from the basidium. Each spore may give rise to a new mycelium.

Chapter 42

THE ANIMAL KINGDOM: INVERTEBRATES

You already have some familiarity with many phyla of the animal kingdom from our previous consideration of their vital processes—their nutrient procurement, gas exchange, internal transport, excretion, coordination, and development. In our earlier discussions we paid particular attention to representatives of the Coelenterata, Platyhelminthes (flatworms), Mollusca, Annelida (segmented worms), Arthropoda, Echinodermata, and Chordata (especially vertebrates). We shall again concentrate on these large and important groups, but shall also briefly mention some smaller phyla we previously ignored. Our aim in both this chapter and the next is to bring together the various phases of animal life discussed in other chapters and to suggest both the immense diversity within the kingdom and possible evolutionary patterns. This chapter will deal with all the animal phyla except Chordata; the next chapter will examine the chordates, which include two small invertebrate subphyla and the vertebrates.

A formal outline of the classification of the animal kingdom used in this book is given in the Appendix. We must stress that this classification, though widely used, is not accepted by all biologists. Much is uncertain regarding the evolution of animal groups, and biologists vary in their interpretation of the data.

PORIFERA (THE SPONGES)

Sponges are aquatic, mostly marine animals (Fig. 42.1). Though the larvae are ciliated and free-swimming, the adults are always sessile and are usually attached to rocks or shells or other submerged objects. They are multicel-

42.1 Colonial sponges
Water, carrying microscopic food particles and oxygen, flows into the central body cavity of each of the sponges through the numerous pores in their body walls, and is discharged through the large oscula.

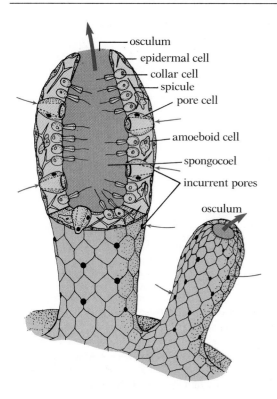

osculum
epidermal cell
collar cell
spicule
pore cell
amoeboid cell
spongocoel
incurrent pores
osculum

42.2 Detailed view of two colonial sponges
One sponge is shown partially sectioned, the other in
exterior view. The blue arrows indicate the path of
water through the sponges. These sponges are of the
asconoid type, with a simple tubular body. The
majority of sponges (syconoid or leuconoid types)
have complexly folded walls in which water passes
through a network of channels, but their basic body
plan may be considered an elaboration of the
asconoid plan.

lular, but show few of the features ordinarily associated with multicellular
animals. For example, they have no digestive system, no nervous system,[1]
and no circulatory system. In fact, they have no organs of any kind, and
even their tissues are not well defined. They thus represent a very low level
of organization.

The body of a sponge is rather like a perforated sac. Its wall is composed
of three layers: an outer layer of flattened epidermal cells, a gelatinous
middle layer with wandering amoeboid cells, and an inner layer of flagellated cells (Fig. 42.2). These last are unusual in that the base of the flagellum is encircled by a delicate collar; such cells are called ***collar cells*** (or
choanocytes).

The wall of a sponge is perforated by numerous pores, each surrounded
by a single pore cell. Water currents flow through the pores into the central
cavity (spongocoel) and out through a larger opening (osculum) at the end
of the body; the flow is enhanced by the beating of the flagella of the collar
cells. Microscopic food particles brought in by the water currents adhere to
the collar cells and are engulfed; the food may be digested in food vacuoles
of the collar cells themselves or passed to the amoeboid cells for digestion.
The water currents also bring oxygen to the cells and carry away carbon
dioxide and nitrogenous wastes (largely ammonia).

Sponges characteristically possess an internal skeleton secreted by the
amoeboid cells. This skeleton is composed of crystalline ***spicules*** or proteinaceous fibers or both. The spicules are made of calcium carbonate or
siliceous material (chiefly silicic acid); their chemical composition and
their shape are the basis for sponge classification. The fibrous skeletons of
the bath sponges (*Spongia*) are cleaned and sold for many uses. A living
bath sponge, which looks rather like a piece of raw liver, bears little resemblance to the familiar commercial object.

Among the free-living flagellated Protista are some collared organisms
(Fig. 42.3) that closely resemble the collar cells of sponges—cells found in
no other organisms. For this reason many biologists think that the Porifera
evolved from collared flagellates. Other biologists disagree, pointing out
that larval sponges have no collar cells, and suggest that sponges arose instead from a hollow free-swimming colonial flagellate. In any case, it
seems likely that sponges arose from the Protista independently of the other
multicellular animals. The phylum Porifera would then stand as an evolutionary development entirely separate from the rest of the animal kingdom,
implying that multicellular animals evolved at least twice from the protists.

Because the Porifera differ so greatly from other multicellular animals,
and probably arose independently, they are often regarded as constituting a
separate subkingdom, the Parazoa.

THE RADIATE PHYLA

The two radiate phyla—Cnidaria and Ctenophora—comprise radially symmetrical animals with bodies at a relatively simple level of construction.
These animals have definite tissue layers, but no distinct internal organs, no
head, and no central nervous system, though they possess nerve nets (see

---

[1] In some sponges a few cells of the body wall possess elongate processes that may have special
conductile properties.

Fig. 18.10, p. 459). There is a digestive cavity, but it has only one opening, which must serve as both mouth and anus—that is, it is a gastrovascular cavity. Tentacles are usually present. There is no coelom or other internal space between the wall of the digestive cavity and the outer body wall.

It was once thought that the bodies of these animals consist of only two layers of cells—an outer epidermis (ectoderm) and an inner gastrodermis (endoderm)—but it is now known that a third layer called mesoglea (mesoderm) usually occurs between these two (see Fig. 10.9, p. 251), just as it does in higher multicellular animals. However, this mesodermal layer is not as well developed in the radiate phyla as in higher phyla. It is usually gelatinous and has a few scattered cells, which may be amoeboid or fibrous.

CNIDARIA (OR COELENTERATA)

The phylum Cnidaria (also called Coelenterata) contains a variety of aquatic organisms, among them the hydras mentioned so frequently in earlier chapters, jellyfish, sea anemones, and corals. The hydras live in fresh water, but most other coelenterates are marine.

The coelenterate body shows some cell specialization and division of labor. Thus the outer epidermis contains sensory-nerve cells, gland cells, special cells that produce nematocysts, small interstitial cells, and epitheliomuscular cells. These last are the main structural elements of the epidermis and consist of a columnar cell body with several contractile basal extensions.

The contractile elements of coelenterates, then, are parts of cells that also have other important functions; there are no cells specialized exclusively for contraction—no separate muscle cells. Note also that these contractile elements are ectodermal, not mesodermal as in most higher animals. The gastrodermis also contains contractile elements; these are basal extensions of cells whose cell bodies constitute the bulk of the lining of the gastrovascular cavity and function in digestion. Here again a single cell performs two functions that in higher animals are performed by separate elements. In short, there is some division of labor among cells in coelenterates, but it is never as complete as in most bilateral multicellular animals; and most functions performed by tissues derived from mesoderm in other animals are performed by ectodermal or endodermal cells in coelenterates.

The phylum Cnidaria is divided into four classes: Hydrozoa, Scyphozoa, Anthozoa, and Cubozoa. The Cubozoa, or sea wasps, has only 20 species, and will not be discussed separately.

Class Hydrozoa   The best-known members of this class are the freshwater hydras. We have discussed their feeding (p. 251), gas exchange (p. 280), nervous control (p. 458), and locomotion (p. 527). Little more need be said about them here. In many ways, however, hydras are not typical members of their class. Many hydrozoans are colonial (Fig. 42.4) and have a complex

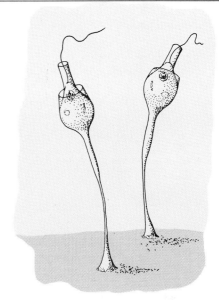

42.3   Two choanoflagellates (*Codosiga*)
These single-celled organisms strikingly resemble the collar cells of sponges, shown in Figure 42.2. They have traditionally been classified as protozoan animals by zoologists and as colorless members of the algal division Chrysophyta by botanists.

42.4   A branching colonial hydrozoan, *Plumularia setacea*
The colony is composed of a stalk with many lateral branches each bearing six to eight polyps.

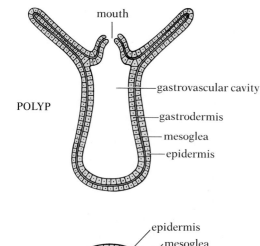

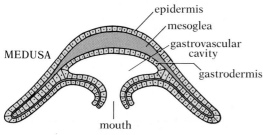

42.5 Diagram contrasting polyp and medusa
The basic structure of these two forms is the same. A medusa is like a flattened polyp turned upside down.

42.7 A hydrozoan medusa, *Gonionemus*
Gonionemus spends most of its life as a medusa—a weakly swimming bell-like creature with numerous tentacles.

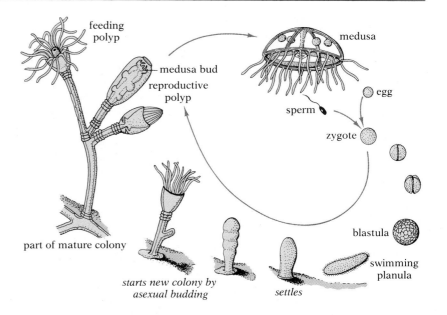

42.6 Life cycle of a colonial hydrozoan (*Obelia*)
Since the medusas are of separate sexes, the eggs and sperm are produced by different individuals. See text for description.

life cycle, in which a sedentary hydralike ***polyp*** stage alternates with a free-swimming jellyfishlike ***medusa*** stage (Fig. 42.5). By contrast, hydras are solitary and have only a polyp stage (which is not completely sedentary).

Let us examine *Obelia* as an example of a colonial hydrozoan (Fig. 42.6). Much of the life of *Obelia* is passed as a sedentary branching colony of polyps. The colony arises from an individual hydralike polyp by asexual budding; the buds fail to separate, and the new polyps remain attached by hollow stemlike connections. The gastrovascular cavities of all the polyps are interconnected via the cavity in the stems. The cells lining the stem cavity have long flagella that circulate the fluid in the cavity. Partly digested food can be passed from one polyp to another in this moving fluid. Both the stems and the polyps (mouth and tentacles excepted) are enclosed in a hard chitinous case secreted by the ectoderm. Rings or joints in the case at intervals along the stems permit some flexibility for the colony.

A mature *Obelia* colony consists of two kinds of polyps: feeding polyps with tentacles and nematocysts, and reproductive polyps without tentacles, which regularly bud off tiny transparent free-swimming medusas. The reproductive polyps are nourished with food captured by the feeding polyps and passed to them through the common gastrovascular cavity in the connecting stems. In view of this division of labor, and of the structural continuity between the polyps, it might be argued that the so-called colony is not really a colony, but a complex individual.

Each medusa is umbrella-shaped or bell-shaped, with numerous tentacles hanging from the margin of the bell (Fig. 42.7). A tube with a mouth at its end hangs from the middle of the undersurface of the bell (where the

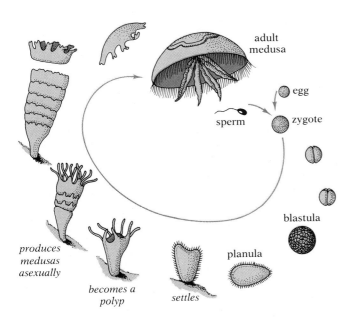

42.8   Life cycle of a jellyfish (*Aurelia*)
The polyps are shown much enlarged. Since the medusas are of separate sexes, the eggs and sperm are produced by different individuals. See text for description.

clapper of a real bell would be). The medusas are the dispersal and sexual stage in the life cycle. They swim feebly by alternately contracting and relaxing the contractile cells in the bell, but much of their movement is a matter of drifting with the water currents.

Certain cells in mature medusas undergo meiosis and give rise to either sperm or eggs, which are released into the surrounding water. Fertilization takes place, and the resulting zygote develops into a hollow blastula. Gastrulation does not take place by the process of invagination described in Chapter 32. Instead, endodermal cells proliferated by the wall (ectoderm) of the blastula wander into the blastocoel until they completely fill it. This solid gastrula then develops into an elongate ciliated larva called a ***planula***. The planula eventually settles to the bottom, attaches by one end to some object, and develops a mouth and tentacles at the other end, becoming a polyp that gives rise to a new colony. The life cycle is thus completed. Note that the alternation of polyp and medusa stages in a coelenterate like *Obelia* differs from the alternation of generations in plants in that both polyp and medusa are diploid; as in all multicellular animals, the only haploid stage in the life cycle is the gametes.

Class Scyphozoa   The scyphozoans are the true jellyfish. In these animals the medusa is the dominant and conspicuous stage in the life cycle, and the polyp stage is restricted to a small larva. This larva, which develops from the planula, promptly produces medusas by budding (Fig. 42.8). Scyphozoan medusas resemble the hydrozoan medusas already described except that they are usually much larger and have long oral arms (endodermal tentacles) arising from the margin of the mouth; the marginal tentacles may be much reduced, as in *Aurelia* (Fig. 42.8), or they may be quite long, as in *Pelagia* (Fig. 42.9).

42.9   A jellyfish with oral arms and long marginal tentacles (*Pelagia*)

42.10   A sea anemone, *Stephanauge*
The animal's mouth is clearly visible within the ring of tentacles.

42.11   A soft coral, *Dendronephythya*
These anthozoans do not deposit a calcareous skeleton, but form numerous spinelike spicules which give structure to their otherwise fleshy bodies.

Class Anthozoa   The class Anthozoa includes sedentary polypoid forms such as sea anemones (Figs. 42.10 and 6.31, p. 161), sea fans, and hard and soft corals (Fig. 42.11). All are marine. There is no trace of a medusa stage in their life cycle. They are the most advanced members of the Cnidaria, and their body structure is much more complex than that of simple polyps like the hydras. They possess a tubular pharynx leading into the gastrovascular cavity, which is divided into numerous radiating compartments by longitudinal septa; their mesoderm (mesoglea) is much thicker than that of other coelenterates and is often elaborated into a fibrous connective tissue; and their muscles are much better developed.

The hard corals—anthozoans that secrete a hard limy skeleton—have played a very important role in the geologic history of the earth, particularly in tropical oceans. As their skeletons have accumulated over the ages, corals have formed many reefs, atolls, and islands, especially in the South Pacific (see p. 978). The Great Barrier Reef, a coral ridge many kilometers wide that extends more than 1,600 km along the eastern coast of Australia, is a particularly impressive example of the way these lowly animals can change the face of the earth and create a favorable environment for an exceptionally diverse ecosystem. Moreover, most of the large oil deposits of the world are thought to be derived from the decay and burial of vast quantities of organic matter in enormous coral reefs.

CTENOPHORA

Like the coelenterates, members of the phylum Ctenophora, known as comb jellies or sea walnuts, are radial animals with a saclike body composed of epidermis, mesoglea, and gastrodermis; and like the coelenterates, they have a digestive cavity of the gastrovascular type and lack a coelom and definitive organ systems. Unlike the coelenterates, however, they have independent mesodermal muscles, lack nematocysts (the tentacles, when present, may have adhesive cells instead), do not have the polymorphic life cycle so common in coelenterates, and characteristically have eight rows of ciliary plates (combs) that run across the surface of the transparent body from the upper pole to the lower pole like the lines of longitude on a globe (Fig. 42.12). The cilia in the eight rows beat in unison and enable the animal to swim feebly. Most ctenophores float near the surface of the sea, chiefly near shore. They are carried about by currents and tides, and large numbers of them may be blown into bays during storms or swept ashore by high waves.

ORIGIN OF THE EUMETAZOA

The gastraea and planuloid hypotheses   The subkingdom Eumetazoa includes essentially all multicellular animals except the Porifera, or sponges. (In older classifications, this subkingdom was called the Metazoa, and the mesozoan phylum was excluded.) Speculation concerning the origin of the Eumetazoa has long centered on the radiate phyla just discussed—not only because the radiates, particularly the hydrozoan coelenterates, seem to be among the simplest eumetazoans, but also because their saclike, essentially two-layered bodies strikingly resemble embryonic gastrulas. Now, a spherical colonial flagellate like *Volvox* (see p. 1065) certainly resembles an embryonic blastula. As long ago as 1874 Ernst Haeckel suggested that the ancestor of multicellular animals was a hollow-sphere colonial flagellate similar to *Volvox* (though presumably without cell walls or chlorophyll), and that the coelenterates arose from this hypothetical ancestor, called a ***blastaea***, by a process of invaginating gastrulation. The higher animals would then have arisen from the early two-layered (diploblastic) ***gastraea*** ancestor of the modern coelenterates by assuming a creeping mode of life and slowly becoming bilateral (Fig. 42.13). According to this hypothesis, the blastula and gastrula stages in the embryonic development of higher animals are recapitulations of early steps in the evolution of these animals from colonial flagellates.

But gastrulation in coelenterates, it was soon pointed out, does not take place by invagination. Instead, as we have already seen, the endoderm arises by inwandering of cells produced at the inner surface of the ectoderm, and a solid gastrula, which develops into a planula larva, is formed. Both the hollowing out of the interior to form the gastrovascular cavity and the breaking-through of a mouth occur later, when the larva develops into a polyp. Thus it seems likely that the invagination type of gastrulation was a later evolutionary development and not the original method of formation of endoderm. Consequently, although the idea of a hollow blastaea ancestor as the starting point was retained in later versions of Haeckel's hypothesis, the idea of a gastraea stage was abandoned and a ***planuloid*** stage hypothesized instead (Fig. 42.14).

The plakula hypothesis and the rediscovery of *Trichoplax* Neither Haeckel's gastraea hypothesis nor its alternative, positing a planuloid stage, explains why gastrulation evolved in the first place, or why the endodermis

42.12 A common ctenophore, *Pleurobrachia*
Note the rows of ciliary plates, as well as the very long antennae, which emerge from sheaths that extend deep into the body.

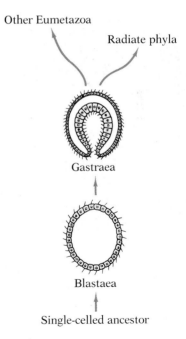

42.13 Diagram of the origin of the Eumetazoa according to Haeckel's blastaea-gastraea hypothesis

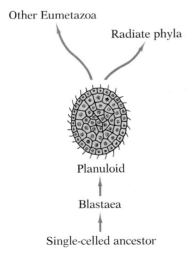

42.14 Diagram of the origin of the Eumetazoa according to the planuloid hypothesis

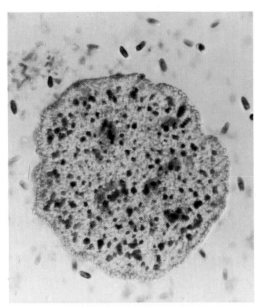

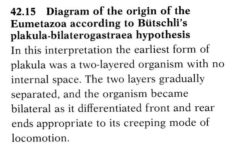

42.15 Diagram of the origin of the Eumetazoa according to Bütschli's plakula-bilaterogastraea hypothesis
In this interpretation the earliest form of plakula was a two-layered organism with no internal space. The two layers gradually separated, and the organism became bilateral as it differentiated front and rear ends appropriate to its creeping mode of locomotion.

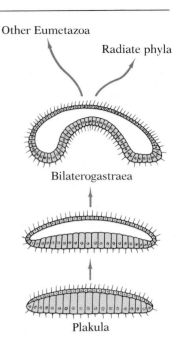

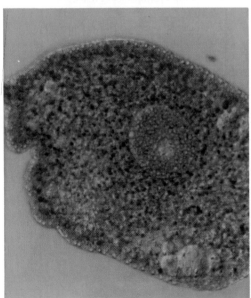

42.16 Photographs of *Trichoplax adhaerens*
Top: The animal has a flat platelike appearance. Of irregular shape, it has no constant symmetry. The many spherical particles in its surface layer are globules of a fatty material. Bottom: An animal containing a large egg cell.

became specialized for nutritional activities. To meet these difficulties, Otto Bütschli of the University of Heidelberg suggested in 1884 that the earliest eumetazoan may have been a ***plakula***, a two-layered, flattened creature creeping about on the sea bottom (Fig. 42.15). Since the ventral cell layer of such an organism would be the layer that would come in contact with food particles most often, it would be reasonable to expect that it should evolve nutritive specializations, and that the dorsal cell layer should evolve protective and perhaps locomotory ones.

Now, if food items were captured beneath the plakula, and if the ventral epithelium was charged with digesting the food, it would have been advantageous, as Bütschli pointed out, for the organism to elevate part of its body and form a temporary digestive chamber from which neither the food nor the digestive enzymes secreted onto it could easily escape. In this manner, then, the plakula might sometimes have become a temporary gastraea. Since Bütschli regarded the plakula as a creeping rather than a swimming organism, he went one step further and suggested that bilateral symmetry[2] would have been more likely than radial symmetry; hence he envisioned a ***bilaterogastraea*** (Fig. 42.15). In other words, he thought that bilateral symmetry was a very early characteristic of the Eumetazoa and that the radial symmetry of such groups as the Cnidaria and Ctenophora evolved secondarily.

When Bütschli first put forward his ideas, he noted that just one year earlier a living organism had been found that corresponded in many ways with his hypothetical plakula. That organism was a tiny animal discovered in

---

[2] Bilateral symmetry is the property of having two similar sides. A bilaterally symmetrical animal has definite dorsal (upper) and ventral (lower) surfaces and definite anterior (head) and posterior (tail) ends.

seawater aquaria by F. E. Schulze of the University of Graz in Austria. The organism, which Schulze named *Trichoplax adhaerens,* was a flattened, two-layered creature that crept about on the bottom. Despite its marked resemblance to Bütschli's hypothetical plakula, *Trichoplax* was soon forgotten, because it was presumed (on very questionable evidence) to be merely a larva of a coelenterate.

There the plakula hypothesis and *Trichoplax* rested, in oblivion, until 1969, when K. G. Grell of the University of Tübingen in Germany rediscovered *Trichoplax* on some algae sent to him from the Red Sea (Fig. 42.16). He soon found that, just as Bütschli had speculated eighty-five years earlier, *Trichoplax* does, in fact, often rear up to form a temporary digestive cavity (Fig. 42.17). Moreover, he showed that this simple organism, with only two principal tissue layers and no organs (Fig. 42.18), is not a larva, because it is capable of reproduction—asexual when conditions are not crowded, sexual when they are.

Grell has proposed a new phylum, Placozoa, for *Trichoplax.* In his estimation *Trichoplax* may be the most primitive multicellular animal known. If it is, then the evolution of the eumetazoans may indeed have followed the pattern proposed by Bütschli so long ago (Fig. 42.19). But because *Trichoplax* has no bilateral symmetry, Grell is uncertain whether the evolutionary stage following the plakula was a gastraea or a bilaterogastraea.

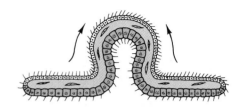

42.17 Formation of a temporary digestive chamber by *Trichoplax adhaerens*
The animal elevates part of its body to form a cavity. It thus becomes, in effect, a temporary gastraea.

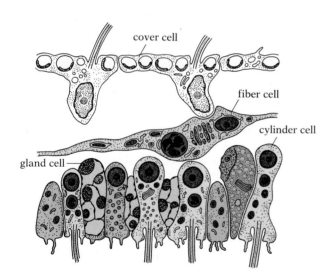

cover cell

fiber cell

cylinder cell

gland cell

42.18 The histology of *Trichoplax adhaerens*
The thin dorsal epithelium, which is transparent, consists of cells of a single type, called cover cells by Grell; each bears a flagellum, and some incorporate large spherical vesicles containing fatty material. The much thicker ventral epithelium contains flagellated cylinder cells and nonflagellated gland cells. In the space between the two epithelia are some fiber cells, which probably function in locomotion.

42.19 Diagram of the origin of the Eumetazoa according to Grell's Placozoa version of the plakula hypothesis
Grell suggests that the Placozoa may be ancestral to the sponges too, but since the evidence for such a filiation is not so convincing, this relationship is not diagramed here.

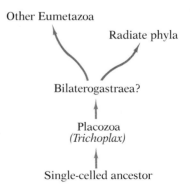

Other Eumetazoa

Radiate phyla

Bilaterogastraea?

Placozoa
(Trichoplax)

Single-celled ancestor

42.20   An aquatic turbellarian, *Prosthecereus vittatus*

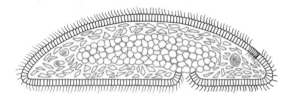

42.21   An acoel flatworm
There is a ventral mouth but no digestive cavity. The entire interior of the animal is filled with an almost solid mass of tissue (color).

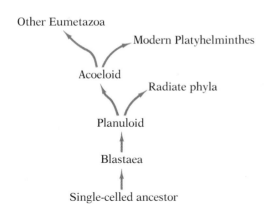

42.22   Diagram of the origin of the bilateral Eumetazoa according to the blastaea-planuloid-acoeloid hypothesis

The ciliate hypothesis   The idea of eumetazoan origin from flagellates via blastaea and planuloid steps has been widely accepted among biologists, and Grell's new version of the plakula hypothesis is likely to win broad support as well. But other proposals have been put forward, the most notable of which is that multicellular animals arose from multinucleate ciliates, not from flagellates.

According to this so-called ciliate hypothesis, the bodies of the earliest eumetazoans were syncytial (coenocytic)—that is, they contained many nuclei, each controlling the cytoplasm around it, but had no cellular partitions. Formation of cellular membranes around each nucleus and its associated cytoplasm would have produced a typical multicellular animal, probably resembling some of the simplest flatworms (see Fig. 42.21). Note that, since this hypothesis starts out with a bilaterally symmetrical ciliate, it assumes, like Bütschli's, that the first eumetazoans were bilateral and that the radial symmetry of the Cnidaria and Ctenophora is secondary.

THE ACOELOMATE BILATERIA

There are three phyla—the Platyhelminthes, the Gnathostomulida, and the Nemertea—that contain what biologists generally regard as the most primitive bilaterally symmetrical animals. In each phylum the body is composed of three well-developed tissue layers—ectoderm, mesoderm, and endoderm—and the mesoderm is a solid mass that fills what once was the embryonic blastocoel. In other words, there is no coelom—no cavity between the digestive tract and the body wall (see Fig. 42.32A). For this reason, these phyla are known as the acoelomate bilateria.

PLATYHELMINTHES (THE FLATWORMS)

The flatworms, as their name implies, are dorsoventrally flattened, elongate animals.[3] Their digestive cavity (not always present) resembles that of coelenterates. It is a gastrovascular cavity with a single opening that must serve as both mouth and anus. However, there is a muscular pharynx leading into the cavity, and the cavity itself is often profusely branched, especially in the free-living species (see p. 253 and Fig. 10.11). As in coelenterates, the amount of extracellular digestion is limited, most of the food particles being phagocytized and digested intracellularly by the cells of the wall of the gastrovascular cavity. Respiratory and circulatory systems are absent.[4] However, there is a flame-cell excretory system (see p. 364 and

---

[3] The term "worm" is applied to a great variety of unrelated animals. It is a descriptive, not a taxonomic, term that denotes possession of a slender elongate body, usually without legs or with very short ones.

[4] A so-called "lymphatic system," present in some flukes, may represent a primitive circulatory system.

Fig. 14.12), and there are well-developed reproductive organs (usually both male and female in each individual).

That both the excretory system, with its flame bulbs and tubules, and the reproductive organs should be present signifies that the flatworms have advanced beyond the tissue level of construction seen in the radiate phyla to an organ level of construction. The more extensive development of mesoderm, leading to greater division of labor, was probably a major factor in making this advance possible. Mesodermal muscles are well developed. Several longitudinal nerve cords running the length of the body and a tiny "brain" ganglion located in the head constitute a central nervous system (see p. 460 and Fig. 18.12).

The phylum is divided into three classes: Turbellaria, Trematoda, and Cestoda. The last two are entirely parasitic.

Class Turbellaria The members of this class, of which the freshwater planarians often mentioned in earlier chapters are examples, are free-living flatworms ranging from microscopic size to a length of several centimeters. The body is clothed by an epidermal layer, which is usually ciliated (at least in part). Although a few turbellarians live on land, most are aquatic (the majority marine) (Fig. 42.20).

Turbellarians usually have a gastrovascular cavity, but most members of one small order, the Acoela, do not (Fig. 42.21). For a variety of reasons (not just the absence of a digestive cavity), some biologists have considered the Acoela the most primitive bilateral animals, and have suggested that a primitive *acoeloid* organism might well have arisen from a planuloid ancestor. In their view, both the more complex flatworms and the other eumetazoan phyla probably evolved from such an acoeloid organism. This version of the early evolution of the Eumetazoa is diagramed in Figure 42.22.

Proponents of the ciliate hypothesis of eumetazoan origin also assume an acoeloid stage at the root of the Eumetazoa, but they derive the acoeloid directly from a multinucleate ciliate, not from a planuloid. They point out that acoels are about the same size as some ciliates, are ciliated, and sometimes have poorly developed cellular partitions. They think the primitive acoeloid organism was the ancestor of the radiate phyla as well as of the bilateral phyla. Their ideas are diagramed in Figure 42.23.

It is unclear how the acoeloid condition, if it is primitive, could be related to Grell's Placozoa hypothesis. However, some biologists regard the Acoela as secondarily simple, not primitive; if this view is correct, the Acoela pose no problem for Grell's hypothesis.

Class Trematoda (the flukes) The flukes are parasitic flatworms. They lack cilia, and in place of the cellular epidermis of their turbellarian ancestors they have a thick cuticle secreted by the cells below. This cuticle is highly resistant to enzyme action and is thus an important adaptation to a parasitic way of life. Flukes characteristically possess suckers, usually two or more, by which they attach themselves to their host (Fig. 42.24). They have a two-branched gastrovascular cavity, which does not ramify throughout the body like that of turbellarians. A large proportion of their bodies is occupied by reproductive organs, including two or more large testes, an ovary, a long much-coiled uterus in which eggs are stored prior to laying, and yolk glands.

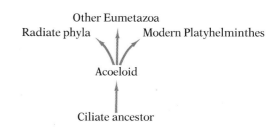

42.23 Diagram of the origin of the Eumetazoa according to the ciliate-acoeloid hypothesis

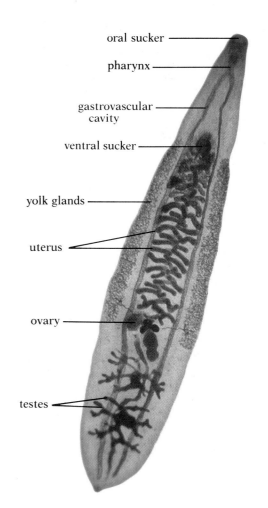

42.24 A Chinese liver fluke, *Opisthorchis sinensis*

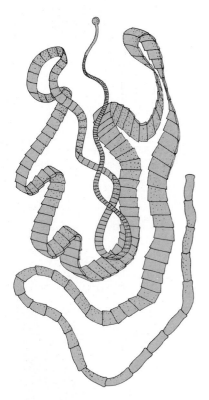

42.25 A tapeworm
The body is composed of a small head and neck, followed by a large number of segments called proglottids. As the proglottids ripen, they break off and pass with the host's feces to the outside. New proglottids are produced just back of the neck.

The members of one order of flukes are ectoparasites (external parasites) on the gills or skin of freshwater and marine fishes. A few of these flukes sometimes wander into the body openings of their hosts, and it may have been from such a beginning that the endoparasitism (internal parasitism) typical of the members of the other two orders arose.

The endoparasitic flukes often have very complicated life cycles involving two to four different kinds of hosts. The blood fluke, *Schistosoma japonicum*, common in China, Japan, Taiwan, and the Philippines, is a species with two hosts. The adult blood fluke inhabits blood vessels near the intestine of a human host. When ready to lay its eggs, it pushes its way into one of the very small blood vessels in the wall of the intestine. There it deposits so many eggs that the vessel ruptures, discharging the eggs into the intestinal cavity, from which they are carried to the exterior in the feces. If there is a modern sewage system, that is the end of the story.

But in many Asiatic countries, human feces are regularly used as fertilizer. Thus the eggs get into water in rice fields, irrigation canals, or rivers, where they hatch into tiny ciliated larvae. A larva swims about until it finds a snail of a certain species; it dies if it cannot soon locate the correct species. When it finds such a snail, it bores into the body of the snail and feeds on its tissues. It then reproduces asexually, and the new individuals thus produced leave the snail and swim about until they come in contact with the skin of a human, such as a farmer wading in a rice paddy or a child swimming in a pond. They attach themselves to the skin and digest their way through it and into a blood vessel. Carried by the blood to the heart and lungs, they eventually reach the vessels of the intestine, where they settle down, mature, and lay eggs, thus initiating a new cycle.

Schistosomas cause a serious disease called **schistosomiasis**, which is characterized initially by a cough, rash, and body pains, followed by severe dysentery and anemia. The disease so saps the strength of its victims that they become weak and emaciated and often die of other diseases to which their weakened condition makes them susceptible. Schistosomiasis is one of the most widespread and debilitating human diseases in the world today, but because it is confined to the warmer, less-developed regions of the earth, most inhabitants of North America and Europe are hardly aware of its existence.

A fluke with three hosts is the Chinese liver fluke, *Opisthorchis sinensis*. The adult lives in the liver of a human host, where it lays its eggs. The eggs pass into the intestine with the bile and are carried to the exterior in the feces. If they get into water and are eaten by a certain species of snail, they hatch, and the larvae bore into the lymph spaces of the snail, where they reproduce asexually. The new individuals thus produced live their entire lives in the snail, but they too reproduce asexually, and their progeny leave the snail, burrow through the skin of a fish, and encyst in the fish's muscles. If a person eats the raw or insufficiently cooked fish, his digestive enzymes weaken the walls of the cysts, and the flukes emerge in his intestine, migrate up the bile duct to the liver, and settle down to start a new cycle. All three hosts—snail, fish, and human—are necessary for completion of the reproductive cycle of this fluke.

Class Cestoda (the tapeworms) Adult tapeworms (Fig. 42.25) live as internal parasites of vertebrates, almost always in the intestine. However, the

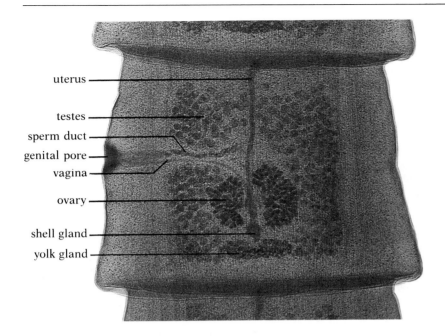

uterus

testes

sperm duct

genital pore

vagina

ovary

shell gland

yolk gland

life cycle usually involves one or two intermediate hosts, which may be invertebrate or vertebrate, depending on the species. The life cycle of the beef tapeworm, in which the intermediate host is a cow and the final host is a human being, was outlined on page 881.

Tapeworms exhibit many special adaptations for their parasitic way of life. Like the flukes, they have a resistant cuticle instead of the epidermis of their free-living ancestors. And they have neither mouth nor digestive tract. Bathed by the food in their host's intestine, they absorb predigested nutrients across their general body surface. Diffusion, probably augmented by active transport, suffices to provision all the cells, because none is far from the surface.

The head of a tapeworm is a small knoblike structure called a *scolex*, which usually bears suckers, and often also hooks, by which the worm attaches to the wall of the host's intestine (Fig. 42.26, right). Immediately behind the scolex is a neck region, which is followed by a very long ribbon-like body (beef tapeworms occasionally grow to 23 m, fish tapeworms to 18 m, and pork tapeworms to 8 m). This long body is usually divided by transverse constrictions into a series of segments called *proglottids* (Fig. 42.26, left).

Each proglottid is essentially a reproductive sac, containing both male and female organs. Sperm cells, usually from a more anterior proglottid of the same animal, enter the genital pore and fertilize the egg cells, which are then combined with yolk from a yolk gland and enclosed in a shell. The fertilized eggs, already undergoing development, are stored in a uterus, which may become so engorged with eggs that it occupies most of the volume of a mature proglottid. Eventually all the sexual organs except the uterus degenerate, and the proglottid, now "ripe," detaches from the worm and passes out of the host's body with the feces. As ripe proglottids are released from the end of the worm, new ones are produced just back of the neck. A

42.26   Scolex and proglottids of the dog tapeworm, *Taenia pisiformis*
Left: A mature proglottid. Right: Scolex, neck, and some of the proglottids. Note the hooks and suckers on the scolex.

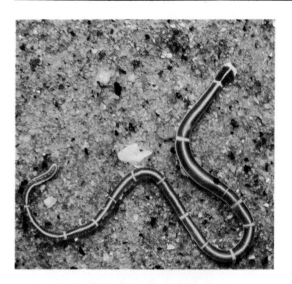

42.27 A nemertine worm, *Tubulanus annulatus*

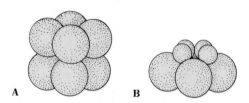

42.28 Radial and spiral cleavage patterns
(A) Radial cleavage, characteristic of deuterostomes.
The cells of the two layers are arranged directly
above each other. (B) Spiral cleavage, characteristic
of protostomes. The cells in the upper layer are
located in the angles between the cells of the lower layer.

single ripe proglottid may contain more than 100,000 eggs, and the annual output of one worm may be more than 600 million.

If an appropriate intermediate host eats food contaminated with feces containing tapeworm eggs, its enzymes digest the shells of the eggs. The embryos thus released bore through the wall of the host's intestine, enter a blood vessel, and are carried by the blood to the muscles, where they encyst. If a person eats the raw or "rare" meat of this intermediate host (for example, beef, pork, or fish), the walls of the cysts are digested away; the young tapeworms attach to the intestinal wall and, nourished by an abundant supply of food, begin to grow and produce eggs, thus starting a new cycle.

NEMERTEA (THE PROBOSCIS WORMS)

The members of the phylum Nemertea are long slender worms (Fig. 42.27) characterized by a very long eversible, muscular proboscis—that is, one that turns itself inside out as it is extruded—enclosed in a tubular cavity at the anterior end of the body. This proboscis, which is used in capturing prey and also in defense, is often two or more times the length of the worm's body and is somewhat coiled when enclosed in its sheath. The worms are common along both the Atlantic and Pacific shores of the United States. They are usually found sheltered under stones, shells, or seaweeds, or burrowing in the sand or mud in shallow water.

Nemertines resemble turbellarian flatworms (their probable ancestors) in their nervous systems and in many other ways, such as in having a ciliated epidermis, a solid mesoderm, and a flame-cell excretory system. But they differ from them in two important characteristics not encountered in the animals considered thus far. First, they have a ***complete digestive system***—one that has two openings, a mouth and an anus. Such a system makes possible specialization of sequentially arranged chambers for different functions and thus permits assembly-line processing of food, as we saw in Chapter 10. Second, they have a simple blood circulatory system, which presumably facilitates transport of materials from one part of the body to another.

GNATHOSTOMULIDA (THE JAW WORMS)

The members of the phylum Gnathostomulida are small, free-living marine worms, first recognized in 1956. These hermaphroditic worms live in detritus-rich sand and possess specialized, paired jaws. They may be more closely related to the pseudocoelomate phylum Gastrotricha, discussed later, than to the other coelenterates.

DIVERGENCE OF THE PROTOSTOMIA AND DEUTEROSTOMIA

We saw in Chapter 31 that the embryonic cavity called the archenteron, formed during gastrulation, becomes the digestive tract of an adult animal. But the archenteron has only one opening to the outside, the blastopore. In animals like coelenterates and flatworms, where the digestive tract is a gastrovascular cavity, the blastopore becomes the combined mouth and anus. Embryologists have shown that in nemertines, however, the site of the embryonic blastopore becomes the mouth and that the anus is an entirely new opening. This is also the case in many other animals, including nematode worms, molluscs, and annelids. But in a few phyla, among them two large and important ones—the Echinodermata and the Chordata—the situation is reversed: the embryonic blastopore becomes the anus, and the mouth is the new opening.

This fundamental difference in embryonic development suggests that a major split occurred in the animal kingdom soon after the origin of a bilateral ancestor. One evolutionary line led to all the phyla in which the blastopore becomes the mouth; these phyla are often called the **Protostomia** (from the Greek *protos*, "first," and *stoma*, "mouth"). The other evolutionary line led to the phyla in which the blastopore becomes the anus and a new mouth is formed: these phyla are called the **Deuterostomia** (from the Greek *deuteros*, "second," or "later," and *stoma*).

As might be expected if the Protostomia and Deuterostomia diverged at a very early stage of their evolution, they differ in a number of other fundamental characters besides the mode of formation of mouth and anus. A further essential difference between them is that the early cleavage stages are usually determinate in protostomes and indeterminate in deuterostomes—that is, the developmental fates of the first few cells of a protostome embryo are usually already at least partly determined, and if these cells are separated not one of them can form a complete individual, whereas the fates of the first few cells of a deuterostome embryo are not determined, and each cell, if separated, can develop into a normal individual (there can be identical twinning). Furthermore, the two groups exhibit strikingly different patterns of cleavage; the early cleavages in protostomes are usually oblique to the polar axis[5] of the embryo and thus give rise to a spiral arrangement of cells, whereas the early cleavages in deuterostomes are either parallel or at right angles to the polar axis and thus give rise to a so-called radial arrangement of cells (Fig. 42.28). The basic larval types are also different in the two groups, as we shall see later in this chapter.

Another fundamental difference between the protostome and deuterostome phyla is seen in the method of origin of the mesoderm in the embryo. The mesoderm of the radiate phyla arises from inwandering cells derived from the ectoderm. A small amount of mesoderm forms this way in the protostomes also, though not usually in the deuterostomes. Most of the mesoderm in protostomes and all of it in deuterostomes is derived from endoderm instead of ectoderm. However, in protostomes this mesoderm arises as a solid ingrowth of cells from a single initial cell located near the blastopore (Fig. 42.29B), whereas in deuterostomes (except vertebrates) it arises by a saclike outfolding of the gut wall, as we saw in Chapter 32.[6]

Still another difference, correlated with the preceding one, has to do with the method of formation of the coelom, if one is present. A true coelom

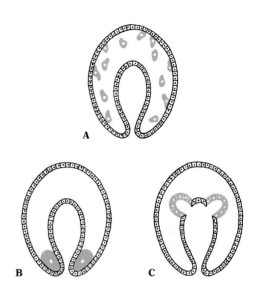

42.29 Different modes of origin of mesoderm
(A) In the radiate phyla, mesoderm (colored cells) arises from inwandering cells derived from the ectoderm. A small amount of the mesoderm of the protostome phyla arises in this way also. (B) In most protostome phyla, a single initial cell (not shown) divides to form two primordial mesodermal cells. The bulk of the mesoderm arises from divisions of these cells, which are located near the blastopore, at the junction between the ectoderm and the endoderm. (C) In the deuterostome phyla, the mesoderm arises as a pair of pouches from the endodermal wall of the archenteron.

---

[5] The polar axis runs between the animal and vegetal poles.

[6] There are actually a variety of other ways in which mesoderm may arise from endoderm, but embryologists usually interpret them as variants of one or the other of the two processes described here.

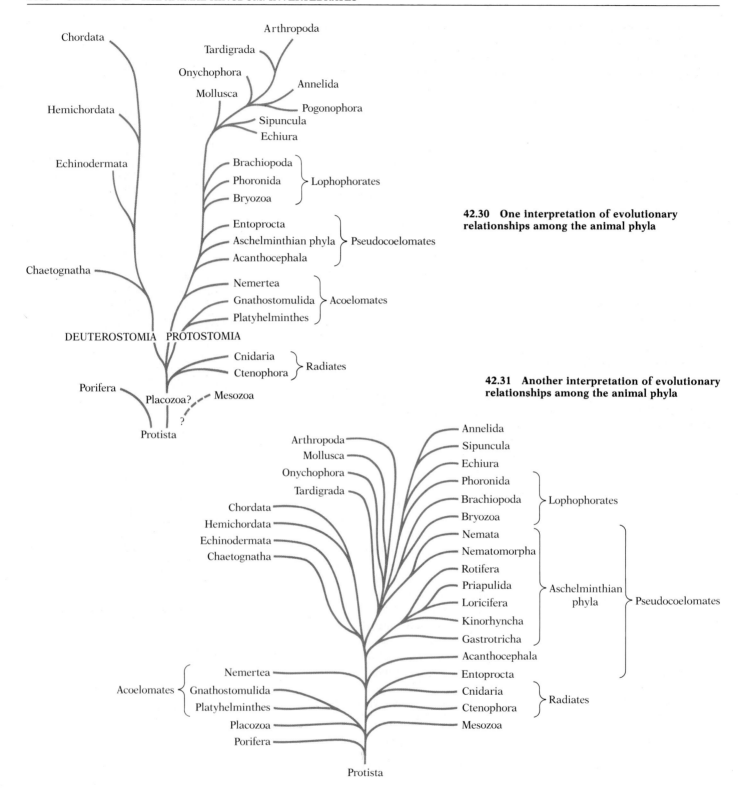

42.30 One interpretation of evolutionary relationships among the animal phyla

42.31 Another interpretation of evolutionary relationships among the animal phyla

is defined as a cavity enclosed entirely by mesoderm and located between the digestive tract and the body wall. In the coelomate protostomes this cavity usually arises as a split in the initially solid mass of mesoderm. In the deuterostomes, by contrast, the coelom arises as the cavity in the mesodermal sacs as they evaginate from the wall of the archenteron.[7]

To repeat, the Protostomia and the Deuterostomia differ most conspicuously in the fate of the embryonic blastopore, in the determinateness and pattern of the initial cleavages, in the mode of origin of the mesoderm and of the coelom (if one is present), and in type of larva. They also differ in many nondevelopmental traits. To give but one example, the visual receptor cells of the protostomes generally have a receptor organelle of specialized microvilli (forming a rhabdom), whereas the analogous organelle in the deuterostomes is composed of specialized cilia.

The differences, however, are not quite so clear-cut as we may have implied. Most of the contrasting characters are subject to exceptions, and some are far less distinct in the animals themselves than descriptions and diagrams tend to suggest. Indeed, some authors have seriously questioned whether the protostome–deuterostome dichotomy is real or imagined. As we have said before, ambiguities of this sort arise in large measure because of the frequent difficulty in distinguishing apparent and real similarities between species—evolutionary analogy versus evolutionary homology. Features are analogous when they evolved separately to serve the same function, such as the strikingly similar wings of bats and birds, the flippers of whales and seals, and the eyes of cephalopods and vertebrates; features are homologous when they have a common evolutionary origin, as the wings of eagles, hummingbirds, and penguins. Mistakes in distinguishing analogy from homology can lead to misleading phylogenetic trees. With the gathering of more extensive comparative data on nonmorphological and nonembryological characters, as well as the determination of nucleotide sequences of equivalent genes in different species, ideas about the relationships between the animal phyla undergo marked changes.

Figure 42.30, drawn in the traditional form of a phylogenetic tree, assumes the reality of the split between the protostomes and the deuterostomes; it shows one of many possible interpretations of evolutionary relationships among the animal phyla. Figure 42.31 shows another interpretation.

THE PSEUDOCOELOMATE PROTOSTOMIA

In several protostome phyla the body cavity is functionally analogous to a coelom but differs from a true coelom, which is entirely enclosed by mesoderm, in being partly bounded by ectoderm and endoderm. Such a cavity is called a ***pseudocoelom*** (Fig. 42.32). It is actually only the remnant of the embryonic blastocoel.

Three phyla of pseudocoelomate protostomes are recognized by many biologists: Acanthocephala, Entoprocta, and Aschelminthes. But the current

---

[7] A coelom that arises as a split in an initially solid mass of mesoderm is called a schizocoelom. One that forms as the cavity in a pouch of mesoderm is called an enterocoelom. The coelomate protostomes are sometimes called the schizocoelous phyla, and the deuterostomes the enterocoelous phyla.

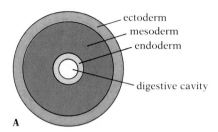

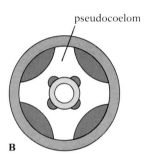

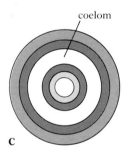

42.32 Diagrams of acoelomate, pseudocoelomate, and coelomate body types
(A) Acoelomate body. There is no body cavity; the entire space between the ectoderm and endoderm is filled by a solid mass of mesoderm.
(B) Pseudocoelomate body. There is a functional body cavity, but it is not entirely bounded by mesoderm.
(C) Coelomate body. The body cavity is completely bounded by mesoderm.

trend is to divide the organisms traditionally classified as one phylum, Aschelminthes, into six separate phyla. For simplicity, we shall discuss these phyla as a group.

Adult Acanthocephala are endoparasites in the digestive tracts of vertebrates; the larvae live in invertebrates. These animals are often called spiny-headed worms because they have a proboscis armed with rows of recurved hooks. They have no digestive tract.

The Entoprocta are tiny sessile, mostly marine animals that live attached by a stalk to rocks, shells, pilings, or animals such as crabs or sponges. Most are colonial. Their digestive tract is U-shaped, and both mouth and anus open inside a circle of ciliated tentacles. They feed on small plankton such as diatoms and protozoans.

THE ASCHELMINTHIAN PHYLA

The largest and most important pseudocoelomate phyla are those of the aschelminthes group. These phyla include an array of generally small, wormlike animals, without a definitely delimited head, that have a straight or slightly curved complete digestive tract, and a cuticle. There is no respiratory or circulatory system. A flame-cell excretory system occurs in most phyla, but not in nematodes, which have a special type of excretory system unique to them. We shall discuss only two of the phyla: Rotifera and Nemata.

Rotifera   The rotifers—or wheel animalcules, as they are commonly called—are microscopic, usually free-living aquatic animals with a crown of cilia at the anterior end (Fig. 42.33). The cilia are generally arranged in a circle, and when beating they often give the appearance of a rotating wheel; hence the name of the class. When feeding, rotifers attach themselves by a tapering posterior "foot," and the beating cilia draw a current of water into the mouth. In this manner, very small protozoans and algae are swept into a complicated muscular pharynx, where they are ground up by seven hard jawlike structures. This complex pharynx, a distinctive feature of the Rotifera, is called a mastax.

The freshwater rotifers are extremely abundant. Anyone examining a drop of water for Protozoa under a microscope is likely to see one or more of these interesting animals. In fact, most of them are no larger than protozoans, and it is often difficult, when encountering them for the first time, to realize that they are multicellular.

Nemata (the roundworms)   Nematode worms have round elongate bodies that usually taper nearly to a point at both ends (Fig. 42.34). Unlike flatworms, they have no cilia. The body is enclosed in a tough cuticle (Fig. 42.35). Just under the epidermal layer of the body wall are bundles of longitudinal muscles; there are no circular muscles. The lack of circular muscles and the stiff cuticle severely limit the types of movement possible for the worms, and they usually thrash about in what appears to be a random and inefficient manner.

The wall of the digestive tract consists of a single layer of endodermal cells; there is usually no muscle layer around the intestine, except some-

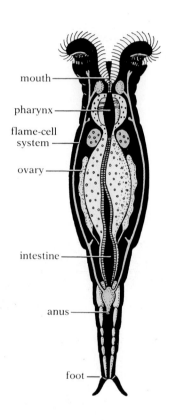

mouth

pharynx

flame-cell system

ovary

intestine

anus

foot

42.33   Section of a rotifer

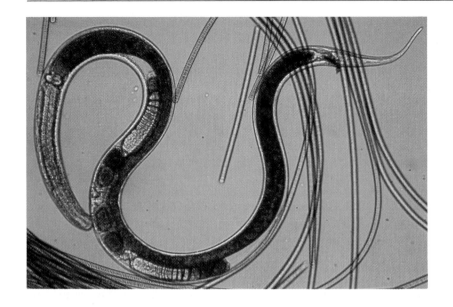

42.34   A living nematode worm viewed among Cyanobacteria (*Oscillatrosia*), its food

times at its posterior end. Between the intestine and the body wall is a fluid-filled cavity. As you can see from Figure 42.35, this cavity is bounded internally by the endodermal wall of the intestine, and it is bounded externally in part by the bands of mesodermal muscle and in part (between the muscle bands) by the ectodermal layer of the body wall. Since the cavity is not entirely enclosed by mesoderm, it is not a coelom, but a pseudocoelom.

Nematodes are extremely abundant, and occur in almost every type of habitat. Of the many free-living in soil or water, most are very tiny, often microscopic. A single spadeful of garden soil may contain a million or more, and a bucket of water from a pond usually contains comparable numbers. Many other nematodes are internal parasites of both plants and animals; these also are often small, but some may attain a length of one meter. So numerous and widespread are nematodes that N. A. Cobb has written:

> If all the matter in the universe except the nematodes were swept away, our world would still be dimly recognizable, and if, as disembodied spirits, we could then investigate it, we should find its mountains, hills, vales, rivers, lakes, and oceans represented by a film of nematodes. The location of towns would be decipherable, since for every massing of human beings there would be a corresponding massing of certain nematodes. Trees would still stand in ghostly rows representing our streets and highways. The location of the various plants and animals would still be decipherable, and, had we sufficient knowledge, in many cases even their species could be determined by an examination of their erstwhile nematode parasites.[8]

Nematodes parasitic on cultivated plants cause an annual loss of millions of dollars. Others parasitic on human beings cause some serious diseases.

[8] Cited in R. Buchsbaum, *Animals without Backbones* (2nd ed.; University of Chicago Press, 1948), pp. 156–57.

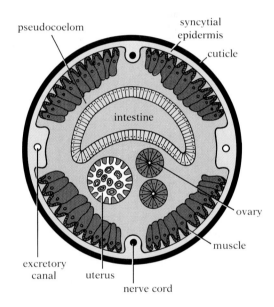

42.35   Diagrammatic cross section of a nematode worm

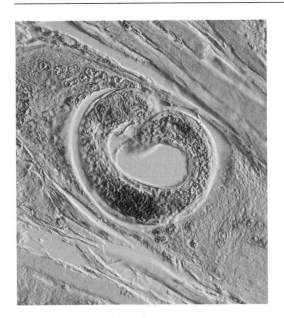

42.36 **Photograph of *Trichinella spiralis* encysted in muscle**

42.37 **Phoronid worms, *Phoronis hippocrepia***

Trichinella spiralis, for example, causes the disease called ***trichinosis,*** often contracted by eating insufficiently cooked pork. Adult *Trichinella* worms inhabit the small intestine of numerous species of mammals, among them hogs. Impregnated females bore through the wall of the intestine and deposit young larvae (which hatched from the eggs while still inside the uterus of the female) in the lymphatic vessels of their host. The larvae are carried by the lymph and blood to all parts of the body. They then bore out of the vessels, eventually entering every organ and tissue. However, only those that bore into skeletal muscles (especially the muscles of the diaphragm, ribs, tongue, and eyes) survive. In the muscles they grow in size (to about one millimeter) and then curl up and encyst (Fig. 42.36); the thick wall of the cyst is formed by the host's tissues. If insufficiently cooked pork containing such cysts is eaten by a human being, the walls of the cysts are digested away and the worms complete their development in the host's intestine. The adult worms then deposit larvae in the lymph vessels in the wall of the intestine, and the larvae move through the host's body as they do through a hog's, eventually encysting in muscles.

Most of the damage of trichinosis occurs during the migration of the larvae, when half a billion or more may simultaneously bore through the body after one infection. Symptoms include excruciating muscular pains, fever, anemia, weakness, and sometimes localized swellings. Some victims die, and those that do not may sustain permanent muscular damage. Prevention of the disease is simple: pork must be thoroughly cooked to kill the encysted larvae. Fortunately, *Trichinella* is exceedingly rare in developed countries.

Among other nematodes that parasitize humans are (1) *Ascaris,* a large worm (up to 30 cm long), which lives in the digestive tract, lays eggs that pass to the outside with the host's feces, and infects new hosts when vegetables grown in soil contaminated with feces are eaten without adequate washing; (2) hookworms, tiny worms, widespread in warm climates, that cause a severely debilitating disease usually contracted by going barefoot on soil contaminated with feces containing eggs; (3) pinworms, likewise tiny worms, common in schoolchildren, who usually get infected by putting unclean fingers with eggs on them in their mouths; (4) filaria worms, which are spread by the bite of certain mosquitoes in tropical and subtropical areas. Filaria worms live in the lymphatic system, where they may accumulate in such numbers that they block the flow of lymph, causing accumulation of fluid and often enormous swelling (elephantiasis) of the infected part of the body (see Fig. 13.19, p. 338).

THE COELOMATE PROTOSTOMIA

All the protostome phyla except the ones discussed above possess true coeloms that in most groups arise as a split in an initially solid mass of mesoderm. All have a complete digestive tract, and most have well-developed circulatory, excretory, and nervous systems.

THE LOPHOPHORATE PHYLA

There are three small phyla, Phoronida, Bryozoa, and Brachiopoda, that resemble one another in having a ***lophophore***—a fold, usually horseshoe-shaped, that encircles the mouth and bears numerous ciliated tentacles. The lophophore is a feeding device; its tentacular cilia create water currents that sweep plankton and tiny particles of detritus into a groove leading to the mouth.

All members of the lophophorate phyla are aquatic, and most are marine. Adults are usually sessile and secrete a protective case, tube, or shell around themselves, but the larvae are ciliated and free-swimming. The digestive tract is U-shaped in Phoronida and Bryozoa and in some Brachiopoda; the anus lies outside the crown of tentacles.

The phylum Phoronida contains only about 15 species of wormlike animals that inhabit a tube of their own secretion (Fig. 42.37). They are found either buried in the sand or attached to rocks, shells, or other objects in shallow seas.

The Bryozoa (or Ectoprocta) are often called moss animals. They are very tiny (usually less than half a millimeter long), colonial, sessile animals enclosed in a case open only at the lophophore end (Fig. 42.38). Unlike most other coelomate protostomes, they lack both excretory and circulatory systems. They superficially resemble entoprocts, which were formerly included in the same phylum with them, but entoprocts have a pseudocoelom instead of a coelom and their anus is inside the ring of tentacles,[9] which is not a true lophophore.

The Brachiopoda, shelled animals that superficially resemble molluscs, are often called lamp shells because they are shaped rather like an old Roman oil lamp (Fig. 42.39). They are usually permanently attached to the ocean bottom by a fleshy stalk. There are only about 260 living species of brachiopods, but more than 30,000 fossil species are known; they were among the most common animals in the Paleozoic seas.

The relationships of the lophophorate phyla to the other phyla are very poorly understood.

MOLLUSCA

The phylum Mollusca is the second-largest in the animal kingdom; it contains nearly 50,000 living species and 35,000 fossil species. Among the best-known molluscs are snails and slugs, clams and oysters, squids and octopuses.

The various groups of molluscs may differ considerably in outward appearance, but most have fundamentally similar body plans.[10] The soft body consists of three principal parts: (1) a large ventral muscular *foot*, which can be extruded from the shell (if one is present) and functions in locomotion; (2) a *visceral mass* above the foot, which contains the digestive system, the excretory organs (nephridia), the heart, and other internal organs; and (3) a heavy fold of tissue called the *mantle*, which covers the visceral mass and which in most species contains glands that secrete a shell. The mantle

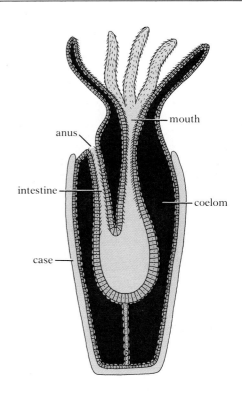

42.38   Section through the body of a bryozoan, showing the U-shaped digestive tract
The anus is located outside the ring of tentacles.

42.39   A brachiopod, *Terebratulina*
The shape of this northern lamp shell from Maine is reminiscent of old Roman lamps. Brachiopods usually attach themselves to the substratum by means of a stalk.

---

[9] The name "Entoprocta" means internal anus—an anus inside the crown of tentacles—and the name "Ectoprocta" means external anus—outside the crown of tentacles.

[10] The generalized body plan described here characterizes all living Mollusca except members of class Solenogastres, which are very primitive wormlike molluscs having no head, mantle, foot, shell, or nephridia.

42.40   A chiton
Note the series of plates composing the chiton's shell.

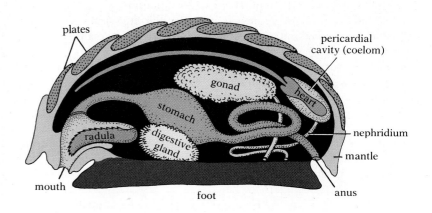

42.41   Lateral view of a chiton

42.42   A marine gastropod mollusc, *Trivia monaca*
The animal moves along the substratum by means of the large muscular foot, here plainly seen. Notice also the head region, bearing prominent antennae, eyes, and a long siphon through which water is brought to the gills.

often overhangs the sides of the visceral mass, thus enclosing a ***mantle cavity***, which frequently accommodates gills (see Fig. 42.47).

Molluscs have an open circulatory system—that is, during part of each circuit the blood is in large open sinuses where it bathes the tissues directly. Blood drains from the sinuses into vessels that run out into the gills, where the blood is oxygenated. From the gills, the blood goes to the heart, which pumps it into vessels that lead it back to the sinuses; a typical circuit, then, is heart–sinuses–gills–heart.

Most marine molluscs pass through one or more ciliated free-swimming larval stages, but freshwater and land snails complete the corresponding developmental stages while still in the egg and hatch as miniature editions of the adult.

The Mollusca are now divided into eight classes: Caudofoveata, Solenogastres, Polyplacophora, Monoplacophora, Gastropoda, Scaphopoda, Bivalva, and Cephalopoda. The first three classes were formerly lumped together and called Amphineura; exclusively marine, these animals are generally regarded as the most primitive living members of the Mollusca, and are thus a suitable place to begin study of the basic molluscan body plan. We shall examine the best-known class here.

Class Polyplacophora (chitons)   The chitons have an ovoid bilaterally symmetrical body with a shell consisting of eight serially arranged dorsal plates (Fig. 42.40). They have an anterior mouth and a posterior anus (Fig. 42.41). The coelom is reduced to a small cavity surrounding the heart.[11]

Chitons lead a sluggish, nearly sessile life. They creep about on the surface of rocks in shallow water, rasping off fragments of algae with a horny toothed organ called a ***radula***. Their broad flat foot can develop tremendous suction, and when disturbed they clamp down so tenaciously to the rock that they can hardly be pried loose.

---

[11] The lumina of the gonads and nephridia are also thought to be remnants of the coelom.

42.43   Two nudibranch gastropods, *Aplysia californica* (left), releasing "ink," and *Flabellinopsis iodinea* (right)
As the designation "nudibranch" suggests, the animals have no shells.

Class Monoplacophora   Members of this class have long been known as fossils, and it was thought that all had been extinct for about 350 million years. In 1952, however, ten living specimens (genus *Neopilina*) were dredged from a deep trench in the Pacific Ocean off the coast of Costa Rica.

These specimens sparked a lively debate on the ancestry of the Mollusca, because they show some internal segmentation, a characteristic seen in no other members of the phylum. Since it was already known that the early cleavage pattern and larval type of molluscs show striking similarities to the corresponding developmental stages in the segmented worms (Annelida), the segmentation of *Neopilina* led many biologists to conclude that the ancestral molluscs were segmented animals, perhaps primitive annelids. Many other biologists, however, are convinced that the segmentation of *Neopilina* is secondary, not primitive, and that the original molluscan body was unsegmented. Whichever view is correct, it seems clear that the Mollusca and the Annelida are fairly closely related.

Class Gastropoda (the snails and their relatives)   Most gastropods have a coiled shell (Fig. 42.42). In some cases, however, the coiling is minimal. Some species, such as *Aplysia* and the other nudibranchs, have lost the shell (Fig. 42.43).

The early larva in gastropods is bilateral, but, as it develops, the digestive tract bends downward and forward until the anus comes to lie close to the mouth. Then the entire visceral mass rotates through an angle of 180 degrees, coming to lie dorsal to the head in the anterior part of the body. Most of the visceral organs on one side (usually the left) atrophy, and growth proceeds asymmetrically, producing the characteristic spiral.

Except for the peculiar twisting and coiling of their bodies, gastropods are thought to be rather like the ancestral molluscs. They have a distinct head with well-developed sense organs, and most have a well-developed ra-

42.44 Shells of some representative gastropods
(A) Salt-marsh snail (*Melampus*). (B) Auger shell
(*Terebra*). (C) Moon shell (*Polinices*). (D) Oyster drill
(*Urosalpinx*). (E) Abalone (*Haliotis*), dorsal and
ventral views. (F) Californian keyhole limpet
(*Diodora*), lateral and dorsal views. (G) Periwinkle
(*Littorina*). (H) Boat shell (*Crepidula*), dorsal and
ventral views. (I) Olive shell (*Oliva*). (J) Channeled
conch (*Busycon*).

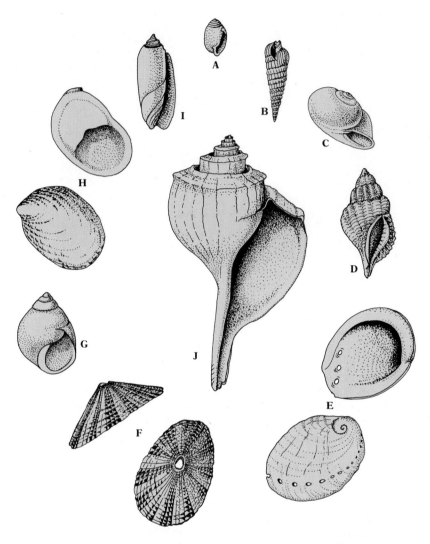

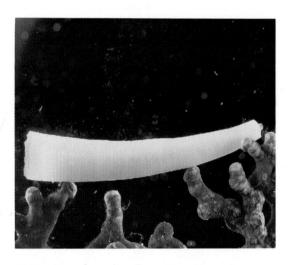

42.45 A scaphopod mollusc, *Dentalium vulgare*
The shell is a long tapering tube open at both ends.

dula and feed on bits of plant or animal tissue that they grate, rasp, or brush loose with this organ.

Gastropods occur in a great variety of habitats. The majority are marine, and their often large and decorative shells are among the most prized finds on a beach (Fig. 42.44), but there are also many freshwater species and some that live on land. The land snails (see Fig. 6.34, p. 163) are among the few groups of fully terrestrial invertebrates. In most of them, the gills have disappeared, but the mantle cavity has become very highly vascularized and functions as a lung. Such snails are said to be pulmonate. Some pulmonate snails have secondarily returned to the water and must periodically come to the surface to obtain air.

Class Scaphopoda (the tusk shells) Scaphopods have a long tubular shell, open at both ends (Fig. 42.45). One end is usually smaller than the

42.46 A scallop (*Chlamys opercularis*), a representative bivalve
The shell is composed of two hinged valves. Note the numerous small eyes around the edges of the mantle.

other, and the shell thus has a tusklike or toothlike appearance. All scaphopods are marine, living buried in mud or sand. The living animals are seldom seen, but the shells of dead ones can sometimes be found washed up on beaches.

Class Bivalva (the bivalve molluscs) As the term "bivalve" indicates, these animals have a two-part shell (Fig. 42.46). The two parts, or valves, are usually similar in shape and size and are hinged on one side (the animal's dorsum) (Fig. 42.47). The animals open and shut them by means of large muscles. Among the more common bivalves are clams, oysters, scallops, cockles, file shells, and mussels. Most lead rather sedentary lives as adults, though scallops sometimes swim about by rapidly opening and shutting the valves of their shells.

The bivalves, which have no radula, are filter feeders, straining tiny food particles from the water flowing across their gills. The process is described on p. 256.

Class Cephalopoda (squids, octopuses, and their relatives) Many of the cephalopods bear little outward resemblance to other molluscs. Unlike their sedentary relatives, they are often specialized for rapid locomotion and a predatory way of life—for killing and eating large prey such as fish and crabs. Though fossil cephalopods often have large shells, these are much reduced or absent in most modern forms. (The chambered nautilus, a modern form with a well-developed shell, is a familiar exception; see Figure 19.16, p. 501.) The body is elongate, with a large and well-developed head encircled by long tentacles.

Some species attain large size, often being several meters long. The giant squids (*Architeuthis*) of the North Atlantic are the largest living invertebrates; the biggest recorded individual was 17 m long (including the tenta-

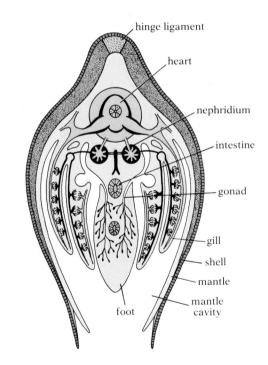

42.47 Cross section of a clam

42.48 A swimming octopus, *Octopus vulgaris*
Note the large suckers on the tentacles.

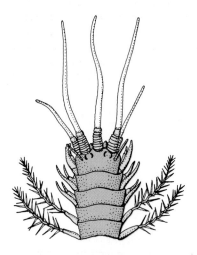

42.49   Head and first two gill-bearing segments of a polychaete worm (*Diopatra*)

cles) and weighed approximately 2 tons. Octopuses (Fig. 42.48) never grow anywhere near this size (except in Hollywood).

Cephalopods, particularly squids (see Fig. 11.11, p. 282), have convergently evolved many similarities to vertebrates. For example, squids have internal cartilaginous supports analogous to the vertebrate skeleton, and they even have a cartilaginous braincase rather like a skull. Furthermore, they have an exceedingly well-developed nervous system with a large and complex brain. Perhaps the most striking of all the squids' similarities to vertebrates are their large camera-type image-forming eyes, which work almost exactly the way ours do.

ANNELIDA

The Annelida, or segmented worms, have received attention in various connections throughout this book. We have considered their digestive system (p. 254), gas exchange (p. 281), closed circulatory system (p. 319), nephridia (p. 365), nervous system (p. 462), and their hydrostatic skeleton, muscle arrangement, and methods of locomotion (p. 528). The discussion here will therefore be brief.

The phylum is usually divided into three classes: Polychaeta, Oligochaeta, and Hirudinoidea.

Class Polychaeta   Polychaetes are marine annelids with a well-defined head bearing eyes and antennae (Fig. 42.49). Each of the numerous serially arranged body segments usually bears a pair of lateral appendages called *parapodia* that function in both locomotion and gas exchange (see Fig. 11.10, p. 281, and Fig. 20.4, p. 529). There are numerous stiff setae (bristles) on the parapodia (the name "Polychaeta" means many setae).

Some polychaetes swim or crawl about actively; others are more sedentary, usually living in tubes they construct in the mud or sand of the ocean bottom. These tubes may be simple mucus-lined burrows, membranous structures, or elaborately constructed dwellings composed of sand grains cemented together. The tubes of some species are straight, while those of others are U-shaped and have two openings (Fig. 42.50). The beating parapodia keep water currents flowing through the tubes; these currents bring oxygen, and in some cases food particles, to the worm. Many of the tube dwellers are beautiful animals, often colored bright red, pink, or green; some are iridescent. Among the most beautiful are the fanworms and peacock worms, which have a crown of colorful, much-branched fanlike or featherlike processes that they wave in the water at the entrance to their tubes (Fig. 42.51).

All the segments of the body are usually much alike. The coelom of each is partly separated from the coeloms of adjacent segments by membranous intersegmental partitions; the partitions of many polychaetes are not complete, however, and in some species they have been entirely lost. Each segment generally has its own ventral ganglion and its own pair of nephridia.[12]

The sexes are separate in the majority of species. In primitive polychaetes most segments produce gametes, but in more advanced species gamete

---

[12] In a few species of polychaetes, there is only one pair of nephridia for the whole animal.

production is restricted to a few specialized segments. The gametes are usually shed into the coelom and leave the body through the nephridia. Fertilization is external. In many species development includes a ciliated free-swimming larval stage called a ***trochophore*** (see Fig. 42.77).

Class Oligochaeta The class Oligochaeta contains the earthworms and many freshwater species. They differ from polychaetes in that they lack a well-developed head and parapodia, have few setae (the name "Oligochaeta" means few setae), usually combine male and female organs in the same individual, and usually have more complete intersegmental partitions. We have described most of the important characteristics of earthworms in earlier chapters (see Fig. 10.12, p. 254; Fig. 10.13, p. 255; Fig. 13.1, p. 319; Fig. 14.13, p. 366; Fig. 17.1, p. 435; and Fig. 18.14, p. 462).

Class Hirudinoidea (the leeches) The leeches, which probably evolved from oligochaetes, are the most specialized annelids. Their body is dorsoventrally flattened and often tapered at both ends. The first and last segments are modified to form suckers, of which the posterior one is much the larger; these suckers are used in locomotion. Leeches show almost no internal segmentation; the intersegmental partitions have been completely lost except in a few very primitive species.

Some leeches are predaceous, capturing invertebrate prey such as worms, snails, and insect larvae and swallowing them whole. More familiar are the bloodsuckers, which attack a variety of vertebrate and invertebrate

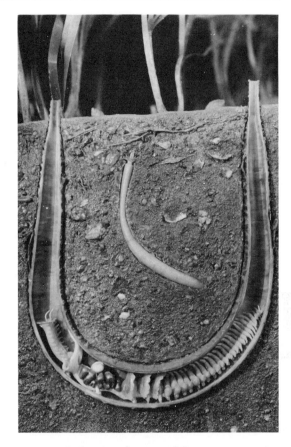

42.50 Polychaete and sipunculid
The parchment worm (*Chaetopterus pergamentacus*) is a polychaete that lives in a U-shaped tube (here shown in section). Between the arms of the tube is a sipunculid (*Phascolosoma gouldii*); the Sipuncula are a small phylum of worms related to the Annelida.

42.51 Fanworms, *Bispira voluticornis*
The featherlike processes of these fanworms move with the surrounding water and circulate food and gases to the worms.

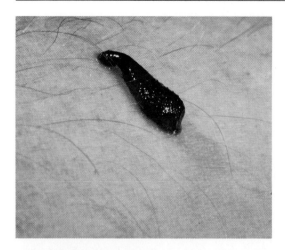

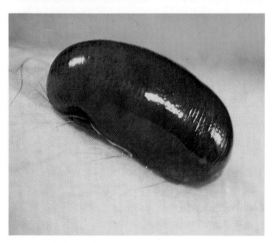

42.52 A terrestrial leech of Australia, sucking blood from a person's arm
Top: The leech has just begun feeding. Bottom: Somewhat later its body has become engorged with blood.

hosts (Fig. 42.52). When a leech of this type attacks a host, it selects a thin area of the host's integument, attaches itself by its posterior sucker, applies the anterior sucker very tightly to the skin, and either painlessly slits the skin with small bladelike jaws or dissolves an opening by means of enzymes. It then secretes into the wound a substance (hirudin) that prevents coagulation of the blood, and begins to suck the blood, usually consuming an enormous quantity at one feeding and then not feeding again for a fairly long time (some leeches have been known to go unfed for more than a year without apparent harm).

ONYCHOPHORA

There are only about 65 living species of this small phylum, all restricted to tropical regions or to the temperate parts of the Southern Hemisphere (Australia, New Zealand, South Africa, and the Andes). They are mostly confined to very moist habitats on land, living beneath leaves, logs, or stones in forests, and are active at night.

Looking rather like caterpillars, onychophorans have a segmented wormlike body with from 14 to 43 pairs of short unjointed legs (Fig. 42.53). These animals are of special interest because they have a combination of annelid and arthropod characters and are regarded as an early evolutionary offshoot from the line leading to the arthropods from an ancient annelidlike ancestor (see Fig. 42.56). They have a thin flexible permeable cuticle more like the cuticle of annelids than the exoskeleton of arthropods, and, like the annelids, they have a pair of nephridia in each segment. However, they resemble arthropods in having claws and an open circulatory system. The tracheal respiratory system of modern onychophorans probably evolved independently of the tracheae of terrestrial arthropods.

ARTHROPODA

The phylum Arthropoda is by far the largest of the phyla. Nearly a million species have been described, and there are doubtless hundreds of thousands more yet to be discovered. Probably more than 80 percent of all animal species on earth belong to this phylum.

Arthropods are characterized by jointed chitinous exoskeletons and jointed legs. The exoskeleton, which is secreted by the epidermis, functions both as a point of attachment for muscles and as a protective armor, but it imposes limitations on growth and must be periodically molted if the animal is to undergo much increase in size (Fig. 42.54; see also Figs. 31.19 and 31.20, pp. 802–4). The arthropod cuticle is not restricted to the exterior surface of the body; long rod-shaped processes that often project from the surface deep into the interior of the animal function as bases for muscle attachment, and both the anterior and the posterior portions of the digestive tract (and the tracheae of land arthropods) are lined with cuticle. These internal extensions of the exoskeleton are also shed at each molt.

Together with their elaborate exoskeleton, arthropods have evolved a complex musculature quite unlike that of most other invertebrates. It comprises not only longitudinal and circular bands, as in so many invertebrates, but also separate muscles that, running in myriad directions, make

42.53 An onychophoran
Note the numerous short unjointed legs and the prominent antennae.

possible an extensive repertoire of movements. Most of the muscles are striated.

The nervous system is very well developed. Of a similar organization as the annelid nervous system, it consists of a dorsal brain and a ventral double nerve cord (see Fig. 18.14, p. 462). In primitive arthropods there are ganglia in each segment, but in many groups the ganglia have moved forward and fused into larger ganglionic masses. Sensory organs are many and varied. Like the nervous system, discussed more extensively in Chapter 18 (p. 462), hormonal control, too, is well developed in arthropods (see p. 403).

As we have seen (p. 320, Fig. 13.2), arthropods have an open circulatory system. There is usually an elongate dorsal vessel called the heart, which pumps the blood forward into arteries (the extent of these arteries varies greatly among the various groups of arthropods). From the arteries, the blood goes into open sinuses, where it bathes the tissues directly. Eventually the blood returns to the posterior portion of the heart.

The body spaces through which the blood moves constitute the *hemocoel*—not a true coelom but a cavity derived from the embryonic blastocoel. Though arthropods almost certainly descended from an annelidlike ancestor with well-developed coelomic cavities, and though such cavities develop in arthropod embryos, they are not retained as the functional body cavity in the adult.[13]

In most aquatic arthropods (excluding secondarily aquatic ones), excretion of nitrogenous wastes (primarily ammonia) is principally by way of the gills. Aquatic species usually also have special saclike glands, located near or in the head, that play a minor role in excretion; these glands (usually called coxal glands or green glands) have their own ducts leading to the

42.54 A molting centiped
Its new yellow-orange exoskeleton glistening, the animal is backing out of its old exoskeleton.

[13] The cavity of the gonads (and that of the excretory ducts in some arthropods) may be a remnant of the true coelom.

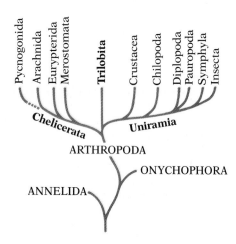

42.55   Diagram of possible relationships between the Annelida, the Onychophora, and the major subphyla and classes of the Arthropoda

42.56   A fossil of a trilobite
Note the two longitudinal furrows that partition the animal's body into a median lobe and two lateral lobes. It is this tripartite arrangement that suggested the name "Trilobita."

outside. The excretory organs in most groups of terrestrial arthropods are Malpighian tubules (see Fig. 14.14, p. 366).

The sexes are usually separate. Fertilization is internal in all terrestrial and in most aquatic forms.

It is generally held that arthropods evolved either from a polychaete annelid or from the ancestor of the polychaetes. The arthropod body plan may be viewed as an elaboration and specialization of the segmented body of that annelid ancestor. The evidence indicates that the first arthropods had long wormlike bodies composed of many nearly identical segments, each bearing a pair of legs. All the legs were alike. Among the host of modifications of this ancestral body plan that have arisen in the various groups of arthropods during the millions of years of their evolution, four tendencies stand out: (1) reduction in the total number of segments; (2) grouping of segments into distinct body regions, such as a head and trunk, or a head, thorax, and abdomen; (3) increasing cephalization—that is, incorporation of more segments into the head and concentration of nervous control and sensory perception in or just behind the head; (4) specialization of the legs of some segments for a variety of functions other than locomotion, and complete loss of legs from many other segments.

The arthropods are now divided into four subphyla: Trilobita (a large group now extinct), Uniramia (a group that includes the insects), Chelicerata (a smaller group that includes the arachnids), and Pentastomida (a tiny group of parasites that will not be discussed separately). Figure 42.55 shows one interpretation of the relationships among the arthropod subphyla and classes.

Subphylum Trilobita   Arthropods were very abundant in the Paleozoic seas, and fossils from that era are plentiful. Particularly common in rocks of the first half of the Paleozoic are the fossils of an extinct group, the Trilobita (Fig. 42.56). The fossils show a usually oval and flattened shape and three body regions: a head, apparently composed of four fused segments, that bore a pair of slender antennae and, often, compound eyes; a thorax consisting of a variable number of separate segments; and an abdomen (pygidium), composed of several fused segments. It is not to this tripartite division, however, that the name "Trilobita" refers, but to a division of the body into a median lobe and two lateral lobes by two prominent longitudinal furrows running along the dorsum.

Trilobites, though they were surely different from the first arthropods, exhibiting specializations of their own such as the longitudinal furrows and the fusion of the abdominal segments, nevertheless seem to approach the hypothetical arthropod ancestor more closely than any other known group. One primitive character stands out—the lack of specialization and structural differentiation of the appendages. The fossils show that every segment bore a pair of legs, and that all these legs, including those of the four head segments, were nearly identical. There were thus no appendages specialized as mouthparts. In both Chelicerata and Uniramia, by contrast, the tendency toward specialization of some appendages and loss of others is quite evident; thus in both, the appendages of the most anterior segments have been modified as mouthparts and no longer function in locomotion.

42.57 A whipscorpion, *Mastigoproctus giganteus*
Whipscorpions have six pairs of appendages: a pair of fanglike chelicerae (not visible in the photograph); a pair of stout toothed pincerlike pedipalps; and four pairs of legs (the long and slender first pair have a sensory-tactile function and are not used in walking). The posterior knob with its "whip" has slits through which the animal can spray a poisonous secretion.

Subphylum Chelicerata   The chelicerate body is usually divided into two regions: a cephalothorax (prosoma) and an abdomen. There are no antennae. The appendages corresponding to the first pair of postoral legs in ancestral arthropods and trilobites are modified as mouthparts called ***chelicerae***, which may be either pincerlike or fanglike. The cephalothorax usually bears five other pairs of appendages besides the chelicerae; in some groups these are all walking legs, while in others only the last four pairs are legs, the first pair being modified as feeding devices called ***pedipalps***, which are often much longer than the chelicerae (Fig. 42.57). The legs of the abdominal segments have been either lost or modified into respiratory or sexual structures.

The subphylum Chelicerata includes four classes. One (Eurypterida) consists entirely of animals extinct since the Paleozoic era, and the members of another (Pycnogonida, the sea spiders) are very rare marine animals.

Members of a third class (Merostomata) are familiar to anyone who has spent some time on the Atlantic beaches of North America (or the coast of Asia from Japan and Korea to Malaysia and Indonesia). These are the horseshoe crabs (Fig. 42.58), which are not really crabs at all but living relicts of an ancient chelicerate class most members of which have been extinct for millions of years. The most common species is named *Limulus polyphemus*.

Members of the fourth class of chelicerates—Arachnida—are familiar to everyone. These are the spiders, ticks, mites, daddy longlegs, scorpions, whipscorpions (see Fig. 42.57), and their relatives. Though the various groups of arachnids differ structurally in many ways, most have two body regions: a cephalothorax and an abdomen (these are not distinguishable in

42.58   A horseshoe crab

ticks, mites, or daddy longlegs). There are often simple eyes on the cephalothorax but never any compound eyes or antennae. The cephalothorax bears six pairs of appendages: a pair of chelicerae, a pair of pedipalps, and four pairs of walking legs (Fig. 42.59). In most groups prey is seized and torn apart by the pedipalps. The chelicerae may also function in manipulating prey, or they may be modified as poison fangs, as in spiders. The abdomen of arachnids may be long, as in scorpions, or short, as in spiders. In some (including spiders), the bases of one or two pairs of abdominal appendages are retained as much-modified lungs; in others, no trace of abdominal appendages remains. In addition to the lungs, many arachnids have tracheae, and some respire by means of tracheae only. Several arachnid groups possess glands that secrete silk.

Subphylum Uniramia (or Mandibulata)   The members of this subphylum differ from chelicerates in having antennae and in having ***mandibles***

42.59   A spider, *Araneus marmoreus*
The four pairs of legs characteristic of the Arachnida are easy to make out. Note how the animal places its first pair of legs on strands of its web; it can detect vibrations when prey touch the web, and can often even distinguish, by the type of vibration, what sort of prey it is.

42.60   A crab and a prawn, representatives of the larger, better-known crustaceans
Top: The crab *Neolithodes grimaldii*. Bottom: A freshwater prawn.

A

B

C

D

42.61   Some representatives of the smaller crustaceans

(A) A marine amphipod, *Gammarus*. (B) The water flea, *Daphnia*. (C) A group of sow bugs (also called wood lice or pill bugs), one of the very few terrestrial crustaceans. (D) A freshwater crustacean, *Asellus aquaticus*.

instead of chelicerae as their first pair of mouthparts. Mandibles are modified from the basal segment (coxa) of the ancestral legs, and function in biting and chewing (though in some species they are secondarily modified for piercing and sucking). They are never clawlike or pincerlike, as chelicerae often are. In most mandibulates there are two additional pairs of mouthparts called **maxillae**.

The subphylum comprises six classes. We shall briefly describe four of them here: Crustacea, Chilopoda, Diplopoda, and Insecta.

Class Crustacea   Some representatives of this class, such as crayfish, lobsters, shrimps, and crabs (Fig. 42.60), are well known to most people. But there are many other species of Crustacea, which bear little superficial resemblance to these familiar animals; among them are fairy shrimps, water fleas, brine shrimps, sand hoppers, barnacles, and sow bugs; many of these are very small odd-looking creatures (Fig. 42.61).

42.62 Goose barnacles, *Lepas*
These animals are sedentary and secrete a protective shell. Most species of barnacles lack the long stalk so prominent in the goose barnacles.

Crustacea characteristically have two pairs of antennae, a pair of mandibles, and two pairs of maxillae. But the rest of the appendages vary greatly from group to group, and whatever could be said about those of one group, such as crayfish and lobsters, would have little relevance to those of other groups. In fact, the Crustacea are an enormously diverse assemblage of animals that can hardly be characterized in any simple way. Some have a cephalothorax and an abdomen; others have a head and a trunk, or a head, thorax, and abdomen, or even a unified body. Most are free-living, but some are parasitic. Most are active swimmers, but some, like barnacles, secrete a shell and are sessile (Fig. 42.62). The majority are marine, but there are many freshwater species, and a few, such as sow bugs, are terrestrial and have a simple tracheal system. We could go on listing divergences, but the point of the amazing diversity of this group has been made. This is a class in which the basic arrangement of a segmented body with numerous jointed appendages has been modified and exploited in countless ways as the members of the class have diverged into different habitats and adopted different modes of life.

Class Chilopoda Members of this class are called centipeds (or hundred-legged worms). Their body is divided into two regions, a head and a trunk (see Fig. 6.37, p. 164). The trunk is elongate and often somewhat flattened. The head bears a single pair of antennae and three pairs of mouthparts (mandibles and two pairs of maxillae). The animals are carnivorous, and the legs of the first trunk segment are modified as large poison claws. Each of the other trunk segments bears a single pair of walking legs. All centipeds are terrestrial and respire by means of tracheae. Excretion is by Malpighian tubules. The genital ducts open at the rear of the body.

Class Diplopoda These animals are called millipeds (or thousand-legged worms). They superficially resemble centipeds and in fact were once placed with them (and with two other smaller groups) in a class called Myriapoda. However, it is now known that centipeds and millipeds are not closely related, and each is placed in a separate class.

The milliped body is divided into a head and a trunk (Fig. 42.63). The head bears a pair of antennae but only two pairs of mouthparts (mandibles, and a pair of maxillae fused to form a platelike underlip). The animals have no poison claws, and are not carnivorous, feeding largely on decaying organic matter of various types. Each of the first four or five trunk segments (depending on the species) bears a single pair of legs, but each of the other segments bears two pairs of legs (and also two pairs of spiracles); it is clear that each of the double-legged segments is formed by the fusion of two segments. Respiration is by tracheae, and excretion is by Malpighian tubules. The genital ducts open anteriorly, on the second segment. In most milliped orders the legs (one or both pairs) of the seventh segment in the males are highly modified and function as organs for inserting sperm into the female reproductive tract (see Fig. 34.10, p. 892).

42.63 A milliped
Each segment (except a few at the front and rear) bears two pairs of legs.

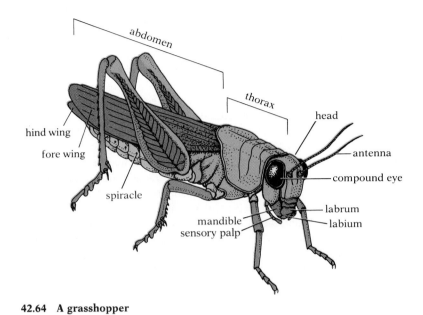

42.64 A grasshopper

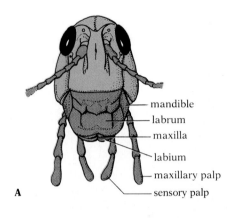

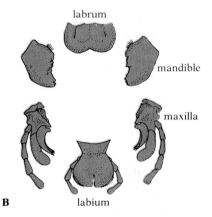

42.65 The mouthparts of a grasshopper
(A) Front view of head, with mouthparts in place.
(B) The mouthparts removed from the head, but kept in their proper relative positions. The mandibles and probably the labrum (upper lip) are derived from the basal segments of ancestral legs; all the other segments of those legs have been lost. The maxillae and labium (lower lip) retain more of the segments of the legs from which they are derived; the basal segments are enlarged, but the distal segments form slender leglike structures called palps, which bear many sensory receptors.

Class Insecta This is an enormous group of diverse animals that occupy almost every conceivable habitat on land and in fresh water. If numbers are the criterion by which to judge biological success, then the insects are the most successful group of animals that has ever lived; there are more species of insects than of all other animal groups combined. But there is one restriction on their dominant role: they do not occur in the sea (although a few species walk on the ocean surface or live in brackish water); the role played by insects on land is played in the sea by Crustacea.

There are a few insect fossils from the Devonian, but it was in the Carboniferous and Permian periods that insects took their place as one of the dominant groups of animals (see Table 37.1, p. 1012). By the end of the Paleozoic era, many of the modern orders had appeared, and the number of species was enormous. A second great period of evolutionary radiation began in the Cretaceous and continues to the present time; this second radiation is correlated with the rise of flowering plants.

The insect body is divided into three regions: a head, a thorax, and an abdomen (Fig. 42.64). The head segments are completely fused, and in adults their boundaries cannot be distinguished. The head bears numerous sensory receptors, usually including compound eyes, one pair of antennae, and three pairs of mouthparts derived from ancestral legs. The mouthparts include a pair of mandibles, a pair of maxillae, and a lower lip, or ***labium***, formed by fusion of the two second maxillae (Fig. 42.65). The upper lip, or ***labrum***, which has not traditionally been classified as a mouthpart, may also be derived from ancestral legs.

The thorax is composed of three segments, each of which bears a pair of walking legs. In many insects (but not all), the second and third thoracic

42.66 A desert locust in flight
Like most flying insects except the flies, the locust—also called the grasshopper—has two pairs of wings. The front wings are leathery and, when the animal is at rest, provide a protective covering for the fragile pleated rear wings, which are the ones important for flight.

segments each bear a pair of wings, and the animals are vigorous fliers (Fig. 42.66).

The abdomen is composed of a variable number of segments (12 or fewer). Abdominal segments are devoid of legs, but highly modified remnants of the ancestral appendages may be present at the posterior end, where they function in mating and egg laying.[14]

We have already discussed the insects at considerable length in other chapters and need not repeat ourselves here. See the discussion of the insect tracheal system (p. 292, Figs. 11.25–11.27), the circulatory system (p. 320, Fig. 13.2), the Malpighian excretory organs (p. 366, Fig. 14.14), hormonal control (p. 405, Fig. 16.1), nervous control and sensory perception (p. 462, Fig. 18.14; p. 500, Figs. 19.13 and 19.14), the exoskeleton and muscles (p. 530, Fig. 20.6B, D), behavior (Chapters 21 and 22), and development (p. 800, Figs. 31.15; 31.19, and 31.20).

The insects are classified in approximately 28 orders (the exact number depends on the authority cited). The following are among the more familiar:

THYSANURA Bristletails and silverfish. Small, primitive, wingless; chewing mouthparts; long tail-like appendages on rear of abdomen. Incomplete metamorphosis. Common in houses, particularly in kitchens and bathrooms; sometimes damage books in libraries.

ODONATA Dragonflies and damselflies (Fig. 42.67A). Medium to large, rapid-flying, predaceous on other insects. Two pairs of long membranous wings; chewing mouthparts; very large compound eyes. Immature stages (nymphs) in fresh water; incomplete metamorphosis.

ORTHOPTERA Grasshoppers, crickets (Fig. 42.67B). Usually two pairs of wings—the coarse-textured fore wings, which are narrower than the hind wings, are not used in flight, but function (when animal is at rest) as covers for the folded fanlike hind wings; chewing mouthparts. Gradual metamorphosis.

ISOPTERA Termites. Highly social insects. Wings briefly present only on members of the reproductive castes; usually chewing mouthparts. Gradual metamorphosis.

HEMIPTERA True bugs (Fig. 42.67C). Usually two pairs of wings—basal half of fore wings thick and leathery, distal half membranous, hind wings membranous; piercing-sucking mouthparts. Gradual metamorphosis.

ANOPLURA Sucking lice (Fig. 42.67D). External parasites. Wingless; piercing-sucking mouthparts; legs and claws adapted for clinging to host. Gradual metamorphosis.

COLEOPTERA Beetles (Fig. 42.67E). Two pairs of wings—fore wings hard, meeting along middorsal line, forming a protective case for the

---

[14] A few primitive insects retain vestiges of appendages on many abdominal segments. These may have a sensory function.

folded membranous hind wings when animal is at rest; chewing mouthparts. Complete metamorphosis.

LEPIDOPTERA Moths and butterflies. Two pairs of large scale-covered wings; chewing mouthparts in larvae, sucking (but not piercing) mouthparts in adults. Complete metamorphosis.

DIPTERA True flies: mosquitoes, gnats, midges, house flies, horse flies, etc. (Fig. 42.67F). One pair of membranous wings; highly modified hind wings acting as tiny balancing organs; piercing-sucking or sponging mouthparts. Complete metamorphosis.

SIPHONAPTERA Fleas (Fig. 42.67G). Intermittent ectoparasites. Small, body laterally compressed; no wings; piercing-sucking mouthparts; long legs, adapted for jumping. Complete metamorphosis.

HYMENOPTERA Sawflies, ants, bees, wasps (Fig. 42.67H). Usually two pairs of membranous wings, interlocked in flight; chewing or chewing-lapping mouthparts; thorax and abdomen connected by a very narrow waist. Complete metamorphosis.

A more complete listing of the insect orders is given in the Appendix.

THE DEUTEROSTOMIA

The four phyla of Deuterostomia mark the transition from the invertebrate animals to the vertebrates. Two phyla are clearly invertebrate: Echinodermata, whose most familiar representatives are the sea stars (starfish), and Chaetognatha, a very small phylum consisting of exclusively marine species known as arrow worms. A third phylum, Chordata, contains all of the vertebrates; the next chapter is devoted to this group. The fourth phylum, Hemichordata, falls somewhere in between the vertebrates and invertebrates. We discuss this small but evolutionarily important phylum in this chapter.[15]

ECHINODERMATA

The echinoderms are exclusively marine, mostly bottom-dwelling animals. They are common in all seas and at all depths from the intertidal zone to the ocean deeps. Included in this distinctive phylum are the sea stars, brittle stars, sea urchins, sand dollars, sea cucumbers, and sea lilies.

The adults are radially symmetric, but the larvae are bilateral, and it is generally held that echinoderms evolved from bilateral ancestors. The radial symmetry probably arose as an adaptation to a sessile way of life. Most of the modern echinoderms (with the exception of sea lilies) move about slowly and are thus not completely sessile, but the ancient echinoderms apparently were.

Almost all members of this phylum possess an internal skeleton com-

[15] Until recently, another small phylum of marine worms—Pogonophora (beard worms)—was placed in the Deuterostomia, but the discovery that part of their body (broken off all older specimens) is segmented and bears numerous setae has prompted the transfer of this phylum to the Protostomia near the Annelida.

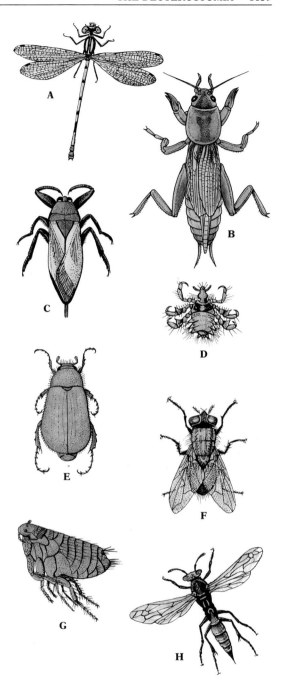

42.67 Some representatives of the major insect orders
(A) Damselfly (order Odonata). (B) Mole-cricket (Orthoptera). (C) Bug (Hemiptera). (D) Louse (Anoplura). (E) Beetle (Coleoptera). (F) Fly (Diptera). (G) Flea (Siphonaptera). (H) Wasp (Hymenoptera).

posed of numerous calcareous plates embedded in the body wall. These plates may be separate, or they may be fused to form a rigid boxlike structure. The skeleton often bears many bumps or spines, particularly noticeable in sea urchins, that project from the surface of the animal (see Fig. 42.71). It is this characteristic that gives the animals the name "Echinodermata" (from the Greek *echino*, "spiny," and *derma*, "skin").

Echinoderms have a well-developed coelom in which the various internal organs are suspended. The complete digestive system is the most prominent of the organ systems. There is no special excretory system, and the blood circulatory system, though present, is poorly developed. The nervous system is radially organized, consisting of nerve networks that connect to ringlike ganglionated nerve cords running around the body of the animal (there are often three of these cords); there is no brain.

A characteristic unique in echinoderms is their **water-vascular system**. This is a system of tubes (usually called canals) filled with watery fluid. Water can enter the system through a sievelike plate, called a madreporite, on the surface of the animal. A tube from this plate leads to a ring canal that encircles the esophagus (Fig. 42.68). Five radial canals branch off the ring canal and run along symmetrically spaced grooves or bands on the surface of the animal. Many short side branches from the radial canals lead to hollow **tube feet** that project to the exterior. Each tube foot is a thin-walled hollow cylinder, with a sucker on its end. At the base of each tube foot is a muscular ampulla containing fluid. When the ampulla contracts, the fluid, prevented by a valve from flowing into the radial canal, is forced into the tube foot, which is thereby extended. The foot attaches to the substratum by

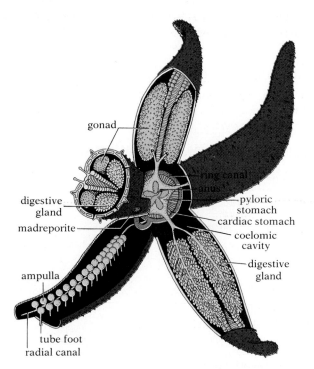

42.68 Dissection of a sea star (dorsal view)

gonad

ring canal

anus

pyloric stomach

cardiac stomach

coelomic cavity

digestive gland

digestive gland

madreporite

ampulla

tube foot

radial canal

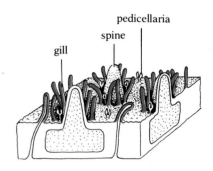

42.69 Sea star
The tiny skin gills are protected by the spines and by the pincerlike pedicellariae, which repel (or capture) small animals that might otherwise settle on the surface of the sea star.

its sucker, and then longitudinal muscles in its wall contract, shortening it and pulling the animal forward (while forcing the water back into the ampulla). This cycle of events, repeated rapidly by the many tube feet of an animal like a sea star, enables it to move slowly. The tube feet may also enable it to hold tightly to a rock or other object by applying suction, or to pull open the valves of the shell of a clam or oyster, on which the sea star will feed.

The sexes are usually separate. Eggs and sperm are shed into the surrounding water, where fertilization occurs. Cleavage is radial and indeterminate. The larva, which is ciliated and free-swimming, has a complete digestive tract (see Fig. 31.18, p. 801); the anus is derived from the embryonic blastopore, the mouth being a new opening.

Class Stelleroidea (or Asteroidea, the sea stars and their relatives)   The body of a sea star (starfish) consists of a central *disc* and usually of five rays, or *arms*,[16] each with a groove bearing rows of tube feet running along the middle of its lower surface. The outer surface of the animal is studded with many short spines and numerous tiny skin gills, which are thin fingerlike evaginations of the body wall that protrude to the outside between the plates of the endoskeleton (Fig. 42.69). The cavity of each skin gill is continuous with the general coelom. Scattered between the spines and skin gills are often numerous small jawlike structures called pedicellariae, which are used for protection and for capturing very small animals. The madreporite is on the upper surface (but not in the center; in this respect the radial symmetry of the animal is not perfect).

---

[16] Though most species of sea stars have five arms, some have more.

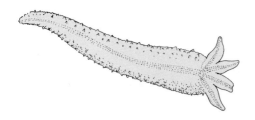

42.70 A detached arm of a sea star regenerating a new body

The mouth is located in the center of the lower surface of the disc and the anus in the center of the upper surface (thus the lower surface is the morphological anterior end of the animal, and the upper surface is the morphological posterior end). The digestive tract of a sea star is straight and very short, consisting of a short esophagus, a broad stomach that fills most of the interior of the disc, and a very short intestine. The stomach is divided by a constriction into two parts: a large eversible part (cardiac stomach) at the esophageal (lower) end, and a smaller noneversible part (pyloric stomach) at the intestinal (upper) end. Attached to the noneversible part are five pairs of large digestive glands; each pair of glands lies in the coelomic cavity of one of the arms (Fig. 42.68).

When the sea star feeds, it pushes the lower part of the stomach out through the mouth, turning it inside out and placing it over food material such as the soft body of a clam or oyster. The stomach secretes digestive enzymes onto the food, and digestion begins. The partly digested food is then taken into the upper part of the stomach and into the digestive glands, where digestion is completed and the products are absorbed.

Sea stars have amazing regenerative abilities. Even a single detached arm can regenerate an entire new animal (Fig. 42.70).

In addition to sea stars, the Asteroidea include brittle stars, serpent stars, and basket stars (Fig. 42.71). The organisms have five arms like sea stars, but the arms are longer, much more slender and flexible, often branched, and grooveless. The body disc is relatively small. The tube feet have no ampullae and are not used in locomotion, which is by rapid lashing of the arms. There is a large stomach, but no intestine and no anus. Gas exchange

42.71 A brittle star (*Ophiopholis aculeata*) and a basket star (*Gorgonocephalus eucnemis*)

Left: The body disc of the brittle star is relatively small, and the arms are long and slender. Right: The arms of the basket star branch repeatedly to produce a mass of coils resembling tentacles.

42.72 A sea urchin
Sea urchins and their relatives have numerous large spines on their hard boxlike skeletons.

is by invaginated pouches in the periphery of the disc. Because of such differences, these organisms were formerly assigned to a separate class (Ophiuroidea).

Other echinoderm classes   Although members of the three other echinoderm classes often show little superficial resemblance to sea stars and brittle stars, their structure is fundamentally similar.

Class Echinoidea   These are the sea urchins, sand dollars, and heart urchins. They have no arms, but, like sea stars, they do have five bands of tube feet. The body is spherical, or flattened and oval, and is covered with long spines (Fig. 42.72). The plates of the endoskeleton are fused to form a rigid box or case. There is a complex chewing aparatus just inside the mouth, and the intestine is long and coiled. Gas exchange is by small but highly branched gills or by modified tube feet.

Class Holothuroidea   The sea cucumbers (Fig. 42.73) differ from the other echinoderms in having a much reduced endoskeleton and a leathery body. Also unlike the members of the classes discussed above, they lie on their sides rather than on the oral surface. The mouth is surrounded by tentacles attached to the water-vascular system. There is a very long coiled intestine, and gas exchange is usually by complexly branched respiratory trees attached to the cloaca (see p. 284, Fig. 11.15).

42.73 A sea cucumber (*Pseudocolochirus*) in the Coral Sea
Notice the rows of tube feet. There are five of these rows, which correspond to the five arms of a sea star.

42.74 A fossil sea lily

Class Crinoidea This is the oldest and most primitive of the living classes of echinoderms. The sea lilies, as most Crinoidea are commonly called, are attached to the substratum by a long stalk and are thus sessile (Fig. 42.74). They have long feathery arms (often branched) around the mouth, which is on the upper side; the sea lilies thus differ from the Stelleroidea and Echinoidea in that their morphological anterior end is directed upward and their morphological posterior end downward. Some modern crinoids—the feather stars—lack a stalk and are not sessile.

HEMICHORDATA

The hemichordates, many of which belong to a class called acorn worms (Fig. 42.75), are entirely marine. Often found living in U-shaped burrows in sand or mud along the coast,[17] acorn worms are fairly large, ranging from 6½ to 43 cm in length. Their bodies consist of an anterior conical proboscis (thought by some to resemble an acorn—hence their name), a collar, and a long trunk (Fig. 42.76). The mouth is situated ventrally, at the junction between the proboscis and the collar.

A particularly important feature in the hemichordates is a series of ***pharyngeal slits*** in the wall of the pharynx. Water drawn into the mouth is forced back into the pharynx and out through these slits into branchial sacs,

---

[17] Some acorn worms live under rocks or shells instead of burrowing; and not all of the burrowers make U-shaped tubes. There are two classes of hemichordates that are not acorn worms.

42.75   An acorn worm, *Balanoglossus*
The short rounded proboscis is partially enveloped by the collar behind it. Only a part of the long trunk is shown.

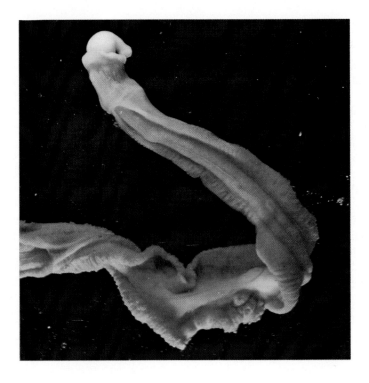

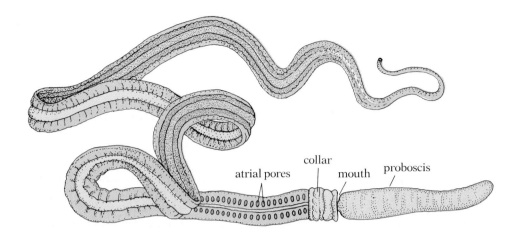

42.76 An adult acorn worm
This particular genus (*Saccoglossus*) has a more elongated proboscis than *Balanoglossus*, shown in Figure 42.75.

which open to the exterior via atrial pores. Oxygen is removed from the in-drawn water, and carbon dioxide released into it, by blood in beds of capillaries in the septa between the slits. Food particles carried by the water into the pharynx do not pass through the slits, but instead move posteriorly into the esophagus. The pharynx, then, acts as a strainer or filter, separating food particles from the water.

Another important characteristic of hemichordates is the occurrence during development of a ciliated larval stage that strikingly resembles the larvae of some echinoderms.

THE RELATIONSHIPS BETWEEN ECHINODERMS, HEMICHORDATES, AND CHORDATES

It may seem strange that the Echinodermata form the major phylum generally considered most closely related to our own phylum, the Chordata. After all, sea stars, sea urchins, and sea cucumbers don't look at all like vertebrates. But as we saw earlier when we discussed the differences between the Protostomia and Deuterostomia, certain characteristics seem to link echinoderms, hemichordates, and chordates and set them apart from all the protostome phyla. These characteristics include formation of the anus from the embryonic blastopore, radial and indeterminate cleavage, origin of the mesoderm as pouches, and formation of the coelom as the cavities in the mesodermal pouches.

The Hemichordata have long held special interest for zoologists because their apparent affinities to both the Echinodermata and the Chordata seem to provide additional evidence of a relationship between those two large and important phyla. The ciliated larvae of hemichordates are so much like those of some echinoderms that they were mistaken for echinoderms when first discovered. This larval type, sometimes called a ***dipleurula*** (Fig. 42.77), is found only in the echinoderms and hemichordates.[18] It has a band

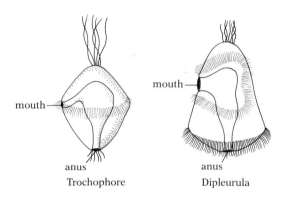

42.77 Trochophore and dipleurula larval types
The band of cilia of the trochophore is located anterior to the mouth, whereas the corresponding band of the dipleurula encircles the mouth.

[18] The larva of a hemichordate is called a tornaria, and the echinoderm larva it resembles is called a bipinnaria (Fig. 31.18, p. 801). Both are of the dipleurula type.

TABLE 42.1 *A comparison of some of the major animal phyla*

Characteristic \ Phylum	Cnidaria	Platyhelminthes	Aschelminthian phyla	Mollusca	Annelida	Arthropoda	Echinodermata	Hemichordata	Chordata
Symmetry	Radial	Bilateral					Secondarily radial	Bilateral	
Cleavage	Determinate						Indeterminate		
Body cavity	None		Pseudocoelom	Coelom much reduced	Coelom	Hemocoel (coelom, degenerate)	Coelom		
Digestive tract	Gastrovascular cavity		Complete, with mouth from blastopore				Complete, with anus from blastopore		
Circulatory system	Absent			Open	Closed	Open	A special type; often poorly developed	Open	Closed (except in tunicates)
Ciliated larva	Planula	Trochophorelike in some	None or a unique type	Trochophore		None	Dipleurula		None
Segmentation	Absent	Absent or correlated with reproduction	Absent	Absent (except in *Neopilina*)	Present		Absent		Present

of cilia that forms a ring encircling the mouth. It thus differs from the trochophore larva found in many protostomes (including some turbellarian Platyhelminthes, the lophophorate phyla, Mollusca, and Annelida), which has a band of cilia encircling the body anterior to the mouth. The similar larvae of hemichordates and echinoderms, as well as the similarities in early embryology mentioned above, indicate that these two groups must stem from a common ancestor. In view of the complicated metamorphosis that in echinoderms produces a radial adult from a bilateral larva, it seems likely that echinoderms have deviated greatly from the ancestral type and that hemichordates are probably nearer that ancestral type.

The most obvious resemblance of hemichordates to chordates are their pharyngeal slits, which are found in all chordates but nowhere else in the animal kingdom. The hemichordates also have a dorsal nerve cord that is sometimes hollow and resembles the dorsal hollow nerve cord characteristic of chordates. Because of these resemblances, the hemichordates were regarded for many years as primitive members of the phylum Chordata. Though they are now generally regarded as a separate phylum, which may actually be closer to the echinoderms than to the chordates, recognition of their ties with both Chordata and Echinodermata has helped clarify the phylogenetic relationship between these two major groups. Note that there is no suggestion here that chordates evolved from echinoderms, but simply that the two groups diverged from a common ancestor at some remote time.

Some of the important characteristics of the major animal phyla are compared in Table 42.1.

THE ANIMAL KINGDOM: CHORDATES

This entire chapter is devoted to a single phylum: Chordata. The Chordata include all the vertebrates—animals which, like ourselves, have backbones. But not all chordates are vertebrates; in fact, the members of two of the three chordate subphyla are invertebrates.

The phylum Chordata is customarily divided into Tunicata (or Urochordata), Cephalochordata, and Vertebrata. Members of these subphyla share three important characteristics: (1) All have, at least during embryonic development, a structure called a ***notochord*** (hence the name "Chordata"). This is a flexible supportive rod running longitudinally through the dorsum of the animal just ventral to the nerve cord. (2) All have pharyngeal slits (or pouches) at some stage in their development. (These slits are often called "gill slits," which may be misleading since they may have originated as feeding devices, and have become modified for hearing, reproduction, or gas exchange—in which case they are properly called gills—over the course of evolution.) (3) All have a dorsal hollow nerve cord distinct from but lying just dorsal to the notochord. Members of the Tunicata and the Cephalochordata have no backbones, and so are invertebrates.

INVERTEBRATE CHORDATA

SUBPHYLUM TUNICATA (THE TUNICATES)

In the best-known class of tunicates (sometimes called sea squirts), the adults are sessile marine animals that little resemble other chordates ex-

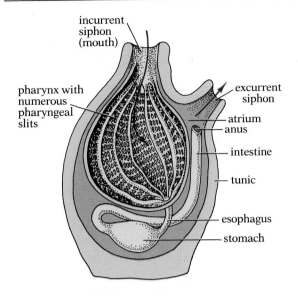

incurrent siphon (mouth)

pharynx with numerous pharyngeal slits

excurrent siphon

atrium

anus

intestine

tunic

esophagus

stomach

43.1 The adult tunicate

Above: The arrows show the path of respiratory water, which is drawn into the pharynx through the incurrent siphon, passes through the pharyngeal slits into the atrium, and then exits through the excurrent siphon. Oxygen is absorbed from the water across the walls of the pharyngeal slits, which thus serve as gills. Food particles drawn into the pharynx with the water do not pass through the slits, but instead are carried through the pharynx into the esophagus. Right: Pharyngeal slits and incurrent and excurrent siphons are clearly visible in this photo of adult tunicates.

cept in having pharyngeal slits.[1] Water taken in through the mouth (also called the incurrent siphon) goes into a large pharynx, and then filters through the pharyngeal slits into a chamber called the atrium, from which it passes to the exterior through the excurrent siphon (Fig. 43.1). The pharyngeal slits function in both gas exchange and feeding, acting as a strainer for removing small food particles from the water flowing through them. The food particles become caught in a layer of mucus in a ciliated groove of the pharynx, called the endostyle, and are carried by the mucus into the esophagus, which leads to the stomach. According to one hypothesis, the versatile pharyngeal slits, which are so distinctive a trait of chordates, first evolved as an adaptation for this sort of filter feeding and only later came to function in gas exchange as well.

Larval tunicates, which are motile, show much more resemblance to other chordates. With their elongate bilaterally symmetrical bodies and long tails, they look rather like tadpoles. They possess a well-developed dorsal hollow nerve cord and a notochord beneath it in the tail region (Fig. 43.2). When the larvae settle down and undergo metamorphosis to the adult form, the notochord and most of the nerve cord are lost.

Some biologists hold that the tunicates and higher chordates descended from a common ancestor that was free-swimming and resembled a modern tunicate larva. If this is so, then the sessile structure of modern adult tunicates is a later specialization. An alternative hypothesis is that the common ancestor was sessile, more like adult tunicates, and that vertebrates evolved from its motile larva; in other words, in the line leading to the vertebrates, the larval stage increased in importance and duration, until finally it could reproduce without undergoing metamorphosis, and the ancestral sessile stage dropped out of the life cycle entirely.

SUBPHYLUM CEPHALOCHORDATA (THE LANCELETS)

There are about 30 species of these small marine animals. Though capable of swimming, they spend most of their time buried tail down in sand in

[1] Members of two smaller classes of tunicates are free-swimming planktonic organisms.

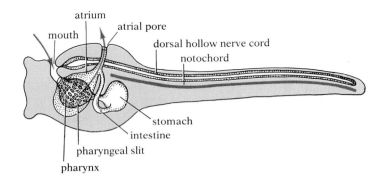

43.2   A larval tunicate
The arrows indicate the path of inflowing and outflowing water.

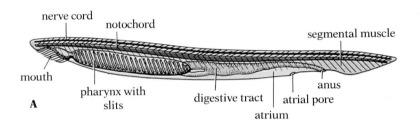

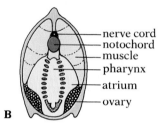

43.3   An adult lancelet (amphioxus)
(A) Longitudinal view. (B) Cross section, showing the relationship between the pharynx, the pharyngeal slits, and the atrium.

shallow water, with only their anterior end exposed. They are filter feeders, taking in water through the mouth and straining it in the pharynx in the same manner as the tunicates. The water passes through pharyngeal slits into a large chamber, the atrium, and thence to the exterior through an atrial pore. Oxygen is removed from the water as it passes through the slits; additional oxygen is obtained through gas exchange across the rest of the body surface. Food particles do not pass through the slits, but are carried posteriorly into the digestive tract.

The genus of lancelets most commonly studied is *Branchiostoma*, usually called amphioxus. The body of a typical specimen is about 5 cm long, translucent, and shaped rather like a fish (Fig. 43.3). Both the dorsal hollow

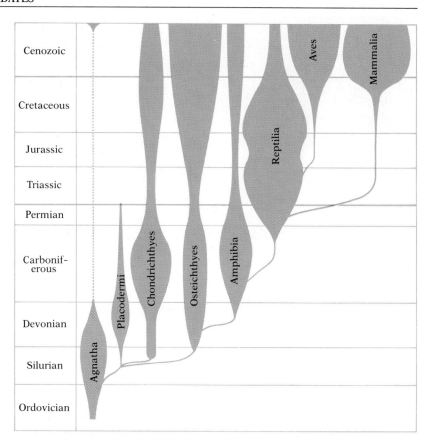

43.4 Evolution of the vertebrate classes

nerve cord and the notochord are well developed and are retained through life. A feature not seen in tunicates but characteristic of both cephalochordates and vertebrates is segmentation. In lancelets this segmentation is most noticeable in the muscles, which are in V-shaped, segmentally arranged bundles.

VERTEBRATE CHORDATA

As the name "Vertebrata" implies, the animals in this subphylum are characterized by an endoskeleton that includes a backbone composed of a series of vertebrae. The vertebrae develop around the notochord, which in most vertebrates is present in the embryo only. As in amphioxus, some parts of the vertebrate body are segmented longitudinally—in particular, the vertebrae and the main body musculature.

We discussed the anatomy, physiology, behavior, and development of vertebrates at length in other parts of this book. Here we shall be concerned primarily with the evolutionary history of the group.

CLASS AGNATHA (JAWLESS FISH)

The vertebrates are one of the few major animal groups poorly represented among the Cambrian fossils. Although some bony scales are preserved

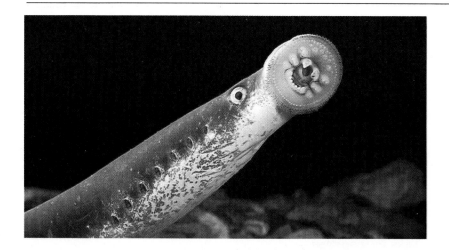

43.5 The head and pharyngeal region of a lamprey
The animal has a large oral sucker instead of jaws, and seven prominent external gill openings.

from the Precambrian period, the oldest complete vertebrate fossils are from the Ordovician period, which began some 500 million years ago (see Table 37.1, p. 1012). Those first vertebrate fossils are of bizarre fishlike animals covered by thick plates of bony material. They lacked an important character found in all later vertebrates—jaws. Furthermore, none of them had true paired fins. These ancient fish constitute the class Agnatha (which means jawless). Most were probably filter feeders, straining food material from mud and water flowing through their gill systems in more or less the same manner as tunicates and amphioxus.

The Agnatha continued as an important group through the Silurian period, sharing the seas with the already abundant sponges, coelenterates, brachiopods (which were far more numerous then than now), molluscs (particularly gastropods and cephalopods), trilobites and eurypterids, and echinoderms. But by the end of the Silurian the Agnatha had begun to decline, and they disappear from the fossil record by the end of the Devonian (Fig. 43.4).

A few peculiar species of jawless vertebrates living today, the lampreys (Fig. 43.5) and the hagfish, are generally classified as Agnatha, although they diverged from a common, cartilaginous ancestor;[2] they are also different from their Paleozoic armored relatives in that they have a soft body without either armor or scales, and their jawless mouth is modified as a round sucker that is lined with many horny teeth and accommodates a rasping tongue.[3] These parasites feed by attaching themselves by their sucker to other fish, rasping a hole in the skin of the prey, and sucking blood and other body fluids. The lampreys have a larval filter-feeding stage that strikingly resembles amphioxus.

The decline of the ancient Agnatha coincided with the rise of three classes of fish: the Placodermi (an armored fish), the Chondrichthyes (cartilaginous fish), and the Osteichthyes (bony fish). All evolved from a common ancestor which had hinged jaws (Fig. 43.6). The acquisition of hinged

---

[2] Lampreys and hagfish are so different from the Paleozoic Agnatha, called ostracoderm, that some biologists erect a separate class for them.

[3] Horny materials are reinforced with protein, while bony materials are reinforced with minerals—usually calcium phosphate ($CaPO_4$).

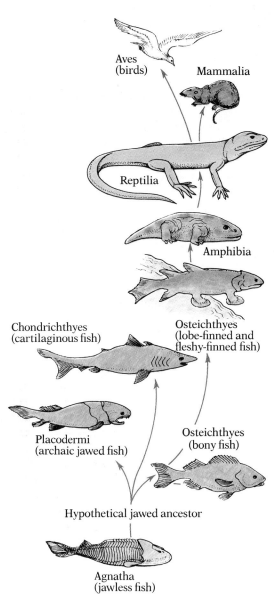

43.6 Evolution of the vertebrate classes

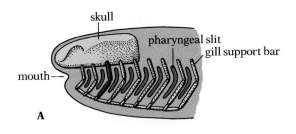

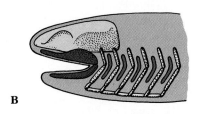

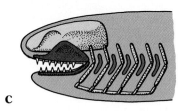

43.7 Evolution of the hinged jaws of vertebrates
(A) The earliest vertebrates had no jaws. The
structures (dark brown) that in their descendants
would become jaws were gill support bars. (B) A pair
of gill support bars has been modified into weak jaws
(the two most anterior support bars, shown in A,
were lost). (C) The jaws have become larger and
stronger.

jaws was one of the most important events in the history of vertebrates. It
made possible a revolution in the method of feeding and hence in the entire
mode of life of early fish. They became more active and wide-ranging animals, usually with paired fins. Many became ferocious predators. Even
those that remained mud feeders were evidently adaptively superior to the
ecologically similar agnaths, which they gradually replaced.

Anatomical and embryological studies have convinced biologists that the
hinged jaws of the placoderms developed from a set of gill support bars
(Fig. 43.7). Notice that hinged jaws arose independently in two important
animal groups, the arthropods and the vertebrates, but that, although they
are functionally analogous structures, they arose in entirely different ways
—in the one case from ancestral legs and in the other from skeletal elements in the wall of the pharyngeal region.

CLASS PLACODERMI

The Placodermi were an important group of armored fish during the
Devonian, but most became extinct by the end of that period; a few survived
until the Permian, when they too disappeared (Fig. 43.8). Some researchers
consider this second group as a separate class, the Ancanthodia.

CLASS CHONDRICHTHYES (CARTILAGINOUS FISH)

The modern Chondrichthyes, consisting of sharks, skates, rays, and their
relatives (Fig. 43.9), have cartilaginous skeletons. Though a cartilaginous
skeleton might at first be taken as a primitive trait, it is not thought to be
one in Chondrichthyes. Primitive sharks had bony internal skeletons, and
loss of the bone must be regarded as an evolutionary specialization.

Chondrichthyes have neither swim bladders nor lungs. Osmoregulation
in the subclass Elasmobranchii is unusual, involving retention of high concentrations of urea in the body fluids (see p. 361). Fertilization is internal,
and the eggs have tough leathery shells. Most species are predaceous, but a
few are plankton feeders.

CLASS OSTEICHTHYES (BONY FISH)

The last class of jawed fish—the Osteichthyes—includes most modern fish.
This is a large class, whose members are the dominant vertebrates in both
fresh water and the oceans, as they have been since the Devonian—the so-
called Age of Fish (see Table 37.1, p. 1012). More than 18,000 species are
known, and many remain to be discovered, particularly in the tropics and
deeper parts of the oceans. According to some biologists, the total number
of living species may be as high as 30,000. A tremendously varied lot, they
range from organisms a centimeter long when mature to giants more than
6 m long. They assume a host of different shapes, many of them bizarre and
grotesque to our eyes. Some are sluggish and sedentary, while others can
swim at speeds as great as 80 km an hour. Almost every type of food is used
by some species of fish.

The earliest members of this class probably lived in fresh water. In addition to gills, they had lungs, which they probably used as supplementary

43.8 Reconstruction of an extinct placoderm
Notice the hinged jaws.

gas-exchange devices when the water was stagnant and acidic, making it difficult to dispose of CO_2. As we saw in Chapter 11, the ventral lungs have been modified into a dorsal swim bladder in most modern bony fish (see Fig. 11.23, p. 290); they rely for gas exchange almost exclusively on their gills (see Fig. 11.13, p. 283). But there are still a few living relict species with lungs.

Soon after the Osteichthyes arose, the class split into two divergent groups. One underwent great evolutionary radiation, giving rise to nearly all the bony fish alive today. The other radiated considerably in the Paleozoic, but today is represented by only six relict species—five species of lungfish (one in Australia, three in Africa, and one in South America) and one species of "lobe-finned" fish known only from deep waters off the southeast coast of Africa. The "fleshy-finned" fish of that period are now totally extinct. Despite its rarity in the present fauna, this group of Osteichthyes is of special evolutionary interest, because it is thought to have been ancestral to the land vertebrates.

Let us look more closely at the ancient group of bony fish known as lobe-finned fish. This group has long been known from fossils, but until 1939 it was thought to have been entirely extinct for some 75 million years. In that year a specimen was caught off the east coast of South Africa; since then, additional specimens of this living fossil, called coelacanths (genus name *Latimeria*), have been caught and studied (Fig. 43.10). The coelacanths are not the particular lobe-fins thought to be the ancestors of land vertebrates —indeed, the majority of researchers favor a fleshy-finned ancestor—but they resemble those ancestral forms in many ways.

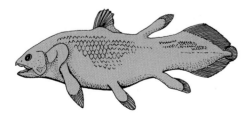

43.10 *Latimeria*, a modern lobe-finned fish

43.9 A mako shark (top) and a manta ray (bottom)
The rays have flattened bodies adapted for life on the bottom. They swim by "flying" through the water, using the thin lateral parts of their bodies as "wings." The manta ray has a further adaptation for its more mobile lifestyle: fleshy appendages on both sides of its mouth direct the microscopic food it eats into its mouth as it moves through the water.

43.11   The movement of vertebrates onto land
Above: A Devonian lobe-finned fish (*Eusthenopteron*), which probably pulled itself out of the water onto mud flats and sandbars. Right: An early amphibian (*Ichthyostega*). Its legs were better suited for locomotion on land than the lobe fins of *Eusthenopteron*, but it too probably spent most of its time in the water.

In addition to lungs, both the lobe-fins and the fleshy-fins had another important preadaptation for life on land—the large fleshy bases of their paired pectoral and pelvic fins. At times, especially during droughts, lobe-fins living in fresh water probably used these leglike fins to pull themselves onto sandbars and mud flats (Fig. 43.11), or to crawl to a new pond or stream when the one they were in dried up.

Now, by the Devonian period the land had already been colonized by plants, but was still nearly devoid of animal life (there is a fossil of what may have been a land scorpion from the Silurian, and the first insects and millipeds appeared in the Devonian, but they did not become common until the Carboniferous). Hence any animal that could survive on land would have had a whole new range of habitats open to it without competition. Any fish that had appendages better suited for land locomotion than those of their fellows would have been able to exploit these habitats more fully; through selection pressure exerted over millions of years, the fins of these first vertebrates to walk (or rather crawl) on land would slowly have evolved into legs. Thus by the end of the Devonian, with a host of other adaptations for life on land evolving at the same time, one group of ancient fleshy-finned (or lobe-finned) fish must have evolved into the first amphibians.

CLASSES AMPHIBIA AND REPTILIA

Numerous fossils indicate that, as would be expected, the first amphibians were still quite fishlike (Fig. 43.11, right). In fact, they probably spent most

43.12 Two modern amphibians
Left: A gold-lined frog (*Rana erythraea*) from Malaya.
Right: A banded salamander (*Salamandra salamandra*).

of their time in the water. But as they progressively exploited the ecological opportunities open to them on land, they slowly became a large and diverse group. So numerous were they during the Carboniferous that that period is often called the Age of Amphibians, just as the period before it, the Devonian, is called the Age of Fish. The amphibians were still abundant in the Permian, but during that period they slowly declined as the members of a new class, the Reptilia, partially replaced them.

The end of the Permian, which also marked the end of the Paleozoic era, was a time of great change, both geological and biological. The ancestral Appalachian Mountains were built up; the last trilobites and the last placoderms disappeared; the once common brachiopods declined; and older types of corals, molluscs, echinoderms, crustaceans, and fish were replaced by more modern representatives of those groups. This so-called Permo-Triassic crisis also witnessed the extinction of most groups of amphibians. By the end of the Triassic, the only members of this class that survived were the immediate ancestors of the modern Amphibia—the salamanders (order Caudata), the rare, wormlike apodes (order Gymnophiona), and the frogs and toads (order Anura) (Fig. 43.12).

The first reptiles had evolved from primitive amphibians by the late Carboniferous. The class expanded during the Permian, replacing its amphibian predecessors in most terrestrial niches, and became a huge and dominant group during the Mesozoic era, which is often called the Age of Reptiles.

One might well wonder why the reptiles were so effectively able to displace the once-dominant amphibians. There were doubtless many reasons, but surely one of the most compelling was that the reptiles, unlike amphibians, were terrestrial in the fullest sense of the word. Amphibians continued to use external fertilization and to lay fishlike eggs (Fig. 43.13)—eggs that

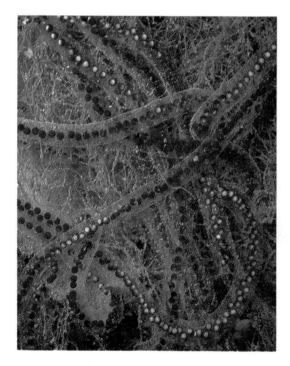

43.13 Eggs of a toad (*Bufo boreus*) in Napa Creek, California

43.14 Baby lizards hatching from their eggs
The shells of reptilian eggs are usually leathery, not
brittle like birds' eggs, as can be seen here from the
way the shells have buckled and bent.

had no amnion or shell and hence had to be deposited either in water or in
very moist places on land, lest they dry up. Larval development remained
aquatic. Amphibians were thus largely bound to the ancestral freshwater
environment by the necessities of their mode of reproduction, and the Mesozoic, during which the reptiles gained ascendancy, was generally much
dryer and warmer, greatly reducing the extent of reliable freshwater habitat suitable for reproduction. Furthermore, even adult amphibians probably had thin moist skin and were in danger of desiccation if conditions
became very dry.[4] Reptiles, on the other hand, used internal fertilization,
laid amniotic shelled eggs (Fig. 43.14), had no larval stage, and had dry,
scaly, relatively impermeable skin. Evolution of the amniotic egg—often
called the "land egg"—which provides a fluid-filled chamber in which the
embryo may develop even when the egg itself is in a dry place, was an advance as important in the conquest of land as the evolution of legs by the
Amphibia.

The Reptilia had many other characteristics that made them better suited
for some modes of terrestrial life than the Amphibia. The legs of the ancient
amphibians were small, weak, attached far up on the sides of the body, and
oriented laterally; hence they were unable to support much weight, and the
belly of the animal often dragged on the ground; walking was doubtless
slow and labored, as it is in salamanders today, while swimming was quite
rapid. The legs of reptiles were usually larger and stronger and could thus
support more weight and effect more rapid locomotion; in many (though
not all) species they were also attached lower on the sides and oriented
more vertically, so that the animal's body cleared the ground—an almost
essential requirement for terrestrial life. Whereas the lungs of amphibians
were well adapted to an aquatic or semiaquatic life, those of reptiles were
larger, and greater rib musculature made their ventilation well adapted to a
more terrestrial life. The amphibian heart was three-chambered (two atria
and one ventricle)—an important adaption when gas exchange can take
place both in the lungs and in the skin: since oxygenated blood may enter
the heart via the atrium serving the pulmonary circulation, or via the
atrium serving the systemic circulation (which includes the skin), or both,
some mixing is adaptive. The heart of reptiles, on the other hand, was four-
chambered (though the partition between the ventricles was seldom complete); hence there was less chance of mixing oxygenated blood from its
sole source, the pulmonary system, with the systemic circulation's returning deoxygenated blood (see Fig. 13.7, p. 326).

The class Reptilia is represented in our modern fauna by members of
four groups: turtles (order Testudines), crocodiles and alligators (order
Crocodylia), lizards and snakes (order Lepidosauria), and the tuatara
(order Rhynchocephalia) (Fig. 43.15). The tuatara (*Sphenodon punctatum*),
which is found only on a few islands off the coast of New Zealand, is the sole
surviving member of its ancient order. Members of the other three orders
are fairly abundant, totaling about 6,500 living species.

All the living reptiles except the crocodilians are directly descended
from an important Permian group called the stem reptiles or root reptiles

---

[4] All modern amphibians have thin moist skin that functions as a respiratory organ (in addition to the gills and/or lungs), but this may not have been true of all ancient amphibians.

A

B

C

43.15 Representatives of the main groups of modern reptiles
(A) A tuatara (*Sphenodon*). (B) A lizard (*Cordylas*).
(C) A snake (*Vipera*). (D) A turtle (*Pseudemys*).
(E) A crocodile (*Crocodylus*).

D

E

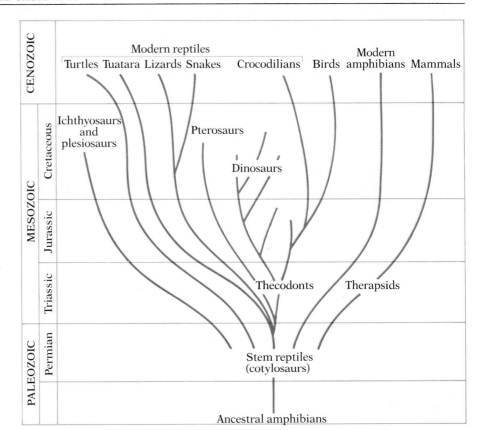

43.16 Evolution of the reptilian groups

43.17 Restoration of a late Permian therapsid reptile from South Africa
Fossil evidence indicates that the mammals evolved from the therapsids. Some investigators think the advanced therapsids had hair (not shown here).

(cotylosaurs) (Fig. 43.16). This group also gave rise to several other lineages, including one (therapsids) (Fig. 43.17) that ultimately led to the mammals, two (ichthyosaurs and plesiosaurs) that returned to the aquatic environment (Fig. 43.18), and one (thecodonts) that in its turn gave rise to crocodilians, the flying reptiles called pterosaurs, the great assemblage of reptiles called dinosaurs, and the birds. The dinosaurs were extremely abundant and varied during the Jurassic and Cretaceous periods (Figs. 43.19 and 43.20).

Recent research has led some investigators to propose a radical revision in the long-established conception of dinosaurs. They think that, unlike modern reptiles, which are cold-blooded, these animals may have been warm-blooded. Their bones sometimes show indications of the extensive vascularization characteristic of warm-blooded animals, and it appears that at least some of them may have had heat-conserving body coverings. For example, the particular dinosaurs from which the birds arose probably had feathers as insulation before the first true birds appeared. Other investigators, however, consider the idea that the dinosaurs were characteristically warm-blooded unconvincing.

By the end of the Cretaceous (which was also the end of the Mesozoic era), all the plesiosaurs and pterosaurs had disappeared. The dinosaurs, too,

43.18   Plesiosaurs (left) and ichthyosaurs (right)

43.19   *Triceratops* and *Tyrannosaurus*, two ancient dinosaurs   *Triceratops* (left) was a herbivore. *Tyrannosaurus,* was a giant carnivore, about 47 feet long and 19 feet high.

43.20   *Brontosaurus,* a giant amphibious dinosaur
Adults probably weighed as much as 25–35 tons.

had disappeared, except for one group of specialized modern descendants, the birds. Besides birds and mammals, only members of the four groups of modern reptiles remained as representatives of this once enormous class.

The decline of the dinosaurs was not necessarily as sudden as is often supposed; it may have taken tens of millions of years. But it was a dramatic event in the history of life on earth nonetheless. Why previously successful animals should have died out on such a scale has never been satisfactorily explained. Extinction was not limited to reptiles; many invertebrates, such as a widespread and abundant group of shelled cephalopods (ammonites), also disappeared. Yet many other groups living in the same sorts of habitats not only did not become extinct but did not even undergo significant change.

Recently, interest in certain mass extinctions has been greatly stimulated by the work of Walter Alvarez and his associates at the University of California at Berkeley. They discovered that the concentration of specific elements, particularly iridium, is 30 times higher than normal in a thin layer of sediments at the Cretaceous/Cenozoic boundary, and they argue that the only plausible source for this element is extraterrestrial. According to their calculations, if a 10-km asteroid or comet struck the earth and exploded, enough iridium would have dispersed into the atmosphere to account for the sudden worldwide distribution of iridium in the sediments.

An asteroid collision or comet might well have contributed to mass extinctions: if either fell into the ocean, an enormous tidal wave perhaps 8 km high would have swept around the world and devastated coastal life. An impact on land might well have darkened the skies with so much dust as to reduce photosynthesis below the level necessary to support most plants and animals. There is even some evidence of a massive fire at the time of one of the extinctions; the amount of what is apparently soot would have blackened the skies for a very long time. Moreover, dust or soot would have had a major effect on temperature, perhaps reflecting the sun's warming rays and plunging the earth into a prolonged and extremely severe winter.

All the large-scale extinctions in the fossil record need not have had an extraterrestrial cause: widespread volcanic activity could also produce enough dust and smoke to alter dramatically world climate and reduce photosynthesis.

An asteroid collision, comet impact, or cycle of intense volcanic activity would almost certainly have created an evolutionary bottleneck of some sort, and would have initiated a series of widespread changes in the tempo and direction of evolution. Though an increasing number of scientists take these hypotheses quite seriously, much about the mass extinctions and their evolutionary impact remains to be explained.

CLASS AVES (BIRDS)

By the late Triassic or early Jurassic, at least two lineages of reptiles, descended from the thecodonts, had developed the power of flight. One of these lineages, the pterosaurs, included animals with wings consisting of a large membrane of skin stretched between the body and the enormously elongated arm and fourth finger; some species had wingspreads as great as 8 m. The pterosaurs were common for a time, but eventually became extinct. The other lineage developed wings of an entirely different sort, in which many long feathers, derived from scales, were attached to the modified forelimbs. This line eventually became sufficiently different from the other reptiles to be designated as a separate class—Aves—the birds. Not all authorities think the birds deserve a class of their own. Some would place them in the Reptilia as surviving dinosaurs; others would erect a new class that would include both the dinosaurs and the birds.

The oldest known fossil bird (*Archaeopteryx*), from the middle Jurassic, still had many reptilian characters including teeth and a long jointed tail. Neither of these traits is present in modern birds, which have a beak instead of teeth and only a tiny remnant of the ancestral tail bones (the tail of a modern bird consists only of feathers).

Along with wings, birds evolved a host of other adaptations for their very active way of life. One of the most important was warm-bloodedness (endothermy)—the ability to maintain a high and constant metabolic rate, and hence great activity, despite fluctuations in environmental temperature. An anatomical feature that helped make possible the metabolic efficiency necessary for endothermy was the complete separation of the two ventricles of the heart; birds have completely four-chambered hearts (Fig. 13.7, p. 326). The insulation against heat loss provided by the body feathers plays an important role in temperature regulation; in modern birds all the scales except those of the feet are modified as feathers. Among other adaptations for flight are light hollow bones and an extensive system of air sacs attached to the lungs (Fig. 11.20, p. 288, and Fig. 11.21, p. 289). Birds also have very keen senses of vision, hearing, and equilibrium.

The newly hatched young of birds are usually not yet capable of complete temperature regulation, and they cannot fly. In many species, in fact, they are featherless, blind, and virtually helpless. Accordingly, most birds exhibit elaborate nest-building and parental-care behavior (Fig. 43.21).

CLASS MAMMALIA

Both birds and mammals evolved from reptiles that were probably at least partly endothermic, and both became highly successful groups of organisms. But the two groups did not arise from the same ancestral reptilian stock. The line leading to the mammals split off from the stem reptiles early in the Permian (Fig. 43.16), while that leading to the birds probably diverged from the branch leading to the dinosaurs in the Triassic.

The mammals themselves, it should be clearly understood, did not appear in the Permian. But the Permian saw the rise of the therapsid reptiles, some of which became very mammal-like (Fig. 43.17); they may even have had hair. Precisely at what point therapsids ceased and mammals began, it is impossible to say; there was no sudden transformation of reptile into mammal, no dramatic event to mark the appearance of the first member of our class. Hence no attempt is made in our review below of some of the characters that distinguish modern mammals from stem reptiles to specify when each of these characters appeared.

Mammals have a four-chambered heart and are endothermic. They have a muscular diaphragm, which increases breathing efficiency. There is increased separation (by the secondary palate) of the respiratory and alimen-

43.21 A cedar waxwing feeding its young
Many baby birds have brightly colored mouth linings, which act as releasers of the parental feeding response.

43.22 Duck-billed platypus, an egg-laying mammal
Though the platypus is well adapted for aquatic life, it lays its eggs on land.

tary passages. The body is usually covered with an insulating layer of hair. The limbs are oriented ventrally and lift the body high off the ground. The lower jaw is composed of only one bone (compared with six or more in most reptiles), and the teeth are differentiated for a variety of functions. There are three bones in the middle ear (compared with one in reptiles and birds). The brain, particularly the neocortex, is much larger than in reptiles, and behavior is more easily modifiable by experience. No eggs are laid (except in monotremes); embryonic development occurs in the uterus of the mother, and the young are born alive. (These reproductive characteristics appear sporadically in other vertebrate classes—live-bearing fish, for example.) After birth, the young are nourished on milk secreted by the mammary glands of the mother.

As indicated above, there is one small group of mammals—the monotremes—that are fundamentally different from all other members of the class. They lay eggs; yet they secrete milk. In many other ways they are a curious blend of reptilian traits, mammalian traits, and traits peculiar to themselves. It seems clear that they were a very early offshoot of the mammalian lineage and were not ancestral to the other mammals. Some biologists think they should be considered mammal-like reptiles rather than reptilelike mammals. The only living monotremes are the echidna (or spiny anteater) and the duck-billed platypus (Fig. 43.22); both are found in Australia, echidnas also occurring in New Guinea.

The main stem of mammalian evolution split into two parts very early, one leading to the marsupials and the other to the placentals. The characteristic difference between them is that marsupial embryos remain in the uterus for a relatively short time and then complete their development while attached to a nipple in an abdominal pouch of the mother (Fig. 43.23), whereas placental embryos complete their development in the uterus.

The living placental mammals are classified in approximately 17 orders, several of which contain species familiar to almost everyone. A few of the most important orders are listed below:

INSECTIVORA   Moles, shrews
CHIROPTERA   Bats
PRIMATA   Lemurs, monkeys, apes, humans
EDENTATA   Sloths, anteaters, armadillos
LAGOMORPHA   Rabbits, hares, pikas
RODENTIA   Rats, mice, squirrels, gophers, beavers, porcupines
ODONTOCETA   Toothed whales, dolphins, porpoises
MYSTICETA   Baleen whales

43.23 A kangaroo with young in pouch
The young kangaroo (called a joey) seen here is hundreds of times larger than when it first entered its mother's pouch; it is no longer attached to a nipple, and it often comes out of the pouch for extended periods.

Carnivora   Cats, dogs, bears, raccoons, weasels, skunks, minks, badgers, otters, hyenas, seals, walruses

Proboscidea   Elephants

Perissodactyla   Odd-toed ungulates (hoofed animals): horses, zebras, tapirs, rhinoceroses

Artiodactyla   Even-toed ungulates: pigs, hippopotamuses, camels, deer, giraffes, antelopes, cattle, sheep, goats, bison

The oldest fossils identified as placental mammalian ones are from the Jurassic. They are of small, probably secretive creatures that are thought to have fed primarily on insects. They remained a relatively unimportant part of the fauna until the end of the Mesozoic. Of the modern orders, Insectivora is closest to this ancient group. The great radiation from the insectivore ancestors dates from the beginning of the Cenozoic era, as the mammals rapidly filled the many niches left open by the demise of the dinosaurs. The Cenozoic, which includes the present, is aptly termed the Age of Mammals.

43.24   A Malay tree shrew
The living tree shrews, which are intermediate in many of their traits between the Insectivora and the Primata, are thought to resemble the early ancestors of the modern Primata. They are not closely related to squirrels (members of the Rodentia), which they resemble superficially.

EVOLUTION OF THE PRIMATES

As members of the mammalian order Primata, we naturally have a special interest in its evolutionary history, and, in particular, in that part of its history that concerns the origin of human beings.

Fossil evidence indicates that the primates arose from an arboreal stock of small shrewlike insectivores very early in the Cenozoic (Fig. 43.24). The group soon split into several evolutionary lines that have had independent histories ever since. Though the modern representatives of these evolutionary lines are a rather heterogeneous lot, most of them share the following characteristics: (1) retention of the clavicle (collarbone), which is greatly reduced or lost in many other mammals; (2) development of a shoulder joint permitting relatively free movements in all directions, and an elbow joint permitting some rotational movement; (3) retention of five functional digits on each foot; (4) enhanced individual mobility of the digits, especially the first digits (thumb and big toe), which are usually opposable; (5) modification of the claws into flattened nails; (6) development of sensitive tactile pads on the digits; (7) abbreviation of the snout or muzzle; (8) elaboration of the visual apparatus and development of binocular vision; (9) expansion of the brain, particularly the cerebral cortex; (10) usually only two mammae; (11) usually only one young per pregnancy. Most of these traits are associated with an arboreal way of life.

In quadrupedal terrestrial mammals the limbs function as props and as instruments of propulsion for running and galloping; they have tended to evolve toward greater stability at the expense of freedom of movement. Think of the forelimbs of a dog or a horse: the clavicles are greatly reduced or lost; the two limbs are positioned close together under the animal, and their movement is restricted largely to one plane, so that they can move easily back and forth, but cannot be spread far to the side like human arms. By contrast, in an animal leaping about in the branches of a tree, the limbs function in grasping, clasping, and swinging; mobility at the shoulder, elbow, and digit joints facilitates such activities, as does attachment of the

43.25 A ring-tailed lemur
The animal has a long snout and a bushy tail—both uncharacteristic of the higher Primata—but its hands and feet have opposable first digits.

limbs (braced by the clavicles) far apart at the sides of the body instead of underneath.

The eyes of many quadrupedal terrestrial mammals, like horses and cows, are located on the sides of the head. As a result, they can survey a very wide total visual field, but the fields of the two eyes overlap only slightly; consequently the animals have little binocular stereoscopic (three-dimensional) vision. But stereoscopic vision aids in localizing nearby objects, and early primates are thought to have been insectivores, searching for and grabbing insects off vegetation in trees. In addition, an animal that jumps from limb to limb must be able to detect very accurately the position of the next limb. Hence the insectivorous and arboreal way of life of the early primates probably led to selection for stereoscopic vision and, consequently, for eyes directed forward rather than laterally. This change, in turn, would have led to the distinctive flattened, forward-directed face of most higher primates.

But hands capable of grasping both small insects and tree limbs, and keen eyes with broadly overlapping fields of vision, would not by themselves have met the requirements of an arboreal, insectivorous way of life. Essential, too, would have been neural and muscular mechanisms capable of very precise eye-hand coordination. This need was doubtless one of the factors that led to the early expansion of the primate brain.

We could continue in this manner, relating other characteristics of primates to the demands of arboreal life, but the point should be that many of the traits most important to us as human beings first evolved because our distant ancestors lived in trees and hunted moving insects to supplement the standard simian diet of leaves and fruit.

THE PROSIMIANS

The living primates are usually classified in two suborders: the prosimians —Stepsirhini (or Prosimii)—and the anthropoids—Haplorhini (or Anthropoidea). The first, the prosimians ("pre-monkeys"), are a miscellaneous group of more or less primitive primates, including the lemurs, aye-ayes, lorises, pottos, and galagos.

The living lemurs and aye-ayes are found only on the island of Madagascar off the east coast of Africa. Their relatives, the lorises, pottos, and galagos, inhabit southern Asia and tropical Africa. Most lemurs are fairly small arboreal animals with a bushier coat than is usual among higher primates. They have fairly long foxlike snouts and bushy tails, and hardly resemble the higher primates (Fig. 43.25). But they have opposable first digits, and the digits are usually provided with flattened nails.

ANTHROPOIDS

The second suborder, the anthropoids (Haplorhini), contains tarsiers, New World monkeys, Old World monkeys, gibbons, and the hominids: apes and humans. By the Oliogcene epoch, after the first members of the anthropoids had evolved from the prosimian stock, the tarsiers arose from an ancestral line that would lead eventually to two or more lines of anthropoids: the New World monkeys (including the marmosets), and the Old World

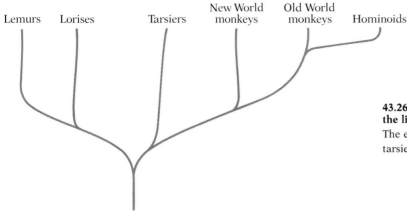

Lemurs   Lorises   Tarsiers   New World monkeys   Old World monkeys   Hominoids

Early prosimians

43.26 Hypothetical phylogenetic relationships of the living groups of Primata
The exact point at which the line leading to the tarsiers diverged is controversial.

monkeys and hominoids. These hypothetical relationships are diagramed in Figure 43.26.

The following is a formal classification of the anthropoid suborder:

Suborder Haplorhini (or Anthropoidea)
    Superfamily Tarsii
        Family Tarsiidae
    Superfamily Platyrrhini (or Ceboidea)
        Family Cebidae, New World monkeys
        Family Callithricidae, marmosets
        Family Callimiconidae, Goeldi's monkey
    Superfamily Catarrhini
        Family Cercopithecidae, Old World monkeys and baboons
        Family Hylobatidae, gibbons
        Family Pongidae, apes
        Family Hominidae, humans

Tarsiers The tarsier, which is a small crepuscular animal found in the Philippines and the East Indies, is a more advanced and specialized prosimian than the lemurs (Fig. 43.27). In some respects it shows more superficial resemblance to monkeys, though it differs in many ways from all other primates. It has a much shorter muzzle than a lemur, and thus has a more distinct face. The eyes are enormous and are directed more completely forward than in lemurs. The hind limbs are long and specialized for leaping. The long tail is naked except at the end.

The monkeys The New World and Old World monkeys differ in far more ways than we can mention here. Three differences, however, are easily seen even on a casual visit to a zoo: (1) Most New World monkeys have a prehensile tail that they use almost like another hand for grasping branches; the tail of Old World monkeys is not prehensile. (2) The nostrils of New World

43.27 A tarsier
Note the distinct face and the large forward-directed eyes.

**43.28   A New World and an Old
World Monkey**
Left: A howler (*Alouatta seniculus*),
one of the New World monkeys, has a
long prehensile tail. Right: The red

patas (*Erythrocebus patas*), one of the
Old World monkeys, lacks a
prehensile tail and has nostrils that
are close together and directed
forward and down.

**43.29   A gibbon (*Hylobates moloch*) from Sunda
Island, Borneo**
Notice the extremely long arms.

monkeys are separated by a wide partition and are thus oriented in a lateral
direction; the nostrils of Old World monkeys are not widely separated and
are directed forward and down. (3) New World monkeys lack the naked
brightly colored areas on the buttocks (ischial callosities) so common in
Old World monkeys.

Among the best-known New World monkeys are capuchins (the traditional organ-grinders' monkeys), howlers (Fig. 43.28), spider monkeys, and
squirrel monkeys. Examples of Old World monkeys are macaques, mandrills, baboons, proboscis monkeys, mona monkeys, and the sacred hanu-
man monkeys of India. One of the macaques, commonly called the rhesus
monkey, has been used extensively in physiological and psychological research; when physiologists or psychologists refer to "the monkey," this is
usually the species they mean.

Gibbons Several species of gibbons are found in Southeast Asia. They are smaller than the apes, with which group they were formerly classified; they are about 3 feet tall when standing. Their arms are exceedingly long, reaching the ground even when the animal is standing erect. The gibbons are amazing arboreal acrobats and spend almost all their time in trees (Fig. 43.29).

The apes The living great apes (Pongidae) fall into three groups: orangutans, gorillas, and chimpanzees. All are fairly large animals that have no tail, a relatively large skull and brain, and very long arms. All have a tendency, when on the ground, to walk semi-erect.

The one living species of orangutan is native to Sumatra and Borneo (Fig. 43.30). Though the orangs are fairly large (males average about 165 pounds), and their movements are slow and deliberate, they nevertheless spend most of their time in trees and only rarely descend to the ground.

Gorillas, of which there are two forms in Africa, are the largest of the apes (Fig. 43.31); wild adult males may weigh as much as 450 pounds (up to 600 pounds in zoos) and stand 6 feet tall. Their arms, while proportionately much longer than those of humans, are not as long as those of gibbons and orangs. Unlike gibbons and orangs, gorillas spend most of their time on the ground. Despite their fierce appearance, they are not usually aggressive.

Chimpanzees, which are native to tropical Africa, have been used extensively in psychological experiments. In general appearance they are the

43.30 An orangutan (*Pongo pygmaeus*) with young

43.31 A young male lowland gorilla (*Gorilla gorilla*) chest thumping

43.32   A chimpanzee using a twig as a tool to get termites out of their nest

most human-looking of the living apes (see Fig. 43.32). They are about the same size as orangs, but their arms are shorter. Although they spend most of their time in trees, they descend to the ground more frequently than orangs, and sometimes even adopt a bipedal position (their usual locomotion, however, is quadrupedal, with the knuckles of the hand used for support). They are quite intelligent and can learn to perform a variety of tasks, such as opening doors and manipulating household gadgets. There has been modest recent success in teaching them to communicate with sign languages and symbols.

THE EVOLUTION OF HUMAN BEINGS

The earliest humans, members of the family Hominidae (not to be confused with the hominids, an informal grouping of apes and humans), almost certainly arose from the same pongid stock that produced the gorillas and chimpanzees. Both paleontological evidence and biochemical and serological data indicate that chimpanzees and humans are more closely related, in terms of recentness of common ancestry, than either is to gorillas, orangutans, or gibbons. Indeed, comparisons of the amino acid sequences of their proteins and base sequences in their DNA have led some investigators to conclude that humans and chimps share about 99 percent of their genes (see table, pp. 916–17). With corroborative paleontological evidence, the amino acid sequencing data suggest that the separation of the line that led to chimpanzees from the hominid line took place no more than 4–6 million years ago. These relationships are shown in Figure 43.33. You will probably notice an anomaly between the standard classification of humans into a family separate from the apes and the evolutionary tree in Figure 43.33. Technically, lineages leading to separate families ought, of logical necessity, to diverge before any divergences leading to separate genera within a family can occur. It follows, therefore, that if orangutans are to be classed in the same family as chimpanzees, and if, as the evidence strongly indicates, the lineages leading to these genera diverged 12 million years ago, then the divergence 4–6 million years ago between the lineages leading to humans and chimpanzees unquestionably places our species in the Pongidae. But taxonomists nevertheless consider humans so different from the apes as to require a separate family. It is doubtful, however, that an objective extraterrestrial taxonomist would find two separate families. Indeed, humans and chimpanzees would probably be placed in the same genus.

The current conception of the common ancestor of modern apes (at least of gorillas and chimpanzees) and humans is founded largely on fossils of several species assigned to the genus *Dryopithecus*, which first appeared some 25 million years ago (during the early Miocene) and ranged widely over Europe, Africa, and Asia.[5] These animals had a skull with a low rounded cranium, moderate supraorbital ridges, and moderate forward projection of face and jaws. The arms were only modestly specialized for brachiation—swinging from branch to branch—and the feet indicate some tendency toward bipedal posture. These generalized features reflect none of the extreme anatomical specializations of any living species of apes.

[5] Included in *Dryopithecus*, as used here, are the African forms sometimes called *Proconsul*.

A few of the many anatomical changes that occurred in the course of evolution from ape ancestor to modern humans are: (1) The jaw became shorter (making the muzzle shorter), and the teeth became smaller. (2) The point of attachment of the skull to the vertebral column shifted from the rear of the braincase to a position under the braincase, the skull thus becoming balanced more on top of the vertebral column (Fig. 43.34). (3) The braincase became much larger, and, as it did, a prominent vertical forehead developed. (4) The eyebrow ridges and other keels on the skull were reduced as the muscles that once attached to them became smaller. (5) The nose became more prominent, with a distinct bridge and tip. (6) The arms (though probably never as long as in the modern apes) became shorter. (7) The feet became flattened, and then an arch developed. (8) The big toe moved back into line with the other toes and ceased being opposable. The various fossil humans are intermediate in these characteristics.

One of the most distinctive traits of human beings is their bipedal locomotion and upright posture. How might this trait have evolved? One possibility immediately suggests itself. When our ancestors moved from the forest to the savanna, their forelimbs, adapted to an arboreal existence, may by that very fact have been preadapted for uses other than locomotion. Not that they didn't serve in locomotion on the savanna: they almost certainly did. But the locomotion may well have been of the kind in which the knuckles rather than the palms are on the ground—the method used by gorillas and chimpanzees today. Knuckle-walking enables a quadrupedal animal like the chimpanzee to carry an object like a tool or weapon in one hand as it walks, using the other arm like a cane. This represents an important advance in transport capabilities.

There is, however, considerable controversy today concerning the selection pressures which led to the evolution of bipedalism. For many years, the assumption that early hominids were primarily hunters led many theorists to believe that having both hands free for the manipulation of weapons might have been the most important selection pressure for bipedalism. Current opinion now favors the view that hunting arose later and that early hominids were mostly herbivorous. Bipedalism is now thought to have arisen incrementally as a series of adaptations to ground feeding on plant food. But adaptation to a terrestrial way of life need not require the evolution of bipedalism: the ancestral baboons also moved into the savanna, and their descendants remain essentially quadrupedal. Because of the current absence of corroborative fossil evidence, all theories about early hominid diet and the selection pressures for bipedalism are necessarily speculative.

The hands of early hominids, having opposable thumbs, must have been preadapted not only for carrying but also for the preparation and manipulation of tools. Both capabilities can be observed in chimpanzees. Jane Goodall of the Gombe Stream Research Centre in Tanzania has reported a host of tool-using practices by these animals: the use of sticks for poking and prying, of blades of grass to collect and eat ants and termites (Fig. 43.32), of leaves for personal grooming, of stones to crack nuts. If such behavior became increasingly important to the early representatives of the hominid line, then there might have been selection for keeping the hands free, and hence for evolution of upright posture.

Another advantage of upright posture would have been the increased

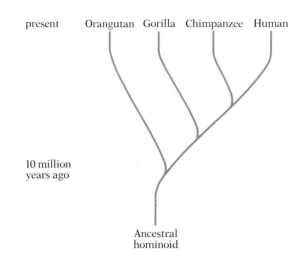

present   Orangutan   Gorilla   Chimpanzee   Human

10 million years ago

Ancestral hominoid

43.33 Hypothetical evolutionary relationships among the living apes and humans

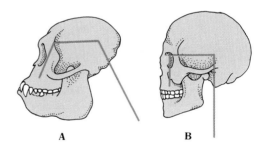

A   B

43.34 Human and monkey skulls compared with respect to their attachment to the vertebral column (A) The color line shows how the vertebral column attaches to the rear of the monkey skull, forming an obtuse angle with the horizontal axis of the cranium; note also the facial angle. (B) In the modern human, the vertebral column joins the skull more ventrally, forming approximately a right angle with the axis of the cranium; the face, too, is oriented nearly at a right angle to the cranial axis.

43.35 A reconstruction of *Australopithecus africanus* standing
Though this primitive human was fully biped, his stance was not as erect as that of our own species.

ease of maintaining surveillance—an important requirement for an animal living on the ground in the open country, if not for a chimpanzee living prevalently on the forest floor.

FOSSIL HUMANS

The first human fossil bones were found in 1856 in Germany by Johann Karl Fuhlrott. They excited lively debate, with Fuhlrott and his supporters maintaining that they were remnants of an ancient manlike organism quite different from modern humans, and his opponents objecting that they were simply the remains of a person who had suffered several deformities. It was many years before enough similar fossils were found to establish the validity of Fuhlrott's claim conclusively.

Many bones of ancient humans have been discovered since 1856. At first, the tendency of anthropologists was to erect both a new genus and a new species for each new find, without regard to biological criteria for erecting such categories. The result was a very long list of names that gave no indication of the probable relationships of the organisms they designated. More recently, however, the modern biological ideas concerning speciation and intraspecific variation have been increasingly applied to the study of fossil humans, and this, together with the discovery of many new fossils, particularly in Africa, has begun to improve our understanding of human evolution and to bring some order out of a growing jumble of information. But much is still unknown, and there is still considerable controversy over how the data should be interpreted; the brief sketch given here is the most likely of a number of possible interpretations.

The first true hominids found in the fossil record are usually assigned to the genus *Australopithecus* (from *australis*, "southern," and *pithecus*, "ape"—"southern" because the first specimens were found in South Africa). The oldest known fossils of *Australopithecus*, which are now usually assigned to the species *Australopithecus afarensis*, are about 4 million years old; it is generally thought that this species must have originated 4 to 6 million years ago, and that the emergence of humans, once associated with the Pleistocene, must have occurred during the Pliocene (see Table 37.1, p. 1012). *A. afarensis* was soon replaced by *Australopithecus africanus* (Fig. 43.35).

Apparently, at least three species of australopithecines lived contemporaneously for at least a million years. *A. africanus* had smaller teeth and a more gracile form than its major contemporary, *A. robustus*, which was distinguished by large bony crests on its skull, and its greater height—40 cm taller on the average (Fig. 43.36A–B). The third species of contemporaneous australopithecine, *A. boisei*, was also smaller than *A. robustus*; all three species are believed to have lived sympatrically.

All australopithecine species, whose fossil remains have been found in South and East Africa, were apparently fully bipedal, though their stance may not have been as upright as that of modern humans. They had large jaws with cheek teeth, but almost no forehead or chin, and their cranial capacity was only about 450–550 cc, compared with 1,200–1,600 (average

A

B

43.36   Skulls of prehistoric humans
(A) *Australopithecus africanus,* skull of a child about
five years old. (B) *A. robustus.* (C) *Homo habilis.*
(D) *H. erectus,* often called Java man.

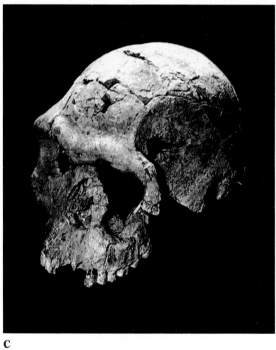

C

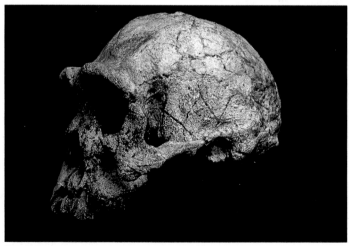

D

about 1,360) for modern humans, and 350–450 for normal chimps (Fig. 43.37). They probably used unworked stones and bits of wood as tools.

After the initial discovery of the australopithecines, a view long prevailed that *A. robustus* and *A. boisei* were evolutionary dead ends—that is, that neither had given rise to any modern descendants. According to this view, *A. africanus* was considered the likely ancestor of our own genus, *Homo*. That interpretation came into serious question, however, when remains of what clearly seemed to have been an early form of *Homo* (*Homo habilis*) were recovered from deposits that also contained bones of *A. africanus* and *A. robustus*. Since then, *A. boisei* remains have also been found with fossils of *Homo habilis*, confirming that the australopithecines must have lived contemporaneously with *Homo*. Evidence collected over the last few years now strongly suggests that *A. afarensis* is in fact the ancestor of both the *Homo* and the *Australopithecus* lines (Fig. 43.38).

Homo habilis (Fig. 43.36C), the first clear representative of the *Homo* line, had a larger cranial capacity than the australopithecines, usually between 650 and 775 cc, and there is convincing evidence that *H. habilis* not only used stones as tools but also chipped and shaped them for various purposes.

A later stage in human evolution is represented by fossils that may be classified as *Homo erectus*[6] (originally described as *Pithecanthropus erectus*, often called Java man) (Fig. 43.36D). Specimens have been found in Asia, Africa, and Europe. This species, which almost certainly descended from *H. habilis*, first appeared about 1.5 million years ago. Its cranial capacity was considerably larger, averaging about 900 cc. However, the facial features remained primitive, with a projecting massive jaw, large teeth, almost no chin, a receding forehead, heavy bony eyebrow ridges, and a broad lowbridged nose. Not only did the members of this species make and use tools, but they also used fire. Casts of the interior of the skulls indicate the presence of the speech areas of the brain; of course we have no way of knowing whether or how language was used.

Modern humans are given the Latin name *Homo sapiens* ("wise man"). Early representatives of this species first appeared about 250,000 years ago. It seems highly likely that *H. sapiens* evolved from *H. erectus*, but it is uncertain where the early stages of this evolutionary transition occurred. The oldest fossils of *H. sapiens* are from England and continental Europe, but the species may well have migrated there from Africa or southern Asia. As we saw in Chapter 22, early humans were probably hunter-gatherers.

During the period from about 100,000 to 40,000 years ago, a very distinctive form of *H. sapiens*, designated *Homo sapiens neanderthalensis*, Neanderthal man, lived throughout most of Europe and also in parts of Asia and Africa.[7] This is the form to which the bones discovered by Fuhlrott belonged. Neanderthals were 5–5½ feet tall, and had a receding forehead, prominent eyebrow ridges, and a receding chin, but their brain was as big

43.37   Reconstruction of head of *Australopithecus africanus*

---

[6] *Homo erectus* includes the forms originally described in *Pithecanthropus*, *Sinanthropus*, *Telanthropus*, and *Atlanthropus*. Some authorities also prefer to regard as early *Homo erectus* the fossils we have called *Homo habilis*.

[7] A few workers still follow the older practice of regarding Neanderthal man as specifically distinct from *Homo sapiens*, and designating this form *Homo neanderthalensis*.

as that of a modern human (perhaps a little bigger). They made many kinds of tools, and they buried their dead, which has been interpreted as showing a capacity on their part for abstract and religious thought. Neanderthals disappeared soon after the modern form of *Homo sapiens* arrived in their range.

THE HUMAN RACES

As we saw in Chapter 34, widespread species often tend to become subdivided into geographic races. Humans are no exception. *Homo sapiens* is an extremely variable species, and regional populations are often recognizably different (or were, prior to the great mobility of the last few centuries). Thus Scandinavians tend to have blue eyes and a fair complexion, while south Europeans tend to have brown eyes and a darker complexion. Eskimos look different from Mohawk Indians, and they in turn look different from Apaches. Pygmies of the Congo are obviously different from their taller neighbors. Many of the differences probably reflect adaptations to different environmental conditions. Thus, for example, the prevalence of darker skin in tropical and subtropical regions may be a protective adaptation against damaging ultraviolet solar radiation.

Races, by definition, are regional populations that differ genetically but have no effective intrinsic isolating mechanisms. There are seldom sharp boundaries between them, and they intergrade over wide areas. Designation of races in most species is therefore an arbitrary matter, and there is no such thing as a "pure" race. An almost unlimited number of races of the snake *Coluber constrictor* could be erected, depending on which of the characters whose distributions are illustrated in Figure 34.2 (p. 885) are chosen for emphasis. The same considerations apply to humans. Some authorities have chosen to recognize as many as 30 races, while others recognize only three: the traditional Caucasoid, Mongoloid, and Negroid. Another widely used system recognizes five: the traditional three, plus American Indians and Australian aborigines.

No one of these systems has any more biological validity than the others, since races, as categories, are largely human inventions. What is biologically real is the geographic variation within the species *Homo sapiens*, a variation that will surely tend to break down as people move about more and more. The main barriers to interbreeding in many parts of the world are now cultural or social rather than geographic, and it seems very unlikely that such barriers will even approach the effectiveness of the original geographic barriers. Hence, whatever races are recognized now, it seems probable that they will become less and less distinct as time goes on. This, too, is a phenomenon that has occurred countless times in other species.

THE INTERACTION OF CULTURAL AND BIOLOGICAL EVOLUTION

One of the most interesting discoveries of modern anthropologists is that early hominids used tools long before their brains were much larger than those of apes. Thus the old idea that a large brain and high intelligence were necessary prerequisites for the use of tools has been discredited. The early hominids' use of tools may, in fact, have been an important factor in leading

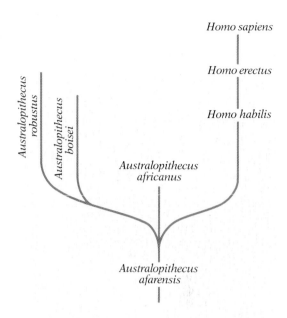

43.38 Hypothetical phylogenetic relationships among the known hominids
There is still no universal consensus upon which to base a phylogenetic tree of hominids. This tree represents one of several possible phylogenetic interpretations.

43.39 A primitive side scraper
This artifact represents a type of stone tool used by
early man, to dress hides, for perhaps half a million
years.

to evolution of higher intelligence. Once the use and making of tools began
(Fig. 43.39), individuals that excelled in these endeavors would surely have
had an advantage over their less talented fellows. There would thus have
been strong selection for neural mechanisms making possible improved
fashioning and use of tools. Thus perhaps, instead of considering culture
the crown of the fully evolved human intelligence, we should regard early
cultural development and increasing intelligence as evolving together synergistically; in a sense, the highly developed brain of modern human beings
may be as much a consequence as a cause of culture.

Cultural evolution can proceed at a far more rapid pace than biological
evolution. Words as units of inheritance are much more effective than
genes in spreading new developments and in giving dominance to new approaches originating with a few talented individuals. But the two types of
evolution continue to be interwoven just as they were in the use of tools;
they may well be even more so in the future.

Human beings, by their unrivaled ability to alter their environment, are
influencing in profound ways the evolution of all species with which they
come in contact. Thus, as we have seen, new strains of bacteria have
evolved in response to the use of antibiotics; many species of insects have
evolved new physiological and behavioral traits as a consequence of the intense selection resulting from the use of insecticides; the clearing of forests
for agricultural purposes has led to drastic decline in the population densities of some species and to increase in others; long-established balances between prey species and their predators and parasites have been destroyed,
often with far-reaching consequences for the entire ecosystem.

While some human actions on the environment have been deliberate,
many have been unintentional. But whether deliberate or unintentional,
they have precipitated a period of wholesale and rapid change rarely
matched since life began. Since disruption of the ecosystem will unavoidably grow as civilization expands, the great challenge to our species is to use
the resources of our knowledge and technology to guide the change in ways
that will benefit both our own species and the other organisms around us.
As Theodosius Dobzhansky has put it, "Creation is not an act but a process;
it did not happen five or six thousand years ago but is going on before our
eyes. Man is not compelled to be a mere spectator; he may become an assistant, a collaborator, a partner in the process of creation."[8]

Not only can we humans influence the evolution of other species, but we
now also have the ability to alter deliberately some aspects of the future
evolution of our own species. Thus modern medicine, by saving people
with gross genetic defects that would once have been fatal, permits perpetuation of genes that natural selection would formerly have eliminated.
Does this mean that we should practice eugenics—deliberately restrict, by
law or by social pressure, the perpetuation of some genetic traits and en-
courage the perpetuation of others? Some thinkers, concerned over what
they see as the inevitable physical "decline" of the species, have urged just
that. Others have pointed out that, although there may indeed be an in-

[8] This and the following quotations are from "Evolution: Implications for religion," by T.
Dobzhansky, in *Changing Man: The Threat and the Promise*, edited by K. Haselden and P. Hefner
(Doubleday, 1968).

crease in traits that would once have been maladaptive, modern human beings live in an environment of their own design, in which those traits are no longer so deleterious. As Dobzhansky says, "man . . . adapts his environments to his genes more frequently and efficiently than his genes to his environments." A decision to restrict reproduction by diabetics, for example, would focus too narrowly on a genetic defect whose symptomatic expression is now largely controllable. Individual diabetics might bear genes for musical ability, or artistic talent, or intellectual acuteness, or compassionate behavior. Moreover, as we saw in an earlier chapter, the Hardy-Weinberg Law dictates that, in the absence of some reproductive advantage, the frequency of once-deleterious genes like the one that causes diabetes will not rise in the population.

Our ability to control our own evolution may eventually no longer depend primarily on regulating reproduction. Now that the genetic code has been deciphered and recombinant DNA techniques developed, the day will surely come when the DNA of genes can deliberately be altered in order to design, at least in part, new human beings. When that day comes, how do we decide what to design? What do we look for in human beings? We might all agree to rid our species of the genes for muscular dystrophy or sickle-cell anemia (at least in areas where there is no malaria), but once the techniques for achieving these apparently worthy ends are mastered, suggestions will surely be put forward for other alterations to which we cannot all agree. As Dobzhansky has asked, "shall we endeavor to breed a race of brawny athletes, or brainy intellectuals, or sensitive esthetes, or some combination of these qualities, or a population containing certain proportions of each kind?" Who shall decide? And who shall control the ones who decide?

More immediately pressing, perhaps, is the problem of regulating the size of human populations now that we have interfered with the action of the many former natural regulating factors. Already some people are asking whether we should abandon the campaigns to eradicate malaria, to cure cancer and heart disease, or to slow the aging process. They point out that the current population problem is the result of major advances in the technology of death control, which societies have generally been as eager to accept as they have been reluctant to accept compensatory birth control. Since no species can continue indefinitely with its birth and death rates unbalanced, and since there is little evidence that most populations of our species will consent soon enough to the efficacious population-control measures that become increasingly necessary as the traditional killers are overcome, perhaps the only solution, short of tolerating mass starvation or bloody wars to control limiting resources, is less death control, heartless as that may seem.

These complex and unnerving questions—at once biological, economic, political, and moral—must be faced, and soon. History provides only unpalatable precedents for population control. The answers the next few generations give to these questions may well have as profound an influence on the future of life as anything that has happened since the first cells materialized in the primordial seas.

APPENDIX: A CLASSIFICATION OF LIVING THINGS

The classification given here is one of many in current use. Some other systems recognize more or fewer divisions and phyla, and combine or divide classes in a variety of other ways; but compared with the large areas of agreement, the differences between the various classifications are minor. Chapters 37 to 43 discuss certain of the points at issue between advocates of different systems.

Botanists have traditionally used the term "division" for the major groups that zoologists have called phyla. In classifications recognizing only two or three kingdoms, this difference in terminology causes little difficulty, because usage can be consistent within each kingdom. But when a kingdom Protista is recognized, as it is here, consistency is achieved only at the expense of violating well-established usage. The Protista contain some plantlike and funguslike groups traditionally called divisions and some animal-like groups traditionally called phyla. These usages we have respected.

Most classes within a division or phylum are listed here, but where there is only one class it is not named. For some classes—Insecta and Mammalia, for example—orders are given too. Except for a few extinct groups of particular evolutionary importance, like Placodermi, only groups with living representatives are included. A few of the better-known genera are mentioned as examples in each of the taxons.

Whenever possible, an estimate (a very rough one) of the number of living species is provided for higher taxons.

KINGDOM MONERA*

SECTION EUBACTERIA (3000)

DIVISION SPIROCHETES. *Treponema, Spirochaeta*

DIVISION CYANOBACTERIA. *Oscillatoria, Nostoc, Gloeocapsa, Microcystis*

DIVISION GRAM-POSITIVE BACTERIA

CLASS CLOSTRIDIA. *Bacillus, Clostridium, Mycoplasma, Staphylococcus, Streptococcus*
CLASS ACTINOMYCES. *Micrococcus, Actinomyces, Streptomyces*

DIVISION GREEN SULFUR BACTERIA. *Chlorobium*

DIVISION PURPLE BACTERIA

CLASS PURPLE SULFUR BACTERIA. *Pseudomonas, Escherichia*
CLASS RHODOPSEUDOMONAS. *Rhodopseudomonas, Rhizobium*
CLASS PURPLE NONSULFUR. *Sphaerotilus, Alcaligenes*
CLASS DESULFOVIBRIO. *Desulfovibrio*

DIVISION PROCHLOROPHYTA. *Prochloron*

SECTION ARCHAEBACTERIA (100)

DIVISION METHANOGENS. *Methanobacterium*

DIVISION HALOPHILES. *Halobacterium*

DIVISION THERMOACIDOPHILES. *Thermoplasma*

KINGDOM PROTISTA

SECTION PROTOPHYTA: Algal protists

DIVISION EUGLENOPHYTA. Euglenoids (800). *Euglena, Eutreptia, Phacus, Colacium*

DIVISION CHRYSOPHYTA

CLASS CHRYSOPHYCEAE. Golden-brown algae (650). *Chrysamoeba, Chromulina, Synura, Mallomonas*

CLASS HAPTOPHYCEAE (or Prymnesiophyceae). Haptophytes and coccolithophores. *Isochrysis, Prymnesium, Phaeocystis, Coccolithus, Hymenomonas*
CLASS XANTHOPHYCEAE. Yellow-green algae (360). *Botrydiopsis, Halosphaera, Tribonema, Botrydium*
CLASS EUSTIGMATOPHYCEAE. Eustigmatophytes. *Pleurochloris, Visheria, Pseudocharaciopsis*
CLASS CHLOROMONADOPHYCEAE. Chloromonads. *Gonyostomum, Reckertia*
CLASS BACILLARIOPHYCEAE. Diatoms (10,000). *Pinnularia, Arachnoidiscus, Triceratium, Pleurosigma*

DIVISION PYRROPHYTA. Dinoflagellates (1,000). *Gonyaulax, Gymnodinium, Ceratium, Gloeodinium*

DIVISION CRYPTOPHYTA. Cryptomonads. *Cryptomonas, Chroomonas, Chilomonas, Hemiselmis*

SECTION PROTOMYCOTA: Fungal protists

DIVISION HYPHOCHYTRIDIOMYCOTA. Hyphochytrids (25). *Rhizidiomyces*

DIVISION CHYTRIDIOMYCOTA. Chytrids (1,000). *Olpidium, Rhizophydium, Diplophlyctis, Cladochytrium*

SECTION GYMNOMYCOTA: Slime molds

DIVISION PLASMODIOPHOROMYCOTA. Plasmodiophores (or endoparasitic slime molds). *Plasmodiophora, Spongospora, Woronina*

DIVISION LABYRINTHULOMYCOTA. Net slime molds. *Labyrinthula*

DIVISION ACRASIOMYCOTA. Cellular slime molds (26). *Dictyostelium, Polysphondylium*

DIVISION MYXOMYCOTA. True slime molds (400). *Physarum, Hemitrichia, Stemonitis*

SECTION PROTOZOA: Animal-like protists

PHYLUM MASTIGOPHORA

CLASS ZOOFLAGELLATA (or Zoomastigina). "Animal" flagellates (5,000). *Trypanosoma, Calonympha, Chilomonas, Trichonympha*
CLASS OPALINATA.[†] Opalinids (200). *Opalina, Zelleriella*

---

* There is no generally accepted classification for bacteria at the higher taxon level. One recent classification divides the bacteria into 17 distinct divisions, some without formal names. Another important classification assigns them to 19 "parts," most without formal names. This one, admittedly incomplete, is based for the most part on similarities in the sequences of one of the ribosomal RNAs, as described in G. E. Fox et al., *Science*, vol. 209, 1980.

---

† The opalinids are sometimes placed in the Ciliata, because they have cilia instead of flagella, but they lack the other diagnostic characters of Ciliata. It must be admitted, however, that they do not fit well in the Mastigophora either.

PHYLUM SARCODINA. Pseudopodal protozoans (11,500)

CLASS RHIZOPODEA. Naked and shelled amoebae, foraminiferans. *Amoeba, Pelomyxa, Entamoeba, Arcella, Globigerina, Textularia*

CLASS ACTINOPODEA. Radiolarians, heliozoans, acantharians. *Aulacantha, Acanthometron, Actinosphaerium, Actinophrys*

PHYLUM SPOROZOA. Sporulation protozoans (3,600)

CLASS TELOSPOREA. *Monocystis, Gregarina, Eineria, Toxoplasma, Plasmodium*

CLASS PIROPLASMEA. *Babesia, Theileria*

PHYLUM CNIDOSPORA. Cnidosporians (1,100)

CLASS MYXOSPOREA. *Myxobolus, Myxidium, Ceratomyxa*

CLASS MICROSPOREA. *Nosema, Haplosporidium, Mrazekia*

PHYLUM CILIATA. Ciliates (6,000). *Paramecium, Stentor, Vorticella, Spirostomum*

KINGDOM PLANTAE

DIVISION CHLOROPHYTA. Green algae (7,000). *Chlamydomonas, Volvox, Ulothrix, Spirogyra, Oedogonium, Ulva*

DIVISION CHAROPHYTA. Stoneworts (300). *Chara, Nitella, Tolypella*

DIVISION PHAEOPHYTA. Brown algae (1,500). *Sargassum, Ectocarpus, Fucus, Laminaria*

DIVISION RHODOPHYTA. Red algae (4,000). *Nemalion, Polysiphonia, Dasya, Chondrus, Batrachospermum*

DIVISION BRYOPHYTA (23,600)

CLASS HEPATICAE. Liverworts. *Marchantia, Conocephalum, Riccia, Porella*

CLASS ANTHOCEROTAE. Hornworts. *Anthoceros*

CLASS MUSCI. Mosses. *Polytrichum, Sphagnum, Mnium*

DIVISION TRACHEOPHYTA. Vascular plants

Subdivision Psilopsida. *Psilotum, Tmesipteris*

Subdivision Lycopsida. Club mosses (1,500). *Lycopodium, Phylloglossum, Selaginella, Isoetes, Stylites*

Subdivision Sphenopsida. Horsetails (25). *Equisetum*

Subdivision Pteropsida. Ferns (10,000). *Polypodium, Osmunda, Dryopteris, Botrychium, Pteridium*

Subdivision Spermopsida. Seed plants

CLASS PTERIDOSPERMAE. Seed ferns. No living representatives

CLASS CYCADAE. Cycads (100). *Zamia*

CLASS GINKGOAE. (1). *Gingko*

CLASS CONIFERAE. Conifers (500). *Pinus, Tsuga, Taxus, Sequoia*

CLASS GNETEAE (70). *Gnetum, Ephedra, Welwitschia*

CLASS ANGIOSPERMAE. Flowering plants

SUBCLASS DICOTYLEDONEAE. Dicots (225,000). *Magnolia, Quercus, Acer, Pisum, Taraxacum, Rosa, Chrysanthemum, Aster, Primula, Ligustrum, Ranunculus*

SUBCLASS MONOCOTYLEDONEAE. Monocots (50,000). *Lilium, Tulipa, Poa, Elymus, Triticum, Zea, Ophyrys, Yucca, Sabal*

KINGDOM FUNGI

DIVISION OOMYCOTA. Water molds, white rusts, downy mildews (400). *Saprolegnia, Phytophthora, Albugo*

DIVISION ZYGOMYCOTA. Conjugation fungi (250)

CLASS ZYGOMYCETES. *Rhizopus, Mucor, Phycomyces, Choanephora, Entomophthora*

CLASS TRICHOMYCETES. *Stachylina*

DIVISION ASCOMYCOTA. Sac fungi (12,000)

CLASS HEMIASCOMYCETES. Yeasts and their relatives. *Saccharomyces, Schizosaccharomyces, Endomyces, Eremascus, Taphrina*

CLASS PLECTOMYCETES. Powdery mildews, fruit molds, etc. *Erysiphe, Podosphaera, Aspergillus, Penicillium, Ceratocystis*

CLASS PYRENOMYCETES. *Sordaria, Neurospora, Chaetomium, Xylaria, Hypoxylon*

CLASS DISCOMYCETES. *Sclerotinia, Trichoscyphella, Rhytisma, Xanthoria, Pyronema*

CLASS LABOULBENIOMYCETES. *Herpomyces, Laboulbenia*

CLASS LOCULOASCOMYCETES. *Cochliobolus, Pyrenophora, Leptosphaeria, Pleospora*

DIVISION BASIDIOMYCOTA. Club fungi (15,000)

CLASS HETEROBASIDIOMYCETES. Rusts and smuts, *Ustilago, Urocystis, Puccinia, Phragmidium, Melampsora*

CLASS HOMOBASIDIOMYCETES. Toadstools, bracket fungi, mushrooms, puffballs, stinkhorns, etc. *Coprinus, Marasmius, Amanita, Agaricus, Lycoperdon, Phallus*

KINGDOM ANIMALIA*

SUBKINGDOM PARAZOA

PHYLUM PORIFERA. Sponges (5,000)

CLASS CALCAREA. Calcareous (Chalky) sponges. *Scypha, Leucosolenia, Sycon, Grantia*

CLASS HEXACTINELLIDA. Glass sponges. *Euplectella, Hyalonema, Monoraphis*

---

* The classification of Animalia has been revised according to *Synopsis and Classification of Living Organisms*, ed. S. P. Parker, McGraw-Hill, New York, 1982, adapted to the five-kingdom system.

Class Demospongiae. *Spongilla, Euspongia, Axinella*
Class Sclerospongiae. Coralline sponges. *Ceratoporella, Stromatospongia*

SUBKINGDOM PHAGOCYTELLOZOA

PHYLUM PLACOZOA   (1). *Trichoplax*

SUBKINGDOM EUMETAZOA

SECTION RADIATA

PHYLUM CNIDARIA   (or Coelenterata)

Class Hydrozoa. Hydrozoans (3,700). *Hydra, Obelia, Gonionemus, Physalia*
Class Cubozoa. Sea wasps (20). *Tripedalia*
Class Scyphozoa. Jellyfish (200). *Aurelia, Pelagia, Cyanea*
Class Anthozoa. Sea anemones and corals (6,100). *Metridium, Pennatula, Gorgonia, Astrangia*

PHYLUM CTENOPHORA.   Comb jellies (90). *Pleurobrachia, Haeckelia*

SECTION PROTOSTOMIA

PHYLUM PLATYHELMINTHES.   Flatworms (10,000)

Class Turbellaria. Free-living flatworms. *Planaria, Dugesia, Leptoplana*
Class Trematoda. Flukes. *Fasciola, Schistosoma, Prosthogonimus*
Class Cestoda. Tapeworms. *Taenia, Dipylidium, Mesocestoides*

PHYLUM GNATHOSTOMULIDA   (100). *Gnathostomula, Haplognathia*

PHYLUM NEMERTEA   (or Rhynchocoela). Proboscis or ribbon worms (650)

Class Anopla. *Tubulanus, Cerebratulus*
Class Enopla. *Amphiporus, Prostoma, Malacobdella*

PHYLUM ACANTHOCEPHALA.   Spiny-headed worms (500). *Echinorhynchus, Gigantorhynchus*

PHYLUM MESOZOA   (50)

Class Rhombozoa. *Dicyema, Pseudicyema, Conocyema*
Class Orthonectida. *Rhopalura*

PHYLUM ROTIFERA.*   Rotifers (1,700). *Asplanchna, Hydatina, Rotaria*

PHYLUM GASTROTRICHA*   (2,000). *Chaetonotus, Macrodasys*

PHYLUM KINORHYNCHA*   (or Echinodera) (100). *Echinoderes, Semnoderes*

PHYLUM NEMATA.*   Roundworms or nematodes (12,000). *Ascaris, Trichinella, Necator, Enterobius, Ancylostoma, Heterodera*

PHYLUM NEMATOMORPHA.*   Horsehair worms (230). *Gordius, Paragordius, Nectonema*

PHYLUM ENTOPROCTA   (150). *Urnatella, Loxosoma, Pedicellina*

PHYLUM LORICIFERA   (3, newly discovered and probably very common)

PHYLUM PRIAPULIDA   (8). *Priapulus, Halicryptus*

PHYLUM BRYOZOA†   (or Ectoprocta). Bryozoans, moss animals (4,000)

Class Gymnolaemata. *Paludicella, Bugula*
Class Phylactolaemata. *Plumatella, Pectinatella*
Class Stenolaemata

PHYLUM PHORONIDA†   (10). *Phoronis, Phoronopsis*

PHYLUM BRACHIOPODA.†   Lamp shells (300)

Class Inarticulata. *Lingula, Glottidia, Discina*
Class Articulata. *Magellania, Neothyris, Terebratula*

PHYLUM MOLLUSCA.   Molluscs

Class Caudofoveata (70). *Chaetoderma*
Class Solenogastres. Solenogasters (180). *Neomenia, Proneomenia*
Class Polyplacophora. Chitons (600). *Chaetopleura, Ischnochiton, Lepidochiton, Amicula*
Class Monoplacophora (8). *Neopilina*
Class Gastropoda. Snails and their allies (univalve molluscs) (25,000). *Helix, Busycon, Crepidula, Haliotis, Littorina, Doris, Limax*
Class Scaphopoda. Tusk shells (350). *Dentalium, Cadulus*
Class Bivalva. Bivalve molluscs (7,500). *Mytilus, Ostrea, Pecten, Mercenaria, Teredo, Tagelus, Unio, Anodonta*
Class Cephalopoda. Squids, octopuses, etc. (600). *Loligo, Octopus, Nautilus*

PHYLUM POGONOPHORA.   Beard worms (100). *Siboglinum, Lamellisabella, Oligobrachia, Polybrachia*

PHYLUM SIPUNCULA   (300). *Sipunculus, Phascolosoma, Dendrostomum*

---

* Formerly considered a class of Phylum Aschelminthes.

† Bryozoa, Phoronida, and Brachiopoda are often referred to as the lophophorate phyla.

PHYLUM ECHIURA   (140)

CLASS ECHIUROINEA. *Echiurus, Ikedella*

CLASS XENOPNEUSTA. *Urechis*

CLASS HETEROMYOTA. *Crisia, Tubulipora*

PHYLUM ANNELIDA.   Segmented worms

CLASS POLYCHAETA (including Archiannelida). Sandworms, tubeworms, etc. (8,000). *Nereis, Chaetopterus, Aphrodite, Diopatra, Arenicola, Hydroides, Sabella*

CLASS OLIGOCHAETA. Earthworms and many freshwater annelids (3,100). *Tubifex, Enchytraeus, Lumbricus, Dendrobaena*

CLASS HIRUDINOIDEA. Leeches (500). *Trachelobdella, Hirudo, Macrobdella, Haemadipsa*

PHYLUM ONYCHOPHORA   (65). *Peripatus. Peripatopsis*

PHYLUM TARDIGRADA.   Water bears (300). *Echiniscus, Macrobiotus*

PHYLUM ARTHROPODA   (at least 2,000,000)

Subphylum Trilobita. No living representatives

Subphylum Chelicerata

CLASS EURYPTERIDA. No living representatives

CLASS MEROSTOMATA. Horseshoe crabs (4). *Limulus*

CLASS ARACHNIDA. Spiders, ticks, mites, scorpions, whipscorpions, daddy longlegs, etc. (55,000; at least 500,000 undiscovered species of mites are thought to exist). *Archaearanea, Latrodectus, Argiope, Centruroides, Chelifer, Mastigoproctus, Phalangium, Ixodes*

CLASS PYCNOGONIDA. Sea spiders (1,000). *Nymphon, Ascorhynchus*

Subphylum Uniramia (or Mandibulata)

CLASS CRUSTACEA (26,000). *Homarus, Cancer, Daphnia, Artemia, Cyclops, Balanus, Porcellio*

CLASS CHILOPODA. Centipeds (2,500). *Scolopendra, Lithobius, Scutigera*

CLASS DIPLOPODA. Millipeds (10,000; another 50,000 species are thought to exist). *Narceus, Apheloria, Polydesmus, Julus, Glomeris*

CLASS PAUROPODA (500). *Pauropus*

CLASS SYMPHYLA (160). *Scutigerella*

CLASS INSECTA. Insects (900,000; another 1,000,000 species are thought to exist)

ORDER THYSANURA. Bristletails, silverfish, firebrats. *Machilis, Lepisma. Thermobia*

ORDER EPHEMERIDA Mayflies. *Hexagenia, Callibaetis, Ephemerella*

ORDER ODONATA. Dragonflies, damselflies. *Archilestes, Lestes, Aeshna, Gomphus*

ORDER ORTHOPTERA. Grasshoppers, crickets, etc. *Schistocerca, Romalea, Nemobius, Megaphasma*

ORDER PHASMATOPTERA. Walking sticks. *Phyllium*

ORDER BLATTARIA. Cockroaches. *Blatta, Periplaneta*

ORDER MANTODEA. Mantids. *Mantis*

ORDER GRYLLOBLATTARIA. *Grylloblatta*

ORDER ISOPTERA. Termites. *Reticulitermes, Kalotermes, Zootermopsis, Nasutitermes*

ORDER DERMAPTERA. Earwigs, *Labia, Forficula, Prolabia*

ORDER EMBIIDINA (or Embiaria or Embioptera). *Oligotoma, Anisembia, Gynembia*

ORDER PLECOPTERA. Stoneflies. *Isoperla, Taeniopteryx, Capnia, Perla*

ORDER ZORAPTERA. *Zorotypus*

ORDER PSOCOPTERA. Book lice. *Ectopsocus, Liposcelis, Trogium*

ORDER MALLOPHAGA. Chewing lice. *Cuclotogaster, Menacanthus, Menopon, Trichodectes*

ORDER ANOPLURA. Sucking lice. *Pediculus, Phthirius, Haematopinus*

ORDER THYSANOPTERA. Thrips. *Heliothrips, Frankliniella, Hercothrips*

ORDER HEMIPTERA. True bugs. *Belostoma, Lygaeus, Notonecta, Cimex, Lygus, Oncopeltus*

ORDER HOMOPTERA. Cicadas, aphids, leafhoppers, scale insects, etc. *Magicicada, Circulifer, Psylla, Aphis, Saissetia*

ORDER NEUROPTERA. Dobsonflies, alderflies, lacewings, mantispids snakeflies, etc. *Corydalus, Hemerobius, Chrysopa, Mantispa, Agulla*

ORDER COLEOPTERA. Beetles, weevils. *Copris, Phyllophaga, Harpalus, Scolytus, Melanotus, Cicindela, Dermestes, Photinus, Coccinella, Tenebrio, Anthonomus. Conotrachelus*

ORDER HYMENOPTERA. Wasps, bees, ants, sawflies. *Cimbex, Vespa, Glypta, Scolia, Bembix, Formica Bombus, Apis*

ORDER STREPSIPTERA. Endoparasites

ORDER MECOPTERA. Scorpionflies. *Panorpa, Boreus, Bittacus*

ORDER SIPHONAPTERA. Fleas. *Pulex, Nosopsyllus, Xenopsylla, Ctenocephalides*

ORDER DIPTERA. True flies, mosquitoes. *Aedes, Asilus, Sarcophaga, Anthomyia, Musca, Chironomus, Tabanus, Tipula, Drosophila*

ORDER TRICHOPTERA. Caddisflies. *Limnephilus, Rhyacophilia, Hydropsyche*

ORDER LEPIDOPTERA. Moths, butterflies. *Tinea, Pyrausta, Malacosoma, Sphinx, Samia, Bombyx, Heliothis, Papilio, Lycaena*

Subphylum Pentastomida

CLASS PENTASTOMIDA. Parasites (60)

SECTION DEUTEROSTOMIA

PHYLUM CHAETOGNATHA.   Arrow worms (70). *Sagitta, Spadella*

PHYLUM ECHINODERMATA

Subphylum Crinozoa

CLASS CRINOIDEA. Crinoids, sea lilies (630). *Antedon, Ptilocrinus, Comactinia*

Subphylum Asterozoa

CLASS STELLEROIDEA. Sea stars, brittle stars (2,600). *Asterias, Ctenodiscus, Luidia, Oreaster, Asteronyx, Amphioplus, Ophiothrix, Ophioderma. Ophiura*

Subphylum Echinozoa

CLASS ECHINOIDEA. Sea urchins, sand dollars, heart urchins (860). *Cidaris, Arbacia, Strongylocentrotus, Echinanthus, Echinarachnius, Moira*

CLASS HOLOTHUROIDEA. Sea cucumbers (900). *Cucumaria, Thyone, Caudina, Synapa*

PHYLUM HEMICHORDATA   (90)

CLASS ENTEROPNEUSTA. Acorn worms. *Saccoglossus, Balanoglossus, Glossobalanus*

CLASS PTEROBRANCHIA. *Rhabdopleura, Cephalodiscus*

CLASS PLANCTOSPHAEROIDEA

PHYLUM CHORDATA.   Chordates

Subphylum Tunicata (or Urochordata). Tunicates (2,000)

CLASS ASCIDIACEA. Ascidians or sea squirts. *Ciona, Clavelina, Molgula, Perophora*

CLASS THALIACEA. *Pyrosoma, Salpa, Doliolum*

CLASS APPENDICULARIA. *Appendicularia, Oikopleura, Fritillaria*

Subphylum Cephalochordata. Lancelets, amphioxus (30). *Branchiostoma, Asymmetron*

Subphylum Vertebrata. Vertebrates

CLASS AGNATHA. Jawless fish (50). *Cephalaspis,* Pteraspis,* Petromyzon, Entosphenus, Myxine, Eptatretus*

CLASS ACANTHODII. No living representatives

CLASS PLACODERMI. No living representatives

CLASS CHONDRICHTHYES. Cartilaginous fish, including sharks and rays (800). *Squalus, Hyporion, Raja, Chimaera*

CLASS OSTEICHTHYES. Bony fish (18,000)

SUBCLASS SARCOPTERYGII

ORDER CERATODIFORMES. Australian lungfish. *Neoceratodus*

ORDER LEPIDOSIRENIFORMES. Lungfish. *Protopterus, Lepidosiren*

SUBCLASS ACTINOPTERYGII. Ray-finned fish. *Amia, Cyprinus, Gadus, Perca, Salmo*

CLASS AMPHIBIA (3,100)

ORDER ANURA. Frogs and toads. *Rana, Hyla, Bufo*

ORDER CAUDATA (or Urodela). Salamanders, *Necturus, Triturus, Plethodon, Ambystoma*

ORDER GYMNOPHIONA (or Apoda). *Ichthyophis, Typhlonectes*

* Extinct.

CLASS REPTILIA (6,500)

ORDER TESTUDINES. Turtles. *Chelydra, Kinosternon, Clemmys, Terrapene*

ORDER RHYNCHOCEPHALIA. Tuatara, *Sphenodon*

ORDER CROCODYLIA. Crocodiles and alligators. *Crocodylus, Alligator*

ORDER LEPIDOSAURIA. Snakes and lizards, *Iguana, Anolis, Sceloporus, Phrynosoma, Natrix, Elaphe, Coluber, Thamnophis, Crotalus*

CLASS AVES. Birds (8,600). *Anas, Larus, Columba, Gallus, Turdus, Dendroica, Sturnus, Passer, Melospiza*

CLASS MAMMALIA. Mammals (4,100)

SUBCLASS PROTOTHERIA

ORDER MONOTREMATA. Egg-laying mammals. *Ornithorhynchus, Tachyglossus*

SUBCLASS THERIA. Marsupial and placental mammals

ORDER METATHERIA (or Marsupialia). Marsupials. *Didelphis, Sarcophilus, Notoryctes, Macropus*

ORDER INSECTIVORA. Insectivores (moles, shrews, etc.). *Scalopus, Sorex, Erinaceus*

ORDER DERMOPTERA. Flying lemurs. *Galeopithecus*

ORDER CHIROPTERA. Bats, *Myotis, Eptesicus, Desmodus*

ORDER PRIMATA. Lemurs, monkeys, apes, humans, *Lemur, Tarsius, Cebus, Macacus, Cynocephalus, Pongo, Pan, Homo*

ORDER EDENTATA. Sloths, anteaters, armadillos, *Bradypus, Myrmecophagus, Dasypus*

ORDER PHOLIDOTA. Pangolin. *Manis*

ORDER LAGOMORPHA. Rabbits, hares, pikas. *Ochotona, Lepus, Sylvilagus, Oryctolagus*

ORDER RODENTIA. Rodents. *Sciurus, Marmota, Dipodomys, Microtus, Peromyscus, Rattus, Mus, Erethizon, Castor*

ORDER ODONTOCETA. Toothed whales, dolphins, porpoises. *Delphinus, Phocaena, Monodon*

ORDER MYSTICETA. Baleen whales. *Balaena*

ORDER CARNIVORA. Carnivores, *Canis, Procyon, Ursus, Mustela, Mephitis, Felis, Hyaena, Eumetopias*

ORDER TUBULIDENTATA. Aardvark. *Orycteropus*

ORDER PROBOSCIDEA. Elephants. *Elephas, Loxodonta*

ORDER HYRACOIEDEA. Hyraxes, conies. *Procavia*

ORDER SIRENIA. Manatees. *Trichechus, Halicore*

ORDER PERISSODACTYLA. Odd-toed ungulates. *Equus, Tapirella, Tapirus, Rhinoceros*

ORDER ARTIODACTYLA. Even-toed ungulates. *Pecari, Sus, Hippopotamus, Camelus, Cervus, Odocoileus, Giraffa, Bison, Ovis, Bos*

SUGGESTED READING

CHAPTER 1

COMROE, J. H., 1977. *Retrospectroscope*. Von Gehr Press, Menlo Park, Calif. *A fascinating study of how important scientific discoveries are made. The author concludes that great advances usually arise out of research directed at wholly unrelated problems.**

DARWIN, C., 1859. *The Origin of Species. Of the many reprints of this classic work, the edition by R. E. Leaky (Hill and Wang, New York, 1979) provides perhaps the best introduction and illustrations.*

GINGERICH, O., 1982. The Galileo affair, *Scientific American* 247 (2). *The ins and outs of the interaction between ecclesiastical politics and Galileo's difficult personality are traced in illuminating detail.*

KOESTLER, A., 1959. *The Sleepwalkers*. Macmillan, New York. *This informal and gossipy account of the Copernican revolution focuses on the personalities and motivations of Copernicus, Galileo (to whom Koestler is unsympathetic), and Kepler. It is a good supplement to Kuhn's book on Copernicus, listed next.*

KUHN, T. S., 1959. *The Copernican Revolution*. Random House, New York.

KUHN, T. S., 1962. *The Structure of Scientific Revolutions*. University of Chicago Press, Chicago. *In this fascinating book Kuhn argues that science works in two ways—that of Normal Science (in which experiments are designed to investigate the dominant theory, or "paradigm") and that of Revolutionary Science (which arises when the dominant theory has accumulated so many anomalies that the field becomes unstable and a replacement is needed).*

TAYLOR, F. S., 1949. *A Short History of Science and Scientific Thought*. W. W. Norton, New York. *An excellent brief history with numerous excerpts from the major writings of important scientists.*

CHAPTER 2

DICKERSON, R. E., and I. GEIS, 1976. *Chemistry, Matter, and the Universe*. W. A. Benjamin, Menlo Park, Calif. *An excellent introduction to chemistry from a biological perspective.*

FRIEDEN, E., 1972. The chemical elements of life, *Scientific American*, 227 (1). *On procedures for determining whether an element is essential to life, with particular emphasis on four elements (fluorine, silicon, tin, and vanadium).*

CHAPTER 3

DOOLITTLE, R. F., 1985. Proteins, *Scientific American* 253 (4). *Reviews the properties of amino acids and the structure of proteins, and discusses the evolution of different modern proteins from common ancestral enzymes.*

KENDREW, J. C., 1961. The three-dimensional structure of a protein molecule, *Scientific American* 205 (6). (Offprint 121) *How the complete folding pattern of myoglobin—the first protein whose conformation was determined—was worked out.*

KOSHLAND, D. E., 1973. Protein shape and biological control, *Scientific American* 229 (4). (Offprint 1280) *On the importance of protein conformation in determining enzymatic activity; how*

---

* Available in paperback.

substances that cause changes in the shape of a protein can regulate its activity.

LEHNINGER, A. L., 1965. *Bioenergetics.* W. A. Benjamin, New York. *Excellent discussion of thermodynamics from a biological perspective.*

STROUD, R. M., 1974. A family of protein-cutting proteins, *Scientific American* 231 (1). (Offprint 1301) *A good discussion of how enzymes like chymotrypsin work.*

STRYER, L., 1981. *Biochemistry,* 2nd ed. W. H. Freeman, San Francisco. *Beautifully produced, clearly written, but highly technical exposition of biochemistry.*

THOMPSON, E. O. P., 1955. The insulin molecule, *Scientific American* 182 (5). (Offprint 42) *On the first determination of the primary structure of a protein—by Frederick Sanger, who labored for ten years before he worked out the amino acid sequence of insulin in 1954.*

CHAPTER 4

BRETSCHER, M. S., 1985. The molecules of the cell membrane, *Scientific American* 253 (4). *Reviews the bilayer plasma membrane and membrane proteins, and the process of endocytosis.*

BROWN, M. S., and J. L. GOLDSTEIN, 1984. How LDL receptors influence cholesterol and atherosclerosis, *Scientific American* 251 (5). (Offprint 1555)

CAPALDI, R. A., 1974. A dynamic model of cell membranes, *Scientific American* 230 (3). (Offprint 1292) *A good discussion of the fluid-mosaic model of membrane structure.*

DAUTRY-VARSAT, A., and H. F. LODISH, 1984. How receptors bring proteins and particles into cells, *Scientific American* 250 (5). (Offprint 1550) *The life cycle of coated pits.*

LODISH, H. F., and J. E. ROTHMAN, 1979. The assembly of cell membranes, *Scientific American* 240 (1). (Offprint 1415) *A good discussion of how the membrane grows and of how and why its two sides differ.*

LURIA, S. E., 1975. Colicins and the energetics of cell membranes, *Scientific American* 233 (6). (Offprint 1332) *How antibiotics synthesized by bacteria are used to study active transport in membranes.*

ROTHMAN, J. E., and J. LENARD, 1977. Membrane asymmetry, *Science* 195, 743–53. *Why some proteins are found on only one side of the membrane.*

SATIR, P., 1975. The final steps in secretion, *Scientific American* 233 (4). (Offprint 1328) *How the membrane of a secretory vesicle interacts with the plasma membrane during exocytosis.*

SHARON, N., 1980. Carbohydrates, *Scientific American* 243 (5). (Offprint 1483) *On the role of carbohydrates in the life of the cell, with particular emphasis on the membrane carbohydrates that are involved in cell recognition.*

SINGER, S. J., and G. NICOLSON, 1972. The fluid-mosaic model of the structure of cell membranes, *Science* 175, 720–31. *The original presentation of the fluid-mosaic hypothesis.*

UNWIN, N., and R. HENDERSON, 1984. The structure of proteins in biological membranes, *Scientific American* 250 (2). (Offprint 1547)

CHAPTER 5

ALBERTS, B., D. BRAY, J. LEWIS, M. RAFF, K. ROBERTS, and J. D. WATSON, 1983. *Molecular Biology of the Cell.* Garland, New York. *This massive tome is the most complete and up-to-date summary of cell biology available.*

DEDUVE, C., 1963. The lysosome, *Scientific American* 208 (5). (Offprint 156) *The discoverer of lysosomes describes his work.*

DEDUVE, C., 1983. Microbodies in the living cell, *Scientific American* 248 (5). (Offprint 1538) *On the class of specialized enzymatic organelles, such as peroxisomes, that are not produced by the Golgi.*

DUSTIN, P., 1980. Microtubules, *Scientific American* 243 (2). (Offprint 1477) *An excellent summary of the formation of microtubules and the diverse roles they play in the cell.*

JENSEN, W. A., 1970. *The Plant Cell,* 2nd ed. Wadsworth, Belmont, Calif. *Short and easy to read.**

MARGULIS, L., 1981. *Symbiosis in Cell Evolution.* W. H. Freeman, San Francisco. *The leading advocate of the endosymbiotic hypothesis argues her case.*

NEUTRA, M., and C. P. LEBLOND, 1969. The Golgi apparatus, *Scientific American* 220 (2). (Offprint 1134) *How radiography was used in working out the function of the Golgi apparatus.*

PORTER, K. R., and J. B. TUCKER, 1981. The ground substance of the living cell, *Scientific American* 244 (3). (Offprint 1494) *Describes the cytoskeleton and the evidence for a microtrabecular lattice.*

ROTHMAN, J. E., 1985. The compartmental organization of the Golgi apparatus. *Scientific American,* 253 (3). *An up-to-date analysis of the fine structure of this important organelle.*

SIMONS, K., H. GAROFF, and A. HELENIUS, 1982. How an animal virus gets into and out of its host cell, *Scientific American* 246 (2). (Offprint 1511) *A fascinating account of how Semliki Forest virus subverts the membrane system of its many vertebrate hosts.*

SWANSON, C. P., 1977. *The Cell,* 4th ed. Prentice-Hall, Englewood Cliffs, N.J. *Readable summary of basic cell biology.**

WEBER, K., and M. OSBORN, 1985. The molecules of the cell matrix, *Scientific American* 253 (4). *Reviews the structure and function of microfilaments and microtubules.*

WESSELLS, N. K., 1971. How living cells change shape, *Scientific American* 225 (4). (Offprint 1233) *The role of microtubules and microfilaments in cell movement.*

CHAPTER 6

ALBERSHEIM, P., 1975. The walls of growing plant cells, *Scientific American* 232 (4). (Offprint 1320) *Some insights into the special properties of cell walls, derived from study of the arrangements of the various polysaccharides they contain.*

GROSS, J., 1961. Collagen, *Scientific American* 204 (5). (Offprint 88) *On the molecular structure of the most abundant protein in the human body.*

* Available in paperback.

SHARON, N., 1977. Lectins, *Scientific American* 236 (6). (Offprint 1360) *On the proteins that are believed to help bind plant cells together.*

STAEHELIN, L. A., and B. E. HULL, 1978. Junctions between living cells, *Scientific American* 238 (5). (Offprint 1388) *A freeze-etch exploration of cellular junctions.*

WOESE, C. R., 1981. Archaebacteria, *Scientific American* 244 (6). (Offprint 1516) *Discusses the evolution of procaryotes and eucaryotes in relation to the Archaebacteria.*

CHAPTER 7

CLOUD, P., 1983. The biosphere, *Scientific American* 249 (3). *On the combined evolution of the earth, life, and the atmosphere, with particular emphasis on the role of oxygen concentration.*

EDMOND, J. M., and K. V. DAMM, 1983. Hot springs on the ocean floor, *Scientific American* 248 (4). *A description of the deep-sea vents where chemosynthetic bacteria thrive in 250°C water, while other species feed on the bacteria.*

HINKLE, P. C., and R. E. McCARTY, 1978. How cells make ATP, *Scientific American* 238 (3). (Offprint 1383) *A difficult but rewarding explanation of how ATP is made.*

LEHNINGER, A. L., 1965. *Bioenergetics*, W. A. Benjamin, New York. *A brief, relatively elementary treatment of energy transformations in organisms.*

RACKER, E., 1968. The membrane of the mitochondrion, *Scientific American* 218 (2). (Offprint 1101)

STRYER, L., 1981. *Biochemistry*, 2nd ed. W. H. Freeman, San Francisco. *Traces in great detail the biochemical pathways of respiration, and their regulation.*

CHAPTER 8

BASSHAM, J. A., 1962. The path of carbon in photosynthesis, *Scientific American* 206 (6). (Offprint 122) *An old but informative account of how the Calvin cycle was worked out.*

BJÖRKMAN, O., and J. BERRY, 1973. High-efficiency photosynthesis, *Scientific American* 229 (4). (Offprint 1281) *The photosynthetic pathway and leaf anatomy of a group of C_4 plants.*

GOVINDJEE and R. GOVINDJEE, 1974. The primary events of photosynthesis, *Scientific American* 231 (6). (Offprint 1310) *Summary account of photosynthesis, including some intermediate steps for which evidence is scant.*

LEVINE, R. P., 1969. The mechanism of photosynthesis, *Scientific American* 221 (6). (Offprint 1163) *A clear explanation of the light reactions.*

MILLER, K. R., 1979. The photosynthetic membrane, *Scientific American* 241 (4). (Offprint 1448) *An excellent discussion relating the chemiosmotic theory of chloroplast function to the structure of thylakoid membranes as shown by freeze-etch microscopy.*

STOECKENIUS, W., 1976. The purple membrane of salt-loving bacteria, *Scientific American* 234 (6). (Offprint 1340) *On a unique photosynthetic mechanism based on a pigment very similar to the vi-*

sual pigments of animals rather than on chlorophyll. Provides food for thought about the evolution of our visual sensitivity.

CHAPTER 9

DENISON, W. C., 1973. Life in tall trees, *Scientific American* 228 (6). *On the interaction between lichens and the trees on which they live.*

EPSTEIN, E., 1973. Roots, *Scientific American* 228 (5). (Offprint 1271) *The mechanisms by which roots take up nutrients from the soil.*

GALSTON, A. W., P. J. DAVIES, and R. L. SLATER, 1980. *The Life of the Green Plant*, 3rd ed. Prentice-Hall, Englewood Cliffs, N.J. *Good short book on plant physiology.**

HESLOP-HARRISON, Y., 1978. Carnivorous plants, *Scientific American* 238 (2). (Offprint 1382)

RAY, P. M., 1972. *The Living Plant*, 2nd ed. Holt, Rinehart & Winston, New York. *Short elementary text on plant physiology.**

CHAPTER 10

CARLSON, A. J., V. JOHNSON, and H. M. CAVERT, 1961. *The Machinery of the Body*, 5th ed. University of Chicago Press, Chicago. *Very clearly written text on human physiology (see esp. Chapters 7 and 8).*

DAVENPORT, H. W., 1972. Why the stomach does not digest itself, *Scientific American* 226 (1). (Offprint 1240)

deDUVE, C., 1963. The lysosome, *Scientific American* 208 (5). (Offprint 156) *Nobel Prize winner deDuve describes this organelle, which he discovered.*

HARPSTEAD, D. D., 1971. High-lysine corn, *Scientific American* 225 (2). (Offprint 1229) *About an attempt to breed corn with an amino acid profile more suitable for our species.*

JARNICK, J., C. H. NOLLER, and C. I. RHYKERD, 1976. The cycles of plant and animal nutrition, *Scientific American* 235 (3). *The movement of nutrients through plants, animals, and the nonliving environment, with a look at some of the agricultural practices that affect this cycling.*

KRETCHMER, N., 1972. Lactose and lactase, *Scientific American* 227 (4). (Offprint 1259) *On differences in human tolerance of milk sugar.*

SCHMIDT-NIELSEN, K., 1970. *Animal Physiology*, 3rd ed. Prentice-Hall, Englewood Cliffs, N.J. *Well-written elementary text on animal physiology.**

YOUNG, V. R., and N. S. SCRIMSHAW, 1971. The physiology of starvation, *Scientific American* 225 (4). (Offprint 1232) *On the remarkable biochemical responses of a starving person's body, which permit relatively long-lasting mineral nutrition of the most important organs, particularly the brain.*

CHAPTER 11

AVERY, M. E., N. S. WANG, and H. W. TAEUSCH, 1973. The lung of the newborn infant, *Scientific American* 228 (4). *Fascinating ac-*

count of the changes that take place at birth to prepare the infant lung for breathing.

FEDER, M. E., and W. W. BURGGREN, 1985. Skin breathing in vertebrates, *Scientific American* 253 (5). *On how some vertebrates supplement or even replace lungs or gills in obtaining oxygen and eliminating carbon dioxide.*

SCHMIDT-NIELSEN, K., 1971. How birds breathe, *Scientific American* 225 (6). (Offprint 1238) *On the remarkable phenomenon of unidirectional flow in the respiratory system of birds.*

SCHMIDT-NIELSEN, K., 1972. *How Animals Work.* Cambridge University Press, New York. *A short, clearly written book that pays particular attention to gas exchange in animals.**

CHAPTER 12

BIDDULPH, O., and S. BIDDULPH, 1959. The circulatory system of plants, *Scientific American* 200 (2). (Offprint 53) *Excellent discussion of the physiology of phloem by two investigators who carried out some of the early radioactive-tracer studies of mineral transport.*

RAY, P. M., 1972. *The Living Plant*, 2nd ed. Holt, Rinehart & Winston, New York. *Short elementary text on plant physiology (see esp. Chapters 5 and 7).**

ZIMMERMAN, M. H., 1963. How sap moves in trees, *Scientific American* 208 (3). (Offprint 154)

CHAPTER 13

ADOLPH, E. F., 1967. The heart's pacemaker, *Scientific American* 216 (3). (Offprint 1067) *How nodal tissue regulates the fundamental rhythm of the heart.*

KILGOUR, F. G., 1952. William Harvey, *Scientific American* 186 (6). *The story of Harvey's classic work on the circulatory system.*

MAYERSON, H. S., 1963. The lymphatic system, *Scientific American* 208 (6). (Offprint 158)

PERUTZ, M. F., 1964. The hemoglobin molecule, *Scientific American* 211 (5). (Offprint 196) *The research that led to the discovery of the structure of hemoglobin described by the major figure in that work.*

PERUTZ, M. F., 1978. Hemoglobin structure and respiratory transport, *Scientific American* 239 (6). (Offprint 1413)

WOOD, J. E., 1968. The venous system, *Scientific American* 218 (1). (Offprint 1093) *How the constriction and dilation of veins help determine the distribution of blood in the human body.*

ZWEIFACH, B. W., 1959. The microcirculation of the blood, *Scientific American* 200 (1). (Offprint 64) *On capillaries, arterioles, and venules.*

CHAPTER 14

SCHMIDT-NIELSEN, K., 1959. Salt glands, *Scientific American* 200 (1). *The salt-secreting glands of marine birds and turtles are described.*

SCHMIDT-NIELSEN, K., 1959. The physiology of the camel, *Scientific American* 201 (6). (Offprint 1096) *Fascinating discussion of the special adaptations that enable camels to survive and prosper in the desert.*

SCHMIDT-NIELSEN, K., 1983. *Animal Physiology: Adaptation and Environment*, 3rd ed. Cambridge University Press, New York.

SCHMIDT-NIELSEN, K., and B. SCHMIDT-NIELSEN, 1953. The desert rat, *Scientific American* 189 (1). (Offprint 1050) *The extraordinary osmoregulatory abilities that allow the desert rat to survive without ever drinking.*

SMITH, H. W., 1953. The kidney, *Scientific American* 188 (1). (Offprint 37)

SMITH, H. W., 1953. *From Fish to Philosopher.* Little, Brown, Boston. (Paperback edition by Doubleday Anchor Books, 1961.) *A little classic on vertebrate evolution in terms of osmoregulation and excretion, enlivened by liberal doses of personal philosophy.**

CHAPTER 15

ALBERSHEIM, P., and A. G. DARVILL, 1985. Oligosaccharins, *Scientific American* 253 (3). *On a newly discovered plant hormone.*

GALSTON, A. W., and P. J. DAVIES, 1970. *Control Mechanisms in Plant Development.* Prentice-Hall, Englewood Cliffs, N.J. *A brief elementary text, which is a good introduction, though slightly out-of-date.**

VAN OVERBEEK, J., 1968. The control of plant growth, *Scientific American* 219 (1). (Offprint 1111)

CHAPTER 16

BERRIDGE, M. J., 1985. The molecular basis of communication within the cell, *Scientific American* 253 (4). *Reviews the operation of second messengers.*

CARMICHAEL, S. W., and H. WINKLER, 1985. The adrenal chromaffin cell, *Scientific American* 253 (2). *On the cellular and molecular bases of adrenalin secretion.*

GARDNER, L. I., 1972. Deprivation dwarfism, *Scientific American* 227 (1). (Offprint 1253) *Inadequate secretion of pituitary hormones, especially growth hormone, as a probable cause of stunted growth in some emotionally deprived children.*

GILLIE, R. B., 1971. Endemic goiter, *Scientific American* 224 (6). *The intriguing history of this disorder; its current distribution as primarily a disease of the poor.*

GUILLEMIN, R., and R. BURGUS, 1972. The hormones of the hypothalamus, *Scientific American* 227 (5). (Offprint 1260) *The discovery of the hypothalamic releasing hormones and their role in regulating the anterior pituitary.*

LEVINE, S., 1971. Stress and behavior, *Scientific American* 224 (1). (Offprint 532) *The major role played in learning by the pituitary and adrenal hormones that regulate responses to stress.*

McEWEN, B. S., 1976. Interactions between hormones and nerve tissue, *Scientific American* 235 (1). (Offprint 1341) *The influence of the steroid hormones on the infant's development of brain circuits that control later behavior.*

* Available in paperback.

NOTKINS, A. L., 1979. The causes of diabetes, *Scientific American* 241 (5). (Offprint 1450)

O'MALLEY, B. W., and W. T. SCHRADER, 1976. The receptors of steroid hormones, *Scientific American* 234 (2). (Offprint 1334) *The mechanism by which steroid hormones are thought to act on their target cells.*

SCHNEIDERMAN, H. A., and L. I. GILBERT, 1964. Control of growth and development in insects, *Science* 143, 325–33.

SNYDER, S. H., 1985. The molecular basis of communication between cells, *Scientific American* 253 (4). *An excellent review of hormones and local chemical mediators, and of the elaborate feedback system for controlling hormone levels.*

WURTMAN, R. J., and J. AXELROD, 1965. The pineal gland, *Scientific American* 213 (1). (Offprint 1015) *The long search for the function of the pineal.*

CHAPTER 17

EPEL, D., 1977. The program of fertilization, *Scientific American* 237 (5). (Offprint 1372) *A discussion of the initial events of fertilization.*

CHAPTER 18

ADLER, J., 1976. Chemotaxis behavior of bacteria, *Scientific American* 234 (4).

DiCARA, L. V., 1970. Learning in the autonomic nervous system, *Scientific American* 222 (1). (Offprint 525) *The modifiability by learning of even such "involuntary" functions as heartbeat and intestinal contraction.*

EVARTS, E. V., 1979. Brain mechanisms of movement, *Scientific American* 241 (3). (Offprint 1443) *How the brain and the muscles interact.*

IVERSEN, L. L., 1979. The chemistry of the brain, *Scientific American* 241 (3). (Offprint 1441) *A look at the workings of the roughly 30 different transmitter chemicals in the brain.*

JULIEN, R. M., 1975. *A Primer of Drug Action.* W. H. Freeman, San Francisco. *An accurate and easy-to-understand discussion of psychoactive drugs.*

KANDEL, E. R., 1976. *Cellular Basis of Behavior.* W. H. Freeman, San Francisco. *A lucid description of Kandel's work on* Aplysia.

KANDEL, E. R., 1979. Small systems of neurons, *Scientific American* 241 (3). (Offprint 1438) *A discussion of the circuits in* Aplysia *that are involved in gill-withdrawal behavior.*

KATZ, B., 1966. *Nerve, Muscle, and Synapse.* McGraw-Hill, New York. *A short but occasionally difficult book on how nerves and synapses work.*

KEYNES, R. D. 1958. The nerve impulse and the squid, *Scientific American* 199 (6). (Offprint 58) *The role of the giant axon of the squid in the development of the modern understanding of impulse conduction.*

LESTER, H. A., 1977. The response to acetylcholine, *Scientific American* 236 (2). (Offprint 1352) *How the receptors for acetylcholine are thought to work.*

LLINÁS, R. R., 1982. Calcium in synaptic transmission. *Scientific American* 247 (4). (Offprint 1523)

NICHOLLS, J. G., and D. VAN ESSEN, 1974. The nervous system of the leech, *Scientific American* 230 (1). (Offprint 1287) *Mapping the circuits in a relatively simple animal.*

SHEPHERD, G. M., 1978. Microcircuits in the nervous system, *Scientific American* 238 (2). (Offprint 1380) *A difficult but important article on dendritic interactions and other unconventional circuits.*

SNYDER, S. H., 1977. Opiate receptors and internal opiates, *Scientific American* 237 (3). (Offprint 1354) *How morphine and some morphinelike substances normally synthesized by certain nerve cells exert their effects on the brain.*

STEVENS, C. F., 1979. The neuron, *Scientific American* 241 (3). (Offprint 1437) *An excellent description of how nerve cells work.*

WILLOWS, A. O. D., 1971. Giant brain cells in mollusks, *Scientific American* 224 (2). (Offprint 1212) *How cells large enough to be identified as individuals interact to produce behavior.*

CHAPTER 19

AMOORE, J. E., J. W. JOHNSTON, and M. RUBIN, 1964. The stereochemical theory of odor, *Scientific American* 208 (2). (Offprint 297) *The explanation of olfaction first developed by Amoore in 1952.*

COWAN, W. M. 1979. The development of the brain, *Scientific American* 241 (3). (Offprint 1440) *Some clues about how the various elements of the brain develop, and how their axons find the correct targets.*

FRENCH, J. D. 1957. The reticular formation, *Scientific American* 196 (5). (Offprint 66) *On the network that arouses and focuses attention.*

GAMOW, R. I., and J. F. HARRIS, 1973. The infrared receptors of snakes, *Scientific American* 228 (5). (Offprint 1272) *How pit vipers can see in the dark.*

GOULD, J. L., 1982. *Ethology: The Mechanisms and Evolution of Behavior.* W. W. Norton, New York. *Discusses animal senses from a behavioral perspective, and provides a less technical treatment of hearing than Michelson.*

HELLER, H. C., L. I. CRAWSHAW, and H. T. HAMMEL, 1978. The thermostat of vertebrate animals, *Scientific American* 239 (2). (Offprint 1398) *How the hypothalamus measures and regulates body temperature.*

HUBEL, D. H. 1963. The visual cortex of the brain, *Scientific American* 209 (5). (Offprint 168) *A description of the classic experiments on feature detectors and maps in the visual cortex of cats.*

HUBEL, D. H., and T. N. WIESEL, 1979. Brain mechanisms of vision, *Scientific American* 241 (3). (Offprint 1442) *More about the processing of visual information in the cortex.*

LEVINE, J. S., and E. F. MacNICHOL, 1982. Color vision in fishes, *Scientific American* 246 (2). (Offprint 1512) *How the choice and arrangement of cones within the retina suit each fish species' needs.*

LISSMANN, H. W., 1963. Electric location by fishes. *Scientific Ameri-*

can 208 (3). (Offprint 152) *How weakly electric fish use their electric fields to sense the world around them.*

MacNichol, E. F., 1964. Three-pigment color vision, *Scientific American* 211 (6). (Offprint 197) *The experimental basis for the three-pigment model.*

Melzack, R., 1961. The perception of pain, *Scientific American* 204 (2). (Offprint 457) *A fascinating discussion of the subject.*

Michael, C. R. 1969. Retinal processing of visual images, *Scientific American* 220 (5). (Offprint 1143) *How the retinas of frogs and of ground squirrels process what they see.*

Michelson, A., 1979. Insect ears as mechanical systems, *American Scientist* 67, 696–706. *A semitechnical look at how three kinds of ears work in insects.*

Miller, W. H., F. Ratliff, and H. K. Hartline, 1961. How cells receive stimuli, *Scientific American* 205 (3). (Offprint 99) *The story of the discovery of lateral inhibition in the horseshoe crab.*

Nauta, W. J. H., and M. Feirtag, 1979. The organization of the brain, *Scientific American* 241 (3). (Offprint 1439) *An excellent discussion of vertebrate—especially human—brain anatomy, with special emphasis on the major pathways of information flow within the brain.*

Newman, E. A., and P. H. Hartline, 1982. The infrared "vision" of snakes, *Scientific American* 246 (3). *A look at the neural basis of the ability of pit vipers to see in the dark.*

Olds, J., 1956. Pleasure centers in the brain, *Scientific American* 195 (4). (Offprint 30) *A fascinating description of the discovery of the places in the brain which, when stimulated, produce highly pleasurable sensations.*

Parker, D. E., 1979. The vestibular apparatus, *Scientific American* 243 (5). (Offprint 1484) *How the otoliths and semicircular canals work.*

Pettigrew, J. D., 1972. The neurophysiology of binocular vision, *Scientific American* 227 (2). (Offprint 1255) *How one layer in visual cortex is wired to determine whether objects are closer or farther away than whatever is currently in focus.*

Ratliff, F., 1972. Contour and contrast, *Scientific American* 226 (6). (Offprint 543) *How lateral inhibition accounts for many optical illusions, particularly those exploited by artists.*

Von Békésy, G., 1957. The ear, *Scientific American* 197 (2). (Offprint 44) *How the ear works, with special emphasis on frequency discrimination.*

Wald, G., 1950. Eye and camera, *Scientific American* 183 (2). (Offprint 46) *A comparison of photographic cameras with the camera eye.*

Werblin, F. S., 1973. The control of sensitivity in the retina, *Scientific American* 228 (1). (Offprint 1264) *The circuitry underlying more complex forms of visual adaptation.*

CHAPTER 20

Carlson, A. J., V. Johnson, and H. M. Cavert, 1961. *The Machinery of the Body*, 5th ed. University of Chicago Press, Chicago. *Exceptionally clear elementary treatment of human physiology.*

Chapter 10 contains a good summary of muscle physiology at the nonmolecular level.

Cohen, C., 1975. The protein switch of muscle contraction, *Scientific American* 233 (5). (Offprint 1329) *A more detailed account than given in this text of the way calcium, troponin, and tropomyosin interact to control muscle contraction,*

Huxley, H. E., 1958. The contraction of muscle, *Scientific American* 199 (5). (Offprint 19) *A description by Huxley of his sliding-filament model, written shortly after he developed it.*

Huxley, H. E., 1965. The mechanism of muscular contraction, *Scientific American* 213 (6). (Offprint 1026)

Lazarides, E., and J. P. Revel, 1979. The molecular basis of cell movement, *Scientific American* 240 (5). (Offprint 1427)

Murray, J. M., and A. Weber, 1974. The cooperative action of muscle proteins, *Scientific American* 230 (2). (Offprint 1290) *The structure and mode of action of the four major proteins that make up the microfilaments of muscle,*

Satir, P., 1974. How cilia move, *Scientific American* 231 (4). (Offprint 1304)

Smith, D. S., 1965. The flight muscles of insects, *Scientific American* 212 (6). (Offprint 1014) *The special arrangements of muscles enabling insect wings to beat hundreds of times per second.*

Wessells, N. K., 1971. How living cells change shape, *Scientific American* 225 (4). (Offprint 1233) *On the way many nonmuscle cells, especially embryonic ones, change shape or travel from one place to another by means of microfilaments and microtubules.*

CHAPTER 21

Dilger, W. C., 1962. The behavior of lovebirds, *Scientific American* 206 (1). (Offprint 1049) *A comparative study of the courtship and nest-building behavior of several species of lovebirds and their hybrids. This classic study offers many insights into the evolution of behavior,*

Emlen, S. T., 1975. The stellar-orientation system of a migratory bird, *Scientific American* 233 (2). (Offprint 1327) *Summarizes elegant planetarium experiments on the star-compass strategy of navigation.*

Ewert, J-P., 1974. The neural basis of visually guided behavior, *Scientific American* 230 (3). (Offprint 1293) *How toads recognize and capture prey.*

Gould, J. L., 1982. *Ethology: The Mechanisms and Evolution of Behavior.* W. W. Norton, New York. *An introductory textbook on animal behavior,*

Hailman, J. P., 1969. How an instinct is "learned," *Scientific American* 221 (6). (Offprint 1165) *The classic reappraisal of how gull chicks know what to peck at and learn to recognize ther parents.*

Hasler, A. D., A. T. Scholz, and R. M. Horrall, 1978. Olfactory imprinting and homing in salmon, *American Scientist* 66, 347–55.

Hess, E. H., 1958. "Imprinting" in animals, *Scientific American* 198 (3). (Offprint 416)

Hess, E. H., 1972. "Imprinting" in a natural laboratory, *Scientific American* 227 (2). (Offprint 546)

* Available in paperback.

KEETON, W. T., 1974. The mystery of pigeon homing, *Scientific American.* 231 (6). (Offprint 1311)

LEVI, H. W. 1978. Orb-weaving spiders and their webs, *American Scientist* 66, 734–42. *A fascinating description of how spiders go about building their webs.*

LORENZ, K. Z., 1952. *King Solomon's Ring.* Crowell, New York. *Delightful accounts of animal behavior by the father of modern ethology.**

MENAKER, M., 1972. Nonvisual light reception, *Scientific American* 226 (3). (Offprint 1243) *The role of the pineal in synchronizing circadian rhythms.*

ROEDER, K. D., 1954. Moths and ultrasound, *Scientific American* 212 (4). (Offprint 1009) *How moths detect and evade bats.*

SMITH, N. G., 1967. Visual isolation in gulls, *Scientific American* 217 (4). (Offprint 1084) *The role of eye color in imprinting in gulls.*

TINBERGEN, N., 1952. The curious behavior of the stickleback, *Scientific American* 187 (6). (Offprint 414)

TRUMAN, J. W., 1973. How moths turn on, *American Scientist* 61, 700–706. *How hormones regulate the behavior of moths.*

WILLIAMS, T. C., and J. M. WILLIAMS, 1978. Oceanic mass migration of land birds, *Scientific American* 239 (4). (Offprint 1411) *How small songbirds get to their wintering ranges in South America by flying out over the Atlantic.*

WILSON, E. O., 1963. Pheromones, *Scientific American* 208 (5). (Offprint 157)

WYLLIE, I., 1981. *The Cuckoo.* Universe, New York.

CHAPTER 22

BENNET-CLARK, H. C., and A. W. EWING, 1970. The love song of the fruit fly, *Scientific American* 223 (1). (Offprint 1183) *A study of the communication system of* Drosophila *in which males signal by means of species-specific patterns of sound bursts.*

BENTLEY, D., and R. R. HOY, 1974. The neurobiology of cricket song. *Scientific American* 231 (2). (Offprint 1302) *An examination of the maturation of a neural circuit, and its genetic basis.*

BERTRAM, B. C. R., 1975. The social system of lions, *Scientific American* 232 (5).

BOWER, T. G. R., 1966. The visual world of infants, *Scientific American* 215 (6). (Offprint 502) *Together with the article listed next, an excellent study of how newborns and infants process what they see and hear.*

BOWER, T. G. R., 1971. The object in the world of the infant, *Scientific American* 225 (4). (Offprint 539)

EATON, G. G., 1976. The social order of Japanese macaques, *Scientific American* 235 (4). (Offprint 1345)

EIBL-EIBESFELDT, I., 1961. The fighting behavior of animals, *Scientific American* 205 (6). (Offprint 470) *The ritualization of aggression in intraspecific conflict.*

EIMAS, P., 1985. The perception of speech in early infancy, *Scientific American* 252 (1). *On the innate ability of infants to recognize consonants.*

GAZZANIGA, M. S., 1967. The split brain in man, *Scientific American* 218 (2). Offprint 508) *Classic experiments demonstrating that the two hemispheres of the human brain are specialized for different sorts of processing.*

GESCHWIND, N., 1979. Specializations of the human brain, *Scientific American* 241 (3). (Offprint 1444) *Studies of patients with localized brain lesions demonstrate that specific parts of the brain have highly specialized jobs.*

GUHL, A. M. 1956. The social order of chickens, *Scientific American* 194 (2). (Offprint 471) *How pecking orders are established and function.*

LIGON, J. D., and S. H. LIGON, 1982. The cooperative breeding behavior of the green woodhoopoe, *Scientific American* 247 (1). *A good example of how altruism turns out to enhance fitness.*

THORNHILL, R., 1980. Sexual selection in the black-tipped hangingfly, *Scientific American* 242 (6). (Offprint 1473) *The fascinating evolution of deception and female choice in a species whose mating ritual includes the offering of "presents."*

WASHBURN, S. L., and I. DeVORE, 1961. The social life of baboons, *Scientific American* 204 (6). (Offprint 614)

CHAPTER 23

ALBERTS, B., et al., 1983. *Molecular Biology of the Cell.* Garland, New York. *Contains a brief but up-to-date discussion of cell division in molecular terms.*

GIBOR, A., 1966. *Acetabularia:* A useful giant cell, *Scientific American* 215 (5). (Offprint 1057) *On some of the classic experiments in which this unicellular alga was used to study the role of the nucleus in controlling differentiation.*

MAZIA, D., 1974. The cell cycle, *Scientific American* 230 (1). (Offprint 1288) *The stages of interphase and mitosis proper; experiments conducted to ascertain the characteristics of these stages and the controls governing them.*

SWANSON, C. P., and P. L. WEBSTER, 1977. *The Cell,* 4th ed. Prentice-Hall, Englewood Cliffs, N.J. *Small book covering many aspects of cell biology at an intermediate level. Chapters 7 and 8 contain clear, well-illustrated descriptions of mitosis and meiosis.*

CHAPTER 24

GOODENOUGH, U., 1978. *Genetics,* 2nd ed. Holt, Rinehart & Winston, New York. *One of several very good introductory genetics texts currently available.*

STERN, C., and E. R. SHERWOOD, 1966. *The Origin of Genetics: A Mendel Sourcebook.* W. H. Freeman, San Francisco. *Provides translations of Mendel's papers and letters, and other early papers in genetics from Mendel's time through the rediscovery of his work. Also modern analyses of why Mendel reported on only a few of the strains he tested.*

CHAPTER 25

CLARKE, C. A., 1968. The prevention of "rhesus" babies, *Scientific American* 219 (5). (Offprint 1126) *Immunological techniques*

* Available in paperback.

that make it possible to prevent Rh disease, which may develop when an Rh⁻ woman is pregnant with an Rh⁺ child.

FRIEDMANN, T., 1971. Prenatal diagnosis of genetic disease, *Scientific American* 225 (5). (Offprint 1234) *The technique of amniocentesis in diagnosing some debilitating genetic diseases at an early stage in fetal development.*

GARDNER, E. J., 1981. *Principles of Genetics*, 6th ed. Wiley, New York. *Clearly written elementary text with good practice problems.*

GOODENOUGH, U., 1978. *Genetics*, 2nd ed. Holt, Rinehart & Winston, New York.

GUILLERY, R. W., 1975. Visual pathways in albinos, *Scientific American* 230 (5). *On the effects of the genes for albinism, including the resulting alterations in the wiring of the visual system of Siamese cats.*

MCKUSICK, V. A., 1971. The mapping of human chromosomes, *Scientific American* 224 (4). (Offprint 1220)

MITTWOCH, U., 1963. Sex differences in cells, *Scientific American* 209 (1). (Offprint 161) *The discovery of Barr bodies and other differences between male and female cells, which permit easy distinction between them.*

CHAPTER 26

BAUER, W. R., F. H. C. CRICK, and J. H. WHITE, 1980. Supercoiled DNA, *Scientific American* 243 (1). (Offprint 1474)

CRICK, F. H. C., 1954. The structure of the hereditary material, *Scientific American* 292 (4). (Offprint 5) *On the discovery of the structure of DNA.*

HOWARD-FLANDERS, P., 1981. Inducible repair of DNA, *Scientific American* 245 (5). (Offprint 1503) *On how cells recognize when the DNA has been damaged, how they switch on genes for repair enzymes, and how the enzymes work.*

UPTON, A. C., 1982. The biological effects of low-level ionizing radiation, *Scientific American* 246 (2). (Offprint 1509) *A superb summary of how DNA is damaged by radiation.*

WANG, J. C., 1982. DNA topoisimoerases, *Scientific American* 247 (1). (Offprint 1520) *On the enzymes responsible for untangling DNA during replication.*

WATSON, J. D., 1980. *The Double Helix*. A Norton Critical Edition, ed. Gunther S. Stent. W. W. Norton, New York. *A fascinating account of the discovery of the structure of DNA by one of the two discoverers. Watson spares neither himself nor others in giving a rare behind-the-scenes look at the dynamics of research in a very competitive field. This edition of the original 1968 book includes articles, relevant to the discovery, by other scientists.*

WATSON, J. D., 1976. *Molecular Biology of the Gene*, 3rd ed. W. A. Benjamin, Menlo Park, Calif. *A particularly well-written text on molecular genetics. Makes even difficult topics easy to understand.*

YUAN, R., and D. L. HAMILTON, 1982. Restriction and modification of DNA by a complex protein, *American Scientist* 70, 61–69. *On how some endonucleases can cut DNA at a specific site if it is fully unmethylated, or finish methylating (and thereby protect) the same site if it is partially methylated.*

CHAPTER 27

CHAMBON, P., 1981. Split genes, *Scientific American* 244 (5). (Offprint 1496) *On the organization of introns and exons.*

CRICK, F. H. C., 1962. The genetic code, *Scientific American* 207 (4). (Offprint 123) *Describes Crick's demonstration that the codon is three bases long.*

DARNELL, J. E., 1983. The processing of RNA, *Scientific American* 249 (4). (Offprint 1543) *An excellent summary of the topic.*

DARNELL, J. E., 1985. RNA, *Scientific American* 253 (4). *Reviews transcription, processing, translation, and transcriptional control.*

JUDSON, H., 1980. *The Eighth Day of Creation*. Simon & Schuster, New York.

LAKE, J. A., 1981. The ribosome, *Scientific American* 245 (2). (Offprint 1501) *On the three-dimensional structure of the ribosome and the details of translation.*

RICH, A., and S. H. KIM, 1978. The three-dimensional structure of transfer RNA, *Scientific American* 238 (1). (Offprint 1377) *How the three-dimensional structure of tRNA was determined, and how that structure helps explain how tRNA works.*

CHAPTER 28

AHARONWITZ, Y., and G. COHEN, 1981. Microbal production of pharmaceuticals, *Scientific American* 245 (3). *On how recombinant DNA techniques are used to make microbes produce antibiotics, hormones, and other drugs. There is also an explanation of how antibiotics work to destroy bacteria, which suggests how plasmid genes may confer resistance.*

BROWN, D., 1973. The isolation of genes, *Scientific American* 229 (2). (Offprint 1278) *How a particular mRNA can be used to locate and purify the gene that codes for it.*

BUTLER, P. J. G., and A. KLUG, 1978. The assembly of a virus, *Scientific American* 239 (5). (Offprint 1412) *How the various components of tobacco-mosaic virus assemble themselves in the host cell.*

CAMPBELL, A. M., 1976. How viruses insert their DNA into the DNA of the host cell, *Scientific American* 235 (6). (Offprint 1347)

CHILTON, M-D., 1983. A vector for introducing new genes into plants. *Scientific American* 248 (6). (Offprint 1539) *On bacteria (as opposed to viruses) that transduce host cells.*

CLOWES, R. C., 1973. The molecule of infectious drug resistance, *Scientific American* 228 (4). (Offprint 1269) *Experiments demonstrating that the bacterial genes for antibiotic resistance are carried on plasmids and can be transmitted from one bacterium to another.*

DEVORET, R., 1979. Bacterial tests for potential carcinogens, *Scientific American* 241 (2). (Offprint 1433) *The close relationship between mutations and cancer, and how to measure mutagenicity.*

FIDDES, J. C., 1977. The nucleotide sequence of a viral DNA, *Scientific American* 237 (6). (Offprint 1374) *How the base sequence of the entire genome of the bacterial virus φχ174 was worked out.*

GILBERT, W., and L. VILLA-KOMAROFF, 1980. Useful proteins from recombinant bacteria, *Scientific American* 242 (4). (Offprint

* Available in paperback.

1466) *How recombinant methods can be used to create insulin-producing bacteria.*

GILLHAM, N. W., 1978. *Organelle Heredity.* Raven Press, New York.

GRIVELL, L. A., 1983. Mitochondrial DNA, *Scientific American* 248 (3). (Offprint 1535) *On the procaryotelike organization of mitochondrial genes, and their unique modification of the genetic code.*

HOLLAND, J. J., 1974. Slow, inapparent, and recurrent viruses, *Scientific American* 230 (2). (Offprint 1289) *On viruses that cause degenerative diseases of humans without revealing their presence by the usual symptoms of infection.*

HOPWOOD, D. A., 1981. The genetic programming of industrial micro-organisms, *Scientific American* 245 (3). *An excellent summary of how recombinant DNA techniques work, and what they have already accomplished.*

KAPLAN, M. M., and R. G. WEBSTER, 1977. The epidemiology of influenza, *Scientific American* 237 (6). (Offprint 1375) *On how genetic recombination between human and animal strains of the influenza virus is probably responsible for the appearance of new subtypes of the virus.*

NOVICK, R. P., 1980. Plasmids, *Scientific American* 243 (6). (Offprint 1486)

WEINBERG, R. A., 1985. The molecules of life, *Scientific American* 253 (4). *A good, very brief review of recombinant DNA techniques.*

CHAPTER 29

BISHOP, J. M., 1982. Oncogenes, *Scientific American* 246 (3). (Offprint 1513) *An illuminating look at the relationship between cancer genes carried by viruses and the similar, noncancerous genes in normal cells.*

CROCE, C. M., and G. KLEIN, 1985. Chromosome translocations and human cancer, *Scientific American* 252 (3). (Offprint 1558) *A clear discussion of the translocations involved in Burkitt's lymphoma.*

FELSENFELD, G., 1985. DNA, *Scientific American* 253 (4). *The role of DNA structure in the regulation of gene expression.*

HUNTER, T., 1984., The proteins of oncogenes, *Scientific American* 251 (2). (Offprint 1553)

MANIATIS, T., and M. PTASHNE, 1976. A DNA operator-repressor system, *Scientific American* 234 (1). (Offprint 1333) *A fine account of how operators, promoters, repressors, and other gene-control elements work.*

NICOLSON, G. L., 1979. Cancer metastasis, *Scientific American* 240 (3). (Offprint 1422)

NOMURA, M., 1984. The control of ribosome synthesis, *Scientific American* 250 (1). (Offprint 1546)

PTASHNE, M., A. D. JOHNSON, and C. O. PABO, 1982. A genetic switch in a bacterial virus, *Scientific American* 247 (5). (Offprint 1526) *An excellent description of the details of the lytic/lysogenic switch of lambda virus.*

RAFFERTY, K. A., 1973. Herpes viruses and cancer, *Scientific American* 229 (4). *On the possibility that these common viruses are causal agents of some types of human cancer.*

RUDDLE, F. H., and R. S. KUCHERLAPATI, 1974. Hybrid cells and human

genes, *Scientific American* 231 (1). (Offprint 1300) *The technique of fusing human cells with cells from other mammals in order to map human genes and study their regulation.*

STEIN, G. S., J. S. STEIN, and L. J. KLEINSMITH, 1975. Chromosomal proteins and gene regulation, *Scientific American* 232 (2). (Offprint 1315) *The possible roles of histones and nonhistone proteins in determining which genes will be switched on.*

WEINBERG, R. A., 1983. A molecular basis of cancer, *Scientific American* 249 (5). (Offprint 1544) *A good discussion of oncogenes and the discovery that a single-base change can transform a prepared cell into a cancer cell.*

CHAPTER 30

BEAUCHAMP, G. K., K. YAMAZAKI, and E. A. BOYSE, 1985. The chemosensory recognition of genetic individuality, *Scientific American* 253 (1). *On how MHC complex is also responsible for individual-specific odors which some animals use for individual recognition.*

BUISSERET, P. D., 1982. Allergy, *Scientific American* 247 (2). (Offprint 1522) *A clear explanation of how the immune system's overreaction to harmless antigens can create annoying and even dangerous allergies.*

CAPRA, J. D., and A. B. EDMUNDSON, 1977. The antibody combining site, *Scientific American* 236 (1). (Offprint 1350) *On the evolution of antibodies and the mechanism of their reaction with antigens.*

COHEN, S. N., and J. A. SHAPIRO, 1980. Transposable genetic elements, *Scientific American* 242 (2). (Offprint 1460)

COLLIER, R. J., and D. A. KAPLAN, 1984. Immunotoxins, *Scientific American* 251 (1). (Offprint 1552) *On attempts to bind toxins to monoclonal antibodies specific for tumor cells.*

COOPER, M. D., and A. R. LAWTON. 1974. The development of the immune system, *Scientific American* 231 (5). (Offprint 1306)

CUNNINGHAM, B. A., 1977. The structure and function of histocompatibility antigens, *Scientific American* 237 (4). (Offprint 1369) *On the role of the MHC antigens on the surface of normal cells, including transplant rejection on the one hand and defense against cancer and infection on the other.*

DONELSON, J. E., and M. J. TURNER, 1985. How the trypanosome changes its coat, *Scientific American* 252 (2). (Offprint 1557) *A discussion of how the parasite evades the immune system.*

DOOLITTLE, R. F., 1985. Proteins, *Scientific American* 253 (4). *Reviews the properties of amino acids and the structure of proteins, and discusses the evolution of different modern proteins from common ancestral enzymes.*

EDELSON, R. L., and J. M. FINK, 1985. The immunologic function of skin, *Scientific American* 252 (6). *On how cells in the skin interact with T cells.*

FEDOROFF, N. V., 1984. Transposable genetic elements in maize, *Scientific American* 250 (6). (Offprint 1551) *Modern analysis of the mobile genetic elements originally discovered by McClintock.*

GALLO, R. C., 1987. The AIDS virus, *Scientific American* 256 (1). *An excellent account of the biology of the AIDS virus and its devastating effects on the immune system.**

---

* Available in paperback.

* For more recent reviews of the biology and epidemiology of AIDS, see the special issue of *Science* (Feb. 5, 1988, vol. 239).

GOSDON, G. N., 1985. Molecular approaches to malaria vaccines, *Scientific American* 252 (5). *An example of how haptens are used in designing vaccines.*

HENLE, W., G. HENLE, and E. T. LENNETTE, 1979. The Epstein-Barr virus, *Scientific American* 241 (1). (Offprint 1431) *A wide-ranging discussion that pulls together immunology, research with monoclonal antibodies, viral biochemistry, and disease statistics to link Epstein-Barr virus with Burkitt's lymphoma.*

HOOD, L. E., I. L. WEISSMAN, and W. B. WOOD, 1978. *Immunology.* Benjamin/Cummings, Menlo Park, Calif. *An excellent summary.*

LEDER, P., 1982. The genetics of antibody diversity, *Scientific American* 246 (5). (Offprint 1518) *On how the exons of antibody genes are combined to create enormous diversity.*

LERNER, R. A., 1983. Synthetic vaccines, *Scientific American* 248 (2). (Offprint 1533) *On how a knowledge of antigen–antibody interaction enables researchers to synthesize a single one of the antigenic determinants of a virus or bacterium and use the resulting harmless chemical as an effective and safe vaccine.*

MAYER, M. M., 1973. The complement system, *Scientific American* 229 (5). (Offprint 1283) *On the way an intricate set of enzymes works with antibodies to make novel channels in the membranes of foreign cells, thereby destroying them.*

MILSTEIN, C., 1980. Monoclonal antibodies, *Scientific American* 243 (4). (Offprint 1479)

NOTKINS, A. L., and H. KOPROWSKI, 1973. How the immune response to a virus can cause disease, *Scientific American* 228 (1). (Offprint 1263)

OLD, L. J., 1977. Cancer immunology, *Scientific American* 236 (5). (Offprint 1358) *On the distinctive antigens on the surfaces of cancer cells and the problem of mobilizing the immune system to combat cancer.*

ROSE, N. R., 1981. Autoimmune diseases, *Scientific American* 244 (2). (Offprint 1491) *What happens when the immune system attacks an organism's own cells.*

TONEGAWA, S., 1985. The molecules of the immune system, *Scientific American* 253 (4). *An excellent review of antibody structure, antigen binding, and B- and T-cell function. Does not discuss the interactions between T cells, B cells, macrophages, and the other elements of the immune system.*

CHAPTER 31

EPEL, D., 1977. The program of fertilization, *Scientific American* 237 (5). (Offprint 1372) *The numerous changes that occur in an egg cell as soon as a sperm cell reaches it.*

GORDON, R., and A. G. JACOBSON, 1978. The shaping of tissues in embryos, *Scientific American* 238 (6). (Offprint 1391)

HAYFLICK, L., 1980. The cell biology of human aging, *Scientific American* 242 (1). (Offprint 1457)

CHAPTER 32

BONNER, J. T., 1983. Chemical signals of social amoebae, *Scientific American* 248 (4). (Offprint 1537) *On the chemical signals that cause social amoebae to aggregate to form a multicellular organism, which may provide a model for the action of morphogens.*

BRACHET, J., 1974. *Introduction to Molecular Embryology.* Springer, Heidelberg. *A clearly written small book.**

BRYANT, P. J., S. V. BRYANT, and V. FRENCH, 1977. Biological regeneration and pattern formation, *Scientific American* 237 (1). (Offprint 1363) *On basic principles of the organization and growth of complex structures in animals.*

COWAN, W. M., 1979. The development of the brain, *Scientific American* 241 (3). (Offprint 1440) *At the peak of brain growth, hundreds of thousands of neurons are added each minute, and yet they are wired together correctly.*

EBERT, J. D., and I. SUSSEX, 1970. *Interacting Systems in Development,* 2nd ed. Holt, Rinehart & Winston, New York. *An excellent short book that gives equal time to plants and animals.**

EDELMAN, G. M., 1984. Cell-adhesion molecules: A molecular basis for animal form, *Scientific American* 250 (4). (Offprint 1549) *On the likely molecular basis of cell-to-cell adhesion and changes in adhesion during embryonic development.*

GARCIA-BELLIDO, A., P. A. LAWRENCE, and G. MORATA, 1979. Compartments in animal development, *Scientific American* 241 (1). (Offprint 1432) *An excellent discussion of imaginal discs and insect development.*

GEHRING, W. J., 1985. The molecular basis of development, *Scientific American* 253 (4). *Focuses exclusively on* Drosophila *development, with a nice discussion of homeotic mutations.*

GIERER, A., 1974. Hydra as a model for the development of biological form, *Scientific American* 231 (6). (Offprint 1309) *On the physicochemical basis of pattern development.*

GOODMAN, C. S., and M. J. BASTIANI, 1985. How embryonic nerve cells recognize one another. *Scientific American* 251 (6). (Offprint 1556) *An excellent description of how axons of invertebrates employ the stepping-stone strategy, following first gradients and then one preexisting axon after another to reach their targets.*

GURDON, J. B., 1968. Transplanted nuclei and cell differentiation, *Scientific American,* 219 (6). (Offprint 1128)

LEVI-MONTALCINI, R., and P. CALISSANO, 1979. The nerve-growth factor, *Scientific American* 240 (6). (Offprint 1430) *About the best-understood molecule important in creating chemical gradients for axon growth and development.*

STENT, G. S., and D. A. WEISBLAT, 1982. The development of a simple nervous system, *Scientific American* 246 (1). (Offprint 1508) *On how the nervous system of the leech is organized and wired up during development.*

WESSELLS, N. K., 1977. *Tissue Interactions and Development.* Benjamin/Cummings, Menlo Park, Calif. *A superb little book, which deals only with animals.**

WOLPERT, L., 1978. Pattern formation in biological development, *Scientific American* 239 (4). (Offprint 1409) *Wolpert's model for pattern development, based on studies of the chick wing.*

CHAPTER 33

BISHOP, J. A., and L. M. COOK, 1975. Moths, melanism and clean air, *Scientific American* 232 (1). (Offprint 1314) *The lessening of air*

* Available in paperback.

pollution in Britain and the diminishing frequency of melanics in some moth populations.

CLARKE, B., 1975. The causes of biological diversity, *Scientific American* 233 (2). (Offprint 1326)

DAWKINS, R., 1976. *The Selfish Gene.* Oxford University Press, New York. *A well-written exposition of the controversial idea that the gene, not the individual organism, is the unit of selection, the organism being merely the container, the robot vehicle, of its selfish genes.*

FUTUYMA, D. J., 1979. *Evolutionary Biology.* Sinauer Associates, Sunderland, Mass.

GRANT, V., 1951. The fertilization of flowers, *Scientific American* 184 (6). (Offprint 12) *The special adaptations of flowers that help ensure their pollination.*

KETTLEWELL, H. B. D., 1959. Darwin's missing evidence, *Scientific American* 200 (3). (Offprint 842) *The story of the industrial melanism of the peppered moth in England.*

LEWONTIN, R. C., 1978. Adaptation. *Scientific American* 239 (3). (Offprint 1408) *An excellent discussion of the process of adaptation, emphasizing that most features are compromises between different selection pressures, and that chance plays a role in evolution when more than one solution to a problem is possible.*

MAYR, E., 1978. Evolution, *Scientific American* 239 (3). (Offprint 1400) *A nice history of evolutionary thought.*

CHAPTER 34

AYALA, F. J., 1978. The mechanisms of evolution *Scientific American* 239 (3). (Offprint 1407) *On the large amount of hidden genetic variation in a species, and its consequences for speciation and taxonomy.*

CLARKE, B., 1975. The causes of biological diversity, *Scientific American* 233 (2). (Offprint 1326)

FUTUYMA, D. J., 1979. *Evolutionary Biology.* Sinauer Associates, Sunderland, Mass.

GOULD, S. J., 1985. *The Flamingo's Smile: Reflections in Natural History.* W. W. Norton, New York. *A compellingly written collection of essays on evolutionary theory and other biological topics by the coauthor of the theory of punctuated equilibrium.*

GRANT, P. R., 1981. Speciation and the adaptive radiation of Darwin's finches, *American Scientist* 69, 653–63. *A well-written and modern analysis of speciation.*

GRANT, V., 1977. *Organic Evolution.* W. H. Freeman, San Francisco. *Excellent treatment, with special emphasis on speciation.*

KIMURA, M., 1979. The neutral theory of molecular evolution, *Scientific American* 241 (5). (Offprint 1451) *On the idea that most single-base mutations are neutral, and therefore provide an evolutionary "clock."*

LACK, D., 1947. *Darwin's Finches.* Cambridge University Press, New York. *Clear account of the classic example of island speciation.*

LOWENSTEIN, J. M., 1985. Molecular approaches to the identification of species, *American Scientist* 73, 541–47. *An excellent introduction to immunological methods of taxonomy.*

* Available in paperback.

RUSSELL, D. A., 1982. The mass extinction of the late Mesozoic, *Scientific American* 246 (1). (Offprint 1507) *On the theory that an asteroid impact may have been responsible for the bout of mass extinctions that took place 65 million years ago.*

SCHOENER, T. W., 1982. The controversy over interspecific competition, *American Scientist* 70, 586–95.

STEBBINS, G. L., and F. J. AYALA, 1985. The evolution of Darwinism, *Scientific American* 253 (1). *A modern summary of evolutionary theory, with particular attention to the claims for punctuated equilibrium.*

WILSON, A. C., 1985. The molecular basis of evolution, *Scientific American* 253 (4). *On methods for tracing evolution through similarities in nucleotide or amino acid sequences.*

CHAPTER 35

BEDDINGTON, J. R., and R. M. MAY, 1982. The harvesting of interacting species in a natural ecosystem, *Scientific American* 247 (5). (Offprint 1525)

BELL, R. H. V., 1971. A grazing ecosystem in the Serengeti, *Scientific American* 225 (1). (Offprint 1228) *The synchronization of the migrations of ungulates across the plains of Tanzania with the growth of certain grasses—a striking example of the precision with which organisms mesh within an ecosystem.*

COOPER, C. F., 1961. The ecology of fire, *Scientific American* 204 (4). (Offprint 1099) *How some biotic communities depend on periodic burning for their continued existence, and why efforts to eliminate fires may be threatening some of our finest forests.*

HORN, H. S., 1975. Forest succession, *Scientific American* 232 (5). (Offprint 1321)

MAY, R. M., 1978. The evolution of ecological systems, *Scientific American* 239 (3). (Offprint 1404)

MAY, R. M., 1983. Parasitic infections as regulators of animal populations, *American Scientist* 71, 36–45.

WAGNER, R. H., 1978. *Environment and Man,* 3rd ed. W. W. Norton, New York. *An excellent treatment of the pressing ecological problems facing civilization today.*

CHAPTER 36

BORMANN, F. H., and G. E. LIKENS, 1970. The nutrient cycles of an ecosystem, *Scientific American* 223 (4). (Offprint 1202) *On the studies of the Hubbard Brook Forest discussed in the text.*

BRILL, W. J., 1977. Biological nitrogen fixation, *Scientific American* 236 (3). (Offprint 922) *How certain bacteria, the Cyanobacteria among them, act as the major suppliers of fixed nitrogen for the rest of the living world.*

DIETZ, R. S., and J. C. HOLDEN, 1970. The breakup of Pangaea, *Scientific American* 223 (4). (Offprint 892)

GATES, D. M., 1971. The flow of energy in the biosphere, *Scientific American* 225 (3). (Offprint 664) *On the radiant energy the earth receives from the sun—and the relatively small percentage trapped by green plants and made available to biotic communities.*

Gosz, J. R., R. T. Holmes, G. E. Likens, and F. H. Bormann, 1978. The flow of energy in a forest ecosystem, *Scientific American* 238 (3). (Offprint 1384)

Hallam, A., 1972. Continental drift and the fossil record, *Scientific American* 227 (5). (Offprint 903)

Likens, G. E., R. F. Wright, J. N. Galloway, and T. J. Butler, 1979. Acid rain, *Scientific American* 241 (4). (Offprint 941)

Myers, N., 1984. *The Primary Source: Tropical Forests and Our Future*. W. W. Norton, New York.

Richards, P. W., 1973. The tropical rain forest, *Scientific American* 229 (6). (Offprint 1286) *On the human threat to its survival.*

Went, F. W., 1949. The plants of Krakatoa, *Scientific American* 181 (3). *The reinvasion of the island of Krakatoa by plants after all life on it had been destroyed by a volcanic eruption.*

CHAPTER 37

Bogorad, L., 1975. Evolution of organelles and eukaryotic genomes, *Science* 188, 891–98. *An alternative to the endosymbiont hypothesis of the origin of eucaryotic cells.*

Dickerson, R. E., 1978. Chemical evolution and the origin of life, *Scientific American* 233 (3). (Offprint 1401)

Eigen, M., W. Gardiner, P. Schuster, and R. Winkler-Oswatitsch, 1981. The origin of genetic information, *Scientific American* 244 (4). (Offprint 1495)

Glaessner, M. F., 1961. Pre-Cambrian animals, *Scientific American* 204 (3). (Offprint 837) *Some of the fascinating Precambrian invertebrate fossils found in Australia.*

Margulis, L., 1971. Symbiosis and evolution, *Scientific American* 225 (2). (Offprint 1230) *Clear exposition of the endosymbiont hypothesis of the origin of eucaryotic cells by one of its principal proponents.*

Miller, S. L., and L. E. Orgel, 1974. *The Origins of Life on Earth.* Prentice-Hall, Englewood Cliffs, N.J. *Easy-to-read introduction to the subject. Miller's experiments of 1953 initiated the modern era of research on the beginnings of life.*

Oparin, A. I., 1969. *Genesis and Evolutionary Development of Life.* Academic Press, New York. *Excellent summary by one of the founding fathers of this field of biology.*

Schopf, J. W., 1978. The evolution of the earliest cells, *Scientific American* 239 (3). (Offprint 1402) *The fossil evidence for the appearance of cellular life on earth; the special metabolic characteristics of bacteria that enabled them to prosper under conditions that shut out most higher forms of life.*

Uzzel, T., and C. Spolsky, 1974. Mitochondria and plastids as endosymbionts: A revival of special creation? *American Scientist* 62, 334–43. *A critique of the endosymbiont hypothesis, and a proposal for an alternative model of the origin of eucaryotic cells.*

Vidal, G., 1984. The oldest eucaryotic cells, *Scientific American* 250 (2).

Whittaker, R. H., 1969. New concepts of kingdoms of organisms, *Science* 163, 150–60. *The five-kingdom system proposed and explained.*

Whittaker, R. H., and L. Margulis, 1978. Protist classification and the kingdoms of organisms, *Biosystems* 10, 3–18. *Alternative ways of applying the five-kingdom system.*

Woese, C. R., 1981. Archaebacteria, *Scientific American* 244 (6). (Offprint 1516) *A clear exposition of the case for considering Archaebacteria a separate kingdom.*

CHAPTER 38

Adler, J., 1976. The sensing of chemicals by bacteria, *Scientific American* 234 (4). (Offprint 1337)

Berg, H. C., 1975. How bacteria swim, *Scientific American* 233 (2). *The structure and mode of action of bacterial flagella.*

Burke, D. C., 1977. The status of interferon, *Scientific American* 236 (4). (Offprint 1356)

Butler, J. G., and A. Klug, 1978. The assembly of a virus, *Scientific American* 239 (5). (Offprint 1412)

Campbell, A. M., 1976. How viruses insert their DNA into the DNA of the host cell, *Scientific American* 235 (6). (Offprint 1347)

Costerton, J. W., G. G. Geesey, and K-J. Cheng, 1978. How bacteria stick, *Scientific American* 238 (1). (Offprint 1379) *On the surface molecules of bacteria that enable these to adhere to host cells—and on the potential for development of a new kind of antibiotic to attack those molecules.*

Diener, T. O., 1981. Viroids, *Scientific American* 244 (1). (Offprint 1488)

Echlin, P., 1966. The blue-green algae, *Scientific American* 214 (6).

Luria, S. E., J. E. Darnell, D. Baltimore, and A. Campbell, 1978. *General Virology*, 3rd ed. Wiley, New York. *Excellent textbook on all aspects of the biology of viruses.*

Prusiner, S. B., 1984. Prions, *Scientific American* 251 (4). (Offprint 1554)

Sharon, N., 1969. The bacterial cell wall, *Scientific American* 220 (5). *On the peculiar structure of bacterial walls, and the way many antibiotics like penicillin block their synthesis.*

Stanier, R. Y., E. A. Adelberg, and J. L. Ingraham, 1976. *The Microbial World*, 4th ed. Prentice-Hall, Englewood Cliffs, N.J. *Excellent general microbiology text.*

Stoeckenius, W., 1976. The purple membrane of salt-loving bacteria, *Scientific American* 234 (6). (Offprint 1340) *Rhodopsin as the light-trapping pigment of a newly discovered kind of photosynthesis carried out by certain Archaebacteria.*

Walsby, A. E., 1977. The gas vacuoles of blue-green algae, *Scientific American* 237 (2). (Offprint 1367) *How Cyanobacteria regulate their buoyancy.*

Woese, C. R., 1981. Archaebacteria, *Scientific American* 244 (6). (Offprint 1516)

CHAPTER 39

Bold, H. C., and M. J. Wynne, 1978. *Introduction to the Algae.* Prentice-Hall, Englewood Cliffs, N.J. *Comprehensive, rather technical text that includes the protistan algae.*

* Available in paperback.

BONNER, J. T., 1969. Hormones in social amoebae and mammals, *Scientific American* 220 (6). (Offprint 1145) *The role of cyclic AMP in the communication system of the cellular slime molds.*

BONNER, J. T., 1983. Chemical signals of social amoebae, *Scientific American* 248 (4). (Offprint 1537)

GRELL, K. G., 1973. *Protozoology*, 2nd ed. Springer, Heidelberg. *Excellent comprehensive treatment of the Protozoa.*

CHAPTER 40

BANKS, H. P., 1970. *Evolution and Plants of the Past*. Wadsworth, Belmont, Calif. *Excellent short book on fossil plants, with special emphasis on the evolutionary relationships of the tracheophyte groups.**

BOLD, H. C., and M. J. WYNNE, 1978. *Introduction to the Algae*. Prentice-Hall, Englewood Cliffs, N.J. *Thorough, rather technical treatment of all the algal groups.*

GRANT, V., 1951. The fertilization of flowers, *Scientific American* 184 (6). (Offprint 12) *The special adaptations of flowers that help ensure their pollination.*

JENSEN, W. A., and F. B. SALISBURY, 1972. *Botany: An Ecological Approach*. Wadsworth, Belmont, Calif. *Very readable general text written from an evolutionary and ecological point of view.*

RAVEN, P. H., R. F. EVERT, and H. CURTIS, 1976. *Biology of Plants*, 2nd ed. Worth, New York. *A well-written broad coverage of botany.*

CHAPTER 41

ABMADJIAN, V., 1963. The fungi of lichens, *Scientific American* 208 (2).

COOKE, R. C., 1978. *Fungi, Man and His Environment*. Longman, London. *Short but fascinating treatment of the biology of fungi, with emphasis on the many ways these organisms affect human beings.*

EMERSON, R., 1952. Molds and men, *Scientific American* 186 (1). (Offprint 115) *The diversity of the fungi, and their many effects on human lives.*

LITTEN, W., 1975. The most poisonous mushrooms, *Scientific American* 232 (3). *The members of the genus* Amanita *and the highly toxic compound they produce.*

RAVEN, P. H., R. F. EVERT, and H. CURTIS, 1976. *Biology of Plants*, 2nd ed. Worth, New York. *Includes an excellent treatment of the fungi.*

WEBSTER, J., 1980. *Introduction to Fungi*, 2nd ed. Cambridge University Press, New York. *Rather technical treatment of the various groups of fungi.**

CHAPTER 42

BARNES, R. D., 1974. *Invertebrate Zoology*, 3rd ed. Saunders, Philadelphia. *Rather technical, comprehensive treatment of all the invertebrate groups.*

* Available in paperback.

BORROR, D. J., D. M. DeLong, and C. A. TRIPLEHORN, 1976. *An Introduction to the Study of Insects*, 4th ed. Holt, Rinehart & Winston, New York. *Thorough coverage of all aspects of insect biology.*

BUCHSBAUM, R. 1948. *Animals Without Backbones*, 2nd ed. University of Chicago Press, Chicago. *One of the most readable and fascinating discussions of the invertebrates ever written. Not technical.**

BUCHSBAUM, R., and L. J. MILNE, 1960. *The Lower Animals: Living Invertebrates of the World*. Doubleday, Garden City, N.Y. *One of the "Living Animals of the World" books. Like the others in the series —on amphibians, birds, fish, insects, mammals, and reptiles— beautifully illustrated, well written, and nontechnical.*

GOREAU, T. F., N. I. GOREAU, and T. J. GOREAU, 1979. Corals and coral reefs, *Scientific American* 241 (2). (Offprint 1434)

KLOTS, A. B., and E. B. KLOTS, 1959. *Living Insects of the World*. Doubleday, Garden City, N.Y.

LAPAN, E. A., and H. J. MOROWITZ, 1972. The Mesozoa, *Scientific American* 277 (6). *An excellent account of these tiny invertebrates.*

CHAPTER 43

BAKKER, R. T., 1975. Dinosaur renaissance, *Scientific American* 232 (4). (Offprint 916) *Reasons for believing that the dinosaurs were warm-blooded and that the birds descended from them.*

COCHRAN, D. M., 1961. *Living Amphibians of the World*. Doubleday, Garden City, N.Y.

GILLIARD, E. T., 1958. *Living Birds of the World*. Doubleday, Garden City, N.Y.

HERALD, E. S., 1961. *Living Fishes of the World*. Doubleday, Garden City, N.Y.

LEAKEY, R. E., and R. LEWIN, 1978. The hominids of East Turkana, *Scientific American* 239 (2). (Offprint 709) *On some of the most recently discovered human fossils.*

PILBEAM, D., 1984. The descent of hominoids and hominids, *Scientific American* 250 (3). *An excellent modern summary of current ideas about human evolution.*

SANDERSON, I. T., 1955. *Living Mammals of the World*. Doubleday, Garden City, N.Y.

SCHMIDT, K. P., and R. F. INGER, 1957. *Living Reptiles of the World*. Doubleday, Garden City, N.Y.

SIMONS, E. L., 1977. Ramapithecus, *Scientific American* 236 (5). (Offprint 695) *On the discovery of many new fossils of this earliest member of the hominid family, and the light they shed on human evolution.*

TELEKI, G., 1973. The omnivorous chimpanzee, *Scientific American* 228 (1). (Offprint 682) *Evidence that chimpanzees, formerly thought entirely herbivorous, sometimes hunt and kill other mammals for food.*

CREDITS

1.1 Courtesy NASA. **1.3** California Institute of Technology Archives. **1.4** SCALA/Art Resource, New York. **1.6** (A) SCALA/Art Resource, New York; (B) Musées Nationaux, Paris. **1.7** By courtesy of the Trustees, The British Museum (Natural History). **1.8** The Master and Fellows of Trinity College, Cambridge. **1.9** Courtesy Rare Book Division, The New York Public Library, Astor, Lenox and Tilden Foundations. **1.10** The Royal Society, London. **1.11** Warder Collection. **1.12** Rijksmuseum, Amsterdam. **1.13** Warder Collection. **1.14** American Museum of Natural History, New York. **1.15** (A) Sovfoto; (B) F. Erize, Bruce Coleman Inc. **1.16** Library of the Museum of Natural History, Paris. **1.18** The New York Public Library, Astor, Lenox and Tilden Foundations. **1.19** Leonard Lee Rue III, Bruce Coleman Inc. **1.20** Photographs by L. Nilsson; from L. Nilsson, *Behold Man*, English translation © 1974, Albert Bonniers Förlag, Stockholm, and Little, Brown and Co. (Canada) Ltd. **1.21** Modified from *The Illustrated Origin of Species* by Charles Darwin, abridged and introduced by Richard E. Leakey, 1979; courtesy Hill and Wang, division of Farrar, Straus & Giroux, Inc. **1.22** By permission to the Syndics of Cambridge University Library. **1.23** National Portrait Gallery, London. **1.24** (center) Kenneth W. Fink, Bruce Coleman Inc.; (other photographs) courtesy Louise B. Van der Meid. **facing p. 1** The Bettmann Archive, Inc.

PART I   THE CHEMICAL AND CELLULAR BASIS OF LIFE

(A) An artist's conception of the primordial environment in which the first chemosynthetic organisms are thought to have arisen. [Painting by Peter Sawyer, photograph by Chip Clark for the National Museum of Natural History, Washington, D.C.]   **(B)** A field of sunflowers. The "invention" of the photosynthetic pathway that produces oxygen as a by-product was an important factor in the evolution of biological diversity. [Courtesy Charlie E. Rogers]   **(C)** A reproductive spore of a yellow-green alga with a full complement of subcellular structures. × 8,400. [Photograph by Gordon Leedale, Photo Researchers, Inc., © Biophoto Associates]   **(D)** Cell movement in a hamster cell. [Courtesy Jean-Paul Revel, California Institute of Technology]

2.22 W. H. Amos, Bruce Coleman Inc. **2.24** (B) W. A. Bentley and W. J. Humphreys, *Snow Crystals*, Dover, New York, 1962.

3.10 From R. G. Kessell and R. H. Kardon, *Tissues and Organs: A Text-Atlas of Scanning Electron Microscopy*, W. H. Freeman, San Francisco, copyright © 1970. **3.11** Courtesy Kenneth Lorenzen, University of California, Davis. **3.20** (C) Redrawn from A. L. Lehninger, *Biochemistry*, 2nd ed., Worth Publishers, New York, 1975, p. 129; based on G. H. Haggis et al., *Introduction to Molecular Biology*, Wiley, New York, 1964. **3.21** (B, C) Redrawn from H. D. Springall, *The Structural Chemistry of Proteins*, Butterworth & Co. and Academic Press, New York, 1954. **3.22** Tony Brain, Science Photo Library. **3.23** Adapted by permission from *The Structure and Action of Proteins* by Richard E. Dickerson and Irving Geis, W. A. Benjamin, Inc., Menlo Park, Calif., Publisher; copyright © 1969 by Dickerson and Geis. **3.24** Adapted by permission from *The Structure and Action of Proteins* by Richard E. Dickerson and Irving Geis, W. A. Benjamin, Inc., Menlo Park, Calif., Publisher; copyright © 1969 by Dickerson and Geis. **3.25** Adapted by permission from *The Structure and Action of Proteins* by Richard E. Dickerson and Irving Geis, W. A. Benjamin, Inc., Menlo Park, Calif., Publisher; copyright © 1969 by Dickerson and Geis. **3.41** Modified from A. L. Lehninger, *Biochemistry*, 2nd ed., Worth Publishers, New York, 1975, p. 233. **p. 63** Courtesy J. A. Bassham and M. Calvin, University of California, Berkeley.

4.3 Courtesy Jeremy D. Pickett-Heaps, University of Colorado. **4.9** Courtesy J. F. Hoffman, Yale University Medical School. **4.12** Courtesy J. David Robertson, Duke University. **4.15** Courtesy Daniel Branton, Harvard University. **4.22** (B) Courtesy Dorothy F. Bainton, University of California, San Francisco. **4.23** (B) J. Ross, J. Olmstead, and J. Rosenbaum, *Tissue and Cell*, vol. 7, 1975. **4.24** M. M. Perry and A. B. Gilbert, *J. Cell Sci.*, vol. 39, 1979, by copyright permission of the Rockefeller University Press. **4.26** (B) V. Herzog, H. Sies, and F. Miller, *J. Cell Biol.*, vol. 70, 1976, by copyright permission of the Rockefeller University Press. **4.28** Courtesy Eva Frei and R. D. Preston, University of Leeds. **4.29** (B) H. Latta, W. Johnson, and T. Stanley, *J. Ultrastruct. Res.*, vol. 51, 1975.

5.1 Courtesy N. B. Gilula, Baylor College of Medicine. **5.2** Courtesy A. H. Sparrow and R. F. Smith, Brookhaven National Laboratory. **5.3** (A) Courtesy Barbara Hamkalo, University of California, Irvine; (B) courtesy Victoria Foe. **5.4** W. G. Whaley, H. H. Mollenhauer, and J. H. Leech, *Am J. Bot.*, vol. 47, 1960. **5.5** Courtesy Daniel Branton, Harvard University. **5.6** Courtesy K. R. Porter, University of Colorado. **5.7** Micrograph courtesy D. S. Friend, University of California, San Francisco. **5.8** Micrograph by C. J. Flickinger, *J. Cell Biol.*, vol. 49, 1971, by copyright permission of the Rockefeller University

Press. **5.10** Courtesy D. S. Friend, University of California, San Francisco. **5.12** Micrograph by S. E. Frederick and E. H. Newcomb, *J. Cell Biol.* vol. 43, 1969. **5.13** Courtesy D. S. Friend, University of California, San Francisco. **5.14** (top) Micrograph by W. P. Wergin, courtesy E. H. Newcomb, University of Wisconsin; (bottom) M. C. Ledbetter, Photo Researchers, Inc. **5.15** Courtesy M. C. Ledbetter, Brookhaven National Laboratory. **5.18** (left) H. Kim, L. I. Binder, and J. L. Rosenbaum, *J. Cell Biol.*, vol. 80, 1979, by copyright permission of the Rockefeller University Press; (right) courtesy D. W. Fawcett, Harvard Medical School. **5.21** Micrograph from J. Heuser and S. R. Salpeter, *J. Cell Bio.*, vol. 82, 1979, by copyright permission of the Rockefeller University Press. **5.22** M. McGill, D. P. Highfield, T. M. Monahan, and B. R. Brinkley, *J. Ultrastruct. Res.*, vol. 57, 1976. **5.23** (A) Courtesy R. W. Linck, Harvard Medical School, and D. T. Woodrum. **5.24** Courtesy E. R. Dirksen, University of California, Los Angeles. **5.25** Micrograph by K. Roberts, John Innes Institute, Norwich, England; from B. Alberts et al., *Molecular Biology of the Cell*, Garland Press, New York, 1983. **5.26** Modified from E. B. Wilson, *The Cell in Development and Heredity*, Macmillan, New York, 1925. **5.27** (mitochondrion) Courtesy K. R. Porter, University of Colorado; (lysosome) courtesy A. B. Novikoff, Albert Einstein College of Medicine; (all other micrographs) courtesy N. B. Gilula, Baylor College of Medicine. **5.28** (clockwise: 1) Courtesy W. G. Whaley et al., *J. Biophys. Biochem. Cytol.*, (now *J. Cell Biol.*), vol. 5, 1959; by copyright permission of the Rockefeller University Press; (2, 4, 5, 6) courtesy M. C. Ledbetter, Brookhaven National Laboratory; (3) W. G. Whaley et al., *Am J. Bot.*, vol. 47, 1960. **5.29** Courtesy A. Ryter, Institut Pasteur, Paris. **5.30** Courtesy J. Griffith, School of Medicine, University of North Carolina, Chapel Hill.

6.1 Modified from L. A. Staehelin and B. E. Hull, *Sci. Am.*, May 1978; copyright © 1978 by Scientific American, Inc.; all rights reserved. **6.2** Ed Reschke, Peter Arnold Inc. **6.3** Courtesy Thomas Eisner, Cornell University. **6.4** Courtesy Nels R. Lersten, Iowa State University. **6.5** (left) Courtesy Victor B. Eichler; (right) courtesy Thomas Eisner, Cornell University. **6.8** Courtesy Ed Reschke. **6.10** Courtesy Ed Reschke. **6.11** Courtesy Ed Reschke. **6.12** (top) Photograph by D. Claugher, by courtesy of the Trustees, The British Museum (Natural History); (bottom) courtesy Jerome Gross, Massachusetts General Hospital. **6.14** (left) Courtesy J. Dlugosz and L. J. Gathercole, University of Bristol; (right) A. Jurand, *J. Embryol. Exp. Morph.*, vol. 28, 1972. **6.16** Photograph courtesy Ed Reschke. **6.17** Courtesy M. Wurtz, University of Basel. **6.18** Tony Brain, Science Photo Library. **6.19** (top) H. Chaumeton, Nature; (bottom) Elliot Scientific Corp. **6.20** (A) Photograph by Ginny Fonte, from H. C. Berg, *Sci. Am.*, August 1975; copyright © 1975 by Scientific American, Inc.; all rights reserved; (B) courtesy T. E. Adams. **6.21** From W. J. Jones, J. A. Leigh, F. Mayer, C. R. Woese, and R. S. Wolfe, *Arch. Microbiol.*, vol. 136, 1983; copyright © 1983 by Springer-Verlag. **6.22** H. Chaumeton, Nature. **6.23** Courtesy E. V. Gravé. **6.24** Courtesy E. V. Gravé. **6.25** H. Chaumeton, Nature. **6.26** H. Chaumeton, Nature. **6.27** H. Chaumeton, Nature. **6.28** Adrian Davies, Bruce Coleman Inc. **6.29** (left) R. P. Carr, and (right) Jane Burton, Bruce Coleman Inc. **6.30** (left) W. H. Amos, Bruce Coleman Inc.; (right) courtesy D. G. Allen. **6.31** R. N. Mariscal, Bruce Coleman Inc. **6.32** Oxford Scientific Films. **6.33** Courtesy T. E. Adams. **6.34** Courtesy J. H. Carmichael, Jr. **6.35** Oxford Scientific Films, Bruce Coleman Inc. **6.36** E. R. Degginger, Bruce Coleman Inc. **6.37** (left) David Hughes, and (right) R. L. Donne, Bruce Coleman Inc. **6.38** Jeff Rotman, Peter Arnold Inc. **6.39** Ferrero, Nature. **6.40** Co Rentmeester, *Life* magazine, copyright © 1970 by Time Inc.

8.11 Micrograph by W. P. Wergin, courtesy E. H. Newcomb, University of Wisconsin. **8.17** Courtesy Raymond Chollet, University of Nebraska.

PART II THE BIOLOGY OF ORGANISMS

(A) The many seed-bearing, bottle-shaped fruit of the dandelion ripen and are spread by the wind. [Photograph by D. Claugher, by courtesy of the Trustees, The British Museum (Natural History)] **(B)** *Didinium nasutum*, a carnivorous protozoan, consumes a paramecium. ✕ 400. [Gary W. Grimes and Steven W. L'Hernault, Taurus Photos] **(C)** The human eye, a camera eye of a type common to vertebrates, produces an image which is abstracted by specialized nerve cells and sent to the brain for further processing. [Photograph by L. Nilsson; from L. Nilsson, *Behold Man*, English translation © 1974, Albert Bonniers, Förlag, Stockholm, and Little, Brown and Co. (Canada) Ltd., p. 219]

(D) The behavior of social animals like these African lions, as well as the behavior of the more solitary creatures of the world, is shaped by an interplay of genes, environment, and evolution. [Yann Arthus-Bertrand/Jacana]

9.3 Courtesy W. E. Loomis, Iowa State University. **9.4** Courtesy Ed Reschke. **9.7** Carolina Biological Supply Co. **9.10** Breck P. Kent, Earth Scenes. **9.11** (A) Courtesy Hans W. Paerl, University of North Carolina; (B) Hans W. Paerl and Kathleen K. Gallucci, *Science*, vol. 227, 1985; copyright © 1985 by the American Association for the Advancement of Science. **9.12** Courtesy E. R. Degginger. **9.13** (top) Oxford Scientific Films; (bottom) D. Lyons, Bruce Coleman Inc. **9.14** Courtesy E. R. Degginger.

10.1 Courtesy R. Farquharson, UNICEF. **10.2** From the *Vitamin Manual*, courtesy The Upjohn Company. **10.3** From the *Vitamin Manual*, courtesy The Upjohn Company. **10.5** David Pramer, *Science*, vol. 144, 1964; copyright © 1964 by the American Association for the Advancement of Science. **10.6** Courtesy K. G. Grell, University of Tübingen. **10.10** Modified from W. D. Russell-Hunter, *A Biology of Lower Invertebrates*, Macmillan Publishing Co., N.Y. 1968. **10.13** Turtox/Cambosco, Macmillan Science Co., Inc. **10.15** (A, B) Courtesy Thomas Eisner; (C) courtesy William A. Watkins. **10.18** Adapted from an original painting by Frank H. Netter, M. D., from *The CIBA Collection of Medical Illustrations*, copyright © 1959 by CIBA Pharmaceutical Company, division of CIBA-GEIGY Corporation. **10.23** Adapted from Norman Kretchmer, *Sci. Am.*, October 1972; copyright © 1972 by Scientific American, Inc.; all rights reserved. **10.24** From Warren Andrew, *Textbook of Comparative Histology*, Oxford, The University Press, 1959. **10.25** Courtesy Susumu Ito, Harvard Medical School.

11.2 Courtesy Thomas Eisner, Cornell University. **11.3** Courtesy J. H. Troughton, Department of Scientific and Industrial Research, Wellington, New Zealand. **11.7** (left) Courtesy Thomas Eisner, Cornell University; (right) G. R. Roberts. **11.10** Modified from Ralph Buchsbaum, *Animals Without Backbones*, by permission of the University of Chicago Press, copyright © 1948 by the University of Chicago. **11.12** Courtesy J. H. Carmichael, Jr. **11.16** Jane Burton, Bruce Coleman Inc. **11.22** Courtesy H-R. Duncker, Justus Liebig University, Geissen, West Germany. **11.23** Modified from *The Vertebrate Body*, 5th ed., by A. S. Romer and T. S. Parsons; copyright © 1977 by W. B. Saunders Co.; reprinted by permission of CBS College Publishing. **11.24** Modified from R. Margaria et al., *J. Appl. Physiol.*, vol. 18, 1963. **11.25** Courtesy Roman Vishniac. **11.27** (left) Photograph by D. Claugher, by courtesy of the Trustees, The British Museum (Natural History); (right) from Warren Andrew, *Textbook of Comparative Histology*, Oxford, The University Press, 1959. **11.28** Kim Taylor, Bruce Coleman Inc.

12.3 Courtesy Thomas Eisner, Cornell University. **12.4** Courtesy Nels R. Lersten, Iowa State University. **12.5** Courtesy Ed Reschke. **12.7** Courtesy Ed Reschke. **12.9** Photograph by B. Bracegirdle; from W. Krommenhoek, J. Sebus, and G. J. van Esch, *Biological Structures*, copyright © 1979 by L. C. G. Malmberg B. V., The Netherlands. **12.10** Courtesy Thomas Eisner, Cornell University. **12.11** Modified from V. A. Greulach and J. E. Adams, *Plants: An Introduction to Modern Botany*, Wiley, New York, 1962. **12.14** (A) Courtesy J. H. Troughton, Department of Scientific and Industrial Research, Wellington, New Zealand; (B, C) courtesy B. G. Butterfield, Canterbury University, and B. A. Meylan, Department of Scientific and Industrial Research, Wellington, New Zealand. **12.15** Courtesy Thomas Eisner, Cornell University. **12.18** Courtesy Thomas Eisner, Cornell University. **12.19** Laura Riley, Bruce Coleman Inc. **12.21** O. Biddulph et al., *Plant Physiol.*, vol. 31, 1958. **12.23** M. H. Zimmerman, *Science*, vol. 133, 1961; copyright © 1961 by the American Association for the Advancement of Science.

13.5 Micrograph from R. G. Kessel and R. H. Kardon, *Tissues and Organs: A Text-Atlas of Scanning Electron Microscopy*, W. H. Freeman, San Francisco, copyright © 1979. **13.7** Modified from B. S. Guttman and J. W. Hopkins III, *Understanding Biology*; copyright © 1983 by Harcourt Brace Jovanovich, Inc.; used by permission of the publisher. **13.13** (left) Courtesy Thomas Eisner, Cornell University; (right) photograph by L. Nilsson; from L. Nilsson, *Behold Man*, English translation copyright © 1974 by Albert Bonniers Förlag, Stockholm, and Little, Brown and Co. Ltd. **13.14** Courtesy D. W. Fawcett, Harvard Medical School. **13.18** Manfred Krage, Peter Arnold Inc. **13.20** Courtesy Turtox/Cambosco, Macmillan Science Co., Inc. **13.21** Courtesy Eila Kairinen, Gillette Research Institute. **13.22** Courtesy Dorothea Zucker-Franklin, New York University Medical Center. **13.23** Courtesy K. R.

vol. 130, 1959; copyright © 1959 by the American Association for the Advancement of Science. **21.22** Based on data from Peter Marler, Rockefeller University. **21.24** Warren Gurst and Genny Gurst, Tom Stack and Associates. **21.26** (A) From B. Elsner, in *Animal Migration, Navigation, and Homing,* edited by K. Schmidt-Koenig and W. T. Keeton, Springer-Verlag, New York, 1978; (B) from M. Michener and C. Walcott, *J. Exp. Biol.,* vol. 47, 1967. **21.27** After S. T. Emlen. **21.28** After W. T. Keeton. **21.29** After W. T. Keeton. **21.30** From W. T. Keeton, *Proc. Natl. Acad. Sci. U.S.A.,* vol. 68, 1971. **21.31** From K. Schmidt-Koenig and C. Walcott, *Animal Behav.,* vol. 26, 1978. **21.32** Based on data from USGS aeromagnetic surveys and from Charles Walcott.

22.1 Based on J. E. Lloyd, *Misc. Publ. Mus. Zool. Univ. Mich.,* vol. 130, 1966; and A. D. Carlson and J. Copeland, *Am. Sci.,* vol. 66, 1978, reprinted by permission of *American Scientist,* Journal of Sigma Xi, the Scientific Research Society. **22.2** D. R. Bentley, *Science,* vol. 174, 1971; copyright © 1971 by the American Association for the Advancement of Science. **22.3** From L. P. Brower, J. V. Z. Brower, and F. P. Cranston, *Zoologica,* vol. 50, 1965; used by permission of the New York Zoological Society. **22.4** Courtesy J. E. Lloyd, University of Florida, and A. Cary. **22.5** Courtesy R. Thornhill, University of New Mexico. **22.6** (B) Redrawn from K. von Frisch, *Bees: Their Vision, Chemical Senses, and Language,* copyright © 1950 by Cornell University; used by permission of Cornell University Press; photograph courtesy Kenneth Lorenzen, University of California, Davis. **22.7** Based on data from R. Boch, *Z. Vergl. Physiol.,* vol. 40, 1957. **22.8** From K. Lorenz, *Zool. Anz.,* vol. 17, 1953. **22.9** Courtesy T. Seeley, Yale University. **22.10** Courtesy Philip Green. **22.11** Courtesy L. David Mech. **22.12** Photograph by Hans Reinhard, Bruce Coleman Inc.; drawing based on H. N. Southern, *J. Zool.,* vol. 162, 1970. **22.13** Based on H. N. Southern, *J. Zool.,* vol. 162, 1970. **22.14** Courtesy R. D. Estes, San Diego State University. **22.15** Redrawn by permission from E. O. Wilson, *Sociobiology,* Harvard University Press, Cambridge, Mass., 1975. **22.17** Courtesy John Sparks, BBC (Natural History). **22.18** Stouffer Productions/Animals Animals. **22.20** George D. Lepp, Bio-Tec Images. **22.21** Roberto Bunge, Ardea London Ltd. **22.22** John Wightman, Ardea London Ltd. **22.23** Courtesy Kenneth Lorenzen, University of California, Davis. **22.24** Courtesy E. S. Ross. **22.25** Courtesy L. T. Nash, Arizona State University. **22.26** Redrawn with permission of Macmillan Publishing Co., Inc., from David R. Pilbeam, *The Ascent of Man: An Introduction to Human Evolution;* copyright © 1972 by David R. Pilbeam. **22.27** (A) J. Tanaka, Anthro-Photo; (B) M. Shostak, Anthro-Photo. **22.28** R. Lee, Anthro-Photo. **22.29** Reprinted with permission of the author and the publishers from D. G. Freedman, *Human Infancy: An Evolutionary Perspective,* Lawrence Erlbaum Associates, Hillsdale, N. J., 1975.

PART III   THE PERPETUATION OF LIFE

(A) Mitosis in several cells of the meristematic tissues of an onion root. × 1,375. [From W. Krommenhoek, J. Sebus, and G. J. van Esch, *Biological Structures,* copyright © 1979 by L. C. G. Malmberg B. V., The Netherlands] **(B)** Free ribosomal subunits are brought together to form a ribosome, the cellular structure upon which proteins are manufactured. **(C)** Gene expression may be influenced by specialized regulator proteins which mask DNA and prevent transcription, as in the case of this *cro* repressor bound to DNA in the bacterial virus lambda. [Computer model and photograph courtesy Brian W. Matthews and Douglas H. Ohlendorf, University of Oregon] **(D)** Antibodies can be cloned, labeled radioactively, and used to help mark specific molecules in cells; the blue structures in this photograph of a fibroblast cell are actin molecules which have been bound by monoclonal antibodies. [Courtesy J. V. Small and G. Rinnerthaler, Institute of Molecular Biology, Austria]

23.2 A. Kerstitch, Sea of Cortez Enterprises. **23.5** Courtesy R. G. E. Murray, University of Western Ontario. **23.6** Micrograph by W. Engler, courtesy of G. F. Bahr, *Fed. Proc., Fed. Am. Soc. Exp. Biol.,* vol. 34, 1975. **23.7** Courtesy M. W. Shaw, University of Michigan. **23.9** Courtesy A. S. Bajer, University of Oregon. **23.12** Courtesy A. S. Bajer, University of Oregon. **23.13** Modified from B. Alberts et al., *Molecular Biology of the Cell,* Garland Press, New York, 1983. **23.14** From M. S. Fuller, *Mycologia,* vol. 60, 1968. **23.15** From H. W. Beams and R. G. Kessel, *Am. Sci.,* vol. 64, 1976; reprinted by permission of *American Scientist,* Journal of Sigma Xi, the Scientific Research Society. **23.16** Adapted from C. J. Avers, *Cell Biology,* copyright © 1976 by Litton Educational Publishing, Inc.; used by permission of Wadsworth, Inc.

23.17 Courtesy A. S. Bajer, University of Oregon. **23.18** From W. G. Whaley et al., *Am. J. Bot.,* vol. 47, 1960. **23.20** (C) From D. von Wettstein, *Proc. Natl. Acad. Sci. U.S.A.,* vol. 68, 1971. **23.22** Courtesy James Kezer, University of Oregon. **23.25** Courtesy A. S. Bajer, University of Oregon. **23.29** Courtesy Thomas Eisner, Cornell University. **23.31** Photograph by L. Nilsson; from L. Nilsson, *Behold Man,* English translation © 1974, Albert Bonniers Förlag, Stockholm, and Little, Brown and Co. (Canada) Ltd.

24.8 Courtesy Ralph Somes, Jr., University of Connecticut. **24.10** Modified from Francisco J. Ayala and John A. Kiger, Jr., *Modern Genetics,* Benjamin Cummings, Menlo Park, Calif., 1980. **24.13** Courtesy S. B. Moore. **24.14** Modified from A. M. Winchester, *Genetics,* 5th ed., Houghton Mifflin, Boston, 1977.

25.1 From a photograph by C. D. Mueller in A. M. Winchester, *Genetics,* 5th ed., Houghton Mifflin, Boston, 1977. **25.2** Courtesy Marion I. Barnhart, Wayne State University School of Medicine, Detroit, Michigan. **25.3** Modified from Sewall Wright, *Genetics,* vol. 6, 1921. **25.5** Courtesy M. L. Barr, *Can. Cancer Conf.,* vol. 2, Academic Press, New York, 1957. **25.6** Courtesy R. A. Boolootian. **25.7** K. R. Dronamraju, in E. J. Gardner, *Principles of Genetics,* copyright © 1975 by John Wiley & Sons, Inc.; reprinted with their permission. **25.10** Science Software Systems, Inc., West Los Angeles. **25.11** Micrograph courtesy Joseph G. Gall, Carnegie Institution, Baltimore. **25.12** Modified from *Biological Science: An Ecological Approach,* 4th ed., Houghton Mifflin, Boston, 1982; used by permission of the publisher and the Biological Sciences Curriculum Study. **25.15** Courtesy Francisco J. Ayala, University of California, Davis. **25.16** Courtesy Ted Bingham, University of Wisconsin.

26.3 Courtesy Lee D. Simon, Waksman Institute, Rutgers University. **26.11** Redrawn from M. S. Meselson and F. W. Stahl, *Proc. Natl. Acad. Sci. U.S.A.,* vol. 44, 1958.

27.7 Adapted from R. Gupta, J. M. Lanter, and C. R. Woese, *Science,* vol. 221, 1983; copyright © 1983 by the American Association for the Advancement of Science. **27.9** Micrograph from O. L. Miller, Jr., B. A. Hamkalo, and C. A. Thomas, Jr., *Science,* vol. 169, 1970; copyright © 1970 by the American Association for the Advancement of Science. **27.10** Photograph courtesy Nigel Unwin, Stanford University School of Medicine. **27.11** Modified from B. Alberts et al., *Molecular Biology of the Cell,* Garland Press, New York, 1983.

28.1 Courtesy A. C. Arnberg, Biochemical Laboratory, State University, Groningen, The Netherlands. **28.2** M. R. Hanson et al., *Mol. Gen. Genet.,* vol. 132, 1974. **28.3** Courtesy S. N. Cohen, Stanford University **28.4** Courtesy L. G. Caro, University of Geneva, and R. Curtiss, University of Alabama.

29.4 Courtesy Steven Henikoff, Fred Hutchinson Cancer Research Center, Seattle, Washington. **29.6** From J. G. Gall, *Methods Cell Physiol.* (edited by D. M. Prescott), vol. 2, Academic Press, New York, 1966. **29.7** Photograph courtesy Michael Ashburner. **29.8** Photograph courtesy O. L. Miller, Jr., and B. R. Beatty, *J. Cell Physiol.,* vol. 74, 1969. **29.9** Courtesy Gunter Albrecht-Buehler, Cold Spring Harbor Laboratory, and Frank Solomon, Massachusetts Institute of Technology. **29.10** K. Porter, G. Fonte, and Weiss, *Cancer Res.,* vol. 34, 1974. **29.11** Curve for retinoblastoma based on data from H. W. Hethcote and A. G. Knudson, *Proc. Natl. Acad. Sci. U.S.A.,* vol. 75, 1978; curves for prostate and skin cancer based on data from Japanese Cancer Association, *Cancer Mortality and Morbidity Statistics,* Japanese Scientific Press, Tokyo, 1981.

30.10 Courtesy M. W. Shaw, University of Michigan. **30.15** Micrograph by S. N. Cohen, from S. N. Cohen and J. A. Shapiro, *Sci. Am.,* February 1980, copyright © 1980 by Scientific American, Inc.; all rights reserved. **p. 777** Courtesy Center for Disease Control, Atlanta. **p. 785** Courtesy Steven T. Brentano.

31.1 Micrograph by P. Sundstrom, Gamma-Liaison. **31.7** Courtesy R. G. Kessel and C. Y. Shih, *Scanning Electron Microscopy in Biology,* Springer-Verlag, New York, 1974. **31.10** Oxford Scientific Films. **31.11** Photographs by L. Nilsson; from L. Nilsson, *Behold Man,* English translation © 1974, Albert Bonniers Förlag, Stockholm, and Little, Brown and Co. (Canada) Ltd. **31.12** Redrawn from G. J. Romanes, *Darwin and After Darwin,* Open Court

PART IV   THE BIOLOGY OF POPULATIONS AND COMMUNITIES

tesy U. S. Department of Agriculture. **36.10** Courtesy D. W. Schindler, *Science*, vol. 184, 1974; copyright © 1974 by the American Association for the Advancement of Science. **36.12** Courtesy P. L. Ames. **36.13** Michael Gallagher, Bruce Coleman Inc. **36.14** Tom McHugh, Photo Researchers, Inc. **36.15** Redrawn from G. E. Likens et al., *Ecol. Monogr.*, vol. 40, 1970; copyright © 1970 by the Ecological Society of America. **36.16** Copyright by the State Organization of Antiquities and Heritage, Iraq. **36.17** Oxford Scientific Films, Earth Scenes. **36.19** C. J. Tucker, J. R. G. Townshend and T. E. Goff, *Science* vol. 227, 1985. **36.20** Photographs courtesy E. S. Ross. **36.22** (left) Leonard Lee Rue III, Bruce Coleman Inc.; (right) courtesy E. R. Degginger. **36.23** (left) Wolfgang Bayer, Bruce Coleman Inc.; (right) Paolo Koch, Photo Researchers, Inc. **36.24** Zig Leszczynski, Earth Scenes. **36.25** (left) Ferrero, Nature; (right) N. de Vore III, Bruce Coleman Inc. **36.26** Courtesy E. R. Degginger. **36.27** (A) Courtesy E. R. Degginger; (B, C) courtesy E. S. Ross. **36.29** Bob and Clara Calhoun, Bruce Coleman Inc. **36.31** After Schwarzbach, 1950. **36.32** (A) Courtesy Howard Hall; (B) courtesy James D. Jordan; (C) courtesy E. H. Newcomb and T. D. Pugh, University of Wisconsin/BPS. **36.34** Redrawn from John Napier, *The Roots of Mankind*, copyright © 1970, Smithsonian Institution Press, Washington, D.C.; used by permission. **36.36** Ferrero, Nature. **36.40** (A) Redrawn from R. H. MacArthur and E. O. Wilson, *The Theory of Island Biogeography*, copyright © 1967 by Princeton University Press; used by permission; (B) redrawn by C. J. Krebs, *Ecology*, 2nd ed., Harper & Row, New York , 1978; after Preston. **36.43** Redrawn from C. J. Krebs, *Ecology*, 2nd ed., Harper & Row, 1978; after Williams. **36.44** (A) Photograph by C. C. Reijnvaan from W. M. Doctors van Leeuwen, *Ann. Jard. Bot. Buitenzorg*, 1936; (B) courtesy Stephen Self, University of Texas, Arlington.

PART V THE GENESIS AND DIVERSITY OF ORGANISMS

(A) An *E. coli* bacterium infected by T4 bacteriophage. [Courtesy Jonathan King, Massachusetts Institute of Technology] **(B)** A parasol fungus. [Hans Reinhard, Bruce Coleman Inc.] **(C)** Meadow flowers, mainly bellflowers, bird's-foot trefoil, and pink thistle. [Hans Reinhard, Bruce Coleman Inc.] **(D)** Giraffe, zebra, and elephant congregate at a watering hole in Etosha Pan, South West Africa. [Clem Haegner, Bruce Coleman Inc.]

37.1 Photographs courtesy NASA. **37.2** Modified from R. E.Dickerson, *Sci. Am.*, September 1978; copyright © 1978 by Scientific American, Inc.; all rights reserved. **37.3** Courtesy Sigurgeir Jónasson. **37.5** Courtesy Sidney W. Fox, University of Miami, and Steven Brooke Studios, Coral Gables, Florida. **37.6** Photo courtesy T. O. Diener, U.S. Department of Agriculture. **37.8** (A) Sovfoto; (B, C) courtesy NASA. **37.9** Courtesy J. W. Schopf, University of California, Los Angeles. **37.10** Courtesy M. F. Glaessner, University of Adelaide. **37.11** (A) Courtesy R. E. Lee, University of the Witwatersrand; (B) courtesy M. Jost, Michigan State University. **37.12** D. A. Stetler and W. M. Laetsch, *Am. J. Bot.*, vol. 56, 1969. **37.13** W. J. Larsen, *J. Cell Biol.*, vol. 47, 1970, by copyright permission of the Rockefeller University Press. **37.14** Courtesy I. B. Dawid, National Institutes of Health, Bethesda, Md., and D. R. Wolstenholme, University of Utah. **37.15** Courtesy E. V. Gravé.

38.1 (A) Courtesy M. Wurtz, University of Basel; (B) courtesy Virus Laboratory, University of California, Berkeley; (C) courtesy M. Gomersall, McGill University. **38.3** M. A. Gonda et al., *Science*, vol. 227, 1985; copyright © 1985 by the American Association for the Advancement of Science. **38.4** (bottom) From K. Corbett, *Virology*. vol. 22, 1964; reprinted by permission of Academic Press, Inc., New York. **38.6** From T. O. Diener, *Am. Sci.*, vol. 71, 1983. **38.8** H. W. Jannasch and C. O. Wirsen, *BioScience*, vol. 29, 1979; copyright © 1979 by the American Institute of Biological Sciences. **38.9** (left) Courtesy David Scharf, Peter Arnold Inc.; (right) Center for Disease Control, Atlanta, Ga. **38.10** Turtox/Cambosco, Macmillan Science Co., Inc. **38.11** Courtesy Z. Skobe, Forsyth Dental Center/BPS. **38.12** Courtesy M. Gomersall, McGill University. **38.13** S. Kimoto and J. C. Russ, *Am. Sci.*, vol. 57, 1969. **38.14** G. B. Chapman, *J. Bacteriol.*, vol. 71, 1956. **38.15** From W. Krommenhoek, J. Sebus, and G. J. van Esch, *Biological Structures*, copyright © 1979 by L. C. G. Malmboj B.V., The Netherlands. **38.16** From R. C. Johnson, M. P. Walsh, B. Ely, and L. Shapiro, *J. Bacteriol.*, vol. 138, 1979. **38.17** Courtesy David Chase, Veterans Hospital, Sepulveda, Calif. **38.18** From R. Kavenhoff and O. A. Ryder, *Chromosoma*, vol. 55, 1976, by courtesy of the authors and L. P. Velten. **38.19** Courtesy Elliot Scientific Corp. **38.20** Courtesy Elliot Scientific

Corp. **38.21** (A) T. E. Adams, Bruce Coleman Inc.; (B, C) courtesy J. M. Kingsbury. **38.22** Courtesy R. Malcolm Brown, Jr., University of Texas. **38.23** Courtesy Rosemarie Rippka, Institut Pasteur, Paris. **38.24** Courtesy Zell A. McGee and E. N. Robinson, Jr., University of Utah.

39.2 Courtesy of E. V. Gravé. **39.3** Photograph courtesy Ed Reschke. **39.4** Courtesy E. V. Gravé. **39.5** Oxford Scientific Films, Bruce Coleman Inc. **39.6** Courtesy E. V. Gravé. **39.9** (A) Courtesy E. V. Gravé; (B) H. Chaumeton, Nature; (C) courtesy Roman Vishniac. **39.10** Courtesy Thomas Eisner, Cornell University. **39.11** Manfred Kage, Peter Arnold Inc. **39.13** Ray Simons, Photo Researchers, Inc. **39.16** (A, B) Courtesy K. B. Raper, *Proc. Am. Philos. Soc.*, vol. 104, 1960; (C) courtesy Roman Vishniac; (D) courtesy David Francis, University of Delaware; (E) from W. F. Loomis, *Dictyostelium Discoideum: A Developmental System*, Academic Press, New York, 1975. **39.17** Courtesy E. V. Gravé. **39.19** Turtox/Cambosco, Macmillan Science Co., Inc. **39.20** (left) H. Chaumeton, Nature; (middle and right) Turtox/Cambosco, Macmillan Science Co., Inc. **39.21** Courtesy E. V. Gravé.

40.3 F. Sauer, Nature. **40.4** Courtesy Ed Reschke. **40.5** Courtesy Roman Vishniac. **40.6** Courtesy Roman Vishniac. **40.8** Courtesy Roman Vishniac. **40.11** Courtesy T. E. Adams. **40.15** Anne Wertheim, Bruce Coleman Inc. **40.16** Courtesy Douglas Faulkner. **40.18** (right) R. P. Carr, Bruce Coleman Inc. **40.20** H. Chaumeton, Nature. **40.21** Modified from H. J. Fuller and O. Tippo, *College Botany*, Holt Rinehart & Winston, Inc., New York, 1954. **40.24** Jane Burton, Bruce Coleman, Inc. **40.25** Adrian Davies, Bruce Coleman Inc. **40.27** Adrian Davies, Bruce Coleman Inc. **40.28** Photo courtesy E. S. Ross. **40.29** Courtesy E. R. Degginger. **40.31** Portion of group in Carnegie Museum, Pittsburgh; used by permission. **40.32** Courtesy Field Museum of Natural History, Chicago. **40.33** Courtesy E. S. Ross. **40.34** John Shaw, Bruce Coleman Inc. **40.35** Courtesy G. R. Roberts. **40.36** Ray Simons, Photo Researchers, Inc. **40.37** Modified from H. J. Fuller and O. Tippo, *College Botany*, Holt, Rinehart & Winston, Inc., New York, 1954. **40.39** Redrawn from H. N. Andrews, *Science*, vol. 142, 1963; copyright © 1963 by the American Association for the Advancement of Science. **40.40** Courtesy E. S. Ross. **40.41** Courtesy Roman Vishniac. **40.42** From W. Krommenhoek, J. Sebus, and G. J. van Esch, *Biological Structures*, copyright © 1979 by L. C. G. Malmboj B.V., The Netherlands. **40.43** (A, B) Courtesy Thomas Eisner, Cornell University; (C) Courtesy E. S. Ross. **40.44** Modified from H. J. Fuller and O. Tippo, *College Botany*, Holt, Rinehart & Winston, Inc., New York, 1954. **40.48** Turtox/Cambosco, Macmillan Science Co., Inc. **40.50** Courtesy E. S. Ross. **40.52** (left) Jane Burton, and (right) R. P. Carr, Bruce Coleman, Inc. **40.53** Adapted from H. J. Fuller and O. Tippo, *College Botany*, rev. ed.; copyright © 1949, 1954, by Holt, Rinehart & Winston, Inc., CBS College Publishing.

41.1 Courtesy Pfizer Inc. **41.4** M. P. L. Fogden, Bruce Coleman, Inc. **41.5** From L. W. Sharp, *Fundamentals of Cytology*, McGraw-Hill Book Co., New York, copyright © 1943; used by permission. **41.8** (A) Jane Burton, Bruce Coleman Inc; (B) L. West, Bruce Coleman Inc.; (C, D) V. Ahmadjian and J. B. Jacobs, *Nature*, vol. 289, 1981 © 1981, Macmillan Journals Ltd. **41.9** (left) Jane Burton, Bruce Coleman Inc.; (right) H. Chaumeton, Nature. **41.10** From L. W. Sharp, *Fundamentals of Cytology*, McGraw-Hill Book Co., New York, copyright © 1943, used by permission.

42.1 Courtesy Jeff Rotman. **42.4** H. Chaumeton, Nature. **42.7** Carolina Biological Supply Co. **42.10** H. Chaumeton, Nature. **42.11** Courtesy Howard Hall. **42.15** After K. G. Grell, University of Tübingen (1974). **42.16** Courtesy K. G. Grell, University of Tübingen. **42.18** After K. G. Grell, University of Tübingen. **42.20** H. Chaumeton, Nature. **42.26** Courtesy Ed Reschke. **42.27** H. Chaumeton, Nature. **42.31** Based on a phylogenetic tree drawn by R. P. Higgins. **42.34** Courtesy T. E. Adams. **42.36** H. Chaumeton, Nature. **42.37** H. Chaumeton, Nature. **42.39** Fred Bavendam, Peter Arnold Inc. **42.40** Oxford Scientific Films. **42.42** H. Chaumeton, Nature. **42.43** Courtesy Howard Hall. **42.44** Based in part on drawings by Louise G. Kingsbury. **42.45** H. Chaumeton, Nature. **42.46** H. Chaumeton, Nature. **42.47** Modified from W. Stempell, *Zoologie im Grundriss*, Borntraeger, 1926. **42.48** H. Chaumeton, Nature. **42.49** Modified from R. D. Barnes, *Invertebrate Zoology*, W. B. Saunders, 1963. **42.50** Courtesy American Museum of Natural History. **42.51** H. Chaumeton, Nature. **42.52** Oxford Scientific Films. **42.53** Courtesy E. S. Ross. **42.54** Jane Burton, Bruce Coleman Inc.

GLOSSARY

The Glossary gives brief definitions of the most important recurrent terms used in the text, excluding taxonomic designations. For fuller definitions, consult the index, where italicized page numbers refer you to explanations of key terms in context.

Of the basic units of measurement, some are tabulated on p. A28, others have their own alphabetical entries.

Interalphabetized with the vocabulary are the main prefixes and combining forms used in biology. You will notice that, while they are generally of Greek or Latin origin, many of them have acquired a new meaning in biology (examples: *blasto-, -cyte, caryo-, -plasm*). Familiarity with these forms will make it easier for you to learn and remember the numerous terms in which they are incorporated.

TABLE 1 *Standard prefixes of the metric system*

kilo- (k)	1,000	10^3
deci- (d)	0.1	10^{-1}
centi- (c)	0.01	10^{-2}
milli- (m)	0.001	10^{-3}
micro- (μ)	0.000001	10^{-6}
nano- (n)	0.000000001	10^{-9}

TABLE 2 *Common units of length, weight, and liquid capacity*

kilometer (km)	1,000 m	0.62137 mile
meter (m)		39.37 inches
centimeter (cm)	0.01 m	0.39 inch
millimeter (mm)	0.001 m	0.039 inch
micrometer* (μm)	10^{-6} m	
nanometer (nm)	10^{-9} m	
angstrom† (Å)	10^{-10} m	
kilogram (kg)	1,000 g	2.2 pounds
gram (g)		0.035 ounce
milligram (mg)	0.001 g	
microgram (μg)	10^{-6} g	
liter (l)	1,000 cm³	1.057 quarts
milliliter (ml)	0.001 l	

*Formerly called micron.
†No longer used; nanometer used instead.

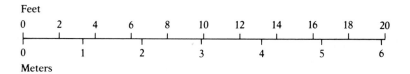

Feet / Meters

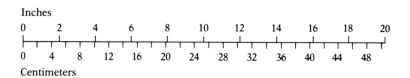

Inches / Centimeters

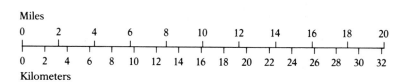

Miles / Kilometers

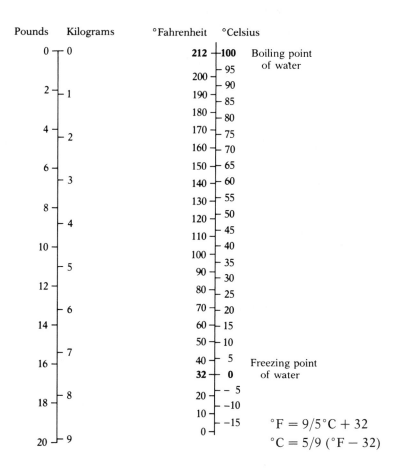

$$°F = 9/5°C + 32$$
$$°C = 5/9 \ (°F - 32)$$

a- Without, lacking.

ab- Away from, off.

abdomen [L belly]   In mammals, the portion of the trunk posterior to the thorax, containing most of the viscera except heart and lungs. In other animals, the posterior portion of the body.

absolute zero   The temperature ($-273°C$) at which all thermal agitation ceases. The lowest possible temperature.

acellular   Not constructed on a cellular basis.

acid [L *acidus* sour]   A substance that increases the concentration of hydrogen ions when dissolved in water. It has a pH lower than 7.

ACTH   *See* adrenocorticotropic hormone.

action potential   *See* potential.

active site   In an enzyme, the portion of the molecule that reacts with a substrate molecule.

active transport   Movement of a substance across a membrane by a process requiring expenditure of energy by the cell.

ad- Next to, at, toward.

adaptation   In evolution, any genetically controlled characteristic that increases an organism's fitness, usually by helping the organism to survive and reproduce in the environment it inhabits. In neurobiology, the process that results in a short-lasting decline in responsiveness of a sensory neuron after repeated firing; *cf.* habituation, sensitization.

adenosine diphosphate (ADP)   A doubly phosphorylated organic compound that can be further phosphorylated to form ATP.

adenosine triphosphate (ATP)   A triply phosphorylated organic compound that functions as "energy currency" for organisms.

adipose [L *adeps* fat]   Fatty.

ADP   *See* adenosine diphosphate.

adrenal [L *renes* kidneys]   An endocrine gland of vertebrates located near the kidney.

adrenalin   A hormone produced by the adrenal medulla that stimulates "fight-or-flight" reactions.

adrenocorticotropic hormone (ACTH)   A hormone produced by the pituitary that stimulates the adrenal cortex.

adsorb [L *sorbēre* to suck up]   Hold on a surface.

advanced   New, unlike the ancestral condition.

aerobic [L *aer* air]   With oxygen.

alcohol   Any of a class of organic compounds in which one or more —OH groups are attached to a carbon backbone.

alkaline   Having a pH of more than 7. *See* base.

all-, allo- [Gk *allos* other]   Other, different.

allele   Any of several alternative gene forms at a given chromosomal locus.

allopatric [L *patria* homeland]   Having different ranges.

all-or-none   The property of responding maximally or not at all.

allosteric   Of an enzyme: one that can exist in two or more conformations. *Allosteric control:* control of the activity of an allosteric enzyme by determination of the particular conformation it will assume.

altruism   The willingness of an individual to sacrifice its fitness for the benefit of another. *Reciprocal altruism:* the performance of a favor by one individual in the expectation of a favor in return, as when two animals groom each other.

alveolus [L little hollow]   A small cavity, especially one of the microscopic cavities that are the functional units of lungs.

amino acid   An organic acid carrying an amino group ($-NH_2$); the building-block compound of proteins.

amnion [Gk caul]   An extraembryonic membrane that forms a fluid-filled sac containing the embryo in reptiles, birds, and mammals.

amoeboid [Gk *amoibē* change]   Amoebalike in the tendency to change shape by protoplasmic flow.

amylase [L *amylum* starch]   A starch-digesting enzyme.

an- Without.

anabolism [Gk *ana-* upward; *metabolē* change]   The biosynthetic building-up aspects of metabolism.

anaerobic [L *aer* air]   Without oxygen.

analogous   Of characters in different organisms: similar in function and often in superficial structure but of different evolutionary origins.

anemia   A condition in which the blood has lower than normal amounts of hemoglobin or red blood corpuscles.

angio-, -angium [Gk *angeion* vessel]   Container, receptacle.

anion   A negatively charged ion.

anterior   Toward the front end.

antheridium [Gk *anthos* flower]   Male reproductive organ of a plant; produces sperm cells.

antibody   A protein, produced by the B lymphocytes of the immune system, that binds to a particular antigen.

antigen   A substance, usually a protein or polysaccharide, that activates an organism's immune system.

anus [L ring]   Opening at the posterior end of the digestive tract, through which indigestible wastes are expelled.

aorta   The main artery of the systemic circulation.

apical   At, toward, or near the apex, or tip, of a structure such as a plant shoot.

apo- Away from.

apoplast   The network cell walls and intercellular spaces within a plant body; permits extensive extracellular movement of water within the plant.

aposematic [Gk *sēma* sign] Serving as a warning, with reference particularly to colors and structures that signal possession of defensive devices.

arch- [Gk *archein* to begin] Primitive, original.

archegonium [Gk *archegonos* the first of a race] Female reproductive organ of a higher plant; produces egg cells.

archenteron [Gk *enteron* intestine] The cavity in an early embryo that becomes the digestive cavity.

arteriole A small artery.

artery A blood vessel that carries blood away from the heart.

articulation A joint between bones. Articulating surfaces are those formed between bones and joints.

artifact A by-product of scientific manipulation rather than an inherent part of the thing observed.

ascus [Gk *askos* bag] The elongate spore sac of a fungus of the Ascomycota group.

asexual Without sex.

atmosphere (atm) (unit of pressure) The normal pressure of air at sea level: 101,325 newtons per square meter (approx. 14.7 pounds per square inch).

atom [Gk *atomos* indivisible] The smallest unit of an element, not divisible by ordinary chemical means.

atomic mass unit (amu) *See* dalton.

atomic weight The average weight of an atom of an element relative to ^{12}C, an isotope of carbon with six neutrons in the nucleus. The atomic weight of ^{12}C has arbitrarily been fixed as 12.

ATP *See* adenosine triphosphate.

auto- Self, same.

autonomic nervous system A portion of the vertebrate nervous system, comprising motor neurons that innervate internal organs and are not normally under direct voluntary control.

autosome [Gk *sōma* body] Any chromosome other than a sex chromosome.

autotrophic [Gk *trophē* food] Capable of manufacturing organic nutrients from inorganic raw materials.

auxin [Gk *auxein* to grow] Any of a class of plant hormones that promote cell elongation and can diffuse into a decapitated plant from an agar block, causing the plant to bend in the dark (Went test).

axon [Gk *axōn* axis] A fiber of a nerve cell that conducts impulses away from the cell body and can release transmitter substance.

bacteriophage [Gk *phagein* to eat] A virus that attacks bacteria, *abbrev.* phage.

basal At, near, or toward the base (the point of attachment) of a structure such as a limb.

basal body A structure, identical to the centriole, found at the base of cilia and eucaryotic flagella; consists of nine triplet microtubules arranged in a circle.

base (or alkali) A substance that increases the concentration of hydroxyl ions when dissolved in water. It has a pH higher than 7.

basidium The spore-bearing structure of Basidiomycota (club fungi).

bi- Two.

bilateral symmetry The property of having two similar sides, with definite upper and lower surfaces and definite anterior and posterior ends.

binary fision Reproduction by the division of a cell into two essentially equal parts by a nonmitotic process.

bio- [Gk *bios* life] Life, living.

biogenesis [Gk *genesis* source] Origin of living organisms from other living organisms.

biological magnification Increasing concentration of relatively stable chemicals as they are passed up a food chain from initial consumers to top predators.

biomass The total weight of all the organisms, or of a designated group of organisms, in a given area.

biome A large climatic region with characteristic sorts of plants and animals.

biotic Pertaining to life.

blasto- [Gk *blastos* bud] Embryo.

blastocoel [Gk *koilos* hollow] The cavity of a blastula.

blastospore [Gk *poros* passage] The opening from the cavity of the archenteron to the exterior in a gastrula.

blastula An early embryonic stage in animals, preceding the delimitation of the three principal tissue layers; frequently spherical and hollow.

B lymphocyte *See* lymphocyte.

buffer A substance that binds H$^+$ ions when their concentration rises and releases them when their concentration falls, thereby minimizing fluctuations in the pH of a solution.

caecum [L *caecus* blind] A blind diverticulum of the digestive tract.

calorie [L *calor* heat] The quantity of energy, in the form of heat, required to raise the temperature of one gram of pure water one degree from 14.5 to 15.5°C. The nutritionists' Calorie (capitalized) is 1,000 calories, or one kilocalorie.

cambium [L *cambiare* to exchange] The principal lateral meristem of vascular plants; gives rise to most secondary tissue.

cAMP *See* cyclic adenosine monophosphate.

capillarity [L *capillus* hair] The tendency of aqueous liquids to rise in narrow tubes with hydrophilic surfaces.

capillary [L *capillus*] A tiny blood vessel with walls one cell thick, across which exchange of materials between blood and the tissues takes place; receives blood from arteries and carries it to veins. Also, a similar vessel of the lymphatic system.

carbohydrate Any of a class of organic compounds composed of carbon, hydrogen, and oxygen in a ratio of about two hydrogens and one oxygen for each carbon; examples are sugar, starch, cellulose.

carbon fixation The process by which CO_2 is incorporated into organic compounds, primarily glucose; energy usually comes from the ATP and $NADP_{re}$ generated by photophosphorylation, and the metabolic pathway utilizing this energy is usually the Calvin cycle.

carboxyl group The —COOH group characteristic of organic acids.

cardiac [Gk *kardia* heart] Pertaining to the heart.

carnivore [L *carnis* of flesh; *vorare* to devour] An organism that feeds on animals.

carotenoid [L *carota* carrot] Any of a group of red, orange, and yellow accessory pigments of plants, found in plastids.

carrying capacity The maximum population that a given environment can support indefinitely.

cartilage A specialized type of dense fibrous connective tissue with a rubbery intercellular matrix.

caryo- [Gk *karyon* kernel] Nucleus.

Casparian strip A waterproof thickening in the radial and end walls of endodermal cells of plants.

cata- Down.

catabolism [Gk *katabolē* a throwing down] The degradational breaking-down aspects of metabolism, by which living things extract energy from food.

catalysis [Gk *katalyein* to dissolve] Acceleration of a chemical reaction by a substance that is not itself permanently changed by the reaction.

catalyst A substance that produces catalysis.

cation A positively charged ion.

caudal [L *cauda* tail] Pertaining to the tail.

cell cycle The cycle of cellular events from one mitosis through the next. Four stages are recognized, of which the last—distribution of genetic material to the two daughter nuclei—is mitosis proper.

cell sap *See* sap.

cellulose [L *cellula* cell] A complex polysaccharide that is a major constituent of most plant cell walls.

centi- [L *centum* hundred] One hundredth.

central nervous system A portion of the nervous system that contains interneurons and exerts some control over the rest of the nervous system. In vertebrates, the brain and the spinal cord.

centri- [L *centrum* center] Center.

centrifugation [L *fugere* to flee] The spinning of a mixture at very high speeds to separate substances of different densities.

centriole A cylindrical cytoplasmic organelle located just outside the nucleus of animal cells and the cells of some lower plants; associated with the spindle during mitosis and meiosis.

centromere [Gk *meros* part] A special region on a chromosome that produces kinetochore microtubules during mitosis or meiosis.

cephalization [Gk *kephalē* head] Localization of neural coordinating centers and sense organs at the anterior end of the body.

cerebellum (L small brain) A part of the hindbrain of vertebrates that controls muscular coordination.

cerebrum [L brain] Part of the forebrain of vertebrates, the chief coordination center of the nervous system.

channel *See* membrane channel.

character Any structure, functional attribute, behavioral trait, or other characteristic of an organism.

character displacement The rapid divergent evolution in sympatric species of characters that minimize competition and/or hybridization between them.

chemiosmotic gradient The combined electrostatic and osmotic-concentration gradient generated by the electron-transport chains of mitochondria and chloroplasts; the energy in this gradient is used, for the most part, to synthesize ATP.

chemosynthesis Autotrophic synthesis of organic materials, energy for which is derived from inorganic molecules.

chitin [Gk *chitōn* tunic] Polysaccharide that forms part of the hard exoskeleton of insects, crustaceans, and other invertebrates; also occurs in the cell walls of fungi.

chlorophyll [Gk *chlōros* greenish yellow; *phyllon* leaf] The green pigment of plants necessary for photosynthesis.

chloroplast A plastid containing chlorophyll.

chrom-, -chrome [Gk *chrōma* color] Colored; pigment.

chromatid A single chromosomal strand.

chromatin The mixture of DNA and protein (mostly histones in the form of nucleosome cores) that comprises eucaryotic nuclear chromosomes.

chromatography Process of separating substances by adsorption on media for which they have different affinities.

chromosome [Gk *sōma* body] A filamentous structure in the cell nucleus (or nucleoid), mitochondria, and chloroplasts, along which the genes are located.

cilium [L eyelid] A short hairlike locomotory organelle on the surface of a cell (*pl* cilia).

cisterna [L cistern] A cavity, sac, or other enclosed space serving as a reservoir.

classical conditioning *See* conditioning.

cleavage Division of a zygote or of the cells of an early embryo.

climax (ecological) A relatively stable stage reached in some ecological successions.

cline [Gk *klinein* to lean] Gradual variation, correlated with geography, in a character of a species.

cloaca [L sewer] Common chamber that receives materials from the digestive, excretory, and reproductive systems.

clone [Gk *klōn* twig]   A group of cells or organisms derived asexually from a single ancestor and hence genetically identical.

co-   With, together.

codon   The unit of genetic coding, three nucleotides long, specifying an amino acid or an instruction to terminate translation.

coel-, -coel [Gk *koilos* hollow]   Hollow, cavity; chamber.

coelom   A body cavity bounded entirely by mesoderm.

coenocytic [Gk *koinos* common]   Having more than one nucleus in a single mass of cytoplasm.

coenzyme   A nonproteinaceous organic molecule that plays an accessory role, but a necessary one, in the catalytic action of an enzyme.

coevolution   Two or more organisms evolving, each in response to the other.

coleoptile [Gk *koleon* sheath, *ptilon* feather]   A sheath around the young shoot of grasses.

collagen   A fibrous protein; the most abundant protein in mammals.

collenchyma [Gk *kolla* glue]   A supportive tissue in plants in which the cells usually have thickenings at the angles of the walls.

colloid [Gk *kolla*]   A stable suspension of particles that, though larger than in a true solution, do not settle out.

colon   The large intestine.

com-   Together.

commensalism [L *mensa* table]   A symbiosis in which one party is benefited and the other party receives neither benefit nor harm.

community   In ecology, a unit composed of all the populations living in a given area.

competition   In ecology, utilization by two or more individuals, or by two or more populations, of the same limited resource; an interaction in which both parties are harmed.

condensation reaction   A reaction joining two compounds with resultant formation of water.

conditioning   Associative learning. *Classical conditioning:* the association of a novel stimulus with an innately recognized stimulus. *Operant conditioning:* learning of a novel behavior as a result of reward or punishment; trial-and-error learning.

conformation (of a protein) [L *conformatio* symmetrical forming]   The three-dimensional pattern according to which the polypeptide chains of a protein coil (secondary structure), fold (tertiary structure), and—if there is more than one chain—fit together (quarternary structure).

conjugation [L *jugare* to join, marry]   Process of genetic recombination between two organisms (e.g., bacteria, algae) through a cytoplasmic bridge between them.

connective tissue   A type of animal tissue whose cells are embedded in an extensive intercellular matrix; connects, supports, or surrounds other tissues and organs.

contractile vacuole   An excretory and/or osmoregulatory vacuole in some cells, which, by contracting, ejects fluids from the cell.

cooperativity   The phenomenon of enhanced reactivity of the remaining binding sites of a protein as a result of the binding of substrate at one site.

cork [L *cortex* bark]   A waterproof tissue, derived from the cork cambium, that forms at the outer surfaces of the older stems and roots of woody plants; the outer bark or periderm.

corpus luteum [L yellow body]   A yellowish structure in the ovary, formed from the follicle after ovulation, that secretes estrogen and progesterone (*pl.* corpora lutea).

cortex [L bark]   In plants, tissue between the epidermis and the vascular cylinder of stems and roots. In animals, the outer barklike tissue of some organs, as *cerebral cortex, adrenal cortex,* etc.

cotyledon [Gk *kotylē* cup]   A "seed leaf," a food-digesting and -storing part of a plant embryo.

covalent bond   A chemical bond resulting from the sharing of a pair of electrons.

crossing over   Exchange of parts between two homologous chromosomes.

cross section   *See* section.

cryptic [Gk *kryptos* hidden]   Concealing.

cuticle [L *cutis* skin]   A waxy layer on the outer surface of leaves, insects, etc.

cyclic, adenosine monophosphate (cyclic AMP or cAMP)   Compound, synthesized in living cells from ATP, that functions as an intracellular mediator of hormonal action; also plays a part in neural transmission and some other kinds of cellular control systems.

cyst [Gk *kystis* bladder, bag]   (1) A saclike abnormal growth. (2) Capsule that certain organisms secrete around themselves and that protects them during resting stages.

-cyte, cyto- [Gk *kytos* container]   Cell.

cytochrome   Any of a group of iron-containing enzymes important in electron transport during respiration or photophosphorylation.

cytokinesis [Gk *kinēsis* motion]   Division of the cytoplasm of a cell.

cytoplasm   All of a cell except the nucleus.

cytosol   The relatively fluid, less structured part of the cytoplasm of a cell, excluding organelles and membranous structures.

dalton   A unit of mass equal to one twelfth the atomic weight of ^{12}C, or 1.66024×10^{-24} gram. Formerly called atomic mass unit (amu).

deamination   Removal of an amino group.

deciduous [L *decidere* to fall off]   Shedding leaves each year.

dehydration reaction   A condensation reaction.

deme [Gk *dēmos* population]   A local unit of population of any one species.

dendr-, dendro- [Gk *dendron* tree]   Tree; branching.

dendrite A short unsheathed fiber of a nerve cell—often spiny, usually branched and tapering—that receives many synapses and carries excitation and inhibition toward the cell body.

deoxyribonucleic acid (DNA) A nucleic acid found in most viruses, all bacteria, chloroplasts, mitochondria, and the nuclei of eucaryotic cells, characterized by the presence of a deoxyribose sugar in each nucleotide; the genetic material of all organisms except the RNA viruses.

-derm [Gk *derma* skin] Skin, covering; tissue layer.

di- Two.

dicot A member of a subclass of the angiosperms, or flowering plants, characterized by the presence of two cotyledons in the embryo, a netlike system of veins in the leaves, and flower petals in fours or fives; *cf.* monocot. *Herbaceous dicot:* a perennial whose aboveground parts die annually. *Woody dicot:* a perennial whose aboveground parts—trunk and branches—remain alive and grow annually.

differentiation The process of developmental change from an immature to a mature form, especially in a cell.

diffusion The movement of dissolved or suspended particles from one place to another as a result of their heat energy (thermal agitation).

digestion Hydrolysis of complex nutrient compounds into their building-block units.

diploid [Gk *diploos* double] Having two of each type of chromosome.

disaccharide A double sugar, one composed of two simple sugars.

distal [L *distare* to stand apart] Situated away from some reference point (usually the main part of the body).

diverticulum [L *devertere* to turn aside] A blind sac branching off a cavity or canal.

DNA *See* deoxyribonucleic acid.

dominant (1) Of an allele: exerting its full phenotypic effect despite the presence of another allele of the same gene, whose phenotypic expression it blocks or masks. *Dominant phenotype, dominant character:* one caused by a dominant allele. (2) Of an individual: occupying a high position in the social hierarchy.

dormancy [L *dormire* to sleep] The state of being inactive, quiescent. In plants, particularly seeds and buds, a period in which growth is arrested until environmental conditions become more favorable.

dorsal [L *dorsum* back] Pertaining to the back.

drive *See* motivation.

duodenum [From a Latin phrase meaning 12 *(duodecin)* finger's-breadths long] The first portion of the small intestine of vertebrates, into which ducts from the pancreas and gallbladder empty.

ecosystem [Gk *oikos* habitation] The sum of physical features and organisms occurring in a given area.

ecto- Outside, external.

ectoderm The outermost tissue layer of an animal embryo. Also, tissue derived from the embryonic ectoderm.

ectothermic *See* poikilothermic.

effector The part of an organism that produces a response, e.g., muscle, cilium, flagellum.

egg An egg cell or female gamete. Also a structure in which embryonic development takes place, especially in birds and reptiles; consists of an egg cell, various membranes, and often a shell.

electrochemical gradient Combined electrostatic and osmotic-concentration gradient, such as the chemiosmotic gradient of mitochondria and chloroplasts.

electron A negatively charged primary subatomic particle.

electronegativity The formal measure of an atom's attraction for free electrons. Atoms with few electron vacancies in their outer shell tend to be more electronegative than those with many. In covalent bonds, the shared electrons are, on average, nearer the more electronegative atom; this asymmetry, in part, gives rise to the polarity of certain molecules.

electronic charge unit The charge of one electron, or 1.6021×10^{-19} coulomb.

electron-transport chain A series of enzymes found in the inner membrane of mitochondria and (with somewhat different components) in the thylakoid membrane of chloroplasts. The chain accepts high-energy electrons and uses their energy to create a chemiosmotic gradient across the membrane in which it is located.

electrostatic force The attraction (also called *electrostatic attraction*) between particles with opposite charges, as between a proton and an electron, or between H^+ and OH^-; and the repulsion between particles with like charges, as between two H^+ ions.

electrostatic gradient The free-energy gradient created by a difference in charge between two points, generally the two sides of a membrane.

elimination (or defecation) The release of unabsorbed wastes from the digestive tract. *Cf.* excretion.

embryo A plant or animal in an early stage of development; generally still contained within the seed, egg, or uterus.

emulsion [L *emulsus* milked out] Suspension, usually as fine droplets, of one liquid in another.

-enchyma [Gk *parenchein* to pour in beside] Tissue.

end-, endo- Within, inside; requiring.

endergonic [Gk *ergon* work] Energy-absorbing; endothermic.

endocrine [Gk *krinein* to separate] Pertaining to ductless glands that produce hormones.

endocytosis The process by which the cell membrane forms an invagination which becomes a vesicle, trapping extracellular material which is then transported within the cell; in general, the invagination is triggered by the binding of membrane receptors to specific substances used by the cell.

endoderm The innermost tissue layer of an animal embryo.

endodermis A plant tissue, especially prominent in roots, that surrounds the vascular cylinder; all endodermal cells have Casparian strips.

endonuclease An enzyme that breaks bonds within nucleic acids, as opposed to an exonuclease, which can digest only a terminal group. *Restriction endonuclease:* an enzyme that breaks bonds only within a specific sequence of bases.

endoplasmic reticulum [L *reticulum* network] A system of membrane-bounded channels in the cytoplasm.

endoskeleton An internal skeleton.

endosperm [Gk *sperma* seed] A nutritive material in seeds.

endosymbiotic hypothesis Hypothesis that certain eucaryotic organelles—in particular mitochondria and chloroplasts—originated as free-living procaryotes that took up mutalistic residence in the ancestors of modern eucaryotes.

endothermic In thermodynamics, energy-absorbing (endergonic). In physiology, warm-blooded (homeothermic).

entropy Measure of the disorder of a system.

enzyme [Gk *zymē* leaven] A compound, usually a protein, that acts as a catalyst.

epi- Upon, outer.

epicotyl A portion of the axis of a plant embryo above the point of attachment of the cotyledons; forms most of the shoot.

epidermis [Gk *derma* skin] The outermost portion of the skin or body wall of an animal.

episome [Gk *sōma* body] Genetic element at times free in the cytoplasm, at other times integrated into a chromosome.

epithelium An animal tissue that forms the covering or lining of all free body surfaces, both external and internal.

equilibrium constant The ratio of products of a reaction to the reactants after the reaction has been allowed to proceed until there is no further change in these concentrations.

erythrocyte [Gk *erythros* red] A red blood cell, i.e., a blood cell containing hemoglobin.

esophagus [Gk *phagein* to eat] An anterior part of the digestive tract; in mammals it leads from the pharynx to the stomach.

estrogen [L *oestrus* frenzy] Any of a group of vertebrate female sex hormones.

estrous cycles [L *oestrus*] In female mammals, the higher primates excepted, a recurrent series of physiological and behavioral changes connected with reproduction.

estuary That portion of a river that is close enough to the sea to be influenced by marine tides.

eu- [Gk *eus* good] Most typical, true.

eucaryotic cell A cell containing a distinct membrane-bounded nucleus, characteristic of all organisms except bacteria.

evaginated [L *vagina* sheath] Folded or protruded outward.

eversible [L *evertere* to turn out] Capable of being turned inside out.

evolution [L *evolutio* unrolling] Change in the genetic makeup of a population with time.

ex-, exo- Out of, outside; producing.

excretion Release of metabolic wastes and excess water. *Cf.* elimination.

exergonic [Gk *ergon* work] Energy-releasing; exothermic.

exocytosis The process by which an intracellular vesicle fuses with the cell membrane, expelling its contents into the surrounding medium.

exon A part of a primary transcript (and the corresponding part of a gene) that is ultimately either translated (in the case of mRNA) or utilized in a final product, such as tRNA.

exoskeleton An external skeleton.

extrinsic External to, not a basic part of; as in *extrinsic isolating mechanism.*

fauna The animals of a given area or period.

feature detector A circuit in the nervous system that responds to a specific type of feature, such as a vertically moving spot or a particular auditory time delay.

feces [L *faeces* dregs] Indigestible wastes discharged from the digestive tract.

feedback The process by which a control mechanism is regulated through the very effects it brings about. *Positive feedback:* the process by which a small effect is amplified, as when a depolarization triggers an action potential. *Negative feedback* (or feedback inhibition): the process by which a control mechanism is activated to restore conditions to their original state.

fermentation Anaerobic production of alcohol, lactic acid, or some similar compound from carbohydrate via the glycolytic pathway.

fertilization Fusion of nuclei of egg and sperm.

fetus [L *fetus* pregnant] An embryo in its later development, still in the egg or uterus.

fitness The probable genetic contribution of an individual (or allele or genotype) to succeeding generations. *Inclusive fitness:* the sum of an individual's personal fitness plus the fitness of that individual's relatives devalued in proportion to their genetic distance from the individual.

fixation (1) Conversion of a substance into a biologically more usable form, as the conversion of CO_2 into carbohydrate by photosynthetic plants or the incorporation of N_2 into more complex molecules by nitrogen-fixing bacteria. (2) Process of treating living tissue for microscopic examination.

flagellum [L whip] A long hairlike locomotory organelle on the surface of a cell.

flora The plants of a given area or period.

follicle [L *follis* bag] A jacket of cells around an egg cell in an ovary.

follicle-stimulating hormone (FSH) A gonadotropic hormone of the anterior pituitary that stimulates growth of follicles in the ovaries of females and function of the seminiferous tubules in males.

food chain Sequence of organisms, including producers, consumers, and decomposers, through which energy and materials may move in a community.

foot-candle Unit of illumination; the illumination of a surface produced by one standard candle at a distance of one foot; *cf.* lambert.

founder effect The difference between the gene pool of a population as a whole and that of a newly isolated population of that species.

free energy Usable energy in a chemical system; energy available for producing change.

fruit A mature ovary or cluster of ovaries (sometimes with additional structures associated with the ovary).

fruiting body A spore-bearing structure (e.g., the aboveground portion of a mushroom).

FSH *See* follicle-stimulating hormone.

gamete [Gk *gametē(s)* wife, husband] A sexual reproductive cell that must usually fuse with another such cell before development begins; an egg or sperm.

gametophyte [Gk *phyton* plant] A haploid plant that can produce gametes.

ganglion [Gk tumor] A structure containing a group of cell bodies of neurons (*pl.* ganglia).

gastr-, gastro- [Gk *gastēr* belly] Stomach; ventral; resembling the stomach.

gastrovascular cavity An often branched digestive cavity, with only one opening to the outside, that conveys nutrients throughout the body; found only in animals without circulatory system.

gastrula A two-layered, later three-layered, animal embryonic stage.

gastrulation The process by which a blastula develops into a gastrula, usually by an involution of cells.

gated channel A membrane channel that can open or close in response to a signal, generally a change in the electrostatic gradient or the binding of a hormone, transmitter, or other molecular signal.

gel Colloid in which the suspended particles form a relatively orderly arrangement; *cf.* sol.

-gen; -geny [Gk *genos* birth, race] Producing; production, generation.

gene [Gk *genos*] The unit of inheritance; usually a portion of a DNA molecule that codes for some product such as a protein, tRNA, or rRNA.

gene amplification Any of the strategies that give rise to multiple copies of certain genes, thus facilitating the rapid synthesis of a product (such as rRNA for ribosomes) for which the demand is great.

gene flow The movement of genes from one part of a population to another, or from one population to another, via gametes.

gene pool The sum total of all the genes of all the individuals in a population.

gene regulation Any of the strategies by which the rate of expression of a gene can be regulated, as by controlling the rate of transcription.

generator potential *See* potential.

genetic drift Change in the gene pool as a result of chance and not as a result of selection, mutation, or migration.

genome The cell's total complement of DNA: in eucaryotes, the nuclear and organelle chromosomes; in procaryotes, the major chromosome, episomes, and plasmids. In viruses and viroids, the total complement of DNA or RNA.

genotype The particular combination of genes present in the cells of an individual.

germ cell A sexual reproductive cell; an egg or sperm.

gibberellin A plant hormone one of whose effects is stem elongation in some dwarf plants.

gill An evaginated area of the body wall of an animal, specialized for gas exchange.

gizzard A chamber of an animal's digestive tract specialized for grinding food.

glucose [Gk *glykys* sweet] A six-carbon sugar; plays a central role in cellular metabolism.

glycocalyx The layer of protein and carbohydrates just outside the plasma membrane of an animal cell; in general, the proteins are anchored in the membrane, and the carbohydrates are bound to the proteins.

glycogen [Gk *glykys*] A polysaccharide that serves as the principal storage form of carbohydrate in animals.

glycolysis [Gk *glykys*] Anaerobic catabolism of carbohydrates to pyruvic acid.

Golgi apparatus Membranous subcellular structure that plays a role in storage and modification particularly of secretory products.

gonadotropic Stimulatory to the gonads.

gonadotropin A hormone stimulatory to the gonads, a gonadotropic hormone.

gonads [Gk *gonos* seed] The testes or ovaries.

gram molecule *See* mole.

granum [L grain] A stacklike grouping of photosynthetic membranes in a chloroplast (*pl.* grana).

guard cell A specialized epidermal cell that regulates the size of stoma of a leaf.

habit [L *habitus* disposition] In biology, the characteristic form or mode of growth of an organism.

habitat [L *it lives*]   The kind of place where a given organism normally lives.

habituation   The process that results in a long-lasting decline in the receptiveness of interneurons (primarily) to the input from sensory neurons or other interneurons; *cf.* sensitization, adaptation.

haploid   [Gk *haploos* single] Having only one of each type of chromosome.

hem-, hemat-, hemo- [Gk *haima* blood]   Blood.

hematopoiesis [Gk *poiĕsis* making]   The formation of blood.

hemoglobin   A red iron-containing pigment in the blood that functions in oxygen transport.

hepatic [Gk *hĕpar* liver]   Pertaining to the liver.

herbaceous [L *herbaceus* grassy]   Having a stem that remains soft and succulent; not woody.

herbaceous dicot   *See* dicot.

herbivore [L *herba* grass; *vorare* to devour]   An animal that eats plants.

Hertz   A unit of frequency (as of sound waves) equal to one cycle per second.

hetero- [Gk *heteros* other]   Other, different.

heterogamy [Gk *gamos* marriage]   The condition of producing gametes of two or more different types.

heterotrophic [Gk *trophĕ* food]   Incapable of manufacturing organic compounds from inorganic raw materials, therefore requiring organic nutrients from the environment.

heterozygous [Gk *zygōtos* yoked]   Having two different alleles of a given gene.

Hg [L *hydrargyrum* mercury]   The symbol for mercury. Pressure is often expressed in *mm Hg*—the pressure exerted by a column of mercury whose height is measured in millimeters (at 0° C, 1 mm Hg = 133.3 newtons per square meter).

hilum   Region where blood vessels, nerves, ducts, enter an organ.

hist- [Gk *histos* web]   Tissue.

histology   The structure and arrangement of the tissues of organisms; the study of these.

histone   One of a class of basic proteins serving as structural elements of eucaryotic chromosomes.

homeo-, homo- [Gk *homoios* like]   Like, similar.

homeostasis   The tendency in an organism toward maintenance of physiological and psychological stability.

homeothermic [Gk *thermĕ* heat]   Capable of self-regulation of body temperature; warm-blooded, endothermic.

home range   An area within which an animal tends to confine all or nearly all its activities for a long period of time.

homologous   Of chromosomes: bearing genes for the same characters. Of characters in different organisms: inherited from a common ancestor.

homozygous [Gk *zygōtos* yoked]   Having two copies of the same allele of a given gene.

hormone [Gk *horman* to set in motion]   A control chemical secreted in one part of the body that affects other parts of the body.

hybrid   In evolutionary biology, a cross between two species. In genetics, a cross between two genetic types.

hydr-, hydro- [Gk *hydōr* water]   Water; fluid; hydrogen.

hydration   Formation of a sphere of water around an electrically charged particle.

hydrocarbon   Any compound containing only carbon and hydrogen.

hydrogen bond   A weak chemical bond formed when two polar molecules, at least one of which usually consists of a hydrogen bonded to a more electronegative atom (usually oxygen or nitrogen), are attracted electrostatically.

hydrolysis [Gk *lysis* loosing]   Breaking apart of a molecule by addition of water.

hydrostatic [Gk *statikos* causing to stand]   Pertaining to the pressure and equilibrium of fluids.

hydroxyl ion   The OH^- ion.

hyper-   Over, overmuch; more.

hypertonic   Of a solution (or colloidal suspension): tending to gain water from some reference solution (or colloidal suspension) separated from it by a selectively permeable membrane—usually because it has a higher osmotic concentration than the reference solution.

hypertrophy [Gk *trophĕ* food]   Abnormal enlargement, excessive growth.

hypha [Gk *hyphĕ* web]   A fungal filament.

hypo-   Under, lower; less.

hypocotyl   The portion of the axis of a plant embryo below the point of attachment of the cotyledons; forms the base of the shoot and the root.

hypothalamus [Gk *thalamos* inner chamber]   Part of the posterior portion of the vertebrate forebrain, containing important centers of the autonomic nervous system and centers of emotion.

hypotonic   Of a solution (or colloidal suspension): tending to lose water to some reference solution (or colloidal suspension) separated from it by a selectively permeable membrane—usually because it has a lower osmotic concentration than the reference solution.

imprinting   A kind of associative learning in which an animal rapidly learns during a particular critical period to recognize an object, individual, or location in the absence of overt reward; distinguished from most other associative learning in that it is retained indefinitely, being difficult or impossible to reverse.

inducer   In embryology, a substance that stimulates differentiation of cells or development of a particular structure. In genetics, a substance that activates particular genes.

inorganic compound A chemical compound not based on carbon.

in situ [L in place] In its natural or original position.

instinct Heritable, genetically specified neural circuitry that guides and directs behavior.

insulin [L *insula* island] A hormone produced by the β islet cells in the pancreas that helps regulate carbohydrate metabolism, especially conversion of glucose into glycogen.

integument [L *integere* to cover] A coat, skin, shell, rind, or other protective surface structure.

inter- Between (e.g., *interspecific*, between two or more different species).

interneuron A neuron that receives input from and synapses on other neurons, as distinguished from a sensory neuron (which receives sensory information) and a motor neuron (which synapses on an effector).

intra- Within (e.g., *intraspecific*, within a single species).

intrinsic Inherent in, a basic part of; as in *intrinsic isolating mechanism*.

intron A part of a primary transcript (and the corresponding part of a gene) that lies between exons, and is removed before the RNA becomes functional.

invaginated [L *vagina* sheath] Folded or protruded inward.

invertebrate [L *vertebra* joint] Lacking a backbone, hence an animal without bones.

in vitro [L in glass] Not in the living organism, in the laboratory.

in vivo [L in the living] In the living organism.

ion An electrically charged atom.

ionic bond A chemical bond formed by the electrostatic attraction between two oppositely charged ions.

iso- Equal, uniform.

isogamy [Gk *gamos* marriage] The condition of producing gametes of only one type, with no distinction existing between male and female.

isolating mechanism An obstacle to interbreeding, either extrinsic, such as a geographical barrier, or intrinsic, such as structural or behavioral incompatibility.

isotonic Of a solution (or colloidal suspension): tending neither to gain nor to lose water when separated from some reference solution (or colloidal suspension) by a selectively permeable membrane—usually because it has the same osmotic concentration as the reference solution.

isotope [Gk *topos* place] An atom differing from another atom of the same element in the number of neutrons in its nucleus.

kilo- A thousand.

kin-, kino- [Gk *kinēma* motion] Motion, action.

lactic acid A three-carbon organic acid produced in animals and some microorganisms by fermentation.

lambert In metric system, unit of brightness of a light source; approximately equivalent to 929 foot-candles.

lamella [L thin plate] A thin platelike structure; a fairly straight intracellular membrane.

larva [L ghost, mask] Immature form of some animals that undergo radical transformation to attain the adult form.

lateral Pertaining to the side.

lateral inhibition Process by which adjacent sensory cells or their targets interact to inhibit one another when excited, the result being an exaggerated contrast; neural basis of feature-detector circuits.

lenticel [L *lenticella* small lentil] A porous region in the periderm of a woody stem through which gases can move.

leukocyte [Gk *leukos* white] A white blood cell; *cf.* lymphocyte, macrophage.

LH *See* luteinizing hormone.

ligament [L *ligare* to bind] A type of connective tissue linking two bones in a joint.

ligase An enzyme that catalyzes the bonding between adjacent nucleotides in DNA and RNA.

lignin [L *lignum* wood] An organic compound in wood that makes cellulose harder and more brittle.

linkage The presence of two or more genes on the same chromosome, which, in the absence of crossing over, causes the characters they control to be inherited together.

lip- [Gk *lipos* fat] Fat or fatlike.

lipase A fat-digesting enzyme.

lipid Any of a variety of compounds insoluble in water but soluble in ethers and alcohols; includes fats, oils, waxes, phospholipids, and steroids.

locus [L place] In genetics, a particular location on a chromosome, hence often used synonymously with gene (*pl.* loci).

lumen [L light, opening] The space or cavity within a tube or sac (*pl.* lumina).

lung An internal chamber specialized for gas exchange in an animal.

luteinizing hormone (LH) A gonadotropic hormone of the pituitary that stimulates conversion of a follicle into corpus luteum and secretion of progesterone by the corpus luteum; also stimulates secretion of sex hormone by the testes.

lymph [L *lympha* water] A fluid derived from tissue fluid and transported in special lymph vessels to the blood.

lymphocyte A white blood cell that responds to the presence of a foreign antigen. *B lymphocyte:* a cell that upon stimulation by an antigen secretes antibodies. *T lymphocyte:* a cell that attacks infected cells and modulates the activity of B lymphocytes.

-lysis, lyso- [Gk *lysis* loosing] Loosening, decomposition.

lysogenic Of bacteria: carrying bacteriophage capable of lysing, i.e., destroying, other bacterial cells.

lysosome A subcellular organelle that stores digestive enzymes.

macro- Large.

macrophage A phagocytic white blood cell that ingests material —particularly viruses, bacteria, and clumped toxins—bound by circulating antibodies.

Malpighian tubule An excretory diverticulum of the digestive tract in insects and some other arthropods.

mast cell Cells that are specialized for the secretion of histamine and other local chemical mediators as part of the immune response.

matrix [L *mater* mother] A mass in which something is embedded, e.g., the intercellular substance of a tissue.

medulla [L marrow, innermost part] (1) The inner portion of an organ, e.g., *adrenal medulla.* (2) The *medulla oblongata*, a portion of the vertebrate hindbrain that connects with the spinal cord.

medusa [*after* Medusa, mythological monster with snaky locks] The free-swimming stage in the life cycle of a coelenterate.

mega- Large. Female.

megaspore A spore that will germinate into a female plant.

meiosis [Gk *meiōsis* diminution] A process of nuclear division in which the number of chromosomes is reduced by half.

membrane A structure, formed mainly by a double layer of phospholipids, which surrounds cells and organelles.

membrane channel A pore in a membrane through which certain molecules may pass.

membrane pump A permease that uses energy, usually from ATP, to move substances across the membrane against their osmotic-concentration or electrostatic gradients.

meristematic tissue [Gk *meristos* divisible] A plant tissue that functions primarily in production of new cells by mitosis.

meso- Middle.

mesoderm The middle tissue layer of an animal embryo.

mesophyll [Gk *phyllon* leaf] The parenchymatous middle tissue layers of a leaf.

meta- Posterior, later; change in.

metabolism [Gk *metabolē* change] The sum of the chemical reactions within a cell (or a whole organism), including the energy-releasing breakdown of molecules (catabolism) and the synthesis of complex molecules and new protoplasm (anabolism).

metamorphosis [Gk *morphē* form] Transformation of an immature animal into an adult. More generally, change in the form of an organ or structure.

micro- Small. Male. In units of measurement, one millionth.

microfilament A long, thin structure, usually formed from the protein actin; when associated with myosin filaments, as in muscles, microfilaments are involved in movement.

microorganism A microscopic organism, especially a bacterium, virus, or protozoan.

microspore A spore that will germinate into a male plant.

microtubule A long, hollow structure formed from the protein tubulin; found in cilia, eucaryotic flagella, basal bodies/centrioles, and the cytoplasm.

middle lamella A layer of substance deposited between the walls of adjacent plant cells.

milli- One thousandth.

mineral In biology, any naturally occurring inorganic substance, excluding water.

mitochondrion [Gk *mitos* thread; *chondrion* small grain] Subcellular organelle in which aerobic respiration takes place.

mitosis [Gk *mitos*] Process of nuclear division in which complex movements of chromosomes along a spindle result in two new nuclei with the same number of chromosomes as the original nucleus.

modulator A control chemical that stabilizes an allosteric enzyme in one of its alternative conformations.

mold Any of many fungi that produce a cottony or furry growth.

mole The amount of a substance that has a weight in grams numerically equal to the molecular weight of the substance. One mole of a substance contains 6.023×10^{23} molecules of that substance; hence one mole of a substance will always contain the same number of molecules as a mole of any other substance.

molecular weight The weight of a molecule calculated as the sum of the atomic weights of its constituent atoms.

molecule A chemical unit consisting of two or more atoms bonded together.

mono- One.

monocot A member of a subclass of angiosperms, or flowering plants, characterized by the presence of a single cotyledon in the embryo, parallel veins in the leaves, and flower petals in threes; *cf.* dicot.

-morph, morpho- [Gk *morphē* form] Form, structure.

morphogenesis The establishment of shape and pattern in an organism.

morphology The form and structure of organisms or parts of organisms; the study of these.

motivation The internal state of an animal that is the immediate cause of its behavior; drive.

motor neuron A neuron, leading away from the central nervous system, that synapses on and controls an effector.

motor program A coordinated, relatively stereotyped series of muscle movements performed as a unit, either innate (as the movements of swallowing) or learned (as in speech); also, the neural circuitry underlying such behavior; *cf.* reflex.

mouthparts Structures or appendages near the mouth used in manipulating food.

mucosa Any membrane that secretes mucus (a slimy protective substance), e.g., the membrane lining the stomach and intestine.

muscle [L *musculus* small mouse, muscle]   A contractile tissue of animals.

mutation [L *mutatio* change]   Any relatively stable heritable change in the genetic material.

mutualism   A symbiosis in which both parties benefit.

mycelium [Gk *mykēs* fungus]   A mass of hyphae forming the body of a fungus.

myo- [Gk *mys* mouse, muscle]   Muscle.

NAD   *See* nicotinamide adenine dinucleotide.

NADP   *See* nicotinamide adenine dinucleotide phosphate.

nano- [L *nanus* dwarf]   One billionth.

natural selection   Differential reproduction in nature, leading to an increase in the frequency of some genes or gene combinations and to a decrease in the frequency of others.

navigation   The initiation and/or maintenance of movement toward a goal.

negative feedback   *See* feedback.

nematocyst [Gk *nēma* thread; *kystis* bag]   A specialized stinging cell in coelenterates; contains a hairlike structure that can be ejected.

neo-   New.

neocortex   Portion of the cerebral cortex in mammals, of relatively recent evolutionary origin; often greatly expanded in the higher primates and dominant over other parts of the brain.

nephr- [Gk *nephros* kidney]   Kidney.

nephridium   An excretory organ consisting of an open bulb and a tubule leading to the exterior; found in many invertebrates, such as segmented worms.

nephron   The functional unit of a vertebrate kidney, consisting of Bowman's capsule, convoluted tubule, and loop of Henle.

nerve [L. *nervus* sinew, nerve]   A bundle of neuron fibers (axons).

nerve net   A nervous system without any central control, as in coelenterates.

neuron [Gk nerve, sinew]   A nerve cell.

neutron   An electrically neutral subatomic particle with approximately the same mass as a proton.

niche   The functional role and position of an organism in the ecosystem; the way an organism makes its living, including, for an animal, not only what it eats, but when, where, and how it obtains food, where it lives, etc.

nicotinamide adenine dinucleotide (NAD)   An organic compound that functions as an electron acceptor, especially in respiration.

nicotinamide adenine dinucleotide phosphate (NADP)   An organic compound that functions as an electron acceptor, especially in biosynthesis.

nitrogen fixation   Incorporation of nitrogen from the atmosphere into substances more generally usable by organisms.

node (of plant) [L *nodus* knot]   Point on a stem where a leaf or bud is (or was) attached.

notochord [Gk *nōtos* back; *chordē* string]   In the lower chordates and in the embryos of the higher vertebrates, a flexible supportive rod running longitudinally through the back just ventral to the nerve cord.

nucleic acid   Any of several organic acids that are polymers of nucleotides and function in transmission of hereditary traits, in protein synthesis, and in control of cellular activities.

nucleoid   A region, not bounded by a membrane, where the chromosome is located in a procaryotic cell.

nucleolus   A dense body within the nucleus, usually attached to one of the chromosomes; consists of multiple copies of the genes for certain kinds of rRNA.

nucleosome   A complex consisting of several histone proteins, which together form a "spool," and chromosomal DNA, which is wrapped around the spool.

nucleotide   A chemical entity consisting of a five-carbon sugar with a phosphate group and a purine or pyrimidine attached; building-block unit of nucleic acids.

nucleus (of cell) [L kernel]   A large membrane-bounded organelle containing the chromosomes.

nutrient [L *nutrire* to nourish]   A food substance usable in metabolism as a source of energy or of building material.

nymph [Gk *nymphē* bride, nymph]   Immature stage of insect that undergoes gradual metamorphosis.

olfaction [L *olfacere* to smell]   The sense of smell.

omnivorous [L *omnis* all; *vorare* to devour]   Eating a variety of foods, including both plants and animals.

oncogene   A gene that causes one of the biochemical changes that lead to cancer.

ontogeny [Gk *ōn* being]   The course of development of an individual organism.

oo- [Gk *ōion* egg]   Egg.

oogamy   A type of heterogamy in which the female gametes are large nonmotile egg cells.

oogonium   Unjacketed female reproductive organ of a thallophyte plant.

operant conditioning   *See* conditioning.

operator   A region on the DNA to which a control substance can bind, thereby altering the rate of transcription.

oral [L *oris* of the mouth]   Relating to the mouth.

organ [Gk *organon* tool]   A body part usually composed of several tissues grouped together into a structural and functional unit.

organelle   A well-defined subcellular structure.

organic compound   A chemical compound containing carbon.

organism   An individual living thing.

orientation The act of turning or moving in relation to some external feature, such as a source of light.

osmol Measure of osmotic concentration; the total number of moles of osmotically active particles per liter of solvent.

osmoregulation Regulation of the osmotic concentration of body fluids in such a manner as to keep them relatively constant despite changes in the external medium.

osmosis [Gk *ōsmos* thrust] Movement of a solvent (usually water in biology) through a selectively permeable membrane.

osmotic potential The free energy of water molecules in a solution or colloid under conditions of constant temperature and pressure; since this free energy decreases as the proportion of osmotically active particles rises, a measure of the tendency of the solution or colloid to lose water.

osmotic pressure The pressure that must be exerted on a solution or colloid to keep it in equilibrium with pure water when it is separated from the water by a selectively permeable membrane; hence a measure of the tendency of the solution or colloid to take in water.

ov-, ovi- [L *ovum* egg] Egg.

ovary Female reproductive organ in which egg cells are produced.

ovulation Release of an egg from the ovary.

ovule A plant structure, composed of an integument, sporangium, and megagametophyte, that develops into a seed after fertilization.

ovum A mature egg cell (*pl.* ova).

oxidation Energy-releasing process involving removal of electrons from a substance; in biological systems, generally by the removal of hydrogen (or sometimes the addition of oxygen).

pancreas In vertebrates, a large glandular organ located near the stomach that secretes digestive enzymes into the duodenum and also produces hormones.

papilla [L nipple] A small nipplelike protuberance.

para- Alongside of.

parapodium [Gk *podion* little foot] One of the paired segmentally arranged lateral flaplike protuberances of polychaete worms.

parasitism [Gk *parasitos* eating with another] A symbiosis in which one party benefits at the expense of the other.

parasympathetic nervous system One of the two parts of the autonomic nervous system.

parathyroids Small endocrine glands of vertebrates located near the thyroid.

parenchyma A plant tissue composed of thin-walled, loosely packed, relatively unspecialized cells.

parthenogenesis [Gk *parthenos* virgin] Production of offspring without fertilization.

pathogen [Gk *pathos* suffering] A disease-causing organism.

pectin A complex polysaccharide that cross-links the cellulose fibrils in a plant cell wall and is a major constituent of the middle lamella.

pellicle [L *pellis* skin] A thin skin or membrane.

pepsin [Gk *pepsis* digestion] A protein-digesting enzyme of the stomach.

peptide bond A bond between two amino acids resulting from a condensation reaction between the amino group of one acid and the acidic group of the other.

perennial A plant that lives for several years, as compared to annuals and biennials, which live for one and two years respectively.

peri- Surrounding.

pericycle A layer of cells inside the endodermis but outside the phloem of roots and stems.

periderm The corky outer bark of older stems and roots.

peristalsis [Gk *stalsis* contraction] Alternating waves of contraction and relaxation passing along a tubular structure such as the digestive tract.

permeable [L *permeare* to go through] Of a membrane: permitting other substances to pass through.

permease [L *permeare*] A protein that allows molecules to move across a membrane; *cf.* gated channel, membrane channel, membrane pump.

petiole [L *pediculus* small foot] The stalk of a leaf.

PGAL *See* phosphoglyceraldehyde.

pH Symbol for the logarithm of the reciprocal of the hydrogen ion concentration; hence a measure of acidity. A pH of 7 is neutral; lower values are acidic, higher values alkaline (basic).

phage *See* bacteriophage.

phagocytosis [Gk *phagein* to eat] The active engulfing of particles by a cell.

pharynx Part of the digestive tract between the oral cavity and the esophagus; in vertebrates, also part of the respiratory passage.

phenotype [Gk *phainein* to show] The physical manifestation of a genetic trait.

pheromone [Gk *pherein* to carry + hormone] A substance that, secreted by one organism, influences the behavior or physiology of other organisms of the same species when they sense its odor.

phloem [Gk *phloios* bark] A plant vascular tissue that transports organic materials; the inner bark.

-phore [Gk *pherein* to carry] Carrier.

phosphoglyceraldehyde (PGAL) A three-carbon phosphorylated carbohydrate, important in both photosynthesis and glycolysis.

phospholipid A compound composed of glycerol, fatty acids, a phosphate group, and often a nitrogenous group.

phosphorylation Addition of a phosphate group.

photo- [Gk *phōs* light] Light.

photon A discrete unit of radiant energy.

photoperiodism A response by an organism to the duration and timing of the light and dark conditions.

photophosphorylation The process by which energy from light is used to convert ADP into ATP.

photosynthesis Autotrophic synthesis of organic materials in which the source of energy is light; *cf.* photophosphorylation.

-phyll [Gk *phyllon* leaf] Leaf.

phylogeny [Gk *phylē* tribe] Evolutionary history of an organism.

physiology [Gk *physis* nature] The life processes and functions of organisms; the study of these.

-phyte, phyto- [Gk *phyton* plant] Plant.

phytochrome A protein pigment of plants sensitive to red and far-red light.

pinocytosis [Gk *pinein* to drink] The active engulfing by cells of liquid or of very small particles.

pistil The female reproductive organ of a flower, composed of one or more megasporophylls.

pith A tissue (usually parenchyma) located in the center of a stem (rarely a root), internal to the xylem.

pituitary An endocrine gland located near the brain of vertebrates; known as the master gland because it secretes hormones that regulate the action of other endocrine glands.

placenta [Gk *plax* flat surface] An organ in mammals, made up of fetal and maternal components, that aids in exchange of materials between the fetus and the mother.

plasm-, plasmo-, -plasm [Gk *plasma* something formed or molded] Formed material; plasma; cytoplasm.

plasma Blood minus the cells and platelets.

plasma membrane The outer membrane of a cell.

plasmid A small circular piece of DNA free in the cytoplasm of a bacterial or yeast cell and replicated independently of the cell's chromosome.

plasmodesma [Gk *desma* bond] A connection between adjacent plant cells through tiny openings in the cell walls (*pl.* plasmodesmata).

plasmolysis Shrinkage of a plant cell away from its wall when in a hypertonic medium.

plastid Relatively large organelle in plant cells that functions in photosynthesis and/or nutrient storage.

pleiotropic [Gk *pleiōn* more] Of a gene: having more than one phenotypic effect.

poikilothermic [Gk *poikilos* various; *thermē* heat] Incapable of precise self-regulation of body temperature, dependent on environmental temperature; cold-blooded, ectothermic.

polar molecule A molecule with oppositely charged sections; the charges, which are far weaker than the charges on ions, arise from differences in electronegativity between the constituent atoms.

pollen grain [L *pollen* flour dust] A microgametophyte of a seed plant.

poly- Many.

polymer [Gk *meros* part] A large molecule consisting of a chain of small molecules bonded together by condensation reactions or similar reactions.

polymerase An enzyme complex that catalyzes the polymerization of nucleotides; examples are DNA polymerase, which is involved in replication, and RNA polymerase, which is involved in transcription.

polymorphism [Gk *morphē* form] The simultaneous occurrence of several discontinuous phenotypes in a population.

polyp [Gk *polypous* many-footed] The sedentary stage in the life cycle of a coelenterate.

polypeptide chain A chain of amino acids linked together by peptide bonds.

polyploid Having more than two complete sets of chromosomes.

polysaccharide Any carbohydrate that is a polymer of simple sugars.

population In ecology, a group of individuals belonging to the same species.

portal system [L *porta* gate] A blood circuit in which two beds of capillaries are connected by a vein (e.g., *hepatic portal system*).

positive feedback *See* feedback.

posterior Toward the hind end.

potential Short for *potential difference:* the difference in electrical charge between two points. *Resting p.:* a relatively steady potential difference across a cell membrane, particularly of a nonfiring nerve cell or a relaxed muscle cell. *Action p.:* a sharp change in the potential difference across the membrane of a nerve or muscle cell that is propagated along the cell; in nerves, identified with the nerve impulse. *Generator p.:* a change in the potential difference across the membrane of a sensory cell that, if it reaches a threshold level, may trigger an action potential along the associated neural pathway.

predation [L *praedatio* plundering] The feeding of free-living organisms on other organisms.

presumptive Describing the developmental fate of a tissue that is not yet differentiated. Presumptive neural tissue, for example, is destined to become part of the nervous system once it has differentiated.

primary transcript Newly synthesized RNA—generally mRNA—before the introns are removed.

primitive [L *primus* first] Old, like the ancestral condition.

primordium [L *primus; ordiri* to begin] Rudiment, earliest stage of development.

pro- Before.

proboscis [Gk *boskein* to feed] A long snout; an elephant's trunk. In invertebrates, an elongate, sometimes eversible process originating in or near the mouth that often serves in feeding.

procaryotic cell A type of cell that lacks a membrane-bounded nucleus; found only in bacteria.

progesterone [L *gestare* to carry] One of the principal female sex hormones of vertebrates.

promoter   The region of DNA to which the transcription complex binds.

prot-, proto-   First, primary.

protease   A protein-digesting enzyme.

protein   A long polypeptide chain.

proteolytic   Protein-digesting.

proton   A positively charged primary subatomic particle.

proto-oncogene   A gene that can, after certain sorts of mutation or translocation, or after mutation or translocation in associated control regions, become an oncogene and cause one of the changes leading to cancer.

protoplasm   Living substance, the material of cells.

provirus   Viral nucleic acid integrated into the genetic material of a host cell.

proximal   Near some reference point (often the main part of the body).

pseudo-   False; temporary.

pseudocoelom   A functional body cavity not entirely enclosed by mesoderm.

pseudogene   An untranscribed region of the DNA that closely resembles a gene.

pseudopod, pseudopodium [L *podium* foot]   A transitory cytoplasmic protrusion of an amoeba or an amoeboid cell.

pulmonary [L *pulmones* lungs]   Relating to the lungs.

purine   Any of several double-ringed nitrogenous bases important in nucleotides.

pyloric [Gk *pylōros* gatekeeper]   Referring to the junction between the stomach and the intestine.

pyrimidine   Any of several single-ringed nitrogenous bases important in nucleotides.

pyruvic acid   A three-carbon compound produced by glycolysis.

race   A subspecies.

radial symmetry   A type of symmetry in which the body parts are arranged regularly around a central line (in animals, running through the oral-anal axis) rather than on the two sides of a plane.

radiation   As an evolutionary phenomenon, divergence of members of a single lineage into different niches or adaptive zones.

receptor   In cell biology, a region, often the exposed part of a membrane protein, that binds a substance but does not catalyze a reaction in the chemical it binds; the membrane protein frequently has another region that, as a result of the binding, undergoes an allosteric change and so becomes catalytically active.

recessive   Of an allele: not expressing its phenotype in the presence of another allele of the same gene, therefore expressing it only in homozygous individuals. *Recessive character, recessive phenotype:* one caused by a recessive allele.

reciprocal altruism   See altruism.

recombination   In genetics, a novel arrangement of alleles resulting from sexual reproduction and from crossing over (or, in procaryotes and eucaryotic organelles, from conjugation). In gene evolution, a novel arrangement of exons resulting from a variety of processes that duplicate and transport segments of the chromosomes within the genome; these processes include transposition, unequal crossing over, and chromosomal breakage and fusion.

rectum [L *rectus* straight]   The terminal portion of the intestine.

redox reaction [*from red*uction-*ox*idation]   A reaction involving reduction and oxidation, which inevitably occur together; *cf.* reduction, oxidation.

reduction   Energy-storing process involving addition of electrons to a substance; in biological systems, generally by the addition of hydrogen (or sometimes the removal of oxygen).

reflex [L *reflexus* bent back]   An automatic act consisting, in its pure form, of a single simple response to a single stimulus, as when a tap on the knee elicits a knee jerk. Distinguished from a motor program, which involves a coordinated response of several muscles.

reflex arc   A functional unit of the nervous system, involving the entire pathway from receptor cell to effector.

reinforcement (psychological)   Reward for a particular behavior.

releaser   *See* sign stimulus.

renal [L *renes* kidneys]   Pertaining to the kidney.

respiration [L *respiratio* breathing out]   (1) The release of energy by oxidation of fuel molecules. (2) The taking in of O_2 and release of CO_2; breathing.

resting potential   *See* potential.

restriction endonuclease   *See* endonuclease.

reticulum [L little net]   A network.

retina   The tissue in the rear of the eye that contains the sensory cells of vision.

retrovirus   An RNA virus that, by means of a special enzyme (reverse transcriptase), makes a DNA copy of its genome which is then incorporated into the host's genome.

rhizoid [Gk *rhiza* root]   Rootlike structure.

ribonucleic acid (RNA)   Nucleic acid characterized by the presence of a ribose sugar in each nucleotide. The primary classes of RNA are mRNA (messenger RNA, which carries the instructions specifying the order of amino acids in new proteins from the genes to the ribosomes where protein synthesis takes place), rRNA (ribosomal RNA, which is incorporated into ribosomes), and tRNA (transfer RNA, which carries amino acids to the ribosomes as part of protein synthesis).

ribosome   A small cytoplasmic organelle that functions in protein synthesis.

RNA   *See* ribonucleic acid.

salt   Any of a class of generally ionic compounds that may be formed by reaction of an acid and a base, e.g., table salt, NaCl.

sap   Water and dissolved materials moving in the xylem; less commonly, solutions moving in the phloem. *Cell sap:* the fluid content of a plant-cell vacuole

saprophyte [Gk *sapros* rotten] A heterotrophic plant or bacterium that lives on dead organic material.

sarcomere [Gk *sarx* flesh; *meros* part] The region of a skeletal-muscle myofibril extending from one Z line to the next; the functional unit of skeletal-muscle contraction.

sclerenchyma [Gk *scleros* hard] A plant supportive tissue composed of cells with thick secondary walls.

section *Cross* or *transverse s.:* section at right angles to the longest axis. *Longitudinal s.:* section parallel to the longest axis. *Radial s.:* longitudinal section along a radius. *Sagittal s.:* vertical longitudinal section along the midline of a bilaterally symmetrical animal.

seed A plant reproductive entity consisting of an embryo and stored food enclosed in a protective coat.

segmentation The subdivision of an organism into more or less equivalent serially arranged units.

selection pressure In a population, the force for genetic change resulting from natural selection.

semipermeable Permeable only to solvent (usually water); less strictly: selectively permeable, i.e., permeable to some substances but not to others.

sensitization The process by which an unexpected stimulus alerts an animal, reducing or eliminating any preexisting habituation; *cf.* adaptation, habituation.

sensory neuron A neuron, leading toward the central nervous system, that receives input from a receptor cell or is itself responsive to sensory stimulation.

septum [L barrier] A partition or wall (*pl.* septa).

sessile [L *sessilis* of sitting, low] Of animals, sedentary. Of plants, without a stalk.

sex-linked Of genes: located on the X chromosome.

sexual dimorphism Morphological differences between the two sexes of a species, as in the size of tails of peacocks as compared to peahens.

sexual selection Selection for morphology or behavior directly related to attracting or winning mates. *Male-contest sexual selection:* selection for morphology or behavior that enables a male to win fights or contests for access to females, gaining a high position in a dominance hierarchy, for example, or possession of a territory. *Female-choice sexual selection:* selection for morphology or behavior that enables a male to attract females directly.

shoot A stem with its leaves, flowers, etc.

sieve element A conductile cell of the phloem.

sign stimulus (or releaser) A simple cue that orients or triggers specific innate behavior.

sinus [L curve, hollow] (1) A channel for the passage of blood lacking the characteristics of a true blood vessel. (2) A hollow within bone or another tissue (e.g. the air-filled sinuses of some of the facial bones).

sol Colloid in which the suspended particles are dispersed at random; *cf.* gel.

solute Substance dissolved in another (the solvent).

solution [L *solutio* loosening] A homogeneous molecular mixture of two or more substances.

solvent Medium in which one or more substances (the solute) are dissolved.

-soma, somat-, -some [Gk *soma* body] Body, entity.

somatic Pertaining to the body; to all cells except the germ cells; to the body wall. *Somatic nervous system:* a portion of the nervous system that is at least potentially under control of the will; *cf.* autonomic nervous system.

specialized Adapted to a special, usually rather narrow, function or way of life.

speciation The process of formation of new species.

species [L kind] The largest unit of population within which effective gene flow occurs or could occur.

sperm [Gk *sperma* seed] A male gamete.

sphincter [Gk *sphinkter* band] A ring-shaped muscle that can close a tubular structure by contracting.

spindle A microtubular structure with which the chromosomes are associated in mitosis and meiosis.

sporangium A plant structure that produces spores.

spore [Gk *spora* seed] An asexual reproductive cell, often a resting stage adapted to resist unfavorable environmental conditions.

sporophyll [Gk *phyllon* leaf] A modified leaf that bears spores.

sporophyte [Gk *phyton* plant] A diploid plant that produces spores.

stamen [L thread] A male sexual part of a flower; a microsporophyll of a flowering plant.

starch A glucose polymer, the principal polysaccharide storage product of vascular plants.

stele [Gk *stele* upright slab] The vascular cylinder in the center of a root or stem, bounded externally by the endodermis.

stereo- [Gk *stereos* solid] Solid; three-dimensional.

steroid Any of a number of complex, often biologically important compounds (e.g., some hormones and vitamins), composed of four interlocking rings of carbon atoms.

stimulus Any environmental factor that is detected by a receptor.

stoma [Gk mouth] An opening, regulated by guard cells, in the epidermis of a leaf or other plant part (*pl.* stomata).

stroma [Gk *stroma* bed, mattress] The ground substance within such organelles as chloroplasts and mitochondria.

subspecies A genetically distinctive geographic subunit of a species.

substrate (1) The base on which an organism lives, e.g., soil. (2) In chemical reactions, a substance acted upon, as by an enzyme.

succession In ecology, progressive change in the plant and animal life of an area.

sucrose A double sugar composed of a unit of glucose and a unit of fructose; table sugar.

suspension A heterogeneous mixture in which the particles of one substance are kept dispersed by agitation.

sym-, syn- Together.

symbiosis [Gk *bios* life] The living together of two organisms in an intimate relationship.

sympathetic nervous system One of the two parts of the autonomic nervous system.

sympatric [L *patria* homeland] Having the same range.

symplast In a plant, the system constituted by the cytoplasm of cells interconnected by plasmodesmata.

synapse [Gk *haptein* to fasten] A juncture between two neurons.

synapsis The pairing of homologous chromosomes during meiosis.

synergistic [Gk *ergon* work] Acting together with another substance or organ to achieve or enhance a given effect.

systemic circulation The part of the circulatory system supplying body parts other than the gas-exchange surfaces.

-tactic Referring to a taxis.

taxis A simple continuously oriented movement in animals (e.g., phototaxis, geotaxis) (*pl.* taxes).

taxonomy [Gk *taxis* arrangement] The classification of organisms on the basis of their evolutionary relationships.

tendon [L *tendere* to stretch] A type of connective tissue attaching muscle to bone.

territory A particular area defended by an individual against intrusion by other individuals, particularly of the same species.

testis Primary male sex organ in which sperm are produced (*pl.* testes).

thalamus [Gk *thalamos* inner chamber] Part of the rear portion of the vertebrate forebrain, a center for integration of sensory impulses.

thallus [Gk *thallos* young shoot] A plant body exhibiting relatively little tissue differentiation and lacking true roots, stems, and leaves.

thorax [Gk *thōrax* breastplate] In mammals, the part of the trunk anterior to the diaphragm, which partitions it from the abdomen. In insects, the body region between the head and the abdomen, bearing the walking legs and wings.

thymus [Gk *thymos* warty excrescence] Glandular organ that plays an important role in the development of immunologic capabilities in vertebrates.

thyroid [Gk *thyreoeidēs* shield-shaped] An endocrine gland of vertebrates located in the neck region.

thyroxin A hormone, produced by the thyroid, that stimulates a speedup of metabolism.

tissue [L *texere* to weave] An aggregate of cells, usually similar in both structure and function, that are bound together by intercellular material.

T lymphocyte *See* lymphocyte.

toxin A proteinaceous substance produced by one organism that is poisonous to another.

trachea In vertebrates, the part of the respiratory system running from the pharynx into the thorax; the "windpipe." In land arthropods, an air duct running from an opening in the body wall to the tissues.

tracheid An elongate thick-walled tapering conductile cell of the xylem.

trans- Across; beyond.

transcription In genetics, the synthesis of RNA from a DNA template.

transduction [L *ducere* to lead] In genetics, the transfer of genetic material from one host cell to another by a virus. In neurobiology, the translation of a stimulus like light or sound into an electrical change in a receptor cell.

transformation The incorporation by bacteria of fragments of DNA released into the medium from dead cells.

translation In genetics, the synthesis of a polypeptide from an mRNA template.

translocation In botany, the movement of organic materials from one place to another within the plant body, primarily through the phloem. In genetics, the exchange of parts between nonhomologous chromosomes.

transpiration Release of water vapor from the aerial parts of a plant, primarily through the stomata.

transposition The movement of DNA from one position in the genome to another. *Transposon:* a mobile segment of DNA, usually encoding the enzymes necessary to effect its own movement.

-trophic [Gk *trophē* food] Nourishing; stimulatory.

tropic hormone A hormone produced by one endocrine gland that stimulates another endocrine gland.

tropism [Gk *tropos* turn] A turning response to a stimulus, primarily by differential growth patterns in plants.

turgid [L *turgidus* swollen] Swollen with fluid.

turgor pressure [L *turgēre* to be swollen] The pressure exerted by the contents of a cell against the cell membrane or cell wall.

tympanic membrane [Gk *tympanon* drum] A membrane of the ear that picks up vibrations from the air and transmits them to other parts of the ear; the eardrum.

urea The nitrogenous waste product of mammals and some other vertebrates, formed in the liver by combination of ammonia and carbon dioxide.

ureter The duct carrying urine from the kidney to the bladder in higher vertebrates.

urethra The duct leading from the bladder to the exterior in higher vertebrates.

uric acid An insoluble nitrogenous waste product of most land arthropods, reptiles, and birds.

uterus In mammals, the chamber of the female reproductive tract in which the embryo undergoes much of its development; the womb.

vaccine [L *vacca* cow] Drug containing an antigen, administered to induce active immunity in the patient.

vacuole [L *vacuus* empty] A membrane-bounded vesicle or chamber in a cell.

valence A measure of the bonding capacity of an atom, which is determined by the number of electrons in the outer shell.

vascular tissue [L *vasculum* small vessel] Tissue concerned with internal transport, such as xylem and phloem in plants and blood and lymph in animals.

vaso- [L *vas* vessel] Blood vessel.

vector [L *vectus* carried] Transmitter of pathogens.

vegetative Of plant cells and organs: not specialized for reproduction. Of reproduction: asexual. Of bodily functions: involuntary.

vein [L *vena* blood vessel] A blood vessel that transports blood toward the heart.

vena cava [L hollow vein] One of the two large veins that return blood to the heart from the systemic circulation of vertebrates.

ventral [L *venter* belly] Pertaining to the belly or underparts.

vessel element A highly specialized cell of the xylem, with thick secondary walls and extensively perforated end walls.

villus [L shaggy hair] A highly vascularized fingerlike process from the intestinal lining or from the surface of some other structure (e.g., a chorionic villus of the placenta) (*pl.* villi).

virus [L slime, poison] A submicroscopic noncellular, obligatorily parasitic entity, composed of a protein shell and a nucleic acid core, that exhibits some properties normally associated with living organisms, including the ability to mutate and to evolve.

viscera [L] The internal organs, especially those of the great central body cavity.

vitamin [L *vita* life] An organic compound, necessary in small quantities, that a given organism cannot synthesize for itself and must obtain prefabricated in the diet.

woody dicot *See* dicot.

X chromosome The female sex chromosome.

xylem [Gk *xylon* wood] A vascular tissue that transports water and dissolved minerals upward through the plant body.

Y chromosome The male sex chromosome.

yolk Stored food material in an egg.

zoo- [Gk *zōion* animal] Animal; motile.

zoospore A ciliated or flagellated plant spore.

zygote [Gk *zygōtos* yoked] A fertilized egg cell.

zymogen [Gk *zymē* leaven] An inactive precursor of an enzyme.

INDEX

Page numbers in **boldface** refer to illustrations; those in *italics* identify definitions or the main treatment of subjects mentioned in several parts of the book.